Theory and Applications of Special Functions for Scientists and Engineers

Xiao-Jun Yang

Theory and Applications of Special Functions for Scientists and Engineers

Springer

Xiao-Jun Yang
School of Mathematics and State Key
Laboratory for Geomechanics and Deep
Underground Engineering
China University of Mining and Technology
Xuzhou, Jiangsu, China

ISBN 978-981-33-6333-5 ISBN 978-981-33-6334-2 (eBook)
https://doi.org/10.1007/978-981-33-6334-2

Mathematics Subject Classification: 33C05, 33C20, 33E12, 44A20, 44A05

This Springer imprint is published by the registered company Springer Nature Singapore Pte Ltd.
The registered company address is: 152 Beach Road, #21-01/04 Gateway East, Singapore 189721, Singapore

To my family, parents, brother, sister, wife,
and my daughters

Preface

The main target of this monograph is to provide the detailed investigations to the newly established special functions involving the Mittag-Leffler, Wiman, Prabhakar, Miller–Ross, Rabotnov, Lorenzo–Hartley, Sonine, Wright, and Kohlrausch–Williams–Watts functions, Gauss hypergeometric series, and Clausen hypergeometric series. The integral transform operators based on the theory of the Wright and Kohlrausch–Williams–Watts functions may be used to solve the complex problems with power-law behaviors in the light of nature complexity. The topics are important and interesting for scientists and engineers to represent the complex phenomena arising in mathematical physics, engineering, and other applied sciences.

The monograph is divided into seven chapters, which are discussed as follows.

Chapter 1 introduces the special functions such as Euler gamma function, Pochhammer symbols, Euler beta function, extended Euler gamma function, extended Euler beta function, Gauss hypergeometric series, and Clausen hypergeometric series as well as calculus operators with respect to monotone function containing the power-law calculus, scaling-law calculus, and complex topology calculus as well as calculus operators with respect to logarithmic and exponential functions.

Chapter 2 investigates the Wright function, Wright's generalized hypergeometric function, supertrigonometric and superhyperbolic functions via Wright function, and Wright's generalized hypergeometric function. The integral representations for the supertrigonometric and superhyperbolic functions are addressed in detail. Some integral transforms via Dunkl transform based on the calculus with respect to power-law function are proposed.

Chapter 3 provides the theory of the Mittag-Leffler function, supertrigonometric functions, and superhyperbolic functions. The integral representations for the Mittag-Leffler function and related functions are addressed, and the general fractional calculus operators are also discussed in detail. The truncated Mittag-Leffler, supertrigonometric, and superhyperbolic functions are considered, and some mathematical models are considered to explain the power-law behaviors in material science.

Chapter 4 shows the theory of the Wiman function, supertrigonometric functions, and superhyperbolic functions. The integral representations for the Wiman function and related functions are addressed, and the general fractional calculus operators are also discussed in detail. The truncated Wiman, supertrigonometric, and superhyperbolic functions are considered, and the integral equations as well as mathematical models related to Wiman function are also presented in detail.

Chapter 5 addresses the theory of Prabhakar function and proposes the supertrigonometric and superhyperbolic functions via Prabhakar function. The Laplace transforms for the new special functions and integral representations for the supertrigonometric and superhyperbolic functions are discussed in detail. The truncated Prabhakar, supertrigonometric, and superhyperbolic functions are proposed, and the general fractional calculus involving the Prabhakar function is considered. The integral equations and mathematical models related to Prabhakar function are also presented.

Chapter 6 presents the Sonine functions, Rabotnov fractional exponential function, Miller–Ross function, and Lorenzo–Hartley functions. The Laplace and Mellin transforms of them are given, and the integral representations for the supertrigonometric and superhyperbolic functions are also presented in detail. The formulas related to the Mittag-Leffler functions and Wright hypergeometric functions are also considered.

Chapter 7 illustrates the Kohlrausch–Williams–Watts function and integral representations. The subtrigonometric functions, subhyperbolic functions, supertrigonometric functions, and superhyperbolic functions are discussed in detail. Moreover, the Fourier-type series, Fourier-type integral transforms, Laplace-type integral transforms, and Mellin-type integral transforms are also proposed.

Xuzhou, China
August 20, 2020

Xiao-Jun Yang

Acknowledgments

I am particularly indebted to Professor George E. Andrews, Professor H. M. Srivastava, Professor Song Jiang, Professor Bo-Ming Yu, Professor Wolfgang Sprößig, Professor Jeffrey Hoffstein, Professor Nouzha El Yacoubi, Professor Mourad Ismail, Professor Tom H. Koornwinder, Professor Michel L. Lapidus, Professor Norbert Hounkonnou, Professor Simeon Oka, Professor Roland W. Lewis, Professor Manuel López Pellicer, Professor Michael Reissig, Professor George A. Anastassiou, Professor Jin-De Cao, Professor Alain Miranville, Professor Minvydas Ragulskis, Professor Miguel A. F. Sanjuan, Professor Tasawar Hayat, Professor Mahmoud Abdel-Aty, Professor Stefano Galatolo, Professor Dumitru Mihalache, Professor Martin Bohner, Professor Thiab Taha, Professor Salvatore Capozziello, Professor André Keller, Professor Martin Ostoja-Starzewski, Professor Vukman Bakic, Professor J. A. Tenreiro Machado, Professor Dumitru Baleanu, Professor Sohail Nadeem, Professor Mauro Bologna, Professor Hui-Ming Wang, Professor Long Jiang, Professor Sheng-Bo Li, Professor Carlo Cattani, Professor Semyon Yakubovich, Professor Ayman S. Abdel-Khalik, Professor Irene Maria Sabadini, Professor Mario Di Paola, Professor Mokhtar Kirane, Professor Giuseppe Failla, Professor Reza Saadati, Professor Yusif Gasimov, and Professor Ivanka Stamova.

My special thanks go to Professor He-Ping Xie, Professor Feng Gao, Professor Fu-Bao Zhou, Professor Hong-Wen Jing, Professor Xiao-Zhao Li, Professor Yang Ju, Professor Hong-Wei Zhou, and Professor Ming-Zhong Gao, and the financial support of the Yue-Qi Scholar of the China University of Mining and Technology (grant no. 04180004). It is my pleasure to thank my students who support me to handle the Latex files. Finally, I also wish to express my special thanks to Springer staff, especially, Daniel Wang and Dimana Tzvetkova, for their cooperation in the production process of this book.

Contents

About the Author

Xiao-Jun Yang, PhD, is Professor of Applied Mathematics and Mechanics at China University of Mining and Technology, Xuzhou, China. His scientific interests include mathematical physics, fractional calculus and applications, fractals, mechanics, analytic number theory, integral transforms, and special functions. He is a recipient of the Atanasije Stojkovič Medal, Belgrade, Serbia (2021). Professor Yang was awarded the Abel Award (Istanbul, Turkey, 2020) for his achievements in the area of fractional calculus and its applications. He was also awarded the Obada-Prize, Cairo, Egypt (2019). He is a recipient of the Young Scientist Award (2019) for contributions in developing local fractional calculus at ICCMAS-2019, Istanbul, Turkey, and the Springer Distinguished Researcher Award (2019) at ICMMAAC-2019, Jaipur, India. He is a highly cited researcher (2021, 2020, and 2019, Clarivate Analytics) in mathematics and Elsevier Most Cited Chinese Researcher in Mathematics (2017, 2018, 2019, and 2020). Professor Yang is one of the scientific committee members of 10th edition of the Pan African Congress of Mathematicians. He is the author and co-author of seven monographs for Elsevier, Springer Nature, CRC, World Science, and Asian Academic, and co-editor of one edited book for De Gruyter.

Chapter 1
Preliminaries

Abstract In this chapter, we investigate the special functions and operator calculus. At first, the Euler gamma function, Pochhammer symbols, Euler beta function, extended Euler gamma function, and extended Euler beta function are introduced. Then, the Gauss hypergeometric series, Clausen hypergeometric series, supertrigonometric and superhyperbolic functions, and Laplace and Mellin transforms are presented. Finally, the calculus operators with respect to monotone function are discussed and the mathematical models in applied sciences are also reported in detail.

1.1 The Euler Gamma Function, Pochhammer Symbols, Euler Beta Function, and Related Functions

In this section, we present the Euler gamma function, Pochhammer symbols, Euler beta function, extended Euler gamma function, and extended Euler beta function.

1.1.1 The Euler Gamma Function

In this part, we introduce the Euler gamma function.

Let $\mathbb{C}$, $\mathbb{R}$, $\mathbb{Z}$, $\mathbb{N}$ be the sets of the complex numbers, real numbers, integrals, and natural numbers, respectively.

Let $\mathbb{Z}^+$, $\mathbb{R}_+$, $\mathbb{Z}^-$, and $\mathbb{R}_-$ be the sets of the positive integrals, positive real numbers, and negative integral numbers, and negative real numbers.

Let $\mathbb{Z}_0^- = \mathbb{Z}^- \cup 0$ and $\mathbb{N}_0 = \mathbb{N} \cup 0$.

Let $Re\,(x)$ denote the real part of x if $x \in \mathbb{C}$.

X.-J. Yang, *Theory and Applications of Special Functions for Scientists and Engineers*, https://doi.org/10.1007/978-981-33-6334-2_1

Definition 1.1 (Euler [1]) The gamma function due to Euler is defined as

$$\Gamma(z) = \int_0^\infty e^{-t} t^{z-1} dt, \tag{1.1}$$

where $Re(z) > 0$ and $z \in \mathbb{C}$.

The formula was discovered by Euler in 1729 (see [1], p.1), and the notation $\Gamma(z)$ was introduced by Legendre in 1814 (see [2], p.476).

Theorem 1.1 (Weierstrassian Product [3]) *If* $z \in \mathbb{C}\backslash\mathbb{Z}_0^-$ *with* $\mathbb{Z}_0^- =: \{0, -1, -2, \cdots\}$ *and* $\gamma := \lim_{n\to\infty}\left(\sum_{k=1}^{n} \frac{1}{k} - \log n\right)$ *is the Euler constant, the Gamma function was given as [3]*

$$\Gamma(z) = \frac{e^{-\gamma z}}{z} \prod_{k=1}^{\infty} \left(\left(1 + \frac{z}{k}\right)^{-1} e^{\frac{z}{k}}\right). \tag{1.2}$$

Moreover, $\Gamma(z)$ *is analytic except at the points* $z \in \mathbb{Z}_0^-$, *where it has simple poles [4].*

The formula for the Weierstrassian product was discovered by Weierstrass in 1856 [3] and by Newman in 1848 [5], respectively, and the proofs were published by Hölder [6], Moore [7], and Baines [8].

Definition 1.2 (Euler [1]) Let $Re(z) > 0$ and $z \in \mathbb{C}$. Then the Euler's functional equation states

$$\Gamma(z+1) = z\Gamma(z). \tag{1.3}$$

The result is the Euler's functional equation discovered by Euler in 1729 [9] and reported by Weierstrass [3], Brunel [10], Gronwall [11], and Olver [12].

Theorem 1.2 (Euler [1]) *If* $z \in \mathbb{N}_0$, *then we have*

$$\Gamma(z+1) = z!. \tag{1.4}$$

The result is the Euler's functional equation discovered by Euler in 1729 [1, 2] and discussed by Weierstrass [4], Brunel [10], and Gronwall [11].

Theorem 1.3 (Euler [1])

$$\Gamma\left(\frac{1}{2}\right) = \sqrt{\pi}. \tag{1.5}$$

This work was discovered by Euler in 1729 [11] and discussed in Bell [13], Luke [14], and Bendersky [15].

Theorem 1.4 (Euler) *If $n, j \in \mathbb{N}$, then we have*

$$\prod_{j=1}^{n} \Gamma\left(\frac{1-j}{n}\right) = (2\pi)^{\frac{n-1}{2}} n^{\frac{1}{2}}. \tag{1.6}$$

The result was reviewed by Gronwall in 1916 [11].

Theorem 1.5 (Winckler [16]) *If $z \in \mathbb{C}$ and $g, k, j, k, l, m, n \in \mathbb{N}$, then we have*

$$\frac{\prod_{j=0}^{n-1} \Gamma\left(hz + \frac{hj}{n}\right)}{\prod_{l=0}^{m-1} \Gamma\left(gz + \frac{gl}{m}\right)} = \left(\frac{h}{g}\right)^{hgz + \frac{hg-h-g}{2}} (2\pi)^{\frac{h-g}{2}}. \tag{1.7}$$

The result was discovered by Winckler in 1856 [16] and reviewed by Gronwall in 1916 [11].

Theorem 1.6 (Schlömilch [17] and Newman [5]) *If $z \in \mathbb{C}$ and $k \in \mathbb{N}$, then we have*

$$\frac{1}{\Gamma(z)} = e^{\gamma z} z \prod_{k=1}^{\infty} \left(1 + \frac{z}{k}\right) e^{-\frac{z}{k}}. \tag{1.8}$$

The result was discovered by Schlömilch in 1844 [17] and by Newman [18].

Theorem 1.7 (Whittaker [19]) *If $Re(z) > 0$, $z \in \mathbb{C}$, and $k \in \mathbb{N}$, then we have*

$$\int_{0}^{\infty} e^{-kt} t^{z-1} dt = \frac{\Gamma(z)}{k^z}. \tag{1.9}$$

The result was first reported by Whittaker in 1902 (see [19], p.184) and further reported by Whittaker and Watson in 1920 [20].

Theorem 1.8 (Whittaker [19]) *If $\alpha, \beta \in \mathbb{C}$, $Re(\alpha) > 0$, and $Re(\beta) > 0$, then we have*

$$\int_{0}^{\frac{\pi}{2}} \cos^{\alpha-1} t \sin^{\beta-1} t dt = \frac{1}{2} \frac{\Gamma\left(\frac{\alpha}{2}\right) \Gamma\left(\frac{\alpha}{2}\right)}{\Gamma\left(\frac{\alpha+\beta}{2}\right)}. \tag{1.10}$$

The result was first defined by Whittaker in 1902 (see [19], p.191) and further reported by Whittaker and Watson in 1920 [20].

Theorem 1.9 (Titchmarsh [21]) *If $\alpha, \beta \in \mathbb{C}$, $Re(\alpha) > 0$, $Re(\beta) > 0$, and $Re(\alpha + \beta) > 1$, then we have*

$$\int_{-\infty}^{\infty} \frac{1}{\Gamma(\alpha + t)\,\Gamma(\beta - t)} dt = \frac{2^{\alpha+\beta-1}}{\Gamma(\alpha + \beta - 1)}. \tag{1.11}$$

The result was first reported in the Titchmarsh's monograph [21].

Theorem 1.10 (Titchmarsh [21]) *If $Re(\alpha) > -1$, $Re(\beta) > -1$, and $Re(\alpha + \beta) > -1$, then we have*

$$f \prod_{k=1}^{\infty} \frac{k(\alpha + \beta + k)}{(\alpha + k)(\beta + k)} = \frac{\Gamma(\alpha + 1)\,\Gamma(\beta + 1)}{\Gamma(\alpha + \beta + 1)}. \tag{1.12}$$

The result was first presented in the Titchmarsh's monograph [21].

Theorem 1.11 (Titchmarsh [21]) *If $z \in \mathbb{C}$ and $k, n \in \mathbb{N}$, then we have*

$$\prod_{k=1}^{n} \left(1 - \frac{z}{k^n}\right) = \left(-\prod_{k=1}^{n} \Gamma\left(-e^{2\pi i \frac{k-1}{n}} z^{\frac{1}{n}}\right)\right)^{-1}. \tag{1.13}$$

The result was first reported in the Titchmarsh's monograph [21].

Theorem 1.12 (Euler [22]) *Let $z \in \mathbb{C}$ and $Re(z) > 0$. Then we have the Euler's completion formula as follows:*

$$\Gamma(z)\,\Gamma(1 - z) = \frac{\pi}{\sin(\pi z)} \tag{1.14}$$

and

$$\sin(\pi z) = \pi z \prod_{k=1}^{n} \left(1 - \frac{z^2}{k^2}\right). \tag{1.15}$$

The result is the Euler's completion formula due to Euler [22].

For more details of the results, readers refer to Weierstrass [4], Manocha and Srivastava [23], Luke [14], Bell [13], Godefroy [24] and Tannery [25].

Theorem 1.13 (Legendre [2], p.485)

*The **Legendre duplication formula** states*

$$\Gamma(2z)\,\Gamma\left(\frac{1}{2}\right) = 2^{2z-1}\Gamma(z)\,\Gamma\left(z + \frac{1}{2}\right), \tag{1.16}$$

where $z \in \mathbb{C} \backslash \mathbb{Z}_0^-$.

The Legendre's duplication formula was first discovered by Legendre in 1809 (see [2], p.477). For more details of the Legendre duplication formula, readers refer to Gronwall [11], Andrews et al. [26], and Manocha and Srivastava [23].

Theorem 1.14 (Gauss [27]) *If* $z \in \mathbb{C}\backslash\left\{0, -\frac{j}{m}\right\}$ *with* $j < m$ *and* $j, m \in \mathbb{N}$, *then we have*

$$\Gamma(mz) = (2\pi)^{\frac{1-m}{2}} m^{mz-\frac{1}{2}} \prod_{j=1}^{m} \Gamma\left(z + \frac{j-1}{m}\right). \tag{1.17}$$

The result is the Gauss' multiplication formula due to Gauss [27]. For more details of the Gauss' multiplication formula, readers refer to Winckler [28], Gronwall [11], Manocha and Srivastava [23], and Andrews et al. [26].

Theorem 1.15 (Weierstrass [3]) *If* $z \in \mathbb{C}/\mathbb{Z}_0^-$, *then we have*

$$\begin{aligned} &\Gamma\left(\tfrac{1}{2} - z\right)\Gamma\left(\tfrac{1}{2}+z\right) \\ &= \pi sec\,(\pi z) \\ &= \tfrac{\pi}{\cos(\pi z)} \\ &= \tfrac{2\pi}{e^{i\pi z}+e^{-i\pi z}}. \end{aligned} \tag{1.18}$$

The result was discovered by Weierstrass [4] and reported by Bell in 1968 [13] and by Luke in 1969 [14].

Theorem 1.16 $\Gamma\left(\frac{1}{2}+iz\right)\Gamma\left(\frac{1}{2}-iz\right) = \frac{\pi}{\cosh(\pi z)} = \frac{2\pi}{e^{\pi z}+e^{-\pi z}}$, *(19)*

$$\left|\Gamma\left(\frac{1}{2}+iz\right)\right|^2 = \frac{2\pi}{e^{\pi z}+e^{-\pi z}} \tag{1.19}$$

and

$$\Gamma(iz)\,\Gamma(-iz) = \frac{\pi}{-iz\sin(\pi zi)} = \frac{2\pi}{z\left(e^{\pi z}-e^{-\pi z}\right)}, \tag{1.20}$$

where $|z| \to \infty$.

The results were reported by different researchers, for example, Lerch [29], Godefroy [30], Stieltjes [31], Bateman [32], and Andrews et al. [26].

Theorem 1.17

$$\prod_{j=1}^{n-1}\Gamma\left(\frac{j}{n}\right)\Gamma\left(1-\frac{j}{n}\right) = \frac{(2\pi)^{n-1}}{n}, \tag{1.21}$$

$$\Gamma\left(-n+\frac{1}{2}\right)=(-1)^n\frac{2^n\sqrt{\pi}}{(2n-1)!},\tag{1.22}$$

$$\Gamma\left(n+\frac{1}{2}\right)=\frac{(2n-1)!\sqrt{\pi}}{2^n},\tag{1.23}$$

$$\Gamma(n+z)\,\Gamma(n-z)=\frac{\pi z}{\sin(\pi z)}\left((n-1)!\right)^2\prod_{j=1}^{n-1}\Gamma\left(1-\frac{z^2}{j}\right)\tag{1.24}$$

and

$$\Gamma\left(n+\frac{1}{2}+z\right)\Gamma\left(n+\frac{1}{2}-z\right)=\frac{\left(\Gamma\left(n+\frac{1}{2}\right)\right)^2}{\cos(\pi z)}\prod_{j=1}^{n}\Gamma\left(1-\frac{4z^2}{(2j-1)^2}\right),\tag{1.25}$$

where $n\in\mathbb{N}$ *and* $z\in\mathbb{C}\backslash\mathbb{Z}_0^-$.

The results were reported by Weierstrass in 1856 [4].

Let us introduce the Temme function which is related to the ratio of two gamma functions [35].

Definition 1.3 The Temme function is defined as

$$\Gamma^*(z)=\frac{\Gamma(z)}{\sqrt{2\pi}\,z^{z-\frac{1}{2}}e^{-z}},\tag{1.26}$$

where $z\in\mathbb{C}$ and $Re(z)>0$.

The result was defined by in Temme's book (see [35], p.66).

Theorem 1.18 *Let* $z,a,b\in\mathbb{C}$, $Re(z)>0$, $Re(z+a)>0$, *and* $Re(z+b)>0$. *Then the ratio of two gamma functions is shown as follows:*

$$\frac{\Gamma(z+a)}{\Gamma(z+b)}=z^{a-b}\frac{\Gamma^*(z+a)}{\Gamma^*(z+b)}Q(z,a,b),\tag{1.27}$$

where

$$Q(z,a,b)=\left(1+\frac{a}{z}\right)^{a-\frac{1}{2}}\left(1+\frac{b}{z}\right)^{\frac{1}{2}-b}e^{z\left[\ln\left(1+\frac{a}{z}\right)-\frac{a}{z}-\ln\left(1+\frac{b}{z}\right)+\frac{b}{z}\right]}.\tag{1.28}$$

The result was discovered by Temme in the book (see [35], p.67).

In an alternative manner, we have (see [35], p.67)

$$\frac{\Gamma(z+a)}{\Gamma(z+b)} = \frac{1}{\Gamma(b-a)} \int_0^1 t^{z+a-1} (1-t)^{b-a-1} \, dt, \tag{1.29}$$

where $a, b, z \in \mathbb{C}$ and $Re\,(b-a) > 0$.

The result was reported by Temme in the book (see [35], p.67).

Here, we introduce the interested formula reported in the book (see [35], p.72) as follows:

Theorem 1.19 *Let $a, b, z \in \mathbb{C}$, $Re\,(a) > 0$, $Re\,(b) > 0$, and $Re\,(z) > 0$.*
Then we have

$$\int_0^\infty t^{z-1} e^{-at^b} dt = \frac{1}{b} \Gamma\left(\frac{z}{b}\right) a^{-\frac{z}{b}}. \tag{1.30}$$

The result was reported by Temme (see [35], p.72).

There are some special cases of (1.29) as follows:

$$\int_0^\infty t^{z-1} e^{-at} dt = a^{-z} \Gamma(z), \tag{1.31}$$

$$\int_0^\infty e^{-t^b} dt = \Gamma\left(\frac{1}{b} + 1\right), \tag{1.32}$$

$$\int_0^\infty e^{-at^b} dt = \frac{1}{b} \Gamma\left(\frac{z}{b}\right) a^{-\frac{z}{b}}, \tag{1.33}$$

$$\int_0^\infty e^{-at^2} dt = \frac{1}{2} \Gamma\left(\frac{z}{2}\right) a^{-\frac{z}{2}}, \tag{1.34}$$

$$\int_0^\infty t^{z-1} e^{-at^2} dt = \frac{1}{2} \Gamma\left(\frac{z}{2}\right) a^{-\frac{z}{2}}, \tag{1.35}$$

$$\int_0^\infty t^{z-1} e^{-t^2} dt = \frac{1}{2} \Gamma\left(\frac{z}{2}\right) \tag{1.36}$$

and

$$\int_0^\infty t^{z-1} e^{-t^b} dt = \frac{1}{b} \Gamma\left(\frac{z}{b}\right), \tag{1.37}$$

where $a, b, z \in \mathbb{C}$, $Re\,(a) > 0$, $Re\,(b) > 0$, and $Re\,(z) > 0$.

There are useful formulas as follows [35]:

$$\Gamma\left(n+\frac{1}{2}\right) = \frac{(2n)!\sqrt{\pi}}{2^{2n} n!}, \tag{1.38}$$

$$\Gamma\left(-n+\frac{1}{2}\right) = (-1)^n \sqrt{\pi} 2^{2n} \frac{n!}{(2n)!}, \tag{1.39}$$

$$\Gamma\left(\frac{1}{2} - z\right) \Gamma\left(z+\frac{1}{2}\right) = \frac{\pi}{\cos(\pi z)}, \; z - \frac{1}{2} \notin \mathbb{Z}, \tag{1.40}$$

$$\Gamma\left(\frac{1}{2} - xi\right) \Gamma\left(xi+\frac{1}{2}\right) = \frac{\pi}{\cosh(\pi x)}, \tag{1.41}$$

$$\int_0^\infty t^{z-1} \sin t dt = \Gamma(z) \sin \frac{\pi z}{2} \tag{1.42}$$

and

$$\int_0^\infty t^{z-1} \cos t dt = \Gamma(z) \cos \frac{\pi z}{2}, \tag{1.43}$$

where $z \in \mathbb{C}$, $Re\,(z) > 0$, $n \in \mathbb{N}$ and $x \geq 0$.

Making use of (1.30) and taking $a = i = \sqrt{-1}$, we have

$$\int_0^\infty t^{z-1} e^{-it} dt = i^{-z} \Gamma(z), \tag{1.44}$$

$$\begin{aligned} \int_0^\infty t^{z-1} e^{-it} dt &= \int_0^\infty t^{z-1} (\cos t - i \sin t)\, dt \\ &= \int_0^\infty t^{z-1} \cos t dt - i \int_0^\infty t^{z-1} \sin t dt \end{aligned} \tag{1.45}$$

and

$$\begin{aligned} i^{-z}\Gamma(z) &= e^{-\frac{z\pi i}{2}}\Gamma(z) \\ &= \left(\cos\frac{z\pi}{2} - \sin\frac{z\pi}{2}\right)\Gamma(z), \end{aligned} \tag{1.46}$$

where $(i\theta)^z = |\theta|\, e^{-\frac{z\pi i sgn(z)}{2}}$, $z \in \mathbb{C}$ and $Re(z) > 0$, such that we obtain (1.42) and (1.43).

1.1.2 The Pochhammer Symbols and Related Formulas

We now introduce the Pochhammer symbols and related theorems.

Definition 1.4 (Pochhammer [34])
The Pochhammer symbol is defined as [34]

$$\begin{aligned} (\alpha)_k &= \prod_{n=1}^{k}(\alpha + n - 1) \\ &= \frac{\Gamma(\alpha+k)}{\Gamma(\alpha)} \\ &= \begin{cases} 1\ (k = 0) \\ \alpha(\alpha+1)\cdots(\alpha+k-1)\ (k \in \mathbb{N}_0) \end{cases} \end{aligned} \tag{1.47}$$

and

$$(\alpha)_0 = 1, \tag{1.48}$$

where $\alpha \in \mathbb{C}$ and $k, n \in \mathbb{N}$.

The Pochhammer symbol was first suggested by Pochhammer in 1870 [34].

The notation was first used by Pochhammer in 1870 [34] and Weierstrass noticed in 1856 that [4]

$$\Gamma(\alpha + k) = \alpha(\alpha+1)\cdots(\alpha+k-1)\,\Gamma(\alpha)\ (k \in \mathbb{N}_0). \tag{1.49}$$

For more information, readers may refer to the monograph [33].

Moreover, there is (see[4, 11, 36])

$$\lim_{k\to\infty}(\alpha)_k = \frac{1}{\Gamma(\alpha)}, \tag{1.50}$$

where $\alpha \in \mathbb{C}\backslash\mathbb{C}_0^-$ and $k \in \mathbb{N}$.

Suppose that $\alpha = -n$ and $n \in \mathbb{N}_0$, then there is (see [36], p.3)

$$(\alpha)_k = \begin{cases} (-n)_k \,, n \geq k \\ -, n < k. \end{cases} \tag{1.51}$$

Theorem 1.20 (Euler [1]) *If $z \in \mathbb{C}\backslash\mathbb{Z}_0^-$, then we have*

$$\Gamma(z) = \lim_{n \to \infty} \frac{n^z}{(z)_{n+1}}. \tag{1.52}$$

The result was discovered by Euler in 1729 [9], reported by Weierstrass in 1856 [3], and discussed by Gronwall in 1916 [11].

Theorem 1.21 *There exist*

$$(\alpha)_k \, (\alpha + k)_n = (\alpha)_{n+k} \,, \tag{1.53}$$

and

$$(\alpha + k)_{m-k} = \frac{(\alpha)_m}{(\alpha)_k}, \tag{1.54}$$

where $\alpha \in \mathbb{C}\backslash\mathbb{Z}_0^-$ and $n, k \in \mathbb{N}$.

The first formula of the results was reported by Rainville in 1960 (see [37], p.59) and the second formula was suggested by Slater in 1966 (see [36], p.31).

There are some useful formulas as follows:

$$(-z)_n = (-1)^n \, (z - n + 1)_n, \tag{1.55}$$

$$(z)_{2n} = 2^{2n} \left(\frac{z}{2}\right)_n \left(\frac{z}{2} + \frac{1}{2}\right)_n, \tag{1.56}$$

$$(z)_{2n+1} = 2^{2n+1} \left(\frac{z}{2}\right)_{n+1} \left(\frac{z}{2} + \frac{1}{2}\right)_n, \tag{1.57}$$

where $z \in \mathbb{C}$ and $n \in \mathbb{N}$.

Theorem 1.22 *Let $j, k, m, n \in \mathbb{N}_0$, $k \leq n$ and $\alpha, \beta \in \mathbb{C}\backslash\mathbb{Z}$, then we have*

$$(1)_n = n!, \tag{1.58}$$

$$\binom{\alpha}{n} = \frac{\alpha(\alpha-1)\cdots(\alpha-k-1)}{n!} = \frac{\Gamma(\alpha+1)}{n!\Gamma(\alpha-n+1)} = \frac{1}{n!(\alpha+1)_n} = \frac{(-1)^n(-\alpha)_n}{n!}, \tag{1.59}$$

$$\frac{\Gamma(\alpha+1)}{\Gamma(\alpha-n+1)} = (-1)^n(-\alpha)_n, \tag{1.60}$$

$$\frac{1}{(m-n)!} = \frac{(-1)^n(-m)_n}{m!}, \tag{1.61}$$

$$\frac{\Gamma(\alpha-n)}{\Gamma(\alpha)} = \frac{(-1)^n}{(1-\alpha)_n}, \tag{1.62}$$

$$\frac{\Gamma(\alpha-n)}{\Gamma(\alpha)} = (\alpha)_{-n} = \frac{(-1)^n}{(1-\alpha)_n}, \tag{1.63}$$

$$\frac{n!}{(\alpha)_{n+1}} - \frac{n!}{(\alpha+1)_{n+1}} = \frac{(n+1)!}{(\alpha)_{n+2}}, \tag{1.64}$$

$$\frac{(\alpha)_n}{(\beta)_n} - \frac{(\alpha)_{n+1}}{(\beta)_{n+1}} = \frac{(\alpha)_n}{(\beta)_{n+1}}(\alpha-\beta), \tag{1.65}$$

$$(\alpha)_{n-k} = \frac{(-1)^k}{(1-\alpha)_{-n}(1-\alpha-n)_k} = \frac{(-1)^k(\alpha)_n}{(1-\alpha-n)_k}, \tag{1.66}$$

$$(1)_{n-k} = (n-k)! = \frac{(-1)^k(1)_n}{(-n)_k} == \frac{(-1)^k n!}{(-n)_k}, \tag{1.67}$$

$$(\alpha)_{mn} = m^{mn}\prod_{j=1}^{m}\left(\frac{\alpha+j-1}{m}\right)_n, \tag{1.68}$$

and

$$(-n)_k = \begin{cases} \frac{(-1)^k n!}{(n-k)!}, & (0 \le k \le n) \\ 0, & (k > n) \end{cases}. \tag{1.69}$$

For more details of the results, readers refer to the works [3, 11, 13, 14, 23, 36].

Theorem 1.23 (Stirling [38])

$$\begin{aligned}&\frac{1}{z-\alpha}\\&=\sum_{k=0}^{\infty}\frac{(\alpha)_k}{(z)_{k+1}}\\&=\sum_{k=0}^{\infty}\frac{\Gamma(z)}{\Gamma(\alpha)}\frac{\Gamma(\alpha+k)}{\Gamma(z+k+1)}\\&=\frac{1}{z}+\frac{\alpha}{z(z+1)}+\frac{\alpha(\alpha+1)}{z(z+1)(z+2)}+\cdots+,\end{aligned}\tag{1.70}$$

where $Re(\alpha)>0$, $Re(z)>0$, $Re(\alpha-z)>0$ *and* $\alpha, z\in\mathbb{C}\backslash\mathbb{Z}_0^-$.

The result was discovered by Stirling in 1730 [38] and reviewed by Gronwall in 1916 [11].

Theorem 1.24 *Let* $h^{(1)}(t)>0$, $h(0)=a$, $h(1)=b$, $x\in\mathbb{C}$ *and* $Re(x)>0$.
Then we have

$$\Gamma(x)=\int_a^b e^{-h(t)}(h(t))^{x-1}h^{(1)}(t)\,dt\,.\tag{1.71}$$

The result was discovered by Yang *et al.* in 2020 when $x\in\mathbb{N}$ [39].

1.1.3 The Euler Beta Function

In this section, we investigate the concept and theorems of the Euler beta function.

Definition 1.5 (Euler [22])
The Euler beta function is defined as

$$B(\alpha,\beta)=\frac{\Gamma(\alpha)\Gamma(\beta)}{\Gamma(\alpha+\beta)}=\int_0^1 t^{\alpha-1}(1-t)^{\beta-1}\,dt,\tag{1.72}$$

where $Re(\alpha)>0$, $Re(\beta)>0$, and $\alpha,\beta\in\mathbb{C}\backslash\mathbb{Z}_0^-$.

The formula (called the Euler integral of the second kind) was first discovered by Euler in 1772 [22] and by Legendre in 1811 (see [40], p.211), and the name of the beta function was introduced for the first time by Binet in 1839 [41]. For more details, see the monograph [33].

It is clear that

$$B(\alpha,1)=\frac{1}{\alpha+1}\tag{1.73}$$

and

$$B(1, \beta) = \frac{1}{1+\beta}, \tag{1.74}$$

where $Re(\alpha) > 0$, $Re(\beta) > 0$, and $\alpha, \beta \in \mathbb{C}\backslash\mathbb{Z}_0^-$.

There are some useful formulas as follows:

$$\int_0^{\frac{\pi}{2}} \sin t^{2\alpha-1} \cos^{2\beta-1} dt = B(\alpha, \beta), \tag{1.75}$$

$$\int_a^b (b-t)^{\alpha-1} (t-a)^{\beta-1} dt = B(\alpha, \beta)(b-a)^{\alpha+\beta-1}, \tag{1.76}$$

where $\alpha, \beta \in \mathbb{C}\backslash\mathbb{Z}_0^-$, $a, b \in \mathbb{R}$, $Re(\alpha) > 0$, $Re(\beta) > 0$ and $b > a$.

In the special case of $a = 0$ and $b = x$ in (1.76), we have

$$\int_0^x (x-t)^{\alpha-1} t^{\beta-1} dt = B(\alpha, \beta) x^{\alpha+\beta-1}, \tag{1.77}$$

where $\alpha, \beta \in \mathbb{C}\backslash\mathbb{Z}_0^-$, $x > 0$ and $Re(\alpha) > 0$, $Re(\beta) > 0 \quad x > 0$.

Theorem 1.25 (Whittaker and Watson [20])

$$\Gamma(\alpha, n) = \lim_{n\to\infty} n^{\alpha} B(\alpha, n), \tag{1.78}$$

where $\alpha \in \mathbb{C}\backslash\mathbb{Z}_0^-$ *and* $n \in \mathbb{N}$.

The result was reported by Whittaker and Watson (see [20], p.254).

Theorem 1.26 (Euler [22])

$$B(\alpha, \beta)(\alpha+\beta, c) = B(\beta, c)(\beta+c, \alpha). \tag{1.79}$$

The result was discovered by Euler in 1772 [22] and further reported by Whittaker and Watson (see [20], p.261).

Theorem 1.27 (Whittaker and Watson [20])

$$B(\alpha, \beta) = \int_0^{\infty} t^{\alpha-1} (1+t)^{-(\alpha+\beta)} dt, \tag{1.80}$$

$$B(\alpha, n+1) = \frac{n!}{(\alpha)_{n+1}} \; (n \in \mathbb{N}_0), \tag{1.81}$$

$$B(\alpha, \beta+1) = \frac{\beta}{\alpha+\beta} B(\alpha, \beta), \tag{1.82}$$

$$B(\alpha, \beta) = B(\alpha+1, \beta) + B(\alpha, \beta+1), \tag{1.83}$$

$$B(\alpha, \beta+1) = \frac{\beta}{\alpha} B(\alpha+1, \beta), \tag{1.84}$$

and

$$B(\alpha, \beta) = B(\beta, \alpha). \tag{1.85}$$

The results were discovered by Whittaker and Watson (see [20], p.254).

Theorem 1.28 (Whittaker and Watson [20])
There is [20]

$$B(n\alpha, n\beta) = n^{-n\beta} \frac{\prod_{k=1}^{n} B\left(\beta + \frac{n-1}{n}, \alpha\right)}{\prod_{k=1}^{n} B((k-1)\alpha, \alpha)}, \tag{1.86}$$

where $Re(\alpha) > 0$, $Re(\beta) > 0$, $n \in \mathbb{N}$ *and* $\alpha, \beta \in \mathbb{C}\backslash\mathbb{Z}_0^-$.

Theorem 1.29 (Hankel [42])

$$\frac{1}{\Gamma(z)} = \frac{1}{2\pi i} \int_L s^{-z} e^s ds, s \in \mathbb{C}, \tag{1.87}$$

where L is the Hankel contour.

The result was discovered by Hankel in 1864 [42].

There is an alternative integral representation as follows:

Theorem 1.30 (Hankel [42])

$$\frac{1}{\Gamma(z)} = \frac{i}{2\pi} \int_{\aleph} (-s)^{-z} e^{-s} ds, s \in \mathbb{C}, \tag{1.88}$$

where $\aleph$ *is the loop contour, which starts at* $0i + \infty$ *encircles the origin and tends to* $0i - \infty$.

The result was discovered by Hankel in 1864 [42].

Definition 1.6

$$\psi(z) = \frac{d \log \Gamma(z)}{dz} = \frac{\Gamma^{(1)}(z)}{\Gamma(z)}, \tag{1.89}$$

where $s \in \mathbb{C}$ and

$$\log \Gamma(z) = \left(z - \frac{1}{2}\right) \log z - s + \log \sqrt{2\pi} + \Sigma(z) \tag{1.90}$$

with the infinite series due to Gudermann [43] $\Sigma(z)$.

The result is the logarithmic derivative of the gamma function [11].

For more information of the Euler gamma function, Pochhammer symbols, and Euler beta function, reader refers to Gronwall [11], Bell [13], Luke [14], Whittaker and Watson [20], Manocha and Srivastava [23], Wang et al. [35], Slater [36], Rainville [37], and Andrews et al. [26].

Theorem 1.31 *Let* $h^{(1)}(t) > 0, h(0) = a, h(\infty) = b, x, y \in \mathbb{C}$, $Re(x) > 0$ *and* $Re(y) > 0.$

Then we have

$$B(x, y) = \int_a^b e^{-h(t)} (h(t))^{x-1} (1 - h(t))^{y-1} h^{(1)}(t)\, dt\,. \tag{1.91}$$

The result was discovered by Yang et al. in 2020 when $x \in\in \mathbb{R}_+$ and $y \in \mathbb{R}_+$ [39].

1.1.4 The Extended Euler Gamma Functions

In this section, we preset the Euler–Chaudhry–Zubair gamma function, which is the extension of the gamma function, and other versions of the gamma-type functions.

Definition 1.7 (Chaudhry and Zubair [44])

The Euler–Chaudhry–Zubair gamma function, denoted by $_I\Gamma_p(z)$, is defined as

$$_I\Gamma_p(z) = \int_0^\infty t^{z-1} e^{-t-\frac{p}{t}} dt, \tag{1.92}$$

where $Re(z) > 0$, $Re(p) > 0$, and $z \in \mathbb{C}$.

The result was discovered by Chaudhry and Zubair in 1994 [44].

Definition 1.8 The Euler-type gamma function, denoted by ${}_{II}\Gamma_p(z)$, is defined as

$$ {}_{II}\Gamma_p(z) = \int_0^\infty t^{z-1} e^{-t-\frac{p}{t(1-t)}} dt, \tag{1.93}$$

where $Re(z) > 0$, $Re(p) > 0$, and $z \in \mathbb{C}$.

In this case, we get

$$\begin{aligned} & {}_{II}\Gamma_p(z) \\ &= \int_0^\infty t^{z-1} e^{-t-\frac{p}{t(1-t)}} dt \\ &= \int_0^\infty t^{z-1} e^{-t-\frac{p}{1-t}-\frac{p}{t}} dt, \end{aligned} \tag{1.94}$$

where $Re(z) > 0$, $Re(p) > 0$, and $z \in \mathbb{C}$.

Definition 1.9 The Euler-type gamma function, denoted by ${}_{III}\Gamma_p(z)$, is defined as

$$ {}_{III}\Gamma_p(z) = \int_0^\infty t^{z-1} e^{-t-\frac{p}{1-t}} dt, \tag{1.95}$$

where $Re(z) > 0$, $Re(p) > 0$, and $z \in \mathbb{C}$.

Definition 1.10 The Euler-type gamma function, denoted by ${}_{IV}\Gamma_{p,q}(z)$, is defined as

$$ {}_{IV}\Gamma_{p,q}(z) = \int_0^\infty t^{z-1} e^{-t-\frac{p}{t}-\frac{q}{1-t}} dt, \tag{1.96}$$

where $Re(z) > 0$, $Re(p) > 0$, $Re(q) > 0$, and $z \in \mathbb{C}$.

Theorem 1.32 *If $Re(z) > 0$, $Re(p) > 0$, $Re(q) > 0$, and $z \in \mathbb{C}$, then we have the following relationships:*

$$ {}_{IV}\Gamma_{p,0}(z) = {}_{I}\Gamma_p(z), \tag{1.97}$$

$$ {}_{IV}\Gamma_{p,p}(z) = {}_{II}\Gamma_p(z), \tag{1.98}$$

$$ {}_{IV}\Gamma_{0,p}(z) = {}_{III}\Gamma_p(z), \tag{1.99}$$

$$ {}_{IV}\Gamma_{0,0}(z) = {}_{I}\Gamma_0(z) = {}_{II}\Gamma_0(z) = {}_{III}\Gamma_0(z) = \Gamma(z). \tag{1.100}$$

Definition 1.11 The Euler-type gamma function, denoted by ${}_{V}\Gamma_p(z)$, is defined as

$$ {}_{V}\Gamma_p(z) = \int_0^\infty t^{z-1} e^{-pt} dt, \tag{1.101}$$

where $Re(z) > 0$, $Re(p) > 0$ and $z \in \mathbb{C}$.

Definition 1.12 The Euler-type gamma function, denoted by ${}_{VI}\Gamma_{p,q,r}(z)$, is defined as

$$ {}_{VI}\Gamma_{p,q,r}(z) = \int_0^\infty t^{z-1} e^{-rt-\frac{p}{t}-\frac{q}{1-t}} dt, \tag{1.102}$$

where $Re(z) > 0$, $Re(p) > 0$, $Re(q) > 0$, $Re(r) > 0$, and $z \in \mathbb{C}$.

Theorem 1.33 *If $Re(z) > 0$, $Re(p) > 0$, $Re(q) > 0$, and $z \in \mathbb{C}$, then we have the following relationships:*

$$ {}_{VI}\Gamma_{0,0,1}(z) = \Gamma(z), \tag{1.103}$$

$$ {}_{VI}\Gamma_{p,q,1}(z) = {}_{IV}\Gamma_{p,q}(z), \tag{1.104}$$

$$ {}_{VI}\Gamma_{0,0,p}(z) = {}_{V}\Gamma_p(z), \tag{1.105}$$

$$ {}_{VI}\Gamma_{0,p,1}(z) = {}_{III}\Gamma_p(z), \tag{1.106}$$

$$ {}_{VI}\Gamma_{p,p,1}(z) = {}_{II}\Gamma_p(z), \tag{1.107}$$

and

$$ {}_{VI}\Gamma_{p,0,1}(z) = {}_{I}\Gamma_p(z), \tag{1.108}$$

where $Re(z) > 0$, $Re(p) > 0$, $Re(q) > 0$, and $z \in \mathbb{C}$.

Definition 1.13 The Euler-type gamma function, denoted by ${}_{VII}\Gamma_{p,r}(z)$, is defined as

$$ {}_{VII}\Gamma_{p,r}(z) = \int_0^\infty t^{z-1} e^{-rt^p}\, dt, \tag{1.109} $$

where $Re(z) > 0$, $Re(p) > 0$, $Re(r) > 0$, and $r, z, p \in \mathbb{C}$.

Theorem 1.34 *If $Re(z) > 0$, $Re(p) > 0$, $Re(r) > 0$, and $r, z, p \in \mathbb{C}$, then we have*

$$ {}_{VII}\Gamma_{p,r}(z) = \frac{r^{-\frac{z}{q}}}{p} \Gamma\left(\frac{z}{p}\right). $$

Corollary 1.1 *If $Re(z) > 0$, $Re(r) > 0$, and $z, r \in \mathbb{C}$, then we have*

$$ {}_{VII}\Gamma_{1,r}(z) = r^{-z} \Gamma(z). \tag{1.110} $$

Corollary 1.2 *If $Re(z) > 0$, $Re(p) > 0$, and $z, p \in \mathbb{C}$, then we have*

$$ {}_{VII}\Gamma_{p,1}(z) = \frac{1}{p} \Gamma\left(\frac{z}{p}\right). \tag{1.111} $$

The results can be obtained from the analytic number theory in Chap. 2.

For more details, see [9, 45–47].

1.1.5 The Extended Euler Beta Functions

In this section, we preset the Euler–Chaudhry–Qadir–Rafique–Zubair beta function [48], which is the extension of the beta function, and other related functions.

Definition 1.14 The Euler-type beta function, denoted by ${}_I B_p(\alpha, \beta)$, is defined as

$$ {}_I B_p(\alpha, \beta) = \int_0^1 t^{\alpha-1} (1-t)^{\beta-1} e^{-\frac{p}{t}}\, dt, \tag{1.112} $$

where $Re(\beta) > Re(\alpha) > 0$, $Re(p) > 0$, and $\alpha, \beta \in \mathbb{C} \backslash \mathbb{Z}_0^-$.

Definition 1.15 (Chaudhry et al. [48])

The Euler–Chaudhry–Qadir–Rafique–Zubair beta function, denoted by ${}_{II}B_p(\alpha, \beta)$, is defined as [48]

$$ {}_{II}B_p(\alpha, \beta) = \int_0^1 t^{\alpha-1} (1-t)^{\beta-1} e^{-\frac{p}{t(1-t)}} dt, \tag{1.113}$$

where $Re(\beta) > Re(\alpha) > 0$, $Re(p) > 0$, and $\alpha, \beta \in \mathbb{C}\backslash\mathbb{Z}_0^-$.

The result was discovered by Chaudhry et al. in 1997 [48].

In this case, we take the form

$$\begin{aligned} &{}_{II}B_p(\alpha, \beta) \\ &= \int_0^1 t^{\alpha-1} (1-t)^{\beta-1} e^{-\frac{p}{t(1-t)}} dt \\ &= \int_0^1 t^{\alpha-1} (1-t)^{\beta-1} e^{-\frac{p}{1-t}-\frac{p}{t}} dt, \end{aligned} \tag{1.114}$$

where $Re(\beta) > Re(\alpha) > 0$, $Re(p) > 0$, and $\alpha, \beta \in \mathbb{C}\backslash\mathbb{Z}_0^-$.

Definition 1.16 The Euler-type beta function, denoted by ${}_{III}B_p(\alpha, \beta)$, is defined as

$$ {}_{III}B_p(\alpha, \beta) = \int_0^1 t^{\alpha-1} (1-t)^{\beta-1} e^{-\frac{p}{1-t}} dt, \tag{1.115}$$

where $Re(\beta) > Re(\alpha) > 0$, $Re(p) > 0$, and $\alpha, \beta \in \mathbb{C}\backslash\mathbb{Z}_0^-$.

Definition 1.17 The Euler–Choi–Rathie–Parmar beta function, denoted by ${}_{IV}B_{p,q}(\alpha, \beta)$, is defined as [49]

$$ {}_{IV}B_{p,q}(\alpha, \beta) = \int_0^1 t^{\alpha-1} (1-t)^{\beta-1} e^{-\frac{p}{t}-\frac{q}{1-t}} dt, \tag{1.116}$$

where $Re(\beta) > Re(\alpha) > 0$, $Re(p) > 0$, $Re(q) > 0$, and $\alpha, \beta \in \mathbb{C}\backslash\mathbb{Z}_0^-$.

The result was discovered by Choi et al. in 2014 [49].

Definition 1.18 The Euler-type beta function, denoted by ${}_{V}B_p(\alpha, \beta)$, is defined as

$$ {}_{V}B_p(\alpha, \beta) = \int_0^1 t^{\alpha-1} (1-t)^{\beta-1} e^{-pt} dt, \tag{1.117} $$

where $Re(\beta) > Re(\alpha) > 0$, $Re(p) > 0$, and $\alpha, \beta \in \mathbb{C}\backslash\mathbb{Z}_0^-$.

Definition 1.19 The Euler-type beta function, denoted by ${}_{VI}B_{p,q,r}(\alpha, \beta)$, is defined as

$$ {}_{VI}B_{p,q,r}(\alpha, \beta) = \int_0^1 t^{\alpha-1} (1-t)^{\beta-1} e^{-rt-\frac{p}{t}-\frac{q}{1-t}} dt, \tag{1.118} $$

where $Re(\beta) > Re(\alpha) > 0$, $Re(p) > 0$, $Re(q) > 0$, $Re(r) > 0$, and $\alpha, \beta \in \mathbb{C}\backslash\mathbb{Z}_0^-$.

Theorem 1.35 *Let $Re(\beta) > Re(\alpha) > 0$, $Re(p) > 0$, $Re(q) > 0$, $Re(r) > 0$ and $\alpha, \beta \in \mathbb{C}\backslash\mathbb{Z}_0^-$. Then we have the following:*

$$ {}_{VI}B_{0,0,p}(\alpha, \beta) = {}_{V}B_p(\alpha, \beta), \tag{1.119} $$

$$ {}_{VI}B_{p,q,0}(\alpha, \beta) = {}_{IV}B_{p,q}(\alpha, \beta), \tag{1.120} $$

$$ {}_{VI}B_{0,q,0}(\alpha, \beta) = {}_{III}B_p(\alpha, \beta), \tag{1.121} $$

$$ {}_{VI}B_{p,p,0}(\alpha, \beta) = {}_{II}B_p(\alpha, \beta), \tag{1.122} $$

$$ {}_{VI}B_{p,0,0}(\alpha, \beta) = {}_{I}B_p(\alpha, \beta). \tag{1.123} $$

Theorem 1.36 (Chaudhry et al. [50])

Suppose that $Re(\beta) > Re(\alpha) > 0$, $Re(\gamma) > 0$, $|\arg(1-z)| < \pi$, $Re(p) > 0$, and $\alpha, \beta \in \mathbb{C}\backslash\mathbb{Z}_0^-$, then we have the following integral representation [50]

$$ F_p(\gamma, \alpha, \beta; z) = \frac{1}{B(\alpha, \beta-\alpha)} \int_0^1 t^{\alpha-1} (1-t)^{\beta-\alpha-1} (1-zt)^{-\gamma} e^{-\frac{p}{t(1-t)}} dt. \tag{1.124} $$

The result was discovered by Chaudhry et al. in 2004 [50].

Theorem 1.37 (Chaudhry–Qadir–Srivastava–Paris Theorem [50])
Suppose that $Re(\beta) > Re(\alpha) > 0$, $Re(p) > 0$, $Re(z) > 0$, *and* $\alpha, \beta \in \mathbb{C}\backslash\mathbb{Z}_0^-$, *then we have the following integral representation [50]*

$$\Phi_p(\alpha, \beta; z) = \frac{1}{B(\alpha, \beta-\alpha)} \int_0^1 t^{\alpha-1}(1-t)^{\beta-\alpha-1} e^{\left(zt - \frac{p}{t(1-t)}\right)} dt. \tag{1.125}$$

The result was discovered by Chaudhry et al in 2004 [50]

As the direct results, we have the following:

Corollary 1.3 *Suppose that* $Re(\beta) > Re(\alpha) > 0$, $Re(p) > 0$, *and* $\alpha, \beta \in \mathbb{C}\backslash\mathbb{Z}_0^-$, *then we have the following integral representation*

$$\Phi_0(\alpha, \beta; z) = \frac{1}{B(\alpha, \beta-\alpha)} \int_0^1 t^{\alpha-1}(1-t)^{\beta-\alpha-1} e^{zt} dt. \tag{1.126}$$

Proof For the above formula with $p = 0$, we get the result. □

Corollary 1.4 *Suppose that* $Re(\beta) > Re(\alpha) > 0$, $Re(p) > 0$, *and* $\alpha, \beta \in \mathbb{C}\backslash\mathbb{Z}_0^-$, *then we have*

$$\Phi_0(\alpha, \beta; z) = \frac{{}_V B_p(\alpha,)}{B(\alpha, \beta-\alpha)}. \tag{1.127}$$

Proof With the definition, we directly obtain the result. □

Corollary 1.5 *Suppose that* $Re(\beta) > Re(\alpha) > 0$, $Re(p) > 0$, $Re(z) > 0$, *and* $\alpha, \beta \in \mathbb{C}\backslash\mathbb{Z}_0^-$, *then we have the following integral representation*

$$\Phi_p(\alpha, \beta; 0) = \frac{{}_{II} B_p(\alpha, \beta)}{B(\alpha, \beta-\alpha)}. \tag{1.128}$$

Proof For $z = 0$, we have the result. □

Theorem 1.38 (Choi et al. [49])
Suppose that $Re(\beta) > Re(\alpha) > 0$, $p > 0$, $q > 0$, *and* $\alpha, \beta \in \mathbb{C}\backslash\mathbb{Z}_0^-$, *then the integral representation*

$$\int_0^1 \int_0^1 {}_{IV} B_{p,q}(\alpha, \beta)\, p^{\kappa-1} q^{\upsilon-1} dp dq = \Gamma(\kappa)\Gamma(\upsilon) B(\alpha+\kappa, \beta+\upsilon) \tag{1.129}$$

holds true.

The result was discovered by Choi *et al* in 2014 [49].

There are the connections among them as follows:

Theorem 1.39 *If $Re(z) > 0$, $Re(p) > 0$, $Re(q) > 0$, and $z \in \mathbb{C}$, then we have the following relationships:*

$$ {}_{VI}B_{0,0,1}(z,0) = B(z,0), \tag{1.130} $$

$$ {}_{VI}B_{p,q,1}(z,0) = {}_{IV}B_{p,q}(z,0), \tag{1.131} $$

$$ {}_{VI}B_{0,0,p}(z,0) = {}_{V}B_{p}(z,0), \tag{1.132} $$

$$ {}_{VI}B_{0,p,1}(z,0) = {}_{III}B_{p}(z,0), \tag{1.133} $$

$$ {}_{VI}B_{p,p,1}(z,0) = {}_{II}B_{p}(z,0), \tag{1.134} $$

$$ {}_{VI}B_{p,0,1}(z,0) = {}_{I}B_{p}(z,0), \tag{1.135} $$

where $Re(z) > 0$, $Re(p) > 0$, $Re(q) > 0$, and $z \in \mathbb{C}$.

Definition 1.20 The Euler-type beta function, denoted by ${}_{VII}B_{p,r}(\alpha,\beta)$, is defined as

$$ {}_{VII}B_{p,r}(\alpha,\beta) = \int_0^1 t^{\alpha-1}(1-t)^{\beta-1} e^{-rt^p} dt, \tag{1.136} $$

where $Re(\beta) > Re(\alpha) > 0$, $Re(p) > 0$, $Re(q) > 0$, and $\alpha, \beta \in \mathbb{C}\backslash\mathbb{Z}_0^-$.

For the detailed comments of the above formulae, see [45–48].

1.2 Gauss Hypergeometric Series and Supertrigonometric and Superhyperbolic Functions

In this section, we present the definition, theorems, and properties for the Gauss hypergeometric series, and the relationships among the hypergeometric supertrigonometric and superhyperbolic functions via Gauss hypergeometric series.

1.2.1 The Gauss Hypergeometric Series

Let us start with the definition of the Gauss hypergeometric series.

Definition 1.21 (Gauss [27])
The Gauss hypergeometric series of form is defined as [27]

$$\begin{aligned} &{}_2F_1\left(a, b; c; z\right) \\ &= 1+\tfrac{ab}{c}z + \tfrac{a(a+1)b(b+1)}{c(c+1)}\tfrac{z^2}{2}+\cdots \\ &= \sum_{n=0}^{\infty} \tfrac{(a)_n (b)_n}{(c)_n}\tfrac{z^n}{n!}, \end{aligned} \tag{1.137}$$

where $a, b, c, z \in \mathbb{C}$, $n \in \mathbb{N}_0$, and $|z| < 1$.
The formula was discovered by Gauss in 1812 [27].

Theorem 1.40 (Euler [51])
If $a, b, c, z \in \mathbb{C}$ and $|z| < 1$, then there exists

$${}_2F_1\left(a, b; c; z\right) = \frac{\Gamma\left(c\right)}{\Gamma\left(b\right)\Gamma\left(c-b\right)}\int_0^1 t^{b-1}\left(1-t\right)^{c-b-1}\left(1-zt\right)^{-a}dt, \tag{1.138}$$

where $Re\left(c\right) > Re\left(b\right) > 0$.
The result was discovered by Euler in 1679 [51].

Theorem 1.41 (Gauss [27])
If $Re\left(c-a-b\right) > 0$, $Re\left(c\right) > Re\left(a\right)$, and $Re\left(c\right) > Re\left(b\right)$, then there exist

$${}_2F_1\left(a, b; c; 1\right) = \sum_{n=0}^{\infty}\frac{(a)_n\left(b\right)_n}{(c)_n}\frac{1}{n!} = \frac{\Gamma\left(c\right)\Gamma\left(c-b-a\right)}{\Gamma\left(c-a\right)\Gamma\left(c-b\right)}. \tag{1.139}$$

For $|z| < 1$ and $a, b, c, z \in \mathbb{C}$ the function ${}_2F_1\left(a, b; c; z\right)$ is analytic except for simple poles at $c = 0$ and $c \in \mathbb{Z}_-$.

The result was discovered by Gauss in 1812 [27].

Theorem 1.42 (Chu-Vandermonde Identity [52, 53])

$${}_2F_1\left(-n, b; c; 1\right) = \frac{(c-b)_n}{(c)_n}. \tag{1.140}$$

The result was discovered by Chu in 1303 [52] and by Vandermonde in 1772 [53], respectively.

Theorem 1.43 (Bateman [54])
If $Re\left(c\right) > Re\left(\mu\right) > 0$ and $|z| < 1$, then there is [54]

$${}_2F_1\left(a, b; c+\mu; z\right) = \frac{\Gamma\left(c+\mu\right)z^{1-(c+\mu)}}{\Gamma\left(c\right)\Gamma\left(\mu\right)}\int_0^z t^{c-1}\left(z-t\right)^{\mu-1}{}_2F_1\left(a, b; c, t\right)dt. \tag{1.141}$$

The result was discovered by Bateman in 1909 [54].

Theorem 1.44 (Bateman [54])
If $Re\,(c) > Re\,(\mu) > 0$ *and* $|z| < 1$, *then there is [54]*

$$ {}_2F_1\,(a, b; c+\mu; z) = \frac{\Gamma\,(c+\mu)}{\Gamma\,(c)\,\Gamma\,(\mu)} \int\limits_0^1 t^{c-1}\,(1-t)^{\mu-1}\;{}_2F_1\,(a, b; c, zt)\,dt. \tag{1.142} $$

This is the special case of Bateman [54].

Theorem 1.45 (Koshliakov [55])
If $Re\,(c) > Re\,(\mu) > 0$ *and* $|z| < 1$, *then there is*

$$ {}_1F_1\,(a, b; c+\mu; z) = \frac{\Gamma\,(c+\mu)}{\Gamma\,(c)\,\Gamma\,(\mu)} \int\limits_0^1 t^{c-1}\,(1-t)^{\mu-1}\;{}_1F_1\,(a, b; c, zt)\,dt. \tag{1.143} $$

This result is known as the Koshliakov's formula [55].

Property 1.1 The hypergeometric series is a solution of the Gauss differential equation of form

$$ z\,(1-z)\,\frac{d^2\varphi}{dz^2} + (c - (a+b+1)\,z)\,\frac{d\varphi}{dz} - ab\varphi = 0, \tag{1.144} $$

where $a, b, z \in \mathbb{C}$, $c \in \mathbb{C} \backslash \mathbb{Z}_0^-$, and $|z| < 1$.

The result was reported by Whittaker and Watson in 1927 [20], and by Bailey in 1935 [56].

Theorem 1.46 (Barnes [57])
Suppose that $Re\,(c) > Re\,(\mu) > 0$, $|z| < 1$, *and* $|\arg\,(-z)| < \pi$, *then we have [57]*

$$ {}_2F_1\,(a, b; c; z) = \frac{\Gamma\,(c)}{\Gamma\,(a)\,\Gamma\,(b)}\,\frac{1}{2\pi i} \int\limits_{-i\infty}^{i\infty} \frac{\Gamma\,(a+s)\,\Gamma\,(b+s)\,\Gamma\,(-s)}{\Gamma\,(c+s)}\,(-z)^s\,ds. \tag{1.145} $$

The result was reported by Barnes in 1908 [57].

Some applications of the ODEs in mechanics of shells were presented by Sun in [58–61].

1.2.2 *The Hypergeometric Supertrigonometric Functions via Gauss Superhyperbolic Series*

We now consider the hypergeometric supertrigonometric functions via Gauss superhyperbolic series [39].

Definition 1.22 The hypergeometric supersine via Gauss superhyperbolic series is defined as [39]

$$ {}_2Supersin_1\,(a, b; c; z) = \sum_{n=0}^{\infty} \frac{(a)_{2n+1}\,(b)_{2n+1}}{(c)_{2n+1}} \frac{(-1)^n\, z^{2n+1}}{(2n+1)!}, \tag{1.146} $$

where $a, b, c, z \in \mathbb{C}$.

Definition 1.23 The hypergeometric supercosine via Gauss superhyperbolic series is defined as [39]

$$ {}_2Supercos_1\,(a, b; c; z) = \sum_{n=0}^{\infty} \frac{(a)_{2n}\,(b)_{2n}}{(c)_{2n}} \frac{(-1)^n\, z^{2n}}{(2n)!}, \tag{1.147} $$

where $a, b, c, z \in \mathbb{C}$.

Definition 1.24 The hypergeometric supertangent via Gauss superhyperbolic series is defined as [39]

$$ {}_2Supertan_1\,(a, b; c; z) = \frac{{}_2Supersin_1\,(a, b; c; z)}{{}_2Supercos_1\,(a, b; c; z)}, \tag{1.148} $$

where $a, b, c, z \in \mathbb{C}$.

Definition 1.25 The hypergeometric supercotangent via Gauss superhyperbolic series is defined as [39]

$$ {}_2Supercot_1\,(a, b; c; z) = \frac{{}_2Supercos_1\,(a, b; c; z)}{{}_2Supersin_1\,(a, b; c; z)}, \tag{1.149} $$

where $a, b, c, z \in \mathbb{C}$.

Definition 1.26 The hypergeometric supersecant via Gauss superhyperbolic series is defined as [39]

$$ {}_2Supersec_1\,(a, b; c; z) = \frac{1}{{}_2Supercos_1\,(a, b; c; z)}, \tag{1.150} $$

where $a, b, c, z \in \mathbb{C}$.

Definition 1.27 The hypergeometric supercosecant via Gauss superhyperbolic series is defined as [39]

$$ {}_2Supercosec_1(a,b;c;z) = \frac{1}{{}_2Supersin_1(a,b;c;z)}, \tag{1.151} $$

where $a, b, c, z \in \mathbb{C}$.

The formulae for the hypergeometric supertrigonometric functions via Gauss superhyperbolic series were discovered by Yang in 2020 [39]. In this case, we present [39]

$$ \begin{aligned} &{}_2Supersin_1(a,b;c;\lambda z) \\ &= \frac{1}{2i}\left({}_2F_1(a,b;c;i\lambda z) - {}_2F_1(a,b;c;-i\lambda z)\right) \\ &= \sum_{n=0}^{\infty} \frac{(a)_{2n+1}(b)_{2n+1}}{(c)_{2n+1}} \frac{(-1)^n(\lambda z)^{2n+1}}{(2n+1)!}, \end{aligned} \tag{1.152} $$

$$ \begin{aligned} &{}_2Supercos_1(a,b;c;\lambda z) \\ &= \frac{1}{2}\left({}_2F_1(a,b;c;i\lambda z) + {}_2F_1(a,b;c;-i\lambda z)\right) \\ &= \sum_{n=0}^{\infty} \frac{(a)_{2n}(b)_{2n}}{(c)_{2n}} \frac{(-1)^n(\lambda z)^{2n}}{(2n)!}, \end{aligned} \tag{1.153} $$

$$ {}_2Supertan_1(a,b;c;\lambda z) = \frac{{}_2Supersin_1(a,b;c;\lambda z)}{{}_2Supercos_1(a,b;c;\lambda z)}, \tag{1.154} $$

$$ {}_2Supercot_1(a,b;c;\lambda z) = \frac{{}_2Supercos_1(a,b;c;\lambda z)}{{}_2Supersin_1(a,b;c;\lambda z)}, \tag{1.155} $$

$$ {}_2Supersec_1(a,b;c;\lambda z) = \frac{1}{{}_2Supercos_1(a,b;c;\lambda z)}, \tag{1.156} $$

and

$$ {}_2Supercosec_1(a,b;c;\lambda z) = \frac{1}{{}_2Supersin_1(a,b;c;\lambda z)}, \tag{1.157} $$

where $a, b, c, \lambda, z \in \mathbb{C}$.

Theorem 1.47 *Suppose that $a, b, c, z \in \mathbb{C}$, and $i = \sqrt{-1}$, then we have*

$$ {}_2F_1(a,b;c;iz) = {}_2Supercos_1(a,b;c;z) + i\,{}_2Supersin_1(a,b;c;z). \tag{1.158} $$

The result was discovered by Yang in 2020 [39].

More generally, we have the following formula:

Suppose that $a, b, c, \lambda, z \in \mathbb{C}$, and $i = \sqrt{-1}$, then we have [39]

$$ {}_2F_1(a,b;c;i\lambda z) = {}_2Supercos_1(a,b;c;\lambda z) + i\,{}_2Supersin_1(a,b;c;\lambda z). \tag{1.159} $$

1.2.3 The Hypergeometric Superhyperbolic Functions via Gauss Hypergeometric Series

We now investigate the hypergeometric superhyperbolic functions via Gauss hypergeometric series.

Definition 1.28 The hypergeometric superhyperbolic sine via Gauss hypergeometric series is defined as [39]

$$ {}_2Supersinh_1\,(a,b;c;z)=\sum_{n=0}^{\infty}\frac{(a)_{2n+1}\,(b)_{2n+1}}{(c)_{2n+1}}\frac{z^{2n+1}}{(2n+1)!}, \tag{1.160}$$

where $a, b, c, z \in \mathbb{C}$.

Definition 1.29 The hypergeometric superhyperbolic cosine via Gauss hypergeometric series is defined as [39]

$$ {}_2Supercosh_1\,(a,b;c;z)=\sum_{n=0}^{\infty}\frac{(a)_{2n}\,(b)_{2n}}{(c)_{2n}}\frac{z^{2n}}{(2n)!}, \tag{1.161}$$

where $a, b, c, z \in \mathbb{C}$.

Definition 1.30 The hypergeometric superhyperbolic tangent via Gauss hypergeometric series is defined as [39]

$$ {}_2Supertanh_1\,(a,b;c;z)=\frac{{}_2Supersinh_1\,(a,b;c;z)}{{}_2Supercosh_1\,(a,b;c;z)}, \tag{1.162}$$

where $a, b, c, z \in \mathbb{C}$.

Definition 1.31 The hypergeometric superhyperbolic cotangent via Gauss hypergeometric series is defined as [39]

$$ {}_2Supercoth_1\,(a,b;c;z)=\frac{{}_2Supercosh_1\,(a,b;c;z)}{{}_2Supersinh_1\,(a,b;c;z)}, \tag{1.163}$$

where $a, b, c, z \in \mathbb{C}$.

Definition 1.32 The hypergeometric superhyperbolic secant via Gauss hypergeometric series is defined as [39]

$$ {}_2Supersech_1\,(a,b;c;z)=\frac{1}{{}_2Supercosh_1\,(a,b;c;z)}, \tag{1.164}$$

where $a, b, c, z \in \mathbb{C}$.

Definition 1.33 The hypergeometric superhyperbolic cosecant via Gauss hypergeometric series is defined as [39]

$$ {}_2Supercosech_1\left(a,b;c;z\right)=\frac{1}{{}_2Supersinh_1\left(a,b;c;z\right)}, \tag{1.165}$$

where $a, b, c, z \in \mathbb{C}$.

The results were proposed by Yang in 2020 [39].

So, there exist [39]

$$\begin{aligned} &{}_2Supersinh_1\left(a,b;c;\lambda z\right)\\ &=\tfrac{1}{2}\left({}_2F_1\left(a,b;c;\lambda z\right)-{}_2F_1\left(a,b;c;-\lambda z\right)\right)\\ &=\sum_{n=0}^{\infty}\tfrac{(a)_{2n+1}(b)_{2n+1}}{(c)_{2n+1}}\tfrac{(\lambda z)^{2n+1}}{(2n+1)!}, \end{aligned}\tag{1.166}$$

$$\begin{aligned} &{}_2Supercosh_1\left(a,b;c;\lambda z\right)\\ &=\tfrac{1}{2}\left({}_2F_1\left(a,b;c;\lambda z\right)+{}_2F_1\left(a,b;c;-\lambda z\right)\right)\\ &=\sum_{n=0}^{\infty}\tfrac{(a)_{2n}(b)_{2n}}{(c)_{2n}}\tfrac{(\lambda z)^{2n}}{(2n)!}, \end{aligned}\tag{1.167}$$

$$ {}_2Supertanh_1\left(a,b;c;\lambda z\right)=\frac{{}_2Supersinh_1\left(a,b;c;\lambda z\right)}{{}_2Supercosh_1\left(a,b;c;\lambda z\right)}, \tag{1.168}$$

$$ {}_2Supercoth_1\left(a,b;c;\lambda z\right)=\frac{{}_2Supercosh_1\left(a,b;c;\lambda z\right)}{{}_2Supersinh_1\left(a,b;c;\lambda z\right)}, \tag{1.169}$$

$$ {}_2Supersech_1\left(a,b;c;\lambda z\right)=\frac{1}{{}_2Supercosh_1\left(a,b;c;\lambda z\right)}, \tag{1.170}$$

$$ {}_2Supercosech_1\left(a,b;c;\lambda z\right)=\frac{1}{{}_2Supersinh_1\left(a,b;c;\lambda z\right)}, \tag{1.171}$$

where $a, b, c, \lambda, z \in \mathbb{C}$.

Theorem 1.48 *Suppose that $a, b, c, z \in \mathbb{C}$, then we have [39]*

$$ {}_2F_1\left(a,b;c;z\right)={}_2Supercosh_1\left(a,b;c;z\right)+{}_2Supersinh_1\left(a,b;c;z\right). \tag{1.172}$$

So, we have [39]

$$ {}_2F_1\left(a,b;c;\lambda z\right)={}_2Supercosh_1\left(a,b;c;\lambda z\right)+{}_2Supersinh_1\left(a,b;c;\lambda z\right), \tag{1.173}$$

where $a, b, c, \lambda, z \in \mathbb{C}$.

Theorem 1.49 *Let $a, b, c \in \mathbb{C}$. Then we have [39]*

$$ {}_2Supersinh_1\,(a, b; c; 0) = 0. \tag{1.174}$$

Theorem 1.50 *Let $a, b, c \in \mathbb{C}$. Then we have [39]*

$$ {}_2Supercosh_1\,(a, b; c; 0) = 1. \tag{1.175}$$

Theorem 1.51 *Let $a, b, c \in \mathbb{C}$. Then we have [39]*

$$ {}_2Supertanh_1\,(a, b; c; 0) = 0. \tag{1.176}$$

Theorem 1.52 *Let $a, b, c \in \mathbb{C}$. Then we have [39]*

$$ {}_2Supersec_1\,(a, b; c; 0) = 1. \tag{1.177}$$

The results were discovered by Yang in 2020 [39].

Theorem 1.53 *Suppose that $Re\,(\alpha) > 0$, $Re\,(\beta) > 0$, $Re\,(a_1) > 0$, $Re\,(a_2) > 0$, $Re\,(c_1) > 0$, $s \in \mathbb{N}$, and $|z| < 1$, then we have [39]*

$$\begin{aligned}&\int_0^z t^{\alpha-1}\,(z-t)^{\beta-1}\,{}_2F_1\left(a_1, a_2; c_1; \lambda\,(z-t)^s\right) dt\\&= B\,(\alpha,\beta)\,z^{\alpha+\beta-1} \times {}_{2+s}F_{1+s}\left(\begin{array}{c} a_1, a_2, \frac{\beta}{s}, \frac{\beta+1}{s}, \ldots, \frac{\beta+s-1}{s}; \\ c_1, \frac{\alpha+\beta}{s}, \frac{\alpha+\beta+1}{s}, \ldots, \frac{\alpha+\beta+s-1}{s}; \end{array} \lambda z^s\right),\end{aligned} \tag{1.178}$$

where λ is a constant.

Theorem 1.54 *Suppose that $Re\,(\alpha) > 0$, $Re\,(\beta) > 0$, $Re\,(a_1) > 0$, $Re\,(c_1) > 0, Re\,(c_2) > 0$, $s \in \mathbb{N}$, and $|z| < 1$, then we have [39]*

$$\begin{aligned}&\int_0^z t^{\alpha-1}\,(z-t)^{\beta-1}\,{}_1F_2\left(a_1; c_1, c_2; \lambda\,(z-t)^s\right) dt\\&= B\,(\alpha,\beta)\,z^{\alpha+\beta-1}\,{}_{1+s}F_{2+s}\left(\begin{array}{c} a_1, \frac{\beta}{s}, \frac{\beta+1}{s}, \ldots, \frac{\beta+s-1}{s}; \\ c_1, c_2, \frac{\alpha+\beta}{s}, \frac{\alpha+\beta+1}{s}, \ldots, \frac{\alpha+\beta+s-1}{s}; \end{array} \lambda z^s\right),\end{aligned} \tag{1.179}$$

where λ is a constant.

Theorem 1.55 *Suppose that* $Re(\alpha) > 0$, $Re(\beta) > 0$, $Re(a_1) > 0$, $Re(a_2) > 0$, $Re(c_1) > 0$, *and* $|z| < 1$, *then we have [39]*

$$\begin{aligned} &\int_0^z t^{\alpha-1}(z-t)^{\beta-1}\,{}_2F_1(a_1, a_2; c_1; \lambda(z-t))\,dt \\ &= B(\alpha,\beta)\, z^{\alpha+\beta-1}\,{}_3F_2\left(\begin{matrix} a_1, a_2, \beta; \\ c_1, \alpha+\beta; \end{matrix}\ \lambda z\right), \end{aligned} \tag{1.180}$$

where λ *is a constant and* $\alpha, \beta, a_1, a_2, c_1 \in \mathbb{C}$.

Theorem 1.56 *Suppose that* $Re(\alpha) > 0$, $Re(\beta) > 0$, $Re(a_1) > 0$, $Re(a_2) > 0$, $Re(c_1) > 0$, *and* $s \in \mathbb{N}$, *then we have [39]*

$$\begin{aligned} &\int_{-1}^{1} (t+1)^{\alpha-1}(1-t)^{\beta-1}\,{}_2F_1\left(a_1, a_2; c_1; \lambda(1-t)^s\right) dt \\ &= B(\alpha,\beta)\, 2^{\alpha+\beta-1}\,{}_{2+s}F_{1+s}\left(\begin{matrix} a_1, a_2, \frac{\beta}{s}, \frac{\beta+1}{s}, \cdots, \frac{\beta+s-1}{s}; \\ c_1, \frac{\alpha+\beta}{s}, \frac{\alpha+\beta+1}{s}, \cdots, \frac{\alpha+\beta+s-1}{s}; \end{matrix}\ \lambda 2^s\right), \end{aligned} \tag{1.181}$$

where λ *is a constant and* $\alpha, \beta, a_1, a_2, c_1 \in \mathbb{C}$.

For more details of the applications of the above results, see [39, 62–67].

1.3 Clausen Hypergeometric Series and Supertrigonometric and Superhyperbolic Functions

In this section, we introduce the definition, theorems, and properties for the Clausen hypergeometric series.

1.3.1 The Clausen Hypergeometric Series

We now consider the definition and theorems for the Clausen hypergeometric series.

Definition 1.34 (Clausen [68])

The Clausen hypergeometric series of form is defined as [68]

$$\begin{aligned} &{}_pF_q((a),(c);z) \\ &= {}_pF_q\left(a_1, \cdots, a_p; c_1, \cdots, c_q; z\right) \\ &= {}_pF_q\left(\begin{matrix} a_1, \cdots, a_p \\ c_1, \cdots, c_q \end{matrix}; z\right) \\ &= \sum_{n=0}^{\infty} \frac{(a_1)_n \cdots (a_p)_n}{(c_1)_n \cdots (c_q)_n} \frac{z^n}{n!}, \end{aligned} \tag{1.182}$$

where $a_n, c_n, z \in \mathbb{C}$ and $n, p, q \in \mathbb{N}_0$.

The result, as an extended version of the Gauss hypergeometric series, was discovered by Clausen in 1828 [68].

Theorem 1.57 (Convergences for the Clausen Hypergeometric Series)
The cases of the convergences of the Clausen hypergeometric series (1.182) *hold for* $a_n \in \mathbb{C}\backslash\mathbb{Z}_0^-$*:*

(1) if $p < q$*, then the series converges absolutely for* $z \in \mathbb{C}$*;*
(2) if $p = q + 1$*, then the series converges absolutely for* $|z| < 1$ *and diverges for* $|z| > 1$*, and for* $|z| = 1$ *it converges absolutely for* $Re\left(\sum_{k=1}^{q} c_k - \sum_{k=1}^{p} a_k\right) > 0$*;*
(3) if $p > q + 1$*, then the series converges only for* $z = 0$*.*

The result was reported by Bailey [56], Srivastava and Kashyap [69], Slater [36], Andrews et al. [26], and Rainville [37].

Theorem 1.58 (The Differential Equation for the Clausen Hypergeometric Series)
If $a_n, c_n, z \in \mathbb{C}$, $n, p, q \in \mathbb{N}_0$, *and* $|z| < 1$, *the Clausen hypergeometric series is a solution of the differential equation [69]*

$$(Q(q, \mu, c_n)\varphi)(z) - (P(p, \mu, a_n)\varphi)(z) = 0, \tag{1.183}$$

where

$$\begin{aligned}
&(Q(q, \mu, c_n)\varphi)(z) \\
&= \left(z\tfrac{d}{dz}\right)\prod_{n=1}^{q}\left(\left(z\tfrac{d}{dz}\right)\varphi(z) + (c_n - 1)\varphi(z)\right) \\
&= z\tfrac{d}{dz}\left(\prod_{n=1}^{q}\left(\left(z\tfrac{d}{dz} + (c_n - 1)\right)\varphi\right)(z)\right)
\end{aligned} \tag{1.184}$$

and

$$\begin{aligned}
&(P(p, \mu, a_n)\varphi)(z) \\
&= z\prod_{n=1}^{p}\left(z\tfrac{d\varphi(z)}{dz} + a_n\varphi(z)\right) \\
&= z\prod_{n=1}^{p}\left(\left(z\tfrac{d}{dz} + a_n\right)\varphi\right)(z).
\end{aligned} \tag{1.185}$$

The result was reported by Srivastava and Kashyap [69], Rainville [37], Andrews et al. [26], and Luke [14].

Theorem 1.59 (The Derivative of the Clausen Hypergeometric Series)

If where $a_n, c_n, \lambda, z \in \mathbb{C}$, $n, p, q \in \mathbb{N}_0$, $Re\left(\sum_{k=1}^{q} c_k - \sum_{k=1}^{p} a_k\right) > 0$, *and* $|z| < 1$, *then there is the derivative of the Clausen hypergeometric series as follows [69]:*

$$\frac{d}{dz}\left({}_pF_q\left(a_1, \cdots, a_p; c_1, \cdots, c_q; z\right)\right) = \frac{\prod_{n=1}^{p} a_n}{\prod_{n=1}^{q} c_n}\, {}_pF_q\left((a_1+1), \cdots, (a_p+1); (c_1+1), \cdots, (c_q+1); z\right) \tag{1.186}$$

and

$$\frac{1}{\lambda}\frac{d}{dz}\left({}_pF_q\left(a_1, \cdots, a_p; c_1, \cdots, c_q; \lambda z\right)\right) = \frac{\prod_{n=1}^{p} a_n}{\prod_{n=1}^{q} c_n}\, {}_pF_q\left((a_1+1), \cdots, (a_p+1); (c_1+1), \cdots, (c_q+1); \lambda z\right). \tag{1.187}$$

The result was discussed by Srivastava and Kashyap [69], Rainville [37], Andrews et al. [26], and Luke [14].

Theorem 1.60 (Rainville [37])

If $p \leq q+1$, $Re(a_1) > 0, \cdots, Re(a_p) > 0$, $Re(c_1) > 0, \cdots, Re(c_q) > 0$, $Re\left(\sum_{k=1}^{q} c_k - \sum_{k=1}^{p} a_k\right) > 0$*and* $|z| < 1$, *then there is [37]*

$$ {}_pF_q\left(\begin{matrix} a_1, \cdots, a_p; \\ c_1, \cdots, c_q; \end{matrix}\ \lambda z\right) = \frac{\Gamma(c_1)}{\Gamma(a_1)\Gamma(c_1-a_1)} \int_0^1 t^{a_1-1}(1-t)^{c_1-a_1-1}\ {}_{p-1}F_{q-1}\left(a_2, \cdots, a_p; c_2, \cdots, c_q; \lambda z t\right) dt, \tag{1.188}$$

where λ *is a constant.*

The result was discovered by Rainville in 1960 (see [37], p.85).

Theorem 1.61 (Askey [70], p.19)

If $p \leq q+1$, $Re(a_1) > 0, \cdots, Re(a_p) > 0$, $Re(c_1) > 0, \cdots, Re(c_q) > 0$, $Re\left(\sum_{k=1}^{q} c_k - \sum_{k=1}^{p} a_k\right) > 0$*and* $|z| < 1$, *then there is [70]*

$$ {}_{p+1}F_{q+1}\left(\begin{matrix} a_1, \cdots, a_p, a; \\ c_1, \cdots, c_q, c; \end{matrix}\ \lambda z\right) = \frac{\Gamma(c)}{\Gamma(a)\Gamma(c-a)} \int_0^1 t^{a-1}(1-t)^{c-a-1}\ {}_pF_q\left(a_1, a_2, \cdots, a_p; c_1, c_2, \cdots, c_q; \lambda z t\right) dt, \tag{1.189}$$

where λ *is a constant.*

The result was discovered by Askey [70].

They were further investigated by Andrews et al. (see [26], p.67), Srivastava and Kashyap [69], and Slater [36].

Theorem 1.62 (Rainville [37], p.104)
Suppose that $Re(\alpha) > 0$, $Re(\beta) > 0$, $Re(a_1) > 0, \cdots$, $Re(a_p) > 0$, $Re(c_1) > 0, \cdots$, $Re(c_q) > 0$, $|z| < 1$, $Re\left(\sum_{k=1}^{q} c_k - \sum_{k=1}^{p} a_k\right) > 0$, *and* $\kappa \in \mathbb{N}$, *then there is [37]*

$$\begin{aligned}
&\frac{1}{B(\alpha,\beta)} \int_0^z t^{\alpha-1} (z-t)^{\beta-1} {}_pF_q\left(a_1, \cdots, a_p; c_1, \cdots, c_q; \lambda t^\kappa\right) dt \\
&= z^{\alpha+\beta-1} \times {}_{p+\kappa}F_{q+\kappa}\left(\begin{matrix} a_1, \cdots, a_p, \frac{\alpha}{\kappa}, \frac{\alpha+1}{\kappa}, \cdots, \frac{\alpha+\kappa-1}{\kappa}; \\ c_1, \cdots, c_q, \frac{\alpha+\beta}{\kappa}, \frac{\alpha+\beta+1}{\kappa}, \cdots, \frac{\alpha+\beta+\kappa-1}{\kappa}; \end{matrix} \lambda z^\kappa\right),
\end{aligned} \tag{1.190}$$

where λ *is a constant.*

The result was discovered by Rainville in 1960 (see [37], p.85).

Theorem 1.63 (Rainville [37])
Suppose that $Re(\alpha) > 0$, $Re(\beta) > 0$, $Re(a_1) > 0, \cdots$, $Re(a_p) > 0$, $Re(c_1) > 0, \cdots$, $Re(c_q) > 0$, $p, q \in \mathbb{N}_0$, $\kappa, s \in \mathbb{N}$, *and* $|z| < 1$, *then there is [37]*

$$\begin{aligned}
&\frac{1}{B(\alpha,\beta)} \int_0^z t^{\alpha-1} (z-t)^{\beta-1} {}_pF_q\left(a_1, \cdots, a_p; c_1, \cdots, c_q; \lambda t^\kappa (z-t)^s\right) dt \\
&= z^{\alpha+\beta-1} \\
&\times {}_{p+\kappa+s}F_{q+\kappa+s}\left(\begin{matrix} a_1, \cdots, a_p, \frac{\alpha}{\kappa}, \frac{\alpha+1}{\kappa}, \cdots, \frac{\alpha+\kappa-1}{\kappa}, \frac{\beta}{s}, \frac{\beta+1}{s}, \cdots, \frac{\beta+s-1}{s}; \\ c_1, \cdots, c_q, \frac{\alpha+\beta}{\kappa+s}, \frac{\alpha+\beta+1}{\kappa+s}, \cdots, \frac{\alpha+\beta+\kappa+s-1}{\kappa+s}; \end{matrix} \frac{\kappa^\kappa s^s \lambda z^{\kappa+s}}{(\kappa+s)^{\kappa+s}}\right),
\end{aligned} \tag{1.191}$$

where λ *is a constant.*

The result was discovered by Rainville in 1960 (see [37], p.104).

Theorem 1.64 *Suppose that* $Re(\alpha) > 0$, $Re(\beta) > 0$, $Re(a_1) > 0, \cdots$, $Re(a_p) > 0$, $Re(c_1) > 0, \cdots$, $Re(c_q) > 0$, $p, q \in \mathbb{N}_0$, $s \in \mathbb{N}$, *and* $|z| < 1$, *then there is [39]*

$$\begin{aligned}
&\frac{1}{B(\alpha,\beta)} \int_0^z t^{\alpha-1} (z-t)^{\beta-1} {}_pF_q\left(a_1, \cdots, a_p; c_1, \cdots, c_q; \lambda (z-t)^s\right) dt \\
&= z^{\alpha+\beta-1} \\
&\times {}_{p+s}F_{q+s}\left(\begin{matrix} a_1, \cdots, a_p, \frac{\beta}{s}, \frac{\beta+1}{s}, \cdots, \frac{\beta+s-1}{s}; \\ c_1, \cdots, c_q, \frac{\alpha+\beta}{s}, \frac{\alpha+\beta+1}{s}, \cdots, \frac{\alpha+\beta+s-1}{s}; \end{matrix} \lambda z^s\right),
\end{aligned} \tag{1.192}$$

where λ *is a constant.*

The result was discovered in [39].

Theorem 1.65 *Suppose that* $Re\,(\alpha) > 0$, $Re\,(a_1) > 0, \cdots, Re\,(a_p) > 0$, $Re\,(c_1) > 0, \cdots, Re\,(c_q) > 0$, $|z| < 1$, $Re\left(\sum_{k=1}^{q} c_k - \sum_{k=1}^{p} a_k\right) > 0$, *and* $\kappa \in \mathbb{N}$, *then there is [39]*

$$\begin{aligned}&\frac{1}{B(\alpha,1)}\int_0^z t^{\alpha-1}\,{}_pF_q\left(a_1,\cdots,a_p;c_1,\cdots,c_q;\lambda t^{\kappa}\right)dt\\&=z^{\alpha}\,{}_{p+\kappa}F_{q+\kappa}\left(\begin{matrix}a_1,\cdots,a_p,\frac{\alpha}{\kappa},\frac{\alpha+1}{\kappa},\cdots,\frac{\alpha+\kappa-1}{\kappa};\\c_1,\cdots,c_q,\frac{\alpha+1}{\kappa},\frac{\alpha+2}{\kappa},\cdots,\frac{\alpha+\kappa}{\kappa};\end{matrix}\lambda z^{\kappa}\right),\end{aligned}\tag{1.193}$$

where λ *is a constant.*

The result was discovered in [39].

Theorem 1.66 *Suppose that* $Re\,(\alpha) > 0$, $Re\,(a_1) > 0, \cdots, Re\,(a_p) > 0$, $Re\,(c_1) > 0, \cdots, Re\,(c_q) > 0$, $p, q \in \mathbb{N}_0$, $\kappa, s \in \mathbb{N}$, *and* $|z| < 1$, *then there is [39]*

$$\begin{aligned}&\frac{1}{B(\alpha,1)}\int_0^z t^{\alpha-1}\,{}_pF_q\left(a_1,\cdots,a_p;c_1,\cdots,c_q;\lambda t^{\kappa}(z-t)^s\right)dt\\&=z^{\alpha}\,{}_{p+\kappa+s}F_{q+\kappa+s}\left(\begin{matrix}a_1,\cdots,a_p,\frac{\alpha}{\kappa},\frac{\alpha+1}{\kappa},\cdots,\frac{\alpha+\kappa-1}{\kappa},\frac{1}{s},\frac{2}{s},\cdots,1;\\c_1,\cdots,c_q,\frac{\alpha+1}{\kappa+s},\frac{\alpha+2}{\kappa+s},\cdots,\frac{\alpha+\kappa+s}{\kappa+s};\end{matrix}\frac{\kappa^{\kappa}s^{s}\lambda z^{\kappa+s}}{(\kappa+s)^{\kappa+s}}\right),\end{aligned}\tag{1.194}$$

where λ *is a constant.*

The result was discovered in [39].

Theorem 1.67 *Suppose that* $Re\,(\alpha) > 0$, $Re\,(a_1) > 0, \cdots, Re\,(a_p) > 0$, $Re\,(c_1) > 0, \cdots, Re\,(c_q) > 0$, $p, q \in \mathbb{N}_0$, $s \in \mathbb{N}$, *and* $|z| < 1$, *then there is [39]*

$$\begin{aligned}&\frac{1}{B(\alpha,1)}\int_0^z t^{\alpha-1}\,{}_pF_q\left(a_1,\cdots,a_p;c_1,\cdots,c_q;\lambda(z-t)^s\right)dt\\&=z^{\alpha}\,{}_{p+s}F_{q+s}\left(\begin{matrix}a_1,\cdots,a_p,\frac{1}{s},\frac{2}{s},\cdots,1;\\c_1,\cdots,c_q,\frac{\alpha+1}{s},\frac{\alpha+2}{s},\cdots,\frac{\alpha+s}{s};\end{matrix}\lambda z^{s}\right),\end{aligned}\tag{1.195}$$

where λ *is a constant.*

The result was discovered in [39].

Corollary 1.6 *Suppose that* $Re(\alpha) > 0$, $Re(\beta) > 0$, $Re(a_1) > 0, \cdots, Re(a_p) > 0$, $Re(c_1) > 0, \cdots, Re(c_q) > 0$, $p, q \in \mathbb{N}_0$, *and* $\kappa, s \in \mathbb{N}$, *then there is [39]*

$$\begin{aligned}&\frac{1}{B(\alpha,\beta)}\int_0^1 t^{\alpha-1}(1-t)^{\beta-1}\,{}_pF_q\left(a_1,\cdots,a_p;c_1,\cdots,c_q;\lambda t^{\kappa}(1-t)^{s}\right)dt\\&={}_{p+\kappa+s}F_{q+\kappa+s}\left(\begin{matrix}a_1,\cdots,a_p,\frac{\alpha}{\kappa},\frac{\alpha+1}{\kappa},\cdots,\frac{\alpha+\kappa-1}{\kappa},\frac{\beta}{s},\frac{\beta+1}{s},\cdots,\frac{\beta+s-1}{s};\\c_1,\cdots,c_q,\frac{\alpha+\beta}{\kappa+s},\frac{\alpha+\beta+1}{\kappa+s},\cdots,\frac{\alpha+\beta+\kappa+s-1}{\kappa+s};\end{matrix}\frac{\kappa^{\kappa}s^{s}\lambda}{(\kappa+s)^{\kappa+s}}\right),\end{aligned} \tag{1.196}$$

where λ *is a constant.*

The result was discovered in [39].

Corollary 1.7 *Suppose that* $Re(\alpha) > 0$, $Re(\beta) > 0$, $Re(a_1) > 0, \cdots, Re(a_p) > 0$, $Re(c_1) > 0, \cdots, Re(c_q) > 0$, $Re\left(\sum_{k=1}^{q} c_k - \sum_{k=1}^{p} a_k\right) > 0$, *and* $\kappa \in \mathbb{N}$, *then there is [39]*

$$\begin{aligned}&\frac{1}{B(\alpha,\beta)}\int_0^1 t^{\alpha-1}(1-t)^{\beta-1}\,{}_pF_q\left(a_1,\cdots,a_p;c_1,\cdots,c_q;\lambda t^{\kappa}\right)dt\\&={}_{p+\kappa}F_{q+\kappa}\left(\begin{matrix}a_1,\cdots,a_p,\frac{\alpha}{\kappa},\frac{\alpha+1}{\kappa},\cdots,\frac{\alpha+\kappa-1}{\kappa};\\c_1,\cdots,c_q,\frac{\alpha+\beta}{\kappa},\frac{\alpha+\beta+1}{\kappa},\cdots,\frac{\alpha+\beta+\kappa-1}{\kappa};\end{matrix}\lambda\right),\end{aligned} \tag{1.197}$$

where λ *is a constant.*

The result was discovered in [39].

Corollary 1.8 *Suppose that* $Re(\alpha) > 0$, $Re(\beta) > 0$, $Re(a_1) > 0, \cdots, Re(a_p) > 0$, $Re(c_1) > 0, \cdots, Re(c_q) > 0$, $p, q \in \mathbb{N}_0$, *and* $s \in \mathbb{N}$, *then there is [39]*

$$\begin{aligned}&\frac{1}{B(\alpha,\beta)}\int_0^1 t^{\alpha-1}(1-t)^{\beta-1}\,{}_pF_q\left(a_1,\cdots,a_p;c_1,\cdots,c_q;\lambda(1-t)^{s}\right)dt\\&={}_{p+s}F_{q+s}\left(\begin{matrix}a_1,\cdots,a_p,\frac{\beta}{s},\frac{\beta+1}{s},\cdots,\frac{\beta+s-1}{s};\\c_1,\cdots,c_q,\frac{\alpha+\beta}{s},\frac{\alpha+\beta+1}{s},\cdots,\frac{\alpha+\beta+s-1}{s};\end{matrix}\lambda\right),\end{aligned} \tag{1.198}$$

where λ *is a constant.*

The result was discovered in [39].

Theorem 1.68 (Manocha and Srivastava [23])
Suppose that $Re(\alpha) > 0$, $Re(\beta) > 0$, $Re(a_1) > 0, \cdots, Re(a_p) > 0$, $Re(c_1) > 0, \cdots, Re(c_q) > 0$, *and* $Re\left(\sum_{k=1}^{q} c_k - \sum_{k=1}^{p} a_k\right) > 0$, *then there is [23]*

$$\begin{aligned}&\frac{1}{B(\alpha,\beta)}\int_0^1 t^{\alpha-1}(1-t)^{\beta-1}\,{}_pF_q\left(a_1,\cdots,a_p;c_1,\cdots,c_q;\lambda t\right)dt\\&={}_{p+1}F_{q+1}\left(\begin{matrix}a_1,\cdots,a_p,\alpha;\\c_1,\cdots,c_q,\alpha+\beta;\end{matrix}\lambda\right),\end{aligned}\tag{1.199}$$

where λ *is a constant.*

Theorem 1.69 *Suppose that* $Re(\alpha) > 0$, $Re(\beta) > 0$, $\kappa, s \in \mathbb{N}$, *and* $|z| < 1$, *then there is [39]*

$$\begin{aligned}&\frac{1}{B(\alpha,\beta)}\int_0^z t^{\alpha-1}(z-t)^{\beta-1}e^{-\lambda t^\kappa(z-t)^s}dt = z^{\alpha+\beta-1}\\&\times{}_{\kappa+s}F_{\kappa+s}\left(\begin{matrix}\frac{\alpha}{\kappa},\frac{\alpha+1}{\kappa},\cdots,\frac{\alpha+\kappa-1}{\kappa},\frac{\beta}{s},\frac{\beta+1}{s},\cdots,\frac{\beta+s-1}{s};\\\frac{\alpha+\beta}{\kappa+s},\frac{\alpha+\beta+1}{\kappa+s},\cdots,\frac{\alpha+\beta+\kappa+s-1}{\kappa+s};\end{matrix}-\frac{\kappa^\kappa s^s\lambda z^{\kappa+s}}{(\kappa+s)^{\kappa+s}}\right),\end{aligned}\tag{1.200}$$

where λ *is a constant.*

Theorem 1.70 (Rainville [37])
Suppose that $Re(\alpha) > 0$, $Re(\beta) > 0$, $|z| < 1$, $Re(\beta) > Re(\alpha) > 0$, *and* $\kappa \in \mathbb{N}_0$, *then we have [37]*

$$\begin{aligned}&\frac{1}{B(\alpha,\beta)}\int_0^z t^{\alpha-1}(z-t)^{\beta-1}e^{-\lambda t^\kappa}dt\\&=z^{\alpha+\beta-1}\,{}_\kappa F_\kappa\left(\begin{matrix}\frac{\alpha}{\kappa},\frac{\alpha+1}{\kappa},\cdots,\frac{\alpha+\kappa-1}{\kappa};\\\frac{\alpha+\beta}{\kappa},\frac{\alpha+\beta+1}{\kappa},\cdots,\frac{\alpha+\beta+\kappa-1}{\kappa};\end{matrix}-\lambda z^\kappa\right),\end{aligned}\tag{1.201}$$

where λ *is a constant.*

Theorem 1.71 *If* $Re(\alpha) > 0$, $Re(\beta) > 0$, $s \in \mathbb{N}$, *and* $|z| < 1$, *then we have [39]*

$$\begin{aligned}&\frac{1}{B(\alpha,\beta)}\int_0^z t^{\alpha-1}(z-t)^{\beta-1}e^{-\lambda(z-t)^s}dt = z^{\alpha+\beta-1}\\&\times{}_sF_s\left(\begin{matrix}\frac{\beta}{s},\frac{\beta+1}{s},\cdots,\frac{\beta+s-1}{s};\\\frac{\alpha+\beta}{s},\frac{\alpha+\beta+1}{s},\cdots,\frac{\alpha+\beta+s-1}{s};\end{matrix}-\lambda z^s\right),\end{aligned}\tag{1.202}$$

where λ *is a constant.*

The result was discovered in [39].

Based on the idea of separation of a power series into its even and odd terms discovered by Euler in 1748 [71], we derive new functions via Clausen hypergeometric series [39].

1.3.2 *The Hypergeometric Supertrigonometric Functions via Clausen Superhyperbolic Series*

We now present the hypergeometric supertrigonometric functions via Clausen superhyperbolic series [39].

Definition 1.35 The hypergeometric supersine via Clausen hypergeometric series is defined as [39]

$$ {}_pSupersin_q\left(a_1,\cdots,a_p;c_1,\cdots,c_q;z\right)=\sum_{n=0}^{\infty}\frac{(a_1)_{2n+1}\cdots\left(a_p\right)_{2n+1}}{(c_1)_{2n+1}\cdots\left(c_q\right)_{2n+1}}\frac{(-1)^n z^{2n+1}}{(2n+1)!}, \tag{1.203}$$

where $a_n, c_n \in \mathbb{C}$ and $n, p, q \in \mathbb{N}_0$.

Definition 1.36 The hypergeometric supercosine via Clausen hypergeometric series is defined as [39]

$$ {}_pSupercos_q\left(a_1,\cdots,a_p;c_1,\cdots,c_q;z\right)=\sum_{n=0}^{\infty}\frac{(a_1)_{2n}\cdots\left(a_p\right)_{2n}}{(c_1)_{2n}\cdots\left(c_q\right)_{2n}}\frac{(-1)^n z^{2n}}{(2n)!}, \tag{1.204}$$

where $a_n, c_n \in \mathbb{C}$ and $n, p, q \in \mathbb{N}_0$.

Definition 1.37 The hypergeometric supertangent via Clausen hypergeometric series is defined as [39]

$$ {}_pSupertan_q\left(a_1,\cdots,a_p;c_1,\cdots,c_q;z\right)=\frac{{}_pSupersin_q\left(a_1,\cdots,a_p;c_1,\cdots,c_q;z\right)}{{}_pSupercos_q\left(a_1,\cdots,a_p;c_1,\cdots,c_q;z\right)}, \tag{1.205}$$

where $a_n, c_n \in \mathbb{C}$ and $p, q \in \mathbb{N}_0$.

Definition 1.38 The hypergeometric supercotangent via Clausen hypergeometric series is defined as [39]

$$ {}_pSupercot_q\left(a_1,\cdots,a_p;c_1,\cdots,c_q;z\right)=\frac{{}_pSupercos_q\left(a_1,\cdots,a_p;c_1,\cdots,c_q;z\right)}{{}_pSupersin_q\left(a_1,\cdots,a_p;c_1,\cdots,c_q;z\right)}, \tag{1.206}$$

where $a_n, c_n \in \mathbb{C}$ and $p, q \in \mathbb{N}_0$.

Definition 1.39 The hypergeometric supersecant via Clausen hypergeometric series is defined as [39]

$$ {}_pSupersec_q\left(a_1,\cdots,a_p;c_1,\cdots,c_q;z\right)=\frac{1}{{}_pSupercos_q\left(a_1,\cdots,a_p;c_1,\cdots,c_q;z\right)}, \tag{1.207} $$

where $a_n, c_n \in \mathbb{C}$ and $p, q \in \mathbb{N}_0$.

Definition 1.40 The hypergeometric supercosecant via Clausen hypergeometric series is defined as [39]

$$ {}_pSupercosec_q\left(a_1,\cdots,a_p;c_1,\cdots,c_q;z\right)=\frac{1}{{}_pSupersin_q\left(a_1,\cdots,a_p;c_1,\cdots,c_q;z\right)}, \tag{1.208} $$

where $a_n, c_n \in \mathbb{C}$ and $p, q \in \mathbb{N}_0$.

1.3.3 The Hypergeometric Superhyperbolic Functions via Clausen Superhyperbolic Series

We now investigate the hypergeometric superhyperbolic functions via Clausen superhyperbolic series [39].

Definition 1.41 The hypergeometric superhyperbolic sine via Clausen hypergeometric series is defined as [39]

$$ {}_pSupersinh_q\left(a_1,\cdots,a_p;c_1,\cdots,c_q;z\right)=\sum_{n=0}^{\infty}\frac{(a_1)_{2n+1}\cdots\left(a_p\right)_{2n+1}}{(c_1)_{2n+1}\cdots\left(c_q\right)_{2n+1}}\frac{z^{2n+1}}{(2n+1)!}, \tag{1.209} $$

where $a_n, c_n \in \mathbb{C}$ and $n, p, q \in \mathbb{N}_0$.

Definition 1.42 The hypergeometric superhyperbolic cosine via Clausen hypergeometric series is defined as [39]

$$ {}_pSupercosh_q\left(a_1,\cdots,a_p;c_1,\cdots,c_q;z\right)=\sum_{n=0}^{\infty}\frac{(a_1)_{2n}\cdots\left(a_p\right)_{2n}}{(c_1)_{2n}\cdots\left(c_q\right)_{2n}}\frac{z^{2n}}{(2n)!}, \tag{1.210} $$

where $a_n, c_n \in \mathbb{C}$ and $n, p, q \in \mathbb{N}_0$.

Definition 1.43 The hypergeometric superhyperbolic tangent via Clausen hypergeometric series is defined as [39]

$$\begin{aligned} &{}_pSupertanh_q\left(a_1,\cdots,a_p;c_1,\cdots,c_q;z\right) \\ &= \frac{{}_pSupersinh_q(a_1,\cdots,a_p;c_1,\cdots,c_q;z)}{{}_pSupercosh_q(a_1,\cdots,a_p;c_1,\cdots,c_q;z)}, \end{aligned} \tag{1.211}$$

where $a_n, c_n \in \mathbb{C}$ and $n, p, q \in \mathbb{N}_0$.

Definition 1.44 The hypergeometric superhyperbolic cotangent via Clausen hypergeometric series is defined as [39]

$$\begin{aligned} &{}_pSupercoth_q\left(a_1,\cdots,a_p;c_1,\cdots,c_q;z\right) \\ &= \frac{{}_pSupercosh_q(a_1,\cdots,a_p;c_1,\cdots,c_q;z)}{{}_pSupersinh_q(a_1,\cdots,a_p;c_1,\cdots,c_q;z)}, \end{aligned} \tag{1.212}$$

where $a_n, c_n \in \mathbb{C}$ and $n, p, q \in \mathbb{N}_0$.

Definition 1.45 The hypergeometric superhyperbolic secant via Clausen hypergeometric series is defined as [39]

$$\begin{aligned} &{}_pSupersech_q\left(a_1,\cdots,a_p;c_1,\cdots,c_q;z\right) \\ &= \frac{1}{{}_pSupercosh_q(a_1,\cdots,a_p;c_1,\cdots,c_q;z)}, \end{aligned} \tag{1.213}$$

where $a_n, c_n \in \mathbb{C}$ and $n, p, q \in \mathbb{N}_0$.

Definition 1.46 The hypergeometric superhyperbolic cosecant via Clausen hypergeometric series is defined as [39]

$$\begin{aligned} &{}_qSupercosech_q\left(a_1,\cdots,a_p;c_1,\cdots,c_q;z\right) \\ &= \frac{1}{{}_qSupersinh_q(a_1,\cdots,a_p;c_1,\cdots,c_q;z)}, \end{aligned} \tag{1.214}$$

where $a_n, c_n \in \mathbb{C}$ and $n, p, q \in \mathbb{N}_0$.

Theorem 1.72 (Yang [39])
Suppose that $a_n, c_n \in \mathbb{C}$ and $n, p, q \in \mathbb{N}_0$, then we have [39]

$$\begin{aligned} &{}_pSupersinh_q\left(a_1,\cdots,a_p;c_1,\cdots,c_q;z\right) \\ &= -i\,{}_pSupersin_q\left(a_1,\cdots,a_p;c_1,\cdots,c_q;iz\right), \end{aligned} \tag{1.215}$$

$$\begin{aligned} &{}_pSupercosh_q\left(a_1,\cdots,a_p;c_1,\cdots,c_q;-z\right) \\ &= {}_pSupercos_q\left(a_1,\cdots,a_p;c_1,\cdots,c_q;iz\right), \end{aligned} \tag{1.216}$$

$$\begin{aligned} &{}_pSupertanh_q\left(a_1,\cdots,a_p;c_1,\cdots,c_q;-z\right) \\ &= -i\,{}_pSupertan_q\left(a_1,\cdots,a_p;c_1,\cdots,c_q;iz\right), \end{aligned} \tag{1.217}$$

$$
\begin{aligned}
&{}_pSupercoth_q\left(a_1,\cdots,a_p;c_1,\cdots,c_q;-z\right)\\
&\quad= i\,{}_pSupercot_q\left(a_1,\cdots,a_p;c_1,\cdots,c_q;iz\right),
\end{aligned}
\tag{1.218}
$$

$$
\begin{aligned}
&{}_pSupersech_q\left(a_1,\cdots,a_p;c_1,\cdots,c_q;-z\right)\\
&\quad= {}_pSupersec_q\left(a_1,\cdots,a_p;c_1,\cdots,c_q;iz\right),
\end{aligned}
\tag{1.219}
$$

and

$$
\begin{aligned}
&{}_pSupercosech_q\left(a_1,\cdots,a_p;c_1,\cdots,c_q;-z\right)\\
&\quad= i\,{}_pSupercosec_q\left(a_1,\cdots,a_p;c_1,\cdots,c_q;iz\right).
\end{aligned}
\tag{1.220}
$$

The results were discovered by Yang in 2020 [39].

1.3.4 The Series Representations for the Special Functions

In this part, we report the series representations related to special functions.

Theorem 1.73 *Let $a, c, p, z \in \mathbb{C}$ and $n \in \mathbb{N}_0$.*
Then we have [72]

$$
\begin{aligned}
&(1-z)^{-a}\ {}_2F_1\left(\begin{matrix}\frac{a}{2},\frac{a}{2}+\frac{1}{2}\\ c\end{matrix};-\frac{4pz}{(1-z)^2}\right)\\
&=\sum_{n=0}^{\infty}(a)_n\ {}_2F_1\left(\begin{matrix}a+n,-n\\ c\end{matrix};p\right)\frac{z^n}{n!}.
\end{aligned}
\tag{1.221}
$$

Theorem 1.74 *Let $a, c, p, z \in \mathbb{C}$ and $n \in \mathbb{N}_0$.*
Then we have [72]

$$
\begin{aligned}
&(1-z)^{-a}\ {}_3F_2\left(\begin{matrix}\frac{a}{2},\frac{a}{2}+\frac{1}{2},b\\ c,d\end{matrix};-\frac{4pz}{(1-z)^2}\right)\\
&=\sum_{n=0}^{\infty}(a)_n\ {}_3F_2\left(\begin{matrix}a+n,-n\\ c,d\end{matrix};p\right)\frac{z^n}{n!}.
\end{aligned}
\tag{1.222}
$$

Theorem 1.75 *Let $a, b, c, p \in \mathbb{C}$, $\alpha, \beta, \gamma, q \in \mathbb{C}$, $z \in \mathbb{C}$ and $n \in \mathbb{N}_0$.*
Then we have [72]

$$
\begin{aligned}
&{}_2F_1\left(\begin{matrix}a,b\\ c\end{matrix};pz\right)\cdot{}_2F_1\left(\begin{matrix}\alpha,\beta\\ \gamma\end{matrix};qz\right)\\
&=\sum_{n=0}^{\infty}\frac{(a)_n(b)_n}{(c)_n}\,{}_4F_3\left(\begin{matrix}\alpha,\beta,1-c-n,-n\\ \beta,1-a-n,1-b-n\end{matrix};-\frac{q}{p}\right)\frac{(pz)^n}{n!}.
\end{aligned}
\tag{1.223}
$$

Theorem 1.76 *Let $a, b, p \in \mathbb{C}$, $\alpha, \beta, q \in \mathbb{C}$, $z \in \mathbb{C}$ and $n \in \mathbb{N}_0$.*

Then we have [72]

$$
\begin{aligned}
&{}_2F_0\left(\begin{matrix} a, b \\ - \end{matrix}; pz\right) \cdot {}_2F_0\left(\begin{matrix} \alpha, \beta \\ - \end{matrix}; qz\right) \\
&= \sum_{n=0}^{\infty} (a)_n\,(b)_n\; {}_3F_2\left(\begin{matrix} \alpha, \beta, -n \\ 1-a-n, 1-b-n \end{matrix}; -\tfrac{q}{p}\right) \tfrac{(pz)^n}{n!}\,.
\end{aligned} \tag{1.224}
$$

Theorem 1.77 *Let $a, c, p \in \mathbb{C}$, $\alpha, \beta, q \in \mathbb{C}$, $z \in \mathbb{C}$ and $n \in \mathbb{N}_0$.*
Then we have [72]

$$
\begin{aligned}
&{}_1F_1\left(\begin{matrix} a \\ c \end{matrix}; pz\right) \cdot {}_1F_1\left(\begin{matrix} \alpha \\ \beta \end{matrix}; qz\right) \\
&= \sum_{n=0}^{\infty} \tfrac{(a)_n}{(c)_n}\; {}_3F_2\left(\begin{matrix} \alpha, 1-c-n, -n \\ \beta, 1-a-n \end{matrix}; -\tfrac{q}{p}\right) \tfrac{(pz)^n}{n!}\,.
\end{aligned} \tag{1.225}
$$

Theorem 1.78 *Let $a, c, p, q \in \mathbb{C}$, $z \in \mathbb{C}$ and $n \in \mathbb{N}_0$.*
Then we have [72]

$$
\begin{aligned}
&{}_1F_1\left(\begin{matrix} a \\ c \end{matrix}; pz\right) \cdot e^{qz} \\
&= \sum_{n=0}^{\infty} \tfrac{(a)_n}{(c)_n}\; {}_2F_1\left(\begin{matrix} 1-c-n, -n \\ 1-a-n \end{matrix}; -\tfrac{q}{p}\right) \tfrac{(pz)^n}{n!}\,.
\end{aligned} \tag{1.226}
$$

Theorem 1.79 *Let $\alpha, \beta, p, q \in \mathbb{C}$, $z \in \mathbb{C}$ and $n \in \mathbb{N}_0$.*
Then we have [72]

$$
\begin{aligned}
&{}_0F_1\left(-; \alpha; pz\right) \cdot {}_0F_1\left(-; \beta; qz\right) \\
&= \sum_{n=0}^{\infty} \tfrac{1}{(\alpha)_n}\; {}_2F_1\left(\begin{matrix} 1-\alpha-n, -n \\ \beta \end{matrix}; \tfrac{q}{p}\right) \tfrac{(pz)^n}{n!}\,.
\end{aligned} \tag{1.227}
$$

Theorem 1.80 *Let $a, a_i, c_j \in \mathbb{C}$, $\lambda, z \in \mathbb{C}$, $i, j, p, q \in \mathbb{N}$ and $n \in \mathbb{N}_0$.*
Then we have [72]

$$
\begin{aligned}
&(1-z)^{-a}\; {}_pF_q\left(\begin{matrix} a_1, \cdots, a_p \\ c_1, \cdots, c_q \end{matrix}; \lambda z\right) \\
&= \sum_{n=0}^{\infty} (a)_n\; {}_{p+1}F_{q+1}\left(\begin{matrix} -n, a_1, \cdots, a_p \\ 1-a-n, c_1, \cdots, c_q \end{matrix}; \lambda\right) \tfrac{z^n}{n!}\,.
\end{aligned} \tag{1.228}
$$

Theorem 1.81 *Let $a, a_i, c_j \in \mathbb{C}$, $\lambda, z \in \mathbb{C}$, $i, j, p, q \in \mathbb{N}$ and $n \in \mathbb{N}_0$.*

Then we have [72]

$$
\begin{aligned}
&(1-z)^{-a}\ {}_{p+2}F_q\left(\begin{matrix}\frac{a}{2},\frac{a}{2}+\frac{1}{2},\cdots,a_p\\ c_1,\cdots,c_q\end{matrix};-\frac{\lambda z}{(1-z)^2}\right)\\
&=\sum_{n=0}^{\infty}(a)_n\ {}_{p+2}F_q\left(\begin{matrix}-n,a+n,a_1,\cdots,a_p\\ c_1,\cdots,c_q\end{matrix};\lambda\right)\frac{z^n}{n!}.
\end{aligned}
\tag{1.229}
$$

Theorem 1.82 *Let $a, a_i, c_j \in \mathbb{C}$, $\lambda, z \in \mathbb{C}$, $i, j, p, q \in \mathbb{N}$ and $n \in \mathbb{N}_0$. Then we have [72]*

$$
\begin{aligned}
&(1-z)^{-a}\ {}_{p+1}F_q\left(\begin{matrix}a,a_1,\cdots,a_p\\ c_1,\cdots,c_q\end{matrix};-\frac{\lambda z}{1-z}\right)\\
&=\sum_{n=0}^{\infty}(a)_n\ {}_{p+1}F_q\left(\begin{matrix}-n,a_1,\cdots,a_p\\ c_1,\cdots,c_q\end{matrix};\lambda\right)\frac{z^n}{n!}.
\end{aligned}
\tag{1.230}
$$

The results were discovered by Burchnall and Chaundy in 1940 [72].

Theorem 1.83 *Let $a, a_i, c_j \in \mathbb{C}$, $\lambda, z \in \mathbb{C}$, $i, j, p, q \in \mathbb{N}$ and $n \in \mathbb{N}_0$. Then we have*

$$
\begin{aligned}
&(1-z)^{-\lambda}\cdot {}_{p+1}F_q\left(\begin{matrix}a_1,\cdots,a_p\\ c_1,\cdots,c_q\end{matrix};-\frac{\lambda z}{1-z}\right)\\
&=\sum_{n=0}^{\infty}(\lambda)_n\cdot {}_{p+1}F_q\left(\begin{matrix}a_1,\cdots,a_p\\ c_1,\cdots,c_q\end{matrix};1\right)\cdot\frac{z^n}{n!}.
\end{aligned}
\tag{1.231}
$$

The result was discovered by Srivastava and Panda in 1973 [69] and the special case reported by Burchnall and Chaundy in 1940 [72].

For more details, see Fasenmyer [73], Nørlund [74], Fields and Wimp [75], Srivastava [76], Weisner [77], Beals and Wong [78], Burchnall and Chaundy [79], Srivastava [80], Srivastava and Buschman [81], Srivastava [82], Abdul-Halim and Al-Salam [83], Carlitz [84], Watson [85], Whipple [86], and Darling [87].

Property 1.2 Let $a_i, c_j \in \mathbb{C}$, $\mu, h, \kappa, \lambda, z \in \mathbb{C}$, $i, j, p, q \in \mathbb{N}$ and $m \in \mathbb{N}_0$. Then we have

$$
\begin{aligned}
&\frac{(\mu+h)_m}{(\mu+\kappa)_m}\ {}_pSupersin_q\left(\begin{matrix}a_1,\cdots,a_p\\ c_1,\cdots,c_q\end{matrix};z\right)\\
&=\frac{(h)_m}{(\kappa)_m}\ {}_{p+2}Supersin_{q+2}\left(\begin{matrix}a_1+n,\cdots,a_p,h+m,\kappa\\ c_1+n,\cdots,c_q,h,\kappa+m\end{matrix};z\right).
\end{aligned}
\tag{1.232}
$$

Proof In view of the relations

$$
\begin{aligned}
&\frac{(\mu+h)_m}{(\mu+\kappa)_m}\ {}_pSupersin_q\left(\begin{matrix}a_1,\cdots,a_p\\ c_1,\cdots,c_q\end{matrix};z\right)\\
&=\frac{(\mu+h)_m}{(\mu+\kappa)_m}\left(\frac{1}{2i}\left({}_pF_q\left(\begin{matrix}a_1,\cdots,a_p\\ c_1,\cdots,c_q\end{matrix};iz\right)-{}_pF_q\left(\begin{matrix}a_1,\cdots,a_p\\ c_1,\cdots,c_q\end{matrix};-iz\right)\right)\right)
\end{aligned}
\tag{1.233}
$$

and

$$\frac{(h)_m}{(\kappa)_m}\, {}_{p+2}Supersin_{q+2}\left(\begin{matrix} a_1+n,\cdots,a_p,h+m,\kappa \\ c_1+n,\cdots,c_q,h,\kappa+m \end{matrix};z\right)$$
$$=\frac{(h)_m}{(\kappa)_m}\left(\frac{1}{2i}\left(\begin{matrix} {}_{p+2}F_{q+2}\left(\begin{matrix} a_1+n,\cdots,a_p,h+m,\kappa \\ c_1+n,\cdots,c_q,h,\kappa+m \end{matrix};iz\right) \\ -\,{}_{p+2}F_{q+2}\left(\begin{matrix} a_1+n,\cdots,a_p,h+m,\kappa \\ c_1+n,\cdots,c_q,h,\kappa+m \end{matrix};-iz\right)\end{matrix}\right)\right), \tag{1.234}$$

we obtain the result, where

$$ {}_{p}Supersin_{q}\left(\begin{matrix} a_1,\cdots,a_p \\ c_1,\cdots,c_q \end{matrix};z\right)$$
$$=\frac{1}{2i}\left({}_{p}F_{q}\left(\begin{matrix} a_1,\cdots,a_p \\ c_1,\cdots,c_q \end{matrix};iz\right)+{}_{p}F_{q}\left(\begin{matrix} a_1,\cdots,a_p \\ c_1,\cdots,c_q \end{matrix};-iz\right)\right) \tag{1.235}$$

and

$$ {}_{p+2}Supersin_{q+2}\left(\begin{matrix} a_1+n,\cdots,a_p,h+m,\kappa \\ c_1+n,\cdots,c_q,h,\kappa+m \end{matrix};z\right)$$
$$=\frac{1}{2i}\left(\begin{matrix} {}_{p+2}F_{q+2}\left(\begin{matrix} a_1+n,\cdots,a_p,h+m,\kappa \\ c_1+n,\cdots,c_q,h,\kappa+m \end{matrix};iz\right) \\ -\,{}_{p+2}F_{q+2}\left(\begin{matrix} a_1+n,\cdots,a_p,h+m,\kappa \\ c_1+n,\cdots,c_q,h,\kappa+m \end{matrix};-iz\right)\end{matrix}\right). \tag{1.236}$$

□

Property 1.3 Let $a_i, c_j \in \mathbb{C}$, $\mu, h, \kappa, \lambda, z \in \mathbb{C}$, $i, j, p, q \in \mathbb{N}$, and $m \in \mathbb{N}_0$. Then we have

$$\frac{(\mu+h)_m}{(\mu+\kappa)_m}\, {}_{p}Supercos_{q}\left(\begin{matrix} a_1,\cdots,a_p \\ c_1,\cdots,c_q \end{matrix};z\right)$$
$$=\frac{(h)_m}{(\kappa)_m}\, {}_{p+2}Supercos_{q+2}\left(\begin{matrix} a_1+n,\cdots,a_p,h+m,\kappa \\ c_1+n,\cdots,c_q,h,\kappa+m \end{matrix};z\right). \tag{1.237}$$

Proof Making use of the relations

$$\frac{(\mu+h)_m}{(\mu+\kappa)_m}\, {}_{p}Supercos_{q}\left(\begin{matrix} a_1,\cdots,a_p \\ c_1,\cdots,c_q \end{matrix};z\right)$$
$$=\frac{(\mu+h)_m}{(\mu+\kappa)_m}\left(\frac{1}{2}\left({}_{p}F_{q}\left(\begin{matrix} a_1,\cdots,a_p \\ c_1,\cdots,c_q \end{matrix};iz\right)+{}_{p}F_{q}\left(\begin{matrix} a_1,\cdots,a_p \\ c_1,\cdots,c_q \end{matrix};-iz\right)\right)\right) \tag{1.238}$$

and

$$\frac{(h)_m}{(\kappa)_m}\,{}_{p+2}Supercos_{q+2}\left(\begin{matrix} a_1+n,\cdots,a_p,h+m,\kappa \\ c_1+n,\cdots,c_q,h,\kappa+m \end{matrix};z\right)$$
$$=\frac{(h)_m}{(\kappa)_m}\left(\frac{1}{2}\left(\begin{matrix} {}_{p+2}F_{q+2}\left(\begin{matrix} a_1+n,\cdots,a_p,h+m,\kappa \\ c_1+n,\cdots,c_q,h,\kappa+m \end{matrix};iz\right) \\ +\,{}_{p+2}F_{q+2}\left(\begin{matrix} a_1+n,\cdots,a_p,h+m,\kappa \\ c_1+n,\cdots,c_q,h,\kappa+m \end{matrix};-iz\right) \end{matrix}\right)\right), \tag{1.239}$$

we have the result, where

$${}_pSupercos_q\left(\begin{matrix} a_1,\cdots,a_p \\ c_1,\cdots,c_q \end{matrix};z\right)$$
$$=\frac{1}{2}\left({}_pF_q\left(\begin{matrix} a_1,\cdots,a_p \\ c_1,\cdots,c_q \end{matrix};iz\right)+{}_pF_q\left(\begin{matrix} a_1,\cdots,a_p \\ c_1,\cdots,c_q \end{matrix};-iz\right)\right) \tag{1.240}$$

and

$${}_{p+2}Supercos_{q+2}\left(\begin{matrix} a_1+n,\cdots,a_p,h+m,\kappa \\ c_1+n,\cdots,c_q,h,\kappa+m \end{matrix};z\right)$$
$$=\frac{1}{2}\left(\begin{matrix} {}_{p+2}F_{q+2}\left(\begin{matrix} a_1+n,\cdots,a_p,h+m,\kappa \\ c_1+n,\cdots,c_q,h,\kappa+m \end{matrix};iz\right) \\ +\,{}_{p+2}F_{q+2}\left(\begin{matrix} a_1+n,\cdots,a_p,h+m,\kappa \\ c_1+n,\cdots,c_q,h,\kappa+m \end{matrix};-iz\right) \end{matrix}\right). \tag{1.241}$$

□

Property 1.4 Let $a, c, z \in \mathbb{C}$ and $n \in \mathbb{N}_0$.
Then we have

$${}_2Supersin_1\left(\begin{matrix} a,-n \\ 1-c-n \end{matrix};z\right)=\frac{(c+a)_n}{(c)_n}\,{}_2Supersin_1\left(\begin{matrix} a,-n \\ c+a \end{matrix};1-z\right). \tag{1.242}$$

Property 1.5 Let $a, c, z \in \mathbb{C}$ and $n \in \mathbb{N}_0$.
Then we have

$${}_2Supercos_1\left(\begin{matrix} a,-n \\ 1-c-n \end{matrix};z\right)=\frac{(c+a)_n}{(c)_n}\,{}_2Supercos_1\left(\begin{matrix} a,-n \\ c+a \end{matrix};1-z\right). \tag{1.243}$$

Property 1.6 Let $a, b, c, d \in \mathbb{C}$ and $n \in \mathbb{N}_0$.

$$ {}_3Supersin_2\left(\begin{matrix} a, b, -n \\ c, 1-d-n \end{matrix}; 1\right) = \frac{(d+a)_n}{(c)_n}\, {}_3Supersin_2\left(\begin{matrix} a, c-b, -n \\ c, d+a \end{matrix}; 1\right). \tag{1.244} $$

Property 1.7 Let $a, b, c, d \in \mathbb{C}$ and $n \in \mathbb{N}_0$.

$$ {}_3Supercos_2\left(\begin{matrix} a, b, -n \\ c, 1-d-n \end{matrix}; 1\right) = \frac{(d+a)_n}{(c)_n}\, {}_3Supercos_2\left(\begin{matrix} a, c-b, -n \\ c, d+a \end{matrix}; 1\right). \tag{1.245} $$

Property 1.8 Let $a, c, z \in \mathbb{C}$ and $n \in \mathbb{N}_0$.

$$ {}_2Supersin_1\left(\begin{matrix} a, -n \\ c \end{matrix}; z\right) = \frac{(a)_n}{(c)_n}(1-z)^n\, {}_2Supersin_1\left(\begin{matrix} c-a, -n \\ 1-a-n \end{matrix}; \frac{1}{1-z}\right). \tag{1.246} $$

Property 1.9 Let $a, c, z \in \mathbb{C}$ and $n \in \mathbb{N}_0$.

$$ {}_2Supercos_1\left(\begin{matrix} a, -n \\ c \end{matrix}; z\right) = \frac{(a)_n}{(c)_n}(1-z)^n\, {}_2Supercos_1\left(\begin{matrix} c-a, -n \\ 1-a-n \end{matrix}; \frac{1}{1-z}\right). \tag{1.247} $$

Theorem 1.84 *Let $a, c, p, z \in \mathbb{C}$ and $n \in \mathbb{N}_0$.*
Then we have

$$ \begin{aligned} &\frac{1}{2i}\left[(1-iz)^{-a}\cdot {}_2F_1\left(\begin{matrix} \frac{a}{2}, \frac{a}{2}+\frac{1}{2} \\ c \end{matrix}; -\frac{4ipz}{(1-iz)^2}\right) - (1+iz)^{-a}\cdot {}_2F_1\left(\begin{matrix} \frac{a}{2}, \frac{a}{2}+\frac{1}{2} \\ c \end{matrix}; \frac{4ipz}{(1+iz)^2}\right)\right] \\ &= \sum_{n=0}^{\infty}(-1)^n (a)_{2n+1}\ {}_2F_1\left(\begin{matrix} a+2n+1, -(2n+1) \\ c \end{matrix}; p\right)\frac{z^{2n+1}}{(2n+1)!}, \end{aligned} \tag{1.248} $$

$$ \begin{aligned} &\frac{1}{2}\left[(1-iz)^{-a}\cdot {}_2F_1\left(\begin{matrix} \frac{a}{2}, \frac{a}{2}+\frac{1}{2} \\ c \end{matrix}; -\frac{4ipz}{(1-iz)^2}\right) + (1+iz)^{-a}\cdot {}_2F_1\left(\begin{matrix} \frac{a}{2}, \frac{a}{2}+\frac{1}{2} \\ c \end{matrix}; \frac{4ipz}{(1+iz)^2}\right)\right] \\ &= \sum_{n=0}^{\infty}(-1)^n (a)_{2n}\ {}_2F_1\left(\begin{matrix} a+2n, -2n \\ c \end{matrix}; p\right)\frac{z^{2n}}{(2n)!}, \end{aligned} \tag{1.249} $$

$$\frac{1}{2}\left[(1-z)^{-a}\cdot {}_2F_1\left(\begin{matrix}\frac{a}{2},\frac{a}{2}+\frac{1}{2}\\ c\end{matrix};-\frac{4pz}{(1-z)^2}\right)-(1+z)^{-a}\cdot {}_2F_1\left(\begin{matrix}\frac{a}{2},\frac{a}{2}+\frac{1}{2}\\ c\end{matrix};\frac{4pz}{(1+z)^2}\right)\right]$$
$$=\sum_{n=0}^{\infty}(a)_{2n+1}\ {}_2F_1\left(\begin{matrix}a+2n+1,-(2n+1)\\ c\end{matrix};p\right)\frac{z^{2n+1}}{(1+2n)!} \tag{1.250}$$

and

$$\frac{1}{2}\left[(1-z)^{-a}\cdot {}_2F_1\left(\begin{matrix}\frac{a}{2},\frac{a}{2}+\frac{1}{2}\\ c\end{matrix};-\frac{4pz}{(1-z)^2}\right)+(1+iz)^{-a}\cdot {}_2F_1\left(\begin{matrix}\frac{a}{2},\frac{a}{2}+\frac{1}{2}\\ c\end{matrix};\frac{4pz}{(1+z)^2}\right)\right]$$
$$=\sum_{n=0}^{\infty}(a)_{2n}\ {}_2F_1\left(\begin{matrix}a+2n,-2n\\ c\end{matrix};p\right)\frac{z^{2n}}{(2n)!}\,. \tag{1.251}$$

Proof With use of the expressions

$$(1-iz)^{-a}\cdot {}_2F_1\left(\begin{matrix}\frac{a}{2},\frac{a}{2}+\frac{1}{2}\\ c\end{matrix};-\frac{4ipz}{(1-iz)^2}\right)=\sum_{n=0}^{\infty}(a)_n\cdot {}_2F_1\left(\begin{matrix}a+n,-n\\ c\end{matrix};p\right)\frac{(iz)^n}{n!}\,, \tag{1.252}$$

$$(1+iz)^{-a}\cdot {}_2F_1\left(\begin{matrix}\frac{a}{2},\frac{a}{2}+\frac{1}{2}\\ c\end{matrix};\frac{4ipz}{(1+iz)^2}\right)=\sum_{n=0}^{\infty}(a)_n\ {}_2F_1\left(\begin{matrix}a+n,-n\\ c\end{matrix};p\right)\frac{(-iz)^n}{n!}\,, \tag{1.253}$$

$$(1-z)^{-a}\cdot {}_2F_1\left(\begin{matrix}\frac{a}{2},\frac{a}{2}+\frac{1}{2}\\ c\end{matrix};-\frac{4pz}{(1-z)^2}\right)=\sum_{n=0}^{\infty}(a)_n\cdot {}_2F_1\left(\begin{matrix}a+n,-n\\ c\end{matrix};p\right)\frac{z^n}{n!} \tag{1.254}$$

and

$$(1+z)^{-a}\cdot {}_2F_1\left(\begin{matrix}\frac{a}{2},\frac{a}{2}+\frac{1}{2}\\ c\end{matrix};\frac{4pz}{(1+z)^2}\right)=\sum_{n=0}^{\infty}(a)_n\ {}_2F_1\left(\begin{matrix}a+n,-n\\ c\end{matrix};p\right)\frac{(-z)^n}{n!}\,, \tag{1.255}$$

we show that

$$\begin{aligned}&(1-iz)^{-a}\cdot {}_2F_1\left(\begin{matrix}\frac{a}{2},\frac{a}{2}+\frac{1}{2}\\ c\end{matrix};-\frac{4ipz}{(1-iz)^2}\right)-(1+iz)^{-a}\cdot {}_2F_1\left(\begin{matrix}\frac{a}{2},\frac{a}{2}+\frac{1}{2}\\ c\end{matrix};\frac{4ipz}{(1+iz)^2}\right)\\&=\sum_{n=0}^{\infty}(a)_n\ {}_2F_1\left(\begin{matrix}a+n,-n\\ c\end{matrix};p\right)\frac{(iz)^n}{n!}-\sum_{n=0}^{\infty}(a)_n\ {}_2F_1\left(\begin{matrix}a+n,-n\\ c\end{matrix};p\right)\frac{(-iz)^n}{n!}\\&=2i\sum_{n=0}^{\infty}(-1)^n\,(a)_{2n+1}\ {}_2F_1\left(\begin{matrix}a+2n+1,-(2n+1)\\ c\end{matrix};p\right)\frac{z^{2n+1}}{(2n+1)!},\end{aligned}\tag{1.256}$$

$$\begin{aligned}&(1-iz)^{-a}\cdot {}_2F_1\left(\begin{matrix}\frac{a}{2},\frac{a}{2}+\frac{1}{2}\\ c\end{matrix};-\frac{4ipz}{(1-iz)^2}\right)+(1+iz)^{-a}\cdot {}_2F_1\left(\begin{matrix}\frac{a}{2},\frac{a}{2}+\frac{1}{2}\\ c\end{matrix};\frac{4ipz}{(1+iz)^2}\right)\\&=\sum_{n=0}^{\infty}(a)_n\ {}_2F_1\left(\begin{matrix}a+n,-n\\ c\end{matrix};p\right)\frac{(iz)^n}{n!}+\sum_{n=0}^{\infty}(a)_n\ {}_2F_1\left(\begin{matrix}a+n,-n\\ c\end{matrix};p\right)\frac{(-iz)^n}{n!}\\&=2\sum_{n=0}^{\infty}(-1)^n\,(a)_{2n}\ {}_2F_1\left(\begin{matrix}a+2n,-2n\\ c\end{matrix};p\right)\frac{z^{2n}}{(2n)!},\end{aligned}\tag{1.257}$$

$$\begin{aligned}&(1-z)^{-a}\cdot {}_2F_1\left(\begin{matrix}\frac{a}{2},\frac{a}{2}+\frac{1}{2}\\ c\end{matrix};-\frac{4pz}{(1-z)^2}\right)-(1+z)^{-a}\cdot {}_2F_1\left(\begin{matrix}\frac{a}{2},\frac{a}{2}+\frac{1}{2}\\ c\end{matrix};\frac{4pz}{(1+z)^2}\right)\\&=\sum_{n=0}^{\infty}(a)_n\ {}_2F_1\left(\begin{matrix}a+n,-n\\ c\end{matrix};p\right)\frac{z^n}{n!}-\sum_{n=0}^{\infty}(a)_n\ {}_2F_1\left(\begin{matrix}a+n,-n\\ c\end{matrix};p\right)\frac{(-z)^n}{n!}\\&=2\sum_{n=0}^{\infty}(a)_{2n+1}\ {}_2F_1\left(\begin{matrix}a+2n+1,-(2n+1)\\ c\end{matrix};p\right)\frac{z^{2n+1}}{(1+2n)!}\end{aligned}\tag{1.258}$$

and

$$\begin{aligned}&(1-z)^{-a}\cdot {}_2F_1\left(\begin{matrix}\frac{a}{2},\frac{a}{2}+\frac{1}{2}\\ c\end{matrix};-\frac{4pz}{(1-z)^2}\right)+(1+iz)^{-a}\cdot {}_2F_1\left(\begin{matrix}\frac{a}{2},\frac{a}{2}+\frac{1}{2}\\ c\end{matrix};\frac{4pz}{(1+z)^2}\right)\\&=\sum_{n=0}^{\infty}(a)_n\ {}_2F_1\left(\begin{matrix}a+n,-n\\ c\end{matrix};p\right)\frac{z^n}{n!}+\sum_{n=0}^{\infty}(a)_n\ {}_2F_1\left(\begin{matrix}a+n,-n\\ c\end{matrix};p\right)\frac{(-z)^n}{n!}\\&=2\sum_{n=0}^{\infty}(a)_{2n}\ {}_2F_1\left(\begin{matrix}a+2n,-2n\\ c\end{matrix};p\right)\frac{z^{2n}}{(2n)!}.\end{aligned}\tag{1.259}$$

□

Theorem 1.85 *Let $a, b, c, d, p, z \in \mathbb{C}$ and $n \in \mathbb{N}_0$.*
Then we have

$$\frac{1}{2i}\left[(1-iz)^{-a}\ {}_3F_2\left(\begin{matrix}\frac{a}{2}, \frac{a}{2}+\frac{1}{2}, b\\ c, d\end{matrix}; -\frac{4ipz}{(1-z)^2}\right) - (1+iz)^{-a}\ {}_3F_2\left(\begin{matrix}\frac{a}{2}, \frac{a}{2}+\frac{1}{2}, b\\ c, d\end{matrix}; \frac{4ipz}{(1+iz)^2}\right)\right]$$
$$= \sum_{n=0}^{\infty} (-1)^n (a)_{2n+1}\ {}_3F_2\left(\begin{matrix}a+2n+1, -(2n+1)\\ c, d\end{matrix}; p\right) \frac{z^{2n+1}}{(2n+1)!}, \tag{1.260}$$

$$\frac{1}{2}\left[(1-iz)^{-a}\ {}_3F_2\left(\begin{matrix}\frac{a}{2}, \frac{a}{2}+\frac{1}{2}, b\\ c, d\end{matrix}; -\frac{4ipz}{(1-z)^2}\right) + (1+iz)^{-a}\ {}_3F_2\left(\begin{matrix}\frac{a}{2}, \frac{a}{2}+\frac{1}{2}, b\\ c, d\end{matrix}; \frac{4ipz}{(1+iz)^2}\right)\right]$$
$$= \sum_{n=0}^{\infty} (-1)^n (a)_{2n}\ {}_3F_2\left(\begin{matrix}a+2n, -2n\\ c, d\end{matrix}; p\right) \frac{z^{2n}}{(2n)!}, \tag{1.261}$$

$$\frac{1}{2}\left[(1-z)^{-a}\ {}_3F_2\left(\begin{matrix}\frac{a}{2}, \frac{a}{2}+\frac{1}{2}, b\\ c, d\end{matrix}; -\frac{4pz}{(1-z)^2}\right) - (1+z)^{-a}\ {}_3F_2\left(\begin{matrix}\frac{a}{2}, \frac{a}{2}+\frac{1}{2}, b\\ c, d\end{matrix}; \frac{4pz}{(1+z)^2}\right)\right]$$
$$= \sum_{n=0}^{\infty} (a)_{2n+1}\ {}_3F_2\left(\begin{matrix}a+2n+1, -(2n+1)\\ c, d\end{matrix}; p\right) \frac{z^{2n+1}}{(2n+1)!}, \tag{1.262}$$

and

$$\frac{1}{2}\left[(1-z)^{-a}\ {}_3F_2\left(\begin{matrix}\frac{a}{2}, \frac{a}{2}+\frac{1}{2}, b\\ c, d\end{matrix}; -\frac{4pz}{(1-z)^2}\right) + (1+z)^{-a}\ {}_3F_2\left(\begin{matrix}\frac{a}{2}, \frac{a}{2}+\frac{1}{2}, b\\ c, d\end{matrix}; \frac{4pz}{(1+iz)^2}\right)\right]$$
$$= \sum_{n=0}^{\infty} (a)_{2n}\ {}_3F_2\left(\begin{matrix}a+2n, -2n\\ c, d\end{matrix}; p\right) \frac{z^{2n}}{(2n)!}. \tag{1.263}$$

Proof In view of the representations

$$(1-iz)^{-a}\ {}_3F_2\left(\begin{matrix}\frac{a}{2}, \frac{a}{2}+\frac{1}{2}, b\\ c, d\end{matrix}; -\frac{4ipz}{(1-z)^2}\right) = \sum_{n=0}^{\infty} (a)_n\ {}_3F_2\left(\begin{matrix}a+n, -n\\ c, d\end{matrix}; p\right) \frac{(iz)^n}{n!}, \tag{1.264}$$

$$(1+iz)^{-a}\ {}_3F_2\left(\begin{matrix}\frac{a}{2}, \frac{a}{2}+\frac{1}{2}, b\\ c, d\end{matrix}; \frac{4ipz}{(1+iz)^2}\right) = \sum_{n=0}^{\infty} (a)_n\ {}_3F_2\left(\begin{matrix}a+n, -n\\ c, d\end{matrix}; p\right) \frac{(-iz)^n}{n!}, \tag{1.265}$$

$$(1-z)^{-a}\ {}_3F_2\left(\begin{matrix}\frac{a}{2}, \frac{a}{2}+\frac{1}{2}, b\\ c, d\end{matrix}; -\frac{4pz}{(1-z)^2}\right) = \sum_{n=0}^{\infty} (a)_n\ {}_3F_2\left(\begin{matrix}a+n, -n\\ c, d\end{matrix}; p\right) \frac{z^n}{n!} \tag{1.266}$$

and

$$(1+z)^{-a}\ {}_3F_2\left(\begin{matrix}\frac{a}{2},\frac{a}{2}+\frac{1}{2},b\\ c,d\end{matrix};\frac{4pz}{(1+z)^2}\right)$$
$$=\sum_{n=0}^{\infty}(a)_n\ {}_3F_2\left(\begin{matrix}a+n,-n\\ c,d\end{matrix};p\right)\frac{(-z)^n}{n!}, \tag{1.267}$$

we give

$$(1-iz)^{-a}\ {}_3F_2\left(\begin{matrix}\frac{a}{2},\frac{a}{2}+\frac{1}{2},b\\ c,d\end{matrix};-\frac{4ipz}{(1-z)^2}\right)-(1+iz)^{-a}\ {}_3F_2\left(\begin{matrix}\frac{a}{2},\frac{a}{2}+\frac{1}{2},b\\ c,d\end{matrix};\frac{4ipz}{(1+iz)^2}\right)$$
$$=\sum_{n=0}^{\infty}(a)_n\ {}_3F_2\left(\begin{matrix}a+n,-n\\ c,d\end{matrix};p\right)\frac{(iz)^n}{n!}-\sum_{n=0}^{\infty}(a)_n\ {}_3F_2\left(\begin{matrix}a+n,-n\\ c,d\end{matrix};p\right)\frac{(-iz)^n}{n!}$$
$$=2i\sum_{n=0}^{\infty}(-1)^n(a)_{2n+1}\ {}_3F_2\left(\begin{matrix}a+2n+1,-(2n+1)\\ c,d\end{matrix};p\right)\frac{z^{2n+1}}{(2n+1)!}, \tag{1.268}$$

$$(1-iz)^{-a}\ {}_3F_2\left(\begin{matrix}\frac{a}{2},\frac{a}{2}+\frac{1}{2},b\\ c,d\end{matrix};-\frac{4ipz}{(1-z)^2}\right)+(1+iz)^{-a}\ {}_3F_2\left(\begin{matrix}\frac{a}{2},\frac{a}{2}+\frac{1}{2},b\\ c,d\end{matrix};\frac{4ipz}{(1+iz)^2}\right)$$
$$=\sum_{n=0}^{\infty}(a)_n\ {}_3F_2\left(\begin{matrix}a+n,-n\\ c,d\end{matrix};p\right)\frac{(iz)^n}{n!}+\sum_{n=0}^{\infty}(a)_n\ {}_3F_2\left(\begin{matrix}a+n,-n\\ c,d\end{matrix};p\right)\frac{(-iz)^n}{n!}$$
$$=2\sum_{n=0}^{\infty}(-1)^n(a)_{2n}\ {}_3F_2\left(\begin{matrix}a+2n,-2n\\ c,d\end{matrix};p\right)\frac{z^{2n}}{(2n)!}, \tag{1.269}$$

$$(1-z)^{-a}\ {}_3F_2\left(\begin{matrix}\frac{a}{2},\frac{a}{2}+\frac{1}{2},b\\ c,d\end{matrix};-\frac{4pz}{(1-z)^2}\right)-(1+z)^{-a}\ {}_3F_2\left(\begin{matrix}\frac{a}{2},\frac{a}{2}+\frac{1}{2},b\\ c,d\end{matrix};\frac{4pz}{(1+z)^2}\right)$$
$$=\sum_{n=0}^{\infty}(a)_n\ {}_3F_2\left(\begin{matrix}a+n,-n\\ c,d\end{matrix};p\right)\frac{z^n}{n!}-\sum_{n=0}^{\infty}(a)_n\ {}_3F_2\left(\begin{matrix}a+n,-n\\ c,d\end{matrix};p\right)\frac{(-z)^n}{n!}$$
$$=2\sum_{n=0}^{\infty}(a)_{2n+1}\ {}_3F_2\left(\begin{matrix}a+2n+1,-(2n+1)\\ c,d\end{matrix};p\right)\frac{z^{2n+1}}{(2n+1)!}, \tag{1.270}$$

and

$$(1-z)^{-a}\ {}_3F_2\left(\begin{matrix}\frac{a}{2},\frac{a}{2}+\frac{1}{2},b\\ c,d\end{matrix};-\frac{4pz}{(1-z)^2}\right)+(1+z)^{-a}\ {}_3F_2\left(\begin{matrix}\frac{a}{2},\frac{a}{2}+\frac{1}{2},b\\ c,d\end{matrix};\frac{4pz}{(1+iz)^2}\right)$$
$$=\sum_{n=0}^{\infty}(a)_n\ {}_3F_2\left(\begin{matrix}a+n,-n\\ c,d\end{matrix};p\right)\frac{z^n}{n!}+\sum_{n=0}^{\infty}(a)_n\ {}_3F_2\left(\begin{matrix}a+n,-n\\ c,d\end{matrix};p\right)\frac{(-z)^n}{n!}$$
$$=2\sum_{n=0}^{\infty}(a)_{2n}\ {}_3F_2\left(\begin{matrix}a+2n,-2n\\ c,d\end{matrix};p\right)\frac{z^{2n}}{(2n)!}. \tag{1.271}$$

□

Theorem 1.86 *Let $a, b, c, p \in \mathbb{C}$, $\alpha, \beta, \gamma, q \in \mathbb{C}$, $z \in \mathbb{C}$ and $n \in \mathbb{N}_0$. Then we have*

$$\frac{1}{2i}\left[{}_2F_1\left(\begin{matrix}a, b\\ c\end{matrix}; ipz\right)\cdot {}_2F_1\left(\begin{matrix}\alpha, \beta\\ \gamma\end{matrix}; iqz\right) - {}_2F_1\left(\begin{matrix}a, b\\ c\end{matrix}; -ipz\right)\cdot {}_2F_1\left(\begin{matrix}\alpha, \beta\\ \gamma\end{matrix}; -iqz\right)\right]$$
$$= \sum_{n=0}^{\infty}(-1)^n \frac{(a)_{2n+1}(b)_{2n+1}}{(c)_{2n+1}}\, {}_4F_3\left(\begin{matrix}\alpha, \beta, -c-2n, -(2n+1)\\ \beta, -a-2n, 1-b-(2n+1)\end{matrix}; -\frac{q}{p}\right)\frac{(pz)^{2n+1}}{(2n+1)!}, \tag{1.272}$$

$$\frac{1}{2}\left[{}_2F_1\left(\begin{matrix}a, b\\ c\end{matrix}; ipz\right)\cdot {}_2F_1\left(\begin{matrix}\alpha, \beta\\ \gamma\end{matrix}; iqz\right) + {}_2F_1\left(\begin{matrix}a, b\\ c\end{matrix}; -ipz\right)\cdot {}_2F_1\left(\begin{matrix}\alpha, \beta\\ \gamma\end{matrix}; -iqz\right)\right]$$
$$= \sum_{n=0}^{\infty}(-1)^n \frac{(a)_{2n}(b)_{2n}}{(c)_{2n}}\, {}_4F_3\left(\begin{matrix}\alpha, \beta, 1-c-2n, -2n\\ \beta, 1-a-2n, 1-b-2n\end{matrix}; -\frac{q}{p}\right)\frac{(pz)^{2n}}{(2n)!}, \tag{1.273}$$

$$\frac{1}{2}\left[{}_2F_1\left(\begin{matrix}a, b\\ c\end{matrix}; pz\right)\cdot {}_2F_1\left(\begin{matrix}\alpha, \beta\\ \gamma\end{matrix}; qz\right) - {}_2F_1\left(\begin{matrix}a, b\\ c\end{matrix}; -pz\right)\cdot {}_2F_1\left(\begin{matrix}\alpha, \beta\\ \gamma\end{matrix}; -qz\right)\right]$$
$$= \sum_{n=0}^{\infty}\frac{(a)_{2n+1}(b)_{2n+1}}{(c)_{2n+1}}\, {}_4F_3\left(\begin{matrix}\alpha, \beta, -c-2n, -(2n+1)\\ \beta, -a-2n, 1-b-(2n+1)\end{matrix}; -\frac{q}{p}\right)\frac{(pz)^{2n+1}}{(2n+1)!} \tag{1.274}$$

and

$$\frac{1}{2}\left[{}_2F_1\left(\begin{matrix}a, b\\ c\end{matrix}; pz\right)\cdot {}_2F_1\left(\begin{matrix}\alpha, \beta\\ \gamma\end{matrix}; qz\right) + {}_2F_1\left(\begin{matrix}a, b\\ c\end{matrix}; -pz\right)\cdot {}_2F_1\left(\begin{matrix}\alpha, \beta\\ \gamma\end{matrix}; -qz\right)\right]$$
$$= \sum_{n=0}^{\infty}\frac{(a)_{2n}(b)_{2n}}{(c)_{2n}}\, {}_4F_3\left(\begin{matrix}\alpha, \beta, 1-c-2n, -2n\\ \beta, 1-a-2n, 1-b-2n\end{matrix}; -\frac{q}{p}\right)\frac{(pz)^{2n}}{(2n)!}. \tag{1.275}$$

Proof In view of the relations

$$\begin{aligned}&{}_2F_1\left(\begin{matrix}a, b\\ c\end{matrix}; ipz\right)\cdot {}_2F_1\left(\begin{matrix}\alpha, \beta\\ \gamma\end{matrix}; iqz\right)\\ &= \sum_{n=0}^{\infty}\frac{(a)_n(b)_n}{(c)_n}\, {}_4F_3\left(\begin{matrix}\alpha, \beta, 1-c-n, -n\\ \beta, 1-a-n, 1-b-n\end{matrix}; -\frac{q}{p}\right)\frac{(ipz)^n}{n!},\end{aligned} \tag{1.276}$$

$$\begin{aligned}&{}_2F_1\left(\begin{matrix}a, b\\ c\end{matrix}; -ipz\right)\cdot {}_2F_1\left(\begin{matrix}\alpha, \beta\\ \gamma\end{matrix}; -iqz\right)\\ &= \sum_{n=0}^{\infty}\frac{(a)_n(b)_n}{(c)_n}\, {}_4F_3\left(\begin{matrix}\alpha, \beta, 1-c-n, -n\\ \beta, 1-a-n, 1-b-n\end{matrix}; -\frac{q}{p}\right)\frac{(-ipz)^n}{n!},\end{aligned} \tag{1.277}$$

$$\begin{aligned}&{}_2F_1\left(\begin{matrix}a, b\\ c\end{matrix}; pz\right)\cdot {}_2F_1\left(\begin{matrix}\alpha, \beta\\ \gamma\end{matrix}; qz\right)\\ &= \sum_{n=0}^{\infty}\frac{(a)_n(b)_n}{(c)_n}\, {}_4F_3\left(\begin{matrix}\alpha, \beta, 1-c-n, -n\\ \beta, 1-a-n, 1-b-n\end{matrix}; -\frac{q}{p}\right)\frac{(pz)^n}{n!},\end{aligned} \tag{1.278}$$

and

$$
\begin{aligned}
&{}_2F_1\left(\begin{matrix}a,b\\c\end{matrix};-pz\right)\cdot {}_2F_1\left(\begin{matrix}\alpha,\beta\\\gamma\end{matrix};-qz\right)\\
&=\sum_{n=0}^{\infty}\frac{(a)_n(b)_n}{(c)_n}\,{}_4F_3\left(\begin{matrix}\alpha,\beta,1-c-n,-n\\\beta,1-a-n,1-b-n\end{matrix};-\frac{q}{p}\right)\frac{(-pz)^n}{n!},
\end{aligned}
\tag{1.279}
$$

we have that

$$
\begin{aligned}
&{}_2F_1\left(\begin{matrix}a,b\\c\end{matrix};ipz\right)\cdot {}_2F_1\left(\begin{matrix}\alpha,\beta\\\gamma\end{matrix};iqz\right)-{}_2F_1\left(\begin{matrix}a,b\\c\end{matrix};-ipz\right)\cdot {}_2F_1\left(\begin{matrix}\alpha,\beta\\\gamma\end{matrix};-iqz\right)\\
&=\sum_{n=0}^{\infty}\frac{(a)_n(b)_n}{(c)_n}\,{}_4F_3\left(\begin{matrix}\alpha,\beta,1-c-n,-n\\\beta,1-a-n,1-b-n\end{matrix};-\frac{q}{p}\right)\frac{(ipz)^n}{n!}\\
&-\sum_{n=0}^{\infty}\frac{(a)_n(b)_n}{(c)_n}\,{}_4F_3\left(\begin{matrix}\alpha,\beta,1-c-n,-n\\\beta,1-a-n,1-b-n\end{matrix};-\frac{q}{p}\right)\frac{(-ipz)^n}{n!}\\
&=2i\sum_{n=0}^{\infty}(-1)^n\frac{(a)_{2n+1}(b)_{2n+1}}{(c)_{2n+1}}\,{}_4F_3\left(\begin{matrix}\alpha,\beta,1-c-(2n+1),-(2n+1)\\\beta,1-a-(2n+1),1-b-(2n+1)\end{matrix};-\frac{q}{p}\right)\frac{(pz)^{2n+1}}{(2n+1)!}\\
&=2i\sum_{n=0}^{\infty}(-1)^n\frac{(a)_{2n+1}(b)_{2n+1}}{(c)_{2n+1}}\,{}_4F_3\left(\begin{matrix}\alpha,\beta,-c-2n,-(2n+1)\\\beta,-a-2n,1-b-(2n+1)\end{matrix};-\frac{q}{p}\right)\frac{(pz)^{2n+1}}{(2n+1)!},
\end{aligned}
\tag{1.280}
$$

$$
\begin{aligned}
&{}_2F_1\left(\begin{matrix}a,b\\c\end{matrix};ipz\right)\cdot {}_2F_1\left(\begin{matrix}\alpha,\beta\\\gamma\end{matrix};iqz\right)+{}_2F_1\left(\begin{matrix}a,b\\c\end{matrix};-ipz\right)\cdot {}_2F_1\left(\begin{matrix}\alpha,\beta\\\gamma\end{matrix};-iqz\right)\\
&=\sum_{n=0}^{\infty}\frac{(a)_n(b)_n}{(c)_n}\,{}_4F_3\left(\begin{matrix}\alpha,\beta,1-c-n,-n\\\beta,1-a-n,1-b-n\end{matrix};-\frac{q}{p}\right)\frac{(ipz)^n}{n!}\\
&+\sum_{n=0}^{\infty}\frac{(a)_n(b)_n}{(c)_n}\,{}_4F_3\left(\begin{matrix}\alpha,\beta,1-c-n,-n\\\beta,1-a-n,1-b-n\end{matrix};-\frac{q}{p}\right)\frac{(-ipz)^n}{n!}\\
&=2\sum_{n=0}^{\infty}(-1)^n\frac{(a)_{2n}(b)_{2n}}{(c)_{2n}}\,{}_4F_3\left(\begin{matrix}\alpha,\beta,1-c-2n,-2n\\\beta,1-a-2n,1-b-2n\end{matrix};-\frac{q}{p}\right)\frac{(pz)^{2n}}{(2n)!},
\end{aligned}
\tag{1.281}
$$

$$
\begin{aligned}
&{}_2F_1\left(\begin{matrix}a,b\\c\end{matrix};pz\right)\cdot {}_2F_1\left(\begin{matrix}\alpha,\beta\\\gamma\end{matrix};qz\right)-{}_2F_1\left(\begin{matrix}a,b\\c\end{matrix};-pz\right)\cdot {}_2F_1\left(\begin{matrix}\alpha,\beta\\\gamma\end{matrix};-qz\right)\\
&=\sum_{n=0}^{\infty}\frac{(a)_n(b)_n}{(c)_n}\,{}_4F_3\left(\begin{matrix}\alpha,\beta,1-c-n,-n\\\beta,1-a-n,1-b-n\end{matrix};-\frac{q}{p}\right)\frac{(pz)^n}{n!}\\
&-\sum_{n=0}^{\infty}\frac{(a)_n(b)_n}{(c)_n}\,{}_4F_3\left(\begin{matrix}\alpha,\beta,1-c-n,-n\\\beta,1-a-n,1-b-n\end{matrix};-\frac{q}{p}\right)\frac{(-pz)^n}{n!}\\
&=2\sum_{n=0}^{\infty}\frac{(a)_{2n+1}(b)_{2n+1}}{(c)_{2n+1}}\,{}_4F_3\left(\begin{matrix}\alpha,\beta,1-c-(2n+1),-(2n+1)\\\beta,1-a-(2n+1),1-b-(2n+1)\end{matrix};-\frac{q}{p}\right)\frac{(pz)^{2n+1}}{(2n+1)!}\\
&=2\sum_{n=0}^{\infty}\frac{(a)_{2n+1}(b)_{2n+1}}{(c)_{2n+1}}\,{}_4F_3\left(\begin{matrix}\alpha,\beta,-c-2n,-(2n+1)\\\beta,-a-2n,1-b-(2n+1)\end{matrix};-\frac{q}{p}\right)\frac{(pz)^{2n+1}}{(2n+1)!}
\end{aligned}
\tag{1.282}
$$

and

$$
\begin{aligned}
&{}_2F_1\left(\begin{matrix}a,b\\c\end{matrix};pz\right)\cdot{}_2F_1\left(\begin{matrix}\alpha,\beta\\\gamma\end{matrix};qz\right)+{}_2F_1\left(\begin{matrix}a,b\\c\end{matrix};-pz\right)\cdot{}_2F_1\left(\begin{matrix}\alpha,\beta\\\gamma\end{matrix};-qz\right)\\
&=\sum_{n=0}^{\infty}\frac{(a)_n(b)_n}{(c)_n}\,{}_4F_3\left(\begin{matrix}\alpha,\beta,1-c-n,-n\\\beta,1-a-n,1-b-n\end{matrix};-\frac{q}{p}\right)\frac{(pz)^n}{n!}\\
&+\sum_{n=0}^{\infty}\frac{(a)_n(b)_n}{(c)_n}\,{}_4F_3\left(\begin{matrix}\alpha,\beta,1-c-n,-n\\\beta,1-a-n,1-b-n\end{matrix};-\frac{q}{p}\right)\frac{(-pz)^n}{n!}\\
&=2\sum_{n=0}^{\infty}\frac{(a)_{2n}(b)_{2n}}{(c)_{2n}}\,{}_4F_3\left(\begin{matrix}\alpha,\beta,1-c-2n,-2n\\\beta,1-a-2n,1-b-2n\end{matrix};-\frac{q}{p}\right)\frac{(pz)^{2n}}{(2n)!}\,.
\end{aligned}
\tag{1.283}
$$

□

Theorem 1.87 *Let $a, b, p \in \mathbb{C}$, $\alpha, \beta, q \in \mathbb{C}$, $z \in \mathbb{C}$ and $n \in \mathbb{N}_0$.*
Then we have

$$
\begin{aligned}
&\frac{1}{2i}\left[{}_2F_0\left(\begin{matrix}a,b\\-\end{matrix};ipz\right)\cdot{}_2F_0\left(\begin{matrix}\alpha,\beta\\-\end{matrix};iqz\right)-{}_2F_0\left(\begin{matrix}a,b\\-\end{matrix};-ipz\right)\cdot{}_2F_0\left(\begin{matrix}\alpha,\beta\\-\end{matrix};-iqz\right)\right]\\
&=\sum_{n=0}^{\infty}(-1)^n\,(a)_{2n+1}\,(b)_{2n+1}\,{}_3F_2\left(\begin{matrix}\alpha,\beta,-(2n+1)\\-a-2n,-b-2n\end{matrix};-\frac{q}{p}\right)\frac{(pz)^{2n+1}}{(2n+1)!}\,,
\end{aligned}
\tag{1.284}
$$

$$
\begin{aligned}
&\frac{1}{2}\left[{}_2F_0\left(\begin{matrix}a,b\\-\end{matrix};ipz\right)\cdot{}_2F_0\left(\begin{matrix}\alpha,\beta\\-\end{matrix};iqz\right)+{}_2F_0\left(\begin{matrix}a,b\\-\end{matrix};-ipz\right)\cdot{}_2F_0\left(\begin{matrix}\alpha,\beta\\-\end{matrix};-iqz\right)\right]\\
&=\sum_{n=0}^{\infty}(-1)^n\,(a)_{2n}\,(b)_{2n}\,{}_3F_2\left(\begin{matrix}\alpha,\beta,-2n\\1-a-2n,1-b-2n\end{matrix};-\frac{q}{p}\right)\frac{(pz)^{2n}}{(2n)!}\,,
\end{aligned}
\tag{1.285}
$$

$$
\begin{aligned}
&\frac{1}{2}\left[{}_2F_0\left(\begin{matrix}a,b\\-\end{matrix};pz\right)\cdot{}_2F_0\left(\begin{matrix}\alpha,\beta\\-\end{matrix};qz\right)-{}_2F_0\left(\begin{matrix}a,b\\-\end{matrix};-pz\right)\cdot{}_2F_0\left(\begin{matrix}\alpha,\beta\\-\end{matrix};-qz\right)\right]\\
&=\sum_{n=0}^{\infty}(a)_{2n+1}\,(b)_{2n+1}\,{}_3F_2\left(\begin{matrix}\alpha,\beta,-(2n+1)\\-a-2n,-b-2n\end{matrix};-\frac{q}{p}\right)\frac{(pz)^{2n+1}}{(2n+1)!}\,,
\end{aligned}
\tag{1.286}
$$

and

$$
\begin{aligned}
&\frac{1}{2}\left[{}_2F_0\left(\begin{matrix}a,b\\-\end{matrix};pz\right)\cdot{}_2F_0\left(\begin{matrix}\alpha,\beta\\-\end{matrix};qz\right)+{}_2F_0\left(\begin{matrix}a,b\\-\end{matrix};-pz\right)\cdot{}_2F_0\left(\begin{matrix}\alpha,\beta\\-\end{matrix};-qz\right)\right]\\
&=\sum_{n=0}^{\infty}(a)_{2n}\,(b)_{2n}\,{}_3F_2\left(\begin{matrix}\alpha,\beta,-2n\\1-a-2n,1-b-2n\end{matrix};-\frac{q}{p}\right)\frac{(pz)^{2n}}{(2n)!}\,.
\end{aligned}
\tag{1.287}
$$

Proof With the relations

$$
\begin{aligned}
&{}_2F_0\left(\begin{matrix}a,b\\ -\end{matrix};ipz\right)\cdot{}_2F_0\left(\begin{matrix}\alpha,\beta\\ -\end{matrix};iqz\right)\\
&=\sum_{n=0}^{\infty}(a)_n\,(b)_n\;{}_3F_2\left(\begin{matrix}\alpha,\beta,-n\\ 1-a-n,1-b-n\end{matrix};-\frac{q}{p}\right)\frac{(ipz)^n}{n!},
\end{aligned}
\tag{1.288}
$$

$$
\begin{aligned}
&{}_2F_0\left(\begin{matrix}a,b\\ -\end{matrix};-ipz\right)\cdot{}_2F_0\left(\begin{matrix}\alpha,\beta\\ -\end{matrix};-iqz\right)\\
&=\sum_{n=0}^{\infty}(a)_n\,(b)_n\;{}_3F_2\left(\begin{matrix}\alpha,\beta,-n\\ 1-a-n,1-b-n\end{matrix};-\frac{q}{p}\right)\frac{(-ipz)^n}{n!},
\end{aligned}
\tag{1.289}
$$

$$
\begin{aligned}
&{}_2F_0\left(\begin{matrix}a,b\\ -\end{matrix};pz\right)\cdot{}_2F_0\left(\begin{matrix}\alpha,\beta\\ -\end{matrix};qz\right)\\
&=\sum_{n=0}^{\infty}(a)_n\,(b)_n\;{}_3F_2\left(\begin{matrix}\alpha,\beta,-n\\ 1-a-n,1-b-n\end{matrix};-\frac{q}{p}\right)\frac{(pz)^n}{n!},
\end{aligned}
\tag{1.290}
$$

and

$$
\begin{aligned}
&{}_2F_0\left(\begin{matrix}a,b\\ -\end{matrix};-pz\right)\cdot{}_2F_0\left(\begin{matrix}\alpha,\beta\\ -\end{matrix};-qz\right)\\
&=\sum_{n=0}^{\infty}(a)_n\,(b)_n\;{}_3F_2\left(\begin{matrix}\alpha,\beta,-n\\ 1-a-n,1-b-n\end{matrix};-\frac{q}{p}\right)\frac{(-pz)^n}{n!},
\end{aligned}
\tag{1.291}
$$

we show that

$$
\begin{aligned}
&{}_2F_0\left(\begin{matrix}a,b\\ -\end{matrix};ipz\right)\cdot{}_2F_0\left(\begin{matrix}\alpha,\beta\\ -\end{matrix};iqz\right)-{}_2F_0\left(\begin{matrix}a,b\\ -\end{matrix};-ipz\right)\cdot{}_2F_0\left(\begin{matrix}\alpha,\beta\\ -\end{matrix};-iqz\right)\\
&=\sum_{n=0}^{\infty}(a)_n\,(b)_n\;{}_3F_2\left(\begin{matrix}\alpha,\beta,-n\\ 1-a-n,1-b-n\end{matrix};-\frac{q}{p}\right)\frac{(ipz)^n}{n!}\\
&-\sum_{n=0}^{\infty}(a)_n\,(b)_n\;{}_3F_2\left(\begin{matrix}\alpha,\beta,-n\\ 1-a-n,1-b-n\end{matrix};-\frac{q}{p}\right)\frac{(-ipz)^n}{n!}\\
&=2i\sum_{n=0}^{\infty}(-1)^n\,(a)_{2n+1}\,(b)_{2n+1}\;{}_3F_2\left(\begin{matrix}\alpha,\beta,-(2n+1)\\ 1-a-(2n+1),1-b-(2n+1)\end{matrix};-\frac{q}{p}\right)\frac{(pz)^{2n+1}}{(2n+1)!}\\
&=2i\sum_{n=0}^{\infty}(-1)^n\,(a)_{2n+1}\,(b)_{2n+1}\;{}_3F_2\left(\begin{matrix}\alpha,\beta,-(2n+1)\\ -a-2n,-b-2n\end{matrix};-\frac{q}{p}\right)\frac{(pz)^{2n+1}}{(2n+1)!},
\end{aligned}
\tag{1.292}
$$

$$
\begin{aligned}
&{}_2F_0\left(\begin{matrix}a,b\\ -\end{matrix};ipz\right)\cdot{}_2F_0\left(\begin{matrix}\alpha,\beta\\ -\end{matrix};iqz\right)+{}_2F_0\left(\begin{matrix}a,b\\ -\end{matrix};-ipz\right)\cdot{}_2F_0\left(\begin{matrix}\alpha,\beta\\ -\end{matrix};-iqz\right)\\
&=\sum_{n=0}^{\infty}(a)_n\,(b)_n\;{}_3F_2\left(\begin{matrix}\alpha,\beta,-n\\ 1-a-n,1-b-n\end{matrix};-\tfrac{q}{p}\right)\tfrac{(ipz)^n}{n!}\\
&+\sum_{n=0}^{\infty}(a)_n\,(b)_n\;{}_3F_2\left(\begin{matrix}\alpha,\beta,-n\\ 1-a-n,1-b-n\end{matrix};-\tfrac{q}{p}\right)\tfrac{(-ipz)^n}{n!}\\
&=2\sum_{n=0}^{\infty}(-1)^n\,(a)_{2n}\,(b)_{2n}\;{}_3F_2\left(\begin{matrix}\alpha,\beta,-2n\\ 1-a-2n,1-b-2n\end{matrix};-\tfrac{q}{p}\right)\tfrac{(pz)^{2n}}{(2n)!},
\end{aligned}
\tag{1.293}
$$

$$
\begin{aligned}
&{}_2F_0\left(\begin{matrix}a,b\\ -\end{matrix};pz\right)\cdot{}_2F_0\left(\begin{matrix}\alpha,\beta\\ -\end{matrix};qz\right)-{}_2F_0\left(\begin{matrix}a,b\\ -\end{matrix};-pz\right)\cdot{}_2F_0\left(\begin{matrix}\alpha,\beta\\ -\end{matrix};-qz\right)\\
&=\sum_{n=0}^{\infty}(a)_n\,(b)_n\;{}_3F_2\left(\begin{matrix}\alpha,\beta,-n\\ 1-a-n,1-b-n\end{matrix};-\tfrac{q}{p}\right)\tfrac{(pz)^n}{n!}\\
&-\sum_{n=0}^{\infty}(a)_n\,(b)_n\;{}_3F_2\left(\begin{matrix}\alpha,\beta,-n\\ 1-a-n,1-b-n\end{matrix};-\tfrac{q}{p}\right)\tfrac{(-pz)^n}{n!}\\
&=2\sum_{n=0}^{\infty}(a)_{2n+1}\,(b)_{2n+1}\;{}_3F_2\left(\begin{matrix}\alpha,\beta,-(2n+1)\\ 1-a-(2n+1),1-b-(2n+1)\end{matrix};-\tfrac{q}{p}\right)\tfrac{(pz)^{2n+1}}{(2n+1)!}\\
&=2\sum_{n=0}^{\infty}(a)_{2n+1}\,(b)_{2n+1}\;{}_3F_2\left(\begin{matrix}\alpha,\beta,-(2n+1)\\ -a-2n,-b-2n\end{matrix};-\tfrac{q}{p}\right)\tfrac{(pz)^{2n+1}}{(2n+1)!},
\end{aligned}
\tag{1.294}
$$

and

$$
\begin{aligned}
&{}_2F_0\left(\begin{matrix}a,b\\ -\end{matrix};pz\right)\cdot{}_2F_0\left(\begin{matrix}\alpha,\beta\\ -\end{matrix};qz\right)+{}_2F_0\left(\begin{matrix}a,b\\ -\end{matrix};-pz\right)\cdot{}_2F_0\left(\begin{matrix}\alpha,\beta\\ -\end{matrix};-qz\right)\\
&=\sum_{n=0}^{\infty}(a)_n\,(b)_n\;{}_3F_2\left(\begin{matrix}\alpha,\beta,-n\\ 1-a-n,1-b-n\end{matrix};-\tfrac{q}{p}\right)\tfrac{(pz)^n}{n!}\\
&+\sum_{n=0}^{\infty}(a)_n\,(b)_n\;{}_3F_2\left(\begin{matrix}\alpha,\beta,-n\\ 1-a-n,1-b-n\end{matrix};-\tfrac{q}{p}\right)\tfrac{(-pz)^n}{n!}\\
&=2\sum_{n=0}^{\infty}(a)_{2n}\,(b)_{2n}\;{}_3F_2\left(\begin{matrix}\alpha,\beta,-2n\\ 1-a-2n,1-b-2n\end{matrix};-\tfrac{q}{p}\right)\tfrac{(pz)^{2n}}{(2n)!}.
\end{aligned}
\tag{1.295}
$$

□

Theorem 1.88 *Let $a, c, p \in \mathbb{C}$, $\alpha, \beta, q \in \mathbb{C}$, $z \in \mathbb{C}$, and $n \in \mathbb{N}_0$.*
Then we have

$$
\begin{aligned}
&\tfrac{1}{2i}\left[{}_1F_1\left(\begin{matrix}a\\ c\end{matrix};ipz\right)\cdot{}_1F_1\left(\begin{matrix}\alpha\\ \beta\end{matrix};iqz\right)-{}_1F_1\left(\begin{matrix}a\\ c\end{matrix};-ipz\right)\cdot{}_1F_1\left(\begin{matrix}\alpha\\ \beta\end{matrix};-iqz\right)\right]\\
&=\sum_{n=0}^{\infty}\tfrac{(-1)^n(a)_{2n+1}}{(c)_{2n+1}}\;{}_3F_2\left(\begin{matrix}\alpha,-c-2n,-(2n+1)\\ \beta,-a-2n\end{matrix};-\tfrac{q}{p}\right)\tfrac{(pz)^{2n+1}}{(2n+1)!},
\end{aligned}
\tag{1.296}
$$

$$\frac{1}{2}\left[{}_1F_1\left(\begin{matrix}a\\c\end{matrix};ipz\right)\cdot{}_1F_1\left(\begin{matrix}\alpha\\\beta\end{matrix};iqz\right)+{}_1F_1\left(\begin{matrix}a\\c\end{matrix};-ipz\right)\cdot{}_1F_1\left(\begin{matrix}\alpha\\\beta\end{matrix};-iqz\right)\right]$$
$$=\sum_{n=0}^{\infty}\frac{(-1)^n(a)_{2n}}{(c)_{2n}}\,{}_3F_2\left(\begin{matrix}\alpha,1-c-2n,-2n\\\beta,1-a-2n\end{matrix};-\frac{q}{p}\right)\frac{(pz)^{2n}}{(2n)!},\tag{1.297}$$

$$\frac{1}{2}\left[{}_1F_1\left(\begin{matrix}a\\c\end{matrix};pz\right)\cdot{}_1F_1\left(\begin{matrix}\alpha\\\beta\end{matrix};qz\right)-{}_1F_1\left(\begin{matrix}a\\c\end{matrix};-pz\right)\cdot{}_1F_1\left(\begin{matrix}\alpha\\\beta\end{matrix};-qz\right)\right]$$
$$=\sum_{n=0}^{\infty}\frac{(a)_{2n+1}}{(c)_{2n+1}}\,{}_3F_2\left(\begin{matrix}\alpha,-c-2n,-(2n+1)\\\beta,-a-2n\end{matrix};-\frac{q}{p}\right)\frac{(pz)^{2n+1}}{(2n+1)!},\tag{1.298}$$

$$\frac{1}{2}\left[{}_1F_1\left(\begin{matrix}a\\c\end{matrix};pz\right)\cdot{}_1F_1\left(\begin{matrix}\alpha\\\beta\end{matrix};qz\right)+{}_1F_1\left(\begin{matrix}a\\c\end{matrix};-pz\right)\cdot{}_1F_1\left(\begin{matrix}\alpha\\\beta\end{matrix};-qz\right)\right]$$
$$=\sum_{n=0}^{\infty}\frac{(a)_{2n}}{(c)_{2n}}\,{}_3F_2\left(\begin{matrix}\alpha,1-c-2n,-2n\\\beta,1-a-2n\end{matrix};-\frac{q}{p}\right)\frac{(pz)^{2n}}{(2n)!}.\tag{1.299}$$

Proof Making use of the relations

$$\begin{aligned}&{}_1F_1\left(\begin{matrix}a\\c\end{matrix};ipz\right)\cdot{}_1F_1\left(\begin{matrix}\alpha\\\beta\end{matrix};iqz\right)\\&=\sum_{n=0}^{\infty}\frac{(a)_n}{(c)_n}\,{}_3F_2\left(\begin{matrix}\alpha,1-c-n,-n\\\beta,1-a-n\end{matrix};-\frac{q}{p}\right)\frac{(ipz)^n}{n!},\end{aligned}\tag{1.300}$$

$$\begin{aligned}&{}_1F_1\left(\begin{matrix}a\\c\end{matrix};-ipz\right)\cdot{}_1F_1\left(\begin{matrix}\alpha\\\beta\end{matrix};-iqz\right)\\&=\sum_{n=0}^{\infty}\frac{(a)_n}{(c)_n}\,{}_3F_2\left(\begin{matrix}\alpha,1-c-n,-n\\\beta,1-a-n\end{matrix};-\frac{q}{p}\right)\frac{(-ipz)^n}{n!},\end{aligned}\tag{1.301}$$

$$\begin{aligned}&{}_1F_1\left(\begin{matrix}a\\c\end{matrix};pz\right)\cdot{}_1F_1\left(\begin{matrix}\alpha\\\beta\end{matrix};qz\right)\\&=\sum_{n=0}^{\infty}\frac{(a)_n}{(c)_n}\,{}_3F_2\left(\begin{matrix}\alpha,1-c-n,-n\\\beta,1-a-n\end{matrix};-\frac{q}{p}\right)\frac{(pz)^n}{n!},\end{aligned}\tag{1.302}$$

and

$$\begin{aligned}&{}_1F_1\left(\begin{matrix}a\\c\end{matrix};-pz\right)\cdot{}_1F_1\left(\begin{matrix}\alpha\\\beta\end{matrix};-qz\right)\\&=\sum_{n=0}^{\infty}\frac{(a)_n}{(c)_n}\,{}_3F_2\left(\begin{matrix}\alpha,1-c-n,-n\\\beta,1-a-n\end{matrix};-\frac{q}{p}\right)\frac{(-pz)^n}{n!},\end{aligned}\tag{1.303}$$

we have that

$$
\begin{aligned}
&{}_1F_1\left(\begin{matrix}a\\c\end{matrix};ipz\right)\cdot{}_1F_1\left(\begin{matrix}\alpha\\\beta\end{matrix};iqz\right)-{}_1F_1\left(\begin{matrix}a\\c\end{matrix};-ipz\right)\cdot{}_1F_1\left(\begin{matrix}\alpha\\\beta\end{matrix};-iqz\right)\\
&=\sum_{n=0}^{\infty}\tfrac{(a)_n}{(c)_n}\,{}_3F_2\left(\begin{matrix}\alpha,1-c-n,-n\\\beta,1-a-n\end{matrix};-\tfrac{q}{p}\right)\tfrac{(ipz)^n}{n!}\\
&-\sum_{n=0}^{\infty}\tfrac{(a)_n}{(c)_n}\,{}_3F_2\left(\begin{matrix}\alpha,1-c-n,-n\\\beta,1-a-n\end{matrix};-\tfrac{q}{p}\right)\tfrac{(-ipz)^n}{n!}\\
&=2i\sum_{n=0}^{\infty}\tfrac{(-1)^n(a)_{2n+1}}{(c)_{2n+1}}\,{}_3F_2\left(\begin{matrix}\alpha,1-c-(2n+1),-(2n+1)\\\beta,1-a-(2n+1)\end{matrix};-\tfrac{q}{p}\right)\tfrac{(pz)^{2n+1}}{(2n+1)!}\\
&=2i\sum_{n=0}^{\infty}\tfrac{(-1)^n(a)_{2n+1}}{(c)_{2n+1}}\,{}_3F_2\left(\begin{matrix}\alpha,-c-2n,-(2n+1)\\\beta,-a-2n\end{matrix};-\tfrac{q}{p}\right)\tfrac{(pz)^{2n+1}}{(2n+1)!},
\end{aligned}
\tag{1.304}
$$

$$
\begin{aligned}
&{}_1F_1\left(\begin{matrix}a\\c\end{matrix};ipz\right)\cdot{}_1F_1\left(\begin{matrix}\alpha\\\beta\end{matrix};iqz\right)+{}_1F_1\left(\begin{matrix}a\\c\end{matrix};-ipz\right)\cdot{}_1F_1\left(\begin{matrix}\alpha\\\beta\end{matrix};-iqz\right)\\
&=\sum_{n=0}^{\infty}\tfrac{(a)_n}{(c)_n}\,{}_3F_2\left(\begin{matrix}\alpha,1-c-n,-n\\\beta,1-a-n\end{matrix};-\tfrac{q}{p}\right)\tfrac{(ipz)^n}{n!}\\
&+\sum_{n=0}^{\infty}\tfrac{(a)_n}{(c)_n}\,{}_3F_2\left(\begin{matrix}\alpha,1-c-n,-n\\\beta,1-a-n\end{matrix};-\tfrac{q}{p}\right)\tfrac{(-ipz)^n}{n!}\\
&=2\sum_{n=0}^{\infty}\tfrac{(-1)^n(a)_{2n}}{(c)_{2n}}\,{}_3F_2\left(\begin{matrix}\alpha,1-c-2n,-2n\\\beta,1-a-2n\end{matrix};-\tfrac{q}{p}\right)\tfrac{(pz)^{2n}}{(2n)!},
\end{aligned}
\tag{1.305}
$$

$$
\begin{aligned}
&{}_1F_1\left(\begin{matrix}a\\c\end{matrix};pz\right)\cdot{}_1F_1\left(\begin{matrix}\alpha\\\beta\end{matrix};qz\right)-{}_1F_1\left(\begin{matrix}a\\c\end{matrix};-pz\right)\cdot{}_1F_1\left(\begin{matrix}\alpha\\\beta\end{matrix};-qz\right)\\
&=\sum_{n=0}^{\infty}\tfrac{(a)_n}{(c)_n}\,{}_3F_2\left(\begin{matrix}\alpha,1-c-n,-n\\\beta,1-a-n\end{matrix};-\tfrac{q}{p}\right)\tfrac{(pz)^n}{n!}\\
&-\sum_{n=0}^{\infty}\tfrac{(a)_n}{(c)_n}\,{}_3F_2\left(\begin{matrix}\alpha,1-c-n,-n\\\beta,1-a-n\end{matrix};-\tfrac{q}{p}\right)\tfrac{(-pz)^n}{n!}\\
&=2\sum_{n=0}^{\infty}\tfrac{(a)_{2n+1}}{(c)_{2n+1}}\,{}_3F_2\left(\begin{matrix}\alpha,1-c-(2n+1),-(2n+1)\\\beta,1-a-(2n+1)\end{matrix};-\tfrac{q}{p}\right)\tfrac{(pz)^{2n+1}}{(2n+1)!}\\
&=2\sum_{n=0}^{\infty}\tfrac{(a)_{2n+1}}{(c)_{2n+1}}\,{}_3F_2\left(\begin{matrix}\alpha,-c-2n,-(2n+1)\\\beta,-a-2n\end{matrix};-\tfrac{q}{p}\right)\tfrac{(pz)^{2n+1}}{(2n+1)!},
\end{aligned}
\tag{1.306}
$$

$$\begin{aligned}&{}_1F_1\left(\begin{matrix}a\\c\end{matrix};pz\right)\cdot{}_1F_1\left(\begin{matrix}\alpha\\\beta\end{matrix};qz\right)+{}_1F_1\left(\begin{matrix}a\\c\end{matrix};-pz\right)\cdot{}_1F_1\left(\begin{matrix}\alpha\\\beta\end{matrix};-qz\right)\\&=\sum_{n=0}^{\infty}\frac{(a)_n}{(c)_n}\,{}_3F_2\left(\begin{matrix}\alpha,1-c-n,-n\\\beta,1-a-n\end{matrix};-\frac{q}{p}\right)\frac{(pz)^n}{n!}\\&+\sum_{n=0}^{\infty}\frac{(a)_n}{(c)_n}\,{}_3F_2\left(\begin{matrix}\alpha,1-c-n,-n\\\beta,1-a-n\end{matrix};-\frac{q}{p}\right)\frac{(-pz)^n}{n!}\\&=2\sum_{n=0}^{\infty}\frac{(a)_{2n}}{(c)_{2n}}\,{}_3F_2\left(\begin{matrix}\alpha,1-c-2n,-2n\\\beta,1-a-2n\end{matrix};-\frac{q}{p}\right)\frac{(pz)^{2n}}{(2n)!}.\end{aligned}\tag{1.307}$$

□

Theorem 1.89 *Let $\alpha,\beta,p,q\in\mathbb{C}$, $z\in\mathbb{C}$, and $n\in\mathbb{N}_0$.*
Then we have

$$\begin{aligned}&\frac{1}{2i}\left[{}_0F_1(-;\alpha;ipz)\cdot{}_0F_1(-;\beta;iqz)-{}_0F_1(-;\alpha;-ipz)\cdot{}_0F_1(-;\beta;-iqz)\right]\\&=\sum_{n=0}^{\infty}\frac{(-1)^n}{(\alpha)_{2n+1}}\,{}_2F_1\left(\begin{matrix}-\alpha-2n,-(2n+1)\\\beta\end{matrix};\frac{q}{p}\right)\frac{(pz)^{2n+1}}{(2n+1)!},\end{aligned}\tag{1.308}$$

$$\begin{aligned}&\frac{1}{2}\left[{}_0F_1(-;\alpha;ipz)\cdot{}_0F_1(-;\beta;iqz)+{}_0F_1(-;\alpha;-ipz)\cdot{}_0F_1(-;\beta;-iqz)\right]\\&=\sum_{n=0}^{\infty}\frac{(-1)^n}{(\alpha)_{2n}}\,{}_2F_1\left(\begin{matrix}1-\alpha-2n,-2n\\\beta\end{matrix};\frac{q}{p}\right)\frac{(pz)^{2n}}{(2n)!},\end{aligned}\tag{1.309}$$

$$\begin{aligned}&\frac{1}{2}\left[{}_0F_1(-;\alpha;pz)\cdot{}_0F_1(-;\beta;qz)-{}_0F_1(-;\alpha;-pz)\cdot{}_0F_1(-;\beta;-qz)\right]\\&=\sum_{n=0}^{\infty}\frac{(-1)^n}{(\alpha)_{2n+1}}\,{}_2F_1\left(\begin{matrix}-\alpha-2n,-(2n+1)\\\beta\end{matrix};\frac{q}{p}\right)\frac{(pz)^{2n+1}}{(2n+1)!},\end{aligned}\tag{1.310}$$

and

$$\begin{aligned}&\frac{1}{2}\left[{}_0F_1(-;\alpha;pz)\cdot{}_0F_1(-;\beta;qz)+{}_0F_1(-;\alpha;-pz)\cdot{}_0F_1(-;\beta;-qz)\right]\\&=\sum_{n=0}^{\infty}\frac{1}{(\alpha)_{2n}}\,{}_2F_1\left(\begin{matrix}1-\alpha-2n,-2n\\\beta\end{matrix};\frac{q}{p}\right)\frac{(pz)^{2n}}{(2n)!}.\end{aligned}\tag{1.311}$$

Proof With use of the relations

$$\begin{aligned}&{}_0F_1(-;\alpha;ipz)\cdot{}_0F_1(-;\beta;iqz)\\&=\sum_{n=0}^{\infty}\frac{1}{(\alpha)_n}\,{}_2F_1\left(\begin{matrix}1-\alpha-n,-n\\\beta\end{matrix};\frac{q}{p}\right)\frac{(ipz)^n}{n!},\end{aligned}\tag{1.312}$$

$$\begin{aligned}&{}_0F_1(-;\alpha;-ipz)\cdot{}_0F_1(-;\beta;-iqz)\\&=\sum_{n=0}^{\infty}\frac{1}{(\alpha)_n}\,{}_2F_1\left(\begin{matrix}1-\alpha-n,-n\\\beta\end{matrix};\frac{q}{p}\right)\frac{(-ipz)^n}{n!},\end{aligned}\tag{1.313}$$

$$
\begin{aligned}
&{}_0F_1\left(-;\alpha;pz\right)\cdot{}_0F_1\left(-;\beta;qz\right)\\
&=\sum_{n=0}^{\infty}\frac{1}{(\alpha)_n}\,{}_2F_1\left(\begin{matrix}1-\alpha-n,-n\\ \beta\end{matrix};\frac{q}{p}\right)\frac{(pz)^n}{n!},
\end{aligned}
\tag{1.314}
$$

and

$$
\begin{aligned}
&{}_0F_1\left(-;\alpha;-pz\right)\cdot{}_0F_1\left(-;\beta;-qz\right)\\
&=\sum_{n=0}^{\infty}\frac{1}{(\alpha)_n}\,{}_2F_1\left(\begin{matrix}1-\alpha-n,-n\\ \beta\end{matrix};\frac{q}{p}\right)\frac{(-pz)^n}{n!},
\end{aligned}
\tag{1.315}
$$

we show that

$$
\begin{aligned}
&{}_0F_1\left(-;\alpha;ipz\right)\cdot{}_0F_1\left(-;\beta;iqz\right)-{}_0F_1\left(-;\alpha;-ipz\right)\cdot{}_0F_1\left(-;\beta;-iqz\right)\\
&=\sum_{n=0}^{\infty}\frac{1}{(\alpha)_n}\,{}_2F_1\left(\begin{matrix}1-\alpha-n,-n\\ \beta\end{matrix};\frac{q}{p}\right)\frac{(ipz)^n}{n!}\\
&-\sum_{n=0}^{\infty}\frac{1}{(\alpha)_n}\,{}_2F_1\left(\begin{matrix}1-\alpha-n,-n\\ \beta\end{matrix};\frac{q}{p}\right)\frac{(-ipz)^n}{n!}\\
&=2i\sum_{n=0}^{\infty}\frac{(-1)^n}{(\alpha)_{2n+1}}\,{}_2F_1\left(\begin{matrix}1-\alpha-(2n+1),-(2n+1)\\ \beta\end{matrix};\frac{q}{p}\right)\frac{(pz)^{2n+1}}{(2n+1)!}\\
&=2i\sum_{n=0}^{\infty}\frac{(-1)^n}{(\alpha)_{2n+1}}\,{}_2F_1\left(\begin{matrix}-\alpha-2n,-(2n+1)\\ \beta\end{matrix};\frac{q}{p}\right)\frac{(pz)^{2n+1}}{(2n+1)!},
\end{aligned}
\tag{1.316}
$$

$$
\begin{aligned}
&{}_0F_1\left(-;\alpha;ipz\right)\cdot{}_0F_1\left(-;\beta;iqz\right)+{}_0F_1\left(-;\alpha;-ipz\right)\cdot{}_0F_1\left(-;\beta;-iqz\right)\\
&=\sum_{n=0}^{\infty}\frac{1}{(\alpha)_n}\,{}_2F_1\left(\begin{matrix}1-\alpha-n,-n\\ \beta\end{matrix};\frac{q}{p}\right)\frac{(ipz)^n}{n!}\\
&+\sum_{n=0}^{\infty}\frac{1}{(\alpha)_n}\,{}_2F_1\left(\begin{matrix}1-\alpha-n,-n\\ \beta\end{matrix};\frac{q}{p}\right)\frac{(-ipz)^n}{n!}\\
&=2\sum_{n=0}^{\infty}\frac{(-1)^n}{(\alpha)_{2n}}\,{}_2F_1\left(\begin{matrix}1-\alpha-2n,-2n\\ \beta\end{matrix};\frac{q}{p}\right)\frac{(pz)^{2n}}{(2n)!}
\end{aligned}
\tag{1.317}
$$

$$
\begin{aligned}
&{}_0F_1\left(-;\alpha;pz\right)\cdot{}_0F_1\left(-;\beta;qz\right)-{}_0F_1\left(-;\alpha;-pz\right)\cdot{}_0F_1\left(-;\beta;-qz\right)\\
&=\sum_{n=0}^{\infty}\frac{1}{(\alpha)_n}\,{}_2F_1\left(\begin{matrix}1-\alpha-n,-n\\ \beta\end{matrix};\frac{q}{p}\right)\frac{(pz)^n}{n!}\\
&-\sum_{n=0}^{\infty}\frac{1}{(\alpha)_n}\,{}_2F_1\left(\begin{matrix}1-\alpha-n,-n\\ \beta\end{matrix};\frac{q}{p}\right)\frac{(-pz)^n}{n!}\\
&=2\sum_{n=0}^{\infty}\frac{(-1)^n}{(\alpha)_{2n+1}}\,{}_2F_1\left(\begin{matrix}1-\alpha-(2n+1),-(2n+1)\\ \beta\end{matrix};\frac{q}{p}\right)\frac{(pz)^{2n+1}}{(2n+1)!}\\
&=2\sum_{n=0}^{\infty}\frac{(-1)^n}{(\alpha)_{2n+1}}\,{}_2F_1\left(\begin{matrix}-\alpha-2n,-(2n+1)\\ \beta\end{matrix};\frac{q}{p}\right)\frac{(pz)^{2n+1}}{(2n+1)!},
\end{aligned}
\tag{1.318}
$$

and

$$\begin{aligned}
&{}_0F_1(-;\alpha;pz)\cdot{}_0F_1(-;\beta;qz)+{}_0F_1(-;\alpha;-pz)\cdot{}_0F_1(-;\beta;-qz)\\
&=\sum_{n=0}^{\infty}\frac{1}{(\alpha)_n}\,{}_2F_1\left(\begin{matrix}1-\alpha-n,-n\\ \beta\end{matrix};\frac{q}{p}\right)\frac{(pz)^n}{n!}\\
&+\sum_{n=0}^{\infty}\frac{1}{(\alpha)_n}\,{}_2F_1\left(\begin{matrix}1-\alpha-n,-n\\ \beta\end{matrix};\frac{q}{p}\right)\frac{(-pz)^n}{n!}\\
&=2\sum_{n=0}^{\infty}\frac{1}{(\alpha)_{2n}}\,{}_2F_1\left(\begin{matrix}1-\alpha-2n,-2n\\ \beta\end{matrix};\frac{q}{p}\right)\frac{(pz)^{2n}}{(2n)!}\,.
\end{aligned}\tag{1.319}$$

□

Theorem 1.90 *Let $a, a_i, c_j \in \mathbb{C}$, $\lambda, z \in \mathbb{C}$, $i, j, p, q \in \mathbb{N}$, and $n \in \mathbb{N}_0$. Then we have*

$$\begin{aligned}
&\frac{1}{2i}\left[(1-iz)^{-a}\;{}_pF_q\left(\begin{matrix}a_1,\cdots,a_p\\ c_1,\cdots,c_q\end{matrix};i\lambda z\right)-(1+iz)^{-a}\;{}_pF_q\left(\begin{matrix}a_1,\cdots,a_p\\ c_1,\cdots,c_q\end{matrix};-i\lambda z\right)\right]\\
&=\sum_{n=0}^{\infty}(-1)^n\,(a)_{2n+1}\;{}_{p+1}F_{q+1}\left(\begin{matrix}-(2n+1),a_1,\cdots,a_p\\ -a-2n,c_1,\cdots,c_q\end{matrix};\lambda\right)\frac{z^{2n+1}}{(2n+1)!}\,,
\end{aligned}\tag{1.320}$$

$$\begin{aligned}
&\frac{1}{2}\left[(1-iz)^{-a}\;{}_pF_q\left(\begin{matrix}a_1,\cdots,a_p\\ c_1,\cdots,c_q\end{matrix};i\lambda z\right)+(1+iz)^{-a}\;{}_pF_q\left(\begin{matrix}a_1,\cdots,a_p\\ c_1,\cdots,c_q\end{matrix};-i\lambda z\right)\right]\\
&=\sum_{n=0}^{\infty}(-1)^n\,(a)_{2n}\;{}_{p+1}F_{q+1}\left(\begin{matrix}-2n,a_1,\cdots,a_p\\ 1-a-2n,c_1,\cdots,c_q\end{matrix};\lambda\right)\frac{z^{2n}}{(2n)!}\,,
\end{aligned}\tag{1.321}$$

$$\begin{aligned}
&\frac{1}{2}\left[(1-z)^{-a}\;{}_pF_q\left(\begin{matrix}a_1,\cdots,a_p\\ c_1,\cdots,c_q\end{matrix};\lambda z\right)-(1+z)^{-a}\;{}_pF_q\left(\begin{matrix}a_1,\cdots,a_p\\ c_1,\cdots,c_q\end{matrix};-\lambda z\right)\right]\\
&=\sum_{n=0}^{\infty}(a)_{2n+1}\;{}_{p+1}F_{q+1}\left(\begin{matrix}-(2n+1),a_1,\cdots,a_p\\ -a-2n,c_1,\cdots,c_q\end{matrix};\lambda\right)\frac{z^{2n+1}}{(2n+1)!}
\end{aligned}\tag{1.322}$$

and

$$\begin{aligned}
&\frac{1}{2}\left[(1-z)^{-a}\;{}_pF_q\left(\begin{matrix}a_1,\cdots,a_p\\ c_1,\cdots,c_q\end{matrix};\lambda z\right)+(1+z)^{-a}\;{}_pF_q\left(\begin{matrix}a_1,\cdots,a_p\\ c_1,\cdots,c_q\end{matrix};-\lambda z\right)\right]\\
&=\sum_{n=0}^{\infty}(a)_{2n}\;{}_{p+1}F_{q+1}\left(\begin{matrix}-2n,a_1,\cdots,a_p\\ 1-a-2n,c_1,\cdots,c_q\end{matrix};\lambda\right)\frac{z^{2n}}{(2n)!}\,.
\end{aligned}\tag{1.323}$$

Proof With the relations

$$(1-iz)^{-a}\ {}_pF_q\left(\begin{matrix}a_1,\cdots,a_p\\c_1,\cdots,c_q\end{matrix};i\lambda z\right)=\sum_{n=0}^{\infty}(a)_n\ {}_{p+1}F_{q+1}\left(\begin{matrix}-n,a_1,\cdots,a_p\\1-a-n,c_1,\cdots,c_q\end{matrix};\lambda\right)\frac{(iz)^n}{n!},\tag{1.324}$$

$$(1+iz)^{-a}\ {}_pF_q\left(\begin{matrix}a_1,\cdots,a_p\\c_1,\cdots,c_q\end{matrix};-i\lambda z\right)=\sum_{n=0}^{\infty}(a)_n\ {}_{p+1}F_{q+1}\left(\begin{matrix}-n,a_1,\cdots,a_p\\1-a-n,c_1,\cdots,c_q\end{matrix};\lambda\right)\frac{(-iz)^n}{n!},\tag{1.325}$$

$$(1-z)^{-a}\ {}_pF_q\left(\begin{matrix}a_1,\cdots,a_p\\c_1,\cdots,c_q\end{matrix};\lambda z\right)=\sum_{n=0}^{\infty}(a)_n\ {}_{p+1}F_{q+1}\left(\begin{matrix}-n,a_1,\cdots,a_p\\1-a-n,c_1,\cdots,c_q\end{matrix};\lambda\right)\frac{z^n}{n!},\tag{1.326}$$

and

$$(1+z)^{-a}\ {}_pF_q\left(\begin{matrix}a_1,\cdots,a_p\\c_1,\cdots,c_q\end{matrix};-\lambda z\right)=\sum_{n=0}^{\infty}(a)_n\ {}_{p+1}F_{q+1}\left(\begin{matrix}-n,a_1,\cdots,a_p\\1-a-n,c_1,\cdots,c_q\end{matrix};\lambda\right)\frac{(-z)^n}{n!},\tag{1.327}$$

we show that

$$\begin{aligned}&(1-iz)^{-a}\ {}_pF_q\left(\begin{matrix}a_1,\cdots,a_p\\c_1,\cdots,c_q\end{matrix};i\lambda z\right)-(1+iz)^{-a}\ {}_pF_q\left(\begin{matrix}a_1,\cdots,a_p\\c_1,\cdots,c_q\end{matrix};-i\lambda z\right)\\&=\sum_{n=0}^{\infty}(a)_n\ {}_{p+1}F_{q+1}\left(\begin{matrix}-n,a_1,\cdots,a_p\\1-a-n,c_1,\cdots,c_q\end{matrix};\lambda\right)\frac{(iz)^n}{n!}\\&-\sum_{n=0}^{\infty}(a)_n\ {}_{p+1}F_{q+1}\left(\begin{matrix}-n,a_1,\cdots,a_p\\1-a-n,c_1,\cdots,c_q\end{matrix};\lambda\right)\frac{(-iz)^n}{n!}\\&=2i\sum_{n=0}^{\infty}(-1)^n(a)_{2n+1}\ {}_{p+1}F_{q+1}\left(\begin{matrix}-(2n+1),a_1,\cdots,a_p\\-a-2n,c_1,\cdots,c_q\end{matrix};\lambda\right)\frac{z^{2n+1}}{(2n+1)!},\end{aligned}\tag{1.328}$$

$$\begin{aligned}&(1-iz)^{-a}\ {}_pF_q\left(\begin{matrix}a_1,\cdots,a_p\\c_1,\cdots,c_q\end{matrix};i\lambda z\right)+(1+iz)^{-a}\ {}_pF_q\left(\begin{matrix}a_1,\cdots,a_p\\c_1,\cdots,c_q\end{matrix};-i\lambda z\right)\\&=\sum_{n=0}^{\infty}(a)_n\ {}_{p+1}F_{q+1}\left(\begin{matrix}-n,a_1,\cdots,a_p\\1-a-n,c_1,\cdots,c_q\end{matrix};\lambda\right)\frac{(iz)^n}{n!}\\&+\sum_{n=0}^{\infty}(a)_n\ {}_{p+1}F_{q+1}\left(\begin{matrix}-n,a_1,\cdots,a_p\\1-a-n,c_1,\cdots,c_q\end{matrix};\lambda\right)\frac{(-iz)^n}{n!}\\&=2\sum_{n=0}^{\infty}(-1)^n(a)_{2n}\ {}_{p+1}F_{q+1}\left(\begin{matrix}-2n,a_1,\cdots,a_p\\1-a-2n,c_1,\cdots,c_q\end{matrix};\lambda\right)\frac{z^{2n}}{(2n)!},\end{aligned}\tag{1.329}$$

$$
\begin{aligned}
&(1-z)^{-a}\ {}_pF_q\left(\begin{matrix} a_1,\cdots,a_p \\ c_1,\cdots,c_q \end{matrix};\lambda z\right) - (1+z)^{-a}\ {}_pF_q\left(\begin{matrix} a_1,\cdots,a_p \\ c_1,\cdots,c_q \end{matrix};-\lambda z\right) \\
&= \sum_{n=0}^{\infty} (a)_n\ {}_{p+1}F_{q+1}\left(\begin{matrix} -n,a_1,\cdots,a_p \\ 1-a-n,c_1,\cdots,c_q \end{matrix};\lambda\right) \tfrac{z^n}{n!} \\
&- \sum_{n=0}^{\infty} (a)_n\ {}_{p+1}F_{q+1}\left(\begin{matrix} -n,a_1,\cdots,a_p \\ 1-a-n,c_1,\cdots,c_q \end{matrix};\lambda\right) \tfrac{(-z)^n}{n!} \\
&= 2\sum_{n=0}^{\infty} (a)_{2n+1}\ {}_{p+1}F_{q+1}\left(\begin{matrix} -(2n+1),a_1,\cdots,a_p \\ -a-2n,c_1,\cdots,c_q \end{matrix};\lambda\right) \tfrac{z^{2n+1}}{(2n+1)!},
\end{aligned}
\tag{1.330}
$$

and

$$
\begin{aligned}
&(1-z)^{-a}\ {}_pF_q\left(\begin{matrix} a_1,\cdots,a_p \\ c_1,\cdots,c_q \end{matrix};\lambda z\right) + (1+z)^{-a}\ {}_pF_q\left(\begin{matrix} a_1,\cdots,a_p \\ c_1,\cdots,c_q \end{matrix};-\lambda z\right) \\
&= \sum_{n=0}^{\infty} (a)_n\ {}_{p+1}F_{q+1}\left(\begin{matrix} -n,a_1,\cdots,a_p \\ 1-a-n,c_1,\cdots,c_q \end{matrix};\lambda\right) \tfrac{z^n}{n!} \\
&+ \sum_{n=0}^{\infty} (a)_n\ {}_{p+1}F_{q+1}\left(\begin{matrix} -n,a_1,\cdots,a_p \\ 1-a-n,c_1,\cdots,c_q \end{matrix};\lambda\right) \tfrac{(-z)^n}{n!} \\
&= 2\sum_{n=0}^{\infty} (a)_{2n}\ {}_{p+1}F_{q+1}\left(\begin{matrix} -2n,a_1,\cdots,a_p \\ 1-a-2n,c_1,\cdots,c_q \end{matrix};\lambda\right) \tfrac{z^{2n}}{(2n)!}.
\end{aligned}
\tag{1.331}
$$

□

Theorem 1.91 *Let* $a, a_i, c_j \in \mathbb{C}$, $\lambda, z \in \mathbb{C}$, $i, j, p, q \in \mathbb{N}$, *and* $n \in \mathbb{N}_0$.
Then we have

$$
\begin{aligned}
&(1-iz)^{-a}\ {}_{p+2}F_q\left(\begin{matrix} \frac{a}{2},\frac{a}{2}+\frac{1}{2},\cdots,a_p \\ c_1,\cdots,c_q \end{matrix};-\tfrac{i\lambda z}{(1-iz)^2}\right) \\
&-(1+iz)^{-a}\ {}_{p+2}F_q\left(\begin{matrix} \frac{a}{2},\frac{a}{2}+\frac{1}{2},\cdots,a_p \\ c_1,\cdots,c_q \end{matrix};\tfrac{i\lambda z}{(1+iz)^2}\right) \\
&= 2i\sum_{n=0}^{\infty} (-1)^n (a)_{2n+1}\ {}_{p+2}F_q\left(\begin{matrix} -(2n+1),a+2n+1,a_1,\cdots,a_p \\ c_1,\cdots,c_q \end{matrix};\lambda\right) \tfrac{z^{2n+1}}{(2n+1)!},
\end{aligned}
\tag{1.332}
$$

$$
\begin{aligned}
&(1-iz)^{-a}\ {}_{p+2}F_q\left(\begin{matrix} \frac{a}{2},\frac{a}{2}+\frac{1}{2},\cdots,a_p \\ c_1,\cdots,c_q \end{matrix};-\tfrac{i\lambda z}{(1-iz)^2}\right) \\
&-(1+iz)^{-a}\ {}_{p+2}F_q\left(\begin{matrix} \frac{a}{2},\frac{a}{2}+\frac{1}{2},\cdots,a_p \\ c_1,\cdots,c_q \end{matrix};\tfrac{i\lambda z}{(1+iz)^2}\right) \\
&= 2\sum_{n=0}^{\infty} (-1)^n (a)_{2n}\ {}_{p+2}F_q\left(\begin{matrix} -2n,a+2n,a_1,\cdots,a_p \\ c_1,\cdots,c_q \end{matrix};\lambda\right) \tfrac{z^{2n}}{(2n)!},
\end{aligned}
\tag{1.333}
$$

$$\begin{aligned}&(1-z)^{-a}\ {}_{p+2}F_q\left(\begin{matrix}\frac{a}{2},\frac{a}{2}+\frac{1}{2},\cdots,a_p\\c_1,\cdots,c_q\end{matrix};-\frac{\lambda z}{(1-z)^2}\right)\\&-(1+z)^{-a}\ {}_{p+2}F_q\left(\begin{matrix}\frac{a}{2},\frac{a}{2}+\frac{1}{2},\cdots,a_p\\c_1,\cdots,c_q\end{matrix};\frac{\lambda z}{(1+z)^2}\right)\\&=2\sum_{n=0}^{\infty}(a)_{2n+1}\ {}_{p+2}F_q\left(\begin{matrix}-(2n+1),a+2n+1,a_1,\cdots,a_p\\c_1,\cdots,c_q\end{matrix};\lambda\right)\frac{z^{2n+1}}{(2n+1)!},\end{aligned} \tag{1.334}$$

and

$$\begin{aligned}&(1-z)^{-a}\ {}_{p+2}F_q\left(\begin{matrix}\frac{a}{2},\frac{a}{2}+\frac{1}{2},\cdots,a_p\\c_1,\cdots,c_q\end{matrix};-\frac{\lambda z}{(1-z)^2}\right)\\&-(1+z)^{-a}\ {}_{p+2}F_q\left(\begin{matrix}\frac{a}{2},\frac{a}{2}+\frac{1}{2},\cdots,a_p\\c_1,\cdots,c_q\end{matrix};\frac{\lambda z}{(1+z)^2}\right)\\&=2\sum_{n=0}^{\infty}(a)_{2n}\ {}_{p+2}F_q\left(\begin{matrix}-2n,a+2n,a_1,\cdots,a_p\\c_1,\cdots,c_q\end{matrix};\lambda\right)\frac{z^{2n}}{(2n)!}.\end{aligned} \tag{1.335}$$

Proof Due to the relations

$$\begin{aligned}&(1-iz)^{-a}\ {}_{p+2}F_q\left(\begin{matrix}\frac{a}{2},\frac{a}{2}+\frac{1}{2},\cdots,a_p\\c_1,\cdots,c_q\end{matrix};-\frac{i\lambda z}{(1-iz)^2}\right)\\&=\sum_{n=0}^{\infty}(a)_n\ {}_{p+2}F_q\left(\begin{matrix}-n,a+n,a_1,\cdots,a_p\\c_1,\cdots,c_q\end{matrix};\lambda\right)\frac{(iz)^n}{n!},\end{aligned} \tag{1.336}$$

$$\begin{aligned}&(1+iz)^{-a}\ {}_{p+2}F_q\left(\begin{matrix}\frac{a}{2},\frac{a}{2}+\frac{1}{2},\cdots,a_p\\c_1,\cdots,c_q\end{matrix};\frac{i\lambda z}{(1+iz)^2}\right)\\&=\sum_{n=0}^{\infty}(a)_n\ {}_{p+2}F_q\left(\begin{matrix}-n,a+n,a_1,\cdots,a_p\\c_1,\cdots,c_q\end{matrix};\lambda\right)\frac{(-iz)^n}{n!},\end{aligned} \tag{1.337}$$

$$\begin{aligned}&(1-z)^{-a}\ {}_{p+2}F_q\left(\begin{matrix}\frac{a}{2},\frac{a}{2}+\frac{1}{2},\cdots,a_p\\c_1,\cdots,c_q\end{matrix};-\frac{\lambda z}{(1-z)^2}\right)\\&=\sum_{n=0}^{\infty}(a)_n\ {}_{p+2}F_q\left(\begin{matrix}-n,a+n,a_1,\cdots,a_p\\c_1,\cdots,c_q\end{matrix};\lambda\right)\frac{z^n}{n!},\end{aligned} \tag{1.338}$$

$$\begin{aligned}&(1+z)^{-a}\ {}_{p+2}F_q\left(\begin{matrix}\frac{a}{2},\frac{a}{2}+\frac{1}{2},\cdots,a_p\\c_1,\cdots,c_q\end{matrix};\frac{\lambda z}{(1+z)^2}\right)\\&=\sum_{n=0}^{\infty}(a)_n\ {}_{p+2}F_q\left(\begin{matrix}-n,a+n,a_1,\cdots,a_p\\c_1,\cdots,c_q\end{matrix};\lambda\right)\frac{(-z)^n}{n!},\end{aligned} \tag{1.339}$$

we have that

$$\begin{aligned}
&(1-iz)^{-a}\ {}_{p+2}F_q\left(\begin{matrix}\frac{a}{2},\frac{a}{2}+\frac{1}{2},\cdots,a_p\\ c_1,\cdots,c_q\end{matrix};-\frac{i\lambda z}{(1-iz)^2}\right)\\
&-(1+iz)^{-a}\ {}_{p+2}F_q\left(\begin{matrix}\frac{a}{2},\frac{a}{2}+\frac{1}{2},\cdots,a_p\\ c_1,\cdots,c_q\end{matrix};\frac{i\lambda z}{(1+iz)^2}\right)\\
&=\sum_{n=0}^{\infty}(a)_n\ {}_{p+2}F_q\left(\begin{matrix}-n,a+n,a_1,\cdots,a_p\\ c_1,\cdots,c_q\end{matrix};\lambda\right)\frac{(iz)^n}{n!}\\
&-\sum_{n=0}^{\infty}(a)_n\ {}_{p+2}F_q\left(\begin{matrix}-n,a+n,a_1,\cdots,a_p\\ c_1,\cdots,c_q\end{matrix};\lambda\right)\frac{(-iz)^n}{n!}\\
&=2i\sum_{n=0}^{\infty}(-1)^n(a)_{2n+1}\ {}_{p+2}F_q\left(\begin{matrix}-(2n+1),a+2n+1,a_1,\cdots,a_p\\ c_1,\cdots,c_q\end{matrix};\lambda\right)\frac{z^{2n+1}}{(2n+1)!},
\end{aligned}\tag{1.340}$$

$$\begin{aligned}
&(1-iz)^{-a}\ {}_{p+2}F_q\left(\begin{matrix}\frac{a}{2},\frac{a}{2}+\frac{1}{2},\cdots,a_p\\ c_1,\cdots,c_q\end{matrix};-\frac{i\lambda z}{(1-iz)^2}\right)\\
&-(1+iz)^{-a}\ {}_{p+2}F_q\left(\begin{matrix}\frac{a}{2},\frac{a}{2}+\frac{1}{2},\cdots,a_p\\ c_1,\cdots,c_q\end{matrix};\frac{i\lambda z}{(1+iz)^2}\right)\\
&=\sum_{n=0}^{\infty}(a)_n\ {}_{p+2}F_q\left(\begin{matrix}-n,a+n,a_1,\cdots,a_p\\ c_1,\cdots,c_q\end{matrix};\lambda\right)\frac{(iz)^n}{n!}\\
&-\sum_{n=0}^{\infty}(a)_n\ {}_{p+2}F_q\left(\begin{matrix}-n,a+n,a_1,\cdots,a_p\\ c_1,\cdots,c_q\end{matrix};\lambda\right)\frac{(-iz)^n}{n!}\\
&=2\sum_{n=0}^{\infty}(-1)^n(a)_{2n}\ {}_{p+2}F_q\left(\begin{matrix}-2n,a+2n,a_1,\cdots,a_p\\ c_1,\cdots,c_q\end{matrix};\lambda\right)\frac{z^{2n}}{(2n)!},
\end{aligned}\tag{1.341}$$

$$\begin{aligned}
&(1-z)^{-a}\ {}_{p+2}F_q\left(\begin{matrix}\frac{a}{2},\frac{a}{2}+\frac{1}{2},\cdots,a_p\\ c_1,\cdots,c_q\end{matrix};-\frac{\lambda z}{(1-z)^2}\right)\\
&-(1+z)^{-a}\ {}_{p+2}F_q\left(\begin{matrix}\frac{a}{2},\frac{a}{2}+\frac{1}{2},\cdots,a_p\\ c_1,\cdots,c_q\end{matrix};\frac{\lambda z}{(1+z)^2}\right)\\
&=\sum_{n=0}^{\infty}(a)_n\ {}_{p+2}F_q\left(\begin{matrix}-n,a+n,a_1,\cdots,a_p\\ c_1,\cdots,c_q\end{matrix};\lambda\right)\frac{z^n}{n!}\\
&-\sum_{n=0}^{\infty}(a)_n\ {}_{p+2}F_q\left(\begin{matrix}-n,a+n,a_1,\cdots,a_p\\ c_1,\cdots,c_q\end{matrix};\lambda\right)\frac{(-z)^n}{n!}\\
&=2\sum_{n=0}^{\infty}(a)_{2n+1}\ {}_{p+2}F_q\left(\begin{matrix}-(2n+1),a+2n+1,a_1,\cdots,a_p\\ c_1,\cdots,c_q\end{matrix};\lambda\right)\frac{z^{2n+1}}{(2n+1)!},
\end{aligned}\tag{1.342}$$

and

$$
\begin{aligned}
&(1-z)^{-a}\ {}_{p+2}F_q\left(\begin{matrix}\frac{a}{2},\frac{a}{2}+\frac{1}{2},\cdots,a_p\\ c_1,\cdots,c_q\end{matrix};-\frac{\lambda z}{(1-z)^2}\right)\\
&-(1+z)^{-a}\ {}_{p+2}F_q\left(\begin{matrix}\frac{a}{2},\frac{a}{2}+\frac{1}{2},\cdots,a_p\\ c_1,\cdots,c_q\end{matrix};\frac{\lambda z}{(1+z)^2}\right)\\
&=\sum_{n=0}^{\infty}(a)_n\ {}_{p+2}F_q\left(\begin{matrix}-n,a+n,a_1,\cdots,a_p\\ c_1,\cdots,c_q\end{matrix};\lambda\right)\frac{z^n}{n!}\\
&-\sum_{n=0}^{\infty}(a)_n\ {}_{p+2}F_q\left(\begin{matrix}-n,a+n,a_1,\cdots,a_p\\ c_1,\cdots,c_q\end{matrix};\lambda\right)\frac{(-z)^n}{n!}\\
&=2\sum_{n=0}^{\infty}(a)_{2n}\ {}_{p+2}F_q\left(\begin{matrix}-2n,a+2n,a_1,\cdots,a_p\\ c_1,\cdots,c_q\end{matrix};\lambda\right)\frac{z^{2n}}{(2n)!}\,.
\end{aligned}
\tag{1.343}
$$

□

Theorem 1.92 *Let $a, a_i, c_j \in \mathbb{C}$, $\lambda, z \in \mathbb{C}$, $i, j, p, q \in \mathbb{N}$, and $n \in \mathbb{N}_0$. Then we have*

$$
\begin{aligned}
&\frac{1}{2i}\left[(1-iz)^{-a}\ {}_{p+1}F_q\left(\begin{matrix}a,a_1,\cdots,a_p\\ c_1,\cdots,c_q\end{matrix};-\frac{i\lambda z}{1-iz}\right)-(1+iz)^{-a}\ {}_{p+1}F_q\left(\begin{matrix}a,a_1,\cdots,a_p\\ c_1,\cdots,c_q\end{matrix};\frac{i\lambda z}{1+iz}\right)\right]\\
&=\sum_{n=0}^{\infty}(-1)^n(a)_{2n+1}\ {}_{p+1}F_q\left(\begin{matrix}-(2n+1),a_1,\cdots,a_p\\ c_1,\cdots,c_q\end{matrix};\lambda\right)\frac{z^{2n+1}}{(2n+1)!},
\end{aligned}
\tag{1.344}
$$

$$
\begin{aligned}
&\frac{1}{2}\left[(1-iz)^{-a}\ {}_{p+1}F_q\left(\begin{matrix}a,a_1,\cdots,a_p\\ c_1,\cdots,c_q\end{matrix};-\frac{i\lambda z}{1-iz}\right)+(1+iz)^{-a}\ {}_{p+1}F_q\left(\begin{matrix}a,a_1,\cdots,a_p\\ c_1,\cdots,c_q\end{matrix};\frac{i\lambda z}{1+iz}\right)\right]\\
&=\sum_{n=0}^{\infty}(-1)^n(a)_{2n}\ {}_{p+1}F_q\left(\begin{matrix}-(2n+1),a_1,\cdots,a_p\\ c_1,\cdots,c_q\end{matrix};\lambda\right)\frac{z^{2n}}{(2n)!},
\end{aligned}
\tag{1.345}
$$

$$
\begin{aligned}
&\frac{1}{2}\left[(1-z)^{-a}\ {}_{p+1}F_q\left(\begin{matrix}a,a_1,\cdots,a_p\\ c_1,\cdots,c_q\end{matrix};-\frac{\lambda z}{1-z}\right)-(1+z)^{-a}\ {}_{p+1}F_q\left(\begin{matrix}a,a_1,\cdots,a_p\\ c_1,\cdots,c_q\end{matrix};\frac{\lambda z}{1+z}\right)\right]\\
&=\sum_{n=0}^{\infty}(a)_{2n+1}\ {}_{p+1}F_q\left(\begin{matrix}-(2n+1),a_1,\cdots,a_p\\ c_1,\cdots,c_q\end{matrix};\lambda\right)\frac{z^{2n+1}}{(2n+1)!},
\end{aligned}
\tag{1.346}
$$

and

$$
\begin{aligned}
&\frac{1}{2}\left[(1-z)^{-a}\ {}_{p+1}F_q\left(\begin{matrix}a,a_1,\cdots,a_p\\ c_1,\cdots,c_q\end{matrix};-\frac{\lambda z}{1-z}\right)+(1+z)^{-a}\ {}_{p+1}F_q\left(\begin{matrix}a,a_1,\cdots,a_p\\ c_1,\cdots,c_q\end{matrix};\frac{\lambda z}{1+z}\right)\right]\\
&=\sum_{n=0}^{\infty}(a)_{2n}\ {}_{p+1}F_q\left(\begin{matrix}-(2n+1),a_1,\cdots,a_p\\ c_1,\cdots,c_q\end{matrix};\lambda\right)\frac{z^{2n}}{(2n)!}\,.
\end{aligned}
\tag{1.347}
$$

Proof Under the consideration of the relations

$$(1-iz)^{-a}\ {}_{p+1}F_q\left(\begin{matrix}a, a_1, \cdots, a_p\\ c_1, \cdots, c_q\end{matrix}; -\frac{i\lambda z}{1-iz}\right) = \sum_{n=0}^{\infty}(a)_n\ {}_{p+1}F_q\left(\begin{matrix}-n, a_1, \cdots, a_p\\ c_1, \cdots, c_q\end{matrix}; \lambda\right)\frac{(iz)^n}{n!}, \tag{1.348}$$

$$(1+iz)^{-a}\ {}_{p+1}F_q\left(\begin{matrix}a, a_1, \cdots, a_p\\ c_1, \cdots, c_q\end{matrix}; \frac{i\lambda z}{1+iz}\right) = \sum_{n=0}^{\infty}(a)_n\ {}_{p+1}F_q\left(\begin{matrix}-n, a_1, \cdots, a_p\\ c_1, \cdots, c_q\end{matrix}; \lambda\right)\frac{(-iz)^n}{n!}, \tag{1.349}$$

$$(1-z)^{-a}\ {}_{p+1}F_q\left(\begin{matrix}a, a_1, \cdots, a_p\\ c_1, \cdots, c_q\end{matrix}; -\frac{\lambda z}{1-z}\right) = \sum_{n=0}^{\infty}(a)_n\ {}_{p+1}F_q\left(\begin{matrix}-n, a_1, \cdots, a_p\\ c_1, \cdots, c_q\end{matrix}; \lambda\right)\frac{z^n}{n!}, \tag{1.350}$$

and

$$(1+z)^{-a}\ {}_{p+1}F_q\left(\begin{matrix}a, a_1, \cdots, a_p\\ c_1, \cdots, c_q\end{matrix}; \frac{\lambda z}{1+z}\right) = \sum_{n=0}^{\infty}(a)_n\ {}_{p+1}F_q\left(\begin{matrix}-n, a_1, \cdots, a_p\\ c_1, \cdots, c_q\end{matrix}; \lambda\right)\frac{(-z)^n}{n!}, \tag{1.351}$$

we present that

$$\begin{aligned}&(1-iz)^{-a}\ {}_{p+1}F_q\left(\begin{matrix}a, a_1, \cdots, a_p\\ c_1, \cdots, c_q\end{matrix}; -\frac{i\lambda z}{1-iz}\right)\\ &-(1+iz)^{-a}\ {}_{p+1}F_q\left(\begin{matrix}a, a_1, \cdots, a_p\\ c_1, \cdots, c_q\end{matrix}; \frac{i\lambda z}{1+iz}\right)\\ &=\sum_{n=0}^{\infty}(a)_n\ {}_{p+1}F_q\left(\begin{matrix}-n, a_1, \cdots, a_p\\ c_1, \cdots, c_q\end{matrix}; \lambda\right)\frac{(iz)^n}{n!}\\ &-\sum_{n=0}^{\infty}(a)_n\ {}_{p+1}F_q\left(\begin{matrix}-n, a_1, \cdots, a_p\\ c_1, \cdots, c_q\end{matrix}; \lambda\right)\frac{(-iz)^n}{n!}\\ &=2i\sum_{n=0}^{\infty}(-1)^n(a)_{2n+1}\ {}_{p+1}F_q\left(\begin{matrix}-(2n+1), a_1, \cdots, a_p\\ c_1, \cdots, c_q\end{matrix}; \lambda\right)\frac{z^{2n+1}}{(2n+1)!},\end{aligned} \tag{1.352}$$

$$
\begin{aligned}
&(1-iz)^{-a}\ {}_{p+1}F_q\left(\begin{matrix} a, a_1, \cdots, a_p \\ c_1, \cdots, c_q \end{matrix}; -\tfrac{i\lambda z}{1-iz}\right) \\
&+(1+iz)^{-a}\ {}_{p+1}F_q\left(\begin{matrix} a, a_1, \cdots, a_p \\ c_1, \cdots, c_q \end{matrix}; \tfrac{i\lambda z}{1+iz}\right) \\
&= \sum_{n=0}^{\infty} (a)_n\ {}_{p+1}F_q\left(\begin{matrix} -n, a_1, \cdots, a_p \\ c_1, \cdots, c_q \end{matrix}; \lambda\right) \tfrac{(iz)^n}{n!} \\
&+ \sum_{n=0}^{\infty} (a)_n\ {}_{p+1}F_q\left(\begin{matrix} -n, a_1, \cdots, a_p \\ c_1, \cdots, c_q \end{matrix}; \lambda\right) \tfrac{(-iz)^n}{n!} \\
&= 2\sum_{n=0}^{\infty} (-1)^n (a)_{2n}\ {}_{p+1}F_q\left(\begin{matrix} -(2n+1), a_1, \cdots, a_p \\ c_1, \cdots, c_q \end{matrix}; \lambda\right) \tfrac{z^{2n}}{(2n)!},
\end{aligned}
\tag{1.353}
$$

$$
\begin{aligned}
&(1-z)^{-a}\ {}_{p+1}F_q\left(\begin{matrix} a, a_1, \cdots, a_p \\ c_1, \cdots, c_q \end{matrix}; -\tfrac{\lambda z}{1-z}\right) \\
&-(1+z)^{-a}\ {}_{p+1}F_q\left(\begin{matrix} a, a_1, \cdots, a_p \\ c_1, \cdots, c_q \end{matrix}; \tfrac{\lambda z}{1+z}\right) \\
&= \sum_{n=0}^{\infty} (a)_n\ {}_{p+1}F_q\left(\begin{matrix} -n, a_1, \cdots, a_p \\ c_1, \cdots, c_q \end{matrix}; \lambda\right) \tfrac{z^n}{n!} \\
&- \sum_{n=0}^{\infty} (a)_n\ {}_{p+1}F_q\left(\begin{matrix} -n, a_1, \cdots, a_p \\ c_1, \cdots, c_q \end{matrix}; \lambda\right) \tfrac{(-z)^n}{n!} \\
&= 2\sum_{n=0}^{\infty} (a)_{2n+1}\ {}_{p+1}F_q\left(\begin{matrix} -(2n+1), a_1, \cdots, a_p \\ c_1, \cdots, c_q \end{matrix}; \lambda\right) \tfrac{z^{2n+1}}{(2n+1)!},
\end{aligned}
\tag{1.354}
$$

and

$$
\begin{aligned}
&(1-z)^{-a}\ {}_{p+1}F_q\left(\begin{matrix} a, a_1, \cdots, a_p \\ c_1, \cdots, c_q \end{matrix}; -\tfrac{\lambda z}{1-z}\right) \\
&+(1+z)^{-a}\ {}_{p+1}F_q\left(\begin{matrix} a, a_1, \cdots, a_p \\ c_1, \cdots, c_q \end{matrix}; \tfrac{\lambda z}{1+z}\right) \\
&= \sum_{n=0}^{\infty} (a)_n\ {}_{p+1}F_q\left(\begin{matrix} -n, a_1, \cdots, a_p \\ c_1, \cdots, c_q \end{matrix}; \lambda\right) \tfrac{z^n}{n!} \\
&+ \sum_{n=0}^{\infty} (a)_n\ {}_{p+1}F_q\left(\begin{matrix} -n, a_1, \cdots, a_p \\ c_1, \cdots, c_q \end{matrix}; \lambda\right) \tfrac{(-z)^n}{n!} \\
&= 2\sum_{n=0}^{\infty} (a)_{2n}\ {}_{p+1}F_q\left(\begin{matrix} -(2n+1), a_1, \cdots, a_p \\ c_1, \cdots, c_q \end{matrix}; \lambda\right) \tfrac{z^{2n}}{(2n)!}.
\end{aligned}
\tag{1.355}
$$

□

Theorem 1.93 *Let $a, a_i, c_j \in \mathbb{C}$, $\lambda, z \in \mathbb{C}$, $i, j, p, q \in \mathbb{N}$, and $n \in \mathbb{N}_0$. Then we have*

$$\frac{1}{2i}\left[(1-iz)^{-\lambda}\cdot {}_{p+1}F_q\left(\begin{matrix}a_1,\cdots,a_p\\c_1,\cdots,c_q\end{matrix};-\tfrac{i\lambda z}{1-iz}\right)-(1+iz)^{-\lambda}\cdot {}_{p+1}F_q\left(\begin{matrix}a_1,\cdots,a_p\\c_1,\cdots,c_q\end{matrix};\tfrac{i\lambda z}{1+iz}\right)\right]$$
$$=\sum_{n=0}^{\infty}(-1)^n(\lambda)_{2n+1}\cdot {}_{p+1}F_q\left(\begin{matrix}a_1,\cdots,a_p\\c_1,\cdots,c_q\end{matrix};1\right)\cdot\tfrac{z^{2n+1}}{(2n+1)!},\tag{1.356}$$

$$\frac{1}{2}\left[(1-iz)^{-\lambda}\cdot {}_{p+1}F_q\left(\begin{matrix}a_1,\cdots,a_p\\c_1,\cdots,c_q\end{matrix};-\tfrac{i\lambda z}{1-iz}\right)+(1+iz)^{-\lambda}\cdot {}_{p+1}F_q\left(\begin{matrix}a_1,\cdots,a_p\\c_1,\cdots,c_q\end{matrix};\tfrac{i\lambda z}{1+iz}\right)\right]$$
$$=\sum_{n=0}^{\infty}(-1)^n(\lambda)_{2n}\cdot {}_{p+1}F_q\left(\begin{matrix}a_1,\cdots,a_p\\c_1,\cdots,c_q\end{matrix};1\right)\cdot\tfrac{z^{2n}}{(2n)!},\tag{1.357}$$

$$\frac{1}{2}\left[(1-z)^{-\lambda}\cdot {}_{p+1}F_q\left(\begin{matrix}a_1,\cdots,a_p\\c_1,\cdots,c_q\end{matrix};-\tfrac{\lambda z}{1-iz}\right)-(1+z)^{-\lambda}\cdot {}_{p+1}F_q\left(\begin{matrix}a_1,\cdots,a_p\\c_1,\cdots,c_q\end{matrix};\tfrac{\lambda z}{1+z}\right)\right]$$
$$=\sum_{n=0}^{\infty}(\lambda)_{2n+1}\cdot {}_{p+1}F_q\left(\begin{matrix}a_1,\cdots,a_p\\c_1,\cdots,c_q\end{matrix};1\right)\cdot\tfrac{z^{2n+1}}{(2n+1)!},\tag{1.358}$$

and

$$\frac{1}{2}\left[(1-z)^{-\lambda}\cdot {}_{p+1}F_q\left(\begin{matrix}a_1,\cdots,a_p\\c_1,\cdots,c_q\end{matrix};-\tfrac{\lambda z}{1-iz}\right)+(1+z)^{-\lambda}\cdot {}_{p+1}F_q\left(\begin{matrix}a_1,\cdots,a_p\\c_1,\cdots,c_q\end{matrix};\tfrac{\lambda z}{1+z}\right)\right]$$
$$=\sum_{n=0}^{\infty}(\lambda)_{2n}\cdot {}_{p+1}F_q\left(\begin{matrix}a_1,\cdots,a_p\\c_1,\cdots,c_q\end{matrix};1\right)\cdot\tfrac{z^{2n}}{(2n)!}.\tag{1.359}$$

Proof In view of the relations

$$\begin{aligned}&(1-iz)^{-\lambda}\cdot {}_{p+1}F_q\left(\begin{matrix}a_1,\cdots,a_p\\c_1,\cdots,c_q\end{matrix};-\tfrac{i\lambda z}{1-iz}\right)\\&=\sum_{n=0}^{\infty}(\lambda)_n\cdot {}_{p+1}F_q\left(\begin{matrix}a_1,\cdots,a_p\\c_1,\cdots,c_q\end{matrix};1\right)\cdot\frac{(iz)^n}{n!},\end{aligned}\tag{1.360}$$

$$\begin{aligned}&(1+iz)^{-\lambda}\cdot {}_{p+1}F_q\left(\begin{matrix}a_1,\cdots,a_p\\c_1,\cdots,c_q\end{matrix};\tfrac{i\lambda z}{1+iz}\right)\\&=\sum_{n=0}^{\infty}(\lambda)_n\cdot {}_{p+1}F_q\left(\begin{matrix}a_1,\cdots,a_p\\c_1,\cdots,c_q\end{matrix};1\right)\cdot\frac{(-iz)^n}{n!},\end{aligned}\tag{1.361}$$

$$\begin{aligned}&(1-z)^{-\lambda}\cdot {}_{p+1}F_q\left(\begin{matrix}a_1,\cdots,a_p\\c_1,\cdots,c_q\end{matrix};-\tfrac{\lambda z}{1-z}\right)\\&=\sum_{n=0}^{\infty}(\lambda)_n\cdot {}_{p+1}F_q\left(\begin{matrix}a_1,\cdots,a_p\\c_1,\cdots,c_q\end{matrix};1\right)\cdot\tfrac{z^n}{n!}.\end{aligned}\tag{1.362}$$

$$\begin{aligned}&(1+z)^{-\lambda}\cdot {}_{p+1}F_q\left(\begin{matrix}a_1,\cdots,a_p\\c_1,\cdots,c_q\end{matrix};\tfrac{\lambda z}{1+z}\right)\\&=\sum_{n=0}^{\infty}(\lambda)_n\cdot {}_{p+1}F_q\left(\begin{matrix}a_1,\cdots,a_p\\c_1,\cdots,c_q\end{matrix};1\right)\cdot\tfrac{(-z)^n}{n!}\end{aligned}\tag{1.363}$$

we give

$$\begin{aligned}&(1-iz)^{-\lambda}\cdot {}_{p+1}F_q\left(\begin{matrix}a_1,\cdots,a_p\\c_1,\cdots,c_q\end{matrix};-\tfrac{i\lambda z}{1-iz}\right)\\&-(1+iz)^{-\lambda}\cdot {}_{p+1}F_q\left(\begin{matrix}a_1,\cdots,a_p\\c_1,\cdots,c_q\end{matrix};\tfrac{i\lambda z}{1+iz}\right)\\&=\sum_{n=0}^{\infty}(\lambda)_n\cdot {}_{p+1}F_q\left(\begin{matrix}a_1,\cdots,a_p\\c_1,\cdots,c_q\end{matrix};1\right)\cdot\tfrac{(iz)^n}{n!}\\&-\sum_{n=0}^{\infty}(\lambda)_n\cdot {}_{p+1}F_q\left(\begin{matrix}a_1,\cdots,a_p\\c_1,\cdots,c_q\end{matrix};1\right)\cdot\tfrac{(-iz)^n}{n!}\\&=2i\sum_{n=0}^{\infty}(-1)^n\,(\lambda)_{2n+1}\cdot {}_{p+1}F_q\left(\begin{matrix}a_1,\cdots,a_p\\c_1,\cdots,c_q\end{matrix};1\right)\cdot\tfrac{z^{2n+1}}{(2n+1)!},\end{aligned}\tag{1.364}$$

$$\begin{aligned}&(1-iz)^{-\lambda}\cdot {}_{p+1}F_q\left(\begin{matrix}a_1,\cdots,a_p\\c_1,\cdots,c_q\end{matrix};-\tfrac{i\lambda z}{1-iz}\right)\\&+(1+iz)^{-\lambda}\cdot {}_{p+1}F_q\left(\begin{matrix}a_1,\cdots,a_p\\c_1,\cdots,c_q\end{matrix};\tfrac{i\lambda z}{1+iz}\right)\\&=\sum_{n=0}^{\infty}(\lambda)_n\cdot {}_{p+1}F_q\left(\begin{matrix}a_1,\cdots,a_p\\c_1,\cdots,c_q\end{matrix};1\right)\cdot\tfrac{(iz)^n}{n!}\\&+\sum_{n=0}^{\infty}(\lambda)_n\cdot {}_{p+1}F_q\left(\begin{matrix}a_1,\cdots,a_p\\c_1,\cdots,c_q\end{matrix};1\right)\cdot\tfrac{(-iz)^n}{n!}\\&=2\sum_{n=0}^{\infty}(-1)^n\,(\lambda)_{2n}\cdot {}_{p+1}F_q\left(\begin{matrix}a_1,\cdots,a_p\\c_1,\cdots,c_q\end{matrix};1\right)\cdot\tfrac{z^{2n}}{(2n)!},\end{aligned}\tag{1.365}$$

$$\begin{aligned}&(1-z)^{-\lambda}\cdot {}_{p+1}F_q\left(\begin{matrix}a_1,\cdots,a_p\\c_1,\cdots,c_q\end{matrix};-\tfrac{\lambda z}{1-iz}\right)\\&-(1+z)^{-\lambda}\cdot {}_{p+1}F_q\left(\begin{matrix}a_1,\cdots,a_p\\c_1,\cdots,c_q\end{matrix};\tfrac{\lambda z}{1+z}\right)\\&=\sum_{n=0}^{\infty}(\lambda)_n\cdot {}_{p+1}F_q\left(\begin{matrix}a_1,\cdots,a_p\\c_1,\cdots,c_q\end{matrix};1\right)\cdot\tfrac{z^n}{n!}\\&-\sum_{n=0}^{\infty}(\lambda)_n\cdot {}_{p+1}F_q\left(\begin{matrix}a_1,\cdots,a_p\\c_1,\cdots,c_q\end{matrix};1\right)\cdot\tfrac{(-z)^n}{n!}\\&=2\sum_{n=0}^{\infty}(\lambda)_{2n+1}\cdot {}_{p+1}F_q\left(\begin{matrix}a_1,\cdots,a_p\\c_1,\cdots,c_q\end{matrix};1\right)\cdot\tfrac{z^{2n+1}}{(2n+1)!},\end{aligned}\tag{1.366}$$

and

$$\begin{aligned}
&(1-z)^{-\lambda} \cdot {}_{p+1}F_q\left(\begin{matrix} a_1, \cdots, a_p \\ c_1, \cdots, c_q \end{matrix}; -\tfrac{\lambda z}{1-iz}\right) \\
&+(1+z)^{-\lambda} \cdot {}_{p+1}F_q\left(\begin{matrix} a_1, \cdots, a_p \\ c_1, \cdots, c_q \end{matrix}; \tfrac{\lambda z}{1+z}\right) \\
&= \sum_{n=0}^{\infty} (\lambda)_n \cdot {}_{p+1}F_q\left(\begin{matrix} a_1, \cdots, a_p \\ c_1, \cdots, c_q \end{matrix}; 1\right) \cdot \tfrac{z^n}{n!} \\
&+ \sum_{n=0}^{\infty} (\lambda)_n \cdot {}_{p+1}F_q\left(\begin{matrix} a_1, \cdots, a_p \\ c_1, \cdots, c_q \end{matrix}; 1\right) \cdot \tfrac{(-z)^n}{n!} \\
&= 2\sum_{n=0}^{\infty} (\lambda)_{2n} \cdot {}_{p+1}F_q\left(\begin{matrix} a_1, \cdots, a_p \\ c_1, \cdots, c_q \end{matrix}; 1\right) \cdot \tfrac{z^{2n}}{(2n)!}.
\end{aligned} \tag{1.367}$$

□

1.4 The Laplace and Mellin Transforms

In this section, we give the investigation of the Laplace and Mellin transforms.

1.4.1 The Laplace Transforms for the Special Functions

We now begin with the definition and properties of the Laplace transforms.

Definition 1.47 The Laplace transform of the function is defined as [88]

$$L[f(t)] = f(s) = \int_0^{\infty} e^{-st} f(t)\, dt \tag{1.368}$$

and the corresponding inverse Laplace transform as [88]

$$f(t) = L^{-1}[f(s)] = \frac{1}{2\pi i} \lim_{T\to\infty} \int_{r-iT}^{r+iT} e^{st} f(s)\, ds, \tag{1.369}$$

where $r \in \mathbb{R}$ and $s = r + iT$.

The result was discovered by Laplace in 1812 [88] and its extended version is showed in Chap. 7.

The properties of Laplace transform are given in Table 1.1 [39].

The Laplace transforms of the functions are given in Table 1.2 [39].

Table 1.1 The properties of the Laplace transform

The functions	Laplace transforms
$f(t) + g(t)$	$f(s) + g(s)$
$tf(t)$	$-f^{(1)}(s)$
$t^{\kappa} f(t)$	$(-1)^{\kappa} f^{(\kappa)}(s)$
$f^{(1)}(t)$	$sf(s) - f(0)$
$f^{(\kappa)}(t)$	$s^{\kappa} f(s) - \sum_{j=1}^{\kappa} s^{\kappa-j} f^{(j-1)}(0)$
$\frac{f(t)}{t}$	$\int_{s}^{\infty} f(\rho)\, d\rho$
$\int_{0}^{t} f(t)\, dt$	$\frac{f(s)}{s}$
$f(t-\lambda)$	$e^{-\lambda s} f(s)$
$e^{\lambda t} f(t)$	$f(s-\lambda)$
$f(at)(\lambda > 0)$	$f(s/\lambda)/\lambda$
$\int_{0}^{t} f(\tau)\, g(t-\tau)\, d\tau$	$f(s)\, g(s)$

Table 1.2 Laplace transforms of the functions

The functions	Laplace transforms of the functions
$e^{-\lambda t}$	$s^{-1}\left(1 + \lambda s^{-1}\right)^{-1}$
$\vartheta(\tau)$	s^{-1}
$\delta(\tau)$	1
$t^{-\nu}/\Gamma(1-\nu)$	s^{ν}
$t^{\nu}/\Gamma(1+\nu)$	$s^{-\nu}$
t^{-1}	s
t	s^{-1}
$t^{1/\kappa}/\Gamma(1+1/\kappa)$	$s^{-(1+1/\kappa)}$
$t^{\nu} e^{-\lambda t}/\Gamma(1+\nu)$	$(s+\lambda)^{-(\nu+1)}$
$\sin(\lambda t)$	$\frac{\lambda}{s^2+\lambda^2}$
$\cos(\lambda t)$	$\frac{s}{s^2+\lambda^2}$
$\sinh(\lambda t)$	$\frac{\lambda}{s^2-\lambda^2}$
$\cosh(\lambda t)$	$\frac{s}{s^2-\lambda^2}$
$t^{\nu} \sinh(-\lambda t)/\Gamma(1+\nu)$	$\frac{1}{2}\left[(s+\lambda)^{-(\nu+1)} + (s-\lambda)^{-(\nu+1)}\right]$
$t^{\nu} \cosh(-\lambda t)/\Gamma(1+\nu)$	$\frac{1}{2}\left[(s+\lambda)^{-(\nu+1)} - (s-\lambda)^{-(\nu+1)}\right]$
$t^{\nu} \sin(-\lambda t)/\Gamma(1+\nu)$	$\frac{1}{2}\left[(s+i\lambda)^{-(\nu+1)} + (s-i\lambda)^{-(\nu+1)}\right]$
$t^{\nu} \cos(\lambda t)/\Gamma(1+\nu)$	$\frac{1}{2}\left[(s+i\lambda)^{-(\nu+1)} - (s-i\lambda)^{-(\nu+1)}\right]$

The Laplace transform of the function $t_p^{\lambda-1} F_q(a_1, \cdots, a_p; c_1, \cdots, c_q; zt)$ is given as follows:

$$
\begin{aligned}
& L\left[t_p^{\lambda-1} F_q(a_1, \cdots, a_p; c_1, \cdots, c_q; zt)\right] \\
&= L\left[t_p^{\lambda-1} F_q\left(\begin{matrix} a_1, \cdots, a_p \\ c_1, \cdots, c_q \end{matrix}; zt\right)\right] \\
&= \int_0^\infty e^{-st}\left(t_p^{\lambda-1} F_q\left(\begin{matrix} a_1, \cdots, a_p \\ c_1, \cdots, c_q \end{matrix}; zt\right)\right) dt \\
&= \frac{\Gamma(\lambda)}{s^\lambda}\, {}_{p+1}F_q\left(\begin{matrix} \lambda, a_1, \cdots, a_p \\ c_1, \cdots, c_q \end{matrix}; \frac{z}{s}\right).
\end{aligned}
\tag{1.370}
$$

The convergent cases are as follows:

(1) $Re(\lambda) > 0, p \le q$;
(2) $Re(s) > 0$ if $p < q$;
(3) $Re(s) > Re(z)$ if $p = q$.

In this case, we have the Laplace transform as follows [89]:

$$
\begin{aligned}
& L^{-1}\left[\frac{\Gamma(\lambda)}{s^\lambda}\, {}_{p+1}F_q\left(\begin{matrix} \lambda, a_1, \cdots, a_p \\ c_1, \cdots, c_q \end{matrix}; \frac{z}{s}\right)\right] \\
&= \frac{t^\lambda}{\Gamma(\lambda)}\, {}_pF_{q+1}\left(\begin{matrix} \lambda, a_1, \cdots, a_p \\ c_1, \cdots, c_q \end{matrix}; zt\right),
\end{aligned}
\tag{1.371}
$$

where $Re(\lambda) > 0$ and $p \le q + 1$.

For the details of the Laplace transforms of the functions, see [90].

1.4.2 The Mellin Transforms for the Special Functions

We now begin with the definition and properties of the Mellin transforms of the special functions.

Definition 1.48 The Mellin transform of the function is defined as [91]

$$
M[f(t)] := f(\varpi) = \int_0^\infty t^{\varpi-1} f(t)\, dt,
\tag{1.372}
$$

and the corresponding inverse Mellin transform is as follows [91]:

$$
f(t) = M^{-1}[f(\varpi)] = \frac{1}{2\pi i} \lim_{T\to\infty} \int_{r-iT}^{r+iT} t^{-\varpi} f(\varpi)\, d\varpi,
\tag{1.373}
$$

where $r \in R$ and $\varpi = r + iT$.

Table 1.3

The functions	Mellin transforms
$f(t)+g(t)$	$f(\varpi)+g(\varpi)$
$tf(t)$	$f(\varpi+1)$
$t^{\kappa}f(t)$	$f(\varpi+\kappa)$
$f^{(\kappa)}(t)$	$\frac{\Gamma(\kappa+1-\varpi)}{\Gamma(1-\varpi)}f(\varpi-\kappa)$
$\left(t\frac{d}{dt}\right)^{\kappa}f(t)$	$(-\varpi)^{\kappa}f(\varpi)$
$\left(\frac{d}{dt}t\right)^{\kappa}f(t)$	$(1-\varpi)^{\kappa}f(\varpi)$
$\frac{f(t)}{t}$	$f(\varpi-1)$
$\int_0^t f(t)\,dt$	$-\varpi^{-1}f(\varpi)$
$\int_t^{\infty} f(t)\,dt$	$\varpi^{-1}f(\varpi)$
$f\left(t^{\lambda}\right)$	$\lambda^{-1}f(\varpi/\lambda)$
$f\left(t^{-\lambda}\right)$	$\lambda^{-1}f(-\varpi/\lambda)$
$t^{\gamma}f\left(at^{\lambda}\right)$	$\lambda^{-1}a^{-(\gamma+\varpi)/\lambda}f((\gamma+\varpi)/\lambda)$
$t^{\gamma}f\left(at^{-\lambda}\right)$	$\lambda^{-1}a^{(\gamma+\varpi)/\lambda}f(-(\gamma+\varpi)/\lambda)$
$f(at)\,(\lambda>0)$	$a^{-\varpi}f(\varpi)$
$t^{\gamma}\int_0^{\infty}\tau^{\lambda}f(t\tau)\,g(\tau)\,d\tau$	$f(\varpi+\gamma)\,g(1-\varpi-\gamma+\lambda)$
$t^{\gamma}\int_0^{\infty}\tau^{\lambda}f(t/\tau)\,g(\tau)\,d\tau$	$f(\varpi+\gamma)\,g(1+\varpi+\gamma+\lambda)$
$\int_0^{\infty}\tau^{\lambda}f(t\tau)\,g(\tau)\,d\tau$	$f(\varpi+1)\,g(\lambda-\varpi)$
$\int_0^{\infty}\tau^{\lambda}f(t/\tau)\,g(\tau)\,d\tau$	$f(\varpi+1)\,g(2+\varpi+\gamma)$
$t^{\gamma}\int_0^{\infty}f(t\tau)\,g(\tau)\,d\tau$	$f(\varpi+\gamma)\,g(-(\varpi+\gamma))$
$t^{\gamma}\int_0^{\infty}f(t/\tau)\,g(\tau)\,d\tau$	$f(\varpi+\gamma)\,g(2+\varpi+\gamma)$
$\int_0^{\infty}f(t\tau)\,g(\tau)\,d\tau$	$f(\varpi+1)\,g(1-\varpi)$
$\int_0^{\infty}f(t/\tau)\,g(\tau)\,d\tau$	$f(\varpi+1)\,g(3+\varpi)$
$\int_0^t f(\tau)\,g\left(\frac{t}{\tau}\right)\frac{d\tau}{\tau}$	$f(\varpi)\,g(\varpi)$

The result was discovered by Mellin in 1896 [91] and its extended version is showed in Chap. 7.

The properties of Mellin transform are given in Table 1.3 [92].

The Mellin transforms of the functions are given in Table 1.4 [92].

Table 1.4

The functions	Mellin transforms of the functions
$(t/a)^{\nu}\ (t < a)$	$(\nu + \varpi)^{-1} a^{\varpi}$
$(t/a)^{\nu}\ (t > a)$	$(\nu + \varpi)^{-1} a^{\varpi}$
$t^{\nu}\ (t < a)$	$(\nu + \varpi)^{-1} a^{\nu+\varpi}$
$t^{\nu}\ (t > a)$	$-(\nu + \varpi)^{-1} a^{\nu+\varpi}$
$1\ (t < a)$	$\varpi^{-1} a^{\varpi}$
$-1\ (t > a)$	$\varpi^{-1} a^{\varpi}$
$e^{-\lambda t}$	$\Gamma(\varpi)\lambda^{-\varpi}$
$\sin t$	$\Gamma(\varpi)\sin\left(\frac{1}{2}\pi\varpi\right)$
$\cos t$	$\Gamma(\varpi)\cos\left(\frac{1}{2}\pi\varpi\right)$
The unit step function $\vartheta(\tau - a)\ (a > 0)$	$-\varpi^{-1} a^{\varpi}$
The unit step function $\vartheta(a - \tau)\ (a < 0)$	$\varpi^{-1} a^{\varpi}$

Barnes Theorem [57]

If $|\arg(-z)| < \pi$, then there is [57]

$$\frac{1}{2\pi i}\int_{-i\infty}^{i\infty}\frac{\Gamma(a+s)\Gamma(b+s)\Gamma(-s)}{\Gamma(c+s)}(-t)^{\varpi}\,d\varpi = \frac{\Gamma(a)\Gamma(b)}{\Gamma(c)}\,{}_2F_1\left(\begin{matrix}a,b\\c\end{matrix};t\right). \tag{1.374}$$

The result was discovered by Barnes in 1908 [57].

The Mellin transform of the function ${}_pF_q\left(a_1,\cdots,a_p;c_1,\cdots,c_q;-zt\right)$ is given as follows [39, 69]:

$$\begin{aligned}
&M\left[{}_pF_q\left(a_1,\cdots,a_p;c_1,\cdots,c_q;-zt\right)\right]\\
&= M\left[{}_pF_q\left(\begin{matrix}a_1,\cdots,a_p\\c_1,\cdots,c_q\end{matrix};-zt\right)\right]\\
&= \int_0^{\infty} t^{\varpi-1}_{\ p}F_q\left(\begin{matrix}a_1,\cdots,a_p\\c_1,\cdots,c_q\end{matrix};-zt\right)dt\\
&= \frac{\prod_{i=1}^{p}\Gamma(a_i-\varpi)\prod_{i=j}^{q}\Gamma(c_j)}{\prod_{i=1}^{p}\Gamma(a_i)\prod_{i=j}^{q}\Gamma(c_j-\varpi)}\Gamma(\omega)z^{-\varpi},
\end{aligned} \tag{1.375}$$

where $z \neq 0$ and $0 < Re(\varpi) < \min\limits_{1\le i\le p} Re(a_i)$.

The convergent cases are as follows:

(1) $p = q - 1, z > 0$;
(2) $p = q, |\arg(z)| \le \frac{\pi}{2}$;
(3) $p = q + 1, |\arg(z)| < \frac{\pi}{2}$.

The result was discovered by Srivastava and Kashyap in 1982 [69] and further reported in [39].

1.5 Calculus with Respect to Monotone Functions

In this section, we introduce the calculus with respect to monotone functions [92].

1.5.1 The Newton–Leibniz Calculus

Let us recall the Newton's and Leibniz's calculus as follows.

Definition 1.49 (Newton [93]; Leibniz [94])
The Newton–Leibniz derivative with respect to the variable t is defined as [93, 94]

$$D^{(1)}\varphi(t) = \frac{d\varphi(t)}{dt} = \lim_{\Delta t \to 0} \frac{\varphi(t+\Delta t) - \varphi(t)}{\Delta t}. \tag{1.376}$$

Definition 1.50 (Newton [93]; Leibniz [94])
The Newton–Leibniz integral with respect to the variable t is defined as [93, 94]

$${}_aI_t^{(1)}\psi(t) = \int_a^t \psi(t)\,dt = \lim_{n \to \infty} \sum_{k=1}^{n} \psi(t_k)\left(\frac{b-a}{n}\right), \tag{1.377}$$

where $a, b \in \mathbb{R}$ and $t_k = a + \frac{k(b-a)}{n}$, $k \in \mathbb{R}$ and $n \in \mathbb{N}$.

For the history of the calculus, see [95, 96].

The relationship between (1.376) and (1.377), the Newton–Leibniz formulae, reads [93, 94]

$$\psi(t) = \frac{d}{dt} \int_a^t \psi(t)\,dt \tag{1.378}$$

and

$$\psi(t) - \psi(a) = \int_a^t \frac{d}{dt}\psi(t)\,dt. \tag{1.379}$$

1.5.2 Calculus with Respect to Monotone Function

This idea of the calculus with respect to monotone functions (the so-called general calculus) goes back to the Leibniz [97], Stieltjes [98], and Lebesgue [99] tasks. The derivative with respect to monotone function was derivated from the chain rules for the Newton–Leibniz derivative for the composite functions by Leibniz in 1676 [97]. The integral with respect to monotone function is derivated from the rule for the change of variables for definite integrals was proposed by Stieltjes in 1894 [98] under the consideration of the Riemann sense [100] and Lebesgue [99] in 1902, which is called the Lebesgue–Stieltjes integral in Lebesgue sense [101, 102] and which is called the Riemann–Stieltjes integral in the Riemann sense [103–108]. For the change of the variable within the Lebesgue integral, see [109].

1.5.2.1 The Leibniz Derivative

Let us consider the Leibniz derivative (the so-called derivative with respect to monotone function) and Lebesgue–Stieltjes integral (the so-called integral with respect to monotone function as follows [110, 111].

Let $g(t)$ be a monotone increasing function for $t \in \mathbb{R}$.

Definition 1.51 The Leibniz derivative of the function $\phi(t)$ (also called the derivative with respect to another function) is defined as [110, 111]

$$D_{t,g(\cdot)}^{(1)}\phi(t) = \frac{d\varphi(g(t))}{dg(t)} = \frac{1}{g^{(1)}(t)} \lim_{\Delta t \to 0} \frac{\phi(t+\Delta t) - \phi(t)}{\Delta t} = \frac{1}{g^{(1)}(t)} \frac{d\phi(t)}{dt}, \tag{1.380}$$

where $\phi(t) = (\varphi \circ g)(t) = \varphi(g(t))$ and $g^{(1)}(t) > 0$.

The result was discovered by Leibniz in 1676 [99] and rediscovered as the differential operator related to the inverse operator of the Lebesgue–Stieltjes integral by [110, 111].

The physical interpretation of the derivative with respect to another function is the slope of the tangent line with respect to monotone function$g(t)$ [111].

The comparison of the geometric interpretations of the Newton–Leibniz derivative and derivative with respect to another function is illustrated in Fig. 1.1.

Definition 1.52 The Leibniz derivative of high-order with respect to another function is defined as [111]

$$D_{t,g(\cdot)}^{(n)}\psi_g(t) = \left(\frac{1}{g^{(1)}(t)}\frac{d}{dt}\right)^n \psi_g(t), \tag{1.381}$$

where $\psi(t) = (\varphi \circ g)(t) = \varphi(g(t))$ and $n \in \mathbb{N}$.

The result was discovered by Yang in 2019 [111].

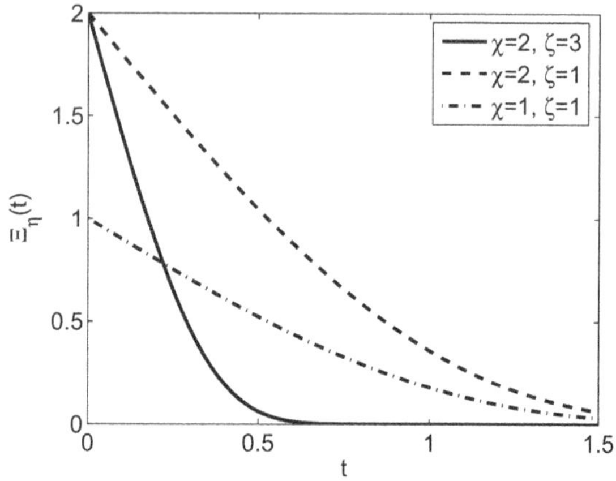

Fig. 1.1 The plots of $\Xi_\eta(t)$ when $\chi = 2$, $\zeta = 3$, $\chi = 2$, $\zeta = 1$, $\chi = 1$, and $\zeta = 1$, respectively

Definition 1.53 The Leibniz differential with respect to another function, denoted by $d\phi(t)$, is defined as

$$d\phi(t) = g^{(1)}(t)\, D^{(1)}_{t,g(\cdot)}\phi(t)\, dt, \tag{1.382}$$

where $g(t)$ is a monotone function for $t \in \mathbb{R}$ and $\phi(t) = (\varphi \circ g)(t) = \varphi(g(t))$.

The theorems for the Leibniz derivative are given as follows:

Theorem 1.94 (The Sum and Difference Rules for the Leibniz Derivative) *Suppose that $D^{(1)}_{t,g(\cdot)}\phi_1(t)$ and $D^{(1)}_{t,g(\cdot)}\phi_2(t)$ exist and $g^{(1)}(t) > 0$, then we have*

$$D^{(1)}_{t,g(\cdot)}(\phi_1(t) + \phi_2(t)) = D^{(1)}_{t,g(\cdot)}\phi_1(t) + D^{(1)}_{t,g(\cdot)}\phi_2(t) \tag{1.383}$$

and

$$D^{(1)}_{t,g(\cdot)}(\phi_1(t) - \phi_2(t)) = D^{(1)}_{t,g(\cdot)}\phi_1(t) - D^{(1)}_{t,g(\cdot)}\phi_2(t). \tag{1.384}$$

Theorem 1.95 (The Constant Multiple Rule for the Leibniz Derivative) *Suppose that $D^{(1)}_{t,g(\cdot)}\phi(t)$ exists, $g^{(1)}(t) > 0$ and l is a constant, then we have*

$$D^{(1)}_{t,g(\cdot)}(l\phi(t)) = l D^{(1)}_{t,g(\cdot)}\phi(t). \tag{1.385}$$

Theorem 1.96 (The Product Rule for the Leibniz Derivative) *Suppose that $D^{(1)}_{t,g(\cdot)}\phi_1(t)$ and $D^{(1)}_{t,g(\cdot)}\phi_2(t)$ exist and $g^{(1)}(t) > 0$, then we have*

$$D^{(1)}_{t,g(\cdot)}(\phi_1(t) \cdot \phi_2(t)) = \phi_2(t)\, D^{(1)}_{t,g(\cdot)}\phi_1(t) + \phi_1(t)\, D^{(1)}_{t,g(\cdot)}\phi_2(t). \tag{1.386}$$

Theorem 1.97 (The Quotient Rule for the Leibniz Derivative)
Suppose that $D^{(1)}_{t,g(\cdot)}\phi_1(t)$ *and* $D^{(1)}_{t,g(\cdot)}\phi_2(t)$ *exist,* $g^{(1)}(t) > 0$ *and* $\phi_2(t) \neq 0$, *then we have*

$$D^{(1)}_{t,g(\cdot)}\left(\frac{\phi_1(t)}{\phi_2(t)}\right) = \frac{\phi_2(t)\, D^{(1)}_{t,g(\cdot)}\phi_1(t) - \phi_1(t)\, D^{(1)}_{t,g(\cdot)}\phi_2(t)}{\phi_2(t)\cdot\phi_2(t)}. \tag{1.387}$$

Theorem 1.98 (The Chain Rule for the Leibniz Derivative)
Suppose that $\frac{d\phi(w)}{dw}$ *and* $D^{(1)}_{t,g(\cdot)}w(t)$ *exist and* $g^{(1)}(t) > 0$, *then we have*

$$D^{(1)}_{t,g(\cdot)}(\phi(w(t))) = \frac{d\phi(w)}{dw}\cdot D^{(1)}_{t,g(\cdot)}w(t). \tag{1.388}$$

The properties of the Leibniz derivative are given as follows:

$$D^{(1)}_{t,g(\cdot)}1 = 0, \tag{1.389}$$

$$D^{(1)}_{t,g(\cdot)}g(t) = 1, \tag{1.390}$$

$$D^{(1)}_{t,g(\cdot)}g^n(t) = ng^{n-1}(t), \tag{1.391}$$

$$D^{(1)}_{t,g(\cdot)}e^{g(t)} = e^{g(t)}, \tag{1.392}$$

$$D^{(1)}_{t,g(\cdot)}\ln g(t) = \frac{1}{g(t)}, \tag{1.393}$$

$$D^{(1)}_{t,g(\cdot)}\log_s g(t) = \frac{1}{(\ln s)\, g(t)}, \tag{1.394}$$

$$D^{(1)}_{t,g(\cdot)}s^{g(t)} = (\ln s)\, s^{g(t)}, \tag{1.395}$$

$$D^{(1)}_{t,g(\cdot)}e^{\psi_g(t)} = e^{\psi_g(t)} D^{(1)}_{t,g(\cdot)}\psi_g(t), \tag{1.396}$$

$$D^{(1)}_{t,g(\cdot)}\left|\psi_g(t)\right| = \frac{\psi_g(t)}{\left|\psi_g(t)\right|} D^{(1)}_{t,g(\cdot)}\psi_g(t),\ \psi_g(t) \neq 0, \tag{1.397}$$

$$D^{(1)}_{t,g(\cdot)}\ln\psi_g(t) = \frac{D^{(1)}_{t,g(\cdot)}\psi_g(t)}{\psi_g(t)}, \tag{1.398}$$

$$D^{(1)}_{t,g(\cdot)}\log_s\psi_g(t) = \frac{D^{(1)}_{t,g(\cdot)}\psi_g(t)}{(\ln s)\,\psi_g(t)}, \tag{1.399}$$

and

$$D_{t,g(\cdot)}^{(1)} s^{\psi_g(t)} = \left[(\ln s)\, s^{\psi_g(t)} \right] \cdot D_{t,g(\cdot)}^{(1)} \psi_g(t) . \tag{1.400}$$

Definition 1.54 Let $g^{(1)}(t) > 0$, where $t \in (a,b)$, and $\psi_g(t) = (\psi \circ g)(t) = \psi(g(t))$.

- If there is an open interval containing γ on which $\psi_g(t) = (\psi \circ g)(t) = \psi(g(t))$is a maximum with respect to another function $g(\gamma)$, then $\psi_g(t) = (\psi \circ g)(t) = \psi(g(t))$is called a relative maximum with respect to another function $g(\gamma)$.
- If there is an open interval containing γ on which $\psi_g(t) = (\psi \circ g)(t) = \psi(g(t))$is a minimum with respect to another function, then $\psi_g(t) = (\psi \circ g)(t) = \psi(g(t))$is called a relative minimum with respect to another function $g(\gamma)$.
- If $\psi_g(t) = (\psi \circ g)(t) = \psi(g(t))$ has a relative minimum or relative maximum with respect to another function $g(t)$ at $t = \gamma$, then γ is a critical number with respect to another function $g(t)$.

The relative maximum and relative minimum with respect to another function $g(t)$ are sometimes called local maximum and local minimum with respect to another function $g(t)$, respectively.

Theorem 1.99 *Let $g^{(1)}(t) > 0$, where $t \in (a,b)$ with $a < b$, $\psi_g(t) = (\psi \circ g)(t) = \psi(g(t))$ and $D_{t,g(\cdot)}^{(1)} \psi_g(\gamma) = 0$ for $\gamma \in (a,b)$. There we have the following cases:*

- *If $D_{t,g(\cdot)}^{(1)} \psi_\eta(t) > 0$ for $t \in (a,\gamma)$ and $D_{t,g(\cdot)}^{(1)} \psi_\eta(t) < 0$ for $t \in (\gamma, b)$, then $\psi_g(t)$ has a local maximum with respect to another function $g(t)$ at $t = \gamma$.*
- *If $D_{t,g(\cdot)}^{(1)} \psi_g(t) < 0$ for $t \in (a,\gamma)$ and $D_{t,g(\cdot)}^{(1)} \psi_g(t) > 0$ for $t \in (\gamma, b)$, then $\psi_g(t)$ has a local minimum with respect to another function $g(t)$ at $t = \gamma$.*

Theorem 1.100 *Let $g^{(1)}(t) > 0$, where $t \in (a,b)$ with $a < b$, and $\psi_g(t) = (\psi \circ g)(t) = \psi(g(t))$.*

Suppose that $D_{t,g(\cdot)}^{(1)} \psi_g(t)$ exists at $t \in (a,b)$with $a < b$, then we have the following cases:

- *If $D_{t,g(\cdot)}^{(2)} \psi_g(t) \geq 0$ for $t \in (a,b)$, then we have $\psi_g(t)$ is the convex function with respect to another function $g(t)$, defined by*

$$\psi(p g_1(t) + (1-p) g_2(t)) \leq p \psi(g_1(t)) + (1-p) \psi(g_2(t)), \tag{1.401}$$

where $p \in (0,1)$, and both $g_1(t)$ and $g_2(t)$ are two function with $g_1^{(1)}(t) > 0$ and $g_2^{(1)}(t) > 0$ for $t \in (a,b)$.

- *If* $D^{(2)}_{t,g(\cdot)}\psi_g(t) < 0$ *for* $t \in (a,b)$, *then we have* $\psi_g(t)$, *which is the concave function with respect to another function* $g(t)$, *defined by*

$$\psi(pg_1(t) + (1-p)g_2(t)) > p\psi(g_1(t)) + (1-p)\psi(g_2(t)), \tag{1.402}$$

where $p \in (0,1)$, *and both* $g_1(t)$ *and* $g_2(t)$ *are two function with* $g_1^{(1)}(t) > 0$ *and* $g_2^{(1)}(t) > 0$ *for* $t \in (a,b)$.

1.5.2.2 The Riemann–Stieltjes Integrals

Let us consider the concepts, properties, and theorems of the Riemann–Stieltjes integrals.

The integral with respect to monotone function is derivated from the rule for the change of variables for definite integrals was proposed by Stieltjes in 1894 [98] and Lebesgue [99] in 1902, which is called the Lebesgue–Stieltjes integral [102], was structured by Stieltjes in 1894 involving the integrals of the Riemann type, which is called the Riemann–Stieltjes integral in the Riemann sense [103–105], and reconsidered as the differential operator related to the inverse operator of the Leibniz derivative by Yang in 2019 [110, 111].

Definition 1.55 The Riemann–Stieltjes integral is defined as

$$\begin{aligned} &{}_aI^{(1)}_{b,g(\cdot)}\psi(t) \\ &= \lim_{n\to\infty}\sum_{k=1}^{n}\Omega(g_k(t))\left(\tfrac{g(b)-g(a)}{n}\right) \\ &= \int_a^b \Omega(g(t))\,dg(t) \\ &= \int_a^b \psi(t)\,g^{(1)}(t)\,dt \end{aligned}, \tag{1.403}$$

where $a, b \in \mathbb{R}$, $\psi(t) = (\Omega \circ g)(t) = \Omega(g(t))$ and

$$g_k(t) = g(a) + \frac{k(g(b) - g(a))}{n},$$

with $n, k \in \mathbb{N}$.

The result was discovered by Stieltjes in 1894 [98] with the sense of Riemann [100] and rediscovered as the differential operator related to the inverse operator of the Leibniz derivative by Yang in 2019 [110, 111].

The Riemann–Stieltjes integral

$$\lim_{n\to\infty}\sum_{k=1}^{n}\Omega\left(g_k\left(t\right)\right)\left(\frac{g\left(b\right)-g\left(a\right)}{n}\right)=\int_{a}^{b}\Omega\left(g\left(t\right)\right)dg\left(t\right), \tag{1.404}$$

where $a, b \in \mathbb{R}$, $\psi\left(t\right)=\left(\Omega\circ g\right)\left(t\right)=\Omega\left(g\left(t\right)\right)$ and $g_k\left(t\right)=g\left(a\right)+\frac{k\left(g\left(b\right)-g\left(a\right)\right)}{n}$ with $n, k \in \mathbb{N}$, was shown by Carmichael in 1919 [112] and discussed by Russell in 1978 [113].

It is clear that the expression

$$\int_{a}^{b}\Omega\left(g\left(t\right)\right)dg\left(t\right)=\int_{a}^{b}\psi\left(t\right)g^{(1)}\left(t\right)dt \tag{1.405}$$

was shown by Widder in 1947 [114].

For more details of the physical interpretation of the Riemann-Stieltjes integral which is the slope of the tangent line with respect to monotone functiong (t), see [92, 111, 114, 115].

The existence of the Riemann–Stieltjes integral was discussed by Bliss [116], Carmichael [117], Aleksander and Rolewicz [118], Russell [119], and Nakasho et al. [120].

As is shown above, the Riemann–Stieltjes integral is the functional integral with respect to monotone function.

In fact, with the aid of the idea of Lebesgue type [99] in 1902, one structures the following measure integral with respect to monotone function as follows:

Definition 1.56 Let $g\left(t\right)$ be a monotone function $-\infty\leq a<t<b\leq\infty$ and let $g\left(t\right)$ and $\psi\left(t\right)$ be the measure functions.

The measure integral with respect to monotone function is defined as

$$\left(M\right){}_{a}I_{t,g(\cdot)}^{(1)}\psi\left(t\right)=\int_{a}^{t}\psi\left(t\right)g^{(1)}\left(t\right)d\mu\left(t\right), \tag{1.406}$$

where $\mu\left(t\right)$ is the Lebesgue measure [99].

The theorems for the Riemann–Stieltjes integral are given as follows:

Theorem 1.101 (The Sum and Difference Rules for the Riemann–Stieltjes Integral)

Suppose that ${}_{a}I_{t,g(\cdot)}^{(1)}\phi_1\left(t\right)$ *and* ${}_{a}I_{t,g(\cdot)}^{(1)}\phi_2\left(t\right)$ *exist and* $g^{(1)}\left(t\right)>0$, *then we have*

$${}_{a}I_{t,g(\cdot)}^{(1)}\left(\phi_1\left(t\right)+\phi_2\left(t\right)\right)={}_{a}I_{t,g(\cdot)}^{(1)}\phi_1\left(t\right)+{}_{a}I_{t,g(\cdot)}^{(1)}\phi_2\left(t\right) \tag{1.407}$$

and

$$ {}_aI_{t,g(\cdot)}^{(1)}\left(\phi_1(t)-\phi_2(t)\right)={}_aI_{t,g(\cdot)}^{(1)}\phi_1(t)-{}_aI_{t,g(\cdot)}^{(1)}\phi_2(t). \tag{1.408} $$

The result was discovered by Widder in 1947 [114] and rediscovered by Reitano in 2017 [121] and by Yang in 2019 [111].

Theorem 1.102 (The Constant Multiple Rule for the Riemann–Stieltjes Integral) *Suppose that ${}_aI_{t,g(\cdot)}^{(1)}\phi(t)$ exists, $g^{(1)}(t)>0$, and l is a constant, then we have*

$$ {}_aI_{t,g(\cdot)}^{(1)}\left(l\phi(t)\right)=l\,{}_aI_{t,g(\cdot)}^{(1)}\phi(t). \tag{1.409} $$

The result was discovered by Widder in 1947 [114] and reported by Yang in 2019 [111].

Theorem 1.103 *Suppose that ${}_aI_{t,g(\cdot)}^{(1)}\phi(t)$ exists and $g^{(1)}(t)>0$, then we have*

$$ {}_aI_{t,g(\cdot)}^{(1)}\phi(t)=-{}_tI_{a,g(\cdot)}^{(1)}\phi(t). \tag{1.410} $$

Theorem 1.104 (The Additivity Rule for the Riemann–Stieltjes Integral) *Suppose that ${}_aI_{t,g(\cdot)}^{(1)}\phi(t)$ exists, $g^{(1)}(t)>0$ and $a<c<b$, then we have*

$$ {}_aI_{b,g(\cdot)}^{(1)}\phi(t)={}_aI_{c,g(\cdot)}^{(1)}\phi(t)+{}_cI_{b,g(\cdot)}^{(1)}\phi(t). \tag{1.411} $$

The result was discovered by Widder in 1947 [114].

Property 1.10 Suppose that ${}_aI_{t,g(\cdot)}^{(1)}\phi_1(t)$ and ${}_aI_{t,g(\cdot)}^{(1)}\phi_2(t)$ exist, $\phi_1(t)\le\phi_2(t)$ and $g^{(1)}(t)>0$, then we have

$$ {}_aI_{b,g(\cdot)}^{(1)}\phi_1(t)\le{}_aI_{b,g(\cdot)}^{(1)}\phi_2(t). \tag{1.412} $$

The result was discovered by Widder in 1947 [114].

Property 1.11 Suppose that ${}_aI_{t,g(\cdot)}^{(1)}\phi(t)$ exists, $g^{(1)}(t)>0$ and $a<t<b$, then we have

$$ \left|{}_aI_{b,g(\cdot)}^{(1)}\phi(t)\right|\le{}_aI_{b,g(\cdot)}^{(1)}\left|\phi(t)\right|. \tag{1.413} $$

The result was discovered by Widder in 1947 [114].

Property 1.12 Suppose that ${}_aI_{t,g(\cdot)}^{(1)}\phi(t)$ exists, $g^{(1)}(t)>0$ and $a\le t\le b$, then we have

$$ \left|{}_aI_{b,g(\cdot)}^{(1)}\phi(t)\right|\le\left|g(b)-g(b)\right|\max_{a\le t\le b}\left|\phi(t)\right|. \tag{1.414} $$

The result was discovered by Widder in 1947 [114].

Theorem 1.105 (The First Fundamental Theorem of the Riemann–Stieltjes Integral)
If $D_{t,g(\cdot)}^{(1)}\psi(t)$ *exists and* $g^{(1)}(t) > 0$, *then we have*

$$\psi(t) - \psi(a) = {}_aI_{t,g(\cdot)}^{(1)}\left(D_{t,g(\cdot)}^{(1)}\psi(t)\right). \tag{1.415}$$

Theorem 1.106 (The Mean Value Theorem for the Riemann–Stieltjes Integral)
If $D_{t,g(\cdot)}^{(1)}\psi(t)$ *exists and* $g^{(1)}(t) > 0$, *then we have*

$${}_aI_{t,g(\cdot)}^{(1)}\psi(t) = \psi(l)(g(t) - g(a)), \tag{1.416}$$

where $a < l < t$.
The result was discovered by Widder in 1947 [114] and by Yang in 2019 [111].

Theorem 1.107 (The Second Fundamental Theorem of the Riemann–Stieltjes Integral)
If $D_{t,g(\cdot)}^{(1)}\psi(t)$ *exists and* $g^{(1)}(t) > 0$, *then we have*

$$\psi(t) = D_{t,g(\cdot)}^{(1)}\left({}_aI_{t,g(\cdot)}^{(1)}\psi(t)\right). \tag{1.417}$$

Theorem 1.108 (The Net Change Theorem for the Riemann–Stieltjes Integral)
If $D_{t,g(\cdot)}^{(1)}\psi(t)$ *exists and* $g^{(1)}(t) > 0$, *then we have*

$$\psi(b) - \psi(a) = {}_aI_{b,g(\cdot)}^{(1)}\left(D_{t,g(\cdot)}^{(1)}\psi(t)\right). \tag{1.418}$$

Theorem 1.109 (The Integration by Parts for the Riemann–Stieltjes Integral)
Suppose that $D_{t,g(\cdot)}^{(1)}\phi_1(t)$ *and* $D_{t,g(\cdot)}^{(1)}\phi_2(t)$ *exist and* $g^{(1)}(t) > 0$, *then we have*

$$\begin{aligned}&{}_aI_{t,g(\cdot)}^{(1)}\left(\phi_2(t)\,D_{t,g(\cdot)}^{(1)}\phi_1(t)\right)\\ &=\phi_1(t)\cdot\phi_2(t) - \phi_1(a)\cdot\phi_2(a) - {}_aI_{t,g(\cdot)}^{(1)}\left(\phi_1(t)\,D_{t,g(\cdot)}^{(1)}\phi_2(t)\right).\end{aligned} \tag{1.419}$$

The result was discovered by Widder in 1947 [114] and by Yang in 2019 [111].

The similar version was discovered by Bray [122] and Ben Dushnik [123].
The properties of the Riemann–Stieltjes integral are given as follows:

$${}_aI_{b,g(\cdot)}^{(1)}1 = \int_a^b g^{(1)}(t)\,dt = g(b) - g(a), \tag{1.420}$$

$$ {}_aI_{b,g(\cdot)}^{(1)}\left(ng^{n-1}(t)\right) = \int_a^b ng^{n-1}(t)\, g^{(1)}(t)\, dt = g^n(b) - g^n(a), \tag{1.421} $$

$$ {}_aI_{b,g(\cdot)}^{(1)}\left(e^{g(t)}\right) = \int_a^b e^{g(t)} g^{(1)}(t)\, dt = e^{g(b)} - e^{g(a)}, \tag{1.422} $$

$$ {}_aI_{b,g(\cdot)}^{(1)}\left(\frac{1}{g(t)}\right) = \int_a^b \frac{g^{(1)}(t)}{g(t)} dt = \ln g(b) - \ln g(a), \tag{1.423} $$

$$ {}_aI_{b,g(\cdot)}^{(1)}\left(\frac{1}{(\ln s)} \cdot \frac{1}{g(t)}\right) = \int_a^b \frac{1}{(\ln s)} \cdot \frac{g^{(1)}(t)}{g(t)} dt = \log_s g(b) - \log_s g(a), \tag{1.424} $$

and

$$ {}_aI_{b,g(\cdot)}^{(1)}\left((\ln s)\, s^{g(t)}\right) = \int_a^b (\ln s)\, s^{g(t)} g^{(1)}(t)\, dt = s^{g(b)} - s^{g(a)}, \tag{1.425} $$

where $\ln s \neq 0$.

For more details of the calculus with respect to another function, see [92, 111].

Definition 1.57 Let $g^{(1)}(t) > 0$, where $t \in (a, +\infty)$ with $a \in \mathbb{R}$, and $\psi_g(t) = \psi(g(t))$.

The Riemann–Stieltjes integral is defined as

$$ {}_aI_{t,g(\cdot)}^{(1)}\psi_g(t) = \int_a^t \psi_g(t)\, g^{(1)}(t)\, dt. \tag{1.426} $$

For more details of the Riemann–Stieltjes integral, which was used to the stochastic integrals, see [124].

We now consider the indefinite Riemann–Stieltjes integral.

Definition 1.58 The indefinite Riemann–Stieltjes integral is defined as

$$ I_{t,g(\cdot)}^{(1)}\psi_g(t) = \int \psi_g(t)\, g^{(1)}(t)\, dt. \tag{1.427} $$

Equation (1.427) is called the antidifferentiation with respect to another function.

In this case, one has

$$\frac{1}{g^{(1)}(t)}\frac{d}{dt}\left(\int \psi_g(t)\,g^{(1)}(t)\,dt\right) = \psi_g(t)\,, \tag{1.428}$$

where

$$\int \psi_g(t)\,g^{(1)}(t)\,dt = \Phi_g(t) + C \tag{1.429}$$

with the constant C.

As a direct result, one has the following result:

Suppose that $\psi_{1,g}(t) = \psi_1(g(t))$ and $\psi_{2,g}(t) = \psi_2(g(t))$ are continuous and $g^{(1)}(t) > 0$ for $t \in \mathbb{R}$, then we have

$$\begin{aligned} &\int \psi_{2,g}(t)\,D^{(1)}_{t,g(\cdot)}\psi_{1,g}(t)\,g^{(1)}(t)\,dt \\ &= \psi_{2,g}(t)\,\psi_{1,g}(t) - \int \psi_{1,g}(t)\,D^{(1)}_{t,g(\cdot)}\psi_{2,g}(t)\,g^{(1)}(t)\,dt. \end{aligned} \tag{1.430}$$

Equation (1.430) is called the integration by parts for the indefinite Riemann–Stieltjes integral.

The properties for the indefinite Riemann–Stieltjes integral are given as follows:

$$I^{(1)}_{t,g(\cdot)}1 = \int g^{(1)}(t)\,dt = g(t) + C\,, \tag{1.431}$$

$$I^{(1)}_{t,g(\cdot)}\left(ng^{n-1}(t)\right) = \int ng^{n-1}(t)\,g^{(1)}(t)\,dt = g^n(t) + C\,, \tag{1.432}$$

$$I^{(1)}_{t,g(\cdot)}\left(e^{g(t)}\right) = \int e^{g(t)}g^{(1)}(t)\,dt = e^{g(t)} + C\,, \tag{1.433}$$

$$I^{(1)}_{t,g(\cdot)}\left(\frac{1}{g(t)}\right) = \int \frac{g^{(1)}(t)}{g(t)}dt = \ln \psi_g(t) + C\,, \tag{1.434}$$

$$I^{(1)}_{t,g(\cdot)}\left(\frac{1}{(\ln s)}\cdot\frac{1}{g(t)}\right) = \int \frac{1}{(\ln s)}\cdot\frac{g^{(1)}(t)}{g(t)}dt = \log_s g(t) + C\,, \tag{1.435}$$

$$I^{(1)}_{t,g(\cdot)}\left((\ln s)\,s^{g(t)}\right) = \int (\ln s)\,s^{g(t)}g^{(1)}(t)\,dt = s^{g(t)} + C\,, \tag{1.436}$$

$$\begin{aligned} &I^{(1)}_{t,g(\cdot)}\left(e^{\psi_g(t)}D^{(1)}_{t,g(\cdot)}\psi_g(t)\right) = \int e^{\psi_g(t)}D^{(1)}_{t,g(\cdot)}\psi_g(t)\,g^{(1)}(t)\,dt \\ &= e^{\psi_g(t)} + C \end{aligned}, \tag{1.437}$$

$$\begin{aligned} &I^{(1)}_{t,g(\cdot)}\left(\frac{\psi_g(t)}{|\psi_g(t)|}D^{(1)}_{t,g(\cdot)}\psi_g(t)\right) = \int \frac{\psi_g(t)}{|\psi_g(t)|}D^{(1)}_{t,g(\cdot)}\psi_g(t)\,g^{(1)}(t)\,dt \\ &= \left|\psi_g(t)\right| + C \end{aligned}, \tag{1.438}$$

$$I^{(1)}_{t,g(\cdot)}\left(\frac{D^{(1)}_{t,g(\cdot)}\psi_g(t)}{\psi_g(t)}\right) = \int \frac{D^{(1)}_{t,g(\cdot)}\psi_g(t)}{\psi_g(t)} g^{(1)}(t)\,dt = \ln \psi_g(t) + C, \tag{1.439}$$

$$I^{(1)}_{t,g(\cdot)}\left(\frac{D^{(1)}_{t,g(\cdot)}\psi_g(t)}{(\ln s)\psi_g(t)}\right) = \int \frac{D^{(1)}_{t,g(\cdot)}\psi_g(t)}{(\ln s)\psi_g(t)} g^{(1)}(t)\,dt = \log_s \psi_g(t) + C, \tag{1.440}$$

and

$$\begin{aligned} &I^{(1)}_{t,g(\cdot)}\left(\left[(\ln s)\, s^{\psi_g(t)}\right] \cdot D^{(1)}_{t,g(\cdot)}\psi_g(t)\right) = \int \left[(\ln s)\, s^{\psi_g(t)}\right] \cdot D^{(1)}_{t,g(\cdot)}\psi_g(t)\, g^{(1)}(t)\,dt \\ &= s^{\psi_g(t)} + C \end{aligned}, \tag{1.441}$$

where C is a constant.

We now present the concepts and properties for the improper Riemann–Stieltjes integral.

Let $g^{(1)}(t) > 0$, where $t \in (a, b)$ with $-\infty \leq a < b \leq \infty$, and $\psi_g(t) = \psi(g(t))$.

We have the following:

- The Riemann–Stieltjes integral in (a, ∞) is defined as follows:

$$\begin{aligned} {}_aI^{(1)}_{\infty,g(\cdot)}\psi_g(t) &= \int_a^\infty \psi_g(t)\, g^{(1)}(t)\,dt \\ &= \lim_{b\to\infty} \int_a^b \psi_g(t)\, g^{(1)}(t)\,dt \end{aligned}. \tag{1.442}$$

- The Riemann–Stieltjes integral in $(-\infty, b)$ is defined as follows:

$$\begin{aligned} {}_{-\infty}I^{(1)}_{b,g(\cdot)}\psi_g(t) &= \int_{-\infty}^b \psi_g(t)\, g^{(1)}(t)\,dt \\ &= \lim_{a\to-\infty} \int_a^b \psi_g(t)\, g^{(1)}(t)\,dt \end{aligned}. \tag{1.443}$$

- The Riemann–Stieltjes integral in $(-\infty, \infty)$ is defined as follows:

$$\begin{aligned} &{}_{-\infty}I^{(1)}_{\infty,g(\cdot)}\psi_\eta(t) = \int_{-\infty}^\infty \psi_g(t)\, g^{(1)}(t)\,dt \\ &= \int_c^\infty \psi_g(t)\, g^{(1)}(t)\,dt + \int_{-\infty}^c \psi_g(t)\, g^{(1)}(t)\,dt \end{aligned}, \tag{1.444}$$

  where c is the constant.

As a typical example, one has

$$\int_{0}^{\infty} e^{-g(t)}\eta^{(1)}(t)\,dt = \lim_{b\to\infty}\int_{0}^{b} e^{-g(t)}g^{(1)}(t)\,dt = \lim_{b\to\infty}\left[-e^{-g(t)}\right]_{0}^{b} = e^{-g(0)}, \tag{1.445}$$

where $\lim_{b\to\infty} e^{-g(b)} = 0$.

There is the version of the improper Riemann–Stieltjes integral of the form [125]

$$\mu(z) = \int_{0}^{\infty} e^{zt}dg(t)\ \ (Re(z)\geq 0), \tag{1.446}$$

which can be written as

$$\mu(z) = \int_{0}^{\infty} e^{zt}g^{(1)}(t)\,dt \tag{1.447}$$

due to the integral equation

$$\int_{0}^{\infty} e^{zt}g^{(1)}(t)\,dt = \int_{0}^{\infty} e^{zt}dg(t)\ \ (Re(z)\geq 0), \tag{1.448}$$

which derived from [114]

$$\int_{a}^{b} e^{zt}dg(t) = \int_{a}^{b} e^{zt}g^{(1)}(t)\,dt. \tag{1.449}$$

1.5.3 The Special Integral Equations

We now consider a new family of the special integral equations as follows:

The integral equation with respect to another function is defined as follows:

$$\lambda\int_{a}^{t}\Omega_{g}(t-\tau)\,\psi_{g}(\tau)\,g^{(1)}(\tau)\,d\tau = \Phi_{g}(t), \tag{1.450}$$

or

$$\psi_g(t) - \lambda \int_a^t \Omega_g(t-\tau)\,\psi_g(\tau)\,g^{(1)}(\tau)\,d\tau = \Phi_g(t), \tag{1.451}$$

where λ is the constant, $\Omega_g(t-\tau) = \Omega(g(t) - g(\tau))$ is the kernel with respect to another function, $\psi_g(t) = \psi(g(t))$ and $\Phi_g(t) = \Phi_g(g(t))$.

The integral equation with respect to another function is defined as follows:

$$\lambda \int_a^b \Omega_g(t-\tau)\,\psi_g(\tau)\,g^{(1)}(\tau)\,d\tau = \Phi_g(t), \tag{1.452}$$

or

$$\psi_g(t) - \lambda \int_a^b \Omega_g(t-\tau)\,\psi_g(\tau)\,g^{(1)}(\tau)\,d\tau = F_g(t), \tag{1.453}$$

where λ is the constant, $\Omega_g(t-\tau) = \Omega(g(t) - g(\tau))$ is the kernel with respect to another function, $\psi_g(t) = \psi(g(t))$ and $\Phi_g(t) = \Phi_g(g(t))$.

In particular, we have the following comments:

- **The Volterra integral equation** [Volterra [126]]

  When $\eta(t) = t$, one gets the Volterra integral equation as follows [128]:

$$\lambda \int_a^t \Omega(t,\tau)\,\psi(\tau)\,d\tau = \Phi(t), \tag{1.454}$$

  or

$$\psi(t) - \lambda \int_a^t \Omega(t,\tau)\,\psi(\tau)\,d\tau = \Phi(t), \tag{1.455}$$

  where λ is the constant.

  The integral equations were discovered by Volterra in 1897 [126].

- **The Fredholm integral equation** [Fredholm [127]]

  When $\eta(t) = t$, one get the Fredholm integral equation as follows [129]

$$\lambda \int_a^b \Omega(t,\tau)\,\psi(\tau)\,d\tau = \Phi(t), \tag{1.456}$$

or

$$\psi(t) - \lambda \int_a^b \Omega(t, \tau)\, \psi(\tau)\, d\tau = F(t), \tag{1.457}$$

where λ is the constant.

The integral equations were discovered by Fredholm in 1903 [127].

Based on the above results, we obtain the following integral equation with respect to another function and its solution as follows:

For $a > 0$ and $0 < \alpha < 1$, the integral equation with respect to another function is defined as follows:

$$\int_a^t \frac{\psi_g(\tau)}{(g(t) - g(\tau))^\alpha} g^{(1)}(\tau)\, d\tau = \Phi_g(t), \tag{1.458}$$

with the solution given by:

$$\psi_g(t) = \frac{\sin(\alpha\pi)}{\pi} \frac{1}{g^{(1)}(t)} \frac{d}{dt} \int_a^t \frac{\Phi_g(\tau)}{(g(t) - g(\tau))^{1-\alpha}} g^{(1)}(\tau)\, d\tau \tag{1.459}$$

or

$$\psi_g(t) = \frac{\sin(\alpha\pi)}{\pi} \left(\frac{\Phi_g(a)}{(g(t) - g(a))^{1-\alpha}} + \int_a^t \frac{1}{(g(t) - g(\tau))^{1-\alpha}} \frac{d\Phi_g(\tau)}{d\tau} d\tau \right), \tag{1.460}$$

where $\psi_g(t) = \left(\psi_g \circ g\right)(t) = \psi(g(t))$ and $\Phi_g(t) = \left(\Phi_g \circ g\right)(t) = \Phi_g(g(t))$.

Let us discuss the special cases of the integral equations:

- **Osler fractional derivative** [Osler [128]]

  The differential operator with respect to another function can be written as follows:

$$\psi_g(t) = \frac{1}{\Gamma(1+\alpha)} \frac{1}{g^{(1)}(t)} \frac{d}{dt} \int_a^t \frac{\Phi_g(\tau)}{(g(t) - g(\tau))^{1-\alpha}} g^{(1)}(\tau)\, d\tau, \tag{1.461}$$

  where $a > 0$ and $0 < \alpha < 1$.

  The Osler fractional derivative was discovered by Osler in 1970 [128, 129].

- **Osler fractional integral** [Osler [128]]

  The integral operator with respect to another function can be presented as follows:

$$\mathbb{M}_g(t) = \frac{1}{\Gamma(1-\alpha)} \int_a^t \frac{\psi_g(\tau)}{(g(t) - g(\tau))^\alpha} g^{(1)}(\tau)\, d\tau, \tag{1.462}$$

  where $a > 0$ and $0 < \alpha < 1$.

  The Osler fractional integral operator was discovered by Osler in 1970 [128].

- **Almeida fractional derivative** [Almeida [130]]

  The differential operator with respect to another function can be given as follows:

$$\Theta_g(t) = \frac{1}{\Gamma(1+\alpha)} \int_a^t \frac{1}{(g(t) - g(\tau))^{1-\alpha}} \frac{d\Phi_g(\tau)}{d\tau} d\tau, \tag{1.463}$$

  where $a > 0$ and $0 < \alpha < 1$.

  The Almeida fractional derivative was discovered by Almeida in 2017 [130].

- **Samko–Kilbas–Marichev fractional derivative** [Samko et al. [131]]

  The differential operator with respect to another function can be written as follows:

$$\psi_g(t) = \frac{1}{\Gamma(1+\alpha)} \frac{1}{g^{(1)}(t)} \frac{d}{dt} \int_{-\infty}^t \frac{\Phi_g(\tau)}{(g(t) - g(\tau))^{1-\alpha}} g^{(1)}(\tau)\, d\tau, \tag{1.464}$$

  where $0 < \alpha < 1$.

  The Samko–Kilbas–Marichev fractional derivative was discovered by Samko, Kilbas, and Marichev in 1987 [131].

- **Samko–Kilbas–Marichev fractional integral** [Samko et al. [131]]

  The integral operator with respect to another function can be presented as follows:

$$\mathbb{M}_g(t) = \frac{1}{\Gamma(1-\alpha)} \int_{-\infty}^t \frac{\psi_g(\tau)}{(g(t) - g(\tau))^\alpha} g^{(1)}(\tau)\, d\tau, \tag{1.465}$$

  where $0 < \alpha < 1$.

  The Samko–Kilbas–Marichev fractional integral was discovered by Samko, Kilbas and Marichev in 1987 [131].

- **Liouville type fractional derivative** [Yang [132]]
  The differential operator with respect to another function can be given as follows:

$$\Theta_g(t) = \frac{1}{\Gamma(1+\alpha)} \int_{-\infty}^{t} \frac{1}{(g(t)-g(\tau))^{1-\alpha}} \frac{d\Phi_g(\tau)}{d\tau} d\tau, \tag{1.466}$$

where $0 < \alpha < 1$.

The Liouville-type fractional derivative operator was discovered by Yang, Gao, and Jun in 2020 [92].

In fact, one has the following comments:

- **The Abel integral equation** [Abel [133]]
  When $\eta(t) = t$, one gets the Abel integral equation as follows [134]

$$\int_{a}^{t} \frac{\psi(\tau)}{(t-a)^{\alpha}} d\tau = \Phi(t), \tag{1.467}$$

with the solution as follows:

$$\psi(t) = \frac{\sin(\alpha\pi)}{\pi} \frac{d}{dt} \int_{a}^{t} \frac{\Phi(\tau)}{(t-\tau)^{1-\alpha}} d\tau \tag{1.468}$$

or

$$\psi(t) = \frac{\sin(\alpha\pi)}{\pi} \left(\frac{\Phi(a)}{(t-a)^{1-\alpha}} + \int_{a}^{t} \frac{1}{(t-\tau)^{1-\alpha}} \frac{d\Phi(\tau)}{d\tau} d\tau \right), \tag{1.469}$$

where $a > 0$ and $0 < \alpha < 1$.

The integral equation was discovered by Abel in 1881 [133].

- **Riemann–Liouville fractional derivative** [Liouville [134], Riemann [135]]
  The differential operator of the fractional order can be written as follows:

$$\psi(t) = \frac{1}{\Gamma(1+\alpha)} \frac{d}{dt} \int_{a}^{t} \frac{\Phi(\tau)}{(t-\tau)^{1-\alpha}} d\tau, \tag{1.470}$$

where $a > 0$ and $0 < \alpha < 1$.

The Riemann–Liouville fractional derivative operator was discovered by Liouville in 1832 with the complement functions [134] and by Riemann in 1847 with the complement function [135].

- **Riemann–Liouville fractional integral** [Liouville [134], Riemann [135]]

The integral operator of the fractional order can be presented as follows:

$$\mathbb{M}(t) = \frac{1}{\Gamma(1-\alpha)} \int_a^t \frac{\psi(\tau)}{(t-\tau)^{\alpha}} d\tau, \tag{1.471}$$

where $a > 0$ and $0 < \alpha < 1$.

The Riemann–Liouville fractional integral was discovered by Liouville in 1832 with the complement functions [134] and by Riemann in 1847 with the complement function [135].

Today, the derivatives of arbitrary order (the so-called Riemann–Liouville definition of fractional integral), was suggested by Holmgren in 1866 [136]. There is another view that Sonine first presented the definition of the Riemann–Liouville operator [137]. We now consider the two equations by Krug in 1890 represented in the forms [138, 139].

- **Liouville-Sonine-Caputo fractional derivative** [Liouville [134], Sonine [140], Caputo [141]]

The differential operator of the fractional order can be given as follows:

$$\Theta(t) = \frac{1}{\Gamma(1+\alpha)} \int_a^t \frac{1}{(t-\tau)^{1-\alpha}} \frac{d\Phi(\tau)}{d\tau} d\tau, \tag{1.472}$$

where $a > 0$ and $0 < \alpha < 1$.

The Liouville-Sonine-Caputo fractional derivative operator was discovered by Liouville in 1832 with the complement functions [134], by Sonine in 1872 [140], and by Caputo in 1967 [141]. We remind that the differential operator was rediscovered by Smit and De Vries in 1970 [142].

- **Liouville-Weyl fractional derivative** [Liouville [134], Weyl [143]]

The differential operator of the fractional order can be written as follows:

$$\psi(t) = \frac{1}{\Gamma(1+\alpha)} \frac{d}{dt} \int_{-\infty}^t \frac{\Phi(\tau)}{(t-\tau)^{1-\alpha}} d\tau, \tag{1.473}$$

where $0 < \alpha < 1$.

The Liouville-Weyl fractional derivative operator was discovered by Liouville in 1832 [134] and by Weyl in [143].

- **Liouville-Weyl fractional integral** [Liouville [134], Weyl [143]]

The integral operator can be presented as follows:

$$\mathbb{M}(t) = \frac{1}{\Gamma(1-\alpha)} \int_{-\infty}^t \frac{\psi(\tau)}{(t-\tau)^{\alpha}} d\tau, \tag{1.474}$$

where $0 < \alpha < 1$.

The Liouville-Weyl fractional integral operator was discovered by Liouville in 1832 [134] and by Weyl in [143].

- **Liouville fractional derivative** [Liouville [134]]

The differential operator with respect to another function can be given as follows:

$$\Theta(t) = \frac{1}{\Gamma(1+\alpha)} \int_{-\infty}^{t} \frac{1}{(t-\tau)^{1-\alpha}} \frac{d\Phi(\tau)}{d\tau} d\tau, \tag{1.475}$$

where $0 < \alpha < 1$.

The Liouville fractional derivative operator was discovered by Liouville in 1832 [134], reconsidered by Gerasimov in 1948 [144] and by Dzhrbashian and Nersesyan in 1968 [145].

Moreover, we give the comments for the other special integral equations as follows:

- **Chen fractional integral** [Chen [146]]

The fractional integral can be given as

$$\frac{1}{\Gamma(\alpha)} \int_{a}^{t} \frac{\psi(\tau)}{|t-\tau|^{1-\alpha}} d\tau = \Phi(t), \tag{1.476}$$

where $a > 0$ and $0 < \alpha < 1$.

The Chen fractional integral operator was discovered by Chen in 1961 [146]

- **Chen fractional derivative of first type** [Chen [146]]

The fractional derivative can be given as

$$\frac{1}{\Gamma(1-\alpha)} \frac{d}{dt} \int_{a}^{t} \frac{\psi(\tau)}{|t-\tau|^{\alpha}} d\tau = \Phi(t), \tag{1.477}$$

where $a > 0$ and $0 < \alpha < 1$.

The Chen fractional derivative operator of first type was discovered by Chen in 1961 [146]

- **Chen fractional derivative of second type** [Chen [146]]

The fractional derivative can be given as

$$\frac{1}{\Gamma(1-\alpha)} \int_{a}^{t} \frac{1}{|t-\tau|^{\alpha}} (\psi(\tau) - \psi(a)) d\tau = \Phi(t), \tag{1.478}$$

where $a > 0$ and $0 < \alpha < 1$.

The Chen fractional derivative of second type was discovered by Chen in 1961 [146].

In this case, one has the following:

Definition 1.59 The Chen-type fractional integral can be given as

$$\frac{1}{\Gamma(\alpha)}\int_{-\infty}^{t}\frac{\psi(\tau)}{|t-\tau|^{1-\alpha}}d\tau=\Phi(t), \tag{1.479}$$

where $0<\alpha<1$.

Definition 1.60 The Chen-type fractional derivative can be given as

$$\frac{1}{\Gamma(1-\alpha)}\frac{d}{dt}\int_{-\infty}^{t}\frac{\psi(\tau)}{|t-\tau|^{\alpha}}d\tau=\Phi(t), \tag{1.480}$$

where $a>0$ and $0<\alpha<1$.

- **Agrawal fractional integral** [Agrawal [147]]
  The fractional integral can be given as

$$\Phi(t)=\frac{1}{2\Gamma(\alpha)}\left(\int_{a}^{t}\frac{\psi(\tau)}{|t-\tau|^{\alpha}}d\tau+\int_{t}^{b}\frac{\psi(\tau)}{|t-\tau|^{\alpha}}d\tau\right), \tag{1.481}$$

where $b>a>0$ and $0<\alpha<1$.

The Agrawal fractional integral operator was discovered by Agrawal in 2007 [147].

- **Agrawal fractional derivative of first type** [Agrawal [147]]
  The fractional derivative can be given as

$$\frac{1}{\Gamma(1-\alpha)}\frac{d}{dt}\int_{a}^{b}\frac{\psi(\tau)}{|t-\tau|^{\alpha}}d\tau=\Phi(t), \tag{1.482}$$

where $b>a>0$ and $0<\alpha<1$.

The Agrawal fractional derivative operator of first type was discovered by Agrawal in 2007 [147].

- **Agrawal fractional derivative of second type** [Agrawal [147]]
  The fractional derivative can be given as

$$\frac{1}{\Gamma(1-\alpha)}\int_{a}^{b}\frac{1}{|t-\tau|^{\alpha}}\frac{d\psi(\tau)}{d\tau}d\tau=\Phi(t), \tag{1.483}$$

where $b>a>0$ and $0<\alpha<1$.

The Agrawal fractional derivative of second type was discovered by Agrawal in 2007 [147]

As the special generalization of the above results, one has the following:

Definition 1.61 The Chen-type integral equation with respect to another function is defined as follows:

$$\frac{1}{\Gamma(\alpha)}\int_a^t \frac{\psi_g(\tau)}{|g(t)-g(\tau)|^{\alpha}} g^{(1)}(\tau)\,d\tau = \Phi_g(t), \tag{1.484}$$

where $b > a > 0$, $0 < \alpha < 1$, $\psi_g(t) = \left(\psi_g \circ g\right)(t) = \psi(g(t))$ and $\Phi_g(t) = \left(\Phi_g \circ g\right)(t) = \Phi_g(g(t))$.

Definition 1.62 The Chen-type differential equation with respect to another function is defined as follows:

$$\frac{1}{\Gamma(1-\alpha)}\frac{1}{g^{(1)}(t)}\frac{d}{dt}\int_a^t \frac{\psi_g(\tau)}{|g(t)-g(\tau)|^{\alpha}} g^{(1)}(\tau)\,d\tau = \Phi_g(t), \tag{1.485}$$

where $b > a > 0$, $0 < \alpha < 1$, $\psi_g(t) = \left(\psi_g \circ g\right)(t) = \psi(g(t))$ and $\Phi_g(t) = \left(\Phi_g \circ g\right)(t) = \Phi_g(g(t))$.

Definition 1.63 The Chen-type differential equation with respect to another function is defined as follows:

$$\frac{1}{\Gamma(1-\alpha)}\frac{1}{g^{(1)}(t)}\frac{d}{dt}\int_a^t \frac{\psi_g(\tau)-\psi_g(a)}{|g(t)-g(\tau)|^{\alpha}} g^{(1)}(\tau)\,d\tau = \Phi_g(t), \tag{1.486}$$

where $b > a > 0$, $0 < \alpha < 1$, $\psi_g(t) = \left(\psi_g \circ g\right)(t) = \psi(g(t))$ and $\Phi_g(t) = \left(\Phi_g \circ g\right)(t) = \Phi_g(g(t))$.

Definition 1.64 The Agrawal-type fractional integral with respect to another function is defined as:

$$\Phi(t) = \frac{1}{2\Gamma(\alpha)}\left(\int_a^t \frac{\psi_g(\tau)}{|g(t)-g(\tau)|^{\alpha}} g^{(1)}(\tau)\,d\tau + \int_t^b \frac{\psi_g(\tau)}{|g(t)-g(\tau)|^{\alpha}} g^{(1)}(\tau)\,d\tau\right), \tag{1.487}$$

where $b > a > 0$, $0 < \alpha < 1$, $\psi_g(t) = \left(\psi_g \circ g\right)(t) = \psi(g(t))$ and $\Phi_g(t) = \left(\Phi_g \circ g\right)(t) = \Phi_g(g(t))$.

Definition 1.65 The Agrawal-type fractional derivative with respect to another function is defined as:

$$\Phi(t) = \frac{1}{\Gamma(1-\alpha)} \int_a^b \frac{\psi_g(\tau)}{|g(t) - g(\tau)|^\alpha} \frac{d\psi(\tau)}{d\tau} d\tau, \tag{1.488}$$

where $b > a > 0$, $0 < \alpha < 1$, $\psi_g(t) = (\psi_g \circ g)(t) = \psi(g(t))$ and $\Phi_g(t) = (\Phi_g \circ g)(t) = \Phi_g(g(t))$.

Definition 1.66 The Agrawal-type fractional derivative with respect to another function can be given as

$$\Phi(t) = \frac{1}{\Gamma(1-\alpha)} \frac{1}{g^{(1)}(t)} \frac{d}{dt} \int_a^b \frac{\psi_g(\tau)}{|g(t) - g(\tau)|^\alpha} g^{(1)}(\tau) d\tau, \tag{1.489}$$

where $b > a > 0$, $0 < \alpha < 1$, $\psi_g(t) = (\psi_g \circ g)(t) = \psi(g(t))$ and $\Phi_g(t) = (\Phi_g \circ g)(t) = \Phi_g(g(t))$.

The Kokilashvili Integral Equation [148]
The Kokilashvili integral equation can be given as follows:

$$\int_{-\infty}^{\infty} \frac{\psi(\tau)}{|g(t) - g(\tau)|^{1-\alpha}} dt = \psi(t), \tag{1.490}$$

where $+\infty > t > -\infty$ and $1 > \alpha > 0$.

The result was discovered by Kokilashvili in 1969 [148].

The Sato Integral Equation [149]
The Sato integral equation can be given as follows [149]:

$$\int_0^t \frac{\psi(\tau)}{(g(t) - g(\tau))^{1-\alpha}} dt = \psi(t), \tag{1.491}$$

where $t > 0$ and $1 > \alpha > 0$.

Let us consider the anomalous relaxation with derivatives with respect to another function.

The anomalous relaxation problem involving derivatives with respect to another function is given as

$$D_{t,g(\cdot)}^{(1)} \Xi_\eta(t) + \Xi_g(t) = 0, \tag{1.492}$$

subjected to the initial condition

$$\Xi_g(0) = \chi, \tag{1.493}$$

where χ is a constant.

In this case, we have

$$\frac{D^{(1)}_{t,g(\cdot)} \Xi_g(t)}{\Xi_g(t)} = -1, \tag{1.494}$$

which, by the general definite integral with respect to another function, leads to

$$\int_0^t \frac{D^{(1)}_{t,g(\cdot)} \Xi_g(t)}{\Xi_g(t)} g^{(1)}(t)\, dt = \ln \Xi_g(t) - \ln \Xi_g(0) = g(0) - g(t). \tag{1.495}$$

It follows from Eq. (1.495) that

$$\ln \frac{\Xi_g(t)}{\Xi_g(0)} = g(0) - g(t), \tag{1.496}$$

which leads to

$$\Xi_g(t) = \Xi_g(0)\, e^{g(0)-g(t)} = \Xi_g(0)\, e^{g(0)-g(t)}. \tag{1.497}$$

In this case, we have the following:

(1) Taking $g(t) = e^{\zeta t}$ with $t \in \mathbb{R}$and$\zeta \in \mathbb{R}$, one has from Eq. (1.497) that

$$\Xi_\eta(t) = \chi e^{1-e^{\zeta t}}, \tag{1.498}$$

where e^{-e^t} is the transcendental function (see [150], p.346) and the plots of the function $\Xi_\eta(t)$ are given in Fig. 1.1.

(2) Taking $g(t) = \ln(t+1)$ with $t > 0$, we present

$$\Xi_g(t) = \frac{\chi}{t+1} \tag{1.499}$$

and the plots of the function $\Xi_g(t)$ are shown in Fig. 1.2.

Fig. 1.2 The plots of the function $\Xi_g(t)$ when $\chi = 3$, $\chi = 2$, and $\chi = 1$, respectively

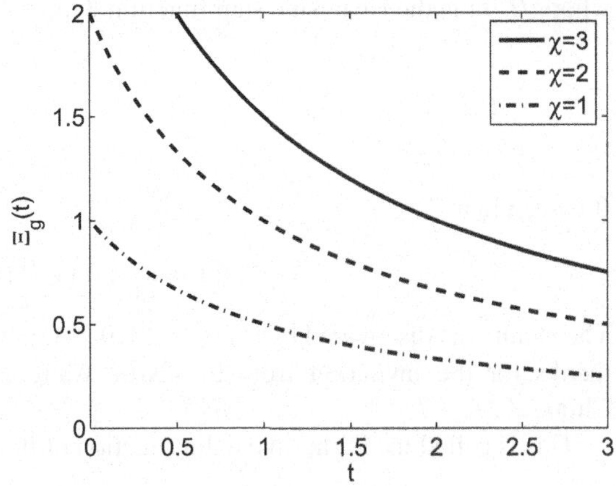

1.5.4 *Generalized Functions and Anomalous Linear Viscoelasticity via Derivative with Respect to Another Function*

We now consider the Heaviside step function, Dirac delta function, and Dirac-like delta functions.

1.5.4.1 Dirac Delta Function

The Dirac delta function with $t \in \mathbb{R}$, denoted by $\delta_D(t)$, is defined as [151]

$$\delta_D(t) = D^{(1)} H(t), \tag{1.500}$$

where $H(t)$ is the Heaviside step function with [152]

$$H(t) = \int_{-\infty}^{\infty} \delta_D(t)\, dt = 1. \tag{1.501}$$

The result was discovered by Dirac in 1926 [151, 152].

1.5.4.2 Dirac-Like Delta Function of First Type

In a similar manner, the Dirac-like delta function of first type, denoted by $\delta_{g(\cdot)}(t)$ with $t \in \mathbb{R}$, is defined as

$$\delta_{g(\cdot)}(t) = D^{(1)}_{t,g(\cdot)} H(t), \tag{1.502}$$

where $H(t)$ is the Heaviside step function with

$$H(t) = {}_{-\infty}I_{\infty,g(\cdot)}^{(1)}\delta_{g(\cdot)}(t) = \int_{-\infty}^{\infty} \delta_{g(\cdot)}(t)\, g^{(1)}(t)\, dt = 1. \tag{1.503}$$

It is seen that

$$\delta_D(t) = \delta_{g(\cdot)}(t)\, g^{(1)}(t). \tag{1.504}$$

The result was discovered by Yang in 2019 [132]. The author would like to express thanks for the invitation from Professor Wang Hui-Ming, Zhejiang University, China.

This is called the Dirac-like delta function of first type.

1.5.4.3 The Anomalous Linear Viscoelasticity via Derivative with Respect to Another Function

We now present the mathematical models for the anomalous linear viscoelasticity based on the derivative with respect to another function as follows.

Let ϖ, θ, $\varepsilon(t)$, $\sigma(t)$, and t be the elastic modulus, viscosity, strain, stress, and time, respectively.

Newton-Like Dashpot Element [132]
The Newton-like dashpot element containing the derivative with respect to another function is given as

$$\sigma(t) = \theta D_{t,g(\cdot)}^{(1)}\varepsilon(t). \tag{1.505}$$

The result was discovered by Yang in 2019 [110].

Maxwell-Like Model [153]
The Maxwell-like model containing the derivative with respect to another function can be given as follows [153]

$$D_{t,g(\cdot)}^{(1)}\varepsilon(t) = \frac{1}{\theta}\sigma(t) + \frac{1}{\varpi}D_{t,g(\cdot)}^{(1)}\sigma(t). \tag{1.506}$$

The result was discovered by Yang, Gao, and Jing in 2019 [153].

Kelvin–Voigt-Like Model [153]
The Kelvin–Voigt-like model containing the general derivative with respect to another function can be presented as follows [153]:

$$\sigma(t) = \theta\varepsilon(t) + \varpi D_{t,g(\cdot)}^{(1)}\varepsilon(t). \tag{1.507}$$

The result was discovered by Yang, Gao, and Jing in 2019 [153].

1.6 Derivative and Integral with Respect to Power-Law Function

We now introduce the calculus with respect to power-law function, which is called the power-law calculus.

1.6.1 The Derivative with Respect to Power-Law Function

Definition 1.67 The derivative with respect to power-law function is defined as [92, 110, 153]

$$ {}^{P}D_t^{(1)}\Theta(t) = \frac{1}{\alpha t^{\alpha-1}}\frac{d\Theta(t)}{dt}, \tag{1.508}$$

where $\alpha \in \mathbb{R}$.

The operator is called the power-law derivative.

Definition 1.68 The derivative of higher order with respect to power-law function is defined as [92, 110, 153]

$$ {}^{P}D_t^{(1)}\Theta(t) = \left(\frac{1}{\alpha t^{\alpha-1}}\frac{d}{dt}\right)^n \Theta(t), \tag{1.509}$$

where $n \in \mathbb{N}$.

We notice the derivative with respect to Hausdorff measure was proposed by Chen Wen [154] and the comment was given in [92].

The theorems for the derivative with respect to power-law function are given as follows:

Theorem 1.110 (The Sum and Difference Rules for the Derivative with Respect to Power-Law Function)
Suppose that ${}^{P}D_t^{(1)}\phi_1(t)$ *and* ${}^{P}D_t^{(1)}\phi_2(t)$ *exist, then we have*

$$ {}^{P}D_t^{(1)}(\phi_1(t)+\phi_2(t)) = {}^{P}D_t^{(1)}\phi_1(t) + {}^{P}D_t^{(1)}\phi_2(t) \tag{1.510}$$

and

$$ {}^{P}D_t^{(1)}(\phi_1(t)-\phi_2(t)) = {}^{P}D_t^{(1)}\phi_1(t) - {}^{P}D_t^{(1)}\phi_2(t). \tag{1.511}$$

Theorem 1.111 (The Constant Multiple Rule for the Derivative with Respect to Power-Law Function)

Suppose that ${}^{P}D_t^{(1)}\phi(t)$ *exists, and* l *is a constant, then we have*

$$ {}^{P}D_t^{(1)}(l\phi(t)) = l\,{}^{P}D_t^{(1)}\phi(t). \tag{1.512}$$

Theorem 1.112 (The Product Rule for the Derivative with Respect to Power-Law Function)

Suppose that ${}^{P}D_t^{(1)}\phi_1(t)$ *and* ${}^{P}D_t^{(1)}\phi_2(t)$ *exist, then we have*

$$ {}^{P}D_t^{(1)}(\phi_1(t)\cdot\phi_2(t)) = \phi_2(t)\,{}^{P}D_t^{(1)}\phi_1(t) + \phi_1(t)\,{}^{P}D_t^{(1)}\phi_2(t). \tag{1.513}$$

Theorem 1.113 (The Quotient Rule for the Derivative with Respect to Power-Law Function)

Suppose that ${}^{P}D_t^{(1)}\phi_1(t)$ *and* ${}^{P}D_t^{(1)}\phi_2(t)$ *exist, and* $\phi_2(t) \neq 0$, *then we have*

$$ {}^{P}D_t^{(1)}\left(\frac{\phi_1(t)}{\phi_2(t)}\right) = \frac{\phi_2(t)\,{}^{P}D_t^{(1)}\phi_1(t) - \phi_1(t)\,{}^{P}D_t^{(1)}\phi_2(t)}{\phi_2(t)\cdot\phi_2(t)}. \tag{1.514}$$

Theorem 1.114 (The Chain Rule for the Derivative with Respect to Power-Law Function)

Suppose that $\frac{d\phi(w)}{dw}$ *and* ${}^{P}D_t^{(1)}w(t)$ *exist, then we have*

$$ {}^{P}D_t^{(1)}(\phi(w(t))) = \frac{d\phi(w)}{dw}\cdot {}^{P}D_t^{(1)}w(t). \tag{1.515}$$

The properties of the derivative with respect to power-law function are given as follows:

$$ {}^{P}D_t^{(1)}1 = 0, \tag{1.516}$$

$$ {}^{P}D_t^{(1)}t^{\alpha} = 1, \tag{1.517}$$

$$ {}^{P}D_t^{(1)}t^{n\alpha} = nt^{(n-1)\alpha}, \tag{1.518}$$

$$ {}^{P}D_t^{(1)}e^{t^{\alpha}} = e^{t^{\alpha}}, \tag{1.519}$$

$$ {}^{P}D_t^{(1)}(\alpha\ln t) = \frac{1}{t^{\alpha}}, \tag{1.520}$$

$$ {}^{P}D_t^{(1)}\left(\alpha\log_s t\right) = \frac{1}{(\ln s)\,t^{\alpha}}, \tag{1.521}$$

$$ {}^{P}D_t^{(1)} s^{t^\alpha} = (\ln s)\, s^{t^\alpha}, \tag{1.522} $$

$$ {}^{P}D_t^{(1)} e^{\psi(t^\alpha)} = e^{\psi(t^\alpha)}\, {}^{P}D_t^{(1)} \psi\left(t^\alpha\right), \tag{1.523} $$

$$ {}^{P}D_t^{(1)} \left|\psi\left(t^\alpha\right)\right| = \frac{\psi\left(t^\alpha\right)}{\left|\psi\left(t^\alpha\right)\right|}\, {}^{P}D_t^{(1)} \psi\left(t^\alpha\right), \psi\left(t^\alpha\right) \neq 0, \tag{1.524} $$

$$ {}^{P}D_t^{(1)} \ln \psi\left(t^\alpha\right) = \frac{{}^{P}D_t^{(1)} \psi\left(t^\alpha\right)}{\psi\left(t^\alpha\right)}, \tag{1.525} $$

$$ {}^{P}D_t^{(1)} \log_s \psi\left(t^\alpha\right) = \frac{{}^{P}D_t^{(1)} \psi\left(t^\alpha\right)}{(\ln s)\, \psi\left(t^\alpha\right)}, \tag{1.526} $$

and

$$ {}^{P}D_t^{(1)} s^{\psi(t^\alpha)} = \left[(\ln s)\, s^{\psi(t^\alpha)}\right] \cdot {}^{P}D_t^{(1)} \psi\left(t^\alpha\right). \tag{1.527} $$

1.6.2 *The Integral with Respect to Power-Law Function*

We now consider the integral with respect to power-law function.

1.6.2.1 The Integral with Respect to Power-Law Function

Definition 1.69 The integral with respect to power-law function is defined as [132]

$$ {}_{a}^{P}I_t^{(1)} \Phi(\tau) = \alpha \int_a^t \Phi(t)\, t^{\alpha-1} dt. \tag{1.528} $$

Here, Eq. (1.528) is called the power-law integral.

The theorems for the integral with respect to power-law function are given as follows:

Theorem 1.115 (The Sum and Difference Rules for the Definite Integral with Respect to Power-Law Function)

Suppose that ${}_{a}^{P}I_t^{(1)} \phi_1(t)$ *and* ${}_{a}^{P}I_t^{(1)} \phi_2(t)$ *exist, then we have*

$$ {}_{a}^{P}I_t^{(1)} \left(\phi_1(t) + \phi_2(t)\right) = {}_{a}^{P}I_t^{(1)} \phi_1(t) + {}_{a}^{P}I_t^{(1)} \phi_2(t) \tag{1.529} $$

and

$$ {}_{a}^{P}I_{t}^{(1)}\left(\phi_1(t)-\phi_2(t)\right)={}_{a}^{P}I_{t}^{(1)}\phi_1(t)-{}_{a}^{P}I_{t}^{(1)}\phi_2(t). \tag{1.530}$$

Theorem 1.116 (The Constant Multiple Rule for the Definite Integral with Respect to Power-Law Function)
Suppose that ${}_{a}^{P}I_{t}^{(1)}\phi(t)$ *exists, and* l *is a constant, then we have*

$$ {}_{a}^{P}I_{t}^{(1)}\left(l\phi(t)\right)=l\,{}_{a}^{P}I_{t}^{(1)}\phi(t). \tag{1.531}$$

Theorem 1.117 *Suppose that* ${}_{a}^{P}I_{t}^{(1)}\phi(t)$ *exists, then we have*

$$ {}_{a}^{P}I_{t}^{(1)}\phi(t)=-{}_{t}^{P}I_{a}^{(1)}\phi(t). \tag{1.532}$$

Theorem 1.118 *Suppose that* ${}_{a}^{P}I_{t}^{(1)}\phi(t)$ *exists and* $a<c<b$, *then we have*

$$ {}_{a}^{P}I_{b}^{(1)}\phi(t)={}_{a}^{P}I_{c}^{(1)}\phi(t)+{}_{c}^{P}I_{b}^{(1)}\phi(t). \tag{1.533}$$

Theorem 1.119 (The First Fundamental Theorem of the Definite Integral with Respect to Power-Law Function)
If ${}^{P}D_{t}^{(1)}\psi(t)$ *exists, then we have*

$$ \psi(t)-\psi(a)={}_{a}^{P}I_{t}^{(1)}\left({}^{P}D_{t}^{(1)}\psi(t)\right). \tag{1.534}$$

Theorem 1.120 (The Mean Value Theorem for the Definite Integral with Respect to Power-Law Function)
If ${}^{P}D_{t}^{(1)}\psi(t)$ *exists, then we have*

$$ {}_{a}^{P}I_{t}^{(1)}\psi(t)=\psi(l)\left(t^{\alpha}-a^{\alpha}\right). \tag{1.535}$$

Theorem 1.121 (The Second Fundamental Theorem of the Definite Integral with Respect to Power-Law Function)
If ${}^{P}D_{t}^{(1)}\psi(t)$ *exists, then we have*

$$ \psi(t)={}^{P}D_{t}^{(1)}\left({}_{a}^{P}I_{t}^{(1)}\psi(t)\right). \tag{1.536}$$

Theorem 1.122 (The Net Change Theorem for the Definite Integral with Respect to Power-Law Function)
If ${}^{P}D_{t}^{(1)}\psi(t)$ *exists, then we have*

$$ \psi(b)-\psi(a)={}_{a}^{P}I_{t}^{(1)}\left({}^{P}D_{t}^{(1)}\psi(t)\right). \tag{1.537}$$

Theorem 1.123 (The Integration by Parts for the Definite Integral with Respect to Power-Law Function)
Suppose that ${}^{P}D_t^{(1)}\phi_1(t)$ *and* ${}^{P}D_t^{(1)}\phi_2(t)$ *exist, then we have*

$$\begin{aligned}&{}_a^P I_t^{(1)}\left(\phi_2(t)\,{}^{P}D_t^{(1)}\phi_1(t)\right)\\&=\phi_1(t)\cdot\phi_2(t)-\phi_1(a)\cdot\phi_2(a)-{}_a^P I_t^{(1)}\left(\phi_1(t)\,{}^{P}D_t^{(1)}\phi_2(t)\right).\end{aligned} \tag{1.538}$$

The properties of the integral with respect to power-law function are given as follows:

$${}_a^P I_b^{(1)}1=b^{\alpha}-a^{\alpha}, \tag{1.539}$$

$${}_a^P I_b^{(1)}nb^{(n-1)\alpha}=b^{n\alpha}-a^{n\alpha}, \tag{1.540}$$

$${}_a^P I_b^{(1)}e^{t^{\alpha}}=e^{b^{\alpha}}-e^{a^{\alpha}}, \tag{1.541}$$

$${}_a^P I_b^{(1)}\left(\frac{1}{t^{\alpha}}\right)=\alpha\ln\frac{b}{a}, \tag{1.542}$$

$${}_a^P I_b^{(1)}\left(\frac{1}{(\ln s)}\cdot\frac{1}{t^{\alpha}}\right)=\alpha\left(\log_s b-\log_s a\right), \tag{1.543}$$

and

$${}_a^P I_b^{(1)}\left((\ln s)\,s^{t^{\alpha}}\right)=s^{b^{\alpha}}-s^{a^{\alpha}}, \tag{1.544}$$

where $\ln s\neq 0$.

1.6.2.2 The Dirac-Like Delta Function of Second Type

The Dirac-like delta function of second type, denoted by $\delta_1(t)$ with $t\in\mathbb{R}$, is defined as

$$\delta_1(t)={}_P D_t^{(1)}H(t), \tag{1.545}$$

where $H(t)$ is the Heaviside step function with

$$H(t)={}_{-\infty}^{P}I_{\infty}^{(1)}\delta_1(t)=\alpha\int_{-\infty}^{\infty}\delta_1(t)\,t^{\alpha-1}dt=1. \tag{1.546}$$

It is seen that

$$\delta_D(t) = \alpha t^{\alpha-1}\delta_1(t). \tag{1.547}$$

The result was discovered by Yang in 2019 [132].

Definition 1.70 The indefinite integral with respect to power-law function is defined as

$${}^{P}I_t^{(1)}\psi(t) = \alpha\int \psi(t)\, t^{\alpha-1}dt. \tag{1.548}$$

Equation (1.548) is called the antidifferentiation with respect to power-law function.

So, we present that

$${}^{P}D_t^{(1)}\left({}^{P}I_t^{(1)}\psi(t)\right) = \left({}^{P}D_t^{(1)} \circ {}^{P}I_t^{(1)}\right)\psi(t) = \psi(t), \tag{1.549}$$

where

$${}^{P}I_t^{(1)}\psi(t) = \Phi(t) + C \tag{1.550}$$

with the constant C.

So, there is

$${}^{P}D_t^{(1)}\left({}^{P}I_t^{(1)}\psi(t)\right) = \left({}^{P}D_t^{(1)} \circ {}^{P}I_t^{(1)}\right)\psi(t) = {}^{P}I\psi(t), \tag{1.551}$$

where ${}^{P}I$ is the unit operator.

In this case, we get the following results for the integration by parts for the indefinite integral with respect to power function.

If $\Lambda_1(t) = (\psi_1 \circ (t^\alpha))(t) = \psi_1(t^\alpha)$ and $\Lambda_2(t) = (\psi_2 \circ (t^\alpha))(t) = \psi_2(t^\alpha)$ are continuous for $t \in \mathbb{R}$, then we have

$${}^{P}I_t^{(1)}\left(\Lambda_2(t)\, {}^{P}D_t^{(1)}\Lambda_1(t)\right) = \Lambda_2(t)\Lambda_1(t) - {}^{P}I_t^{(1)}\left(\Lambda_1(t)\, {}^{P}D_t^{(1)}\Lambda_2(t)\right). \tag{1.552}$$

Equation (1.552) is called the integration by parts for the indefinite integral with respect to power function.

The properties for the indefinite integral with respect to power function are given as follows:

$${}^{P}I_t^{(1)}1 = t^\alpha + C, \tag{1.553}$$

$${}^{P}I_t^{(1)}\left(nt^{(n-1)\alpha}\right) = t^{n\alpha} + C, \tag{1.554}$$

$$ {}^{P}I_t^{(1)} e^{t^\alpha} = e^{t^\alpha} + C, \tag{1.555}$$

$$ {}^{P}I_t^{(1)} \left(\frac{1}{t^\alpha}\right) = \alpha \ln (t) + C, \tag{1.556}$$

$$ {}^{P}I_t^{(1)} \left(\frac{1}{(\ln s)} \cdot \frac{1}{t^\alpha}\right) = \alpha \log_s (t) + C, \tag{1.557}$$

$$ {}^{P}I_t^{(1)} \left((\ln s)\, s^{t^\alpha}\right) = s^{t^\alpha} + C, \tag{1.558}$$

$$ {}^{P}I_t^{(1)} \left(e^{\Lambda(t)}\, {}^{P}D_t^{(1)} \Lambda (t)\right) = e^{\Lambda(t)} + C, \tag{1.559}$$

$$ {}^{P}I_t^{(1)} \left(\frac{\Lambda (t)}{|\Lambda (t)|}\, {}^{P}D_t^{(1)} \Lambda (t)\right) = |\Lambda (t)| + C, \tag{1.560}$$

$$ {}^{P}I_t^{(1)} \left(\frac{{}^{P}D_t^{(1)} \Lambda (t)}{\Lambda (t)}\right) = \ln \Lambda (t) + C, \tag{1.561}$$

$$ {}^{P}I_t^{(1)} \left(\frac{{}^{P}D_t^{(1)} \Lambda (t)}{(\ln s)\, \Lambda (t)}\right) = \log_s \Lambda (t) + C, \tag{1.562}$$

and

$$ {}^{P}I_t^{(1)} \left(\left[(\ln s)\, s^{\Lambda(t)}\right] \cdot {}^{P}D_t^{(1)} \Lambda (t)\right) = s^{\Lambda(t)} + C, \tag{1.563}$$

where C is a constant, and $e^{-\kappa t^\beta}$ is the Kohlrausch–Williams–Watts function (see Chap. 7).

The plots of the Kohlrausch–Williams–Watts function $e^{-\kappa t^\beta}$ are illustrated in Fig. 1.3.

We now present the concepts and properties for the improper integral with respect to power function as follows:

Let $\Lambda (t) = (\psi \circ (t^\alpha)) (t) = \psi (t^\alpha)$, where $-\infty \le a < t < b \le \infty$. Then we have the following:

- The integral with respect to power function in (a, ∞) is defined as follows:

$$\begin{aligned} &{}_{a}^{P}I_{\infty}^{(1)} \psi_g (t) \\ &= \alpha \int_a^\infty \Lambda (t)\, t^{\alpha-1} dt \\ &= \lim_{b\to\infty} \alpha \int_a^\infty \Lambda (t)\, t^{\alpha-1} dt. \end{aligned} \tag{1.564}$$

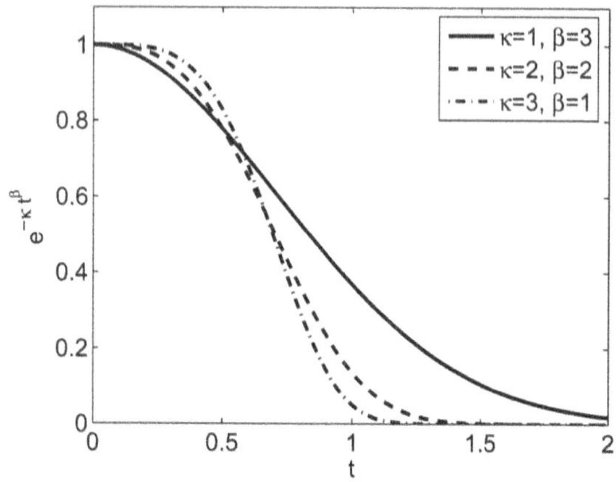

Fig. 1.3 The plots of $e^{-\kappa t^{\beta}}$ with the different values κ and β

- The integral in $(-\infty, b)$ is defined as follows:

$$\begin{aligned} & {}^{P}_{-\infty}I_{b}^{(1)}\psi_{g}(t) \\ & = \alpha \int_{-\infty}^{b} \Lambda(t)\, t^{\alpha-1} dt \\ & = \lim_{a \to -\infty} \alpha \int_{a}^{\infty} \Lambda(t)\, t^{\alpha-1} dt. \end{aligned} \tag{1.565}$$

- The integral with respect to power function in $(-\infty, \infty)$ is defined as follows:

$$\begin{aligned} & {}^{P}_{-\infty}I_{\infty}^{(1)}\Lambda(t) \\ & = \beta \int_{-\infty}^{\infty} \Lambda(t)\, t^{\alpha-1} dt \\ & = \alpha \left(\int_{c}^{\infty} \Lambda(t)\, t^{\alpha-1} dt + \int_{-\infty}^{c} \Lambda(t)\, t^{\alpha-1} dt \right), \\ & = {}^{P}_{-\infty}I_{c}^{(1)}\Lambda(t) + {}^{P}_{c}I_{\infty}^{(1)}\Lambda(t) \end{aligned} \tag{1.566}$$

where c is the constant.

We now present a typical application to the anomalous relaxation within the derivative with respect to power function.

The anomalous relaxation problem within the derivative with respect to power function is presented as

$${}^{P}D_{t}^{(1)}\Xi(t) + \Xi(t) = 0, \tag{1.567}$$

subjected to the initial condition

$$\Xi(0) = \chi, \tag{1.568}$$

where χ is a constant.

So, we give

$$\frac{{}^{P}D_t^{(1)}\Xi(t)}{\Xi(t)} = -1, \tag{1.569}$$

which, by the scaling-law integral, leads to

$${}_0^P I_t^{(1)}\left(\frac{{}^{P}D_t^{(1)}\Xi(t)}{\Xi(t)}\right) = \ln\Xi(t) - \ln\Xi(0) = -t^{\alpha}. \tag{1.570}$$

It follows from Eq. (1.570) that

$$\ln\frac{\Xi(t)}{\Xi(0)} = -t^{\alpha}, \tag{1.571}$$

which leads to

$$\Xi(t) = \Xi(0)\,e^{-t^{\alpha}} = \chi e^{-t^{\alpha}}, \tag{1.572}$$

where $e^{-t^{\alpha}}$ is the Kohlrausch–Williams–Watts function (see Chap. 7).

The plots of the solution of the anomalous relaxation problem for $\chi = 2, \alpha = 3$, $\alpha = 2$, and $\alpha = 1$ are given in Fig. 1.4.

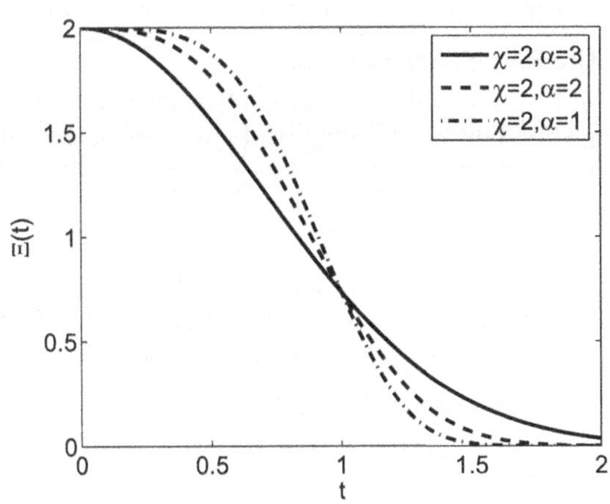

Fig. 1.4 The solution of the anomalous relaxation problem with the different values $\chi = 2, \alpha = 3, \alpha = 2$ and $\alpha = 1$

1.6.3 *The Scaling-Law Calculus*

In this section, we introduce the scaling-law calculus, which consists of the scaling-law derivative (the so-called derivative with respect to scaling-law function) and scaling-law integral (the so-called integral with respect to scaling-law function). Based on the scaling-law calculus, we consider the anomalous relaxation with complex topology.

Let us recall the scaling-law functions as follows:

Mandelbrot Scaling-Law Function
The Mandelbrot scaling-law function is defined as follows [155]

$$g(t) = \kappa t^{1-D}, \tag{1.573}$$

where $\kappa \in \mathbb{R}_+$, $t \in \mathbb{R}_+$ and D is the fractal dimension with $D \in \mathbb{R}_+$.

The result was discovered by Mandelbrot in 1967 [155].

Richardson Scaling-Law Function
The Richardson empirical law (the so-called Richardson scaling-law function, Richardson empirical law, Richardson law, and Richardson-exponent) is defined as

$$g(t) = \kappa t^{D}, \tag{1.574}$$

where $\kappa \in \mathbb{R}_+$, $t \in \mathbb{R}_+$ and $D \in \mathbb{R}$ is the scaling exponent.

The result was discovered by Richardson in 1926 [156].

The Korcak Scaling-Law
In 1938, Korcak [157] presented the scaling-law function (the so-called Korcak Scaling-law, Korcak empirical law, Korcak law, and Korcak-exponent) given as

$$g(t) = \kappa t^{-D}, \tag{1.575}$$

where $\kappa \in \mathbb{R}_+$, $t \in \mathbb{R}_+$ and D is the scaling exponent for $D \in \mathbb{R}_+$.

The result was discovered by Korcak in 1938 [157], applied by Turcotte in 1997 [158], and reviewed by Imre and Novotny in 2016 [159].

The Scaling Law in Life
The scaling-law function in life is presented as follows [160]:

$$g(t) = \kappa t^{D}, \tag{1.576}$$

where κ is the normalization constant for $\kappa \in \mathbb{R}_+$, $t \in \mathbb{R}_+$, and $D \in \mathbb{R}$ is the scaling exponent.

The result was discovered by West, Brown, and Enquist in 1999 [160].

For more details of the connection between the fractals and scaling-law function, see [161].

The Complex Topology by Using the Scaling-Law Function
We now consider the complex topology by using the scaling-law function.

There is the complex topology by using the scaling-law function, which is given as

$$g(t) = \kappa t^{\beta}, \tag{1.577}$$

where κ is the normalization constant for $\kappa \in \mathbb{R}_+$, β is the scaling exponent for $\beta \in \mathbb{R}$, and t is the radius for $t \in \mathbb{R}$.

Let the scaling-law function $g(t)$ be monotone for $t \in \mathbb{R}$ or $t \in \mathbb{R}_+$.

1.6.3.1 The Scaling-Law Derivatives

We now introduce the definitions, properties, and theorems for the scaling-law derivatives.

Definition 1.71 The scaling-law derivative of the function $\phi(t)$ is defined as

$${}^{SL}D_t^{(1)}\phi(t) := D_{t,\kappa t^{\beta}}^{(1)}\phi(t) = \frac{1}{\beta\kappa t^{\beta-1}}\frac{d\phi(t)}{dt}, \tag{1.578}$$

where $\phi(t) = \left(\varphi \circ \left(\kappa t^{\beta}\right)\right)(t) = \varphi\left(\kappa t^{\beta}\right)$, κ is the normalization constant, and β is the scaling exponent.

Definition 1.72 The scaling-law derivative of the function $\phi(t)$ of order n is defined as

$${}^{SL}D_t^{(n)}\phi(t) = \left(\frac{1}{\beta\kappa t^{\beta-1}}\frac{d}{dt}\right)^n \phi(t), \tag{1.579}$$

where $\phi(t) = \left(\varphi \circ \left(\kappa t^{\beta}\right)\right)(t) = \varphi\left(\kappa t^{\beta}\right)$, $n \in \mathbb{N}$, κ is the normalization constant, and β is the scaling exponent.

Definition 1.73 The scaling-law differential of the function $\phi(t)$, denoted by $d\phi(t)$, is defined as

$$d\phi(t) = \beta\kappa t^{\beta-1} D_{t,\kappa t^{\beta}}^{(1)}\phi(t)\, dt, \tag{1.580}$$

where $\phi(t) = \left(\varphi \circ \left(\kappa t^{\beta}\right)\right)(t) = \varphi\left(\kappa t^{\beta}\right)$, κ is the normalization constant, and β is the scaling exponent.

The theorems for the scaling-law derivative are given as follows:

Theorem 1.124 (The Sum and Difference Rules for the Scaling-Law Derivative) *Suppose that* ${}^{SL}D_t^{(1)}\phi_1(t)$ *and* ${}^{SL}D_t^{(1)}\phi_2(t)$ *exist, then we have*

$$ {}^{SL}D_t^{(1)}\left(\phi_1(t)+\phi_2(t)\right)={}^{SL}D_t^{(1)}\phi_1(t)+{}^{SL}D_t^{(1)}\phi_2(t) \tag{1.581}$$

and

$$ {}^{SL}D_t^{(1)}\left(\phi_1(t)-\phi_2(t)\right)={}^{SL}D_t^{(1)}\phi_1(t)-{}^{SL}D_t^{(1)}\phi_2(t). \tag{1.582}$$

Theorem 1.125 (The Constant Multiple Rule for the Scaling-Law Derivative) *Suppose that* ${}^{SL}D_t^{(1)}\phi(t)$ *exists and* l *is a constant, then we have*

$$ {}^{SL}D_t^{(1)}\left(l\phi(t)\right)=l\,{}^{SL}D_t^{(1)}\phi(t). \tag{1.583}$$

Theorem 1.126 (The Product Rule for the Scaling-Law Derivative) *Suppose that* ${}^{SL}D_t^{(1)}\phi_1(t)$ *and* ${}^{SL}D_t^{(1)}\phi_2(t)$ *exist, then we have*

$$ {}^{SL}D_t^{(1)}\left(\phi_1(t)\cdot\phi_2(t)\right)=\phi_2(t)\,{}^{SL}D_t^{(1)}\phi_1(t)+\phi_1(t)\,{}^{SL}D_t^{(1)}\phi_2(t). \tag{1.584}$$

Theorem 1.127 (The Quotient Rule for the Scaling-Law Derivative) *Suppose that* ${}^{SL}D_t^{(1)}\phi_1(t)$ *and* ${}^{SL}D_t^{(1)}\phi_2(t)$ *exist, and* $\phi_2(t)\neq 0$, *then we have*

$$ {}^{SL}D_t^{(1)}\left(\frac{\phi_1(t)}{\phi_2(t)}\right)=\frac{\phi_2(t)\,{}^{SL}D_t^{(1)}\phi_1(t)-\phi_1(t)\,{}^{SL}D_t^{(1)}\phi_2(t)}{\phi_2(t)\cdot\phi_2(t)}. \tag{1.585}$$

Theorem 1.128 (The Chain Rule for the Scaling-Law Derivative) *Suppose that* $\frac{d\phi(w)}{dw}$ *and* ${}^{SL}D_t^{(1)}w(t)$ *exist, then we have*

$$ {}^{SL}D_t^{(1)}\left(\phi(w(t))\right)=\frac{d\phi(w)}{dw}\cdot{}^{SL}D_t^{(1)}w(t). \tag{1.586}$$

The properties of the scaling-law derivative are given as follows:

$$ {}^{SL}D_t^{(1)}1=0, \tag{1.587}$$

$$ {}^{SL}D_t^{(1)}\left(\kappa t^{\beta}\right)=1, \tag{1.588}$$

$$ {}^{SL}D_t^{(1)}\left(\kappa t^{\beta}\right)^n=n\kappa^{n-1}t^{(n-1)\beta}, \tag{1.589}$$

$$ {}^{SL}D_t^{(1)}e^{\kappa t^{\beta}}=e^{\kappa t^{\beta}}, \tag{1.590}$$

$$ {}^{SL}D_t^{(1)} \ln\left(\kappa t^\beta\right) = \frac{1}{\kappa t^\beta}, \tag{1.591} $$

$$ {}^{SL}D_t^{(1)} \log_s\left(\kappa t^\beta\right) = \frac{1}{(\ln s)\,\kappa t^\beta}, \tag{1.592} $$

$$ {}^{SL}D_t^{(1)} s^{\kappa t^\beta} = (\ln s)\, s^{\kappa t^\beta}, \tag{1.593} $$

$$ {}^{SL}D_t^{(1)} e^{\psi(\kappa t^\beta)} = e^{\psi(\kappa t^\beta)}\, {}^{SL}D_t^{(1)} \psi\left(\kappa t^\beta\right), \tag{1.594} $$

$$ {}^{SL}D_t^{(1)} \left|\psi\left(\kappa t^\beta\right)\right| = \frac{\psi\left(\kappa t^\beta\right)}{\left|\psi\left(\kappa t^\beta\right)\right|}\, {}^{SL}D_t^{(1)} \psi\left(\kappa t^\beta\right), \ \psi\left(\kappa t^\beta\right) \neq 0, \tag{1.595} $$

$$ {}^{SL}D_t^{(1)} \ln \psi\left(\kappa t^\beta\right) = \frac{{}^{SL}D_t^{(1)} \psi\left(\kappa t^\beta\right)}{\psi\left(\kappa t^\beta\right)}, \tag{1.596} $$

$$ {}^{SL}D_t^{(1)} \log_s \psi\left(\kappa t^\beta\right) = \frac{{}^{SL}D_t^{(1)} \psi\left(\kappa t^\beta\right)}{(\ln s)\, \psi\left(\kappa t^\beta\right)}, \tag{1.597} $$

and

$$ {}^{SL}D_t^{(1)} s^{\psi(\kappa t^\beta)} = \left[(\ln s)\, s^{\psi(\kappa t^\beta)}\right] \cdot {}^{SL}D_t^{(1)} \psi\left(\kappa t^\beta\right), \tag{1.598} $$

where $e^{-\kappa t^\beta}$ is the Kohlrausch–Williams–Watts function (see Chap. 7).

1.6.3.2 The Scaling-Law Integrals

We now investigate the definitions, properties, and theorems for the scaling-law integrals.

Definition 1.74 The scaling-law integral of the function $\psi(t)$ is defined as

$$ {}^{SL}_{a}I_t^{(1)} \psi(t) := {}_{a}I^{(1)}_{t,\kappa t^\beta} \psi(t) = \beta\kappa \int_a^t \psi(t)\, t^{\beta-1} dt. \tag{1.599} $$

where $t > a$, $\phi(t) = \left(\varphi \circ \left(\kappa t^\beta\right)\right)(t) = \varphi\left(\kappa t^\beta\right)$, κ is the normalization constant, and β is the scaling exponent.

As a special case, one has

$$ {}^{SL}_{0}I_t^{(1)} \psi(t) = \beta\kappa \int_0^t \psi(t)\, t^{\beta-1} dt, \tag{1.600} $$

where $t > 0$, $\phi(t) = \left(\varphi \circ \left(\kappa t^{\beta}\right)\right)(t) = \varphi\left(\kappa t^{\beta}\right)$, κ is the normalization constant, and β is the scaling exponent. The theorems for the scaling-law integral are given as follows:

Theorem 1.129 (The Sum and Difference Rules for the Scaling-Law Integral) *Suppose that* ${}^{SL}_{a}I_t^{(1)}\phi_1(t)$ *and* ${}^{SL}_{a}I_t^{(1)}\phi_2(t)$ *exist, where* κ*is the normalization constant and* β *is the scaling exponent, then we have*

$$ {}^{SL}_{a}I_t^{(1)}\left(\phi_1(t) + \phi_2(t)\right) = {}^{SL}_{a}I_t^{(1)}\phi_1(t) + {}^{SL}_{a}I_t^{(1)}\phi_2(t) \tag{1.601} $$

and

$$ {}^{SL}_{a}I_t^{(1)}\left(\phi_1(t) - \phi_2(t)\right) = {}^{SL}_{a}I_t^{(1)}\phi_1(t) - {}^{SL}_{a}I_t^{(1)}\phi_2(t). \tag{1.602} $$

Theorem 1.130 (The Constant Multiple Rule for the Scaling-Law Integral) *Suppose that* ${}^{SL}_{a}I_t^{(1)}\phi(t)$ *exists, where* κ*is the normalization constant and* β *is the scaling exponent, and* l *is a constant, then we have*

$$ {}^{SL}_{a}I_t^{(1)}\left(l\phi(t)\right) = l\,{}^{SL}_{a}I_t^{(1)}\phi(t). \tag{1.603} $$

Theorem 1.131 *Suppose that* ${}^{SL}_{a}I_t^{(1)}\phi(t)$ *exists, where* κ*is the normalization constant and* β *is the scaling exponent, then we have*

$$ {}^{SL}_{a}I_t^{(1)}\phi(t) = -{}^{SL}_{t}I_a^{(1)}\phi(t). \tag{1.604} $$

Theorem 1.132 (The First Fundamental Theorem of the Scaling-Law Integral) *If* ${}^{SL}D_t^{(1)}\psi(t)$ *exists, where* κ*is the normalization constant and* β *is the scaling exponent,, then we have*

$$ \psi(t) - \psi(a) = {}^{SL}_{a}I_t^{(1)}\left({}^{SL}D_t^{(1)}\psi(t)\right). \tag{1.605} $$

Theorem 1.133 (The Mean Value Theorem for the Scaling-Law Integral) *If* ${}^{SL}D_t^{(1)}\psi(t)$ *exists, where* κ*is the normalization constant and* β *is the scaling exponent, then we have*

$$ {}^{SL}_{a}I_t^{(1)}\psi(t) = \kappa\psi(l)\left(t^{\beta} - a^{\beta}\right), \tag{1.606} $$

where $a < l < t$.

Theorem 1.134 (The Second Fundamental Theorem of the Scaling-Law Integral)
If ${}^{SL}D_t^{(1)}\psi(t)$ *exists, where* κ *is the normalization constant and* β *is the scaling exponent, then we have*

$$\psi(t) = {}^{SL}D_t^{(1)}\left({}_a^{SL}I_t^{(1)}\psi(t)\right). \tag{1.607}$$

Theorem 1.135 (The Net Change Theorem for the Scaling-Law Integral)
If ${}^{SL}D_t^{(1)}\psi(t)$ *exists, where* κ *is the normalization constant and* β *is the scaling exponent, then we have*

$$\psi(b) - \psi(a) = {}_a^{SL}I_b^{(1)}\left({}^{SL}D_t^{(1)}\psi(t)\right). \tag{1.608}$$

Theorem 1.136 (The Integration by Parts for the Scaling-Law Integral)
Suppose that ${}^{SL}D_t^{(1)}\phi_1(t)$ *and* ${}^{SL}D_t^{(1)}\phi_2(t)$ *exist where* κ *is the normalization constant and* β *is the scaling exponent, then we have*

$$\begin{aligned}&{}_a^{SL}I_t^{(1)}\left(\phi_2(t)\,{}^{SL}D_t^{(1)}\phi_1(t)\right)\\&=\phi_1(t)\cdot\phi_2(t) - \phi_1(a)\cdot\phi_2(a) - {}_a^{SL}I_t^{(1)}\left(\phi_1(t)\,{}^{SL}D_t^{(1)}\phi_2(t)\right).\end{aligned} \tag{1.609}$$

The properties of the scaling-law integral are given as follows:

$${}_a^{SL}I_b^{(1)}1=\kappa\left(b^\beta - a^\beta\right), \tag{1.610}$$

$${}_a^{SL}I_b^{(1)}\left(n\kappa^{n-1}t^{n\beta-1}\right) = \kappa^n\left(b^{n\beta} - a^{n\beta}\right), \tag{1.611}$$

$${}_a^{SL}I_b^{(1)}e^{\kappa t^\beta} = e^{\kappa b^\beta} - e^{\kappa a^\beta}, \tag{1.612}$$

$${}_a^{SL}I_b^{(1)}\left(\frac{1}{\kappa t^\beta}\right) = \beta\ln\frac{b}{a}, \tag{1.613}$$

$${}_a^{SL}I_b^{(1)}\left(\frac{1}{(\ln s)}\left(\frac{1}{\kappa t^\beta}\right)\right) = \beta\log_s\frac{b}{a}, \tag{1.614}$$

and

$${}_a^{SL}I_b^{(1)}\left((\ln s)\,s^{\kappa t^\beta}\right) = s^{\kappa b^\beta} - s^{\kappa a^\beta}, \tag{1.615}$$

where $\ln s \neq 0$, κ *is the normalization constant,* β *is the scaling exponent, and* $e^{-\kappa t^\beta}$ *is the Kohlrausch–Williams–Watts function (see Chap. 7).*

We now consider the indefinite scaling-law integral as follows.

Definition 1.75 The indefinite scaling-law integral of the function is defined as

$$ {}^{SL}I_t^{(1)}\psi(t) = \beta\kappa \int \psi(t)\, t^{\beta-1} dt, \tag{1.616} $$

where $\kappa$is the normalization constant and β is the scaling exponent.

Equation (1.616) is called the antidifferentiation for the scaling-law derivative.

In this case, we have

$$ {}^{SL}D_t^{(1)}\left({}^{SL}I_t^{(1)}\psi(t)\right) = \left({}^{SL}D_t^{(1)} \circ {}^{SL}I_t^{(1)}\right)\psi(t) = \psi(t), \tag{1.617} $$

where

$$ {}^{SL}I_t^{(1)}\psi(t) = \Phi(t) + C \tag{1.618} $$

with the constant C.

It follows that

$$ {}^{SL}D_t^{(1)}\left({}^{SL}I_t^{(1)}\psi(t)\right) = \left({}^{SL}D_t^{(1)} \circ {}^{SL}I_t^{(1)}\right)\psi(t) = {}^{SL}I\psi(t), \tag{1.619} $$

where ${}^{SL}I$ is the unit operator, $\kappa$is the normalization constant, and β is the scaling exponent.

As a direct result, one has the following result for the integration by parts for the indefinite scaling-law integral.

Suppose that $\Lambda_1(t) = \left(\psi_1 \circ \left(\kappa t^\beta\right)\right)(t) = \psi_1\left(\kappa t^\beta\right)$ and $\Lambda_2(t) = \left(\psi_2 \circ \left(\kappa t^\beta\right)\right)(t) = \psi_2\left(\kappa t^\beta\right)$ are continuous for $t \in \mathbb{R}$, where κ is the normalization constant and β is the scaling exponent, then we have

$$ {}^{SL}I_t^{(1)}\left(\Lambda_2(t)\, {}^{SL}D_t^{(1)}\Lambda_1(t)\right) = \Lambda_2(t)\Lambda_1(t) - {}^{SL}I_t^{(1)}\left(\Lambda_1(t)\, {}^{SL}D_t^{(1)}\Lambda_2(t)\right). \tag{1.620} $$

Equation (1.620) is called the integration by parts for the indefinite scaling-law integral.

The properties for the indefinite scaling-law integral are given as follows:

$$ {}^{SL}I_t^{(1)} 1 = \kappa t^\beta + C, \tag{1.621} $$

$$ {}^{SL}I_t^{(1)}\left(n\kappa^{n-1} t^{(n-1)\beta}\right) = \kappa^n t^{n\beta} + C, \tag{1.622} $$

$$ {}^{SL}I_t^{(1)} e^{\kappa t^\beta} = e^{\kappa t^\beta} + C, \tag{1.623} $$

$$ {}^{SL}I_t^{(1)}\left(\frac{1}{\kappa t^{\beta}}\right) = \ln\left(\kappa t^{\beta}\right) + C, \tag{1.624}$$

$$ {}^{SL}I_t^{(1)}\left(\frac{1}{(\ln s)}\cdot\frac{1}{\kappa t^{\beta}}\right) = \log_s\left(\kappa t^{\beta}\right) + C, \tag{1.625}$$

$$ {}^{SL}I_t^{(1)}\left((\ln s)\, s^{\kappa t^{\beta}}\right) = s^{\kappa t^{\beta}} + C, \tag{1.626}$$

$$ {}^{SL}I_t^{(1)}\left(e^{\Lambda(t)}\, {}^{SL}D_t^{(1)}\Lambda(t)\right) = e^{\Lambda(t)} + C, \tag{1.627}$$

$$ {}^{SL}I_t^{(1)}\left(\frac{\Lambda(t)}{|\Lambda(t)|}\, {}^{SL}D_t^{(1)}\Lambda(t)\right) = |\Lambda(t)| + C, \tag{1.628}$$

$$ {}^{SL}I_t^{(1)}\left(\frac{{}^{SL}D_t^{(1)}\Lambda(t)}{\Lambda(t)}\right) = \ln\Lambda(t) + C, \tag{1.629}$$

$$ {}^{SL}I_t^{(1)}\left(\frac{{}^{SL}D_t^{(1)}\Lambda(t)}{(\ln s)\,\Lambda(t)}\right) = \log_s\Lambda(t) + C, \tag{1.630}$$

and

$$ {}^{SL}I_t^{(1)}\left(\left[(\ln s)\, s^{\Lambda(t)}\right]\cdot {}^{SL}D_t^{(1)}\Lambda(t)\right) = s^{\Lambda(t)} + C, \tag{1.631}$$

where C is a constant, $\Lambda(t) = \left(\psi \circ \left(\kappa t^{\beta}\right)\right)(t) = \psi\left(\kappa t^{\beta}\right)$, κ is the normalization constant, β is the scaling exponent, and $e^{-\kappa t^{\beta}}$ is the Kohlrausch–Williams–Watts function (see Chap. 7).

1.6.3.3 The Improper Scaling-Law Integral and Scaling-Law Integral Equations

We now present the concepts and properties for the improper scaling-law integral as follows:

Let $\Lambda(t) = \left(\psi \circ \left(\kappa t^{\beta}\right)\right)(t) = \psi\left(\kappa t^{\beta}\right)$ where $-\infty \le a < t < b \le \infty$, $\kappa$is the normalization constant, and β is the scaling exponent.

We have the following:

- The scaling-law integral in (a, ∞) is defined as follows:

$$ \begin{aligned} &{}_a^{SL}I_{\infty}^{(1)}\psi_g(t) \\ &= \kappa\beta \int_a^{\infty} \Lambda(t)\, t^{\beta-1} dt \\ &= \lim_{b\to\infty} \kappa\beta \int_a^{\infty} \Lambda(t)\, t^{\beta-1} dt \end{aligned}, \tag{1.632}$$

where κ is the normalization constant and β is the scaling exponent.

- The scaling-law integral in $(-\infty, b)$ is defined as follows:

$$\begin{aligned} &{}_{-\infty}^{SL}I_b^{(1)}\psi_g(t) \\ &= \kappa\beta \int_{-\infty}^{b} \Lambda(t)\, t^{\beta-1}dt \\ &= \lim_{a\to-\infty} \kappa\beta \int_{a}^{\infty} \Lambda(t)\, t^{\beta-1}dt \end{aligned}, \tag{1.633}$$

where κ is the normalization constant and β is the scaling exponent.

- The scaling-law integral in $(-\infty, \infty)$ is defined as follows:

$$\begin{aligned} &{}_{-\infty}^{SL}I_\infty^{(1)}\Lambda(t) \\ &= \kappa\beta \int_{-\infty}^{\infty} \Lambda(t)\, t^{\beta-1}dt \\ &= \kappa\beta \left(\int_{c}^{\infty} \Lambda(t)\, t^{\beta-1}dt + \int_{-\infty}^{c} \Lambda(t)\, t^{\beta-1}dt \right) \\ &= {}_{-\infty}^{SL}I_c^{(1)}\Lambda(t) + {}_{c}^{SL}I_\infty^{(1)}\Lambda(t) \end{aligned}, \tag{1.634}$$

where c is the constant, κ is the normalization constant, and β is the scaling exponent.

As a typical example, one has

$${}_{0}^{SL}I_\infty^{(1)}\left(e^{-\kappa t^\beta}\right) = \lim_{b\to\infty} {}_{0}^{SL}I_b^{(1)}\left(e^{-\kappa t^\beta}\right) = \lim_{b\to\infty}\left[-e^{-\kappa t^\beta}\right]_0^b = 1, \tag{1.635}$$

where $e^{-\kappa t^\beta}$ is the Kohlrausch–Williams–Watts function (see Chap. 7), κ is the normalization constant, and β is the scaling exponent.

We now consider a new family of the scaling-law integral equations as follows: The scaling-law integral equation is defined as follows:

$$\lambda\kappa\beta \int_{a}^{t} \Pi(t-\tau)\,\Lambda(\tau)\,\tau^{\beta-1}d\tau = \Phi(t), \tag{1.636}$$

or

$$\Lambda(t) - \lambda\kappa\beta \int_{a}^{t} \Pi(t-\tau)\,\Lambda(\tau)\,\tau^{\beta-1}d\tau = \Phi(t), \tag{1.637}$$

where λ is the constant, $\Pi(t-\tau) = \left(\Omega \circ \left(\kappa t^\beta\right)\right)(t-\tau) = \Omega\left(\kappa t^\beta - \kappa\tau^\beta\right)$, $\Lambda(t) = \left(\psi \circ \left(\kappa t^\beta\right)\right)(t) = \psi\left(\kappa t^\beta\right)$ and $\Phi(t) = \left(F \circ \left(\kappa t^\beta\right)\right)(t) = F\left(\kappa t^\beta\right)$.

The scaling-law integral equation is defined as follows:

$$\lambda\kappa\beta\int_a^b \Pi(t-\tau)\Lambda(\tau)\tau^{\beta-1}d\tau=\Phi(t), \tag{1.638}$$

or

$$\Lambda(t)-\lambda\kappa\beta\int_a^b \Pi(t-\tau)\Lambda(\tau)\tau^{\beta-1}d\tau=\Phi(t), \tag{1.639}$$

where λ is the constant, $\Pi(t-\tau)=\left(\Omega\circ\left(\kappa t^{\beta}\right)\right)(t-\tau)=\Omega\left(\kappa t^{\beta}-\kappa\tau^{\beta}\right)$, $\Lambda(t)=\left(\psi\circ\left(\kappa t^{\beta}\right)\right)(t)=\psi\left(\kappa t^{\beta}\right)$ and $\Phi(t)=\left(F\circ\left(\kappa t^{\beta}\right)\right)(t)=F\left(\kappa t^{\beta}\right)$.

1.6.3.4 The Anomalous Relaxation

Let us consider the applications of the scaling-law calculus to mathematical problems in applied sciences.

The anomalous relaxation problem involving general derivatives with respect to another function is given as

$${}^{SL}D_t^{(1)}\Xi(t)+\Xi(t)=0, \tag{1.640}$$

subjected to the initial condition

$$\Xi(0)=\chi, \tag{1.641}$$

where χ is a constant.

In this case, we have

$$\frac{{}^{SL}D_t^{(1)}\Xi(t)}{\Xi(t)}=-1, \tag{1.642}$$

which, by the scaling-law integral, leads to

$${}^{SL}_{0}I_t^{(1)}\left(\frac{{}^{SL}D_t^{(1)}\Xi(t)}{\Xi(t)}\right)=\ln\Xi(t)-\ln\Xi(0)=-\kappa t^{\beta}. \tag{1.643}$$

It follows from Eq. (1.643) that

$$\ln\frac{\Xi(t)}{\Xi(0)}=-\kappa t^{\beta}, \tag{1.644}$$

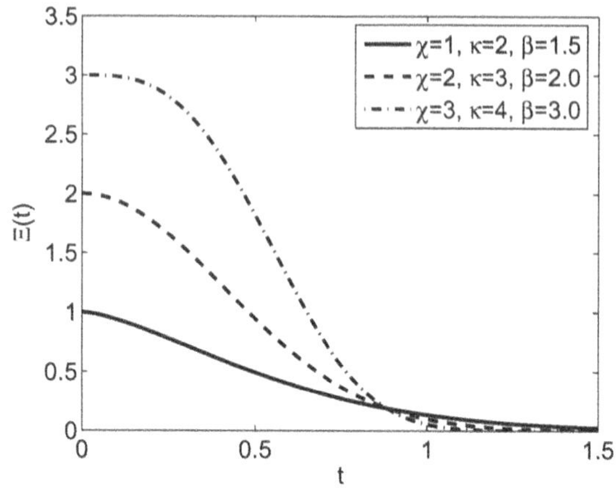

Fig. 1.5 The plots of the anomalous relaxation problem

which leads to

$$\Xi(t) = \Xi(0)\, e^{-\kappa t^{\beta}} = \chi e^{-\kappa t^{\beta}}, \tag{1.645}$$

where $e^{-\kappa t^{\beta}}$ is the Kohlrausch–Williams–Watts function (see Chap. 7).

The plots of the anomalous relaxation problem are given in Fig. 1.5.

The results are important for implementing the non-conventional theory of the relaxation in complex materials, where there are the time-scaling-law behaviors [162].

1.6.3.5 The Anomalous Linear Viscoelasticity

We now present the potential applications of the scaling-law calculus to the anomalous linear viscoelasticity.

Let$\varepsilon(t), \sigma(t), \omega, \theta, t, G(t), J(t), J(t), \Delta\sigma_i, \Delta\varepsilon_i, \sigma_0$, and ε_0 be the strain, stress, elastic modulus, viscosity, time, relaxation modulus, creep compliance, discrete stress increment applied at time $t = \tau_i$, discrete strain increment applied at time $t = \tau_i$, constant strain, and constant strain, respectively.

The Newton-like dashpot element with the scaling-law derivative is given as [132]

$$\sigma(t) = \theta\, {}^{SL}D_t^{(1)} \varepsilon(t). \tag{1.646}$$

The Maxwell-like element with the scaling-law derivative can be given as follows [132]:

$${}^{SL}D_t^{(1)} \varepsilon(t) = \frac{1}{\theta}\sigma(t) + \frac{1}{\omega}\, {}^{SL}D_t^{(1)} \sigma(t). \tag{1.647}$$

The Kelvin–Voigt-like element with the scaling-law derivative can be presented as follows [132]:

$$\sigma(t) = \theta\varepsilon(t) + \omega^{SL}D_t^{(1)}\varepsilon(t). \tag{1.648}$$

The result was discovered by Yang in 2019 [132]. The author would like to express thanks for the invitation from Professor Wang Hui-Ming, Zhejiang University, China.

They are important for implementing the non-conventional theory of the solid mechanics.

To describe the scaling-law behaviors of the complex materials, we present the Boltzmann type superposition principle within the scaling-law calculus.

We set up

$$\varepsilon(t) = \sigma_0 J(t) + \beta\kappa \int_0^\infty \sigma(t-\tau)\,\tau^{\beta-1SL}D_t^{(1)}J(\tau)\,d\tau, \tag{1.649}$$

which can be written as follows:

$$\varepsilon(t) = \beta\kappa \int_{-\infty}^{t} J(t-\tau)\,\tau^{\beta-1SL}D_t^{(1)}\sigma(\tau)\,d\tau. \tag{1.650}$$

Thus, we obtain

$$\varepsilon(t) = \sum_{i=1}^{n} \Delta\sigma_i J(t-\tau_i). \tag{1.651}$$

In an analogous manner, we give

$$\sigma(t) = \varepsilon_0 G(t) + \beta\kappa \int_0^\infty \varepsilon(t-\tau)\,\tau^{\beta-1SL}D_t^{(1)}G(\tau)\,d\tau, \tag{1.652}$$

which can be written as follows:

$$\sigma(t) = \beta\kappa \int_{-\infty}^{t} G(t-\tau)\,\tau^{\beta-1SL}D_t^{(1)}\varepsilon(\tau)\,d\tau. \tag{1.653}$$

In this case, we have

$$\sigma(t) = \sum_{i=1}^{n} \Delta\varepsilon_i G(t - \tau_i). \tag{1.654}$$

The result was discovered by Yang in 2019 [111].

1.6.3.6 The Dirac-Like Delta Function of Third Type

The Dirac-like delta function of third type, denoted by $\delta_2(t)$ with $t \in \mathbb{R}$, is defined as [132]

$$\delta_2(t) = {}_{SL}D_t^{(1)} H(t), \tag{1.655}$$

where $H(t)$ is the Heaviside step function with

$$H(t) = {}_{-\infty}^{SL}I_{\infty}^{(1)} \delta_2(t) = \beta\kappa \int_{-\infty}^{\infty} \delta_2(t) t^{\beta-1} dt = 1. \tag{1.656}$$

It is seen that

$$\delta_D(t) = \beta\kappa t^{\beta-1} \delta_2(t). \tag{1.657}$$

The result was discovered by Yang in 2019 [132].

1.6.4 The Special Formulas via Scaling-Law Calculus

We now consider the special formulas based on the scaling-law calculus.

Theorem 1.137 *Let* $Re(z) > 0$ *and* $z \in \mathbb{C}$*. Then we have*

$$\int_0^{\infty} e^{-\kappa t^{\beta}} t^{\beta z - 1} dt = \frac{\Gamma(z)}{\beta \kappa^z},$$

where $\beta \in \mathbb{R}_+$ *and* $\kappa \in \mathbb{R}_+$.

Proof Taking $x = \kappa t^{\beta}$, we have

$$\Gamma(z) = \int_0^{\infty} e^{-x} x^{z-1} dx. \tag{1.658}$$

such that

$$
\begin{aligned}
&\Gamma(z) \\
&= \int_0^\infty e^{-\kappa t^\beta} \left(\kappa t^\beta\right)^{z-1} \beta \kappa t^{\beta-1} dt \\
&= \beta \kappa^z \int_0^\infty e^{-\kappa t^\beta} t^{\beta z-1} dt.
\end{aligned}
\tag{1.659}
$$

Thus, we get the result. □

Theorem 1.138 *Let $Re(z) > 0$ and $z \in \mathbb{C}$. Then we have*

$$\int_0^\infty e^{-\kappa t} t^{z-1} dt = \frac{\Gamma(z)}{\kappa^z}, \tag{1.660}$$

where $\kappa \in \mathbb{R}_+$.

Theorem 1.139 *Let $Re(z) > 0$ and $z \in \mathbb{C}$. Then we have*

$$\int_0^\infty e^{-t^\beta} t^{\beta z-1} dt = \frac{\Gamma(z)}{\beta}, \tag{1.661}$$

where $\beta \in \mathbb{R}_+$.

Theorem 1.140 *Let $Re(z) > 0$ and $z \in \mathbb{C}$. Then we have*

$$\int_0^\infty e^{-t^2} t^{2z-1} dt = \frac{\Gamma(z)}{2}. \tag{1.662}$$

Theorem 1.141 *Let $Re(z) > 0$ and $z \in \mathbb{C}$. Then we have*

$$\int_0^\infty e^{-\kappa t^2} t^{2z-1} dt = \frac{\Gamma(z)}{2\kappa^z}, \tag{1.663}$$

where $\kappa \in \mathbb{R}_+$.

Theorem 1.142 *Let $Re(z) > 0$ and $z \in \mathbb{C}$. Then we have*

$$\int_0^\infty e^{-\kappa t^3} t^{3z-1} dt = \frac{\Gamma(z)}{3\kappa^z},$$

where $\kappa \in \mathbb{R}_+$.

Theorem 1.143 *Let $\kappa \in \mathbb{R}_+$. Then we have*

$$\int_0^\infty e^{-\kappa t^2} t dt = \frac{1}{2\kappa^z}. \tag{1.664}$$

Theorem 1.144 *Let $Re(z) > 0$ and $z \in \mathbb{C}$. Then we have*

$$\int_0^\infty e^{-\kappa t^4} t^{4z-1} dt = \frac{\Gamma(z)}{4\kappa^z}, \tag{1.665}$$

where $\kappa \in \mathbb{R}_+$.

Theorem 1.145 *Let $\kappa \in \mathbb{R}_+$. Then we have*

$$\int_0^\infty e^{-\kappa t^4} dt = \frac{\Gamma\left(\frac{1}{4}\right)}{4\sqrt[4]{\kappa}}. \tag{1.666}$$

Theorem 1.146 *Let $\kappa \in \mathbb{R}_+$. Then we have*

$$\int_0^\infty e^{-\kappa t^4} t dt = \frac{1}{4}\sqrt{\frac{\pi}{\kappa}}. \tag{1.667}$$

Theorem 1.147

$$\int_0^\infty e^{-t^4} t dt = \frac{\sqrt{\pi}}{4}. \tag{1.668}$$

Theorem 1.148 *Let $Re(z) > 0$ and $z \in \mathbb{C}$. Then we have*

$$\int_0^\infty e^{-\kappa t^3} t^{3z-1} dt = \frac{\Gamma(z)}{3\kappa^z},$$

where $\kappa \in \mathbb{R}_+$.

Theorem 1.149 *Let $\kappa \in \mathbb{R}_+$. Then we have*

$$\int_0^\infty e^{-\kappa t^3} dt = \frac{\Gamma\left(\frac{1}{3}\right)}{3\sqrt[3]{\kappa}}. \tag{1.669}$$

Theorem 1.150 *Let $\kappa \in \mathbb{R}_+$. Then we have*

$$\int_0^\infty e^{-t^3} dt = \frac{\Gamma\left(\frac{1}{3}\right)}{3}. \tag{1.670}$$

Theorem 1.151 *Let $\kappa \in \mathbb{R}_+$. Then we have*

$$\int_0^\infty e^{-\kappa t^3} t dt = \frac{\Gamma\left(\frac{2}{3}\right)}{3\sqrt[3]{\kappa^2}}. \tag{1.671}$$

Theorem 1.152

$$\int_0^\infty e^{-t^3} t dt = \frac{\Gamma\left(\frac{2}{3}\right)}{3}. \tag{1.672}$$

Theorem 1.153 *Let $Re(z) > 0$ and $z \in \mathbb{C}$. Then we have*

$$\int_0^\infty e^{-\kappa\sqrt{t}} t^{\frac{z}{2}-1} dt = \frac{2\Gamma(z)}{\kappa^z}, \tag{1.673}$$

where $\kappa \in \mathbb{R}_+$.

Theorem 1.154 *Let $Re(z) > 0$ and $z \in \mathbb{C}$. Then we have*

$$\int_0^\infty e^{-\kappa t\sqrt{t}} t^{\frac{3z}{2}-1} dt = \frac{2\Gamma(z)}{3\kappa^z}, \tag{1.674}$$

where $\kappa \in \mathbb{R}_+$.

Theorem 1.155

$$\int_0^\infty e^{-2t^5} t^{\frac{3}{2}} dt = \frac{1}{5}\sqrt{\frac{\pi}{2}}. \tag{1.675}$$

Theorem 1.156

$$\int_0^\infty e^{-2t^6} t^5 dt = \frac{1}{6}\sqrt{\frac{\pi}{2}}. \tag{1.676}$$

Theorem 1.157

$$\int_0^\infty e^{-2t^7} t^{\frac{5}{2}} dt = \frac{1}{7}\sqrt{\frac{\pi}{2}}. \tag{1.677}$$

Theorem 1.158

$$\int_0^\infty e^{-2t^8} t^3 dt = \frac{1}{8}\sqrt{\frac{\pi}{2}}. \tag{1.678}$$

Theorem 1.159

$$\int_0^\infty e^{-2t^9} t^{\frac{7}{2}} dt = \frac{1}{9}\sqrt{\frac{\pi}{2}}. \tag{1.679}$$

Theorem 1.160

$$\int_0^\infty e^{-2t^{10}} t^4 dt = \frac{1}{10}\sqrt{\frac{\pi}{2}}. \tag{1.680}$$

Theorem 1.161

$$\int_0^\infty e^{-3t^2} dt = \frac{1}{2}\sqrt{\frac{\pi}{3}}. \tag{1.681}$$

Theorem 1.162

$$\int_0^\infty e^{-3t^3} t^{\frac{1}{2}} dt = \frac{1}{3}\sqrt{\frac{\pi}{3}}. \tag{1.682}$$

Theorem 1.163

$$\int_0^\infty e^{-3t^4} t dt = \frac{1}{4}\sqrt{\frac{\pi}{3}}. \tag{1.683}$$

Theorem 1.164

$$\int_0^\infty e^{-3t^5} t^{\frac{3}{2}} dt = \frac{1}{5}\sqrt{\frac{\pi}{3}}. \tag{1.684}$$

Theorem 1.165

$$\int_0^\infty e^{-3t^6} t^2 dt = \frac{1}{6}\sqrt{\frac{\pi}{3}}. \tag{1.685}$$

Theorem 1.166

$$\int_0^\infty e^{-3t^7} t^{\frac{5}{2}} dt = \frac{1}{7}\sqrt{\frac{\pi}{3}}. \tag{1.686}$$

Theorem 1.167

$$\int_0^\infty e^{-3t^8} t^3 dt = \frac{1}{8}\sqrt{\frac{\pi}{3}}. \tag{1.687}$$

Theorem 1.168

$$\int_0^\infty e^{-3t^9} t^{\frac{7}{2}} dt = \frac{1}{9}\sqrt{\frac{\pi}{3}}. \tag{1.688}$$

Theorem 1.169

$$\int_0^\infty e^{-3t^{10}} t^4 dt = \frac{1}{10}\sqrt{\frac{\pi}{3}}. \tag{1.689}$$

Theorem 1.170

$$\int_0^\infty e^{-4t^3} t^{\frac{1}{2}} dt = \frac{1}{3}\sqrt{\frac{\pi}{4}}. \tag{1.690}$$

Theorem 1.171

$$\int_0^\infty e^{-4t^4} t\,dt = \frac{1}{4}\sqrt{\frac{\pi}{4}}. \tag{1.691}$$

Theorem 1.172

$$\int_0^\infty e^{-4t^5} t^{\frac{3}{2}}dt = \frac{1}{5}\sqrt{\frac{\pi}{4}}. \tag{1.692}$$

Theorem 1.173

$$\int_0^\infty e^{-4t^6} t^2 dt = \frac{1}{6}\sqrt{\frac{\pi}{4}}. \tag{1.693}$$

Theorem 1.174

$$\int_0^\infty e^{-4t^7} t^{\frac{5}{2}}dt = \frac{1}{7}\sqrt{\frac{\pi}{4}}. \tag{1.694}$$

Theorem 1.175

$$\int_0^\infty e^{-4t^8} t^3 dt = \frac{1}{8}\sqrt{\frac{\pi}{4}}. \tag{1.695}$$

Theorem 1.176

$$\int_0^\infty e^{-5t^3} dt = \frac{\Gamma\left(\frac{1}{3}\right)}{3\sqrt[3]{5}}. \tag{1.696}$$

Theorem 1.177

$$\int_0^\infty e^{-6t^3} dt = \frac{\Gamma\left(\frac{1}{3}\right)}{3\sqrt[3]{6}}. \tag{1.697}$$

Theorem 1.178

$$\int_0^\infty e^{-7t^3}dt = \frac{\Gamma\left(\frac{1}{3}\right)}{3\sqrt[3]{7}}. \tag{1.698}$$

Theorem 1.179

$$\int_0^\infty e^{-8t^3}dt = \frac{\Gamma\left(\frac{1}{3}\right)}{3\sqrt[3]{8}}. \tag{1.699}$$

Theorem 1.180

$$\int_0^\infty e^{-9t^3}dt = \frac{\Gamma\left(\frac{1}{3}\right)}{3\sqrt[3]{9}}. \tag{1.700}$$

Theorem 1.181

$$\int_0^\infty e^{-10t^3}dt = \frac{\Gamma\left(\frac{1}{3}\right)}{3\sqrt[3]{10}}. \tag{1.701}$$

Theorem 1.182

$$\int_0^\infty e^{-4t^3}tdt = \frac{\Gamma\left(\frac{2}{3}\right)}{3\sqrt[3]{16}}. \tag{1.702}$$

Theorem 1.183

$$\int_0^\infty e^{-5t^3}tdt = \frac{\Gamma\left(\frac{2}{3}\right)}{3\sqrt[3]{25}}. \tag{1.703}$$

Theorem 1.184

$$\int_0^\infty e^{-6t^3}tdt = \frac{\Gamma\left(\frac{2}{3}\right)}{3\sqrt[3]{36}}. \tag{1.704}$$

Theorem 1.185 *Let $\kappa \in \mathbb{R}_+$. Then we have*

$$\int\limits_0^\infty e^{-7t^3} t dt = \frac{\Gamma\left(\frac{2}{3}\right)}{3\sqrt[3]{49}}. \tag{1.705}$$

Theorem 1.186

$$\int\limits_0^\infty e^{-8t^3} t dt = \frac{\Gamma\left(\frac{2}{3}\right)}{3\sqrt[3]{64}}. \tag{1.706}$$

Theorem 1.187

$$\int\limits_0^\infty e^{-9t^3} t dt = \frac{\Gamma\left(\frac{2}{3}\right)}{3\sqrt[3]{81}}. \tag{1.707}$$

Theorem 1.188

$$\int\limits_0^\infty e^{-10t^3} t dt = \frac{\Gamma\left(\frac{2}{3}\right)}{3\sqrt[3]{100}}. \tag{1.708}$$

Theorem 1.189 *Let $\kappa \in \mathbb{R}_+$. Then we have*

$$\int\limits_0^\infty e^{-\kappa t\sqrt{t}} t^2 dt = \frac{2}{3\kappa^2}. \tag{1.709}$$

Theorem 1.190

$$\int\limits_0^\infty e^{-2t\sqrt{t}} t^2 dt = \frac{1}{6}. \tag{1.710}$$

Theorem 1.191

$$\int\limits_0^\infty e^{-3t\sqrt{t}} t^2 dt = \frac{2}{27}. \tag{1.711}$$

Theorem 1.192

$$\int_0^\infty e^{-4t\sqrt{t}}t^2dt = \frac{1}{24}. \tag{1.712}$$

Theorem 1.193

$$\int_0^\infty e^{-5t\sqrt{t}}t^2dt = \frac{2}{75}. \tag{1.713}$$

Theorem 1.194

$$\int_0^\infty e^{-6t\sqrt{t}}t^2dt = \frac{1}{54}. \tag{1.714}$$

Theorem 1.195

$$\int_0^\infty e^{-7t\sqrt{t}}t^2dt = \frac{2}{147}. \tag{1.715}$$

Theorem 1.196

$$\int_0^\infty e^{-8t\sqrt{t}}t^2dt = \frac{1}{96}. \tag{1.716}$$

Theorem 1.197

$$\int_0^\infty e^{-9t\sqrt{t}}t^2dt = \frac{2}{243}. \tag{1.717}$$

Theorem 1.198

$$\int_0^\infty e^{-10t\sqrt{t}}t^2dt = \frac{1}{150}. \tag{1.718}$$

Theorem 1.199 *Suppose that* $z, q, p \in \mathbb{C}$, $Re\,(q) > 0$, $Re\,(z) > 0$, *and* $Re\,(p) > 0$, *then there is*

$$\int_0^\infty e^{-p\kappa^q t^{\beta q}} t^{\beta z-1} dt = \frac{1}{\beta\kappa^z q p^{\frac{z}{q}}} \Gamma\left(\frac{z}{q}\right), \tag{1.719}$$

where $\beta \in \mathbb{R}_+$ *and* $\kappa \in \mathbb{R}_+$.

Proof Taking $x = \kappa t^\beta$, we have

$$\frac{1}{q p^{\frac{z}{q}}} \Gamma\left(\frac{z}{q}\right) = \int_0^\infty e^{-px^q} x^{z-1} dx \tag{1.720}$$

such that

$$\begin{aligned} &\frac{1}{q p^{\frac{z}{q}}} \Gamma\left(\frac{z}{q}\right) \\ &= \int_0^\infty e^{-p(\kappa t^\beta)^q} \left(\kappa t^\beta\right)^{z-1} \beta\kappa t^{\beta-1} dt \\ &= \beta\kappa^z \int_0^\infty t^{z\beta-1} \cdot \exp\left(-p\kappa^q t^{\beta q}\right) dt. \end{aligned} \tag{1.721}$$

Thus, we get the result. □

Theorem 1.200 *Suppose that* $p, q \in \mathbb{C}$, $Re\,(q) > 0$ *and* $Re\,(p) > 0$, *then there is*

$$\int_0^\infty e^{-p\kappa^q t^{\beta q}} t^{\beta-1} dt = \frac{1}{2\beta\kappa\sqrt{p}} \Gamma\left(\frac{1}{q}\right), \tag{1.722}$$

where $\beta \in \mathbb{R}_+$ *and* $\kappa \in \mathbb{R}_+$.

Theorem 1.201 *Suppose that* $z, q \in \mathbb{C}$, $Re\,(q) > 0$, *and* $Re\,(z) > 0$, *then there is*

$$\int_0^\infty e^{-\kappa^q t^{\beta q}} t^{\beta z-1} dt = \frac{\Gamma\,(z)}{\beta q \kappa^z}, \tag{1.723}$$

where $\beta \in \mathbb{R}_+$ *and* $\kappa \in \mathbb{R}_+$.

Theorem 1.202 *Suppose that* $z \in \mathbb{C}$ *and* $Re\,(z) > 0$, *then there is*

$$\int_0^\infty e^{-\kappa^2 t^{2\beta}} t^{\beta z-1} dt = \frac{\Gamma\,(z)}{2\beta\kappa^z}, \tag{1.724}$$

where $\beta \in \mathbb{R}_+$ *and* $\kappa \in \mathbb{R}_+$.

Theorem 1.203 *Suppose that* $\beta \in \mathbb{R}_+$ *and* $\kappa \in \mathbb{R}_+$, *then there is*

$$\int_0^\infty t^{\beta(1+i)-1} \cdot e^{-\kappa^2 t^{2\beta}} dt = \frac{\Gamma(1+i)}{2\beta\kappa^{1+i}}. \tag{1.725}$$

Theorem 1.204 *Suppose that* $\beta \in \mathbb{R}_+$ *and* $\kappa \in \mathbb{R}_+$, *then there is*

$$\int_0^\infty t^{\beta(2+5i)-1} \cdot e^{-\kappa^2 t^{2\beta}} dt = \frac{\Gamma(2+5i)}{2\beta\kappa^{2+5i}}. \tag{1.726}$$

Theorem 1.205 *Suppose that* $\beta \in \mathbb{R}_+$ *and* $\kappa \in \mathbb{R}_+$, *then there is*

$$\int_0^\infty e^{-\kappa^2 t^{2\beta}} t^{\beta-1} dt = \frac{1}{2\beta\kappa}. \tag{1.727}$$

Theorem 1.206 *Suppose that* $\beta \in \mathbb{R}_+$, *then there is*

$$\int_0^\infty e^{-4t^{2\beta}} t^{\beta-1} dt = \frac{1}{4\beta}. \tag{1.728}$$

Theorem 1.207

$$\int_0^\infty e^{-4t^4} t dt = \frac{1}{8}. \tag{1.729}$$

Theorem 1.208

$$\int_0^\infty e^{-4t^6} t^2 dt = \frac{1}{12}. \tag{1.730}$$

Theorem 1.209

$$\int_0^\infty e^{-4t^8} t^3 dt = \frac{1}{16}. \tag{1.731}$$

Theorem 1.210

$$\int_0^\infty e^{-4t^{10}} t^4 dt = \frac{1}{20}. \tag{1.732}$$

Theorem 1.211

$$\int_0^\infty e^{-4t^{10}} t^5 dt = \frac{1}{24}. \tag{1.733}$$

Theorem 1.212 *Suppose that* $\kappa \in \mathbb{R}_+$, *then there is*

$$\int_0^\infty e^{-\kappa^2 t^6} t^2 dt = \frac{1}{6\kappa}. \tag{1.734}$$

Theorem 1.213

$$\int_0^\infty e^{-9t^6} t^2 dt = \frac{1}{18}. \tag{1.735}$$

Theorem 1.214

$$\int_0^\infty e^{-16t^6} t^2 dt = \frac{1}{24}. \tag{1.736}$$

Theorem 1.215

$$\int_0^\infty e^{-25t^6} t^2 dt = \frac{1}{30}. \tag{1.737}$$

Theorem 1.216

$$\int_0^\infty e^{-36t^6} t^2 dt = \frac{1}{36}. \tag{1.738}$$

Theorem 1.217

$$\int_0^\infty e^{-49t^6} t^2 dt = \frac{1}{42}. \tag{1.739}$$

Theorem 1.218

$$\int_0^\infty e^{-64t^6} t^2 dt = \frac{1}{48}. \tag{1.740}$$

Theorem 1.219

$$\int_0^\infty e^{-81t^6} t^2 dt = \frac{1}{54}. \tag{1.741}$$

Theorem 1.220

$$\int_0^\infty e^{-100t^6} t^2 dt = \frac{1}{60}. \tag{1.742}$$

Theorem 1.221 *Suppose that $\beta \in \mathbb{R}_+$ and $\kappa \in \mathbb{R}_+$, then there is*

$$\int_0^\infty t^{2\beta-1} \exp\left(-\kappa^3 t^{3\beta}\right) dt = \frac{1}{3\beta\kappa^2}. \tag{1.743}$$

Theorem 1.222 *Suppose that $\kappa \in \mathbb{R}_+$, then there is*

$$\int_0^\infty e^{-\kappa^3 t^{12}} t^7 dt = \frac{1}{12\kappa^2}. \tag{1.744}$$

Theorem 1.223 *Suppose that $\kappa \in \mathbb{R}_+$, then there is*

$$\int_0^\infty e^{-\kappa^3 t^{15}} t^9 dt = \frac{1}{15\kappa^2}. \tag{1.745}$$

Theorem 1.224

$$\int_0^\infty e^{-27t^{15}} t^9 dt = \frac{1}{135}. \tag{1.746}$$

Theorem 1.225

$$\int_0^\infty e^{-64t^{15}} t^9 dt = \frac{1}{240}. \tag{1.747}$$

Theorem 1.226

$$\int_0^\infty e^{-125t^{15}} t^9 dt = \frac{1}{375}. \tag{1.748}$$

Theorem 1.227

$$\int_0^\infty e^{-216t^{15}} t^9 dt = \frac{1}{540}. \tag{1.749}$$

Theorem 1.228

$$\int_0^\infty e^{-343t^{15}} t^9 dt = \frac{1}{735}. \tag{1.750}$$

Theorem 1.229

$$\int_0^\infty e^{-\kappa^3 t^{18}} t^{11} dt = \frac{1}{18\kappa^2}. \tag{1.751}$$

Theorem 1.230

$$\int_0^\infty e^{-t^{18}} t^{11} dt = \frac{1}{18}. \tag{1.752}$$

Theorem 1.231

$$\int_0^\infty e^{-8t^{18}} t^{11} dt = \frac{1}{72}. \tag{1.753}$$

Theorem 1.232

$$\int_0^\infty e^{-27t^{18}} t^{11} dt = \frac{1}{162}. \tag{1.754}$$

Theorem 1.233

$$\int_0^\infty e^{-96t^{18}} t^{11} dt = \frac{1}{432}. \tag{1.755}$$

Theorem 1.234

$$\int_0^\infty e^{-125t^{18}} t^{11} dt = \frac{1}{600}. \tag{1.756}$$

Theorem 1.235 *Suppose that $\beta \in \mathbb{R}_+$ and $\kappa \in \mathbb{R}_+$, then there is*

$$\int_0^\infty e^{-\kappa^2 t^{2\beta}} t^{3\beta-1} dt = \frac{3}{\beta\kappa^3}, \tag{1.757}$$

where $\beta \in \mathbb{R}_+$ and $\kappa \in \mathbb{R}_+$.

Theorem 1.236 *Suppose that $\beta \in \mathbb{R}_+$ and $\kappa \in \mathbb{R}_+$, then there is*

$$\int_0^\infty e^{-\kappa^2 t^{2\beta}} t^{4\beta-1} dt = \frac{12}{\beta\kappa^4}. \tag{1.758}$$

Theorem 1.237 *Suppose that $\beta \in \mathbb{R}_+$ and $\kappa \in \mathbb{R}_+$, then there is*

$$\int_0^\infty e^{-\kappa^2 t^{2\beta}} t^{\frac{\beta}{2}-1} dt = \frac{1}{2\beta}\sqrt{\frac{\pi}{\kappa}}. \tag{1.759}$$

Theorem 1.238 *Suppose that $\beta \in \mathbb{R}_+$ and $\kappa \in \mathbb{R}_+$, then there is*

$$\int_0^\infty e^{-\kappa^2 t^{2\beta}} t^{\frac{3\beta}{2}-1} dt = \frac{3\sqrt{\pi}}{4\beta \sqrt[\frac{3}{2}]{\kappa}}. \tag{1.760}$$

Theorem 1.239 *Suppose that $\beta \in \mathbb{R}_+$ and $\kappa \in \mathbb{R}_+$, then there is*

$$\int_0^\infty e^{-\kappa^2 t^{2\beta}} t^{\frac{5\beta}{2}-1} dt = \frac{15\sqrt{\pi}}{8\beta \sqrt[\frac{5}{2}]{\kappa}}. \tag{1.761}$$

Theorem 1.240 *Suppose that $p \in \mathbb{C}$ and $Re\,(p) > 0$, then there is*

$$\int_0^\infty t^{\beta-1} \exp\left(-p\kappa^2 t^{2\beta}\right) dt = \frac{1}{2\beta\kappa}\sqrt{\frac{\pi}{p}}, \tag{1.762}$$

where $\beta \in \mathbb{R}_+$ and $\kappa \in \mathbb{R}_+$.

Theorem 1.241 *Suppose that $p \in \mathbb{C}$ and $Re\,(p) > 0$, then there is*

$$\int_0^\infty t^{\beta-1} \exp\left(-p \cdot \kappa^3 \cdot t^{3\beta}\right) dt = \frac{1}{2\beta\kappa\sqrt{p}}\Gamma\left(\frac{1}{3}\right), \tag{1.763}$$

where $\beta \in \mathbb{R}_+$ and $\kappa \in \mathbb{R}_+$.

Theorem 1.242 *Suppose that $p \in \mathbb{C}$ and $Re\,(p) > 0$, then there is*

$$\int_0^\infty t^{\beta-1} \exp\left(-p \cdot \sqrt[\frac{\beta}{2}]{t} \cdot \sqrt{\kappa}\right) dt = \frac{1}{2\beta\kappa\sqrt{p}}, \tag{1.764}$$

where $\beta \in \mathbb{R}_+$ and $\kappa \in \mathbb{R}_+$.

Theorem 1.243 *Suppose that $p \in \mathbb{C}$ and $Re\,(p) > 0$, then there is*

$$\int_0^\infty t^{\beta-1} \exp\left(-p \cdot \sqrt[\frac{\beta}{2}]{t} \cdot \sqrt[\frac{1}{2}]{\kappa}\right) dt = \frac{1}{\beta\kappa\sqrt{p}}, \tag{1.765}$$

where $\beta \in \mathbb{R}_+$ and $\kappa \in \mathbb{R}_+$.

Theorem 1.244 *Suppose that $q, p \in \mathbb{C}$, $Re\,(q) > 0$ and $Re\,(p) > 0$, then there is*

$$\int_0^\infty e^{-p\kappa^q t^{\beta q}} t^{2\beta-1} dt = \frac{1}{\beta\kappa^2 q p^{\frac{2}{q}}} \Gamma\left(\frac{2}{q}\right), \tag{1.766}$$

where $\beta \in \mathbb{R}_+$ and $\kappa \in \mathbb{R}_+$.

Theorem 1.245 *Suppose that $q \in \mathbb{C}$ and $Re\,(q) > 0$, then there is*

$$\int_0^\infty e^{-p\kappa^3 t^{3\beta}} t^{2\beta-1} dt = \frac{1}{3\beta\kappa^2 p^{\frac{2}{3}}} \Gamma\left(\frac{2}{3}\right), \tag{1.767}$$

where $\beta \in \mathbb{R}_+$ and $\kappa \in \mathbb{R}_+$.

Theorem 1.246 *Suppose that $q, p \in \mathbb{C}$, $Re\,(q) > 0$ and $Re\,(p) > 0$, then there is*

$$\int_0^\infty e^{-p\kappa^q t^{\beta q}} t^{3\beta-1} dt = \frac{1}{\beta\kappa^3 q p^{\frac{3}{q}}} \Gamma\left(\frac{3}{q}\right), \tag{1.768}$$

where $\beta \in \mathbb{R}_+$ and $\kappa \in \mathbb{R}_+$.

Theorem 1.247 *Suppose that $p \in \mathbb{C}$ and $Re\,(p) > 0$, then there is*

$$\int_0^\infty e^{-p\kappa^2 t^{2\beta}} t^{3\beta-1} dt = \frac{3\sqrt{\pi}}{4\beta\kappa^3 p^{\frac{3}{2}}}, \tag{1.769}$$

where $\beta \in \mathbb{R}_+$ and $\kappa \in \mathbb{R}_+$.

Theorem 1.248 *Suppose that $p \in \mathbb{C}$ and $Re\,(p) > 0$, then there is*

$$\int_0^\infty e^{-4pt^{2\beta}} t^{3\beta-1} dt = \frac{3\sqrt{\pi}}{32\beta p^{\frac{3}{2}}}, \tag{1.770}$$

where $\beta \in \mathbb{R}_+$.

Theorem 1.249 *Suppose that $\beta \in \mathbb{R}_+$, then there is*

$$\int_0^\infty e^{-18t^{2\beta}} t^{3\beta-1} dt = \frac{3\sqrt{\pi}}{32\beta 3^{\frac{3}{2}}}. \tag{1.771}$$

Theorem 1.250 *Suppose that $\beta \in \mathbb{R}_+$, then there is*

$$\int_0^\infty e^{-20t^{2\beta}} t^{3\beta-1} dt = \frac{3\sqrt{\pi}}{32\beta 5^{\frac{3}{2}}}. \tag{1.772}$$

Theorem 1.251 *Suppose that $\beta \in \mathbb{R}_+$, then there is*

$$\int_0^\infty e^{-28t^{2\beta}} t^{3\beta-1} dt = \frac{3\sqrt{\pi}}{32\beta 7^{\frac{3}{2}}}. \tag{1.773}$$

1.6.5 Other Calculus Operators with Respect to Monotone Functions

In this section, we introduce the calculus operators with respect to monotone functions, which are called a family of the general calculus operators.

1.6.5.1 The Calculus with Respect to Exponential Function

Definition 1.76 The derivative with respect to exponential function is defined as [110]

$${}^{e}D_t^{(1)} \Theta(t) =: D_{t,e^{\lambda t}}^{(1)} \Theta(t) = \frac{1}{\lambda e^{\lambda t}} \frac{d\Theta(t)}{dt}. \tag{1.774}$$

Definition 1.77 The integral with respect to exponential function is defined as [110]

$${}_{a}^{e}I_t^{(1)} \Theta(t) =: {}_{a}I_{t,e^{\lambda t}}^{(1)} \Theta(t) = \int_a^t \Theta(\tau) \lambda e^{\lambda t} dt. \tag{1.775}$$

Definition 1.78 The derivative of higher order with respect to another function is defined as

$${}^{e}D_t^{(n)} \Theta(t) = \left(\frac{1}{\lambda e^{\lambda t}} \frac{d}{dt} \right)^n \Theta(t). \tag{1.776}$$

Their relationships between (1.774) and (1.775) are given as follows:

$$\Theta(t) = {}^{e}D_t^{(1)} \left({}_{a}^{e}I_t^{(1)} \Theta(t) \right) \tag{1.777}$$

and

$$\Theta(\tau) - \Theta(a) = {}_a^e I_t^{(1)} \left({}^e D_t^{(1)} \Theta(t) \right). \tag{1.778}$$

1.6.5.2 The Calculus with Respect to Logarithmic Function

We now present the derivatives with respect to logarithmic function as follows:

Definition 1.79 The derivative with respect to logarithmic function is defined as [110]

$$^L D_t^{(1)} \Theta(t) =: D_{t,\ln t}^{(1)} \Theta(t) = t \frac{d\Theta(t)}{dt}. \tag{1.779}$$

Definition 1.80 The integral with respect to logarithmic function is defined as [110, 153]

$${}_a^L I_t^{(1)} \Theta(t) =: {}_a I_{t,\ln t}^{(1)} \Theta(t) = \int_a^t \Theta(t) \frac{dt}{\tau}. \tag{1.780}$$

Definition 1.81 The derivative of higher order with respect to logarithmic function is defined as [131]

$$^L D_t^{(n)} \Theta(t) = \left(t \frac{d}{dt} \right)^n \Theta(t). \tag{1.781}$$

Their relationships between (1.779) and (1.780) can be written as follows [131]:

$$\begin{aligned} \Theta(t) &= {}^L D_t^{(1)} \left({}_a^L I_t^{(1)} \Theta(t) \right) \\ &= \left(t \tfrac{d}{dt} \right) \int_a^t \Theta(t) \tfrac{dt}{t} \\ &= \left(t \tfrac{d}{dt} \right) \int_a^t \Theta(t) \tfrac{dt}{t} \end{aligned} \tag{1.782}$$

and

$$\Theta(t) - \Theta(0) = {}_a^L I_t^{(1)} \left({}^L D_t^{(1)} \Theta(t) \right). \tag{1.783}$$

1.6.5.3 The Calculus with Respect to Complex Topology

We now consider the mapping complex topology by using the scaling-law function and the calculus with respect to complex topology.

The complex topology by using the scaling-law function is defined as

$$g(t) = \kappa t^{\beta} + c, \tag{1.784}$$

where $\kappa$is the normalization constant for $\kappa \in \mathbb{R}_{+}$, β is the scaling exponent for $\beta \in \mathbb{R}$, t is the radius for $t \in \mathbb{R}$, and $c$is the moving term for $c \in \mathbb{R}$.

Definition 1.82 The derivative with respect to complex topology is defined as

$${}^{T}D_{t}^{(1)}\Theta(t) =: D_{t,\kappa t^{\beta}+c}^{(1)}\Theta(t) = \frac{1}{\left(\kappa t^{\beta}+c\right)^{(1)}}\frac{d\Theta(t)}{dt}, \tag{1.785}$$

where $\Theta(t) = \left(\varphi \circ \left(\kappa t^{\beta}+c\right)\right)(t) = \varphi\left(\kappa t^{\beta}+c\right)$, κ is the normalization constant, β is the scaling exponent, t is the radius, and c is the moving term.

Definition 1.83 The derivative of higher order with respect to complex topology is defined as

$${}^{T}D_{t}^{(n)}\Theta(t) = \left(\frac{1}{\left(\kappa t^{\beta}+c\right)^{(1)}}\frac{d}{dt}\right)^{n}\Theta(t), \tag{1.786}$$

where $\Theta(t) = \left(\varphi \circ \left(\kappa t^{\beta}+c\right)\right)(t) = \varphi\left(\kappa t^{\beta}+c\right)$, κ is the normalization constant, β is the scaling exponent, t is the radius, and c is the moving term.

Their relationships between (1.785) and (1.786) can be written as follows [131]:

$$\begin{aligned}
&\Theta(t) \\
&= {}^{T}D_{t}^{(1)}\left({}_{a}^{T}I_{t}^{(1)}\Theta(t)\right) \\
&= \tfrac{1}{\left(\kappa t^{\beta}+c\right)^{(1)}}\tfrac{d}{dt}\int_{a}^{t}\Theta(t)\left(\kappa t^{\beta}+c\right)^{(1)}dt
\end{aligned} \tag{1.787}$$

and

$$\Theta(t) - \Theta(0) = {}_{a}^{T}I_{t}^{(1)}\left({}^{T}D_{t}^{(1)}\Theta(t)\right). \tag{1.788}$$

We now present the properties for the derivative with respect to complex topology as follows:

$${}^{T}D_{t}^{(1)}1 = 0, \tag{1.789}$$

$${}^{T}D_{t}^{(1)}\left(\kappa t^{\beta}+c\right) = 1, \tag{1.790}$$

$${}^{T}D_{t}^{(1)}\left(\kappa t^{\beta}+c\right)^{n} = n\left(\kappa t^{\beta}+c\right)^{n-1}, \tag{1.791}$$

$$ {}^{T}D_t^{(1)} e^{\kappa t^{\beta}+c} = e^{\kappa t^{\beta}+c}, \tag{1.792} $$

$$ {}^{T}D_t^{(1)} \ln\left(\kappa t^{\beta}+c\right) = \frac{1}{\left(\kappa t^{\beta}+c\right)}, \tag{1.793} $$

$$ {}^{T}D_t^{(1)} \log_s\left(\kappa t^{\beta}+c\right) = \frac{1}{(\ln s)\left(\kappa t^{\beta}+c\right)}, \tag{1.794} $$

$$ {}^{T}D_t^{(1)} s^{\kappa t^{\beta}+c} = (\ln s)\, s^{\kappa t^{\beta}+c}, \tag{1.795} $$

$$ {}^{T}D_t^{(1)} e^{\psi(t)} = e^{\psi(t)}\, {}^{T}D_t^{(1)} \psi(t), \tag{1.796} $$

$$ {}^{T}D_t^{(1)} |\psi(t)| = \frac{\psi(t)}{|\psi(t)|}\, {}^{T}D_t^{(1)} \psi(t),\ \psi(t) \neq 0, \tag{1.797} $$

$$ {}^{T}D_t^{(1)} \ln\psi(t) = \frac{{}^{T}D_t^{(1)} \psi(t)}{\psi(t)}, \tag{1.798} $$

$$ {}^{T}D_t^{(1)} \log_s \psi(t) = \frac{{}^{T}D_t^{(1)} \psi(t)}{(\ln s)\,\psi(t)}, \tag{1.799} $$

and

$$ {}^{T}D_t^{(1)} s^{\psi(t)} = \left[(\ln s)\, s^{\psi(t)}\right] \cdot {}^{T}D_t^{(1)} \psi(t), \tag{1.800} $$

where $\psi(t) = \left(\Lambda \circ \left(\kappa t^{\beta}+c\right)\right)(t) = \Lambda\left(\kappa t^{\beta}+c\right)$, κ is the normalization constant, β is the scaling exponent, t is the radius, and c is the moving term.

Definition 1.84 The integral with respect to complex topology is defined as

$$ {}_{a}^{T}I_t^{(1)} \Theta(t) =: {}_{a}I_{t,\kappa t^{\beta}+c}^{(1)} \Theta(t) = \int_a^t \Theta(t) \left(\kappa t^{\beta}+c\right)^{(1)} dt, \tag{1.801} $$

where $\Theta(t) = \left(\varphi \circ \left(\kappa t^{\beta}+c\right)\right)(t) = \varphi\left(\kappa t^{\beta}+c\right)$, κ is the normalization constant, β is the scaling exponent, t is the radius, and c is the moving term.

The properties of the integral with respect to complex topology are given as follows:

$$ {}_{a}^{T}I_t^{(1)} 1 = \kappa\left(b^{\beta} - a^{\beta}\right), \tag{1.802} $$

$$ {}_{a}^{T}I_t^{(1)} \left(n\left(\kappa t^{\beta}+c\right)^{n-1}\right) = \left(\kappa b^{\beta}+c\right)^{n} - \left(\kappa a^{\beta}+c\right)^{n}, \tag{1.803} $$

$$ {}_{a}^{T}I_{t}^{(1)}\left(e^{\kappa t^{\beta}+c}\right)=e^{\kappa b^{\beta}+c}-e^{\kappa a^{\beta}+c}, \tag{1.804}$$

$$ {}_{a}^{T}I_{t}^{(1)}\left(\frac{1}{\kappa t^{\beta}+c}\right)=\ln\left(\kappa b^{\beta}+c\right)-\ln\left(\kappa a^{\beta}+c\right), \tag{1.805}$$

$$ {}_{a}^{T}I_{t}^{(1)}\left(\frac{1}{(\ln s)}\cdot\frac{1}{\kappa t^{\beta}+c}\right)=\log_{s}\left(\kappa b^{\beta}+c\right)-\log_{s}\left(\kappa a^{\beta}+c\right), \tag{1.806}$$

and

$$ {}_{a}^{T}I_{t}^{(1)}\left((\ln s)\,s^{\kappa t^{\beta}+c}\right)=s^{\kappa b^{\beta}+c}-s^{\kappa a^{\beta}+c}, \tag{1.807}$$

where $\ln s \neq 0$, κ is the normalization constant, β is the scaling exponent, t is the radius, c is the moving term, and $e^{-\kappa t^{\beta}}$ is the Kohlrausch–Williams–Watts function (see Chap. 7).

Definition 1.85 The indefinite integral with respect to complex topology is defined as

$$ {}^{T}I_{t}^{(1)}\Theta(t)=\int\Theta(t)\left(\kappa t^{\beta}+c\right)^{(1)}dt, \tag{1.808}$$

where $\psi(t)=\left(\Lambda\circ\left(\kappa t^{\beta}+c\right)\right)(t)=\Lambda\left(\kappa t^{\beta}+c\right)$, κ is the normalization constant, β is the scaling exponent, t is the radius, and c is the moving term.

Equation (1.808) is called the antidifferentiation with respect to complex topology.

In this case, we give

$$ {}^{T}D_{t}^{(1)}\left({}^{T}I_{t}^{(1)}\psi(t)\right)=\psi(t), \tag{1.809}$$

where

$$ {}^{T}I_{t}^{(1)}\psi(t)=\Phi(t)+C \tag{1.810}$$

with the constant C.

So, we have the following result:

Suppose that $\psi_1(t)=\Lambda_1\left(\kappa t^{\beta}+c\right)$ and $\psi_2(t)=\Lambda_2\left(\kappa t^{\beta}+c\right)$ are continuous, then there is

$$ {}^{T}I_{t}^{(1)}\left(\psi_2(t)\,{}^{T}D_{t}^{(1)}\psi_1(t)\right)=\psi_2(t)\,\psi_1(t)-{}^{T}I_{t}^{(1)}\left(\psi_1(t)\,{}^{T}D_{t}^{(1)}\psi_2(t)\right). \tag{1.811}$$

Equation (1.811) is called the integration by parts for the indefinite integral with respect to complex topology.

The properties for the indefinite integral with respect to complex topology are given as follows:

$$ {}^{T}I_t^{(1)}1 = \kappa t^{\beta} + c + C, \tag{1.812} $$

$$ {}^{T}I_t^{(1)}\left(n\left(\kappa t^{\beta}+c\right)^{n-1}\right) = \left(\kappa t^{\beta}+c\right)^{n} + C, \tag{1.813} $$

$$ {}^{T}I_t^{(1)}\left(e^{\kappa t^{\beta}+c}\right) = e^{\kappa t^{\beta}+c} + C, \tag{1.814} $$

$$ {}^{T}I_t^{(1)}\left(\frac{1}{\kappa t^{\beta}+c}\right) = \ln\left(\kappa t^{\beta}+c\right) + C, \tag{1.815} $$

$$ {}^{T}I_t^{(1)}\left(\frac{1}{(\ln s)}\cdot\frac{1}{\kappa t^{\beta}+c}\right) = \log_s\left(\kappa t^{\beta}+c\right) + C, \tag{1.816} $$

$$ {}^{T}I_t^{(1)}\left((\ln s)\, s^{\kappa t^{\beta}+c}\right) = s^{\kappa t^{\beta}+c} + C, \tag{1.817} $$

$$ {}^{T}I_t^{(1)}\left(e^{\psi(t)}\, {}^{T}D_t^{(1)}\psi(t)\right) = e^{\psi(t)} + C, \tag{1.818} $$

$$ {}^{T}I_t^{(1)}\left(\frac{\psi(t)}{|\psi(t)|}\, {}^{T}D_t^{(1)}\psi(t)\right) = |\psi(t)| + C, \tag{1.819} $$

$$ {}^{T}I_t^{(1)}\left(\frac{{}^{T}D_t^{(1)}\psi(t)}{\psi(t)}\right) = \ln\psi(t) + C, \tag{1.820} $$

$$ {}^{T}I_t^{(1)}\left(\frac{{}^{T}D_t^{(1)}\psi(t)}{(\ln s)\,\psi(t)}\right) = \log_s\psi(t) + C, \tag{1.821} $$

and

$$ {}^{T}I_t^{(1)}\left(\left[(\ln s)\, s^{\psi(t)}\right]\cdot {}^{T}D_t^{(1)}\psi(t)\right) = s^{\psi(t)} + C, \tag{1.822} $$

where $\ln s \neq 0$, κ is the normalization constant, β is the scaling exponent, t is the radius, c is the moving term, and $e^{-\kappa t^{\beta}}$ is the Kohlrausch–Williams–Watts function (see Chap. 7).

We now consider the anomalous relaxation within the derivative with respect to complex topology.

The anomalous relaxation problem within the derivative with respect to complex topology n is given as

$$ {}^{T}D_t^{(1)}\Xi(t) + \Xi(t) = 0, \tag{1.823} $$

subjected to the initial condition

$$\Xi(0) = \chi, \tag{1.824}$$

where χ is a constant.

So, we give

$$\frac{{}^{T}D_t^{(1)}\Xi(t)}{\Xi(t)} = -1, \tag{1.825}$$

which, by the scaling-law integral, leads to

$${}_0^T I_t^{(1)}\left(\frac{{}^{T}D_t^{(1)}\Xi(t)}{\Xi(t)}\right) = \ln\Xi(t) - \ln\Xi(0) = -\left(\kappa t^{\beta} + c\right). \tag{1.826}$$

It follows from Eq. (1.826) that

$$\ln\frac{\Xi(t)}{\Xi(0)} = -\left(\kappa t^{\beta} + c\right), \tag{1.827}$$

which leads to

$$\Xi(t) = \Xi(0)\, e^{-\left(\kappa t^{\beta}+c\right)} = \chi e^{-\left(\kappa t^{\beta}+c\right)}, \tag{1.828}$$

where $e^{-\kappa t^{\beta}}$ is the Kohlrausch–Williams–Watts function (see Chap. 7).

The plot of the solution for the anomalous relaxation problem $\Xi(t)$ is shown in Fig. 1.6.

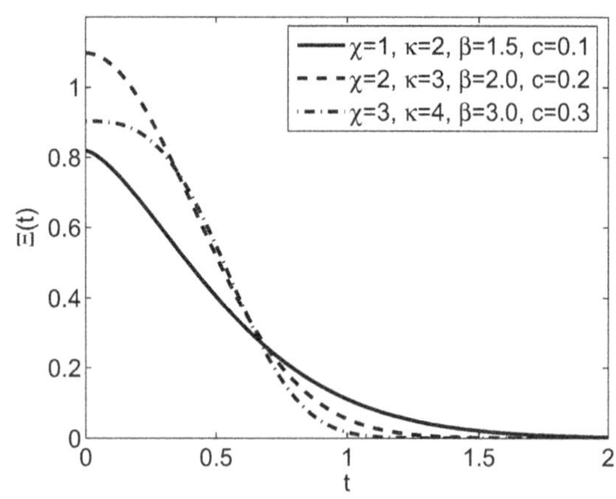

Fig. 1.6 The plot of the solution for the anomalous relaxation problem

We now present the concepts for the improper integrals with respect to complex topology as follows:

- The integral with respect to complex topology in (a, ∞) is defined as follows:

$$
\begin{aligned}
&{}_{a}^{T}I_{\infty}^{(1)}\Theta(t) \\
&= \int_{a}^{\infty}\Theta(t)\left(\kappa t^{\beta}+c\right)^{(1)}dt \\
&= \lim_{b\to\infty}\int_{a}^{b}\Theta(t)\left(\kappa t^{\beta}+c\right)^{(1)}dt,
\end{aligned}
\tag{1.829}
$$

  where $\Theta(t) = \left(\varphi \circ \left(\kappa t^{\beta}+c\right)\right)(t) = \varphi\left(\kappa t^{\beta}+c\right)$, κ is the normalization constant, β is the scaling exponent, t is the radius, and c is the moving term.
- The integral with respect to complex topology in $(-\infty, b)$ is defined as follows:

$$
\begin{aligned}
&{}_{-\infty}^{T}I_{b}^{(1)}\Theta(t) \\
&= \int_{-\infty}^{b}\Theta(t)\left(\kappa t^{\beta}+c\right)^{(1)}dt \\
&= \lim_{a\to\infty}\int_{a}^{b}\Theta(t)\left(\kappa t^{\beta}+c\right)^{(1)}dt,
\end{aligned}
\tag{1.830}
$$

  where $\Theta(t) = \left(\varphi \circ \left(\kappa t^{\beta}+c\right)\right)(t) = \varphi\left(\kappa t^{\beta}+c\right)$, κ is the normalization constant, β is the scaling exponent, t is the radius, and c is the moving term.
- The integral with respect to complex topology in $(-\infty, \infty)$ is defined as follows:

$$
\begin{aligned}
&{}_{-\infty}^{T}I_{\infty}^{(1)}\Theta(t) \\
&= \int_{-\infty}^{\infty}\Theta(t)\left(\kappa t^{\beta}+c\right)^{(1)}dt \\
&= \int_{-\infty}^{c}\Theta(t)\left(\kappa t^{\beta}+c\right)^{(1)}dt + \int_{c}^{\infty}\Theta(t)\left(\kappa t^{\beta}+c\right)^{(1)}dt,
\end{aligned}
\tag{1.831}
$$

  where c is the constant, $\psi(t) = \left(\varphi \circ \left(\kappa t^{\beta}+c\right)\right)(t) = \varphi\left(\kappa t^{\beta}+c\right)$, κ is the normalization constant, β is the scaling exponent, t is the radius, and c is the moving term.

According to the above results, the topics on the calculus operators with respect to monotone functions are still open from the point of view of the operator calculus due to the complex topology.

Chapter 2
Wright Function and Integral Transforms via Dunkl Transform

Abstract In this chapter, we investigate the concepts and theorems for the Wright's generalized hypergeometric function and Wright function and introduce the supertrigonometric and superhyperbolic functions related to them. The integral representations for the supertrigonometric and superhyperbolic functions are addressed in detail. The Dunkl transform, new integral transforms based on the calculus with respect to power-law function, and the Wright functions are proposed.

2.1 The Special Functions Related to Wright Function and Integral Representations

2.1.1 The Wright's Generalized Hypergeometric Function

We now begin with the definitions and theorems for the Wright's generalized hypergeometric function.

To formulate the result we use the following notation:

$$\Delta = \sum_{j=1}^{p} A_j - \sum_{k=1}^{q} B_k, \tag{2.1}$$

$$\delta = \prod_{j=1}^{p} |A_j|^{-A_j} \prod_{k=1}^{q} |B_k|^{-B_k} \tag{2.2}$$

and

$$h = \sum_{k=1}^{q} \beta_k - \sum_{j=1}^{p} \alpha_j + \frac{p-q}{2}, \tag{2.3}$$

where $j, k \in \mathbb{N}$, $\alpha_j, \beta_k \in \mathbb{C}$, and $A_j, B_k \in \mathbb{R}_+$.

X.-J. Yang, *Theory and Applications of Special Functions for Scientists and Engineers*, https://doi.org/10.1007/978-981-33-6334-2_2

Definition 2.1 The Wright's generalized hypergeometric series is defined as [163, 164]

$$
\begin{aligned}
{}_p\varphi_q\,(z) &= {}_p\varphi_q\left(\begin{matrix}(\alpha_1, A_1), \cdots, (\alpha_p, A_p) \\ (\beta_1, B_1), \cdots, (\beta_q, B_q)\end{matrix}; z\right) \\
&= {}_p\varphi_q\left(\begin{matrix}(\alpha_p, A_p)_{1,p} \\ (\beta_q, B_q)_{1,q}\end{matrix}; z\right) \\
&= \sum_{u=0}^{\infty} \frac{\left\{\prod_{j=1}^{p} \Gamma(\alpha_j + A_j u)\right\}}{\left\{\prod_{k=1}^{p} \Gamma(\beta_k + B_k u)\right\}} \frac{z^u}{u!},
\end{aligned} \tag{2.4}
$$

where $z \in \mathbb{C}$, $j, k, u \in \mathbb{N}$, $\alpha_j, \beta_k \in \mathbb{C}$, and $A_j, B_k \in \mathbb{R}_+$.

The result was discovered by Wright in 1935 [163] and further investigated by Wright in 1940 [164].

The Wright's generalized hypergeometric function is defined by the following series representation [163, 164]:

$$
{}_p\varphi_q\left(\begin{matrix}(\alpha_1, A_1), \cdots, (\alpha_p, A_p) \\ (\beta_1, B_1), \cdots, (\beta_q, B_q)\end{matrix}; z\right) = \sum_{u=0}^{\infty} \frac{\left\{\prod_{j=1}^{p} \Gamma\left(\alpha_j + A_j u\right)\right\}}{\left\{\prod_{k=1}^{p} \Gamma\left(\beta_k + B_k u\right)\right\}} \frac{z^u}{u!}, \tag{2.5}
$$

where $z \in \mathbb{C}$, $j, k, u \in \mathbb{N}$, $\alpha_j, \beta_k \in \mathbb{C}$, $A_j, B_k \in \mathbb{R}_+$, and $\Delta > -1$.

The result was discovered by Wright in 1935 [163, 164] and further investigated in [165–169].

Theorem 2.1 *Let $z \in \mathbb{C}$, $j, k, u \in \mathbb{N}$, $\alpha_j, \beta_k \in \mathbb{C}$, and $A_j, B_k \in \mathbb{R}_+$.*
Then we have the cases:

(1a) *If $\Delta + 1 > 0$, then the series in (2.4) is absolutely convergent for $z \in \mathbb{C}$;*
(2a) *If $\Delta + 1 = 0$, then the series in (2.4) is absolutely convergent for all values of $|z| < \delta$, and of*
$|z| = \delta$ and $Re\,(h) > \frac{1}{2}$.

The result was discovered by Kibas in 2005 [170] and further reported in [171].

Definition 2.2 The Wright function is defined by the following series representation:

$$
\psi\,(\beta, B, z) = {}_0\varphi_1\left(\begin{matrix}- \\ (B, \beta)\end{matrix}; z\right) = \sum_{k=0}^{\infty} \frac{1}{\Gamma\,(\beta + Bk)} \frac{z^k}{k!}, \tag{2.6}
$$

where $z \in \mathbb{C}$, $\beta \in \mathbb{C}$, and $B \in \mathbb{R}$.

The result was discovered by Wright in 1933 [172].

Theorem 2.2 *The Wright function for $\beta > -1$ and $B \in \mathbb{C}$ ($B \in \mathbb{Z}_0^-$ if $\beta = 0$) is an entire function of finite order p and type σ given by $p = \frac{1}{1+\beta}$ and $\sigma = (1+\beta)\,|\beta|^{-\frac{\beta}{1+\beta}}$.*

The result was discovered by Luchko in 2019 [173].

Theorem 2.3 (Wright Theorem [163]) *The Wright function is an entire function of z for all values of the parameters $\beta > -1$ and $B \in \mathbb{C}$.*

The result was discovered by Wright in 1935 [163, 174], rediscovered by Kilbas, Saigo, and Trujillo in 2002 [165], and further reported by Luchko in 2019 [173].

Definition 2.3 The Bessel–Maitland function (also called the Wright generalized Bessel function) is defined as [170, 175, 176]

$$\Theta\left(\alpha, A, z\right) = {}_0\varphi_1\left(\begin{matrix} - \\ (\alpha+1, B) \end{matrix}; -z\right) = \sum_{k=0}^{\infty} \frac{1}{\Gamma\left(\alpha+1+Ak\right)} \frac{(-z)^k}{k!}, \tag{2.7}$$

where $z \in \mathbb{C}$, $\alpha \in \mathbb{C}$, and $A \in \mathbb{R}$.

Theorem 2.4 *Let $z \in \mathbb{C}$, $j, k, u \in \mathbb{N}$, $\alpha_j, \beta_k \in \mathbb{C}$ and $A_j, B_k \in \mathbb{R}_+$. Then the Wright's generalized hypergeometric function is an entire function of z.*

The result was discovered by Kibas in 2005 [170].

Theorem 2.5 *Let $B \in \mathbb{R}$ and $\beta \in \mathbb{C}$.*

(1A) *If $B+1 > 0$, then the series in (2.6) is absolutely convergent for $z \in \mathbb{C}$.*
(2A) *If $B+1 = 0$, then the series in (2.6) is absolutely convergent for all values of $|z| < 1$, and of $|z| = 1$ and $Re\,(h) > \frac{1}{2}$.*

The result was discovered by Kibas in 2005 [170].

Theorem 2.6 *Let $B > -1$ and $\beta \in \mathbb{C}$. Then the Wright function $\psi\left(\beta, B, z\right)$ is an entire function of z.*

The result was discovered by Kibas in 2005 [170].

Theorem 2.7 *Let $z \in \mathbb{C}$, $j, k, u \in \mathbb{N}$, $\alpha_j, \beta_k \in \mathbb{C}$, and $A_j, B_k \in \mathbb{R}_+$.*
Then we have [177]

$${}_p\varphi_q\left(\begin{matrix} (\alpha_1, 1), \cdots, (\alpha_p, 1) \\ (\beta_1, 1), \cdots, (\beta_q, 1) \end{matrix}; z\right) = \frac{\prod_{j=1}^{p} \Gamma\left(\alpha_j\right)}{\prod_{k=1}^{p} \Gamma\left(\beta_k\right)} \, {}_pF_q\left(\begin{matrix} \alpha_1, \cdots, \alpha_p \\ \beta_1, \cdots, \beta_q \end{matrix}; z\right), \tag{2.8}$$

where $\Delta + 1 \geq 0$.

The result was discovered by Srivastava and Karlsson in 1985 (see [177], p.21).

2.1.2 The Integral Representations via Wright's Generalized Hypergeometric Function

Theorem 2.8 (Kilbas Theorem [170]) *Let $\alpha, \gamma \in \mathbb{C}$, $Re(\alpha) > 0$, $Re(\gamma) > 0$, $\lambda \in \mathbb{C}$, $\beta > 0$, $j, k, u \in \mathbb{N}$, $\alpha_j, \beta_k \in \mathbb{C}$, and $A_j, B_k \in \mathbb{R}_+$. Then we have*

$$\begin{aligned}&\tfrac{1}{\Gamma(\alpha)} \int_0^t (t-\tau)^{\alpha} \tau^{\gamma-1} {}_p\varphi_q \left(\begin{matrix} (\alpha_1, A_1), \cdots, (\alpha_p, A_p) \\ (\beta_1, B_1), \cdots, (\beta_q, B_q) \end{matrix} ; \lambda\tau^{\beta} \right) d\tau \\ &= t^{\gamma+\alpha-1} {}_{p+1}\varphi_{q+1} \left(\begin{matrix} (\alpha_1, A_1), \cdots, (\alpha_p, A_p), (\gamma, \beta) \\ (\beta_1, B_1), \cdots, (\beta_q, B_q), (\gamma+\alpha, \beta) \end{matrix} ; \lambda t^{\beta} \right).\end{aligned} \tag{2.9}$$

The result was discovered by Kibas in 2005 [170].

As the special case, we have the following property:

Property 2.1 Let $\gamma \in \mathbb{C}$, $Re(\gamma) > 0$, $\lambda \in \mathbb{C}$, $\beta > 0$, $j, k, u \in \mathbb{N}$, $\alpha_j, \beta_k \in \mathbb{C}$, and $A_j, B_k \in \mathbb{R}_+$. Then we have

$$\begin{aligned}&\int_0^t (t-\tau) \tau^{\gamma-1} {}_p\varphi_q \left(\begin{matrix} (\alpha_1, A_1), \cdots, (\alpha_p, A_p) \\ (\beta_1, B_1), \cdots, (\beta_q, B_q) \end{matrix} ; \lambda\tau^{\beta} \right) d\tau \\ &= t^{\gamma} {}_{p+1}\varphi_{q+1} \left(\begin{matrix} (\alpha_1, A_1), \cdots, (\alpha_p, A_p), (\gamma, \beta) \\ (\beta_1, B_1), \cdots, (\beta_q, B_q), (\gamma+1, \beta) \end{matrix} ; \lambda t^{\beta} \right).\end{aligned} \tag{2.10}$$

The result was discovered by Raina in 2005 [178].

Property 2.2 Let $\lambda \in \mathbb{C}$, $\beta > 0$, $j, k, u \in \mathbb{N}$, $\alpha_j, \beta_k \in \mathbb{C}$, and $A_j, B_k \in \mathbb{R}_+$. Then we have

$$\begin{aligned}&\int_0^t (t-\tau) {}_p\varphi_q \left(\begin{matrix} (\alpha_1, A_1), \cdots, (\alpha_p, A_p) \\ (\beta_1, B_1), \cdots, (\beta_q, B_q) \end{matrix} ; \lambda\tau^{\beta} \right) d\tau \\ &= t^{\gamma} {}_{p+1}\varphi_{q+1} \left(\begin{matrix} (\alpha_1, A_1), \cdots, (\alpha_p, A_p), (1, \beta) \\ (\beta_1, B_1), \cdots, (\beta_q, B_q), (2, \beta) \end{matrix} ; \lambda t^{\beta} \right).\end{aligned} \tag{2.11}$$

Proof When $\gamma = 1$ in (2.10), we obtain the result. □

Property 2.3 Let $\gamma \in \mathbb{C}$, $Re(\gamma) > 0$, $\lambda \in \mathbb{C}$, $\beta > 0$, $j, k, u \in \mathbb{N}$, $\alpha_j, \beta_k \in \mathbb{C}$, and $A_j, B_k \in \mathbb{R}_+$. Then we have

$$\begin{aligned}&\int_0^1 (1-\tau) \tau^{\gamma-1} {}_p\varphi_q \left(\begin{matrix} (\alpha_1, A_1), \cdots, (\alpha_p, A_p) \\ (\beta_1, B_1), \cdots, (\beta_q, B_q) \end{matrix} ; \lambda\tau^{\beta} \right) d\tau \\ &= {}_{p+1}\varphi_{q+1} \left(\begin{matrix} (\alpha_1, A_1), \cdots, (\alpha_p, A_p), (\gamma, \beta) \\ (\beta_1, B_1), \cdots, (\beta_q, B_q), (\gamma+1, \beta) \end{matrix} ; \lambda \right).\end{aligned} \tag{2.12}$$

Proof When $t = 1$ in (2.10), we obtain the result. □

Property 2.4 Let $\alpha, \gamma \in \mathbb{C}$, $Re(\alpha) > 0$, $Re(\gamma) > 0$, $\lambda \in \mathbb{C}$, $\beta > 0$, $j, k, u \in \mathbb{N}$, $\alpha_j, \beta_k \in \mathbb{C}$, and $A_j, B_k \in \mathbb{R}_+$. Then we have

$$\begin{aligned} &\frac{1}{\Gamma(\alpha)} \int_0^1 (1-\tau)^{\alpha} \tau^{\gamma-1} {}_p\varphi_q \left(\begin{matrix} (\alpha_1, A_1), \cdots, (\alpha_p, A_p) \\ (\beta_1, B_1), \cdots, (\beta_q, B_q) \end{matrix} ; \lambda\tau^{\beta} \right) d\tau \\ &= {}_{p+1}\varphi_{q+1} \left(\begin{matrix} (\alpha_1, A_1), \cdots, (\alpha_p, A_p), (\gamma, \beta) \\ (\beta_1, B_1), \cdots, (\beta_q, B_q), (\gamma+\alpha, \beta) \end{matrix} ; \lambda \right). \end{aligned} \tag{2.13}$$

Proof When $t = 1$ in (2.10), we obtain the result. □

Property 2.5 Let $\alpha, \gamma \in \mathbb{C}$, $Re(\alpha) > 0$, $Re(\gamma) > 0$, $\beta > 0$, $j, k, u \in \mathbb{N}$, $\alpha_j, \beta_k \in \mathbb{C}$, and $A_j, B_k \in \mathbb{R}_+$. Then we have

$$\begin{aligned} &\frac{1}{\Gamma(\alpha)} \int_0^1 (1-\tau)^{\alpha} \tau^{\gamma-1} {}_p\varphi_q \left(\begin{matrix} (\alpha_1, A_1), \cdots, (\alpha_p, A_p) \\ (\beta_1, B_1), \cdots, (\beta_q, B_q) \end{matrix} ; \tau^{\beta} \right) d\tau \\ &= {}_{p+1}\varphi_{q+1} \left(\begin{matrix} (\alpha_1, A_1), \cdots, (\alpha_p, A_p), (1, \beta) \\ (\beta_1, B_1), \cdots, (\beta_q, B_q), (1+\alpha, \beta) \end{matrix} ; 1 \right). \end{aligned} \tag{2.14}$$

Proof When $\lambda = 1$ in (2.10), we obtain the result. □

Property 2.6 Let $\gamma \in \mathbb{C}$, $Re(\gamma) > 0$, $\lambda \in \mathbb{C}$, $\beta > 0$, $j, k, u \in \mathbb{N}$, $\alpha_j, \beta_k \in \mathbb{C}$, and $A_j, B_k \in \mathbb{R}_+$. Then we have

$$\begin{aligned} &\int_0^1 \tau^{\gamma-1} {}_p\varphi_q \left(\begin{matrix} (\alpha_1, A_1), \cdots, (\alpha_p, A_p) \\ (\beta_1, B_1), \cdots, (\beta_q, B_q) \end{matrix} ; \tau^{\beta} \right) d\tau \\ &= {}_{p+1}\varphi_{q+1} \left(\begin{matrix} (\alpha_1, A_1), \cdots, (\alpha_p, A_p), (1, \beta) \\ (\beta_1, B_1), \cdots, (\beta_q, B_q), (2, \beta) \end{matrix} ; 1 \right). \end{aligned} \tag{2.15}$$

Proof When $\alpha = 1$ in (2.10), we obtain the result. □

Property 2.7 Let $\lambda \in \mathbb{C}$, $\beta > 0$, $j, k, u \in \mathbb{N}$, $\alpha_j, \beta_k \in \mathbb{C}$, and $A_j, B_k \in \mathbb{R}_+$. Then we have

$$\begin{aligned} &\int_0^1 {}_p\varphi_q \left(\begin{matrix} (\alpha_1, A_1), \cdots, (\alpha_p, A_p) \\ (\beta_1, B_1), \cdots, (\beta_q, B_q) \end{matrix} ; \lambda\tau^{\beta} \right) d\tau \\ &= {}_{p+1}\varphi_{q+1} \left(\begin{matrix} (\alpha_1, A_1), \cdots, (\alpha_p, A_p), (1, \beta) \\ (\beta_1, B_1), \cdots, (\beta_q, B_q), (2, \beta) \end{matrix} ; \lambda \right). \end{aligned} \tag{2.16}$$

Proof When $\gamma = 1$ in (2.10), we obtain the result. □

Theorem 2.9 *Let $\gamma \in \mathbb{C}$, $n \in \mathbb{N}$, $Re(\gamma) > 0$, $\lambda \in \mathbb{C}$, $\beta > 0$, $j, k, u \in \mathbb{N}$, $\alpha_j, \beta_k \in \mathbb{C}$, and $A_j, B_k \in \mathbb{R}_+$. Then we have*

$$\begin{aligned}&\frac{\partial^n}{\partial t^n}\left\{t^{\gamma-1}\ {}_p\varphi_q\left(\begin{matrix}(\alpha_1, A_1), \cdots, (\alpha_p, A_p)\\(\beta_1, B_1), \cdots, (\beta_q, B_q)\end{matrix}; \lambda t^\beta\right)\right\}\\&= t^{\gamma-n-1}\ {}_{p+1}\varphi_{q+1}\left(\begin{matrix}(\alpha_1, A_1), \cdots, (\alpha_p, A_p), (\gamma, \beta)\\(\beta_1, B_1), \cdots, (\beta_q, B_q), (\gamma-n, \beta)\end{matrix}; \lambda t^\beta\right).\end{aligned} \tag{2.17}$$

The result was discovered by Raina in 2005 [178].

As the special cases, we derive the following properties:

Property 2.8 Let $n \in \mathbb{N}$, $\lambda \in \mathbb{C}$, $\beta > 0$, $j, k, u \in \mathbb{N}$, $\alpha_j, \beta_k \in \mathbb{C}$, and $A_j, B_k \in \mathbb{R}_+$. Then we have

$$\begin{aligned}&\frac{\partial^n}{\partial t^n}\left\{{}_p\varphi_q\left(\begin{matrix}(\alpha_1, A_1), \cdots, (\alpha_p, A_p)\\(\beta_1, B_1), \cdots, (\beta_q, B_q)\end{matrix}; \lambda t^\beta\right)\right\}\\&= t^{-n}\ {}_{p+1}\varphi_{q+1}\left(\begin{matrix}(\alpha_1, A_1), \cdots, (\alpha_p, A_p), (1, \beta)\\(\beta_1, B_1), \cdots, (\beta_q, B_q), (1-n, \beta)\end{matrix}; \lambda t^\beta\right).\end{aligned} \tag{2.18}$$

Proof When $\gamma = 1$ in (2.17), we have the result. □

Property 2.9 Let $\lambda \in \mathbb{C}$, $\beta > 0$, $j, k, u \in \mathbb{N}$, $\alpha_j, \beta_k \in \mathbb{C}$, and $A_j, B_k \in \mathbb{R}_+$. Then we have

$$\begin{aligned}&\frac{\partial}{\partial t}\left\{{}_p\varphi_q\left(\begin{matrix}(\alpha_1, A_1), \cdots, (\alpha_p, A_p)\\(\beta_1, B_1), \cdots, (\beta_q, B_q)\end{matrix}; \lambda t^\beta\right)\right\}\\&= t^{-1}\ {}_{p+1}\varphi_{q+1}\left(\begin{matrix}(\alpha_1, A_1), \cdots, (\alpha_p, A_p), (1, \beta)\\(\beta_1, B_1), \cdots, (\beta_q, B_q), (0, \beta)\end{matrix}; \lambda t^\beta\right).\end{aligned} \tag{2.19}$$

Proof When $n = 1$ in (2.17), we obtain the result. □

Theorem 2.10 *Let $\alpha, \gamma, \lambda, \varsigma \in \mathbb{C}$, $Re(\alpha) > 0$, $Re(\gamma) > 0$, $\beta > 0$, and $B > -1$. Then we have*

$$\begin{aligned}&\frac{1}{\Gamma(\alpha)}\int_0^t (t-\tau)^\alpha\, \tau^{\gamma-1}\psi\left(\varsigma, B, \lambda\tau^\beta\right) d\tau\\&= t^{\gamma+\alpha-1}\ {}_1\varphi_2\left(\begin{matrix}(\gamma, \beta)\\(\varsigma, B), (\gamma+\alpha, \beta)\end{matrix}; \lambda t^\beta\right).\end{aligned} \tag{2.20}$$

The result was discovered by Kibas in 2005 [170].

In a similar way, one has the following property:

Property 2.10 Let $\gamma, \lambda, \varsigma \in \mathbb{C}$, $Re(\gamma) > 0$, $\beta > 0$, and $B > -1$. Then we have

$$\int_0^t \tau^{\gamma-1}\psi\left(\varsigma, B, \lambda\tau^\beta\right) d\tau = t^\gamma\ {}_1\varphi_2\left(\begin{matrix}(\gamma, \beta)\\(\varsigma, B), (\gamma+1, \beta)\end{matrix}; \lambda t^\beta\right). \tag{2.21}$$

Proof When $\alpha = 1$ in (2.20), we obtain the result. □

Property 2.11 Let $\gamma, \lambda, \varsigma \in \mathbb{C}$, $n \in \mathbb{N}$, $Re(\gamma) > 0$, $\beta > 0$, and $B > -1$. Then we have

$$\frac{\partial^n}{\partial t^n}\left[t^{\gamma-1}\psi\left(\varsigma, B, \lambda t^{\beta}\right)\right] = t^{\gamma}\ {}_1\varphi_2\left(\begin{matrix}(\gamma, \beta) \\ (\varsigma, B), (\gamma - n, \beta)\end{matrix}; \lambda t^{\beta}\right). \tag{2.22}$$

The result was discovered by Raina in 2005 [178].

Property 2.12 Let $\gamma, \lambda, \varsigma \in \mathbb{C}$, $Re(\gamma) > 0$, $\beta > 0$, and $B > -1$. Then we have

$$\int_0^1 \tau^{\gamma-1}\psi\left(\varsigma, B, \lambda\tau^{\beta}\right)d\tau = {}_1\varphi_2\left(\begin{matrix}(\gamma, \beta) \\ (\varsigma, B), (\gamma + 1, \beta)\end{matrix}; \lambda\right). \tag{2.23}$$

Proof When $\alpha = 1$ and $t = 1$ in (2.20), we obtain the result. □

Property 2.13 Let $\alpha, \gamma, \lambda, \varsigma \in \mathbb{C}$, $Re(\alpha) > 0$, $Re(\gamma) > 0$, $\beta > 0$, and $B > -1$. Then we have

$$\frac{1}{\Gamma(\alpha)}\int_0^1 (1-\tau)^{\alpha}\tau^{\gamma-1}\psi\left(\varsigma, B, \lambda\tau^{\beta}\right)d\tau = {}_1\varphi_2\left(\begin{matrix}(\gamma, \beta) \\ (\varsigma, B), (\gamma + \alpha, \beta)\end{matrix}; \lambda\right). \tag{2.24}$$

Proof When $t = 1$ in (2.20), we obtain the result. □

It is not difficult to derive the following properties:

Property 2.14 Let $\alpha, \gamma, \lambda, \varsigma \in \mathbb{C}$, $Re(\alpha) > 0$, $Re(\gamma) > 0$, $\beta > 0$, and $B > -1$. Then we have

$$\frac{1}{\Gamma(\alpha)}\int_0^1 (1-\tau)^{\alpha}\psi\left(\varsigma, B, \lambda\tau^{\beta}\right)d\tau = {}_1\varphi_2\left(\begin{matrix}(1, \beta) \\ (\varsigma, B), (1 + \alpha, \beta)\end{matrix}; \lambda\right). \tag{2.25}$$

Proof When $t = 1$ in (2.24), we obtain the result. □

Property 2.15 Let $\lambda, \varsigma \in \mathbb{C}$, $\beta > 0$, and $B > -1$. Then we have

$$\int_0^t \psi\left(\varsigma, B, \lambda\tau^{\beta}\right)d\tau = {}_1\varphi_2\left(\begin{matrix}(1, \beta) \\ (\varsigma, B), (2, \beta)\end{matrix}; \lambda t^{\beta}\right). \tag{2.26}$$

Proof When $\gamma = 1$ in (2.21), we obtain the result. □

Property 2.16 Let $\lambda, \varsigma \in \mathbb{C}$, $\beta > 0$, and $B > -1$. Then we have

$$\int_0^1 \psi\left(\varsigma, B, \lambda\tau^{\beta}\right)d\tau = {}_1\varphi_2\left(\begin{matrix}(1, \beta) \\ (\varsigma, B), (2, \beta)\end{matrix}; \lambda\right). \tag{2.27}$$

Proof When $t = 1$ in (2.26), we obtain the result. □

The following special cases are as follows:

- The Virchenko function, denoted as ${}_1\varphi_2\left(\begin{matrix}(1,1)\\(\varsigma,B),(\alpha,\beta)\end{matrix};t\right)$, was introduced by Virchenko in 1999 [179].
- The Djzrbashian function, denoted as ${}_1\varphi_2\left(\begin{matrix}(1,1)\\(\varsigma,B),(\alpha,\beta)\end{matrix};t\right)$, was introduced by Djzrbashian in 1960 [180].

Theorem 2.11 (Luchko Theorem [181]) *The cases of the zeros of the Wright function are given as follows:*

(A 1) $\beta > 0$*: All zeros with large enough absolute values are simple and located on the negative real semi-axis;*
(A 2) $\beta = 0$*: The Wright function is the exponential function with a constant factor (equal to zero if* $B \in \mathbb{Z}_0^-$ *) and has no zeros;*
(A 3) $-\frac{1}{3} < \beta < 0$*: All zeros with large enough absolute values are simple and located on the positive real semi-axis;*
(A 4) $\beta = -\frac{1}{2}$, $B \in \mathbb{Z}_0^-$ *or* $\beta = -\frac{1}{2}$, $B = -\frac{1}{2} - n$, $n \in \mathbb{Z}_0^-$*: The Wright function has exactly* $2n + 1$ *or* $2n$ *zeros, respectively;*
(A 5) $-1 < \beta < -\frac{1}{3}$ *(excluding the case 4): All zeros with large enough absolute values are simple and located in the neighborhoods of the rays* $\arg z = \pm\frac{1}{2}\pi\,(-1 - 3\beta)$.

The result was discovered by Luchko in 2000 [181, 182].

2.1.3 The Integral Transforms for the Generalized Wright Functions

Theorem 2.12 (Pogb́ny–Srivastava Theorem [183]) *The Laplace transform of the function* $t^{\gamma-1}\,{}_p\varphi_q\left(\begin{matrix}(\alpha_1,A_1),\cdots,(\alpha_p,A_p)\\(\beta_1,B_1),\cdots,(\beta_q,B_q)\end{matrix};\lambda t\right)$ *is given as follows [183, 184]:*

$$\begin{aligned}
&L\left[t^{\gamma-1}\,{}_p\varphi_q\left(\begin{matrix}(\alpha_1,A_1),\cdots,(\alpha_p,A_p)\\(\beta_1,B_1),\cdots,(\beta_q,B_q)\end{matrix};\lambda t\right)\right]\\
&= \int_0^\infty e^{-st}t^{\gamma-1}\,{}_p\varphi_q\left(\begin{matrix}(\alpha_1,A_1),\cdots,(\alpha_p,A_p)\\(\beta_1,B_1),\cdots,(\beta_q,B_q)\end{matrix};\lambda t\right)dt\\
&= s^{-\gamma}\,{}_{p+1}\varphi_q\left(\begin{matrix}(\gamma,1),(\alpha_1,A_1),\cdots,(\alpha_p,A_p)\\(\beta_1,B_1),\cdots,(\beta_q,B_q)\end{matrix};\frac{\lambda}{s}\right),
\end{aligned} \tag{2.28}$$

where $\lambda \in \mathbb{C}$, $Re\,(s) > 0$, $Re\,(\gamma) > 0$, $j,k,u \in \mathbb{N}$, $\alpha_j, \beta_k > 0$, *and* $A_j, B_k > 0$.

The result was discovered by Pogбny and Srivastava in 2009 [183] and reported by Mehrez in 2018 [184].

As a special case, we have the following property:

Property 2.17 The Laplace transform of the generalized Wright function $_p\varphi_q\left(\begin{matrix}(\alpha_1,A_1),\cdots,(\alpha_p,A_p)\\(\beta_1,B_1),\cdots,(\beta_q,B_q)\end{matrix};\lambda t\right)$ is given as follows:

$$\begin{aligned}&L\left[{}_p\varphi_q\left(\begin{matrix}(\alpha_1,A_1),\cdots,(\alpha_p,A_p)\\(\beta_1,B_1),\cdots,(\beta_q,B_q)\end{matrix};\lambda t\right)\right]\\&=\int_0^\infty e^{-st}\,{}_p\varphi_q\left(\begin{matrix}(\alpha_1,A_1),\cdots,(\alpha_p,A_p)\\(\beta_1,B_1),\cdots,(\beta_q,B_q)\end{matrix};\lambda t\right)dt\\&=s^{-1}\,{}_{p+1}\varphi_q\left(\begin{matrix}(\gamma,1),(\alpha_1,A_1),\cdots,(\alpha_p,A_p)\\(\beta_1,B_1),\cdots,(\beta_q,B_q)\end{matrix};\frac{\lambda}{s}\right),\end{aligned} \tag{2.29}$$

where $\lambda\in\mathbb{C}$, $Re(s)>0$, $j,k,u\in\mathbb{N}$, $\alpha_j,\beta_k>0$, and $A_j,B_k>0$.

Theorem 2.13 *The Mellin transform of the generalized Wright function is given as follows:*

$$\begin{aligned}&M\left[{}_p\varphi_q\left(\begin{matrix}(\alpha_1,A_1),\cdots,(\alpha_p,A_p)\\(\beta_1,B_1),\cdots,(\beta_q,B_q)\end{matrix};-\lambda t\right)\right]\\&=\int_0^\infty t^{\varpi-1}\,{}_p\varphi_q\left(\begin{matrix}(\alpha_1,A_1),\cdots,(\alpha_p,A_p)\\(\beta_1,B_1),\cdots,(\beta_q,B_q)\end{matrix};-\lambda t\right)dt\\&=\frac{\Gamma(\varpi)\prod\limits_{i=1}^{p}\Gamma(\alpha_p-A_p\varpi)}{\prod\limits_{i=1}^{q}\Gamma(\beta_q-B_q\varpi)}\lambda^{-\varpi},\end{aligned} \tag{2.30}$$

where $\lambda\in\mathbb{C}$, $Re(s)>0$, $j,k,u\in\mathbb{N}$, $\alpha_j,\beta_k>0$, *and* $A_j,B_k>0$, *and the Mellin transform of the function is denoted as*

$$M[f(t)]=f(\varpi)=\int_0^\infty t^{\varpi-1}f(t)\,dt. \tag{2.31}$$

The result was discovered by Kilbas, Saigo, and Trujillo in 2002 [165] and further reported by Mathai and Haubold in 2010 [185].

Theorem 2.14 *The Laplace transform of the function* $t^{\gamma-1}\psi(\beta, B, \lambda z)$ *is given as follows:*

$$\begin{aligned} &L\left[t^{\gamma-1}\psi(\beta, B, \lambda z)\right] \\ &= \int_0^{\infty} e^{-st} t^{\gamma-1}\psi(\beta, B, \lambda z)\, dt \\ &= s^{-\gamma}\, {}_1\varphi_1\left(\begin{matrix}(\gamma, 1)\\(\beta, B)\end{matrix}; \frac{\lambda}{s}\right), \end{aligned} \tag{2.32}$$

where $\lambda \in \mathbb{C}$, $Re(s) > 0$, $Re(\gamma) > 0$, $\beta > 0$, *and* $B > 0$.

The case of $\lambda = \pm 1$ in (2.32) was proposed by Gerenflo *et al.* in 1999 [186]

Theorem 2.15 *The Mellin transform of the Wright function was given as follows:*

$$\begin{aligned} &M\left[\psi(\beta, B, -\lambda z)\right] \\ &= \int_0^{\infty} t^{\varpi-1}\psi(\beta, B, -\lambda z)\, dt \\ &= \frac{\Gamma(\varpi)}{\Gamma(\beta - B\varpi)}\lambda^{-\varpi}, \end{aligned} \tag{2.33}$$

where $\lambda \in \mathbb{C}$, $Re(s) > 0$, $\beta > 0$, *and* $B > 0$.

The case of $\gamma \in \mathbb{N}$ and $\lambda = 1$ in (2.32) was proposed by Gerenflo *et al.* in 1999 [186], and the case of $\gamma \in \mathbb{C}$ and $\lambda = 1$ in (2.32) was proposed by Ansari in 2019 [187].

Theorem 2.16 *The Wright function for* $\beta > -1$ *and* $B \in \mathbb{C}$ $(B \in \mathbb{Z}_0^-$ *if* $\beta = 0)$ *is an entire function of finite order* p *and type* σ *given by* $p = \frac{1}{1+\beta}$ *and* $\sigma = (1+\beta)\,|\beta|^{-\frac{\beta}{1+\beta}}$.

The result was discovered by Luchko in 2019 [173].

Theorem 2.17 (Wright Theorem [163]) *The Wright function is an entire function of* z *for all values of the parameters* $\beta > -1$ *and* $B \in \mathbb{C}$.

The result was discovered by Wright in 1935 [163, 174], rediscovered by Kilbas, Saigo, and Trujillo in 2002 [165] and further reported by Luchko in 2019 [173].

More generally,

$$\begin{aligned} &{}_p\varphi_q(\lambda z) \\ &= {}_p\varphi_q\left(\begin{matrix}(\alpha_1, A_1), \cdots, (\alpha_p, A_p)\\(\beta_1, B_1), \cdots, (\beta_q, B_q)\end{matrix}; \lambda z\right) \\ &= {}_p\varphi_q\left(\begin{matrix}(\alpha_p, A_p)_{1,p}\\(\beta_q, B_q)_{1,q}\end{matrix}; \lambda z\right) \\ &= \sum_{k=0}^{\infty}\frac{\left\{\prod_{j=1}^{p}\Gamma(\alpha_j + A_j k)\right\}}{\left\{\prod_{j=1}^{p}\Gamma(\beta_j + B_j k)\right\}}\frac{(\lambda z)^k}{k!} \end{aligned} \tag{2.34}$$

and

$$
\begin{aligned}
&{}_p\varphi_q(-\lambda z) \\
&= {}_p\varphi_q\left(\begin{matrix}(\alpha_1, A_1), \cdots, (\alpha_p, A_p) \\ (\beta_1, B_1), \cdots, (\beta_q, B_q)\end{matrix}; -\lambda z\right) \\
&= {}_p\varphi_q\left(\begin{matrix}(\alpha_p, A_p)_{1,p} \\ (\beta_q, B_q)_{1,q}\end{matrix}; -\lambda z\right) \\
&= \sum_{k=0}^{\infty} \frac{\left\{\prod\limits_{j=1}^{p} \Gamma(\alpha_j + A_j k)\right\}}{\left\{\prod\limits_{j=1}^{p} \Gamma(\beta_j + B_j k)\right\}} \frac{(-\lambda z)^k}{k!},
\end{aligned}
\tag{2.35}
$$

where $z, \lambda \in \mathbb{C}$, $\alpha_j, \beta_j \in \mathbb{C}$, $A_j, B_j \in \mathbb{R}_+$, and $\sum\limits_{j=1}^{p} A_j - \sum\limits_{j=1}^{q} B_j = \Delta > -1$.

Let us consider the series representations

$$
{}_p\varphi_q(i\lambda z) = \sum_{k=0}^{\infty} \frac{\left\{\prod\limits_{j=1}^{p} \Gamma(\alpha_j + A_j k)\right\}}{\left\{\prod\limits_{j=1}^{p} \Gamma(\beta_j + B_j k)\right\}} \frac{(i\lambda z)^k}{k!} \tag{2.36}
$$

and

$$
{}_p\varphi_q(-i\lambda z) = \sum_{k=0}^{\infty} \frac{\left\{\prod\limits_{j=1}^{p} \Gamma(\alpha_j + A_j k)\right\}}{\left\{\prod\limits_{j=1}^{p} \Gamma(\beta_j + B_j k)\right\}} \frac{(-i\lambda z)^k}{k!} \tag{2.37}
$$

where $i = \sqrt{-1}$, $z, \lambda \in \mathbb{C}$, $\alpha_j, \beta_j \in \mathbb{C}$, $A_j, B_j \in \mathbb{R}_+$, and $\sum\limits_{j=1}^{p} A_j - \sum\limits_{j=1}^{q} B_j = \Delta > -1$.

2.1.4 The Supertrigonometric Functions via Wright's Generalized Hypergeometric Function

In this part, based on the above formulas, we propose the definitions for the supertrigonometric functions via Wright's generalized hypergeometric function.

Definition 2.4 The supersine via Wright's generalized hypergeometric function is defined as

$$
\begin{aligned}
& {}_p sin_q\ (z) \\
& = \frac{1}{2i}\left({}_p\varphi_q\ (iz) - {}_p\varphi_q\ (-iz)\right) \\
& = \sum_{k=0}^{\infty} (-1)^k \frac{\left\{\prod_{j=1}^{p} \Gamma(\alpha_j + A_j(2k+1))\right\}}{\left\{\prod_{j=1}^{p} \Gamma(\beta_j + B_j(2k+1))\right\}} \frac{z^{2k+1}}{(2k+1)!},
\end{aligned}
\tag{2.38}
$$

where $i = \sqrt{-1}$, $z \in \mathbb{C}$, $j, k, u \in \mathbb{N}$, $\alpha_j, \beta_k \in \mathbb{C}$, $A_j, B_k \in \mathbb{R}_+$.

Theorem 2.18 *Let $z \in \mathbb{C}$, $j, k, u \in \mathbb{N}$, $\alpha_j, \beta_k \in \mathbb{C}$, and $A_j, B_k \in \mathbb{R}_+$.*
Then we have the cases:

(1a) *If $\Delta + 1 > 0$, then the series in (2.38) is absolutely convergent for $z \in \mathbb{C}$;*
(2a) *If $\Delta + 1 = 0$, then the series in (2.38) is absolutely convergent for all values of $|z| < \delta$, and of $|z| = \delta$ and $Re\ (h) > \frac{1}{2}$.*

Definition 2.5 The supercosine via Wright's generalized hypergeometric function is defined as

$$
\begin{aligned}
& {}_p cos_q\ (z) \\
& = \frac{1}{2}\left({}_p\varphi_q\ (iz) + {}_p\varphi_q\ (-iz)\right) \\
& = \sum_{k=0}^{\infty} (-1)^k \frac{\left\{\prod_{j=1}^{p} \Gamma(\alpha_j + 2kA_j)\right\}}{\left\{\prod_{j=1}^{p} \Gamma(\beta_j + 2kB_j)\right\}} \frac{z^{2k}}{(2k)!}
\end{aligned}
\tag{2.39}
$$

where $i = \sqrt{-1}$, $z \in \mathbb{C}$, $j, k, u \in \mathbb{N}$, $\alpha_j, \beta_k \in \mathbb{C}$ and $A_j, B_k \in \mathbb{R}_+$.

Theorem 2.19 *Let $z \in \mathbb{C}$, $j, k, u \in \mathbb{N}$, $\alpha_j, \beta_k \in \mathbb{C}$, and $A_j, B_k \in \mathbb{R}_+$.*
Then we have the cases:

(1a) *If $\Delta + 1 > 0$, then the series in (2.39) is absolutely convergent for $z \in \mathbb{C}$;*
(2a) *If $\Delta + 1 = 0$, then the series in (2.39) is absolutely convergent for all values of $|z| < \delta$, and of $|z| = \delta$ and $Re\ (h) > \frac{1}{2}$.*

Definition 2.6 The supertangent via Wright's generalized hypergeometric function is defined as

$$
\begin{aligned}
&{}_p tan_q\,(z)\\
&= \frac{{}_p sin_q\,(z)}{{}_p cos_q\,(z)}\\
&= \frac{1}{i}\frac{{}_p\varphi_q\,(iz) - {}_p\varphi_q\,(-iz)}{{}_p\varphi_q\,(iz) + {}_p\varphi_q\,(-iz)}\\
&= \frac{\sum\limits_{k=0}^{\infty}(-1)^k \frac{\left\{\prod\limits_{j=1}^{p}\Gamma\left(\alpha_j + A_j(2k+1)\right)\right\}}{\left\{\prod\limits_{j=1}^{p}\Gamma\left(\beta_j + B_j(2k+1)\right)\right\}}\frac{z^{2k+1}}{(2k+1)!}}{\sum\limits_{k=0}^{\infty}(-1)^k \frac{\left\{\prod\limits_{j=1}^{p}\Gamma\left(\alpha_j + 2kA_j\right)\right\}}{\left\{\prod\limits_{j=1}^{p}\Gamma\left(\beta_j + 2kB_j\right)\right\}}\frac{z^{2k}}{(2k)!}},
\end{aligned}
\tag{2.40}
$$

where $i = \sqrt{-1}$, $z \in \mathbb{C}$, $j, k, u \in \mathbb{N}$, α_j, $\beta_k \in \mathbb{C}$, and A_j, $B_k \in \mathbb{R}_+$.

Definition 2.7 The supercotangent via Wright's generalized hypergeometric function is defined as

$$
\begin{aligned}
&{}_p cot_q\,(z)\\
&= \frac{{}_p cos_q\,(z)}{{}_p sin_q\,(z)}\\
&= i\,\frac{{}_p\varphi_q\,(iz) + {}_p\varphi_q\,(-iz)}{{}_p\varphi_q\,(iz) - {}_p\varphi_q\,(-iz)}\\
&= \frac{\sum\limits_{k=0}^{\infty}(-1)^k \frac{\left\{\prod\limits_{j=1}^{p}\Gamma\left(\alpha_j + 2kA_j\right)\right\}}{\left\{\prod\limits_{j=1}^{p}\Gamma\left(\beta_j + 2kB_j\right)\right\}}\frac{z^{2k}}{(2k)!}}{\sum\limits_{k=0}^{\infty}(-1)^k \frac{\left\{\prod\limits_{j=1}^{p}\Gamma\left(\alpha_j + A_j(2k+1)\right)\right\}}{\left\{\prod\limits_{j=1}^{p}\Gamma\left(\beta_j + B_j(2k+1)\right)\right\}}\frac{z^{2k+1}}{(2k+1)!}},
\end{aligned}
\tag{2.41}
$$

where $i = \sqrt{-1}$, $z \in \mathbb{C}$, $j, k, u \in \mathbb{N}$, α_j, $\beta_k \in \mathbb{C}$, and A_j, $B_k \in \mathbb{R}_+$.

Definition 2.8 The supersecant via Wright's generalized hypergeometric function is defined as

$$
\begin{aligned}
&{}_p sec_q\ (z) \\
&= \frac{1}{{}_p cos_q\ (z)} \\
&= \frac{2}{{}_p\varphi_q\ (iz) + {}_p\varphi_q\ (-iz)} \\
&= \frac{1}{\sum\limits_{k=0}^{\infty} (-1)^k \frac{\left\{\prod\limits_{j=1}^{p} \Gamma\left(\alpha_j + 2kA_j\right)\right\}}{\left\{\prod\limits_{j=1}^{p} \Gamma\left(\beta_j + 2kB_j\right)\right\}} \frac{z^{2k}}{(2k)!}},
\end{aligned}
\tag{2.42}
$$

where $i = \sqrt{-1}$, $z \in \mathbb{C}$, $j, k, u \in \mathbb{N}$, α_j, $\beta_k \in \mathbb{C}$, and A_j, $B_k \in \mathbb{R}_+$.

Definition 2.9 The supercosecant via Wright's generalized hypergeometric function is defined as

$$
\begin{aligned}
&{}_p cosec_q\ (z) \\
&= \frac{1}{{}_p sin_q\ (z)} \\
&= \frac{2i}{{}_p\varphi_q\ (iz) - {}_p\varphi_q\ (-iz)} \\
&= \frac{1}{\sum\limits_{k=0}^{\infty} (-1)^k \frac{\left\{\prod\limits_{j=1}^{p} \Gamma\left(\alpha_j + A_j(2k+1)\right)\right\}}{\left\{\prod\limits_{j=1}^{p} \Gamma\left(\beta_j + B_j(2k+1)\right)\right\}} \frac{z^{2k+1}}{(2k+1)!}},
\end{aligned}
\tag{2.43}
$$

where $i = \sqrt{-1}$, $z \in \mathbb{C}$, $j, k, u \in \mathbb{N}$, α_j, $\beta_k \in \mathbb{C}$ and A_j, $B_k \in \mathbb{R}_+$.

Theorem 2.20 *Let* $i = \sqrt{-1}$, $z \in \mathbb{C}$, $\alpha_j, \beta_j \in \mathbb{C}$, $A_j, B_j \in \mathbb{R}_+$, *and* $\sum\limits_{j=1}^{p} A_j - \sum\limits_{j=1}^{q} B_j = \Delta > -1$.

Then we have

$$
{}_p\varphi_q\ (iz) = {}_p cos_q\ (z) + i\ {}_p sin_q\ (z) \tag{2.44}
$$

and

$$
{}_p\varphi_q\ (-iz) = {}_p cos_q\ (z) - i\ {}_p sin_q\ (z). \tag{2.45}
$$

Definition 2.10 The supersine ${}_p sin_q\ (z)$, supercosine ${}_p cos_q\ (z)$, supertangent ${}_p tan_q\ (z)$, supercotangent ${}_p cot_q\ (z)$, supersecant ${}_p sec_q\ (z)$, and supercosecant ${}_p cosec_q\ (z)$ are called as the supertrigonometric functions via Wright's generalized hypergeometric function.

It is not difficult to show the following series representations:

$$
\begin{aligned}
&{}_p sin_q\ (\lambda z) \\
&= \tfrac{1}{2i}\left({}_p\varphi_q\ (i\lambda z) - {}_p\varphi_q\ (-i\lambda z)\right) \\
&= \sum_{k=0}^{\infty} (-1)^k \frac{\left\{\prod\limits_{j=1}^{p} \Gamma(\alpha_j + A_j(2k+1))\right\}}{\left\{\prod\limits_{j=1}^{p} \Gamma(\beta_j + B_j(2k+1))\right\}} \frac{(\lambda z)^{2k+1}}{(2k+1)!}
\end{aligned}
\tag{2.46}
$$

$$
\begin{aligned}
&{}_p cos_q\ (\lambda z) \\
&= \tfrac{1}{2}\left({}_p\varphi_q\ (i\lambda z) + {}_p\varphi_q\ (-i\lambda z)\right) \\
&= \sum_{k=0}^{\infty} (-1)^k \frac{\left\{\prod\limits_{j=1}^{p} \Gamma(\alpha_j + 2kA_j)\right\}}{\left\{\prod\limits_{j=1}^{p} \Gamma(\beta_j + 2kB_j)\right\}} \frac{(\lambda z)^{2k}}{(2k)!}
\end{aligned}
\tag{2.47}
$$

$$
\begin{aligned}
&{}_p tan_q\ (\lambda z) \\
&= \frac{{}_p sin_q\ (\lambda z)}{{}_p cos_q\ (\lambda z)} \\
&= \frac{1}{i} \frac{{}_p\varphi_q\ (i\lambda z) - {}_p\varphi_q\ (-i\lambda z)}{{}_p\varphi_q\ (i\lambda z) + {}_p\varphi_q\ (-i\lambda z)} \\
&= \frac{\sum\limits_{k=0}^{\infty} (-1)^k \frac{\left\{\prod\limits_{j=1}^{p} \Gamma\left(\alpha_j + A_j(2k+1)\right)\right\}}{\left\{\prod\limits_{j=1}^{p} \Gamma\left(\beta_j + B_j(2k+1)\right)\right\}} \frac{(\lambda z)^{2k+1}}{(2k+1)!}}{\sum\limits_{k=0}^{\infty} (-1)^k \frac{\left\{\prod\limits_{j=1}^{p} \Gamma\left(\alpha_j + 2kA_j\right)\right\}}{\left\{\prod\limits_{j=1}^{p} \Gamma\left(\beta_j + 2kB_j\right)\right\}} \frac{(\lambda z)^{2k}}{(2k)!}},
\end{aligned}
\tag{2.48}
$$

$$
\begin{aligned}
&{}_p cot_q\ (\lambda z) \\
&= \frac{{}_p cos_q\ (\lambda z)}{{}_p sin_q\ (\lambda z)} \\
&= i \frac{{}_p\varphi_q\ (i\lambda z) + {}_p\varphi_q\ (-i\lambda z)}{{}_p\varphi_q\ (i\lambda z) - {}_p\varphi_q\ (-i\lambda z)} \\
&= \frac{\sum\limits_{k=0}^{\infty} (-1)^k \frac{\left\{\prod\limits_{j=1}^{p} \Gamma\left(\alpha_j + 2kA_j\right)\right\}}{\left\{\prod\limits_{j=1}^{p} \Gamma\left(\beta_j + 2kB_j\right)\right\}} \frac{(\lambda z)^{2k}}{(2k)!}}{\sum\limits_{k=0}^{\infty} (-1)^k \frac{\left\{\prod\limits_{j=1}^{p} \Gamma\left(\alpha_j + A_j(2k+1)\right)\right\}}{\left\{\prod\limits_{j=1}^{p} \Gamma\left(\beta_j + B_j(2k+1)\right)\right\}} \frac{(\lambda z)^{2k+1}}{(2k+1)!}},
\end{aligned}
\tag{2.49}
$$

$$\begin{aligned}
&{}_p sec_q\,(\lambda z)\\
&= \frac{1}{{}_p cos_q\,(\lambda z)}\\
&= \frac{2}{{}_p\varphi_q\,(i\lambda z) + {}_p\varphi_q\,(-i\lambda z)}\\
&= \frac{1}{\sum\limits_{k=0}^{\infty}(-1)^k \frac{\left\{\prod\limits_{j=1}^{p}\Gamma\left(\alpha_j+2kA_j\right)\right\}}{\left\{\prod\limits_{j=1}^{p}\Gamma\left(\beta_j+2kB_j\right)\right\}}\frac{(\lambda z)^{2k}}{(2k)!}}
\end{aligned} \tag{2.50}$$

and

$$\begin{aligned}
&{}_p cosec_q\,(\lambda z)\\
&= \frac{1}{{}_p sin_q\,(\lambda z)}\\
&= \frac{2i}{{}_p\varphi_q\,(i\lambda z) - {}_p\varphi_q\,(-i\lambda z)}\\
&= \frac{1}{\sum\limits_{k=0}^{\infty}(-1)^k \frac{\left\{\prod\limits_{j=1}^{p}\Gamma\left(\alpha_j+A_j(2k+1)\right)\right\}}{\left\{\prod\limits_{j=1}^{p}\Gamma\left(\beta_j+B_j(2k+1)\right)\right\}}\frac{(\lambda z)^{2k+1}}{(2k+1)!}},
\end{aligned} \tag{2.51}$$

where $i = \sqrt{-1}$, $z, \lambda \in \mathbb{C}$, $j, k, u \in \mathbb{N}$, $\alpha_j, \beta_k \in \mathbb{C}$, and $A_j, B_k \in \mathbb{R}_+$.

Theorem 2.21 *Let $z \in \mathbb{C}$, $\Delta + 1 \geq 0$, $j, k, u \in \mathbb{N}$, $\alpha_j, \beta_k \in \mathbb{C}$, and $A_j, B_k \in \mathbb{R}_+$. Then we have*

$${}_p sin_q\left(\begin{matrix}(\alpha_1,1),\cdots,(\alpha_p,1)\\(\beta_1,1),\cdots,(\beta_q,1)\end{matrix};z\right) = \frac{\prod\limits_{j=1}^{p}\Gamma(\alpha_j)}{\prod\limits_{k=1}^{p}\Gamma(\beta_k)}\,{}_p Supersin_q\left(\begin{matrix}\alpha_1,\cdots,\alpha_p\\\beta_1,\cdots,\beta_q\end{matrix};z\right), \tag{2.52}$$

and

$${}_p cos_q\left(\begin{matrix}(\alpha_1,1),\cdots,(\alpha_p,1)\\(\beta_1,1),\cdots,(\beta_q,1)\end{matrix};-\lambda z\right) = \frac{\prod\limits_{j=1}^{p}\Gamma(\alpha_j)}{\prod\limits_{k=1}^{p}\Gamma(\beta_k)}\,{}_p cos_q\left(\begin{matrix}\alpha_1,\cdots,\alpha_p\\\beta_1,\cdots,\beta_q\end{matrix};-\lambda z\right), \tag{2.53}$$

Proof On setting

$$ {}_p\varphi_q\left(\begin{matrix}(\alpha_1,1),\cdots,(\alpha_p,1)\\(\beta_1,1),\cdots,(\beta_q,1)\end{matrix};i\lambda z\right)=\frac{\prod_{j=1}^{p}\Gamma(\alpha_j)}{\prod_{k=1}^{p}\Gamma(\beta_k)}\,{}_pF_q\left(\begin{matrix}\alpha_1,\cdots,\alpha_p\\\beta_1,\cdots,\beta_q\end{matrix};i\lambda z\right), \tag{2.54}$$

and

$$ {}_p\varphi_q\left(\begin{matrix}(\alpha_1,1),\cdots,(\alpha_p,1)\\(\beta_1,1),\cdots,(\beta_q,1)\end{matrix};-i\lambda z\right)=\frac{\prod_{j=1}^{p}\Gamma(\alpha_j)}{\prod_{k=1}^{p}\Gamma(\beta_k)}\,{}_pF_q\left(\begin{matrix}\alpha_1,\cdots,\alpha_p\\\beta_1,\cdots,\beta_q\end{matrix};-i\lambda z\right), \tag{2.55}$$

where $z\in\mathbb{C}$, $j,k,u\in\mathbb{N}$, $\alpha_j,\beta_k\in\mathbb{C}$, and $A_j,B_k\in\mathbb{R}_+$, $\Delta+1\geq 0$, we have the result. □

For more information of the Wright's generalized hypergeometric function, see [188, 189].

We now consider the integral representations via Wright's generalized hypergeometric function as follows.

Theorem 2.22 *Let $\gamma\in\mathbb{C}$, $n\in\mathbb{N}$, $Re(\gamma)>0$, $\lambda\in\mathbb{C}$, $\beta>0$, $j,k,u\in\mathbb{N}$, $\alpha_j,\beta_k\in\mathbb{C}$, and $A_j,B_k\in\mathbb{R}_+$. Then we have*

$$\begin{aligned}&\frac{\partial^n}{\partial t^n}\left\{t^{\gamma-1}\,{}_psin_q\left(\begin{matrix}(\alpha_1,A_1),\cdots,(\alpha_p,A_p)\\(\beta_1,B_1),\cdots,(\beta_q,B_q)\end{matrix};\lambda t^\beta\right)\right\}\\&=t^{\gamma-n-1}\,{}_{p+1}sin_{q+1}\left(\begin{matrix}(\alpha_1,A_1),\cdots,(\alpha_p,A_p),(\gamma,\beta)\\(\beta_1,B_1),\cdots,(\beta_q,B_q),(\gamma-n,\beta)\end{matrix};\lambda t^\beta\right)\end{aligned} \tag{2.56}$$

and

$$\begin{aligned}&\frac{\partial^n}{\partial t^n}\left\{t^{\gamma-1}\,{}_pcos_q\left(\begin{matrix}(\alpha_1,A_1),\cdots,(\alpha_p,A_p)\\(\beta_1,B_1),\cdots,(\beta_q,B_q)\end{matrix};\lambda t^\beta\right)\right\}\\&=t^{\gamma-n-1}\,{}_{p+1}cos_{q+1}\left(\begin{matrix}(\alpha_1,A_1),\cdots,(\alpha_p,A_p),(\gamma,\beta)\\(\beta_1,B_1),\cdots,(\beta_q,B_q),(\gamma-n,\beta)\end{matrix};\lambda t^\beta\right).\end{aligned} \tag{2.57}$$

Proof Making the expressions

$$\begin{aligned}&\frac{\partial^n}{\partial t^n}\left\{t^{\gamma-1}\ {}_p\varphi_q\left(\begin{matrix}(\alpha_1,A_1),\cdots,(\alpha_p,A_p)\\(\beta_1,B_1),\cdots,(\beta_q,B_q)\end{matrix};i\lambda t^\beta\right)\right\}\\&=t^{\gamma-n-1}\ {}_{p+1}\varphi_{q+1}\left(\begin{matrix}(\alpha_1,A_1),\cdots,(\alpha_p,A_p),(\gamma,\beta)\\(\beta_1,B_1),\cdots,(\beta_q,B_q),(\gamma-n,\beta)\end{matrix};i\lambda t^\beta\right)\end{aligned}\tag{2.58}$$

and

$$\begin{aligned}&\frac{\partial^n}{\partial t^n}\left\{t^{\gamma-1}\ {}_p\varphi_q\left(\begin{matrix}(\alpha_1,A_1),\cdots,(\alpha_p,A_p)\\(\beta_1,B_1),\cdots,(\beta_q,B_q)\end{matrix};-i\lambda t^\beta\right)\right\}\\&=t^{\gamma-n-1}\ {}_{p+1}\varphi_{q+1}\left(\begin{matrix}(\alpha_1,A_1),\cdots,(\alpha_p,A_p),(\gamma,\beta)\\(\beta_1,B_1),\cdots,(\beta_q,B_q),(\gamma-n,\beta)\end{matrix};-i\lambda t^\beta\right),\end{aligned}\tag{2.59}$$

we have

$$\begin{aligned}&\frac{\partial^n}{\partial t^n}\left\{t^{\gamma-1}\ {}_psin_q\left(\begin{matrix}(\alpha_1,A_1),\cdots,(\alpha_p,A_p)\\(\beta_1,B_1),\cdots,(\beta_q,B_q)\end{matrix};\lambda t^\beta\right)\right\}\\&=t^{\gamma-n-1}\ {}_{p+1}sin_{q+1}\left(\begin{matrix}(\alpha_1,A_1),\cdots,(\alpha_p,A_p),(\gamma,\beta)\\(\beta_1,B_1),\cdots,(\beta_q,B_q),(\gamma-n,\beta)\end{matrix};\lambda t^\beta\right)\end{aligned}\tag{2.60}$$

and

$$\begin{aligned}&\frac{\partial^n}{\partial t^n}\left\{t^{\gamma-1}\ {}_pcos_q\left(\begin{matrix}(\alpha_1,A_1),\cdots,(\alpha_p,A_p)\\(\beta_1,B_1),\cdots,(\beta_q,B_q)\end{matrix};\lambda t^\beta\right)\right\}\\&=t^{\gamma-n-1}\ {}_{p+1}cos_{q+1}\left(\begin{matrix}(\alpha_1,A_1),\cdots,(\alpha_p,A_p),(\gamma,\beta)\\(\beta_1,B_1),\cdots,(\beta_q,B_q),(\gamma-n,\beta)\end{matrix};\lambda t^\beta\right).\end{aligned}\tag{2.61}$$

□

Theorem 2.23 *Let* $\gamma\in\mathbb{C}$, $Re(\gamma)>0$, $\lambda\in\mathbb{C}$, $\beta>0$, $j,k,u\in\mathbb{N}$, $\alpha_j,\beta_k\in\mathbb{C}$, *and* $A_j, B_k\in\mathbb{R}_+$. *Then we have*

$$\begin{aligned}&\int_0^t(t-\tau)\,\tau^{\gamma-1}\ {}_psin_q\left(\begin{matrix}(\alpha_1,A_1),\cdots,(\alpha_p,A_p)\\(\beta_1,B_1),\cdots,(\beta_q,B_q)\end{matrix};\lambda\tau^\beta\right)d\tau\\&=t^{\gamma}\ {}_{p+1}sin_{q+1}\left(\begin{matrix}(\alpha_1,A_1),\cdots,(\alpha_p,A_p),(\gamma,\beta)\\(\beta_1,B_1),\cdots,(\beta_q,B_q),(\gamma+1,\beta)\end{matrix};\lambda t^\beta\right)\end{aligned}\tag{2.62}$$

and

$$\begin{aligned}&\int_0^t(t-\tau)\,\tau^{\gamma-1}\ {}_pcos_q\left(\begin{matrix}(\alpha_1,A_1),\cdots,(\alpha_p,A_p)\\(\beta_1,B_1),\cdots,(\beta_q,B_q)\end{matrix};\lambda\tau^\beta\right)d\tau\\&=t^{\gamma}\ {}_{p+1}cos_{q+1}\left(\begin{matrix}(\alpha_1,A_1),\cdots,(\alpha_p,A_p),(\gamma,\beta)\\(\beta_1,B_1),\cdots,(\beta_q,B_q),(\gamma+1,\beta)\end{matrix};\lambda t^\beta\right).\end{aligned}\tag{2.63}$$

Proof With the aid of (2.10), (2.58), and (2.59), we directly obtain the results. □

Corollary 2.1 *Let $\lambda \in \mathbb{C}$, $\beta > 0$, $j, k, u \in \mathbb{N}$, $\alpha_j, \beta_k \in \mathbb{C}$, and $A_j, B_k \in \mathbb{R}_+$. Then we have*

$$\begin{aligned} &\int_0^t (t-\tau)\ {}_p sin_q \left(\begin{matrix} (\alpha_1, A_1), \cdots, (\alpha_p, A_p) \\ (\beta_1, B_1), \cdots, (\beta_q, B_q) \end{matrix} ; \lambda\tau^\beta \right) d\tau \\ &= t\ {}_{p+1} sin_{q+1} \left(\begin{matrix} (\alpha_1, A_1), \cdots, (\alpha_p, A_p), (1, \beta) \\ (\beta_1, B_1), \cdots, (\beta_q, B_q), (2, \beta) \end{matrix} ; \lambda t^\beta \right) \end{aligned} \tag{2.64}$$

and

$$\begin{aligned} &\int_0^t (t-\tau)\ {}_p cos_q \left(\begin{matrix} (\alpha_1, A_1), \cdots, (\alpha_p, A_p) \\ (\beta_1, B_1), \cdots, (\beta_q, B_q) \end{matrix} ; \lambda\tau^\beta \right) d\tau \\ &= t\ {}_{p+1} cos_{q+1} \left(\begin{matrix} (\alpha_1, A_1), \cdots, (\alpha_p, A_p), (1, \beta) \\ (\beta_1, B_1), \cdots, (\beta_q, B_q), (2, \beta) \end{matrix} ; \lambda t^\beta \right). \end{aligned} \tag{2.65}$$

Proof Taking $\gamma = 1$ in (2.62) and (2.63), we directly present the results. □

Corollary 2.2 *Let $\gamma \in \mathbb{C}$, $Re(\gamma) > 0$, $\lambda \in \mathbb{C}$, $\beta > 0$, $j, k, u \in \mathbb{N}$, $\alpha_j, \beta_k \in \mathbb{C}$, and $A_j, B_k \in \mathbb{R}_+$. Then we have*

$$\begin{aligned} &\int_0^1 \tau^{\gamma-1}\ {}_p sin_q \left(\begin{matrix} (\alpha_1, A_1), \cdots, (\alpha_p, A_p) \\ (\beta_1, B_1), \cdots, (\beta_q, B_q) \end{matrix} ; \lambda\tau^\beta \right) d\tau \\ &= {}_{p+1} sin_{q+1} \left(\begin{matrix} (\alpha_1, A_1), \cdots, (\alpha_p, A_p), (\gamma, \beta) \\ (\beta_1, B_1), \cdots, (\beta_q, B_q), (\gamma+1, \beta) \end{matrix} ; \lambda \right) \end{aligned} \tag{2.66}$$

and

$$\begin{aligned} &\int_0^1 \tau^{\gamma-1}\ {}_p cos_q \left(\begin{matrix} (\alpha_1, A_1), \cdots, (\alpha_p, A_p) \\ (\beta_1, B_1), \cdots, (\beta_q, B_q) \end{matrix} ; \lambda\tau^\beta \right) d\tau \\ &= {}_{p+1} cos_{q+1} \left(\begin{matrix} (\alpha_1, A_1), \cdots, (\alpha_p, A_p), (\gamma, \beta) \\ (\beta_1, B_1), \cdots, (\beta_q, B_q), (\gamma+1, \beta) \end{matrix} ; \lambda \right). \end{aligned} \tag{2.67}$$

Proof Taking $t = 1$ in (2.62) and (2.63), we directly have the results. □

Corollary 2.3 *Let $\lambda \in \mathbb{C}$, $\beta > 0$, $j, k, u \in \mathbb{N}$, $\alpha_j, \beta_k \in \mathbb{C}$, and $A_j, B_k \in \mathbb{R}_+$. Then we have*

$$\begin{aligned} &\int_0^1 {}_p sin_q \left(\begin{matrix} (\alpha_1, A_1), \cdots, (\alpha_p, A_p) \\ (\beta_1, B_1), \cdots, (\beta_q, B_q) \end{matrix} ; \lambda\tau^\beta \right) d\tau \\ &= {}_{p+1} sin_{q+1} \left(\begin{matrix} (\alpha_1, A_1), \cdots, (\alpha_p, A_p), (1, \beta) \\ (\beta_1, B_1), \cdots, (\beta_q, B_q), (2, \beta) \end{matrix} ; \lambda \right) \end{aligned} \tag{2.68}$$

and

$$\int_0^1 {}_p cos_q \left(\begin{matrix} (\alpha_1, A_1), \cdots, (\alpha_p, A_p) \\ (\beta_1, B_1), \cdots, (\beta_q, B_q) \end{matrix}; \lambda\tau^\beta \right) d\tau$$
$$= {}_{p+1}cos_{q+1} \left(\begin{matrix} (\alpha_1, A_1), \cdots, (\alpha_p, A_p), (1, \beta) \\ (\beta_1, B_1), \cdots, (\beta_q, B_q), (2, \beta) \end{matrix}; \lambda \right). \tag{2.69}$$

Proof Taking $t = 1$ and $\gamma = 1$ in (2.62) and (2.63), we directly have the results. □

Theorem 2.24 *Let* $\alpha, \gamma \in \mathbb{C}$, $Re(\alpha) > 0$, $Re(\gamma) > 0$, $\lambda \in \mathbb{C}$, $\beta > 0$, $j, k, u \in \mathbb{N}$, $\alpha_j, \beta_k \in \mathbb{C}$, *and* $A_j, B_k \in \mathbb{R}_+$. *Then we have*

$$\frac{1}{\Gamma(\alpha)} \int_0^t (t-\tau)^\alpha \tau^{\gamma-1} {}_p sin_q \left(\begin{matrix} (\alpha_1, A_1), \cdots, (\alpha_p, A_p) \\ (\beta_1, B_1), \cdots, (\beta_q, B_q) \end{matrix}; \lambda\tau^\beta \right) d\tau$$
$$= t^{\gamma+\alpha-1} {}_{p+1}sin_{q+1} \left(\begin{matrix} (\alpha_1, A_1), \cdots, (\alpha_p, A_p), (\gamma, \beta) \\ (\beta_1, B_1), \cdots, (\beta_q, B_q), (\gamma+\alpha, \beta) \end{matrix}; \lambda t^\beta \right) \tag{2.70}$$

and

$$\frac{1}{\Gamma(\alpha)} \int_0^t (t-\tau)^\alpha \tau^{\gamma-1} {}_p cos_q \left(\begin{matrix} (\alpha_1, A_1), \cdots, (\alpha_p, A_p) \\ (\beta_1, B_1), \cdots, (\beta_q, B_q) \end{matrix}; \lambda\tau^\beta \right) d\tau$$
$$= t^{\gamma+\alpha-1} {}_{p+1}cos_{q+1} \left(\begin{matrix} (\alpha_1, A_1), \cdots, (\alpha_p, A_p), (\gamma, \beta) \\ (\beta_1, B_1), \cdots, (\beta_q, B_q), (\gamma+\alpha, \beta) \end{matrix}; \lambda t^\beta \right). \tag{2.71}$$

Proof Making use of (2.9), we have

$$\frac{1}{\Gamma(\alpha)} \int_0^t (t-\tau)^\alpha \tau^{\gamma-1} {}_p\varphi_q \left(\begin{matrix} (\alpha_1, A_1), \cdots, (\alpha_p, A_p) \\ (\beta_1, B_1), \cdots, (\beta_q, B_q) \end{matrix}; i\lambda\tau^\beta \right) d\tau$$
$$= t^{\gamma+\alpha-1} {}_{p+1}\varphi_{q+1} \left(\begin{matrix} (\alpha_1, A_1), \cdots, (\alpha_p, A_p), (\gamma, \beta) \\ (\beta_1, B_1), \cdots, (\beta_q, B_q), (\gamma+\alpha, \beta) \end{matrix}; i\lambda t^\beta \right) \tag{2.72}$$

and

$$\frac{1}{\Gamma(\alpha)} \int_0^t (t-\tau)^\alpha \tau^{\gamma-1} {}_p\varphi_q \left(\begin{matrix} (\alpha_1, A_1), \cdots, (\alpha_p, A_p) \\ (\beta_1, B_1), \cdots, (\beta_q, B_q) \end{matrix}; -i\lambda\tau^\beta \right) d\tau$$
$$= t^{\gamma+\alpha-1} {}_{p+1}\varphi_{q+1} \left(\begin{matrix} (\alpha_1, A_1), \cdots, (\alpha_p, A_p), (\gamma, \beta) \\ (\beta_1, B_1), \cdots, (\beta_q, B_q), (\gamma+\alpha, \beta) \end{matrix}; -i\lambda t^\beta \right), \tag{2.73}$$

which reduces to the results. □

Corollary 2.4 *Let $\alpha, \gamma \in \mathbb{C}$, $Re(\alpha) > 0$, $Re(\gamma) > 0$, $\lambda \in \mathbb{C}$, $\beta > 0$, $j, k, u \in \mathbb{N}$, $\alpha_j, \beta_k \in \mathbb{C}$, and $A_j, B_k \in \mathbb{R}_+$. Then we have*

$$\begin{aligned}&\frac{1}{\Gamma(\alpha)}\int_0^1 (1-\tau)^{\alpha}\tau^{\gamma-1}\,{}_p sin_q\left(\begin{matrix}(\alpha_1, A_1), \cdots, (\alpha_p, A_p)\\ (\beta_1, B_1), \cdots, (\beta_q, B_q)\end{matrix}; \lambda\tau^{\beta}\right)d\tau\\ &= {}_{p+1}sin_{q+1}\left(\begin{matrix}(\alpha_1, A_1), \cdots, (\alpha_p, A_p), (\gamma, \beta)\\ (\beta_1, B_1), \cdots, (\beta_q, B_q), (\gamma+\alpha, \beta)\end{matrix}; \lambda\right)\end{aligned} \tag{2.74}$$

and

$$\begin{aligned}&\frac{1}{\Gamma(\alpha)}\int_0^1 (1-\tau)^{\alpha}\tau^{\gamma-1}\,{}_p cos_q\left(\begin{matrix}(\alpha_1, A_1), \cdots, (\alpha_p, A_p)\\ (\beta_1, B_1), \cdots, (\beta_q, B_q)\end{matrix}; \lambda\tau^{\beta}\right)d\tau\\ &= {}_{p+1}cos_{q+1}\left(\begin{matrix}(\alpha_1, A_1), \cdots, (\alpha_p, A_p), (\gamma, \beta)\\ (\beta_1, B_1), \cdots, (\beta_q, B_q), (\gamma+\alpha, \beta)\end{matrix}; \lambda\right).\end{aligned} \tag{2.75}$$

Proof Taking $t = 1$ in (2.74) and (2.75), we directly have the results. □

Corollary 2.5 *Let $\alpha \in \mathbb{C}$, $Re(\alpha) > 0$, $\lambda \in \mathbb{C}$, $\beta > 0$, $j, k, u \in \mathbb{N}$, $\alpha_j, \beta_k \in \mathbb{C}$, and $A_j, B_k \in \mathbb{R}_+$. Then we have*

$$\begin{aligned}&\frac{1}{\Gamma(\alpha)}\int_0^t (t-\tau)^{\alpha}\,{}_p sin_q\left(\begin{matrix}(\alpha_1, A_1), \cdots, (\alpha_p, A_p)\\ (\beta_1, B_1), \cdots, (\beta_q, B_q)\end{matrix}; \lambda\tau^{\beta}\right)d\tau\\ &= t^{\alpha}\,{}_{p+1}sin_{q+1}\left(\begin{matrix}(\alpha_1, A_1), \cdots, (\alpha_p, A_p), (1, \beta)\\ (\beta_1, B_1), \cdots, (\beta_q, B_q), (1+\alpha, \beta)\end{matrix}; \lambda t^{\beta}\right)\end{aligned} \tag{2.76}$$

and

$$\begin{aligned}&\frac{1}{\Gamma(\alpha)}\int_0^t (t-\tau)^{\alpha}\,{}_p cos_q\left(\begin{matrix}(\alpha_1, A_1), \cdots, (\alpha_p, A_p)\\ (\beta_1, B_1), \cdots, (\beta_q, B_q)\end{matrix}; \lambda\tau^{\beta}\right)d\tau\\ &= t^{\alpha}\,{}_{p+1}cos_{q+1}\left(\begin{matrix}(\alpha_1, A_1), \cdots, (\alpha_p, A_p), (1, \beta)\\ (\beta_1, B_1), \cdots, (\beta_q, B_q), (1+\alpha, \beta)\end{matrix}; \lambda t^{\beta}\right).\end{aligned} \tag{2.77}$$

Proof Taking $\gamma = 1$ in (2.74) and (2.75), we directly have the results. □

Corollary 2.6 *Let $\alpha \in \mathbb{C}$, $Re(\alpha) > 0$, $\lambda \in \mathbb{C}$, $\beta > 0$, $j, k, u \in \mathbb{N}$, $\alpha_j, \beta_k \in \mathbb{C}$, and $A_j, B_k \in \mathbb{R}_+$. Then we have*

$$\begin{aligned}&\frac{1}{\Gamma(\alpha)}\int_0^1 (1-\tau)^{\alpha}\,{}_p sin_q\left(\begin{matrix}(\alpha_1, A_1), \cdots, (\alpha_p, A_p)\\ (\beta_1, B_1), \cdots, (\beta_q, B_q)\end{matrix}; \lambda\tau^{\beta}\right)d\tau\\ &= {}_{p+1}sin_{q+1}\left(\begin{matrix}(\alpha_1, A_1), \cdots, (\alpha_p, A_p), (1, \beta)\\ (\beta_1, B_1), \cdots, (\beta_q, B_q), (1+\alpha, \beta)\end{matrix}; \lambda\right)\end{aligned} \tag{2.78}$$

and

$$\frac{1}{\Gamma(\alpha)}\int_0^1 (1-\tau)^{\alpha}\ {}_pcos_q\left(\begin{matrix}(\alpha_1, A_1), \cdots, (\alpha_p, A_p)\\(\beta_1, B_1), \cdots, (\beta_q, B_q)\end{matrix}; \lambda\tau^{\beta}\right)d\tau \\ = {}_{p+1}cos_{q+1}\left(\begin{matrix}(\alpha_1, A_1), \cdots, (\alpha_p, A_p), (1, \beta)\\(\beta_1, B_1), \cdots, (\beta_q, B_q), (1+\alpha, \beta)\end{matrix}; \lambda\right). \tag{2.79}$$

Proof Taking $t = 1$ in (2.74) and (2.75), we directly have the results. □

2.1.5 *The Superhyperbolic Functions via Wright's Generalized Hypergeometric Function*

In this part, we introduce the definitions for the superhyperbolic functions via Wright's generalized hypergeometric function.

Definition 2.11 The superhyperbolic sine via Wright's generalized hypergeometric function is defined as

$$\begin{aligned}&{}_psinh_q(z)\\&=\frac{1}{2}\left({}_p\varphi_q(z) - {}_p\varphi_q(-z)\right)\\&=\sum_{k=0}^{\infty}\frac{\left\{\prod_{j=1}^{p}\Gamma(\alpha_j+A_j(2k+1))\right\}}{\left\{\prod_{j=1}^{p}\Gamma(\beta_j+B_j(2k+1))\right\}}\frac{z^{2k+1}}{(2k+1)!}\end{aligned} \tag{2.80}$$

where $z \in \mathbb{C}$, $j, k, u \in \mathbb{N}$, $\alpha_j, \beta_k \in \mathbb{C}$, and $A_j, B_k \in \mathbb{R}_+$.

Theorem 2.25 *Let* $z \in \mathbb{C}$, $j, k, u \in \mathbb{N}$, $\alpha_j, \beta_k \in \mathbb{C}$, *and* $A_j, B_k \in \mathbb{R}_+$.
Then we have the cases:

(1a) If $\Delta + 1 > 0$*, then the series in (2.80) is absolutely convergent for* $z \in \mathbb{C}$*;*
(2a) If $\Delta + 1 = 0$*, then the series in (2.80) is absolutely convergent for all values of* $|z| < \delta$*, and of*
$|z| = \delta$ *and* $Re(h) > \frac{1}{2}$.

Definition 2.12 The superhyperbolic cosine via Wright's generalized hypergeometric function is defined as

$$\begin{aligned}&{}_pcosh_q(z)\\&=\frac{1}{2}\left({}_p\varphi_q(z) + {}_p\varphi_q(-z)\right)\\&=\sum_{k=0}^{\infty}\frac{\left\{\prod_{j=1}^{p}\Gamma(\alpha_j+2kA_j)\right\}}{\left\{\prod_{j=1}^{p}\Gamma(\beta_j+2kB_j)\right\}}\frac{z^{2k}}{(2k)!},\end{aligned} \tag{2.81}$$

where $z \in \mathbb{C}$, $j, k, u \in \mathbb{N}$, $\alpha_j, \beta_k \in \mathbb{C}$, and $A_j, B_k \in \mathbb{R}_+$.

Theorem 2.26 *Let $z \in \mathbb{C}$, $j, k, u \in \mathbb{N}$, $\alpha_j, \beta_k \in \mathbb{C}$, and $A_j, B_k \in \mathbb{R}_+$.*
Then we have the cases:

(1a) *If $\Delta + 1 > 0$, then the series in (2.81) is absolutely convergent for $z \in \mathbb{C}$;*
(2a) *If $\Delta + 1 = 0$, then the series in (2.81) is absolutely convergent for all values of $|z| < \delta$, and of $|z| = \delta$ and $Re(h) > \frac{1}{2}$.*

Definition 2.13 The superhyperbolic tangent via Wright's generalized hypergeometric function is defined as

$$
\begin{aligned}
&{}_p tanh_q(z) \\
&= \frac{{}_p sinh_q(z)}{{}_p cosh_q(z)} \\
&= \frac{{}_p\varphi_q(z) - {}_p\varphi_q(-z)}{{}_p\varphi_q(z) + {}_p\varphi_q(-z)} \\
&= \frac{\sum\limits_{k=0}^{\infty} \left\{ \frac{\prod\limits_{j=1}^{p} \Gamma(\alpha_j + A_j(2k+1))}{\prod\limits_{j=1}^{p} \Gamma(\beta_j + B_j(2k+1))} \right\} \frac{z^{2k+1}}{(2k+1)!}}{\sum\limits_{k=0}^{\infty} \left\{ \frac{\prod\limits_{j=1}^{p} \Gamma(\alpha_j + 2kA_j)}{\prod\limits_{j=1}^{p} \Gamma(\beta_j + 2kB_j)} \right\} \frac{z^{2k}}{(2k)!}},
\end{aligned}
\tag{2.82}
$$

where $z \in \mathbb{C}$, $j, k, u \in \mathbb{N}$, $\alpha_j, \beta_k \in \mathbb{C}$, and $A_j, B_k \in \mathbb{R}_+$.

Definition 2.14 The superhyperbolic cotangent via Wright's generalized hypergeometric function is defined as

$$
\begin{aligned}
&{}_p coth_q(z) \\
&= \frac{{}_p cosh_q(z)}{{}_p sinh_q(z)} \\
&= \frac{{}_p\varphi_q(z) + {}_p\varphi_q(-z)}{{}_p\varphi_q(z) - {}_p\varphi_q(-z)} \\
&= \frac{\sum\limits_{k=0}^{\infty} \left\{ \frac{\prod\limits_{j=1}^{p} \Gamma(\alpha_j + 2kA_j)}{\prod\limits_{j=1}^{p} \Gamma(\beta_j + 2kB_j)} \right\} \frac{z^{2k}}{(2k)!}}{\sum\limits_{k=0}^{\infty} \left\{ \frac{\prod\limits_{j=1}^{p} \Gamma(\alpha_j + A_j(2k+1))}{\prod\limits_{j=1}^{p} \Gamma(\beta_j + B_j(2k+1))} \right\} \frac{z^{2k+1}}{(2k+1)!}},
\end{aligned}
\tag{2.83}
$$

where $z \in \mathbb{C}$, $j, k, u \in \mathbb{N}$, $\alpha_j, \beta_k \in \mathbb{C}$, and $A_j, B_k \in \mathbb{R}_+$.

Definition 2.15 The superhyperbolic secant via Wright's generalized hypergeometric function is defined as

$$\begin{aligned} {}_p sech_q\,(z) &= \frac{1}{{}_p cosh_q\,(z)} \\ &= \frac{2}{{}_p\varphi_q\,(z) + {}_p\varphi_q\,(-z)} \\ &= \frac{1}{\sum_{k=0}^{\infty} \frac{\left\{\prod_{j=1}^{p} \Gamma(\alpha_j + 2kA_j)\right\}}{\left\{\prod_{j=1}^{p} \Gamma(\beta_j + 2kB_j)\right\}} \frac{z^{2k}}{(2k)!}}, \end{aligned} \tag{2.84}$$

where $z \in \mathbb{C}$, $j, k, u \in \mathbb{N}$, $\alpha_j, \beta_k \in \mathbb{C}$, and $A_j, B_k \in \mathbb{R}_+$.

Definition 2.16 The superhyperbolic cosecant via Wright's generalized hypergeometric function is defined as

$$\begin{aligned} {}_p cosech_q\,(z) &= \frac{1}{{}_p sinh_q\,(z)} \\ &= \frac{2}{{}_p\varphi_q\,(z) - {}_p\varphi_q\,(-z)} \\ &= \frac{1}{\sum_{k=0}^{\infty} \frac{\left\{\prod_{j=1}^{p} \Gamma(\alpha_j + A_j(2k+1))\right\}}{\left\{\prod_{j=1}^{p} \Gamma(\beta_j + B_j(2k+1))\right\}} \frac{z^{2k+1}}{(2k+1)!}}, \end{aligned} \tag{2.85}$$

where $z \in \mathbb{C}$, $j, k, u \in \mathbb{N}$, $\alpha_j, \beta_k \in \mathbb{C}$, and $A_j, B_k \in \mathbb{R}_+$.

Theorem 2.27 *Let $z \in \mathbb{C}$, $\alpha_j, \beta_j \in \mathbb{C}$, $A_j, B_j \in \mathbb{R}_+$, and $\sum_{j=1}^{p} A_j - \sum_{j=1}^{q} B_j = \Delta > -1$.*

Then we have

$${}_p\varphi_q\,(z) = {}_p cosh_q\,(z) + {}_p sinh_q\,(z) \tag{2.86}$$

and

$${}_p\varphi_q\,(-z) = {}_p cosh_q\,(z) - {}_p sinh_q\,(z). \tag{2.87}$$

Definition 2.17 The superhyperbolic sine ${}_p sinh_q\,(z)$, superhyperbolic cosine ${}_p cosh_q\,(z)$, superhyperbolic tangent ${}_p tanh_q\,(z)$, superhyperbolic cotangent ${}_p coth_q\,(z)$, superhyperbolic secant ${}_p sech_q\,(z)$, and superhyperbolic cosecant ${}_p cosec_q\,(z)$ are called as the superhyperbolic functions via Wright's generalized hypergeometric function.

More generally,

$$
\begin{aligned}
&{}_p sinh_q\ (\lambda z)\\
&= \tfrac{1}{2}\left({}_p\varphi_q\ (\lambda z) - {}_p\varphi_q\ (-\lambda z)\right)\\
&= \sum_{k=0}^{\infty} \frac{\left\{\prod_{j=1}^{p} \Gamma(\alpha_j + A_j(2k+1))\right\}}{\left\{\prod_{j=1}^{p} \Gamma(\beta_j + B_j(2k+1))\right\}} \frac{(\lambda z)^{2k+1}}{(2k+1)!},
\end{aligned}
\tag{2.88}
$$

$$
\begin{aligned}
&{}_p cosh_q\ (\lambda z)\\
&= \tfrac{1}{2}\left({}_p\varphi_q\ (\lambda z) + {}_p\varphi_q\ (-\lambda z)\right)\\
&= \sum_{k=0}^{\infty} \frac{\left\{\prod_{j=1}^{p} \Gamma(\alpha_j + 2kA_j)\right\}}{\left\{\prod_{j=1}^{p} \Gamma(\beta_j + 2kB_j)\right\}} \frac{(\lambda z)^{2k}}{(2k)!},
\end{aligned}
\tag{2.89}
$$

$$
\begin{aligned}
&{}_p tanh_q\ (\lambda z)\\
&= \frac{{}_p sinh_q\ (\lambda z)}{{}_p cosh_q\ (\lambda z)}\\
&= \frac{{}_p\varphi_q\ (\lambda z) - {}_p\varphi_q\ (-\lambda z)}{{}_p\varphi_q\ (\lambda z) + {}_p\varphi_q\ (-\lambda z)}\\
&= \frac{\sum_{k=0}^{\infty} \frac{\left\{\prod_{j=1}^{p} \Gamma\left(\alpha_j + A_j(2k+1)\right)\right\}}{\left\{\prod_{j=1}^{p} \Gamma\left(\beta_j + B_j(2k+1)\right)\right\}} \frac{(\lambda z)^{2k+1}}{(2k+1)!}}{\sum_{k=0}^{\infty} \frac{\left\{\prod_{j=1}^{p} \Gamma\left(\alpha_j + 2kA_j\right)\right\}}{\left\{\prod_{j=1}^{p} \Gamma\left(\beta_j + 2kB_j\right)\right\}} \frac{(\lambda z)^{2k}}{(2k)!}},
\end{aligned}
\tag{2.90}
$$

$$
\begin{aligned}
&{}_p coth_q\ (\lambda z)\\
&= \frac{{}_p cosh_q\ (\lambda z)}{{}_p sinh_q\ (\lambda z)}\\
&= \frac{{}_p\varphi_q\ (\lambda z) + {}_p\varphi_q\ (-\lambda z)}{{}_p\varphi_q\ (i\lambda z) - {}_p\varphi_q\ (-\lambda z)}\\
&= \frac{\sum_{k=0}^{\infty} \frac{\left\{\prod_{j=1}^{p} \Gamma\left(\alpha_j + 2kA_j\right)\right\}}{\left\{\prod_{j=1}^{p} \Gamma\left(\beta_j + 2kB_j\right)\right\}} \frac{(\lambda z)^{2k}}{(2k)!}}{\sum_{k=0}^{\infty} \frac{\left\{\prod_{j=1}^{p} \Gamma\left(\alpha_j + A_j(2k+1)\right)\right\}}{\left\{\prod_{j=1}^{p} \Gamma\left(\beta_j + B_j(2k+1)\right)\right\}} \frac{(\lambda z)^{2k+1}}{(2k+1)!}},
\end{aligned}
\tag{2.91}
$$

$$
\begin{aligned}
&{}_p sech_q\,(\lambda z)\\
&= \frac{1}{{}_p cosh_q\,(\lambda z)}\\
&= \frac{2}{{}_p\varphi_q\,(\lambda z) + {}_p\varphi_q\,(-\lambda z)}\\
&= \frac{1}{\sum\limits_{k=0}^{\infty} \frac{\left\{\prod\limits_{j=1}^{p} \Gamma\left(\alpha_j + 2kA_j\right)\right\}}{\left\{\prod\limits_{j=1}^{p} \Gamma\left(\beta_j + 2kB_j\right)\right\}} \frac{(\lambda z)^{2k}}{(2k)!}}
\end{aligned}
\tag{2.92}
$$

and

$$
\begin{aligned}
&{}_p cosech_q\,(\lambda z)\\
&= \frac{1}{{}_p sinh_q\,(\lambda z)}\\
&= \frac{2i}{{}_p\varphi_q\,(\lambda z) - {}_p\varphi_q\,(-\lambda z)}\\
&= \frac{1}{\sum\limits_{k=0}^{\infty} \frac{\left\{\prod\limits_{j=1}^{p} \Gamma\left(\alpha_j + A_j(2k+1)\right)\right\}}{\left\{\prod\limits_{j=1}^{p} \Gamma\left(\beta_j + B_j(2k+1)\right)\right\}} \frac{(\lambda z)^{2k+1}}{(2k+1)!}},
\end{aligned}
\tag{2.93}
$$

where $z \in \mathbb{C}$, $j, k, u \in \mathbb{N}$, $\alpha_j, \beta_k \in \mathbb{C}$, and $A_j, B_k \in \mathbb{R}_+$.

Theorem 2.28 *Let $\gamma \in \mathbb{C}$, $n \in \mathbb{N}$, $Re(\gamma) > 0$, $\lambda \in \mathbb{C}$, $\beta > 0$, $j, k, u \in \mathbb{N}$, $\alpha_j, \beta_k \in \mathbb{C}$, and $A_j, B_k \in \mathbb{R}_+$. Then we have*

$$
\begin{aligned}
&\frac{\partial^n}{\partial t^n}\left\{t^{\gamma-1}\,{}_p sinh_q\left(\begin{matrix}(\alpha_1, A_1), \cdots, (\alpha_p, A_p)\\ (\beta_1, B_1), \cdots, (\beta_q, B_q)\end{matrix}; \lambda t^{\beta}\right)\right\}\\
&= t^{\gamma-n-1}\,{}_{p+1} sinh_{q+1}\left(\begin{matrix}(\alpha_1, A_1), \cdots, (\alpha_p, A_p), (\gamma, \beta)\\ (\beta_1, B_1), \cdots, (\beta_q, B_q), (\gamma-n, \beta)\end{matrix}; \lambda t^{\beta}\right)
\end{aligned}
\tag{2.94}
$$

and

$$
\begin{aligned}
&\frac{\partial^n}{\partial t^n}\left\{t^{\gamma-1}\,{}_p cosh_q\left(\begin{matrix}(\alpha_1, A_1), \cdots, (\alpha_p, A_p)\\ (\beta_1, B_1), \cdots, (\beta_q, B_q)\end{matrix}; \lambda t^{\beta}\right)\right\}\\
&= t^{\gamma-n-1}\,{}_{p+1} cosh_{q+1}\left(\begin{matrix}(\alpha_1, A_1), \cdots, (\alpha_p, A_p), (\gamma, \beta)\\ (\beta_1, B_1), \cdots, (\beta_q, B_q), (\gamma-n, \beta)\end{matrix}; \lambda t^{\beta}\right).
\end{aligned}
\tag{2.95}
$$

Proof In view of the expressions

$$\frac{\partial^n}{\partial t^n}\left\{t^{\gamma-1}\,{}_p\varphi_q\left(\begin{matrix}(\alpha_1,A_1),\cdots,(\alpha_p,A_p)\\(\beta_1,B_1),\cdots,(\beta_q,B_q)\end{matrix};\lambda t^\beta\right)\right\}$$
$$=t^{\gamma-n-1}\,{}_{p+1}\varphi_{q+1}\left(\begin{matrix}(\alpha_1,A_1),\cdots,(\alpha_p,A_p),(\gamma,\beta)\\(\beta_1,B_1),\cdots,(\beta_q,B_q),(\gamma-n,\beta)\end{matrix};\lambda t^\beta\right) \tag{2.96}$$

and

$$\frac{\partial^n}{\partial t^n}\left\{t^{\gamma-1}\,{}_p\varphi_q\left(\begin{matrix}(\alpha_1,A_1),\cdots,(\alpha_p,A_p)\\(\beta_1,B_1),\cdots,(\beta_q,B_q)\end{matrix};-\lambda t^\beta\right)\right\}$$
$$=t^{\gamma-n-1}\,{}_{p+1}\varphi_{q+1}\left(\begin{matrix}(\alpha_1,A_1),\cdots,(\alpha_p,A_p),(\gamma,\beta)\\(\beta_1,B_1),\cdots,(\beta_q,B_q),(\gamma-n,\beta)\end{matrix};-\lambda t^\beta\right), \tag{2.97}$$

we have

$$\frac{\partial^n}{\partial t^n}\left\{t^{\gamma-1}\,{}_psinh_q\left(\begin{matrix}(\alpha_1,A_1),\cdots,(\alpha_p,A_p)\\(\beta_1,B_1),\cdots,(\beta_q,B_q)\end{matrix};\lambda t^\beta\right)\right\}$$
$$=t^{\gamma-n-1}\,{}_{p+1}sinh_{q+1}\left(\begin{matrix}(\alpha_1,A_1),\cdots,(\alpha_p,A_p),(\gamma,\beta)\\(\beta_1,B_1),\cdots,(\beta_q,B_q),(\gamma-n,\beta)\end{matrix};\lambda t^\beta\right) \tag{2.98}$$

and

$$\frac{\partial^n}{\partial t^n}\left\{t^{\gamma-1}\,{}_pcosh_q\left(\begin{matrix}(\alpha_1,A_1),\cdots,(\alpha_p,A_p)\\(\beta_1,B_1),\cdots,(\beta_q,B_q)\end{matrix};\lambda t^\beta\right)\right\}$$
$$=t^{\gamma-n-1}\,{}_{p+1}cosh_{q+1}\left(\begin{matrix}(\alpha_1,A_1),\cdots,(\alpha_p,A_p),(\gamma,\beta)\\(\beta_1,B_1),\cdots,(\beta_q,B_q),(\gamma-n,\beta)\end{matrix};\lambda t^\beta\right). \tag{2.99}$$

□

Theorem 2.29 *Let* $\gamma\in\mathbb{C}$, $Re(\gamma)>0$, $\lambda\in\mathbb{C}$, $\beta>0$, $j,k,u\in\mathbb{N}$, $\alpha_j,\beta_k\in\mathbb{C}$, *and* $A_j,B_k\in\mathbb{R}_+$. *Then we have*

$$\int_0^t \tau^{\gamma-1}\,{}_psinh_q\left(\begin{matrix}(\alpha_1,A_1),\cdots,(\alpha_p,A_p)\\(\beta_1,B_1),\cdots,(\beta_q,B_q)\end{matrix};\lambda\tau^\beta\right)d\tau$$
$$=t^{\gamma}\,{}_{p+1}sinh_{q+1}\left(\begin{matrix}(\alpha_1,A_1),\cdots,(\alpha_p,A_p),(\gamma,\beta)\\(\beta_1,B_1),\cdots,(\beta_q,B_q),(\gamma+1,\beta)\end{matrix};\lambda t^\beta\right) \tag{2.100}$$

and

$$\int_0^t \tau^{\gamma-1}\,{}_pcosh_q\left(\begin{matrix}(\alpha_1,A_1),\cdots,(\alpha_p,A_p)\\(\beta_1,B_1),\cdots,(\beta_q,B_q)\end{matrix};\lambda\tau^\beta\right)d\tau$$
$$=t^{\gamma}\,{}_{p+1}cosh_{q+1}\left(\begin{matrix}(\alpha_1,A_1),\cdots,(\alpha_p,A_p),(\gamma,\beta)\\(\beta_1,B_1),\cdots,(\beta_q,B_q),(\gamma+1,\beta)\end{matrix};\lambda t^\beta\right). \tag{2.101}$$

Proof From (2.89), (2.96), and (2.97), we directly have the results. □

Corollary 2.7 *Let $\lambda \in \mathbb{C}$, $\beta > 0$, $j, k, u \in \mathbb{N}$, $\alpha_j, \beta_k \in \mathbb{C}$, and $A_j, B_k \in \mathbb{R}_+$. Then we have*

$$\begin{aligned} &\int_0^t {}_p sinh_q \left(\begin{matrix} (\alpha_1, A_1), \cdots, (\alpha_p, A_p) \\ (\beta_1, B_1), \cdots, (\beta_q, B_q) \end{matrix}; \lambda\tau^\beta \right) d\tau \\ &= t\ {}_{p+1} sinh_{q+1} \left(\begin{matrix} (\alpha_1, A_1), \cdots, (\alpha_p, A_p), (1, \beta) \\ (\beta_1, B_1), \cdots, (\beta_q, B_q), (2, \beta) \end{matrix}; \lambda t^\beta \right) \end{aligned} \tag{2.102}$$

and

$$\begin{aligned} &\int_0^t {}_p cosh_q \left(\begin{matrix} (\alpha_1, A_1), \cdots, (\alpha_p, A_p) \\ (\beta_1, B_1), \cdots, (\beta_q, B_q) \end{matrix}; \lambda\tau^\beta \right) d\tau \\ &= t\ {}_{p+1} cosh_{q+1} \left(\begin{matrix} (\alpha_1, A_1), \cdots, (\alpha_p, A_p), (1, \beta) \\ (\beta_1, B_1), \cdots, (\beta_q, B_q), (2, \beta) \end{matrix}; \lambda t^\beta \right). \end{aligned} \tag{2.103}$$

Proof Taking $\gamma = 1$ in (2.100) and (2.101), we directly present the results. □

Corollary 2.8 *Let $\gamma \in \mathbb{C}$, $Re(\gamma) > 0$, $\lambda \in \mathbb{C}$, $\beta > 0$, $j, k, u \in \mathbb{N}$, $\alpha_j, \beta_k \in \mathbb{C}$, and $A_j, B_k \in \mathbb{R}_+$. Then we have*

$$\begin{aligned} &\int_0^1 \tau^{\gamma-1}\ {}_p sinh_q \left(\begin{matrix} (\alpha_1, A_1), \cdots, (\alpha_p, A_p) \\ (\beta_1, B_1), \cdots, (\beta_q, B_q) \end{matrix}; \lambda\tau^\beta \right) d\tau \\ &= {}_{p+1} sinh_{q+1} \left(\begin{matrix} (\alpha_1, A_1), \cdots, (\alpha_p, A_p), (\gamma, \beta) \\ (\beta_1, B_1), \cdots, (\beta_q, B_q), (\gamma+1, \beta) \end{matrix}; \lambda \right) \end{aligned} \tag{2.104}$$

and

$$\begin{aligned} &\int_0^1 \tau^{\gamma-1}\ {}_p cosh_q \left(\begin{matrix} (\alpha_1, A_1), \cdots, (\alpha_p, A_p) \\ (\beta_1, B_1), \cdots, (\beta_q, B_q) \end{matrix}; \lambda\tau^\beta \right) d\tau \\ &= {}_{p+1} cosh_{q+1} \left(\begin{matrix} (\alpha_1, A_1), \cdots, (\alpha_p, A_p), (\gamma, \beta) \\ (\beta_1, B_1), \cdots, (\beta_q, B_q), (\gamma+1, \beta) \end{matrix}; \lambda \right). \end{aligned} \tag{2.105}$$

Proof Taking $t = 1$ in (2.100) and (2.101), we directly have the results. □

Corollary 2.9 *Let $\lambda \in \mathbb{C}$, $\beta > 0$, $j, k, u \in \mathbb{N}$, $\alpha_j, \beta_k \in \mathbb{C}$, and $A_j, B_k \in \mathbb{R}_+$. Then we have*

$$\begin{aligned} &\int_0^1 {}_p sinh_q \left(\begin{matrix} (\alpha_1, A_1), \cdots, (\alpha_p, A_p) \\ (\beta_1, B_1), \cdots, (\beta_q, B_q) \end{matrix}; \lambda\tau^\beta \right) d\tau \\ &= {}_{p+1} sinh_{q+1} \left(\begin{matrix} (\alpha_1, A_1), \cdots, (\alpha_p, A_p), (1, \beta) \\ (\beta_1, B_1), \cdots, (\beta_q, B_q), (2, \beta) \end{matrix}; \lambda \right) \end{aligned} \tag{2.106}$$

and

$$\int_0^1 {}_pcosh_q\left(\begin{matrix}(\alpha_1,A_1),\cdots,(\alpha_p,A_p)\\(\beta_1,B_1),\cdots,(\beta_q,B_q)\end{matrix};\lambda\tau^\beta\right)d\tau \\ = {}_{p+1}cosh_{q+1}\left(\begin{matrix}(\alpha_1,A_1),\cdots,(\alpha_p,A_p),(1,\beta)\\(\beta_1,B_1),\cdots,(\beta_q,B_q),(2,\beta)\end{matrix};\lambda\right). \tag{2.107}$$

Proof Taking $t = 1$ and $\gamma = 1$ in (2.100) and (2.101), we directly have the results. □

Theorem 2.30 *Let $\alpha, \gamma \in \mathbb{C}$, $Re(\alpha) > 0$, $Re(\gamma) > 0$, $\lambda \in \mathbb{C}$, $\beta > 0$, $j, k, u \in \mathbb{N}$, $\alpha_j, \beta_k \in \mathbb{C}$, and $A_j, B_k \in \mathbb{R}_+$. Then we have*

$$\frac{1}{\Gamma(\alpha)}\int_0^t (t-\tau)^\alpha\tau^{\gamma-1} {}_psinh_q\left(\begin{matrix}(\alpha_1,A_1),\cdots,(\alpha_p,A_p)\\(\beta_1,B_1),\cdots,(\beta_q,B_q)\end{matrix};\lambda\tau^\beta\right)d\tau \\ = t^{\gamma+\alpha-1} {}_{p+1}sinh_{q+1}\left(\begin{matrix}(\alpha_1,A_1),\cdots,(\alpha_p,A_p),(\gamma,\beta)\\(\beta_1,B_1),\cdots,(\beta_q,B_q),(\gamma+\alpha,\beta)\end{matrix};\lambda t^\beta\right) \tag{2.108}$$

and

$$\frac{1}{\Gamma(\alpha)}\int_0^t (t-\tau)^\alpha\tau^{\gamma-1} {}_pcosh_q\left(\begin{matrix}(\alpha_1,A_1),\cdots,(\alpha_p,A_p)\\(\beta_1,B_1),\cdots,(\beta_q,B_q)\end{matrix};\lambda\tau^\beta\right)d\tau \\ = t^{\gamma+\alpha-1} {}_{p+1}cosh_{q+1}\left(\begin{matrix}(\alpha_1,A_1),\cdots,(\alpha_p,A_p),(\gamma,\beta)\\(\beta_1,B_1),\cdots,(\beta_q,B_q),(\gamma+\alpha,\beta)\end{matrix};\lambda t^\beta\right). \tag{2.109}$$

Proof Making use of (2.88), we have

$$\frac{1}{\Gamma(\alpha)}\int_0^t (t-\tau)^\alpha\tau^{\gamma-1} {}_p\varphi_q\left(\begin{matrix}(\alpha_1,A_1),\cdots,(\alpha_p,A_p)\\(\beta_1,B_1),\cdots,(\beta_q,B_q)\end{matrix};\lambda\tau^\beta\right)d\tau \\ = t^{\gamma+\alpha-1} {}_{p+1}\varphi_{q+1}\left(\begin{matrix}(\alpha_1,A_1),\cdots,(\alpha_p,A_p),(\gamma,\beta)\\(\beta_1,B_1),\cdots,(\beta_q,B_q),(\gamma+\alpha,\beta)\end{matrix};\lambda t^\beta\right) \tag{2.110}$$

and

$$\frac{1}{\Gamma(\alpha)}\int_0^t (t-\tau)^\alpha\tau^{\gamma-1} {}_p\varphi_q\left(\begin{matrix}(\alpha_1,A_1),\cdots,(\alpha_p,A_p)\\(\beta_1,B_1),\cdots,(\beta_q,B_q)\end{matrix};-\lambda\tau^\beta\right)d\tau \\ = t^{\gamma+\alpha-1} {}_{p+1}\varphi_{q+1}\left(\begin{matrix}(\alpha_1,A_1),\cdots,(\alpha_p,A_p),(\gamma,\beta)\\(\beta_1,B_1),\cdots,(\beta_q,B_q),(\gamma+\alpha,\beta)\end{matrix};-\lambda t^\beta\right) \tag{2.111}$$

which reduces the results. □

Corollary 2.10 *Let $\alpha, \gamma \in \mathbb{C}$, $Re(\alpha) > 0$, $Re(\gamma) > 0$, $\lambda \in \mathbb{C}$, $\beta > 0$, $j, k, u \in \mathbb{N}$, $\alpha_j, \beta_k \in \mathbb{C}$, and $A_j, B_k \in \mathbb{R}_+$. Then we have*

$$\begin{aligned}&\frac{1}{\Gamma(\alpha)}\int_0^1 (1-\tau)^{\alpha}\tau^{\gamma-1}\,{}_p sinh_q\left(\begin{matrix}(\alpha_1, A_1), \cdots, (\alpha_p, A_p)\\(\beta_1, B_1), \cdots, (\beta_q, B_q)\end{matrix}; \lambda\tau^{\beta}\right)d\tau\\&= {}_{p+1}sinh_{q+1}\left(\begin{matrix}(\alpha_1, A_1), \cdots, (\alpha_p, A_p), (\gamma, \beta)\\(\beta_1, B_1), \cdots, (\beta_q, B_q), (\gamma+\alpha, \beta)\end{matrix}; \lambda\right)\end{aligned} \tag{2.112}$$

and

$$\begin{aligned}&\frac{1}{\Gamma(\alpha)}\int_0^1 (1-\tau)^{\alpha}\tau^{\gamma-1}\,{}_p cosh_q\left(\begin{matrix}(\alpha_1, A_1), \cdots, (\alpha_p, A_p)\\(\beta_1, B_1), \cdots, (\beta_q, B_q)\end{matrix}; \lambda\tau^{\beta}\right)d\tau\\&= {}_{p+1}cosh_{q+1}\left(\begin{matrix}(\alpha_1, A_1), \cdots, (\alpha_p, A_p), (\gamma, \beta)\\(\beta_1, B_1), \cdots, (\beta_q, B_q), (\gamma+\alpha, \beta)\end{matrix}; \lambda\right).\end{aligned} \tag{2.113}$$

Proof Taking $t = 1$ in (2.108) and (2.109), we directly have the results. □

Corollary 2.11 *Let $\alpha \in \mathbb{C}$, $Re(\alpha) > 0$, $\lambda \in \mathbb{C}$, $\beta > 0$, $j, k, u \in \mathbb{N}$, $\alpha_j, \beta_k \in \mathbb{C}$, and $A_j, B_k \in \mathbb{R}_+$. Then we have*

$$\begin{aligned}&\frac{1}{\Gamma(\alpha)}\int_0^t (t-\tau)^{\alpha}\,{}_p sinh_q\left(\begin{matrix}(\alpha_1, A_1), \cdots, (\alpha_p, A_p)\\(\beta_1, B_1), \cdots, (\beta_q, B_q)\end{matrix}; \lambda\tau^{\beta}\right)d\tau\\&= t^{\alpha}\,{}_{p+1}sinh_{q+1}\left(\begin{matrix}(\alpha_1, A_1), \cdots, (\alpha_p, A_p), (1, \beta)\\(\beta_1, B_1), \cdots, (\beta_q, B_q), (1+\alpha, \beta)\end{matrix}; \lambda t^{\beta}\right)\end{aligned} \tag{2.114}$$

and

$$\begin{aligned}&\frac{1}{\Gamma(\alpha)}\int_0^t (t-\tau)^{\alpha}\,{}_p cosh_q\left(\begin{matrix}(\alpha_1, A_1), \cdots, (\alpha_p, A_p)\\(\beta_1, B_1), \cdots, (\beta_q, B_q)\end{matrix}; \lambda\tau^{\beta}\right)d\tau\\&= t^{\alpha}\,{}_{p+1}cosh_{q+1}\left(\begin{matrix}(\alpha_1, A_1), \cdots, (\alpha_p, A_p), (1, \beta)\\(\beta_1, B_1), \cdots, (\beta_q, B_q), (1+\alpha, \beta)\end{matrix}; \lambda t^{\beta}\right).\end{aligned} \tag{2.115}$$

Proof Taking $\gamma = 1$ in (2.108) and (2.109), we directly have the results. □

Corollary 2.12 *Let $\alpha \in \mathbb{C}$, $Re(\alpha) > 0$, $\lambda \in \mathbb{C}$, $\beta > 0$, $j, k, u \in \mathbb{N}$, $\alpha_j, \beta_k \in \mathbb{C}$, and $A_j, B_k \in \mathbb{R}_+$. Then we have*

$$\begin{aligned}&\frac{1}{\Gamma(\alpha)}\int_0^1 (1-\tau)^{\alpha}\,{}_p sinh_q\left(\begin{matrix}(\alpha_1, A_1), \cdots, (\alpha_p, A_p)\\(\beta_1, B_1), \cdots, (\beta_q, B_q)\end{matrix}; \lambda\tau^{\beta}\right)d\tau\\&= {}_{p+1}sinh_{q+1}\left(\begin{matrix}(\alpha_1, A_1), \cdots, (\alpha_p, A_p), (1, \beta)\\(\beta_1, B_1), \cdots, (\beta_q, B_q), (1+\alpha, \beta)\end{matrix}; \lambda\right)\end{aligned} \tag{2.116}$$

and

$$\begin{aligned}&\frac{1}{\Gamma(\alpha)}\int_0^1 (1-\tau)^{\alpha}\ {}_p cosh_q\left(\begin{matrix}(\alpha_1,A_1),\cdots,(\alpha_p,A_p)\\(\beta_1,B_1),\cdots,(\beta_q,B_q)\end{matrix};\lambda\tau^{\beta}\right)d\tau\\&=\ {}_{p+1}cosh_{q+1}\left(\begin{matrix}(\alpha_1,A_1),\cdots,(\alpha_p,A_p),(1,\beta)\\(\beta_1,B_1),\cdots,(\beta_q,B_q),(1+\alpha,\beta)\end{matrix};\lambda\right).\end{aligned}\tag{2.117}$$

Proof Taking $t=1$ in (2.114) and (2.115), we directly have the results. □

Theorem 2.31 *The Laplace transforms of the functions*

$$t^{\gamma-1}\ {}_p sin_q\left(\begin{matrix}(\alpha_1,A_1),\cdots,(\alpha_p,A_p)\\(\beta_1,B_1),\cdots,(\beta_q,B_q)\end{matrix};\lambda t\right)\quad,$$

$$t^{\gamma-1}\ {}_p cos_q\left(\begin{matrix}(\alpha_1,A_1),\cdots,(\alpha_p,A_p)\\(\beta_1,B_1),\cdots,(\beta_q,B_q)\end{matrix};\lambda t\right)$$

$t^{\gamma-1}\ {}_p sin_q\left(\begin{matrix}(\alpha_1,A_1),\cdots,(\alpha_p,A_p)\\(\beta_1,B_1),\cdots,(\beta_q,B_q)\end{matrix};\lambda t\right)$ *and* $t^{\gamma-1}\ {}_p cos_q\left(\begin{matrix}(\alpha_1,A_1),\cdots,(\alpha_p,A_p)\\(\beta_1,B_1),\cdots,(\beta_q,B_q)\end{matrix};\lambda t\right)$
are given as follows:

$$\begin{aligned}&L\left[t^{\gamma-1}\ {}_p sin_q\left(\begin{matrix}(\alpha_1,A_1),\cdots,(\alpha_p,A_p)\\(\beta_1,B_1),\cdots,(\beta_q,B_q)\end{matrix};\lambda t\right)\right]\\&=\int_0^{\infty}e^{-st}t^{\gamma-1}\ {}_p sin_q\left(\begin{matrix}(\alpha_1,A_1),\cdots,(\alpha_p,A_p)\\(\beta_1,B_1),\cdots,(\beta_q,B_q)\end{matrix};\lambda t\right)dt\\&=s^{-\gamma}\ {}_{p+1}sin_q\left(\begin{matrix}(\gamma,1),(\alpha_1,A_1),\cdots,(\alpha_p,A_p)\\(\beta_1,B_1),\cdots,(\beta_q,B_q)\end{matrix};\frac{\lambda}{s}\right),\end{aligned}\tag{2.118}$$

$$\begin{aligned}&L\left[t^{\gamma-1}\ {}_p cos_q\left(\begin{matrix}(\alpha_1,A_1),\cdots,(\alpha_p,A_p)\\(\beta_1,B_1),\cdots,(\beta_q,B_q)\end{matrix};\lambda t\right)\right]\\&=\int_0^{\infty}e^{-st}t^{\gamma-1}\ {}_p cos_q\left(\begin{matrix}(\alpha_1,A_1),\cdots,(\alpha_p,A_p)\\(\beta_1,B_1),\cdots,(\beta_q,B_q)\end{matrix};\lambda t\right)dt\\&=s^{-\gamma}\ {}_{p+1}cos_q\left(\begin{matrix}(\gamma,1),(\alpha_1,A_1),\cdots,(\alpha_p,A_p)\\(\beta_1,B_1),\cdots,(\beta_q,B_q)\end{matrix};\frac{\lambda}{s}\right),\end{aligned}\tag{2.119}$$

$$\begin{aligned}&L\left[t^{\gamma-1}\ {}_p sinh_q\left(\begin{matrix}(\alpha_1,A_1),\cdots,(\alpha_p,A_p)\\(\beta_1,B_1),\cdots,(\beta_q,B_q)\end{matrix};\lambda t\right)\right]\\&=\int_0^{\infty}e^{-st}t^{\gamma-1}\ {}_p sinh_q\left(\begin{matrix}(\alpha_1,A_1),\cdots,(\alpha_p,A_p)\\(\beta_1,B_1),\cdots,(\beta_q,B_q)\end{matrix};\lambda t\right)dt\\&=s^{-\gamma}\ {}_{p+1}sinh_q\left(\begin{matrix}(\gamma,1),(\alpha_1,A_1),\cdots,(\alpha_p,A_p)\\(\beta_1,B_1),\cdots,(\beta_q,B_q)\end{matrix};\frac{\lambda}{s}\right),\end{aligned}\tag{2.120}$$

$$L\left[t^{\gamma-1}\ {}_pcosh_q\left(\begin{matrix}(\alpha_1,A_1),\cdots,(\alpha_p,A_p)\\(\beta_1,B_1),\cdots,(\beta_q,B_q)\end{matrix};\lambda t\right)\right]$$
$$=\int_0^\infty e^{-st}t^{\gamma-1}\ {}_pcosh_q\left(\begin{matrix}(\alpha_1,A_1),\cdots,(\alpha_p,A_p)\\(\beta_1,B_1),\cdots,(\beta_q,B_q)\end{matrix};\lambda t\right)dt \tag{2.121}$$
$$=s^{-\gamma}\ {}_{p+1}cosh_q\left(\begin{matrix}(\gamma,1),(\alpha_1,A_1),\cdots,(\alpha_p,A_p)\\(\beta_1,B_1),\cdots,(\beta_q,B_q)\end{matrix};\tfrac{\lambda}{s}\right),$$

where $\lambda\in\mathbb{C}$, $Re(s)>0$, $Re(\gamma)>0$, $j,k,u\in\mathbb{N}$, $\alpha_j,\beta_k>0$, *and* $A_j,B_k>0$.

Proof With the help of the integral relations

$$L\left[t^{\gamma-1}\ {}_p\varphi_q\left(\begin{matrix}(\alpha_1,A_1),\cdots,(\alpha_p,A_p)\\(\beta_1,B_1),\cdots,(\beta_q,B_q)\end{matrix};i\lambda t\right)\right]$$
$$=\int_0^\infty e^{-st}t^{\gamma-1}\ {}_p\varphi_q\left(\begin{matrix}(\alpha_1,A_1),\cdots,(\alpha_p,A_p)\\(\beta_1,B_1),\cdots,(\beta_q,B_q)\end{matrix};i\lambda t\right)dt \tag{2.122}$$
$$=s^{-\gamma}\ {}_{p+1}\varphi_q\left(\begin{matrix}(\gamma,1),(\alpha_1,A_1),\cdots,(\alpha_p,A_p)\\(\beta_1,B_1),\cdots,(\beta_q,B_q)\end{matrix};i\tfrac{\lambda}{s}\right),$$

$$L\left[t^{\gamma-1}\ {}_p\varphi_q\left(\begin{matrix}(\alpha_1,A_1),\cdots,(\alpha_p,A_p)\\(\beta_1,B_1),\cdots,(\beta_q,B_q)\end{matrix};-i\lambda t\right)\right]$$
$$=\int_0^\infty e^{-st}t^{\gamma-1}\ {}_p\varphi_q\left(\begin{matrix}(\alpha_1,A_1),\cdots,(\alpha_p,A_p)\\(\beta_1,B_1),\cdots,(\beta_q,B_q)\end{matrix};-i\lambda t\right)dt \tag{2.123}$$
$$=s^{-\gamma}\ {}_{p+1}\varphi_q\left(\begin{matrix}(\gamma,1),(\alpha_1,A_1),\cdots,(\alpha_p,A_p)\\(\beta_1,B_1),\cdots,(\beta_q,B_q)\end{matrix};\tfrac{-i\lambda}{s}\right),$$

$$L\left[t^{\gamma-1}\ {}_p\varphi_q\left(\begin{matrix}(\alpha_1,A_1),\cdots,(\alpha_p,A_p)\\(\beta_1,B_1),\cdots,(\beta_q,B_q)\end{matrix};\lambda t\right)\right]$$
$$=\int_0^\infty e^{-st}t^{\gamma-1}\ {}_p\varphi_q\left(\begin{matrix}(\alpha_1,A_1),\cdots,(\alpha_p,A_p)\\(\beta_1,B_1),\cdots,(\beta_q,B_q)\end{matrix};\lambda t\right)dt \tag{2.124}$$
$$=s^{-\gamma}\ {}_{p+1}\varphi_q\left(\begin{matrix}(\gamma,1),(\alpha_1,A_1),\cdots,(\alpha_p,A_p)\\(\beta_1,B_1),\cdots,(\beta_q,B_q)\end{matrix};\tfrac{\lambda}{s}\right)$$

and

$$L\left[t^{\gamma-1}\ {}_p\varphi_q\left(\begin{matrix}(\alpha_1,A_1),\cdots,(\alpha_p,A_p)\\(\beta_1,B_1),\cdots,(\beta_q,B_q)\end{matrix};-\lambda t\right)\right]$$
$$=\int_0^\infty e^{-st}t^{\gamma-1}\ {}_p\varphi_q\left(\begin{matrix}(\alpha_1,A_1),\cdots,(\alpha_p,A_p)\\(\beta_1,B_1),\cdots,(\beta_q,B_q)\end{matrix};-\lambda t\right)dt \tag{2.125}$$
$$=s^{-\gamma}\ {}_{p+1}\varphi_q\left(\begin{matrix}(\gamma,1),(\alpha_1,A_1),\cdots,(\alpha_p,A_p)\\(\beta_1,B_1),\cdots,(\beta_q,B_q)\end{matrix};-\tfrac{\lambda}{s}\right),$$

we obtain the desired results. □

Theorem 2.32 *Let $\alpha, \gamma \in \mathbb{C}$, $Re(\alpha) > 0$, $Re(\gamma) > 0$, $\lambda \in \mathbb{C}$, $\beta > 0$, $j, k, u \in \mathbb{N}$, $\alpha_j, \beta_k \in \mathbb{C}$, and $A_j, B_k \in \mathbb{R}_+$. Then we have*

$$\begin{aligned}
&\tfrac{1}{\Gamma(\alpha)} \int_0^2 (2-\tau)^{\alpha} \tau^{\gamma-1} {}_p\varphi_q \left(\begin{matrix} (\alpha_1, A_1), \cdots, (\alpha_p, A_p) \\ (\beta_1, B_1), \cdots, (\beta_q, B_q) \end{matrix} ; \lambda \tau^{\beta} \right) d\tau \\
&= \tfrac{1}{\Gamma(\alpha)} \int_{-1}^1 (1-\tau)^{\alpha} (1+\tau)^{\gamma-1} {}_p\varphi_q \left(\begin{matrix} (\alpha_1, A_1), \cdots, (\alpha_p, A_p) \\ (\beta_1, B_1), \cdots, (\beta_q, B_q) \end{matrix} ; \lambda (1+\tau)^{\beta} \right) d\tau \\
&= 2^{\gamma+\alpha-1} {}_{p+1}\varphi_{q+1} \left(\begin{matrix} (\alpha_1, A_1), \cdots, (\alpha_p, A_p), (\gamma, \beta) \\ (\beta_1, B_1), \cdots, (\beta_q, B_q), (\gamma+\alpha, \beta) \end{matrix} ; \lambda 2^{\beta} \right).
\end{aligned} \tag{2.126}$$

Proof Taking $t = 2$ in (2.125), we obtain the result. □

Theorem 2.33 *Let $\alpha, \gamma \in \mathbb{C}$, $Re(\alpha) > 0$, $Re(\gamma) > 0$, $\lambda \in \mathbb{C}$, $\beta > 0$, $j, k, u \in \mathbb{N}$, $\alpha_j, \beta_k \in \mathbb{C}$, and $A_j, B_k \in \mathbb{R}_+$. Then we have*

$$\begin{aligned}
&\tfrac{1}{\Gamma(\alpha)} \int_0^2 (2-\tau)^{\alpha} \tau^{\gamma-1} {}_p sin_q \left(\begin{matrix} (\alpha_1, A_1), \cdots, (\alpha_p, A_p) \\ (\beta_1, B_1), \cdots, (\beta_q, B_q) \end{matrix} ; \lambda \tau^{\beta} \right) d\tau \\
&= \tfrac{1}{\Gamma(\alpha)} \int_{-1}^1 (1-\tau)^{\alpha} (1+\tau)^{\gamma-1} {}_p sin_q \\
&\quad \times \left(\begin{matrix} (\alpha_1, A_1), \cdots, (\alpha_p, A_p) \\ (\beta_1, B_1), \cdots, (\beta_q, B_q) \end{matrix} ; \lambda (1+\tau)^{\beta} \right) d\tau \\
&= 2^{\gamma+\alpha-1} {}_{p+1} sin_{q+1} \left(\begin{matrix} (\alpha_1, A_1), \cdots, (\alpha_p, A_p), (\gamma, \beta) \\ (\beta_1, B_1), \cdots, (\beta_q, B_q), (\gamma+\alpha, \beta) \end{matrix} ; \lambda 2^{\beta} \right).
\end{aligned} \tag{2.127}$$

Proof Taking $t = 2$ in (2.126), we get the result. □

Theorem 2.34 *Let $\alpha, \gamma \in \mathbb{C}$, $Re(\alpha) > 0$, $Re(\gamma) > 0$, $\lambda \in \mathbb{C}$, $\beta > 0$, $j, k, u \in \mathbb{N}$, $\alpha_j, \beta_k \in \mathbb{C}$, and $A_j, B_k \in \mathbb{R}_+$. Then we have*

$$\begin{aligned}
&\tfrac{1}{\Gamma(\alpha)} \int_0^2 (2-\tau)^{\alpha} \tau^{\gamma-1} {}_p cos_q \left(\begin{matrix} (\alpha_1, A_1), \cdots, (\alpha_p, A_p) \\ (\beta_1, B_1), \cdots, (\beta_q, B_q) \end{matrix} ; \lambda \tau^{\beta} \right) d\tau \\
&= \tfrac{1}{\Gamma(\alpha)} \int_{-1}^1 (1-\tau)^{\alpha} (1+\tau)^{\gamma-1} {}_p cos_q \\
&\quad \times \left(\begin{matrix} (\alpha_1, A_1), \cdots, (\alpha_p, A_p) \\ (\beta_1, B_1), \cdots, (\beta_q, B_q) \end{matrix} ; \lambda (1+\tau)^{\beta} \right) d\tau \\
&= 2^{\gamma+\alpha-1} {}_{p+1} cos_{q+1} \left(\begin{matrix} (\alpha_1, A_1), \cdots, (\alpha_p, A_p), (\gamma, \beta) \\ (\beta_1, B_1), \cdots, (\beta_q, B_q), (\gamma+\alpha, \beta) \end{matrix} ; \lambda 2^{\beta} \right).
\end{aligned} \tag{2.128}$$

Proof Taking $t = 2$ in (2.126), we give the result. □

Theorem 2.35 *Let* $\alpha, \gamma \in \mathbb{C}$, $Re(\alpha) > 0$, $Re(\gamma) > 0$, $\lambda \in \mathbb{C}$, $\beta > 0$, $j, k, u \in \mathbb{N}$, $\alpha_j, \beta_k \in \mathbb{C}$, *and* $A_j, B_k \in \mathbb{R}_+$. *Then we have*

$$\begin{aligned}
&\frac{1}{\Gamma(\alpha)} \int_0^2 (2-\tau)^\alpha \tau^{\gamma-1} {}_p sinh_q \left(\begin{matrix} (\alpha_1, A_1), \cdots, (\alpha_p, A_p) \\ (\beta_1, B_1), \cdots, (\beta_q, B_q) \end{matrix} ; \lambda \tau^\beta \right) d\tau \\
&= \frac{1}{\Gamma(\alpha)} \int_{-1}^1 (1-\tau)^\alpha (1+\tau)^{\gamma-1} {}_p sinh_q \\
&\times \left(\begin{matrix} (\alpha_1, A_1), \cdots, (\alpha_p, A_p) \\ (\beta_1, B_1), \cdots, (\beta_q, B_q) \end{matrix} ; \lambda (1+\tau)^\beta \right) d\tau \\
&= 2^{\gamma+\alpha-1} {}_{p+1} sinh_{q+1} \left(\begin{matrix} (\alpha_1, A_1), \cdots, (\alpha_p, A_p), (\gamma, \beta) \\ (\beta_1, B_1), \cdots, (\beta_q, B_q), (\gamma+\alpha, \beta) \end{matrix} ; \lambda 2^\beta \right).
\end{aligned} \tag{2.129}$$

Proof Taking $t = 2$ in (2.126), we have the result. □

Theorem 2.36 *Let* $\alpha, \gamma \in \mathbb{C}$, $Re(\alpha) > 0$, $Re(\gamma) > 0$, $\lambda \in \mathbb{C}$, $\beta > 0$, $j, k, u \in \mathbb{N}$, $\alpha_j, \beta_k \in \mathbb{C}$, *and* $A_j, B_k \in \mathbb{R}_+$. *Then we have*

$$\begin{aligned}
&\frac{1}{\Gamma(\alpha)} \int_0^2 (2-\tau)^\alpha \tau^{\gamma-1} {}_p cosh_q \left(\begin{matrix} (\alpha_1, A_1), \cdots, (\alpha_p, A_p) \\ (\beta_1, B_1), \cdots, (\beta_q, B_q) \end{matrix} ; \lambda \tau^\beta \right) d\tau \\
&= \frac{1}{\Gamma(\alpha)} \int_{-1}^1 (1-\tau)^\alpha (1+\tau)^{\gamma-1} {}_p cosh_q \\
&\times \left(\begin{matrix} (\alpha_1, A_1), \cdots, (\alpha_p, A_p) \\ (\beta_1, B_1), \cdots, (\beta_q, B_q) \end{matrix} ; \lambda (1+\tau)^\beta \right) d\tau \\
&= 2^{\gamma+\alpha-1} {}_{p+1} cosh_{q+1} \left(\begin{matrix} (\alpha_1, A_1), \cdots, (\alpha_p, A_p), (\gamma, \beta) \\ (\beta_1, B_1), \cdots, (\beta_q, B_q), (\gamma+\alpha, \beta) \end{matrix} ; \lambda 2^\beta \right).
\end{aligned} \tag{2.130}$$

Proof Taking $t = 2$ in (2.126), we give the result. □

Theorem 2.37 *Let* $\alpha \in \mathbb{C}$, $Re(\alpha) > 0$, $\lambda \in \mathbb{C}$, $\beta > 0$, $j, k, u \in \mathbb{N}$, $\alpha_j, \beta_k \in \mathbb{C}$, *and* $A_j, B_k \in \mathbb{R}_+$. *Then we have*

$$\begin{aligned}
&\frac{1}{\Gamma(\alpha)} \int_0^2 (2-\tau)^\alpha {}_p\varphi_q \left(\begin{matrix} (\alpha_1, A_1), \cdots, (\alpha_p, A_p) \\ (\beta_1, B_1), \cdots, (\beta_q, B_q) \end{matrix} ; \lambda \tau^\beta \right) d\tau \\
&= \frac{1}{\Gamma(\alpha)} \int_{-1}^1 (1-\tau)^\alpha {}_p\varphi_q \left(\begin{matrix} (\alpha_1, A_1), \cdots, (\alpha_p, A_p) \\ (\beta_1, B_1), \cdots, (\beta_q, B_q) \end{matrix} ; \lambda (1+\tau)^\beta \right) d\tau \\
&= 2^{\gamma+\alpha-1} {}_{p+1}\varphi_{q+1} \left(\begin{matrix} (\alpha_1, A_1), \cdots, (\alpha_p, A_p), (1, \beta) \\ (\beta_1, B_1), \cdots, (\beta_q, B_q), (1+\alpha, \beta) \end{matrix} ; \lambda 2^\beta \right).
\end{aligned} \tag{2.131}$$

Proof Taking $\gamma = 1$ in (2.126), we obtain the result. □

Theorem 2.38 *Let* $\alpha \in \mathbb{C}$, $Re(\alpha) > 0$, $\lambda \in \mathbb{C}$, $\beta > 0$, $j, k, u \in \mathbb{N}$, $\alpha_j, \beta_k \in \mathbb{C}$, *and* $A_j, B_k \in \mathbb{R}_+$. *Then we have*

$$\begin{aligned} &\frac{1}{\Gamma(\alpha)} \int_0^2 (2-\tau)^{\alpha} {}_p sin_q \left(\begin{matrix} (\alpha_1, A_1), \cdots, (\alpha_p, A_p) \\ (\beta_1, B_1), \cdots, (\beta_q, B_q) \end{matrix} ; \lambda \tau^{\beta} \right) d\tau \\ &= \frac{1}{\Gamma(\alpha)} \int_{-1}^1 (1-\tau)^{\alpha} {}_p sin_q \left(\begin{matrix} (\alpha_1, A_1), \cdots, (\alpha_p, A_p) \\ (\beta_1, B_1), \cdots, (\beta_q, B_q) \end{matrix} ; \lambda (1+\tau)^{\beta} \right) d\tau \\ &= 2^{\gamma+\alpha-1} {}_{p+1} sin_{q+1} \left(\begin{matrix} (\alpha_1, A_1), \cdots, (\alpha_p, A_p), (1, \beta) \\ (\beta_1, B_1), \cdots, (\beta_q, B_q), (1+\alpha, \beta) \end{matrix} ; \lambda 2^{\beta} \right). \end{aligned} \tag{2.132}$$

Proof Taking $\gamma = 1$ in (2.131), we get the result. □

Theorem 2.39 *Let* $\alpha \in \mathbb{C}$, $Re(\alpha) > 0$, $\lambda \in \mathbb{C}$, $\beta > 0$, $j, k, u \in \mathbb{N}$, $\alpha_j, \beta_k \in \mathbb{C}$, *and* $A_j, B_k \in \mathbb{R}_+$. *Then we have*

$$\begin{aligned} &\frac{1}{\Gamma(\alpha)} \int_0^2 (2-\tau)^{\alpha} {}_p cos_q \left(\begin{matrix} (\alpha_1, A_1), \cdots, (\alpha_p, A_p) \\ (\beta_1, B_1), \cdots, (\beta_q, B_q) \end{matrix} ; \lambda \tau^{\beta} \right) d\tau \\ &= \frac{1}{\Gamma(\alpha)} \int_{-1}^1 (1-\tau)^{\alpha} {}_p cos_q \left(\begin{matrix} (\alpha_1, A_1), \cdots, (\alpha_p, A_p) \\ (\beta_1, B_1), \cdots, (\beta_q, B_q) \end{matrix} ; \lambda (1+\tau)^{\beta} \right) d\tau \\ &= 2^{\gamma+\alpha-1} {}_{p+1} cos_{q+1} \left(\begin{matrix} (\alpha_1, A_1), \cdots, (\alpha_p, A_p), (1, \beta) \\ (\beta_1, B_1), \cdots, (\beta_q, B_q), (1+\alpha, \beta) \end{matrix} ; \lambda 2^{\beta} \right). \end{aligned} \tag{2.133}$$

Proof Taking $\gamma = 1$ in (2.131), we give the result. □

Theorem 2.40 *Let* $\alpha \in \mathbb{C}$, $Re(\alpha) > 0$, $\lambda \in \mathbb{C}$, $\beta > 0$, $j, k, u \in \mathbb{N}$, $\alpha_j, \beta_k \in \mathbb{C}$ *and* $A_j, B_k \in \mathbb{R}_+$. *Then we have*

$$\begin{aligned} &\frac{1}{\Gamma(\alpha)} \int_0^2 (2-\tau)^{\alpha} {}_p sinh_q \left(\begin{matrix} (\alpha_1, A_1), \cdots, (\alpha_p, A_p) \\ (\beta_1, B_1), \cdots, (\beta_q, B_q) \end{matrix} ; \lambda \tau^{\beta} \right) d\tau \\ &= \frac{1}{\Gamma(\alpha)} \int_{-1}^1 (1-\tau)^{\alpha} {}_p sinh_q \left(\begin{matrix} (\alpha_1, A_1), \cdots, (\alpha_p, A_p) \\ (\beta_1, B_1), \cdots, (\beta_q, B_q) \end{matrix} ; \lambda (1+\tau)^{\beta} \right) d\tau \\ &= 2^{\gamma+\alpha-1} {}_{p+1} sinh_{q+1} \left(\begin{matrix} (\alpha_1, A_1), \cdots, (\alpha_p, A_p), (1, \beta) \\ (\beta_1, B_1), \cdots, (\beta_q, B_q), (1+\alpha, \beta) \end{matrix} ; \lambda 2^{\beta} \right). \end{aligned} \tag{2.134}$$

Proof Taking $\gamma = 1$ in (2.131), we have the result. □

Theorem 2.41 *Let $\alpha \in \mathbb{C}$, $Re(\alpha) > 0$, $\lambda \in \mathbb{C}$, $\beta > 0$, $j, k, u \in \mathbb{N}$, $\alpha_j, \beta_k \in \mathbb{C}$ and $A_j, B_k \in \mathbb{R}_+$. Then we have*

$$\begin{aligned} &\frac{1}{\Gamma(\alpha)} \int_0^2 (2-\tau)^\alpha \ {}_p cosh_q \left(\begin{matrix} (\alpha_1, A_1), \cdots, (\alpha_p, A_p) \\ (\beta_1, B_1), \cdots, (\beta_q, B_q) \end{matrix} ; \lambda\tau^\beta \right) d\tau \\ &= \frac{1}{\Gamma(\alpha)} \int_{-1}^1 (1-\tau)^\alpha \ {}_p cosh_q \left(\begin{matrix} (\alpha_1, A_1), \cdots, (\alpha_p, A_p) \\ (\beta_1, B_1), \cdots, (\beta_q, B_q) \end{matrix} ; \lambda (1+\tau)^\beta \right) d\tau \\ &= 2^{\gamma+\alpha-1} \ {}_{p+1} cosh_{q+1} \left(\begin{matrix} (\alpha_1, A_1), \cdots, (\alpha_p, A_p), (1, \beta) \\ (\beta_1, B_1), \cdots, (\beta_q, B_q), (1+\alpha, \beta) \end{matrix} ; \lambda 2^\beta \right). \end{aligned} \tag{2.135}$$

Proof Taking $\gamma = 1$ in (2.131), we give the result. □

Theorem 2.42 *Let $\alpha \in \mathbb{C}$, $Re(\alpha) > 0$, $\lambda \in \mathbb{C}$, $j, k, u \in \mathbb{N}$, $\alpha_j, \beta_k \in \mathbb{C}$ and $A_j, B_k \in \mathbb{R}_+$. Then we have*

$$\begin{aligned} &\frac{1}{\Gamma(\alpha)} \int_0^2 (2-\tau)^\alpha \ {}_p\varphi_q \left(\begin{matrix} (\alpha_1, A_1), \cdots, (\alpha_p, A_p) \\ (\beta_1, B_1), \cdots, (\beta_q, B_q) \end{matrix} ; \lambda\tau \right) d\tau \\ &= \frac{1}{\Gamma(\alpha)} \int_{-1}^1 (1-\tau)^\alpha \ {}_p\varphi_q \left(\begin{matrix} (\alpha_1, A_1), \cdots, (\alpha_p, A_p) \\ (\beta_1, B_1), \cdots, (\beta_q, B_q) \end{matrix} ; \lambda (1+\tau) \right) d\tau \\ &= 2^{\gamma+\alpha-1} \ {}_{p+1}\varphi_{q+1} \left(\begin{matrix} (\alpha_1, A_1), \cdots, (\alpha_p, A_p), (1, 1) \\ (\beta_1, B_1), \cdots, (\beta_q, B_q), (1+\alpha, 1) \end{matrix} ; 2\lambda \right). \end{aligned} \tag{2.136}$$

Proof Taking $\gamma = 1$ and $\gamma = 1$ in (2.126), we obtain the result. □

Theorem 2.43 *Let $\alpha \in \mathbb{C}$, $Re(\alpha) > 0$, $\lambda \in \mathbb{C}$, $j, k, u \in \mathbb{N}$, $\alpha_j, \beta_k \in \mathbb{C}$ and $A_j, B_k \in \mathbb{R}_+$. Then we have*

$$\begin{aligned} &\frac{1}{\Gamma(\alpha)} \int_0^2 (2-\tau)^\alpha \ {}_p sin_q \left(\begin{matrix} (\alpha_1, A_1), \cdots, (\alpha_p, A_p) \\ (\beta_1, B_1), \cdots, (\beta_q, B_q) \end{matrix} ; \lambda\tau^\beta \right) d\tau \\ &= \frac{1}{\Gamma(\alpha)} \int_{-1}^1 (1-\tau)^\alpha \ {}_p sin_q \left(\begin{matrix} (\alpha_1, A_1), \cdots, (\alpha_p, A_p) \\ (\beta_1, B_1), \cdots, (\beta_q, B_q) \end{matrix} ; \lambda (1+\tau)^\beta \right) d\tau \\ &= 2^{\gamma+\alpha-1} \ {}_{p+1} sin_{q+1} \left(\begin{matrix} (\alpha_1, A_1), \cdots, (\alpha_p, A_p), (1, \beta) \\ (\beta_1, B_1), \cdots, (\beta_q, B_q), (1+\alpha, \beta) \end{matrix} ; \lambda 2^\beta \right). \end{aligned} \tag{2.137}$$

Proof Taking $\gamma = 1$ and $\gamma = 1$ in (2.136), we get the result. □

Theorem 2.44 *Let $\alpha \in \mathbb{C}$, $Re(\alpha) > 0$, $\lambda \in \mathbb{C}$, $j, k, u \in \mathbb{N}$, $\alpha_j, \beta_k \in \mathbb{C}$, and $A_j, B_k \in \mathbb{R}_+$. Then we have*

$$\begin{aligned} &\frac{1}{\Gamma(\alpha)} \int_0^2 (2-\tau)^\alpha \ {}_p cos_q \left(\begin{matrix} (\alpha_1, A_1), \cdots, (\alpha_p, A_p) \\ (\beta_1, B_1), \cdots, (\beta_q, B_q) \end{matrix} ; \lambda\tau \right) d\tau \\ &= \frac{1}{\Gamma(\alpha)} \int_{-1}^1 (1-\tau)^\alpha \ {}_p cos_q \left(\begin{matrix} (\alpha_1, A_1), \cdots, (\alpha_p, A_p) \\ (\beta_1, B_1), \cdots, (\beta_q, B_q) \end{matrix} ; \lambda(1+\tau) \right) d\tau \\ &= 2^{\gamma+\alpha-1} \ {}_{p+1} cos_{q+1} \left(\begin{matrix} (\alpha_1, A_1), \cdots, (\alpha_p, A_p), (1,1) \\ (\beta_1, B_1), \cdots, (\beta_q, B_q), (1+\alpha, 1) \end{matrix} ; 2\lambda \right). \end{aligned} \tag{2.138}$$

Proof Taking $\gamma = 1$ and $\gamma = 1$ in (2.136), we give the result. □

Theorem 2.45 *Let $\alpha \in \mathbb{C}$, $Re(\alpha) > 0$, $\lambda \in \mathbb{C}$, $j, k, u \in \mathbb{N}$, $\alpha_j, \beta_k \in \mathbb{C}$, and $A_j, B_k \in \mathbb{R}_+$. Then we have*

$$\begin{aligned} &\frac{1}{\Gamma(\alpha)} \int_0^2 (2-\tau)^\alpha \ {}_p sinh_q \left(\begin{matrix} (\alpha_1, A_1), \cdots, (\alpha_p, A_p) \\ (\beta_1, B_1), \cdots, (\beta_q, B_q) \end{matrix} ; \lambda\tau \right) d\tau \\ &= \frac{1}{\Gamma(\alpha)} \int_{-1}^1 (1-\tau)^\alpha \ {}_p sinh_q \left(\begin{matrix} (\alpha_1, A_1), \cdots, (\alpha_p, A_p) \\ (\beta_1, B_1), \cdots, (\beta_q, B_q) \end{matrix} ; \lambda(1+\tau) \right) d\tau \\ &= 2^{\gamma+\alpha-1} \ {}_{p+1} sinh_{q+1} \left(\begin{matrix} (\alpha_1, A_1), \cdots, (\alpha_p, A_p), (1,1) \\ (\beta_1, B_1), \cdots, (\beta_q, B_q), (1+\alpha, 1) \end{matrix} ; 2\lambda \right). \end{aligned} \tag{2.139}$$

Proof Taking $\gamma = 1$ and $\gamma = 1$ in (2.136), we have the result. □

Theorem 2.46 *Let $\alpha \in \mathbb{C}$, $Re(\alpha) > 0$, $\lambda \in \mathbb{C}$, $j, k, u \in \mathbb{N}$, $\alpha_j, \beta_k \in \mathbb{C}$, and $A_j, B_k \in \mathbb{R}_+$. Then we have*

$$\begin{aligned} &\frac{1}{\Gamma(\alpha)} \int_0^2 (2-\tau)^\alpha \ {}_p cosh_q \left(\begin{matrix} (\alpha_1, A_1), \cdots, (\alpha_p, A_p) \\ (\beta_1, B_1), \cdots, (\beta_q, B_q) \end{matrix} ; \lambda\tau \right) d\tau \\ &= \frac{1}{\Gamma(\alpha)} \int_{-1}^1 (1-\tau)^\alpha \ {}_p cosh_q \left(\begin{matrix} (\alpha_1, A_1), \cdots, (\alpha_p, A_p) \\ (\beta_1, B_1), \cdots, (\beta_q, B_q) \end{matrix} ; \lambda(1+\tau) \right) d\tau \\ &= 2^{\gamma+\alpha-1} \ {}_{p+1} cosh_{q+1} \left(\begin{matrix} (\alpha_1, A_1), \cdots, (\alpha_p, A_p), (1,1) \\ (\beta_1, B_1), \cdots, (\beta_q, B_q), (1+\alpha, 1) \end{matrix} ; 2\lambda \right). \end{aligned} \tag{2.140}$$

Proof Taking $\gamma = 1$ and $\gamma = 1$ in (2.136), we give the result. □

2.1.6 The Supertrigonometric Functions via Wright Function

In this part, we introduce the definitions for the supertrigonometric functions via Wright function.

Definition 2.18 The supersine via Wright function is defined as

$$\begin{aligned}
&{}_0sin_1\left(\begin{matrix} - \\ (B,\beta) \end{matrix};z\right)\\
&=\frac{1}{2i}\left(\psi\left(\beta,B,iz\right)-\psi\left(\beta,B,-iz\right)\right)\\
&=\frac{1}{2i}\left({}_0\varphi_1\left(\begin{matrix} - \\ (B,\beta) \end{matrix};iz\right)-{}_0\varphi_1\left(\begin{matrix} - \\ (B,\beta) \end{matrix};-iz\right)\right)\\
&=\sum_{k=0}^{\infty}\frac{(-1)^k}{\Gamma(\beta+B(2k+1))}\frac{z^{2k+1}}{(2k+1)!}
\end{aligned} \tag{2.141}$$

where $z\in\mathbb{C}$, $\beta\in\mathbb{C}$, and $B\in\mathbb{R}$.

Definition 2.19 The supercosine via Wright function is defined as

$$\begin{aligned}
&{}_0cos_1\left(\begin{matrix} - \\ (B,\beta) \end{matrix};z\right)\\
&=\frac{1}{2}\left(\psi\left(\beta,B,iz\right)+\psi\left(\beta,B,-iz\right)\right)\\
&=\frac{1}{2}\left({}_0\varphi_1\left(\begin{matrix} - \\ (B,\beta) \end{matrix};iz\right)-{}_0\varphi_1\left(\begin{matrix} - \\ (B,\beta) \end{matrix};-iz\right)\right)\\
&=\sum_{k=0}^{\infty}\frac{(-1)^k}{\Gamma(\beta+2kB)}\frac{z^{2k}}{(2k)!}
\end{aligned} \tag{2.142}$$

where $z\in\mathbb{C}$, $\beta\in\mathbb{C}$, and $B\in\mathbb{R}$.

Definition 2.20 The supertangent via Wright function is defined as

$$\begin{aligned}
&{}_0tan_1\left(\begin{matrix} - \\ (B,\beta) \end{matrix};z\right)\\
&=\frac{{}_0sin_1\left(\begin{matrix} - \\ (B,\beta) \end{matrix};z\right)}{{}_0cos_1\left(\begin{matrix} - \\ (B,\beta) \end{matrix};z\right)}\\
&=\frac{1}{i}\frac{\psi(\beta,B,iz)-\psi(\beta,B,-iz)}{\psi(\beta,B,iz)+\psi(\beta,B,-iz)}\\
&=\frac{\sum_{k=0}^{\infty}\frac{(-1)^k}{\Gamma(\beta+(2k+1)B)}\frac{z^{2k+1}}{(2k+1)!}}{\sum_{k=0}^{\infty}\frac{(-1)^k}{\Gamma(\beta+2kB)}\frac{z^{2k}}{(2k)!}},
\end{aligned} \tag{2.143}$$

where $z\in\mathbb{C}$, $\beta\in\mathbb{C}$, $i=\sqrt{-1}$, and $B\in\mathbb{R}$.

Definition 2.21 The supercotangent via Wright function is defined as

$$
\begin{aligned}
&{}_0cot_1\left(\begin{matrix} - \\ (B,\beta)\end{matrix};z\right) \\
&= \frac{{}_0cos_1\left(\begin{matrix} - \\ (B,\beta)\end{matrix};z\right)}{{}_0sin_1\left(\begin{matrix} - \\ (B,\beta)\end{matrix};z\right)} \\
&= i\frac{\psi(\beta,B,iz)+\psi(\beta,B,-iz)}{\psi(\beta,B,iz)-\psi(\beta,B,-iz)} \\
&= \frac{\sum\limits_{k=0}^{\infty}\frac{(-1)^k}{\Gamma(\beta+2kB)}\frac{z^{2k+1}}{(2k)!}}{\sum\limits_{k=0}^{\infty}\frac{(-1)^k}{\Gamma(\beta+(2k+1)B)}\frac{z^{2k+1}}{(2k+1)!}},
\end{aligned}
\tag{2.144}
$$

where $z \in \mathbb{C}$, $\beta \in \mathbb{C}$, $i = \sqrt{-1}$, and $B \in \mathbb{R}$.

Definition 2.22 The supersecant via Wright function is defined as

$$
\begin{aligned}
&{}_0sec_1\left(\begin{matrix} - \\ (B,\beta)\end{matrix};z\right) \\
&= \frac{1}{{}_0cos_1\left(\begin{matrix} - \\ (B,\beta)\end{matrix};z\right)} \\
&= \frac{2}{\psi(\beta,B,iz)+\psi(\beta,B,-iz)},
\end{aligned}
\tag{2.145}
$$

where $z \in \mathbb{C}$, $\beta \in \mathbb{C}$, $i = \sqrt{-1}$, and $B \in \mathbb{R}$.

Definition 2.23 The supercosecant via Wright function is defined as

$$
\begin{aligned}
&{}_0cosec_1\left(\begin{matrix} - \\ (B,\beta)\end{matrix};z\right) \\
&= \frac{1}{{}_0sin_1\left(\begin{matrix} - \\ (B,\beta)\end{matrix};z\right)} \\
&= \frac{2i}{\psi(\beta,B,iz)-\psi(\beta,B,-iz)},
\end{aligned}
\tag{2.146}
$$

where $z \in \mathbb{C}$, $\beta \in \mathbb{C}$, $i = \sqrt{-1}$, and $B \in \mathbb{R}$.

Theorem 2.47 *Let $z \in \mathbb{C}$, $\beta \in \mathbb{C}$, $i = \sqrt{-1}$, and $B \in \mathbb{R}$. Then we have*

$$
\psi(\beta,B,iz) = {}_0cos_1\left(\begin{matrix} - \\ (B,\beta)\end{matrix};z\right) + i\,{}_0sin_1\left(\begin{matrix} - \\ (B,\beta)\end{matrix};z\right)
\tag{2.147}
$$

and

$$\psi(\beta, B, -iz) = {}_0cos_1\left(\begin{matrix} - \\ (B,\beta) \end{matrix}; z\right) - i\; {}_0sin_1\left(\begin{matrix} - \\ (B,\beta) \end{matrix}; z\right). \tag{2.148}$$

Proof Making use of the expression

$$\psi(\beta, B, iz) = {}_0\varphi_1\left(\begin{matrix} - \\ (B,\beta) \end{matrix}; iz\right) = \sum_{k=0}^{\infty} \frac{1}{\Gamma(\beta + Bk)} \frac{(iz)^k}{k!} \tag{2.149}$$

and

$$\psi(\beta, B, -iz) = {}_0\varphi_1\left(\begin{matrix} - \\ (B,\beta) \end{matrix}; -iz\right) = \sum_{k=0}^{\infty} \frac{1}{\Gamma(\beta + Bk)} \frac{(-iz)^k}{k!}, \tag{2.150}$$

where $z, \lambda \in \mathbb{C}$, $\beta \in \mathbb{C}$, and $B \in \mathbb{R}$, we obtain the desired results. □

More generally, if $z \in \mathbb{C}$, $\beta \in \mathbb{C}$, $i = \sqrt{-1}$, and $B \in \mathbb{R}$, then we have

$$\psi(\beta, B, i\lambda z) = {}_0cos_1\left(\begin{matrix} - \\ (B,\beta) \end{matrix}; \lambda z\right) + i\; {}_0sin_1\left(\begin{matrix} - \\ (B,\beta) \end{matrix}; \lambda z\right) \tag{2.151}$$

and

$$\psi(\beta, B, -i\lambda z) = {}_0cos_1\left(\begin{matrix} - \\ (B,\beta) \end{matrix}; \lambda z\right) - i\; {}_0sin_1\left(\begin{matrix} - \\ (B,\beta) \end{matrix}; \lambda z\right), \tag{2.152}$$

where

$$\psi(\beta, B, i\lambda z) = {}_0\varphi_1\left(\begin{matrix} - \\ (B,\beta) \end{matrix}; i\lambda z\right) = \sum_{k=0}^{\infty} \frac{1}{\Gamma(\beta + Bk)} \frac{(i\lambda z)^k}{k!} \tag{2.153}$$

and

$$\psi(\beta, B, -i\lambda z) = {}_0\varphi_1\left(\begin{matrix} - \\ (B,\beta) \end{matrix}; -i\lambda z\right) = \sum_{k=0}^{\infty} \frac{1}{\Gamma(\beta + Bk)} \frac{(-i\lambda z)^k}{k!}. \tag{2.154}$$

Theorem 2.48 *Let $\alpha, \gamma, \lambda, \varsigma \in \mathbb{C}$, $Re(\alpha) > 0$, $Re(\gamma) > 0$, $\beta > 0$, and $B > -1$. Then we have*

$$\begin{aligned} &\tfrac{1}{\Gamma(\alpha)} \int_0^t (t-\tau)^{\alpha} \tau^{\gamma-1}\, {}_0sin_1\left(\begin{matrix} - \\ (B,\varsigma) \end{matrix}; \lambda\tau^{\beta}\right) d\tau \\ &= t^{\gamma+\alpha-1}\, {}_1sin_2\left(\begin{matrix} (\gamma,\beta) \\ (\varsigma, B), (\gamma+\alpha,\beta) \end{matrix}; \lambda t^{\beta}\right) \end{aligned} \tag{2.155}$$

and

$$\begin{aligned}&\frac{1}{\Gamma(\alpha)}\int_0^t (t-\tau)^{\alpha}\,\tau^{\gamma-1}\,{}_0cos_1\left(\begin{matrix}-\\(B,\varsigma)\end{matrix};\lambda\tau^{\beta}\right)d\tau\\&=t^{\gamma+\alpha-1}\,{}_1cos_2\left(\begin{matrix}(\gamma,\beta)\\(\varsigma,B),(\gamma+\alpha,\beta)\end{matrix};\lambda t^{\beta}\right).\end{aligned}\tag{2.156}$$

Proof Making use of the expression

$$\begin{aligned}&\frac{1}{\Gamma(\alpha)}\int_0^t (t-\tau)^{\alpha}\,\tau^{\gamma-1}\psi\left(\varsigma,B,i\lambda\tau^{\beta}\right)d\tau\\&=t^{\gamma+\alpha-1}\,{}_1\varphi_2\left(\begin{matrix}(\gamma,\beta)\\(\varsigma,B),(\gamma+\alpha,\beta)\end{matrix};i\lambda t^{\beta}\right)\end{aligned}\tag{2.157}$$

and

$$\begin{aligned}&\frac{1}{\Gamma(\alpha)}\int_0^t (t-\tau)^{\alpha}\,\tau^{\gamma-1}\psi\left(\varsigma,B,-i\lambda\tau^{\beta}\right)d\tau\\&=t^{\gamma+\alpha-1}\,{}_1\varphi_2\left(\begin{matrix}(\gamma,\beta)\\(\varsigma,B),(\gamma+\alpha,\beta)\end{matrix};-i\lambda t^{\beta}\right)\end{aligned}\tag{2.158}$$

we have that

$$\begin{aligned}&\frac{1}{\Gamma(\alpha)}\int_0^t (t-\tau)^{\alpha}\,\tau^{\gamma-1}\,{}_0sin_1\left(\begin{matrix}-\\(B,\varsigma)\end{matrix};\lambda\tau^{\beta}\right)d\tau\\&=t^{\gamma+\alpha-1}\,{}_1sin_2\left(\begin{matrix}(\gamma,\beta)\\(\varsigma,B),(\gamma+\alpha,\beta)\end{matrix};\lambda t^{\beta}\right)\end{aligned}\tag{2.159}$$

and

$$\begin{aligned}&\frac{1}{\Gamma(\alpha)}\int_0^t (t-\tau)^{\alpha}\,\tau^{\gamma-1}\,{}_0cos_1\left(\begin{matrix}-\\(B,\varsigma)\end{matrix};\lambda\tau^{\beta}\right)d\tau\\&=t^{\gamma+\alpha-1}\,{}_1cos_2\left(\begin{matrix}(\gamma,\beta)\\(\varsigma,B),(\gamma+\alpha,\beta)\end{matrix};\lambda t^{\beta}\right),\end{aligned}\tag{2.160}$$

with the aid of the given conditions. □

Corollary 2.13 *Let $\gamma,\lambda,\varsigma\in\mathbb{C}$, $Re(\gamma)>0$, and $B>-1$. Then we have*

$$\begin{aligned}&\frac{1}{\Gamma(\alpha)}\int_0^t (t-\tau)^{\alpha}\,\tau^{\gamma-1}\,{}_0sin_1\left(\begin{matrix}-\\(B,\varsigma)\end{matrix};\lambda\tau\right)d\tau\\&=t^{\gamma+\alpha-1}\,{}_1sin_2\left(\begin{matrix}(\gamma,1)\\(\varsigma,B),(\gamma+\alpha,1)\end{matrix};\lambda t\right)\end{aligned}\tag{2.161}$$

and

$$\begin{aligned}&\frac{1}{\Gamma(\alpha)} y_0^t(t-\tau)^{\alpha}\,\tau^{\gamma-1}\,{}_0cos_1\left(\begin{matrix}-\\(B,\varsigma)\end{matrix};\lambda\tau\right)d\tau\\&=t^{\gamma+\alpha-1}\,{}_1cos_2\left(\begin{matrix}(\gamma,1)\\(\varsigma,B),(\gamma+\alpha,1)\end{matrix};\lambda t\right).\end{aligned}\tag{2.162}$$

Proof Taking $\beta=1$ in (2.161) and (2.162), we obtain the results. □

Corollary 2.14 *Let $\alpha,\lambda,\varsigma\in\mathbb{C}$, $Re\,(\alpha)>0$, $\beta>0$, and $B>-1$. Then we have*

$$\begin{aligned}&\frac{1}{\Gamma(\alpha)}\int_0^t(t-\tau)^{\alpha}\,{}_0sin_1\left(\begin{matrix}-\\(B,\varsigma)\end{matrix};\lambda\tau^{\beta}\right)d\tau\\&=t^{\alpha}\,{}_1sin_2\left(\begin{matrix}(1,\beta)\\(\varsigma,B),(1+\alpha,\beta)\end{matrix};\lambda t^{\beta}\right)\end{aligned}\tag{2.163}$$

and

$$\begin{aligned}&\frac{1}{\Gamma(\alpha)}\int_0^t(t-\tau)^{\alpha}\,{}_0cos_1\left(\begin{matrix}-\\(B,\varsigma)\end{matrix};\lambda\tau^{\beta}\right)d\tau\\&=t^{\alpha}\,{}_1cos_2\left(\begin{matrix}(1,\beta)\\(\varsigma,B),(1+\alpha,\beta)\end{matrix};\lambda t^{\beta}\right).\end{aligned}\tag{2.164}$$

Proof Taking $\gamma=1$ in (2.163) and (2.164), we obtain the results. □

Theorem 2.49 *Let $\alpha,\gamma,\lambda,\varsigma\in\mathbb{C}$, $Re\,(\alpha)>0$, $Re\,(\gamma)>0$, $\beta>0$, and $B>-1$. Then we have*

$$\begin{aligned}&\frac{1}{\Gamma(\alpha)}\int_0^1(1-\tau)^{\alpha}\,\tau^{\gamma-1}\,{}_0sin_1\left(\begin{matrix}-\\(B,\varsigma)\end{matrix};\lambda\tau^{\beta}\right)d\tau\\&={}_1sin_2\left(\begin{matrix}(\gamma,\beta)\\(\varsigma,B),(\gamma+\alpha,\beta)\end{matrix};\lambda\right)\end{aligned}\tag{2.165}$$

and

$$\begin{aligned}&\frac{1}{\Gamma(\alpha)}\int_0^1(1-\tau)^{\alpha}\,\tau^{\gamma-1}\,{}_0cos_1\left(\begin{matrix}-\\(B,\varsigma)\end{matrix};\lambda\tau^{\beta}\right)d\tau\\&={}_1cos_2\left(\begin{matrix}(\gamma,\beta)\\(\varsigma,B),(\gamma+\alpha,\beta)\end{matrix};\lambda\right).\end{aligned}\tag{2.166}$$

Proof Taking $t=1$ in (2.165) and (2.166), we obtain the results. □

Corollary 2.15 *Let $\gamma,\lambda,\varsigma\in\mathbb{C}$, $Re\,(\gamma)>0$, $\beta>0$, and $B>-1$. Then we have*

$$\begin{aligned}&\frac{1}{\Gamma(\alpha)}\int_0^1(1-\tau)^{\alpha}\,\tau^{\gamma-1}\,{}_0sin_1\left(\begin{matrix}-\\(B,\varsigma)\end{matrix};\lambda\tau\right)d\tau\\&={}_1sin_2\left(\begin{matrix}(\gamma,1)\\(\varsigma,B),(\gamma+\alpha,1)\end{matrix};\lambda\right)\end{aligned}\tag{2.167}$$

and

$$\begin{aligned}&\frac{1}{\Gamma(\alpha)}\int_0^1(1-\tau)^{\alpha}\,\tau^{\gamma-1}\,{}_0cos_1\left(\begin{matrix}-\\(B,\varsigma)\end{matrix};\lambda\tau\right)d\tau\\&={}_1cos_2\left(\begin{matrix}(\gamma,1)\\(\varsigma,B),(\gamma+\alpha,1)\end{matrix};\lambda\right).\end{aligned}\tag{2.168}$$

Proof Taking $t=1$ and $\beta=1$ in (2.167) and (2.168), we obtain the results. □

Corollary 2.16 *Let* $\alpha,\lambda,\varsigma\in\mathbb{C}$, $Re(\alpha)>0$, $\beta>0$, *and* $B>-1$. *Then we have*

$$\begin{aligned}&\frac{1}{\Gamma(\alpha)}\int_0^1(1-\tau)^{\alpha}\,{}_0sin_1\left(\begin{matrix}-\\(B,\varsigma)\end{matrix};\lambda\tau^{\beta}\right)d\tau\\&={}_1sin_2\left(\begin{matrix}(1,\beta)\\(\varsigma,B),(1+\alpha,\beta)\end{matrix};\lambda\right)\end{aligned}\tag{2.169}$$

and

$$\begin{aligned}&\frac{1}{\Gamma(\alpha)}\int_0^1(1-\tau)^{\alpha}\,{}_0cos_1\left(\begin{matrix}-\\(B,\varsigma)\end{matrix};\lambda\tau^{\beta}\right)d\tau\\&={}_1cos_2\left(\begin{matrix}(1,\beta)\\(\varsigma,B),(1+\alpha,\beta)\end{matrix};\lambda\right).\end{aligned}\tag{2.170}$$

Proof Taking $t=1$ and $\gamma=1$ in (2.169) and (2.170), we obtain the results. □

2.1.7 The Superhyperbolic Functions via Wright Function

In this part, we introduce the definitions for the superhyperbolic functions via Wright function.

Definition 2.24 The superhyperbolic sine via Wright function is defined as

$$\begin{aligned}&{}_0sinh_1\left(\begin{matrix}-\\(B,\beta)\end{matrix};z\right)\\&=\frac{1}{2}\left(\psi(\beta,B,z)-\psi(\beta,B,-z)\right)\\&=\frac{1}{2}\left({}_0\varphi_1\left(\begin{matrix}-\\(B,\beta)\end{matrix};z\right)-{}_0\varphi_1\left(\begin{matrix}-\\(B,\beta)\end{matrix};-z\right)\right)\\&=\sum_{k=0}^{\infty}\frac{1}{\Gamma(\beta+B(2k+1))}\frac{z^{2k+1}}{(2k+1)!},\end{aligned}\tag{2.171}$$

where $z\in\mathbb{C}$, $\beta\in\mathbb{C}$, and $B\in\mathbb{R}$.

Definition 2.25 The superhyperbolic cosine via Wright function is defined as

$$
\begin{aligned}
&{}_0cosh_1\left(\begin{matrix}-\\(B,\beta)\end{matrix};z\right)\\
&=\frac{1}{2}\left(\psi\left(\beta,B,z\right)+\psi\left(\beta,B,-z\right)\right)\\
&=\frac{1}{2}\left({}_0\varphi_1\left(\begin{matrix}-\\(B,\beta)\end{matrix};z\right)-{}_0\varphi_1\left(\begin{matrix}-\\(B,\beta)\end{matrix};-z\right)\right)\\
&=\sum_{k=0}^{\infty}\frac{1}{\Gamma(\beta+2kB)}\frac{z^{2k}}{(2k)!}
\end{aligned}
\tag{2.172}
$$

where $z\in\mathbb{C}$, $\beta\in\mathbb{C}$, and $B\in\mathbb{R}$.

Definition 2.26 The superhyperbolic tangent via Wright function is defined as

$$
\begin{aligned}
&{}_0tanh_1\left(\begin{matrix}-\\(B,\beta)\end{matrix};z\right)\\
&=\frac{{}_0sinh_1\left(\begin{matrix}-\\(B,\beta)\end{matrix};z\right)}{{}_0cosh_1\left(\begin{matrix}-\\(B,\beta)\end{matrix};z\right)}\\
&=\frac{\psi(\beta,B,z)-\psi(\beta,B,-z)}{\psi(\beta,B,z)+\psi(\beta,B,-z)}\\
&=\frac{\sum_{k=0}^{\infty}\frac{1}{\Gamma(\beta+(2k+1)B)}\frac{z^{2k+1}}{(2k+1)!}}{\sum_{k=0}^{\infty}\frac{1}{\Gamma(\beta+2kB)}\frac{z^{2k}}{(2k)!}},
\end{aligned}
\tag{2.173}
$$

where $z\in\mathbb{C}$, $\beta\in\mathbb{C}$, and $B\in\mathbb{R}$.

Definition 2.27 The superhyperbolic cotangent via Wright function is defined as

$$
\begin{aligned}
&{}_0coth_1\left(\begin{matrix}-\\(B,\beta)\end{matrix};z\right)\\
&=\frac{{}_0cosh_1\left(\begin{matrix}-\\(B,\beta)\end{matrix};z\right)}{{}_0sinh_1\left(\begin{matrix}-\\(B,\beta)\end{matrix};z\right)}\\
&=\frac{\psi(\beta,B,z)+\psi(\beta,B,-z)}{\psi(\beta,B,z)-\psi(\beta,B,-z)}\\
&=\frac{\sum_{k=0}^{\infty}\frac{1}{\Gamma(\beta+2kB)}\frac{z^{2k+1}}{(2k)!}}{\sum_{k=0}^{\infty}\frac{1}{\Gamma(\beta+(2k+1)B)}\frac{z^{2k+1}}{(2k+1)!}},
\end{aligned}
\tag{2.174}
$$

where $z\in\mathbb{C}$, $\beta\in\mathbb{C}$, and $B\in\mathbb{R}$.

Definition 2.28 The superhyperbolic secant via Wright function is defined as

$$
\begin{aligned}
&{}_0sech_1\left(\begin{matrix}-\\(B,\beta)\end{matrix};z\right)\\
&=\frac{1}{{}_0cosh_1\left(\begin{matrix}-\\(B,\beta)\end{matrix};z\right)}\\
&=\frac{2}{\psi(\beta,B,iz)+\psi(\beta,B,-iz)},
\end{aligned}
\tag{2.175}
$$

where $z \in \mathbb{C}$, $\beta \in \mathbb{C}$, and $B \in \mathbb{R}$.

Definition 2.29 The superhyperbolic cosecant via Wright function is defined as

$$
\begin{aligned}
&{}_0cosech_1\left(\begin{matrix}-\\(B,\beta)\end{matrix};z\right)\\
&=\frac{1}{{}_0sinh_1\left(\begin{matrix}-\\(B,\beta)\end{matrix};z\right)}\\
&=\frac{2i}{\psi(\beta,B,iz)-\psi(\beta,B,-iz)},
\end{aligned}
\tag{2.176}
$$

where $z \in \mathbb{C}$, $\beta \in \mathbb{C}$, and $B \in \mathbb{R}$.

Theorem 2.50 *Let $z \in \mathbb{C}$, $\beta \in \mathbb{C}$, and $B \in \mathbb{R}$. Then we have*

$$
\psi(\beta, B, z) = {}_0cosh_1\left(\begin{matrix}-\\(B,\beta)\end{matrix};z\right) + {}_0sinh_1\left(\begin{matrix}-\\(B,\beta)\end{matrix};z\right)
\tag{2.177}
$$

and

$$
\psi(\beta, B, -z) = {}_0cosh_1\left(\begin{matrix}-\\(B,\beta)\end{matrix};z\right) - {}_0sinh_1\left(\begin{matrix}-\\(B,\beta)\end{matrix};z\right).
\tag{2.178}
$$

Proof Setting

$$
\psi(\beta, B, z) = {}_0\varphi_1\left(\begin{matrix}-\\(B,\beta)\end{matrix};z\right) = \sum_{k=0}^{\infty}\frac{1}{\Gamma(\beta+Bk)}\frac{z^k}{k!}
\tag{2.179}
$$

and

$$
\psi(\beta, B, -z) = {}_0\varphi_1\left(\begin{matrix}-\\(B,\beta)\end{matrix};-z\right) = \sum_{k=0}^{\infty}\frac{1}{\Gamma(\beta+Bk)}\frac{(-z)^k}{k!},
\tag{2.180}
$$

where $z, \lambda \in \mathbb{C}$, $\beta \in \mathbb{C}$, and $B \in \mathbb{R}$, we obtain the desired results. □

For $z \in \mathbb{C}$, $\beta \in \mathbb{C}$, and $B \in \mathbb{R}$ we have

$$\psi\left(\beta, B, \lambda z\right) = {}_0cos_1\left(\begin{matrix} - \\ (B, \beta) \end{matrix}; \lambda z\right) + {}_0sin_1\left(\begin{matrix} - \\ (B, \beta) \end{matrix}; \lambda z\right) \tag{2.181}$$

and

$$\psi\left(\beta, B, -\lambda z\right) = {}_0cos_1\left(\begin{matrix} - \\ (B, \beta) \end{matrix}; \lambda z\right) - {}_0sin_1\left(\begin{matrix} - \\ (B, \beta) \end{matrix}; \lambda z\right), \tag{2.182}$$

where

$$\psi\left(\beta, B, \lambda z\right) = {}_0\varphi_1\left(\begin{matrix} - \\ (B, \beta) \end{matrix}; \lambda z\right) = \sum_{k=0}^{\infty} \frac{1}{\Gamma\left(\beta + Bk\right)} \frac{(\lambda z)^k}{k!} \tag{2.183}$$

and

$$\psi\left(\beta, B, -\lambda z\right) = {}_0\varphi_1\left(\begin{matrix} - \\ (B, \beta) \end{matrix}; -\lambda z\right) = \sum_{k=0}^{\infty} \frac{1}{\Gamma\left(\beta + Bk\right)} \frac{(-\lambda z)^k}{k!}. \tag{2.184}$$

Theorem 2.51 *Let $\alpha, \gamma, \lambda, \varsigma \in \mathbb{C}$, $Re\left(\alpha\right) > 0$, $Re\left(\gamma\right) > 0$, $\beta > 0$, and $B > -1$. Then we have*

$$\begin{aligned} &\tfrac{1}{\Gamma(\alpha)} \int_0^t \left(t - \tau\right)^{\alpha} \tau^{\gamma-1} {}_0sinh_1\left(\begin{matrix} - \\ (B, \varsigma) \end{matrix}; \lambda \tau^{\beta}\right) d\tau \\ &= t^{\gamma+\alpha-1} {}_1sinh_2\left(\begin{matrix} (\gamma, \beta) \\ (\varsigma, B), (\gamma + \alpha, \beta) \end{matrix}; \lambda t^{\beta}\right) \end{aligned} \tag{2.185}$$

and

$$\begin{aligned} &\tfrac{1}{\Gamma(\alpha)} \int_0^t \left(t - \tau\right)^{\alpha} \tau^{\gamma-1} {}_0cosh_1\left(\begin{matrix} - \\ (B, \varsigma) \end{matrix}; \lambda \tau^{\beta}\right) d\tau \\ &= t^{\gamma+\alpha-1} {}_1cosh_2\left(\begin{matrix} (\gamma, \beta) \\ (\varsigma, B), (\gamma + \alpha, \beta) \end{matrix}; \lambda t^{\beta}\right). \end{aligned} \tag{2.186}$$

Proof With use of the expressions

$$\begin{aligned} &\tfrac{1}{\Gamma(\alpha)} y_0^t \left(t - \tau\right)^{\alpha} \tau^{\gamma-1} \psi\left(\varsigma, B, \lambda \tau^{\beta}\right) d\tau \\ &= t^{\gamma+\alpha-1} {}_1\varphi_2\left(\begin{matrix} (\gamma, \beta) \\ (\varsigma, B), (\gamma + \alpha, \beta) \end{matrix}; \lambda t^{\beta}\right) \end{aligned} \tag{2.187}$$

and

$$\frac{1}{\Gamma(\alpha)} \int_0^t (t-\tau)^{\alpha} \tau^{\gamma-1} \psi \left(\varsigma, B, -\lambda \tau^{\beta}\right) d\tau = t^{\gamma+\alpha-1} {}_1\varphi_2 \left(\begin{matrix} (\gamma, \beta) \\ (\varsigma, B), (\gamma+\alpha, \beta) \end{matrix} ; -\lambda t^{\beta} \right) \tag{2.188}$$

we have that

$$\frac{1}{\Gamma(\alpha)} \int_0^t (t-\tau)^{\alpha} \tau^{\gamma-1} {}_0sinh_1 \left(\begin{matrix} - \\ (B, \varsigma) \end{matrix} ; \lambda \tau^{\beta} \right) d\tau = t^{\gamma+\alpha-1} {}_1sinh_2 \left(\begin{matrix} (\gamma, \beta) \\ (\varsigma, B), (\gamma+\alpha, \beta) \end{matrix} ; \lambda t^{\beta} \right) \tag{2.189}$$

and

$$\frac{1}{\Gamma(\alpha)} y_0^t (t-\tau)^{\alpha} \tau^{\gamma-1} {}_0cosh_1 \left(\begin{matrix} - \\ (B, \varsigma) \end{matrix} ; \lambda \tau^{\beta} \right) d\tau = t^{\gamma+\alpha-1} {}_1cosh_2 \left(\begin{matrix} (\gamma, \beta) \\ (\varsigma, B), (\gamma+\alpha, \beta) \end{matrix} ; \lambda t^{\beta} \right) \tag{2.190}$$

with the aid of the given conditions. □

Corollary 2.17 *Let $\gamma, \lambda, \varsigma \in \mathbb{C}$, $Re(\gamma) > 0$, and $B > -1$. Then we have*

$$\frac{1}{\Gamma(\alpha)} \int_0^t (t-\tau)^{\alpha} \tau^{\gamma-1} {}_0sinh_1 \left(\begin{matrix} - \\ (B, \varsigma) \end{matrix} ; \lambda \tau \right) d\tau = t^{\gamma+\alpha-1} {}_1sinh_2 \left(\begin{matrix} (\gamma, 1) \\ (\varsigma, B), (\gamma+\alpha, 1) \end{matrix} ; \lambda t \right) \tag{2.191}$$

and

$$\frac{1}{\Gamma(\alpha)} \int_0^t (t-\tau)^{\alpha} \tau^{\gamma-1} {}_0cosh_1 \left(\begin{matrix} - \\ (B, \varsigma) \end{matrix} ; \lambda \tau \right) d\tau = t^{\gamma+\alpha-1} {}_1cosh_2 \left(\begin{matrix} (\gamma, 1) \\ (\varsigma, B), (\gamma+\alpha, 1) \end{matrix} ; \lambda t \right). \tag{2.192}$$

Proof Taking $\beta = 1$ in (2.191) and (2.192), we obtain the results. □

Corollary 2.18 *Let $\alpha, \lambda, \varsigma \in \mathbb{C}$, $Re(\alpha) > 0$, $\beta > 0$, and $B > -1$. Then we have*

$$\frac{1}{\Gamma(\alpha)} \int_0^t (t-\tau)^{\alpha} {}_0sinh_1 \left(\begin{matrix} - \\ (B, \varsigma) \end{matrix} ; \lambda \tau^{\beta} \right) d\tau = t^{\alpha} {}_1sinh_2 \left(\begin{matrix} (1, \beta) \\ (\varsigma, B), (1+\alpha, \beta) \end{matrix} ; \lambda t^{\beta} \right) \tag{2.193}$$

and

$$\begin{aligned}&\tfrac{1}{\Gamma(\alpha)}\int_0^t (t-\tau)^{\alpha}\ {}_0cosh_1\left(\begin{matrix}-\\(B,\varsigma)\end{matrix};\lambda\tau^{\beta}\right)d\tau\\ &\quad = t^{\alpha}\ {}_1cosh_2\left(\begin{matrix}(1,\beta)\\(\varsigma,B),(1+\alpha,\beta)\end{matrix};\lambda t^{\beta}\right).\end{aligned} \tag{2.194}$$

Proof Taking $\gamma=1$ in (2.193) and (2.194), we obtain the results. □

Theorem 2.52 *Let* $\alpha,\gamma,\lambda,\varsigma\in\mathbb{C}$, $Re(\alpha)>0$, $Re(\gamma)>0$, $\beta>0$, *and* $B>-1$. *Then we have*

$$\begin{aligned}&\tfrac{1}{\Gamma(\alpha)}\int_0^1 (1-\tau)^{\alpha}\tau^{\gamma-1}\ {}_0sinh_1\left(\begin{matrix}-\\(B,\varsigma)\end{matrix};\lambda\tau^{\beta}\right)d\tau\\ &\quad = {}_1sinh_2\left(\begin{matrix}(\gamma,\beta)\\(\varsigma,B),(\gamma+\alpha,\beta)\end{matrix};\lambda\right)\end{aligned} \tag{2.195}$$

and

$$\begin{aligned}&\tfrac{1}{\Gamma(\alpha)}\int_0^1 (1-\tau)^{\alpha}\tau^{\gamma-1}\ {}_0cosh_1\left(\begin{matrix}-\\(B,\varsigma)\end{matrix};\lambda\tau^{\beta}\right)d\tau\\ &\quad = {}_1cosh_2\left(\begin{matrix}(\gamma,\beta)\\(\varsigma,B),(\gamma+\alpha,\beta)\end{matrix};\lambda\right).\end{aligned} \tag{2.196}$$

Proof Taking $t=1$ in (2.195) and (2.196), we obtain the results. □

Corollary 2.19 *Let* $\gamma,\lambda,\varsigma\in\mathbb{C}$, $Re(\gamma)>0$, $\beta>0$, *and* $B>-1$. *Then we have*

$$\begin{aligned}&\tfrac{1}{\Gamma(\alpha)}\int_0^1 (1-\tau)^{\alpha}\tau^{\gamma-1}\ {}_0sinh_1\left(\begin{matrix}-\\(B,\varsigma)\end{matrix};\lambda\tau\right)d\tau\\ &\quad = {}_1sinh_2\left(\begin{matrix}(\gamma,1)\\(\varsigma,B),(\gamma+\alpha,1)\end{matrix};\lambda\right)\end{aligned} \tag{2.197}$$

and

$$\begin{aligned}&\tfrac{1}{\Gamma(\alpha)}\int_0^1 (1-\tau)^{\alpha}\tau^{\gamma-1}\ {}_0cosh_1\left(\begin{matrix}-\\(B,\varsigma)\end{matrix};\lambda\tau\right)d\tau\\ &\quad = {}_1cosh_2\left(\begin{matrix}(\gamma,1)\\(\varsigma,B),(\gamma+\alpha,1)\end{matrix};\lambda\right).\end{aligned} \tag{2.198}$$

Proof Taking $t=1$ and $\beta=1$ in (2.197) and (2.198), we obtain the results. □

Corollary 2.20 *Let $\alpha, \lambda, \varsigma \in \mathbb{C}$, $Re(\alpha) > 0$, $\beta > 0$, and $B > -1$. Then we have*

$$\begin{aligned}&\tfrac{1}{\Gamma(\alpha)}\int_0^1 (1-\tau)^{\alpha}\ {}_0sinh_1\left(\begin{matrix}-\\(B,\varsigma)\end{matrix};\lambda\tau^{\beta}\right)d\tau\\&= {}_1sinh_2\left(\begin{matrix}(1,\beta)\\(\varsigma,B),(1+\alpha,\beta)\end{matrix};\lambda\right)\end{aligned} \tag{2.199}$$

and

$$\begin{aligned}&\tfrac{1}{\Gamma(\alpha)}\int_0^1 (1-\tau)^{\alpha}\ {}_0cosh_1\left(\begin{matrix}-\\(B,\varsigma)\end{matrix};\lambda\tau^{\beta}\right)d\tau\\&= {}_1cosh_2\left(\begin{matrix}(1,\beta)\\(\varsigma,B),(1+\alpha,\beta)\end{matrix};\lambda\right).\end{aligned} \tag{2.200}$$

Proof Taking $t = 1$ and $\gamma = 1$ in (2.199) and (2.200), we obtain the results. □

Theorem 2.53 *The Laplace transform of the function $t^{\gamma-1}\psi(\beta, B, \lambda t) = t^{\gamma-1}\ {}_0\varphi_1\left(\begin{matrix}-\\(B,\varsigma)\end{matrix};\lambda t\right)$ is given as*

$$\begin{aligned}&L\left[t^{\gamma-1}\ {}_0\varphi_1\left(\begin{matrix}-\\(B,\varsigma)\end{matrix};\lambda t\right)\right]\\&= \int_0^{\infty} e^{-st}t^{\gamma-1}\ {}_0\varphi_1\left(\begin{matrix}-\\(B,\varsigma)\end{matrix};\lambda t\right)dt\\&= s^{-\gamma}\ {}_0\varphi_1\left(\begin{matrix}(\gamma,1)\\(B,\varsigma)\end{matrix};\tfrac{\lambda}{s}\right),\end{aligned} \tag{2.201}$$

where $\gamma, \lambda, \varsigma \in \mathbb{C}$, $Re(\gamma) > 0$, $\beta > 0$, and $B > -1$.

Proof The result is the special case of the expression (2.201). □

Theorem 2.54 *The Laplace transforms of the functions*

$t^{\gamma-1}\ {}_0sin_1\left(\begin{matrix}-\\(B,\varsigma)\end{matrix};\lambda t\right)$, $t^{\gamma-1}\ {}_0cos_1\left(\begin{matrix}-\\(B,\varsigma)\end{matrix};\lambda t\right)$, $t^{\gamma-1}\ {}_0sin_1\left(\begin{matrix}-\\(B,\varsigma)\end{matrix};\lambda t\right)$,

and $t^{\gamma-1}\ {}_0cos_1\left(\begin{matrix}-\\(B,\varsigma)\end{matrix};\lambda t\right)$ are given as follows:

$$\begin{aligned}&L\left[t^{\gamma-1}\ {}_0sin_1\left(\begin{matrix}-\\(B,\varsigma)\end{matrix};\lambda t\right)\right]\\&= \int_0^{\infty} e^{-st}t^{\gamma-1}\ {}_0sin_1\left(\begin{matrix}-\\(B,\varsigma)\end{matrix};\lambda t\right)dt\\&= s^{-\gamma}\ {}_1sin_1\left(\begin{matrix}(\gamma,1)\\(B,\varsigma)\end{matrix};\tfrac{\lambda}{s}\right),\end{aligned} \tag{2.202}$$

$$
\begin{aligned}
&L\left[t^{\gamma-1}\,{}_0cos_1\left(\begin{matrix} - \\ (B,\varsigma)\end{matrix};\lambda t\right)\right] \\
&= \int_0^\infty e^{-st}t^{\gamma-1}\,{}_0cos_1\left(\begin{matrix} - \\ (B,\varsigma)\end{matrix};\lambda t\right)dt \\
&= s^{-\gamma}\,{}_1cos_1\left(\begin{matrix} (\gamma,1) \\ (B,\varsigma)\end{matrix};\tfrac{\lambda}{s}\right),
\end{aligned}
\tag{2.203}
$$

$$
\begin{aligned}
&L\left[t^{\gamma-1}\,{}_0sinh_1\left(\begin{matrix} - \\ (B,\varsigma)\end{matrix};\lambda t\right)\right] \\
&= \int_0^\infty e^{-st}t^{\gamma-1}\,{}_0sinh_1\left(\begin{matrix} - \\ (B,\varsigma)\end{matrix};\lambda t\right)dt \\
&= s^{-\gamma}\,{}_1sinh_1\left(\begin{matrix} (\gamma,1) \\ (B,\varsigma)\end{matrix};\tfrac{\lambda}{s}\right)
\end{aligned}
\tag{2.204}
$$

and

$$
\begin{aligned}
&L\left[t^{\gamma-1}\,{}_0cosh_1\left(\begin{matrix} - \\ (B,\varsigma)\end{matrix};\lambda t\right)\right] \\
&= y_0^\infty e^{-st}t^{\gamma-1}\,{}_0cosh_1\left(\begin{matrix} - \\ (B,\varsigma)\end{matrix};\lambda t\right)dt \\
&= s^{-\gamma}\,{}_1cosh_1\left(\begin{matrix} (\gamma,1) \\ (B,\varsigma)\end{matrix};\tfrac{\lambda}{s}\right),
\end{aligned}
\tag{2.205}
$$

where $\gamma, \lambda, \varsigma \in \mathbb{C}$, $Re(\gamma) > 0$, $\beta > 0$, *and* $B > -1$.

Proof With the aid of the expressions

$$
\begin{aligned}
&L\left[t^{\gamma-1}\,{}_0\varphi_1\left(\begin{matrix} - \\ (B,\varsigma)\end{matrix};i\lambda t\right)\right] \\
&= \int_0^\infty e^{-st}t^{\gamma-1}\,{}_0\varphi_1\left(\begin{matrix} - \\ (B,\varsigma)\end{matrix};i\lambda t\right)dt \\
&= s^{-\gamma}\,{}_0\varphi_1\left(\begin{matrix} (\gamma,1) \\ (B,\varsigma)\end{matrix};i\tfrac{\lambda}{s}\right),
\end{aligned}
\tag{2.206}
$$

$$
\begin{aligned}
&L\left[t^{\gamma-1}\,{}_0\varphi_1\left(\begin{matrix} - \\ (B,\varsigma)\end{matrix};-i\lambda t\right)\right] \\
&= \int_0^\infty e^{-st}t^{\gamma-1}\,{}_0\varphi_1\left(\begin{matrix} - \\ (B,\varsigma)\end{matrix};-i\lambda t\right)dt \\
&= s^{-\gamma}\,{}_0\varphi_1\left(\begin{matrix} (\gamma,1) \\ (B,\varsigma)\end{matrix};-i\tfrac{\lambda}{s}\right),
\end{aligned}
\tag{2.207}
$$

$$
\begin{aligned}
&L\left[t^{\gamma-1}{}_0\varphi_1\left(\begin{matrix}-\\(B,\varsigma)\end{matrix};\lambda t\right)\right]\\
&=\int_0^\infty e^{-st}t^{\gamma-1}{}_0\varphi_1\left(\begin{matrix}-\\(B,\varsigma)\end{matrix};\lambda t\right)dt \qquad (2.208)\\
&=s^{-\gamma}{}_0\varphi_1\left(\begin{matrix}(\gamma,1)\\(B,\varsigma)\end{matrix};\frac{\lambda}{s}\right),
\end{aligned}
$$

and

$$
\begin{aligned}
&L\left[t^{\gamma-1}{}_0\varphi_1\left(\begin{matrix}-\\(B,\varsigma)\end{matrix};-\lambda t\right)\right]\\
&=\int_0^\infty e^{-st}t^{\gamma-1}{}_0\varphi_1\left(\begin{matrix}-\\(B,\varsigma)\end{matrix};-\lambda t\right)dt \qquad (2.209)\\
&=s^{-\gamma}{}_0\varphi_1\left(\begin{matrix}(\gamma,1)\\(B,\varsigma)\end{matrix};-\frac{\lambda}{s}\right),
\end{aligned}
$$

we easily obtain the results. □

As the special cases, we have the following corollaries:

Corollary 2.21 *Let $\gamma, \lambda, \varsigma \in \mathbb{C}$, $Re(\gamma) > 0$, $\beta > 0$, and $B > -1$. Then we have*

$$\int_0^\infty e^{-t}t^{\gamma-1}{}_0sin_1\left(\begin{matrix}-\\(B,\varsigma)\end{matrix};\lambda t\right)dt = {}_1sin_1\left(\begin{matrix}(\gamma,1)\\(B,\varsigma)\end{matrix};\lambda\right), \tag{2.210}$$

$$\int_0^\infty e^{-t}t^{\gamma-1}{}_0cos_1\left(\begin{matrix}-\\(B,\varsigma)\end{matrix};\lambda t\right)dt = {}_1cos_1\left(\begin{matrix}(\gamma,1)\\(B,\varsigma)\end{matrix};\lambda\right), \tag{2.211}$$

$$\int_0^\infty e^{-t}t^{\gamma-1}{}_0sinh_1\left(\begin{matrix}-\\(B,\varsigma)\end{matrix};\lambda t\right)dt = {}_1sinh_1\left(\begin{matrix}(\gamma,1)\\(B,\varsigma)\end{matrix};\lambda\right), \tag{2.212}$$

and

$$\int_0^\infty e^{-t}t^{\gamma-1}{}_0cosh_1\left(\begin{matrix}-\\(B,\varsigma)\end{matrix};\lambda t\right)dt = {}_1cosh_1\left(\begin{matrix}(\gamma,1)\\(B,\varsigma)\end{matrix};\lambda\right), \tag{2.213}$$

Proof On setting $s = 1$ in the expressions (2.202)–(2.205) we have the desired results. □

Corollary 2.22 *Let $\gamma, \lambda, \varsigma \in C$, $Re(\gamma) > 0$, $\beta > 0$, and $B > -1$. Then we have*

$$\begin{aligned} &\int_0^\infty e^{-t} t^{\gamma-1} {}_0\varphi_1 \left(\begin{matrix} - \\ (B, \varsigma) \end{matrix} ; \lambda t \right) dt \\ &= {}_0\varphi_1 \left(\begin{matrix} (\gamma, 1) \\ (B, \varsigma) \end{matrix} ; \lambda \right), \end{aligned} \tag{2.214}$$

where $\gamma, \lambda, \varsigma \in \mathbb{C}$, $Re(\gamma) > 0$, $\beta > 0$, and $B > -1$.

Proof The result is the special case of the expression (2.201) when $s = 1$. □

Corollary 2.23 *Let $\lambda, \varsigma \in \mathbb{C}$, $\beta > 0$, and $B > -1$.*
Then we have

$$\int_0^\infty e^{-t} {}_0sin_1 \left(\begin{matrix} - \\ (B, \varsigma) \end{matrix} ; \lambda t \right) dt = {}_1sin_1 \left(\begin{matrix} (1, 1) \\ (B, \varsigma) \end{matrix} ; \lambda \right), \tag{2.215}$$

$$\int_0^\infty e^{-t} {}_0cos_1 \left(\begin{matrix} - \\ (B, \varsigma) \end{matrix} ; \lambda t \right) dt = {}_1cos_1 \left(\begin{matrix} (1, 1) \\ (B, \varsigma) \end{matrix} ; \lambda \right), \tag{2.216}$$

$$y_0^\infty e^{-t} {}_0sinh_1 \left(\begin{matrix} - \\ (B, \varsigma) \end{matrix} ; \lambda t \right) dt = {}_1sinh_1 \left(\begin{matrix} (1, 1) \\ (B, \varsigma) \end{matrix} ; \lambda \right), \tag{2.217}$$

and

$$\int_0^\infty e^{-t} {}_0cosh_1 \left(\begin{matrix} - \\ (B, \varsigma) \end{matrix} ; \lambda t \right) dt = {}_1cosh_1 \left(\begin{matrix} (1, 1) \\ (B, \varsigma) \end{matrix} ; \lambda \right). \tag{2.218}$$

Proof On setting $\gamma = 1$ in the expressions (2.210)–(2.213) we have the desired results. □

Corollary 2.24 *Let $\gamma, \lambda, \varsigma \in \mathbb{C}$, $Re(\gamma) > 0$, $\beta > 0$, and $B > -1$. Then we have*

$$\begin{aligned} &y_0^\infty e^{-t} t^{\gamma-1} {}_0\varphi_1 \left(\begin{matrix} - \\ (B, \varsigma) \end{matrix} ; \lambda t \right) dt \\ &= {}_0\varphi_1 \left(\begin{matrix} (\gamma, 1) \\ (B, \varsigma) \end{matrix} ; \lambda \right). \end{aligned} \tag{2.219}$$

Proof The result is the special case of the expression (2.209) when $\gamma = 1$. □

It is not difficult to derive the properties for the new special functions.

Property 2.18 Let $\varsigma \in \mathbb{C}$, $\beta > 0$, and $B > -1$. Then we have

$${}_0sin_1 \left(\begin{matrix} - \\ (B, \varsigma) \end{matrix} ; -\lambda t \right) = - {}_0sin_1 \left(\begin{matrix} - \\ (B, \varsigma) \end{matrix} ; t \right), \tag{2.220}$$

$$ {}_0cos_1\left(\begin{matrix} - \\ (B,\varsigma) \end{matrix};-\lambda t\right) = {}_0cos_1\left(\begin{matrix} - \\ (B,\varsigma) \end{matrix};t\right), \tag{2.221} $$

$$ {}_0tan_1\left(\begin{matrix} - \\ (B,\varsigma) \end{matrix};-t\right) = -\,{}_0tan_1\left(\begin{matrix} - \\ (B,\varsigma) \end{matrix};t\right), \tag{2.222} $$

$$ {}_0cot_1\left(\begin{matrix} - \\ (B,\varsigma) \end{matrix};-t\right) = -{}_0cot_1\left(\begin{matrix} - \\ (B,\varsigma) \end{matrix};t\right), \tag{2.223} $$

$$ {}_0sec_1\left(\begin{matrix} - \\ (B,\varsigma) \end{matrix};-t\right) = {}_0sec_1\left(\begin{matrix} - \\ (B,\varsigma) \end{matrix};t\right), \tag{2.224} $$

and

$$ {}_0cosec_1\left(\begin{matrix} - \\ (B,\varsigma) \end{matrix};-t\right) = -\,{}_0cosec_1\left(\begin{matrix} - \\ (B,\varsigma) \end{matrix};t\right). \tag{2.225} $$

Proof With the definition of the Wright function we obtain the results. □

Property 2.19 Let $\varsigma \in \mathbb{C}$, $\beta > 0$, and $B > -1$. Then we have

$$ {}_0sinh_1\left(\begin{matrix} - \\ (B,\varsigma) \end{matrix};-\lambda t\right) = -\,{}_0sinh_1\left(\begin{matrix} - \\ (B,\varsigma) \end{matrix};t\right), \tag{2.226} $$

$$ {}_0cosh_1\left(\begin{matrix} - \\ (B,\varsigma) \end{matrix};-\lambda t\right) = {}_0cosh_1\left(\begin{matrix} - \\ (B,\varsigma) \end{matrix};t\right), \tag{2.227} $$

$$ {}_0tanh_1\left(\begin{matrix} - \\ (B,\varsigma) \end{matrix};-t\right) = -\,{}_0tanh_1\left(\begin{matrix} - \\ (B,\varsigma) \end{matrix};t\right), \tag{2.228} $$

$$ {}_0coth_1\left(\begin{matrix} - \\ (B,\varsigma) \end{matrix};-t\right) = -{}_0coth_1\left(\begin{matrix} - \\ (B,\varsigma) \end{matrix};t\right), \tag{2.229} $$

$$ {}_0sech_1\left(\begin{matrix} - \\ (B,\varsigma) \end{matrix};-t\right) = {}_0sech_1\left(\begin{matrix} - \\ (B,\varsigma) \end{matrix};t\right), \tag{2.230} $$

and

$$ {}_0cosech_1\left(\begin{matrix} - \\ (B,\varsigma) \end{matrix};-t\right) = -\,{}_0cosech_1\left(\begin{matrix} - \\ (B,\varsigma) \end{matrix};t\right). \tag{2.231} $$

Proof With the definition of the Wright function we obtain the results. □

Property 2.20 Let $\varsigma \in \mathbb{C}$, $\beta > 0$, and $B > -1$. Then we have

$$ {}_0sinh_1\left(\begin{matrix} - \\ (B, \varsigma) \end{matrix} ; t \right) = -i \, {}_0sin_1\left(\begin{matrix} - \\ (B, \varsigma) \end{matrix} ; it \right), \tag{2.232} $$

$$ {}_0cosh_1\left(\begin{matrix} - \\ (B, \varsigma) \end{matrix} ; t \right) = {}_0cos_1\left(\begin{matrix} - \\ (B, \varsigma) \end{matrix} ; it \right), \tag{2.233} $$

$$ {}_0tanh_1\left(\begin{matrix} - \\ (B, \varsigma) \end{matrix} ; t \right) = -i \, {}_0tan_1\left(\begin{matrix} - \\ (B, \varsigma) \end{matrix} ; it \right), \tag{2.234} $$

$$ {}_0coth_1\left(\begin{matrix} - \\ (B, \varsigma) \end{matrix} ; t \right) = i {}_0cot_1\left(\begin{matrix} - \\ (B, \varsigma) \end{matrix} ; it \right), \tag{2.235} $$

$$ {}_0sech_1\left(\begin{matrix} - \\ (B, \varsigma) \end{matrix} ; -t \right) = {}_0sec_1\left(\begin{matrix} - \\ (B, \varsigma) \end{matrix} ; it \right), \tag{2.236} $$

and

$$ {}_0cosech_1\left(\begin{matrix} - \\ (B, \varsigma) \end{matrix} ; t \right) = i {}_0cosec_1\left(\begin{matrix} - \\ (B, \varsigma) \end{matrix} ; it \right). \tag{2.237} $$

Proof With the definition of the Wright function we obtain the results. □

More generally, we show the following properties.

Property 2.21 Let $t \in \mathbb{C}$, $j, k, u \in \mathbb{N}$, $\alpha_j, \beta_k \in \mathbb{C}$, and $A_j, B_k \in \mathbb{R}_+$. Then we have

$$ {}_psin_q\left(\begin{matrix} (\alpha_1, A_1), \cdots, (\alpha_p, A_p) \\ (\beta_1, B_1), \cdots, (\beta_q, B_q) \end{matrix} ; -t \right) = - {}_psin_q\left(\begin{matrix} (\alpha_1, A_1), \cdots, (\alpha_p, A_p) \\ (\beta_1, B_1), \cdots, (\beta_q, B_q) \end{matrix} ; t \right), \tag{2.238} $$

$$ {}_pcos_q\left(\begin{matrix} (\alpha_1, A_1), \cdots, (\alpha_p, A_p) \\ (\beta_1, B_1), \cdots, (\beta_q, B_q) \end{matrix} ; -t \right) = {}_pcos_q\left(\begin{matrix} (\alpha_1, A_1), \cdots, (\alpha_p, A_p) \\ (\beta_1, B_1), \cdots, (\beta_q, B_q) \end{matrix} ; t \right), \tag{2.239} $$

$$ {}_ptan_q\left(\begin{matrix} (\alpha_1, A_1), \cdots, (\alpha_p, A_p) \\ (\beta_1, B_1), \cdots, (\beta_q, B_q) \end{matrix} ; -t \right) = - {}_ptan_q\left(\begin{matrix} (\alpha_1, A_1), \cdots, (\alpha_p, A_p) \\ (\beta_1, B_1), \cdots, (\beta_q, B_q) \end{matrix} ; t \right), \tag{2.240} $$

$$ {}_pcot_q\left(\begin{matrix} (\alpha_1, A_1), \cdots, (\alpha_p, A_p) \\ (\beta_1, B_1), \cdots, (\beta_q, B_q) \end{matrix} ; -t \right) = - {}_pcot_q\left(\begin{matrix} (\alpha_1, A_1), \cdots, (\alpha_p, A_p) \\ (\beta_1, B_1), \cdots, (\beta_q, B_q) \end{matrix} ; t \right), \tag{2.241} $$

$$ {}_p sec_q \left(\begin{matrix} (\alpha_1, A_1), \cdots, (\alpha_p, A_p) \\ (\beta_1, B_1), \cdots, (\beta_q, B_q) \end{matrix} ; -t \right) = {}_p sec_q \left(\begin{matrix} (\alpha_1, A_1), \cdots, (\alpha_p, A_p) \\ (\beta_1, B_1), \cdots, (\beta_q, B_q) \end{matrix} ; t \right), \tag{2.242} $$

and

$$ {}_p cosec_q \left(\begin{matrix} (\alpha_1, A_1), \cdots, (\alpha_p, A_p) \\ (\beta_1, B_1), \cdots, (\beta_q, B_q) \end{matrix} ; -t \right) $$
$$ = - {}_p cosec_q \left(\begin{matrix} (\alpha_1, A_1), \cdots, (\alpha_p, A_p) \\ (\beta_1, B_1), \cdots, (\beta_q, B_q) \end{matrix} ; t \right). \tag{2.243} $$

Proof With the definition of the Wright's generalized hypergeometric function we obtain the results. □

Property 2.22 Let $t \in \mathbb{C}$, $j, k, u \in \mathbb{N}$, $\alpha_j, \beta_k \in \mathbb{C}$, and $A_j, B_k \in \mathbb{R}_+$. Then we have

$$ {}_p sinh_q \left(\begin{matrix} (\alpha_1, A_1), \cdots, (\alpha_p, A_p) \\ (\beta_1, B_1), \cdots, (\beta_q, B_q) \end{matrix} ; -t \right) $$
$$ = - {}_p sinh_q \left(\begin{matrix} (\alpha_1, A_1), \cdots, (\alpha_p, A_p) \\ (\beta_1, B_1), \cdots, (\beta_q, B_q) \end{matrix} ; t \right), \tag{2.244} $$

$$ {}_p cosh_q \left(\begin{matrix} (\alpha_1, A_1), \cdots, (\alpha_p, A_p) \\ (\beta_1, B_1), \cdots, (\beta_q, B_q) \end{matrix} ; -t \right) $$
$$ = {}_p cosh_q \left(\begin{matrix} (\alpha_1, A_1), \cdots, (\alpha_p, A_p) \\ (\beta_1, B_1), \cdots, (\beta_q, B_q) \end{matrix} ; t \right), \tag{2.245} $$

$$ {}_p tanh_q \left(\begin{matrix} (\alpha_1, A_1), \cdots, (\alpha_p, A_p) \\ (\beta_1, B_1), \cdots, (\beta_q, B_q) \end{matrix} ; -t \right) $$
$$ = - {}_p tanh_q \left(\begin{matrix} (\alpha_1, A_1), \cdots, (\alpha_p, A_p) \\ (\beta_1, B_1), \cdots, (\beta_q, B_q) \end{matrix} ; t \right), \tag{2.246} $$

$$ {}_p coth_q \left(\begin{matrix} (\alpha_1, A_1), \cdots, (\alpha_p, A_p) \\ (\beta_1, B_1), \cdots, (\beta_q, B_q) \end{matrix} ; -t \right) $$
$$ = - {}_p coth_q \left(\begin{matrix} (\alpha_1, A_1), \cdots, (\alpha_p, A_p) \\ (\beta_1, B_1), \cdots, (\beta_q, B_q) \end{matrix} ; t \right), \tag{2.247} $$

$$ {}_p sech_q \left(\begin{matrix} (\alpha_1, A_1), \cdots, (\alpha_p, A_p) \\ (\beta_1, B_1), \cdots, (\beta_q, B_q) \end{matrix} ; -t \right) $$
$$ = {}_p sech_q \left(\begin{matrix} (\alpha_1, A_1), \cdots, (\alpha_p, A_p) \\ (\beta_1, B_1), \cdots, (\beta_q, B_q) \end{matrix} ; t \right), \tag{2.248} $$

and

$$
{}_p cosech_q \left(\begin{matrix} (\alpha_1, A_1), \cdots, (\alpha_p, A_p) \\ (\beta_1, B_1), \cdots, (\beta_q, B_q) \end{matrix} ; -t \right)
$$

$$
= -{}_p cosech_q \left(\begin{matrix} (\alpha_1, A_1), \cdots, (\alpha_p, A_p) \\ (\beta_1, B_1), \cdots, (\beta_q, B_q) \end{matrix} ; t \right). \tag{2.249}
$$

Proof With the definition of the Wright's generalized hypergeometric function we obtain the results. □

Property 2.23 Let $t \in \mathbb{C}$, $j, k, u \in \mathbb{N}$, $\alpha_j, \beta_k \in \mathbb{C}$, and $A_j, B_k \in \mathbb{R}_+$. Then we have

$$
{}_p sinh_q \left(\begin{matrix} (\alpha_1, A_1), \cdots, (\alpha_p, A_p) \\ (\beta_1, B_1), \cdots, (\beta_q, B_q) \end{matrix} ; t \right)
$$

$$
= -i\, {}_p sin_q \left(\begin{matrix} (\alpha_1, A_1), \cdots, (\alpha_p, A_p) \\ (\beta_1, B_1), \cdots, (\beta_q, B_q) \end{matrix} ; it \right), \tag{2.250}
$$

$$
{}_p cosh_q \left(\begin{matrix} (\alpha_1, A_1), \cdots, (\alpha_p, A_p) \\ (\beta_1, B_1), \cdots, (\beta_q, B_q) \end{matrix} ; t \right) = {}_p cos_q \left(\begin{matrix} (\alpha_1, A_1), \cdots, (\alpha_p, A_p) \\ (\beta_1, B_1), \cdots, (\beta_q, B_q) \end{matrix} ; it \right), \tag{2.251}
$$

$$
{}_p tanh_q \left(\begin{matrix} (\alpha_1, A_1), \cdots, (\alpha_p, A_p) \\ (\beta_1, B_1), \cdots, (\beta_q, B_q) \end{matrix} ; t \right)
$$

$$
= -i\, {}_p tan_q \left(\begin{matrix} (\alpha_1, A_1), \cdots, (\alpha_p, A_p) \\ (\beta_1, B_1), \cdots, (\beta_q, B_q) \end{matrix} ; it \right), \tag{2.252}
$$

$$
{}_p coth_q \left(\begin{matrix} (\alpha_1, A_1), \cdots, (\alpha_p, A_p) \\ (\beta_1, B_1), \cdots, (\beta_q, B_q) \end{matrix} ; t \right)
$$

$$
= i\, {}_p cot_q \left(\begin{matrix} (\alpha_1, A_1), \cdots, (\alpha_p, A_p) \\ (\beta_1, B_1), \cdots, (\beta_q, B_q) \end{matrix} ; it \right), \tag{2.253}
$$

$$
{}_p sech_q \left(\begin{matrix} (\alpha_1, A_1), \cdots, (\alpha_p, A_p) \\ (\beta_1, B_1), \cdots, (\beta_q, B_q) \end{matrix} ; t \right)
$$

$$
= {}_p sec_q \left(\begin{matrix} (\alpha_1, A_1), \cdots, (\alpha_p, A_p) \\ (\beta_1, B_1), \cdots, (\beta_q, B_q) \end{matrix} ; it \right), \tag{2.254}
$$

and

$$
{}_p cosech_q \left(\begin{matrix} (\alpha_1, A_1), \cdots, (\alpha_p, A_p) \\ (\beta_1, B_1), \cdots, (\beta_q, B_q) \end{matrix} ; t \right)
$$
$$
= i \, {}_p cosech_q \left(\begin{matrix} (\alpha_1, A_1), \cdots, (\alpha_p, A_p) \\ (\beta_1, B_1), \cdots, (\beta_q, B_q) \end{matrix} ; it \right). \tag{2.255}
$$

Proof With the definition of the Wright's generalized hypergeometric function we obtain the results. □

Definition 2.30 The shifted Wright's generalized hypergeometric series is defined as

$$
\begin{aligned}
&{}_p\mathbb{S}_q(z) \\
&= {}_p\mathbb{S}_q \left(\begin{matrix} (\alpha_1, A_1; \beta_1, B_1), \cdots, (\alpha_p, A_p; \beta_p, B_p) \\ (\gamma_1, C_1; \vartheta_1, D_1), \cdots, (\gamma_q, C_q; \vartheta_q, D_q) \end{matrix} ; z \right) \\
&= {}_p\mathbb{S}_q \left(\begin{matrix} (\alpha_p, A_p; \beta_p, B_p)_{1,p} \\ (\gamma_q, C_q; \vartheta_q, D_q)_{1,q} \end{matrix} ; z \right) \\
&= \sum_{k=0}^{\infty} \frac{\left\{ \prod_{m=1}^{p} B(\alpha_m + A_m k; \beta_m + B_m k) \right\}}{\left\{ \prod_{n=1}^{q} B(\gamma_n + C_n k; \vartheta_n + D_n k) \right\}} \frac{z^k}{k!} \\
&= \sum_{k=0}^{\infty} \frac{\prod_{m=1}^{p} (\Gamma(\alpha_m + A_m k)\Gamma(\beta_m + B_m k)) \prod_{n=1}^{q} \Gamma[(\gamma_n + \vartheta_n) + (C_n + D_n)k]}{\prod_{m=1}^{p} \Gamma[(\beta_m + \alpha_m) + (B_m + A_m)k] \prod_{n=1}^{q} (\Gamma(\gamma_n + C_n k)\Gamma(\vartheta_n + D_n k))} \frac{z^k}{k!},
\end{aligned} \tag{2.256}
$$

where $k \in \mathbb{N} \cup 0$, $m, n \in \mathbb{N}$, $\alpha_m, \beta_m, \gamma_n, \vartheta_n, z \in \mathbb{C}$, and $A_m, B_m, C_n, D_n \in \mathbb{R}_+$.

There are some special cases as follows:

$$
{}_0\mathbb{S}_0(z) = e^z, \tag{2.257}
$$

$$
\begin{aligned}
&{}_1\mathbb{S}_0(z) \\
&= {}_1\mathbb{S}_0 \left(\begin{matrix} (\alpha, A; \beta, B) \\ - \end{matrix} ; z \right) \\
&= \sum_{k=0}^{\infty} B(\alpha + Ak; \beta + Bk) \frac{z^k}{k!} \\
&= \sum_{k=0}^{\infty} \frac{\Gamma(\alpha + Ak)\Gamma(\beta + Bk)}{\Gamma[(\beta + \alpha) + (B + B)k]} \frac{z^k}{k!} \\
&= {}_2\varphi_1 \left(\begin{matrix} (\alpha, B), (\beta, B) \\ (\beta + \alpha, B + B) \end{matrix} ; z \right),
\end{aligned} \tag{2.258}
$$

$$\begin{aligned}
&{}_0\mathbb{S}_1(z) \\
&= {}_0\mathbb{S}_1\left(\begin{matrix} - \\ (\alpha, A; \beta, B) \end{matrix}; z\right) \\
&= \sum_{k=0}^{\infty} \frac{1}{B(\alpha+Ak;\beta+Bk)} \frac{z^k}{k!} \\
&= \sum_{k=0}^{\infty} \frac{\Gamma[(\beta+\alpha)+(B+B)k]}{\Gamma(\alpha+Ak)\Gamma(\beta+Bk)} \frac{z^k}{k!} \\
&= {}_1\varphi_2\left(\begin{matrix} (\beta+\alpha, B+B) \\ (\alpha, B), (\beta, B) \end{matrix}; z\right),
\end{aligned} \tag{2.259}$$

$$\begin{aligned}
&{}_1\mathbb{S}_1(z) \\
&= {}_1\mathbb{S}_1\left(\begin{matrix} (\alpha, A; \beta, B) \\ (\gamma, C; \vartheta, D) \end{matrix}; z\right) \\
&= \sum_{k=0}^{\infty} \frac{B(\alpha+Ak;\beta+Bk)}{B(\gamma+Ck;\vartheta+Dk)} \frac{z^k}{k!} \\
&= \sum_{k=0}^{\infty} \frac{\Gamma(\alpha+Ak)\Gamma(\beta+Bk)\Gamma[(\gamma+\vartheta)+(C+D)k]}{\Gamma[(\beta+\alpha)+(B+B)k]\Gamma(\gamma+Ck)\Gamma(\vartheta+Dk)} \frac{z^k}{k!} \\
&= {}_3\varphi_3\left(\begin{matrix} (\alpha, B), (\beta, B), (\gamma+\vartheta, C+D) \\ (\gamma, C), (\vartheta, D), (\beta+\alpha, B+B) \end{matrix}; z\right),
\end{aligned} \tag{2.260}$$

where $k \in \mathbb{N} \cup 0$, $\alpha, \beta, \gamma, \vartheta, z \in \mathbb{C}$, and $A, B, C, D \in \mathbb{R}_+$.

2.2 The Truncated Wright's Generalized Hypergeometric Function

Definition 2.31 The Wright's generalized hypergeometric series is defined as

$$\left[{}_p\varphi_q\right]^n(z) = \sum_{u=0}^{n} \frac{\left\{\prod_{j=1}^{p} \Gamma\left(\alpha_j + A_j u\right)\right\}}{\left\{\prod_{k=1}^{q} \Gamma\left(\beta_k + B_k u\right)\right\}} \frac{z^u}{u!}, \tag{2.261}$$

where $z \in \mathbb{C}$, $n, j, k, u, p, q \in \mathbb{N}$, $\alpha_j, \beta_k \in \mathbb{C}$, and $A_j, B_k \in \mathbb{R}_+$.

Definition 2.32 The truncated supersine via Wright's generalized hypergeometric function is defined as

$$\begin{aligned}
&\left[{}_p sin_q\right]^n(z)\\
&= \frac{1}{2i}\left(\left[{}_p\varphi_q\right]^n(iz) - \left[{}_p\varphi_q\right]^n(-iz)\right)\\
&= \sum_{k=0}^{n}(-1)^k \frac{\left\{\prod_{j=1}^{p}\Gamma(\alpha_j+A_j(2k+1))\right\}}{\left\{\prod_{j=1}^{q}\Gamma(\beta_j+B_j(2k+1))\right\}}\frac{z^{2k+1}}{(2k+1)!},
\end{aligned} \tag{2.262}$$

where $i=\sqrt{-1}$, $z\in\mathbb{C}$, $n,j,k,p,q\in\mathbb{N}$, $\alpha_j,\beta_k\in\mathbb{C}$, $A_j,B_k\in\mathbb{R}_+$.

Definition 2.33 The truncated supercosine via Wright's generalized hypergeometric function is defined as

$$\begin{aligned}
&\left[{}_p cos_q\right]^n(z)\\
&= \frac{1}{2}\left(\left[{}_p\varphi_q\right]^n(iz) + \left[{}_p\varphi_q\right]^n(-iz)\right)\\
&= \sum_{k=0}^{n}(-1)^k \frac{\left\{\prod_{j=1}^{p}\Gamma(\alpha_j+2kA_j)\right\}}{\left\{\prod_{j=1}^{q}\Gamma(\beta_j+2kB_j)\right\}}\frac{z^{2k}}{(2k)!}
\end{aligned} \tag{2.263}$$

where $i=\sqrt{-1}$, $z\in\mathbb{C}$, $n,j,k,p,q\in\mathbb{N}$, $\alpha_j,\beta_k\in\mathbb{C}$, and $A_j,B_k\in\mathbb{R}_+$.

Definition 2.34 The truncated superhyperbolic sine via Wright's generalized hypergeometric function is defined as

$$\begin{aligned}
&\left[{}_p sinh_q\right]^n(z)\\
&= \frac{1}{2}\left(\left[{}_p\varphi_q\right]^n(z) - \left[{}_p\varphi_q\right]^n(-z)\right)\\
&= \sum_{k=0}^{n} \frac{\left\{\prod_{j=1}^{p}\Gamma(\alpha_j+A_j(2k+1))\right\}}{\left\{\prod_{j=1}^{q}\Gamma(\beta_j+B_j(2k+1))\right\}}\frac{z^{2k+1}}{(2k+1)!},
\end{aligned} \tag{2.264}$$

where $z\in\mathbb{C}$, $n,j,k,p,q\in\mathbb{N}$, $\alpha_j,\beta_k\in\mathbb{C}$, and $A_j,B_k\in\mathbb{R}_+$.

Definition 2.35 The truncated superhyperbolic cosine via Wright's generalized hypergeometric function is defined as

$$\begin{aligned}
&\left[{}_p cosh_q\right]^n(z)\\
&= \frac{1}{2}\left(\left[{}_p\varphi_q\right]^n(z) + \left[{}_p\varphi_q\right]^n(-z)\right)\\
&= \sum_{k=0}^{n} \frac{\left\{\prod_{j=1}^{p}\Gamma(\alpha_j+2kA_j)\right\}}{\left\{\prod_{j=1}^{q}\Gamma(\beta_j+2kB_j)\right\}}\frac{z^{2k}}{(2k)!},
\end{aligned} \tag{2.265}$$

where $z\in\mathbb{C}$, $n,j,k,p,q\in\mathbb{N}$, $\alpha_j,\beta_k\in\mathbb{C}$, and $A_j,B_k\in\mathbb{R}_+$.

For more interesting tasks, see Stankovic [190], Paris [191], Dziok and Raina [192], Mainardi and Pagnini [193], Virchenko et al. [194], Aouf and Dziok [195], Raina and Sharma [196], Virchenko [197], and so on.

2.3 The Integral Transforms via Dunkl Transform

In this section, we investigate and propose the results via the Dunkl transform.

2.3.1 The Dunkl Transform

In this part, we propose the results associated to the Dunkl transform.

Definition 2.36 The Dunkl operator, denoted by $\widehat{\mathbb{T}_\alpha} f(t)$, is defined as

$$\widehat{\mathbb{T}_\alpha} f(t) = \frac{df(t)}{dt} + \frac{2\alpha+1}{t}\left[\frac{f(t)-f(-t)}{2}\right], \tag{2.266}$$

where $\alpha \geq \frac{1}{2}$.

The result was discovered by Dunkl in 1989 [198] and reported by de Jeu [199] and by Soltani [200].

Theorem 2.55 *The Dunkl equation is given as follows:*

$$\widehat{\mathbb{T}_\alpha} f(t) = \lambda f(t) \tag{2.267}$$

subjected to the initial conditions

$$f(0) = 1 \tag{2.268}$$

with the solution given as

$$\begin{aligned} &f(t) \\ &= \epsilon_\alpha(\lambda t) \\ &= \Im_\alpha(\lambda t) + \tfrac{\lambda t}{2(\alpha+1)} \Im_{\alpha+1}(\lambda t) \end{aligned}, \tag{2.269}$$

where

$$\Im_\alpha(\lambda t) = \Gamma(1+\alpha) \sum_{k=0}^{\infty} \frac{1}{\Gamma(k+\alpha+1)} \frac{(\lambda t/2)^{2k}}{k!} \tag{2.270}$$

for $\lambda \in \mathbb{C}$.

The result was discovered by Dunkl in 1989 [198] and by Soltani [200].

It is clear that

$$\begin{aligned}
&\Im_\alpha(\lambda t)+\tfrac{\lambda t}{2(\alpha+1)}\Im_{\alpha+1}(\lambda t)\\
&=\Gamma(1+\alpha)\left(\sum_{k=0}^{\infty}\tfrac{1}{\Gamma(k+\alpha+1)}\tfrac{(\lambda t/2)^{2k}}{k!}+\sum_{k=0}^{\infty}\tfrac{1}{\Gamma(k+\alpha+2)}\tfrac{(\lambda t/2)^{2k+1}}{k!}\right)\\
&=\Gamma(1+\alpha)\left({}_0\varphi_1\left(\begin{matrix}-\\(\alpha+1,1)\end{matrix};\tfrac{\lambda^2t^2}{4}\right)+\tfrac{\lambda t}{2}\,{}_0\varphi_1\left(\begin{matrix}-\\(\alpha+2,1)\end{matrix};\tfrac{\lambda^2t^2}{4}\right)\right),
\end{aligned}\tag{2.271}$$

where the Wright function is denoted as

$${}_0\varphi_1\left(\begin{matrix}-\\(\beta,\zeta)\end{matrix};t\right)=\sum_{k=0}^{\infty}\frac{1}{\Gamma(\zeta k+\beta)}\frac{t^k}{k!}.\tag{2.272}$$

For $\lambda\in\mathbb{C}$ we have

$$\begin{aligned}
&\in_\alpha(i\lambda t)\\
&=\Gamma(1+\alpha)\left({}_0\varphi_1\left(\begin{matrix}-\\(\alpha+1,1)\end{matrix};-\tfrac{\lambda^2t^2}{4}\right)+\tfrac{i\lambda t}{2}\,{}_0\varphi_1\left(\begin{matrix}-\\(\alpha+2,1)\end{matrix};-\tfrac{\lambda^2t^2}{4}\right)\right)
\end{aligned}\tag{2.273}$$

and

$$\begin{aligned}
&\in_\alpha(-i\lambda t)\\
&=\Gamma(1+\alpha)\left({}_0\varphi_1\left(\begin{matrix}-\\(\alpha+1,1)\end{matrix};-\tfrac{\lambda^2t^2}{4}\right)-\tfrac{i\lambda t}{2}\,{}_0\varphi_1\left(\begin{matrix}-\\(\alpha+2,1)\end{matrix};-\tfrac{\lambda^2t^2}{4}\right)\right)
\end{aligned}\tag{2.274}$$

such that

$$\begin{aligned}
&{}_1\mathbb{S}cos_1(\lambda t)\\
&=\tfrac{\in_\alpha(i\lambda t)+\in_\alpha(-i\lambda t)}{2}\\
&=\Gamma(1+\alpha)\ {}_0\varphi_1\left(\begin{matrix}-\\(\alpha+1,1)\end{matrix};-\tfrac{\lambda^2t^2}{4}\right)
\end{aligned}\tag{2.275}$$

and

$$\begin{aligned}
&{}_1\mathbb{S}sin_1(\lambda t)\\
&=\tfrac{\in_\alpha(i\lambda t)-\in_\alpha(-i\lambda t)}{2i}\\
&=\tfrac{\lambda t}{2}\ {}_0\varphi_1\left(\begin{matrix}-\\(\alpha+2,1)\end{matrix};\tfrac{\lambda^2t^2}{4}\right),
\end{aligned}\tag{2.276}$$

where

$$\in_{\alpha}(i\lambda t) = {}_1\mathbb{S}cos_1(\lambda t) + i\,{}_1\mathbb{S}sin_1(\lambda t) \tag{2.277}$$

and

$$\in_{\alpha}(-i\lambda t) = {}_1\mathbb{S}cos_1(\lambda t) - i\,{}_1\mathbb{S}sin_1(\lambda t)\,. \tag{2.278}$$

Similarly, it is shown that

$$\begin{aligned}&\in_{\alpha}(\lambda t)\\&= \Gamma(1+\alpha)\left({}_0\varphi_1\left(\begin{matrix}-\\(\alpha+1,1)\end{matrix};\frac{\lambda^2t^2}{4}\right) + \frac{\lambda t}{2}\,{}_0\varphi_1\left(\begin{matrix}-\\(\alpha+2,1)\end{matrix};\frac{\lambda^2t^2}{4}\right)\right)\end{aligned} \tag{2.279}$$

and

$$\begin{aligned}&\in_{\alpha}(-\lambda t)\\&= \Gamma(1+\alpha)\left({}_0\varphi_1\left(\begin{matrix}-\\(\alpha+1,1)\end{matrix};\frac{\lambda^2t^2}{4}\right) - \frac{\lambda t}{2}\,{}_0\varphi_1\left(\begin{matrix}-\\(\alpha+2,1)\end{matrix};\frac{\lambda^2t^2}{4}\right)\right)\end{aligned} \tag{2.280}$$

with

$$\begin{aligned}&{}_1\mathbb{S}cosh_1(\lambda t)\\&= \frac{\in_{\alpha}(\lambda t)+\in_{\alpha}(-\lambda t)}{2}\\&= \Gamma(1+\alpha)\;{}_0\varphi_1\left(\begin{matrix}-\\(\alpha+1,1)\end{matrix};\frac{\lambda^2t^2}{4}\right)\end{aligned} \tag{2.281}$$

and

$$\begin{aligned}&{}_1\mathbb{S}sinh_1(\lambda t)\\&= \frac{\in_{\alpha}(\lambda t)-\in_{\alpha}(-\lambda t)}{2}\\&= \frac{\Gamma(1+\alpha)\lambda t}{2}\;{}_0\varphi_1\left(\begin{matrix}-\\(\alpha+2,1)\end{matrix};\frac{\lambda^2t^2}{4}\right)\end{aligned} \tag{2.282}$$

since

$$\in_{\alpha}(\lambda t) = {}_1\mathbb{S}cosh_1(\lambda t) + {}_1\mathbb{S}sinh_1(\lambda t) \tag{2.283}$$

and

$$\in_{\alpha}(-\lambda t) = {}_1\mathbb{S}cosh_1(\lambda t) - {}_1\mathbb{S}sinh_1(\lambda t)\,. \tag{2.284}$$

Definition 2.37 The Dunkl–Sonine operator is defined as

$$\widehat{S}_{\alpha,\beta}(f)(t) = \frac{\Gamma(\beta+1)}{\Gamma(\beta-\alpha)\Gamma(\alpha+1)} \int_{-1}^{1} f(t\tau)(1+\tau)\left(1-\tau^2\right)^{\beta-\alpha-1} |\tau|^{2\alpha+1} d\tau, \tag{2.285}$$

where $\beta > \alpha$.

The result was investigated by Soltani in 2008 [200].

When $\beta - \alpha = 1$, one has

$$\widehat{G}_{\alpha}(f)(t) = \widehat{S}_{\alpha,\alpha+1}(f)(t) = (\alpha+1)\int_{-1}^{1} f(t\tau)(1+\tau)|\tau|^{2\alpha+1} d\tau \tag{2.286}$$

which leads to

$$\widehat{G}_{\alpha}(f)(t) = \widehat{S}_{\alpha,\alpha+1}(f)(t) = (\alpha+1)\left(\frac{1}{t}\right)^{2\alpha+2} \int_{-t}^{t} f(\tau)\left(1+\frac{\tau}{t}\right)|\tau|^{2\alpha+1} d\tau. \tag{2.287}$$

Definition 2.38 Let us define the unit operator $\widehat{I_{\alpha}}$ such that

$$\widehat{I_{\alpha}} f(t) = \left[\left(\widehat{H_{\alpha}} \circ \widehat{\mathbb{T}_{\alpha}}\right) f\right](t) = \widehat{H_{\alpha}}\left(\widehat{\mathbb{T}_{\alpha}} f(t)\right), \tag{2.288}$$

where $\widehat{H_{\alpha}} f(t)$ is the inverse operator of the Dunkl operator with $\widehat{H_{\alpha}} f(t) = \widehat{\mathbb{T}_{\alpha}}^{-1} f(t)$.

Definition 2.39 Let $\alpha \geq \frac{1}{2}$. The Dunkl transform of a function $f(t)$ on $\mathbb{R}$ is defined as follows:

$$\begin{aligned} &\wp_{\alpha}[f(t)] \\ &= f(\varsigma) \\ &= \int_{-\infty}^{\infty} \in_{\alpha}(-i\varsigma t) f(t) |t|^{2\alpha+1} dt \end{aligned}, \tag{2.289}$$

and the inverse Dunkl transform on $\mathbb{R}$ as

$$\begin{aligned} &\wp_{\alpha}^{-1}[f(\varsigma)] \\ &= f(t) \\ &= \hbar_{\alpha}^{2} \int_{-\infty}^{\infty} \in_{\alpha}(i\varsigma t) f(\varsigma) |\varsigma|^{2\alpha+1} d\varsigma \end{aligned}, \tag{2.290}$$

where

$$\hbar_{\alpha} = \frac{1}{2^{\alpha+1}\Gamma(\alpha+1)}. \tag{2.291}$$

The result was discovered by Dunkl in 1989 [198] and reported by de Jeu [199] and by Soltani [200].

Let

$$\mathrm{U}_{\beta}\left(\lambda t\right) ={}_{\beta-\frac{1}{2}}\left(i\lambda t\right) + \frac{\lambda t}{2\beta+1}{}_{\beta+\frac{1}{2}}\left(i\lambda t\right) \tag{2.292}$$

and

$$\in_{\alpha}\left(\lambda t\right) = \Im_{\alpha}\left(\lambda t\right) + \frac{\lambda t}{2\left(\alpha+1\right)}\Im_{\alpha+1}\left(\lambda t\right), \tag{2.293}$$

which leads to

$$\mathrm{U}_{\beta}\left(i\lambda t\right) ={}_{\beta-\frac{1}{2}}\left(-\lambda t\right) + \frac{i\lambda t}{2\beta+1}{}_{\beta+\frac{1}{2}}\left(-\lambda t\right) \tag{2.294}$$

$$\in_{\alpha}\left(i\lambda t\right) = \Im_{\alpha}\left(i\lambda t\right) + \frac{i\lambda t}{2\left(\alpha+1\right)}\Im_{\alpha+1}\left(i\lambda t\right), \tag{2.295}$$

$$\mathrm{U}_{\beta}\left(-i\lambda t\right) ={}_{\beta-\frac{1}{2}}\left(\lambda t\right) - \frac{i\lambda t}{2\beta+1}{}_{\beta+\frac{1}{2}}\left(\lambda t\right) \tag{2.296}$$

and

$$\in_{\alpha}\left(-i\lambda t\right) = \Im_{\alpha}\left(-i\lambda t\right) - \frac{i\lambda t}{2\left(\alpha+1\right)}\Im_{\alpha+1}\left(-i\lambda t\right), \tag{2.297}$$

where

$${}_{\beta-\frac{1}{2}}\left(i\lambda t\right) = \Gamma\left(\beta+\frac{1}{2}\right)\sum_{n=0}^{\infty}\frac{1}{n!}\left(\frac{\lambda t}{2}\right)^{2n}\frac{(-1)^n}{\Gamma\left(n+\beta+\frac{1}{2}\right)}, \tag{2.298}$$

$$\in_{\alpha}\left(\lambda t\right) = \Gamma\left(\alpha+1\right)\sum_{n=0}^{\infty}\frac{1}{n!}\frac{\left(\frac{\lambda t}{2}\right)^{2n}}{\Gamma\left(n+\alpha+1\right)}, \tag{2.299}$$

$${}_{\beta+\frac{1}{2}}\left(i\lambda t\right) = \Gamma\left(\beta+\frac{3}{2}\right)\sum_{n=0}^{\infty}\frac{1}{n!}\left(\frac{\lambda t}{2}\right)^{2n}\frac{(-1)^n}{\Gamma\left(n+\beta+\frac{3}{2}\right)}, \tag{2.300}$$

and

$$\in_{\alpha+1}(\lambda t) = \Gamma(\alpha+2)\sum_{n=0}^{\infty}\frac{1}{n!}\frac{\left(\frac{\lambda t}{2}\right)^{2n}}{\Gamma(n+\alpha+2)}, \tag{2.301}$$

for $\beta = \alpha + \frac{1}{2}$.

Thus, we have

$$\mathrm{U}_{\beta}(i\lambda t) = \in_{\alpha}(i\lambda t) \tag{2.302}$$

and

$$\mathrm{U}_{\beta}(-i\lambda t) = \in_{\alpha}(-i\lambda t) \tag{2.303}$$

so that

$$\bar{\lambda}_{\alpha}\int_{-\infty}^{\infty}\mathrm{U}_{\beta}(-i\varsigma t)\, f(t)\,|t|^{2\alpha}\,dt = \hbar_{\alpha}\int_{-\infty}^{\infty}\in_{\alpha}(-i\varsigma t)\, f(t)\,|t|^{2\alpha+1}\,dt, \tag{2.304}$$

$$\bar{\lambda}_{\alpha}\int_{-\infty}^{\infty}\mathrm{U}_{\beta}(i\varsigma t)\, f(t)\,|\varsigma|^{2\beta}\,dt = \hbar_{\alpha}\int_{-\infty}^{\infty}\in_{\alpha}(i\varsigma t)\, f(t)\,|\varsigma|^{2\alpha+1}\,dt, \tag{2.305}$$

$${}_{\beta-\frac{1}{2}}(-\lambda t) = \Im_{\alpha}(i\lambda t) \tag{2.306}$$

and

$${}_{\beta-\frac{1}{2}}(-\lambda t) = \Im_{\alpha}(i\lambda t), \tag{2.307}$$

where

$$\hbar_{\alpha} = \frac{1}{2^{\alpha+1}\Gamma(\alpha+1)} \tag{2.308}$$

and

$$\bar{\lambda}_{\alpha} = \frac{1}{2^{\beta+\frac{1}{2}}\Gamma\left(\beta+\frac{1}{2}\right)}. \tag{2.309}$$

Therefore, we have the alternative forms of the Dunkl transform [200, 201].

As a special case, we have the following theorems.

Theorem 2.56 (Dunkl Theorem) *If* $\alpha \geq \frac{1}{2}$ *and* $\wp_{\alpha}[f(t)] = \widehat{\mathbb{T}_{\alpha}} f(\varsigma)$, *then there is*

$$\wp_{\alpha}\left[\widehat{\mathbb{T}_{\alpha}} f(t)\right] = i\varsigma f(\varsigma), \tag{2.310}$$

where

$$\hbar_\alpha = \frac{1}{2^{\alpha+1}\Gamma(\alpha+1)}. \tag{2.311}$$

The result was discovered by Dunkl in 1989 [198] and reported by de Jeu [199] and by Soltani [200].

Theorem 2.57 *If* $\alpha \geq \frac{1}{2}$ *and* $\wp_\alpha[f(t)] = \widehat{\mathbb{T}_\alpha} f(\varsigma)$, *then there is*

$$\wp_\alpha\left[\widehat{H_\alpha} f(t)\right] = \frac{1}{i\varsigma} f(\varsigma), \tag{2.312}$$

where

$$\hbar_\alpha = \frac{1}{2^{\alpha+1}\Gamma(\alpha+1)}. \tag{2.313}$$

Proof Making use of

$$\wp_\alpha\left[\widehat{\mathbb{T}_\alpha} f(t)\right] = i\varsigma f(\varsigma) \tag{2.314}$$

we have

$$\wp_\alpha\left[\widehat{H_\alpha}\left(\widehat{\mathbb{T}_\alpha} f(t)\right)\right] = f(\varsigma) \tag{2.315}$$

and

$$\wp_\alpha\left[\widehat{I_\alpha} f(t)\right] = \wp_\alpha[f(t)] = f(\varsigma) \tag{2.316}$$

such that

$$\wp_\alpha\left[\widehat{H_\alpha}\left(\widehat{\mathbb{T}_\alpha} f(t)\right)\right] = \wp_\alpha\left[\widehat{I_\alpha} f(t)\right]. \tag{2.317}$$

Thus, we reduce to the result. □

Theorem 2.58 (Dunkl Transform Theorem) *If* $\alpha \geq \frac{1}{2}$, $\wp_\alpha[f(t)] = f(\varsigma)$, *and* $\wp_\alpha^{-1}[f(\varsigma)] = f(t)$, *then there exist*

$$\wp_\alpha^{-1}[\wp_\alpha[f(t)]] = f(t) \tag{2.318}$$

and

$$\wp_\alpha\left[\wp_\alpha^{-1}[f(\varsigma)]\right] = f(\varsigma), \tag{2.319}$$

where

$$\hbar_\alpha = \frac{1}{2^{\alpha+1}\Gamma(\alpha+1)}. \tag{2.320}$$

Proof According to the definition of the Dunkl transform, we directly reduce to the result. □

The result was discovered by Dunkl in 1989 [198] and reported by Soltani [200].

Corollary 2.25 *If* $\alpha \geq \frac{1}{2}$, $\wp_\alpha[f(t)] = f(\varsigma)$, *and* $\wp_\alpha^{-1}[f(\varsigma)] = f(t)$, *then there exist*

$$\hbar_\alpha \int_{-\infty}^{\infty} |\varsigma|^{2\alpha+1} \in_\alpha (i\varsigma\tau) \left(\hbar_\alpha \int_{-\infty}^{\infty} |\tau|^{2\alpha+1} \in_\alpha (-i\varsigma\tau)\, d\tau\right) d\varsigma = 1 \tag{2.321}$$

and

$$\hbar_\alpha \int_{-\infty}^{\infty} |\tau|^{2\alpha+1} \in_\alpha (-i\varsigma\tau) \left(\hbar_\alpha \int_{-\infty}^{\infty} |\varsigma|^{2\alpha+1} \in_\alpha (i\varsigma\tau)\, d\varsigma\right) d\tau = 1, \tag{2.322}$$

where

$$\hbar_\alpha = \frac{1}{2^{\alpha+1}\Gamma(\alpha+1)}. \tag{2.323}$$

Proof In view of

$$\wp_\alpha^{-1}[\wp_\alpha[f(t)]] = f(t) \tag{2.324}$$

and

$$\wp_\alpha\left[\wp_\alpha^{-1}[f(\varsigma)]\right] = f(\varsigma), \tag{2.325}$$

and taking $f(t) = 1$ and $f(\varsigma) = 1$, we obtain the results, respectively. □

Theorem 2.59 *If* $\alpha \geq \frac{1}{2}$, *then we have*

$$\int_{-\infty}^{\infty} |\varsigma|^{2\alpha+1} \in_\alpha (i\varsigma\tau)\, d\varsigma = l\tau^{-2(\alpha+1)} \tag{2.326}$$

and

$$\int_{-\infty}^{\infty} |\tau|^{2\alpha+1} \in_\alpha (-i\varsigma\tau)\, d\tau = l\tau^{-2(\alpha+1)}, \tag{2.327}$$

where $\varsigma \in \mathbb{C}$ *and*

$$\ell = 2^{2(\alpha+1)}\Gamma^2(1+\alpha). \tag{2.328}$$

Proof Making use of Eqs. (2.273) and (2.274) and considering that

$$y_{-\infty}^{\infty}|\varsigma|^{2\alpha+1}\frac{\varsigma\tau}{2}\;{}_0\varphi_1\left(\begin{matrix}-\\(\alpha+2,1)\end{matrix};-\frac{\varsigma^2\tau^2}{4}\right)d\varsigma=0, \tag{2.329}$$

and

$$\begin{aligned}&\int_{-\infty}^{\infty}|\varsigma|^{2\alpha+1}\;{}_0\varphi_1\left(\begin{matrix}-\\(\alpha+1,1)\end{matrix};-\frac{\varsigma^2\tau^2}{4}\right)d\varsigma\\&=2\int_{0}^{\infty}\varsigma^{2\alpha+1}\;{}_0\varphi_1\left(\begin{matrix}-\\(\alpha+1,1)\end{matrix};-\frac{\varsigma^2\tau^2}{4}\right)d\varsigma,\end{aligned} \tag{2.330}$$

we have

$$\begin{aligned}&\int_{-\infty}^{\infty}|\varsigma|^{2\alpha+1}\,\epsilon_\alpha\,(i\varsigma t)\,d\varsigma\\&=\Gamma\,(1+\alpha)\int_{-\infty}^{\infty}|\varsigma|^{2\alpha+1}\;{}_0\varphi_1\left(\begin{matrix}-\\(\alpha+1,1)\end{matrix};-\frac{\varsigma^2\tau^2}{4}\right)d\varsigma\\&\quad+i\Gamma\,(1+\alpha)\int_{-\infty}^{\infty}|\varsigma|^{2\alpha+1}\frac{\varsigma\tau}{2}\;{}_0\varphi_1\left(\begin{matrix}-\\(\alpha+2,1)\end{matrix};-\frac{\varsigma^2\tau^2}{4}\right)d\varsigma\\&=\Gamma\,(1+\alpha)\int_{-\infty}^{\infty}|\varsigma|^{2\alpha+1}\;{}_0\varphi_1\left(\begin{matrix}-\\(\alpha+1,1)\end{matrix};-\frac{\varsigma^2\tau^2}{4}\right)d\varsigma\\&=\Gamma\,(1+\alpha)\int_{-\infty}^{\infty}\varsigma^{2\alpha+1}\;{}_0\varphi_1\left(\begin{matrix}-\\(\alpha+1,1)\end{matrix};-\frac{\varsigma^2\tau^2}{4}\right)d\varsigma\\&=2\Gamma\,(1+\alpha)\int_{0}^{\infty}\varsigma^{2\alpha+1}\;{}_0\varphi_1\left(\begin{matrix}-\\(\alpha+1,1)\end{matrix};-\frac{\varsigma^2\tau^2}{4}\right)d\varsigma\\&=2^{2(\alpha+1)}\Gamma^2\,(1+\alpha)\,\tau^{-2(\alpha+1)}\end{aligned} \tag{2.331}$$

and

$$\begin{aligned}&y_{-\infty}^{\infty}|\varsigma|^{2\alpha+1}\,\epsilon_\alpha\,(-i\varsigma t)\,d\varsigma\\&=\Gamma\,(1+\alpha)\int_{-\infty}^{\infty}|\varsigma|^{2\alpha+1}\;{}_0\varphi_1\left(\begin{matrix}-\\(\alpha+1,1)\end{matrix};-\frac{\varsigma^2\tau^2}{4}\right)d\varsigma\\&\quad-i\Gamma\,(1+\alpha)\int_{-\infty}^{\infty}|\varsigma|^{2\alpha+1}\frac{\varsigma\tau}{2}\;{}_0\varphi_1\left(\begin{matrix}-\\(\alpha+2,1)\end{matrix};-\frac{\varsigma^2\tau^2}{4}\right)d\varsigma\\&=\Gamma\,(1+\alpha)\int_{-\infty}^{\infty}|\varsigma|^{2\alpha+1}\;{}_0\varphi_1\left(\begin{matrix}-\\(\alpha+1,1)\end{matrix};-\frac{\varsigma^2\tau^2}{4}\right)d\varsigma\\&=2\Gamma\,(1+\alpha)\,y_{0}^{\infty}\varsigma^{2\alpha+1}\;{}_0\varphi_1\left(\begin{matrix}-\\(\alpha+1,1)\end{matrix};-\frac{\varsigma^2\tau^2}{4}\right)d\varsigma\\&=2^{2(\alpha+1)}\Gamma^2\,(1+\alpha)\,\tau^{-2(\alpha+1)}\end{aligned} \tag{2.332}$$

since

$$\int_0^\infty \varsigma^{2\alpha+1}\ {}_0\varphi_1\left(\begin{matrix} - \\ (\alpha+1,1)\end{matrix};-\lambda\varsigma^2\right)d\varsigma=\frac{\lambda^{-(\alpha+1)}\Gamma(\alpha+1)}{2}. \tag{2.333}$$

□

Definition 2.40 Let $\Theta(\tau)=\left(f\circ\tau^\beta\right)(\tau)=f\left(\tau^\beta\right)$, where $\tau\in\mathbb{R}$, $\beta>0$, $\varsigma^{\beta-1}>0$, and $\tau^{\beta-1}>0$, and let $\alpha\geq\frac{1}{2}$.

The Dunkl type transform of a function $\Theta(\tau)$ on $\mathbb{R}$ is defined as follows:

$$\begin{aligned}&\widehat{\wp}_{\alpha,\beta}[\Theta(\tau)]\\&=\Theta(\varsigma)\\&=\beta\int_{-\infty}^{\infty}\in_\alpha\left(-i\varsigma^\beta\tau^\beta\right)\Theta(\tau)|\tau|^{2\alpha\beta-1}d\tau,\end{aligned} \tag{2.334}$$

and the inverse Dunkl transform on $\mathbb{R}$ as

$$\begin{aligned}&\widehat{\wp}^{-1}_{\alpha,\beta}[\Theta(\varsigma)]\\&=\Theta(\tau)\\&=\hbar_\alpha^2\beta\int_{-\infty}^{\infty}\in_\alpha\left(i\varsigma^\beta\tau^\beta\right)\Theta(\varsigma)|\varsigma|^{2\alpha\beta-1}d\varsigma,\end{aligned} \tag{2.335}$$

where

$$\hbar_\alpha=\frac{1}{2^{\alpha+1}\Gamma(\alpha+1)}. \tag{2.336}$$

Theorem 2.60 *If* $\alpha\geq\frac{1}{2}$, $\tau\in\mathbb{R}$, $\beta>0,\varsigma^{\beta-1}>0$, $\tau^{\beta-1}>0$, $\widehat{\wp}_{\alpha,\beta}[\Theta(\tau)]=\Theta(\varsigma)$, *and* $\widehat{\wp}^{-1}_{\alpha,\beta}[\Theta(\varsigma)]=\Theta(\tau)$, *then there exist*

$$\widehat{\wp}^{-1}_{\alpha,\beta}\left[\widehat{\wp}_{\alpha,\beta}[\Theta(\tau)]\right]=\Theta(\tau) \tag{2.337}$$

and

$$\widehat{\wp}_{\alpha,\beta}\left[\widehat{\wp}^{-1}_{\alpha,\beta}[\Theta(\varsigma)]\right]=\Theta(\varsigma). \tag{2.338}$$

Proof Let ς^β and τ^β take the place of ς and τin (2.318) and (2.319), respectively, and we have the results. □

Theorem 2.61 *If* $\alpha\geq\frac{1}{2}$, $\tau\in\mathbb{R}$, $\beta>0,\varsigma^{\beta-1}>0$, $\tau^{\beta-1}>0$, $\widehat{\wp}_{\alpha,\beta}[\Theta(\tau)]=\Theta(\varsigma)$, *and* $\widehat{\wp}^{-1}_{\alpha,\beta}[\Theta(\varsigma)]=\Theta(\tau)$, *then there exist*

$$\widehat{\wp}^{-1}_{\alpha,\beta}\left[\widehat{\wp}_{\alpha,\beta}[1]\right]=1 \tag{2.339}$$

and

$$\widehat{\wp}_{\alpha,\beta}\left[\widehat{\wp}^{-1}_{\alpha,\beta}[1]\right] = 1. \tag{2.340}$$

Proof Making use of (2.337) and (2.338), we reduce the results. □

Definition 2.41 Let $\Theta(\tau) = \left(f \circ \tau^{\beta}\right)(\tau) = f\left(\tau^{\beta}\right)$, where $\tau \in \mathbb{R}$, $\beta > 0$, $\varsigma^{\beta-1} > 0$, and $\tau^{\beta-1} > 0$, and let $\alpha \geq \frac{1}{2}$.

The Dunkl type transform of a function $\Theta(\tau)$ on $\mathbb{R}$ is defined as follows:

$$\begin{aligned} &\widehat{\mathbb{S}}_{\alpha,\beta}\left[\Theta(\tau)\right] \\ &\quad = \Theta(\varsigma) \\ &\quad = \hbar_{\alpha}\beta \int_{-\infty}^{\infty} \in_{\alpha}\left(-i\varsigma^{\beta}\tau^{\beta}\right)\Theta(\tau)\left|\tau\right|^{2\alpha\beta-1}d\tau, \end{aligned} \tag{2.341}$$

and the inverse Dunkl transform on $\mathbb{R}$ as

$$\begin{aligned} &\widehat{\mathbb{S}}^{-1}_{\alpha,\beta}\left[\Theta(\varsigma)\right] \\ &\quad = \Theta(\tau) \\ &\quad = \hbar_{\alpha}\beta y_{-\infty}^{\infty} \in_{\alpha}\left(i\varsigma^{\beta}\tau^{\beta}\right)\Theta(\varsigma)\left|\varsigma\right|^{2\alpha\beta-1}d\varsigma, \end{aligned} \tag{2.342}$$

where

$$\hbar_{\alpha} = \frac{1}{2^{\alpha+1}\Gamma(\alpha+1)}. \tag{2.343}$$

Theorem 2.62 *If* $\alpha \geq \frac{1}{2}$, $\tau \in \mathbb{R}$, $\beta > 0, \varsigma^{\beta-1} > 0$, $\tau^{\beta-1} > 0$, $\widehat{\mathbb{S}}_{\alpha,\beta}\left[\Theta(\tau)\right] = \Theta(\varsigma)$, *and* $\widehat{\mathbb{S}}^{-1}_{\alpha,\beta}\left[\Theta(\varsigma)\right] = \Theta(\tau)$, *then there exist*

$$\widehat{\mathbb{S}}^{-1}_{\alpha,\beta}\left[\widehat{\mathbb{S}}_{\alpha,\beta}\left[\Theta(\tau)\right]\right] = \Theta(\tau) \tag{2.344}$$

and

$$\widehat{\mathbb{S}}_{\alpha,\beta}\left[\widehat{\mathbb{S}}^{-1}_{\alpha,\beta}\left[\Theta(\varsigma)\right]\right] = \Theta(\varsigma). \tag{2.345}$$

Proof With the aid of (2.318), (2.319), (2.341), and (2.342), we obtain the results. □

Theorem 2.63 *If* $\alpha \geq \frac{1}{2}$, $\tau \in \mathbb{R}$, $\beta > 0, \varsigma^{\beta-1} > 0$*and* $\tau^{\beta-1} > 0$, $\widehat{\mathbb{S}}_{\alpha,\beta}\left[\Theta(\tau)\right] = \Theta(\varsigma)$, *and* $\widehat{\mathbb{S}}^{-1}_{\alpha,\beta}\left[\Theta(\varsigma)\right] = \Theta(\tau)$, *then there exist*

$$\widehat{\mathbb{S}}^{-1}_{\alpha,\beta}\left[\widehat{\mathbb{S}}_{\alpha,\beta}[1]\right] = 1 \tag{2.346}$$

and

$$\widehat{\mathbb{S}}_{\alpha,\beta}\left[\widehat{\mathbb{S}}_{\alpha,\beta}^{-1}[1]\right] = 1. \tag{2.347}$$

Proof In view of (2.345) and (2.346) we reduce to results. □

Definition 2.42 Let $\Pi(\tau) = \left(f \circ \tau^{\beta}\right)(\tau) = f\left(\tau^{\beta}\right)$, where $\tau \in \mathbb{R}$, $\beta > 0$, and $\tau^{\beta-1} > 0$, and let $\alpha \geq \frac{1}{2}$.

The Dunkl type transform of a function $\Pi(\tau)$ on $\mathbb{R}$ is defined as follows:

$$\begin{aligned}&\overline{\overline{\wp}}_{\alpha}^{\beta}[\Pi(\tau)]\\&= \Pi(\varsigma)\\&= \beta\int_{-\infty}^{\infty} \in_{\alpha}\left(-i\varsigma\tau^{\beta}\right)\Pi(\tau)\,|\tau|^{2\alpha\beta-1}\,dt,\end{aligned} \tag{2.348}$$

and the inverse Dunkl transform on $\mathbb{R}$ as

$$\begin{aligned}&\left(\overline{\overline{\wp}}_{\alpha}^{\beta}\right)^{-1}[\Pi(\varsigma)]\\&= \Pi(\tau)\\&= \hbar_{\alpha}^{2}\int_{-\infty}^{\infty} \in_{\alpha}\left(i\varsigma\tau^{\beta}\right) f(\varsigma)\,|\varsigma|^{2\alpha+1}\,d\varsigma,\end{aligned} \tag{2.349}$$

where

$$\hbar_{\alpha} = \frac{1}{2^{\alpha+1}\Gamma(\alpha+1)}. \tag{2.350}$$

Theorem 2.64 *If* $\alpha \geq \frac{1}{2}$, $\tau \in \mathbb{R}$, $\beta > 0$, $\tau^{\beta-1} > 0$, $\overline{\overline{\wp}}_{\alpha}^{\beta}[\Pi(\tau)] = \Pi(\varsigma)$, *and* $\left(\overline{\overline{\wp}}_{\alpha}^{\beta}\right)^{-1}[\Pi(\varsigma)] = \Pi(\tau)$, *then there exist*

$$\left(\overline{\overline{\wp}}_{\alpha}^{\beta}\right)^{-1}\left[\overline{\overline{\wp}}_{\alpha}^{\beta}[\Pi(\tau)]\right] = \Pi(\tau) \tag{2.351}$$

and

$$\overline{\overline{\wp}}_{\alpha}^{\beta}\left[\left(\overline{\overline{\wp}}_{\alpha}^{\beta}\right)^{-1}[\Pi(\varsigma)]\right] = \Pi(\varsigma). \tag{2.352}$$

Proof With the aid of (2.318), (2.319), (2.348), and (2.349) we reduce to the results. □

Theorem 2.65 *If* $\alpha \geq \frac{1}{2}$, $\tau \in \mathbb{R}$, $\beta > 0$, $\tau^{\beta-1} > 0$, $\overline{\overline{\wp}}_{\alpha}^{\beta}[\Pi(\tau)] = \Pi(\varsigma)$, *and* $\left(\overline{\overline{\wp}}_{\alpha}^{\beta}\right)^{-1}[\Pi(\varsigma)] = \Pi(\tau)$, *then there exist*

$$\left(\overline{\overline{\wp}}_{\alpha}^{\beta}\right)^{-1}\left[\overline{\overline{\wp}}_{\alpha}^{\beta}[1]\right] = 1 \tag{2.353}$$

and

$$\overline{\overline{\wp}}_{\alpha}^{\beta}\left[\left(\overline{\overline{\wp}}_{\alpha}^{\beta}\right)^{-1}[1]\right]=1. \tag{2.354}$$

Proof Making use of (2.353) and (2.354) we reduce to the results. □

Definition 2.43 Let $\tau \in \mathbb{R}$, $\beta > 0$ and $\varsigma^{\beta-1} > 0$, and let $\alpha \geq \frac{1}{2}$.
The Dunkl type transform of a function $\Pi(\tau)$ on $\mathbb{R}$ is defined as follows:

$$\begin{aligned}&\overline{\overline{Y}}_{\alpha}^{\beta}[\Pi(\tau)]\\&=\Pi(\varsigma)\\&=\int_{-\infty}^{\infty}\epsilon_{\alpha}\left(-i\varsigma^{\beta}\tau\right)\Pi(\tau)|\tau|^{2\alpha+1}dt\end{aligned}, \tag{2.355}$$

and the inverse Dunkl transform on $\mathbb{R}$ as

$$\begin{aligned}&\left(\overline{\overline{Y}}_{\alpha}^{\beta}\right)^{-1}[\Pi(\varsigma)]\\&=\Pi(\tau)\\&=\beta\hbar_{\alpha}^{2}\int_{-\infty}^{\infty}\epsilon_{\alpha}\left(i\varsigma^{\beta}\tau\right)f(\varsigma)|\varsigma|^{2\alpha\beta-1}d\varsigma\end{aligned}, \tag{2.356}$$

where

$$\hbar_{\alpha}=\frac{1}{2^{\alpha+1}\Gamma(\alpha+1)}. \tag{2.357}$$

Theorem 2.66 *If* $\alpha \geq \frac{1}{2}$, $\tau \in \mathbb{R}$, $\beta > 0$, $\tau^{\beta-1} > 0$, $\overline{\overline{Y}}_{\alpha}^{\beta}[\Pi(\tau)] = \Pi(\varsigma)$ *and* $\left(\overline{\overline{Y}}_{\alpha}^{\beta}\right)^{-1}[\Pi(\varsigma)] = \Pi(\tau)$, *then there exist*

$$\left(\overline{\overline{Y}}_{\alpha}^{\beta}\right)^{-1}\left[\overline{\overline{Y}}_{\alpha}^{\beta}[\Pi(\tau)]\right]=\Pi(\tau) \tag{2.358}$$

and

$$\overline{\overline{Y}}_{\alpha}^{\beta}\left[\left(\overline{\overline{Y}}_{\alpha}^{\beta}\right)^{-1}[\Pi(\varsigma)]\right]=\Pi(\varsigma). \tag{2.359}$$

Proof With the use of (2.318), (2.319), (2.355), and (2.356) we obtain the results. □

Theorem 2.67 *If* $\alpha \geq \frac{1}{2}$, $\tau \in \mathbb{R}$, $\beta > 0$, $\tau^{\beta-1} > 0$, $\overline{\overline{Y}}_{\alpha}^{\beta} [\Pi(\tau)] = \Pi(\varsigma)$, *and* $\left(\overline{\overline{Y}}_{\alpha}^{\beta}\right)^{-1} [\Pi(\varsigma)] = \Pi(\tau)$, *then there exist*

$$\left(\overline{\overline{Y}}_{\alpha}^{\beta}\right)^{-1} \left[\overline{\overline{Y}}_{\alpha}^{\beta} [1]\right] = 1 \tag{2.360}$$

and

$$\overline{\overline{Y}}_{\alpha}^{\beta} \left[\left(\overline{\overline{Y}}_{\alpha}^{\beta}\right)^{-1} [1]\right] = 1. \tag{2.361}$$

Proof In view of (2.358) and (2.359) we deduce the results. □

Definition 2.44 Let $\Theta(\tau) = (f \circ \tau^{\alpha})(\tau) = f(\tau^{\alpha})$, where $\tau \in \mathbb{R}, \alpha > 0, \varsigma^{\alpha-1} > 0$ and $\tau^{\alpha-1} > 0$, and let $\alpha \geq \frac{1}{2}$.

The Dunkl type transform of a function $\Theta(\tau)$ on $\mathbb{R}$ is defined as follows:

$$\begin{aligned} &\widehat{\wp}_{\alpha,\alpha}[\Theta(\tau)] \\ &= \Theta(\varsigma) \\ &= \alpha \int_{-\infty}^{\infty} \in_{\alpha}(-i\varsigma^{\alpha}\tau^{\alpha})\, \Theta(\tau)\, |\tau|^{2\alpha^2-1}\, d\tau \end{aligned}, \tag{2.362}$$

and the inverse Dunkl transform on $\mathbb{R}$ as

$$\begin{aligned} &\widehat{\wp}_{\alpha,\alpha}^{-1}[\Theta(\varsigma)] \\ &= \Theta(\tau) \\ &= \alpha \hbar_{\alpha}^{2} \int_{-\infty}^{\infty} \in_{\alpha}(i\varsigma^{\alpha}\tau^{\alpha})\, \Theta(\varsigma)\, |\varsigma|^{2\alpha^2-1}\, d\varsigma \end{aligned}, \tag{2.363}$$

where

$$\hbar_{\alpha} = \frac{1}{2^{\alpha+1}\Gamma(\alpha+1)}. \tag{2.364}$$

Theorem 2.68 *If* $\alpha \geq \frac{1}{2}$, $\tau \in \mathbb{R}$, $\varsigma^{\alpha-1} > 0$, $\tau^{\alpha-1} > 0$, $\widehat{\wp}_{\alpha,\alpha}[\Theta(\tau)] = \Theta(\varsigma)$ *and* $\widehat{\wp}_{\alpha,\alpha}^{-1}[\Theta(\varsigma)] = \Theta(\tau)$, *then there exist*

$$\widehat{\wp}_{\alpha,\alpha}^{-1}\left[\widehat{\wp}_{\alpha,\alpha}[\Theta(\tau)]\right] = \Theta(\tau) \tag{2.365}$$

and

$$\widehat{\wp}_{\alpha,\alpha}\left[\widehat{\wp}_{\alpha,\alpha}^{-1}[\Theta(\varsigma)]\right] = \Theta(\varsigma). \tag{2.366}$$

Proof Making use of $\alpha = \beta$ in (2.355) and (2.356), we deduce the results. □

Theorem 2.69 *If* $\alpha \geq \frac{1}{2}$, $\tau \in \mathbb{R}$, $\varsigma^{\alpha-1} > 0$, $\tau^{\alpha-1} > 0$, $\widehat{\wp}_{\alpha,\alpha}[\Theta(\tau)] = \Theta(\varsigma)$, *and* $\widehat{\wp}_{\alpha,\alpha}^{-1}[\Theta(\varsigma)] = \Theta(\tau)$, *then there exist*

$$\widehat{\wp}_{\alpha,\alpha}^{-1}\left[\widehat{\wp}_{\alpha,\alpha}[1]\right] = 1 \tag{2.367}$$

and

$$\widehat{\wp}_{\alpha,\alpha}\left[\widehat{\wp}_{\alpha,\alpha}^{-1}[1]\right] = 1. \tag{2.368}$$

Proof Making use of $\Theta(\tau) = 1$ and $\Theta(\varsigma) = 1$ in (2.365) and (2.366), respectively, we obtain the results. □

Definition 2.45 Let $\Theta(\tau) = f(\tau^{\alpha})$, $\tau \in \mathbb{R}$, $\alpha > 0$, $\tau^{\alpha-1} > 0$, and $\varsigma^{\alpha-1} > 0$.
The Dunkl type transform of a function $\Theta(\tau)$ on $\mathbb{R}$ is defined as follows:

$$\begin{aligned}\widehat{\mathbb{F}}_{\alpha,\alpha}&[\Theta(\tau)]\\ &= \Theta(\varsigma)\\ &= \alpha\hbar_{\alpha}\int_{-\infty}^{\infty} \in_{\alpha}(-i\varsigma^{\alpha}\tau^{\alpha})\,\Theta(\tau)\,|\tau|^{2\alpha^2-1}\,d\tau,\end{aligned} \tag{2.369}$$

and the inverse Dunkl transform on $\mathbb{R}$ as

$$\begin{aligned}\widehat{\mathbb{F}}_{\alpha,\alpha}^{-1}&[\Theta(\varsigma)]\\ &= \Theta(\tau)\\ &= \alpha\hbar_{\alpha}\int_{-\infty}^{\infty} \in_{\alpha}(i\varsigma^{\alpha}\tau^{\alpha})\,\Theta(\varsigma)\,|\varsigma|^{2\alpha^2-1}\,d\varsigma,\end{aligned} \tag{2.370}$$

where

$$\hbar_{\alpha} = \frac{1}{2^{\alpha+1}\Gamma(\alpha+1)}. \tag{2.371}$$

Theorem 2.70 *If* $\alpha \geq \frac{1}{2}$, $\tau \in \mathbb{R}$, $\tau^{\alpha-1} > 0$, $\varsigma^{\alpha-1} > 0$, $\widehat{\mathbb{F}}_{\alpha,\alpha}[\Theta(\tau)] = \Theta(\varsigma)$, *and* $\widehat{\mathbb{F}}_{\alpha,\alpha}^{-1}[\Theta(\varsigma)] = \Theta(\tau)$, *then there exist*

$$\widehat{\mathbb{F}}_{\alpha,\alpha}^{-1}\left[\widehat{\mathbb{F}}_{\alpha,\alpha}[\Theta(\tau)]\right] = \Theta(\tau) \tag{2.372}$$

and

$$\widehat{\mathbb{F}}_{\alpha,\alpha}\left[\widehat{\mathbb{F}}_{\alpha,\alpha}^{-1}[\Theta(\varsigma)]\right] = \Theta(\varsigma). \tag{2.373}$$

Proof With the aid of (2.318) and (2.319) we obtain the results. □

Theorem 2.71 *If* $\alpha \geq \frac{1}{2}$, $\tau \in \mathbb{R}$, $\tau^{\alpha-1} > 0$, $\varsigma^{\alpha-1} > 0$, $\widehat{\mathbb{F}}_{\alpha,\alpha}[\Theta(\tau)] = \Theta(\varsigma)$, *and* $\widehat{\mathbb{F}}^{-1}_{\alpha,\alpha}[\Theta(\varsigma)] = \Theta(\tau)$, *then there exist*

$$\widehat{\mathbb{F}}^{-1}_{\alpha,\alpha}\left[\widehat{\wp}_{\alpha,\alpha}[1]\right] = 1 \tag{2.374}$$

and

$$\widehat{\mathbb{F}}_{\alpha,\alpha}\left[\widehat{\wp}^{-1}_{\alpha,\alpha}[1]\right] = 1. \tag{2.375}$$

Proof On setting $\Theta(\tau) = 1$ and $\Theta(\varsigma) = 1$ in (2.372) and (2.373), respectively, we reduce to the results. □

2.3.2 New Integral Transforms of First Type

Taking $\zeta = i\varsigma$ in the Dunkl transform, we have the following integral representation.

Definition 2.46 Let $\alpha \geq \frac{1}{2}$.

The transform of a function $f(t)$ on $\mathbb{R}$ is defined as follows:

$$\begin{aligned} &\mathbb{B}_\alpha[f(t)] \\ &= f(\zeta) \\ &= \int_{-\infty}^{\infty} \in_\alpha(-\zeta t)\, f(t)\, |t|^{2\alpha+1}\, dt, \end{aligned} \tag{2.376}$$

and the inverse transform on $\mathbb{R}$ as

$$\begin{aligned} &\mathbb{B}_\alpha^{-1}[f(\zeta)] \\ &= f(t) \\ &= \hbar_\alpha^2 \lim_{T\to\infty} \int_{-iT}^{+iT} \in_\alpha(\zeta t)\, f(\zeta)\, |\zeta|^{2\alpha+1}\, d\zeta, \end{aligned} \tag{2.377}$$

where

$$\hbar_\alpha = \frac{1}{2^{\alpha+1}\Gamma(\alpha+1)}. \tag{2.378}$$

Note: When $\alpha = \frac{1}{2}$, one obtains the Laplace transform.

Theorem 2.72 *If* $\alpha \geq \frac{1}{2}$ *and* $\mathbb{B}_\alpha[f(t)] = f(\zeta)$ *and* $\mathbb{B}_\alpha^{-1}[f(\zeta)] = f(t)$, *then there exist*

$$\mathbb{B}_\alpha^{-1}[\mathbb{B}_\alpha[f(t)]] = f(t) \tag{2.379}$$

and

$$\mathbb{B}_\alpha\left[\mathbb{B}_\alpha^{-1}\left[f(\zeta)\right]\right] = f(\zeta), \tag{2.380}$$

where

$$\hbar_\alpha = \frac{1}{2^{\alpha+1}\Gamma(\alpha+1)}. \tag{2.381}$$

Proof Setting $\zeta = i\varsigma$ in (2.318) and (2.319), we obtain the result. □

Theorem 2.73 *If* $\alpha \geq \frac{1}{2}$ *and* $\mathbb{B}_\alpha\left[f(t)\right] = f(\zeta)$, *then there is*

$$\mathbb{B}_\alpha\left[\widehat{\mathbb{T}_\alpha} f(t)\right] = \zeta f(\zeta). \tag{2.382}$$

Proof Setting $\zeta = i\varsigma$ in the Dunkl transform, we obtain the result. □

Theorem 2.74 *If* $\alpha \geq \frac{1}{2}$ *and* $\mathbb{B}_\alpha\left[f(t)\right] = f(\zeta)$, *then there is*

$$\mathbb{B}_\alpha\left[\widehat{H_\alpha} f(t)\right] = \frac{1}{\zeta} f(\zeta). \tag{2.383}$$

Proof Setting $\zeta = i\varsigma$ in the Dunkl transform, we show the result. □

Corollary 2.26 *If* $\alpha \geq \frac{1}{2}$ *and* $\mathbb{B}_\alpha\left[f(t)\right] = f(\zeta)$ *and* $\mathbb{B}_\alpha^{-1}\left[f(\zeta)\right] = f(t)$, *then there exist*

$$\hbar_\alpha \int_{-\infty}^{\infty} |\zeta|^{2\alpha+1} \in_\alpha (\zeta t) \left(\hbar_\alpha \int_{-\infty}^{\infty} |\tau|^{2\alpha+1} \in_\alpha (-\zeta\tau)\, d\tau\right) d\zeta = 1 \tag{2.384}$$

and

$$\hbar_\alpha \int_{-\infty}^{\infty} |\tau|^{2\alpha+1} \in_\alpha (-\zeta\tau) \left(\hbar_\alpha \int_{-\infty}^{\infty} |\varsigma|^{2\alpha+1} \in_\alpha (\zeta t)\, d\varsigma\right) d\tau = 1, \tag{2.385}$$

where

$$\hbar_\alpha = \frac{1}{2^{\alpha+1}\Gamma(\alpha+1)}. \tag{2.386}$$

Proof Taking $\zeta = i\varsigma$ in the Dunkl transform, there is

$$|\zeta| = |i\varsigma| = |\varsigma| \tag{2.387}$$

such that we obtain the result. □

Definition 2.47 Let $\alpha \geq \frac{1}{2}$.

The transform of a function $f(t)$ on $\mathbb{R}$ is defined as follows:

$$\begin{aligned} &\mathbb{B}_\alpha [f(t)] \\ &= f(\zeta) \\ &= \hbar_\alpha \int_{-\infty}^{\infty} \in_\alpha (-\zeta t) f(t) |t|^{2\alpha+1} dt \end{aligned}, \tag{2.388}$$

and the inverse transform on $\mathbb{R}$ as

$$\begin{aligned} &\mathbb{B}_\alpha^{-1} [f(\zeta)] \\ &= f(t) \\ &= \hbar_\alpha \lim_{T\to\infty} \int_{-iT}^{+iT} \in_\alpha (\zeta t) f(\zeta) |\zeta|^{2\alpha+1} d\zeta \end{aligned}, \tag{2.389}$$

where

$$\hbar_\alpha = \frac{1}{2^{\alpha+1}\Gamma(\alpha+1)}. \tag{2.390}$$

When $\alpha = \frac{1}{2}$, one obtains the Laplace transform.

Theorem 2.75 *If* $\alpha \geq \frac{1}{2}$ *and* $\mathbb{B}_\alpha [f(t)] = f(\zeta)$ *and* $\mathbb{B}_\alpha^{-1} [f(\zeta)] = f(t)$*, then there exist*

$$\mathbb{B}_\alpha^{-1} [\mathbb{B}_\alpha [f(t)]] = f(t) \tag{2.391}$$

and

$$\mathbb{B}_\alpha \left[\mathbb{B}_\alpha^{-1} [f(\zeta)]\right] = f(\zeta). \tag{2.392}$$

Proof Making use of (2.379) and (2.380), we get the results. □

Definition 2.48 Let $\Theta(t) = \left(f \circ t^\beta\right)(t) = f\left(t^\beta\right)$, where $\tau \in \mathbb{R}$, $\beta > 0$and $\tau^{\beta-1} > 0$ and let $\alpha \geq \frac{1}{2}$.

The transform of a function $\Theta(t)$ on $\mathbb{R}$ is defined as follows:

$$\begin{aligned} &\widehat{\mathbb{B}}_{\alpha,\beta} [\Theta(t)] \\ &= \Theta(\zeta) \\ &= \beta \int_{-\infty}^{\infty} \in_\alpha \left(-\zeta^\beta t^\beta\right) \Theta(t) |t|^{2\alpha\beta-1} dt, \end{aligned} \tag{2.393}$$

and the inverse transform on $\mathbb{R}$ as

$$\begin{aligned} &\widehat{\mathbb{B}}_{\alpha,\beta}^{-1} [\Theta(\zeta)] \\ &= \Theta(t) \\ &= \beta\hbar_\alpha^2 \lim_{T\to\infty} \int_{-iT}^{+iT} \in_\alpha \left(\zeta^\beta t^\beta\right) \Theta(\zeta) |\zeta|^{2\alpha\beta-1} d\zeta, \end{aligned} \tag{2.394}$$

where

$$\hbar_\alpha = \frac{1}{2^{\alpha+1}\Gamma(\alpha+1)}. \tag{2.395}$$

Theorem 2.76 *If* $\alpha \geq \frac{1}{2}, \beta > 0$, $\tau^{\beta-1} > 0$, $\widehat{\mathbb{B}}_{\alpha,\beta}[\Theta(t)] = \Theta(\zeta)$, *and* $\widehat{\mathbb{B}}^{-1}_{\alpha,\beta}[\Theta(\zeta)] = \Theta(t)$, *then there exist*

$$\widehat{\mathbb{B}}^{-1}_{\alpha,\beta}\left[\widehat{\mathbb{B}}_{\alpha,\beta}[\Theta(t)]\right] = \Theta(t) \tag{2.396}$$

and

$$\widehat{\mathbb{B}}_{\alpha,\beta}\left[\widehat{\mathbb{B}}^{-1}_{\alpha,\beta}[\Theta(\zeta)]\right] = \Theta(\zeta). \tag{2.397}$$

Proof In view of (2.376) and (2.377), we reduce to the results. □

Definition 2.49 Let $\Theta(t) = \left(f \circ t^\beta\right)(t) = f\left(t^\beta\right)$, where $\tau \in \mathbb{R}$, $\alpha \geq \frac{1}{2}$, $\beta > 0$, and $\tau^{\beta-1} > 0$.

Let $\alpha \geq \frac{1}{2}$.

The transform of a function $\Theta(t)$ on $\mathbb{R}$ is defined as follows:

$$\begin{aligned}&\widehat{\mathbb{C}}_{\alpha,\beta}[\Theta(t)]\\&= \Theta(\zeta)\\&= \beta\hbar_\alpha \textstyle\int_{-\infty}^{\infty} \in_\alpha\left(-\zeta^\beta t^\beta\right)\Theta(t)\, t^{2\alpha\beta-1}dt,\end{aligned} \tag{2.398}$$

and the inverse transform on $\mathbb{R}$ as

$$\begin{aligned}&\widehat{\mathbb{C}}^{-1}_{\alpha,\beta}[\Theta(\zeta)]\\&= \Theta(t)\\&= \beta\hbar_\alpha \lim_{T\to\infty} \textstyle\int_{-iT}^{+iT} \in_\alpha\left(\zeta^\beta t^\beta\right)\Theta(\zeta)\, \zeta^{2\alpha\beta-1}d\zeta,\end{aligned} \tag{2.399}$$

where

$$\hbar_\alpha = \frac{1}{2^{\alpha+1}\Gamma(\alpha+1)}. \tag{2.400}$$

Theorem 2.77 *If* $\alpha \geq \frac{1}{2}, \beta > 0$, $\tau^{\beta-1} > 0$, $\widehat{\mathbb{C}}_{\alpha,\beta}[\Theta(t)] = \Theta(\zeta)$, *and* $\widehat{\mathbb{C}}^{-1}_{\alpha,\beta}[\Theta(\zeta)] = \Theta(t)$, *then there exist*

$$\widehat{\mathbb{C}}^{-1}_{\alpha,\beta}\left[\widehat{\mathbb{C}}_{\alpha,\beta}[\Theta(t)]\right] = \Theta(t) \tag{2.401}$$

and

$$\widehat{\mathbb{C}}_{\alpha,\beta}\left[\widehat{\mathbb{C}}_{\alpha,\beta}^{-1}\left[\Theta\left(\zeta\right)\right]\right]=\Theta\left(\zeta\right). \tag{2.402}$$

Proof With the aid of (2.396) and (2.397), we deduce the results. □

Definition 2.50 Let $\Theta(t) = f\left(t^{\beta}\right)$, where $\tau \in \mathbb{R}$, $\beta > 0$, and $\tau^{\beta-1} > 0$, and let $\alpha \geq \frac{1}{2}$.

The transform of a function $\Theta(t)$ on $\mathbb{R}$ is defined as follows:

$$\begin{aligned}&\widehat{\mathbb{X}}_{\alpha,\beta}\left[\Theta\left(t\right)\right]\\&=\Theta\left(\zeta\right)\\&=\beta\int_{-\infty}^{\infty}\in_{\alpha}\left(-\zeta t^{\beta}\right)\Theta\left(t\right)\left|t\right|^{2\alpha\beta-1}dt\end{aligned}, \tag{2.403}$$

and the inverse transform on $\mathbb{R}$ as

$$\begin{aligned}&\widehat{\mathbb{X}}_{\alpha,\beta}^{-1}\left[\Theta\left(\zeta\right)\right]\\&=\Theta\left(t\right)\\&=\hbar_{\alpha}^{2}\lim_{T\to\infty}\int_{-iT}^{+iT}\in_{\alpha}\left(\zeta t^{\beta}\right)\Theta\left(\zeta\right)\left|\zeta\right|^{2\alpha+1}d\zeta\end{aligned}, \tag{2.404}$$

where

$$\hbar_{\alpha}=\frac{1}{2^{\alpha+1}\Gamma\left(\alpha+1\right)}. \tag{2.405}$$

Theorem 2.78 *If* $\alpha \geq \frac{1}{2}, \beta > 0$, $t^{\beta-1} > 0$, $\widehat{\mathbb{X}}_{\alpha,\beta}\left[\Theta(t)\right] = \Theta(\zeta)$, *and* $\widehat{\mathbb{X}}_{\alpha,\beta}^{-1}\left[\Theta(\zeta)\right] = \Theta(t)$, *then there exist*

$$\widehat{\mathbb{X}}_{\alpha,\beta}^{-1}\left[\widehat{\mathbb{X}}_{\alpha,\beta}\left[\Theta\left(t\right)\right]\right]=\Theta\left(t\right) \tag{2.406}$$

and

$$\widehat{\mathbb{X}}_{\alpha,\beta}\left[\widehat{\mathbb{X}}_{\alpha,\beta}^{-1}\left[\Theta\left(\zeta\right)\right]\right]=\Theta\left(\zeta\right). \tag{2.407}$$

Proof Making use of (2.379) and (2.380) and taking place of t by t^{β}, we have the results. □

Definition 2.51 Let $t \in \mathbb{R}$, $\beta > 0$, and $\alpha \geq \frac{1}{2}$.

The transform of a function $\Theta(t)$ on $\mathbb{R}$ is defined as follows:

$$\begin{aligned}&\widehat{\mathbb{Y}}_{\alpha,\beta}\left[\Theta\left(t\right)\right]\\&=\Theta\left(\zeta\right)\\&=\int_{-\infty}^{\infty}\in_{\alpha}\left(-\zeta^{\beta}t\right)\Theta\left(t\right)\left|t\right|^{2\alpha+1}dt\end{aligned}, \tag{2.408}$$

and the inverse transform on $\mathbb{R}$ as

$$\begin{aligned}&\widehat{\mathbb{Y}}_{\alpha,\beta}^{-1}\left[\Theta\left(\zeta\right)\right]\\&=\Theta\left(t\right)\\&=\beta\hbar_{\alpha}^{2}\lim_{T\to\infty}\int_{-iT}^{+iT}\in_{\alpha}\left(\zeta^{\beta}t\right)\Theta\left(\zeta\right)\left|\zeta\right|^{2\alpha\beta-1}d\zeta\end{aligned}, \tag{2.409}$$

where

$$\hbar_{\alpha}=\frac{1}{2^{\alpha+1}\Gamma\left(\alpha+1\right)}. \tag{2.410}$$

Theorem 2.79 *If* $\alpha\geq\frac{1}{2},\beta>0$, $\widehat{\mathbb{Y}}_{\alpha,\beta}\left[\Theta\left(t\right)\right]=\Theta\left(\zeta\right)$, *and* $\widehat{\mathbb{Y}}_{\alpha,\beta}^{-1}\left[\Theta\left(\zeta\right)\right]=\Theta\left(t\right)$, *then there exist*

$$\widehat{\mathbb{Y}}_{\alpha,\beta}^{-1}\left[\widehat{\mathbb{Y}}_{\alpha,\beta}\left[\Theta\left(t\right)\right]\right]=\Theta\left(t\right) \tag{2.411}$$

and

$$\widehat{\mathbb{Y}}_{\alpha,\beta}\left[\widehat{\mathbb{Y}}_{\alpha,\beta}^{-1}\left[\Theta\left(\zeta\right)\right]\right]=\Theta\left(\zeta\right). \tag{2.412}$$

Proof Making use of (2.379) and (2.380) and taking place of ζ by ζ^{β}, we have the results. □

Definition 2.52 Let $\Theta\left(t\right)=\left(f\circ t^{\alpha}\right)\left(t\right)=f\left(t^{\alpha}\right)$, $\tau\in\mathbb{R}$, $\alpha>0$, and $\tau^{\alpha-1}>0$.
The transform of a function $\Theta\left(t\right)$ on $\mathbb{R}$ is defined as follows:

$$\begin{aligned}&\widehat{X}_{\alpha,\alpha}\left[\Theta\left(t\right)\right]\\&=\Theta\left(\zeta\right)\\&=\alpha\int_{-\infty}^{\infty}\in_{\alpha}\left(-\zeta^{\alpha}t^{\alpha}\right)\Theta\left(t\right)\left|t\right|^{2\alpha^{2}-1}dt,\end{aligned} \tag{2.413}$$

and the inverse transform on $\mathbb{R}$ as

$$\begin{aligned}&\widehat{X}_{\alpha,\alpha}^{-1}\left[\Theta\left(\zeta\right)\right]\\&=\Theta\left(t\right)\\&=\alpha\hbar_{\alpha}^{2}\lim_{T\to\infty}y_{-iT}^{+iT}\in_{\alpha}\left(\zeta^{\alpha}t^{\alpha}\right)\Theta\left(\zeta\right)\left|\zeta\right|^{2\alpha^{2}-1}d\zeta,\end{aligned} \tag{2.414}$$

where

$$\hbar_{\alpha}=\frac{1}{2^{\alpha+1}\Gamma\left(\alpha+1\right)}. \tag{2.415}$$

Theorem 2.80 *If $\alpha \geq \frac{1}{2}$, $\tau^{\alpha-1} > 0$, $\widehat{X}_{\alpha,\alpha}[\Theta(t)] = \Theta(\zeta)$, and $\widehat{X}^{-1}_{\alpha,\alpha}[\Theta(\zeta)] = \Theta(t)$, then there exist*

$$\widehat{X}^{-1}_{\alpha,\alpha}\left[\widehat{X}_{\alpha,\alpha}[\Theta(t)]\right] = \Theta(t) \tag{2.416}$$

and

$$\widehat{X}_{\alpha,\alpha}\left[\widehat{X}^{-1}_{\alpha,\alpha}[\Theta(\zeta)]\right] = \Theta(\zeta). \tag{2.417}$$

Proof On setting $\beta = \alpha$ in (2.406) and (2.407), we have the results. □

Definition 2.53 Let $\Theta(t) = (f \circ t^{\alpha})(t) = f(t^{\alpha})$, $\alpha \geq \frac{1}{2}$, $\tau \in \mathbb{R}$ and $\tau^{\alpha-1} > 0$.
The transform of a function $\Theta(t)$ on $\mathbb{R}$ is defined as follows:

$$\begin{aligned}&\widehat{\mathbb{B}}_{\alpha,\alpha}[\Theta(t)]\\ &\quad= \Theta(\zeta)\\ &\quad= \alpha\hbar_{\alpha}\int_{-\infty}^{\infty}\in_{\alpha}(-\zeta^{\alpha}t^{\alpha})\,\Theta(t)\,|t|^{2\alpha^2-1}\,dt,\end{aligned} \tag{2.418}$$

and the inverse transform on $\mathbb{R}$ as

$$\begin{aligned}&\widehat{\mathbb{B}}^{-1}_{\alpha,\alpha}[\Theta(\zeta)]\\ &\quad= \Theta(t)\\ &\quad= \alpha\hbar_{\alpha}\lim_{T\to\infty}\int_{-iT}^{+iT}\in_{\alpha}(\zeta^{\alpha}t^{\alpha})\,\Theta(\zeta)\,|\zeta|^{2\alpha^2-1}\,d\zeta,\end{aligned} \tag{2.419}$$

where

$$\hbar_{\alpha} = \frac{1}{2^{\alpha+1}\Gamma(\alpha+1)}. \tag{2.420}$$

Theorem 2.81 *If $\alpha \geq \frac{1}{2}$, $\tau^{\alpha-1} > 0$, $\widehat{\mathbb{B}}_{\alpha,\alpha}[\Theta(t)] = \Theta(\zeta)$, and $\widehat{\mathbb{B}}^{-1}_{\alpha,\alpha}[\Theta(\zeta)] = \Theta(t)$, then there exist*

$$\widehat{\mathbb{B}}^{-1}_{\alpha,\alpha}\left[\widehat{\mathbb{B}}_{\alpha,\alpha}[\Theta(t)]\right] = \Theta(t) \tag{2.421}$$

and

$$\widehat{\mathbb{B}}_{\alpha,\alpha}\left[\widehat{\mathbb{B}}^{-1}_{\alpha,\alpha}[\Theta(\zeta)]\right] = \Theta(\zeta). \tag{2.422}$$

Proof Taking $\beta = \alpha$ in (2.416) and (2.417), we deduce the results. □

Theorem 2.82 *If $\alpha \geq \frac{1}{2}$, then we have*

$$\int_{-\infty}^{\infty}|t|^{2\alpha+1}\in_{\alpha}(-\varsigma t)\,dt = \ell\,(i\varsigma)^{-2(\alpha+1)}, \tag{2.423}$$

where $\varsigma \in \mathbb{R}$ *and*

$$\ell = 2^{2(\alpha+1)}\Gamma^2\,(1+\alpha)\,. \tag{2.424}$$

Proof With the help of the representation

$$\begin{aligned}&\in_\alpha(\lambda t)\\&=\Gamma\,(1+\alpha)\left({}_0\varphi_1\left(\begin{matrix}-\\(\alpha+1,1)\end{matrix};\tfrac{\lambda^2t^2}{4}\right)+\tfrac{\lambda t}{2}\,{}_0\varphi_1\left(\begin{matrix}-\\(\alpha+2,1)\end{matrix};\tfrac{\lambda^2t^2}{4}\right)\right)\end{aligned} \tag{2.425}$$

we have that

$$\begin{aligned}&\textstyle\int_{-\infty}^{\infty}|t|^{2\alpha+1}\in_\alpha(-\varsigma t)\,dt\\&=\Gamma\,(1+\alpha)\textstyle\int_{-\infty}^{\infty}|t|^{2\alpha+1}\;{}_0\varphi_1\left(\begin{matrix}-\\(\alpha+1,1)\end{matrix};\tfrac{\varsigma^2t^2}{4}\right)dt\\&\quad-\Gamma\,(1+\alpha)\textstyle\int_{-\infty}^{\infty}\tfrac{\varsigma t}{2}|t|^{2\alpha+1}\;{}_0\varphi_1\left(\begin{matrix}-\\(\alpha+2,1)\end{matrix};\tfrac{\varsigma^2t^2}{4}\right)dt\\&=\Gamma\,(1+\alpha)\textstyle\int_{-\infty}^{\infty}|t|^{2\alpha+1}\;{}_0\varphi_1\left(\begin{matrix}-\\(\alpha+1,1)\end{matrix};\tfrac{\varsigma^2t^2}{4}\right)dt\\&=2\Gamma\,(1+\alpha)\textstyle\int_0^{\infty}t^{2\alpha+1}\;{}_0\varphi_1\left(\begin{matrix}-\\(\alpha+1,1)\end{matrix};\tfrac{\varsigma^2t^2}{4}\right)dt\\&=2^{2(\alpha+1)}\Gamma^2\,(1+\alpha)\,(i\varsigma)^{-2(\alpha+1)}\end{aligned} \tag{2.426}$$

since there exist

$$y_{-\infty}^{\infty}\frac{\varsigma t}{2}\,|t|^{2\alpha+1}\;{}_0\varphi_1\left(\begin{matrix}-\\(\alpha+2,1)\end{matrix};\tfrac{\varsigma^2t^2}{4}\right)dt=0 \tag{2.427}$$

$$\begin{aligned}&\int_{-\infty}^{\infty}|t|^{2\alpha+1}\;{}_0\varphi_1\left(\begin{matrix}-\\(\alpha+1,1)\end{matrix};\tfrac{\varsigma^2t^2}{4}\right)dt\\&\quad=2\int_0^{\infty}|t|^{2\alpha+1}\;{}_0\varphi_1\left(\begin{matrix}-\\(\alpha+1,1)\end{matrix};\tfrac{\varsigma^2t^2}{4}\right)dt\end{aligned} \tag{2.428}$$

and

$$\int_0^{\infty}t^{2\alpha+1}\;{}_0\varphi_1\left(\begin{matrix}-\\(\alpha+1,1)\end{matrix};\tfrac{\varsigma^2t^2}{4}\right)dt=2^{2\alpha+1}\Gamma\,(1+\alpha)\,(i\varsigma)^{-2(\alpha+1)}\,. \tag{2.429}$$

□

2.3.3 New Integral Transforms of Second Type

Taking $\zeta = i\varsigma$ in the Dunkl transform, we give the following integral representation.

Definition 2.54 The new integral transform of a function $f(t)$ on $\mathbb{R}_+$ is defined as follows:

$$\Im_\alpha [f(t)] = f(\xi) = \int_0^\infty f(t) \in_\alpha (-\xi t) t^{2\alpha+1} dt, \tag{2.430}$$

and the inverse transform on $\mathbb{R}_+$ as

$$\Im_\alpha^{-1} [f(\xi)] = f(t) = \hbar_\alpha^2 \lim_{T\to\infty} \int_{-iT}^{+iT} f(\xi) \in_\alpha (\xi t) |\xi|^{2\alpha+1} d\xi, \tag{2.431}$$

where $\xi = iT$, $\hbar_\alpha = \frac{1}{2^{\alpha+1}\Gamma(\alpha+1)}$ and $\alpha \geq \frac{1}{2}$.

When $\alpha = \frac{1}{2}$, one obtains the Laplace transform.

Theorem 2.83 *If $\alpha \geq \frac{1}{2}$ and $\Im_\alpha [f(t)] = f(\xi)$, then there is*

$$\Im_\alpha \left[\Im_\alpha^{-1} [f(\xi)]\right] = f(\xi) \tag{2.432}$$

and

$$\Im_\alpha^{-1} [\Im_\alpha [f(t)]] = f(t), \tag{2.433}$$

where $\hbar_\alpha = \frac{1}{2^{\alpha+1}\Gamma(\alpha+1)}$.

Proof With the aid of (2.376) and (2.377), we directly give the result. □

Theorem 2.84 *If $\alpha \geq \frac{1}{2}$ and $\Im_\alpha [f(t)] = f(\xi)$, then there is*

$$\Im_\alpha \left[\widehat{\mathbb{T}_\alpha} f(t)\right] = \xi f(\xi), \tag{2.434}$$

where

$$\hbar_\alpha = \frac{1}{2^{\alpha+1}\Gamma(\alpha+1)}. \tag{2.435}$$

Proof Making use of the Dunkl theorem and taking $\xi = i\varsigma$, we obtain the result. □

Theorem 2.85 *If $\alpha \geq \frac{1}{2}$ and $\Im_\alpha [f(t)] = f(\varsigma)$, then there is*

$$\wp_\alpha \left[\widehat{H_\alpha} f(t)\right] = \frac{1}{\xi} f(\xi), \tag{2.436}$$

where

$$\hbar_\alpha = \frac{1}{2^{\alpha+1}\Gamma(\alpha+1)}. \tag{2.437}$$

Proof Making use of the Dunkl theorem and taking $\xi = i\varsigma$, we give the result. □

Corollary 2.27 *If* $\alpha \geq \frac{1}{2}$, *then there exist*

$$\hbar_\alpha \int_{-i\infty}^{i\infty} |\xi|^{2\alpha+1} \in_\alpha (\xi t) \left(\hbar_\alpha \int_0^\infty |\tau|^{2\alpha+1} \in_\alpha (-\xi t)\, d\tau \right) d\xi = 1 \tag{2.438}$$

and

$$\hbar_\alpha \int_0^\infty \tau^{2\alpha+1} \in_\alpha (-\xi t) \left(\hbar_\alpha \int_{-i\infty}^{i\infty} |\xi|^{2\alpha+1} \in_\alpha (\zeta t)\, d\varsigma \right) d\tau = 1, \tag{2.439}$$

where

$$\hbar_\alpha = \frac{1}{2^{\alpha+1}\Gamma(\alpha+1)}. \tag{2.440}$$

Proof Taking $f(\xi) = 1$ and $f(t) = 1$ in (2.432) and (2.433), we give the results. □

Definition 2.55 Let $\alpha \geq \frac{1}{2}$ and $\hbar_\alpha = \frac{1}{2^{\alpha+1}\Gamma(\alpha+1)}$.

The new integral transform of a function $f(t)$ on $\mathbb{R}_+$ is defined as follows:

$$\begin{aligned} &\mathbb{L}_\alpha [f(t)] \\ &= f(\xi) \\ &= \hbar_\alpha y_0^\infty \in_\alpha (-\xi t) f(t) t^{2\alpha+1} dt \end{aligned}, \tag{2.441}$$

and the inverse transform on $\mathbb{R}_+$ as

$$\begin{aligned} &\mathbb{L}_\alpha^{-1} [f(\xi)] \\ &= f(t) \\ &= \hbar_\alpha \lim_{T\to\infty} \int_{-iT}^{+iT} \in_\alpha (\xi t) f(\xi) |\xi|^{2\alpha+1} d\xi \end{aligned}, \tag{2.442}$$

where $\xi = iT$.

Theorem 2.86 *If* $\alpha \geq \frac{1}{2}$ *and* $\mathbb{L}_\alpha [f(t)] = f(\zeta)$ *and* $\mathbb{L}_\alpha^{-1} [f(\zeta)] = f(t)$, *then there exist*

$$\mathbb{L}_\alpha^{-1} [\mathbb{L}_\alpha [f(t)]] = f(t) \tag{2.443}$$

and

$$\mathbb{L}_{\alpha}\left[\mathbb{L}_{\alpha}^{-1}\left[f\left(\zeta\right)\right]\right]=f\left(\zeta\right). \tag{2.444}$$

Proof Making use of (2.432) and (2.433), we obtain the results. □

Definition 2.56 Let $\Theta(t) = \left(f \circ t^{\beta}\right)(t) = f\left(t^{\beta}\right)$, where $\tau \in \mathbb{R}$, $\beta > 0$, and $\tau^{\beta-1} > 0$ and let $\alpha \geq \frac{1}{2}$.

The transform of a function $\Theta(t)$ on $\mathbb{R}$ is defined as follows:

$$\begin{aligned}&\widehat{\mathbb{T}}_{\alpha,\beta}\left[\Theta\left(t\right)\right]\\&=\Theta\left(\zeta\right)\\&=\beta y_{0}^{\infty}\in_{\alpha}\left(-\zeta^{\beta}t^{\beta}\right)\Theta\left(t\right)\left|t\right|^{2\alpha\beta-1}dt,\end{aligned} \tag{2.445}$$

and the inverse transform on $\mathbb{R}$ as

$$\begin{aligned}&\widehat{\mathbb{T}}_{\alpha,\beta}^{-1}\left[\Theta\left(\zeta\right)\right]\\&=\Theta\left(t\right)\\&=\beta\hbar_{\alpha}^{2}\lim_{T\to\infty}\int_{-iT}^{+iT}\in_{\alpha}\left(\zeta^{\beta}t^{\beta}\right)\Theta\left(\zeta\right)\left|\zeta\right|^{2\alpha\beta-1}d\zeta,\end{aligned} \tag{2.446}$$

where

$$\hbar_{\alpha}=\frac{1}{2^{\alpha+1}\Gamma\left(\alpha+1\right)}. \tag{2.447}$$

Theorem 2.87 *If* $\alpha \geq \frac{1}{2}, \beta > 0$, $\tau^{\beta-1} > 0$, $\widehat{\mathbb{T}}_{\alpha,\beta}\left[\Theta(t)\right] = \Theta(\zeta)$, *and* $\widehat{\mathbb{T}}_{\alpha,\beta}^{-1}\left[\Theta(\zeta)\right] = \Theta(t)$, *then there exist*

$$\widehat{\mathbb{T}}_{\alpha,\beta}^{-1}\left[\widehat{\mathbb{T}}_{\alpha,\beta}\left[\Theta\left(t\right)\right]\right]=\Theta\left(t\right) \tag{2.448}$$

and

$$\widehat{\mathbb{T}}_{\alpha,\beta}\left[\widehat{\mathbb{T}}_{\alpha,\beta}^{-1}\left[\Theta\left(\zeta\right)\right]\right]=\Theta\left(\zeta\right). \tag{2.449}$$

Proof Taking place of t by t^{β}in (2.443) and taking place of ζ by ζ^{β}in (2.444), we have the results. □

Definition 2.57 Let $\Theta(t) = \left(f \circ t^{\beta}\right)(t) = f\left(t^{\beta}\right)$, where $\tau \in \mathbb{R}$, $\alpha \geq \frac{1}{2}$, $\beta > 0$, and $\tau^{\beta-1} > 0$, and let

$$\alpha\geq\frac{1}{2}.$$

The transform of a function $\Theta(t)$ on $\mathbb{R}$ is defined as follows:

$$\begin{aligned}&\mathbb{T}_{\alpha,\beta}\left[\Theta(t)\right]\\&=\Theta(\zeta)\\&=\beta\hbar_\alpha\int_0^\infty \in_\alpha\left(-\zeta^\beta t^\beta\right)\Theta(t)\,t^{2\alpha\beta-1}dt,\end{aligned} \tag{2.450}$$

and the inverse transform on $\mathbb{R}$ as

$$\begin{aligned}&\mathbb{T}^{-1}_{\alpha,\beta}\left[\Theta(\zeta)\right]\\&=\Theta(t)\\&=\beta\hbar_\alpha\lim_{T\to\infty}\int_{-iT}^{+iT}\in_\alpha\left(\zeta^\beta t^\beta\right)\Theta(\zeta)\,|\zeta|^{2\alpha\beta-1}\,d\zeta\end{aligned}, \tag{2.451}$$

where

$$\hbar_\alpha=\frac{1}{2^{\alpha+1}\Gamma(\alpha+1)}. \tag{2.452}$$

Theorem 2.88 *If* $\alpha \geq \frac{1}{2}, \beta > 0,\ \tau^{\beta-1} > 0,\ \mathbb{T}_{\alpha,\beta}\left[\Theta(t)\right] = \Theta(\zeta)$, *and* $\mathbb{T}^{-1}_{\alpha,\beta}\left[\Theta(\zeta)\right]=\Theta(t)$, *then there exist*

$$\mathbb{T}^{-1}_{\alpha,\beta}\left[\mathbb{T}_{\alpha,\beta}\left[\Theta(t)\right]\right]=\Theta(t) \tag{2.453}$$

and

$$\mathbb{T}_{\alpha,\beta}\left[\mathbb{T}^{-1}_{\alpha,\beta}\left[\Theta(\zeta)\right]\right]=\Theta(\zeta). \tag{2.454}$$

Proof Taking place of t by t^βin (2.443) and taking place of ζ by ζ^βin (2.444), we have the results. □

Definition 2.58 Let $\Theta(t) = \left(f\circ t^\beta\right)(t) = f\left(t^\beta\right)$, where $\tau\in\mathbb{R}$, $\beta>0$, and $\tau^{\beta-1}>0$, and let $\alpha\geq\frac{1}{2}$.

The transform of a function $\Theta(t)$ on $\mathbb{R}$ is defined as follows:

$$\begin{aligned}&\widehat{\mathbb{E}}_{\alpha,\beta}\left[\Theta(t)\right]\\&=\Theta(\zeta)\\&=\beta\int_0^\infty \in_\alpha\left(-\zeta t^\beta\right)\Theta(t)\,|t|^{2\alpha\beta-1}\,dt\end{aligned}, \tag{2.455}$$

and the inverse transform on $\mathbb{R}$ as

$$\begin{aligned}&\widehat{\mathbb{E}}^{-1}_{\alpha,\beta}\left[\Theta(\zeta)\right]\\&=\Theta(t)\\&=\hbar^2_\alpha\lim_{T\to\infty}\int_{-iT}^{+iT}\in_\alpha\left(\zeta t^\beta\right)\Theta(\zeta)\,|\zeta|^{2\alpha+1}\,d\zeta\end{aligned}, \tag{2.456}$$

where

$$\hbar_\alpha = \frac{1}{2^{\alpha+1}\Gamma(\alpha+1)}. \tag{2.457}$$

Theorem 2.89 *If* $\alpha \geq \frac{1}{2}, \beta > 0, \tau^{\beta-1} > 0, \widehat{\mathbb{E}}_{\alpha,\beta}[\Theta(t)] = \Theta(\zeta)$, *and* $\widehat{\mathbb{E}}^{-1}_{\alpha,\beta}[\Theta(\zeta)] = \Theta(t)$, *then there exist*

$$\widehat{\mathbb{E}}^{-1}_{\alpha,\beta}\left[\widehat{\mathbb{E}}_{\alpha,\beta}[\Theta(t)]\right] = \Theta(t) \tag{2.458}$$

and

$$\widehat{\mathbb{E}}_{\alpha,\beta}\left[\widehat{\mathbb{E}}^{-1}_{\alpha,\beta}[\Theta(\zeta)]\right] = \Theta(\zeta). \tag{2.459}$$

Proof Taking place of t by t^βin (2.443) and (2.444), we derive the results. □

Definition 2.59 Let $\alpha \geq \frac{1}{2}$, $\tau \in \mathbb{R}$, and $\beta > 0$.

The transform of a function $\Theta(t)$ on $\mathbb{R}$ is defined as follows:

$$\begin{aligned}&\widehat{\mathbb{F}}_{\alpha,\beta}[\Theta(t)]\\ &= \Theta(\zeta)\\ &= \int_0^\infty \in_\alpha\left(-\zeta^\beta t\right)\Theta(t)\,t^{2\alpha+1}dt,\end{aligned} \tag{2.460}$$

and the inverse transform on $\mathbb{R}$ as

$$\begin{aligned}&\widehat{\mathbb{F}}^{-1}_{\alpha,\beta}[\Theta(\zeta)]\\ &= \Theta(t)\\ &= \beta\hbar_\alpha^2 \lim_{T\to\infty}\int_{-iT}^{+iT} \in_\alpha\left(\zeta^\beta t\right)\Theta(\zeta)\,t^{2\alpha\beta-1}d\zeta,\end{aligned} \tag{2.461}$$

where

$$\hbar_\alpha = \frac{1}{2^{\alpha+1}\Gamma(\alpha+1)}. \tag{2.462}$$

Theorem 2.90 *If* $\alpha \geq \frac{1}{2}, \beta > 0, \widehat{\mathbb{F}}_{\alpha,\beta}[\Theta(t)] = \Theta(\zeta)$, *and* $\widehat{\mathbb{F}}^{-1}_{\alpha,\beta}[\Theta(\zeta)] = \Theta(t)$, *then there exist*

$$\widehat{\mathbb{F}}^{-1}_{\alpha,\beta}\left[\widehat{\mathbb{F}}_{\alpha,\beta}[\Theta(t)]\right] = \Theta(t) \tag{2.463}$$

and

$$\widehat{\mathbb{F}}_{\alpha,\beta}\left[\widehat{\mathbb{F}}^{-1}_{\alpha,\beta}[\Theta(\zeta)]\right] = \Theta(\zeta). \tag{2.464}$$

Proof Taking place of t by t^{β} in (2.443) and (2.444), we obtain the results. □

Definition 2.60 Let $\Theta(t) = (f \circ t^{\alpha})(t) = f(t^{\alpha})$, where $\tau \in \mathbb{R}$, $\alpha > 0$, and $\tau^{\alpha-1} > 0$, and let $\alpha \geq \frac{1}{2}$.

The transform of a function $\Theta(t)$ on $\mathbb{R}$ is defined as follows:

$$\begin{aligned} &\widehat{\mathbb{J}}_{\alpha,\alpha}[\Theta(t)] \\ &= \Theta(\zeta) \\ &= \alpha y_0^{\infty} \in_{\alpha} (-\zeta^{\alpha} t^{\alpha}) \Theta(t) t^{2\alpha^2-1} dt \end{aligned}, \tag{2.465}$$

and the inverse transform on $\mathbb{R}$ as

$$\begin{aligned} &\widehat{\mathbb{J}}_{\alpha,\alpha}^{-1}[\Theta(\zeta)] \\ &= \Theta(t) \\ &= \alpha \hbar_{\alpha}^{2} \lim_{T \to \infty} \int_{-iT}^{+iT} \in_{\alpha} (\zeta^{\alpha} t^{\alpha}) \Theta(\zeta) \zeta^{2\alpha^2-1} d\zeta \end{aligned}, \tag{2.466}$$

where

$$\hbar_{\alpha} = \frac{1}{2^{\alpha+1} \Gamma(\alpha+1)}. \tag{2.467}$$

Theorem 2.91 *If* $\alpha \geq \frac{1}{2}$, $\tau^{\alpha-1} > 0$, $\widehat{\mathbb{J}}_{\alpha,\alpha}[\Theta(t)] = \Theta(\zeta)$, *and* $\widehat{\mathbb{J}}_{\alpha,\alpha}^{-1}[\Theta(\zeta)] = \Theta(t)$, *then there exist*

$$\widehat{\mathbb{J}}_{\alpha,\alpha}^{-1}\left[\widehat{\mathbb{J}}_{\alpha,\alpha}[\Theta(t)]\right] = \Theta(t) \tag{2.468}$$

and

$$\widehat{\mathbb{J}}_{\alpha,\alpha}\left[\widehat{\mathbb{J}}_{\alpha,\alpha}^{-1}[\Theta(\zeta)]\right] = \Theta(\zeta). \tag{2.469}$$

Proof Taking place of ζ by ζ^{β} in (2.160) and (2.161) and $\alpha = \beta$, we have the results. □

Definition 2.61 Let $\Theta(t) = (f \circ t^{\alpha})(t) = f(t^{\alpha})$, where $\alpha \geq \frac{1}{2}$, $\tau \in \mathbb{R}$, and $\tau^{\alpha-1} > 0$.

The transform of a function $\Theta(t)$ on $\mathbb{R}$ is defined as follows:

$$\begin{aligned} &\mathbb{J}_{\alpha,\alpha}[\Theta(t)] \\ &= \Theta(\zeta) \\ &= \alpha \int_0^{\infty} \in_{\alpha} (-\zeta^{\alpha} t^{\alpha}) \Theta(t) t^{2\alpha^2-1} dt \end{aligned}, \tag{2.470}$$

and the inverse transform on $\mathbb{R}$ as

$$\begin{aligned}&\mathbb{J}_{\alpha,\alpha}^{-1}[\Theta(\zeta)]\\&=\Theta(t)\\&=\alpha\hbar_\alpha^2\lim_{T\to\infty}\int_{-iT}^{+iT}E_\alpha(\zeta^\alpha t^\alpha)\Theta(\zeta)\zeta^{2\alpha^2-1}d\zeta\end{aligned}, \tag{2.471}$$

where

$$\hbar_\alpha=\frac{1}{2^{\alpha+1}\Gamma(\alpha+1)}. \tag{2.472}$$

Theorem 2.92 *If* $\alpha\geq\frac{1}{2}$, $\tau^{\alpha-1}>0$, $\mathbb{J}_{\alpha,\alpha}[\Theta(t)]=\Theta(\zeta)$, *and* $\mathbb{J}_{\alpha,\alpha}^{-1}[\Theta(\zeta)]=\Theta(t)$, *then there exist*

$$\mathbb{J}_{\alpha,\alpha}^{-1}\left[\mathbb{J}_{\alpha,\alpha}[\Theta(t)]\right]=\Theta(t) \tag{2.473}$$

and

$$\mathbb{J}_{\alpha,\alpha}\left[\mathbb{J}_{\alpha,\alpha}^{-1}[\Theta(\zeta)]\right]=\Theta(\zeta). \tag{2.474}$$

Proof Taking place of ζ by ζ^β in (2.443) and (2.444) and $\alpha=\beta$, we have the results. □

Definition 2.62 The operator, denoted by $\widehat{\mathbb{T}_{\alpha,\beta}}f(t)$, is defined as

$$\widehat{\mathbb{T}_{\alpha,\beta}}f(t)=\frac{1}{\beta t^{\beta-1}}\frac{df(t)}{dt}+\frac{2\alpha+1}{t^\beta}\left[\frac{f(t)-f(-t)}{2}\right], \tag{2.475}$$

where $\alpha\geq\frac{1}{2}$, $t^{\beta-1}>0$, and $\beta>0$.

Theorem 2.93 *There is the equation*

$$\widehat{\mathbb{T}_{\alpha,\beta}}f(t)=\lambda f(t) \tag{2.476}$$

subjected to the initial conditions

$$f(0)=1 \tag{2.477}$$

with the solution given as

$$\begin{aligned}&f(t)\\&=\epsilon_\alpha\left(\lambda t^\beta\right)\\&=\Im_\alpha\left(\lambda t^\beta\right)+\tfrac{\lambda t^\beta}{2(\alpha+1)}\Im_{\alpha+1}\left(\lambda t^\beta\right)\end{aligned}, \tag{2.478}$$

where

$$\Im_{\alpha}(\lambda t) = \Gamma(1+\alpha) \sum_{k=0}^{\infty} \frac{1}{\Gamma(k+\alpha+1)} \frac{(\lambda t/2)^{2k}}{k!} \tag{2.479}$$

for $\lambda \in \mathbb{C}$.

Proof With the aid of the Dunkl equation and operator, we obtain the result. □

2.3.4 New Integral Transforms of Third Type

Based on the above results, we give the concepts and theorems for the new integral transforms.

Motivations
Let $\alpha \geq \frac{1}{2}$ and $\hbar_{\alpha} = \frac{1}{2^{\alpha+1}\Gamma(\alpha+1)}$.

With the aid of

$$\in_{\alpha}(i\lambda t) = {}_1\mathbb{S}cos_1(\lambda t) + i\, {}_1\mathbb{S}sin_1(\lambda t) \tag{2.480}$$

and

$$\in_{\alpha}(-i\lambda t) = {}_1\mathbb{S}cos_1(\lambda t) - i\, {}_1\mathbb{S}sin_1(\lambda t), \tag{2.481}$$

where

$$\begin{aligned} &{}_1\mathbb{S}cos_1(\lambda t) \\ &= \frac{\in_{\alpha}(i\lambda t) + \in_{\alpha}(-i\lambda t)}{2} \\ &= \Gamma(1+\alpha)\ {}_0\varphi_1\left(\begin{matrix} - \\ (\alpha+1, 1) \end{matrix}; -\frac{\lambda^2 t^2}{4}\right) \end{aligned} \tag{2.482}$$

and

$$\begin{aligned} &{}_1\mathbb{S}sin_1(\lambda t) \\ &= \frac{\in_{\alpha}(i\lambda t) - \in_{\alpha}(-i\lambda t)}{2i} \\ &= \frac{\lambda t}{2}\ {}_0\varphi_1\left(\begin{matrix} - \\ (\alpha+2, 1) \end{matrix}; -\frac{\lambda^2 t^2}{4}\right), \end{aligned} \tag{2.483}$$

we have that

$$
\begin{aligned}
&\wp_\alpha^{-1}[f(\varsigma)] \\
&= \hbar_\alpha^2 y_{-\infty}^{\infty}\left[\in_\alpha(i\varsigma\tau)|\varsigma|^{2\alpha+1}\right]\times\left[\int_{-\infty}^{\infty}\in_\alpha(i\varsigma\tau)f(\tau)|\tau|^{2\alpha+1}d\tau\right]d\varsigma \\
&= \hbar_\alpha^2\int_{-\infty}^{\infty}\begin{array}{l}\left[\left({}_1\mathbb{S}cos_1(\varsigma\tau)+i\,{}_1\mathbb{S}sin_1(\varsigma\tau)\right)|\varsigma|^{2\alpha+1}\right] \\ \times\left[\int_{-\infty}^{\infty}{}_1\mathbb{S}cos_1(\varsigma\tau)f(\tau)|\tau|^{2\alpha+1}d\tau\right]d\varsigma\end{array} \\
&\quad -i\hbar_\alpha^2\int_{-\infty}^{\infty}\begin{array}{l}\left[\left({}_1\mathbb{S}cos_1(\varsigma\tau)+i\,{}_1\mathbb{S}sin_1(\varsigma\tau)\right)|\varsigma|^{2\alpha+1}\right] \\ \times\left[\int_{-\infty}^{\infty}{}_1\mathbb{S}sin_1(\varsigma\tau)f(\tau)|\tau|^{2\alpha+1}d\tau\right]d\varsigma\end{array} \\
&= \hbar_\alpha^2\int_{-\infty}^{\infty}{}_1\mathbb{S}cos_1(\varsigma\tau)|\varsigma|^{2\alpha+1}y_{-\infty}^{\infty}{}_1\mathbb{S}cos_1(\varsigma\tau)f(\tau)|\tau|^{2\alpha+1}d\tau d\varsigma \\
&\quad +i\hbar_\alpha^2\int_{-\infty}^{\infty}{}_1\mathbb{S}sin_1(\varsigma\tau)|\varsigma|^{2\alpha+1}\int_{-\infty}^{\infty}{}_1\mathbb{S}cos_1(\varsigma\tau)f(\tau)|\tau|^{2\alpha+1}d\tau d\varsigma \\
&\quad -i\hbar_\alpha^2\int_{-\infty}^{\infty}{}_1\mathbb{S}cos_1(\varsigma\tau)|\varsigma|^{2\alpha+1}\int_{-\infty}^{\infty}{}_1\mathbb{S}sin_1(\varsigma\tau)f(\tau)|\tau|^{2\alpha+1}d\tau d\varsigma \\
&\quad +\hbar_\alpha^2\int_{-\infty}^{\infty}{}_1\mathbb{S}sin_1(\varsigma\tau)|\varsigma|^{2\alpha+1}\int_{-\infty}^{\infty}{}_1\mathbb{S}sin_1(\varsigma\tau)f(\tau)|\tau|^{2\alpha+1}d\tau d\varsigma \\
&= \hbar_\alpha^2\int_{-\infty}^{\infty}{}_1\mathbb{S}cos_1(\varsigma\tau)|\varsigma|^{2\alpha+1}\left[\int_{-\infty}^{\infty}{}_1\mathbb{S}cos_1(\varsigma\tau)f(\tau)|\tau|^{2\alpha+1}d\tau\right]d\varsigma \\
&\quad +\hbar_\alpha^2 y_{-\infty}^{\infty}{}_1\mathbb{S}sin_1(\varsigma\tau)|\varsigma|^{2\alpha+1}\left[\int_{-\infty}^{\infty}{}_1\mathbb{S}sin_1(\varsigma\tau)f(\tau)|\tau|^{2\alpha+1}d\tau\right]d\varsigma.
\end{aligned}
\tag{2.484}
$$

Thus, we get two integral equations as follows:

$$
f(\tau)=\hbar_\alpha^2 y_{-\infty}^{\infty}{}_1\mathbb{S}sin_1(\varsigma\tau)|\varsigma|^{2\alpha+1}\left[\int_{-\infty}^{\infty}{}_1\mathbb{S}sin_1(\varsigma\tau)f(\tau)|\tau|^{2\alpha+1}d\tau\right]d\varsigma
\tag{2.485}
$$

and

$$
f(\tau)=\hbar_\alpha^2\int_{-\infty}^{\infty}{}_1\mathbb{S}cos_1(\varsigma\tau)|\varsigma|^{2\alpha+1}\left[\int_{-\infty}^{\infty}{}_1\mathbb{S}cos_1(\varsigma\tau)f(\tau)|\tau|^{2\alpha+1}d\tau\right]d\varsigma,
\tag{2.486}
$$

which reduces to

$$
f(\tau)=2\hbar_\alpha^2\int_0^{\infty}{}_1\mathbb{S}sin_1(\varsigma\tau)\varsigma^{2\alpha+1}\left[\int_{-\infty}^{\infty}{}_1\mathbb{S}sin_1(\varsigma\tau)f(\tau)|\tau|^{2\alpha+1}d\tau\right]d\varsigma,
\tag{2.487}
$$

$$
f(\tau)=2\hbar_\alpha^2\int_0^{\infty}{}_1\mathbb{S}cos_1(\varsigma\tau)\varsigma^{2\alpha+1}\left[\int_{-\infty}^{\infty}{}_1\mathbb{S}cos_1(\varsigma\tau)f(\tau)|\tau|^{2\alpha+1}d\tau\right]d\varsigma,
\tag{2.488}
$$

$$
f(\tau)=4\hbar_\alpha^2\int_0^{\infty}{}_1\mathbb{S}sin_1(\varsigma\tau)\varsigma^{2\alpha+1}\left[y_0^{\infty}{}_1\mathbb{S}sin_1(\varsigma\tau)f(\tau)\tau^{2\alpha+1}d\tau\right]d\varsigma,
\tag{2.489}
$$

and

$$f(\tau) = 4\hbar_\alpha^2 \int_0^\infty {}_1\mathbb{S}cos_1(\varsigma\tau)\,\varsigma^{2\alpha+1}\left[\int_0^\infty {}_1\mathbb{S}cos_1(\varsigma\tau)\,f(\tau)\,\tau^{2\alpha+1}d\tau\right]d\varsigma, \tag{2.490}$$

where

$$\hbar_\alpha = \frac{1}{2^{\alpha+1}\Gamma(\alpha+1)}. \tag{2.491}$$

Similarly, with the aid of the expression

$$\wp_\alpha[f(t)] = \hbar_\alpha^2 \int_{-\infty}^\infty \epsilon_\alpha(i\varsigma\tau)\left[\int_{-\infty}^\infty \epsilon_\alpha(i\varsigma\tau)\,f(\varsigma)\,|\varsigma|^{2\alpha+1}d\varsigma\right]|\tau|^{2\alpha+1}d\tau \tag{2.492}$$

we obtain

$$f(\varsigma) = \hbar_\alpha^2 \int_{-\infty}^\infty {}_1\mathbb{S}sin_1(\varsigma\tau)\,|\tau|^{2\alpha+1}\left[\int_{-\infty}^\infty {}_1\mathbb{S}sin_1(\varsigma\tau)\,f(\varsigma)\,|\varsigma|^{2\alpha+1}d\varsigma\right]d\tau \tag{2.493}$$

and

$$f(\varsigma) = \hbar_\alpha^2 \int_{-\infty}^\infty {}_1\mathbb{S}cos_1(\varsigma\tau)\,|\tau|^{2\alpha+1}\left[\int_{-\infty}^\infty {}_1\mathbb{S}cos_1(\varsigma\tau)\,f(\varsigma)\,|\varsigma|^{2\alpha+1}d\varsigma\right]d\tau, \tag{2.494}$$

which reduces to

$$f(\varsigma) = 2\hbar_\alpha^2 \int_0^\infty {}_1\mathbb{S}sin_1(\varsigma\tau)\,\tau^{2\alpha+1}\left[\int_{-\infty}^\infty {}_1\mathbb{S}sin_1(\varsigma\tau)\,f(\varsigma)\,|\varsigma|^{2\alpha+1}d\varsigma\right]d\tau, \tag{2.495}$$

$$f(\varsigma) = 2\hbar_\alpha^2 y_0^\infty\ {}_1\mathbb{S}cos_1(\varsigma\tau)\,\tau^{2\alpha+1}\left[\int_{-\infty}^\infty {}_1\mathbb{S}cos_1(\varsigma\tau)\,f(\varsigma)\,|\varsigma|^{2\alpha+1}d\tau\right]d\tau, \tag{2.496}$$

$$f(\varsigma) = 4\hbar_\alpha^2 \int_0^\infty {}_1\mathbb{S}sin_1(\varsigma\tau)\,\tau^{2\alpha+1}\left[\int_0^\infty {}_1\mathbb{S}sin_1(\varsigma\tau)\,f(\varsigma)\,\varsigma^{2\alpha+1}d\varsigma\right]d\tau, \tag{2.497}$$

and

$$f(\varsigma) = 4\hbar_\alpha^2 \int_0^\infty {}_1\mathbb{S}cos_1(\varsigma\tau)\,\varsigma^{2\alpha+1}\left[\int_0^\infty {}_1\mathbb{S}cos_1(\varsigma\tau)\,f(\tau)\,\tau^{2\alpha+1}d\tau\right]d\varsigma. \tag{2.498}$$

2.3.4.1 New Integral Transforms of First Type Containing the Wright Function

We now consider the new integral transforms of first type containing the Wright function as follows.

Definition 2.63 Let $f(t)$ be an odd function, $\alpha \geq \frac{1}{2}$, and $\hbar_\alpha = \frac{1}{2^{\alpha+1}\Gamma(\alpha+1)}$.
The new integral transform of a function $f(t)$ is defined as follows:

$$\begin{aligned}
&\aleph_\alpha[f(t)] \\
&= f(\theta) \\
&= \textstyle\int_0^\infty {}_1\mathbb{S}sin_1(\theta t)\,f(t)\,t^{2\alpha+1}dt \\
&= \textstyle\int_0^\infty \left[\frac{\theta t}{2}\,{}_0\varphi_1\left(\begin{matrix}-\\(\alpha+2,1)\end{matrix};-\frac{\theta^2t^2}{4}\right)\right] f(t)\,t^{2\alpha+1}dt,
\end{aligned} \tag{2.499}$$

and the inverse integral transform as

$$\begin{aligned}
&\aleph_\alpha^{-1}[f(\theta)] \\
&= f(t) \\
&= 4\hbar_\alpha^2 \textstyle\int_0^\infty {}_1\mathbb{S}sin_1(\theta t)\,\theta^{2\alpha+1}f(\theta)\,d\theta \\
&= 4\hbar_\alpha^2 \textstyle\int_0^\infty \left[\frac{\theta t}{2}\,{}_0\varphi_1\left(\begin{matrix}-\\(\alpha+2,1)\end{matrix};-\frac{\theta^2t^2}{4}\right)\right]\theta^{2\alpha+1}f(\theta)\,d\theta,
\end{aligned} \tag{2.500}$$

where

$$\hbar_\alpha = \frac{1}{2^{\alpha+1}\Gamma(\alpha+1)}. \tag{2.501}$$

As a direct result, we give the following theorem.

Theorem 2.94 *Let $f(t)$ be an odd function, $\alpha \geq \frac{1}{2}$, and $\hbar_\alpha = \frac{1}{2^{\alpha+1}\Gamma(\alpha+1)}$. Then we have*

$$\begin{aligned}
&f(t) \\
&= 4\hbar_\alpha^2 y_0^\infty\, {}_1\mathbb{S}sin_1(\theta t)\,\theta^{2\alpha+1}\left(\textstyle\int_0^\infty {}_1\mathbb{S}sin_1(\theta t)\,f(t)\,t^{2\alpha+1}dt\right)d\theta \\
&= 4\hbar_\alpha^2 \textstyle\int_0^\infty \begin{array}{l}\left[\frac{\theta t}{2}\,{}_0\varphi_1\left(\begin{matrix}-\\(\alpha+2,1)\end{matrix};-\frac{\theta^2t^2}{4}\right)\right]\theta^{2\alpha+1} \\ \times\left(\int_0^\infty \left[\frac{\theta t}{2}\,{}_0\varphi_1\left(\begin{matrix}-\\(\alpha+2,1)\end{matrix};-\frac{\theta^2t^2}{4}\right)\right] f(t)\,t^{2\alpha+1}dt\right)d\theta\end{array}
\end{aligned} \tag{2.502}$$

and

$$
\begin{aligned}
&f(\theta)\\
&= 4\hbar_\alpha^2 \int_0^\infty {}_1\mathbb{S}sin_1(\theta t)\, t^{2\alpha+1} \left(\int_0^\infty {}_1\mathbb{S}sin_1(\theta t)\, f(\theta)\, \theta^{2\alpha+1} d\theta\right) dt\\
&= 4\hbar_\alpha^2 \int_0^\infty \begin{array}{l} \left[\frac{\theta t}{2}\; {}_0\varphi_1\left(\begin{array}{c} - \\ (\alpha+2,1) \end{array}; -\frac{\theta^2 t^2}{4}\right)\right] t^{2\alpha+1} \\ \times \left(\int_0^\infty \left[\frac{\theta t}{2}\; {}_0\varphi_1\left(\begin{array}{c} - \\ (\alpha+2,1) \end{array}; -\frac{\theta^2 t^2}{4}\right)\right] f(\omega)\, \theta^{2\alpha+1} d\theta\right) dt \end{array}.
\end{aligned}
\tag{2.503}
$$

Proof With the aid of (2.489) and (2.493), we directly obtain the results. □

Definition 2.64 Let $f(t)$ be an odd function, $\alpha \geq \frac{1}{2}$, and $\hbar_\alpha = \frac{1}{2^{\alpha+1}\Gamma(\alpha+1)}$. The new integral transform of a function $f(t)$ is defined as follows:

$$
\begin{aligned}
&\widehat{\aleph}_\alpha[f(t)]\\
&= f(\theta)\\
&= \hbar_\alpha \int_0^\infty {}_1\mathbb{S}sin_1(\theta t)\, f(t)\, t^{2\alpha+1} dt\\
&= \hbar_\alpha \int_0^\infty \left[\frac{\theta t}{2}\; {}_0\varphi_1\left(\begin{array}{c} - \\ (\alpha+2,1) \end{array}; -\frac{\theta^2 t^2}{4}\right)\right] f(t)\, t^{2\alpha+1} dt,
\end{aligned}
\tag{2.504}
$$

and the inverse integral transform as

$$
\begin{aligned}
&\widehat{\aleph}_\alpha^{-1}[f(\theta)]\\
&= f(t)\\
&= \hbar_\alpha y_0^\infty {}_1\mathbb{S}sin_1(\theta t)\, \theta^{2\alpha+1} f(\theta)\, d\theta\\
&= \hbar_\alpha \int_0^\infty \left[\frac{\theta t}{2}\; {}_0\varphi_1\left(\begin{array}{c} - \\ (\alpha+2,1) \end{array}; -\frac{\theta^2 t^2}{4}\right)\right] \theta^{2\alpha+1} f(\theta)\, d\theta,
\end{aligned}
\tag{2.505}
$$

where

$$
\hbar_\alpha = \frac{1}{2^{\alpha+1}\Gamma(\alpha+1)}. \tag{2.506}
$$

Theorem 2.95 *Let $f(t)$ be an odd function, $\alpha \geq \frac{1}{2}$, $\hbar_\alpha = \frac{1}{2^{\alpha+1}\Gamma(\alpha+1)}$, $\widehat{\aleph}_\alpha[f(t)] = f(\theta)$, and $\widehat{\aleph}_\alpha^{-1}[f(\theta)] = f(t)$.*

Then we have

$$
\widehat{\aleph}_\alpha^{-1}\left[\widehat{\aleph}_\alpha[f(t)]\right] = f(t) \tag{2.507}
$$

and

$$
\widehat{\aleph}_\alpha\left[\widehat{\aleph}_\alpha^{-1}[f(\theta)]\right] = f(\theta). \tag{2.508}
$$

Proof With the aid of (2.502) and (2.503), we reduce to the results. □

Definition 2.65 Let $f(t)$ be an odd function, $\alpha \geq \frac{1}{2}$, and $\hbar_\alpha = \frac{1}{2^{\alpha+1}\Gamma(\alpha+1)}$.
The new integral transform of a function $f(t)$ is defined as follows:

$$\begin{aligned}&\widehat{\mathbb{G}}_\alpha [f(t)] \\ &= f(\theta) \\ &= \hbar_\alpha \int_{-\infty}^{\infty} {}_1\mathbb{S}sin_1(\theta t)\, f(t)\, |t|^{2\alpha+1}\, dt \\ &= \hbar_\alpha \int_{-\infty}^{\infty} \left[\frac{\theta t}{2}\ {}_0\varphi_1 \left(\begin{matrix} - \\ (\alpha+2,1) \end{matrix} ; -\frac{\theta^2 t^2}{4} \right) \right] f(t)\, |t|^{2\alpha+1}\, dt,\end{aligned} \tag{2.509}$$

and the inverse integral transform as

$$\begin{aligned}&\widehat{\mathbb{G}}_\alpha^{-1} [f(\theta)] \\ &= f(t) \\ &= \hbar_\alpha \int_{-\infty}^{\infty} {}_1\mathbb{S}sin_1(\theta t)\, |\theta|^{2\alpha+1}\, f(\theta)\, d\theta \\ &= \hbar_\alpha \int_{-\infty}^{\infty} \left[\frac{\theta t}{2}\ {}_0\varphi_1 \left(\begin{matrix} - \\ (\alpha+2,1) \end{matrix} ; -\frac{\theta^2 t^2}{4} \right) \right] |\theta|^{2\alpha+1}\, f(\theta)\, d\theta,\end{aligned} \tag{2.510}$$

where

$$\hbar_\alpha = \frac{1}{2^{\alpha+1}\Gamma(\alpha+1)}. \tag{2.511}$$

Theorem 2.96 *Let $f(t)$ be an odd function, $\alpha \geq \frac{1}{2}$, $\hbar_\alpha = \frac{1}{2^{\alpha+1}\Gamma(\alpha+1)}$, $\widehat{\mathbb{G}}_\alpha [f(t)] = f(\theta)$, and $\widehat{\mathbb{G}}_\alpha^{-1} [f(\theta)] = f(t)$.*
Then we have

$$\widehat{\mathbb{G}}_\alpha^{-1} \left[\widehat{\mathbb{G}}_\alpha [f(t)] \right] = f(t) \tag{2.512}$$

and

$$\widehat{\mathbb{G}}_\alpha \left[\widehat{\mathbb{G}}_\alpha^{-1} [f(\theta)] \right] = f(\theta). \tag{2.513}$$

Proof With the aid of (2.485) and (2.493), we reduce to the results. □

The alternative forms of the above results are given as follows:
Let $f(t)$ be an odd function, $\alpha \geq \frac{1}{2}$, and $\hbar_\alpha = \frac{1}{2^{\alpha+1}\Gamma(\alpha+1)}$.
Then we have

$$f(t) = \hbar_\alpha^2 \int_{-\infty}^{\infty} {}_1\mathbb{S}sin_1(\varsigma t)\, |\varsigma|^{2\alpha+1} \int_{-\infty}^{\infty} {}_1\mathbb{S}sin_1(\varsigma t)\, f(t)\, |t|^{2\alpha+1}\, dt d\varsigma \tag{2.514}$$

and

$$f(\varsigma) = \hbar_\alpha^2 \int_{-\infty}^{\infty} {}_1\mathbb{S}sin_1(\varsigma t)\,|t|^{2\alpha+1}\left[\int_{-\infty}^{\infty} {}_1\mathbb{S}sin_1(\varsigma t)\, f(\varsigma)\,|\varsigma|^{2\alpha+1}\, d\varsigma\right] dt. \tag{2.515}$$

Definition 2.66 Let $\beta > 0$, $t^{\beta-1} > 0$, $\alpha \geq \frac{1}{2}$, and $\hbar_\alpha = \frac{1}{2^{\alpha+1}\Gamma(\alpha+1)}$.
The new integral transform of a function $\Theta(t)$ is defined as follows:

$$\begin{aligned}
&\widehat{\mathbb{K}}_{\alpha,\beta}[\Theta(t)] \\
&= \Theta(\theta) \\
&= \beta\hbar_\alpha \textstyle\int_{-\infty}^{\infty} {}_1\mathbb{S}sin_1\left(\theta^\beta t^\beta\right)\Theta(t)\,|t|^{2\alpha\beta-1}\, dt \\
&= \beta\hbar_\alpha \textstyle\int_{-\infty}^{\infty} \left[\frac{\theta^\beta t^\beta}{2}\ {}_0\varphi_1\left(\begin{matrix} - \\ (\alpha+2,1) \end{matrix}; -\frac{\theta^{2\beta}t^{2\beta}}{4}\right)\right]\Theta(t)\,|t|^{2\alpha\beta-1}\, dt,
\end{aligned} \tag{2.516}$$

and the inverse integral transform as

$$\begin{aligned}
&\widehat{\mathbb{K}}^{-1}_{\alpha,\beta}[\Theta(\theta)] \\
&= \Theta(t) \\
&= \beta\hbar_\alpha \textstyle\int_{-\infty}^{\infty} {}_1\mathbb{S}sin_1\left(\theta^\beta t^\beta\right)|\theta|^{2\alpha\beta-1}\,\Theta(\theta)\, d\theta \\
&= \beta\hbar_\alpha \textstyle\int_{-\infty}^{\infty} \left[\frac{\theta^\beta t^\beta}{2}\ {}_0\varphi_1\left(\begin{matrix} - \\ (\alpha+2,1) \end{matrix}; -\frac{\theta^{2\beta}t^{2\beta}}{4}\right)\right]|\theta|^{2\alpha\beta-1}\,\Theta(\theta)\, d\theta.
\end{aligned} \tag{2.517}$$

Theorem 2.97 *Let $\beta > 0$, $t^{\beta-1} > 0$, $\alpha \geq \frac{1}{2}$, and $\hbar_\alpha = \frac{1}{2^{\alpha+1}\Gamma(\alpha+1)}$.*
Then we have

$$\widehat{\mathbb{K}}^{-1}_{\alpha,\beta}\left[\widehat{\mathbb{K}}_{\alpha,\beta}[f(t)]\right] = \Theta(t) \tag{2.518}$$

and

$$\widehat{\mathbb{K}}_{\alpha,\beta}\left[\widehat{\mathbb{K}}^{-1}_{\alpha,\beta}[f(\theta)]\right] = \Theta(\theta). \tag{2.519}$$

Proof Taking the place of θ and t by θ^β and t^β in (2.514) and (2.515), respectively, we obtain the results. □

Definition 2.67 Let $\beta > 0$, $t^{\beta-1} > 0$, $\alpha \geq \frac{1}{2}$, and $\hbar_\alpha = \frac{1}{2^{\alpha+1}\Gamma(\alpha+1)}$.

The new integral transform of a function $\Theta(t)$ is defined as follows:

$$\begin{aligned}
&\widehat{\mathbb{P}}_{\alpha,\beta}\left[\Theta(t)\right] \\
&= \Theta(\theta) \\
&= 2\beta\hbar_\alpha \int_0^\infty {}_1\mathbb{S}sin_1\left(\theta^\beta t^\beta\right)\Theta(t)\,t^{2\alpha\beta-1}dt \\
&= 2\beta\hbar_\alpha \int_0^\infty \left[\frac{\theta^\beta t^\beta}{2}\;{}_0\varphi_1\left(\begin{matrix} - \\ (\alpha+2,1)\end{matrix};-\frac{\theta^{2\beta}t^{2\beta}}{4}\right)\right]\Theta(t)\,|t|^{2\alpha\beta-1}\,dt,
\end{aligned} \tag{2.520}$$

and the inverse integral transform as

$$\begin{aligned}
&\widehat{\mathbb{P}}^{-1}_{\alpha,\beta}\left[\Theta(\theta)\right] \\
&= \Theta(t) \\
&= 2\beta\hbar_\alpha \int_0^\infty {}_1\mathbb{S}sin_1\left(\theta^\beta t^\beta\right)|\theta|^{2\alpha\beta-1}\,\Theta(\theta)\,d\theta \\
&= 2\beta\hbar_\alpha \int_0^\infty \left[\frac{\theta^\beta t^\beta}{2}\;{}_0\varphi_1\left(\begin{matrix} - \\ (\alpha+2,1)\end{matrix};-\frac{\theta^{2\beta}t^{2\beta}}{4}\right)\right]\theta^{2\alpha\beta-1}\Theta(\theta)\,d\theta.
\end{aligned} \tag{2.521}$$

Theorem 2.98 *Let* $\beta > 0$, $t^{\beta-1} > 0$, $\theta^{\beta-1} > 0$, $\alpha \geq \frac{1}{2}$, $\hbar_\alpha = \frac{1}{2^{\alpha+1}\Gamma(\alpha+1)}$, $\widehat{\mathbb{P}}_{\alpha,\beta}\left[\Theta(t)\right] = \Theta(\theta)$, *and* $\widehat{\mathbb{P}}^{-1}_{\alpha,\beta}\left[\Theta(\theta)\right] = \Theta(t)$.

Then we have

$$\widehat{\mathbb{P}}^{-1}_{\alpha,\beta}\left[\widehat{\mathbb{P}}_{\alpha,\beta}\left[\Theta(t)\right]\right] = \Theta(t) \tag{2.522}$$

and

$$\widehat{\mathbb{P}}_{\alpha,\beta}\left[\widehat{\mathbb{P}}^{-1}_{\alpha,\beta}\left[\Theta(\theta)\right]\right] = \Theta(\theta). \tag{2.523}$$

Proof Taking the place of θ and t by θ^β and t^β in (2.514) and (2.515), respectively, we obtain the results. □

Definition 2.68 Let $\Theta(t) = f(t^\alpha)$ be an odd function, $t^{\alpha-1} > 0$, $\alpha \geq \frac{1}{2}$ and $\hbar_\alpha = \frac{1}{2^{\alpha+1}\Gamma(\alpha+1)}$.

The new integral transform of a function $\Theta(t)$ is defined as follows:

$$\begin{aligned}
&\widehat{\mathbb{K}}_{\alpha,\alpha}\left[\Theta(t)\right] \\
&= \Theta(\theta) \\
&= \alpha\hbar_\alpha \int_{-\infty}^\infty {}_1\mathbb{S}sin_1\left(\theta^\alpha t^\alpha\right)\Theta(t)\,|t|^{2\alpha^2-1}\,dt \\
&= \alpha\hbar_\alpha \int_{-\infty}^\infty \left[\frac{\theta^\alpha t^\alpha}{2}\;{}_0\varphi_1\left(\begin{matrix} - \\ (\alpha+2,1)\end{matrix};-\frac{\theta^{2\alpha}t^{2\alpha}}{4}\right)\right]\Theta(t)\,|t|^{2\alpha^2-1}\,dt,
\end{aligned} \tag{2.524}$$

and the inverse integral transform as

$$\begin{aligned}
&\widehat{\mathbb{K}}_{\alpha,\alpha}^{-1}\left[\Theta\left(\theta\right)\right]\\
&=\Theta\left(t\right)\\
&=\alpha\hbar_{\alpha}\int_{-\infty}^{\infty}{}_{1}\mathbb{S}sin_{1}\left(\theta^{\alpha}t^{\alpha}\right)\left|\theta\right|^{2\alpha^{2}-1}\Theta\left(\theta\right)d\theta\\
&=\alpha\hbar_{\alpha}\int_{-\infty}^{\infty}\left[\frac{\theta^{\alpha}t^{\alpha}}{2}\;{}_{0}\varphi_{1}\left(\begin{matrix}-\\(\alpha+2,1)\end{matrix};-\frac{\theta^{2\alpha}t^{2\alpha}}{4}\right)\right]\left|\theta\right|^{2\alpha^{2}-1}\Theta\left(\theta\right)d\theta.
\end{aligned} \tag{2.525}$$

Theorem 2.99 *Let $\Theta\left(t\right)$ be an odd function, $\alpha\geq\frac{1}{2}$, $t^{\alpha-1}>0$, $\hbar_{\alpha}=\frac{1}{2^{\alpha+1}\Gamma(\alpha+1)}$, $\widehat{\mathbb{K}}_{\alpha,\alpha}\left[\Theta\left(t\right)\right]=\Theta\left(\theta\right)$, and $\widehat{\mathbb{K}}_{\alpha,\alpha}^{-1}\left[\Theta\left(\theta\right)\right]=\Theta\left(t\right)$.*

Then we have

$$\widehat{\mathbb{K}}_{\alpha,\alpha}^{-1}\left[\widehat{\mathbb{K}}_{\alpha,\alpha}\left[\Theta\left(t\right)\right]\right]=\Theta\left(t\right) \tag{2.526}$$

and

$$\widehat{\mathbb{K}}_{\alpha,\alpha}\left[\widehat{\mathbb{K}}_{\alpha,\alpha}^{-1}\left[\Theta\left(\theta\right)\right]\right]=\Theta\left(\theta\right). \tag{2.527}$$

Proof Making use of (2.516) and (2.517) and taking $\beta=\alpha$, we obtain the results.

□

Definition 2.69 Let $\Theta\left(t\right)$ be an odd function, $\alpha\geq\frac{1}{2}$, $t^{\alpha-1}>0$, $\theta^{\alpha-1}>0$, and $\hbar_{\alpha}=\frac{1}{2^{\alpha+1}\Gamma(\alpha+1)}$.

The new integral transform of a function $\Theta\left(t\right)$ is defined as follows:

$$\begin{aligned}
&\widehat{\mathbb{P}}_{\alpha,\alpha}\left[\Theta\left(t\right)\right]\\
&=\Theta\left(\theta\right)\\
&=2\alpha\hbar_{\alpha}\int_{0}^{\infty}{}_{1}\mathbb{S}sin_{1}\left(\theta^{\alpha}t^{\alpha}\right)\Theta\left(t\right)\left|t\right|^{2\alpha^{2}-1}dt\\
&=2\alpha\hbar_{\alpha}y_{0}^{\infty}\left[\frac{\theta^{\alpha}t^{\alpha}}{2}\;{}_{0}\varphi_{1}\left(\begin{matrix}-\\(\alpha+2,1)\end{matrix};-\frac{\theta^{2\alpha}t^{2\alpha}}{4}\right)\right]\Theta\left(t\right)t^{2\alpha^{2}-1}dt,
\end{aligned} \tag{2.528}$$

and the inverse integral transform as

$$\begin{aligned}
&\widehat{\mathbb{P}}_{\alpha,\alpha}^{-1}\left[\Theta\left(\theta\right)\right]\\
&=\Theta\left(t\right)\\
&=2\alpha\hbar_{\alpha}\int_{0}^{\infty}{}_{1}\mathbb{S}sin_{1}\left(\theta^{\alpha}t^{\alpha}\right)\theta^{2\alpha^{2}-1}\Theta\left(\theta\right)d\theta\\
&=2\alpha\hbar_{\alpha}\int_{0}^{\infty}\left[\frac{\theta^{\alpha}t^{\alpha}}{2}\;{}_{0}\varphi_{1}\left(\begin{matrix}-\\(\alpha+2,1)\end{matrix};-\frac{\theta^{2\alpha}t^{2\alpha}}{4}\right)\right]\theta^{2\alpha^{2}-1}\Theta\left(\theta\right)d\theta.
\end{aligned} \tag{2.529}$$

Theorem 2.100 *Let $\Theta\left(t\right)$ be an odd function, $\alpha\geq\frac{1}{2}$, $t^{\alpha-1}>0$, $\theta^{\alpha-1}>0$, $\hbar_{\alpha}=\frac{1}{2^{\alpha+1}\Gamma(\alpha+1)}$, $\widehat{\mathbb{P}}_{\alpha,\alpha}\left[\Theta\left(t\right)\right]=\Theta\left(\theta\right)$, and $\widehat{\mathbb{P}}_{\alpha,\alpha}^{-1}\left[\Theta\left(\theta\right)\right]=\Theta\left(t\right)$.*

Then we have

$$\widehat{\mathbb{P}}_{\alpha,\alpha}^{-1}\left[\widehat{\mathbb{P}}_{\alpha,\alpha}\left[\Theta\left(t\right)\right]\right]=\Theta\left(t\right) \tag{2.530}$$

and

$$\widehat{\mathbb{P}}_{\alpha,\alpha}\left(\widehat{\mathbb{P}}_{\alpha,\alpha}^{-1}\left[\Theta\left(\theta\right)\right]\right)=\Theta\left(\theta\right). \tag{2.531}$$

Proof Making use of (2.524) and (2.525), we have the results. □

Definition 2.70 Let $f(t)$ is an odd function.

The operator, denoted by $\widehat{\mathbb{O}_\alpha} f(t)$, is defined as

$$\widehat{\mathbb{O}_\alpha} f\left(t\right)=\left(\widehat{\mathbb{T}_\alpha}\cdot\widehat{\mathbb{T}_\alpha}\right)f\left(t\right)=\left(\widehat{\mathbb{T}_\alpha}\right)^2 f\left(t\right), \tag{2.532}$$

where $\alpha \geq \frac{1}{2}$.

Theorem 2.101 *We have the equation*

$$\widehat{\mathbb{O}_\alpha} f\left(t\right)=\lambda f\left(t\right) \tag{2.533}$$

subjected to the initial conditions

$$f\left(0\right)=0 \tag{2.534}$$

with the solution

$$\begin{aligned} f\left(t\right) & \\ &= {}_1\mathbb{S}sinh_1\left(\lambda t\right) \\ &= \frac{\epsilon_\alpha(\lambda t)-\epsilon_\alpha(-\lambda t)}{2} \\ &= \frac{\Gamma(1+\alpha)\lambda t}{2}\ {}_0\varphi_1\left(\begin{matrix} - \\ (\alpha+2,1)\end{matrix};\frac{\lambda^2 t^2}{4}\right)\end{aligned} \tag{2.535}$$

for $\lambda \in \mathbb{C}$.

Proof Making use of the Dunkl equation, we obtain the result. □

2.3.4.2 New Integral Transforms of Second Type Containing Wright Function

We now consider the new integral transforms of second type containing the Wright function as follows.

Definition 2.71 Let $f(t)$ be an even function, $\alpha \geq \frac{1}{2}$, and $\hbar_\alpha = \frac{1}{2^{\alpha+1}\Gamma(\alpha+1)}$.
The new integral transform of the function $f(t)$ is defined as follows:

$$\begin{aligned} &\mathbb{G}_\alpha[f(t)] \\ &= f(\vartheta) \\ &= \hbar_\alpha \int_0^\infty {}_1\mathbb{S}cos_1(\vartheta t) f(t) t^{2\alpha+1} dt \\ &= \frac{1}{2^{\alpha+1}} \int_0^\infty {}_0\varphi_1\left(\begin{matrix} - \\ (\alpha+1,1) \end{matrix}; -\frac{\vartheta^2 t^2}{4}\right) f(t) t^{2\alpha+1} dt, \end{aligned} \tag{2.536}$$

and the inverse integral transform as

$$\begin{aligned} &\mathbb{G}_\alpha^{-1}[f(t)] \\ &= f(t) \\ &= 4h_\alpha \int_0^\infty {}_1\mathbb{S}cos_1(\vartheta t)\, \vartheta^{2\alpha+1} d\vartheta \\ &= \frac{1}{2^{\alpha-1}} \int_0^\infty {}_0\varphi_1\left(\begin{matrix} - \\ (\alpha+1,1) \end{matrix}; -\frac{\vartheta^2 t^2}{4}\right) \vartheta^{2\alpha+1} d\vartheta. \end{aligned} \tag{2.537}$$

Theorem 2.102 *Let $f(\tau)$ be an even function, $\alpha \geq \frac{1}{2}$, and $\hbar_\alpha = \frac{1}{2^{\alpha+1}\Gamma(\alpha+1)}$.*
Then we have

$$f(t) = 4\hbar_\alpha^2 \int_0^\infty {}_1\mathbb{S}cos_1(\varsigma t)\, \varsigma^{2\alpha+1} \left[\int_0^\infty {}_1\mathbb{S}cos_1(\varsigma t) f(t) t^{2\alpha+1} dt\right] d\varsigma \tag{2.538}$$

and

$$f(\varsigma) = 4\hbar_\alpha^2 \int_0^\infty {}_1\mathbb{S}cos_1(\varsigma t)\, t^{2\alpha+1} \left[\int_0^\infty {}_1\mathbb{S}cos_1(\varsigma t) f(\varsigma)\, \varsigma^{2\alpha+1} d\varsigma\right] dt. \tag{2.539}$$

Proof In view of (2.490) and (2.498), we have the desired results. □

As alternative representations, we have the following:
Let $f(\tau)$ be an even function, $\alpha \geq \frac{1}{2}$, and $\hbar_\alpha = \frac{1}{2^{\alpha+1}\Gamma(\alpha+1)}$, and let $\mathbb{G}_\alpha[f(t)] = f(\vartheta)$ and $\mathbb{G}_\alpha^{-1}[f(t)] = f(t)$.
Then we have

$$\mathbb{G}_\alpha^{-1}[\mathbb{G}_\alpha[f(t)]] = f(t) \tag{2.540}$$

and

$$\mathbb{G}_\alpha\left[\mathbb{G}_\alpha^{-1}[f(\vartheta)]\right] = f(\vartheta). \tag{2.541}$$

Definition 2.72 Let $f(\tau)$ be an even function, $\alpha \geq \frac{1}{2}$, and $\hbar_\alpha = \frac{1}{2^{\alpha+1}\Gamma(\alpha+1)}$.

The new integral transform of a function $f(t)$ is defined as follows:

$$\begin{aligned} &\mathbb{D}_\alpha[f(t)] \\ &= f(\vartheta) \\ &= \hbar_\alpha \int_{-\infty}^{\infty} {}_1\mathbb{S}cos_1(\vartheta t) f(t) |t|^{2\alpha+1} dt \\ &= \frac{1}{2^{\alpha+1}} \int_{-\infty}^{\infty} {}_0\varphi_1 \left(\begin{matrix} - \\ (\alpha+1, 1) \end{matrix} ; -\frac{\vartheta^2 t^2}{4} \right) f(t) |t|^{2\alpha+1} dt, \end{aligned} \tag{2.542}$$

and the inverse integral transform as

$$\begin{aligned} &\mathbb{D}_\alpha^{-1}[f(t)] \\ &= f(t) \\ &= \hbar_\alpha \int_{-\infty}^{\infty} {}_1\mathbb{S}cos_1(\vartheta t) |\vartheta|^{2\alpha+1} d\vartheta \\ &= \frac{1}{2^{\alpha+1}} \int_{-\infty}^{\infty} {}_0\varphi_1 \left(\begin{matrix} - \\ (\alpha+1, 1) \end{matrix} ; -\frac{\vartheta^2 t^2}{4} \right) |\vartheta|^{2\alpha+1} d\vartheta. \end{aligned} \tag{2.543}$$

Theorem 2.103 *Let $f(\tau)$ be an even function, $\alpha \geq \frac{1}{2}$, and $\hbar_\alpha = \frac{1}{2^{\alpha+1}\Gamma(\alpha+1)}$, and let $\mathbb{D}_\alpha[f(t)] = f(\vartheta)$ and $\mathbb{D}_\alpha^{-1}[f(t)] = f(t)$.*

Then we have

$$\mathbb{D}_\alpha^{-1}[\mathbb{D}_\alpha[f(t)]] = f(t) \tag{2.544}$$

and

$$\mathbb{D}_\alpha\left[\mathbb{D}_\alpha^{-1}[f(\vartheta)]\right] = f(\vartheta). \tag{2.545}$$

Proof Making use of (2.538) and (2.539), we obtain the results. □

Definition 2.73 Let $f(t)$ be an even function, $\alpha \geq \frac{1}{2}$, and $\hbar_\alpha = \frac{1}{2^{\alpha+1}\Gamma(\alpha+1)}$.

The new integral transform of a function $f(t)$ is defined as follows:

$$\begin{aligned} &G_\alpha[f(t)] \\ &= f(\vartheta) \\ &= 2\hbar_\alpha \int_0^{\infty} {}_1\mathbb{S}cos_1(\vartheta t) f(t) t^{2\alpha+1} dt \\ &= \frac{1}{2^\alpha} \int_0^{\infty} {}_0\varphi_1 \left(\begin{matrix} - \\ (\alpha+1, 1) \end{matrix} ; -\frac{\vartheta^2 t^2}{4} \right) f(t) t^{2\alpha+1} dt, \end{aligned} \tag{2.546}$$

and the inverse integral transform as

$$\begin{aligned} &G_{\alpha}^{-1}\left[f(t)\right] \\ &= f(t) \\ &= 2h_{\alpha}\int_0^{\infty} {}_1\mathbb{S}cos_1\left(\vartheta t\right)\vartheta^{2\alpha+1}d\vartheta \\ &= \frac{1}{2^{\alpha}}\int_0^{\infty} {}_0\varphi_1\left(\begin{matrix} - \\ (\alpha+1,1)\end{matrix};-\frac{\vartheta^2 t^2}{4}\right)\vartheta^{2\alpha+1}d\vartheta. \end{aligned} \tag{2.547}$$

Theorem 2.104 *Let $f(\tau)$ be an even function, $\alpha \geq \frac{1}{2}$, and $\hbar_{\alpha} = \frac{1}{2^{\alpha+1}\Gamma(\alpha+1)}$, and let $G_{\alpha}\left[f(t)\right] = f(\vartheta)$ and $G_{\alpha}^{-1}\left[f(t)\right] = f(t)$.*

Then we have

$$G_{\alpha}^{-1}\left[G_{\alpha}\left[f(t)\right]\right] = f(t) \tag{2.548}$$

and

$$G_{\alpha}\left[G_{\alpha}^{-1}\left[f(\vartheta)\right]\right] = f(\vartheta). \tag{2.549}$$

Proof With the use of (4.255) and (4.256), we deduce the results. □

Definition 2.74 Let $f(t)$ be an even function, $\beta > 0$, $t^{\beta-1} > 0$, $\vartheta^{\beta-1} > 0$, $\alpha \geq \frac{1}{2}$, and $\hbar_{\alpha} = \frac{1}{2^{\alpha+1}\Gamma(\alpha+1)}$.

The new integral transform of a function $f(t)$ is defined as follows:

$$\begin{aligned} &\widehat{G}_{\alpha,\beta}\left[f(t)\right] \\ &= f(\vartheta) \\ &= 2\beta\hbar_{\alpha}\int_0^{\infty} {}_1\mathbb{S}cos_1\left(\vartheta^{\beta}t^{\beta}\right)f(t)\,t^{2\alpha\beta-1}dt \\ &= \frac{\beta}{2^{\alpha}}\int_0^{\infty} {}_0\varphi_1\left(\begin{matrix} - \\ (\alpha+1,1)\end{matrix};-\frac{\vartheta^{2\beta}t^{2\beta}}{4}\right)f(t)\,t^{2\alpha\beta-1}dt, \end{aligned} \tag{2.550}$$

and the inverse integral transform as

$$\begin{aligned} &\widehat{G}_{\alpha,\beta}^{-1}\left[f(\vartheta)\right] \\ &= f(t) \\ &= 2\beta h_{\alpha}\int_0^{\infty} {}_1\mathbb{S}cos_1\left(\vartheta^{\beta}t^{\beta}\right)f(\vartheta)\,\vartheta^{2\alpha\beta-1}d\vartheta \\ &= \frac{\beta}{2^{\alpha}}\int_0^{\infty} {}_0\varphi_1\left(\begin{matrix} - \\ (\alpha+1,1)\end{matrix};-\frac{\theta^{2\beta}t^{2\beta}}{4}\right)\vartheta^{2\alpha\beta-1}d\vartheta. \end{aligned} \tag{2.551}$$

Theorem 2.105 *Let $f(t)$ be an even function, $\beta > 0$, $t^{\beta-1} > 0$, $\vartheta^{\beta-1} > 0$, $\alpha \geq \frac{1}{2}$, and $\hbar_{\alpha} = \frac{1}{2^{\alpha+1}\Gamma(\alpha+1)}$, and let $\widehat{G}_{\alpha,\beta}\left[f(t)\right] = f(\vartheta)$ and $\widehat{G}_{\alpha,\beta}^{-1}\left[f(\vartheta)\right] = f(t)$.*

Then we have

$$\widehat{G}^{-1}_{\alpha,\beta}\left[\widehat{G}_{\alpha,\beta}\left[f(t)\right]\right] = f(t) \tag{2.552}$$

and

$$\widehat{G}_{\alpha,\beta}\left[\widehat{G}^{-1}_{\alpha,\beta}\left[f(\vartheta)\right]\right] = f(\vartheta). \tag{2.553}$$

Proof Taking the place of ϑ and t by ϑ^{β} and t^{β} in (2.538) and (2.539), respectively, we obtain the results. □

Definition 2.75 Let $f(t)$ be an even function, $\beta > 0$, $t^{\beta-1} > 0$, $\vartheta^{\beta-1} > 0$, $\alpha \geq \frac{1}{2}$, and $\hbar_{\alpha} = \frac{1}{2^{\alpha+1}\Gamma(\alpha+1)}$.

The new integral transform of a function $f(t)$ is defined as follows:

$$\begin{aligned}
&\widehat{\mathbb{D}}_{\alpha,\beta}\left[f(t)\right] \\
&= f(\vartheta) \\
&= \beta\hbar_{\alpha}\int_{-\infty}^{\infty} {}_1\mathbb{S}cos_1\left(\vartheta^{\beta}t^{\beta}\right) f(t)\,|t|^{2\alpha\beta-1}\,dt \\
&= \frac{\beta}{2^{\alpha+1}}\int_{-\infty}^{\infty} {}_0\varphi_1\left(\begin{matrix} - \\ (\alpha+1,1) \end{matrix};-\frac{\vartheta^{2\beta}t^{2\beta}}{4}\right) f(t)\,|t|^{2\alpha\beta-1}\,dt,
\end{aligned} \tag{2.554}$$

and the inverse integral transform as

$$\begin{aligned}
&\widehat{\mathbb{D}}^{-1}_{\alpha,\beta}\left[f(t)\right] \\
&= f(t) \\
&= \beta h_{\alpha}\int_{-\infty}^{\infty} {}_1\mathbb{S}cos_1\left(\vartheta^{\beta}t^{\beta}\right) f(\vartheta)\,|\vartheta|^{2\alpha\beta-1}\,d\vartheta \\
&= \frac{\beta}{2^{\alpha+1}}\int_{-\infty}^{\infty} {}_0\varphi_1\left(\begin{matrix} - \\ (\alpha+1,1) \end{matrix};-\frac{\vartheta^{2\beta}t^{2\beta}}{4}\right) |\vartheta|^{2\alpha\beta-1}\,d\vartheta.
\end{aligned} \tag{2.555}$$

Theorem 2.106 *Let $f(t)$ be an even function, $\beta > 0$, $t^{\beta-1} > 0$, $\vartheta^{\beta-1} > 0$, $\alpha \geq \frac{1}{2}$, and $\hbar_{\alpha} = \frac{1}{2^{\alpha+1}\Gamma(\alpha+1)}$, and let $\widehat{\mathbb{D}}_{\alpha,\beta}\left[f(t)\right] = f(\vartheta)$ and $\widehat{\mathbb{D}}^{-1}_{\alpha,\beta}\left[f(\vartheta)\right] = f(t)$.*

Then we have

$$\widehat{\mathbb{D}}^{-1}_{\alpha,\beta}\left[\widehat{\mathbb{D}}_{\alpha,\beta}\left[f(t)\right]\right] = f(t) \tag{2.556}$$

and

$$\widehat{\mathbb{D}}_{\alpha,\beta}\left[\widehat{\mathbb{D}}^{-1}_{\alpha,\beta}\left[f(\vartheta)\right]\right] = f(\vartheta). \tag{2.557}$$

Proof Making use of (2.550) and (2.551), we get the results. □

Definition 2.76 Let $f(t)$ be an even function, $t^{\alpha-1} > 0$, $\vartheta^{\alpha-1} > 0$, $\alpha \geq \frac{1}{2}$, and $\hbar_{\alpha} = \frac{1}{2^{\alpha+1}\Gamma(\alpha+1)}$.

The new integral transform of a function $f(t)$ is defined as follows:

$$\begin{aligned}
&\widehat{\mathbb{U}}_{\alpha,\alpha}[f(t)] \\
&= f(\vartheta) \\
&= \alpha\hbar_\alpha \int_{-\infty}^{\infty} {}_1\mathbb{S}cos_1\left(\vartheta^\alpha t^\alpha\right) f(t)\,|t|^{2\alpha^2-1}\,dt \\
&= \frac{\alpha}{2^{\alpha+1}} y_{-\infty}^{\infty}\, {}_0\varphi_1\left(\begin{matrix} - \\ (\alpha+1,1) \end{matrix};-\frac{\vartheta^{2\alpha}t^{2\alpha}}{4}\right) f(t)\,|t|^{2\alpha^2-1}\,dt,
\end{aligned} \tag{2.558}$$

and the inverse integral transform as

$$\begin{aligned}
&\widehat{\mathbb{U}}_{\alpha,\alpha}^{-1}[f(t)] \\
&= f(t) \\
&= \alpha h_\alpha \int_{-\infty}^{\infty} {}_1\mathbb{S}cos_1\left(\vartheta^\beta t^\beta\right) f(\vartheta)\,|\vartheta|^{2\alpha^2-1}\,d\vartheta \\
&= \frac{\alpha}{2^{\alpha+1}} y_{-\infty}^{\infty}\, {}_0\varphi_1\left(\begin{matrix} - \\ (\alpha+1,1) \end{matrix};-\frac{\vartheta^{2\alpha}t^{2\alpha}}{4}\right) |\vartheta|^{2\alpha^2-1}\,d\vartheta.
\end{aligned} \tag{2.559}$$

Theorem 2.107 *Let $f(t)$ be an even function, $t^{\alpha-1}>0$, $\vartheta^{\alpha-1}>0$, $\alpha \geq \frac{1}{2}$, and $\hbar_\alpha = \frac{1}{2^{\alpha+1}\Gamma(\alpha+1)}$, and let $\widehat{\mathbb{U}}_{\alpha,\alpha}[f(t)] = f(\vartheta)$ and $\widehat{\mathbb{U}}_{\alpha,\alpha}^{-1}[f(\vartheta)] = f(t)$.*

Then we have

$$\widehat{\mathbb{U}}_{\alpha,\alpha}^{-1}\left[\widehat{\mathbb{U}}_{\alpha,\alpha}[f(t)]\right] = f(t) \tag{2.560}$$

and

$$\widehat{\mathbb{U}}_{\alpha,\alpha}\left[\widehat{\mathbb{U}}_{\alpha,\alpha}^{-1}[f(\vartheta)]\right] = f(\vartheta). \tag{2.561}$$

Proof With the use of (2.555) and (2.557), we derive the desired results. □

Definition 2.77 Let $f(t)$ be an even function, $\beta>0$, $t^{\beta-1}>0$, $\alpha \geq \frac{1}{2}$, and $\hbar_\alpha = \frac{1}{2^{\alpha+1}\Gamma(\alpha+1)}$.

The new integral transform of a function $f(t)$ is defined as follows:

$$\begin{aligned}
&\widehat{V}_{\alpha,\beta}[f(t)] \\
&= f(\vartheta) \\
&= 2\beta\hbar_\alpha \int_0^{\infty} {}_1\mathbb{S}cos_1\left(\vartheta t^\beta\right) f(t)\,t^{2\alpha\beta-1}dt \\
&= \frac{\beta}{2^\alpha} \int_0^{\infty} {}_0\varphi_1\left(\begin{matrix} - \\ (\alpha+1,1) \end{matrix};-\frac{\vartheta^2 t^{2\beta}}{4}\right) f(t)\,t^{2\alpha\beta-1}dt,
\end{aligned} \tag{2.562}$$

and the inverse integral transform as

$$\begin{aligned}
&\widehat{V}_{\alpha,\beta}^{-1}\left[f(\vartheta)\right] \\
&= f(t) \\
&= 2h_\alpha \int_0^\infty {}_1\mathbb{S}cos_1\left(\vartheta t^\beta\right) f(\vartheta)\, \vartheta^{2\alpha+1} d\vartheta \\
&= \frac{1}{2^\alpha}\int_0^\infty {}_0\varphi_1\left(\begin{matrix} - \\ (\alpha+1,1)\end{matrix}; -\frac{\vartheta^2 t^{2\beta}}{4}\right)\vartheta^{2\alpha+1} d\vartheta.
\end{aligned} \tag{2.563}$$

Theorem 2.108 *Let $f(t)$ be an even function, $\beta > 0$, $t^{\beta-1} > 0$, $\alpha \geq \frac{1}{2}$, and $\hbar_\alpha = \frac{1}{2^{\alpha+1}\Gamma(\alpha+1)}$, and let $\widehat{V}_{\alpha,\beta}\left[f(t)\right] = f(\vartheta)$ and $\widehat{V}_{\alpha,\beta}^{-1}\left[f(\vartheta)\right] = f(t)$.*
Then we have

$$\widehat{V}_{\alpha,\beta}^{-1}\left[\widehat{V}_{\alpha,\beta}\left[f(t)\right]\right] = f(t) \tag{2.564}$$

and

$$\widehat{V}_{\alpha,\beta}\left[\widehat{V}_{\alpha,\beta}^{-1}\left[f(\vartheta)\right]\right] = f(\vartheta). \tag{2.565}$$

Proof Taking the place of t by t^β in (2.538) and (2.539), respectively, we get the results. □

Definition 2.78 Let $f(t)$ be an even function, $\beta > 0$, $\vartheta^{\beta-1} > 0$, $\alpha \geq \frac{1}{2}$, and $\hbar_\alpha = \frac{1}{2^{\alpha+1}\Gamma(\alpha+1)}$.
The new integral transform of a function $f(t)$ is defined as follows:

$$\begin{aligned}
&\widehat{\mathbb{V}}_{\alpha,\beta}\left[f(t)\right] \\
&= f(\vartheta) \\
&= 2\hbar_\alpha \int_0^\infty {}_1\mathbb{S}cos_1\left(\vartheta^\beta t\right) f(t)\, t^{2\alpha+1} dt \\
&= \frac{1}{2^\alpha}\int_0^\infty {}_0\varphi_1\left(\begin{matrix} - \\ (\alpha+1,1)\end{matrix}; -\frac{\vartheta^{2\beta} t^2}{4}\right) f(t)\, t^{2\alpha+1} dt,
\end{aligned} \tag{2.566}$$

and the inverse integral transform as

$$\begin{aligned}
&\widehat{\mathbb{V}}_{\alpha,\beta}^{-1}\left[f(\vartheta)\right] \\
&= f(t) \\
&= 2\beta h_\alpha \int_0^\infty {}_1\mathbb{S}cos_1\left(\vartheta^\beta t\right) f(\vartheta)\, \vartheta^{2\alpha\beta-1} d\vartheta \\
&= \frac{\beta}{2^\alpha}\int_0^\infty {}_0\varphi_1\left(\begin{matrix} - \\ (\alpha+1,1)\end{matrix}; -\frac{\vartheta^{2\beta} t^2}{4}\right)\vartheta^{2\alpha\beta-1} d\vartheta.
\end{aligned} \tag{2.567}$$

Theorem 2.109 *Let $f(t)$ be an even function, $\beta > 0$, $\vartheta^{\beta-1} > 0$, $\alpha \geq \frac{1}{2}$, and $\hbar_\alpha = \frac{1}{2^{\alpha+1}\Gamma(\alpha+1)}$, and let $\widehat{\mathbb{V}}_{\alpha,\beta}\left[f(t)\right] = f(\vartheta)$ and $\widehat{\mathbb{V}}_{\alpha,\beta}^{-1}\left[f(\vartheta)\right] = f(t)$.*

Then we have

$$\widehat{\mathbb{V}}_{\alpha,\beta}^{-1}\left[\widehat{\mathbb{V}}_{\alpha,\beta}\left[f(t)\right]\right]=f(t) \tag{2.568}$$

and

$$\widehat{\mathbb{V}}_{\alpha,\beta}\left[\widehat{\mathbb{V}}_{\alpha,\beta}^{-1}\left[f(\vartheta)\right]\right]=f(\vartheta). \tag{2.569}$$

Proof Taking the place of ϑ by ϑ^{β} in (2.538) and (2.539), respectively, we have the results. □

Definition 2.79 Let $f(t)$ be an even function, $t^{\alpha-1}>0$, $\alpha\geq\frac{1}{2}$, and $\hbar_{\alpha}=\frac{1}{2^{\alpha+1}\Gamma(\alpha+1)}$.

The new integral transform of a function $f(t)$ is defined as follows:

$$\begin{aligned}
&\widehat{W}_{\alpha,\alpha}\left[f(t)\right]\\
&=f(\vartheta)\\
&=2\alpha\hbar_{\alpha}\int_{0}^{\infty}{}_{1}\mathbb{S}cos_{1}\left(\vartheta t^{\alpha}\right)f(t)\,t^{2\alpha^{2}-1}dt\\
&=\frac{\alpha}{2^{\alpha}}y_{0}^{\infty}\,{}_{0}\varphi_{1}\left(\begin{matrix}-\\(\alpha+1,1)\end{matrix};-\frac{\vartheta^{2}t^{2\alpha}}{4}\right)f(t)\,t^{2\alpha^{2}-1}dt,
\end{aligned} \tag{2.570}$$

and the inverse integral transform as

$$\begin{aligned}
&\widehat{W}_{\alpha,\alpha}^{-1}\left[f(\vartheta)\right]\\
&=f(t)\\
&=2h_{\alpha}\int_{0}^{\infty}{}_{1}\mathbb{S}cos_{1}\left(\vartheta t^{\alpha}\right)f(\vartheta)\,\vartheta^{2\alpha+1}d\vartheta\\
&=\frac{1}{2^{\alpha}}\int_{0}^{\infty}{}_{0}\varphi_{1}\left(\begin{matrix}-\\(\alpha+1,1)\end{matrix};-\frac{\vartheta^{2}t^{2\alpha}}{4}\right)\vartheta^{2\alpha+1}d\vartheta.
\end{aligned} \tag{2.571}$$

Theorem 2.110 *Let $f(t)$ be an even function, $t^{\alpha-1}>0$, $\alpha\geq\frac{1}{2}$, and $\hbar_{\alpha}=\frac{1}{2^{\alpha+1}\Gamma(\alpha+1)}$, and let $\widehat{W}_{\alpha,\alpha}\left[f(t)\right]=f(\vartheta)$ and $\widehat{W}_{\alpha,\alpha}^{-1}\left[f(\vartheta)\right]=f(t)$.*

Then we have

$$\widehat{W}_{\alpha,\alpha}^{-1}\left[\widehat{W}_{\alpha,\alpha}\left[f(t)\right]\right]=f(t) \tag{2.572}$$

and

$$\widehat{W}_{\alpha,\alpha}\left[\widehat{W}_{\alpha,\alpha}^{-1}\left[f(\vartheta)\right]\right]=f(\vartheta). \tag{2.573}$$

Proof Taking the place of t by t^{α} in (2.538) and (2.539), respectively, we have the results. □

Definition 2.80 Let $f(t)$ be an even function, $\vartheta^{\alpha-1} > 0$, $\alpha \geq \frac{1}{2}$, and $\hbar_\alpha = \frac{1}{2^{\alpha+1}\Gamma(\alpha+1)}$.

The new integral transform of a function $f(t)$ is defined as follows:

$$\begin{aligned}
&\widehat{\mathbb{W}}_{\alpha,\alpha}[f(t)] \\
&= f(\vartheta) \\
&= 2\hbar_\alpha \int_0^\infty {}_1\mathbb{S}cos_1(\vartheta^\alpha t) f(t) t^{2\alpha+1} dt \\
&= \frac{1}{2^\alpha}\int_0^\infty {}_0\varphi_1\left(\begin{matrix} - \\ (\alpha+1,1) \end{matrix}; -\frac{\vartheta^{2\alpha}t^2}{4}\right) f(t) t^{2\alpha+1} dt,
\end{aligned} \tag{2.574}$$

and the inverse integral transform as

$$\begin{aligned}
&\widehat{\mathbb{W}}^{-1}_{\alpha,\beta}[f(\vartheta)] \\
&= f(t) \\
&= 2\alpha h_\alpha \int_0^\infty {}_1\mathbb{S}cos_1(\vartheta^\alpha t) f(\vartheta) \vartheta^{2\alpha^2-1} d\vartheta \\
&= \frac{\alpha}{2^\alpha}\int_0^\infty {}_0\varphi_1\left(\begin{matrix} - \\ (\alpha+1,1) \end{matrix}; -\frac{\vartheta^{2\alpha}t^2}{4}\right) \vartheta^{2\alpha^2-1} d\vartheta.
\end{aligned} \tag{2.575}$$

Theorem 2.111 *Let $f(t)$ be an even function, $\vartheta^{\alpha-1} > 0$, $\alpha \geq \frac{1}{2}$, and $\hbar_\alpha = \frac{1}{2^{\alpha+1}\Gamma(\alpha+1)}$, and let $\widehat{\mathbb{W}}_{\alpha,\alpha}[f(t)] = f(\vartheta)$ and $\widehat{\mathbb{W}}^{-1}_{\alpha,\alpha}[f(\vartheta)] = f(t)$.*

Then we have

$$\widehat{\mathbb{W}}^{-1}_{\alpha,\alpha}\left[\widehat{\mathbb{W}}_{\alpha,\alpha}[f(t)]\right] = f(t) \tag{2.576}$$

and

$$\widehat{\mathbb{W}}_{\alpha,\alpha}\left[\widehat{\mathbb{W}}^{-1}_{\alpha,\alpha}[f(\vartheta)]\right] = f(\vartheta). \tag{2.577}$$

Proof Taking the place of ϑ by ϑ^α in (2.538) and (2.539), respectively, we have the results. □

Theorem 2.112 *We have the equation*

$$\widehat{\mathbb{O}}_\alpha f(t) = \lambda f(t) \tag{2.578}$$

subjected to the initial conditions

$$f(0) = 1 \tag{2.579}$$

with the solution

$$\begin{aligned} & f(t) \\ & = {}_1\mathbb{S}cosh_1(\lambda t) \\ & = \frac{\in_\alpha(\lambda t)+\in_\alpha(-\lambda t)}{2} \\ & = \Gamma(1+\alpha)\ {}_0\varphi_1\left(\begin{matrix} - \\ (\alpha+1,1) \end{matrix};\frac{\lambda^2 t^2}{4}\right) \end{aligned} \tag{2.580}$$

for $\lambda \in \mathbb{C}$ *and* $\alpha \geq \frac{1}{2}$.

Proof Making use of the Dunkl equation, we obtain the result. □

Chapter 3
Mittag-Leffler, Supertrigonometric, and Superhyperbolic Functions

Abstract In this chapter, we introduce the theory of the Mittag-Leffler function, supertrigonometric functions, and superhyperbolic functions. The integral representations for the Mittag-Leffler function and related functions are addressed and the general fractional calculus operators are also discussed in detail. The integral equations in the kernels of the Mittag-Leffler function, supertrigonometric functions, and superhyperbolic functions are presented; the truncated Mittag-Leffler, supertrigonometric, and superhyperbolic functions are considered; and some mathematical models are derived to explain the power-law behaviors in material science.

3.1 The Mittag-Leffler Function: History, Definitions, Properties, and Theorems

In this section, we present the results related to the Mittag-Leffler function.

3.1.1 The Mittag-Leffler Function

In 1903, Swedish mathematician Gösta Magnus Mittag-Leffler coined the well-known Mittag-Leffler function [202].

We now start with the definition of the Mittag-Leffler function as follows.

Definition 3.1 (Mittag-Leffler [202]) The Mittag-Leffler function is defined as [202]

$$E_{\alpha}(t) = \sum_{n=0}^{\infty} \frac{t^n}{\Gamma(n\alpha + 1)}, \tag{3.1}$$

where $\alpha, t \in \mathbb{C}$, $n \in \mathbb{N}$, $Re(\alpha) > 0$, and $\Gamma(\cdot)$ is the Euler Gamma function.

X.-J. Yang, *Theory and Applications of Special Functions for Scientists and Engineers*, https://doi.org/10.1007/978-981-33-6334-2_3

This work (the function of Mittag-Leffler of first type) was discovered by Mittag-Leffler in 1903 [202] and further reported by Bateman in 1940 [203].

The special form of Mittag-Leffler function can be written as [203]

$$E_1(-t) = e^{-t}, \tag{3.2}$$

and

$$E_0(-t) = \frac{1}{1+t} \; (|t| < 1), \tag{3.3}$$

where $t \in \mathbb{C}$.

The results were discovered by Pollard in 1948 [203].

There are some results for the Mittag-Leffler function as follows [204, 205]:

$$E_0(t) = \frac{1}{1-t} \; (|t| < 1), \tag{3.4}$$

$$E_2(t) = cosh\left(\sqrt{t}\right), \tag{3.5}$$

and

$$E_2\left(-t^2\right) = cos(t), \tag{3.6}$$

where $t \in \mathbb{C}$.

For more information of the Mittag-Leffler function, see Berberan-Santos [206], Peng and Li [207], Barnes [208], Kilbas and Koroleva [209], Gorenflo et al. [210], Kilbas et al. [211], Valiron [212], Hardy [213], Hille [214], Hille and Tamarkin [215], Cartwright [216], Rademacher [217], and so on.

The Pollard function, denoted as $L_\beta(u)$, is expressed by Pollard [218]

$$e^{-t^\beta} = \int_0^\infty e^{ut} L_\beta(u)\, du, \tag{3.7}$$

where e^{-t^β} is the Kohlrausch–Williams–Watts function (see Chap. 7), which leads to the following integral representation of the form [218]

$$E_\alpha(-t) = \int_0^\infty e^{-u^\beta t} L_\beta(u)\, du, \tag{3.8}$$

where $0 < \beta < 1$.

The result was discovered by Pollard in 1948 [218] and reported by Berberan-Santos in 2005 [206].

So, there exists [219, 220]

$$E_\alpha(-t) = \int_0^\infty e^{-ut} \frac{u^{-\left(1+\frac{1}{\alpha}\right)}}{\beta} L_\beta\left(u^{-\frac{1}{\beta}}\right) du, \tag{3.9}$$

where $0 < \beta < 1$.

Definition 3.2 The function of Mittag-Leffler of second type is defined as [215]

$$E_\alpha\left(t^\alpha\right) = \sum_{n=0}^{\infty} \frac{t^{n\alpha}}{\Gamma(n\alpha+1)}, \tag{3.10}$$

where $\alpha, t \in \mathbb{C}$, $n \in \mathbb{N}$, and $Re(\alpha) > 0$.

The result was discovered by Hille and Tamarkin due to the expression [215]

$$Q(t,\lambda) = \frac{d}{dt} E_{1-\alpha}\left(\lambda\Gamma(1-\alpha)\, t^{1-\alpha}\right)$$

with the Hille–Tamarkin integral equation, which is given as follows:

$$\begin{aligned} &H(t) \\ &= f(t) - \lambda \int_0^0 Q(t-\tau) f(\tau)\, d\tau \\ &= f(t) - \lambda \int_0^0 \left[\tfrac{d}{d\tau} E_{1-\alpha}\left(\lambda\Gamma(1-\alpha)(t-\tau)^{1-\alpha}\right)\right] f(\tau)\, d\tau, \end{aligned} \tag{3.11}$$

where $\lambda \in \mathbb{R}$.

With the aid of the Pollard function, there is the following integral representation of the form [206, 218]

$$E_\alpha\left(-t^\alpha\right) = \int_0^\infty e^{-u^\alpha t^\alpha} L_\alpha(u)\, du, \tag{3.12}$$

where $0 < \alpha < 1$.

Definition 3.3 The function of Mittag-Leffler of third type is defined as

$$E_\alpha\left(t^\beta\right) = \sum_{n=0}^{\infty} \frac{t^{n\beta}}{\Gamma(n\alpha+1)}, \tag{3.13}$$

where $\alpha, \beta, t \in \mathbb{C}$, $n \in \mathbb{N}$, $Re(\alpha) > 0$, $Re(\beta) > 0$ and $\Gamma(\cdot)$ is the Euler Gamma function.

The result was discovered by Pollard in 1948 [218] and reported by Berberan-Santos in 2005 [206].

With the aid of the Pollard function, there is the following integral representation of the form

$$E_\alpha\left(-t^\beta\right) = \int_0^\infty e^{-u^\beta t^\beta} L_\beta(u)\, du, \tag{3.14}$$

where $0 < \beta < 1$.

Definition 3.4 The Mittag-Leffler function of Gauss type is defined as

$$E_\alpha\left(t^{2\beta}\right) = \sum_{n=0}^{\infty} \frac{t^{2n\beta}}{\Gamma(n\alpha + 1)}, \tag{3.15}$$

where $\alpha, \beta, t \in \mathbb{C}$, $n \in \mathbb{N}$, $Re(\alpha) > 0$, $Re(\beta) > 0$, and $\Gamma(\cdot)$ is the Euler Gamma function.

In this case, one has the special cases:

- If $\beta = 2\alpha$ in Eq. (3.15), then there is

$$E_\alpha\left(\lambda t^{2\alpha}\right) = \sum_{n=0}^{\infty} \frac{\lambda^n t^{2n\alpha}}{\Gamma(n\alpha + 1)}, \tag{3.16}$$

  where $\alpha, \lambda, t \in \mathbb{C}$, $n \in \mathbb{N}$, and $Re(\alpha) > 0$.
- If $\beta = 2$ in Eq. (3.15), then there is

$$E_\alpha\left(\lambda t^{2}\right) = \sum_{n=0}^{\infty} \frac{\lambda^n t^{2n}}{\Gamma(n\alpha + 1)}, \tag{3.17}$$

  where $\alpha, t, \lambda \in \mathbb{C}$, $n \in \mathbb{N}$, and $Re(\alpha) > 0$.
- If $\beta = 2\alpha$ in Eq. (3.15), then there is

$$E_{2\alpha}\left(\lambda t^{2\alpha}\right) = \sum_{n=0}^{\infty} \frac{\lambda^n t^{2\alpha n}}{\Gamma(2n\alpha + 1)}, \tag{3.18}$$

  where $\alpha, t, \lambda \in \mathbb{C}$, $n \in \mathbb{N}$, and $Re(\alpha) > 0$.
- If $\beta = 2\alpha$ in Eq. (3.15), then there is

$$E_{2\alpha}\left(-t^{2\alpha}\right) = \sum_{n=0}^{\infty} \frac{(-1)^n t^{2\alpha n}}{\Gamma(2n\alpha + 1)}, \tag{3.19}$$

  where $\alpha, t \in \mathbb{C}$, $n \in \mathbb{N}$ and $Re(\alpha) > 0$.

- If $\alpha = 1$ in Eq. (3.15), then there is

$$e^{\lambda t^2} = \sum_{n=0}^{\infty} \frac{\lambda^n t^{2n}}{\Gamma(n+1)}, \tag{3.20}$$

where $t, \alpha \in \mathbb{C}$, $n \in \mathbb{N}$ and $Re(\alpha) > 0$.

Theorem 3.1 (Phragmén Theorem [221]) *Let $x \geq 0$ and $0 \leq \alpha \leq 1$. The function $E_\alpha(x)$ is an entire function of order $\rho = \frac{1}{\alpha}$ and type $\mu = 1$.*

The result was discovered by Phragmén [221].

The function $E_\alpha(x)$ for $\alpha > 0$ is the simplest entire function of its order, proposed by Phragmén in 1904 [221], and its applications were reported by Buhl in 1925 [222].

Theorem 3.2 (Feller–Pollard Theorem [218]) *The function $E_\alpha(-x)$ is completely monotonic for $x \geq 0$ and $0 \leq \alpha \leq 1$;*

or alternatively

$$(-1)^n \frac{d^n}{dx^n} E_\alpha(-x) > 0. \tag{3.21}$$

The result was conjectured by Feller [218], proved by Pollard in 1948 [218]. There exists an alternative form as follows:

$$\frac{d^n}{dx^n} E_\alpha(x) > 0, \tag{3.22}$$

where $x \in \mathbb{R}$ and $0 \leq \alpha \leq 1$.

It is not difficult to show the following theorem:

Theorem 3.3 (Pollard Theorem [218]) *The function $E_\alpha(x)$ has no real zeros when $0 \leq \alpha \leq 1$ and $x \in \mathbb{R}$.*

The result was discovered in 1948 [218].

The function $E_\alpha(-x)$ was proposed by Bieberbach in 1931 (see [223, p. 273]) and reported by Pollard in 1948 [218] and further studied by Feller in 1949 [224].

The function $E_\alpha(-x^\alpha)$ for $t > 0$ and $0 < \alpha < 1$ was proposed by Goreno and Mainardi in 1997 [224] and further investigated by Mainardi in 2014 [225].

3.1.2 Special Integral Representations

Here, there are the special integral representations as follows [219]:

Theorem 3.4 *Let $\alpha \in \mathbb{R}_+$. Then we have*

$$\int_{-\infty}^{\infty} E_\alpha\left(-t^2\right) dt = \frac{\pi}{\Gamma\left(1 - \frac{\alpha}{2}\right)}. \tag{3.23}$$

Some examples are given as follows:

Example 3.1

$$\int_{-\infty}^{\infty} E_3\left(-t^2\right) dt = \frac{\pi}{\Gamma\left(-\frac{1}{2}\right)}. \tag{3.24}$$

Example 3.2

$$\int_{-\infty}^{\infty} E_5\left(-t^2\right) dt = \frac{\pi}{\Gamma\left(-\frac{3}{2}\right)}. \tag{3.25}$$

Theorem 3.5 *Let $\alpha, \beta \in \mathbb{R}_+$, $\alpha, \beta \notin \mathbb{Z}$, $\beta^{-1} \notin \mathbb{Z}$, and $\alpha \neq \beta$. Then we have*

$$\begin{aligned} &\int_0^\infty E_\alpha\left(-t^\beta\right) dt \\ &= \frac{1}{\beta}\frac{\Gamma\left(\frac{1}{\beta}\right)\Gamma\left(1-\frac{1}{\beta}\right)}{\Gamma\left(1-\frac{\alpha}{\beta}\right)} \\ &= \frac{\pi}{\beta sin\left(\frac{\pi}{\beta}\right)}\frac{1}{\Gamma\left(1-\frac{\alpha}{\beta}\right)}. \end{aligned} \tag{3.26}$$

The results were discovered by Dattoli et al in 2017 [219].

Some examples are given as follows:

Example 3.3

$$\int_0^\infty E_2\left(-t^3\right) dt = \frac{1}{2}\Gamma\left(\frac{5}{3}\right). \tag{3.27}$$

Example 3.4

$$\int_0^\infty E_{\frac{1}{3}}\left(-\sqrt{t}\right) dt = \frac{\pi}{2}\cdot\frac{1}{\Gamma\left(\frac{5}{6}\right)}. \tag{3.28}$$

Example 3.5

$$\int_0^\infty E_{\frac{1}{3}}\left(-\sqrt[3]{t}\right) dt = \frac{\Gamma\left(\frac{4}{3}\right)\Gamma\left(\frac{2}{3}\right)}{\Gamma\left(\frac{8}{9}\right)}. \tag{3.29}$$

Example 3.6

$$\int_0^\infty E_{\frac{1}{2}}\left(-\sqrt{t}\right)dt = \frac{\pi}{2}\cdot\frac{1}{\Gamma\left(\frac{3}{4}\right)}. \tag{3.30}$$

Example 3.7

$$\int_0^\infty E_{\frac{1}{2}}\left(-\sqrt[3]{t}\right)dt = \frac{\Gamma\left(\frac{4}{3}\right)\Gamma\left(\frac{2}{3}\right)}{\Gamma\left(\frac{5}{6}\right)}. \tag{3.31}$$

As a direct result, we have the following:

Let β be the set of the even numbers ∂, then it is denoted by $\beta \in \partial$.

Theorem 3.6 *Let $\alpha, \beta \in \mathbb{R}_+$, $\alpha, \beta \notin \mathbb{Z}$, $\beta^{-1} \notin \mathbb{Z}$, $\alpha \neq \beta$, and $\beta \in \partial$. Then we have*

$$\begin{aligned}&\int_{-\infty}^{\infty} E_\alpha\left(-t^\beta\right)dt\\ &= \frac{2}{\beta}\cdot\frac{\Gamma\left(\frac{1}{\beta}\right)\Gamma\left(1-\frac{1}{\beta}\right)}{\Gamma\left(1-\frac{\alpha}{\beta}\right)}\\ &= \frac{2\pi}{\beta sin\left(\frac{\pi}{\beta}\right)}\frac{1}{\Gamma\left(1-\frac{\alpha}{\beta}\right)}.\end{aligned} \tag{3.32}$$

Some examples are given as follows:

Example 3.8

$$\int_{-\infty}^\infty E_7\left(-t^2\right)dt = \frac{\pi}{\Gamma\left(-\frac{5}{2}\right)}. \tag{3.33}$$

Example 3.9

$$\int_{-\infty}^\infty E_{\frac{1}{2}}\left(-t^4\right)dt = \frac{1}{2}\cdot\frac{\Gamma\left(\frac{1}{4}\right)\Gamma\left(\frac{3}{4}\right)}{\Gamma\left(\frac{7}{8}\right)}. \tag{3.34}$$

Example 3.10

$$\int_{-\infty}^\infty E_1\left(-t^6\right)dt = \frac{1}{3}\Gamma\left(\frac{1}{6}\right). \tag{3.35}$$

Example 3.11

$$\int_{-\infty}^{\infty} E_{\frac{3}{2}}\left(-t^{8}\right) dt = \frac{2}{\beta} \frac{\Gamma\left(\frac{1}{8}\right) \Gamma\left(\frac{7}{8}\right)}{\Gamma\left(\frac{13}{16}\right)}. \tag{3.36}$$

Theorem 3.7 *Let $\alpha, \beta \in \mathbb{R}_{+}$, $\beta \neq n$, $n \in \mathbb{Z}_{+}$, $\lambda \in \mathbb{C}$ and $\frac{\alpha}{\beta} - 1 \notin \mathbb{Z}_{+}$. Then we have*

$$\int_{0}^{\infty} t^{\beta-1} E_{\alpha}\left(-\lambda t^{2\beta}\right) dt = \frac{\pi}{2\beta\sqrt{\lambda}} \cdot \frac{1}{\Gamma\left(1-\frac{\alpha}{2}\right)} \tag{3.37}$$

and

$$\int_{-\infty}^{\infty} |t|^{\beta-1} E_{\alpha}\left(-\lambda t^{2\beta}\right) dt = \frac{\pi}{\beta\sqrt{\lambda}} \cdot \frac{1}{\Gamma\left(1-\frac{\alpha}{2}\right)}. \tag{3.38}$$

Some examples are given as follows:

Example 3.12

$$\int_{0}^{\infty} t^{2} E_{1}\left(-2t^{6}\right) dt = \frac{\sqrt{\pi}}{6\sqrt{2}}. \tag{3.39}$$

Example 3.13

$$\int_{-\infty}^{\infty} |t|^{2} E_{\frac{1}{2}}\left(-3t^{3}\right) dt = \frac{\pi}{3\sqrt{3}} \cdot \frac{1}{\Gamma\left(\frac{3}{4}\right)}. \tag{3.40}$$

Example 3.14

$$\int_{0}^{\infty} t^{2} E_{7}\left(-4t^{6}\right) dt = \frac{\pi}{12} \cdot \frac{1}{\Gamma\left(-\frac{5}{2}\right)}. \tag{3.41}$$

Example 3.15

$$\int_{-\infty}^{\infty} |t|^{2} E_{9}\left(-5t^{3}\right) dt = \frac{\pi}{3\sqrt{5}} \cdot \frac{1}{\Gamma\left(-\frac{7}{2}\right)}. \tag{3.42}$$

Example 3.16

$$\int_{0}^{\infty} t^{2} E_{1}\left(-\frac{1}{2}t^{6}\right) dt = \frac{2}{3} \cdot \sqrt{\frac{\pi}{2}}. \tag{3.43}$$

Example 3.17

$$\int_{-\infty}^{\infty} |t|^2 E_{\frac{1}{2}}\left(-\frac{1}{2}t^3\right) dt = \frac{2\pi}{3\sqrt{2}} \cdot \frac{1}{\Gamma\left(\frac{3}{4}\right)}. \tag{3.44}$$

Example 3.18

$$\int_{0}^{\infty} t^2 E_7\left(-\frac{1}{2}t^6\right) dt = \frac{\pi}{3\sqrt{2}} \cdot \frac{1}{\Gamma\left(-\frac{5}{2}\right)}. \tag{3.45}$$

Example 3.19

$$\int_{-\infty}^{\infty} |t|^2 E_9\left(-\frac{1}{2}t^3\right) dt = \frac{2\pi}{3\sqrt{2}} \cdot \frac{1}{\Gamma\left(-\frac{7}{2}\right)}. \tag{3.46}$$

Theorem 3.8 *Let $\alpha \in \mathbb{R}_+$, $\alpha \neq 2$, and $\lambda \in \mathbb{C}$. Then we have*

$$\int_{0}^{\infty} t^{\alpha-1} E_\alpha\left(-\lambda t^{2\alpha}\right) dt = \frac{\pi}{2\alpha\sqrt{\lambda}} \cdot \frac{1}{\Gamma\left(1-\frac{\alpha}{2}\right)} \tag{3.47}$$

and

$$\int_{-\infty}^{\infty} |t|^{\alpha-1} E_\alpha\left(-\lambda t^{2\alpha}\right) dt = \frac{\pi}{\alpha\sqrt{\lambda}} \cdot \frac{1}{\Gamma\left(1-\frac{\alpha}{2}\right)}. \tag{3.48}$$

We now find the properties of the special integral equations as follows:

Property 3.1 If $\alpha, \beta, \gamma, \lambda \in \mathbb{R}_+$, $\beta \neq 1$ and $\alpha \neq \beta$, then we have

$$\int_{0}^{\infty} t^{\gamma-1} E_\alpha\left(-\lambda t^{\gamma\beta}\right) dt = \frac{1}{\gamma} \cdot \frac{1}{\sqrt[\beta]{\lambda}} \cdot \frac{\Gamma\left(1-\frac{1}{\beta}\right)\Gamma\left(1+\frac{1}{\beta}\right)}{\Gamma\left(1-\frac{\alpha}{\beta}\right)} \; (\alpha \neq \beta). \tag{3.49}$$

Note: The result is also valid for $\lambda \in \mathbb{C}$.

Some examples are given as follows:

Example 3.20

$$\int_{0}^{\infty} t E_{\frac{1}{2}}\left(-2t^3\right) dt = \frac{1}{2} \cdot \frac{1}{\sqrt[3]{2}} \cdot \frac{\Gamma\left(-\frac{1}{2}\right)\Gamma\left(\frac{5}{2}\right)}{\Gamma\left(1-\frac{1}{3}\right)}. \tag{3.50}$$

Example 3.21

$$\int_0^\infty t^2 E_{\frac{2}{3}}\left(-3t^{\frac{9}{2}}\right) dt = \frac{1}{3} \cdot \frac{1}{\sqrt[\frac{3}{2}]{3}} \cdot \frac{\Gamma\left(-\frac{1}{2}\right)\Gamma\left(\frac{5}{2}\right)}{\Gamma\left(\frac{5}{9}\right)}. \tag{3.51}$$

Example 3.22

$$\int_0^\infty t^{\frac{1}{2}} E_7\left(-4t\right) dt = \frac{3}{2} \cdot \frac{1}{\sqrt[\frac{3}{2}]{4}} \cdot \frac{\Gamma\left(-\frac{1}{2}\right)\Gamma\left(\frac{5}{2}\right)}{\Gamma\left(-\frac{11}{3}\right)}. \tag{3.52}$$

Example 3.23

$$\int_0^\infty t^{-\frac{4}{5}} E_{\frac{1}{2}}\left(-5t^{\frac{3}{10}}\right) dt = 5 \cdot \frac{1}{\sqrt[\frac{3}{2}]{5}} \cdot \frac{\Gamma\left(-\frac{1}{2}\right)\Gamma\left(\frac{5}{2}\right)}{\Gamma\left(1-\frac{1}{3}\right)}. \tag{3.53}$$

Example 3.24

$$\int_0^\infty t^{-\frac{3}{4}} E_{\frac{2}{3}}\left(-6t^{\frac{3}{8}}\right) dt = 4 \cdot \frac{1}{\sqrt[\frac{3}{2}]{6}} \cdot \frac{\Gamma\left(-\frac{1}{2}\right)\Gamma\left(\frac{5}{2}\right)}{\Gamma\left(\frac{5}{9}\right)}. \tag{3.54}$$

Example 3.25

$$\int_0^\infty t^{-\frac{1}{2}} E_7\left(-7t^{\frac{3}{4}}\right) dt = 2 \cdot \frac{1}{\sqrt[\frac{3}{2}]{7}} \cdot \frac{\Gamma\left(-\frac{1}{2}\right)\Gamma\left(\frac{5}{2}\right)}{\Gamma\left(-\frac{11}{3}\right)}. \tag{3.55}$$

Theorem 3.9 *Let $\alpha, \beta, \gamma, \lambda \in \mathbb{R}_+$, $\beta \neq 1$, $\alpha \neq \beta$, and $\gamma \in \partial$. Then we have*

$$\int_{-\infty}^\infty |t|^{\gamma-1} E_\alpha\left(-\lambda\,|t|^{\gamma\beta}\right) dt = \frac{2}{\gamma} \cdot \frac{1}{\sqrt[\beta]{\lambda}} \cdot \frac{\Gamma\left(1-\frac{1}{\beta}\right)\Gamma\left(1+\frac{1}{\beta}\right)}{\Gamma\left(1-\frac{\alpha}{\beta}\right)} \,(\alpha \neq \beta)\,. \tag{3.56}$$

Note: The result is also valid for $\lambda \in \mathbb{C}$.

Some examples are given as follows:

Example 3.26

$$\int_{-\infty}^{\infty} |t|\, E_2\left(-2\,|t|^6\right) dt = \frac{2}{2} \cdot \frac{1}{\sqrt[3]{2}} \cdot \frac{\Gamma\left(\frac{2}{3}\right)\Gamma\left(\frac{4}{3}\right)}{\Gamma\left(1-\frac{2}{3}\right)}. \tag{3.57}$$

Example 3.27

$$\int_{-\infty}^{\infty} |t|\, E_1\left(-3\,|t|^6\right) dt = \frac{2}{2} \cdot \frac{1}{\sqrt[3]{3}} \cdot \Gamma\left(\frac{4}{3}\right). \tag{3.58}$$

Example 3.28

$$\int_{-\infty}^{\infty} |t|\, E_{\frac{1}{2}}\left(-4\,|t|^6\right) dt = \frac{2}{2} \cdot \frac{1}{\sqrt[3]{4}} \cdot \frac{\Gamma\left(\frac{2}{3}\right)\Gamma\left(\frac{4}{3}\right)}{\Gamma\left(\frac{5}{6}\right)}. \tag{3.59}$$

Example 3.29

$$\int_{-\infty}^{\infty} |t|\, E_2\left(-5\,|t|^6\right) dt = \frac{2}{2} \cdot \frac{1}{\sqrt[3]{5}} \cdot \frac{\Gamma\left(\frac{2}{3}\right)\Gamma\left(\frac{4}{3}\right)}{\Gamma\left(1-\frac{2}{3}\right)}. \tag{3.60}$$

Example 3.30

$$\int_{-\infty}^{\infty} |t|\, E_1\left(-6\,|t|^6\right) dt = \frac{2}{2} \cdot \frac{1}{\sqrt[3]{6}} \cdot \Gamma\left(\frac{4}{3}\right). \tag{3.61}$$

Example 3.31

$$\int_{-\infty}^{\infty} |t|\, E_{\frac{1}{2}}\left(-7\,|t|^6\right) dt = \frac{2}{2} \cdot \frac{1}{\sqrt[3]{7}} \cdot \frac{\Gamma\left(\frac{2}{3}\right)\Gamma\left(\frac{4}{3}\right)}{\Gamma\left(\frac{5}{6}\right)}. \tag{3.62}$$

Theorem 3.10 *Let $\alpha, \beta, \gamma, \lambda \in \mathbb{R}_+$ and $\beta \in \partial$. Then we have*

$$\int_{-\infty}^{\infty} |t|^{\gamma-1}\, E_\alpha\left(-\lambda\,|t|^{\gamma\beta}\right) dt = \frac{2}{\gamma} \cdot \frac{1}{\sqrt[\beta]{\lambda}} \cdot \frac{\Gamma\left(1-\frac{1}{\beta}\right)\Gamma\left(1+\frac{1}{\beta}\right)}{\Gamma\left(1-\frac{\alpha}{\beta}\right)}\ (\alpha \neq \beta). \tag{3.63}$$

Note: The result is also valid for $\lambda \in \mathbb{C}$.

Some examples are given as follows:

Example 3.32

$$\int_{-\infty}^{\infty} |t|^4 E_{\frac{1}{2}}\left(-2|t|^{10}\right) dt = \frac{2}{5} \cdot \frac{1}{\sqrt[2]{2}} \cdot \frac{\Gamma\left(\frac{1}{2}\right)\Gamma\left(\frac{3}{2}\right)}{\Gamma\left(\frac{3}{4}\right)}. \tag{3.64}$$

Example 3.33

$$\int_{-\infty}^{\infty} |t|^{-\frac{1}{2}} E_1(-3|t|)\, dt = 4 \cdot \frac{1}{\sqrt[2]{3}} \cdot \Gamma\left(\frac{3}{2}\right). \tag{3.65}$$

Theorem 3.11 *If $\alpha, \beta, \lambda \in \mathbb{R}_+$, then we have*

$$\int_0^{\infty} E_\alpha\left(-\lambda t^\beta\right) dt = \frac{1}{\sqrt[\beta]{\lambda}} \cdot \frac{\Gamma\left(1-\frac{1}{\beta}\right)\Gamma\left(1+\frac{1}{\beta}\right)}{\Gamma\left(1-\frac{\alpha}{\beta}\right)} \,(\alpha \neq \beta). \tag{3.66}$$

Note: The result is also valid for $\lambda \in \mathbb{C}$.

Some examples are given as follows:

Example 3.34

$$\int_0^{\infty} t^4 E_{\frac{2}{3}}\left(-2t^{10}\right) dt = \frac{1}{5} \cdot \frac{1}{\sqrt[2]{2}} \cdot \frac{\Gamma\left(\frac{1}{2}\right)\Gamma\left(\frac{3}{2}\right)}{\Gamma\left(\frac{2}{3}\right)}. \tag{3.67}$$

Example 3.35

$$\int_0^{\infty} t^{-\frac{1}{2}} E_{\frac{1}{4}}(-3t)\, dt = 2 \cdot \frac{1}{\sqrt[2]{3}} \cdot \frac{\Gamma\left(\frac{1}{2}\right)\Gamma\left(\frac{3}{2}\right)}{\Gamma\left(\frac{7}{8}\right)}. \tag{3.68}$$

Theorem 3.12 *If $\alpha, \beta, \lambda \in \mathbb{R}_+$, then we have*

$$\int_{-\infty}^{\infty} E_\alpha\left(-\lambda|t|^\beta\right) dt = \frac{2}{\sqrt[\beta]{\lambda}} \cdot \frac{\Gamma\left(1-\frac{1}{\beta}\right)\Gamma\left(1+\frac{1}{\beta}\right)}{\Gamma\left(1-\frac{\alpha}{\beta}\right)} \,(\alpha \neq \beta). \tag{3.69}$$

Note: The result is also valid for $\lambda \in \mathbb{C}$.

Some examples are given as follows:

Example 3.36

$$\int_{-\infty}^{\infty} |t|^4 E_{\frac{2}{3}}\left(-2|t|^{10}\right) dt = \frac{2}{5} \cdot \frac{1}{\sqrt[2]{2}} \cdot \frac{\Gamma\left(\frac{1}{2}\right)\Gamma\left(\frac{3}{2}\right)}{\Gamma\left(\frac{2}{3}\right)}. \tag{3.70}$$

Example 3.37

$$\int_{-\infty}^{\infty} |t|^{-\frac{1}{2}} E_{\frac{1}{4}}\left(-3|t|\right) dt = 4 \cdot \frac{1}{\sqrt[2]{3}} \cdot \frac{\Gamma\left(\frac{1}{2}\right)\Gamma\left(\frac{3}{2}\right)}{\Gamma\left(\frac{7}{8}\right)}. \tag{3.71}$$

Theorem 3.13 *If $\alpha, \gamma \in \mathbb{R}_+$, then we have*

$$\int_0^{\infty} t^{\gamma-1} E_{2\alpha}\left(-t^{2\gamma}\right) dt = \frac{\pi}{2\gamma} \cdot \frac{1}{\Gamma(1-\alpha)} \ (\alpha \neq 1). \tag{3.72}$$

Some examples are given as follows:

Example 3.38

$$\int_0^{\infty} t^2 E_{\frac{1}{4}}\left(-t^6\right) dt = \frac{\pi}{6} \cdot \frac{1}{\Gamma\left(1-\frac{1}{8}\right)}. \tag{3.73}$$

Example 3.39

$$\int_0^{\infty} t^2 E_3\left(-t^6\right) dt = \frac{\pi}{6} \cdot \frac{1}{\Gamma\left(-\frac{1}{2}\right)}. \tag{3.74}$$

Theorem 3.14 *If $\alpha, \gamma, \lambda \in \mathbb{R}_+$, then we have*

$$\int_0^{\infty} t^{\gamma-1} E_{2\alpha}\left(-\lambda t^{2\gamma}\right) dt = \frac{1}{2\gamma} \cdot \frac{\pi}{\sqrt[2]{\lambda}} \cdot \frac{1}{\Gamma(1-\alpha)} \ (\alpha \neq 1). \tag{3.75}$$

Theorem 3.15 *Let $\alpha, \beta, \lambda \in \mathbb{R}_+$ and $\alpha \neq 2$. Then we have*

$$\int_0^{\infty} t^{\frac{\beta}{2}-1} E_{\alpha}\left(-\lambda t^{\beta}\right) dt = \frac{\pi}{\beta\sqrt[2]{\lambda}} \cdot \frac{1}{\Gamma\left(1-\frac{\alpha}{2}\right)}. \tag{3.76}$$

Theorem 3.16 *Let $\alpha, \beta, \lambda \in \mathbb{R}_+$ and $\alpha \neq 2$. Then we have*

$$\int_{-\infty}^{\infty} t^{\frac{\beta}{2}-1} E_\alpha \left(-\lambda |t|^\beta\right) dt = \frac{2\pi}{\beta \sqrt[2]{\lambda}} \cdot \frac{1}{\Gamma\left(1-\frac{\alpha}{2}\right)}, \tag{3.77}$$

where $\beta \in \partial$.

Theorem 3.17 *Let $\alpha, \lambda \in \mathbb{R}_+$ and $\alpha \neq 2$. Then we have*

$$\int_0^{\infty} t^{-\frac{1}{2}} E_\alpha (-\lambda t)\, dt = \frac{\pi}{\sqrt[2]{\lambda}} \cdot \frac{1}{\Gamma\left(1-\frac{\alpha}{2}\right)}. \tag{3.78}$$

Theorem 3.18 *Let $\alpha, \lambda \in \mathbb{R}_+$ and $\alpha \neq 1$. Then we have*

$$\int_0^{\infty} t^{-\frac{1}{2}} E_{2\alpha} (-\lambda t)\, dt = \frac{\pi}{\sqrt[2]{\lambda}} \cdot \frac{1}{\Gamma(1-\alpha)}. \tag{3.79}$$

Theorem 3.19 *Let $\alpha \in \mathbb{R}_+$ and $\alpha \neq 2$. Then we have*

$$\int_0^{\infty} t^{-\frac{1}{2}} E_\alpha (-t)\, dt = \frac{\pi}{\Gamma\left(1-\frac{\alpha}{2}\right)}. \tag{3.80}$$

Theorem 3.20 *Let $\alpha \in \mathbb{R}_+$ and $\alpha \neq 1$. Then we have*

$$\int_0^{\infty} t^{-\frac{1}{2}} E_{2\alpha} (-t)\, dt = \frac{\pi}{\Gamma(1-\alpha)}. \tag{3.81}$$

Theorem 3.21 *Let $\alpha \in \mathbb{R}_+$ and $\alpha \neq \frac{2}{3}$. Then we have*

$$\int_0^{\infty} t^{-\frac{1}{2}} E_{3\alpha} (-t)\, dt = \frac{\pi}{\Gamma\left(1-\frac{3}{2}\alpha\right)}. \tag{3.82}$$

Theorem 3.22 *Let $\alpha, \beta \in \mathbb{R}_+$ and $\alpha \neq 2$. Then we have*

$$\int_0^{\infty} t^{\frac{\beta}{2}-1} E_\alpha \left(-t^\beta\right) dt = \frac{\pi}{\beta} \cdot \frac{1}{\Gamma\left(1-\frac{\alpha}{2}\right)}. \tag{3.83}$$

Theorem 3.23 *Let $\alpha, \beta \in \mathbb{R}_+$ and $\alpha \neq 2$. Then we have*

$$\int_{-\infty}^{\infty} |t|^{\frac{\beta}{2}-1} E_\alpha \left(-|t|^\beta\right) dt = \frac{2\pi}{\beta} \cdot \frac{1}{\Gamma\left(1-\frac{\alpha}{2}\right)}, \tag{3.84}$$

where $\beta \in \partial$.

Theorem 3.24 *Let $\alpha, \beta \in \mathbb{R}_+$ and $\alpha \neq 1$. Then we have*

$$\int_0^{\infty} t^{\frac{\beta}{2}-1} E_{2\alpha}\left(-t^{\beta}\right) dt = \frac{\pi}{\beta} \cdot \frac{1}{\Gamma(1-\alpha)}. \tag{3.85}$$

Theorem 3.25 *Let $\alpha, \beta \in \mathbb{R}_+$ and $\alpha \neq 1$. Then we have*

$$\int_{-\infty}^{\infty} |t|^{\frac{\beta}{2}-1} E_{2\alpha}\left(-|t|^{\beta}\right) dt = \frac{2\pi}{\beta} \cdot \frac{1}{\Gamma(1-\alpha)}, \tag{3.86}$$

where $\beta \in \partial$.

Theorem 3.26 *Let $\alpha, \beta \in \mathbb{R}_+$ and $\alpha \neq \frac{2}{3}$. Then we have*

$$\int_0^{\infty} t^{\frac{\beta}{2}-1} E_{3\alpha}\left(-t^{\beta}\right) dt = \frac{\pi}{\beta} \cdot \frac{1}{\Gamma\left(1-\frac{3\alpha}{2}\right)}. \tag{3.87}$$

Theorem 3.27 *Let $\alpha, \beta \in \mathbb{R}_+$ and $\alpha \neq \frac{2}{3}$. Then we have*

$$\int_{-\infty}^{\infty} |t|^{\frac{\beta}{2}-1} E_{3\alpha}\left(-|t|^{\beta}\right) dt = \frac{2\pi}{\beta} \cdot \frac{1}{\Gamma\left(1-\frac{3\alpha}{2}\right)}, \tag{3.88}$$

where $\beta \in \partial$.

Theorem 3.28 *Let $\alpha, \beta \in \mathbb{R}_+$ and $\alpha \neq \frac{1}{2}$. Then we have*

$$\int_0^{\infty} t^{\frac{\beta}{2}-1} E_{4\alpha}\left(-t^{\beta}\right) dt = \frac{\pi}{\beta} \cdot \frac{1}{\Gamma(1-2\alpha)}. \tag{3.89}$$

Theorem 3.29 *Let $\alpha, \beta \in \mathbb{R}_+$ and $\alpha \neq \frac{1}{2}$. Then we have*

$$\int_{-\infty}^{\infty} |t|^{\frac{\beta}{2}-1} E_{4\alpha}\left(-|t|^{\beta}\right) dt = \frac{2\pi}{\beta} \cdot \frac{1}{\Gamma(1-2\alpha)}, \tag{3.90}$$

where $\beta \in \partial$.

Theorem 3.30 *Let $\alpha, \beta, \lambda \in \mathbb{R}_+$ and $\alpha \neq 2$. Then we have*

$$\int_0^{\infty} t^{\beta-1} E_{\alpha}\left(-\lambda t^{2\beta}\right) dt = \frac{\pi}{2\beta\sqrt[2]{\lambda}} \cdot \frac{1}{\Gamma\left(1-\frac{\alpha}{2}\right)}. \tag{3.91}$$

Theorem 3.31 *Let $\alpha, \beta, \lambda \in \mathbb{R}_+$ and $\alpha \neq 2$. Then we have*

$$\int_{-\infty}^{\infty} |t|^{\beta-1} E_{\alpha}\left(-\lambda t^{2\beta}\right) dt = \frac{\pi}{\beta\sqrt[2]{\lambda}} \cdot \frac{1}{\Gamma\left(1-\frac{\alpha}{2}\right)}. \tag{3.92}$$

Theorem 3.32 *Let* $\alpha, \beta \in \mathbb{R}_+$ *and* $\alpha \neq 1$. *Then we have*

$$\int_0^{\infty} t^{\beta-1} E_{2\alpha}\left(-t^{2\beta}\right) dt = \frac{\pi}{2\beta} \cdot \frac{1}{\Gamma(1-\alpha)}. \tag{3.93}$$

Theorem 3.33 *Let* $\alpha, \beta \in \mathbb{R}_+$ *and* $\alpha \neq 1$. *Then we have*

$$\int_{-\infty}^{\infty} |t|^{\beta-1} E_{2\alpha}\left(-t^{2\beta}\right) dt = \frac{1}{\beta} \cdot \frac{\pi}{\Gamma(1-\alpha)}. \tag{3.94}$$

Theorem 3.34 *Let* $\alpha, \beta \in \mathbb{R}_+$ *and* $\alpha \neq \frac{2}{3}$. *Then we have*

$$\int_0^{\infty} t^{\beta-1} E_{3\alpha}\left(-t^{2\beta}\right) dt = \frac{\pi}{2\beta} \cdot \frac{1}{\Gamma\left(1-\frac{3\alpha}{2}\right)}. \tag{3.95}$$

Theorem 3.35 *Let* $\alpha, \beta \in \mathbb{R}_+$ *and* $\alpha \neq \frac{2}{3}$. *Then we have*

$$\int_{-\infty}^{\infty} |t|^{\beta-1} E_{3\alpha}\left(-t^{2\beta}\right) dt = \frac{\pi}{\beta} \cdot \frac{1}{\Gamma\left(1-\frac{3\alpha}{2}\right)}. \tag{3.96}$$

Theorem 3.36 *Let* $\alpha, \beta \in \mathbb{R}_+$ *and* $\alpha \neq \frac{1}{2}$. *Then we have*

$$\int_0^{\infty} t^{\beta-1} E_{4\alpha}\left(-t^{2\beta}\right) dt = \frac{\pi}{2\beta} \cdot \frac{1}{\Gamma(1-2\alpha)}. \tag{3.97}$$

Theorem 3.37 *Let* $\alpha, \beta \in \mathbb{R}_+$ *and* $\alpha \neq \frac{1}{2}$. *Then we have*

$$\int_{-\infty}^{\infty} |t|^{\beta-1} E_{4\alpha}\left(-t^{2\beta}\right) dt = \frac{\pi}{\beta} \cdot \frac{1}{\Gamma(1-2\alpha)}. \tag{3.98}$$

Theorem 3.38 *Let* $\alpha \in \mathbb{R}_+$ *and* $\alpha \neq 1$. *Then we have*

$$\int_0^{\infty} t^{\alpha-1} E_{2\alpha}\left(-t^{2\alpha}\right) dt = \frac{\pi}{2\alpha} \cdot \frac{1}{\Gamma(1-\alpha)}. \tag{3.99}$$

Theorem 3.39 *Let* $\alpha \in \mathbb{R}_+$ *and* $\alpha \neq 1$. *Then we have*

$$\int_{-\infty}^{\infty} |t|^{\alpha-1} E_{2\alpha}\left(-t^{2\alpha}\right) dt = \frac{1}{\alpha} \cdot \frac{\pi}{\Gamma(1-\alpha)}. \tag{3.100}$$

Theorem 3.40 *Let $\alpha \in \mathbb{R}_{+}$, $\alpha \neq 2$, and $\lambda \in \mathbb{C}$. Then we have*

$$\int_0^{\infty} t^{\frac{\alpha}{2}-1} E_\alpha\left(-\lambda t^\alpha\right) dt = \frac{\pi}{\alpha \sqrt[2]{\lambda}} \cdot \frac{1}{\Gamma\left(1-\frac{\alpha}{2}\right)}. \tag{3.101}$$

Theorem 3.41 *Letα, $\lambda \in \mathbb{R}_{+}$ and $\alpha \neq 2$. Then we have*

$$\int_{-\infty}^{\infty} |t|^{\frac{\alpha}{2}-1} E_\alpha\left(-\lambda t^\alpha\right) dt = \frac{2\pi}{\alpha \sqrt[2]{\lambda}} \cdot \frac{1}{\Gamma\left(1-\frac{\alpha}{2}\right)}, \tag{3.102}$$

where $\alpha \in \partial$.

Theorem 3.42 *If $\alpha, \lambda \in \mathbb{R}_{+}$ and $\alpha \neq \pm\frac{1}{2}$, then we have*

$$\int_0^{\infty} E_\alpha\left(-\lambda t^{2\alpha}\right) dt = \frac{1}{\sqrt[2\alpha]{\lambda}} \cdot \frac{1}{\sqrt{\pi}} \Gamma\left(1-\frac{1}{2\alpha}\right) \Gamma\left(1+\frac{1}{2\alpha}\right). \tag{3.103}$$

Note: The result is also valid for $\lambda \in \mathbb{C}$.

Theorem 3.43 *If $\alpha, \lambda \in \mathbb{R}_{+}$, $\alpha \neq \pm\frac{1}{2}$, and $\beta \in \partial$, then we have*

$$\int_{-\infty}^{\infty} E_\alpha\left(-\lambda t^{2\alpha}\right) dt = \frac{1}{\sqrt[2\alpha]{\lambda}} \cdot \frac{2}{\sqrt{\pi}} \cdot \Gamma\left(1-\frac{1}{2\alpha}\right) \Gamma\left(1+\frac{1}{2\alpha}\right). \tag{3.104}$$

Note: The result is also valid for $\lambda \in \mathbb{C}$.

Theorem 3.44 *If $|z| < 1$, $\alpha \in \mathbb{Z}$, $Re(\alpha) > 0$, then we have*

$$\int_0^{\infty} e^{-t} E_\alpha\left(-z t^\alpha\right) dt = \frac{1}{1-z}. \tag{3.105}$$

The result was discovered by Erd'elyi et al. [226].
Let us consider the following special cases:

Theorem 3.45 *If $\alpha, \lambda \in \mathbb{R}_{+}$, $|z| < 1$, and $\lambda \neq 1$, then we have*

$$\int_0^{\infty} e^{-t} E_\alpha\left(-\lambda^\alpha t^\alpha\right) dt = \frac{1}{1-\lambda^\alpha}. \tag{3.106}$$

Theorem 3.46 *If $\alpha, \beta, \lambda \in \mathbb{R}_{+}$, $|z| < 1$, and $\lambda \neq 1$, then we have*

$$\int_0^{\infty} e^{-t} E_\alpha\left(-\lambda^\beta t^\alpha\right) dt = \frac{1}{1-\lambda^\beta}. \tag{3.107}$$

Theorem 3.47 *If $\alpha, \lambda \in \mathbb{R}_{+}$, $\lambda \neq 0$, and $\lambda \neq 1$, then we have*

$$\int_0^{\infty} e^{-\lambda t} E_\alpha\left(-t^\alpha\right) dt = \frac{1}{\lambda-\lambda^{1-\alpha}}. \tag{3.108}$$

Theorem 3.48 *If $\alpha, \beta, \gamma, \lambda \in \mathbb{R}_+$ and $\lambda \neq 1$, then we have*

$$\int_0^\infty t^{\frac{\beta}{\alpha}-1} e^{-\frac{\gamma}{\lambda} t^{\frac{\beta}{\alpha}}} E_\alpha\left(-\gamma t^\beta\right) dt = \frac{\alpha}{\beta} \cdot \frac{\lambda\gamma}{\gamma^2 - \lambda\gamma^\alpha}. \tag{3.109}$$

Theorem 3.49 *If $\alpha, \gamma, \lambda \in \mathbb{R}_+$ and $\lambda \neq \sqrt[2-\alpha]{\gamma}$, then we have*

$$\int_0^\infty t^3 e^{-\frac{\gamma}{\lambda} t^4} E_\alpha\left(-\gamma t^{4\alpha}\right) dt = \frac{1}{4} \cdot \frac{\lambda\gamma}{\gamma^2 - \lambda\gamma^\alpha}. \tag{3.110}$$

Theorem 3.50 *If $\alpha, \gamma, \lambda \in \mathbb{R}_+$ and $\lambda \neq \sqrt[2-\alpha]{\gamma}$, then we have*

$$\int_0^\infty t^2 e^{-\frac{\gamma}{\lambda} t^3} E_\alpha\left(-\gamma t^{3\alpha}\right) dt = \frac{1}{3} \cdot \frac{\lambda\gamma}{\gamma^2 - \lambda\gamma^\alpha}. \tag{3.111}$$

Theorem 3.51 *If $\alpha, \gamma, \lambda \in \mathbb{R}_+$ and $\lambda \neq \sqrt[2-\alpha]{\gamma}$, then we have*

$$\int_0^\infty t e^{-\frac{\gamma}{\lambda} t^2} E_\alpha\left(-\gamma t^{2\alpha}\right) dt = \frac{1}{2} \cdot \frac{\lambda\gamma}{\gamma^2 - \lambda\gamma^\alpha}. \tag{3.112}$$

Theorem 3.52 *If $\alpha, \gamma, \lambda \in \mathbb{R}_+$ and $\lambda \neq \sqrt[2-\alpha]{\gamma}$, then we have*

$$\int_0^\infty t^{\frac{1}{\alpha}-1} e^{-\frac{\gamma}{\lambda} t^{\frac{1}{\alpha}}} E_\alpha\left(-\gamma t\right) dt = \frac{\alpha\lambda\gamma}{\gamma^2 - \lambda\gamma^\alpha}. \tag{3.113}$$

Theorem 3.53 *If $\beta, \gamma, \lambda \in \mathbb{R}_+$ and $\lambda \neq \sqrt[2-\alpha]{\gamma}$, then we have*

$$\int_0^\infty t^{\beta-1} e^{-\left(\frac{\gamma}{\lambda}+\gamma\right) t^\beta} dt = \frac{1}{\beta} \frac{\lambda\gamma}{\gamma^2 - \lambda\gamma}. \tag{3.114}$$

3.1.3 *The Integral Transforms for the Mittag-Leffler Functions*

The Laplace transform for the Mittag-Leffler function can be given as follows [227]:

$$\begin{aligned} &L\left\{E_\alpha\left(t^\alpha\right)\right\} \\ &= L\left\{\sum_{n=0}^{\infty} \frac{t^{n\alpha}}{\Gamma(n\alpha+1)}\right\} \\ &= \sum_{n=0}^{\infty} \frac{1}{s^{n\alpha+1}} \\ &= \frac{1}{s} \frac{1}{1-s^{-\alpha}} \left(\left|s^{-\alpha}\right| < 1\right) \\ &= s^{-1}\left(1-s^{-\alpha}\right)^{-1} \left(\left|s^{-\alpha}\right| < 1\right) \end{aligned} \tag{3.115}$$

and

$$\begin{aligned} & L\left\{E_\alpha\left(\lambda t^\alpha\right)\right\} \\ & =L\left\{\sum_{n=0}^{\infty} \frac{\lambda^n t^{n\alpha}}{\Gamma(n\alpha+1)}\right\} \\ & =\sum_{n=0}^{\infty} \frac{\lambda^n}{s^{n\alpha+1}} \\ & =\frac{1}{s} \frac{1}{1-\lambda s^{-\alpha}}\left(\left|\lambda s^{-\alpha}\right|<1\right) \\ & =s^{-1}\left(1-\lambda s^{-\alpha}\right)^{-1}\left(\left|\lambda s^{-\alpha}\right|<1\right) . \end{aligned} \tag{3.116}$$

The result was discovered by Humbert in 1953 [227].

We now consider the Mellin–Barnes integral representation for the Mittag-Leffler function.

Theorem 3.54 *If $Re(\alpha) > 0$ and $t \in \mathbb{C}$, then there is the following integral representation as follows:*

$$E_\alpha(t)=\frac{1}{2\pi i} \int_{-i\infty}^{i\infty} \frac{\Gamma(s)\Gamma(1-s)}{\Gamma(1-\alpha s)}(-t)^{-s}\, ds, \tag{3.117}$$

where $|\arg(t)|<\pi$.

The result was discovered by Humbert [227] and reported by Haubold et al. [205] and Gorenflo et al. [210].

The Mittag-Leffler integral representation is given as follows:

$$E_\alpha(t)=\frac{1}{2\pi i} \int_{\widehat{H}} \frac{s^{\alpha-1} e^s}{s^\alpha-t} ds, \tag{3.118}$$

where $\widehat{H}$ is the Hankel path.

The Mellin transform of the Mittag-Leffler function can be given as follows:

$$M\left[E_\alpha(-\lambda t)\right]=\int_0^{\infty} t^{\varpi-1} E_\alpha(-\lambda t)\, dt=\frac{\Gamma(\varpi)\Gamma(1-\varpi)}{\lambda^{\varpi}\Gamma(1-\alpha\varpi)}, \tag{3.119}$$

where $\alpha, \varpi \in \mathbb{C}$ and $Re(\alpha)>0$.

The result was reported by Yang in 2019 [92].

Theorem 3.55 *There is [226]*

$$\frac{d^n}{dt^n} E_n\left(t^n\right)=E_n\left(t^n\right). \tag{3.120}$$

The result was discovered by Erd'elyi et al . [226] and reported by Haubold et al. [205] and Gorenflo et al. [210].

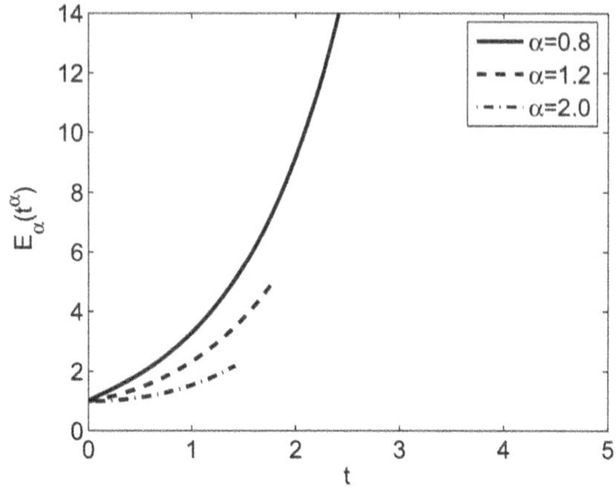

Fig. 3.1 The plots of $E_\alpha(t^\alpha)$ with the different values of the parameter α

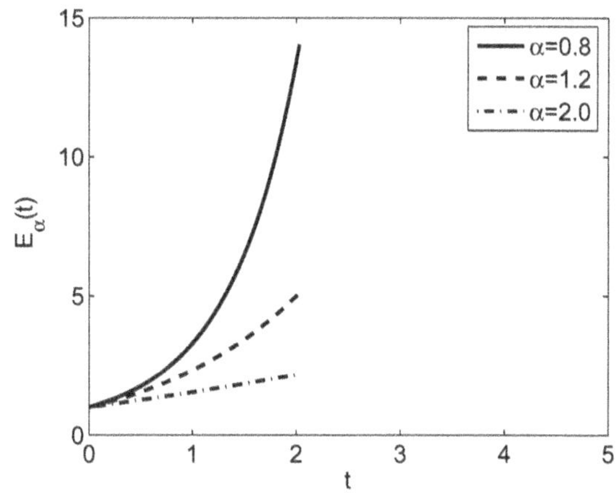

Fig. 3.2 The plots of $E_\alpha(t)$ with the different values of the parameter α

The plots of the functions $E_\alpha(t^\alpha)$, $E_\alpha(t)$, $E_\alpha(-t^\alpha)$, and $E_\alpha(-t)$ with the different values of the parameter α are shown in Figs. 3.1, 3.2, 3.3, and 3.4, respectively.

3.1.4 The Supertrigonometric Functions via Mittag-Leffler Function

In 1988, Bagley introduced the supersine and supercosine functions based on the Mittag-Leffler function [228]. Here we introduce the definitions of the supertrigonometric functions based on the Mittag-Leffler function.

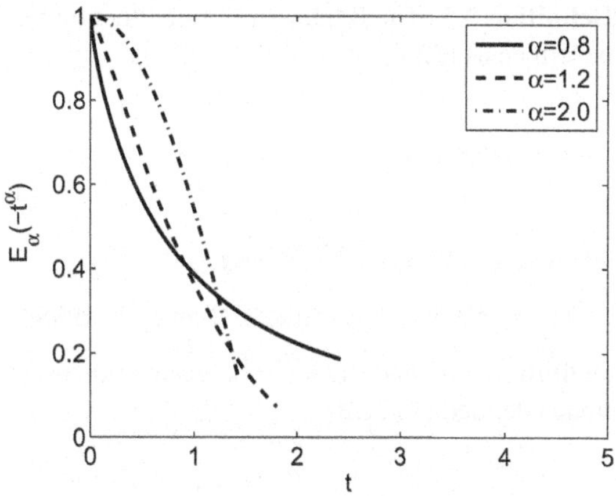

Fig. 3.3 The plots of $E_\alpha(-t^\alpha)$ with the different values of the parameter α

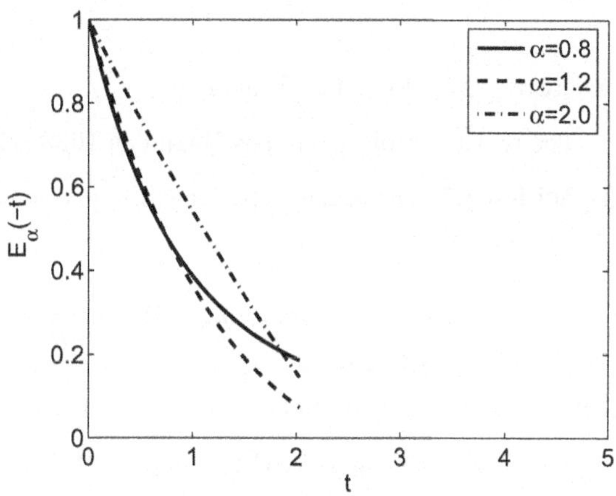

Fig. 3.4 The plots of $E_\alpha(-t)$ with the different values of the parameter α

Let

$$E_\alpha\left(it^\alpha\right) = \sum_{n=0}^{\infty} \frac{i^n t^{n\alpha}}{\Gamma\left(n\alpha + 1\right)} \tag{3.121}$$

and

$$E_\alpha\left(-it^\alpha\right) = \sum_{n=0}^{\infty} \frac{(-i)^n\, t^{n\alpha}}{\Gamma\left(n\alpha + 1\right)}, \tag{3.122}$$

where $\alpha, t \in \mathbb{C}$, $Re\,(\alpha) > 0$, and $i = \sqrt{-1}$.

Definition 3.5 The Bagley supersine function based on the Mittag-Leffler function is defined as [228]

$$supersin_{\alpha}\left(t^{\alpha}\right)=\frac{E_{\alpha}\left(it^{\alpha}\right)-E_{\alpha}\left(-it^{\alpha}\right)}{2i}=\sum_{n=0}^{\infty}\frac{(-1)^{n}\,t^{(2n+1)\alpha}}{\Gamma\left((2n+1)\,\alpha+1\right)}, \tag{3.123}$$

where $\alpha, t \in \mathbb{C}$, $Re\,(\alpha) > 0$, and $i = \sqrt{-1}$.

The result was discovered by Bagley in 1988 [228].

Definition 3.6 The Bagley supercosine function based on the Mittag-Leffler function is defined as [228]

$$supercos_{\alpha}\left(t^{\alpha}\right)=\frac{E_{\alpha}\left(it^{\alpha}\right)+E_{\alpha}\left(-it^{\alpha}\right)}{2}=\sum_{n=0}^{\infty}\frac{(-1)^{n}\,t^{2n\alpha}}{\Gamma\left(2n\alpha+1\right)}, \tag{3.124}$$

where $\alpha, t \in \mathbb{C}$, $Re\,(\alpha) > 0$, and $i = \sqrt{-1}$.

The result was discovered by Bagley in 1988 [228].

Definition 3.7 The Zhang supertangent function based on the Mittag-Leffler function is defined as

$$supertan_{\alpha}\left(t^{\alpha}\right)=\frac{supersin_{\alpha}\left(t^{\alpha}\right)}{supercos_{\alpha}\left(t^{\alpha}\right)}, \tag{3.125}$$

where $t, \alpha \in \mathbb{C}$, and $Re\,(\alpha) > 0$.

The result was discovered by Zhang in 2011 [229].

Definition 3.8 The Zhang supercotangent function based on the Mittag-Leffler function is defined as [229]

$$supercotan_{\alpha}\left(t^{\alpha}\right)=\frac{supercos_{\alpha}\left(t^{\alpha}\right)}{supersin_{\alpha}\left(t^{\alpha}\right)}, \tag{3.126}$$

where $t, \alpha \in \mathbb{C}$, and $Re\,(\alpha) > 0$.

The result was discovered by Zhang in 2011 [229].

Definition 3.9 The supercosecant function based on the Mittag-Leffler function is defined as

$$supercsc_{\alpha}\left(t^{\alpha}\right)=\frac{1}{supersin_{\alpha}\left(t^{\alpha}\right)}\left(supersin_{\alpha}\left(t^{\alpha}\right)\neq 0\right), \tag{3.127}$$

where $t, \alpha \in \mathbb{C}$, and $Re\,(\alpha) > 0$.

Definition 3.10 The supersecant function based on the Mittag-Leffler function is defined as

$$supersec_\alpha\left(t^\alpha\right) = \frac{1}{supercos_\alpha\left(t^\alpha\right)}\left(supercos_\alpha\left(t^\alpha\right) \neq 0\right), \tag{3.128}$$

where $t, \alpha \in \mathbb{C}$, and $Re(\alpha) > 0$.

Due to the above results, we may have that

$$supersin_\alpha(\lambda t^\alpha) = \frac{E_\alpha(i\lambda t^\alpha) - E_\alpha(-i\lambda t^\alpha)}{2i} = \sum_{n=0}^{\infty} \frac{(-\lambda)^n t^{(2n+1)\alpha}}{\Gamma((2n+1)\alpha+1)}, \tag{3.129}$$

$$supercos_\alpha(\lambda t^\alpha) = \frac{E_\alpha(i\lambda t^\alpha) + E_\alpha(-i\lambda t^\alpha)}{2} = \sum_{n=0}^{\infty} \frac{(-\lambda)^n t^{2n\alpha}}{\Gamma(2n\alpha+1)}, \tag{3.130}$$

$$supertan_\alpha\left(\lambda t^\alpha\right) = \frac{supersin_\alpha(\lambda t^\alpha)}{supercos_\alpha(\lambda t^\alpha)}, \tag{3.131}$$

$$supercotan_\alpha\left(\lambda t^\alpha\right) = \frac{supercos_\alpha(\lambda t^\alpha)}{supersin_\alpha(\lambda t^\alpha)}, \tag{3.132}$$

$$\begin{aligned} &supercsc_\alpha(\lambda t^\alpha) \\ &= \frac{1}{\frac{1}{2i}(E_\alpha(i\lambda t^\alpha) - E_\alpha(-i\lambda t^\alpha))} \\ &= \frac{2i}{(E_\alpha(i\lambda t^\alpha) - E_\alpha(-i\lambda t^\alpha))}\left((E_\alpha(i\lambda t^\alpha) - E_\alpha(-i\lambda t^\alpha)) \neq 0\right), \end{aligned} \tag{3.133}$$

and

$$\begin{aligned} &supersec_\alpha(\lambda t^\alpha) \\ &= \frac{1}{\frac{1}{2}(E_\alpha(i\lambda t^\alpha) + E_\alpha(-i\lambda t^\alpha))} \\ &= \frac{2}{E_\alpha(i\lambda t^\alpha) + E_\alpha(-i\lambda t^\alpha)}\left(E_\alpha(i\lambda t^\alpha) + E_\alpha(-i\lambda t^\alpha) \neq 0\right) \end{aligned} \tag{3.134}$$

with

$$E_\alpha\left(i\lambda t^\alpha\right) = \sum_{n=0}^{\infty} \frac{(i\lambda)^n t^{n\alpha}}{\Gamma(n\alpha+1)} \tag{3.135}$$

and

$$E_\alpha\left(-i\lambda t^\alpha\right) = \sum_{n=0}^{\infty} \frac{(-i\lambda)^n t^{n\alpha}}{\Gamma(n\alpha+1)}, \tag{3.136}$$

where $t, \alpha \in \mathbb{C}$, $\lambda \in \mathbb{R}$, $i = \sqrt{-1}$, and $Re(\alpha) > 0$.

Theorem 3.56 *Let $\alpha \in \mathbb{C}$, $i = \sqrt{-1}$, and $Re(\alpha) > 0$. Then we have*

$$E_\alpha(it^\alpha) = supercos_\alpha(t^\alpha) + isupersin_\alpha(t^\alpha). \tag{3.137}$$

Similarly,

$$E_\alpha(-it^\alpha) = supercos_\alpha(t^\alpha) - isupersin_\alpha(t^\alpha), \tag{3.138}$$

where $t, \alpha \in \mathbb{C}$, $i = \sqrt{-1}$ and $Re(\alpha) > 0$.

Here we have

$$E_\alpha(it^\alpha) E_\alpha(-it^\alpha) = supercos_\alpha^2(t^\alpha) + supersin_\alpha^2(t^\alpha), \tag{3.139}$$

where

$$supercos_\alpha^2(t^\alpha) = supercos_\alpha(t^\alpha) supercos_\alpha(t^\alpha) \tag{3.140}$$

and

$$supersin_\alpha^2(t^\alpha) = supersin_\alpha(t^\alpha) supersin_\alpha(t^\alpha). \tag{3.141}$$

Thus, we have

$$E_\alpha(i\lambda t^\alpha) = supercos_\alpha(\lambda t^\alpha) + isupersin_\alpha(\lambda t^\alpha), \tag{3.142}$$

and

$$E_\alpha(-i\lambda t^\alpha) = supercos_\alpha(\lambda t^\alpha) - isupersin_\alpha(\lambda t^\alpha), \tag{3.143}$$

so that

$$E_\alpha(i\lambda t^\alpha) E_\alpha(-i\lambda t^\alpha) = supercos_\alpha^2(\lambda t^\alpha) + supersin_\alpha^2(\lambda t^\alpha), \tag{3.144}$$

with

$$supercos_\alpha^2(\lambda t^\alpha) = supercos_\alpha(\lambda t^\alpha) supercos_\alpha(\lambda t^\alpha) \tag{3.145}$$

and

$$supersin_\alpha^2(\lambda t^\alpha) = \sup ersin_\alpha(\lambda t^\alpha) supersin_\alpha(\lambda t^\alpha), \tag{3.146}$$

where $t, \alpha \in \mathbb{C}$, $\lambda \in \mathbb{R}$, $i = \sqrt{-1}$ and $Re(\alpha) > 0$.

There exists the series via Mittag-Leffler function as follows [230]:

$$\Theta(t) = \frac{a_0}{2} + \sum_{n=0}^{\infty} \left(a_n supercos_\alpha\left(nt^\alpha\right) + b_n supersin_\alpha\left(nt^\alpha\right)\right), \tag{3.147}$$

where a_n and b_n are the coefficients of the series via Mittag-Leffler function.

The result was discovered by Jumarie in 2008 [230].

In this case, we have the following cases:

There are the forms of the series via Mittag-Leffler function [230]

$$\Omega_1(t) = \frac{a_0}{2} + \sum_{n=0}^{\infty} \left(a_n supercos_\alpha\left(nt^\alpha\right)\right), \tag{3.148}$$

and

$$\Omega_2(t) = \sum_{n=0}^{\infty} \left(b_n supersin_\alpha\left(nt^\alpha\right)\right), \tag{3.149}$$

where a_n and b_n are the coefficients of the series via Mittag-Leffler function and $n \in \mathbb{N}$.

3.1.5 The Superhyperbolic Functions via Mittag-Leffler Function

In the section, we present the concept of the superhyperbolic supercosecant, and superhyperbolic supersecant functions based on the Mittag-Leffler function.

Let

$$E_\alpha\left(t^\alpha\right) = \sum_{n=0}^{\infty} \frac{t^{n\alpha}}{\Gamma(n\alpha+1)} \tag{3.150}$$

and

$$E_\alpha\left(-t^\alpha\right) = \sum_{n=0}^{\infty} \frac{(-1)^n t^{n\alpha}}{\Gamma(n\alpha+1)}, \tag{3.151}$$

where $t, \alpha \in \mathbb{C}$ and $Re(\alpha) > 0$.

Definition 3.11 The Bagley superhyperbolic supersine function based on the Mittag-Leffler function is defined as [228]

$$\begin{aligned} &supersinh_\alpha (t^\alpha) \\ &= \frac{E_\alpha(t^\alpha)-E_\alpha(-t^\alpha)}{2} \\ &= \sum_{n=0}^{\infty} \frac{t^{(2n+1)\alpha}}{\Gamma((2n+1)\alpha+1)}, \end{aligned} \tag{3.152}$$

where $t, \alpha \in \mathbb{C}$ and $Re(\alpha) > 0$.

The result was discovered by Bagley in 1988 [228].

Definition 3.12 The Bagley superhyperbolic supercosine function based on the Mittag-Leffler function is defined as [228]

$$\begin{aligned} &supercosh_\alpha (t^\alpha) \\ &= \frac{E_\alpha(t^\alpha)+E_\alpha(-t^\alpha)}{2} \\ &= \sum_{n=0}^{\infty} \frac{t^{2n\alpha}}{\Gamma(2n\alpha+1)}, \end{aligned} \tag{3.153}$$

where $\xi, \alpha \in \mathbb{C}$ and $Re(\alpha) > 0$.

The result was discovered by Bagley in 1988 [228].

Definition 3.13 The Zhang superhyperbolic supertangent function based on the Mittag-Leffler function is defined as [229]

$$supertanh_\alpha \left(t^\alpha\right) = \frac{Supersinh_\alpha (t^\alpha)}{Supercosh_\alpha (t^\alpha)}, \tag{3.154}$$

where $t, \alpha \in \mathbb{C}$ and $Re(\alpha) > 0$.

The result was discovered by Zhang in 2011 [229].

Definition 3.14 The Zhang superhyperbolic supercotangent function based on the Mittag-Leffler function is defined as [229]

$$supercotanh_\alpha \left(t^\alpha\right) = \frac{supercosh_\alpha (t^\alpha)}{supersinh_\alpha (t^\alpha)}, \tag{3.155}$$

where $t, \alpha \in \mathbb{C}$ and $Re(\alpha) > 0$.

The result was discovered by Zhang in 2011 [229].

Definition 3.15 The superhyperbolic supercosecant function based on the Mittag-Leffler function is defined as

$$supercsch_{\alpha}\left(t^{\alpha}\right)=\frac{1}{supersinh_{\alpha}\left(t^{\alpha}\right)}\left(supersinh_{\alpha}\left(t^{\alpha}\right)\neq 0\right), \tag{3.156}$$

where $t, \alpha \in \mathbb{C}$ and $Re\left(\alpha\right) > 0$.

Definition 3.16 The superhyperbolic supersecant function based on the Mittag-Leffler function is defined as

$$supersech_{\alpha}\left(t^{\alpha}\right)=\frac{1}{supercosh_{\alpha}\left(t^{\alpha}\right)}\left(supercosh_{\alpha}\left(t^{\alpha}\right)\neq 0\right), \tag{3.157}$$

where $t, \alpha \in \mathbb{C}$ and $Re\left(\alpha\right) > 0$.

Let

$$E_{\alpha}\left(\lambda t^{\alpha}\right)=\sum_{n=0}^{\infty}\frac{\lambda^{n}t^{n\alpha}}{\Gamma\left(n\alpha+1\right)} \tag{3.158}$$

and

$$E_{\alpha}\left(-\lambda t^{\alpha}\right)=\sum_{n=0}^{\infty}\frac{\left(-\lambda\right)^{n}t^{n\alpha}}{\Gamma\left(n\alpha+1\right)}, \tag{3.159}$$

where $t, \alpha \in \mathbb{C}$, $\lambda \in \mathbb{R}$ and $Re\left(\alpha\right) > 0$.

Then we have that

$$\begin{aligned} &supersinh_{\alpha}\left(\lambda t^{\alpha}\right)\\ &=\frac{E_{\alpha}(\lambda t^{\alpha})-E_{\alpha}(-\lambda t^{\alpha})}{2}\\ &=\sum_{n=0}^{\infty}\frac{\lambda^{n}t^{(2n+1)\alpha}}{\Gamma((2n+1)\alpha+1)} \end{aligned}, \tag{3.160}$$

$$\begin{aligned} &supercosh_{\alpha}\left(\lambda t^{\alpha}\right)\\ &=\frac{E_{\alpha}(\lambda t^{\alpha})+E_{\alpha}(-\lambda t^{\alpha})}{2}\\ &=\sum_{n=0}^{\infty}\frac{\lambda^{n}t^{2n\alpha}}{\Gamma(2n\alpha+1)} \end{aligned}, \tag{3.161}$$

$$supertanh_{\alpha}\left(\lambda t^{\alpha}\right)=\frac{supersinh_{\alpha}\left(\lambda t^{\alpha}\right)}{supercosh_{\alpha}\left(\lambda t^{\alpha}\right)}, \tag{3.162}$$

$$supercotanh_{\alpha}\left(\lambda t^{\alpha}\right)=\frac{supercosh_{\alpha}\left(\lambda t^{\alpha}\right)}{supersinh_{\alpha}\left(\lambda t^{\alpha}\right)}, \tag{3.163}$$

$$\begin{aligned} &supercsch_{\alpha}\left(\lambda t^{\alpha}\right) \\ &\quad = \frac{1}{\frac{1}{2}\left(E_{\alpha}\left(\lambda t^{\alpha}\right) - E_{\alpha}\left(-\lambda t^{\alpha}\right)\right)} \\ &\quad = \frac{2}{\left(E_{\alpha}\left(\lambda t^{\alpha}\right) - E_{\alpha}\left(-\lambda t^{\alpha}\right)\right)} \left(\left(E_{\alpha}\left(\lambda t^{\alpha}\right) - E_{\alpha}\left(-\lambda t^{\alpha}\right)\right) \neq 0\right), \end{aligned} \tag{3.164}$$

and

$$\begin{aligned} &supersech_{\alpha}\left(\lambda t^{\alpha}\right) \\ &\quad = \frac{1}{\frac{1}{2}\left(E_{\alpha}\left(\lambda t^{\alpha}\right) + E_{\alpha}\left(-\lambda t^{\alpha}\right)\right)} \\ &\quad = \frac{2}{E_{\alpha}\left(\lambda t^{\alpha}\right) + E_{\alpha}\left(-\lambda t^{\alpha}\right)} \left(E_{\alpha}\left(\lambda t^{\alpha}\right) + E_{\alpha}\left(-\lambda t^{\alpha}\right) \neq 0\right), \end{aligned} \tag{3.165}$$

where $t, \alpha \in \mathbb{C}$, $\lambda \in \mathbb{R}$, and $Re(\alpha) > 0$.

In this case, we obtain the following results:

Theorem 3.57 *Let $\alpha \in \mathbb{C}$ and $Re(\alpha) > 0$. Then we have*

$$E_{\alpha}\left(t^{\alpha}\right) = supercosh_{\alpha}\left(t^{\alpha}\right) + supersinh_{\alpha}\left(t^{\alpha}\right). \tag{3.166}$$

Similarly,

$$E_{\alpha}\left(-t^{\alpha}\right) = supercosh_{\alpha}\left(t^{\alpha}\right) - supersinh_{\alpha}\left(t^{\alpha}\right), \tag{3.167}$$

where $t, \alpha \in \mathbb{C}$ and $Re(\alpha) > 0$.

Here we have

$$E_{\alpha}\left(t^{\alpha}\right) E_{\alpha}\left(-t^{\alpha}\right) = supercosh_{\alpha}^{2}\left(t^{\alpha}\right) - supersinh_{\alpha}^{2}\left(t^{\alpha}\right), \tag{3.168}$$

where

$$supercosh_{\alpha}^{2}\left(t^{\alpha}\right) = supercosh_{\alpha}\left(t^{\alpha}\right) supercosh_{\alpha}\left(t^{\alpha}\right)$$

and

$$supersinh_{\alpha}^{2}\left(t^{\alpha}\right) = supersinh_{\alpha}\left(t^{\alpha}\right) supersinh_{\alpha}\left(t^{\alpha}\right).$$

Thus, we have

$$E_{\alpha}\left(\lambda t^{\alpha}\right) = supercosh_{\alpha}\left(\lambda t^{\alpha}\right) + supersinh_{\alpha}\left(\lambda t^{\alpha}\right), \tag{3.169}$$

and

$$E_{\alpha}\left(-\lambda t^{\alpha}\right) = supercosh_{\alpha}\left(\lambda t^{\alpha}\right) - supersinh_{\alpha}\left(\lambda t^{\alpha}\right), \tag{3.170}$$

so that

$$E_\alpha\left(\lambda t^\alpha\right)E_\alpha\left(-\lambda t^\alpha\right) = supercosh_\alpha^2\left(\lambda t^\alpha\right) - supersinh_\alpha^2\left(\lambda t^\alpha\right), \tag{3.171}$$

with

$$supercosh_\alpha^2\left(\lambda t^\alpha\right) = supercosh_\alpha\left(\lambda t^\alpha\right)\sup ercosh_\alpha\left(\lambda t^\alpha\right)$$

and

$$supersinh_\alpha^2\left(\lambda t^\alpha\right) = supersinh_\alpha\left(\lambda t^\alpha\right)supersinh_\alpha\left(\lambda t^\alpha\right),$$

where $t, \alpha \in \mathbb{C}$, $\lambda \in \mathbb{R}$ *and* $Re(\alpha) > 0$.

Property 3.2 Let $\alpha \in \mathbb{C}$ and $Re(\alpha) > 0$. Then we have

$$supersin_\alpha\left(-t^\alpha\right) = -supersin_\alpha\left(t^\alpha\right), \tag{3.172}$$

$$supercos_\alpha\left(-t^\alpha\right) = supercos_\alpha\left(t^\alpha\right), \tag{3.173}$$

$$supertan_\alpha\left(-t^\alpha\right) = -supertan_\alpha\left(t^\alpha\right), \tag{3.174}$$

$$supercotan_\alpha\left(-t^\alpha\right) = -supercotan_\alpha\left(t^\alpha\right), \tag{3.175}$$

$$supercsc_\alpha\left(-t^\alpha\right) = -supercsc_\alpha\left(t^\alpha\right), \tag{3.176}$$

$$supersec_\alpha\left(-t^\alpha\right) = supersec_\alpha\left(t^\alpha\right), \tag{3.177}$$

$$supersinh_\alpha\left(-t^\alpha\right) = -supersinh_\alpha\left(t^\alpha\right), \tag{3.178}$$

$$supercosh_\alpha\left(-t^\alpha\right) = supercosh_\alpha\left(t^\alpha\right), \tag{3.179}$$

$$supertanh_\alpha\left(-t^\alpha\right) = -supertanh_\alpha\left(t^\alpha\right), \tag{3.180}$$

$$superanh_\alpha\left(-t^\alpha\right) = -supercotanh_\alpha\left(t^\alpha\right), \tag{3.181}$$

$$supercsch_\alpha\left(-t^\alpha\right) = -supercsch_\alpha\left(t^\alpha\right), \tag{3.182}$$

and

$$supercsch_\alpha\left(-t^\alpha\right) = supercsch_\alpha\left(t^\alpha\right). \tag{3.183}$$

3.1.6 The Pre-supertrigonometric Functions via Mittag-Leffler Function

In this section, we propose the concepts of the pre-supersine, pre-supercosine, pre-supertangent, pre-supercotangent, pre-supercosecant, and pre-supersecant via Mittag-Leffler function.

Let

$$E_\alpha(it) = \sum_{n=0}^{\infty} \frac{i^n t^n}{\Gamma(n\alpha + 1)}, \tag{3.184}$$

and

$$E_\alpha(-it) = \sum_{n=0}^{\infty} \frac{(-i)^n t^n}{\Gamma(n\alpha + 1)}, \tag{3.185}$$

where $t, \alpha \in \mathbb{C}$ and $Re(\alpha) > 0$.

Definition 3.17 The pre-supersine function via Mittag-Leffler function is defined as

$$\begin{aligned} &presin_\alpha(t) \\ &= \tfrac{1}{2i}\left(E_\alpha(it) - E_\alpha(-it)\right) \\ &= \sum_{n=0}^{\infty} \tfrac{(-1)^n t^{2n+1}}{\Gamma((2n+1)\alpha+1)}, \end{aligned} \tag{3.186}$$

where $t, \alpha \in \mathbb{C}$ and $Re(\alpha) > 0$.

The result was discovered by Dattoli et al in 2017 [219].

Definition 3.18 The pre-supercosine function via Mittag-Leffler function is defined as

$$\begin{aligned} &precos_\alpha(t) \\ &= \tfrac{1}{2}\left(E_\alpha(it) + E_\alpha(-it)\right) \\ &= \sum_{n=0}^{\infty} \tfrac{(-1)^n t^{2n}}{\Gamma(2n\alpha+1)} \end{aligned}, \tag{3.187}$$

where $t, \alpha \in \mathbb{C}$ and $Re(\alpha) > 0$.

The result was discovered by Dattoli et al in 2017 [219].

Definition 3.19 The pre-presupertangent via Mittag-Leffler function is defined as

$$pretan_\alpha(t) = \frac{presin_\alpha(t)}{precos_\alpha(t)} \left(precos_\alpha(t) \neq 0\right), \tag{3.188}$$

where $t, \alpha \in \mathbb{C}$ and $Re(\alpha) > 0$.

Definition 3.20 The pre-supercotangent via Mittag-Leffler function is defined as

$$precotan_{\alpha}(\xi) = \frac{precos_{\alpha}(\xi)}{presin_{\alpha}(\xi)} \left(presin_{\alpha,\beta}(\xi) \neq 0\right), \tag{3.189}$$

where $t, \alpha \in \mathbb{C}$ and $Re(\alpha) > 0$.

Definition 3.21 The pre-supercosecant via Mittag-Leffler function is defined as

$$precsc_{\alpha}(t) = \frac{1}{presin_{\alpha}(t)} \left(presin_{\alpha}(t) \neq 0\right), \tag{3.190}$$

where $t, \alpha \in \mathbb{C}$ and $Re(\alpha) > 0$.

Definition 3.22 The pre-supersecant via Mittag-Leffler function is defined as

$$presec_{\alpha}(t) = \frac{1}{precos_{\alpha}(t)} \left(precos_{\alpha}(t) \neq 0\right), \tag{3.191}$$

where $t, \alpha \in \mathbb{C}$ and $Re(\alpha) > 0$.

Let

$$E_{\alpha}(i\lambda t) = \sum_{n=0}^{\infty} \frac{(i\lambda)^n t^n}{\Gamma(n\alpha + 1)}, \tag{3.192}$$

and

$$E_{\alpha}(-i\lambda t) = \sum_{n=0}^{\infty} \frac{(-i\lambda)^n t^n}{\Gamma(n\alpha + 1)}, \tag{3.193}$$

where $t, \alpha \in \mathbb{C}$, $\lambda \in \mathbb{R}$ and $Re(\alpha) > 0$.

The pre-supersine via Mittag-Leffler function is alternatively defined as

$$\begin{aligned} &presin_{\alpha}(\lambda t) \\ &= \sum_{n=0}^{\infty} \frac{(-1)^n \lambda^{2n+1} t^{2n+1}}{\Gamma((2n+1)\alpha+1)} \\ &= \frac{1}{2i}\left(E_{\alpha}(i\lambda t) - E_{\alpha}(-i\lambda t)\right), \end{aligned} \tag{3.194}$$

where $t, \alpha \in \mathbb{C}$, $\lambda \in \mathbb{R}$, and $Re(\alpha) > 0$.

The pre-supercosine via Mittag-Leffler function is alternatively defined as

$$\begin{aligned} &precos_{\alpha}(\lambda t) \\ &= \sum_{n=0}^{\infty} \frac{(-1)^n \lambda^{2n} t^{2n}}{\Gamma(2n\alpha+1)} \\ &= \frac{1}{2}\left(E_{\alpha}(i\lambda\xi) + E_{\alpha}(-i\lambda t)\right), \end{aligned} \tag{3.195}$$

where $t, \alpha \in \mathbb{C}$, $\lambda \in \mathbb{R}$, and $Re(\alpha) > 0$.

The pre-supertangent via Mittag-Leffler function is alternatively defined as

$$\begin{array}{l} pretan_{\alpha}\left(\lambda t\right) = \frac{1}{i}\frac{E_{\alpha}(i\lambda t)-E_{\alpha}(-i\lambda t)}{E_{\alpha}(i\lambda t)+E_{\alpha}(-i\lambda t)}, \\ \left(E_{\alpha}\left(i\lambda t\right)+E_{\alpha}\left(-i\lambda t\right)\neq 0\right) \end{array} \tag{3.196}$$

where $t, \alpha \in \mathbb{C}$, $\lambda \in \mathbb{R}$, and $Re\left(\alpha\right) > 0$.

The pre-supercotangent via Mittag-Leffler function is alternatively defined as

$$\begin{array}{l} precotan_{\alpha}\left(\lambda t\right) = i\frac{E_{\alpha}(i\lambda t)+E_{\alpha}(-i\lambda t)}{E_{\alpha}(i\lambda t)-E_{\alpha}(-i\lambda t)}, \\ \left(E_{\alpha}\left(i\lambda t\right)-E_{\alpha}\left(-i\lambda t\right)\neq 0\right) \end{array} \tag{3.197}$$

where $t, \alpha \in \mathbb{C}$, $\lambda \in \mathbb{R}$, and $Re\left(\alpha\right) > 0$.

The pre-supercosecant via Mittag-Leffler function is alternatively defined as

$$\begin{array}{l} precsc_{\alpha}\left(\lambda t\right) = \frac{2i}{E_{\alpha}(i\lambda t)-E_{\alpha}(-i\lambda t)}, \\ \left(E_{\alpha}\left(i\lambda t\right)-E_{\alpha}\left(-i\lambda t\right)\neq 0\right) \end{array} \tag{3.198}$$

where $t, \alpha \in \mathbb{C}$, $\lambda \in \mathbb{R}$ and $Re\left(\alpha\right) > 0$.

The pre-supersecant via Mittag-Leffler function is alternatively defined as

$$\begin{array}{l} presec_{\alpha}\left(\lambda t\right) = \frac{2}{E_{\alpha}(i\lambda t)+E_{\alpha}(-i\lambda t)}, \\ \left(E_{\alpha}\left(i\lambda t\right)+E_{\alpha}\left(-i\lambda t\right)\neq 0\right) \end{array} \tag{3.199}$$

where $t, \alpha \in \mathbb{C}$, $\lambda \in \mathbb{R}$, and $Re\left(\alpha\right) > 0$.

Here we have

$$E_{\alpha}\left(it\right) = precos_{\alpha}\left(t\right) + ipresin_{\alpha}\left(t\right) \tag{3.200}$$

and

$$E_{\alpha}\left(-it\right) = precos_{\alpha}\left(t\right) - ipresin_{\alpha}\left(t\right), \tag{3.201}$$

where $t, \alpha \in \mathbb{C}$, $\lambda \in \mathbb{R}$, $i = \sqrt{-1}$, and $Re\left(\alpha\right) > 0$.

The results were discovered by Dattoli et al. in 2017 [219].

In this case, there exists the series via Mittag-Leffler function as follows:

$$\Omega\left(t\right) = \frac{a_0}{2} + \sum_{n=0}^{\infty}\left(a_n precos_{\alpha}\left(nt\right) + b_n presin_{\alpha}\left(nt\right)\right), \tag{3.202}$$

where a_n and b_n are the coefficients of the series via Mittag-Leffler function and $n \in \mathbb{N}$.

In this case, we have the following cases:
There are the forms of the series via Mittag-Leffler function

$$\Omega_1(t) = \frac{a_0}{2} + \sum_{n=0}^{\infty} (a_n precos_\alpha(nt)), \tag{3.203}$$

and

$$\Omega_2(t) = \sum_{n=0}^{\infty} (b_n presin_\alpha(nt)), n \in \mathbb{N}, \tag{3.204}$$

where a_n and b_n are the coefficients of the series via Mittag-Leffler function.
In this case, we obtain that

$$E_\alpha(it) E_\alpha(-it) = precos_\alpha^2(t) + presin_\alpha^2(t), \tag{3.205}$$

where

$$precos_\alpha^2(t) = precos_\alpha(t) precos_\alpha(t)$$

and

$$presin_\alpha^2(t) = presin_\alpha(t) presin_\alpha(t).$$

Thus, we have

$$E_\alpha(i\lambda t) = precos_\alpha(\lambda t) + ipresin_\alpha(\lambda t), \tag{3.206}$$

and

$$E_\alpha(-i\lambda t) = precos_\alpha(\lambda t) - ipresin_\alpha(\lambda t), \tag{3.207}$$

so that

$$E_\alpha(i\lambda t) E_\alpha(-i\lambda t) = precos_\alpha^2(\lambda t) + presin_\alpha^2(\lambda t), \tag{3.208}$$

with

$$precos_\alpha^2(\lambda t) = precos_\alpha(\lambda t) precos_\alpha(\lambda t)$$

and

$$presin_\alpha^2(\lambda t) = presin_\alpha(\lambda t) presin_\alpha(\lambda t),$$

where $t, \alpha \in \mathbb{C}$, $\lambda \in \mathbb{R}$, $i = \sqrt{-1}$ and $Re(\alpha) > 0$.

In this case, we have the following:

Theorem 3.58 *If* $t, \alpha \in \mathbb{C}$, $\lambda \in \mathbb{R}$, *and* $Re(\alpha) > 0$, *then there are*

$$supersin_{\alpha}\left(\lambda t^{\alpha}\right) = presin_{\alpha}\left(\lambda t^{\alpha}\right), \tag{3.209}$$

$$supercos_{\alpha}\left(\lambda t^{\alpha}\right) = precos_{\alpha}\left(\lambda t^{\alpha}\right), \tag{3.210}$$

$$supertan_{\alpha}\left(\lambda t^{\alpha}\right) = pretan_{\alpha}\left(\lambda t^{\alpha}\right), \tag{3.211}$$

$$supercotan_{\alpha}\left(\lambda t^{\alpha}\right) = precotan_{\alpha}\left(\lambda t^{\alpha}\right), \tag{3.212}$$

$$supercsc_{\alpha}\left(\lambda t^{\alpha}\right) = precsc_{\alpha}\left(\lambda t^{\alpha}\right), \tag{3.213}$$

and

$$supersec_{\alpha}\left(\lambda t^{\alpha}\right) = presec_{\alpha}\left(\lambda t^{\alpha}\right). \tag{3.214}$$

Definition 3.23 The pre-supersine via Mittag-Leffler function is defined as

$$\begin{aligned} &presin_{\alpha}\left(t^{\beta}\right) \\ &= \frac{1}{2i}\left(E_{\alpha}\left(it^{\beta}\right) - E_{\alpha}\left(-it^{\beta}\right)\right) \\ &= \sum_{n=0}^{\infty} \frac{(-1)^n t^{(2n+1)\beta}}{\Gamma((2n+1)\alpha+1)} \end{aligned}, \tag{3.215}$$

where $t, \alpha, \beta \in \mathbb{C}$, $Re(\alpha) > 0$, and $Re(\beta) > 0$.

Definition 3.24 The pre-supercosine via Mittag-Leffler function is defined as

$$\begin{aligned} &precos_{\alpha}\left(t^{\beta}\right) \\ &= \frac{1}{2}\left(E_{\alpha}\left(it^{\beta}\right) + E_{\alpha}\left(-it^{\beta}\right)\right) \\ &= \sum_{n=0}^{\infty} \frac{(-1)^n t^{2n\beta}}{\Gamma(2n\alpha+1)} \end{aligned}, \tag{3.216}$$

where $\xi, \alpha, \beta \in \mathbb{C}$, $Re(\alpha) > 0$ and $Re(\beta) > 0$.

Definition 3.25 The pre-presupertangent via Mittag-Leffler function is defined as

$$\begin{aligned} &pretan_{\alpha}\left(t^{\beta}\right) = \frac{presin_{\alpha}\left(t^{\beta}\right)}{precos_{\alpha}\left(t^{\beta}\right)} \\ &\left(precos_{\alpha}\left(t^{\beta}\right) \neq 0\right) \end{aligned}, \tag{3.217}$$

where $\xi, \alpha, \beta \in \mathbb{C}$, $Re(\alpha) > 0$ and $Re(\beta) > 0$.

Definition 3.26 The pre-supercotangent via Mittag-Leffler function is defined as

$$\begin{array}{l} precotan_{\alpha}\left(t^{\beta}\right) = \frac{precos_{\alpha}(t^{\beta})}{presin_{\alpha}(t^{\beta})}, \\ \left(presin_{\alpha}\left(t^{\beta}\right) \neq 0\right) \end{array} \tag{3.218}$$

where $\xi, \alpha, \beta \in \mathbb{C}$, $Re\,(\alpha) > 0$ and $Re\,(\beta) > 0$.

Definition 3.27 The pre-supercosecant via Mittag-Leffler function is defined as

$$\begin{array}{l} precsc_{\alpha}\left(t^{\beta}\right) = \frac{1}{presin_{\alpha}(t^{\beta})}, \\ \left(presin_{\alpha}\left(t^{\beta}\right) \neq 0\right) \end{array} \tag{3.219}$$

where $\xi, \alpha, \beta \in \mathbb{C}$, $Re\,(\alpha) > 0$, and $Re\,(\beta) > 0$.

Definition 3.28 The pre-supersecant via Mittag-Leffler function is defined as

$$\begin{array}{l} presec_{\alpha}\left(t^{\beta}\right) = \frac{1}{precos_{\alpha}(t^{\beta})}, \\ \left(precos_{\alpha}\left(t^{\beta}\right) \neq 0\right) \end{array} \tag{3.220}$$

where $\xi, \alpha, \beta \in \mathbb{C}$, $Re\,(\alpha) > 0$ and $Re\,(\beta) > 0$.

Similarly, we define the following functions:
The pre-supersine via Mittag-Leffler function is defined as

$$\begin{array}{l} presin_{\alpha}\left(\lambda t^{\beta}\right) \\ \quad = \frac{1}{2i}\left(E_{\alpha}\left(i\lambda t^{\beta}\right) - E_{\alpha}\left(-i\lambda t^{\beta}\right)\right) \\ \quad = \sum\limits_{n=0}^{\infty} \frac{(-\lambda)^{n} t^{(2n+1)\beta}}{\Gamma((2n+1)\alpha+1)} \end{array}, \tag{3.221}$$

where $t, \alpha, \beta \in \mathbb{C}$, $\lambda \in \mathbb{R}$, $Re\,(\alpha) > 0$, and $Re\,(\beta) > 0$.

The pre-supercosine via Mittag-Leffler function is defined as

$$\begin{array}{l} precos_{\alpha}\left(\lambda t^{\beta}\right) \\ \quad = \frac{1}{2}\left(E_{\alpha}\left(i\lambda t^{\beta}\right) + E_{\alpha}\left(-i\lambda t^{\beta}\right)\right) \\ \quad = \sum\limits_{n=0}^{\infty} \frac{(-\lambda)^{n} t^{2n\beta}}{\Gamma(2n\alpha+1)} \end{array}, \tag{3.222}$$

where $\xi, \alpha, \beta \in \mathbb{C}$, $\lambda \in \mathbb{R}$, $Re\,(\alpha) > 0$, and $Re\,(\beta) > 0$.

The pre-presupertangent via Mittag-Leffler function is defined as

$$\begin{array}{l} pretan_{\alpha}\left(\lambda t^{\beta}\right) = \frac{presin_{\alpha}(\lambda t^{\beta})}{precos_{\alpha}(\lambda t^{\beta})}, \\ \left(precos_{\alpha}\left(\lambda t^{\beta}\right) \neq 0\right) \end{array} \tag{3.223}$$

where $\xi, \alpha, \beta \in \mathbb{C}$, $\lambda \in \mathbb{R}$, $Re\,(\alpha) > 0$, and $Re\,(\beta) > 0$.

The pre-supercotangent via Mittag-Leffler function is defined as

$$precotan_{\alpha}\left(\lambda t^{\beta}\right)=\frac{precos_{\alpha}(\lambda t^{\beta})}{presin_{\alpha}(\lambda t^{\beta})},\quad \left(presin_{\alpha}\left(\lambda t^{\beta}\right)\neq 0\right) \tag{3.224}$$

where $\xi, \alpha, \beta \in \mathbb{C}$, $\lambda \in \mathbb{R}$, $Re(\alpha) > 0$, and $Re(\beta) > 0$.

The pre-supercosecant via Mittag-Leffler function is defined as

$$precsc_{\alpha}\left(\lambda t^{\beta}\right)=\frac{1}{presin_{\alpha}(\lambda t^{\beta})},\quad \left(presin_{\alpha}\left(\lambda t^{\beta}\right)\neq 0\right) \tag{3.225}$$

where $\xi, \alpha, \beta \in \mathbb{C}$, $\lambda \in \mathbb{R}$, $Re(\alpha) > 0$, and $Re(\beta) > 0$.

The pre-supersecant via Mittag-Leffler function is defined as

$$presec_{\alpha}\left(\lambda t^{\beta}\right)=\frac{1}{precos_{\alpha}(\lambda t^{\beta})},\quad \left(precos_{\alpha}\left(\lambda t^{\beta}\right)\neq 0\right) \tag{3.226}$$

where $\xi, \alpha, \beta \in \mathbb{C}$, $\lambda \in \mathbb{R}$, $Re(\alpha) > 0$, and $Re(\beta) > 0$.

In this case, we have the following theorem:

Theorem 3.59 *If $t, \alpha, \beta \in \mathbb{C}$, $i = \sqrt{-1}$, $Re(\alpha) > 0$ and $Re(\beta) > 0$, then there exist*

$$E_{\alpha}\left(it^{\beta}\right)=precos_{\alpha}\left(t^{\beta}\right)+ipresin_{\alpha}\left(t^{\beta}\right) \tag{3.227}$$

and

$$E_{\alpha}\left(-it^{\beta}\right)=precos_{\alpha}\left(t^{\beta}\right)-ipresin_{\alpha}\left(t^{\beta}\right). \tag{3.228}$$

In this case, we give the following results:

Theorem 3.60 *If $\xi, \alpha, \beta \in \mathbb{C}$, $\lambda \in \mathbb{R}$, $i = \sqrt{-1}$, $Re(\alpha) > 0$, and $Re(\beta) > 0$, then there is*

$$E_{\alpha}\left(i\lambda t^{\beta}\right)=precos_{\alpha}\left(\lambda t^{\beta}\right)+ipresin_{\alpha}\left(\lambda t^{\beta}\right) \tag{3.229}$$

and

$$E_{\alpha}\left(-i\lambda t^{\beta}\right)=precos_{\alpha}\left(\lambda t^{\beta}\right)-ipresin_{\alpha}\left(\lambda t^{\beta}\right). \tag{3.230}$$

In this case, we obtain that

$$E_{\alpha}\left(it^{\beta}\right)E_{\alpha}\left(-it^{\beta}\right)=precos_{\alpha}^{2}\left(t^{\beta}\right)+presin_{\alpha}^{2}\left(t^{\beta}\right), \tag{3.231}$$

where

$$precos_{\alpha}^{2}\left(t^{\beta}\right)=precos_{\alpha}\left(t^{\beta}\right)precos_{\alpha}\left(t^{\beta}\right)$$

and

$$presin_{\alpha}^{2}\left(t^{\beta}\right)=presin_{\alpha}\left(t^{\beta}\right)presin_{\alpha}\left(t^{\beta}\right).$$

Thus, we have

$$E_{\alpha}\left(i\lambda t^{\beta}\right)=precos_{\alpha}\left(\lambda t^{\beta}\right)+ipresin_{\alpha}\left(\lambda t^{\beta}\right), \tag{3.232}$$

and

$$E_{\alpha}\left(-i\lambda t^{\beta}\right)=precos_{\alpha}\left(\lambda t^{\beta}\right)-ipresin_{\alpha}\left(\lambda t^{\beta}\right), \tag{3.233}$$

so that

$$E_{\alpha}\left(i\lambda t^{\beta}\right)E_{\alpha}\left(-i\lambda t^{\beta}\right)=precos_{\alpha}^{2}\left(\lambda t^{\beta}\right)+presin_{\alpha}^{2}\left(\lambda t^{\beta}\right), \tag{3.234}$$

with

$$precos_{\alpha}^{2}\left(\lambda t^{\beta}\right)=precos_{\alpha}\left(\lambda t^{\beta}\right)precos_{\alpha}\left(\lambda t^{\beta}\right)$$

and

$$presin_{\alpha}^{2}\left(\lambda t^{\beta}\right)=presin_{\alpha}\left(\lambda t^{\beta}\right)presin_{\alpha}\left(\lambda t^{\beta}\right),$$

where $t,\alpha\in\mathbb{C}$, $\lambda\in\mathbb{R}$, $i=\sqrt{-1}$, *and* $Re(\alpha)>0$.

In this case, there exists the series via Mittag-Leffler function as follows:

$$\Omega(t)=\frac{a_0}{2}+\sum_{n=0}^{\infty}\left(a_n precos_{\alpha}\left(nt^{\beta}\right)+b_n presin_{\alpha}\left(nt^{\beta}\right)\right), \tag{3.235}$$

where a_n and b_n are the coefficients of the series via Mittag-Leffler function and $n\in\mathbb{N}$.

In this case, we have the following cases:

There are the forms of the series via Mittag-Leffler function

$$\Omega_1(t)=\frac{a_0}{2}+\sum_{n=0}^{\infty}\left(a_n precos_{\alpha}\left(nt^{\beta}\right)\right), \tag{3.236}$$

and

$$\Omega_2(t) = \sum_{n=0}^{\infty} \left(b_n presin_\alpha\left(nt^\beta\right)\right), \tag{3.237}$$

where a_n and b_n are the coefficients of the series via Mittag-Leffler function and $n \in \mathbb{N}$.

3.1.7 The Pre-superhyperbolic Functions via Mittag-Leffler Function

In this section, we now consider the concepts of the pre-superhyperbolic supersine, pre-superhyperbolic supercosine, pre-superhyperbolic presupertangent, pre-superhyperbolic presupertangent, pre-superhyperbolic supercotangent, pre-superhyperbolic supercosecant, and pre-superhyperbolic supersecant based on the Mittag-Leffler function.

Let

$$E_\alpha(t) = \sum_{n=0}^{\infty} \frac{t^n}{\Gamma(n\alpha + 1)}, \tag{3.238}$$

and

$$E_\alpha(-t) = \sum_{n=0}^{\infty} \frac{(-1)^n t^n}{\Gamma(n\alpha + 1)}, \tag{3.239}$$

where $t, \alpha \in \mathbb{C}$ and $Re(\alpha) > 0$.

Definition 3.29 The pre-superhyperbolic supersine via Mittag-Leffler function is defined as:

$$\begin{aligned} &presinh_\alpha(t) \\ &= \tfrac{1}{2}\left(E_\alpha(t) - E_\alpha(-t)\right) \\ &= \sum_{n=0}^{\infty} \tfrac{t^{2n+1}}{\Gamma((2n+1)\alpha+1)} \end{aligned}, \tag{3.240}$$

where $t, \alpha \in \mathbb{C}$ and $Re(\alpha) > 0$.

Definition 3.30 The pre-superhyperbolic supercosine via Mittag-Leffler function is defined as:

$$\begin{aligned} & precosh_{\alpha}(t) \\ & \quad = \tfrac{1}{2}\left(E_{\alpha}(t) + E_{\alpha}(-t)\right) \\ & \quad = \sum_{n=0}^{\infty} \tfrac{t^{2n}}{\Gamma(2n\alpha+1)} \end{aligned}, \tag{3.241}$$

where $t, \alpha \in \mathbb{C}$ and $Re(\alpha) > 0$.

Definition 3.31 The pre-superhyperbolic presupertangent via Mittag-Leffler function is defined as:

$$pretanh_{\alpha}(t) = \frac{presinh_{\alpha}(t)}{precosh_{\alpha}(t)}, \tag{3.242}$$

where $t, \alpha \in \mathbb{C}$ and $Re(\alpha) > 0$.

Definition 3.32 The pre-superhyperbolic supercotangent via Mittag-Leffler function is defined as:

$$precotanh_{\alpha}(t) = \frac{precosh_{\alpha}(t)}{presinh_{\alpha}(t)}, \tag{3.243}$$

where $t, \alpha \in \mathbb{C}$ and $Re(\alpha) > 0$.

Definition 3.33 The pre-superhyperbolic supercosecant via Mittag-Leffler function is defined as:

$$precsch_{\alpha}(t) = \frac{1}{presinh_{\alpha}(t)}, \tag{3.244}$$

where $t, \alpha \in \mathbb{C}$ and $Re(\alpha) > 0$.

Definition 3.34 The pre-superhyperbolic supersecant via Mittag-Leffler function is defined as:

$$presech_{\alpha}(t) = \frac{1}{precosh_{\alpha}(t)}, \tag{3.245}$$

where $t, \alpha \in \mathbb{C}$ and $Re(\alpha) > 0$.

Let

$$E_{\alpha}(\lambda t) = \sum_{n=0}^{\infty} \frac{\lambda^{n} t^{n}}{\Gamma(n\alpha+1)}, \tag{3.246}$$

and

$$E_\alpha(-\lambda t) = \sum_{n=0}^{\infty} \frac{(-\lambda)^n t^n}{\Gamma(n\alpha + 1)}, \tag{3.247}$$

where $t, \alpha \in \mathbb{C}$, $\lambda \in \mathbb{R}$, and $Re(\alpha) > 0$.

The pre-superhyperbolic supersine via Mittag-Leffler function is alternatively defined as:

$$\begin{aligned} &presinh_\alpha(\lambda t) \\ &= \sum_{\kappa=0}^{\infty} \frac{\lambda^{2\kappa+1} t^{2\kappa+1}}{\Gamma((2\kappa+1)\alpha+1)} \\ &= \frac{1}{2}\left(E_\alpha(\lambda t) - E_\alpha(-\lambda t)\right) \end{aligned}, \tag{3.248}$$

where $t, \alpha \in \mathbb{C}$, $\lambda \in \mathbb{R}$, and $Re(\alpha) > 0$.

The pre-superhyperbolic supercosine via Mittag-Leffler function is alternatively defined as:

$$\begin{aligned} &precosh_\alpha(\lambda t) \\ &= \sum_{\kappa=0}^{\infty} \frac{\lambda^{2\kappa} t^{2\kappa}}{\Gamma(2\kappa\alpha+1)} \\ &= \frac{1}{2}\left(E_\alpha(\lambda t) + E_\alpha(-\lambda t)\right) \end{aligned}, \tag{3.249}$$

where $t, \alpha \in \mathbb{C}$, $\lambda \in \mathbb{R}$, and $Re(\alpha) > 0$.

The pre-superhyperbolic presupertangent via Mittag-Leffler function is alternatively defined as:

$$\begin{aligned} &pretanh_\alpha(\lambda t) \\ &= \frac{\frac{1}{2}(E_\alpha(\lambda t) - E_\alpha(-\lambda t))}{\frac{1}{2}(E_\alpha(\lambda t) + E_\alpha(-\lambda t))} \\ &= \frac{E_\alpha(\lambda t) - E_\alpha(-\lambda t)}{E_\alpha(\lambda t) + E_\alpha(-\lambda t)}, \end{aligned} \tag{3.250}$$

where $t, \alpha \in \mathbb{C}$, $\lambda \in \mathbb{R}$, and $Re(\alpha) > 0$.

The pre-superhyperbolic supercotangent via Mittag-Leffler function is alternatively defined as:

$$\begin{aligned} &precotanh_\alpha(\lambda t) \\ &= \frac{\frac{1}{2}(E_\alpha(\lambda t) + E_\alpha(\lambda t))}{\frac{1}{2}(E_\alpha(\lambda t) - E_\alpha(-\lambda t))} \\ &= \frac{E_\alpha(\lambda t) + E_\alpha(-\lambda t)}{E_\alpha(\lambda t) - E_\alpha(-\lambda t)}, \end{aligned} \tag{3.251}$$

where $t, \alpha \in \mathbb{C}$, $\lambda \in \mathbb{R}$, and $Re(\alpha) > 0$.

The pre-superhyperbolic supercosecant via Mittag-Leffler function is alternatively defined as:

$$\begin{aligned} &precsch_{\alpha}(\lambda t) \\ &= \frac{1}{\frac{1}{2}(E_{\alpha}(\lambda t)-E_{\alpha}(-\lambda t))} \\ &= \frac{2}{E_{\alpha}(\lambda t)-E_{\alpha}(-\lambda t)}, \end{aligned} \tag{3.252}$$

where $t, \alpha \in \mathbb{C}$, $\lambda \in \mathbb{R}$, and $Re(\alpha) > 0$.

The pre-superhyperbolic supersecant via Mittag-Leffler function is alternatively defined as:

$$\begin{aligned} &presech_{\alpha}(\lambda t) \\ &= \frac{1}{\frac{1}{2}(E_{\alpha}(\lambda t)+E_{\alpha}(-\lambda t))} \\ &= \frac{2}{E_{\alpha}(\lambda t)+E_{\alpha}(-\lambda t)}, \end{aligned} \tag{3.253}$$

where $t, \alpha \in \mathbb{C}$, $\lambda \in \mathbb{R}$, and $Re(\alpha) > 0$.

In this case, we may present

$$E_{\alpha}(t) = precosh_{\alpha}(t) + presinh_{\alpha}(t) \tag{3.254}$$

and

$$E_{\alpha}(-t) = precosh_{\alpha}(t) - presinh_{\alpha}(t) \tag{3.255}$$

which leads to

$$E_{\alpha}(\lambda t) = precosh_{\alpha}(\lambda t) + presinh_{\alpha}(\lambda t) \tag{3.256}$$

and

$$E_{\alpha}(-\lambda t) = precosh_{\alpha}(\lambda t) - presinh_{\alpha}(\lambda t), \tag{3.257}$$

where $t, \alpha \in \mathbb{C}$, $\lambda \in \mathbb{R}$ and $Re(\alpha) > 0$.

Thus,

$$E_{\alpha}(t)\, E_{\alpha}(-t) = precosh_{\alpha}^{2}(t) - presinh_{\alpha}^{2}(t), \tag{3.258}$$

where

$$precosh_{\alpha}^{2}(t) = precosh_{\alpha}(t)\, precosh_{\alpha}(t)$$

and

$$presinh_{\alpha}^{2}(t) = presinh_{\alpha}(t)\, presinh_{\alpha}(t).$$

Thus, we have

$$E_{\alpha}(\lambda t)\, E_{\alpha}(-\lambda t) = precosh_{\alpha}^{2}(\lambda t) + presinh_{\alpha}^{2}(\lambda t), \tag{3.259}$$

where

$$precosh_{\alpha}^{2}(\lambda t) = precosh_{\alpha}(\lambda t)\, precosh_{\alpha}(\lambda t)$$

and

$$presinh_{\alpha}^{2}(\lambda t) = presinh_{\alpha}(\lambda t)\, presinh_{\alpha}(\lambda t),$$

where $t, \alpha \in \mathbb{C}$, $\lambda \in \mathbb{R}$ and $Re(\alpha) > 0$.

In view of above results, we may present the following results:

Theorem 3.61 *If $t, \alpha \in \mathbb{C}$, $\lambda \in \mathbb{R}$ and $Re(\alpha) > 0$, then there exist*

$$supersinh_{\alpha}\left(\lambda t^{\alpha}\right) = presinh_{\alpha}\left(\lambda t^{\alpha}\right), \tag{3.260}$$

$$supercosh_{\alpha}\left(\lambda t^{\alpha}\right) = precosh_{\alpha}\left(\lambda t^{\alpha}\right), \tag{3.261}$$

$$supertanh_{\alpha}\left(\lambda t^{\alpha}\right) = pertanh_{\alpha}\left(\lambda t^{\alpha}\right), \tag{3.262}$$

$$supercotanh_{\alpha}\left(\lambda t^{\alpha}\right) = precotanh_{\alpha}\left(\lambda t^{\alpha}\right), \tag{3.263}$$

$$supercsch_{\alpha}\left(\lambda t^{\alpha}\right) = precsch_{\alpha}\left(\lambda t^{\alpha}\right), \tag{3.264}$$

and

$$supersech_{\alpha}\left(\lambda t^{\alpha}\right) = presech_{\alpha}\left(\lambda t^{\alpha}\right). \tag{3.265}$$

Definition 3.35 The pre-superhyperbolic supersine via Mittag-Leffler function is defined as:

$$\begin{aligned} &presinh_{\alpha}\left(t^{\beta}\right) \\ &\quad = \frac{1}{2}\left(E_{\alpha}\left(t^{\beta}\right) - E_{\alpha}\left(-t^{\beta}\right)\right) \\ &\quad = \sum_{n=0}^{\infty} \frac{t^{(2n+1)\beta}}{\Gamma((2n+1)\alpha+1)} \end{aligned}, \tag{3.266}$$

where $t, \alpha, \beta \in \mathbb{C}$, $Re(\alpha) > 0$, and $Re(\beta) > 0$.

Definition 3.36 The pre-superhyperbolic supercosine via Mittag-Leffler function is defined as:

$$\begin{aligned} &precosh_\alpha\left(t^\beta\right) \\ &= \tfrac{1}{2}\left(E_\alpha\left(t^\beta\right) + E_\alpha\left(-t^\beta\right)\right) \\ &= \sum_{n=0}^{\infty} \tfrac{t^{2n\beta}}{\Gamma(2n\alpha+1)} \end{aligned}, \tag{3.267}$$

where $t, \alpha, \beta \in \mathbb{C}$, $Re(\alpha) > 0$, and $Re(\beta) > 0$.

Definition 3.37 The pre-superhyperbolic presupertangent via Mittag-Leffler function is defined as:

$$pretanh_\alpha\left(t^\beta\right) = \frac{presin_\alpha\left(t^\beta\right)}{precos_\alpha\left(t^\beta\right)}, \tag{3.268}$$

where $t, \alpha, \beta \in \mathbb{C}$, $Re(\alpha) > 0$, and $Re(\beta) > 0$.

Definition 3.38 The pre-superhyperbolic supercotangent via Mittag-Leffler function is defined as:

$$precotanh_\alpha\left(t^\beta\right) = \frac{precos_\alpha\left(t^\beta\right)}{presin_\alpha\left(t^\beta\right)}, \tag{3.269}$$

where $t, \alpha, \beta \in \mathbb{C}$, $Re(\alpha) > 0$, and $Re(\beta) > 0$.

Definition 3.39 The pre-superhyperbolic supercosecant via Mittag-Leffler function is defined as:

$$precsch_\alpha\left(t^\beta\right) = \frac{1}{presin_\alpha\left(t^\beta\right)}, \tag{3.270}$$

where $t, \alpha, \beta \in \mathbb{C}$, $Re(\alpha) > 0$, and $Re(\beta) > 0$.

Definition 3.40 The pre-superhyperbolic supersecant via Mittag-Leffler function is defined as:

$$presech_\alpha\left(t^\beta\right) = \frac{1}{precos_\alpha\left(t^\beta\right)}, \tag{3.271}$$

where $t, \alpha, \beta \in \mathbb{C}$, $Re(\alpha) > 0$, and $Re(\beta) > 0$.

The pre-superhyperbolic supersine via Mittag-Leffler function is defined as:

$$\begin{aligned} & presinh_\alpha\left(\lambda t^\beta\right) \\ & \quad = \tfrac{1}{2}\left(E_\alpha\left(\lambda t^\beta\right) - E_\alpha\left(-\lambda t^\beta\right)\right) \\ & \quad = \sum_{n=0}^{\infty} \frac{\lambda^n t^{(2n+1)\beta}}{\Gamma((2n+1)\alpha+1)} \end{aligned}, \tag{3.272}$$

where $t, \lambda, \alpha, \beta \in \mathbb{C}$, $Re(\alpha) > 0$, and $Re(\beta) > 0$.

The pre-superhyperbolic supercosine via Mittag-Leffler function is defined as:

$$\begin{aligned} & precosh_\alpha\left(\lambda t^\beta\right) \\ & \quad = \tfrac{1}{2}\left(E_\alpha\left(\lambda t^\beta\right) + E_\alpha\left(-\lambda t^\beta\right)\right) \\ & \quad = \sum_{n=0}^{\infty} \frac{\lambda^n t^{2n\beta}}{\Gamma(2n\alpha+1)} \end{aligned}, \tag{3.273}$$

where $t, \alpha, \beta \in \mathbb{C}$, $Re(\alpha) > 0$, and $Re(\beta) > 0$.

The pre-superhyperbolic presupertangent via Mittag-Leffler function is defined as:

$$pretanh_\alpha\left(\lambda t^\beta\right) = \frac{presinh_\alpha\left(\lambda t^\beta\right)}{precosh_\alpha\left(\lambda t^\beta\right)}, \tag{3.274}$$

where $t, \alpha, \beta \in \mathbb{C}$, $Re(\alpha) > 0$, and $Re(\beta) > 0$.

The pre-superhyperbolic supercotangent via Mittag-Leffler function is defined as:

$$precotanh_\alpha\left(\lambda t^\beta\right) = \frac{precos_\alpha\left(\lambda t^\beta\right)}{presin_\alpha\left(\lambda t^\beta\right)}, \tag{3.275}$$

where $t, \alpha, \beta \in \mathbb{C}$, $Re(\alpha) > 0$, and $Re(\beta) > 0$.

The pre-superhyperbolic supercosecant via Mittag-Leffler function is defined as:

$$precsch_\alpha\left(\lambda t^\beta\right) = \frac{1}{presin_\alpha\left(\lambda t^\beta\right)}, \tag{3.276}$$

where $t, \alpha, \beta \in \mathbb{C}$, $Re(\alpha) > 0$, and $Re(\beta) > 0$.

The pre-superhyperbolic supersecant via Mittag-Leffler function is defined as:

$$presech_\alpha\left(\lambda t^\beta\right) = \frac{1}{precos_\alpha\left(\lambda t^\beta\right)}, \tag{3.277}$$

where $t, \alpha, \beta \in \mathbb{C}$, $Re(\alpha) > 0$ and $Re(\beta) > 0$.

If $t, \alpha, \beta \in \mathbb{C}$, $Re(\alpha) > 0$, and $Re(\beta) > 0$, then we have

$$E_\alpha\left(t^\beta\right) = precosh_\alpha\left(t^\beta\right) + presinh_\alpha\left(t^\beta\right) \tag{3.278}$$

and

$$E_\alpha\left(-t^\beta\right) = precosh_\alpha\left(t^\beta\right) - presinh_\alpha\left(t^\beta\right) \tag{3.279}$$

such that

$$E_\alpha\left(t^\beta\right) E_\alpha\left(-t^\beta\right) = precosh_\alpha^2\left(t^\beta\right) - presinh_\alpha^2\left(t^\beta\right), \tag{3.280}$$

where

$$precosh_\alpha^2\left(t^\beta\right) = precosh_\alpha\left(t^\beta\right) precosh_\alpha\left(t^\beta\right)$$

and

$$presinh_\alpha^2\left(t^\beta\right) = presinh_\alpha\left(t^\beta\right) presinh_\alpha\left(t^\beta\right).$$

Similarly, we have the following result:

If $t, \alpha, \beta, \lambda \in \mathbb{C}$, $Re(\alpha) > 0$, and $Re(\beta) > 0$, then we have

$$E_\alpha\left(\lambda t^\beta\right) = precosh_\alpha\left(\lambda t^\beta\right) + presinh_\alpha\left(\lambda t^\beta\right) \tag{3.281}$$

and

$$E_\alpha\left(-\lambda t^\beta\right) = precosh_\alpha\left(\lambda t^\beta\right) - presinh_\alpha\left(\lambda t^\beta\right) \tag{3.282}$$

such that

$$E_\alpha\left(\lambda t^\beta\right) E_\alpha\left(-\lambda t^\beta\right) = precosh_\alpha^2\left(\lambda t^\beta\right) - presinh_\alpha^2\left(\lambda t^\beta\right), \tag{3.283}$$

where

$$precosh_\alpha^2\left(\lambda t^\beta\right) = precosh_\alpha\left(\lambda t^\beta\right) precosh_\alpha\left(\lambda t^\beta\right)$$

and

$$presinh_\alpha^2\left(\lambda t^\beta\right) = presinh_\alpha\left(\lambda t^\beta\right) presinh_\alpha\left(\lambda t^\beta\right).$$

Property 3.3 Let $\alpha \in \mathbb{C}$ and $Re(\alpha) > 0$. Then we have

$$presin_\alpha(-t) = -presin_\alpha(t), \tag{3.284}$$

$$precos_\alpha(-t) = precos_\alpha(t), \tag{3.285}$$

$$pretan_\alpha(-t) = -\,pretan_\alpha(t), \tag{3.286}$$

$$precotan_\alpha(-t) = -\,precotan_\alpha(t), \tag{3.287}$$

$$precsc_\alpha(-t) = -precsc_\alpha(t), \tag{3.288}$$

$$presec_\alpha(-t) = presec_\alpha(t), \tag{3.289}$$

$$presinh_\alpha(-t) = -presinh_\alpha(t), \tag{3.290}$$

$$precosh_\alpha(-t) = precosh_\alpha(t), \tag{3.291}$$

$$pretanh_\alpha(-t) = -\,pretanh_\alpha(t), \tag{3.292}$$

$$precotanh_\alpha(-t) = -\,precotanh_\alpha(t), \tag{3.293}$$

$$precsch_\alpha(-t) = -precsch_\alpha(t), \tag{3.294}$$

$$precsch_\alpha(-t) = precsch_\alpha(t), \tag{3.295}$$

$$presin_\alpha\left(-t^\beta\right) = -presin_\alpha\left(t^\beta\right), \tag{3.296}$$

$$precos_\alpha\left(-t^\beta\right) = precos_\alpha\left(t^\beta\right), \tag{3.297}$$

$$pretan_\alpha\left(-t^\beta\right) = -\,pretan_\alpha\left(t^\beta\right), \tag{3.298}$$

$$precotan_\alpha\left(-t^\beta\right) = -\,precotan_\alpha\left(t^\beta\right), \tag{3.299}$$

$$precsc_\alpha\left(-t^\beta\right) = -precsc_\alpha\left(t^\beta\right), \tag{3.300}$$

$$presec_\alpha\left(-t^\beta\right) = presec_\alpha\left(t^\beta\right), \tag{3.301}$$

$$presinh_\alpha\left(-t^\beta\right) = -presinh_\alpha\left(t^\beta\right), \tag{3.302}$$

$$precosh_\alpha\left(-t^\beta\right) = precosh_\alpha\left(t^\beta\right), \tag{3.303}$$

$$pretanh_\alpha\left(-t^\beta\right) = -\,pretanh_\alpha\left(t^\beta\right), \tag{3.304}$$

$$precotanh_\alpha\left(-t^\beta\right) = -\ precotanh_\alpha\left(t^\beta\right), \tag{3.305}$$

$$precsch_\alpha\left(-t^\beta\right) = -precsch_\alpha\left(t^\beta\right), \tag{3.306}$$

and

$$precsch_\alpha\left(-t^\beta\right) = precsch_\alpha\left(t^\beta\right). \tag{3.307}$$

3.1.8 The Laplace Transforms of the Special Functions via Mittag-Leffler Function

The Laplace transform for the special functions via Mittag-Leffler function can be given as follows [92]:

$$\begin{aligned} &L\left\{supersin_\alpha\left(\lambda t^\alpha\right)\right\} \\ &= L\left\{\sum_{n=0}^{\infty} \frac{(-1)^n \lambda^{2n+1} t^{(2n+1)\alpha}}{\Gamma((2n+1)\alpha+1)}\right\} \\ &= \sum_{n=0}^{\infty} \frac{(-1)^n \lambda^{2n+1}}{s^{(2n+1)\alpha+1}} \\ &= \frac{\lambda}{s^{1+\alpha}} \frac{1}{1+\lambda^2 s^{-2\alpha}} \left(\left|\lambda s^{-\alpha}\right| < 1\right) \\ &= \lambda s^{-(1+\alpha)} \left(1+\lambda^2 s^{-2\alpha}\right)^{-1} \left(\left|\lambda s^{-\alpha}\right| < 1\right), \end{aligned} \tag{3.308}$$

$$\begin{aligned} &L\left\{supercos_\alpha\left(\lambda t^\alpha\right)\right\} \\ &= L\left\{\sum_{n=0}^{\infty} \frac{(-1)^n \lambda^{2n} t^{2n\alpha}}{\Gamma(2n\alpha+1)}\right\} \\ &= \sum_{n=0}^{\infty} \frac{(-1)^n \lambda^{2n}}{s^{2n\alpha+1}} \\ &= \frac{1}{s} \frac{1}{1+\lambda^2 s^{-2\alpha}} \left(\left|\lambda s^{-\alpha}\right| < 1\right) \\ &= s^{-1} \left(1+\lambda^2 s^{-2\alpha}\right)^{-1} \left(\left|\lambda s^{-\alpha}\right| < 1\right), \end{aligned} \tag{3.309}$$

$$\begin{aligned} &L\left\{supersinh_\alpha\left(\lambda t^\alpha\right)\right\} \\ &= L\left\{\sum_{n=0}^{\infty} \frac{\lambda^{2n+1} t^{(2n+1)\alpha}}{\Gamma((2n+1)\alpha+1)}\right\} \\ &= \sum_{n=0}^{\infty} \frac{\lambda^{2n+1}}{s^{(2n+1)\alpha+1}} \\ &= \frac{\lambda}{s^{1+\alpha}} \frac{1}{1-\lambda^2 s^{-2\alpha}} \left(\left|\lambda s^{-\alpha}\right| < 1\right) \\ &= \lambda s^{-(1+\alpha)} \left(1-\lambda^2 s^{-2\alpha}\right)^{-1} \left(\left|\lambda s^{-\alpha}\right| < 1\right), \end{aligned} \tag{3.310}$$

and

$$\begin{aligned}
&L\left\{supercosh_{\alpha}\left(\lambda t^{\alpha}\right)\right\}\\
&=L\left\{\sum_{n=0}^{\infty}\frac{\lambda^{2n}t^{2n\alpha}}{\Gamma(2n\alpha+1)}\right\}\\
&=\sum_{n=0}^{\infty}\frac{\lambda^{2n}}{s^{2n\alpha+1}}\\
&=\frac{1}{s}\frac{1}{1-\lambda^{2}s^{-2\alpha}}\left(\left|\lambda s^{-\alpha}\right|<1\right)\\
&=s^{-1}\left(1-\lambda^{2}s^{-2\alpha}\right)^{-1}\left(\left|\lambda s^{-\alpha}\right|<1\right).
\end{aligned} \tag{3.311}$$

3.2 Analytic Number Theory Involving the Mittag-Leffler Function

In this section, we introduce the analytic number theory involving the Mittag-Leffler function.

3.2.1 *The Basic Formulas Involving the Mittag-Leffler Function*

We now present [92]

$$\vartheta_0(\alpha)=\sum_{\kappa=0}^{\infty}\frac{1}{\Gamma(\kappa\alpha+1)}, \tag{3.312}$$

$$\vartheta_1(\alpha)=\sum_{\kappa=0}^{\infty}\frac{1}{\Gamma(2\kappa\alpha+1)}, \tag{3.313}$$

$$\vartheta_2(\alpha)=\sum_{\kappa=0}^{\infty}\frac{1}{\Gamma((2\kappa+1)\alpha+1)}, \tag{3.314}$$

$$\vartheta_3(\alpha)=\sum_{\kappa=0}^{\infty}\frac{(-1)^{\kappa}}{\Gamma(2\kappa\alpha+1)} \tag{3.315}$$

and

$$\vartheta_4(\alpha)=\sum_{\kappa=0}^{\infty}\frac{(-1)^{\kappa}}{\Gamma((2\kappa+1)\alpha+1)}, \tag{3.316}$$

where $\alpha\in\mathbb{C}$ and $Re(\alpha)>0$.

Theorem 3.62 *[92]*

$$E_{\alpha}(0) = 1, \tag{3.317}$$

where $\alpha \in \mathbb{C}$ *and* $Re(\alpha) > 0$.

Theorem 3.63

$$supersin_{\alpha}(0) = 0, \tag{3.318}$$

where $\alpha \in \mathbb{C}$ *and* $Re(\alpha) > 0$.

Theorem 3.64

$$supercos_{\alpha}(0) = 1, \tag{3.319}$$

where $\alpha \in \mathbb{C}$ *and* $Re(\alpha) > 0$.

Theorem 3.65

$$supertan_{\alpha}(0) = 0, \tag{3.320}$$

where $\alpha \in \mathbb{C}$ *and* $Re(\alpha) > 0$.

Theorem 3.66

$$supercsc_{\alpha}(0) = 1, \tag{3.321}$$

where $\alpha \in \mathbb{C}$ *and* $Re(\alpha) > 0$.

Theorem 3.67

$$supersinh_{\alpha}(0) = 0, \tag{3.322}$$

where $\alpha \in \mathbb{C}$ *and* $Re(\alpha) > 0$.

Theorem 3.68

$$supercosh_{\alpha}(0) = 1, \tag{3.323}$$

where $\alpha \in \mathbb{C}$ *and* $Re(\alpha) > 0$.

Theorem 3.69

$$supertanh_{\alpha}(0) = 0, \tag{3.324}$$

where $\alpha \in \mathbb{C}$ *and* $Re(\alpha) > 0$.

Theorem 3.70

$$supercsch_{\alpha}(0) = 1, \tag{3.325}$$

where $\alpha \in \mathbb{C}$ *and* $Re(\alpha) > 0$.

The plots of the functions $supersin_{\alpha}(t^{\alpha})$, $presin_{\alpha}(t)$, $supercos_{\alpha}(t^{\alpha})$, $precos_{\alpha}(t)$, $supersinh_{\alpha}(t^{\alpha})$, $presinh_{\alpha}(t)$, $supercosh_{\alpha}(t^{\alpha})$, and $precosh_{\alpha}(t)$ with the different values of the parameter α are shown in Figs. 3.5, 3.6, 3.7, 3.8, 3.9, 3.10, 3.11, and 3.12, respectively.

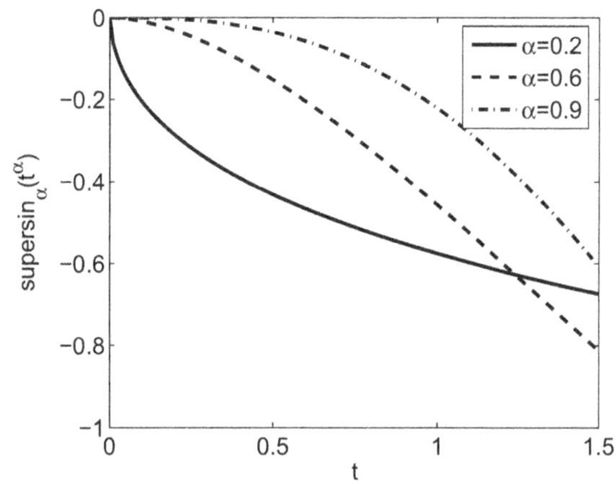

Fig. 3.5 The plots of $supersin_{\alpha}(t^{\alpha})$ with the different values of the parameter α

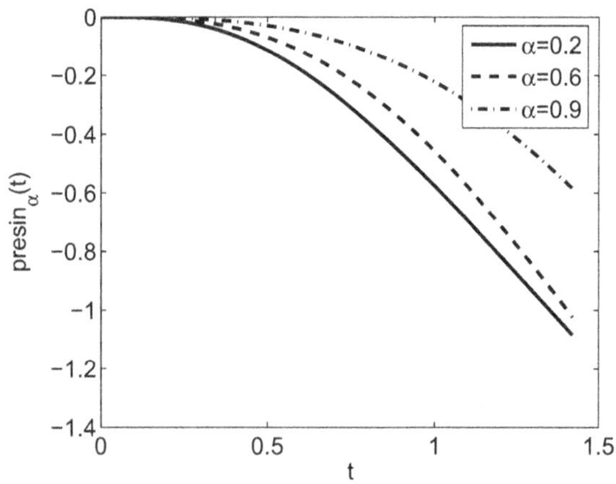

Fig. 3.6 The plots of $presin_{\alpha}(t)$ with the different values of the parameter α

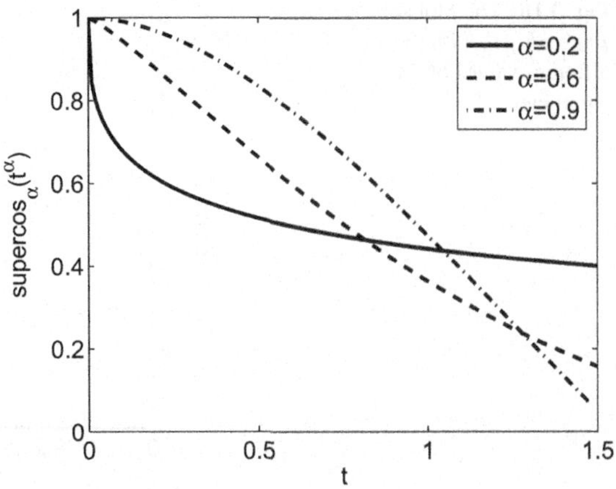

Fig. 3.7 The plots of $supercos_\alpha(t^\alpha)$ with the different values of the parameter α

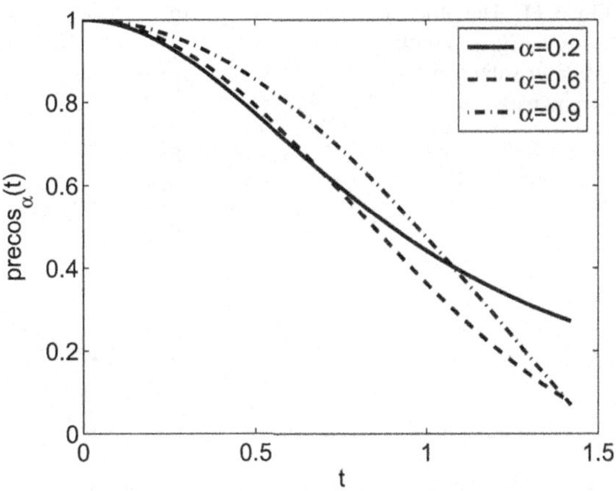

Fig. 3.8 The plots of $precos_\alpha(t)$ with the different values of the parameter α

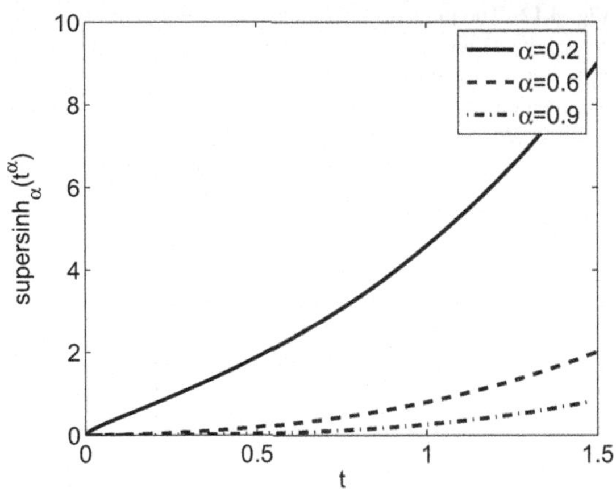

Fig. 3.9 The plots of $supersinh_\alpha(t^\alpha)$ with the different values of the parameter α

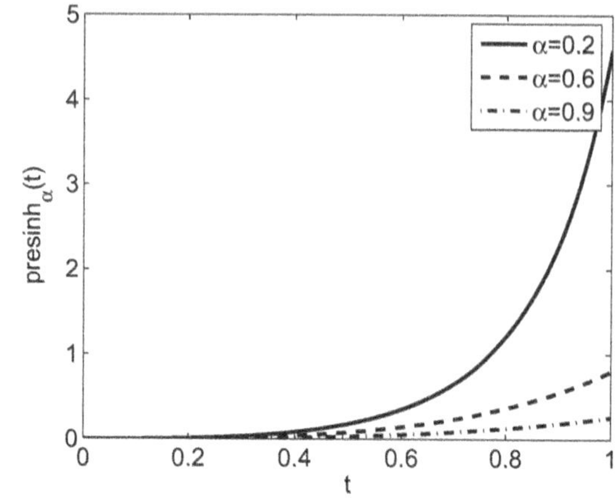

Fig. 3.10 The plots of $presinh_\alpha(t)$ with the different values of the parameter α

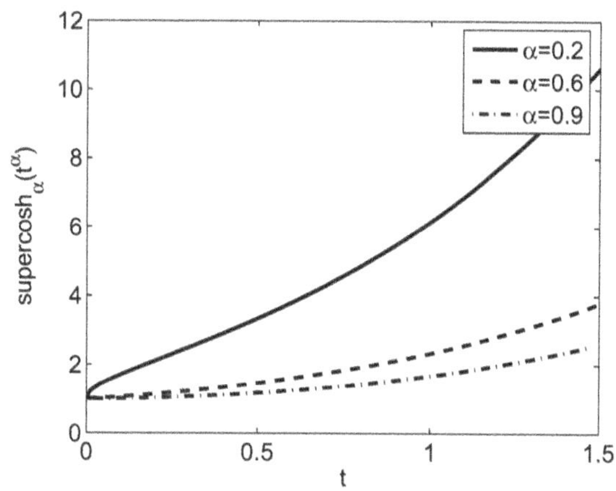

Fig. 3.11 The plots of $supercosh_\alpha(t^\alpha)$ with the different values of the parameter α

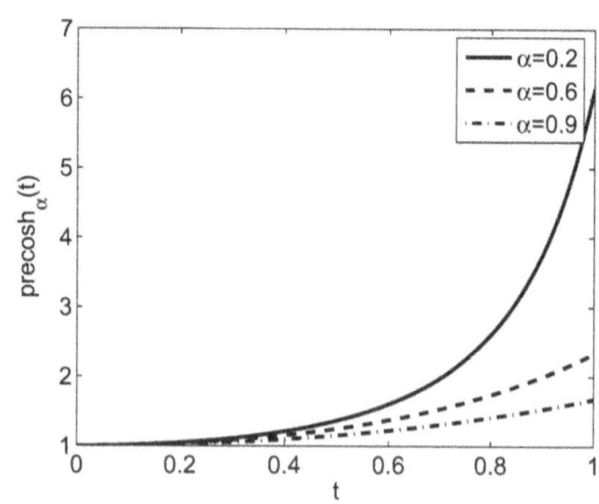

Fig. 3.12 The plots of $precosh_\alpha(t)$ with the different values of the parameter α

3.2.2 The Generalized Hyperbolic Function

The generalized hyperbolic function, as extended version of Mittag-Leffler, is defined as follows:

$$\Xi_{k,r}^{\alpha}(t)=\sum_{n=0}^{\infty}\frac{\alpha^{n}}{\Gamma(\kappa n+r)}t^{n\kappa+r}, \tag{3.326}$$

where $n,r\in\mathbb{N}$ and $\alpha\in\mathbb{C}\backslash\{0\}$.

The result was discovered by Riccati in 1757 [231].

The special cases of the generalized hyperbolic function are presented as follows:

$$\Xi_{k,,0}^{\alpha}(0)=1, \tag{3.327}$$

$$\Xi_{1,0}^{\alpha}(t)=\sum_{n=0}^{\infty}\frac{(\alpha t)^{n}}{\Gamma(n)}=e^{at}, \tag{3.328}$$

$$\Xi_{2,0}^{\alpha}(t)=cosh\left(\sqrt{\alpha}t\right), \tag{3.329}$$

$$\Xi_{2,1}^{\alpha}(t)=\frac{1}{\sqrt{\alpha}}sinh\left(\sqrt{\alpha}t\right), \tag{3.330}$$

and

$$\Xi_{k,0}^{1}(t)=\sum_{n=0}^{\infty}\frac{t^{n\kappa}}{\Gamma(n\kappa)}=t\sum_{n=0}^{\infty}\frac{t^{n\kappa-1}}{\Gamma(n\kappa)}=tE_{k}(t). \tag{3.331}$$

The results were discovered by Muldoon and Ungar in 1996 [232].

For more details of the generalized hyperbolic function, see [233].

3.3 The Special Integral Equations via Mittag-Leffler Function and Related Functions

In this section, we present the theory of the special integral equations via Mittag-Leffler function.

3.3.1 The Integral Equations of Volterra Type

In this part, we give the examples for the integral equations of Volterra type.

Example 3.40 The Volterra–Hille–Tamarkin integral equation is given as follows [215]:

$$H(t) = f(t) - \lambda \int_a^t E_\alpha\left(\gamma (t-\tau)^\alpha\right) f(\tau)\, d\tau \tag{3.332}$$

or

$$H(t) = \lambda \int_a^t E_\alpha\left(\gamma (t-\tau)^\alpha\right) f(\tau)\, d\tau, \tag{3.333}$$

where $\lambda, \gamma \in \mathbb{R}$ and $t > a$.

Example 3.41 The integral equation of Volterra type is given as follows:

$$H(t) = f(t) - \lambda \int_a^t E_\alpha(\gamma (t-\tau)) f(\tau)\, d\tau \tag{3.334}$$

or

$$H(t) = \lambda \int_a^t E_\alpha(\gamma (t-\tau)) f(\tau)\, d\tau, \tag{3.335}$$

where $\lambda, \gamma \in \mathbb{R}$ and $t > a$.

Example 3.42 The integral equation of Volterra type is given as follows:

$$H(t) = f(t) - \lambda \int_a^t E_\alpha\left(\gamma (t-\tau)^\beta\right) f(\tau)\, d\tau \tag{3.336}$$

or

$$H(t) = \lambda \int_a^t E_\alpha\left(\gamma (t-\tau)^\beta\right) f(\tau)\, d\tau, \tag{3.337}$$

where $\lambda, \gamma \in \mathbb{R}$ and $t > a$.

Example 3.43 The integral equation of Volterra type is given as follows:

$$H(t) = f(t) - \lambda \int_a^t presin_\alpha\left(\gamma (t-\tau)^\beta\right) f(\tau)\, d\tau \tag{3.338}$$

or

$$H(t) = \lambda \int_a^t presin_\alpha\left(\gamma (t-\tau)^\beta\right) f(\tau)\, d\tau, \tag{3.339}$$

where $\lambda, \gamma \in \mathbb{R}$ and $t > a$.

Example 3.44 The integral equation of Volterra type is given as follows:

$$H(t) = f(t) - \lambda \int_a^t supersin_\alpha \left(\gamma (t-\tau)^\alpha\right) f(\tau)\, d\tau \tag{3.340}$$

or

$$H(t) = \lambda \int_a^t supersin_\alpha \left(\gamma (t-\tau)^\alpha\right) f(\tau)\, d\tau, \tag{3.341}$$

where $\lambda, \gamma \in \mathbb{R}$ and $t > a$.

Example 3.45 The integral equation of Volterra type is given as follows:

$$H(t) = f(t) - \lambda \int_a^t presin_\alpha (\gamma (t-\tau)) f(\tau)\, d\tau \tag{3.342}$$

or

$$H(t) = \lambda \int_a^t presin_\alpha (\gamma (t-\tau)) f(\tau)\, d\tau, \tag{3.343}$$

where $\lambda, \gamma \in \mathbb{R}$ and $t > a$.

Example 3.46 The integral equation of Volterra type is given as follows:

$$H(t) = f(t) - \lambda \int_a^t presinh_\alpha \left(\gamma (t-\tau)^\beta\right) f(\tau)\, d\tau \tag{3.344}$$

or

$$H(t) = \lambda \int_a^t presinh_\alpha \left(\gamma (t-\tau)^\beta\right) f(\tau)\, d\tau, \tag{3.345}$$

where $\lambda, \gamma \in \mathbb{R}$ and $t > a$.

Example 3.47 The integral equation of Volterra type is given as follows:

$$H(t) = f(t) - \lambda \int_a^t supersinh_\alpha \left(\gamma (t-\tau)^\alpha\right) f(\tau)\, d\tau \tag{3.346}$$

or

$$H(t) = \lambda \int_a^t supersinh_\alpha \left(\gamma (t-\tau)^\alpha\right) f(\tau)\, d\tau, \tag{3.347}$$

where $\lambda, \gamma \in \mathbb{R}$ and $t > a$.

Example 3.48 The integral equation of Volterra type is given as follows:

$$H(t) = f(t) - \lambda \int_a^t presinh_\alpha(\gamma(t-\tau)) f(\tau) d\tau \tag{3.348}$$

or

$$H(t) = \lambda \int_a^t presinh_\alpha(\gamma(t-\tau)) f(\tau) d\tau, \tag{3.349}$$

where $\lambda, \gamma \in \mathbb{R}$ and $t > a$.

Example 3.49 The integral equation of Volterra type is given as follows:

$$H(t) = f(t) - \lambda \int_a^t precos_\alpha\left(\gamma(t-\tau)^\beta\right) f(\tau) d\tau \tag{3.350}$$

or

$$H(t) = \lambda \int_a^t precos_\alpha\left(\gamma(t-\tau)^\beta\right) f(\tau) d\tau, \tag{3.351}$$

where $\lambda, \gamma \in \mathbb{R}$ and $t > a$.

Example 3.50 The integral equation of Volterra type is given as follows:

$$H(t) = f(t) - \lambda \int_a^t precos_\alpha(\gamma(t-\tau)) f(\tau) d\tau \tag{3.352}$$

or

$$H(t) = \lambda \int_a^t precos_\alpha(\gamma(t-\tau)) f(\tau) d\tau, \tag{3.353}$$

where $\lambda, \gamma \in \mathbb{R}$ and $t > a$.

Example 3.51 The integral equation of Volterra type is given as follows:

$$H(t) = f(t) - \lambda \int_a^t supercos_\alpha\left(\gamma(t-\tau)^\alpha\right) f(\tau) d\tau \tag{3.354}$$

or

$$H(t) = \lambda \int_a^t supercos_\alpha\left(\gamma(t-\tau)^\alpha\right) f(\tau) d\tau, \tag{3.355}$$

where $\lambda, \gamma \in \mathbb{R}$ and $t > a$.

Example 3.52 The integral equation of Volterra type is given as follows:

$$H(t) = f(t) - \lambda \int_a^t precosh_\alpha\left(\gamma (t-\tau)^\beta\right) f(\tau)\, d\tau \tag{3.356}$$

or

$$H(t) = \lambda \int_a^t precosh_\alpha\left(\gamma (t-\tau)^\beta\right) f(\tau)\, d\tau, \tag{3.357}$$

where $\lambda, \gamma \in \mathbb{R}$ and $t > a$.

Example 3.53 The integral equation of Volterra type is given as follows:

$$H(t) = f(t) - \lambda \int_a^t precosh_\alpha\left(\gamma (t-\tau)\right) f(\tau)\, d\tau \tag{3.358}$$

or

$$H(t) = \lambda \int_a^t precosh_\alpha\left(\gamma (t-\tau)\right) f(\tau)\, d\tau, \tag{3.359}$$

where $\lambda, \gamma \in \mathbb{R}$ and $t > a$.

Example 3.54 The integral equation of Volterra type is given as follows:

$$H(t) = f(t) - \lambda \int_a^t supercosh_\alpha\left(\gamma (t-\tau)^\alpha\right) f(\tau)\, d\tau \tag{3.360}$$

or

$$H(t) = \lambda \int_a^t supercosh_\alpha\left(\gamma (t-\tau)^\alpha\right) f(\tau)\, d\tau, \tag{3.361}$$

where $\lambda, \gamma \in \mathbb{R}$ and $t > a$.

3.3.2 The Integral Equations of Fredholm Type

In this part, we propose the examples for the integral equations of Fredholm type.

Example 3.55 The integral equation of Fredholm type is given as follows:

$$H(t) = f(t) - \lambda \int_a^b E_\alpha\left(\gamma (t-\tau)^\alpha\right) f(\tau)\, d\tau \tag{3.362}$$

or

$$H(t) = \lambda \int_a^b E_\alpha\left(\gamma (t-\tau)^\alpha\right) f(\tau)\, d\tau, \tag{3.363}$$

where $\lambda, \gamma \in \mathbb{R}$ and $b > \tau > a$.

Example 3.56 The integral equation of Fredholm type is given as follows:

$$H(t) = f(t) - \lambda \int_a^b E_\alpha(\gamma (t-\tau)) f(\tau)\, d\tau \tag{3.364}$$

or

$$H(t) = \lambda \int_a^b E_\alpha(\gamma (t-\tau)) f(\tau)\, d\tau, \tag{3.365}$$

where $\lambda, \gamma \in \mathbb{R}$ and $b > \tau > a$.

Example 3.57 The integral equation of Fredholm type is given as follows:

$$H(t) = f(t) - \lambda \int_a^b E_\alpha\left(\gamma (t-\tau)^\beta\right) f(\tau)\, d\tau \tag{3.366}$$

or

$$H(t) = \lambda \int_a^b E_\alpha\left(\gamma (t-\tau)^\beta\right) f(\tau)\, d\tau, \tag{3.367}$$

where $\lambda, \gamma \in \mathbb{R}$ and $b > \tau > a$.

Example 3.58 The integral equation of Fredholm type is given as follows:

$$H(t) = f(t) - \lambda \int_a^b presin_\alpha\left(\gamma (t-\tau)^\beta\right) f(\tau)\, d\tau \tag{3.368}$$

or

$$H(t) = \lambda \int_a^b presin_\alpha\left(\gamma (t-\tau)^\beta\right) f(\tau)\, d\tau, \tag{3.369}$$

where $\lambda, \gamma \in \mathbb{R}$ and $b > \tau > a$.

Example 3.59 The integral equation of Fredholm type is given as follows:

$$H(t) = f(t) - \lambda \int_a^b supersin_\alpha\left(\gamma (t-\tau)^\alpha\right) f(\tau)\, d\tau \tag{3.370}$$

or

$$H(t) = \lambda \int_a^b supersin_\alpha (\gamma (t - \tau)) f(\tau) d\tau, \tag{3.371}$$

where $\lambda, \gamma \in \mathbb{R}$ and $b > \tau > a$.

Example 3.60 The integral equation of Fredholm type is given as follows:

$$H(t) = f(t) - \lambda \int_a^b presin_\alpha (\gamma (t - \tau)) f(\tau) d\tau \tag{3.372}$$

or

$$H(t) = \lambda \int_a^b presin_\alpha (\gamma (t - \tau)) f(\tau) d\tau, \tag{3.373}$$

where $\lambda, \gamma \in \mathbb{R}$ and $b > \tau > a$.

Example 3.61 The integral equation of Fredholm type is given as follows:

$$H(t) = f(t) - \lambda \int_a^b presinh_\alpha \left(\gamma (t - \tau)^\beta\right) f(\tau) d\tau \tag{3.374}$$

or

$$H(t) = \lambda \int_a^b presinh_\alpha \left(\gamma (t - \tau)^\beta\right) f(\tau) d\tau, \tag{3.375}$$

where $\lambda, \gamma \in \mathbb{R}$ and $b > \tau > a$.

Example 3.62 The integral equation of Fredholm type is given as follows:

$$H(t) = f(t) - \lambda \int_a^b supersinh_\alpha \left(\gamma (t - \tau)^\alpha\right) f(\tau) d\tau \tag{3.376}$$

or

$$H(t) = \lambda \int_a^b supersinh_\alpha (\gamma (t - \tau)) f(\tau) d\tau, \tag{3.377}$$

where $\lambda, \gamma \in \mathbb{R}$ and $b > \tau > a$.

Example 3.63 The integral equation of Fredholm type is given as follows:

$$H(t) = f(t) - \lambda \int_a^b presinh_\alpha (\gamma (t - \tau)) f(\tau) d\tau \tag{3.378}$$

or

$$H(t) = \lambda \int_a^b presinh_\alpha(\gamma(t-\tau)) f(\tau) d\tau, \tag{3.379}$$

where $\lambda, \gamma \in \mathbb{R}$ and $b > \tau > a$.

Example 3.64 The integral equation of Fredholm type is given as follows:

$$H(t) = f(t) - \lambda \int_a^b precos_\alpha\left(\gamma(t-\tau)^\beta\right) f(\tau) d\tau \tag{3.380}$$

or

$$H(t) = \lambda \int_a^b precos_\alpha\left(\gamma(t-\tau)^\beta\right) f(\tau) d\tau, \tag{3.381}$$

where $\lambda, \gamma \in \mathbb{R}$ and $b > \tau > a$.

Example 3.65 The integral equation of Fredholm type is given as follows:

$$H(t) = f(t) - \lambda \int_a^b precos_\alpha(\gamma(t-\tau)) f(\tau) d\tau \tag{3.382}$$

or

$$H(t) = \lambda \int_a^b precos_\alpha(\gamma(t-\tau)) f(\tau) d\tau \tag{3.383}$$

where $\lambda, \gamma \in \mathbb{R}$ and $b > \tau > a$.

Example 3.66 The integral equation of Fredholm type is given as follows:

$$H(t) = f(t) - \lambda \int_a^b supercos_\alpha\left(\gamma(t-\tau)^\alpha\right) f(\tau) d\tau \tag{3.384}$$

or

$$H(t) = \lambda \int_a^b supercos_\alpha\left(\gamma(t-\tau)^\alpha\right) f(\tau) d\tau \tag{3.385}$$

where $\lambda, \gamma \in \mathbb{R}$ and $b > \tau > a$.

Example 3.67 The integral equation of Fredholm type is given as follows:

$$H(t) = f(t) - \lambda \int_a^b precosh_\alpha\left(\gamma(t-\tau)^\beta\right) f(\tau) d\tau \tag{3.386}$$

or

$$H(t) = \lambda \int_a^b precosh_\alpha \left(\gamma (t - \tau)^\beta\right) f(\tau) \, d\tau \tag{3.387}$$

where $\lambda, \gamma \in \mathbb{R}$ and $b > \tau > a$.

Example 3.68 The integral equation of Fredholm type is given as follows:

$$H(t) = f(t) - \lambda \int_a^b precosh_\alpha (\gamma (t - \tau)) f(\tau) \, d\tau \tag{3.388}$$

or

$$H(t) = \lambda \int_a^b precosh_\alpha (\gamma (t - \tau)) f(\tau) \, d\tau \tag{3.389}$$

where $\lambda, \gamma \in \mathbb{R}$ and $b > \tau > a$.

Example 3.69 The integral equation of Fredholm type is given as follows:

$$H(t) = f(t) - \lambda \int_a^b supercosh_\alpha \left(\gamma (t - \tau)^\alpha\right) f(\tau) \, d\tau \tag{3.390}$$

or

$$H(t) = \lambda \int_a^b supercosh_\alpha \left(\gamma (t - \tau)^\alpha\right) f(\tau) \, d\tau \tag{3.391}$$

where $\lambda, \gamma \in \mathbb{R}$ and $b > \tau > a$.

3.4 The Integral Representations for the Special Function via Mittag-Leffler Function

In this section, we give the integral representations for the special function via Mittag-Leffler function.

Theorem 3.71

$$\int_0^\infty e^{-t} supersin_\alpha \left(\lambda t^\alpha\right) dt = -\frac{\lambda}{1 + \lambda^2}, \tag{3.392}$$

$$\int_0^\infty e^{-t} supercos_\alpha \left(\lambda t^\alpha\right) dt = \frac{1}{1 + \lambda^2}, \tag{3.393}$$

$$\int_0^{\infty} e^{-t} supersinh_{\alpha}\left(\lambda t^{\alpha}\right) dt = -\frac{\lambda}{1-\lambda^2}, \tag{3.394}$$

and

$$\int_0^{\infty} e^{-t} supercosh_{\alpha}\left(\lambda t^{\alpha}\right) dt = \frac{1}{1-\lambda^2}, \tag{3.395}$$

where $\lambda, \alpha \in \mathbb{C}$ and $Re(\alpha) > 0$.

Theorem 3.72

$$\int_0^{\infty} t^{\beta-1} e^{-t^{\beta}} presin_{\alpha}\left(\lambda t^{\alpha\beta}\right) dt = -\frac{\lambda}{\beta\left(1+\lambda^2\right)}, \tag{3.396}$$

$$\int_0^{\infty} t^{\beta-1} e^{-t^{\beta}} precos_{\alpha}\left(\lambda t^{\alpha\beta}\right) dt = \frac{1}{\beta\left(1+\lambda^2\right)}, \tag{3.397}$$

$$\int_0^{\infty} t^{\beta-1} e^{-t^{\beta}} presinh_{\alpha}\left(\lambda t^{\alpha\beta}\right) dt = \frac{-\lambda}{\beta\left(1-\lambda^2\right)}, \tag{3.398}$$

and

$$\int_0^{\infty} t^{\beta-1} e^{-t^{\beta}} precosh_{\alpha}\left(\lambda t^{\alpha\beta}\right) dt = \frac{1}{\beta\left(1-\lambda^2\right)}, \tag{3.399}$$

where $\lambda, \alpha \in \mathbb{C}$, $Re(\alpha) > 0$ and $\beta > 0$.

Theorem 3.73

$$\int_0^{\infty} t^{\beta-1} e^{-\theta t^{\beta}} presin_{\alpha}\left(\lambda t^{\alpha\beta}\right) dt = -\frac{\lambda}{\beta\left(\theta^{2\alpha}+\lambda^2\right)}, \tag{3.400}$$

$$\int_0^{\infty} t^{\beta-1} e^{-\theta t^{\beta}} precos_{\alpha}\left(\lambda t^{\alpha\beta}\right) dt = \frac{\theta^{\alpha}}{\beta\left(\theta^{2\alpha}+\lambda^2\right)}, \tag{3.401}$$

$$\int_0^{\infty} t^{\beta-1} e^{-\theta t^{\beta}} presinh_{\alpha}\left(\lambda t^{\alpha\beta}\right) dt = -\frac{\lambda}{\beta\left(\theta^{2\alpha}-\lambda^2\right)}, \tag{3.402}$$

and

$$\int_0^{\infty} t^{\beta-1} e^{-\theta t^{\beta}} precosh_{\alpha}\left(\lambda t^{\alpha\beta}\right) dt = \frac{\theta^{\alpha}}{\beta\left(\theta^{2\alpha}-\lambda^2\right)}, \tag{3.403}$$

where $\lambda, \alpha \in \mathbb{C}$, $Re(\alpha) > 0$, $\theta > 0$ and $\beta > 0$.

Theorem 3.74

$$\int_0^{\infty} t^{\frac{1}{\alpha}-1} e^{-\theta \sqrt[\alpha]{t}} presin_{\alpha}(\lambda t)\, dt = -\frac{\lambda\alpha}{\theta^{2\alpha}+\lambda^2}, \tag{3.404}$$

$$\int_0^{\infty} t^{\frac{1}{\alpha}-1} e^{-\theta \sqrt[\alpha]{t}} precos_{\alpha}(\lambda t)\, dt = \frac{\alpha\theta^{\alpha}}{\theta^{2\alpha}+\lambda^2}, \tag{3.405}$$

$$\int_0^{\infty} t^{\frac{1}{\alpha}-1} e^{-\theta \sqrt[\alpha]{t}} presinh_{\alpha}(\lambda t)\, dt = -\frac{\alpha\lambda}{\theta^{2\alpha}-\lambda^2}, \tag{3.406}$$

and

$$\int_0^{\infty} t^{\frac{1}{\alpha}-1} e^{-\theta \sqrt[\alpha]{t}} precosh_{\alpha}(\lambda t)\, dt = \frac{\alpha\theta^{\alpha}}{\theta^{2\alpha}-\lambda^2}, \tag{3.407}$$

where $\lambda, \alpha \in \mathbb{C}$, $Re(\alpha) > 0$, *and* $\theta > 0$.

3.5 The Fractional Equations via Mittag-Leffler Function and Related Functions

In this section, we present the theorem for the fractional equations as follows.

Theorem 3.75 (Haubold–Mathai–Saxena Theorem [205])
If $Re(\alpha) > 0$, *then there is [215]*

$$G(t) - G_0 = -c^{\alpha}\, {}^{\widehat{RL}}DG(t) \tag{3.408}$$

with the Mittag-Leffler type function solution

$$G(t) = G(0)\, E_{\alpha}\left(-c^{\alpha} t^{\alpha}\right), \tag{3.409}$$

where the Riemann–Liouville fractional derivative operator is given as (see the Chap. 1):

$${}^{\widehat{RL}}D\Phi(t) = \frac{1}{\Gamma(1+\alpha)} \frac{d}{dt} \int_0^t \frac{\Phi(\tau)}{(t-\tau)^{1-\alpha}} d\tau. \tag{3.410}$$

The result was discovered by Haubold, Mathai, and Saxena in 2011 [215]

As the generalized version of the Haubold–Mathai–Saxena Theorem, we have the following results:

Corollary 3.1 *If $Re(\alpha) > 0$, then there is*

$$G(t) - G_0 = -c^{\alpha}\, \widehat{{}^{RL}D_{g(\cdot)}} G(t) \tag{3.411}$$

with the Mittag-Leffler type function solution

$$G(t) = G(0)\, E_{\alpha}\left(-c^{\alpha} (g(t))^{\alpha}\right), \tag{3.412}$$

where the Osler fractional derivative operator is given as (see the Chap. 2):

$$\widehat{{}^{RL}D_{g(\cdot)}} \Phi(t) = \frac{1}{\Gamma(1+\alpha)} \frac{1}{g^{(1)}(t)} \frac{d}{dt} \int_0^t \frac{\Phi(\tau)\, g^{(1)}(\tau)}{(g(t) - g(\tau))^{1-\alpha}} d\tau, \tag{3.413}$$

with $\Phi(t) = (\Phi \circ g)(t) = \Phi(g(t))$, $g^{(1)}(t) > 0$ and $g(0) = 0$.

Corollary 3.2 *If $Re(\alpha) > 0$ and $Re(\beta) > 0$, then there is*

$$G(t) - G_0 = -c^{\alpha}\, \widehat{{}^{RL}D_{g(\cdot)}} G(t) \tag{3.414}$$

with the Mittag-Leffler type function solution

$$G(t) = G(0)\, E_{\alpha}\left(-c^{\alpha} t^{\alpha\beta}\right), \tag{3.415}$$

where the Osler fractional derivative operator is given as (see the Chap. 2):

$$\widehat{{}^{RL}D_{g(\cdot)}} \Phi(t) = \frac{1}{\Gamma(1+\alpha)} \frac{1}{g^{(1)}(t)} \frac{d}{dt} \int_0^t \frac{\Phi(\tau)\, g^{(1)}(\tau)}{(g(t) - g(\tau))^{1-\alpha}} d\tau, \tag{3.416}$$

with $\Phi(t) = (\Phi \circ g)(t) = \Phi(g(t))$ and $g(t) = t^{\beta}$.

Corollary 3.3 *If $Re(\alpha) > 0$, $Re(\beta) > 0$, and $\lambda \in \mathbb{R}$, then there is*

$$G(t) - G_0 = -c^{\alpha}\, \widehat{{}^{RL}D_{g(\cdot)}} G(t) \tag{3.417}$$

with the Mittag-Leffler type function solution

$$G(t) = G(0)\, E_{\alpha}\left(-c^{\alpha} \lambda^{\alpha} t^{\alpha\beta}\right), \tag{3.418}$$

where the Osler fractional derivative operator is given as (see the Chap. 2):

$$\widehat{{}^{RL}D_{g(\cdot)}} \Phi(t) = \frac{1}{\Gamma(1+\alpha)} \frac{1}{g^{(1)}(t)} \frac{d}{dt} \int_0^t \frac{\Phi(\tau)\, g^{(1)}(\tau)}{(g(t) - g(\tau))^{1-\alpha}} d\tau, \tag{3.419}$$

with $\Phi(t) = (\Phi \circ g)(t) = \Phi(g(t))$ and $g(t) = \lambda t^{\beta}$.

Theorem 3.76 (Mainardi Theorem [234]) *If* $Re(\alpha) > 0$*, then there is*

$$G(t) = -c^{\alpha}\widehat{{}^{C}D}G(t) \tag{3.420}$$

with the Mittag-Leffler type function solution

$$G(t) = G(0)\, E_{\alpha}\left(-c^{\alpha}t^{\alpha}\right), \tag{3.421}$$

where the Riemann–Liouville fractional derivative operator is given as (see the Chap. 2):

$$\widehat{{}^{C}D}\Phi(t) = \frac{1}{\Gamma(1+\alpha)} \int_0^t \frac{\Phi^{(1)}(\tau)}{(t-\tau)^{1-\alpha}} d\tau. \tag{3.422}$$

The result was discovered by Mainardi in 2011 [234].

As the generalized version of the Mainardi Theorem, we give the following results:

Corollary 3.4 *If* $Re(\alpha) > 0$*, then there is*

$$G(t) = -c^{\alpha}\widehat{{}^{C}D_{g(\cdot)}}G(t) \tag{3.423}$$

with the Mittag-Leffler type function solution

$$G(t) = G(0)\, E_{\alpha}\left(-c^{\alpha}(g(t))^{\alpha}\right), \tag{3.424}$$

where the Almeida fractional derivative operator is given as (see the Chap. 2):

$$\widehat{{}^{C}D_{g(\cdot)}}\Phi(t) = \frac{1}{\Gamma(1+\alpha)} \int_0^t \frac{g^{(1)}(\tau)}{(g(t)-g(\tau))^{1-\alpha}} \left(\frac{1}{g^{(1)}(\tau)}\frac{d}{d\tau}\right) \Phi(\tau)\, d\tau, \tag{3.425}$$

with $\Phi(t) = (\Phi \circ g)(t) = \Phi(g(t))$*,* $g^{(1)}(t) > 0$ *and* $g(0) = 0$*.*

Corollary 3.5 *If* $Re(\alpha) > 0$ *and* $Re(\beta) > 0$*, then there is*

$$G(t) = -c^{\alpha}\widehat{{}^{C}D_{g(\cdot)}}G(t) \tag{3.426}$$

with the Mittag-Leffler type function solution

$$G(t) = G(0)\, E_{\alpha}\left(-c^{\alpha}t^{\alpha\beta}\right), \tag{3.427}$$

where the Almeida fractional derivative operator is given as (see the Chap. 2):

$$\widehat{{}^{C}D_{g(\cdot)}}\Phi(t) = \frac{1}{\Gamma(1+\alpha)} \int_0^t \frac{g^{(1)}(\tau)}{(g(t)-g(\tau))^{1-\alpha}} \left(\frac{1}{g^{(1)}(\tau)} \frac{d}{d\tau} \right) \Phi(\tau) \, d\tau, \tag{3.428}$$

with $\Phi(t) = (\Phi \circ g)(t) = \Phi(g(t))$ *and* $g(t) = t^{\beta}$.

Corollary 3.6 *If* $Re(\alpha) > 0$, $Re(\beta) > 0$ *and* $\lambda \in \mathbb{R}$, *then there is*

$$G(t) = -c^{\alpha} \widehat{{}^{C}D_{g(\cdot)}} G(t) \tag{3.429}$$

with the Mittag-Leffler type function solution

$$G(t) = G(0)\, E_{\alpha}\left(-c^{\alpha}\lambda^{\alpha}t^{\alpha\beta}\right), \tag{3.430}$$

where the Almeida fractional derivative operator is given as (see the Chap. 2):

$$\widehat{{}^{C}D_{g(\cdot)}}\Phi(t) = \frac{1}{\Gamma(1+\alpha)} \int_0^t \frac{g^{(1)}(\tau)}{(g(t)-g(\tau))^{1-\alpha}} \left(\frac{1}{g^{(1)}(\tau)} \frac{d}{d\tau} \right) \Phi(\tau) \, d\tau, \tag{3.431}$$

with $\Phi(t) = (\Phi \circ g)(t) = \Phi(g(t))$ *and* $g(t) = \lambda t^{\beta}$.

3.6 General Fractional Calculus Operators with Mittag-Leffler Function

In this section, we introduce the general fractional calculus operators with Mittag-Leffler function.

3.6.1 Hille–Tamarkin General Fractional Derivative

Definition 3.41 Let $1 \geq \alpha \geq 0$ and $\lambda \in \mathbb{R}$.

The left-sided Hille–Tamarkin general fractional derivative is defined as [92, 215]

$${}_{Ml}^{RL}D_{a+}^{\alpha,\lambda} f(t) = \frac{d}{dt} \int_a^t E_{\alpha}\left(-\lambda (t-\tau)^{\alpha}\right) f(\tau) \, d\tau, \tag{3.432}$$

and the right-sided Hille–Tamarkin general fractional derivative as

$$ {}_{Ml}^{RL}D_{b-}^{\alpha,\lambda} f(t) = -\frac{d}{dt}\int_t^b E_\alpha\left(-\lambda(\tau-t)^\alpha\right) f(\tau)\, d\tau. \tag{3.433}$$

Definition 3.42 Let $\alpha \geq 0$, $\kappa = [\alpha] + 1$, and $\lambda \in \mathbb{R}$.

The left-sided Hille–Tamarkin general fractional derivative is defined as [92, 214]

$$ {}_{Ml}^{RL}D_{a+}^{\alpha,\kappa,\lambda} f(t) = \frac{d^\kappa}{dt^\kappa}\int_a^t E_\alpha\left(-\lambda(t-\tau)^\alpha\right) f(\tau)\, d\tau, \tag{3.434}$$

and the right-sided Hille–Tamarkin general fractional derivative as

$$ {}_{Ml}^{RL}D_{b-}^{\alpha,\kappa,\lambda} f(t) = (-1)^\kappa \frac{d^\kappa}{dt^\kappa}\int_t^b E_\alpha\left(-\lambda(\tau-t)^\alpha\right) f(\tau)\, d\tau. \tag{3.435}$$

For $\kappa = 1$ and $a = 0$ we have

$$ \begin{aligned} &\tfrac{d}{dt}\textstyle\int_0^t E_\alpha(-\lambda(t-\tau)^\alpha) f(\tau)\, d\tau \\ &= \textstyle\int_0^t E_\alpha(-\lambda(t-\tau)^\alpha) f^{(1)}(\tau)\, d\tau - E_\alpha(-\lambda t^\alpha) f(0), \end{aligned} \tag{3.436}$$

such that the Gorenflo–Mainardi formula is given as [235]

$$ \begin{aligned} &\tfrac{d}{dt}\textstyle\int_0^t E_\alpha(-\lambda(t-\tau)^\alpha) f(\tau)\, d\tau \\ &= \textstyle\int_0^t E_\alpha(-\lambda(t-\tau)^\alpha) f^{(1)}(\tau)\, d\tau + E_\alpha(-\lambda t^\alpha) f(0) \end{aligned} \tag{3.437}$$

due to

$$ \int_0^t E_\alpha\left(-\lambda\tau^\alpha\right) f^{(1)}(t-\tau)\, d\tau = \int_0^t E_\alpha\left(-\lambda(t-\tau)^\alpha\right) f^{(1)}(\tau)\, d\tau. \tag{3.438}$$

Gorenflo and Mainardi did not continue to work the Hille–Tamarkin general fractional derivative [235].

For more details of the Hille–Tamarkin general fractional derivative, see [92] .

Definition 3.43 Let $1 \geq \alpha \geq 0$ and $\lambda \in \mathbb{R}$.

The left-sided Liouville–Sonine–Caputo type general fractional derivative via Mittag-Leffler function is defined as [92]

$$ {}_{Ml}^{LSC}D_{a+}^{\alpha,\lambda} f(t) = \int_a^t E_\alpha\left(-\lambda(t-\tau)^\alpha\right) f^{(1)}(\tau)\, d\tau, \tag{3.439}$$

and the right-sided Liouville–Sonine–Caputo type general fractional derivative via Mittag-Leffler function as

$$ {}_{Ml}^{LSC}D_{b-}^{\alpha,\lambda} f(t) = -\int_t^b E_\alpha\left(-\lambda(\tau-t)^\alpha\right) f^{(1)}(\tau)\,d\tau. \tag{3.440} $$

Definition 3.44 Let $\alpha \geq 0$, $\kappa = [\alpha] + 1$, and $\lambda \in \mathbb{R}$.

The left-sided Liouville–Sonine–Caputo type general fractional derivative via Mittag-Leffler function is defined as [92] .

$$ {}_{Ml}^{LSC}D_{a+}^{\alpha,\kappa,\lambda} f(t) = \int_a^t E_\alpha\left(-\lambda(t-\tau)^\alpha\right) f^{(\kappa)}(\tau)\,d\tau, \tag{3.441} $$

and the right-sided Liouville–Sonine–Caputo type general fractional derivative via Mittag-Leffler function as

$$ {}_{Ml}^{LSC}D_{b-}^{\alpha,\kappa,\lambda} f(t) = (-1)^\kappa \int_t^b E_\alpha\left(-\lambda(\tau-t)^\alpha\right) f^{(\kappa)}(\tau)\,d\tau. \tag{3.442} $$

However, we have to repeat the following cases:

In 2016, the left-sided Liouville–Sonine–Caputo type general fractional derivative via Mittag-Leffler function is defined as [236]

$$ {}_{MlA}^{LSC}D_{a+}^{\alpha,\lambda} f(t) = \frac{\Im(\alpha)}{1-\alpha}\int_a^t E_\alpha\left(-\frac{\alpha}{1-\alpha}(t-\tau)^\alpha\right) f^{(1)}(\tau)\,d\tau, \tag{3.443} $$

and the right-sided Liouville–Sonine–Caputo type general fractional derivative via Mittag-Leffler function as [236]

$$ {}_{MlA}^{LSC}D_{b-}^{\alpha,\lambda} f(t) = \frac{\Im(\alpha)}{1-\alpha}\int_t^b E_\alpha\left(-\frac{\alpha}{1-\alpha}(\tau-t)^\alpha\right) f^{(1)}(\tau)\,d\tau, \tag{3.444} $$

the left-sided general fractional derivative via Mittag-Leffler function is defined as [236]

$$ {}_{MlA}^{RL}D_{a+}^{\alpha,\lambda} f(t) = \frac{\Im(\alpha)}{1-\alpha}\frac{d}{dt}\int_a^t E_\alpha\left(-\frac{\alpha}{1-\alpha}(t-\tau)^\alpha\right) f(\tau)\,d\tau, \tag{3.445} $$

and the right-sided general fractional derivative via Mittag-Leffler function as [236]

$$ {}_{MlA}^{RL}D_{b-}^{\alpha,\kappa,\lambda} f(t) = (-1)^\kappa \frac{\Im(\alpha)}{1-\alpha}\frac{d}{dt}\int_t^b E_\alpha\left(-\frac{\alpha}{1-\alpha}(\tau-t)^\alpha\right) f(\tau)\,d\tau, \tag{3.446} $$

where $\Im(\alpha)$ is the normalization constant with the conditions $\Im(0) = 1$ and $\Im(1) = 1$.

3.6.2 Hille–Tamarkin General Fractional Integrals

As the inverse operator of the Hille–Tamarkin general fractional derivative, we have the following results:

Definition 3.45 Let$\alpha > 0$, $\kappa = [\alpha]+1$, $-\infty < a < b < \infty$ and $\lambda \in \mathbb{R}$.

The left-sided Hille–Tamarkin general fractional integral, as the inverse operator of the Hille–Tamarkin general fractional derivative, is defined as [92]

$$ {}_{MI}^{RL}I_{a+}^{\alpha,\kappa,\lambda} f(t) = \int_a^t \frac{1}{(t-\tau)^{\alpha+1-\kappa}} E_{\alpha,\kappa-\alpha}^{-1}\left(-\lambda(t-\tau)^{\alpha}\right) f(\tau)\, d\tau, \tag{3.447} $$

and the right-sided Hille–Tamarkin general fractional integral as

$$ {}_{MI}^{RL}I_{b-}^{\alpha,\kappa,\lambda} f(t) = (-1)^{\kappa} \int_t^b \frac{1}{(\tau-t)^{\alpha+1-\kappa}} E_{\alpha,\kappa-\alpha}^{-1}\left(-\lambda(\tau-t)^{\alpha}\right) f(\tau)\, d\tau. \tag{3.448} $$

For $\kappa = 1$ the left-sided Hille–Tamarkin general fractional integral, as the inverse operator of the Hille–Tamarkin general fractional derivative, is defined as [92, 237]

$$ \begin{aligned} &{}_{MI}^{RL}I_{a+}^{\alpha,\lambda} f(t) \\ &= \int_a^t \tfrac{1}{(t-\tau)^{\alpha}} E_{\alpha,1-\alpha}^{-1}\left(-\lambda(t-\tau)^{\alpha}\right) f(\tau)\, d\tau \\ &= f(t) + \tfrac{\lambda}{\Gamma(\alpha)} \int_a^t \tfrac{f(\tau)}{(t-\tau)^{1-\alpha}} d\tau, \end{aligned} \tag{3.449} $$

and the right-sided Hille–Tamarkin general fractional integral as

$$ \begin{aligned} &{}_{MI}^{RL}I_{b-}^{\alpha,\lambda} f(t) \\ &= -\int_t^b \tfrac{1}{(\tau-t)^{\alpha}} E_{\alpha,1-\alpha}^{-1}\left(-\lambda(\tau-t)^{\alpha}\right) f(\tau)\, d\tau \\ &= -\left[f(t) + \tfrac{\lambda}{\Gamma(\alpha)} \int_t^b \tfrac{f(\tau)}{(\tau-t)^{1-\alpha}} d\tau\right]. \end{aligned} \tag{3.450} $$

We have the properties of the Hille–Tamarkin general fractional calculus as follows:

Property 3.4 Let $\alpha > 0$, $\kappa = [\alpha]+1$, and $\lambda \in \mathbb{R}$. Then we have [92]

$$ L\left\{{}_{MI}^{RL}D_{0+}^{\alpha,\lambda} f(t)\right\} = \left(1+\lambda s^{-\alpha}\right)^{-1} f(s) - {}_{MI}^{RL}I_{0+}^{\alpha,1,\lambda} f(+0). \tag{3.451} $$

Property 3.5 Let $1 > \alpha > 0$ and $\lambda \in \mathbb{R}$. Then we have [92]

$$ L\left\{{}_{MI}^{RL}D_{0+}^{\alpha,\lambda} f(t)\right\} = \left(1+\lambda s^{-\alpha}\right)^{-1} f(s) - {}_{MI}^{RL}I_{0+}^{\alpha,\lambda} f(+0). \tag{3.452} $$

Property 3.6 Let $\alpha > 0$, $\kappa = [\alpha]+1$, and $\lambda \in \mathbb{R}$. Then we have [92]

$$L\left\{{}_{Ml}^{RL}D_{0+}^{\alpha,\kappa,\lambda} f(t)\right\} = s^{\kappa-1}\left(1+\lambda s^{-\alpha}\right)^{-1} f(s) - \sum_{\kappa=0}^{\kappa-1} s^{\kappa-\mu-1}\left(D^{\mu}\left[{}_{Ml}^{RL}I_{0+}^{\alpha,\kappa,\lambda} f(+0)\right]\right). \tag{3.453}$$

Property 3.7 Let $\alpha > 0$, $\kappa = [\alpha]+1$ and $\lambda \in \mathbb{R}$. Then we have [92]

$$L\left\{{}_{Ml}^{LSC}D_{0+}^{\alpha,\kappa,\lambda} f(t)\right\} = s^{-1}\left(1+\lambda s^{-\alpha}\right)^{-1}\left(s^{\kappa} f(s) - \sum_{j=1}^{\kappa} s^{\kappa-j} f^{(j-1)}(0)\right). \tag{3.454}$$

3.6.3 Liouville–Weyl–Hille–Tamarkin Type General Fractional Calculus

Based on the Liouville-Weyl fractional calculus, the Liouville–Weyl–Hille–Tamarkin type general fractional calculus was proposed in the book [92].

Definition 3.46 Let $1 \geq \alpha \geq 0$ and $\lambda \in \mathbb{R}$.

The left-sided Liouville–Weyl–Hille–Tamarkin type general fractional derivative is defined as

$$ {}_{Mlr}^{RL}D_{+}^{\alpha,\lambda} f(t) = \frac{d}{dt}\int_{-\infty}^{t} E_{\alpha}\left(-\lambda(t-\tau)^{\alpha}\right) f(\tau)\, d\tau, \tag{3.455}$$

and the right-sided Liouville–Weyl–Hille–Tamarkin type general fractional derivative as

$$ {}_{Mlr}^{RL}D_{-}^{\alpha,\lambda} f(t) = -\frac{d}{dt}\int_{t}^{\infty} E_{\alpha}\left(-\lambda(\tau-t)^{\alpha}\right) f(\tau)\, d\tau. \tag{3.456}$$

The result was discovered by Yang in 2019 [92].

Definition 3.47 Let $\alpha \geq 0$, $\kappa = [\alpha]+1$, and $\lambda \in \mathbb{R}$.

The left-sided Liouville–Weyl–Hille–Tamarkin type general fractional derivative is defined as

$$ {}_{Mlr}^{RL}D_{+}^{\alpha,\kappa,\lambda} f(t) = \frac{d^{\kappa}}{dt^{\kappa}}\int_{-\infty}^{t} E_{\alpha}\left(-\lambda(t-\tau)^{\alpha}\right) f(\tau)\, d\tau, \tag{3.457}$$

and the right-sided Liouville–Weyl–Hille–Tamarkin type general fractional derivative as

$$ {}_{Ml}^{RL}D_{-}^{\alpha,\kappa,\lambda} f(t) = (-1)^{\kappa} \frac{d^{\kappa}}{dt^{\kappa}} \int_{t}^{\infty} E_{\alpha}\left(-\lambda(\tau - t)^{\alpha}\right) f(\tau)\, d\tau. \tag{3.458} $$

The result was discovered by Yang in 2019 [92]

Definition 3.48 Let $1 \geq \alpha \geq 0$ and $\lambda \in \mathbb{R}$.

The left-sided Liouville–Sonine–Caputo type general fractional derivative via Mittag-Leffler function is defined as

$$ {}_{Ml}^{LSC}D_{+}^{\alpha,\lambda} f(t) = \int_{-\infty}^{t} E_{\alpha}\left(-\lambda(t - \tau)^{\alpha}\right) f^{(1)}(\tau)\, d\tau, \tag{3.459} $$

and the right-sided Liouville–Sonine–Caputo type general fractional derivative via Mittag-Leffler function as

$$ {}_{Ml}^{LSC}D_{-}^{\alpha,\lambda} f(t) = (-1)^{\kappa} \int_{t}^{\infty} E_{\alpha}\left(-\lambda(\tau - t)^{\alpha}\right) f^{(1)}(\tau)\, d\tau. \tag{3.460} $$

The result was discovered by Yang in 2019 [92]

Definition 3.49 Let $\alpha \geq 0$, $\kappa = [\alpha] + 1$ and $\lambda \in \mathbb{R}$.

The left-sided Liouville–Sonine–Caputo type general fractional derivative via Mittag-Leffler function is defined as

$$ {}_{Ml}^{LSC}D_{+}^{\alpha,\kappa,\lambda} f(t) = \int_{-\infty}^{t} E_{\alpha}\left(-\lambda(t - \tau)^{\alpha}\right) f^{(\kappa)}(\tau)\, d\tau, \tag{3.461} $$

and the right-sided Liouville–Sonine–Caputo type general fractional derivative via Mittag-Leffler function as

$$ {}_{Ml}^{LSC}D_{-}^{\alpha,\kappa,\lambda} f(t) = (-1)^{\kappa} \int_{t}^{\infty} E_{\alpha}\left(-\lambda(\tau - t)^{\alpha}\right) f^{(\kappa)}(\tau)\, d\tau. \tag{3.462} $$

The result was discovered by Yang in 2019 [92].

Definition 3.50 Let $\alpha > 0$, $\kappa = [\alpha]+1$, $-\infty < a < b < \infty$ and $\lambda \in \mathbb{R}$.

The left-sided Liouville–Weyl–Hille–Tamarkin type general fractional integral, as the inverse operator of the Liouville–Weyl–Hille–Tamarkin type general fractional derivative, is defined as

$$ {}_{Ml}^{RL}I_{+}^{\alpha,\kappa,\lambda} f(t) = \int_{-\infty}^{t} \frac{1}{(t - \tau)^{\alpha+1-\kappa}} E_{\alpha,\kappa-\alpha}^{-1}\left(-\lambda(t - \tau)^{\alpha}\right) f(\tau)\, d\tau, \tag{3.463} $$

and the right-sided Liouville–Weyl–Hille–Tamarkin type general fractional integral as

$$ {}_{Ml}^{RL}I_{-}^{\alpha,\kappa,\lambda} f(t) = (-1)^{\kappa} \int_t^{\infty} \frac{1}{(\tau - t)^{\alpha+1-\kappa}} E_{\alpha,\kappa-\alpha}^{-1} \left(-\lambda (\tau - t)^{\alpha}\right) f(\tau)\, d\tau. \tag{3.464} $$

The result was discovered by Yang in 2019 [92].

3.6.4 *Hilfer–Hille–Tamarkin Type General Fractional Derivative with Nonsingular Kernel*

Definition 3.51 Let $0 < \alpha < 1$ and $0 \leq \beta \leq 1$.

The left-sided Hilfer–Hille–Tamarkin type general fractional derivative is defined as

$$ {}_{HHT}D_{a+}^{\alpha,\beta,\lambda} f(t) = \left({}_{ML}I_{a+}^{\beta(1-\alpha),\lambda} \frac{d}{dt} {}_{ML}I_{a+}^{(1-\beta)(1-\alpha),\lambda}\right) f(t), \tag{3.465} $$

where

$$ {}_{ML}I_{a+}^{\beta(1-\alpha),\lambda} f(t) = \int_a^t E_{\alpha}\left(-\lambda (t - \tau)^{\beta(1-\alpha)}\right) f(\tau)\, d\tau, \tag{3.466} $$

and

$$ {}_{ML}I_{a+}^{(1-\beta)(1-\alpha),\lambda} f(t) = \int_a^t E_{\alpha}\left(-\lambda (t - \tau)^{(1-\beta)(1-\alpha)}\right) f(\tau)\, d\tau, \tag{3.467} $$

and the right-sided Hilfer–Hille–Tamarkin type general fractional derivative as

$$ {}_{HHT}D_{b-}^{\alpha,\beta,\lambda} f(t) = -\left({}_{ML}I_{b-}^{\beta(1-\alpha),\lambda} \frac{d}{dt} {}_{ML}I_{b-}^{(1-\beta)(1-\alpha),\lambda}\right) f(t), \tag{3.468} $$

where

$$ {}_{ML}I_{b-}^{\beta(1-\alpha),\lambda} f(t) = \int_t^b E_{\alpha}\left(-\lambda (\tau - t)^{\beta(1-\alpha)}\right) f(\tau)\, d\tau, \tag{3.469} $$

and

$$ {}_{ML}I_{b-}^{(1-\beta)(1-\alpha),\lambda} f(t) = \int_t^b E_{\alpha}\left(-\lambda (\tau - t)^{(1-\beta)(1-\alpha)}\right) f(\tau)\, d\tau. \tag{3.470} $$

The result was discovered by Yang in 2020 [92].

Property 3.8 Suppose that $0 < \alpha < 1$ and $0 \le \beta \le 1$, then we have

$$L\left\{ {}_{HHT}D_{0+}^{\alpha,\beta,\lambda} f(t) \right\}(s) = s^{\alpha} f(s) - s^{1-\beta(1-\alpha)} e^{-\frac{\lambda}{s}} {}_{SH}I_{0+}^{(1-\beta)(1-\alpha),\lambda} f(0). \tag{3.471}$$

Proof

$$\begin{aligned}
& L\left\{ {}_{HHT}D_{0+}^{\alpha,\beta,\lambda} f(t) \right\}(s) \\
& = L\left\{ \left({}_{ML}I_{0+}^{\beta(1-\alpha),\lambda} \frac{d}{dt} {}_{ML}I_{0+}^{(1-\beta)(1-\alpha),\lambda} \right) f(t) \right\} \\
& = s^{-1}\left(1+\lambda s^{-\beta(1-\alpha)}\right)^{-1} \left(s \times s^{-1}\left(1+\lambda s^{-(1-\beta)(1-\alpha)}\right)^{-1} f(s) \right. \\
& \left. - {}_{SH}I_{0+}^{(1-\beta)(1-\alpha),\lambda} f(0) \right) \\
& = s^{-1}\left(1+\lambda s^{-\beta(1-\alpha)}\right)^{-1} \left(\left(1+\lambda s^{-(1-\beta)(1-\alpha)}\right)^{-1} f(s) \right. \\
& \left. - {}_{ML}I_{0+}^{(1-\beta)(1-\alpha),\lambda} f(0) \right).
\end{aligned} \tag{3.472}$$

□

Thus, we give the proof.

The result was discovered by Yang in 2020 [92].

Definition 3.52 Let $0 < \alpha < 1$ and $0 \le \beta \le 1$.

The left-sided Hilfer–Hille–Tamarkin type general fractional derivative in the real line is defined as

$${}_{HHT}D_{+}^{\alpha,\beta,\lambda} f(t) = \left({}_{ML}I_{+}^{\beta(1-\alpha),\lambda} \frac{d}{dt} {}_{ML}I_{+}^{(1-\beta)(1-\alpha),\lambda} \right) f(t), \tag{3.473}$$

where

$${}_{ML}I_{+}^{\beta(1-\alpha),\lambda} f(t) = \int_{-\infty}^{t} E_{\alpha}\left(-\lambda (t-\tau)^{\beta(1-\alpha)}\right) f(\tau)\, d\tau, \tag{3.474}$$

and

$${}_{ML}I_{+}^{(1-\beta)(1-\alpha),\lambda} f(t) = \int_{-\infty}^{t} E_{\alpha}\left(-\lambda (t-\tau)^{(1-\beta)(1-\alpha)}\right) f(\tau)\, d\tau, \tag{3.475}$$

and the right-sided Hilfer–Hille–Tamarkin type general fractional derivative in the real line as

$$_{HHT}D_{-}^{\alpha,\beta,\lambda} f(t) = -\left({}_{ML}I_{-}^{\beta(1-\alpha),\lambda} \frac{d}{dt} {}_{ML}I_{-}^{(1-\beta)(1-\alpha),\lambda} \right) f(t), \tag{3.476}$$

where

$$_{ML}I_{-}^{\beta(1-\alpha),\lambda} f(t) = \int_{t}^{\infty} E_{\alpha}\left(-\lambda(\tau - t)^{\beta(1-\alpha)}\right) f(\tau)\,d\tau, \tag{3.477}$$

and

$$_{ML}I_{-}^{(1-\beta)(1-\alpha),\lambda} f(t) = \int_{t}^{\infty} E_{\alpha}\left(-\lambda(\tau - t)^{(1-\beta)(1-\alpha)}\right) f(\tau)\,d\tau. \tag{3.478}$$

The result was discovered by Yang in 2020 [92].

3.6.5 Hille–Tamarkin General Fractional Derivative with Respect to Another Function

Definition 3.53 Let $h^{(1)}(t) > 0$, $1 \geq \alpha \geq 0$ and $\lambda \in \mathbb{R}$.

The left-sided Hille–Tamarkin general fractional derivative with respect to another function is defined as

$$_{Ml}^{RL}D_{a+,h}^{\alpha,\lambda} f(t) = \left(\frac{1}{h^{(1)}(t)}\frac{d}{dt}\right)\int_{a}^{t} E_{\alpha}\left(-\lambda(h(t) - h(\tau))^{\alpha}\right) h^{(1)}(\tau) f(\tau)\,d\tau, \tag{3.479}$$

and the right-sided Hille–Tamarkin general fractional derivative as

$$_{Ml}^{RL}D_{b-,h}^{\alpha,\lambda} f(t) = \left(-\frac{1}{h^{(1)}(t)}\frac{d}{dt}\right)\int_{t}^{b} E_{\alpha}\left(-\lambda(h(\tau) - h(t))^{\alpha}\right) h^{(1)}(\tau) f(\tau)\,d\tau. \tag{3.480}$$

The result was discovered by Yang in 2020 [92].

Definition 3.54 Let $1 \geq \alpha \geq 0$ and $\lambda \in \mathbb{R}$.

The left-sided Hille–Tamarkin general fractional derivative with power function is defined as

$$_{\Omega}^{RL}D_{a+}^{\alpha,\lambda} f(t) = \left(\frac{1}{t^{\beta-1}}\frac{d}{dt}\right)\int_{a}^{t} E_{\alpha}\left(-\lambda\left(t^{\beta} - \tau^{\beta}\right)^{\alpha}\right) \tau^{\beta-1} f(\tau)\,d\tau, \tag{3.481}$$

and the right-sided Hille–Tamarkin general fractional derivative as

$$ {}_{\Omega}^{RL}D_{b-,h}^{\alpha,\lambda}f(t)=\left(-\frac{1}{t^{\beta-1}}\frac{d}{dt}\right)\int_{t}^{b}E_{\alpha}\left(-\lambda\left(t^{\beta}-\tau^{\beta}\right)^{\alpha}\right)\tau^{\beta-1}f(\tau)\,d\tau. \tag{3.482} $$

Definition 3.55 Let $h^{(1)}(t)>0$, $\alpha\geq 0$, $\kappa=[\alpha]+1$ and $\lambda\in\mathbb{R}$.

The left-sided Hille–Tamarkin general fractional derivative with respect to another function is defined as

$$ {}_{Ml}^{RL}D_{a+,h}^{\alpha,\kappa,\lambda}f(t)=\left(\frac{1}{h^{(1)}(t)}\frac{d}{dt}\right)^{\kappa}\int_{a}^{t}E_{\alpha}\left(-\lambda\left(h(t)-h(\tau)\right)^{\alpha}\right)h^{(1)}(\tau)\,f(\tau)\,d\tau, \tag{3.483} $$

and the right-sided Hille–Tamarkin general fractional derivative with respect to another function as

$$ {}_{Ml}^{RL}D_{b-,h}^{\alpha,\kappa,\lambda}f(t)=\left(\frac{-1}{h^{(1)}(t)}\frac{d}{dt}\right)^{\kappa}\int_{t}^{b}E_{\alpha}\left(-\lambda\left(h(\tau)-h(t)\right)^{\alpha}\right)h^{(1)}(\tau)\,f(\tau)\,d\tau. \tag{3.484} $$

The result was discovered by Yang in 2020 [92].

Definition 3.56 Let $h^{(1)}(t)>0$, $1\geq\alpha\geq 0$ and $\lambda\in\mathbb{R}$.

The left-sided Liouville–Sonine–Caputo type general fractional derivative with respect to another function via Mittag-Leffler function is defined as

$$ {}_{Ml}^{LSC}D_{a+,h}^{\alpha,\lambda}f(t)=\int_{a}^{t}E_{\alpha}\left(-\lambda\left(h(t)-h(\tau)\right)^{\alpha}\right)h^{(1)}(\tau)\left(\frac{1}{h^{(1)}(\tau)}\frac{d}{d\tau}\right)f(\tau)\,d\tau, \tag{3.485} $$

and the right-sided Liouville–Sonine–Caputo type general fractional derivative with respect to another function via Mittag-Leffler function as

$$ {}_{Ml}^{LSC}D_{b-,h}^{\alpha,\lambda}f(t)=-\int_{t}^{b}E_{\alpha}\left(-\lambda\left(h(\tau)-h(t)\right)^{\alpha}\right)h^{(1)}(\tau)\left(\frac{1}{h^{(1)}(\tau)}\frac{d}{d\tau}\right)f(\tau)\,d\tau. \tag{3.486} $$

The result was discovered by Yang in 2020 [92].

Definition 3.57 Let $h^{(1)}(t)>0$, $1\geq\alpha\geq 0$ and $\lambda\in\mathbb{R}$.

The left-sided Liouville–Sonine–Caputo type general fractional derivative with respect to another function via Mittag-Leffler function is defined as

$$ {}_{\Omega}^{LSC}D_{a+}^{\alpha,\lambda}f(t)=\int_{a}^{t}E_{\alpha}\left(-\lambda\left(t^{\beta}-\tau^{\beta}\right)^{\alpha}\right)f^{(1)}(\tau)\,d\tau, \tag{3.487} $$

and the right-sided Liouville–Sonine–Caputo type general fractional derivative with respect to another function via Mittag-Leffler function as

$$ {}_{\Omega}^{LSC}D_{b-}^{\alpha,\lambda} f(t) = -\int_t^b E_\alpha\left(-\lambda\left(\tau^\beta - t^\beta\right)^\alpha\right) f^{(1)}(\tau)\, d\tau. \tag{3.488} $$

Definition 3.58 Let $h^{(1)}(t) > 0$, $\alpha \geq 0$, $\kappa = [\alpha] + 1$ and $\lambda \in \mathbb{R}$.

The left-sided Liouville–Sonine–Caputo type general fractional derivative with respect to another function via Mittag-Leffler function is defined as

$$ \begin{aligned} &{}_{Ml}^{LSC}D_{a+,h}^{\alpha,\kappa,\lambda} f(t) \\ &= \int_a^t E_\alpha\left(-\lambda\left(h(t) - h(\tau)\right)^\alpha\right) h^{(1)}(\tau)\left(\frac{1}{h^{(1)}(\tau)}\frac{d}{d\tau}\right)^\kappa f(\tau)\, d\tau, \end{aligned} \tag{3.489} $$

and the right-sided Liouville–Sonine–Caputo type general fractional derivative with respect to another function via Mittag-Leffler function as

$$ \begin{aligned} &{}_{Ml}^{LSC}D_{b-,h}^{\alpha,\kappa,\lambda} f(t) \\ &= (-1)^\kappa \int_t^b E_\alpha\left(-\lambda\left(h(\tau) - h(t)\right)^\alpha\right) h^{(1)}(\tau)\left(\frac{1}{h^{(1)}(\tau)}\frac{d}{d\tau}\right)^\kappa f(\tau)\, d\tau. \end{aligned} \tag{3.490} $$

The result was discovered by Yang in 2020 [92].

3.6.6 *Hille–Tamarkin General Fractional Integrals with Respect to Another Function*

As the inverse operator of the Hille–Tamarkin general fractional derivative with respect to another function, we have the following results:

Definition 3.59 Let $h^{(1)}(t) > 0$, $\alpha > 0$, $\kappa = [\alpha] + 1$, $-\infty < a < b < \infty$ and $\lambda \in \mathbb{R}$.

The left-sided Hille–Tamarkin general fractional integral with respect to another function, as the inverse operator of the Hille–Tamarkin general fractional derivative with respect to another function, is defined as

$$ {}_{Ml}^{RL}I_{a+,h}^{\alpha,\kappa,\lambda} f(t) = \int_a^t \frac{h^{(1)}(\tau)}{\left(h(t) - h(\tau)\right)^{\alpha+1-\kappa}} E_{\alpha,\kappa-\alpha}^{-1}\left(-\lambda\left(h(t) - h(\tau)\right)^\alpha\right) f(\tau)\, d\tau, \tag{3.491} $$

and the right-sided Hille–Tamarkin general fractional integral with respect to another function as

$$
\begin{aligned}
&{}_{Ml}^{RL}I_{b-,h}^{\alpha,\kappa,\lambda} f(t) \\
&= (-1)^{\kappa} \int_t^b \frac{h^{(1)}(\tau)}{(h(\tau)-h(t))^{\alpha+1-\kappa}} E_{\alpha,\kappa-\alpha}^{-1}\left(-\lambda(h(\tau)-h(t))^{\alpha}\right) f(\tau)\, d\tau.
\end{aligned}
\tag{3.492}
$$

The result was discovered by Yang in 2020 [92].

For $h^{(1)}(t) > 0$ and $\kappa = 1$ the left-sided Hille–Tamarkin general fractional integral with respect to another function, as the inverse operator of the Hille–Tamarkin general fractional derivative with respect to another function, is defined as

$$
{}_{Ml}^{RL}I_{a+,h}^{\alpha,\lambda} f(t) = f(t) + \frac{\lambda}{\Gamma(\alpha)} \int_a^t \frac{h^{(1)}(\tau) f(\tau)}{(h(t)-h(\tau))^{1-\alpha}} d\tau, \tag{3.493}
$$

and the right-sided Hille–Tamarkin general fractional integral with respect to another function as

$$
{}_{Ml}^{RL}I_{b-,h}^{\alpha,\lambda} f(t) = -\left[f(t) + \frac{\lambda}{\Gamma(\alpha)} \int_t^b \frac{h^{(1)}(\tau) f(\tau)}{(h(\tau)-h(t))^{1-\alpha}} d\tau\right]. \tag{3.494}
$$

The result was discovered by Yang in 2020 [92].

The left-sided Hille–Tamarkin general fractional integral with respect to another function is defined as

$$
{}_{\Omega}^{RL}I_{a+}^{\alpha,\lambda} f(t) = f(t) + \frac{\lambda\beta}{\Gamma(\alpha)} \int_a^t \frac{\tau^{\beta-1} f(\tau)}{\left(t^{\beta}-\tau^{\beta}\right)^{1-\alpha}} d\tau, \tag{3.495}
$$

and the right-sided Hille–Tamarkin general fractional integral with respect to another function as

$$
{}_{\Omega}^{RL}I_{b-}^{\alpha,\lambda} f(t) = -\left[f(t) + \frac{\lambda\beta}{\Gamma(\alpha)} \int_t^b \frac{\tau^{\beta-1} f(\tau)}{\left(\tau^{\beta}-t^{\beta}\right)^{1-\alpha}} d\tau\right]. \tag{3.496}
$$

3.6.7 Liouville–Weyl–Hille–Tamarkin Type General Fractional Calculus with Respect to Another Function

We now present the Liouville–Weyl–Hille–Tamarkin type general fractional calculus with respect to another function.

Definition 3.60 Let $h^{(1)}(t) > 0$, $1 \geq \alpha \geq 0$ and $\lambda \in \mathbb{R}$.

The left-sided Liouville–Weyl–Hille–Tamarkin type general fractional derivative with respect to another function is defined as

$$ {}_{Mlr}^{RL}D_{+,h}^{\alpha,\lambda} f(t) = \left(\frac{1}{h^{(1)}(t)}\frac{d}{dt}\right)\int_{-\infty}^{t} E_{\alpha}\left(-\lambda\left(h(t)-h(\tau)\right)^{\alpha}\right)h^{(1)}(\tau) f(\tau)\, d\tau, \tag{3.497} $$

and the right-sided Liouville–Weyl–Hille–Tamarkin type general fractional derivative with respect to another function as

$$ {}_{Mlr}^{RL}D_{-}^{\alpha,\lambda} f(t) = \left(-\frac{1}{h^{(1)}(t)}\frac{d}{dt}\right)\int_{t}^{\infty} E_{\alpha}\left(-\lambda\left(h(\tau)-h(t)\right)^{\alpha}\right)h^{(1)}(\tau) f(\tau)\, d\tau. \tag{3.498} $$

The result was discovered by Yang in 2020 [92].

Definition 3.61 Let $h^{(1)}(t) > 0$, $\alpha \geq 0$, $\kappa = [\alpha] + 1$ and $\lambda \in \mathbb{R}$.

The left-sided Liouville–Weyl–Hille–Tamarkin type general fractional derivative with respect to another function is defined as

$$ {}_{Mlr}^{RL}D_{+,h}^{\alpha,\kappa,\lambda} f(t) = \left(\frac{1}{h^{(1)}(t)}\frac{d}{dt}\right)^{\kappa}\int_{-\infty}^{t} E_{\alpha}\left(-\lambda\left(h(t)-h(\tau)\right)^{\alpha}\right)h^{(1)}(\tau) f(\tau)\, d\tau, \tag{3.499} $$

and the right-sided Liouville–Weyl–Hille–Tamarkin type general fractional derivative with respect to another function as

$$ {}_{Ml}^{RL}D_{-,h}^{\alpha,\kappa,\lambda} f(t) = \left(\frac{-1}{h^{(1)}(t)}\frac{d}{dt}\right)^{\kappa}\int_{t}^{\infty} E_{\alpha}\left(-\lambda\left(h(\tau)-h(t)\right)^{\alpha}\right)h^{(1)}(\tau) f(\tau)\, d\tau. \tag{3.500} $$

The result was discovered by Yang in 2020 [92].

Definition 3.62 Let $h^{(1)}(t) > 0$, $1 \geq \alpha \geq 0$ and $\lambda \in \mathbb{R}$.

The left-sided Liouville–Sonine–Caputo type general fractional derivative with respect to another function via Mittag-Leffler function in the real line is defined as [25]

$$ {}_{Ml}^{LSC}D_{+,h}^{\alpha,\lambda} f(t) = \int_{-\infty}^{t} E_{\alpha}\left(-\lambda\left(h(t)-h(\tau)\right)^{\alpha}\right)h^{(1)}(\tau)\left(\frac{1}{h^{(1)}(\tau)}\frac{d}{d\tau}\right) f(\tau)\, d\tau, \tag{3.501} $$

and the right-sided Liouville–Sonine–Caputo type general fractional derivative with respect to another function via Mittag-Leffler function in the real line as

$$ {}_{Ml}^{LSC}D_{-,h}^{\alpha,\lambda} f(t) = \int_t^{\infty} E_\alpha\left(-\lambda\left(h(\tau) - h(t)\right)^\alpha\right) h^{(1)}(\tau) \left(\frac{-1}{h^{(1)}(\tau)}\frac{d}{d\tau}\right) f(\tau)\, d\tau. \tag{3.502} $$

The result was discovered by Yang in 2020 [92].

Definition 3.63 Let $h^{(1)}(t) > 0$, $\alpha \geq 0$, $\kappa = [\alpha] + 1$ and $\lambda \in \mathbb{R}$.

The left-sided Liouville–Sonine–Caputo type general fractional derivative with respect to another function via Mittag-Leffler function is defined as

$$ {}_{Ml}^{LSC}D_{+,h}^{\alpha,\kappa,\lambda} f(t) = \int_{-\infty}^{t} E_\alpha\left(-\lambda\left(h(t) - h(\tau)\right)^\alpha\right) h^{(1)}(\tau) \left(\frac{1}{h^{(1)}(\tau)}\frac{d}{d\tau}\right)^\kappa f(\tau)\, d\tau, \tag{3.503} $$

and the right-sided Liouville–Sonine–Caputo type general fractional derivative with respect to another function via Mittag-Leffler function as

$$ {}_{Ml}^{LSC}D_{-,h}^{\alpha,\kappa,\lambda} f(t) = \int_t^{\infty} E_\alpha\left(-\lambda\left(h(\tau) - h(t)\right)^\alpha\right) h^{(1)}(\tau) \left(\frac{-1}{h^{(1)}(\tau)}\frac{d}{d\tau}\right)^\kappa f(\tau)\, d\tau. \tag{3.504} $$

The result was discovered by Yang in 2020 [92].

Definition 3.64 Let $h^{(1)}(t) > 0$, $\alpha > 0$, $\kappa = [\alpha]+1$, $-\infty < a < b < \infty$ and $\lambda \in \mathbb{R}$.

The left-sided Liouville–Weyl–Hille–Tamarkin type general fractional integral with respect to another function, as the inverse operator of the Liouville–Weyl–Hille–Tamarkin type general fractional derivative with respect to another function, is defined as

$$ {}_{Ml}^{RL}I_{+,h}^{\alpha,\kappa,\lambda} f(t) = \int_{-\infty}^{t} \frac{h^{(1)}(\tau)}{\left(h(t) - h(\tau)\right)^{\alpha+1-\kappa}} E_{\alpha,\kappa-\alpha}^{-1}\left(-\lambda\left(h(t) - h(\tau)\right)^\alpha\right) f(\tau)\, d\tau, \tag{3.505} $$

and the right-sided Liouville–Weyl–Hille–Tamarkin type general fractional integral with respect to another function as

$$ {}_{Ml}^{RL}I_{-,h}^{\alpha,\kappa,\lambda} f(t) = (-1)^\kappa \int_t^{\infty} \frac{h^{(1)}(\tau)}{\left(h(\tau) - h(t)\right)^{\alpha+1-\kappa}} E_{\alpha,\kappa-\alpha}^{-1}\left(-\lambda\left(h(\tau) - h(t)\right)^\alpha\right) f(\tau)\, d\tau. \tag{3.506} $$

The result was discovered by Yang in 2020 [92].

3.6.8 Hilfer–Hille–Tamarkin Type General Fractional Derivative with Respect to Another Function

We now propose new general fractional derivatives with Mittag-Leffler kernel with respect to another function, which is called the Hilfer–Hille–Tamarkin type general fractional derivative with respect to another function.

Definition 3.65 Let $h^{(1)}(t) > 0$, $0 < \alpha < 1$ and $0 \le \beta \le 1$.

The left-sided Hilfer–Hille–Tamarkin type general fractional derivative with respect to another function is defined as

$$ {}_{HHT}D_{a+,h}^{\alpha,\beta,\lambda} f(t) = \left({}_{ML}I_{a+,h}^{\beta(1-\alpha),\lambda} \left(\frac{1}{h^{(1)}(t)} \frac{d}{dt} \right) {}_{ML}I_{a+,h}^{(1-\beta)(1-\alpha),\lambda} \right) f(t), \tag{3.507} $$

where

$$ {}_{ML}I_{a+,h}^{\beta(1-\alpha),\lambda} f(t) = \int_a^t E_\alpha \left(-\lambda (h(t) - h(\tau))^{\beta(1-\alpha)} \right) h^{(1)}(\tau) f(\tau) d\tau, \tag{3.508} $$

and

$$ {}_{ML}I_{a+,h}^{(1-\beta)(1-\alpha),\lambda} f(t) = \int_a^t E_\alpha \left(-\lambda (h(t) - h(\tau))^{(1-\beta)(1-\alpha)} \right) h^{(1)}(\tau) f(\tau) d\tau, \tag{3.509} $$

and the right-sided Hilfer–Hille–Tamarkin type general fractional derivative with respect to another function as

$$ {}_{HHT}D_{b-,h}^{\alpha,\beta,\lambda} f(t) = -\left({}_{ML}I_{b-}^{\beta(1-\alpha),\lambda} \left(\frac{1}{h^{(1)}(t)} \frac{d}{dt} \right) {}_{ML}I_{b-}^{(1-\beta)(1-\alpha),\lambda} \right) f(t), \tag{3.510} $$

where

$$ {}_{ML}I_{b-,h}^{\beta(1-\alpha),\lambda} f(t) = \int_t^b E_\alpha \left(-\lambda (h(\tau) - h(t))^{\beta(1-\alpha)} \right) h^{(1)}(\tau) f(\tau) d\tau, \tag{3.511} $$

and

$$ {}_{ML}I_{b-,h}^{(1-\beta)(1-\alpha),\lambda} f(t) = \int_t^b E_\alpha \left(-\lambda (h(\tau) - h(t))^{(1-\beta)(1-\alpha)} \right) h^{(1)}(\tau) f(\tau) d\tau. \tag{3.512} $$

The result was discovered by Yang in 2020 [92].

Definition 3.66 Let $h^{(1)}(t) > 0, 0 < \alpha < 1$ and $0 \leq \beta \leq 1$.

The left-sided Hilfer–Hille–Tamarkin type general fractional derivative with respect to another function is defined as

$$ {}_{HHT}D_{+,h}^{\alpha,\beta,\lambda} f(t) = \left({}_{ML}I_{+,h}^{\beta(1-\alpha),\lambda} \left(\frac{1}{h^{(1)}(t)} \frac{d}{dt} \right) {}_{ML}I_{+,h}^{(1-\beta)(1-\alpha),\lambda} \right) f(t), \tag{3.513} $$

where

$$ {}_{ML}I_{+,h}^{\beta(1-\alpha),\lambda} f(t) = \int_{-\infty}^{t} E_{\alpha} \left(-\lambda (h(t) - h(\tau))^{\beta(1-\alpha)} \right) h^{(1)}(\tau) f(\tau) d\tau, \tag{3.514} $$

and

$$ {}_{ML}I_{+,h}^{(1-\beta)(1-\alpha),\lambda} f(t) = \int_{-\infty}^{t} E_{\alpha} \left(-\lambda (h(t) - h(\tau))^{(1-\beta)(1-\alpha)} \right) h^{(1)}(\tau) f(\tau) d\tau, \tag{3.515} $$

and the right-sided Hilfer–Hille–Tamarkin type general fractional derivative with respect to another function as

$$ {}_{HHT}D_{-,h}^{\alpha,\beta,\lambda} f(t) = -\left({}_{ML}I_{-,h}^{\beta(1-\alpha),\lambda} \left(\frac{1}{h^{(1)}(t)} \frac{d}{dt} \right) {}_{ML}I_{-,h}^{(1-\beta)(1-\alpha),\lambda} \right) f(t), \tag{3.516} $$

where

$$ {}_{ML}I_{-,h}^{\beta(1-\alpha),\lambda} f(t) = \int_{t}^{\infty} E_{\alpha} \left(-\lambda (h(\tau) - h(t))^{\beta(1-\alpha)} \right) h^{(1)}(\tau) f(\tau) d\tau, \tag{3.517} $$

and

$$ {}_{ML}I_{-,h}^{(1-\beta)(1-\alpha),\lambda} f(t) = \int_{t}^{\infty} E_{\alpha} \left(-\lambda (h(\tau) - h(t))^{(1-\beta)(1-\alpha)} \right) h^{(1)}(\tau) f(\tau) d\tau. \tag{3.518} $$

The result was discovered by Yang in 2020 [92].

3.7 The Integral Representations Related to Mittag-Leffler Function

We now present the theorems related to the integral representations via Mittag-Leffler function.

Theorem 3.77 *Let* $\alpha \in \mathbb{C}$, $\gamma \in \mathbb{R}_+$, $\left|\lambda s^{-\alpha}\right| < 1$, *and* $Re(\alpha) > 0$.
Then there is

$$\int_0^{\infty} e^{-st^{\gamma}} t^{\gamma-1} E_\alpha\left(\lambda t^{\alpha\gamma}\right) dt = \gamma^{-1} s^{-1}\left(1 - \lambda s^{-\alpha}\right)^{-1}. \tag{3.519}$$

Proof If the variable t takes the place of t^{γ}, we directly deduce the result. □

Theorem 3.78 *Let* $\alpha \in \mathbb{C}$, $\gamma \in \mathbb{R}_+$, $\left|\lambda s^{-\alpha}\right| < 1$, *and* $Re(\alpha) > 0$.
Then there is

$$\int_0^{\infty} e^{-st^{\gamma}} t^{\gamma-1} supersin_\alpha\left(\lambda t^{\alpha\gamma}\right) dt = \gamma^{-1} \lambda s^{-\alpha-1}\left(1 + \lambda^2 s^{-2\alpha}\right)^{-1}. \tag{3.520}$$

Proof Making use of the relation

$$supersin_\alpha\left(\lambda t^{\alpha\gamma}\right) = \frac{1}{2i}\left(E_\alpha\left(i\lambda t^{\alpha\gamma}\right) - E_\alpha\left(-i\lambda t^{\alpha\gamma}\right)\right), \tag{3.521}$$

we have

$$\begin{aligned} &\textstyle\int_0^{\infty} e^{-st^{\gamma}} t^{\gamma-1} supersin_{\alpha,\beta}\left(\lambda t^{\alpha\gamma}\right) dt \\ &= \textstyle\int_0^{\infty} e^{-st^{\gamma}} t^{\gamma-1}\left[\frac{1}{2i}\left(E_\alpha\left(i\lambda t^{\alpha\gamma}\right) - E_\alpha\left(-i\lambda t^{\alpha\gamma}\right)\right)\right] dt \\ &= \frac{\gamma^{-1}s^{-1}}{2i}\left(\left(1 - i\lambda s^{-\alpha}\right)^{-1} - \left(1 + i\lambda s^{-\alpha}\right)^{-1}\right) \\ &= \frac{\gamma^{-1}\lambda s^{-\alpha-1}}{1+\lambda^2 s^{-2\alpha}}. \end{aligned} \tag{3.522}$$

□

Theorem 3.79 *Let* $\alpha \in \mathbb{C}$, $\gamma \in \mathbb{R}_+$, $\left|\lambda s^{-\alpha}\right| < 1$, *and* $Re(\alpha) > 0$.
Then there is

$$\int_0^{\infty} e^{-st^{\gamma}} t^{\gamma-1} supercos_\alpha\left(\lambda t^{\alpha\gamma}\right) dt = \gamma^{-1} s^{-1}\left(1 + \lambda^2 s^{-2\alpha}\right)^{-1}. \tag{3.523}$$

Proof Making use of the relation

$$supercos_\alpha\left(\lambda t^{\alpha\gamma}\right) = \frac{1}{2}\left(E_\alpha\left(i\lambda t^{\alpha\gamma}\right) + E_\alpha\left(-i\lambda t^{\alpha\gamma}\right)\right), \tag{3.524}$$

we have

$$\begin{aligned}
&\int_0^\infty e^{-st^\gamma} t^{\gamma-1} supercos_\alpha\left(\lambda t^{\alpha\gamma}\right) dt \\
&= \int_0^\infty e^{-st^\gamma} t^{\gamma-1} \left[\tfrac{1}{2}\left(E_\alpha\left(i\lambda t^{\alpha\gamma}\right) + E_\alpha\left(-i\lambda t^{\alpha\gamma}\right)\right)\right] dt \\
&= \tfrac{\gamma^{-1}s^{-1}}{2}\left(\left(1 - i\lambda s^{-\alpha}\right)^{-1} + \left(1 + i\lambda s^{-\alpha}\right)^{-1}\right) \\
&= \tfrac{\gamma^{-1}s^{-1}}{1+\lambda^2 s^{-2\alpha}}.
\end{aligned} \tag{3.525}$$

□

Theorem 3.80 *Let* $\alpha \in \mathbb{C}$, $\gamma \in \mathbb{R}_+$, $\left|\lambda s^{-\alpha}\right| < 1$ *and* $Re\left(\alpha\right) > 0$.
Then there is

$$\int_0^\infty e^{-st^\gamma} t^{\gamma-1} supersinh_\alpha\left(\lambda t^{\alpha\gamma}\right) dt = \gamma^{-1}\lambda s^{-\alpha-1}\left(1 - \lambda^2 s^{-2\alpha}\right)^{-1}. \tag{3.526}$$

Proof Making use of the relation

$$supersinh_\alpha\left(\lambda t^{\alpha\gamma}\right) = \frac{1}{2}\left(E_\alpha\left(\lambda t^{\alpha\gamma}\right) - E_\alpha\left(-\lambda t^{\alpha\gamma}\right)\right), \tag{3.527}$$

we have

$$\begin{aligned}
&\int_0^\infty e^{-st^\gamma} t^{\gamma-1} supersinh_\alpha\left(\lambda t^{\alpha\gamma}\right) dt \\
&= \int_0^\infty e^{-st^\gamma} t^{\gamma-1} \left[\tfrac{1}{2}\left(E_\alpha\left(\lambda t^{\alpha\gamma}\right) - E_\alpha\left(-\lambda t^{\alpha\gamma}\right)\right)\right] dt \\
&= \tfrac{\gamma^{-1}}{2}\left(\left(1 - \lambda s^{-\alpha}\right)^{-1} - \left(1 + \lambda s^{-\alpha}\right)^{-1}\right) \\
&= \tfrac{\gamma^{-1}\lambda s^{-\alpha-1}}{1-\lambda^2 s^{-2\alpha}}.
\end{aligned} \tag{3.528}$$

□

Theorem 3.81 *Let* $\alpha \in \mathbb{C}$, $\gamma \in \mathbb{R}_+$, $\left|\lambda s^{-\alpha}\right| < 1$ *and* $Re\left(\alpha\right) > 0$.
Then there is

$$\int_0^\infty e^{-st^\gamma} t^{\gamma-1} supercosh_\alpha\left(\lambda t^{\alpha\gamma}\right) dt = \gamma^{-1}s^{-1}\left(1 - \lambda^2 s^{-2\alpha}\right)^{-1}. \tag{3.529}$$

Proof Making use of the relation

$$supercosh_\alpha\left(\lambda t^{\alpha\gamma}\right) = \frac{1}{2}\left(E_\alpha\left(\lambda t^{\alpha\gamma}\right) + E_\alpha\left(-\lambda t^{\alpha\gamma}\right)\right), \tag{3.530}$$

we have

$$\begin{aligned}
&\int_0^\infty e^{-st^\gamma} t^{\gamma-1} supercosh_{\alpha,\beta}\left(\lambda t^{\alpha\gamma}\right) dt \\
&= \int_0^\infty e^{-st^\gamma} t^{\gamma-1} \left[\tfrac{1}{2}\left(E_\alpha\left(\lambda t^{\alpha\gamma}\right) + E_\alpha\left(-\lambda t^{\alpha\gamma}\right)\right)\right] dt \\
&= \tfrac{\gamma^{-1}}{2}\left(\left(1-\lambda s^{-\alpha}\right)^{-1} + \left(1+\lambda s^{-\alpha}\right)^{-1}\right) \\
&= \tfrac{\gamma^{-1}s^{-1}}{1-\lambda^2 s^{-2\alpha}}.
\end{aligned} \tag{3.531}$$

□

3.8 The Relationship Between Mittag-Leffler Function and Wright's Generalized Hypergeometric Function

In this section, we discuss the relationship between Mittag-Leffler function and Wright's generalized hypergeometric function.

We have

$$E_\alpha\left(t^\alpha\right) = {}_1\varphi_1\left(\begin{matrix}(1,1)\\(1,\alpha)\end{matrix}; t^\alpha\right), \tag{3.532}$$

$$E_\alpha(t) = {}_1\varphi_1\left(\begin{matrix}(1,1)\\(1,\alpha)\end{matrix}; t\right), \tag{3.533}$$

$$E_\alpha\left(t^\vartheta\right) = {}_1\varphi_1\left(\begin{matrix}(1,1)\\(1,\alpha)\end{matrix}; t^\vartheta\right), \tag{3.534}$$

$$supersin_\alpha\left(t^\alpha\right) = {}_1sin_1\left(\begin{matrix}(1,1)\\(1,\alpha)\end{matrix}; t^\alpha\right), \tag{3.535}$$

$$supercos_\alpha\left(t^\alpha\right) = {}_1cos_1\left(\begin{matrix}(1,1)\\(\beta,\alpha)\end{matrix}; t^\alpha\right), \tag{3.536}$$

$$supersinh_\alpha\left(t^\alpha\right) = {}_1sinh_1\left(\begin{matrix}(1,1)\\(1,\alpha)\end{matrix}; t^\alpha\right), \tag{3.537}$$

$$supercosh_\alpha\left(t^\alpha\right) = {}_1cosh_1\left(\begin{matrix}(1,1)\\(1,\alpha)\end{matrix}; t^\alpha\right), \tag{3.538}$$

$$presin_\alpha(t) = {}_1sin_1\left(\begin{matrix}(1,1)\\(1,\alpha)\end{matrix}; t\right), \tag{3.539}$$

$$precos_{\alpha}(t) = {}_1cos_1\left(\begin{matrix}(1,1)\\(1,\alpha)\end{matrix};t\right), \tag{3.540}$$

$$presinh_{\alpha}(t) = {}_1sinh_1\left(\begin{matrix}(1,1)\\(1,\alpha)\end{matrix};t\right), \tag{3.541}$$

$$precosh_{\alpha}(t) = {}_1cosh_1\left(\begin{matrix}(1,1)\\(1,\alpha)\end{matrix};t\right), \tag{3.542}$$

$$presin_{\alpha}\left(t^{\vartheta}\right) = {}_1sin_1\left(\begin{matrix}(1,1)\\(1,\alpha)\end{matrix};t^{\vartheta}\right), \tag{3.543}$$

$$precos_{\alpha}\left(t^{\vartheta}\right) = {}_1cos_1\left(\begin{matrix}(1,1)\\(1,\alpha)\end{matrix};t^{\vartheta}\right), \tag{3.544}$$

$$presinh_{\alpha}\left(t^{\vartheta}\right) = {}_1sinh_1\left(\begin{matrix}(1,1)\\(1,\alpha)\end{matrix};t^{\vartheta}\right), \tag{3.545}$$

$$precosh_{\alpha}\left(t^{\vartheta}\right) = {}_1cosh_1\left(\begin{matrix}(1,1)\\(1,\alpha)\end{matrix};t^{\vartheta}\right), \tag{3.546}$$

where $\alpha, \beta, t \in \mathbb{C}$, $\vartheta \in \mathbb{C}$, $Re(\alpha) > 0$, and $Re(\beta) > 0$.

The expression (3.2) was reported in [238].

From the Chap. 2, we have that

$$ {}_2\varphi_2\left(\begin{matrix}(\alpha_1, A_1), (\alpha_2, A_2)\\(\beta_1, B_1), (\beta_2, B_2)\end{matrix};t\right) = \sum_{k=0}^{\infty}\frac{\Gamma(\alpha_1 + A_1 k)\,\Gamma(\alpha_2 + A_2 k)}{\Gamma(\beta_1 + B_1 k)\,\Gamma(\beta_2 + B_2 k)}\frac{t^k}{k!}, \tag{3.547}$$

$$\begin{aligned}
&{}_2sin_2\left(\begin{matrix}(\alpha_1, A_1), (\alpha_2, A_2)\\(\beta_1, B_1), (\beta_2, B_2)\end{matrix};t^{\alpha}\right)\\
&= \frac{1}{2i}\left({}_2\varphi_2\left(\begin{matrix}(\alpha_1, A_1), (\alpha_2, A_2)\\(\beta_1, B_1), (\beta_2, B_2)\end{matrix};it^{\alpha}\right) - {}_2\varphi_2\left(\begin{matrix}(\alpha_1, A_1), (\alpha_2, A_2)\\(\beta_1, B_1), (\beta_2, B_2)\end{matrix};-it^{\alpha}\right)\right)\\
&= \sum_{k=0}^{\infty}(-1)^k\frac{\Gamma(\alpha_1+A_1(2k+1))\Gamma(\alpha_2+A_2(2k+1))}{\Gamma(\beta_1+B_1(2k+1))\Gamma(\beta_2+B_2(2k+1))}\frac{t^{(2k+1)\alpha}}{(2k+1)!},
\end{aligned} \tag{3.548}$$

$$\begin{aligned}
&{}_2cos_2\left(\begin{matrix}(\alpha_1, A_1), (\alpha_2, A_2)\\(\beta_1, B_1), (\beta_2, B_2)\end{matrix};t^{\alpha}\right)\\
&= \frac{1}{2}\left({}_2\varphi_2\left(\begin{matrix}(\alpha_1, A_1), (\alpha_2, A_2)\\(\beta_1, B_1), (\beta_2, B_2)\end{matrix};it^{\alpha}\right) + {}_2\varphi_2\left(\begin{matrix}(\alpha_1, A_1), (\alpha_2, A_2)\\(\beta_1, B_1), (\beta_2, B_2)\end{matrix};-it^{\alpha}\right)\right)\\
&= \sum_{k=0}^{\infty}(-1)^k\frac{\Gamma(\alpha_1+2kA_1)\Gamma(\alpha_2+2kA_2)}{\Gamma(\beta_1+2kB_1)\Gamma(\beta_2+2kB_2)}\frac{t^{2k\alpha}}{(2k)!},
\end{aligned} \tag{3.549}$$

$$
\begin{aligned}
&{}_2sinh_2\left(\begin{matrix}(\alpha_1, A_1), (\alpha_2, A_2)\\ (\beta_1, B_1), (\beta_2, B_2)\end{matrix}; t^{\alpha}\right)\\
&= \frac{1}{2}\left({}_2\varphi_2\left(\begin{matrix}(\alpha_1, A_1), (\alpha_2, A_2)\\ (\beta_1, B_1), (\beta_2, B_2)\end{matrix}; t^{\alpha}\right) - {}_2\varphi_2\left(\begin{matrix}(\alpha_1, A_1), (\alpha_2, A_2)\\ (\beta_1, B_1), (\beta_2, B_2)\end{matrix}; -t^{\alpha}\right)\right)\\
&= \sum_{k=0}^{\infty} \frac{\Gamma(\alpha_1+A_1(2k+1))\Gamma(\alpha_2+A_2(2k+1))}{\Gamma(\beta_1+B_1(2k+1))\Gamma(\beta_2+B_2(2k+1))} \frac{t^{(2k+1)\alpha}}{(2k+1)!},
\end{aligned}
\tag{3.550}
$$

$$
\begin{aligned}
&{}_2cosh_2\left(\begin{matrix}(\alpha_1, A_1), (\alpha_2, A_2)\\ (\beta_1, B_1), (\beta_2, B_2)\end{matrix}; t^{\alpha}\right)\\
&= \frac{1}{2}\left({}_2\varphi_2\left(\begin{matrix}(\alpha_1, A_1), (\alpha_2, A_2)\\ (\beta_1, B_1), (\beta_2, B_2)\end{matrix}; t^{\alpha}\right) + {}_2\varphi_2\left(\begin{matrix}(\alpha_1, A_1), (\alpha_2, A_2)\\ (\beta_1, B_1), (\beta_2, B_2)\end{matrix}; -t^{\alpha}\right)\right)\\
&= \sum_{k=0}^{\infty} \frac{\Gamma(\alpha_1+2kA_1)\Gamma(\alpha_2+2kA_2)}{\Gamma(\beta_1+2kB_1)\Gamma(\beta_2+2kB_2)} \frac{t^{2k\alpha}}{(2k)!},
\end{aligned}
\tag{3.551}
$$

$$
\begin{aligned}
&{}_2sin_2\left(\begin{matrix}(\alpha_1, A_1), (\alpha_2, A_2)\\ (\beta_1, B_1), (\beta_2, B_2)\end{matrix}; t^{\vartheta}\right)\\
&= \frac{1}{2i}\left({}_2\varphi_2\left(\begin{matrix}(\alpha_1, A_1), (\alpha_2, A_2)\\ (\beta_1, B_1), (\beta_2, B_2)\end{matrix}; it^{\vartheta}\right) - {}_2\varphi_2\left(\begin{matrix}(\alpha_1, A_1), (\alpha_2, A_2)\\ (\beta_1, B_1), (\beta_2, B_2)\end{matrix}; -it^{\vartheta}\right)\right)\\
&= \sum_{k=0}^{\infty} (-1)^k \frac{\Gamma(\alpha_1+A_1(2k+1))\Gamma(\alpha_2+A_2(2k+1))}{\Gamma(\beta_1+B_1(2k+1))\Gamma(\beta_2+B_2(2k+1))} \frac{t^{(2k+1)\vartheta}}{(2k+1)!},
\end{aligned}
\tag{3.552}
$$

$$
\begin{aligned}
&{}_2cos_2\left(\begin{matrix}(\alpha_1, A_1), (\alpha_2, A_2)\\ (\beta_1, B_1), (\beta_2, B_2)\end{matrix}; t^{\vartheta}\right)\\
&= \frac{1}{2}\left({}_2\varphi_2\left(\begin{matrix}(\alpha_1, A_1), (\alpha_2, A_2)\\ (\beta_1, B_1), (\beta_2, B_2)\end{matrix}; it^{\vartheta}\right) + {}_2\varphi_2\left(\begin{matrix}(\alpha_1, A_1), (\alpha_2, A_2)\\ (\beta_1, B_1), (\beta_2, B_2)\end{matrix}; -it^{\vartheta}\right)\right)\\
&= \sum_{k=0}^{\infty} (-1)^k \frac{\Gamma(\alpha_1+2kA_1)\Gamma(\alpha_2+2kA_2)}{\Gamma(\beta_1+2kB_1)\Gamma(\beta_2+2kB_2)} \frac{t^{2k\vartheta}}{(2k)!},
\end{aligned}
\tag{3.553}
$$

$$
\begin{aligned}
&{}_2sinh_2\left(\begin{matrix}(\alpha_1, A_1), (\alpha_2, A_2)\\ (\beta_1, B_1), (\beta_2, B_2)\end{matrix}; t^{\vartheta}\right)\\
&= \frac{1}{2}\left({}_2\varphi_2\left(\begin{matrix}(\alpha_1, A_1), (\alpha_2, A_2)\\ (\beta_1, B_1), (\beta_2, B_2)\end{matrix}; t^{\vartheta}\right) - {}_2\varphi_2\left(\begin{matrix}(\alpha_1, A_1), (\alpha_2, A_2)\\ (\beta_1, B_1), (\beta_2, B_2)\end{matrix}; -t^{\vartheta}\right)\right)\\
&= \sum_{k=0}^{\infty} \frac{\Gamma(\alpha_1+A_1(2k+1))\Gamma(\alpha_2+A_2(2k+1))}{\Gamma(\beta_1+B_1(2k+1))\Gamma(\beta_2+B_2(2k+1))} \frac{t^{(2k+1)\vartheta}}{(2k+1)!},
\end{aligned}
\tag{3.554}
$$

$$
\begin{aligned}
&{}_2cosh_2\left(\begin{matrix}(\alpha_1, A_1),(\alpha_2, A_2)\\(\beta_1, B_1),(\beta_2, B_2)\end{matrix};t^{\vartheta}\right)\\
&=\frac{1}{2}\left({}_2\varphi_2\left(\begin{matrix}(\alpha_1, A_1),(\alpha_2, A_2)\\(\beta_1, B_1),(\beta_2, B_2)\end{matrix};t^{\vartheta}\right)+{}_2\varphi_2\left(\begin{matrix}(\alpha_1, A_1),(\alpha_2, A_2)\\(\beta_1, B_1),(\beta_2, B_2)\end{matrix};-t^{\vartheta}\right)\right)\\
&=\sum_{k=0}^{\infty}\frac{\Gamma(\alpha_1+2kA_1)\Gamma(\alpha_2+2kA_2)}{\Gamma(\beta_1+2kB_1)\Gamma(\beta_2+2kB_2)}\frac{t^{2k\vartheta}}{(2k)!},
\end{aligned}
\tag{3.555}
$$

$$
\begin{aligned}
&{}_2sin_2\left(\begin{matrix}(\alpha_1, A_1),(\alpha_2, A_2)\\(\beta_1, B_1),(\beta_2, B_2)\end{matrix};t\right)\\
&=\frac{1}{2i}\left({}_2\varphi_2\left(\begin{matrix}(\alpha_1, A_1),(\alpha_2, A_2)\\(\beta_1, B_1),(\beta_2, B_2)\end{matrix};it\right)-{}_2\varphi_2\left(\begin{matrix}(\alpha_1, A_1),(\alpha_2, A_2)\\(\beta_1, B_1),(\beta_2, B_2)\end{matrix};-it\right)\right)\\
&=\sum_{k=0}^{\infty}(-1)^k\frac{\Gamma(\alpha_1+A_1(2k+1))\Gamma(\alpha_2+A_2(2k+1))}{\Gamma(\beta_1+B_1(2k+1))\Gamma(\beta_2+B_2(2k+1))}\frac{t^{2k+1}}{(2k+1)!},
\end{aligned}
\tag{3.556}
$$

$$
\begin{aligned}
&{}_2cos_2\left(\begin{matrix}(\alpha_1, A_1),(\alpha_2, A_2)\\(\beta_1, B_1),(\beta_2, B_2)\end{matrix};t\right)\\
&=\frac{1}{2}\left({}_2\varphi_2\left(\begin{matrix}(\alpha_1, A_1),(\alpha_2, A_2)\\(\beta_1, B_1),(\beta_2, B_2)\end{matrix};it\right)+{}_2\varphi_2\left(\begin{matrix}(\alpha_1, A_1),(\alpha_2, A_2)\\(\beta_1, B_1),(\beta_2, B_2)\end{matrix};-it\right)\right)\\
&=\sum_{k=0}^{\infty}(-1)^k\frac{\Gamma(\alpha_1+2kA_1)\Gamma(\alpha_2+2kA_2)}{\Gamma(\beta_1+2kB_1)\Gamma(\beta_2+2kB_2)}\frac{t^{2k}}{(2k)!},
\end{aligned}
\tag{3.557}
$$

$$
\begin{aligned}
&{}_2sinh_2\left(\begin{matrix}(\alpha_1, A_1),(\alpha_2, A_2)\\(\beta_1, B_1),(\beta_2, B_2)\end{matrix};t\right)\\
&=\frac{1}{2}\left({}_2\varphi_2\left(\begin{matrix}(\alpha_1, A_1),(\alpha_2, A_2)\\(\beta_1, B_1),(\beta_2, B_2)\end{matrix};t\right)-{}_2\varphi_2\left(\begin{matrix}(\alpha_1, A_1),(\alpha_2, A_2)\\(\beta_1, B_1),(\beta_2, B_2)\end{matrix};-t\right)\right)\\
&=\sum_{k=0}^{\infty}\frac{\Gamma(\alpha_1+A_1(2k+1))\Gamma(\alpha_2+A_2(2k+1))}{\Gamma(\beta_1+B_1(2k+1))\Gamma(\beta_2+B_2(2k+1))}\frac{t^{2k+1}}{(2k+1)!},
\end{aligned}
\tag{3.558}
$$

and

$$
\begin{aligned}
&{}_2cosh_2\left(\begin{matrix}(\alpha_1, A_1),(\alpha_2, A_2)\\(\beta_1, B_1),(\beta_2, B_2)\end{matrix};t\right)\\
&=\frac{1}{2}\left({}_2\varphi_2\left(\begin{matrix}(\alpha_1, A_1),(\alpha_2, A_2)\\(\beta_1, B_1),(\beta_2, B_2)\end{matrix};t\right)+{}_2\varphi_2\left(\begin{matrix}(\alpha_1, A_1),(\alpha_2, A_2)\\(\beta_1, B_1),(\beta_2, B_2)\end{matrix};-t\right)\right)\\
&=\sum_{k=0}^{\infty}\frac{\Gamma(\alpha_1+2kA_1)\Gamma(\alpha_2+2kA_2)}{\Gamma(\beta_1+2kB_1)\Gamma(\beta_2+2kB_2)}\frac{t^{2k}}{(2k)!},
\end{aligned}
\tag{3.559}
$$

which leads to

$$
{}_2sin_2\left(\begin{matrix}(\alpha_1, A_1),(\beta_1, B_1)\\(\beta_1, B_1),(\beta_2, B_2)\end{matrix};\lambda t\right)={}_1sin_1\left(\begin{matrix}(\alpha_1, A_1)\\(\beta_2, B_2)\end{matrix};\lambda t\right),
\tag{3.560}
$$

$$ {}_2sin_2\left(\begin{matrix}(\alpha_1, A_1), (\beta_1, B_1)\\(\beta_1, B_1), (\beta_2, B_2)\end{matrix};\lambda t\right) = {}_1sin_1\left(\begin{matrix}(\alpha_1, A_1)\\(\beta_2, B_2)\end{matrix};\lambda t\right), \tag{3.561}$$

$$ {}_2sinh_2\left(\begin{matrix}(\alpha_1, A_1), (\beta_1, B_1)\\(\beta_1, B_1), (\beta_2, B_2)\end{matrix};\lambda t\right) = {}_1sinh_1\left(\begin{matrix}(\alpha_1, A_1)\\(\beta_2, B_2)\end{matrix};\lambda t\right), \tag{3.562}$$

and

$$ {}_2cosh_2\left(\begin{matrix}(\alpha_1, A_1), (\beta_1, B_1)\\(\beta_1, B_1), (\beta_2, B_2)\end{matrix};\lambda t\right) = {}_1cosh_1\left(\begin{matrix}(\alpha_1, A_1)\\(\beta_2, B_2)\end{matrix};\lambda t\right), \tag{3.563}$$

where $\lambda \in \mathbb{C}, \alpha_1, \beta_1, \alpha_2, \beta_2 \in \mathbb{C}$ and $A_1, B_1, A_1, B_1 \in \mathbb{R}_+$.

Theorem 3.82 *Let $\varpi, \gamma \in \mathbb{C}$, $Re(\varpi) > 0$, $Re(\gamma) > 0$, $\lambda \in \mathbb{C}$, $\vartheta > 0$ and $\alpha \in \mathbb{R}_+$. Then we have*

$$\begin{aligned}&\tfrac{1}{\Gamma(\varpi)}\textstyle\int_0^t (t-\tau)^{\varpi}\,\tau^{\gamma-1}\,{}_1\varphi_1\left(\begin{matrix}(1,1)\\(1,\alpha)\end{matrix};\lambda\tau^{\vartheta}\right)d\tau\\ &= t^{\gamma+\varpi-1}\,{}_2\varphi_2\left(\begin{matrix}(1,1),(\gamma,\vartheta)\\(1,\alpha),(\gamma+\varpi,\vartheta)\end{matrix};\lambda t^{\vartheta}\right).\end{aligned} \tag{3.564}$$

Proof From the Chap. 2, we obtain the result. □

Property 3.9 Let $\varpi, \gamma \in \mathbb{C}$, $Re(\varpi) > 0$, $Re(\gamma) > 0$, $\lambda \in \mathbb{C}$, $\vartheta > 0$ and $\alpha \in \mathbb{R}_+$. Then we have

$$\begin{aligned}&\frac{1}{\Gamma(\varpi)}\int_0^1 (1-\tau)^{\varpi}\,\tau^{\gamma-1}\,{}_1\varphi_1\left(\begin{matrix}(1,1)\\(1,\alpha)\end{matrix};\lambda\tau^{\vartheta}\right)d\tau\\ &= {}_2\varphi_2\left(\begin{matrix}(1,1),(\gamma,\vartheta)\\(1,\alpha),(\gamma+\varpi,\vartheta)\end{matrix};\lambda\right).\end{aligned} \tag{3.565}$$

Proof When $t = 1$ in (3.35), we give the result. □

Theorem 3.83 *Let $\varpi, \gamma \in \mathbb{C}$, $Re(\varpi) > 0$, $Re(\gamma) > 0$, $\lambda \in \mathbb{C}$, $\vartheta > 0$ and $\alpha \in \mathbb{R}_+$. Then we have*

$$\begin{aligned}&\textstyle\int_{-1}^1 (1-\tau)^{\varpi}(1+\tau)^{\gamma-1}\,{}_1\varphi_1\left(\begin{matrix}(1,1)\\(1,\alpha)\end{matrix};\lambda(1+\tau)^{\vartheta}\right)d\tau\\ &= \textstyle\int_{-1}^1 (1+\tau)^{\varpi}(1-\tau)^{\gamma-1}\,{}_1\varphi_1\left(\begin{matrix}(1,1)\\(1,\alpha)\end{matrix};\lambda(1-\tau)^{\vartheta}\right)d\tau\end{aligned} \tag{3.566}$$

and

$$\begin{aligned}
&\tfrac{1}{\Gamma(\varpi)}\int_0^2 (2-\tau)^{\varpi}\,\tau^{\gamma-1}\;{}_1\varphi_1\left(\begin{matrix}(1,1)\\(1,\alpha)\end{matrix};\lambda\tau^{\vartheta}\right)d\tau\\
&=\tfrac{1}{\Gamma(\varpi)}\int_{-1}^1 (1-\tau)^{\varpi}\,(1+\tau)^{\gamma-1}\;{}_1\varphi_1\left(\begin{matrix}(1,1)\\(1,\alpha)\end{matrix};\lambda\,(1+\tau)^{\vartheta}\right)d\tau\\
&=\tfrac{1}{\Gamma(\varpi)}\int_{-1}^1 (1+\tau)^{\varpi}\,(1-\tau)^{\gamma-1}\;{}_1\varphi_1\left(\begin{matrix}(1,1)\\(1,\alpha)\end{matrix};\lambda\,(1-\tau)^{\vartheta}\right)d\tau\\
&=2^{\gamma+\varpi-1}\;{}_2\varphi_2\left(\begin{matrix}(1,1),(\gamma,\vartheta)\\(1,\alpha),(\gamma+\varpi,\vartheta)\end{matrix};\lambda 2^{\vartheta}\right).
\end{aligned}\tag{3.567}$$

Proof When $t=2$ in (3.35), we obtain the result. □

Theorem 3.84 *Let $\varpi,\gamma\in\mathbb{C}$, $Re(\varpi)>0$, $Re(\gamma)>0$, $\lambda\in\mathbb{C}$, $\vartheta>0$ and $\alpha\in\mathbb{R}_{+}$. Then we have*

$$\begin{aligned}
&\tfrac{1}{\Gamma(\varpi)}\int_0^t (t-\tau)^{\varpi}\,\tau^{\gamma-1} presin_{\alpha}\left(\lambda\tau^{\vartheta}\right)d\tau\\
&=t^{\gamma+\varpi-1}\;{}_2sin_2\left(\begin{matrix}(1,1),(\gamma,\vartheta)\\(1,\alpha),(\gamma+\varpi,\vartheta)\end{matrix};\lambda t^{\vartheta}\right)
\end{aligned}\tag{3.568}$$

and

$$\begin{aligned}
&\tfrac{1}{\Gamma(\varpi)}\int_0^t (t-\tau)^{\varpi}\,\tau^{\gamma-1} precos_{\alpha}\left(\lambda\tau^{\vartheta}\right)d\tau\\
&=t^{\gamma+\varpi-1}\;{}_2cos_2\left(\begin{matrix}(1,1),(\gamma,\vartheta)\\(1,\alpha),(\gamma+\varpi,\vartheta)\end{matrix};\lambda t^{\vartheta}\right).
\end{aligned}\tag{3.569}$$

Proof Making use of (3.564), (3.543), and (3.544), we have

$$\begin{aligned}
&\tfrac{1}{\Gamma(\varpi)}\int_0^t (t-\tau)^{\varpi}\,\tau^{\gamma-1}\;{}_1\varphi_1\left(\begin{matrix}(1,1)\\(1,\alpha)\end{matrix};i\lambda\tau^{\vartheta}\right)d\tau\\
&=t^{\gamma+\varpi-1}\;{}_2\varphi_2\left(\begin{matrix}(1,1),(\gamma,\vartheta)\\(1,\alpha),(\gamma+\varpi,\vartheta)\end{matrix};i\lambda t^{\vartheta}\right)
\end{aligned}\tag{3.570}$$

and

$$\begin{aligned}
&\tfrac{1}{\Gamma(\varpi)}\int_0^t (t-\tau)^{\varpi}\,\tau^{\gamma-1}\;{}_1\varphi_1\left(\begin{matrix}(1,1)\\(1,\alpha)\end{matrix};-i\lambda\tau^{\vartheta}\right)d\tau\\
&=t^{\gamma+\varpi-1}\;{}_2\varphi_2\left(\begin{matrix}(1,1),(\gamma,\vartheta)\\(1,\alpha),(\gamma+\varpi,\vartheta)\end{matrix};-i\lambda t^{\vartheta}\right),
\end{aligned}\tag{3.571}$$

which reduces the desired results. □

Corollary 3.7 *Let $\varpi, \gamma \in \mathbb{C}$, $Re(\varpi) > 0$, $Re(\gamma) > 0$, $\lambda \in \mathbb{C}$, $\vartheta > 0$ and $\alpha \in \mathbb{R}_+$. Then we have*

$$\begin{aligned}&\tfrac{1}{\Gamma(\varpi)} \int_0^1 (1-\tau)^{\varpi} \tau^{\gamma-1} presin_\alpha\left(\lambda\tau^{\vartheta}\right) d\tau \\ &= {}_2sin_2\left(\begin{matrix}(1,1),(\gamma,\vartheta) \\ (1,\alpha),(\gamma+\varpi,\vartheta)\end{matrix};\lambda\right)\end{aligned} \tag{3.572}$$

and

$$\begin{aligned}&\tfrac{1}{\Gamma(\varpi)} \int_0^1 (1-\tau)^{\varpi} \tau^{\gamma-1} precos_\alpha\left(\lambda\tau^{\vartheta}\right) d\tau \\ &= {}_2cos_2\left(\begin{matrix}(1,1),(\gamma,\vartheta) \\ (1,\alpha),(\gamma+\varpi,\vartheta)\end{matrix};\lambda\right).\end{aligned} \tag{3.573}$$

Proof Taking $t = 1$ in (3.568) and (3.569), we directly have the results. □

Theorem 3.85 *Let $\varpi, \gamma \in \mathbb{C}$, $Re(\varpi) > 0$, $Re(\gamma) > 0$, $\lambda \in \mathbb{C}$, $\vartheta > 0$ and $\alpha \in \mathbb{R}_+$. Then we have*

$$\begin{aligned}&\int_{-1}^1 (1-\tau)^{\varpi} (1+\tau)^{\gamma-1} presin_\alpha\left(\lambda(1+\tau)^{\vartheta}\right) d\tau \\ &= \int_{-1}^1 (1+\tau)^{\varpi} (1-\tau)^{\gamma-1} presin_\alpha\left(\lambda(1-\tau)^{\vartheta}\right) d\tau\end{aligned} \tag{3.574}$$

and

$$\begin{aligned}&\tfrac{1}{\Gamma(\varpi)} \int_0^2 (2-\tau)^{\varpi} \tau^{\gamma-1} presin_\alpha\left(\lambda\tau^{\vartheta}\right) d\tau \\ &= \tfrac{1}{\Gamma(\varpi)} \int_{-1}^1 (1-\tau)^{\varpi} (1+\tau)^{\gamma-1} presin_\alpha\left(\lambda(1+\tau)^{\vartheta}\right) d\tau \\ &= \tfrac{1}{\Gamma(\varpi)} \int_{-1}^1 (1+\tau)^{\varpi} (1-\tau)^{\gamma-1} presin_\alpha\left(\lambda(1-\tau)^{\vartheta}\right) d\tau \\ &= 2^{\gamma+\varpi-1}\, {}_2sin_2\left(\begin{matrix}(1,1),(\gamma,\vartheta) \\ (1,\alpha),(\gamma+\varpi,\vartheta)\end{matrix};\lambda 2^{\vartheta}\right).\end{aligned} \tag{3.575}$$

Proof When $t = 2$ in (3.568), we have the result. □

Theorem 3.86 *Let $\varpi, \gamma \in \mathbb{C}$, $Re(\varpi) > 0$, $Re(\gamma) > 0$, $\lambda \in \mathbb{C}$, $\vartheta > 0$ and $\alpha \in \mathbb{R}_+$. Then we have*

$$\begin{aligned}&\int_{-1}^1 (1-\tau)^{\varpi} (1+\tau)^{\gamma-1} precos_\alpha\left(\lambda(1+\tau)^{\vartheta}\right) d\tau \\ &= \int_{-1}^1 (1+\tau)^{\varpi} (1-\tau)^{\gamma-1} precos_\alpha\left(\lambda(1-\tau)^{\vartheta}\right) d\tau\end{aligned} \tag{3.576}$$

and

$$\begin{aligned}&\frac{1}{\Gamma(\varpi)}\int_0^2 (2-\tau)^{\varpi}\,\tau^{\gamma-1} precos_{\alpha}\left(\lambda\tau^{\vartheta}\right)d\tau\\&=\frac{1}{\Gamma(\varpi)}\int_{-1}^1 (1-\tau)^{\varpi}\,(1+\tau)^{\gamma-1}\;precos_{\alpha}\left(\lambda\left(1+\tau\right)^{\vartheta}\right)d\tau\\&=\frac{1}{\Gamma(\varpi)}\int_{-1}^1 (1+\tau)^{\varpi}\,(1-\tau)^{\gamma-1}\;precos_{\alpha}\left(\lambda\left(1-\tau\right)^{\vartheta}\right)d\tau\\&=2^{\gamma+\varpi-1}\;{}_2cos_2\left(\begin{matrix}(1,1),(\gamma,\vartheta)\\(1,\alpha),(\gamma+\varpi,\vartheta)\end{matrix};\lambda 2^{\vartheta}\right).\end{aligned}\tag{3.577}$$

Proof When $t = 2$ in (3.569), we obtain the result. □

Theorem 3.87 *Let $\varpi, \gamma \in \mathbb{C}$, $Re(\varpi) > 0$, $Re(\gamma) > 0$, $\lambda \in \mathbb{C}$, $\vartheta > 0$, and $\alpha \in \mathbb{R}_+$. Then we have*

$$\begin{aligned}&\frac{1}{\Gamma(\varpi)}\int_0^t (t-\tau)^{\varpi}\,\tau^{\gamma-1} presinh_{\alpha}\left(\lambda\tau^{\vartheta}\right)d\tau\\&=t^{\gamma+\varpi-1}\;{}_2sinh_2\left(\begin{matrix}(1,1),(\gamma,\vartheta)\\(1,\alpha),(\gamma+\varpi,\vartheta)\end{matrix};\lambda t^{\vartheta}\right)\end{aligned}\tag{3.578}$$

and

$$\begin{aligned}&\frac{1}{\Gamma(\varpi)}\int_0^t (t-\tau)^{\varpi}\,\tau^{\gamma-1} precosh_{\alpha}\left(\lambda\tau^{\vartheta}\right)d\tau\\&=t^{\gamma+\varpi-1}\;{}_2cosh_2\left(\begin{matrix}(1,1),(\gamma,\vartheta)\\(1,\alpha),(\gamma+\varpi,\vartheta)\end{matrix};\lambda t^{\vartheta}\right).\end{aligned}\tag{3.579}$$

Proof Making use of (3.35), we have

$$\begin{aligned}&\frac{1}{\Gamma(\varpi)}\int_0^t (t-\tau)^{\varpi}\,\tau^{\gamma-1}\;{}_1\varphi_1\left(\begin{matrix}(1,1)\\(1,\alpha)\end{matrix};\lambda\tau^{\vartheta}\right)d\tau\\&=t^{\gamma+\varpi-1}\;{}_2\varphi_2\left(\begin{matrix}(1,1),(\gamma,\vartheta)\\(1,\alpha),(\gamma+\varpi,\vartheta)\end{matrix};\lambda t^{\vartheta}\right)\end{aligned}\tag{3.580}$$

and

$$\begin{aligned}&\frac{1}{\Gamma(\varpi)}\int_0^t (t-\tau)^{\varpi}\,\tau^{\gamma-1}\;{}_1\varphi_1\left(\begin{matrix}(1,1)\\(1,\alpha)\end{matrix};-\lambda\tau^{\vartheta}\right)d\tau\\&=t^{\gamma+\varpi-1}\;{}_2\varphi_2\left(\begin{matrix}(1,1),(\gamma,\vartheta)\\(1,\alpha),(\gamma+\varpi,\vartheta)\end{matrix};-\lambda t^{\vartheta}\right),\end{aligned}\tag{3.581}$$

which reduces the results. □

Corollary 3.8 *Let $\varpi, \gamma \in \mathbb{C}$, $Re(\varpi) > 0$, $Re(\gamma) > 0$, $\lambda \in \mathbb{C}$, $\vartheta > 0$, and $\alpha \in \mathbb{R}_+$. Then we have*

$$\begin{aligned}&\frac{1}{\Gamma(\varpi)}\int_0^1 (1-\tau)^{\varpi}\,\tau^{\gamma-1} presinh_{\alpha}\left(\lambda\tau^{\vartheta}\right)d\tau\\&=\;{}_2sinh_2\left(\begin{matrix}(1,1),(\gamma,\vartheta)\\(1,\alpha),(\gamma+\varpi,\vartheta)\end{matrix};\lambda\right)\end{aligned}\tag{3.582}$$

and

$$\begin{aligned}&\frac{1}{\Gamma(\varpi)}\int_0^1 (1-\tau)^{\varpi}\,\tau^{\gamma-1} precosh_{\alpha}\left(\lambda\tau^{\vartheta}\right)d\tau\\&\quad= {}_2cosh_2\left(\begin{matrix}(1,1),(\gamma,\vartheta)\\(1,\alpha),(\gamma+\varpi,\vartheta)\end{matrix};\lambda\right).\end{aligned}\tag{3.583}$$

Proof Taking $t = 1$ in (3.578) and (3.579), we directly have the results. □

Theorem 3.88 *Let $\varpi, \gamma \in \mathbb{C}$, $Re(\varpi) > 0$, $Re(\gamma) > 0$, $\lambda \in \mathbb{C}$, $\vartheta > 0$, and $\alpha \in \mathbb{R}_+$. Then we have*

$$\begin{aligned}&\int_{-1}^1 (1-\tau)^{\varpi}(1+\tau)^{\gamma-1}\, presinh_{\alpha}\left(\lambda(1+\tau)^{\vartheta}\right)d\tau\\&\quad=\int_{-1}^1 (1+\tau)^{\varpi}(1-\tau)^{\gamma-1}\, presinh_{\alpha}\left(\lambda(1-\tau)^{\vartheta}\right)d\tau\end{aligned}\tag{3.584}$$

and

$$\begin{aligned}&\frac{1}{\Gamma(\varpi)}\int_0^2 (2-\tau)^{\varpi}\,\tau^{\gamma-1} presinh_{\alpha}\left(\lambda\tau^{\vartheta}\right)d\tau\\&\quad=\frac{1}{\Gamma(\varpi)}\int_{-1}^1 (1-\tau)^{\varpi}(1+\tau)^{\gamma-1}\, presinh_{\alpha}\left(\lambda(1+\tau)^{\vartheta}\right)d\tau\\&\quad=\frac{1}{\Gamma(\varpi)}\int_{-1}^1 (1+\tau)^{\varpi}(1-\tau)^{\gamma-1}\, presinh_{\alpha}\left(\lambda(1-\tau)^{\vartheta}\right)d\tau\\&\quad=2^{\gamma+\varpi-1}\,{}_2sinh_2\left(\begin{matrix}(1,1),(\gamma,\vartheta)\\(1,\alpha),(\gamma+\varpi,\vartheta)\end{matrix};\lambda 2^{\vartheta}\right).\end{aligned}\tag{3.585}$$

Proof When $t = 2$ in (3.578), we obtain the result. □

Theorem 3.89 *Let $\varpi, \gamma \in \mathbb{C}$, $Re(\varpi) > 0$, $Re(\gamma) > 0$, $\lambda \in \mathbb{C}$, $\vartheta > 0$, and $\alpha \in \mathbb{R}_+$. Then we have*

$$\begin{aligned}&\int_{-1}^1 (1-\tau)^{\varpi}(1+\tau)^{\gamma-1}\, precosh_{\alpha}\left(\lambda(1+\tau)^{\vartheta}\right)d\tau\\&\quad=\int_{-1}^1 (1+\tau)^{\varpi}(1-\tau)^{\gamma-1}\, precosh_{\alpha}\left(\lambda(1-\tau)^{\vartheta}\right)d\tau\end{aligned}\tag{3.586}$$

and

$$\begin{aligned}&\frac{1}{\Gamma(\varpi)}\int_0^2 (2-\tau)^{\varpi}\,\tau^{\gamma-1} precosh_{\alpha}\left(\lambda\tau^{\vartheta}\right)d\tau\\&\quad=\frac{1}{\Gamma(\varpi)}\int_{-1}^1 (1-\tau)^{\varpi}(1+\tau)^{\gamma-1}\, precosh_{\alpha}\left(\lambda(1+\tau)^{\vartheta}\right)d\tau\\&\quad=\frac{1}{\Gamma(\varpi)}\int_{-1}^1 (1+\tau)^{\varpi}(1-\tau)^{\gamma-1}\, precosh_{\alpha}\left(\lambda(1-\tau)^{\vartheta}\right)d\tau\\&\quad=2^{\gamma+\varpi-1}\,{}_2cosh_2\left(\begin{matrix}(1,1),(\gamma,\vartheta)\\(1,\alpha),(\gamma+\varpi,\vartheta)\end{matrix};\lambda 2^{\vartheta}\right).\end{aligned}\tag{3.587}$$

Proof When $t = 2$ in (3.579), we obtain the result. □

Theorem 3.90 *Let $\varpi, \gamma \in \mathbb{C}$, $Re(\varpi) > 0$, $Re(\gamma) > 0$, $\lambda \in \mathbb{C}$, and $\alpha \in \mathbb{R}_+$. Then we have*

$$\begin{aligned}&\tfrac{1}{\Gamma(\varpi)} \int_0^t (t-\tau)^{\varpi} \tau^{\gamma-1}\ {}_1\varphi_1\left(\begin{matrix}(1,1)\\(1,\alpha)\end{matrix};\lambda\tau\right)d\tau\\&= t^{\gamma+\varpi-1}\ {}_2\varphi_2\left(\begin{matrix}(1,1),(\gamma,1)\\(1,\alpha),(\gamma+\varpi,1)\end{matrix};\lambda t\right).\end{aligned} \tag{3.588}$$

Proof Taking $\vartheta = 1$ in (3.564), we get the result. □

Property 3.10 Let $\varpi, \gamma \in \mathbb{C}$, $Re(\varpi) > 0$, $Re(\gamma) > 0$, $\lambda \in \mathbb{C}$, and $\alpha \in \mathbb{R}_+$. Then we have

$$\frac{1}{\Gamma(\varpi)}\int_0^1 (1-\tau)^{\varpi}\tau^{\gamma-1}\ {}_1\varphi_1\left(\begin{matrix}(1,1)\\(1,\alpha)\end{matrix};\lambda\tau\right)d\tau = {}_2\varphi_2\left(\begin{matrix}(1,1),(\gamma,1)\\(1,\alpha),(\gamma+\varpi,1)\end{matrix};\lambda\right). \tag{3.589}$$

Proof Taking $\vartheta = 1$ in (3.565), we get the result. □

Theorem 3.91 *Let $\varpi, \gamma \in \mathbb{C}$, $Re(\varpi) > 0$, $Re(\gamma) > 0$, $\lambda \in \mathbb{C}$, and $\alpha \in \mathbb{R}_+$. Then we have*

$$\begin{aligned}&\int_{-1}^1 (1-\tau)^{\varpi}(1+\tau)^{\gamma-1}\ {}_1\varphi_1\left(\begin{matrix}(1,1)\\(1,\alpha)\end{matrix};\lambda(1+\tau)\right)d\tau\\&= \int_{-1}^1 (1+\tau)^{\varpi}(1-\tau)^{\gamma-1}\ {}_1\varphi_1\left(\begin{matrix}(1,1)\\(1,\alpha)\end{matrix};\lambda(1-\tau)\right)d\tau\end{aligned} \tag{3.590}$$

and

$$\begin{aligned}&\tfrac{1}{\Gamma(\varpi)}\int_0^2 (2-\tau)^{\varpi}\tau^{\gamma-1}\ {}_1\varphi_1\left(\begin{matrix}(1,1)\\(1,\alpha)\end{matrix};\lambda\tau\right)d\tau\\&= \tfrac{1}{\Gamma(\varpi)}\int_{-1}^1 (1-\tau)^{\varpi}(1+\tau)^{\gamma-1}\ {}_1\varphi_1\left(\begin{matrix}(1,1)\\(1,\alpha)\end{matrix};\lambda(1+\tau)\right)d\tau\\&= \tfrac{1}{\Gamma(\varpi)}\int_{-1}^1 (1+\tau)^{\varpi}(1-\tau)^{\gamma-1}\ {}_1\varphi_1\left(\begin{matrix}(1,1)\\(1,\alpha)\end{matrix};\lambda(1-\tau)\right)d\tau\\&= 2^{\gamma+\varpi-1}\ {}_2\varphi_2\left(\begin{matrix}(1,1),(\gamma,1)\\(1,\alpha),(\gamma+\varpi,1)\end{matrix};2\lambda\right).\end{aligned} \tag{3.591}$$

Proof Taking $\vartheta = 1$ in (3.565) and (3.567), we get the results. □

Theorem 3.92 *Let $\varpi, \gamma \in \mathbb{C}$, $Re(\varpi) > 0$, $Re(\gamma) > 0$, $\lambda \in \mathbb{C}$, and $\alpha \in \mathbb{R}_+$. Then we have*

$$\begin{aligned}&\tfrac{1}{\Gamma(\varpi)}\int_0^t (t-\tau)^{\varpi}\tau^{\gamma-1} presin_{\alpha}(\lambda\tau)\,d\tau\\&= t^{\gamma+\varpi-1}\ {}_2sin_2\left(\begin{matrix}(1,1),(\gamma,1)\\(1,\alpha),(\gamma+\varpi,1)\end{matrix};\lambda t\right)\end{aligned} \tag{3.592}$$

and

$$\begin{aligned} &\tfrac{1}{\Gamma(\varpi)} \int_0^t (t-\tau)^{\varpi}\, \tau^{\gamma-1} precos_{\alpha}(\lambda\tau)\, d\tau \\ &= t^{\gamma+\varpi-1}\, {}_2cos_2 \left(\begin{matrix} (1,1), (\gamma,1) \\ (1,\alpha), (\gamma+\varpi,1) \end{matrix} ; \lambda t \right). \end{aligned} \tag{3.593}$$

Proof Taking $\vartheta = 1$ in (3.568) and (3.569), we get the results. □

Corollary 3.9 *Let $\varpi, \gamma \in \mathbb{C}$, $Re(\varpi) > 0$, $Re(\gamma) > 0$, $\lambda \in \mathbb{C}$, and $\alpha \in \mathbb{R}_+$. Then we have*

$$\begin{aligned} &\tfrac{1}{\Gamma(\varpi)} \int_0^1 (1-\tau)^{\varpi}\, \tau^{\gamma-1} presin_{\alpha}(\lambda\tau)\, d\tau \\ &= {}_2sin_2 \left(\begin{matrix} (1,1), (\gamma,1) \\ (1,\alpha), (\gamma+\varpi,1) \end{matrix} ; \lambda \right) \end{aligned} \tag{3.594}$$

and

$$\begin{aligned} &\tfrac{1}{\Gamma(\varpi)} \int_0^1 (1-\tau)^{\varpi}\, \tau^{\gamma-1} precos_{\alpha}(\lambda\tau)\, d\tau \\ &= {}_2cos_2 \left(\begin{matrix} (1,1), (\gamma,1) \\ (1,\alpha), (\gamma+\varpi,1) \end{matrix} ; \lambda \right). \end{aligned} \tag{3.595}$$

Proof Taking $\vartheta = 1$ in (3.224) and (3.225), we get the results. □

Theorem 3.93 *Let $\varpi, \gamma \in \mathbb{C}$, $Re(\varpi) > 0$, $Re(\gamma) > 0$, $\lambda \in \mathbb{C}$, and $\alpha \in \mathbb{R}_+$. Then we have*

$$\begin{aligned} &\int_{-1}^1 (1-\tau)^{\varpi} (1+\tau)^{\gamma-1}\, presin_{\alpha}(\lambda(1+\tau))\, d\tau \\ &= \int_{-1}^1 (1+\tau)^{\varpi} (1-\tau)^{\gamma-1}\, presin_{\alpha}(\lambda(1-\tau))\, d\tau \end{aligned} \tag{3.596}$$

and

$$\begin{aligned} &\tfrac{1}{\Gamma(\varpi)} \int_0^2 (2-\tau)^{\varpi}\, \tau^{\gamma-1}\, presin_{\alpha}(\lambda\tau)\, d\tau \\ &= \tfrac{1}{\Gamma(\varpi)} \int_{-1}^1 (1-\tau)^{\varpi} (1+\tau)^{\gamma-1}\, presin_{\alpha}(\lambda(1+\tau))\, d\tau \\ &= \tfrac{1}{\Gamma(\varpi)} \int_{-1}^1 (1+\tau)^{\varpi} (1-\tau)^{\gamma-1}\, presin_{\alpha}(\lambda(1-\tau))\, d\tau \\ &= 2^{\gamma+\varpi-1}\, {}_2sin_2 \left(\begin{matrix} (1,1), (\gamma,1) \\ (1,\alpha), (\gamma+\varpi,1) \end{matrix} ; 2\lambda \right). \end{aligned} \tag{3.597}$$

Proof Taking $\vartheta = 1$ in (3.574) and (3.575), we get the results. □

Theorem 3.94 *Let $\varpi, \gamma \in \mathbb{C}$, $Re(\varpi) > 0$, $Re(\gamma) > 0$, $\lambda \in \mathbb{C}$, and $\alpha \in \mathbb{R}_+$. Then we have*

$$\begin{aligned} &\int_{-1}^1 (1-\tau)^{\varpi} (1+\tau)^{\gamma-1}\, precos_{\alpha}(\lambda(1+\tau))\, d\tau \\ &= \int_{-1}^1 (1+\tau)^{\varpi} (1-\tau)^{\gamma-1}\, precos_{\alpha}(\lambda(1-\tau))\, d\tau \end{aligned} \tag{3.598}$$

and

$$\begin{aligned}
&\tfrac{1}{\Gamma(\varpi)}\int_0^2 (2-\tau)^{\varpi}\,\tau^{\gamma-1} precos_{\alpha}(\lambda\tau)\,d\tau \\
&=\tfrac{1}{\Gamma(\varpi)}\int_{-1}^1 (1-\tau)^{\varpi}\,(1+\tau)^{\gamma-1}\; precos_{\alpha}(\lambda(1+\tau))\,d\tau \\
&=\tfrac{1}{\Gamma(\varpi)}\int_{-1}^1 (1+\tau)^{\varpi}\,(1-\tau)^{\gamma-1}\; precos_{\alpha}(\lambda(1-\tau))\,d\tau \\
&=2^{\gamma+\varpi-1}\;{}_2cos_2\left(\begin{matrix}(1,1),(\gamma,1)\\(1,\alpha),(\gamma+\varpi,1)\end{matrix};2\lambda\right).
\end{aligned} \tag{3.599}$$

Proof Taking $\vartheta = 1$ in (3.576) and (3.577), we get the results. □

Theorem 3.95 *Let $\varpi,\gamma\in\mathbb{C}$, $Re(\varpi)>0$, $Re(\gamma)>0$, $\lambda\in\mathbb{C}$, and $\alpha\in\mathbb{R}_+$. Then we have*

$$\begin{aligned}
&\tfrac{1}{\Gamma(\varpi)}\int_0^t (t-\tau)^{\varpi}\,\tau^{\gamma-1} presinh_{\alpha}(\lambda\tau)\,d\tau \\
&=t^{\gamma+\varpi-1}\;{}_2sinh_2\left(\begin{matrix}(1,1),(\gamma,1)\\(1,\alpha),(\gamma+\varpi,1)\end{matrix};\lambda t\right)
\end{aligned} \tag{3.600}$$

and

$$\begin{aligned}
&\tfrac{1}{\Gamma(\varpi)}\int_0^t (t-\tau)^{\varpi}\,\tau^{\gamma-1} precosh_{\alpha}(\lambda\tau)\,d\tau \\
&=t^{\gamma+\varpi-1}\;{}_2cosh_2\left(\begin{matrix}(1,1),(\gamma,1)\\(1,\alpha),(\gamma+\varpi,1)\end{matrix};\lambda t\right).
\end{aligned} \tag{3.601}$$

Proof Taking $\vartheta = 1$ in (3.230) and (3.231), we get the results. □

Corollary 3.10 *Let $\varpi,\gamma\in\mathbb{C}$, $Re(\varpi)>0$, $Re(\gamma)>0$, $\lambda\in\mathbb{C}$, and $\alpha\in\mathbb{R}_+$. Then we have*

$$\begin{aligned}
&\tfrac{1}{\Gamma(\varpi)}\int_0^1 (1-\tau)^{\varpi}\,\tau^{\gamma-1} presinh_{\alpha}(\lambda\tau)\,d\tau \\
&=\;{}_2sinh_2\left(\begin{matrix}(1,1),(\gamma,1)\\(1,\alpha),(\gamma+\varpi,1)\end{matrix};\lambda\right)
\end{aligned} \tag{3.602}$$

and

$$\begin{aligned}
&\tfrac{1}{\Gamma(\varpi)}\int_0^1 (1-\tau)^{\varpi}\,\tau^{\gamma-1} precosh_{\alpha}(\lambda\tau)\,d\tau \\
&=\;{}_2cosh_2\left(\begin{matrix}(1,1),(\gamma,1)\\(1,\alpha),(\gamma+\varpi,1)\end{matrix};\lambda\right).
\end{aligned} \tag{3.603}$$

Proof Taking $\vartheta = 1$ in (3.582) and (3.583), we get the results. □

Theorem 3.96 *Let $\varpi,\gamma\in\mathbb{C}$, $Re(\varpi)>0$, $Re(\gamma)>0$, $\lambda\in\mathbb{C}$, and $\alpha\in\mathbb{R}_+$. Then we have*

$$\begin{aligned}
&\int_{-1}^1 (1-\tau)^{\varpi}\,(1+\tau)^{\gamma-1}\; presinh_{\alpha}(\lambda(1+\tau))\,d\tau \\
&=\int_{-1}^1 (1+\tau)^{\varpi}\,(1-\tau)^{\gamma-1}\; presinh_{\alpha}(\lambda(1-\tau))\,d\tau
\end{aligned} \tag{3.604}$$

and

$$\begin{aligned}&\tfrac{1}{\Gamma(\varpi)}\int_0^2 (2-\tau)^{\varpi}\,\tau^{\gamma-1} presinh_{\alpha}(\lambda\tau)\,d\tau\\ &\quad=\tfrac{1}{\Gamma(\varpi)}\int_{-1}^1 (1-\tau)^{\varpi}\,(1+\tau)^{\gamma-1}\; presinh_{\alpha}(\lambda(1+\tau))\,d\tau\\ &=\tfrac{1}{\Gamma(\varpi)}\int_{-1}^1 (1+\tau)^{\varpi}\,(1-\tau)^{\gamma-1}\; presinh_{\alpha}(\lambda(1-\tau))\,d\tau\\ &\quad=2^{\gamma+\varpi-1}\;{}_2sinh_2\left(\begin{matrix}(1,1),(\gamma,1)\\(1,\alpha),(\gamma+\varpi,1)\end{matrix};2\lambda\right).\end{aligned}\tag{3.605}$$

Proof Taking $\vartheta=1$ in (3.584) and (3.585), we get the results. □

Theorem 3.97 *Let $\varpi,\gamma\in\mathbb{C}$, $Re(\varpi)>0$, $Re(\gamma)>0$, $\lambda\in\mathbb{C}$, and $\alpha\in\mathbb{R}_+$. Then we have*

$$\begin{aligned}&\int_{-1}^1 (1-\tau)^{\varpi}\,(1+\tau)^{\gamma-1}\; precosh_{\alpha}(\lambda(1+\tau))\,d\tau\\ &\quad=\int_{-1}^1 (1+\tau)^{\varpi}\,(1-\tau)^{\gamma-1}\; precosh_{\alpha}(\lambda(1-\tau))\,d\tau\end{aligned}\tag{3.606}$$

and

$$\begin{aligned}&\tfrac{1}{\Gamma(\varpi)}\int_0^2 (2-\tau)^{\varpi}\,\tau^{\gamma-1} precosh_{\alpha}(\lambda\tau)\,d\tau\\ &\quad=\tfrac{1}{\Gamma(\varpi)}\int_{-1}^1 (1-\tau)^{\varpi}\,(1+\tau)^{\gamma-1}\; precosh_{\alpha}(\lambda(1+\tau))\,d\tau\\ &\quad=\tfrac{1}{\Gamma(\varpi)}\int_{-1}^1 (1+\tau)^{\varpi}\,(1-\tau)^{\gamma-1}\; precosh_{\alpha}(\lambda(1-\tau))\,d\tau\\ &\quad=2^{\gamma+\varpi-1}\;{}_2cosh_2\left(\begin{matrix}(1,1),(\gamma,1)\\(1,\alpha),(\gamma+\varpi,1)\end{matrix};2\lambda\right).\end{aligned}\tag{3.607}$$

Proof Taking $\vartheta=1$ in (3.586) and (3.587), we get the results. □

Theorem 3.98 *Let $\varpi\in\mathbb{C}$, $Re(\varpi)>0$, $\lambda\in\mathbb{C}$ and $\alpha\in\mathbb{R}_+$. Then we have*

$$\frac{1}{\Gamma(\varpi)}\int_0^t (t-\tau)^{\varpi}\,\tau^{\beta-1}\;{}_1\varphi_1\left(\begin{matrix}(1,1)\\(1,\alpha)\end{matrix};\lambda\tau^{\alpha}\right)d\tau=t^{\varpi}\;{}_1\varphi_1\left(\begin{matrix}(1,1)\\(1+\varpi,\alpha)\end{matrix};\lambda t^{\alpha}\right).\tag{3.608}$$

Proof With the aid of (3.564) and (3.532), we have

$$\tau^{\beta-1}\;{}_1\varphi_1\left(\begin{matrix}(1,1)\\(1,\alpha)\end{matrix};\lambda\tau^{\alpha}\right)=\tau^{\beta-1}E_{\alpha}\left(\lambda\tau^{\alpha}\right)\tag{3.609}$$

such that we have the desired result, where

$$t^{\varpi}\;{}_2\varphi_2\left(\begin{matrix}(1,1),(1,\alpha)\\(1,\alpha),(1+\varpi,\alpha)\end{matrix};\lambda t^{\alpha}\right)=t^{\varpi}\;{}_1\varphi_1\left(\begin{matrix}(1,1)\\(1+\varpi,\alpha)\end{matrix};\lambda t^{\alpha}\right).$$

□

Theorem 3.99 *Let $\varpi \in \mathbb{C}$, $Re(\varpi) > 0$, $\lambda \in \mathbb{C}$ and $\alpha \in \mathbb{R}_+$. Then we have*

$$\frac{1}{\Gamma(\varpi)} \int_0^t (t-\tau)^{\varpi}\, supersin_\alpha \left(\lambda\tau^\alpha\right) d\tau = t^{\varpi}\ {}_1sin_1 \left(\begin{matrix} (1,1) \\ (1+\varpi,\alpha) \end{matrix} ; \lambda t^\alpha \right). \tag{3.610}$$

Proof With the aid of (3.535) and (3.608), we have the result. □

Theorem 3.100 *Let $\varpi \in \mathbb{C}$, $Re(\varpi) > 0$, $\lambda \in \mathbb{C}$ and $\alpha \in \mathbb{R}_+$. Then we have*

$$\frac{1}{\Gamma(\varpi)} \int_0^t (t-\tau)^{\varpi}\, supercos_\alpha \left(\lambda\tau^\alpha\right) d\tau = t^{\varpi}\ {}_1cos_1 \left(\begin{matrix} (1,1) \\ (1+\varpi,\alpha) \end{matrix} ; \lambda t^\alpha \right). \tag{3.611}$$

Proof With the aid of (3.188) and (3.260), we have the result. □

Theorem 3.101 *Let $\varpi \in \mathbb{C}$, $Re(\varpi) > 0$, $\lambda \in \mathbb{C}$ and $\alpha \in \mathbb{R}_+$. Then we have*

$$\frac{1}{\Gamma(\varpi)} \int_0^t (t-\tau)^{\varpi}\, supersinh_\alpha \left(\lambda\tau^\alpha\right) d\tau = t^{\varpi}\ {}_1sinh_1 \left(\begin{matrix} (1,1) \\ (1+\varpi,\alpha) \end{matrix} ; \lambda t^\alpha \right). \tag{3.612}$$

Proof With the aid of (3.537) and (3.608), we have the result. □

Theorem 3.102 *Let $\varpi \in \mathbb{C}$, $Re(\varpi) > 0$, $\lambda \in \mathbb{C}$ and $\alpha \in \mathbb{R}_+$. Then we have*

$$\frac{1}{\Gamma(\varpi)} \int_0^t (t-\tau)^{\varpi}\, supercosh_\alpha \left(\lambda\tau^\alpha\right) d\tau = t^{\beta+\varpi-1}\ {}_1cosh_1 \left(\begin{matrix} (1,1) \\ (1+\varpi,\alpha) \end{matrix} ; \lambda t^\alpha \right). \tag{3.613}$$

Proof With the aid of (3.538) and (3.608), we have the result. □

Theorem 3.103 *Let $\varpi \in \mathbb{C}$, $Re(\varpi) > 0$, $\lambda \in \mathbb{C}$ and $\alpha \in \mathbb{R}_+$. Then we have*

$$\frac{1}{\Gamma(\varpi)} \int_0^1 (1-\tau)^{\varpi}\, \tau^{\beta-1}\ {}_1\varphi_1 \left(\begin{matrix} (1,1) \\ (1,\alpha) \end{matrix} ; \lambda\tau^\alpha \right) d\tau = {}_1\varphi_1 \left(\begin{matrix} (1,1) \\ (1+\varpi,\alpha) \end{matrix} ; \lambda \right). \tag{3.614}$$

Proof Taking $t = 1$ in (3.6) and (3.532) we have the desired result. □

Theorem 3.104 *Let $\varpi \in \mathbb{C}$, $Re(\varpi) > 0$, $\lambda \in \mathbb{C}$ and $\alpha \in \mathbb{R}_+$. Then we have*

$$\frac{1}{\Gamma(\varpi)} \int_0^1 (1-\tau)^{\varpi}\, supersin_\alpha \left(\lambda\tau^\alpha\right) d\tau = {}_1sin_1 \left(\begin{matrix} (1,1) \\ (1+\varpi,\alpha) \end{matrix} ; \lambda \right). \tag{3.615}$$

Proof Setting $t = 1$ in (3.535) and (3.608), we have the result. □

Theorem 3.105 *Let $\varpi \in \mathbb{C}$, $Re(\varpi) > 0$, $\lambda \in \mathbb{C}$ and $\alpha \in \mathbb{R}_{+}$. Then we have*

$$\frac{1}{\Gamma(\varpi)} \int_0^1 (1-\tau)^{\varpi}\, supercos_{\alpha}\left(\lambda \tau^{\alpha}\right) d\tau = {}_1cos_1\left(\begin{matrix}(1,1)\\(1+\varpi,\alpha)\end{matrix};\lambda\right). \tag{3.616}$$

Proof Setting $t = 1$ in (3.536) and (3.608), we have the result. □

Theorem 3.106 *Let $\varpi \in \mathbb{C}$, $Re(\varpi) > 0$, $\lambda \in \mathbb{C}$ and $\alpha \in \mathbb{R}_{+}$. Then we have*

$$\frac{1}{\Gamma(\varpi)} \int_0^1 (1-\tau)^{\varpi}\, supersinh_{\alpha}\left(\lambda \tau^{\alpha}\right) d\tau = {}_1sinh_1\left(\begin{matrix}(1,1)\\(1+\varpi,\alpha)\end{matrix};\lambda\right). \tag{3.617}$$

Proof Taking $t = 1$ in (3.537) and (3.608), we have the result. □

Theorem 3.107 *Let $\varpi \in \mathbb{C}$, $Re(\varpi) > 0$, $\lambda \in \mathbb{C}$ and $\alpha \in \mathbb{R}_{+}$. Then we have*

$$\frac{1}{\Gamma(\varpi)} \int_0^1 (1-\tau)^{\varpi}\, supercosh_{\alpha}\left(\lambda \tau^{\alpha}\right) d\tau = {}_1cosh_1\left(\begin{matrix}(1,1)\\(1+\varpi,\alpha)\end{matrix};\lambda\right). \tag{3.618}$$

Proof Setting $t = 1$ in (3.538) and (3.608), we have the result. □

Theorem 3.108 *Let $\varpi \in \mathbb{C}$, $Re(\varpi) > 0$, $\lambda \in \mathbb{C}$ and $\alpha \in \mathbb{R}_{+}$. Then we have*

$$\begin{aligned}&\int_{-1}^1 (1-\tau)^{\varpi}\; {}_1\varphi_1\left(\begin{matrix}(1,1)\\(1,\alpha)\end{matrix};\lambda(\tau+1)^{\alpha}\right) d\tau\\&= \int_{-1}^1 (1+\tau)^{\varpi}\; {}_1\varphi_1\left(\begin{matrix}(1,1)\\(1,\alpha)\end{matrix};\lambda(1-\tau)^{\alpha}\right) d\tau\end{aligned} \tag{3.619}$$

and

$$\begin{aligned}&\tfrac{1}{\Gamma(\varpi)} \textstyle\int_0^2 (2-\tau)^{\varpi}\; {}_1\varphi_1\left(\begin{matrix}(1,1)\\(1,\alpha)\end{matrix};\lambda\tau^{\alpha}\right) d\tau\\&= \tfrac{1}{\Gamma(\varpi)} \textstyle\int_{-1}^1 (1-\tau)^{\varpi}\; {}_1\varphi_1\left(\begin{matrix}(1,1)\\(1,\alpha)\end{matrix};\lambda(\tau+1)^{\alpha}\right) d\tau\\&= \tfrac{1}{\Gamma(\varpi)} \textstyle\int_{-1}^1 (1+\tau)^{\varpi}\; {}_1\varphi_1\left(\begin{matrix}(1,1)\\(1,\alpha)\end{matrix};\lambda(1-\tau)^{\alpha}\right) d\tau\\&= 2^{\varpi}\; {}_1\varphi_1\left(\begin{matrix}(1,1)\\(1+\varpi,\alpha)\end{matrix};\lambda 2^{\alpha}\right).\end{aligned} \tag{3.620}$$

Proof Setting $t = 2$ in (3.564) and (3.532), we have

$$\begin{aligned}
&\tfrac{1}{\Gamma(\varpi)}\int_0^2 (2-\tau)^{\varpi}\ {}_1\varphi_1\left(\begin{matrix}(1,1)\\(1,\alpha)\end{matrix};\lambda\tau^{\alpha}\right)d\tau\\
&=\tfrac{1}{\Gamma(\varpi)}\int_{-1}^1 (1-\tau)^{\varpi}\ {}_1\varphi_1\left(\begin{matrix}(1,1)\\(1,\alpha)\end{matrix};\lambda\left(\tau+1\right)^{\alpha}\right)d\tau\\
&=\tfrac{1}{\Gamma(\varpi)}\int_{-1}^1 (1+\tau)^{\varpi}\ {}_1\varphi_1\left(\begin{matrix}(1,1)\\(1,\alpha)\end{matrix};\lambda\left(1-\tau\right)^{\alpha}\right)d\tau\\
&=2^{\varpi}\ {}_1\varphi_1\left(\begin{matrix}(1,1)\\(1+\varpi,\alpha)\end{matrix};\lambda 2^{\alpha}\right)
\end{aligned} \tag{3.621}$$

which reduces that

$$\begin{aligned}
&\frac{1}{\Gamma\left(\varpi\right)}\int_{-1}^1 (1-\tau)^{\varpi}\ {}_1\varphi_1\left(\begin{matrix}(1,1)\\(1,\alpha)\end{matrix};\lambda\left(\tau+1\right)^{\alpha}\right)d\tau\\
&=\int_{-1}^1 (1+\tau)^{\varpi}\ {}_1\varphi_1\left(\begin{matrix}(1,1)\\(1,\alpha)\end{matrix};\lambda\left(1-\tau\right)^{\alpha}\right)d\tau.
\end{aligned} \tag{3.622}$$

□

Theorem 3.109 *Let $\varpi \in \mathbb{C}$, $Re\left(\varpi\right) > 0$, $\lambda \in \mathbb{C}$ and $\alpha \in \mathbb{R}_{+}$. Then we have*

$$\int_{-1}^1 (1-\tau)^{\varpi}\ supersin_{\alpha}\left(\lambda\left(\tau+1\right)^{\alpha}\right)d\tau=\int_{-1}^1 (1+\tau)^{\varpi}\ supersin_{\alpha}\left(\lambda\left(1-\tau\right)^{\alpha}\right)d\tau \tag{3.623}$$

and

$$\begin{aligned}
&\tfrac{1}{\Gamma(\varpi)}\int_0^2 (2-\tau)^{\varpi}\ supersin_{\alpha}\left(\lambda\tau^{\alpha}\right)d\tau\\
&=\tfrac{1}{\Gamma(\varpi)}\int_{-1}^1 (1-\tau)^{\varpi}\ supersin_{\alpha}\left(\lambda\left(\tau+1\right)^{\alpha}\right)d\tau\\
&=\tfrac{1}{\Gamma(\varpi)}\int_{-1}^1 (1+\tau)^{\varpi}\ supersin_{\alpha}\left(\lambda\left(1-\tau\right)^{\alpha}\right)d\tau\\
&=2^{\beta+\varpi-1}\ {}_1sin_1\left(\begin{matrix}(1,1)\\(\beta+\varpi,\alpha)\end{matrix};\lambda 2^{\alpha}\right).
\end{aligned} \tag{3.624}$$

Proof With the aid of (3.55) , (3.619), and (3.620), we have the results. □

Theorem 3.110 *Let $\varpi \in \mathbb{C}$, $Re\left(\varpi\right) > 0$, $\lambda \in \mathbb{C}$, and $\alpha \in \mathbb{R}_{+}$. Then we have*

$$\int_{-1}^1 (1-\tau)^{\varpi}\ supercos_{\alpha}\left(\lambda\left(\tau+1\right)^{\alpha}\right)d\tau=\int_{-1}^1 (1+\tau)^{\varpi}\ supercos_{\alpha}\left(\lambda\left(1-\tau\right)^{\alpha}\right)d\tau \tag{3.625}$$

and

$$\begin{aligned}
&\tfrac{1}{\Gamma(\varpi)}\int_0^2 (2-\tau)^{\varpi}\, supercos_{\alpha}\left(\lambda\tau^{\alpha}\right) d\tau \\
&= \tfrac{1}{\Gamma(\varpi)}\int_{-1}^1 (1-\tau)^{\varpi}\, supercos_{\alpha}\left(\lambda\left(\tau+1\right)^{\alpha}\right) d\tau \\
&= \tfrac{1}{\Gamma(\varpi)}\int_{-1}^1 (1+\tau)^{\varpi}\, supercos_{\alpha}\left(\lambda\left(1-\tau\right)^{\alpha}\right) d\tau \\
&= 2^{\varpi}\ {}_1cos_1\left(\begin{matrix}(1,1)\\(1+\varpi,\alpha)\end{matrix};\lambda 2^{\alpha}\right).
\end{aligned} \tag{3.626}$$

Proof With the aid of (3.536) , (3.619), and (3.620), we have the results. □

Theorem 3.111 *Let $\varpi\in\mathbb{C}$, $Re(\varpi)>0$, $\lambda\in\mathbb{C}$ and $\alpha\in\mathbb{R}_+$. Then we have*

$$\begin{aligned}
&\int_{-1}^1 (1-\tau)^{\varpi}\, supersinh_{\alpha}\left(\lambda\left(\tau+1\right)^{\alpha}\right) d\tau \\
&\quad= \int_{-1}^1 (1+\tau)^{\varpi}\, supersinh_{\alpha}\left(\lambda\left(1-\tau\right)^{\alpha}\right) d\tau
\end{aligned} \tag{3.627}$$

and

$$\begin{aligned}
&\tfrac{1}{\Gamma(\varpi)}\int_0^2 (2-\tau)^{\varpi}\, supersinh_{\alpha}\left(\lambda\tau^{\alpha}\right) d\tau \\
&= \tfrac{1}{\Gamma(\varpi)}\int_{-1}^1 (1-\tau)^{\varpi}\, supersinh_{\alpha}\left(\lambda\left(\tau+1\right)^{\alpha}\right) d\tau \\
&= \tfrac{1}{\Gamma(\varpi)}\int_{-1}^1 (1+\tau)^{\varpi}\, supersinh_{\alpha}\left(\lambda\left(1-\tau\right)^{\alpha}\right) d\tau \\
&= 2^{\varpi}\ {}_1sinh_1\left(\begin{matrix}(1,1)\\(1+\varpi,\alpha)\end{matrix};\lambda 2^{\alpha}\right).
\end{aligned} \tag{3.628}$$

Proof With the aid of (3.537) , (3.619), and (3.620), we have the results. □

Theorem 3.112 *Let $\varpi\in\mathbb{C}$, $Re(\varpi)>0$, $\lambda\in\mathbb{C}$, and $\alpha\in\mathbb{R}_+$. Then we have*

$$\begin{aligned}
&\int_{-1}^1 (1-\tau)^{\varpi}\, supercosh_{\alpha}\left(\lambda\left(\tau+1\right)^{\alpha}\right) d\tau \\
&\quad= \int_{-1}^1 (1+\tau)^{\varpi}\, supercosh_{\alpha}\left(\lambda\left(1-\tau\right)^{\alpha}\right) d\tau
\end{aligned} \tag{3.629}$$

and

$$\begin{aligned}
&\tfrac{1}{\Gamma(\varpi)}\int_0^2 (2-\tau)^{\varpi}\, supercosh_{\alpha}\left(\lambda\tau^{\alpha}\right) d\tau \\
&= \tfrac{1}{\Gamma(\varpi)}\int_{-1}^1 (1-\tau)^{\varpi}\, supercosh_{\alpha}\left(\lambda\left(\tau+1\right)^{\alpha}\right) d\tau \\
&= \tfrac{1}{\Gamma(\varpi)}\int_{-1}^1 (1+\tau)^{\varpi}\, supercosh_{\alpha}\left(\lambda\left(1-\tau\right)^{\alpha}\right) d\tau \\
&= 2^{\varpi}\ {}_1supercosh_1\left(\begin{matrix}(1,1)\\(1+\varpi,\alpha)\end{matrix};\lambda 2^{\alpha}\right).
\end{aligned} \tag{3.630}$$

Proof With the aid of (3.538) , (3.619), and (3.620), we have the results. □

3.9 The Truncated Mittag-Leffler Functions and Related Functions

In this section, we propose the truncated Mittag-Leffler functions and related functions.

Definition 3.67 The truncated Mittag-Leffler function is defined as

$$\{E_\alpha\}^n(t) = \sum_{\kappa=0}^{n} \frac{t^\kappa}{\Gamma(\kappa\alpha+1)}, \tag{3.631}$$

where $t, \alpha \in \mathbb{C}$, $n, \kappa \in \mathbb{N}_0$, and $Re(\alpha) > 0$.

In this case, we have that

$$\{E_\alpha\}^0(t) = 1, \tag{3.632}$$

$$\{E_\alpha\}^1(t) = 1 + \frac{t}{\Gamma(\kappa\alpha+1)}, \tag{3.633}$$

$$\{E_\alpha\}^2(t) = 1 + \frac{t}{\Gamma(\alpha+1)} + \frac{t^2}{\Gamma(2\alpha+1)} \tag{3.634}$$

and

$$\{E_\alpha\}^3(t) = 1 + \frac{t}{\Gamma(\alpha+1)} + \frac{t^2}{\Gamma(2\alpha+1)} + \frac{t^3}{\Gamma(3\alpha+1)}. \tag{3.635}$$

Let

$$\{E_\alpha\}^n(i\lambda t) = \sum_{\kappa=0}^{n} \frac{(i\lambda t)^\kappa}{\Gamma(\kappa\alpha+1)}, \tag{3.636}$$

$$\{E_\alpha\}^n(-i\lambda t) = \sum_{\kappa=0}^{n} \frac{(-i\lambda t)^\kappa}{\Gamma(\kappa\alpha+1)}, \tag{3.637}$$

$$\{E_\alpha\}^n(-\lambda t) = \sum_{\kappa=0}^{n} \frac{(\lambda t)^\kappa}{\Gamma(\kappa\alpha+1)} \tag{3.638}$$

and

$$\{E_\alpha\}^n(-\lambda t) = \sum_{\kappa=0}^{n} \frac{(-\lambda t)^\kappa}{\Gamma(\kappa\alpha+1)}, \tag{3.639}$$

where $i = \sqrt{-1}$, $t, \alpha, \lambda \in \mathbb{C}$, $n, \kappa \in \mathbb{N}_0$, and $Re(\alpha) > 0$.

Moreover, we have that

$$\{E_\alpha\}^n\left(\lambda t^\alpha\right)=\sum_{\kappa=0}^{n}\frac{\lambda^\kappa t^{\kappa\alpha}}{\Gamma(\kappa\alpha+1)} \tag{3.640}$$

or

$$\{E_\alpha\}^n\left(\lambda t^\gamma\right)=\sum_{\kappa=0}^{n}\frac{\lambda^\kappa t^{\kappa\gamma}}{\Gamma(\kappa\alpha+1)}, \tag{3.641}$$

where $i=\sqrt{-1}$, $t,\alpha,\lambda,\gamma\in\mathbb{C}$, $n,\kappa\in\mathbb{N}_0$, and $Re(\alpha)>0$.

Definition 3.68 The truncated pre-supersine function based on the truncated Mittag-Leffler function is defined as

$$\begin{aligned}&\{presin_\alpha\}^n(t)\\&=\tfrac{1}{2i}\left(\{E_\alpha\}^n(it)-\{E_\alpha\}^n(-it)\right)\\&=\sum_{\kappa=0}^{n}\tfrac{(-1)^\kappa t^{2\kappa+1}}{\Gamma((2\kappa+1)\alpha+1)},\end{aligned} \tag{3.642}$$

where $i=\sqrt{-1}$, $t,\alpha\in\mathbb{C}$, $n,\kappa\in\mathbb{N}_0$, and $Re(\alpha)>0$.

In this case, we present

$$\begin{aligned}&\{presin_\alpha\}^0(t)\\&=\tfrac{1}{2i}\left(\{E_\alpha\}^0(it)-\{E_\alpha\}^0(-it)\right)\\&=\tfrac{t}{\Gamma(\alpha+1)},\end{aligned} \tag{3.643}$$

$$\begin{aligned}&\{presin_\alpha\}^1(t)\\&=\tfrac{1}{2i}\left(\{E_\alpha\}^1(it)-\{E_\alpha\}^1(-it)\right)\\&=\tfrac{t}{\Gamma(\alpha+1)}-\tfrac{t^3}{\Gamma(3\alpha+1)},\end{aligned} \tag{3.644}$$

and

$$\begin{aligned}&\{presin_\alpha\}^2(t)\\&=\tfrac{1}{2i}\left(\{E_\alpha\}^2(it)-\{E_\alpha\}^2(-it)\right)\\&=\tfrac{t}{\Gamma(\alpha+1)}-\tfrac{t^3}{\Gamma(3\alpha+1)}+\tfrac{t^5}{\Gamma(5\alpha+1)},\end{aligned} \tag{3.645}$$

where $i=\sqrt{-1}$, $t,\alpha\in\mathbb{C}$, $n,\kappa\in\mathbb{N}_0$, and $Re(\alpha)>0$.

Definition 3.69 The truncated pre-supercosine function based on the truncated Mittag-Leffler function is defined as

$$\begin{aligned}&\{precos_\alpha\}^n(t)\\&=\tfrac{1}{2}\left(\{E_\alpha\}^n(it)+\{E_\alpha\}^n(-it)\right)\\&=\sum_{\kappa=0}^{n}\frac{(-1)^\kappa t^{2\kappa}}{\Gamma(2\kappa\alpha+1)},\end{aligned}\tag{3.646}$$

where $i=\sqrt{-1}$, $t,\alpha\in\mathbb{C}$, $n,\kappa\in\mathbb{N}_0$, and $Re(\alpha)>0$.

Thus,

$$\begin{aligned}&\{precos_\alpha\}^0(t)\\&=\tfrac{1}{2}\left(\{E_\alpha\}^0(it)+\{E_\alpha\}^0(-it)\right)\\&=1,\end{aligned}\tag{3.647}$$

$$\begin{aligned}&\{precos_\alpha\}^1(t)\\&=\tfrac{1}{2}\left(\{E_\alpha\}^1(it)+\{E_\alpha\}^1(-it)\right)\\&=1-\frac{t^2}{\Gamma(2\alpha+1)},\end{aligned}\tag{3.648}$$

and

$$\begin{aligned}&\{precos_\alpha\}^2(t)\\&=\tfrac{1}{2}\left(\{E_\alpha\}^2(it)+\{E_\alpha\}^2(-it)\right)\\&=1-\frac{t^2}{\Gamma(2\alpha+1)}+\frac{t^4}{\Gamma(4\alpha+1)},\end{aligned}\tag{3.649}$$

where $i=\sqrt{-1}$, $t,\alpha\in\mathbb{C}$, $n,\kappa\in\mathbb{N}_0$, and $Re(\alpha)>0$.

Definition 3.70 The truncated superhyperbolic pre-supersine function based on the truncated Mittag-Leffler function is defined as

$$\begin{aligned}&\{presinh_\alpha\}^n(t)\\&=\tfrac{1}{2}\left(\{E_\alpha\}^n(t)-\{E_\alpha\}^n(-t)\right)\\&=\sum_{\kappa=0}^{n}\frac{t^{2\kappa+1}}{\Gamma((2\kappa+1)\alpha+1)},\end{aligned}\tag{3.650}$$

where $i=\sqrt{-1}$, $t,\alpha\in\mathbb{C}$, $n,\kappa\in\mathbb{N}_0$, and $Re(\alpha)>0$.

So, we present

$$\begin{aligned}&\{presinh_\alpha\}^0(t)\\&=\tfrac{1}{2}\left(\{E_\alpha\}^0(t)-\{E_\alpha\}^0(-t)\right)\\&=\frac{t}{\Gamma(\alpha+1)},\end{aligned}\tag{3.651}$$

$$\begin{aligned}&\{presinh_\alpha\}^1(t)\\&=\tfrac{1}{2}\left(\{E_\alpha\}^1(t)-\{E_\alpha\}^1(-t)\right)\\&=\tfrac{t}{\Gamma(\alpha+1)}+\tfrac{t^3}{\Gamma(3\alpha+1)},\end{aligned}\tag{3.652}$$

and

$$\begin{aligned}&\{presinh_\alpha\}^2(t)\\&=\tfrac{1}{2}\left(\{E_\alpha\}^2(t)-\{E_\alpha\}^2(-t)\right)\\&=\tfrac{t}{\Gamma(\alpha+1)}+\tfrac{t^3}{\Gamma(3\alpha+1)}+\tfrac{t^5}{\Gamma(5\alpha+1)},\end{aligned}\tag{3.653}$$

where $i=\sqrt{-1}$, $t,\alpha\in\mathbb{C}$, $n,\kappa\in\mathbb{N}_0$, and $Re(\alpha)>0$.

Definition 3.71 The truncated superhyperbolic pre-supercosine function based on the truncated Mittag-Leffler function is defined as

$$\begin{aligned}&\{precosh_\alpha\}^n(t)\\&=\tfrac{1}{2}\left(\{E_\alpha\}^n(t)+\{E_\alpha\}^n(-t)\right)\\&=\sum_{\kappa=0}^{n}\tfrac{t^{2\kappa}}{\Gamma(2\kappa\alpha+1)},\end{aligned}\tag{3.654}$$

where $i=\sqrt{-1}$, $t,\alpha\in\mathbb{C}$, $n,\kappa\in\mathbb{N}_0$, and $Re(\alpha)>0$.

Thus, we have that

$$\begin{aligned}&\{precosh_\alpha\}^0(t)\\&=\tfrac{1}{2}\left(\{E_\alpha\}^0(t)+\{E_\alpha\}^0(-t)\right)\\&=1,\end{aligned}\tag{3.655}$$

$$\begin{aligned}&\{precosh_\alpha\}^1(t)\\&=\tfrac{1}{2}\left(\{E_\alpha\}^1(t)+\{E_\alpha\}^1(-t)\right)\\&=1+\tfrac{t^2}{\Gamma(2\alpha+1)},\end{aligned}\tag{3.656}$$

and

$$\begin{aligned}&\{precosh_\alpha\}^2(t)\\&=\tfrac{1}{2}\left(\{E_\alpha\}^2(t)+\{E_\alpha\}^2(-t)\right)\\&=1+\tfrac{t^2}{\Gamma(2\alpha+1)}+\tfrac{t^4}{\Gamma(4\alpha+1)},\end{aligned}\tag{3.657}$$

where $i=\sqrt{-1}$, $t,\alpha\in\mathbb{C}$, $n,\kappa\in\mathbb{N}_0$, and $Re(\alpha)>0$.

From the above definitions, we directly derive the following results:

$$\{presin_\alpha\}^n(\lambda t)=\frac{1}{2i}\left(\{E_\alpha\}^n(i\lambda t)-\{E_\alpha\}^n(-i\lambda t)\right),\tag{3.658}$$

$$\{precos_\alpha\}^n(\lambda t) = \frac{1}{2}\left(\{E_\alpha\}^n(i\lambda t) + \{E_\alpha\}^n(-i\lambda t)\right), \tag{3.659}$$

$$\{presinh_\alpha\}^n(\lambda t) = \frac{1}{2}\left(\{E_\alpha\}^n(\lambda t) - \{E_\alpha\}^n(-\lambda t)\right), \tag{3.660}$$

and

$$\{precosh_\alpha\}^n(\lambda t) = \frac{1}{2}\left(\{E_\alpha\}^n(\lambda t) + \{E_\alpha\}^n(-\lambda t)\right), \tag{3.661}$$

where $i = \sqrt{-1}$, $t, \alpha \in \mathbb{C}$, $n, \kappa \in \mathbb{N}_0$, and $Re(\alpha) > 0$.

Definition 3.72 The truncated pre-supersine function based on the truncated Mittag-Leffler function is defined as

$$\begin{aligned} &\{presin_\alpha\}^n(t^\gamma) \\ &= \tfrac{1}{2i}\left(\{E_\alpha\}^n(it^\gamma) - \{E_\alpha\}^n(-it^\gamma)\right) \\ &= \sum_{\kappa=0}^{n} \tfrac{(-1)^\kappa t^{(2\kappa+1)\gamma}}{\Gamma((2\kappa+1)\alpha+1)}, \end{aligned} \tag{3.662}$$

where $i = \sqrt{-1}$, $t, \alpha, \gamma \in \mathbb{C}$, $n, \kappa \in \mathbb{N}_0$, and $Re(\alpha) > 0$.

Thus, we have

$$\begin{aligned} &\{presin_\alpha\}^0(t^\gamma) \\ &= \tfrac{1}{2i}\left(\{E_\alpha\}^0(it^\gamma) - \{E_\alpha\}^0(-it^\gamma)\right) \\ &= \tfrac{t^\gamma}{\Gamma(\alpha+1)}, \end{aligned} \tag{3.663}$$

$$\begin{aligned} &\{presin_\alpha\}^1(t^\gamma) \\ &= \tfrac{1}{2i}\left(\{E_\alpha\}^1(it^\gamma) - \{E_\alpha\}^1(-it^\gamma)\right) \\ &= \tfrac{t^\gamma}{\Gamma(\alpha+1)} - \tfrac{t^{3\gamma}}{\Gamma(3\alpha+1)}, \end{aligned} \tag{3.664}$$

and

$$\begin{aligned} &\{presin_\alpha\}^2(t^\gamma) \\ &= \tfrac{1}{2i}\left(\{E_\alpha\}^2(it^\gamma) - \{E_\alpha\}^2(-it^\gamma)\right) \\ &= \tfrac{t^\gamma}{\Gamma(\alpha+1)} - \tfrac{t^{3\gamma}}{\Gamma(3\alpha+1)} + \tfrac{t^{5\gamma}}{\Gamma(5\alpha+1)}, \end{aligned} \tag{3.665}$$

where $i = \sqrt{-1}$, $t, \alpha, \gamma \in \mathbb{C}$, $n, \kappa \in \mathbb{N}_0$, and $Re(\alpha) > 0$.

Definition 3.73 The truncated pre-supercosine function based on the truncated Mittag-Leffler function is defined as

$$\begin{aligned}&\{precos_\alpha\}^n\left(t^\gamma\right)\\&=\tfrac{1}{2}\left(\{E_\alpha\}^n\left(it^\gamma\right)+\{E_\alpha\}^n\left(-it^\gamma\right)\right)\\&=\sum_{\kappa=0}^{n}\tfrac{(-1)^\kappa t^{2\kappa\gamma}}{\Gamma(2\kappa\alpha+1)},\end{aligned}\tag{3.666}$$

where $i=\sqrt{-1}$, $t,\alpha,\gamma\in\mathbb{C}$, $n,\kappa\in\mathbb{N}_0$, and $Re\,(\alpha)>0$.

Thus,

$$\begin{aligned}&\{precos_\alpha\}^0\left(t^\gamma\right)\\&=\tfrac{1}{2}\left(\{E_\alpha\}^0\left(it^\gamma\right)+\{E_\alpha\}^0\left(-it^\gamma\right)\right)\\&=1,\end{aligned}\tag{3.667}$$

$$\begin{aligned}&\{precos_\alpha\}^1\left(t^\gamma\right)\\&=\tfrac{1}{2}\left(\{E_\alpha\}^1\left(it^\gamma\right)+\{E_\alpha\}^1\left(-it^\gamma\right)\right)\\&=1-\tfrac{t^{2\gamma}}{\Gamma(2\alpha+1)},\end{aligned}\tag{3.668}$$

and

$$\begin{aligned}&\{precos_\alpha\}^2\left(t^\gamma\right)\\&=\tfrac{1}{2}\left(\{E_\alpha\}^2\left(it^\gamma\right)+\{E_\alpha\}^2\left(-it^\gamma\right)\right)\\&=1-\tfrac{t^{2\gamma}}{\Gamma(2\alpha+1)}+\tfrac{t^{4\gamma}}{\Gamma(4\alpha+1)},\end{aligned}\tag{3.669}$$

where $i=\sqrt{-1}$, $t,\alpha,\gamma\in\mathbb{C}$, $n,\kappa\in\mathbb{N}_0$, and $Re\,(\alpha)>0$.

Definition 3.74 The truncated superhyperbolic pre-supersine function based on the truncated Mittag-Leffler function is defined as

$$\begin{aligned}&\{presinh_\alpha\}^n\left(t^\gamma\right)\\&=\tfrac{1}{2}\left(\{E_\alpha\}^n\left(t^\gamma\right)-\{E_\alpha\}^n\left(-t^\gamma\right)\right)\\&=\sum_{\kappa=0}^{n}\tfrac{t^{(2\kappa+1)\gamma}}{\Gamma((2\kappa+1)\alpha+1)},\end{aligned}\tag{3.670}$$

where $i=\sqrt{-1}$, $t,\alpha,\gamma\in\mathbb{C}$, $n,\kappa\in\mathbb{N}_0$, and $Re\,(\alpha)>0$.

So, we have that

$$\begin{aligned}&\{presinh_\alpha\}^0\left(t^\gamma\right)\\&=\tfrac{1}{2}\left(\{E_\alpha\}^0\left(t^\gamma\right)-\{E_\alpha\}^0\left(-t^\gamma\right)\right)\\&=\tfrac{t^\gamma}{\Gamma(\alpha+1)},\end{aligned}\tag{3.671}$$

$$\begin{aligned}&\{presinh_\alpha\}^1(t^\gamma)\\&=\tfrac{1}{2}\left(\{E_\alpha\}^1(t^\gamma)-\{E_\alpha\}^1(-t^\gamma)\right)\\&=\tfrac{t^\gamma}{\Gamma(\alpha+1)}+\tfrac{t^{3\gamma}}{\Gamma(3\alpha+1)},\end{aligned}\tag{3.672}$$

and

$$\begin{aligned}&\{presinh_\alpha\}^2(t^\gamma)\\&=\tfrac{1}{2}\left(\{E_\alpha\}^2(t^\gamma)-\{E_\alpha\}^2(-t^\gamma)\right)\\&=\tfrac{t^\gamma}{\Gamma(\alpha+1)}+\tfrac{t^{3\gamma}}{\Gamma(3\alpha+1)}+\tfrac{t^{5\gamma}}{\Gamma(5\alpha+1)},\end{aligned}\tag{3.673}$$

where $i=\sqrt{-1}$, $t,\alpha,\gamma\in\mathbb{C}$, $n,\kappa\in\mathbb{N}_0$, and $Re(\alpha)>0$.

Definition 3.75 The truncated superhyperbolic pre-supercosine function based on the truncated Mittag-Leffler function is defined as

$$\begin{aligned}&\{precosh_\alpha\}^n(t^\gamma)\\&=\tfrac{1}{2}\left(\{E_\alpha\}^n(t^\gamma)+\{E_\alpha\}^n(-t^\gamma)\right)\\&=\sum_{\kappa=0}^{n}\tfrac{t^{2\kappa\gamma}}{\Gamma(2\kappa\alpha+1)},\end{aligned}\tag{3.674}$$

where $i=\sqrt{-1}$, $t,\alpha,\gamma\in\mathbb{C}$, $n,\kappa\in\mathbb{N}_0$ and $Re(\alpha)>0$.

Thus, we have that

$$\begin{aligned}&\{precosh_\alpha\}^0(t^\gamma)\\&=\tfrac{1}{2}\left(\{E_\alpha\}^0(t^\gamma)+\{E_\alpha\}^0(-t^\gamma)\right)\\&=1,\end{aligned}\tag{3.675}$$

$$\begin{aligned}&\{precosh_\alpha\}^1(t^\gamma)\\&=\tfrac{1}{2}\left(\{E_\alpha\}^1(t^\gamma)+\{E_\alpha\}^1(-t^\gamma)\right)\\&=1+\tfrac{t^{2\gamma}}{\Gamma(2\alpha+1)},\end{aligned}\tag{3.676}$$

and

$$\begin{aligned}&\{precosh_\alpha\}^2(t^\gamma)\\&=\tfrac{1}{2}\left(\{E_\alpha\}^2(t^\gamma)+\{E_\alpha\}^2(-t^\gamma)\right)\\&=1+\tfrac{t^{2\gamma}}{\Gamma(2\alpha+1)}+\tfrac{t^{4\gamma}}{\Gamma(4\alpha+1)},\end{aligned}\tag{3.677}$$

where $i=\sqrt{-1}$, $t,\alpha,\gamma\in\mathbb{C}$, $n,\kappa\in\mathbb{N}_0$, and $Re(\alpha)>0$.

With the above definitions, we can write

$$\{presin_\alpha\}^n\left(\lambda t^\gamma\right)=\frac{1}{2i}\left(\{E_\alpha\}^n\left(i\lambda t^\gamma\right)-\{E_\alpha\}^n\left(-i\lambda t^\gamma\right)\right),\tag{3.678}$$

$$\{precos_{\alpha}\}^{n}\left(\lambda t^{\gamma}\right)=\frac{1}{2}\left(\{E_{\alpha}\}^{n}\left(i\lambda t^{\gamma}\right)+\{E_{\alpha}\}^{n}\left(-i\lambda t^{\gamma}\right)\right), \tag{3.679}$$

$$\{presinh_{\alpha}\}^{n}\left(\lambda t^{\gamma}\right)=\frac{1}{2}\left(\{E_{\alpha}\}^{n}\left(\lambda t^{\gamma}\right)-\{E_{\alpha}\}^{n}\left(-\lambda t^{\gamma}\right)\right), \tag{3.680}$$

and

$$\{precosh_{\alpha}\}^{n}\left(\lambda t^{\gamma}\right)=\frac{1}{2}\left(\{E_{\alpha}\}^{n}\left(\lambda t^{\gamma}\right)+\{E_{\alpha}\}^{n}\left(-\lambda t^{\gamma}\right)\right), \tag{3.681}$$

where $i=\sqrt{-1}$, $t,\alpha,\gamma\in\mathbb{C}$, $n,\kappa\in\mathbb{N}_0$, and $Re(\alpha)>0$.

Definition 3.76 The truncated supersine function based on the truncated Mittag-Leffler function is defined as

$$\begin{aligned}&\{supersin_{\alpha}\}^{n}\left(t^{\alpha}\right)\\&=\tfrac{1}{2i}\left(\{E_{\alpha}\}^{n}\left(it^{\alpha}\right)-\{E_{\alpha}\}^{n}\left(-it^{\alpha}\right)\right)\\&=\sum_{\kappa=0}^{n}\tfrac{(-1)^{\kappa}t^{(2\kappa+1)\alpha}}{\Gamma((2\kappa+1)\alpha+1)},\end{aligned} \tag{3.682}$$

where $i=\sqrt{-1}$, $t,\alpha\in\mathbb{C}$, $n,\kappa\in\mathbb{N}_0$, and $Re(\alpha)>0$.

In this case, we present

$$\begin{aligned}&\{supersin_{\alpha}\}^{0}\left(t^{\alpha}\right)\\&=\tfrac{1}{2i}\left(\{E_{\alpha}\}^{0}\left(it^{\alpha}\right)-\{E_{\alpha}\}^{0}\left(-it^{\alpha}\right)\right)\\&=\tfrac{t^{\alpha}}{\Gamma(\alpha+1)},\end{aligned} \tag{3.683}$$

$$\begin{aligned}&\{supersin_{\alpha}\}^{1}\left(t^{\alpha}\right)\\&=\tfrac{1}{2i}\left(\{E_{\alpha}\}^{1}\left(it^{\alpha}\right)-\{E_{\alpha}\}^{1}\left(-it^{\alpha}\right)\right)\\&=\tfrac{t^{\alpha}}{\Gamma(\alpha+1)}-\tfrac{t^{3\alpha}}{\Gamma(3\alpha+1)},\end{aligned} \tag{3.684}$$

and

$$\begin{aligned}&\{supersin_{\alpha}\}^{2}\left(t^{\alpha}\right)\\&=\tfrac{1}{2i}\left(\{E_{\alpha}\}^{2}\left(it^{\alpha}\right)-\{E_{\alpha}\}^{2}\left(-it^{\alpha}\right)\right)\\&=\tfrac{t^{\alpha}}{\Gamma(\alpha+1)}-\tfrac{t^{3\alpha}}{\Gamma(3\alpha+1)}+\tfrac{t^{5\alpha}}{\Gamma(5\alpha+1)},\end{aligned} \tag{3.685}$$

where $i=\sqrt{-1}$, $t,\alpha\in\mathbb{C}$, $n,\kappa\in\mathbb{N}_0$, and $Re(\alpha)>0$.

Definition 3.77 The truncated supercosine function based on the truncated Mittag-Leffler function is defined as

$$\begin{aligned}&\{supercos_\alpha\}^n(t^\alpha)\\&=\frac{1}{2}\left(\{E_\alpha\}^n(it^\alpha)+\{E_\alpha\}^n(-it^\alpha)\right)\\&=\sum_{\kappa=0}^{n}\frac{(-1)^\kappa t^{2\kappa\alpha}}{\Gamma(2\kappa\alpha+1)},\end{aligned}\tag{3.686}$$

where $i=\sqrt{-1}$, $t,\alpha\in\mathbb{C}$, $n,\kappa\in\mathbb{N}_0$, and $Re(\alpha)>0$.

Thus, we present

$$\begin{aligned}&\{supercos_\alpha\}^0(t^\alpha)\\&=\frac{1}{2}\left(\{E_\alpha\}^0(it^\alpha)+\{E_\alpha\}^0(-it^\alpha)\right)\\&=1,\end{aligned}\tag{3.687}$$

$$\begin{aligned}&\{supercos_\alpha\}^1(t^\alpha)\\&=\frac{1}{2}\left(\{E_\alpha\}^1(it^\alpha)+\{E_\alpha\}^1(-it^\alpha)\right)\\&=1-\frac{t^{2\alpha}}{\Gamma(2\alpha+1)},\end{aligned}\tag{3.688}$$

and

$$\begin{aligned}&\{supercos_\alpha\}^2(t^\alpha)\\&=\frac{1}{2}\left(\{E_\alpha\}^2(it^\alpha)+\{E_\alpha\}^2(-it^\alpha)\right)\\&=1-\frac{t^{2\alpha}}{\Gamma(2\alpha+1)}+\frac{t^{4\alpha}}{\Gamma(4\alpha+1)},\end{aligned}\tag{3.689}$$

where $i=\sqrt{-1}$, $t,\alpha\in\mathbb{C}$, $n,\kappa\in\mathbb{N}_0$, and $Re(\alpha)>0$.

Definition 3.78 The truncated superhyperbolic supersine function based on the truncated Mittag-Leffler function is defined as

$$\begin{aligned}&\{supersinh_\alpha\}^n(t^\alpha)\\&=\frac{1}{2}\left(\{E_\alpha\}^n(t^\alpha)-\{E_\alpha\}^n(-t^\alpha)\right)\\&=\sum_{\kappa=0}^{n}\frac{t^{(2\kappa+1)\alpha}}{\Gamma((2\kappa+1)\alpha+1)},\end{aligned}\tag{3.690}$$

where $i=\sqrt{-1}$, $t,\alpha\in\mathbb{C}$, $n,\kappa\in\mathbb{N}_0$, and $Re(\alpha)>0$.

In this case, we present

$$\begin{aligned}&\{supersinh_\alpha\}^0(t^\alpha)\\&=\frac{1}{2}\left(\{E_\alpha\}^0(t^\alpha)-\{E_\alpha\}^0(-t^\alpha)\right)\\&=\frac{t^\alpha}{\Gamma(\alpha+1)},\end{aligned}\tag{3.691}$$

$$\begin{aligned}
&\{supersinh_\alpha\}^1 (t^\alpha)\\
&= \frac{1}{2}\left(\{E_\alpha\}^1 (t^\alpha) - \{E_\alpha\}^1 (-t^\alpha)\right) \\
&= \frac{t^\alpha}{\Gamma(\alpha+1)} + \frac{t^{3\alpha}}{\Gamma(3\alpha+1)},
\end{aligned} \tag{3.692}$$

and

$$\begin{aligned}
&\{supersinh_\alpha\}^2 (t^\alpha)\\
&= \frac{1}{2}\left(\{E_\alpha\}^2 (t^\alpha) - \{E_\alpha\}^2 (-t^\alpha)\right) \\
&= \frac{t^\alpha}{\Gamma(\alpha+1)} + \frac{t^{3\alpha}}{\Gamma(3\alpha+1)} + \frac{t^{5\alpha}}{\Gamma(5\alpha+1)},
\end{aligned} \tag{3.693}$$

where $i = \sqrt{-1}$, $t, \alpha \in \mathbb{C}$, $n, \kappa \in \mathbb{N}_0$, and $Re(\alpha) > 0$.

Definition 3.79 The truncated superhyperbolic supercosine function based on the truncated Mittag-Leffler function is defined as

$$\begin{aligned}
&\{supercosh_\alpha\}^n (t^\alpha)\\
&= \frac{1}{2}\left(\{E_\alpha\}^n (t^\alpha) + \{E_\alpha\}^n (-t^\alpha)\right) \\
&= \sum_{\kappa=0}^{n} \frac{t^{2\kappa\alpha}}{\Gamma(2\kappa\alpha+1)},
\end{aligned} \tag{3.694}$$

where $i = \sqrt{-1}$, $t, \alpha \in \mathbb{C}$, $n, \kappa \in \mathbb{N}_0$, and $Re(\alpha) > 0$.

Thus, we have that

$$\begin{aligned}
&\{supercosh_\alpha\}^0 (t^\alpha)\\
&= \frac{1}{2}\left(\{E_\alpha\}^0 (t^\alpha) + \{E_\alpha\}^0 (-t^\alpha)\right) \\
&= 1,
\end{aligned} \tag{3.695}$$

$$\begin{aligned}
&\{supercosh_\alpha\}^1 (t^\alpha)\\
&= \frac{1}{2}\left(\{E_\alpha\}^1 (t^\alpha) + \{E_\alpha\}^1 (-t^\alpha)\right) \\
&= 1 + \frac{t^{2\alpha}}{\Gamma(2\alpha+1)},
\end{aligned} \tag{3.696}$$

and

$$\begin{aligned}
&\{supercosh_\alpha\}^2 (t^\alpha)\\
&= \frac{1}{2}\left(\{E_\alpha\}^2 (t^\alpha) + \{E_\alpha\}^2 (-t^\alpha)\right) \\
&= 1 + \frac{t^{2\alpha}}{\Gamma(2\alpha+1)} + \frac{t^{4\alpha}}{\Gamma(4\alpha+1)},
\end{aligned} \tag{3.697}$$

where $i = \sqrt{-1}$, $t, \alpha \in \mathbb{C}$, $n, \kappa \in \mathbb{N}_0$, and $Re(\alpha) > 0$.

With the aid of the above definitions, we may give the following results:

$$\{supersin_\alpha\}^n \left(\lambda t^\alpha\right) = \frac{1}{2i}\left(\{E_\alpha\}^n \left(i\lambda t^\alpha\right) - \{E_\alpha\}^n \left(-i\lambda t^\alpha\right)\right), \tag{3.698}$$

$$\{supercos_{\alpha}\}^{n}\left(\lambda t^{\alpha}\right)=\frac{1}{2}\left(\{E_{\alpha}\}^{n}\left(i\lambda t^{\alpha}\right)+\{E_{\alpha}\}^{n}\left(-i\lambda t^{\alpha}\right)\right), \tag{3.699}$$

$$\{supersinh_{\alpha}\}^{n}\left(\lambda t^{\alpha}\right)=\frac{1}{2}\left(\{E_{\alpha}\}^{n}\left(\lambda t^{\alpha}\right)-\{E_{\alpha}\}^{n}\left(-\lambda t^{\alpha}\right)\right), \tag{3.700}$$

and

$$\{supercosh_{\alpha}\}^{n}\left(\lambda t^{\alpha}\right)=\frac{1}{2}\left(\{E_{\alpha}\}^{n}\left(\lambda t^{\alpha}\right)+\{E_{\alpha}\}^{n}\left(-\lambda t^{\alpha}\right)\right), \tag{3.701}$$

where $i=\sqrt{-1}$, $t,\alpha\in\mathbb{C}$, $n,\kappa\in\mathbb{N}_0$, and $Re\left(\alpha\right)>0$.

3.10 Applications in Anomalous Linear Viscoelasticity

In this section, we now present the mathematical models for the anomalous linear viscoelasticity.

Let ω be the elastic modulus, θ be the viscosity, $\varepsilon\left(t\right)$ be the strain, $\sigma\left(t\right)$ be the stress, and t be the time.

Model The Newton-like dashpot element with the Hille–Tamarkin general fractional derivative is given as [92]

$$\sigma\left(t\right)=\theta\,{}_{Ml}^{RL}D_{a+}^{\alpha,\lambda}\varepsilon\left(t\right). \tag{3.702}$$

Model The Newton-like dashpot element with the Liouville–Sonine–Caputo type general fractional derivative via Mittag-Leffler function is given as [92]

$$\sigma\left(t\right)=\theta\,{}_{Ml}^{LSC}D_{a+}^{\alpha,\lambda}\varepsilon\left(t\right). \tag{3.703}$$

Model The Newton-like dashpot element with the Liouville–Weyl–Hille–Tamarkin type general fractional derivative is given as [92]

$$\sigma\left(t\right)=\theta\,{}_{Mlr}^{RL}D_{+}^{\alpha,\lambda}\varepsilon\left(t\right). \tag{3.704}$$

Model The Newton-like dashpot element with the Liouville–Sonine–Caputo type general fractional derivative via Mittag-Leffler function is given as [92]

$$\sigma\left(t\right)=\theta\,{}_{Ml}^{LSC}D_{+}^{\alpha,\lambda}\varepsilon\left(t\right). \tag{3.705}$$

Model The Newton-like dashpot element with the Hilfer–Hille–Tamarkin type general fractional derivative is given as [92]

$$\sigma\left(t\right)=\theta\,{}_{HHT}D_{a+}^{\alpha,\beta,\lambda}\varepsilon\left(t\right). \tag{3.706}$$

Model The Newton-like dashpot element with the Hille–Tamarkin general fractional derivative with respect to another function is given as

$$\sigma(t) = \theta_{Ml}^{RL} D_{a+,h}^{\alpha,\lambda} \varepsilon(t). \tag{3.707}$$

Model The Newton-like dashpot element with the Hille–Tamarkin general fractional derivative with power function is given as

$$\sigma(t) = \theta_{\Omega}^{RL} D_{a+}^{\alpha,\lambda} \varepsilon(t). \tag{3.708}$$

Model The Newton-like dashpot element with the Liouville–Sonine–Caputo type general fractional derivative with respect to another function via Mittag-Leffler function is given as

$$\sigma(t) = \theta_{Ml}^{LSC} D_{a+,h}^{\alpha,\lambda} \varepsilon(t). \tag{3.709}$$

Model The Newton-like dashpot element with the Liouville–Sonine–Caputo type general fractional derivative with respect to another function via Mittag-Leffler function is given as

$$\sigma(t) = \theta_{\Omega}^{LSC} D_{a+}^{\alpha,\lambda} \varepsilon(t). \tag{3.710}$$

Model The Newton-like dashpot element with the Liouville–Weyl–Hille–Tamarkin type general fractional derivative with respect to another function is given as

$$\sigma(t) = \theta_{Mlr}^{RL} D_{+,h}^{\alpha,\lambda} \varepsilon(t). \tag{3.711}$$

Model The Newton-like dashpot element with the Liouville–Sonine–Caputo type general fractional derivative with respect to another function via Mittag-Leffler function is given as

$$\sigma(t) = \theta_{Ml}^{LSC} D_{+,h}^{\alpha,\lambda} \varepsilon(t). \tag{3.712}$$

Model The Newton-like dashpot element with the Hilfer–Hille–Tamarkin type general fractional derivative with respect to another function is given as

$$\sigma(t) = \theta_{HHT} D_{a+,h}^{\alpha,\beta,\lambda} \varepsilon(t). \tag{3.713}$$

Model The Newton-like dashpot element with the Hilfer–Hille–Tamarkin type general fractional derivative with respect to another function is given as

$$\sigma(t) = \theta_{HHT} D_{+,h}^{\alpha,\beta,\lambda} \varepsilon(t). \tag{3.714}$$

Chapter 4
Wiman, Supertrigonometric, and Superhyperbolic Functions

Abstract In this chapter, we introduce the Wiman function and present the supertrigonometric and superhyperbolic functions via Wiman function. The Laplace transforms and integral representations for the supertrigonometric and superhyperbolic functions are considered. The truncated Wiman, supertrigonometric, and superhyperbolic functions are proposed, and the general fractional calculus via Wiman function is also discussed. The integral equations and some of the mathematical models related to the Wiman function are finally presented.

4.1 The Wiman Function: History, Definitions, Properties, and Theorems

In this section, we introduce the concept, theorems, properties, and corollaries for the Wiman function.

4.1.1 The Wiman Function

In 1905, Anders Wiman was a Swedish mathematician who worked on algebraic geometry and applications of group theory and generalized the Mittag-Leffler function [239], called the two-parametric Mittag-Leffler function, the Mittag-Leffler type function, or the Wiman function.

Definition 4.1 The Wiman function is defined as [239, 240]

$$E_{\alpha,\beta}(t) = \sum_{n=0}^{\infty} \frac{t^n}{\Gamma(n\alpha+\beta)}, \tag{4.1}$$

where $t, \alpha, \beta \in \mathbb{C}$, $Re(\alpha) > 0$, and $Re(\beta) > 0$.

The work was discovered by Wiman in 1905 [239, 240].

X.-J. Yang, *Theory and Applications of Special Functions for Scientists and Engineers*, https://doi.org/10.1007/978-981-33-6334-2_4

It was called the Mittag-Leffler type function by Agarwal in 1953 [241], Humbert in 1953 [227], and Dzherbashyan in [242, 243]. For the generalized function, see [164].

The relation between the Wiman function and the generalized Wright function is given as follows [238]:

$$E_{\alpha,\beta}(t) = {}_1\varphi_1\left(\begin{matrix}(1,1)\\(\beta,\alpha)\end{matrix};t\right), \tag{4.2}$$

where $t, \alpha, \beta \in \mathbb{C}$, $Re(\alpha) > 0$, and $Re(\beta) > 0$.

Theorem 4.1 *Let $\alpha > 0$ and $\beta > 0$.*
Then, we have

$$tE_{\alpha,\alpha+\beta}(t) = E_{\alpha,\beta}(t) - \frac{1}{\Gamma(\beta)}, \tag{4.3}$$

where $t \in \mathbb{C}$.

The work was discovered by Erdelyi et al. in 1953 [32] and reported by Gorenflo, Kilbas, and Rogosin in 1998 [244].

Corollary 4.1 *Then, we have*

$$tE_{\alpha,\alpha+1}(t) = E_{\alpha}(t) - 1, \tag{4.4}$$

where $t \in \mathbb{C}$.

The work was discovered by Gorenflo, Kilbas, and Rogosin in 1998 [244].

Definition 4.2 The function of Wiman type is defined as

$$t^{\beta-1}E_{\alpha,\beta}\left(t^{\alpha}\right) = \sum_{n=0}^{\infty}\frac{t^{n\alpha+\beta-1}}{\Gamma(n\alpha+\beta)}, \tag{4.5}$$

where $t, \alpha, \beta \in \mathbb{C}$, $Re(\alpha) > 0$, and $Re(\beta) > 0$.

The work was discovered by Erdelyi et al. in 1953 [32] and reported by Srivastava in 1994 [245].

Theorem 4.2 *Let $\alpha, \beta \in \mathbb{C}$, $t \in \mathbb{R}$, $n \in \mathbb{N}$, $Re(\alpha) > 0$, and $Re(\beta) > 0$.*
Then, we have

$$\frac{\partial^n}{\partial t^n}\left[t^{\beta-1}E_{\alpha,\beta}\left(t^{\alpha}\right)\right] = t^{\beta-n-1}E_{\alpha,\beta-n}\left(t^{\alpha}\right). \tag{4.6}$$

The work was discovered by Erdelyi et al. in 1953 [32].

Property 4.1 If $t, \alpha \in \mathbb{C}$ and $Re(\alpha) > 0$, then we have

$$E_{\alpha,1}(t) = E_{\alpha}(t). \tag{4.7}$$

The work was discovered by Wiman in 1905 [239, 240].

Theorem 4.3 *Let $\alpha, \beta \in \mathbb{C}$, $t, \lambda \in \mathbb{R}$, $Re(\alpha) > 0$, and $Re(\beta) > 0$.*
Then, we have

$$\int_0^t t^{\beta-1} E_{\alpha,\beta}\left(\lambda t^{\alpha}\right) dt = t^{\beta} E_{\alpha,\beta+1}\left(\lambda t^{\alpha}\right). \tag{4.8}$$

Theorem 4.4 *Let $\alpha, \beta \in \mathbb{C}$, $t, \lambda \in \mathbb{R}$, $Re(\alpha) > 0$, and $Re(\beta) > 0$.*
Then, we have

$$\frac{\partial}{\partial t}\left[t^{\beta-1} E_{\alpha,\beta}\left(\lambda t^{\alpha}\right)\right] = \lambda t^{\beta-2} E_{\alpha,\beta-1}\left(\lambda t^{\alpha}\right). \tag{4.9}$$

The work is the extended version discovered by Erdelyi et al. in 1953 [32].

Theorem 4.5 *Let $\alpha, \beta \in \mathbb{C}$, $\lambda \in \mathbb{R}$, $Re(\alpha) > 0$, and $Re(\beta) > 0$.*
Then, we have

$$\int_0^{\infty} t^{\beta-1} e^{-t} E_{\alpha,\beta}\left(\lambda t^{\alpha}\right) dt = \frac{1}{1-\lambda}. \tag{4.10}$$

The work was discovered in [227] and further reported by Yu and Zhang in 2006 [246].

Theorem 4.6 *Let $\alpha, \beta \in \mathbb{C}$, $\lambda \in \mathbb{R}$, $Re(\alpha) > 0$, and $Re(\beta) > 0$.*
Then, we have

$$\int_{-\infty}^{\infty} |t|^{\beta-1} e^{-|t|} E_{\alpha,\beta}\left(\lambda |t|^{\alpha}\right) dt = \frac{2}{1-\lambda}. \tag{4.11}$$

Theorem 4.7 *Let $\alpha, \beta \in \mathbb{C}$, $t \in \mathbb{R}$, $Re(\alpha) > 0$, and $Re(\beta) > 0$.*
Then, we have

$$E_{\alpha,\beta}(t) = \left(\beta + \alpha t \frac{dt}{dt}\right) E_{\alpha,\beta+1}(t). \tag{4.12}$$

The work was discovered by Gorenflo et al. in 2014 [211].

Theorem 4.8 *Let $\alpha, \beta \in \mathbb{C}$, $t \in \mathbb{R}$, $Re(\alpha) > 0$, and $Re(\beta) > 0$.*

Then, we have

$$E_{1,1}(t) = e^t, \tag{4.13}$$

$$E_{1,2}(t) = \frac{e^t - 1}{t}, \tag{4.14}$$

$$E_{2,1}(t) = cosh\sqrt{t}, \tag{4.15}$$

and

$$E_{2,2}(t) = \frac{sinh\sqrt{t}}{\sqrt{t}}. \tag{4.16}$$

The Laplace transform for the Wiman function is given as

$$\begin{aligned} &L\left\{t^{\beta-1}E_{\alpha,\beta}(\lambda t^{\alpha})\right\} \\ &= \int_0^{\infty} e^{-st}t^{\beta-1}E_{\alpha,\beta}(\lambda t^{\alpha})\,dt \\ &= L\left\{\sum_{n=0}^{\infty}\frac{\lambda^n t^{n\alpha+\beta-1}}{\Gamma(n\alpha+\beta)}\right\} \\ &= \sum_{n=0}^{\infty}\frac{\lambda^n}{s^{n\alpha+\beta}} \\ &= s^{-\beta}\left(1-\lambda s^{-\alpha}\right)^{-1}, \end{aligned} \tag{4.17}$$

where $\alpha, \beta \in \mathbb{C}$, $\left|\lambda s^{-\alpha}\right| < 1$, $Re(\alpha) > 0$, *and* $Re(\beta) > 0$.

The result is the special case of the result discovered by Gorenflo et al. in 2014 [211] and reported by Yang in 2019 [92].

Moreover, the Mellin transform for the Wiman function is given as

$$M\left[E_{\alpha,\beta}(-\lambda t)\right] = \int_0^{\infty} t^{\varpi-1}E_{\alpha,\beta}(-\lambda t)\,dt = \frac{\Gamma(\varpi)\Gamma(1-\varpi)}{\Gamma(\beta-\alpha\varpi)}\lambda^{-\varpi}, \tag{4.18}$$

where $\alpha, \beta, \varpi \in \mathbb{C}$, $\lambda \in \mathbb{R}$, $Re(\alpha) > 0$, $Re(\varpi) > 0$, and $Re(\beta) > 0$.

The result was reported by Yang in 2019 [92].

Theorem 4.9 *Let* $\alpha, \beta \in \mathbb{C}$, $\mu, \lambda \in \mathbb{R}$, $Re(\alpha) > 0$, *and* $Re(\beta) > 0$.

Then, we have

$$\int_0^{\infty} e^{-\mu t}t^{\beta-1}E_{\alpha,\beta}\left(\lambda t^{\alpha}\right)dt = \mu^{-\beta}\left(1-\lambda\mu^{-\alpha}\right)^{-1}. \tag{4.19}$$

Some examples are given as follows.

Example 4.1 Let $\alpha, \beta \in \mathbb{C}$, $\mu, \lambda \in \mathbb{R}$, $Re(\alpha) > 0$, and $Re(\beta) > 0$.
Then, we have

$$\int_{-\infty}^{\infty} e^{-\mu|t|} |t|^{\beta-1} E_{\alpha,\beta}\left(\lambda |t|^{\alpha}\right) dt = 2\mu^{-\beta}\left(1 - \lambda\mu^{-\alpha}\right)^{-1}. \tag{4.20}$$

Example 4.2 Let $\alpha, \beta \in \mathbb{C}$, $\lambda \in \mathbb{R}$, $Re(\alpha) > 0$, and $Re(\beta) > 0$. Then, we have

$$\int_{0}^{\infty} e^{-2t} t^{\beta-1} E_{\alpha,\beta}\left(\lambda t^{\alpha}\right) dt = 2^{-\beta}\left(1 - \lambda 2^{-\alpha}\right)^{-1}. \tag{4.21}$$

Example 4.3 Let $\beta \in \mathbb{C}$, $\mu, \lambda \in \mathbb{R}$, and $Re(\beta) > 0$. Then, we have

$$\int_{0}^{\infty} e^{-\mu t} t^{\beta-1} E_{2,\beta}\left(\lambda t^{2}\right) dt = \mu^{-\beta}\left(1 - \lambda\mu^{-2}\right)^{-1}. \tag{4.22}$$

Theorem 4.10 *Let* $\alpha, \beta \in \mathbb{C}$, $\mu, \lambda \in \mathbb{R}$, $\gamma > 0$, $Re(\alpha) > 0$, *and* $Re(\beta) > 0$. *Then, we have*

$$\int_{0}^{\infty} e^{-\mu t^{\gamma}} t^{\gamma(\beta-1)-2} E_{\alpha,\beta}\left(\lambda t^{\alpha\gamma}\right) dt = \gamma^{-1}\mu^{-\beta}\left(1 - \lambda\mu^{-\alpha}\right)^{-1}. \tag{4.23}$$

Some examples are given as follows.

Example 4.4 Let $\alpha \in \mathbb{C}$, $\lambda \in \mathbb{R}$, $\gamma > 0$, and $Re(\alpha) > 0$. Then, we have

$$\int_{0}^{\infty} e^{-2t^{\gamma}} t^{\gamma(3-1)-2} E_{\alpha,3}\left(\lambda t^{\alpha\gamma}\right) dt = \gamma^{-1} 2^{-3}\left(1 - \lambda 2^{-\alpha}\right)^{-1}. \tag{4.24}$$

Theorem 4.11 *Let* $\alpha, \beta \in \mathbb{C}$, $\lambda \in \mathbb{R}$, $\gamma > 0$, $Re(\alpha) > 0$, *and* $Re(\beta) > 0$. *Then, we have*

$$\int_{0}^{\infty} e^{-t^{\gamma}} t^{\gamma(\beta-1)-2} E_{\alpha,\beta}\left(\lambda t^{\alpha\gamma}\right) dt = \gamma^{-1}\left(1 - \lambda\right)^{-1}. \tag{4.25}$$

Example 4.5 Let $\beta \in \mathbb{C}$, $\lambda \in \mathbb{R}$, $\gamma > 0$, and $Re(\beta) > 0$. Then, we have

$$\int_0^\infty e^{-t^\gamma} t^{\gamma(\beta-1)-2} E_{1,\beta}\left(\lambda t^\gamma\right) dt = \gamma^{-1}(1-\lambda)^{-1}. \tag{4.26}$$

Theorem 4.12 *Let $\alpha, \beta \in \mathbb{C}$, $\lambda \in \mathbb{R}$, $\gamma > 0$, $Re(\alpha) > 0$, and $Re(\beta) > 0$. Then, we have*

$$\int_{-\infty}^\infty e^{-|t|^\gamma} |t|^{\gamma(\beta-1)-2} E_{\alpha,\beta}\left(-|t|^{\alpha\gamma}\right) dt = \gamma^{-1}. \tag{4.27}$$

Example 4.6 Let $\beta \in \mathbb{C}$, $\lambda \in \mathbb{R}$, $\gamma > 0$, and $Re(\beta) > 0$. Then, we have

$$\int_{-\infty}^\infty e^{-|t|^\gamma} |t|^{\gamma(\beta-1)-2} E_{1,\beta}\left(-|t|^{\gamma}\right) dt = \gamma^{-1}. \tag{4.28}$$

Example 4.7 Let $\alpha, \beta \in \mathbb{C}$, $\lambda \in \mathbb{R}$, $\gamma > 0$, $Re(\alpha) > 0$, and $Re(\beta) > 0$. Then, we have

$$\int_{-\infty}^\infty |t|^{-2} e^{-|t|^\gamma} E_{\alpha}\left(-|t|^{\alpha\gamma}\right) dt = \gamma^{-1}. \tag{4.29}$$

Theorem 4.13 *Let $\beta \in \mathbb{C}$, $\mu, \lambda \in \mathbb{R}$, $\gamma > 0$, and $Re(\beta) > 0$. Then, we have*

$$\int_0^\infty e^{-\mu t^\gamma} t^{\gamma(\beta-1)-2} E_{1,\beta}\left(\lambda t^\gamma\right) dt = \gamma^{-1}\mu^{-\beta}\left(1-\lambda\mu^{-1}\right)^{-1}. \tag{4.30}$$

Example 4.8 Let $\beta \in \mathbb{C}$, $t, \mu, \lambda \in \mathbb{R}$, $\gamma > 0$, and $Re(\beta) > 0$. Then, we have

$$\int_0^\infty e^{-\mu t^\gamma} t^{\gamma-2} E_{1,2}\left(\lambda t^\gamma\right) dt = \gamma^{-1}\mu^{-2}\left(1-\lambda\mu^{-1}\right)^{-1}. \tag{4.31}$$

Theorem 4.14 *Let $\lambda \in \mathbb{R}$, $\gamma > 0$, $\alpha > 0$, and $\beta > 0$. Then, we have*

$$\int_0^\infty t^{\gamma(\varpi-1)-2} E_{\alpha,\beta}\left(-\lambda t^\gamma\right) dt = \frac{\lambda^{-\varpi}}{\gamma} \cdot \frac{\Gamma(\varpi)\Gamma(1-\varpi)}{\Gamma(\beta-\alpha\varpi)}. \tag{4.32}$$

Example 4.9 Let $\lambda \in \mathbb{R}$, $\gamma > 0$, $\alpha > 0$, and $\beta > 0$. Then, we have

$$\int_0^\infty t^{2\varpi-4} E_{\alpha,\beta}\left(-\lambda t^2\right) dt = \frac{\lambda^{-\varpi}}{2} \cdot \frac{\Gamma(\varpi)\Gamma(1-\varpi)}{\Gamma(\beta-\alpha\varpi)}. \tag{4.33}$$

Example 4.10 Let $\lambda \in \mathbb{R}$, $\gamma > 0$, and $\beta > 0$. Then, we have

$$\int_0^\infty t^{2\varpi-4} E_{1,\beta}\left(-\lambda t^2\right) dt = \frac{\lambda^{-\varpi}}{2} \cdot \frac{\Gamma(\varpi)\Gamma(1-\varpi)}{\Gamma(\beta-\varpi)}. \tag{4.34}$$

Theorem 4.15 *Let $\gamma > 0$, $\alpha > 0$, and $\beta > 0$. Then, we have*

$$t^\gamma E_{\alpha,\alpha+\beta}\left(t^\gamma\right) = E_{\alpha,\beta}\left(t^\gamma\right) - \frac{1}{\Gamma(\beta)}, \tag{4.35}$$

where $t \in \mathbb{C}$.

Theorem 4.16 *Let $\gamma > 0$ and $\alpha > 0$. Then, we have*

$$t^\gamma E_{\alpha,\alpha+1}\left(t^\gamma\right) = E_\alpha\left(t^\gamma\right) - 1, \tag{4.36}$$

where $t \in \mathbb{C}$.

Theorem 4.17 *Let $\alpha > 0$. Then, we have*

$$t^\alpha E_{\alpha,\alpha+1}\left(t^\alpha\right) = E_\alpha\left(t^\alpha\right) - 1, \tag{4.37}$$

where $t \in \mathbb{C}$.

Theorem 4.18 *Let $\lambda \in \mathbb{R}$, $\gamma > 0$, $\alpha > 0$, and $\beta > 0$. Then, we have*

$$\lambda t^\gamma E_{\alpha,\alpha+\beta}\left(\lambda t^\gamma\right) = E_{\alpha,\beta}\left(\lambda t^\gamma\right) - \frac{1}{\Gamma(\beta)}, \tag{4.38}$$

where $t \in \mathbb{C}$.

Theorem 4.19 *Let $\lambda \in \mathbb{R}$, $\gamma > 0$, and $\alpha > 0$. Then, we have*

$$\lambda t^\gamma E_{\alpha,\alpha+1}\left(\lambda t^\gamma\right) = E_\alpha\left(\lambda t^\gamma\right) - 1, \tag{4.39}$$

where $t \in \mathbb{C}$.

Theorem 4.20 *Let $\lambda \in \mathbb{R}$ and $\alpha > 0$. Then, we have*

$$\lambda t^\alpha E_{\alpha,\alpha+1}\left(\lambda t^\alpha\right) = E_\alpha\left(\lambda t^\alpha\right) - 1, \tag{4.40}$$

where $t \in \mathbb{C}$.

Theorem 4.21 *Let* $\lambda \in \mathbb{R}$, $\gamma > 0$, $\alpha > 0$, *and* $\beta > 0$. *Then, we have*

$$-t^{\gamma} E_{\alpha,\alpha+1}\left(-t^{\gamma}\right) = E_{\alpha}\left(-t^{\gamma}\right) - 1, \tag{4.41}$$

where $t \in \mathbb{C}$.

Theorem 4.22 (Djrbashian [242, 243]) *Let* $x \geq 0$, $\alpha > 0$, *and* $\beta > 0$. *Then,* $E_{\alpha,\beta}(x)$ *is an entire function of order* $\rho = \frac{1}{\alpha}$ *and type* $\mu = 1$.

The result for the parameters $\alpha > 0$ and $\beta > 0$ was discovered by Dzherbashyan in 1966 [242], and the case for the parameters $\alpha > 0$ and $\beta > 0$ was investigated by Gorenflo et al. in 2014 [211].

The history of the functions is investigated as follows:

- The function $E_{\alpha,\beta}(x^{\alpha})$ was proposed by Erdélyi et al. in 1955 (see [226]).
- The function $E_{\alpha,\beta}(-x)$ was proposed by Schneider in 1996 [247] and discussed by Miller and Samko in 1999 [248].
- The function $t^{\beta-1}E_{\alpha,\beta}(-x^{\alpha})$ was first reported by Erdélyi et al. in 1955 [226] and further reported by Dzherbashian in 1966 [242].

The plots of the functions $t^{\beta-1}E_{\alpha,\beta}(t^{\alpha})$, $t^{\beta-1}E_{\alpha,\beta}(t)$, $t^{\beta-1}E_{\alpha,\beta}(-t^{\alpha})$, $t^{\beta-1}E_{\alpha,\beta}(-t)$, $E_{\alpha,\beta}(t^{\alpha})$, $E_{\alpha,\beta}(t)$, $E_{\alpha,\beta}(-t^{\alpha})$, and $E_{\alpha,\beta}(-t)$ with the different values of two parameters α and β are shown in Figs. 4.1, 4.2, 4.3, 4.4, 4.5, 4.6, 4.7, and 4.8, respectively.

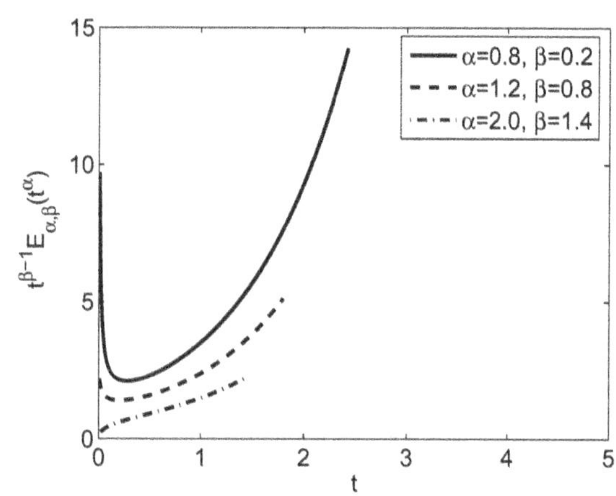

Fig. 4.1 The plots of $t^{\beta-1}E_{\alpha,\beta}(t^{\alpha})$ with the different values of two parameters α and β

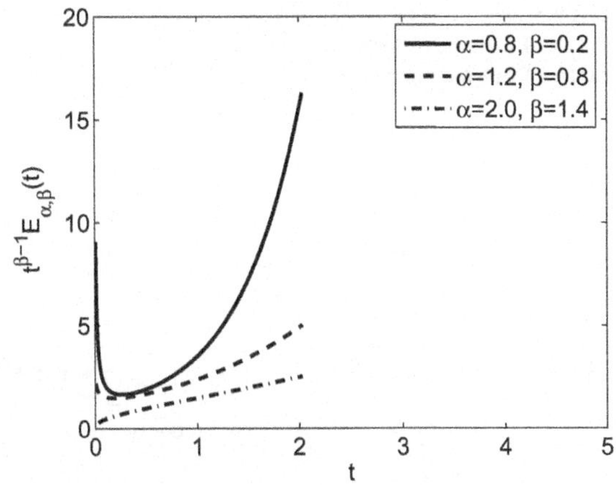

Fig. 4.2 The plots of $t^{\beta-1}E_{\alpha,\beta}(t)$ with the different values of two parameters α and β

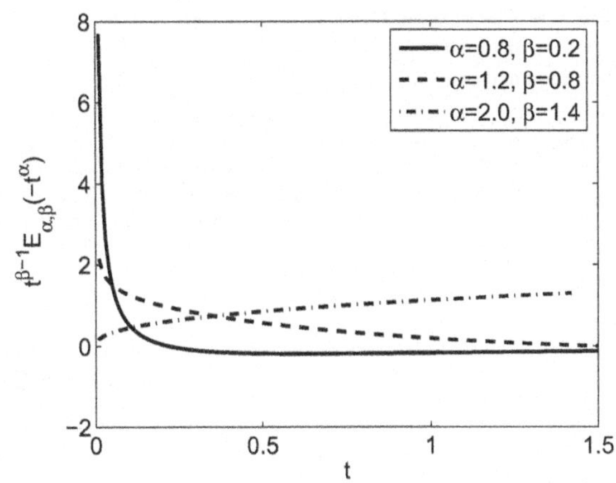

Fig. 4.3 The plots of $t^{\beta-1}E_{\alpha,\beta}(-t^{\alpha})$ with the different values of two parameters α and β

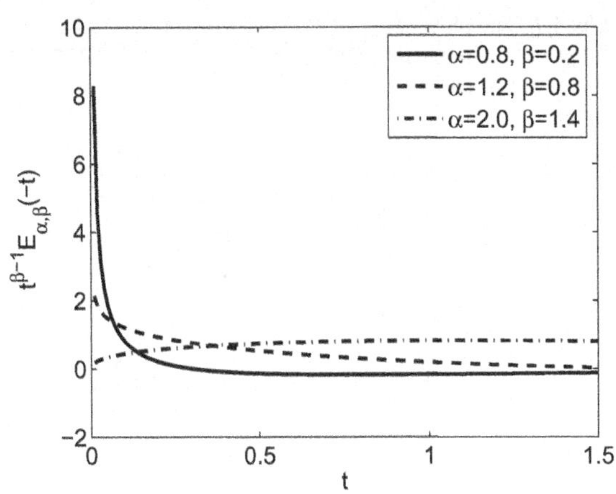

Fig. 4.4 The plots of $t^{\beta-1}E_{\alpha,\beta}(-t)$ with the different values of two parameters α and β

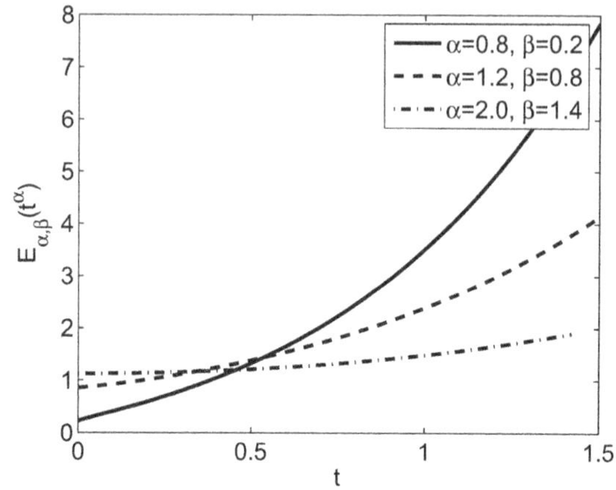

Fig. 4.5 The plots of $E_{\alpha,\beta}(t^{\alpha})$ with the different values of two parameters α and β

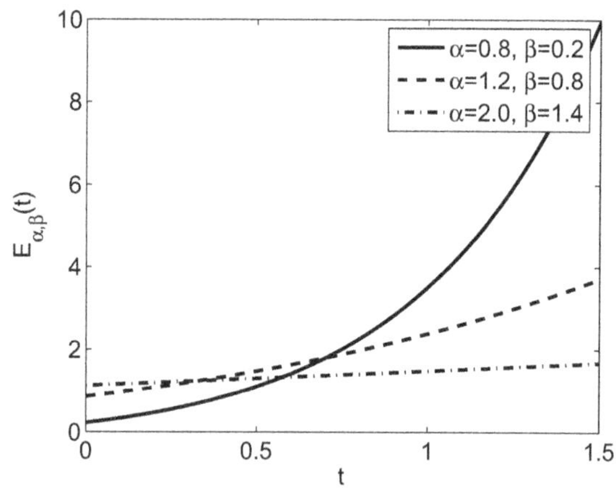

Fig. 4.6 The plots of $E_{\alpha,\beta}(t)$ with the different values of two parameters α and β

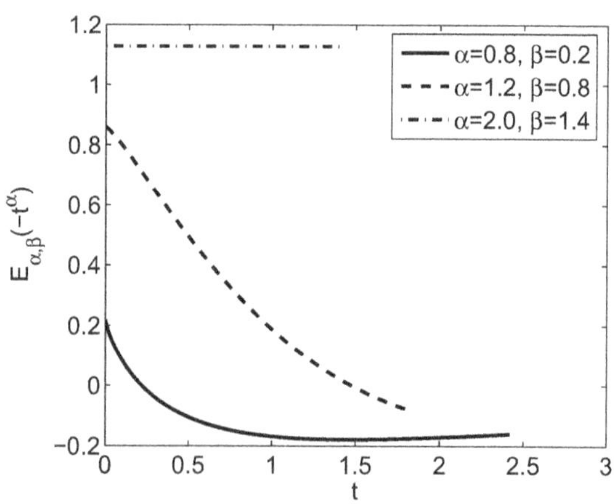

Fig. 4.7 The plots of $E_{\alpha,\beta}(-t^{\alpha})$ with the different values of two parameters α and β

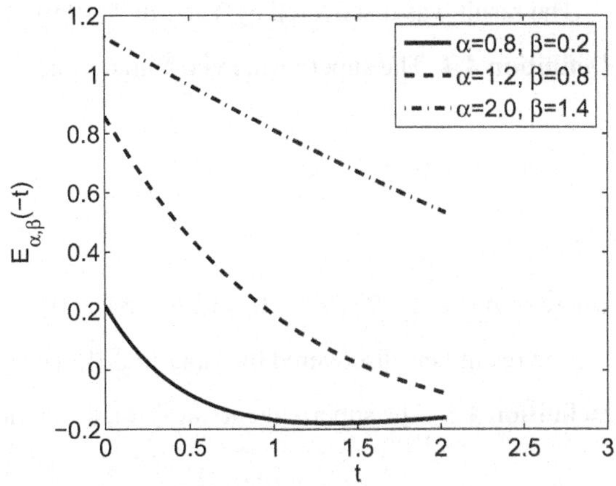

Fig. 4.8 The plots of $E_{\alpha,\beta}(-t)$ with the different values of two parameters α and β

4.1.2 The Supertrigonometric Functions via Wiman Function

Let $\alpha, \beta, t \in \mathbb{C}, \lambda \in \mathbb{C}, Re(\alpha) > 0$, and $Re(\beta) > 0$.

Then, we have

$$t^{\beta-1} E_{\alpha,\beta}\left(\lambda t^{\alpha}\right) = t^{\beta-1} \sum_{\kappa=0}^{\infty} \frac{\lambda^{\kappa} t^{\kappa\alpha}}{\Gamma(\kappa\alpha + \beta)} = \sum_{\kappa=0}^{\infty} \frac{\lambda^{\kappa} t^{\kappa\alpha+\beta-1}}{\Gamma(\kappa\alpha + \beta)}, \tag{4.42}$$

such that

$$t^{\beta-1} E_{\alpha,\beta}\left(it^{\alpha}\right) = t^{\beta-1} \sum_{\kappa=0}^{\infty} \frac{i^{\kappa} t^{\kappa\alpha}}{\Gamma(\kappa\alpha + \beta)} = \sum_{\kappa=0}^{\infty} \frac{i^{\kappa} t^{\kappa\alpha+\beta-1}}{\Gamma(\kappa\alpha + \beta)}, \tag{4.43}$$

and

$$t^{\beta-1} E_{\alpha,\beta}\left(-it^{\alpha}\right) = t^{\beta-1} \sum_{\kappa=0}^{\infty} \frac{(-i)^{\kappa} t^{\kappa\alpha}}{\Gamma(\kappa\alpha + \beta)} = \sum_{\kappa=0}^{\infty} \frac{(-i)^{\kappa} t^{\kappa\alpha+\beta-1}}{\Gamma(\kappa\alpha + \beta)}, \tag{4.44}$$

where $i = \sqrt{-1}$.

Definition 4.3 The supersine via Wiman function is defined as

$$supersin_{\alpha,\beta}\left(t^{\alpha}\right) = \frac{t^{\beta-1}}{2i}\left(E_{\alpha,\beta}\left(it^{\alpha}\right) - E_{\alpha,\beta}\left(-it^{\alpha}\right)\right) = \sum_{n=0}^{\infty} \frac{(-1)^{n} t^{(2n+1)\alpha+\beta-1}}{\Gamma((2n+1)\alpha + \beta)}, \tag{4.45}$$

where $\alpha, \beta, t \in \mathbb{C}, Re(\alpha) > 0$, and $Re(\beta) > 0$.

The result was discovered by Yang in 2019 [92].

Definition 4.4 The supercosine via Wiman function is defined as

$$supercos_{\alpha,\beta}\left(t^{\alpha}\right)=\frac{t^{\beta-1}}{2}\left(E_{\alpha,\beta}\left(it^{\alpha}\right)+E_{\alpha,\beta}\left(-it^{\alpha}\right)\right)=\sum_{n=0}^{\infty}\frac{(-1)^{n}\,t^{2n\alpha+\beta-1}}{\Gamma\left(2n\alpha+\beta\right)},\tag{4.46}$$

where $\alpha, \beta, t \in \mathbb{C}$, $Re\,(\alpha) > 0$, and $Re\,(\beta) > 0$.

The result was discovered by Yang in 2019 [92].

Definition 4.5 The supertangent via Wiman function is defined as

$$supertan_{\alpha,\beta}\left(t^{\alpha}\right)=\frac{supersin_{\alpha,\beta}\left(t^{\alpha}\right)}{supercos_{\alpha,\beta}\left(t^{\alpha}\right)}\left(supercos_{\alpha,\beta}\left(t^{\alpha}\right)\neq 0\right),\tag{4.47}$$

where $\alpha, \beta, t \in \mathbb{C}$, $Re\,(\alpha) > 0$, and $Re\,(\beta) > 0$.

The result was discovered by Yang in 2019 [92].

Definition 4.6 The supercotangent via Wiman function is defined as

$$supercotan_{\alpha,\beta}\left(t^{\alpha}\right)=\frac{supercos_{\alpha,\beta}\left(t^{\alpha}\right)}{supersin_{\alpha,\beta}\left(t^{\alpha}\right)}\left(supersin_{\alpha,\beta}\left(t^{\alpha}\right)\neq 0\right),\tag{4.48}$$

where $\alpha, \beta, t \in \mathbb{C}$, $Re\,(\alpha) > 0$, and $Re\,(\beta) > 0$.

The result was discovered by Yang in 2019 [92].

Definition 4.7 The supercosecant via Wiman function is defined as

$$supercsc_{\alpha,\beta}\left(t^{\alpha}\right)=\frac{1}{supersin_{\alpha,\beta}\left(t^{\alpha}\right)}\left(supersin_{\alpha,\beta}\left(t^{\alpha}\right)\neq 0\right),\tag{4.49}$$

where $\alpha, \beta, t \in \mathbb{C}$, $Re\,(\alpha) > 0$, and $Re\,(\beta) > 0$.

The result was discovered by Yang in 2019 [92].

Definition 4.8 The supersecant via Wiman function is defined as

$$supersec_{\alpha,\beta}\left(t^{\alpha}\right)=\frac{1}{supercos_{\alpha,\beta}\left(t^{\alpha}\right)}\left(supercos_{\alpha,\beta}\left(t^{\alpha}\right)\neq 0\right),\tag{4.50}$$

where $\alpha, \beta, t \in \mathbb{C}$, $Re\,(\alpha) > 0$, and $Re\,(\beta) > 0$.

The result was discovered by Yang in 2019 [92].

Generally, we have that

$$supersin_{\alpha,\beta}\left(\lambda t^{\alpha}\right)=\frac{t^{\beta-1}}{2i}\left(E_{\alpha,\beta}\left(i\lambda t^{\alpha}\right)-E_{\alpha,\beta}\left(-i\lambda t^{\alpha}\right)\right)=\sum_{n=0}^{\infty}\frac{(-\lambda)^{n}t^{(2n+1)\alpha+\upsilon-1}}{\Gamma\left((2n+1)\alpha+\beta\right)}, \tag{4.51}$$

$$supercos_{\alpha,\beta}\left(\lambda t^{\alpha}\right)=\frac{t^{\beta-1}}{2}\left(E_{\alpha,\beta}\left(i\lambda t^{\alpha}\right)+E_{\alpha,\beta}\left(-i\lambda t^{\alpha}\right)\right)=\sum_{n=0}^{\infty}\frac{(-\lambda)^{n}t^{2n\alpha+\upsilon-1}}{\Gamma\left(2n\alpha+\beta\right)}, \tag{4.52}$$

$$supertan_{\alpha,\beta}\left(\lambda t^{\alpha}\right)=\frac{supersin_{\alpha,\beta}\left(\lambda t^{\alpha}\right)}{supercos_{\alpha,\beta}\left(\lambda t^{\alpha}\right)}\left(supercos_{\alpha,\beta}\left(\lambda t^{\alpha}\right)\neq 0\right), \tag{4.53}$$

$$supercotan_{\alpha,\beta}\left(\lambda t^{\alpha}\right)=\frac{supercos_{\alpha,\beta}\left(\lambda t^{\alpha}\right)}{supersin_{\alpha,\beta}\left(\lambda t^{\alpha}\right)}\left(supersin_{\alpha,\beta}\left(\lambda t^{\alpha}\right)\neq 0\right), \tag{4.54}$$

$$supercsc_{\alpha,\beta}\left(\lambda t^{\alpha}\right)=\frac{1}{supersin_{\alpha,\beta}\left(\lambda t^{\alpha}\right)}\left(supersin_{\alpha,\beta}\left(\lambda t^{\alpha}\right)\neq 0\right), \tag{4.55}$$

and

$$supersec_{\alpha,\beta}\left(\lambda t^{\alpha}\right)=\frac{1}{supercos_{\alpha,\beta}\left(\lambda t^{\alpha}\right)}\left(supercos_{\alpha,\beta}\left(\lambda t^{\alpha}\right)\neq 0\right), \tag{4.56}$$

where $\alpha, \beta, t, \lambda \in \mathbb{C}$, $Re(\alpha) > 0$, and $Re(\beta) > 0$.

The results were discovered by Yang in 2019 [92].

As a direct result, we have the following theorem.

Theorem 4.23 *Let $\alpha, \beta, t \in \mathbb{C}$, $Re(\alpha) > 0$, and $Re(\beta) > 0$. Then, we have*

$$t^{\beta-1}E_{\alpha,\beta}\left(it^{\alpha}\right)=supercos_{\alpha,\beta}\left(t^{\alpha}\right)+isupersin_{\alpha,\beta}\left(t^{\alpha}\right) \tag{4.57}$$

and

$$t^{\beta-1}E_{\alpha,\beta}\left(-it^{\alpha}\right)=supercos_{\alpha,\beta}\left(t^{\alpha}\right)-isupersin_{\alpha,\beta}\left(t^{\alpha}\right). \tag{4.58}$$

The result was discovered by Yang in 2019 [92].

Generally, there are

$$t^{\beta-1}E_{\alpha,\beta}\left(i\lambda t^{\alpha}\right)=supercos_{\alpha,\beta}\left(\lambda t^{\alpha}\right)+isupersin_{\alpha,\beta}\left(\lambda t^{\alpha}\right) \tag{4.59}$$

and

$$t^{\beta-1}E_{\alpha,\beta}\left(-i\lambda t^{\alpha}\right)=supercos_{\alpha,\beta}\left(\lambda t^{\alpha}\right)-isupersin_{\alpha,\beta}\left(\lambda t^{\alpha}\right), \tag{4.60}$$

where $\alpha, \beta, t \in \mathbb{C}$, $\lambda \in \mathbb{C}$, $Re(\alpha) > 0$, and $Re(\beta) > 0$.

Thus, we have

$$\left[t^{\beta-1}E_{\alpha,\beta}\left(it^{\alpha}\right)\right]\left[t^{\beta-1}E_{\alpha,\beta}\left(-it^{\alpha}\right)\right]=supercos_{\alpha,\beta}^{2}\left(t^{\alpha}\right)+supersin_{\alpha,\beta}^{2}\left(t^{\alpha}\right), \tag{4.61}$$

which is provided that

$$supercos_{\alpha,\beta}^{2}\left(t^{\alpha}\right)=supercos_{\alpha,\beta}\left(t^{\alpha}\right)supercos_{\alpha,\beta}\left(t^{\alpha}\right) \tag{4.62}$$

and

$$supersin_{\alpha,\beta}^{2}\left(t^{\alpha}\right)=supersin_{\alpha,\beta}\left(t^{\alpha}\right)supersin_{\alpha,\beta}\left(t^{\alpha}\right), \tag{4.63}$$

where $\alpha, \beta, t \in \mathbb{C}$, $Re(\alpha) > 0$, and $Re(\beta) > 0$.

So, there exists

$$\left[t^{\beta-1}E_{\alpha,\beta}\left(i\lambda t^{\alpha}\right)\right]\left[t^{\beta-1}E_{\alpha,\beta}\left(-i\lambda t^{\alpha}\right)\right]=supercos_{\alpha,\beta}^{2}\left(\lambda t^{\alpha}\right)+supersin_{\alpha,\beta}^{2}\left(\lambda t^{\alpha}\right), \tag{4.64}$$

which is provided that

$$supercos_{\alpha,\beta}^{2}\left(\lambda t^{\alpha}\right)=supercos_{\alpha,\beta}\left(t^{\alpha}\right)supercos_{\alpha,\beta}\left(\lambda t^{\alpha}\right) \tag{4.65}$$

and

$$supersin_{\alpha,\beta}^{2}\left(\lambda t^{\alpha}\right)=supersin_{\alpha,\beta}\left(\lambda t^{\alpha}\right)supersin_{\alpha,\beta}\left(\lambda t^{\alpha}\right), \tag{4.66}$$

where $\alpha, \beta, t \in \mathbb{C}$, $\lambda \in \mathbb{C}$, $Re(\alpha) > 0$, and $Re(\beta) > 0$.

The results were discovered by Yang in 2019 [92].

Property 4.2 Let $\alpha, \beta, t \in \mathbb{C}, Re(\alpha) > 0$, and $Re(\beta) > 0$. Then, we have

$$supersin_{\alpha,\beta}\left(-t^{\alpha}\right)=-supersin_{\alpha,\beta}\left(t^{\alpha}\right), \tag{4.67}$$

$$supercos_{\alpha,\beta}\left(-t^{\alpha}\right)=supercos_{\alpha,\beta}\left(t^{\alpha}\right), \tag{4.68}$$

$$supertan_{\alpha,\beta}\left(-t^{\alpha}\right)=-\,supertan_{\alpha,\beta}\left(t^{\alpha}\right), \tag{4.69}$$

$$supercotan_{\alpha,\beta}\left(-t^{\alpha}\right)=-\,supercotan_{\alpha,\beta}\left(t^{\alpha}\right), \tag{4.70}$$

$$supercsc_{\alpha,\beta}\left(-t^{\alpha}\right)=-supercsc_{\alpha,\beta}\left(t^{\alpha}\right), \tag{4.71}$$

and

$$supersec_{\alpha,\beta}\left(-t^{\alpha}\right) = supersec_{\alpha,\beta}\left(t^{\alpha}\right). \tag{4.72}$$

The results were discovered by Yang in 2019 [92].

Property 4.3 (Laplace Transform of the supersine via Wiman Function)
Let $\alpha, \beta \in \mathbb{C}$, $\left|s^{-2\alpha}\right| < 1$, $Re(\alpha) > 0$, and $Re(\beta) > 0$. Then, we have

$$\begin{aligned} &L\left\{supersin_{\alpha,\beta}\left(t^{\alpha}\right)\right\} \\ &= L\left\{\sum_{\kappa=0}^{\infty} \frac{(-1)^{\kappa} t^{(2\kappa+1)\alpha+\beta-1}}{\Gamma((2\kappa+1)\alpha+\beta)}\right\} \\ &= \sum_{\kappa=0}^{\infty} \frac{(-1)^{\kappa}}{s^{(2\kappa+1)\alpha+\beta}} \\ &= s^{-(\beta+\alpha)}\left(1+s^{-2\alpha}\right)^{-1}. \end{aligned} \tag{4.73}$$

The result was discovered and proved by Yang in 2019 [92].

Property 4.4 (Laplace Transform of the Supercosine via Wiman Function)
Let $\alpha, \beta \in \mathbb{C}$, $\left|s^{-2\alpha}\right| < 1$, $Re(\alpha) > 0$, and $Re(\beta) > 0$. Then, we have

$$\begin{aligned} &L\left\{supercos_{\alpha,\beta}\left(t^{\alpha}\right)\right\} \\ &= L\left\{\sum_{\kappa=0}^{\infty} \frac{(-1)^{\kappa} t^{2\kappa\alpha+\beta-1}}{\Gamma(2\kappa\alpha+\beta)}\right\} \\ &= \sum_{\kappa=0}^{\infty} \frac{(-1)^{\kappa}}{s^{2\kappa\alpha+\beta}} \\ &= s^{-\beta}\left(1+s^{-2\alpha}\right)^{-1}. \end{aligned} \tag{4.74}$$

The result was discovered and proved by Yang in 2019 [92].
More generally,

$$\begin{aligned} &L\left\{supersin_{\alpha,\beta}\left(\lambda t^{\alpha}\right)\right\} \\ &= L\left\{\sum_{\kappa=0}^{\infty} \frac{(-1)^{\kappa} \lambda^{2\kappa+1} t^{(2\kappa+1)\alpha+\beta-1}}{\Gamma((2\kappa+1)\alpha+\beta)}\right\} \\ &= \sum_{\kappa=0}^{\infty} \frac{(-1)^{\kappa} \lambda^{2\kappa+1}}{s^{(2\kappa+1)\alpha+\beta}} \\ &= \lambda s^{-(\beta+\alpha)}\left(1+\lambda^{2} s^{-2\alpha}\right)^{-1} \end{aligned} \tag{4.75}$$

and

$$
\begin{aligned}
&L\left\{supercos_{\alpha,\beta}\left(\lambda t^{\alpha}\right)\right\} \\
&= L\left\{\sum_{\kappa=0}^{\infty}\frac{(-1)^{\kappa}\lambda^{2\kappa}t^{2\kappa\alpha+\beta-1}}{\Gamma(2\kappa\alpha+\beta)}\right\} \\
&= \sum_{\kappa=0}^{\infty}\frac{(-1)^{\kappa}\lambda^{2\kappa}}{s^{2\kappa\alpha+\beta}} \\
&= s^{-\beta}\left(1+\lambda^{2}s^{-2\alpha}\right)^{-1},
\end{aligned}
\tag{4.76}
$$

where α, $\beta \in \mathbb{C}$, $\lambda \in \mathbb{C}$, $Re\,(\alpha) > 0$, and $Re\,(\beta) > 0$.

The results were discovered and proved by Yang in 2019 [92].

The plots of the functions $supersin_{\alpha,\beta}(t^{\alpha})$ and $supercos_{\alpha,\beta}(t^{\alpha})$ with the different values of two parameters α and β are given in Figs. 4.9 and 4.10, respectively.

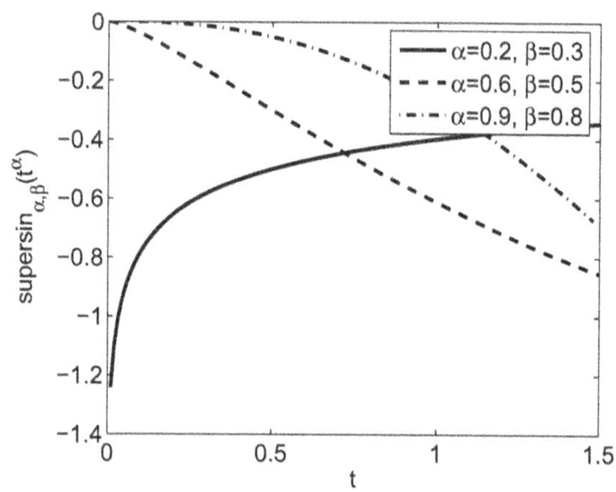

Fig. 4.9 The plots of $supersin_{\alpha,\beta}(t^{\alpha})$ with the different values of two parameters α and β

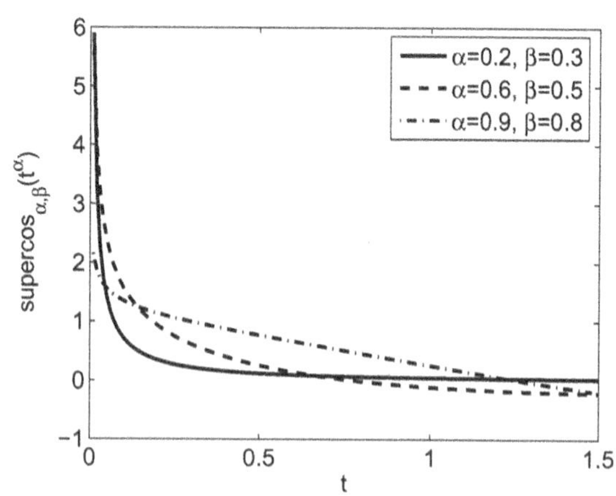

Fig. 4.10 The plots of $supercos_{\alpha,\beta}(t^{\alpha})$ with the different values of two parameters α and β

4.1.3 The Superhyperbolic Functions via Wiman Function

In the section, we present the concept of the superhyperbolic functions via Wiman function.

Let $\alpha, \beta, t \in \mathbb{C}$, $\lambda \in \mathbb{C}$, $Re\,(\alpha) > 0$, and $Re\,(\beta) > 0$.

Then, we have

$$t^{\upsilon-1} E_{\alpha,\beta}\left(\lambda t^{\alpha}\right) = t^{\beta-1} \sum_{\kappa=0}^{\infty} \frac{\lambda^{\kappa} t^{\kappa\alpha}}{\Gamma\left(\kappa\alpha+\beta\right)} = \sum_{\kappa=0}^{\infty} \frac{\lambda^{\kappa} t^{\kappa\alpha+\beta-1}}{\Gamma\left(\kappa\alpha+\beta\right)}, \tag{4.77}$$

such that

$$t^{\beta-1} E_{\alpha,\beta}\left(t^{\alpha}\right) = t^{\beta-1} \sum_{\kappa=0}^{\infty} \frac{t^{\kappa\alpha}}{\Gamma\left(\kappa\alpha+\beta\right)} = \sum_{\kappa=0}^{\infty} \frac{i^{\kappa} t^{\kappa\alpha+\beta-1}}{\Gamma\left(\kappa\alpha+\beta\right)}, \tag{4.78}$$

and

$$t^{\beta-1} E_{\alpha,\beta}\left(-t^{\alpha}\right) = t^{\beta-1} \sum_{\kappa=0}^{\infty} \frac{(-1)^{\kappa} t^{\kappa\alpha}}{\Gamma\left(\kappa\alpha+\beta\right)} = \sum_{\kappa=0}^{\infty} \frac{(-1)^{\kappa} t^{\kappa\alpha+\beta-1}}{\Gamma\left(\kappa\alpha+\beta\right)}. \tag{4.79}$$

Definition 4.9 The superhyperbolic supersine via Wiman function is defined as

$$supersinh_{\alpha,\beta}\left(t^{\alpha}\right) = \frac{t^{\beta-1}}{2}\left(E_{\alpha,\beta}\left(t^{\alpha}\right) - E_{\alpha,\beta}\left(-t^{\alpha}\right)\right) = \sum_{n=0}^{\infty} \frac{t^{(2n+1)\alpha+\beta-1}}{\Gamma\left((2n+1)\,\alpha+\beta\right)}, \tag{4.80}$$

where $\alpha, \beta, t \in \mathbb{C}$, $Re\,(\alpha) > 0$, and $Re\,(\beta) > 0$.

The result was discovered by Yang in 2019 [92].

Definition 4.10 The superhyperbolic supercosine via Wiman function is defined as

$$supercosh_{\alpha,\beta}\left(t^{\alpha}\right) = \frac{t^{\beta-1}}{2}\left(E_{\alpha,\beta}\left(t^{\alpha}\right) + E_{\alpha,\beta}\left(-t^{\alpha}\right)\right) = \sum_{n=0}^{\infty} \frac{t^{2n\alpha+\beta-1}}{\Gamma\left(2n\alpha+\beta\right)}, \tag{4.81}$$

where $\alpha, \beta, t \in \mathbb{C}$, $Re\,(\alpha) > 0$, and $Re\,(\beta) > 0$.

The result was discovered by Yang in 2019 [92].

Definition 4.11 The superhyperbolic supertangent via Wiman function is defined as

$$supertanh_{\alpha,\beta}\left(t^{\alpha}\right)=\frac{supersinh_{\alpha,\beta}\left(t^{\alpha}\right)}{supercosh_{\alpha,\beta}\left(t^{\alpha}\right)}\left(supercosh_{\alpha,\beta}\left(t^{\alpha}\right)\neq 0\right), \tag{4.82}$$

where $\alpha, \beta, t \in \mathbb{C}$, $Re(\alpha) > 0$, and $Re(\beta) > 0$.

The result was discovered by Yang in 2019 [92].

Definition 4.12 The superhyperbolic supercotangent via Wiman function is defined as

$$supercotanh_{\alpha,\beta}\left(t^{\alpha}\right)=\frac{supercosh_{\alpha,\beta}\left(t^{\alpha}\right)}{supersinh_{\alpha,\beta}\left(t^{\alpha}\right)}\left(supersinh_{\alpha,\beta}\left(t^{\alpha}\right)\neq 0\right), \tag{4.83}$$

where $\alpha, \beta, t \in \mathbb{C}$, $Re(\alpha) > 0$, and $Re(\beta) > 0$.

The result was discovered by Yang in 2019 [92].

Definition 4.13 The superhyperbolic supercosecant via Wiman function is defined as

$$supercsch_{\alpha,\beta}\left(t^{\alpha}\right)=\frac{1}{supersinh_{\alpha,\beta}\left(t^{\alpha}\right)}\left(supersinh_{\alpha,\beta}\left(t^{\alpha}\right)\neq 0\right), \tag{4.84}$$

where $\alpha, \beta, t \in \mathbb{C}$, $Re(\alpha) > 0$, and $Re(\beta) > 0$.

The result was discovered by Yang in 2019 [92].

Definition 4.14 The superhyperbolic supersecant via Wiman function is defined as

$$supersech_{\alpha,\beta}\left(t^{\alpha}\right)=\frac{1}{supercosh_{\alpha,\beta}\left(t^{\alpha}\right)}\left(supercosh_{\alpha,\beta}\left(t^{\alpha}\right)\neq 0\right), \tag{4.85}$$

where $\alpha, \beta, t \in \mathbb{C}$, $Re(\alpha) > 0$, and $Re(\beta) > 0$.

The result was discovered by Yang in 2019 [92].
Generally, we have

$$supersinh_{\alpha,\beta}\left(\lambda t^{\alpha}\right)=\frac{t^{\beta-1}}{2}\left(E_{\alpha,\beta}\left(\lambda t^{\alpha}\right)-E_{\alpha,\beta}\left(-\lambda t^{\alpha}\right)\right)=\sum_{n=0}^{\infty}\frac{\lambda^{n}t^{(2n+1)\alpha+\beta-1}}{\Gamma\left((2n+1)\alpha+\beta\right)}, \tag{4.86}$$

$$supercosh_{\alpha,\beta}\left(\lambda t^{\alpha}\right)=\frac{t^{\beta-1}}{2}\left(E_{\alpha,\beta}\left(\lambda t^{\alpha}\right)+E_{\alpha,\beta}\left(-\lambda t^{\alpha}\right)\right)=\sum_{n=0}^{\infty}\frac{\lambda^{n}t^{2n\alpha+\beta-1}}{\Gamma\left(2n\alpha+\beta\right)}, \tag{4.87}$$

$$supertanh_{\alpha,\beta}\left(\lambda t^{\alpha}\right) = \frac{supersinh_{\alpha,\beta}\left(\lambda t^{\alpha}\right)}{supercosh_{\alpha,\beta}\left(\lambda t^{\alpha}\right)}\left(supercosh_{\alpha,\beta}\left(\lambda t^{\alpha}\right) \neq 0\right), \tag{4.88}$$

$$supercotanh_{\alpha,\beta}\left(\lambda t^{\alpha}\right) = \frac{supercosh_{\alpha,\beta}\left(\lambda t^{\alpha}\right)}{supersinh_{\alpha,\beta}\left(\lambda t^{\alpha}\right)}\left(supersinh_{\alpha,\beta}\left(\lambda t^{\alpha}\right) \neq 0\right), \tag{4.89}$$

$$supercsch_{\alpha,\beta}\left(\lambda t^{\alpha}\right) = \frac{1}{supersinh_{\alpha,\beta}\left(\lambda t^{\alpha}\right)}\left(supersinh_{\alpha,\beta}\left(\lambda t^{\alpha}\right) \neq 0\right), \tag{4.90}$$

and

$$supersech_{\alpha,\beta}\left(\lambda t^{\alpha}\right) = \frac{1}{supercosh_{\alpha,\beta}\left(\lambda t^{\alpha}\right)}\left(supercosh_{\alpha,\beta}\left(\lambda t^{\alpha}\right) \neq 0\right), \tag{4.91}$$

where $\alpha, \beta, t, \lambda \in \mathbb{C}$, $Re\left(\alpha\right) > 0$, and $Re\left(\beta\right) > 0$.

The results were discovered by Yang in 2019 [92].

The plots of the functions $supersinh_{\alpha,\beta}(t^{\alpha})$ and $supercosh_{\alpha,\beta}(t^{\alpha})$ with the different values of two parameters α and β are given in Figs. 4.11 and 4.12, respectively.

As a direct result, we have the following theorem.

Theorem 4.24 *Let $\alpha, \beta, t \in \mathbb{C}$, $Re\left(\alpha\right) > 0$, and $Re\left(\beta\right) > 0$. Then, we have*

$$t^{\beta-1}E_{\alpha,\beta}\left(t^{\alpha}\right) = supercosh_{\alpha,\beta}\left(t^{\alpha}\right) + supersinh_{\alpha,\beta}\left(t^{\alpha}\right) \tag{4.92}$$

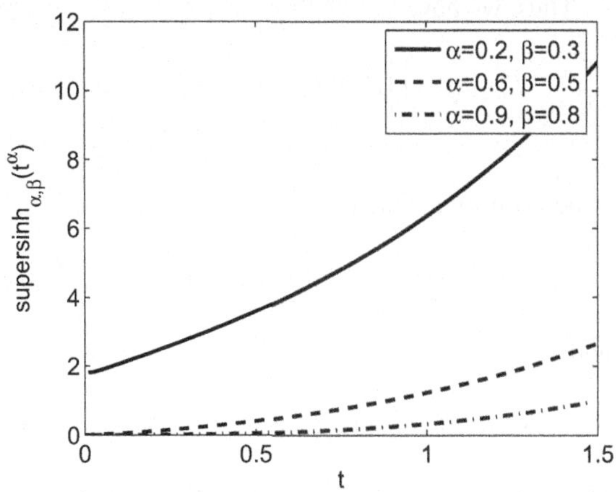

Fig. 4.11 The plots of $supersinh_{\alpha,\beta}(t^{\alpha})$ with the different values of two parameters α and β

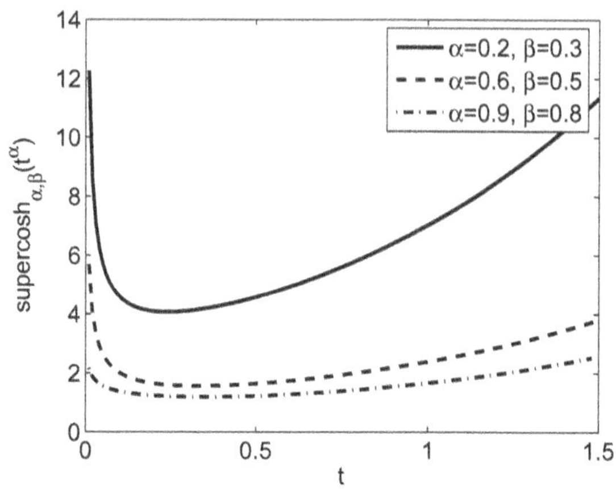

Fig. 4.12 The plots of $supercosh_{\alpha,\beta}(t^{\alpha})$ with the different values of two parameters α and β

and

$$t^{\beta-1}E_{\alpha,\beta}\left(-t^{\alpha}\right)=supercosh_{\alpha,\beta}\left(t^{\alpha}\right)-isupersinh_{\alpha,\beta}\left(t^{\alpha}\right). \tag{4.93}$$

The result was discovered by Yang in 2019 [92].

Generally, there are

$$t^{\beta-1}E_{\alpha,\beta}\left(\lambda t^{\alpha}\right)=supercosh_{\alpha,\beta}\left(\lambda t^{\alpha}\right)+supersinh_{\alpha,\beta}\left(\lambda t^{\alpha}\right) \tag{4.94}$$

and

$$t^{\beta-1}E_{\alpha,\beta}\left(-\lambda t^{\alpha}\right)=supercosh_{\alpha,\beta}\left(\lambda t^{\alpha}\right)-supersinh_{\alpha,\beta}\left(\lambda t^{\alpha}\right), \tag{4.95}$$

where $\alpha, \beta, t \in \mathbb{C}$, $\lambda \in \mathbb{C}$, $Re(\alpha) > 0$, and $Re(\beta) > 0$.

Thus, we have

$$\left[t^{\beta-1}E_{\alpha,\beta}\left(t^{\alpha}\right)\right]\left[t^{\beta-1}E_{\alpha,\beta}\left(-t^{\alpha}\right)\right]=supercosh^{2}_{\alpha,\beta}\left(t^{\alpha}\right)-supersinh^{2}_{\alpha,\beta}\left(t^{\alpha}\right), \tag{4.96}$$

which is provided that

$$supercosh^{2}_{\alpha,\beta}\left(t^{\alpha}\right)=supercosh_{\alpha,\beta}\left(t^{\alpha}\right)supercosh_{\alpha,\beta}\left(t^{\alpha}\right) \tag{4.97}$$

and

$$supersinh^{2}_{\alpha,\beta}\left(t^{\alpha}\right)=supersinh_{\alpha,\beta}\left(t^{\alpha}\right)supersinh_{\alpha,\beta}\left(t^{\alpha}\right), \tag{4.98}$$

where $\alpha, \beta, t \in \mathbb{C}$, $Re(\alpha) > 0$, and $Re(\beta) > 0$.

So, there exists

$$\left[t^{\beta-1}E_{\alpha,\beta}\left(\lambda t^{\alpha}\right)\right]\left[t^{\beta-1}E_{\alpha,\beta}\left(-\lambda t^{\alpha}\right)\right]=supercosh_{\alpha,\beta}^{2}\left(\lambda t^{\alpha}\right)-supersinh_{\alpha,\beta}^{2}\left(\lambda t^{\alpha}\right),\tag{4.99}$$

which is provided that

$$supercosh_{\alpha,\beta}^{2}\left(\lambda t^{\alpha}\right)=supercosh_{\alpha,\beta}\left(t^{\alpha}\right)supercosh_{\alpha,\beta}\left(\lambda t^{\alpha}\right)\tag{4.100}$$

and

$$supersinh_{\alpha,\beta}^{2}\left(\lambda t^{\alpha}\right)=supersinh_{\alpha,\beta}\left(\lambda t^{\alpha}\right)supersinh_{\alpha,\beta}\left(\lambda t^{\alpha}\right),\tag{4.101}$$

where $\alpha, \beta, t \in \mathbb{C}$, $\lambda \in \mathbb{C}$, $Re(\alpha) > 0$, and $Re(\beta) > 0$.

The results were discovered by Yang in 2019 [92].

Property 4.5 Let $\alpha, \beta, t \in \mathbb{C}, Re(\alpha) > 0$, and $Re(\beta) > 0$. Then, we have

$$supersinh_{\alpha,\beta}\left(-t^{\alpha}\right)=-supersinh_{\alpha,\beta}\left(t^{\alpha}\right),\tag{4.102}$$

$$supercosh_{\alpha,\beta}\left(-t^{\alpha}\right)=supercosh_{\alpha,\beta}\left(t^{\alpha}\right),\tag{4.103}$$

$$supertanh_{\alpha,\beta}\left(-t^{\alpha}\right)=-\,supertanh_{\alpha,\beta}\left(t^{\alpha}\right),\tag{4.104}$$

$$supercotanh_{\alpha,\beta}\left(-t^{\alpha}\right)=-\,supercotanh_{\alpha,\beta}\left(t^{\alpha}\right),\tag{4.105}$$

$$supercsch_{\alpha,\beta}\left(-t^{\alpha}\right)=-supercsch_{\alpha,\beta}\left(t^{\alpha}\right),\tag{4.106}$$

and

$$supersech_{\alpha,\beta}\left(-t^{\alpha}\right)=supersech_{\alpha,\beta}\left(t^{\alpha}\right).\tag{4.107}$$

The results were discovered by Yang in 2019 [92].

Property 4.6 (Laplace Transform of the Superhyperbolic Supersine via Wiman Function) Let $\alpha, \beta \in \mathbb{C}$, $\left|s^{-2\alpha}\right| < 1$, $Re(\alpha) > 0$, and $Re(\beta) > 0$. Then, we have

$$\begin{aligned}&L\left\{supersinh_{\alpha,\beta}\left(t^{\alpha}\right)\right\}\\&=L\left\{\sum_{\kappa=0}^{\infty}\frac{t^{(2\kappa+1)\alpha+\beta-1}}{\Gamma((2\kappa+1)\alpha+\beta)}\right\}\\&=\sum_{\kappa=0}^{\infty}\frac{1}{s^{(2\kappa+1)\alpha+\beta}}\\&=s^{-(\beta+\alpha)}\left(1-s^{-2\alpha}\right)^{-1}.\end{aligned}\tag{4.108}$$

The result was discovered and proved by Yang in 2019 [92].

Property 4.7 (Laplace Transform of the Superhyperbolic Supercosine via Wiman Function) Let $\alpha, \beta \in \mathbb{C}$, $\left|s^{-2\alpha}\right| < 1$, $Re(\alpha) > 0$, and $Re(\beta) > 0$. Then, we have

$$\begin{aligned} &L\left\{supercosh_{\alpha,\beta}\left(t^{\alpha}\right)\right\} \\ &= L\left\{\sum_{\kappa=0}^{\infty} \frac{t^{2\kappa\alpha+\beta-1}}{\Gamma(2\kappa\alpha+\beta)}\right\} \\ &= \sum_{\kappa=0}^{\infty} \frac{1}{s^{2\kappa\alpha+\beta}} \\ &= s^{-\beta}\left(1-s^{-2\alpha}\right)^{-1}. \end{aligned} \tag{4.109}$$

The result was discovered and proved by Yang in 2019 [92].

More generally,

$$\begin{aligned} &L\left\{supersinh_{\alpha,\beta}\left(\lambda t^{\alpha}\right)\right\} \\ &= L\left\{\sum_{\kappa=0}^{\infty} \frac{\lambda^{2\kappa+1} t^{(2\kappa+1)\alpha+\beta-1}}{\Gamma((2\kappa+1)\alpha+\beta)}\right\} \\ &= \sum_{\kappa=0}^{\infty} \frac{\lambda^{2\kappa+1}}{s^{(2\kappa+1)\alpha+\beta}} \\ &= \lambda s^{-(\beta+\alpha)}\left(1-\lambda^{2}s^{-2\alpha}\right)^{-1} \end{aligned} \tag{4.110}$$

and

$$\begin{aligned} &L\left\{supercosh_{\alpha,\beta}\left(\lambda t^{\alpha}\right)\right\} \\ &= L\left\{\sum_{\kappa=0}^{\infty} \frac{\lambda^{2\kappa} t^{2\kappa\alpha+\beta-1}}{\Gamma(2\kappa\alpha+\beta)}\right\} \\ &= \sum_{\kappa=0}^{\infty} \frac{\lambda^{2\kappa}}{s^{2\kappa\alpha+\beta}} \\ &= s^{-\beta}\left(1-\lambda^{2}s^{-2\alpha}\right)^{-1}, \end{aligned} \tag{4.111}$$

where $\alpha, \beta \in \mathbb{C}$, $\lambda \in \mathbb{C}$, $Re(\alpha) > 0$, and $Re(\beta) > 0$.

The results were discovered and proved by Yang in 2019 [92].

There exists the series via Wiman function as follows:

$$\Theta(t) = \frac{a_0}{2} + \sum_{n=0}^{\infty}\left(a_n supercosh_{\alpha,\beta}\left(nt^{\alpha}\right) + b_n supersinh_{\alpha,\beta}\left(nt^{\alpha}\right)\right), \tag{4.112}$$

where a_n and b_n are the coefficients of the series via Wiman function.

Alternatively, there are the forms of the series via Wiman function

$$\Omega_1(t) = \sum_{n=0}^{\infty}\left(a_n supersin_{\alpha,\beta}\left(nt^{\alpha}\right)\right), \tag{4.113}$$

and

$$\Omega_2(t) = \frac{a_0}{2} + \sum_{n=0}^{\infty} \left(b_n supercos_{\alpha,\beta}\left(nt^{\alpha}\right)\right), \tag{4.114}$$

where a_n and b_n are the coefficients of the series via Wiman function and $n \in \mathbb{N}$.

As a special case of $\alpha = 1$, we have the supersine via Wiman function given as

$$\begin{aligned} &supersin_{1,\beta}(\lambda t) \\ &= \frac{t^{\beta-1}}{2i}\left(E_{1,\beta}(i\lambda t) - E_{1,\beta}(-i\lambda t)\right) \\ &= \sum_{n=0}^{\infty} \frac{(-\lambda)^n t^{2n+1+\beta-1}}{\Gamma(2n+1+\beta)}, \end{aligned} \tag{4.115}$$

and the supercosine via Wiman function as

$$\begin{aligned} &supercos_{1,\beta}(\lambda t) \\ &= \frac{t^{\beta-1}}{2}\left(E_{1,\beta}(i\lambda t) + E_{1,\beta}(-i\lambda t)\right) \\ &= \sum_{n=0}^{\infty} \frac{(-\lambda)^n t^{2n+\beta-1}}{\Gamma(2n+\beta)}, \end{aligned} \tag{4.116}$$

the superhyperbolic supersine via Wiman function as

$$\begin{aligned} &supersinh_{1,\beta}(\lambda t) \\ &= \frac{t^{\beta-1}}{2}\left(E_{1,\beta}(\lambda t) - E_{1,\beta}(-\lambda t)\right) \\ &= \sum_{n=0}^{\infty} \frac{\lambda^n t^{2n+1+\beta-1}}{\Gamma(2n+1+\beta)}, \end{aligned} \tag{4.117}$$

and the superhyperbolic supercosine via Wiman function as

$$\begin{aligned} &supercosh_{1,\beta}(\lambda t) \\ &= \frac{t^{\beta-1}}{2}\left(E_{1,\beta}(\lambda t^{\alpha}) + E_{1,\beta}(-\lambda t^{\alpha})\right) \\ &= \sum_{n=0}^{\infty} \frac{\lambda^n t^{2n+\beta-1}}{\Gamma(2n+\beta)}, \end{aligned} \tag{4.118}$$

where $\beta \in \mathbb{C}$, $\lambda \in \mathbb{C}$, and $Re(\beta) > 0$.

Thus, we have

$$t^{\beta-1} E_{1,\beta}(\lambda t) = supercosh_{1,\beta}(\lambda t) + supersinh_{1,\beta}(\lambda t) \tag{4.119}$$

and

$$t^{\beta-1} E_{1,\beta}(i\lambda t) = Supercos_{1,\beta}(\lambda t) + i\,Supersin_{1,\beta}(\lambda t) \tag{4.120}$$

such that

$$supersin_{1,\beta}(t) = \frac{t^{\beta-1}}{2i} \cdot \left(E_{1,\beta}(it) - E_{1,\beta}(-it)\right), \tag{4.121}$$

$$supercos_{1,\beta}(t) = \frac{t^{\beta-1}}{2} \cdot \left(E_{1,\beta}(it) + E_{1,\beta}(-it)\right), \tag{4.122}$$

$$supersinh_{1,\beta}(t) = \frac{t^{\beta-1}}{2} \cdot \left(E_{1,\beta}(t) - E_{1,\beta}(-t)\right), \tag{4.123}$$

$$supercosh_{1,\beta}(t) = \frac{t^{\beta-1}}{2} \cdot \left(E_{1,\beta}(t) + E_{1,\beta}(-t)\right), \tag{4.124}$$

$$supersin_{1,\beta}(\lambda t) = \frac{t^{\beta-1}}{2i} \cdot \left(E_{1,\beta}(i\lambda t) - E_{1,\beta}(-i\lambda t)\right), \tag{4.125}$$

$$supercos_{1,\beta}(\lambda t) = \frac{t^{\beta-1}}{2} \cdot \left(E_{1,\beta}(i\lambda t) + E_{1,\beta}(-i\lambda t)\right), \tag{4.126}$$

$$supersinh_{1,\beta}(\lambda t) = \frac{t^{\beta-1}}{2} \cdot \left(E_{1,\beta}(\lambda t) - E_{1,\beta}(-\lambda t)\right), \tag{4.127}$$

and

$$supercosh_{1,\beta}(\lambda t) = \frac{t^{\beta-1}}{2} \cdot \left(E_{1,\beta}(\lambda t) + E_{1,\beta}(-\lambda t)\right), \tag{4.128}$$

where $\beta \in \mathbb{C}$, $\lambda \in \mathbb{C}$, and $Re(\beta) > 0$.

In this case, we have

$$\begin{aligned} & L\left\{t^{\beta-1} E_{1,\beta}(\lambda t)\right\} \\ & = L\left\{\sum_{\kappa=0}^{\infty} \frac{\lambda^{\kappa} t^{\kappa+\beta-1}}{\Gamma(\kappa+\beta)}\right\} \\ & = \sum_{\kappa=0}^{\infty} \frac{\lambda^{\kappa}}{s^{\kappa+\beta}} \\ & = s^{-\beta}\left(1 - \lambda s^{-1}\right) \end{aligned} \tag{4.129}$$

such that

$$\begin{aligned} & L\left\{supersin_{1,\beta}(\lambda t)\right\} \\ & = L\left\{\sum_{\kappa=0}^{\infty} \frac{(-1)^{\kappa} \lambda^{2\kappa+1} t^{2\kappa+\beta}}{\Gamma(2\kappa+1+\beta)}\right\} \\ & = \sum_{\kappa=0}^{\infty} \frac{(-1)^{\kappa} \lambda^{2\kappa+1}}{s^{2\kappa+1+\beta}} \\ & = \lambda s^{-(\beta+1)}\left(1+\lambda^2 s^{-2}\right)^{-1}, \end{aligned} \tag{4.130}$$

$$\begin{aligned} & L\left\{supercos_{1,\beta}(\lambda t)\right\} \\ & = L\left\{\sum_{\kappa=0}^{\infty}\frac{(-1)^{\kappa}\lambda^{2\kappa}t^{2\kappa+\beta-1}}{\Gamma(2\kappa+\beta)}\right\} \\ & = \sum_{\kappa=0}^{\infty}\frac{(-1)^{\kappa}\lambda^{2\kappa}}{s^{2\kappa+\beta}} \\ & = s^{-\beta}\left(1+\lambda^{2}s^{-2}\right)^{-1}, \end{aligned} \tag{4.131}$$

$$\begin{aligned} & L\left\{supersinh_{1,\beta}(\lambda t)\right\} \\ & = L\left\{\sum_{\kappa=0}^{\infty}\frac{\lambda^{2\kappa+1}t^{2\kappa+\beta}}{\Gamma(2\kappa+1+\beta)}\right\} \\ & = \sum_{\kappa=0}^{\infty}\frac{\lambda^{2\kappa+1}}{s^{2\kappa+1+\beta}} \\ & = \lambda s^{-(\beta+1)}\left(1-\lambda^{2}s^{-2}\right)^{-1}, \end{aligned} \tag{4.132}$$

and [25]

$$\begin{aligned} & L\left\{supercosh_{1,\beta}(\lambda t)\right\} \\ & = L\left\{\sum_{\kappa=0}^{\infty}\frac{\lambda^{2\kappa}t^{2\kappa+\beta-1}}{\Gamma(2\kappa+\beta)}\right\} \\ & = \sum_{\kappa=0}^{\infty}\frac{\lambda^{2\kappa}}{s^{2\kappa+\beta}} \\ & = s^{-\beta}\left(1-\lambda^{2}s^{-2}\right)^{-1}, \end{aligned} \tag{4.133}$$

where $\beta \in \mathbb{C}$, $\lambda \in \mathbb{C}$, $\left|\lambda s^{-1}\right| < 1$, and $Re(\beta) > 0$.

The results were discovered and proved by Yang in 2019 [92].

4.1.4 The Pre-supertrigonometric Functions via Wiman Function

Let $\alpha, \beta, t, \lambda \in \mathbb{C}$, $\lambda \in \mathbb{C}$, $Re(\alpha) > 0$, and $Re(\beta) > 0$.

Then, we have

$$E_{\alpha,\beta}(\lambda t) = \sum_{\kappa=0}^{\infty}\frac{\lambda^{\kappa}t^{\kappa}}{\Gamma(\kappa\alpha+\beta)}, \tag{4.134}$$

and

$$E_{\alpha,\beta}(-\lambda t) = \sum_{\kappa=0}^{\infty}\frac{(-\lambda)^{\kappa}t^{\kappa}}{\Gamma(\kappa\alpha+\beta)}, \tag{4.135}$$

such that

$$E_{\alpha,\beta}(it) = \sum_{\kappa=0}^{\infty} \frac{i^{\kappa} t^{\kappa}}{\Gamma(\kappa\alpha + \beta)}, \tag{4.136}$$

and

$$E_{\alpha,\beta}(-it) = \sum_{\kappa=0}^{\infty} \frac{(-i)^{\kappa} t^{\kappa}}{\Gamma(\kappa\alpha + \beta)}, \tag{4.137}$$

where $i = \sqrt{-1}$.

Definition 4.15 The presupersine via Wiman function is defined as

$$presin_{\alpha,\beta}(t) = \frac{1}{2i}\left(E_{\alpha,\beta}(it) - E_{\alpha,\beta}(-it)\right) = \sum_{n=0}^{\infty} \frac{(-1)^n t^{2n+1}}{\Gamma((2n+1)\alpha + \beta)}, \tag{4.138}$$

where $\alpha, \beta, t \in \mathbb{C}$, $Re(\alpha) > 0$, and $Re(\beta) > 0$.

Definition 4.16 The precosine via Wiman function is defined as

$$precos_{\alpha,\beta}(t) = \frac{1}{2}\left(E_{\alpha,\beta}(it) + E_{\alpha,\beta}(-it)\right) = \sum_{n=0}^{\infty} \frac{(-1)^n t^{2n}}{\Gamma(2n\alpha + \beta)}, \tag{4.139}$$

where $\alpha, \beta, t \in \mathbb{C}$, $Re(\alpha) > 0$, and $Re(\beta) > 0$.

Definition 4.17 The pretangent via Wiman function is defined as

$$pretan_{\alpha,\beta}(t) = \frac{presin_{\alpha,\beta}(t)}{precos_{\alpha,\beta}(t)}\left(precos_{\alpha,\beta}(t) \neq 0\right), \tag{4.140}$$

where $\alpha, \beta, t \in \mathbb{C}$, $Re(\alpha) > 0$, and $Re(\beta) > 0$.

Definition 4.18 The precotangent via Wiman function is defined as

$$precotan_{\alpha,\beta}(t) = \frac{precos_{\alpha,\beta}(t)}{presin_{\alpha,\beta}(t)}\left(presin_{\alpha,\beta}(t) \neq 0\right), \tag{4.141}$$

where $\alpha, \beta, t \in \mathbb{C}$, $Re(\alpha) > 0$, and $Re(\beta) > 0$.

Definition 4.19 The precosecant via Wiman function is defined as

$$precsc_{\alpha,\beta}(t) = \frac{1}{presin_{\alpha,\beta}(t)}\left(presin_{\alpha,\beta}(t) \neq 0\right), \tag{4.142}$$

where $\alpha, \beta, t \in \mathbb{C}$, $Re(\alpha) > 0$, and $Re(\beta) > 0$.

Definition 4.20 The presecant via Wiman function is defined as

$$presec_{\alpha,\beta}\left(t^{\alpha}\right) = \frac{1}{precos_{\alpha,\beta}(t)} \left(precos_{\alpha,\beta}(t) \neq 0\right), \tag{4.143}$$

where $\alpha, \beta, t \in \mathbb{C}$, $Re(\alpha) > 0$, and $Re(\beta) > 0$.

Alternatively, there exist

$$presin_{\alpha,\beta}(\lambda t) = \frac{1}{2i}\left(E_{\alpha,\beta}(i\lambda t) - E_{\alpha,\beta}(-i\lambda t)\right) = \sum_{n=0}^{\infty} \frac{(-\lambda)^n t^{2n+1}}{\Gamma((2n+1)\alpha + \beta)}, \tag{4.144}$$

$$precos_{\alpha,\beta}(\lambda t) = \frac{1}{2}\left(E_{\alpha,\beta}(i\lambda t) + E_{\alpha,\beta}(-i\lambda t)\right) = \sum_{n=0}^{\infty} \frac{(-\lambda)^n t^{2n}}{\Gamma(2n\alpha + \beta)}, \tag{4.145}$$

$$pretan_{\alpha,\beta}(\lambda t) = \frac{presin_{\alpha,\beta}(\lambda t)}{precos_{\alpha,\beta}(\lambda t)} \left(precos_{\alpha,\beta}(\lambda t) \neq 0\right), \tag{4.146}$$

$$precotan_{\alpha,\beta}(\lambda t) = \frac{precos_{\alpha,\beta}(\lambda t)}{presin_{\alpha,\beta}(\lambda t)} \left(presin_{\alpha,\beta}(\lambda t) \neq 0\right), \tag{4.147}$$

$$precsc_{\alpha,\beta}(\lambda t) = \frac{1}{presin_{\alpha,\beta}(\lambda t)} \left(presin_{\alpha,\beta}(\lambda t) \neq 0\right), \tag{4.148}$$

and

$$presec_{\alpha,\beta}(\lambda t) = \frac{1}{precos_{\alpha,\beta}(\lambda t)} \left(precos_{\alpha,\beta}(\lambda t) \neq 0\right), \tag{4.149}$$

where $\alpha, \beta, t \in \mathbb{C}$, $Re(\alpha) > 0$, and $Re(\beta) > 0$.

As a direct result, we have the following theorem.

Theorem 4.25 *Let $\alpha, \beta, t \in \mathbb{C}$, $Re(\alpha) > 0$, and $Re(\beta) > 0$. Then, we have*

$$E_{\alpha,\beta}(it) = precos_{\alpha,\beta}(t) + i\,presin_{\alpha,\beta}(t) \tag{4.150}$$

and

$$E_{\alpha,\beta}(-it) = precos_{\alpha,\beta}(t) - i\,presin_{\alpha,\beta}(t). \tag{4.151}$$

Generally, there are

$$E_{\alpha,\beta}(i\lambda t) = precos_{\alpha,\beta}(\lambda t) + ipresin_{\alpha,\beta}(\lambda t) \tag{4.152}$$

and

$$E_{\alpha,\beta}(-i\lambda t) = precos_{\alpha,\beta}(\lambda t) - ipresin_{\alpha,\beta}(\lambda t), \tag{4.153}$$

where $\alpha, \beta, t, \lambda \in \mathbb{C}$, $\lambda \in \mathbb{C}$, $Re(\alpha) > 0$, *and* $Re(\beta) > 0$.

Thus, we have

$$\left[E_{\alpha,\beta}(it)\right]\left[E_{\alpha,\beta}(-it)\right] = precos^2_{\alpha,\beta}(t) + presin^2_{\alpha,\beta}(t), \tag{4.154}$$

which is provided that

$$precos^2_{\alpha,\beta}(t) = precos_{\alpha,\beta}(t) \cdot precos_{\alpha,\beta}(t) \tag{4.155}$$

and

$$presin^2_{\alpha,\beta}(t) = presin_{\alpha,\beta}(t)\, presin_{\alpha,\beta}(t), \tag{4.156}$$

where $\alpha, \beta, t \in \mathbb{C}$, $Re(\alpha) > 0$, and $Re(\beta) > 0$.

Thus, there exists

$$\left[E_{\alpha,\beta}(i\lambda t)\right]\left[E_{\alpha,\beta}(-i\lambda t)\right] = precos^2_{\alpha,\beta}(\lambda t) + presin^2_{\alpha,\beta}(\lambda t), \tag{4.157}$$

which is provided that

$$precos^2_{\alpha,\beta}(\lambda t) = precos_{\alpha,\beta}(t)\, precos_{\alpha,\beta}(\lambda t) \tag{4.158}$$

and

$$presin^2_{\alpha,\beta}(\lambda t) = presin_{\alpha,\beta}\left(\lambda t^{\alpha}\right) presin_{\alpha,\beta}(\lambda t), \tag{4.159}$$

where $\alpha, \beta, t, \lambda \in \mathbb{C}$, $Re(\alpha) > 0$, and $Re(\beta) > 0$.

Property 4.8 Let $\alpha, \beta, t \in \mathbb{C}, Re(\alpha) > 0$, and $Re(\beta) > 0$. Then, we have

$$presin_{\alpha,\beta}(-t) = -presin_{\alpha,\beta}(t), \tag{4.160}$$

$$precos_{\alpha,\beta}(-t) = precos_{\alpha,\beta}(t), \tag{4.161}$$

$$pretan_{\alpha,\beta}(-t) = -\ pretan_{\alpha,\beta}(t), \tag{4.162}$$

$$precotan_{\alpha,\beta}(-t) = -\ precotan_{\alpha,\beta}(t), \tag{4.163}$$

$$precsc_{\alpha,\beta}(-t) = -precsc_{\alpha,\beta}(t), \tag{4.164}$$

and

$$presec_{\alpha,\beta}(-t) = presec_{\alpha,\beta}(t). \tag{4.165}$$

There exists the series via Wiman function as follows:

$$\Theta(t) = \frac{a_0}{2} + \sum_{n=0}^{\infty} \left(a_n precos_{\alpha,\beta}(nt) + b_n presin_{\alpha,\beta}(nt)\right), \tag{4.166}$$

where a_n and b_n are the coefficients of the series via Wiman function.

Alternatively, there are the forms of the series via Wiman function

$$\Omega_1(t) = \sum_{n=0}^{\infty} \left(a_n presin_{\alpha,\beta}(nt)\right), \tag{4.167}$$

and

$$\Omega_2(t) = \frac{a_0}{2} + \sum_{n=0}^{\infty} \left(b_n precos_{\alpha,\beta}(nt)\right), \tag{4.168}$$

where a_n and b_n are the coefficients of the series via Wiman function and $n \in \mathbb{N}$.

4.1.5 The Pre-superhyperbolic Functions via Wiman Function

Let $\alpha, \beta, t \in \mathbb{C}$, $\lambda \in \mathbb{C}$, $Re(\alpha) > 0$, and $Re(\beta) > 0$.

Then, we have

$$E_{\alpha,\beta}(\lambda t) = \sum_{\kappa=0}^{\infty} \frac{\lambda^{\kappa} t^{\kappa}}{\Gamma(\kappa\alpha + \beta)}, \tag{4.169}$$

and

$$E_{\alpha,\beta}(-\lambda t) = \sum_{\kappa=0}^{\infty} \frac{(-\lambda)^{\kappa} t^{\kappa}}{\Gamma(\kappa\alpha + \beta)}, \tag{4.170}$$

such that

$$E_{\alpha,\beta}(t) = \sum_{\kappa=0}^{\infty} \frac{i^{\kappa} t^{\kappa}}{\Gamma(\kappa\alpha + \beta)}, \tag{4.171}$$

and

$$E_{\alpha,\beta}(-t) = \sum_{\kappa=0}^{\infty} \frac{(-1)^{\kappa} t^{\kappa}}{\Gamma(\kappa\alpha + \beta)}, \tag{4.172}$$

where $i = \sqrt{-1}$.

Definition 4.21 The hyperbolic presupersine via Wiman function is defined as

$$presinh_{\alpha,\beta}(t) = \frac{1}{2}\left(E_{\alpha,\beta}(t) - E_{\alpha,\beta}(-t)\right) = \sum_{n=0}^{\infty} \frac{t^{2n+1}}{\Gamma((2n+1)\alpha + \beta)}, \tag{4.173}$$

where $\alpha, \beta, t \in \mathbb{C}$, $Re(\alpha) > 0$, and $Re(\beta) > 0$.

Definition 4.22 The hyperbolic precosine via Wiman function is defined as

$$precosh_{\alpha,\beta}(t) = \frac{1}{2}\left(E_{\alpha,\beta}(t) + E_{\alpha,\beta}(-t)\right) = \sum_{n=0}^{\infty} \frac{t^{2n}}{\Gamma(2n\alpha + \beta)}, \tag{4.174}$$

where $\alpha, \beta, t \in \mathbb{C}$, $Re(\alpha) > 0$, and $Re(\beta) > 0$.

Definition 4.23 The hyperbolic pretangent via Wiman function is defined as

$$pretanh_{\alpha,\beta}(t) = \frac{presinh_{\alpha,\beta}(t)}{precosh_{\alpha,\beta}(t)} \left(precosh_{\alpha,\beta}(t) \neq 0\right), \tag{4.175}$$

where $\alpha, \beta, t \in \mathbb{C}$, $Re(\alpha) > 0$, and $Re(\beta) > 0$.

Definition 4.24 The hyperbolic precotangent via Wiman function is defined as

$$precotanh_{\alpha,\beta}(t) = \frac{precosh_{\alpha,\beta}(t)}{presinh_{\alpha,\beta}(t)} \left(presinh_{\alpha,\beta}(t) \neq 0\right), \tag{4.176}$$

where $\alpha, \beta, t \in \mathbb{C}$, $Re(\alpha) > 0$, and $Re(\beta) > 0$.

Definition 4.25 The hyperbolic precosecant via Wiman function is defined as

$$precsch_{\alpha,\beta}(t) = \frac{1}{presinh_{\alpha,\beta}(t)} \left(presinh_{\alpha,\beta}(t) \neq 0\right), \tag{4.177}$$

where $\alpha, \beta, t \in \mathbb{C}$, $Re(\alpha) > 0$, and $Re(\beta) > 0$.

The hyperbolic presecant via Wiman function is defined as

$$presech_{\alpha,\beta}(t) = \frac{1}{precosh_{\alpha,\beta}(t)} \left(precosh_{\alpha,\beta}(t) \neq 0\right), \tag{4.178}$$

where $\alpha, \beta, t \in \mathbb{C}$, $Re(\alpha) > 0$, and $Re(\beta) > 0$.

Here, we have that

$$presinh_{\alpha,\beta}(\lambda t) = \frac{1}{2}\left(E_{\alpha,\beta}(i\lambda t) - E_{\alpha,\beta}(-i\lambda t)\right) = \sum_{n=0}^{\infty} \frac{\lambda^n t^{2n+1}}{\Gamma((2n+1)\alpha+\beta)}, \tag{4.179}$$

$$precosh_{\alpha,\beta}(\lambda t) = \frac{1}{2}\left(E_{\alpha,\beta}(i\lambda t) + E_{\alpha,\beta}(-i\lambda t)\right) = \sum_{n=0}^{\infty} \frac{\lambda^n t^{2n}}{\Gamma(2n\alpha+\beta)}, \tag{4.180}$$

$$pretanh_{\alpha,\beta}(\lambda t) = \frac{presinh_{\alpha,\beta}(\lambda t)}{precosh_{\alpha,\beta}(\lambda t)} \left(precosh_{\alpha,\beta}(\lambda t) \neq 0\right), \tag{4.181}$$

$$precotanh_{\alpha,\beta}(\lambda t) = \frac{precosh_{\alpha,\beta}(\lambda t)}{presinh_{\alpha,\beta}(\lambda t)} \left(presinh_{\alpha,\beta}(\lambda t) \neq 0\right), \tag{4.182}$$

$$precsch_{\alpha,\beta}(\lambda t) = \frac{1}{presinh_{\alpha,\beta}(\lambda t)} \left(presinh_{\alpha,\beta}(\lambda t) \neq 0\right), \tag{4.183}$$

and

$$presech_{\alpha,\beta}(\lambda t) = \frac{1}{precosh_{\alpha,\beta}(\lambda t)} \left(precosh_{\alpha,\beta}(\lambda t) \neq 0\right), \tag{4.184}$$

where $\alpha, \beta, t, \lambda \in \mathbb{C}$, $Re(\alpha) > 0$, and $Re(\beta) > 0$.

As a direct result, we have the following theorem.

Theorem 4.26 *Let $\alpha, \beta, t \in \mathbb{C}$, $Re(\alpha) > 0$, and $Re(\beta) > 0$. Then, we have*

$$E_{\alpha,\beta}(t) = precosh_{\alpha,\beta}(t) + presinh_{\alpha,\beta}(t) \tag{4.185}$$

and

$$E_{\alpha,\beta}(-t) = precosh_{\alpha,\beta}(t) - presinh_{\alpha,\beta}(t). \tag{4.186}$$

Generally, there are

$$E_{\alpha,\beta}(\lambda t) = precosh_{\alpha,\beta}(\lambda t) + presinh_{\alpha,\beta}(\lambda t) \tag{4.187}$$

and

$$E_{\alpha,\beta}(-\lambda t) = precosh_{\alpha,\beta}(\lambda t) - presinh_{\alpha,\beta}(\lambda t), \tag{4.188}$$

where $\alpha, \beta, t, \lambda \in \mathbb{C}$, $Re(\alpha) > 0$, and $Re(\beta) > 0$.

Thus, we have

$$\left[E_{\alpha,\beta}(t)\right]\left[E_{\alpha,\beta}(-t)\right] = precosh^2_{\alpha,\beta}(t) + presinh^2_{\alpha,\beta}(t), \tag{4.189}$$

which is provided that

$$precosh^2_{\alpha,\beta}(t) = precosh_{\alpha,\beta}(t) \cdot precosh_{\alpha,\beta}(t) \tag{4.190}$$

and

$$presinh^2_{\alpha,\beta}(t) = presinh_{\alpha,\beta}(t)\, presinh_{\alpha,\beta}(t), \tag{4.191}$$

where $\alpha, \beta, t \in \mathbb{C}$, $Re(\alpha) > 0$, and $Re(\beta) > 0$.

So, there exists

$$\left[E_{\alpha,\beta}(\lambda t)\right]\left[E_{\alpha,\beta}(-\lambda t)\right] = precosh^2_{\alpha,\beta}(\lambda t) + presinh^2_{\alpha,\beta}(\lambda t), \tag{4.192}$$

which is provided that

$$precosh^2_{\alpha,\beta}(\lambda t) = precosh_{\alpha,\beta}(t)\, precosh_{\alpha,\beta}(\lambda t) \tag{4.193}$$

and

$$presinh^2_{\alpha,\beta}(\lambda t) = presinh_{\alpha,\beta}\left(\lambda t^{\alpha}\right) presinh_{\alpha,\beta}(\lambda t), \tag{4.194}$$

where $\alpha, \beta, t, \lambda \in \mathbb{C}$, $Re(\alpha) > 0$, and $Re(\beta) > 0$.

Property 4.9 Let $\alpha, \beta, t \in \mathbb{C}, Re(\alpha) > 0$, and $Re(\beta) > 0$. Then, we have

$$presinh_{\alpha,\beta}(-t) = -presinh_{\alpha,\beta}(t), \tag{4.195}$$

$$precosh_{\alpha,\beta}(-t) = precosh_{\alpha,\beta}(t), \tag{4.196}$$

$$pretanh_{\alpha,\beta}(-t) = -\, pretanh_{\alpha,\beta}(t), \tag{4.197}$$

$$precotanh_{\alpha,\beta}(-t) = -\, precotanh_{\alpha,\beta}(t), \tag{4.198}$$

$$precsch_{\alpha,\beta}(-t) = -precsch_{\alpha,\beta}(t), \tag{4.199}$$

and

$$presech_{\alpha,\beta}(-t) = presech_{\alpha,\beta}(t). \tag{4.200}$$

4.1.6 *Some Special Cases via Wiman Function*

4.1.6.1 Case 1: The Supertrigonometric and Superhyperbolic Functions with the Power Law

Let $\alpha, \beta, t \in \mathbb{C}$, $\lambda, \gamma \in \mathbb{C}$, $Re(\alpha) > 0$, and $Re(\beta) > 0$.

Then, we have

$$E_{\alpha,\beta}\left(\lambda t^{\gamma}\right) = \sum_{\kappa=0}^{\infty} \frac{\lambda^{\kappa} t^{\kappa\gamma}}{\Gamma(\kappa\alpha + \beta)}, \tag{4.201}$$

and

$$E_{\alpha,\beta}\left(-\lambda t^{\gamma}\right) = \sum_{\kappa=0}^{\infty} \frac{(-\lambda)^{\kappa} t^{\kappa\gamma}}{\Gamma(\kappa\alpha + \beta)}, \tag{4.202}$$

such that

$$E_{\alpha,\beta}\left(i t^{\gamma}\right) = \sum_{\kappa=0}^{\infty} \frac{i^{\kappa} t^{\kappa\gamma}}{\Gamma(\kappa\alpha + \beta)}, \tag{4.203}$$

and

$$E_{\alpha,\beta}\left(-i t^{\gamma}\right) = \sum_{\kappa=0}^{\infty} \frac{(-i)^{\kappa} t^{\kappa\gamma}}{\Gamma(\kappa\alpha + \beta)}, \tag{4.204}$$

where $i = \sqrt{-1}$.

Definition 4.26 The presupersine with power law via Wiman function is defined as

$$presin_{\alpha,\beta}\left(t^{\gamma}\right) = \frac{1}{2i}\left(E_{\alpha,\beta}\left(i t^{\gamma}\right) - E_{\alpha,\beta}\left(-i t^{\gamma}\right)\right) = \sum_{n=0}^{\infty} \frac{(-1)^{n} t^{(2n+1)\gamma}}{\Gamma((2n+1)\alpha + \beta)}, \tag{4.205}$$

where $\alpha, \beta, t, \gamma \in \mathbb{C}$, $Re(\alpha) > 0$, and $Re(\beta) > 0$.

Definition 4.27 The precosine with power law via Wiman function is defined as

$$precos_{\alpha,\beta}\left(t^{\gamma}\right) = \frac{1}{2}\left(E_{\alpha,\beta}\left(i t^{\gamma}\right) + E_{\alpha,\beta}\left(-i t^{\gamma}\right)\right) = \sum_{n=0}^{\infty} \frac{(-1)^{n} t^{2n\gamma}}{\Gamma(2n\alpha + \beta)}, \tag{4.206}$$

where $\alpha, \beta, t, \gamma \in \mathbb{C}$, $Re(\alpha) > 0$, and $Re(\beta) > 0$.

Definition 4.28 The pretangent with power law via Wiman function is defined as

$$pretan_{\alpha,\beta}\left(t^{\gamma}\right)=\frac{presin_{\alpha,\beta}\left(t^{\gamma}\right)}{precos_{\alpha,\beta}\left(t^{\gamma}\right)}\left(precos_{\alpha,\beta}\left(t^{\gamma}\right)\neq 0\right), \tag{4.207}$$

where $\alpha, \beta, t, \gamma \in \mathbb{C}$, $Re\,(\alpha) > 0$, and $Re\,(\beta) > 0$.

Definition 4.29 The precotangent with power law via Wiman function is defined as

$$precotan_{\alpha,\beta}\left(t^{\gamma}\right)=\frac{precos_{\alpha,\beta}\left(t^{\gamma}\right)}{presin_{\alpha,\beta}\left(t^{\gamma}\right)}\left(presin_{\alpha,\beta}\left(t^{\gamma}\right)\neq 0\right), \tag{4.208}$$

where $\alpha, \beta, t, \gamma \in \mathbb{C}$, $Re\,(\alpha) > 0$, and $Re\,(\beta) > 0$.

Definition 4.30 The precosecant with power law via Wiman function is defined as

$$precsc_{\alpha,\beta}\left(t^{\gamma}\right)=\frac{1}{presin_{\alpha,\beta}\left(t^{\gamma}\right)}\left(presin_{\alpha,\beta}\left(t^{\gamma}\right)\neq 0\right), \tag{4.209}$$

where $\alpha, \beta, t, \gamma \in \mathbb{C}$, $Re\,(\alpha) > 0$, and $Re\,(\beta) > 0$.

Definition 4.31 The presecant with power law via Wiman function is defined as

$$presec_{\alpha,\beta}\left(t^{\gamma}\right)=\frac{1}{precos_{\alpha,\beta}\left(t^{\gamma}\right)}\left(precos_{\alpha,\beta}\left(t^{\gamma}\right)\neq 0\right), \tag{4.210}$$

where $\alpha, \beta, t, \gamma \in \mathbb{C}$, $Re\,(\alpha) > 0$, and $Re\,(\beta) > 0$.

Therefore, we have that

$$presin_{\alpha,\beta}\left(\lambda t^{\gamma}\right)=\frac{1}{2i}\left(E_{\alpha,\beta}\left(i\lambda t^{\gamma}\right)-E_{\alpha,\beta}\left(-i\lambda t^{\gamma}\right)\right)=\sum_{n=0}^{\infty}\frac{(-\lambda)^{n}\,t^{(2n+1)\gamma}}{\Gamma\left((2n+1)\,\alpha+\beta\right)}, \tag{4.211}$$

$$precos_{\alpha,\beta}\left(\lambda t^{\gamma}\right)=\frac{1}{2}\left(E_{\alpha,\beta}\left(i\lambda t^{\gamma}\right)+E_{\alpha,\beta}\left(-i\lambda t^{\gamma}\right)\right)=\sum_{n=0}^{\infty}\frac{(-\lambda)^{n}\,t^{2n\gamma}}{\Gamma\left(2n\alpha+\beta\right)}, \tag{4.212}$$

$$pretan_{\alpha,\beta}\left(\lambda t^{\gamma}\right)=\frac{presin_{\alpha,\beta}\left(\lambda t^{\gamma}\right)}{precos_{\alpha,\beta}\left(\lambda t^{\gamma}\right)}\left(precos_{\alpha,\beta}\left(\lambda t^{\gamma}\right)\neq 0\right), \tag{4.213}$$

$$precotan_{\alpha,\beta}\left(\lambda t^{\gamma}\right)=\frac{precos_{\alpha,\beta}\left(\lambda t^{\gamma}\right)}{presin_{\alpha,\beta}\left(\lambda t^{\gamma}\right)}\left(presin_{\alpha,\beta}\left(\lambda t^{\gamma}\right)\neq 0\right), \tag{4.214}$$

$$precsc_{\alpha,\beta}\left(\lambda t^{\gamma}\right)=\frac{1}{presin_{\alpha,\beta}\left(\lambda t^{\gamma}\right)}\left(presin_{\alpha,\beta}\left(\lambda t^{\gamma}\right)\neq 0\right), \tag{4.215}$$

and

$$presec_{\alpha,\beta}\left(\lambda t^{\gamma}\right) = \frac{1}{precos_{\alpha,\beta}\left(\lambda t^{\gamma}\right)}\left(precos_{\alpha,\beta}\left(\lambda t^{\gamma}\right) \neq 0\right), \tag{4.216}$$

where $\alpha, \beta, t, \gamma \in \mathbb{C}$, $Re(\alpha) > 0$, and $Re(\beta) > 0$.

As a direct result, we have the following theorem.

Theorem 4.27 *Let $\alpha, \beta, t \in \mathbb{C}$, $Re(\alpha) > 0$, and $Re(\beta) > 0$. Then, we have*

$$E_{\alpha,\beta}\left(it^{\gamma}\right) = precos_{\alpha,\beta}\left(t^{\gamma}\right) + ipresin_{\alpha,\beta}\left(t^{\gamma}\right) \tag{4.217}$$

and

$$E_{\alpha,\beta}\left(-it^{\gamma}\right) = precos_{\alpha,\beta}\left(t^{\gamma}\right) - ipresin_{\alpha,\beta}\left(t^{\gamma}\right). \tag{4.218}$$

Generally, there are

$$E_{\alpha,\beta}\left(i\lambda t^{\gamma}\right) = precos_{\alpha,\beta}\left(\lambda t^{\gamma}\right) + ipresin_{\alpha,\beta}\left(\lambda t^{\gamma}\right) \tag{4.219}$$

and

$$E_{\alpha,\beta}\left(-i\lambda t^{\gamma}\right) = precos_{\alpha,\beta}\left(\lambda t^{\gamma}\right) - ipresin_{\alpha,\beta}\left(\lambda t^{\gamma}\right), \tag{4.220}$$

where $\alpha, \beta, t, \gamma, \lambda \in \mathbb{C}$, $Re(\alpha) > 0$, and $Re(\beta) > 0$.

Thus, we have

$$\left[E_{\alpha,\beta}\left(it^{\gamma}\right)\right]\left[E_{\alpha,\beta}\left(-it^{\gamma}\right)\right] = precos^{2}_{\alpha,\beta}\left(t^{\gamma}\right) + presin^{2}_{\alpha,\beta}\left(t^{\gamma}\right), \tag{4.221}$$

which is provided that

$$precos^{2}_{\alpha,\beta}\left(t^{\gamma}\right) = precos_{\alpha,\beta}\left(t^{\gamma}\right) \cdot precos_{\alpha,\beta}\left(t^{\gamma}\right) \tag{4.222}$$

and

$$presin^{2}_{\alpha,\beta}\left(t^{\gamma}\right) = presin_{\alpha,\beta}\left(t^{\gamma}\right) presin_{\alpha,\beta}\left(t^{\gamma}\right), \tag{4.223}$$

where $\alpha, \beta, t, \gamma \in \mathbb{C}$, $Re(\alpha) > 0$, and $Re(\beta) > 0$.

So, there exists

$$\left[E_{\alpha,\beta}\left(i\lambda t^{\gamma}\right)\right]\left[E_{\alpha,\beta}\left(-i\lambda t^{\gamma}\right)\right] = precos^{2}_{\alpha,\beta}\left(\lambda t^{\gamma}\right) + presin^{2}_{\alpha,\beta}\left(\lambda t^{\gamma}\right), \tag{4.224}$$

which is provided that

$$precos^{2}_{\alpha,\beta}\left(\lambda t^{\gamma}\right) = precos_{\alpha,\beta}\left(t^{\gamma}\right) precos_{\alpha,\beta}\left(\lambda t^{\gamma}\right) \tag{4.225}$$

and

$$presin_{\alpha,\beta}^{2}\left(\lambda t^{\gamma}\right) = presin_{\alpha,\beta}\left(\lambda t^{\gamma}\right) presin_{\alpha,\beta}\left(\lambda t^{\gamma}\right), \tag{4.226}$$

where $\alpha, \beta, t, \gamma, \lambda \in \mathbb{C}$, $Re(\alpha) > 0$, and $Re(\beta) > 0$.

Property 4.10 Let $\alpha, \beta, t, \gamma \in \mathbb{C}, Re(\alpha) > 0$, and $Re(\beta) > 0$. Then, we have

$$presin_{\alpha,\beta}\left(-t^{\gamma}\right) = -presin_{\alpha,\beta}\left(t^{\gamma}\right), \tag{4.227}$$

$$precos_{\alpha,\beta}\left(-t^{\gamma}\right) = precos_{\alpha,\beta}\left(t^{\gamma}\right), \tag{4.228}$$

$$pretan_{\alpha,\beta}\left(-t^{\gamma}\right) = -\ pretan_{\alpha,\beta}\left(t^{\gamma}\right), \tag{4.229}$$

$$precotan_{\alpha,\beta}\left(-t^{\gamma}\right) = -\ precotan_{\alpha,\beta}\left(t^{\gamma}\right), \tag{4.230}$$

$$precsc_{\alpha,\beta}\left(-t^{\gamma}\right) = -precsc_{\alpha,\beta}\left(t^{\gamma}\right), \tag{4.231}$$

and

$$presec_{\alpha,\beta}\left(-t^{\gamma}\right) = presec_{\alpha,\beta}\left(t^{\gamma}\right). \tag{4.232}$$

There exists the series via Wiman function as follows:

$$\Theta(t) = \frac{a_0}{2} + \sum_{n=0}^{\infty}\left(a_n precos_{\alpha,\beta}\left(nt^{\gamma}\right) + b_n presin_{\alpha,\beta}\left(nt^{\gamma}\right)\right), \tag{4.233}$$

where a_n and b_n are the coefficients of the series via Wiman function.

Alternatively, there are the forms of the series via Wiman function

$$\Omega_1(t) = \sum_{n=0}^{\infty}\left(a_n presin_{\alpha,\beta}\left(nt^{\gamma}\right)\right), \tag{4.234}$$

and

$$\Omega_2(t) = \frac{a_0}{2} + \sum_{n=0}^{\infty}\left(b_n precos_{\alpha,\beta}\left(nt^{\gamma}\right)\right), \tag{4.235}$$

where a_n and b_n are the coefficients of the series via Wiman function and $n \in \mathbb{N}$.

Let $\alpha, \beta, t, \lambda \in \mathbb{C}, \lambda \in \mathbb{C}$, $Re(\alpha) > 0$, and $Re(\beta) > 0$.

Then, we have

$$E_{\alpha,\beta}\left(\lambda t^{\gamma}\right) = \sum_{\kappa=0}^{\infty}\frac{\lambda^{\kappa}t^{\kappa\gamma}}{\Gamma(\kappa\alpha+\beta)}, \tag{4.236}$$

and

$$E_{\alpha,\beta}\left(-\lambda t^{\gamma}\right)=\sum_{\kappa=0}^{\infty}\frac{(-\lambda)^{\kappa}\,t^{\kappa\gamma}}{\Gamma\left(\kappa\alpha+\beta\right)}, \tag{4.237}$$

such that

$$E_{\alpha,\beta}\left(t^{\gamma}\right)=\sum_{\kappa=0}^{\infty}\frac{i^{\kappa}t^{\kappa\gamma}}{\Gamma\left(\kappa\alpha+\beta\right)}, \tag{4.238}$$

and

$$E_{\alpha,\beta}\left(-t^{\gamma}\right)=\sum_{\kappa=0}^{\infty}\frac{(-1)^{\kappa}\,t^{\kappa\gamma}}{\Gamma\left(\kappa\alpha+\beta\right)}, \tag{4.239}$$

where $i=\sqrt{-1}$.

Definition 4.32 The hyperbolic presupersine with power law via Wiman function is defined as

$$presinh_{\alpha,\beta}\left(t^{\gamma}\right)=\frac{1}{2}\left(E_{\alpha,\beta}\left(t^{\gamma}\right)-E_{\alpha,\beta}\left(-t^{\gamma}\right)\right)=\sum_{n=0}^{\infty}\frac{t^{(2n+1)\gamma}}{\Gamma\left((2n+1)\,\alpha+\beta\right)}, \tag{4.240}$$

where $\alpha,\beta,t,\gamma\in\mathbb{C}$, $Re\,(\alpha)>0$, and $Re\,(\beta)>0$.

Definition 4.33 The hyperbolic precosine with power law via Wiman function is defined as

$$precosh_{\alpha,\beta}\left(t^{\gamma}\right)=\frac{1}{2}\left(E_{\alpha,\beta}\left(t^{\gamma}\right)+E_{\alpha,\beta}\left(-t^{\gamma}\right)\right)=\sum_{n=0}^{\infty}\frac{t^{2n\gamma}}{\Gamma\left(2n\alpha+\beta\right)}, \tag{4.241}$$

where $\alpha,\beta,t,\gamma\in\mathbb{C}$, $Re\,(\alpha)>0$, and $Re\,(\beta)>0$.

Definition 4.34 The hyperbolic pretangent with power law via Wiman function is defined as

$$pretanh_{\alpha,\beta}\left(t^{\gamma}\right)=\frac{presinh_{\alpha,\beta}\left(t^{\gamma}\right)}{precosh_{\alpha,\beta}\left(t^{\gamma}\right)}\left(precosh_{\alpha,\beta}\left(t^{\gamma}\right)\neq 0\right), \tag{4.242}$$

where $\alpha,\beta,t,\gamma\in\mathbb{C}$, $Re\,(\alpha)>0$, and $Re\,(\beta)>0$.

Definition 4.35 The hyperbolic precotangent with power law via Wiman function is defined as

$$precotanh_{\alpha,\beta}\left(t^{\gamma}\right)=\frac{precosh_{\alpha,\beta}\left(t^{\gamma}\right)}{presinh_{\alpha,\beta}\left(t^{\gamma}\right)}\left(presinh_{\alpha,\beta}\left(t^{\gamma}\right)\neq 0\right), \tag{4.243}$$

where $\alpha, \beta, t, \gamma \in \mathbb{C}$, $Re\left(\alpha\right) > 0$, and $Re\left(\beta\right) > 0$.

Definition 4.36 The hyperbolic precosecant with power law via Wiman function is defined as

$$precsch_{\alpha,\beta}\left(t^{\gamma}\right)=\frac{1}{presinh_{\alpha,\beta}\left(t^{\gamma}\right)}\left(presinh_{\alpha,\beta}\left(t^{\gamma}\right)\neq 0\right), \tag{4.244}$$

where $\alpha, \beta, t, \gamma \in \mathbb{C}$, $Re\left(\alpha\right) > 0$, and $Re\left(\beta\right) > 0$.

Definition 4.37 The hyperbolic presecant with power law via Wiman function is defined as

$$presech_{\alpha,\beta}\left(t^{\gamma}\right)=\frac{1}{precosh_{\alpha,\beta}\left(t^{\gamma}\right)}\left(precosh_{\alpha,\beta}\left(t^{\gamma}\right)\neq 0\right), \tag{4.245}$$

where $\alpha, \beta, t, \gamma \in \mathbb{C}$, $Re\left(\alpha\right) > 0$, and $Re\left(\beta\right) > 0$.

Generally, we present

$$presinh_{\alpha,\beta}\left(\lambda t^{\gamma}\right)=\frac{1}{2}\left(E_{\alpha,\beta}\left(\lambda t^{\gamma}\right)-E_{\alpha,\beta}\left(-\lambda t^{\gamma}\right)\right)=\sum_{n=0}^{\infty}\frac{\lambda^{n}t^{(2n+1)\gamma}}{\Gamma\left((2n+1)\alpha+\beta\right)}, \tag{4.246}$$

$$precosh_{\alpha,\beta}\left(\lambda t^{\gamma}\right)=\frac{1}{2}\left(E_{\alpha,\beta}\left(i\lambda t^{\gamma}\right)+E_{\alpha,\beta}\left(-i\lambda t^{\gamma}\right)\right)=\sum_{n=0}^{\infty}\frac{\lambda^{n}t^{2n\gamma}}{\Gamma\left(2n\alpha+\beta\right)}, \tag{4.247}$$

$$pretanh_{\alpha,\beta}\left(\lambda t^{\gamma}\right)=\frac{presinh_{\alpha,\beta}\left(\lambda t^{\gamma}\right)}{precosh_{\alpha,\beta}\left(\lambda t^{\gamma}\right)}\left(precosh_{\alpha,\beta}\left(\lambda t^{\gamma}\right)\neq 0\right), \tag{4.248}$$

$$precotanh_{\alpha,\beta}\left(\lambda t^{\gamma}\right)=\frac{precosh_{\alpha,\beta}\left(\lambda t^{\gamma}\right)}{presinh_{\alpha,\beta}\left(\lambda t^{\gamma}\right)}\left(presinh_{\alpha,\beta}\left(\lambda t^{\gamma}\right)\neq 0\right), \tag{4.249}$$

$$precsch_{\alpha,\beta}\left(\lambda t^{\gamma}\right)=\frac{1}{presinh_{\alpha,\beta}\left(\lambda t^{\gamma}\right)}\left(presinh_{\alpha,\beta}\left(\lambda t^{\gamma}\right)\neq 0\right), \tag{4.250}$$

and

$$presech_{\alpha,\beta}\left(\lambda t^{\gamma}\right) = \frac{1}{precosh_{\alpha,\beta}\left(\lambda t^{\gamma}\right)}\left(precosh_{\alpha,\beta}\left(\lambda t^{\gamma}\right) \neq 0\right), \tag{4.251}$$

where $\alpha, \beta, t, \gamma \in \mathbb{C}$, $Re(\alpha) > 0$, and $Re(\beta) > 0$.

As a direct result, we have the following theorem.

Theorem 4.28 *Let $\alpha, \beta, t \in \mathbb{C}$, $Re(\alpha) > 0$, and $Re(\beta) > 0$. Then, we have*

$$E_{\alpha,\beta}\left(t^{\gamma}\right) = precosh_{\alpha,\beta}\left(t^{\gamma}\right) + presinh_{\alpha,\beta}\left(t^{\gamma}\right) \tag{4.252}$$

and

$$E_{\alpha,\beta}\left(-t^{\gamma}\right) = precosh_{\alpha,\beta}\left(t^{\gamma}\right) - presinh_{\alpha,\beta}\left(t^{\gamma}\right). \tag{4.253}$$

Generally, there are

$$E_{\alpha,\beta}\left(\lambda t^{\gamma}\right) = precosh_{\alpha,\beta}\left(\lambda t^{\gamma}\right) + presinh_{\alpha,\beta}\left(\lambda t^{\gamma}\right) \tag{4.254}$$

and

$$E_{\alpha,\beta}\left(-\lambda t^{\gamma}\right) = precosh_{\alpha,\beta}\left(\lambda t^{\gamma}\right) - presinh_{\alpha,\beta}\left(\lambda t^{\gamma}\right), \tag{4.255}$$

where $\alpha, \beta, t, \gamma \in \mathbb{C}$, $\lambda \in \mathbb{C}$, $Re(\alpha) > 0$, and $Re(\beta) > 0$.

Thus, we have

$$\left[E_{\alpha,\beta}(t)\right]\left[E_{\alpha,\beta}(-t)\right] = precosh^{2}_{\alpha,\beta}\left(t^{\gamma}\right) + presinh^{2}_{\alpha,\beta}\left(t^{\gamma}\right), \tag{4.256}$$

which is provided that

$$precosh^{2}_{\alpha,\beta}\left(t^{\gamma}\right) = precosh_{\alpha,\beta}\left(t^{\gamma}\right) \cdot precosh_{\alpha,\beta}\left(t^{\gamma}\right) \tag{4.257}$$

and

$$presinh^{2}_{\alpha,\beta}\left(t^{\gamma}\right) = presinh_{\alpha,\beta}\left(t^{\gamma}\right) presinh_{\alpha,\beta}\left(t^{\gamma}\right), \tag{4.258}$$

where $\alpha, \beta, t, \gamma \in \mathbb{C}$, $Re(\alpha) > 0$, and $Re(\beta) > 0$.

So, there exists

$$\left[E_{\alpha,\beta}\left(\lambda t^{\gamma}\right)\right]\left[E_{\alpha,\beta}\left(-\lambda t^{\gamma}\right)\right] = precosh^{2}_{\alpha,\beta}\left(\lambda t^{\gamma}\right) + presinh^{2}_{\alpha,\beta}\left(\lambda t^{\gamma}\right), \tag{4.259}$$

which is provided that

$$precosh^2_{\alpha,\beta}\left(\lambda t^{\gamma}\right) = precosh_{\alpha,\beta}\left(\lambda t^{\gamma}\right) precosh_{\alpha,\beta}\left(\lambda t^{\gamma}\right) \tag{4.260}$$

and

$$presinh^2_{\alpha,\beta}\left(\lambda t^{\gamma}\right) = presinh_{\alpha,\beta}\left(\lambda t^{\gamma}\right) presinh_{\alpha,\beta}\left(\lambda t^{\gamma}\right), \tag{4.261}$$

where $\alpha, \beta, t, \gamma \in \mathbb{C}$, $\lambda \in \mathbb{C}$, $Re\,(\alpha) > 0$, and $Re\,(\beta) > 0$.

Property 4.11 Let $\alpha, \beta, t \in \mathbb{C}, Re\,(\alpha) > 0$, and $Re\,(\beta) > 0$. Then, we have

$$presinh_{\alpha,\beta}\left(-t^{\gamma}\right) = -presinh_{\alpha,\beta}\left(t^{\gamma}\right), \tag{4.262}$$

$$precosh_{\alpha,\beta}\left(-t^{\gamma}\right) = precosh_{\alpha,\beta}\left(t^{\gamma}\right), \tag{4.263}$$

$$pretanh_{\alpha,\beta}\left(-t^{\gamma}\right) = -\ pretanh_{\alpha,\beta}\left(t^{\gamma}\right), \tag{4.264}$$

$$precotanh_{\alpha,\beta}\left(-t^{\gamma}\right) = -\ precotanh_{\alpha,\beta}\left(t^{\gamma}\right), \tag{4.265}$$

$$precsch_{\alpha,\beta}\left(-t^{\gamma}\right) = -precsch_{\alpha,\beta}\left(t^{\gamma}\right), \tag{4.266}$$

and

$$presech_{\alpha,\beta}\,(-t) = presech_{\alpha,\beta}\,(t)\,. \tag{4.267}$$

4.1.6.2 Case 2: The Supertrigonometric and Superhyperbolic Functions with the Parameter

Let $\alpha, \beta, t \in \mathbb{C}$, $\lambda \in \mathbb{C}$, $Re\,(\alpha) > 0$, and $Re\,(\beta) > 0$.

Then, we have

$$t^{\beta-1} E_{\alpha,\beta}\,(\lambda t) = t^{\beta-1} \sum_{\kappa=0}^{\infty} \frac{\lambda^{\kappa} t^{\kappa}}{\Gamma\,(\kappa\alpha + \beta)}, \tag{4.268}$$

and

$$t^{\beta-1} E_{\alpha,\beta}\,(-\lambda t) = t^{\beta-1} \sum_{\kappa=0}^{\infty} \frac{(-\lambda)^{\kappa}\, t^{\kappa}}{\Gamma\,(\kappa\alpha + \beta)}, \tag{4.269}$$

such that

$$t^{\beta-1} E_{\alpha,\beta}\,(it) = t^{\beta-1} \sum_{\kappa=0}^{\infty} \frac{i^{\kappa} t^{\kappa}}{\Gamma\,(\kappa\alpha + \beta)}, \tag{4.270}$$

and

$$t^{\beta-1}E_{\alpha,\beta}\left(-it\right)=t^{\beta-1}\sum_{\kappa=0}^{\infty}\frac{(-i)^{\kappa}\,t^{\kappa}}{\Gamma\left(\kappa\alpha+\beta\right)},\tag{4.271}$$

where $i=\sqrt{-1}$.

Definition 4.38 The presupersine via Wiman function is defined as

$$Tresin_{\alpha,\beta}\left(t\right)=\frac{t^{\beta-1}}{2i}\left(E_{\alpha,\beta}\left(it\right)-E_{\alpha,\beta}\left(-it\right)\right)=t^{\beta-1}\sum_{n=0}^{\infty}\frac{(-1)^{n}\,t^{2n+1}}{\Gamma\left(\left(2n+1\right)\alpha+\beta\right)},\tag{4.272}$$

where $\alpha,\beta,t\in\mathbb{C}$, $Re\left(\alpha\right)>0$, and $Re\left(\beta\right)>0$.

Definition 4.39 The precosine via Wiman function is defined as

$$Trecos_{\alpha,\beta}\left(t\right)=\frac{t^{\beta-1}}{2}\left(E_{\alpha,\beta}\left(it\right)+E_{\alpha,\beta}\left(-it\right)\right)=t^{\beta-1}\sum_{n=0}^{\infty}\frac{(-1)^{n}\,t^{2n}}{\Gamma\left(2n\alpha+\beta\right)},\tag{4.273}$$

where $\alpha,\beta,t\in\mathbb{C}$, $Re\left(\alpha\right)>0$, and $Re\left(\beta\right)>0$.

Definition 4.40 The pretangent via Wiman function is defined as

$$Tretan_{\alpha,\beta}\left(t\right)=\frac{Tresin_{\alpha,\beta}\left(t\right)}{Trecos_{\alpha,\beta}\left(t\right)}\left(Trecos_{\alpha,\beta}\left(t\right)\neq 0\right),\tag{4.274}$$

where $\alpha,\beta,t\in\mathbb{C}$, $Re\left(\alpha\right)>0$, and $Re\left(\beta\right)>0$.

Definition 4.41 The precotangent via Wiman function is defined as

$$Trecotan_{\alpha,\beta}\left(t\right)=\frac{Trecos_{\alpha,\beta}\left(t\right)}{Tresin_{\alpha,\beta}\left(t\right)}\left(Tresin_{\alpha,\beta}\left(t\right)\neq 0\right),\tag{4.275}$$

where $\alpha,\beta,t\in\mathbb{C}$, $Re\left(\alpha\right)>0$, and $Re\left(\beta\right)>0$.

Definition 4.42 The precosecant via Wiman function is defined as

$$Trecsc_{\alpha,\beta}\left(t\right)=\frac{1}{Tresin_{\alpha,\beta}\left(t\right)}\left(Tresin_{\alpha,\beta}\left(t\right)\neq 0\right),\tag{4.276}$$

where $\alpha,\beta,t\in\mathbb{C}$, $Re\left(\alpha\right)>0$, and $Re\left(\beta\right)>0$.

Definition 4.43 The presecant via Wiman function is defined as

$$Tresec_{\alpha,\beta}(t) = \frac{1}{Trecos_{\alpha,\beta}(t)} \left(Trecos_{\alpha,\beta}(t) \neq 0\right), \tag{4.277}$$

where $\alpha, \beta, t \in \mathbb{C}$, $Re(\alpha) > 0$, and $Re(\beta) > 0$.

Generally, we give the alternative definitions as follows.

The supersine via Wiman function is defined as

$$Tresin_{\alpha,\beta}(\lambda t) = \frac{t^{\beta-1}}{2i}\left(E_{\alpha,\beta}(i\lambda t) - E_{\alpha,\beta}(-i\lambda t)\right) = t^{\beta-1}\sum_{n=0}^{\infty}\frac{(-\lambda)^n t^{2n+1}}{\Gamma((2n+1)\alpha+\beta)}, \tag{4.278}$$

where $\alpha, \beta, t, \lambda \in \mathbb{C}$, $Re(\alpha) > 0$, and $Re(\beta) > 0$.

Theorem 4.29 *Let $\alpha, \beta, t, \lambda \in \mathbb{C}$, $Re(\alpha) > 0$, and $Re(\beta) > 0$. Then, we have*

$$Tresin_{\alpha,\beta}(\lambda t) = t^{\beta-1} presin_{\alpha,\beta}(\lambda t). \tag{4.279}$$

The precosine via Wiman function is defined as

$$Trecos_{\alpha,\beta}(\lambda t) = \frac{t^{\beta-1}}{2}\left(E_{\alpha,\beta}(i\lambda t) + E_{\alpha,\beta}(-i\lambda t)\right) = t^{\beta-1}\sum_{n=0}^{\infty}\frac{(-\lambda)^n t^{2n}}{\Gamma(2n\alpha+\beta)}, \tag{4.280}$$

where $\alpha, \beta, t, \lambda \in \mathbb{C}$, $Re(\alpha) > 0$, and $Re(\beta) > 0$.

Theorem 4.30 *Let $\alpha, \beta, t, \lambda \in \mathbb{C}$, $Re(\alpha) > 0$, and $Re(\beta) > 0$. Then, we have*

$$Trecos_{\alpha,\beta}(\lambda t) = t^{\beta-1} precos_{\alpha,\beta}(\lambda t). \tag{4.281}$$

The pretangent via Wiman function is defined as

$$Tretan_{\alpha,\beta}(\lambda t) = \frac{Tresin_{\alpha,\beta}(\lambda t)}{Trecos_{\alpha,\beta}(\lambda t)} \left(Trecos_{\alpha,\beta}(\lambda t) \neq 0\right), \tag{4.282}$$

where $\alpha, \beta, t, \lambda \in \mathbb{C}$, $Re(\alpha) > 0$, and $Re(\beta) > 0$.

The precotangent via Wiman function is defined as

$$Trecotan_{\alpha,\beta}(\lambda t) = \frac{Trecos_{\alpha,\beta}(\lambda t)}{Tresin_{\alpha,\beta}(\lambda t)} \left(Tresin_{\alpha,\beta}(\lambda t) \neq 0\right), \tag{4.283}$$

where $\alpha, \beta, t, \lambda \in \mathbb{C}$, $Re(\alpha) > 0$, and $Re(\beta) > 0$.

The precosecant via Wiman function is defined as

$$Trecsc_{\alpha,\beta}(\lambda t) = \frac{1}{Tresin_{\alpha,\beta}(\lambda t)} \left(Tresin_{\alpha,\beta}(\lambda t) \neq 0\right), \tag{4.284}$$

where $\alpha, \beta, t, \lambda \in \mathbb{C}$, $Re(\alpha) > 0$, and $Re(\beta) > 0$.

The presecant via Wiman function is defined as

$$Tresec_{\alpha,\beta}(\lambda t) = \frac{1}{Trecos_{\alpha,\beta}(\lambda t)} \left(Trecos_{\alpha,\beta}(\lambda t) \neq 0\right), \tag{4.285}$$

where $\alpha, \beta, t, \lambda \in \mathbb{C}$, $Re(\alpha) > 0$, and $Re(\beta) > 0$.

As a direct result, we have the following theorem.

Theorem 4.31 *Let $\alpha, \beta, t \in \mathbb{C}$, $Re(\alpha) > 0$, and $Re(\beta) > 0$. Then, we have*

$$t^{\beta-1} E_{\alpha,\beta}(it) = Trecos_{\alpha,\beta}(t) + iTresin_{\alpha,\beta}(t) \tag{4.286}$$

and

$$t^{\beta-1} E_{\alpha,\beta}(-it) = Trecos_{\alpha,\beta}(t) - iTresin_{\alpha,\beta}(t). \tag{4.287}$$

Generally, there are

$$t^{\beta-1} E_{\alpha,\beta}(i\lambda t) = Trecos_{\alpha,\beta}(\lambda t) + iTresin_{\alpha,\beta}(\lambda t) \tag{4.288}$$

and

$$t^{\beta-1} E_{\alpha,\beta}(-i\lambda t) = Trecos_{\alpha,\beta}(\lambda t) - iTresin_{\alpha,\beta}(\lambda t), \tag{4.289}$$

where $\alpha, \beta, t \in \mathbb{C}$, $\lambda \in \mathbb{C}$, $Re(\alpha) > 0$, and $Re(\beta) > 0$.

Thus, we have

$$\left[t^{\beta-1} E_{\alpha,\beta}(it)\right]\left[t^{\beta-1} E_{\alpha,\beta}(-it)\right] = Trecos^2_{\alpha,\beta}(t) + Tresin^2_{\alpha,\beta}(t), \tag{4.290}$$

which is provided that

$$Trecos^2_{\alpha,\beta}(t) = Trecos_{\alpha,\beta}(t) \cdot Trecos_{\alpha,\beta}(t) \tag{4.291}$$

and

$$Tresin^2_{\alpha,\beta}(t) = Tresin_{\alpha,\beta}(t)\, Tresin_{\alpha,\beta}(t), \tag{4.292}$$

where $\alpha, \beta, t \in \mathbb{C}$, $Re(\alpha) > 0$, and $Re(\beta) > 0$.

So, there exists

$$\left[t^{\beta-1}E_{\alpha,\beta}\left(i\lambda t\right)\right]\left[t^{\beta-1}E_{\alpha,\beta}\left(-i\lambda t\right)\right]=Trecos_{\alpha,\beta}^{2}\left(\lambda t\right)+Tresin_{\alpha,\beta}^{2}\left(\lambda t\right), \tag{4.293}$$

which is provided that

$$Trecos_{\alpha,\beta}^{2}\left(\lambda t\right)=Trecos_{\alpha,\beta}\left(t\right)Trecos_{\alpha,\beta}\left(\lambda t\right) \tag{4.294}$$

and

$$Tresin_{\alpha,\beta}^{2}\left(\lambda t\right)=Tresin_{\alpha,\beta}\left(\lambda t\right)Tresin_{\alpha,\beta}\left(\lambda t\right), \tag{4.295}$$

where α, $\beta, t \in \mathbb{C}$, $\lambda \in \mathbb{C}$, $Re\left(\alpha\right) > 0$, and $Re\left(\beta\right) > 0$.

Property 4.12 Let α, $\beta, t \in \mathbb{C}, Re\left(\alpha\right) > 0$, and $Re\left(\beta\right) > 0$. Then, we have

$$Tresin_{\alpha,\beta}\left(-t\right)=-Tresin_{\alpha,\beta}\left(t\right), \tag{4.296}$$

$$Trecos_{\alpha,\beta}\left(-t\right)=Trecos_{\alpha,\beta}\left(t\right), \tag{4.297}$$

$$Tretan_{\alpha,\beta}\left(-t\right)=-\,Tretan_{\alpha,\beta}\left(t\right), \tag{4.298}$$

$$Trecotan_{\alpha,\beta}\left(-t\right)=-\,Trecotan_{\alpha,\beta}\left(t\right), \tag{4.299}$$

$$Trecsc_{\alpha,\beta}\left(-t\right)=-Trecsc_{\alpha,\beta}\left(t\right), \tag{4.300}$$

and

$$Tresec_{\alpha,\beta}\left(-t\right)=Tresec_{\alpha,\beta}\left(t\right). \tag{4.301}$$

There exists the series via Wiman function as follows:

$$\Theta\left(t\right)=\frac{a_0}{2}+\sum_{n=0}^{\infty}\left(a_nTrecos_{\alpha,\beta}\left(nt\right)+b_nTresin_{\alpha,\beta}\left(nt\right)\right), \tag{4.302}$$

where a_n and b_n are the coefficients of the series via Wiman function.

Alternatively, there are the forms of the series via Wiman function

$$\Omega_1\left(t\right)=\sum_{n=0}^{\infty}\left(a_nTresin_{\alpha,\beta}\left(nt\right)\right), \tag{4.303}$$

and

$$\Omega_2(t) = \frac{a_0}{2} + \sum_{n=0}^{\infty} \left(b_n Trecos_{\alpha,\beta}(nt) \right), \tag{4.304}$$

where a_n and b_n are the coefficients of the series via Wiman function and $n \in \mathbb{N}$.

Let $\alpha, \beta, t \in \mathbb{C}$, $\lambda \in \mathbb{C}$, $Re(\alpha) > 0$, and $Re(\beta) > 0$.

Then, we have

$$t^{\beta-1} E_{\alpha,\beta}(\lambda t) = t^{\beta-1} \sum_{\kappa=0}^{\infty} \frac{\lambda^{\kappa} t^{\kappa}}{\Gamma(\kappa\alpha + \beta)}, \tag{4.305}$$

and

$$t^{\beta-1} E_{\alpha,\beta}(-\lambda t) = t^{\beta-1} \sum_{\kappa=0}^{\infty} \frac{(-\lambda)^{\kappa} t^{\kappa}}{\Gamma(\kappa\alpha + \beta)}, \tag{4.306}$$

such that

$$t^{\beta-1} E_{\alpha,\beta}(t) = t^{\beta-1} \sum_{\kappa=0}^{\infty} \frac{i^{\kappa} t^{\kappa}}{\Gamma(\kappa\alpha + \beta)}, \tag{4.307}$$

and

$$t^{\beta-1} E_{\alpha,\beta}(-t) = t^{\beta-1} \sum_{\kappa=0}^{\infty} \frac{(-1)^{\kappa} t^{\kappa}}{\Gamma(\kappa\alpha + \beta)}, \tag{4.308}$$

where $i = \sqrt{-1}$.

Definition 4.44 The hyperbolic presupersine via Wiman function is defined as

$$Tresinh_{\alpha,\beta}(t) = \frac{t^{\beta-1}}{2} \left(E_{\alpha,\beta}(t) - E_{\alpha,\beta}(-t) \right) = t^{\beta-1} \sum_{n=0}^{\infty} \frac{t^{2n+1}}{\Gamma((2n+1)\alpha + \beta)}, \tag{4.309}$$

where $\alpha, \beta, t, \lambda \in \mathbb{C}$, $Re(\alpha) > 0$, and $Re(\beta) > 0$.

Definition 4.45 The hyperbolic precosine via Wiman function is defined as

$$Trecosh_{\alpha,\beta}(t) = \frac{t^{\beta-1}}{2} \left(E_{\alpha,\beta}(t) + E_{\alpha,\beta}(-t) \right) = \sum_{n=0}^{\infty} \frac{t^{2n}}{\Gamma(2n\alpha + \beta)}, \tag{4.310}$$

where $\alpha, \beta, t, \lambda \in \mathbb{C}$, $Re(\alpha) > 0$, and $Re(\beta) > 0$.

Definition 4.46 The hyperbolic pretangent via Wiman function is defined as

$$Tretanh_{\alpha,\beta}(t) = \frac{Tresinh_{\alpha,\beta}(t)}{Trecosh_{\alpha,\beta}(t)} \left(Trecosh_{\alpha,\beta}(t) \neq 0\right), \tag{4.311}$$

where $\alpha, \beta, t, \lambda \in \mathbb{C}$, $Re(\alpha) > 0$, and $Re(\beta) > 0$.

Definition 4.47 The hyperbolic precotangent via Wiman function is defined as

$$Trecotanh_{\alpha,\beta}(t) = \frac{Trecosh_{\alpha,\beta}(t)}{Tresinh_{\alpha,\beta}(t)} \left(Tresinh_{\alpha,\beta}(t) \neq 0\right), \tag{4.312}$$

where $\alpha, \beta, t, \lambda \in \mathbb{C}$, $Re(\alpha) > 0$, and $Re(\beta) > 0$.

Definition 4.48 The hyperbolic precosecant via Wiman function is defined as

$$Trecsch_{\alpha,\beta}(t) = \frac{1}{Tresinh_{\alpha,\beta}(t)} \left(Tresinh_{\alpha,\beta}(t) \neq 0\right), \tag{4.313}$$

where $\alpha, \beta, t, \lambda \in \mathbb{C}$, $Re(\alpha) > 0$, and $Re(\beta) > 0$.

Definition 4.49 The hyperbolic presecant via Wiman function is defined as

$$Tresech_{\alpha,\beta}(t) = \frac{1}{Trecosh_{\alpha,\beta}(t)} \left(Trecosh_{\alpha,\beta}(t) \neq 0\right), \tag{4.314}$$

where $\alpha, \beta, t, \lambda \in \mathbb{C}$, $Re(\alpha) > 0$, and $Re(\beta) > 0$.

Generally, we have

$$Tresinh_{\alpha,\beta}(\lambda t) = \frac{t^{\beta-1}}{2}\left(E_{\alpha,\beta}(\lambda t) - E_{\alpha,\beta}(-\lambda t)\right) = t^{\beta-1}\sum_{n=0}^{\infty}\frac{\lambda^n t^{2n+1}}{\Gamma((2n+1)\alpha+\beta)}, \tag{4.315}$$

$$Trecosh_{\alpha,\beta}(\lambda t) = \frac{t^{\beta-1}}{2}\left(E_{\alpha,\beta}(\lambda t) + E_{\alpha,\beta}(-\lambda t)\right) = t^{\beta-1}\sum_{n=0}^{\infty}\frac{\lambda^n t^{2n}}{\Gamma(2n\alpha+\beta)}, \tag{4.316}$$

$$Tretanh_{\alpha,\beta}(\lambda t) = \frac{Tresinh_{\alpha,\beta}(\lambda t)}{Trecosh_{\alpha,\beta}(\lambda t)} \left(Trecosh_{\alpha,\beta}(\lambda t) \neq 0\right), \tag{4.317}$$

$$Trecotanh_{\alpha,\beta}(\lambda t) = \frac{Trecosh_{\alpha,\beta}(\lambda t)}{Tresinh_{\alpha,\beta}(\lambda t)} \left(Tresinh_{\alpha,\beta}(\lambda t) \neq 0\right), \tag{4.318}$$

$$Trecsch_{\alpha,\beta}(\lambda t) = \frac{1}{Tresinh_{\alpha,\beta}(\lambda t)} \left(Tresinh_{\alpha,\beta}(\lambda t) \neq 0\right), \tag{4.319}$$

and

$$Tresech_{\alpha,\beta}(\lambda t) = \frac{1}{Trecosh_{\alpha,\beta}(\lambda t)} \left(Trecosh_{\alpha,\beta}(\lambda t) \neq 0\right), \tag{4.320}$$

where $\alpha, \beta, t, \lambda \in \mathbb{C}$, $Re(\alpha) > 0$, and $Re(\beta) > 0$.

As a direct result, we have the following theorem.

Theorem 4.32 *Let $\alpha, \beta, t \in \mathbb{C}$, $Re(\alpha) > 0$, and $Re(\beta) > 0$. Then, we have*

$$t^{\beta-1} E_{\alpha,\beta}(t) = Trecosh_{\alpha,\beta}(t) + Tresinh_{\alpha,\beta}(t) \tag{4.321}$$

and

$$t^{\beta-1} E_{\alpha,\beta}(-t) = Trecosh_{\alpha,\beta}(t) - Tresinh_{\alpha,\beta}(t). \tag{4.322}$$

Generally, there are

$$t^{\beta-1} E_{\alpha,\beta}(\lambda t) = Trecosh_{\alpha,\beta}(\lambda t) + Tresinh_{\alpha,\beta}(\lambda t) \tag{4.323}$$

and

$$t^{\beta-1} E_{\alpha,\beta}(-\lambda t) = Trecosh_{\alpha,\beta}(\lambda t) - Tresinh_{\alpha,\beta}(\lambda t), \tag{4.324}$$

where $\alpha, \beta, t \in \mathbb{C}$, $\lambda \in \mathbb{C}$, $Re(\alpha) > 0$, and $Re(\beta) > 0$.

Thus, we have

$$\left[t^{\beta-1} E_{\alpha,\beta}(t)\right]\left[t^{\beta-1} E_{\alpha,\beta}(-t)\right] = Trecosh^2_{\alpha,\beta}(t) + Tresinh^2_{\alpha,\beta}(t), \tag{4.325}$$

which is provided that

$$Trecosh^2_{\alpha,\beta}(t) = Trecosh_{\alpha,\beta}(t)\, Trecosh_{\alpha,\beta}(t) \tag{4.326}$$

and

$$Tresinh^2_{\alpha,\beta}(t) = Tresinh_{\alpha,\beta}(t)\, Tresinh_{\alpha,\beta}(t), \tag{4.327}$$

where $\alpha, \beta, t \in \mathbb{C}$, $Re(\alpha) > 0$, and $Re(\beta) > 0$.

So, there exists

$$\left[t^{\beta-1} E_{\alpha,\beta}(\lambda t)\right]\left[t^{\beta-1} E_{\alpha,\beta}(-\lambda t)\right] = Tprecosh^2_{\alpha,\beta}(\lambda t) + Tresinh^2_{\alpha,\beta}(\lambda t), \tag{4.328}$$

which is provided that

$$Trecosh^{2}_{\alpha,\beta}(\lambda t) = Trecosh_{\alpha,\beta}(t)\, Trecosh_{\alpha,\beta}(\lambda t) \quad (4.329)$$

and

$$Tresinh^{2}_{\alpha,\beta}(\lambda t) = Tresinh_{\alpha,\beta}\left(\lambda t^{\alpha}\right) Tresinh_{\alpha,\beta}(\lambda t), \quad (4.330)$$

where $\alpha, \beta, t \in \mathbb{C}$, $\lambda \in \mathbb{C}$, $Re(\alpha) > 0$, *and* $Re(\beta) > 0$.

Property 4.13 Let $\alpha, \beta, t \in \mathbb{C}, Re(\alpha) > 0$, and $Re(\beta) > 0$. Then, we have

$$Tresinh_{\alpha,\beta}(-t) = -Tresinh_{\alpha,\beta}(t), \quad (4.331)$$

$$Trecosh_{\alpha,\beta}(-t) = Trecosh_{\alpha,\beta}(t), \quad (4.332)$$

$$Tretanh_{\alpha,\beta}(-t) = -\, Tretanh_{\alpha,\beta}(t), \quad (4.333)$$

$$Trecotanh_{\alpha,\beta}(-t) = -\, Trecotanh_{\alpha,\beta}(t), \quad (4.334)$$

$$Trecsch_{\alpha,\beta}(-t) = -Trecsch_{\alpha,\beta}(t), \quad (4.335)$$

and

$$Tresech_{\alpha,\beta}(-t) = Tresech_{\alpha,\beta}(t). \quad (4.336)$$

4.1.6.3 Case 3: The Supertrigonometric and Superhyperbolic Functions with the Power Law and the Parameter

Let $\alpha, \beta, t \in \mathbb{C}$, $\lambda, \gamma \in \mathbb{C}$, $Re(\alpha) > 0$, and $Re(\beta) > 0$.

Then, we have

$$t^{\beta-1} E_{\alpha,\beta}\left(\lambda t^{\gamma}\right) = t^{\beta-1} \sum_{\kappa=0}^{\infty} \frac{\lambda^{\kappa} t^{\kappa\gamma}}{\Gamma(\kappa\alpha+\beta)}, \quad (4.337)$$

and

$$t^{\beta-1} E_{\alpha,\beta}\left(-\lambda t^{\gamma}\right) = t^{\beta-1} \sum_{\kappa=0}^{\infty} \frac{(-\lambda)^{\kappa} t^{\kappa\gamma}}{\Gamma(\kappa\alpha+\beta)}, \quad (4.338)$$

such that

$$t^{\beta-1} E_{\alpha,\beta}\left(it^{\gamma}\right) = t^{\beta-1} \sum_{\kappa=0}^{\infty} \frac{i^{\kappa} t^{\kappa\gamma}}{\Gamma(\kappa\alpha+\beta)}, \quad (4.339)$$

and

$$t^{\beta-1}E_{\alpha,\beta}\left(-it^{\gamma}\right)=t^{\beta-1}\sum_{\kappa=0}^{\infty}\frac{(-i)^{\kappa}\,t^{\kappa\gamma}}{\Gamma\left(\kappa\alpha+\beta\right)},\tag{4.340}$$

where $i=\sqrt{-1}$.

Definition 4.50 The presupersine with the power law t^{γ} and the parameter $t^{\beta-1}$ via Wiman function is defined as

$$Gresin_{\alpha,\beta}\left(t^{\gamma}\right)=\frac{t^{\beta-1}}{2i}\left(E_{\alpha,\beta}\left(it^{\gamma}\right)-E_{\alpha,\beta}\left(-it^{\gamma}\right)\right)=\sum_{n=0}^{\infty}\frac{(-1)^{n}\,t^{(2n+1)\gamma}}{\Gamma\left((2n+1)\,\alpha+\beta\right)},\tag{4.341}$$

where $\alpha,\beta,t,\gamma\in\mathbb{C}$, $Re\,(\alpha)>0$, and $Re\,(\beta)>0$.

Definition 4.51 The precosine with the power law t^{γ} and the parameter $t^{\beta-1}$ via Wiman function is defined as

$$Grecos_{\alpha,\beta}\left(t^{\gamma}\right)=\frac{t^{\beta-1}}{2}\left(E_{\alpha,\beta}\left(it^{\gamma}\right)+E_{\alpha,\beta}\left(-it^{\gamma}\right)\right)=\sum_{n=0}^{\infty}\frac{(-1)^{n}\,t^{2n\gamma}}{\Gamma\left(2n\alpha+\beta\right)},\tag{4.342}$$

where $\alpha,\beta,t,\gamma\in\mathbb{C}$, $Re\,(\alpha)>0$, and $Re\,(\beta)>0$.

Definition 4.52 The pretangent with the power law t^{γ} and the parameter $t^{\beta-1}$ via Wiman function is defined as

$$Gretan_{\alpha,\beta}\left(t^{\gamma}\right)=\frac{Gresin_{\alpha,\beta}\left(t^{\gamma}\right)}{Grecos_{\alpha,\beta}\left(t^{\gamma}\right)}\left(Grecos_{\alpha,\beta}\left(t^{\gamma}\right)\neq 0\right),\tag{4.343}$$

where $\alpha,\beta,t,\gamma\in\mathbb{C}$, $Re\,(\alpha)>0$, and $Re\,(\beta)>0$.

Definition 4.53 The precotangent with the power law t^{γ} and the parameter $t^{\beta-1}$ via Wiman function is defined as

$$Grecotan_{\alpha,\beta}\left(t^{\gamma}\right)=\frac{Grecos_{\alpha,\beta}\left(t^{\gamma}\right)}{Gresin_{\alpha,\beta}\left(t^{\gamma}\right)}\left(Gresin_{\alpha,\beta}\left(t^{\gamma}\right)\neq 0\right),\tag{4.344}$$

where $\alpha,\beta,t,\gamma\in\mathbb{C}$, $Re\,(\alpha)>0$, and $Re\,(\beta)>0$.

Definition 4.54 The precosecant with the power law t^{γ} and the parameter $t^{\beta-1}$ via Wiman function is defined as

$$Grecsc_{\alpha,\beta}\left(t^{\gamma}\right)=\frac{1}{Gresin_{\alpha,\beta}\left(t^{\gamma}\right)}\left(Gresin_{\alpha,\beta}\left(t^{\gamma}\right)\neq 0\right),\tag{4.345}$$

where $\alpha,\beta,t,\gamma\in\mathbb{C}$, $Re\,(\alpha)>0$, and $Re\,(\beta)>0$.

Definition 4.55 The presecant with the power law t^{γ} and the parameter $t^{\beta-1}$ via Wiman function is defined as

$$Gresec_{\alpha,\beta}\left(t^{\gamma}\right) = \frac{1}{Grecos_{\alpha,\beta}\left(t^{\gamma}\right)} \left(Grecos_{\alpha,\beta}\left(t^{\gamma}\right) \neq 0\right), \tag{4.346}$$

where $\alpha, \beta, t, \gamma \in \mathbb{C}$, $Re(\alpha) > 0$, and $Re(\beta) > 0$.

The supersine with the power law t^{γ} and the parameter $t^{\beta-1}$ via Wiman function is defined as

$$\begin{aligned} Gresin_{\alpha,\beta}\left(\lambda t^{\gamma}\right) &= \frac{t^{\beta-1}}{2i}\left(E_{\alpha,\beta}\left(i\lambda t^{\gamma}\right) - E_{\alpha,\beta}\left(-i\lambda t^{\gamma}\right)\right) \\ &= t^{\beta-1}\sum_{n=0}^{\infty}\frac{(-\lambda)^{n}\, t^{(2n+1)\gamma}}{\Gamma\left((2n+1)\alpha+\beta\right)}, \end{aligned} \tag{4.347}$$

where $\alpha, \beta, t, \gamma, \lambda \in \mathbb{C}$, $Re(\alpha) > 0$, and $Re(\beta) > 0$.

Theorem 4.33 *Let* $\alpha, \beta, t \in \mathbb{C}$, $\lambda, \gamma \in \mathbb{C}$, $Re(\alpha) > 0$, *and* $Re(\beta) > 0$.

$$Gresin_{\alpha,\beta}\left(\lambda t^{\gamma}\right) = t^{\beta-1} presin_{\alpha,\beta}\left(\lambda t^{\gamma}\right). \tag{4.348}$$

The precosine with the power law t^{γ} and the parameter $t^{\beta-1}$ via Wiman function is defined as

$$Grecos_{\alpha,\beta}\left(\lambda t^{\gamma}\right) = \frac{t^{\beta-1}}{2}\left(E_{\alpha,\beta}\left(i\lambda t^{\gamma}\right) + E_{\alpha,\beta}\left(-i\lambda t^{\gamma}\right)\right) = t^{\beta-1}\sum_{n=0}^{\infty}\frac{(-\lambda)^{n}\, t^{2n\gamma}}{\Gamma\left(2n\alpha+\beta\right)} \tag{4.349}$$

where $\alpha, \beta, t, \gamma, \lambda \in \mathbb{C}$, $Re(\alpha) > 0$, and $Re(\beta) > 0$.

Theorem 4.34 *Let* $\alpha, \beta, t \in \mathbb{C}$, $\lambda, \gamma \in \mathbb{C}$, $Re(\alpha) > 0$, *and* $Re(\beta) > 0$.

$$Grecos_{\alpha,\beta}\left(\lambda t^{\gamma}\right) = t^{\beta-1} precos_{\alpha,\beta}\left(\lambda t^{\gamma}\right). \tag{4.350}$$

The pretangent with the power law t^{γ} *and the parameter* $t^{\beta-1}$ *via Wiman function is defined as*

$$Gretan_{\alpha,\beta}\left(\lambda t^{\gamma}\right) = \frac{Gresin_{\alpha,\beta}\left(\lambda t^{\gamma}\right)}{Grecos_{\alpha,\beta}\left(\lambda t^{\gamma}\right)} \left(Grecos_{\alpha,\beta}\left(\lambda t^{\gamma}\right) \neq 0\right), \tag{4.351}$$

where $\alpha, \beta, t, \gamma, \lambda \in \mathbb{C}$, $Re(\alpha) > 0$, *and* $Re(\beta) > 0$.

The precotangent with the power law t^{γ} and the parameter $t^{\beta-1}$ via Wiman function is defined as

$$Grecotan_{\alpha,\beta}\left(\lambda t^{\gamma}\right)=\frac{Grecos_{\alpha,\beta}\left(\lambda t^{\gamma}\right)}{Gresin_{\alpha,\beta}\left(\lambda t^{\gamma}\right)}\left(Gresin_{\alpha,\beta}\left(\lambda t^{\gamma}\right)\neq 0\right), \tag{4.352}$$

where $\alpha, \beta, t, \gamma, \lambda \in \mathbb{C}$, $Re(\alpha) > 0$, and $Re(\beta) > 0$.

The precosecant with the power law t^{γ} and the parameter $t^{\beta-1}$ via Wiman function is defined as

$$Grecsc_{\alpha,\beta}\left(\lambda t^{\gamma}\right)=\frac{1}{Gresin_{\alpha,\beta}\left(\lambda t^{\gamma}\right)}\left(Gresin_{\alpha,\beta}\left(\lambda t^{\gamma}\right)\neq 0\right), \tag{4.353}$$

where $\alpha, \beta, t, \gamma, \lambda \in \mathbb{C}$, $Re(\alpha) > 0$, and $Re(\beta) > 0$.

The presecant with the power law t^{γ} and the parameter $t^{\beta-1}$ via Wiman function is defined as

$$Gresec_{\alpha,\beta}\left(\lambda t^{\gamma}\right)=\frac{1}{Grecos_{\alpha,\beta}\left(\lambda t^{\gamma}\right)}\left(Grecos_{\alpha,\beta}\left(\lambda t^{\gamma}\right)\neq 0\right), \tag{4.354}$$

where $\alpha, \beta, t \in \mathbb{C}$, $\lambda, \gamma \in \mathbb{C}$, $Re(\alpha) > 0$, and $Re(\beta) > 0$.

As a direct result, we have the following theorem.

Theorem 4.35 *Let $\alpha, \beta, t, \gamma \in \mathbb{C}$, $Re(\alpha) > 0$, and $Re(\beta) > 0$. Then, we have*

$$t^{\beta-1}E_{\alpha,\beta}\left(it^{\gamma}\right)=Grecos_{\alpha,\beta}\left(t^{\gamma}\right)+iGresin_{\alpha,\beta}\left(t^{\gamma}\right) \tag{4.355}$$

and

$$t^{\beta-1}E_{\alpha,\beta}\left(-it^{\gamma}\right)=Grecos_{\alpha,\beta}\left(t^{\gamma}\right)-iGresin_{\alpha,\beta}\left(t^{\gamma}\right). \tag{4.356}$$

Generally, there are

$$t^{\beta-1}E_{\alpha,\beta}\left(i\lambda t^{\gamma}\right)=Grecos_{\alpha,\beta}\left(\lambda t^{\gamma}\right)+iGresin_{\alpha,\beta}\left(\lambda t^{\gamma}\right) \tag{4.357}$$

and

$$t^{\beta-1}E_{\alpha,\beta}\left(-i\lambda t^{\gamma}\right)=Grecos_{\alpha,\beta}\left(\lambda t^{\gamma}\right)-iGresin_{\alpha,\beta}\left(\lambda t^{\gamma}\right), \tag{4.358}$$

where $\alpha, \beta, t \in \mathbb{C}$, $\lambda, \gamma \in \mathbb{C}$, $Re(\alpha) > 0$, and $Re(\beta) > 0$.

Thus, we have

$$\left[t^{\beta-1}E_{\alpha,\beta}\left(it^{\gamma}\right)\right]\left[t^{\beta-1}E_{\alpha,\beta}\left(-it^{\gamma}\right)\right]=Grecos^{2}_{\alpha,\beta}\left(t^{\gamma}\right)+Gresin^{2}_{\alpha,\beta}\left(t^{\gamma}\right), \tag{4.359}$$

which is provided that

$$Grecos^2_{\alpha,\beta}\left(t^\gamma\right) = Grecos_{\alpha,\beta}\left(t^\gamma\right) \cdot Grecos_{\alpha,\beta}\left(t^\gamma\right) \tag{4.360}$$

and

$$Gresin^2_{\alpha,\beta}\left(t^\gamma\right) = Gresin_{\alpha,\beta}\left(t^\gamma\right) Gresin_{\alpha,\beta}\left(t^\gamma\right), \tag{4.361}$$

where $\alpha, \beta, t, \gamma \in \mathbb{C}$, $Re(\alpha) > 0$, *and* $Re(\beta) > 0$.

So, there exists

$$\left[t^{\beta-1}E_{\alpha,\beta}\left(i\lambda t^\gamma\right)\right]\left[t^{\beta-1}E_{\alpha,\beta}\left(-i\lambda t^\gamma\right)\right] = Grecos^2_{\alpha,\beta}\left(\lambda t^\gamma\right) + Gresin^2_{\alpha,\beta}\left(\lambda t^\gamma\right), \tag{4.362}$$

which is provided that

$$Grecos^2_{\alpha,\beta}\left(\lambda t^\gamma\right) = Grecos_{\alpha,\beta}\left(t^\gamma\right) Grecos_{\alpha,\beta}\left(\lambda t^\gamma\right) \tag{4.363}$$

and

$$Gresin^2_{\alpha,\beta}\left(\lambda t^\gamma\right) = Gresin_{\alpha,\beta}\left(\lambda t^\gamma\right) Gresin_{\alpha,\beta}\left(\lambda t^\gamma\right), \tag{4.364}$$

where $\alpha, \beta, t \in \mathbb{C}$, $\lambda, \gamma \in \mathbb{C}$, $Re(\alpha) > 0$, *and* $Re(\beta) > 0$.

Property 4.14 Let $\alpha, \beta, t, \gamma \in \mathbb{C}, Re(\alpha) > 0$, and $Re(\beta) > 0$. Then, we have

$$Gresin_{\alpha,\beta}\left(-t^\gamma\right) = -Gresin_{\alpha,\beta}\left(t^\gamma\right), \tag{4.365}$$

$$Grecos_{\alpha,\beta}\left(-t^\gamma\right) = Grecos_{\alpha,\beta}\left(t^\gamma\right), \tag{4.366}$$

$$Gretan_{\alpha,\beta}\left(-t^\gamma\right) = -\, Gretan_{\alpha,\beta}\left(t^\gamma\right), \tag{4.367}$$

$$Grecotan_{\alpha,\beta}\left(-t^\gamma\right) = -\, Grecotan_{\alpha,\beta}\left(t^\gamma\right), \tag{4.368}$$

$$Grecsc_{\alpha,\beta}\left(-t^\gamma\right) = -Grecsc_{\alpha,\beta}\left(t^\gamma\right), \tag{4.369}$$

and

$$Gresec_{\alpha,\beta}\left(-t^\gamma\right) = Gresec_{\alpha,\beta}\left(t^\gamma\right). \tag{4.370}$$

Thus, we have

$$Gresin_{\alpha,\beta}\left(-\lambda t^\gamma\right) = -Gresin_{\alpha,\beta}\left(\lambda t^\gamma\right), \tag{4.371}$$

$$Grecos_{\alpha,\beta}\left(-\lambda t^\gamma\right) = Grecos_{\alpha,\beta}\left(\lambda t^\gamma\right), \tag{4.372}$$

$$Gretan_{\alpha,\beta}\left(-\lambda t^{\gamma}\right) = -\ Gretan_{\alpha,\beta}\left(\lambda t^{\gamma}\right), \tag{4.373}$$

$$Grecotan_{\alpha,\beta}\left(-\lambda t^{\gamma}\right) = -\ Grecotan_{\alpha,\beta}\left(\lambda t^{\gamma}\right), \tag{4.374}$$

$$Grecsc_{\alpha,\beta}\left(-\lambda t^{\gamma}\right) = -Grecsc_{\alpha,\beta}\left(\lambda t^{\gamma}\right), \tag{4.375}$$

and

$$Gresec_{\alpha,\beta}\left(-\lambda t^{\gamma}\right) = Gresec_{\alpha,\beta}\left(\lambda t^{\gamma}\right), \tag{4.376}$$

where $\alpha, \beta, t, \gamma, \lambda \in \mathbb{C}, Re\left(\alpha\right) > 0$, and $Re\left(\beta\right) > 0$.

There exists the series via Wiman function as follows:

$$\Theta\left(t\right) = \frac{a_0}{2} + \sum_{n=0}^{\infty}\left(a_n Grecos_{\alpha,\beta}\left(nt^{\gamma}\right) + b_n Gresin_{\alpha,\beta}\left(nt^{\gamma}\right)\right), \tag{4.377}$$

where a_n and b_n are the coefficients of the series via Wiman function.

Alternatively, there are the forms of the series via Wiman function

$$\Omega_1\left(t\right) = \sum_{n=0}^{\infty}\left(a_n Gresin_{\alpha,\beta}\left(nt^{\gamma}\right)\right), \tag{4.378}$$

and

$$\Omega_2\left(t\right) = \frac{a_0}{2} + \sum_{n=0}^{\infty}\left(b_n Grecos_{\alpha,\beta}\left(nt^{\gamma}\right)\right), \tag{4.379}$$

where a_n and b_n are the coefficients of the series via Wiman function and $n \in \mathbb{N}$.

Let $\alpha, \beta, t \in \mathbb{C}$, $\lambda, \gamma \in \mathbb{C}$, $Re\left(\alpha\right) > 0$, and $Re\left(\beta\right) > 0$.

Then, we have

$$t^{\beta-1}E_{\alpha,\beta}\left(\lambda t^{\gamma}\right) = t^{\beta-1}\sum_{\kappa=0}^{\infty}\frac{\lambda^{\kappa}t^{\kappa\gamma}}{\Gamma\left(\kappa\alpha+\beta\right)}, \tag{4.380}$$

and

$$t^{\beta-1}E_{\alpha,\beta}\left(-\lambda t^{\gamma}\right) = t^{\beta-1}\sum_{\kappa=0}^{\infty}\frac{\left(-\lambda\right)^{\kappa}t^{\kappa\gamma}}{\Gamma\left(\kappa\alpha+\beta\right)}, \tag{4.381}$$

such that

$$t^{\beta-1} E_{\alpha,\beta}\left(t^{\gamma}\right) = t^{\beta-1} \sum_{\kappa=0}^{\infty} \frac{i^{\kappa} t^{\kappa\gamma}}{\Gamma\left(\kappa\alpha+\beta\right)}, \tag{4.382}$$

and

$$t^{\beta-1} E_{\alpha,\beta}\left(-t^{\gamma}\right) = t^{\beta-1} \sum_{\kappa=0}^{\infty} \frac{(-1)^{\kappa} t^{\kappa\gamma}}{\Gamma\left(\kappa\alpha+\beta\right)}, \tag{4.383}$$

where $i = \sqrt{-1}$.

Definition 4.56 The hyperbolic presupersine with the power law t^{γ} and the parameter $t^{\beta-1}$ via Wiman function is defined as

$$Gresinh_{\alpha,\beta}\left(t^{\gamma}\right) = \frac{t^{\beta-1}}{2}\left(E_{\alpha,\beta}\left(t^{\gamma}\right) - E_{\alpha,\beta}\left(-t^{\gamma}\right)\right) = t^{\beta-1} \sum_{n=0}^{\infty} \frac{t^{(2n+1)\gamma}}{\Gamma\left((2n+1)\alpha+\beta\right)}, \tag{4.384}$$

where $\alpha, \beta, t, \gamma \in \mathbb{C}$, $Re\left(\alpha\right) > 0$, and $Re\left(\beta\right) > 0$.

Definition 4.57 The hyperbolic precosine with the power law t^{γ} and the parameter $t^{\beta-1}$ via Wiman function is defined as

$$Grecosh_{\alpha,\beta}\left(t^{\gamma}\right) = \frac{t^{\beta-1}}{2}\left(E_{\alpha,\beta}\left(t^{\gamma}\right) + E_{\alpha,\beta}\left(-t^{\gamma}\right)\right) = t^{\beta-1} \sum_{n=0}^{\infty} \frac{t^{2n\gamma}}{\Gamma\left(2n\alpha+\beta\right)}, \tag{4.385}$$

where $\alpha, \beta, t, \gamma \in \mathbb{C}$, $Re\left(\alpha\right) > 0$, and $Re\left(\beta\right) > 0$.

Definition 4.58 The hyperbolic pretangent with the power law t^{γ} and the parameter $t^{\beta-1}$ via Wiman function is defined as

$$Gretanh_{\alpha,\beta}\left(t^{\gamma}\right) = \frac{Gresinh_{\alpha,\beta}\left(t^{\gamma}\right)}{precosh_{\alpha,\beta}\left(t^{\gamma}\right)}\left(Grecosh_{\alpha,\beta}\left(t^{\gamma}\right) \neq 0\right), \tag{4.386}$$

where $\alpha, \beta, t, \gamma \in \mathbb{C}$, $Re\left(\alpha\right) > 0$, and $Re\left(\beta\right) > 0$.

Definition 4.59 The hyperbolic precotangent with the power law t^{γ} and the parameter $t^{\beta-1}$ via Wiman function is defined as

$$Grecotanh_{\alpha,\beta}\left(t^{\gamma}\right) = \frac{Grecosh_{\alpha,\beta}\left(t^{\gamma}\right)}{Gresinh_{\alpha,\beta}\left(t^{\gamma}\right)}\left(Gresinh_{\alpha,\beta}\left(t^{\gamma}\right) \neq 0\right), \tag{4.387}$$

where $\alpha, \beta, t, \gamma \in \mathbb{C}$, $Re\left(\alpha\right) > 0$, and $Re\left(\beta\right) > 0$.

Definition 4.60 The hyperbolic precosecant with the power law t^{γ} and the parameter $t^{\beta-1}$ via Wiman function is defined as

$$Grecsch_{\alpha,\beta}\left(t^{\gamma}\right)=\frac{1}{Gresinh_{\alpha,\beta}\left(t^{\gamma}\right)}\left(Gresinh_{\alpha,\beta}\left(t^{\gamma}\right)\neq 0\right), \tag{4.388}$$

where $\alpha, \beta, t, \gamma \in \mathbb{C}$, $Re(\alpha) > 0$, and $Re(\beta) > 0$.

Definition 4.61 The hyperbolic presecant with the power law t^{γ} and the parameter $t^{\beta-1}$ via Wiman function is defined as

$$Gresech_{\alpha,\beta}\left(t^{\gamma}\right)=\frac{1}{Grecosh_{\alpha,\beta}\left(t^{\gamma}\right)}\left(Grecosh_{\alpha,\beta}\left(t^{\gamma}\right)\neq 0\right), \tag{4.389}$$

where $\alpha, \beta, t, \gamma \in \mathbb{C}$, $Re(\alpha) > 0$, and $Re(\beta) > 0$.

Generally, we present the alternative definitions as follows.

The hyperbolic supersine with the power law t^{γ} and the parameter $t^{\beta-1}$ via Wiman function is defined as

$$\begin{aligned}Gresinh_{\alpha,\beta}\left(\lambda t^{\gamma}\right)&=\frac{t^{\beta-1}}{2}\left(E_{\alpha,\beta}\left(\lambda t^{\gamma}\right)-E_{\alpha,\beta}\left(-\lambda t^{\gamma}\right)\right)\\&=t^{\beta-1}\sum_{n=0}^{\infty}\frac{\lambda^{n}t^{(2n+1)\gamma}}{\Gamma\left((2n+1)\alpha+\beta\right)},\end{aligned} \tag{4.390}$$

where $\alpha, \beta, t, \gamma, \lambda \in \mathbb{C}$, $Re(\alpha) > 0$, and $Re(\beta) > 0$.

The hyperbolic precosine with the power law t^{γ} and the parameter $t^{\beta-1}$ via Wiman function is defined as

$$Grecosh_{\alpha,\beta}\left(\lambda t^{\gamma}\right)=\frac{t^{\beta-1}}{2}\left(E_{\alpha,\beta}\left(\lambda t^{\gamma}\right)+E_{\alpha,\beta}\left(-\lambda t^{\gamma}\right)\right)=t^{\beta-1}\sum_{n=0}^{\infty}\frac{\lambda^{n}t^{2n\gamma}}{\Gamma\left(2n\alpha+\beta\right)}, \tag{4.391}$$

where $\alpha, \beta, t, \gamma, \lambda \in \mathbb{C}$, $Re(\alpha) > 0$, and $Re(\beta) > 0$.

The hyperbolic pretangent with the power law t^{γ} and the parameter $t^{\beta-1}$ via Wiman function is defined as

$$Gretanh_{\alpha,\beta}\left(\lambda t^{\gamma}\right)=\frac{Gresinh_{\alpha,\beta}\left(\lambda t^{\gamma}\right)}{Grecosh_{\alpha,\beta}\left(\lambda t^{\gamma}\right)}\left(Grecosh_{\alpha,\beta}\left(\lambda t^{\gamma}\right)\neq 0\right), \tag{4.392}$$

where $\alpha, \beta, t, \gamma, \lambda \in \mathbb{C}$, $Re(\alpha) > 0$, and $Re(\beta) > 0$.

The hyperbolic precotangent with the power law t^{γ} and the parameter $t^{\beta-1}$ via Wiman function is defined as

$$Grecotanh_{\alpha,\beta}\left(\lambda t^{\gamma}\right)=\frac{Grecosh_{\alpha,\beta}\left(\lambda t^{\gamma}\right)}{Gresinh_{\alpha,\beta}\left(\lambda t^{\gamma}\right)}\left(Gresinh_{\alpha,\beta}\left(\lambda t^{\gamma}\right)\neq 0\right), \tag{4.393}$$

where $\alpha, \beta, t, \gamma, \lambda \in \mathbb{C}$, $Re(\alpha) > 0$, and $Re(\beta) > 0$.

The hyperbolic precosecant with the power law t^{γ} and the parameter $t^{\beta-1}$ via Wiman function is defined as

$$Grecsch_{\alpha,\beta}\left(\lambda t^{\gamma}\right)=\frac{1}{Gresinh_{\alpha,\beta}\left(\lambda t^{\gamma}\right)}\left(Gresinh_{\alpha,\beta}\left(\lambda t^{\gamma}\right)\neq 0\right), \tag{4.394}$$

where $\alpha, \beta, t, \gamma, \lambda \in \mathbb{C}$, $Re(\alpha) > 0$, and $Re(\beta) > 0$.

The hyperbolic presecant with the power law t^{γ} and the parameter $t^{\beta-1}$ via Wiman function is defined as

$$Gresech_{\alpha,\beta}\left(\lambda t^{\gamma}\right)=\frac{1}{Grecosh_{\alpha,\beta}\left(\lambda t^{\gamma}\right)}\left(Grecosh_{\alpha,\beta}\left(\lambda t^{\gamma}\right)\neq 0\right), \tag{4.395}$$

where $\alpha, \beta, t, \gamma, \lambda \in \mathbb{C}$, $Re(\alpha) > 0$, and $Re(\beta) > 0$.

As a direct result, we have the following theorem.

Theorem 4.36 *Let $\alpha, \beta, t, \gamma \in \mathbb{C}$, $Re(\alpha) > 0$, and $Re(\beta) > 0$. Then, we have*

$$t^{\beta-1}E_{\alpha,\beta}\left(t^{\gamma}\right)=Grecosh_{\alpha,\beta}\left(t^{\gamma}\right)+Gresinh_{\alpha,\beta}\left(t^{\gamma}\right) \tag{4.396}$$

and

$$t^{\beta-1}E_{\alpha,\beta}\left(-t^{\gamma}\right)=Grecosh_{\alpha,\beta}\left(t^{\gamma}\right)-Gresinh_{\alpha,\beta}\left(t^{\gamma}\right). \tag{4.397}$$

Generally, there are

$$t^{\beta-1}E_{\alpha,\beta}\left(\lambda t^{\gamma}\right)=Grecosh_{\alpha,\beta}\left(\lambda t^{\gamma}\right)+Gresinh_{\alpha,\beta}\left(\lambda t^{\gamma}\right) \tag{4.398}$$

and

$$t^{\beta-1}E_{\alpha,\beta}\left(-\lambda t^{\gamma}\right)=Grecosh_{\alpha,\beta}\left(\lambda t^{\gamma}\right)-Gresinh_{\alpha,\beta}\left(\lambda t^{\gamma}\right), \tag{4.399}$$

where $\alpha, \beta, t \in \mathbb{C}$, $\lambda, \gamma \in \mathbb{C}$, $Re(\alpha) > 0$, and $Re(\beta) > 0$.

Thus, we have

$$\left[t^{\beta-1}E_{\alpha,\beta}\left(t^{\gamma}\right)\right]\left[t^{\beta-1}E_{\alpha,\beta}\left(-t^{\gamma}\right)\right]=Grecosh^{2}_{\alpha,\beta}\left(t^{\gamma}\right)+Gresinh^{2}_{\alpha,\beta}\left(t^{\gamma}\right), \tag{4.400}$$

which is provided that

$$Grecosh^{2}_{\alpha,\beta}\left(t^{\gamma}\right)=Grecosh_{\alpha,\beta}\left(t^{\gamma}\right)\cdot Grecosh_{\alpha,\beta}\left(t^{\gamma}\right) \tag{4.401}$$

and

$$Gresinh^2_{\alpha,\beta}\left(t^\gamma\right) = Gresinh_{\alpha,\beta}\left(t^\gamma\right) Gresinh_{\alpha,\beta}\left(t^\gamma\right), \tag{4.402}$$

where $\alpha, \beta, t, \gamma \in \mathbb{C}$, $Re(\alpha) > 0$, *and* $Re(\beta) > 0$.

So, there exists

$$\left[t^{\beta-1} E_{\alpha,\beta}\left(\lambda t^\gamma\right)\right]\left[t^{\beta-1} E_{\alpha,\beta}\left(-\lambda t^\gamma\right)\right] = Grecosh^2_{\alpha,\beta}\left(\lambda t^\gamma\right) + Gresinh^2_{\alpha,\beta}\left(\lambda t^\gamma\right), \tag{4.403}$$

which is provided that

$$Grecosh^2_{\alpha,\beta}\left(\lambda t^\gamma\right) = Grecosh_{\alpha,\beta}\left(\lambda t^\gamma\right) Grecosh_{\alpha,\beta}\left(\lambda t^\gamma\right) \tag{4.404}$$

and

$$Gresinh^2_{\alpha,\beta}\left(\lambda t^\gamma\right) = Gresinh_{\alpha,\beta}\left(\lambda t^\gamma\right) Gresinh_{\alpha,\beta}\left(\lambda t^\gamma\right), \tag{4.405}$$

where $\alpha, \beta, t \in \mathbb{C}$, $\lambda, \gamma \in \mathbb{C}$, $Re(\alpha) > 0$, *and* $Re(\beta) > 0$.

Property 4.15 Let $\alpha, \beta, t, \gamma \in \mathbb{C}, Re(\alpha) > 0$, and $Re(\beta) > 0$. Then, we have

$$Gresinh_{\alpha,\beta}\left(-t^\gamma\right) = -Gresinh_{\alpha,\beta}\left(t^\gamma\right), \tag{4.406}$$

$$Grecosh_{\alpha,\beta}\left(-t^\gamma\right) = Grecosh_{\alpha,\beta}\left(t^\gamma\right), \tag{4.407}$$

$$Gretanh_{\alpha,\beta}\left(-t^\gamma\right) = -\ Gretanh_{\alpha,\beta}\left(t^\gamma\right), \tag{4.408}$$

$$Grecotanh_{\alpha,\beta}\left(-t^\gamma\right) = -\ Grecotanh_{\alpha,\beta}\left(t^\gamma\right), \tag{4.409}$$

$$Grecsch_{\alpha,\beta}\left(-t^\gamma\right) = -Grecsch_{\alpha,\beta}\left(t^\gamma\right), \tag{4.410}$$

and

$$Gresech_{\alpha,\beta}\left(-t^\gamma\right) = Gresech_{\alpha,\beta}\left(t^\gamma\right). \tag{4.411}$$

Generally,

$$Gresinh_{\alpha,\beta}\left(-\lambda t^\gamma\right) = -Gresinh_{\alpha,\beta}\left(\lambda t^\gamma\right), \tag{4.412}$$

$$Grecosh_{\alpha,\beta}\left(-\lambda t^\gamma\right) = Grecosh_{\alpha,\beta}\left(\lambda t^\gamma\right), \tag{4.413}$$

$$Gretanh_{\alpha,\beta}\left(-\lambda t^\gamma\right) = -\ Gretanh_{\alpha,\beta}\left(\lambda t^\gamma\right), \tag{4.414}$$

$$Grecotanh_{\alpha,\beta}\left(-\lambda t^{\gamma}\right) = -\ Grecotanh_{\alpha,\beta}\left(\lambda t^{\gamma}\right), \tag{4.415}$$

$$Grecsch_{\alpha,\beta}\left(-\lambda t^{\gamma}\right) = -Grecsch_{\alpha,\beta}\left(\lambda t^{\gamma}\right), \tag{4.416}$$

and

$$Gresech_{\alpha,\beta}\left(-\lambda t^{\gamma}\right) = Gresech_{\alpha,\beta}\left(\lambda t^{\gamma}\right), \tag{4.417}$$

where $\alpha, \beta, t, \gamma, \lambda \in \mathbb{C}, Re\,(\alpha) > 0$, and $Re\,(\beta) > 0$.

4.1.7 *The Special Integral Equations via Viman Function and Related Functions*

In this section, we present the theory of the special integral equations via Viman function.

4.1.7.1 Integral Equations of Volterra Type

Example 4.11 The integral equation of Volterra type is given as follows:

$$H\,(t) = f\,(t) - \lambda \int_a^t (t-\tau)^{\beta-1}\, E_{\alpha,\beta}\left(\lambda\,(t-\tau)^{\alpha}\right) f\,(\tau)\, d\tau \tag{4.418}$$

or

$$H\,(t) = \lambda \int_a^t (t-\tau)^{\beta-1}\, E_{\alpha,\beta}\left(\lambda\,(t-\tau)^{\alpha}\right) f\,(\tau)\, d\tau, \tag{4.419}$$

where $\lambda, \gamma, \alpha, \beta \in \mathbb{R}$ and $t > a$.

Example 4.12 The integral equation of Volterra type is given as follows:

$$H\,(t) = f\,(t) - \lambda \int_a^t supersin_{\alpha,\beta}\left(\lambda\,(t-\tau)^{\alpha}\right) f\,(\tau)\, d\tau \tag{4.420}$$

or

$$H(t) = \lambda \int_a^t supersin_{\alpha,\beta}\left(\lambda (t-\tau)^{\alpha}\right) f(\tau)\, d\tau, \tag{4.421}$$

where $\lambda, \gamma, \alpha, \beta \in \mathbb{R}$ and $t > a$.

Example 4.13 The integral equation of Volterra type is given as follows:

$$H(t) = f(t) - \lambda \int_a^t supercos_{\alpha,\beta}\left(\lambda (t-\tau)^{\alpha}\right) f(\tau)\, d\tau \tag{4.422}$$

or

$$H(t) = \lambda \int_a^t supercos_{\alpha,\beta}\left(\lambda (t-\tau)^{\alpha}\right) f(\tau)\, d\tau, \tag{4.423}$$

where $\lambda, \gamma, \alpha, \beta \in \mathbb{R}$ and $t > a$.

Example 4.14 The integral equation of Volterra type is given as follows:

$$H(t) = f(t) - \lambda \int_a^t supersinh_{\alpha,\beta}\left(\lambda (t-\tau)^{\alpha}\right) f(\tau)\, d\tau \tag{4.424}$$

or

$$H(t) = \lambda \int_a^t supersinh_{\alpha,\beta}\left(\lambda (t-\tau)^{\alpha}\right) f(\tau)\, d\tau, \tag{4.425}$$

where $\lambda, \gamma, \alpha, \beta \in \mathbb{R}$ and $t > a$.

Example 4.15 The integral equation of Volterra type is given as follows:

$$H(t) = f(t) - \lambda \int_a^t supercosh_{\alpha,\beta}\left(\lambda (t-\tau)^{\alpha}\right) f(\tau)\, d\tau \tag{4.426}$$

or

$$H(t) = \lambda \int_a^t supercosh_{\alpha,\beta}\left(\lambda (t-\tau)^{\alpha}\right) f(\tau)\, d\tau, \tag{4.427}$$

where $\lambda, \gamma, \alpha, \beta \in \mathbb{R}$ and $t > a$.

Example 4.16 The integral equation of Volterra type is given as follows:

$$H(t) = f(t) - \lambda \int_a^t (t-\tau)^{\beta-1} E_{\alpha,\beta}\left(\lambda (t-\tau)^{\gamma}\right) f(\tau)\, d\tau \tag{4.428}$$

or

$$H(t) = \lambda \int_a^t (t-\tau)^{\beta-1} E_{\alpha,\beta}\left(\lambda (t-\tau)^{\gamma}\right) f(\tau)\, d\tau \tag{4.429}$$

where $\lambda, \gamma, \alpha, \beta \in \mathbb{R}$ and $t > a$.

Example 4.17 The integral equation of Volterra type is given as follows:

$$H(t) = f(t) - \lambda \int_a^t Gresin_{\alpha,\beta}\left(\lambda (t-\tau)^{\gamma}\right) f(\tau)\, d\tau \tag{4.430}$$

or

$$H(t) = \lambda \int_a^t Gresin_{\alpha,\beta}\left(\lambda (t-\tau)^{\gamma}\right) f(\tau)\, d\tau, \tag{4.431}$$

where $\lambda, \gamma, \alpha, \beta \in \mathbb{R}$ and $t > a$.

Example 4.18 The integral equation of Volterra type is given as follows:

$$H(t) = f(t) - \lambda \int_a^t Grecos_{\alpha,\beta}\left(\lambda (t-\tau)^{\gamma}\right) f(\tau)\, d\tau \tag{4.432}$$

or

$$H(t) = \lambda \int_a^t Grecos_{\alpha,\beta}\left(\lambda (t-\tau)^{\gamma}\right) f(\tau)\, d\tau, \tag{4.433}$$

where $\lambda, \gamma, \alpha, \beta \in \mathbb{R}$ and $t > a$.

Example 4.19 The integral equation of Volterra type is given as follows:

$$H(t) = f(t) - \lambda \int_a^t Gresinh_{\alpha,\beta}\left(\lambda (t-\tau)^{\gamma}\right) f(\tau)\, d\tau \tag{4.434}$$

or

$$H(t) = \lambda \int_{a}^{t} Gresinh_{\alpha,\beta}\left(\lambda (t-\tau)^{\gamma}\right) f(\tau)\, d\tau, \tag{4.435}$$

where $\lambda, \gamma, \alpha, \beta \in \mathbb{R}$ and $t > a$.

Example 4.20 The integral equation of Volterra type is given as follows:

$$H(t) = f(t) - \lambda \int_{a}^{t} Grecosh_{\alpha,\beta}\left(\lambda (t-\tau)^{\gamma}\right) f(\tau)\, d\tau \tag{4.436}$$

or

$$H(t) = \lambda \int_{a}^{t} Grecosh_{\alpha,\beta}\left(\lambda (t-\tau)^{\gamma}\right) f(\tau)\, d\tau, \tag{4.437}$$

where $\lambda, \gamma, \alpha, \beta \in \mathbb{R}$ and $t > a$.

4.1.7.2 Integral Equations of Fredholm Type

Example 4.21 The integral equation of Fredholm type is given as follows:

$$H(t) = f(t) - \lambda \int_{a}^{b} (t-\tau)^{\beta-1} E_{\alpha,\beta}\left(\lambda (t-\tau)^{\alpha}\right) f(\tau)\, d\tau \tag{4.438}$$

or

$$H(t) = \lambda \int_{a}^{b} (t-\tau)^{\beta-1} E_{\alpha,\beta}\left(\lambda (t-\tau)^{\alpha}\right) f(\tau)\, d\tau, \tag{4.439}$$

where $\lambda, \alpha, \beta \in \mathbb{R}$ and $b > \tau > a$.

Example 4.22 The integral equation of Fredholm type is given as follows:

$$H(t) = f(t) - \lambda \int_{a}^{b} supersin_{\alpha,\beta}\left(\lambda (t-\tau)^{\alpha}\right) f(\tau)\, d\tau \tag{4.440}$$

or

$$H(t) = \lambda \int_a^b supersin_{\alpha,\beta}\left(\lambda (t-\tau)^\alpha\right) f(\tau)\, d\tau, \tag{4.441}$$

where $\lambda, \alpha, \beta \in \mathbb{R}$ and $b > \tau > a$.

Example 4.23 The integral equation of Fredholm type is given as follows:

$$H(t) = f(t) - \lambda \int_a^b supercos_{\alpha,\beta}\left(\lambda (t-\tau)^\alpha\right) f(\tau)\, d\tau \tag{4.442}$$

or

$$H(t) = \lambda \int_a^b supercos_{\alpha,\beta}\left(\lambda (t-\tau)^\alpha\right) f(\tau)\, d\tau, \tag{4.443}$$

where $\lambda, \alpha, \beta \in \mathbb{R}$ and $b > \tau > a$.

Example 4.24 The integral equation of Fredholm type is given as follows:

$$H(t) = f(t) - \lambda \int_a^b supersinh_{\alpha,\beta}\left(\lambda (t-\tau)^\alpha\right) f(\tau)\, d\tau \tag{4.444}$$

or

$$H(t) = \lambda \int_a^b supersinh_{\alpha,\beta}\left(\lambda (t-\tau)^\alpha\right) f(\tau)\, d\tau, \tag{4.445}$$

where $\lambda, \alpha, \beta \in \mathbb{R}$ and $b > \tau > a$.

Example 4.25 The integral equation of Fredholm type is given as follows:

$$H(t) = f(t) - \lambda \int_a^b supercosh_{\alpha,\beta}\left(\lambda (t-\tau)^\alpha\right) f(\tau)\, d\tau \tag{4.446}$$

or

$$H(t) = \lambda \int_a^b supercosh_{\alpha,\beta}\left(\lambda (t-\tau)^\alpha\right) f(\tau)\, d\tau, \tag{4.447}$$

where $\lambda, \alpha, \beta \in \mathbb{R}$ and $b > \tau > a$.

Example 4.26 The integral equation of Fredholm type is given as follows:

$$H(t) = f(t) - \lambda \int_a^b (t-\tau)^{\beta-1} E_{\alpha,\beta}\left(\lambda (t-\tau)^{\gamma}\right) f(\tau)\, d\tau \tag{4.448}$$

or

$$H(t) = \lambda \int_a^t (t-\tau)^{\beta-1} E_{\alpha,\beta}\left(\lambda (t-\tau)^{\gamma}\right) f(\tau)\, d\tau, \tag{4.449}$$

where $\lambda, \alpha, \beta, \gamma \in \mathbb{R}$ and $b > \tau > a$.

Example 4.27 The integral equation of Fredholm type is given as follows:

$$H(t) = f(t) - \lambda \int_a^b Gresin_{\alpha,\beta}\left(\lambda (t-\tau)^{\gamma}\right) f(\tau)\, d\tau \tag{4.450}$$

or

$$H(t) = \lambda \int_a^b Gresin_{\alpha,\beta}\left(\lambda (t-\tau)^{\gamma}\right) f(\tau)\, d\tau, \tag{4.451}$$

where $\lambda, \alpha, \beta, \gamma \in \mathbb{R}$ and $b > \tau > a$.

Example 4.28 The integral equation of Fredholm type is given as follows:

$$H(t) = f(t) - \lambda \int_a^b Grecos_{\alpha,\beta}\left(\lambda (t-\tau)^{\gamma}\right) f(\tau)\, d\tau \tag{4.452}$$

or

$$H(t) = \lambda \int_a^b Grecos_{\alpha,\beta}\left(\lambda (t-\tau)^{\gamma}\right) f(\tau)\, d\tau, \tag{4.453}$$

where $\lambda, \gamma \in \mathbb{R}$ and $b > \tau > a$.

Example 4.29 The integral equation of Fredholm type is given as follows:

$$H(t) = f(t) - \lambda \int_a^b Gresinh_{\alpha,\beta}\left(\lambda (t-\tau)^{\gamma}\right) f(\tau)\, d\tau \tag{4.454}$$

or

$$H(t) = \lambda \int_a^b Gresinh_{\alpha,\beta}\left(\lambda(t-\tau)^{\gamma}\right) f(\tau)\, d\tau, \tag{4.455}$$

where $\lambda, \alpha, \beta, \gamma \in \mathbb{R}$ and $b > \tau > a$.

Example 4.30 The integral equation of Fredholm type is given as follows:

$$H(t) = f(t) - \lambda \int_a^b Grecosh_{\alpha,\beta}\left(\lambda(t-\tau)^{\gamma}\right) f(\tau)\, d\tau \tag{4.456}$$

or

$$H(t) = \lambda \int_a^b Grecosh_{\alpha,\beta}\left(\lambda(t-\tau)^{\gamma}\right) f(\tau)\, d\tau, \tag{4.457}$$

where $\lambda, \alpha, \beta, \gamma \in \mathbb{R}$ and $b > \tau > a$.

4.1.8 The Integral Representations Related to Viman Function

We now consider the integral representations related to the Viman function and give some theorems.

Theorem 4.37 *Let* $\alpha, \beta \in \mathbb{C}$, $\gamma \in \mathbb{R}_+$, $\left|\lambda s^{-\alpha}\right| < 1$, $Re(\alpha) > 0$, *and* $Re(\beta) > 0$. *Then, there is*

$$\int_0^{\infty} e^{-st^{\gamma}} t^{\beta\gamma-1} E_{\alpha,\beta}\left(\lambda t^{\alpha\gamma}\right) dt = \gamma^{-1} s^{-\beta}\left(1-\lambda s^{-\alpha}\right)^{-1}. \tag{4.458}$$

Theorem 4.38 *Let* $\alpha, \beta \in \mathbb{C}$, $\gamma \in \mathbb{R}_+$, $\left|\lambda s^{-\alpha}\right| < 1$, $Re(\alpha) > 0$, *and* $Re(\beta) > 0$. *Then, there is*

$$\int_0^{\infty} e^{-st^{\gamma}} t^{\beta\gamma-1} presin_{\alpha,\beta}\left(\lambda t^{\alpha\gamma}\right) dt = \gamma^{-1} \lambda s^{-\alpha-\beta}\left(1+\lambda^2 s^{-2\alpha}\right)^{-1}. \tag{4.459}$$

Proof Making use of the relation

$$presin_{\alpha,\beta}\left(\lambda t^{\alpha\gamma}\right)=\frac{1}{2i}\left(E_{\alpha,\beta}\left(i\lambda t^{\alpha\gamma}\right)-E_{\alpha,\beta}\left(-i\lambda t^{\alpha\gamma}\right)\right), \tag{4.460}$$

we have

$$\begin{aligned}&\int_0^\infty e^{-st^\gamma}t^{\beta\gamma-1}presin_{\alpha,\beta}\left(\lambda t^{\alpha\gamma}\right)dt\\&=\int_0^\infty e^{-st^\gamma}t^{\beta\gamma-1}\left[\tfrac{1}{2i}\left(E_{\alpha,\beta}\left(i\lambda t^{\alpha\gamma}\right)-E_{\alpha,\beta}\left(-i\lambda t^{\alpha\gamma}\right)\right)\right]dt\\&=\tfrac{\gamma^{-1}s^{-\beta}}{2i}\left(\left(1-i\lambda s^{-\alpha}\right)^{-1}-\left(1+i\lambda s^{-\alpha}\right)^{-1}\right)\\&=\tfrac{\gamma^{-1}\lambda s^{-\alpha-\beta}}{1+\lambda^2s^{-2\alpha}}.\end{aligned} \tag{4.461}$$

□

Theorem 4.39 *Let* $\alpha,\beta\in\mathbb{C}$, $\gamma\in\mathbb{R}_+$, $\left|\lambda s^{-\alpha}\right|<1$, $Re\,(\alpha)>0$, *and* $Re\,(\beta)>0$. *Then, there is*

$$\int_0^\infty e^{-st^\gamma}t^{\beta\gamma-1}precos_{\alpha,\beta}\left(\lambda t^{\alpha\gamma}\right)dt=\gamma^{-1}s^{-\beta}\left(1+\lambda^2s^{-2\alpha}\right)^{-1}. \tag{4.462}$$

Proof Making use of the relation

$$precos_{\alpha,\beta}\left(\lambda t^{\alpha\gamma}\right)=\frac{1}{2}\left(E_{\alpha,\beta}\left(i\lambda t^{\alpha\gamma}\right)+E_{\alpha,\beta}\left(-i\lambda t^{\alpha\gamma}\right)\right), \tag{4.463}$$

we have

$$\begin{aligned}&\int_0^\infty e^{-st^\gamma}t^{\beta\gamma-1}precos_{\alpha,\beta}\left(\lambda t^{\alpha\gamma}\right)dt\\&=\int_0^\infty e^{-st^\gamma}t^{\beta\gamma-1}\left[\tfrac{1}{2}\left(E_{\alpha,\beta}\left(i\lambda t^{\alpha\gamma}\right)+E_{\alpha,\beta}\left(-i\lambda t^{\alpha\gamma}\right)\right)\right]dt\\&=\tfrac{\gamma^{-1}s^{-\beta}}{2}\left(\left(1-i\lambda s^{-\alpha}\right)^{-1}+\left(1+i\lambda s^{-\alpha}\right)^{-1}\right)\\&=\tfrac{\gamma^{-1}s^{-\beta}}{1+\lambda^2s^{-2\alpha}}.\end{aligned} \tag{4.464}$$

□

Theorem 4.40 *Let* $\alpha,\beta\in\mathbb{C}$, $\gamma\in\mathbb{R}_+$, $\left|\lambda s^{-\alpha}\right|<1$, $Re\,(\alpha)>0$, *and* $Re\,(\beta)>0$. *Then, there is*

$$\int_0^\infty e^{-st^\gamma}t^{\beta\gamma-1}presinh_{\alpha,\beta}\left(\lambda t^{\alpha\gamma}\right)dt=\gamma^{-1}\lambda s^{-\alpha-\beta}\left(1-\lambda^2s^{-2\alpha}\right)^{-1}. \tag{4.465}$$

Proof Making use of the relation

$$presinh_{\alpha,\beta}\left(\lambda t^{\alpha\gamma}\right) = \frac{1}{2}\left(E_{\alpha,\beta}\left(\lambda t^{\alpha\gamma}\right) - E_{\alpha,\beta}\left(-\lambda t^{\alpha\gamma}\right)\right), \tag{4.466}$$

we have

$$\begin{aligned}
&\int_0^\infty e^{-st^\gamma} t^{\beta\gamma-1} presinh_{\alpha,\beta}\left(\lambda t^{\alpha\gamma}\right) dt \\
&= \int_0^\infty e^{-st^\gamma} t^{\beta\gamma-1} \left[\tfrac{1}{2}\left(E_{\alpha,\beta}\left(\lambda t^{\alpha\gamma}\right) - E_{\alpha,\beta}\left(-\lambda t^{\alpha\gamma}\right)\right)\right] dt \\
&= \tfrac{\gamma^{-1}s^{-\beta}}{2}\left(\left(1-\lambda s^{-\alpha}\right)^{-1} - \left(1+\lambda s^{-\alpha}\right)^{-1}\right) \\
&= \tfrac{\gamma^{-1}\lambda s^{-\alpha-\beta}}{1-\lambda^2 s^{-2\alpha}}.
\end{aligned} \tag{4.467}$$

□

Theorem 4.41 *Let $\alpha, \beta \in \mathbb{C}$, $\gamma \in \mathbb{R}_+$, $\left|\lambda s^{-\alpha}\right| < 1$, $Re(\alpha) > 0$, and $Re(\beta) > 0$. Then, there is*

$$\int_0^\infty e^{-st^\gamma} t^{\beta\gamma-1} precosh_{\alpha,\beta}\left(\lambda t^{\alpha\gamma}\right) dt = \gamma^{-1}s^{-\beta}\left(1-\lambda^2 s^{-2\alpha}\right)^{-1}. \tag{4.468}$$

Proof Making use of the relation

$$precosh_{\alpha,\beta}\left(\lambda t^{\alpha\gamma}\right) = \frac{1}{2}\left(E_{\alpha,\beta}\left(\lambda t^{\alpha\gamma}\right) + E_{\alpha,\beta}\left(-\lambda t^{\alpha\gamma}\right)\right), \tag{4.469}$$

we have

$$\begin{aligned}
&\int_0^\infty e^{-st^\gamma} t^{\beta\gamma-1} precosh_{\alpha,\beta}\left(\lambda t^{\alpha\gamma}\right) dt \\
&= \int_0^\infty e^{-st^\gamma} t^{\beta\gamma-1} \left[\tfrac{1}{2}\left(E_{\alpha,\beta}\left(\lambda t^{\alpha\gamma}\right) + E_{\alpha,\beta}\left(-\lambda t^{\alpha\gamma}\right)\right)\right] dt \\
&= \tfrac{\gamma^{-1}s^{-\beta}}{2}\left(\left(1-\lambda s^{-\alpha}\right)^{-1} + \left(1+\lambda s^{-\alpha}\right)^{-1}\right) \\
&= \tfrac{\gamma^{-1}s^{-\beta}}{1-\lambda^2 s^{-2\alpha}}.
\end{aligned} \tag{4.470}$$

□

4.1.9 The Special Cases Based on the Wiman Function

In this section, we present the special cases based on the Wiman function.

Definition 4.62 The supersine via Wiman function is defined as

$$supersin_{1,\beta}\left(t^{\alpha}\right)=\frac{t^{\beta-1}}{2i}\left(E_{1,\beta}\left(it^{\alpha}\right)-E_{1,\beta}\left(-it^{\alpha}\right)\right)=\sum_{n=0}^{\infty}\frac{(-1)^{n}t^{(2n+1)\alpha}}{\Gamma\left(2n+1+\beta\right)}, \tag{4.471}$$

where $\alpha, \beta, t, \gamma \in \mathbb{C}$, $Re\left(\alpha\right)>0$, and $Re\left(\beta\right)>0$.

The result was discovered by Yang in 2019 [92].

Definition 4.63 The supercosine via Wiman function is defined as

$$supercos_{1,\beta}\left(t^{\alpha}\right)=\frac{t^{\beta-1}}{2}\left(E_{1,\beta}\left(it^{\alpha}\right)+E_{1,\beta}\left(-it^{\alpha}\right)\right)=\sum_{n=0}^{\infty}\frac{(-1)^{n}t^{2n\alpha}}{\Gamma\left(2n+\beta\right)}, \tag{4.472}$$

where $\alpha, \beta, t, \gamma \in \mathbb{C}$, $Re\left(\alpha\right)>0$, and $Re\left(\beta\right)>0$.

The result was discovered by Yang in 2019 [92].

Definition 4.64 The supertangent via Wiman function is defined as

$$supertan_{1,\beta}\left(t^{\alpha}\right)=\frac{supersin_{1,\beta}\left(t^{\alpha}\right)}{supercos_{1,\beta}\left(t^{\alpha}\right)}\left(supercos_{1,\beta}\left(t^{\alpha}\right)\neq 0\right), \tag{4.473}$$

where $\alpha, \beta, t, \gamma \in \mathbb{C}$, $Re\left(\alpha\right)>0$, and $Re\left(\beta\right)>0$.

Definition 4.65 The supercotangent via Wiman function is defined as

$$supercotan_{1,\beta}\left(t^{\alpha}\right)=\frac{supercos_{1,\beta}\left(t^{\alpha}\right)}{supersin_{1,\beta}\left(t^{\alpha}\right)}\left(supersin_{1,\beta}\left(t^{\alpha}\right)\neq 0\right), \tag{4.474}$$

where $\alpha, \beta, t, \gamma \in \mathbb{C}$, $Re\left(\alpha\right)>0$, and $Re\left(\beta\right)>0$.

Definition 4.66 The supercosecant via Wiman function is defined as

$$supercsc_{1,\beta}\left(t^{\alpha}\right)=\frac{1}{supersin_{1,\beta}\left(t^{\alpha}\right)}\left(supersin_{1,\beta}\left(t^{\alpha}\right)\neq 0\right), \tag{4.475}$$

where $\alpha, \beta, t, \gamma \in \mathbb{C}$, $Re\left(\alpha\right)>0$, and $Re\left(\beta\right)>0$.

Definition 4.67 The supersecant via Wiman function is defined as

$$supersec_{1,\beta}\left(t^{\alpha}\right)=\frac{1}{supercos_{1,\beta}\left(t^{\alpha}\right)}\left(supercos_{1,\beta}\left(t^{\alpha}\right)\neq 0\right), \tag{4.476}$$

where $\alpha, \beta, t, \gamma \in \mathbb{C}$, $Re\left(\alpha\right)>0$, and $Re\left(\beta\right)>0$.

Definition 4.68 The hyperbolic supersine via Wiman function is defined as

$$supersinh_{1,\beta}\left(t^{\alpha}\right)=\frac{t^{\beta-1}}{2}\left(E_{1,\beta}\left(t^{\alpha}\right)-E_{1,\beta}\left(-t^{\alpha}\right)\right)=t^{\beta-1}\sum_{n=0}^{\infty}\frac{t^{(2n+1)\alpha}}{\Gamma\left(2n+1+\beta\right)}, \tag{4.477}$$

where $\alpha, \beta, t, \gamma \in \mathbb{C}$, $Re(\alpha) > 0$, and $Re(\beta) > 0$.

The result was discovered by Yang in 2019 [92].

Definition 4.69 The hyperbolic supercosine via Wiman function is defined as

$$supercosh_{1,\beta}\left(t^{\alpha}\right)=\frac{t^{\beta-1}}{2}\left(E_{1,\beta}\left(t^{\alpha}\right)+E_{1,\beta}\left(-t^{\alpha}\right)\right)=t^{\beta-1}\sum_{n=0}^{\infty}\frac{t^{2n\alpha}}{\Gamma\left(2n+\beta\right)}, \tag{4.478}$$

where $\alpha, \beta, t, \gamma \in \mathbb{C}$, $Re(\alpha) > 0$, and $Re(\beta) > 0$.

The result was discovered by Yang in 2019 [92].

Definition 4.70 The hyperbolic supertangent via Wiman function is defined as

$$supertanh_{1,\beta}\left(t^{\alpha}\right)=\frac{supersinh_{1,\beta}\left(t^{\alpha}\right)}{supercosh_{1,\beta}\left(t^{\alpha}\right)}\left(supercosh_{1,\beta}\left(t^{\alpha}\right)\neq 0\right), \tag{4.479}$$

where $\alpha, \beta, t \in \mathbb{C}$, $Re(\alpha) > 0$, and $Re(\beta) > 0$.

Definition 4.71 The hyperbolic supercotangent via Wiman function is defined as

$$supercotanh_{1,\beta}\left(t^{\alpha}\right)=\frac{supercosh_{1,\beta}\left(t^{\alpha}\right)}{supersinh_{1,\beta}\left(t^{\alpha}\right)}\left(supersinh_{1,\beta}\left(t^{\alpha}\right)\neq 0\right), \tag{4.480}$$

where $\alpha, \beta, t \in \mathbb{C}$, $Re(\alpha) > 0$, and $Re(\beta) > 0$.

Definition 4.72 The hyperbolic supercosecant via Wiman function is defined as

$$supercsch_{1,\beta}\left(t^{\alpha}\right)=\frac{1}{supersinh_{1,\beta}\left(t^{\alpha}\right)}\left(supersinh_{1,\beta}\left(t^{\alpha}\right)\neq 0\right), \tag{4.481}$$

where $\alpha, \beta, t \in \mathbb{C}$, $Re(\alpha) > 0$, and $Re(\beta) > 0$.

Definition 4.73 The hyperbolic supersecant via Wiman function is defined as

$$supersech_{1,\beta}\left(t^{\alpha}\right)=\frac{1}{supercosh_{1,\beta}\left(t^{\alpha}\right)}\left(supercosh_{1,\beta}\left(t^{\alpha}\right)\neq 0\right), \tag{4.482}$$

where $\alpha, \beta, t \in \mathbb{C}$, $Re(\alpha) > 0$, and $Re(\beta) > 0$.

Definition 4.74 The presupersine via Wiman function is defined as

$$presin_{1,\beta}(t) = \frac{1}{2i}\left(E_{1,\beta}(it) - E_{1,\beta}(-it)\right) = \sum_{n=0}^{\infty} \frac{(-1)^n t^{2n+1}}{\Gamma(2n+1+\beta)}, \tag{4.483}$$

where $\alpha, \beta, t \in \mathbb{C}$, $Re(\alpha) > 0$, and $Re(\beta) > 0$.

Definition 4.75 The precosine via Wiman function is defined as

$$precos_{1,\beta}(t) = \frac{1}{2}\left(E_{1,\beta}(it) + E_{1,\beta}(-it)\right) = \sum_{n=0}^{\infty} \frac{(-1)^n t^{2n}}{\Gamma(2n+\beta)}, \tag{4.484}$$

where $\alpha, \beta, t \in \mathbb{C}$, $Re(\alpha) > 0$, and $Re(\beta) > 0$.

Definition 4.76 The pretangent via Wiman function is defined as

$$pretan_{1,\beta}(t) = \frac{presin_{1,\beta}(t)}{precos_{1,\beta}(t)} \left(precos_{1,\beta}(t) \neq 0\right), \tag{4.485}$$

where $\alpha, \beta, t \in \mathbb{C}$, $Re(\alpha) > 0$, and $Re(\beta) > 0$.

Definition 4.77 The precotangent via Wiman function is defined as

$$precotan_{1,\beta}(t) = \frac{precos_{1,\beta}(t)}{presin_{1,\beta}(t)} \left(presin_{1,\beta}(t) \neq 0\right), \tag{4.486}$$

where $\alpha, \beta, t \in \mathbb{C}$, $Re(\alpha) > 0$, and $Re(\beta) > 0$.

Definition 4.78 The precosecant via Wiman function is defined as

$$precsc_{1,\beta}(t) = \frac{1}{presin_{1,\beta}(t)} \left(presin_{1,\beta}(t) \neq 0\right), \tag{4.487}$$

where $\alpha, \beta, t \in \mathbb{C}$, $Re(\alpha) > 0$, and $Re(\beta) > 0$.

Definition 4.79 The presecant via Wiman function is defined as

$$presec_{1,\beta}(t) = \frac{1}{precos_{1,\beta}(t)} \left(precos_{1,\beta}(t) \neq 0\right), \tag{4.488}$$

where $\alpha, \beta, t \in \mathbb{C}$, $Re(\alpha) > 0$, and $Re(\beta) > 0$.

Definition 4.80 The hyperbolic presupersine via Wiman function is defined as

$$presinh_{1,\beta}(t) = \frac{1}{2}\left(E_{1,\beta}(t) - E_{1,\beta}(-t)\right) = \sum_{n=0}^{\infty} \frac{t^{2n+1}}{\Gamma(2n+1+\beta)}, \tag{4.489}$$

where $\alpha, \beta, t \in \mathbb{C}$, $Re(\alpha) > 0$, and $Re(\beta) > 0$.

Definition 4.81 The hyperbolic precosine via Wiman function is defined as

$$precosh_{1,\beta}(t) = \frac{1}{2}\left(E_{1,\beta}(t) + E_{1,\beta}(-t)\right) = \sum_{n=0}^{\infty} \frac{t^{2n}}{\Gamma(2n+\beta)}, \tag{4.490}$$

where $\alpha, \beta, t \in \mathbb{C}$, $Re(\alpha) > 0$, and $Re(\beta) > 0$.

Definition 4.82 The hyperbolic pretangent via Wiman function is defined as

$$pretanh_{1,\beta}(t) = \frac{presinh_{1,\beta}(t)}{precosh_{1,\beta}(t)} \left(precosh_{1,\beta}(t) \neq 0\right), \tag{4.491}$$

where $\alpha, \beta, t \in \mathbb{C}$, $Re(\alpha) > 0$, and $Re(\beta) > 0$.

Definition 4.83 The hyperbolic precotangent via Wiman function is defined as

$$precotanh_{1,\beta}(t) = \frac{precosh_{1,\beta}(t)}{presinh_{1,\beta}(t)} \left(presinh_{1,\beta}(t) \neq 0\right), \tag{4.492}$$

where $\alpha, \beta, t \in \mathbb{C}$, $Re(\alpha) > 0$, and $Re(\beta) > 0$.

Definition 4.84 The hyperbolic precosecant via Wiman function is defined as

$$precsch_{1,\beta}(t) = \frac{1}{presinh_{1,\beta}(t)} \left(presinh_{1,\beta}(t) \neq 0\right), \tag{4.493}$$

where $\alpha, \beta, t \in \mathbb{C}$, $Re(\alpha) > 0$, and $Re(\beta) > 0$.

Definition 4.85 The hyperbolic presecant via Wiman function is defined as

$$presech_{1,\beta}(t) = \frac{1}{precosh_{1,\beta}(t)} \left(precosh_{1,\beta}(t) \neq 0\right), \tag{4.494}$$

where $\alpha, \beta, t \in \mathbb{C}$, $Re(\alpha) > 0$, and $Re(\beta) > 0$.

Definition 4.86 The presupersine with the power law t^{γ} and the parameter $t^{\beta-1}$ via Wiman function is defined as

$$Gresin_{1,\beta}\left(t^{\gamma}\right) = \frac{t^{\beta-1}}{2i}\left(E_{1,\beta}\left(it^{\gamma}\right) - E_{1,\beta}\left(-it^{\gamma}\right)\right) = \sum_{n=0}^{\infty} \frac{(-1)^n t^{(2n+1)\gamma}}{\Gamma(2n+1+\beta)}, \tag{4.495}$$

Definition 4.87 The precosine with the power law t^γ and the parameter $t^{\beta-1}$ via Wiman function is defined as

$$Grecos_{1,\beta}\left(t^\gamma\right)=\frac{t^{\beta-1}}{2}\left(E_{1,\beta}\left(it^\gamma\right)+E_{1,\beta}\left(-it^\gamma\right)\right)=\sum_{n=0}^{\infty}\frac{(-1)^n\,t^{2n\gamma}}{\Gamma\left(2n+\beta\right)},\tag{4.496}$$

where $\alpha,\beta,t\in\mathbb{C}$, $Re\left(\alpha\right)>0$, and $Re\left(\beta\right)>0$.

Definition 4.88 The pretangent with the power law t^γ and the parameter $t^{\beta-1}$ via Wiman function is defined as

$$Gretan_{1,\beta}\left(t^\gamma\right)=\frac{Gresin_{1,\beta}\left(t^\gamma\right)}{Grecos_{1,\beta}\left(t^\gamma\right)}\left(Grecos_{1,\beta}\left(t^\gamma\right)\neq 0\right),\tag{4.497}$$

where $\alpha,\beta,t\in\mathbb{C}$, $Re\left(\alpha\right)>0$, and $Re\left(\beta\right)>0$.

Definition 4.89 The precotangent with the power law t^γ and the parameter $t^{\beta-1}$ via Wiman function is defined as

$$Grecotan_{1,\beta}\left(t^\gamma\right)=\frac{Grecos_{1,\beta}\left(t^\gamma\right)}{Gresin_{1,\beta}\left(t^\gamma\right)}\left(Gresin_{1,\beta}\left(t^\gamma\right)\neq 0\right),\tag{4.498}$$

where $\alpha,\beta,t\in\mathbb{C}$, $Re\left(\alpha\right)>0$, and $Re\left(\beta\right)>0$.

Definition 4.90 The precosecant with the power law t^γ and the parameter $t^{\beta-1}$ via Wiman function is defined as

$$Grecsc_{1,\beta}\left(t^\gamma\right)=\frac{1}{Gresin_{1,\beta}\left(t^\gamma\right)}\left(Gresin_{1,\beta}\left(t^\gamma\right)\neq 0\right),\tag{4.499}$$

where $\alpha,\beta,t\in\mathbb{C}$, $Re\left(\alpha\right)>0$, and $Re\left(\beta\right)>0$.

Definition 4.91 The presecant with the power law t^γ and the parameter $t^{\beta-1}$ via Wiman function is defined as

$$Gresec_{1,\beta}\left(t^\gamma\right)=\frac{1}{Grecos_{1,\beta}\left(t^\gamma\right)}\left(Grecos_{1,\beta}\left(t^\gamma\right)\neq 0\right),\tag{4.500}$$

where $\alpha,\beta,t\in\mathbb{C}$, $Re\left(\alpha\right)>0$, and $Re\left(\beta\right)>0$.

Definition 4.92 The hyperbolic presupersine with the power law t^{γ} and the parameter $t^{\beta-1}$ via Wiman function is defined as

$$Gresinh_{1,\beta}\left(t^{\gamma}\right)=\frac{t^{\beta-1}}{2}\left(E_{1,\beta}\left(t^{\gamma}\right)-E_{1,\beta}\left(-t^{\gamma}\right)\right)=t^{\beta-1}\sum_{n=0}^{\infty}\frac{t^{(2n+1)\gamma}}{\Gamma\left(2n+1+\beta\right)}, \tag{4.501}$$

where $\alpha, \beta, t \in \mathbb{C}$, $Re\left(\alpha\right)>0$, and $Re\left(\beta\right)>0$.

Definition 4.93 The hyperbolic precosine with the power law t^{γ} and the parameter $t^{\beta-1}$ via Wiman function is defined as

$$Grecosh_{1,\beta}\left(t^{\gamma}\right)=\frac{t^{\beta-1}}{2}\left(E_{1,\beta}\left(t^{\gamma}\right)+E_{1,\beta}\left(-t^{\gamma}\right)\right)=t^{\beta-1}\sum_{n=0}^{\infty}\frac{t^{2n\gamma}}{\Gamma\left(2n+\beta\right)}, \tag{4.502}$$

where $\alpha, \beta, t \in \mathbb{C}$, $Re\left(\alpha\right)>0$, and $Re\left(\beta\right)>0$.

Definition 4.94 The hyperbolic pretangent with the power law t^{γ} and the parameter $t^{\beta-1}$ via Wiman function is defined as

$$Gretanh_{1,\beta}\left(t^{\gamma}\right)=\frac{Gresinh_{1,\beta}\left(t^{\gamma}\right)}{precosh_{1,\beta}\left(t^{\gamma}\right)}\left(Grecosh_{1,\beta}\left(t^{\gamma}\right)\neq 0\right), \tag{4.503}$$

where $\alpha, \beta, t \in \mathbb{C}$, $Re\left(\alpha\right)>0$, and $Re\left(\beta\right)>0$.

Definition 4.95 The hyperbolic precotangent with the power law t^{γ} and the parameter $t^{\beta-1}$ via Wiman function is defined as

$$Grecotanh_{1,\beta}\left(t^{\gamma}\right)=\frac{Grecosh_{1,\beta}\left(t^{\gamma}\right)}{Gresinh_{1,\beta}\left(t^{\gamma}\right)}\left(Gresinh_{1,\beta}\left(t^{\gamma}\right)\neq 0\right), \tag{4.504}$$

where $\alpha, \beta, t \in \mathbb{C}$, $Re\left(\alpha\right)>0$, and $Re\left(\beta\right)>0$.

Definition 4.96 The hyperbolic precosecant with the power law t^{γ} and the parameter $t^{\beta-1}$ via Wiman function is defined as

$$Grecsch_{1,\beta}\left(t^{\gamma}\right)=\frac{1}{Gresinh_{1,\beta}\left(t^{\gamma}\right)}\left(Gresinh_{1,\beta}\left(t^{\gamma}\right)\neq 0\right), \tag{4.505}$$

where $\alpha, \beta, t \in \mathbb{C}$, $Re\left(\alpha\right)>0$, and $Re\left(\beta\right)>0$.

Definition 4.97 The hyperbolic presecant with the power law t^{γ} and the parameter $t^{\beta-1}$ via Wiman function is defined as

$$Gresech_{1,\beta}\left(t^{\gamma}\right)=\frac{1}{Grecosh_{1,\beta}\left(t^{\gamma}\right)}\left(Grecosh_{1,\beta}\left(t^{\gamma}\right)\neq 0\right), \tag{4.506}$$

where $\alpha, \beta, t \in \mathbb{C}$, $Re\left(\alpha\right)>0$, and $Re\left(\beta\right)>0$.

4.2 The Integral Representations Related to Wiman, Supertrigonometric, and Superhyperbolic Functions

There are

$$E_{\alpha,\beta}\left(t^{\alpha}\right) = {}_{1}\varphi_{1}\left(\begin{matrix}(1,1)\\(\beta,\alpha)\end{matrix};t^{\alpha}\right), \tag{4.507}$$

$$E_{\alpha,\beta}\left(t\right) = {}_{1}\varphi_{1}\left(\begin{matrix}(1,1)\\(\beta,\alpha)\end{matrix};t\right), \tag{4.508}$$

$$E_{\alpha,\beta}\left(t^{\vartheta}\right) = {}_{1}\varphi_{1}\left(\begin{matrix}(1,1)\\(\beta,\alpha)\end{matrix};t^{\vartheta}\right), \tag{4.509}$$

$$supersin_{\alpha,\beta}\left(t^{\alpha}\right) = t^{\beta-1}\,{}_{1}sin_{1}\left(\begin{matrix}(1,1)\\(\beta,\alpha)\end{matrix};t^{\alpha}\right), \tag{4.510}$$

$$supercos_{\alpha,\beta}\left(t^{\alpha}\right) = t^{\beta-1}\,{}_{1}cos_{1}\left(\begin{matrix}(1,1)\\(\beta,\alpha)\end{matrix};t^{\alpha}\right), \tag{4.511}$$

$$supersinh_{\alpha,\beta}\left(t^{\alpha}\right) = t^{\beta-1}\,{}_{1}sinh_{1}\left(\begin{matrix}(1,1)\\(\beta,\alpha)\end{matrix};t^{\alpha}\right), \tag{4.512}$$

$$supercosh_{\alpha,\beta}\left(t^{\alpha}\right) = t^{\beta-1}\,{}_{1}cosh_{1}\left(\begin{matrix}(1,1)\\(\beta,\alpha)\end{matrix};t^{\alpha}\right), \tag{4.513}$$

$$presin_{\alpha,\beta}\left(t\right) = {}_{1}sin_{1}\left(\begin{matrix}(1,1)\\(\beta,\alpha)\end{matrix};t\right), \tag{4.514}$$

$$precos_{\alpha,\beta}\left(t\right) = {}_{1}cos_{1}\left(\begin{matrix}(1,1)\\(\beta,\alpha)\end{matrix};t\right), \tag{4.515}$$

$$presinh_{\alpha,\beta}\left(t\right) = {}_{1}sinh_{1}\left(\begin{matrix}(1,1)\\(\beta,\alpha)\end{matrix};t\right), \tag{4.516}$$

$$precosh_{\alpha,\beta}\left(t\right) = {}_{1}cosh_{1}\left(\begin{matrix}(1,1)\\(\beta,\alpha)\end{matrix};t\right), \tag{4.517}$$

$$presin_{\alpha,\beta}\left(t^{\vartheta}\right) = {}_{1}sin_{1}\left(\begin{matrix}(1,1)\\(\beta,\alpha)\end{matrix};t^{\vartheta}\right), \tag{4.518}$$

$$precos_{\alpha,\beta}\left(t^{\vartheta}\right) = {}_1cos_1\left(\begin{matrix}(1,1)\\(\beta,\alpha)\end{matrix};t^{\vartheta}\right), \tag{4.519}$$

$$presinh_{\alpha,\beta}\left(t^{\vartheta}\right) = {}_1sinh_1\left(\begin{matrix}(1,1)\\(\beta,\alpha)\end{matrix};t^{\vartheta}\right), \tag{4.520}$$

$$precosh_{\alpha,\beta}\left(t^{\vartheta}\right) = {}_1cosh_1\left(\begin{matrix}(1,1)\\(\beta,\alpha)\end{matrix};t^{\vartheta}\right), \tag{4.521}$$

$$Tresin_{\alpha,\beta}(t) = t^{\beta-1}\,{}_1sin_1\left(\begin{matrix}(1,1)\\(\beta,\alpha)\end{matrix};t\right), \tag{4.522}$$

$$Trecos_{\alpha,\beta}(t) = t^{\beta-1}\,{}_1cos_1\left(\begin{matrix}(1,1)\\(\beta,\alpha)\end{matrix};t\right), \tag{4.523}$$

$$Tresinh_{\alpha,\beta}(t) = t^{\beta-1}\,{}_1sinh_1\left(\begin{matrix}(1,1)\\(\beta,\alpha)\end{matrix};t\right), \tag{4.524}$$

$$Trecosh_{\alpha,\beta}(t) = t^{\beta-1}\,{}_1cosh_1\left(\begin{matrix}(1,1)\\(\beta,\alpha)\end{matrix};t\right), \tag{4.525}$$

$$Gresin_{\alpha,\beta}\left(t^{\gamma}\right) = t^{\beta-1}\,{}_1sin_1\left(\begin{matrix}(1,1)\\(\beta,\alpha)\end{matrix};t^{\gamma}\right), \tag{4.526}$$

$$Grecos_{\alpha,\beta}\left(t^{\gamma}\right) = t^{\beta-1}\,{}_1cos_1\left(\begin{matrix}(1,1)\\(\beta,\alpha)\end{matrix};t^{\gamma}\right), \tag{4.527}$$

$$Gresinh_{\alpha,\beta}\left(t^{\gamma}\right) = t^{\beta-1}\,{}_1sinh_1\left(\begin{matrix}(1,1)\\(\beta,\alpha)\end{matrix};t^{\gamma}\right), \tag{4.528}$$

and

$$Grecosh_{\alpha,\beta}\left(t^{\gamma}\right) = t^{\beta-1}\,{}_1cosh_1\left(\begin{matrix}(1,1)\\(\beta,\alpha)\end{matrix};t^{\gamma}\right), \tag{4.529}$$

where

$$\begin{aligned}
&{}_1sin_1\left(\begin{matrix}(1,1)\\(\beta,\alpha)\end{matrix};t^{\alpha}\right)\\
&= \tfrac{1}{2i}\left({}_1\varphi_1\left(\begin{matrix}(1,1)\\(\beta,\alpha)\end{matrix};it^{\alpha}\right) - {}_1\varphi_1\left(\begin{matrix}(1,1)\\(\beta,\alpha)\end{matrix};-it^{\alpha}\right)\right)\\
&= \sum_{n=0}^{\infty}\tfrac{(-1)^n t^{(2n+1)\alpha}}{\Gamma((2n+1)\alpha+\beta)},
\end{aligned} \tag{4.530}$$

$$
\begin{aligned}
&{}_1cos_1\left(\begin{matrix}(1,1)\\(\beta,\alpha)\end{matrix};t^{\alpha}\right)\\
&=\frac{1}{2}\left({}_1\varphi_1\left(\begin{matrix}(1,1)\\(\beta,\alpha)\end{matrix};it^{\alpha}\right)+{}_1\varphi_1\left(\begin{matrix}(1,1)\\(\beta,\alpha)\end{matrix};-it^{\alpha}\right)\right)\\
&=\sum_{n=0}^{\infty}\frac{(-1)^n t^{2n\alpha}}{\Gamma((2n+1)\alpha+\beta)},
\end{aligned}
\tag{4.531}
$$

$$
\begin{aligned}
&{}_1sinh_1\left(\begin{matrix}(1,1)\\(\beta,\alpha)\end{matrix};t^{\alpha}\right)\\
&=\frac{1}{2}\left({}_1\varphi_1\left(\begin{matrix}(1,1)\\(\beta,\alpha)\end{matrix};t^{\alpha}\right)-{}_1\varphi_1\left(\begin{matrix}(1,1)\\(\beta,\alpha)\end{matrix};-t^{\alpha}\right)\right)\\
&=\sum_{n=0}^{\infty}\frac{t^{(2n+1)\alpha}}{\Gamma((2n+1)\alpha+\beta)},
\end{aligned}
\tag{4.532}
$$

$$
\begin{aligned}
&{}_1cosh_1\left(\begin{matrix}(1,1)\\(\beta,\alpha)\end{matrix};t^{\alpha}\right)\\
&=\frac{1}{2}\left({}_1\varphi_1\left(\begin{matrix}(1,1)\\(\beta,\alpha)\end{matrix};t^{\alpha}\right)+{}_1\varphi_1\left(\begin{matrix}(1,1)\\(\beta,\alpha)\end{matrix};-t^{\alpha}\right)\right)\\
&=\sum_{n=0}^{\infty}\frac{t^{2n\alpha}}{\Gamma((2n+1)\alpha+\beta)},
\end{aligned}
\tag{4.533}
$$

$$
\begin{aligned}
&{}_1sin_1\left(\begin{matrix}(1,1)\\(\beta,\alpha)\end{matrix};t\right)\\
&=\frac{1}{2i}\left({}_1\varphi_1\left(\begin{matrix}(1,1)\\(\beta,\alpha)\end{matrix};it\right)-{}_1\varphi_1\left(\begin{matrix}(1,1)\\(\beta,\alpha)\end{matrix};-it\right)\right)\\
&=\sum_{n=0}^{\infty}\frac{(-1)^n t^{2n+1}}{\Gamma((2n+1)\alpha+\beta)},
\end{aligned}
\tag{4.534}
$$

$$
\begin{aligned}
&{}_1cos_1\left(\begin{matrix}(1,1)\\(\beta,\alpha)\end{matrix};t\right)\\
&=\frac{1}{2}\left({}_1\varphi_1\left(\begin{matrix}(1,1)\\(\beta,\alpha)\end{matrix};it\right)+{}_1\varphi_1\left(\begin{matrix}(1,1)\\(\beta,\alpha)\end{matrix};-it\right)\right)\\
&=\sum_{n=0}^{\infty}\frac{(-1)^n t^{2n}}{\Gamma((2n+1)\alpha+\beta)},
\end{aligned}
\tag{4.535}
$$

$$
\begin{aligned}
&{}_1sinh_1\left(\begin{matrix}(1,1)\\(\beta,\alpha)\end{matrix};t\right)\\
&=\frac{1}{2}\left({}_1\varphi_1\left(\begin{matrix}(1,1)\\(\beta,\alpha)\end{matrix};t\right)-{}_1\varphi_1\left(\begin{matrix}(1,1)\\(\beta,\alpha)\end{matrix};-t\right)\right)\\
&=\sum_{n=0}^{\infty}\frac{t^{2n+1}}{\Gamma((2n+1)\alpha+\beta)},
\end{aligned}
\tag{4.536}
$$

$$
\begin{aligned}
&{}_1cosh_1\left(\begin{matrix}(1,1)\\(\beta,\alpha)\end{matrix};t\right)\\
&=\frac{1}{2}\left({}_1\varphi_1\left(\begin{matrix}(1,1)\\(\beta,\alpha)\end{matrix};t\right)+{}_1\varphi_1\left(\begin{matrix}(1,1)\\(\beta,\alpha)\end{matrix};-t\right)\right)\\
&=\sum_{n=0}^{\infty}\frac{t^{2n\alpha}}{\Gamma((2n+1)\alpha+\beta)},
\end{aligned}
\tag{4.537}
$$

$$
\begin{aligned}
&{}_1sin_1\left(\begin{matrix}(1,1)\\(\beta,\alpha)\end{matrix};t^{\vartheta}\right)\\
&=\frac{1}{2i}\left({}_1\varphi_1\left(\begin{matrix}(1,1)\\(\beta,\alpha)\end{matrix};it^{\vartheta}\right)-{}_1\varphi_1\left(\begin{matrix}(1,1)\\(\beta,\alpha)\end{matrix};-it^{\vartheta}\right)\right)\\
&=\sum_{n=0}^{\infty}\frac{(-1)^n t^{(2n+1)\vartheta}}{\Gamma((2n+1)\alpha+\beta)},
\end{aligned}
\tag{4.538}
$$

$$
\begin{aligned}
&{}_1cos_1\left(\begin{matrix}(1,1)\\(\beta,\alpha)\end{matrix};t^{\vartheta}\right)\\
&=\frac{1}{2}\left({}_1\varphi_1\left(\begin{matrix}(1,1)\\(\beta,\alpha)\end{matrix};it^{\vartheta}\right)+{}_1\varphi_1\left(\begin{matrix}(1,1)\\(\beta,\alpha)\end{matrix};-it^{\vartheta}\right)\right)\\
&=\sum_{n=0}^{\infty}\frac{(-1)^n t^{2n\vartheta}}{\Gamma((2n+1)\alpha+\beta)},
\end{aligned}
\tag{4.539}
$$

$$
\begin{aligned}
&{}_1sinh_1\left(\begin{matrix}(1,1)\\(\beta,\alpha)\end{matrix};t^{\vartheta}\right)\\
&=\frac{1}{2}\left({}_1\varphi_1\left(\begin{matrix}(1,1)\\(\beta,\alpha)\end{matrix};t^{\vartheta}\right)-{}_1\varphi_1\left(\begin{matrix}(1,1)\\(\beta,\alpha)\end{matrix};-t^{\vartheta}\right)\right)\\
&=\sum_{n=0}^{\infty}\frac{t^{(2n+1)\vartheta}}{\Gamma((2n+1)\alpha+\beta)},
\end{aligned}
\tag{4.540}
$$

and

$$
\begin{aligned}
&{}_1cosh_1\left(\begin{matrix}(1,1)\\(\beta,\alpha)\end{matrix};t^{\vartheta}\right)\\
&=\frac{1}{2}\left({}_1\varphi_1\left(\begin{matrix}(1,1)\\(\beta,\alpha)\end{matrix};t^{\vartheta}\right)+{}_1\varphi_1\left(\begin{matrix}(1,1)\\(\beta,\alpha)\end{matrix};-t^{\vartheta}\right)\right)\\
&=\sum_{n=0}^{\infty}\frac{t^{2n\vartheta}}{\Gamma((2n+1)\alpha+\beta)},
\end{aligned}
\tag{4.541}
$$

for $\alpha, \beta, t, \gamma \in \mathbb{C}$, $\vartheta \in \mathbb{C}$, $Re(\alpha) > 0$, and $Re(\beta) > 0$.

From Chap. 2, we have that

$$
{}_2\varphi_2\left(\begin{matrix}(\alpha_1, A_1),(\alpha_2, A_2)\\(\beta_1, B_1),(\beta_2, B_2)\end{matrix};\lambda t\right)=\sum_{k=0}^{\infty}\frac{\Gamma(\alpha_1+A_1k)\,\Gamma(\alpha_2+A_2k)}{\Gamma(\beta_1+B_1k)\,\Gamma(\beta_2+B_2k)}\frac{(\lambda t)^k}{k!},
\tag{4.542}
$$

$$
{}_2sin_2\left(\begin{matrix}(\alpha_1, A_1), (\alpha_2, A_2)\\(\beta_1, B_1), (\beta_2, B_2)\end{matrix}; \lambda t^\alpha\right)
= \sum_{k=0}^{\infty} (-\lambda)^k \frac{\Gamma(\alpha_1+A_1(2k+1))\Gamma(\alpha_2+A_2(2k+1))}{\Gamma(\beta_1+B_1(2k+1))\Gamma(\beta_2+B_2(2k+1))} \frac{t^{(2k+1)\alpha}}{(2k+1)!}, \tag{4.543}
$$

$$
{}_2cos_2\left(\begin{matrix}(\alpha_1, A_1), (\alpha_2, A_2)\\(\beta_1, B_1), (\beta_2, B_2)\end{matrix}; \lambda t^\alpha\right)
= \sum_{k=0}^{\infty} (-\lambda)^k \frac{\Gamma(\alpha_1+2kA_1)\Gamma(\alpha_2+2kA_2)}{\Gamma(\beta_1+2kB_1)\Gamma(\beta_2+2kB_2)} \frac{t^{2k\alpha}}{(2k)!}, \tag{4.544}
$$

$$
{}_2sinh_2\left(\begin{matrix}(\alpha_1, A_1), (\alpha_2, A_2)\\(\beta_1, B_1), (\beta_2, B_2)\end{matrix}; \lambda t^\alpha\right)
= \sum_{k=0}^{\infty} \lambda^k \frac{\Gamma(\alpha_1+A_1(2k+1))\Gamma(\alpha_2+A_2(2k+1))}{\Gamma(\beta_1+B_1(2k+1))\Gamma(\beta_2+B_2(2k+1))} \frac{t^{(2k+1)\alpha}}{(2k+1)!}, \tag{4.545}
$$

$$
{}_2cosh_2\left(\begin{matrix}(\alpha_1, A_1), (\alpha_2, A_2)\\(\beta_1, B_1), (\beta_2, B_2)\end{matrix}; \lambda t^\alpha\right)
= \sum_{k=0}^{\infty} \lambda^k \frac{\Gamma(\alpha_1+2kA_1)\Gamma(\alpha_2+2kA_2)}{\Gamma(\beta_1+2kB_1)\Gamma(\beta_2+2kB_2)} \frac{t^{2k\alpha}}{(2k)!}, \tag{4.546}
$$

$$
{}_2sin_2\left(\begin{matrix}(\alpha_1, A_1), (\alpha_2, A_2)\\(\beta_1, B_1), (\beta_2, B_2)\end{matrix}; \lambda t^\vartheta\right)
= \sum_{k=0}^{\infty} (-\lambda)^k \frac{\Gamma(\alpha_1+A_1(2k+1))\Gamma(\alpha_2+A_2(2k+1))}{\Gamma(\beta_1+B_1(2k+1))\Gamma(\beta_2+B_2(2k+1))} \frac{t^{(2k+1)\vartheta}}{(2k+1)!}, \tag{4.547}
$$

$$
{}_2cos_2\left(\begin{matrix}(\alpha_1, A_1), (\alpha_2, A_2)\\(\beta_1, B_1), (\beta_2, B_2)\end{matrix}; \lambda t^\vartheta\right)
= \sum_{k=0}^{\infty} (-\lambda)^k \frac{\Gamma(\alpha_1+2kA_1)\Gamma(\alpha_2+2kA_2)}{\Gamma(\beta_1+2kB_1)\Gamma(\beta_2+2kB_2)} \frac{t^{2k\vartheta}}{(2k)!}, \tag{4.548}
$$

$$
{}_2sinh_2\left(\begin{matrix}(\alpha_1, A_1), (\alpha_2, A_2)\\(\beta_1, B_1), (\beta_2, B_2)\end{matrix}; \lambda t^\vartheta\right)
= \sum_{k=0}^{\infty} \lambda^k \frac{\Gamma(\alpha_1+A_1(2k+1))\Gamma(\alpha_2+A_2(2k+1))}{\Gamma(\beta_1+B_1(2k+1))\Gamma(\beta_2+B_2(2k+1))} \frac{t^{(2k+1)\vartheta}}{(2k+1)!}, \tag{4.549}
$$

$$
{}_2cosh_2\left(\begin{matrix}(\alpha_1, A_1), (\alpha_2, A_2)\\(\beta_1, B_1), (\beta_2, B_2)\end{matrix}; \lambda t^\vartheta\right)
= \sum_{k=0}^{\infty} \lambda^k \frac{\Gamma(\alpha_1+2kA_1)\Gamma(\alpha_2+2kA_2)}{\Gamma(\beta_1+2kB_1)\Gamma(\beta_2+2kB_2)} \frac{t^{2k\vartheta}}{(2k)!}, \tag{4.550}
$$

$$
{}_2sin_2\left(\begin{matrix}(\alpha_1, A_1), (\alpha_2, A_2)\\(\beta_1, B_1), (\beta_2, B_2)\end{matrix}; \lambda t\right) = \sum_{k=0}^{\infty} (-\lambda)^k \frac{\Gamma(\alpha_1+A_1(2k+1))\Gamma(\alpha_2+A_2(2k+1))}{\Gamma(\beta_1+B_1(2k+1))\Gamma(\beta_2+B_2(2k+1))} \frac{t^{2k+1}}{(2k+1)!}, \tag{4.551}
$$

$$
{}_2cos_2\left(\begin{matrix}(\alpha_1, A_1), (\alpha_2, A_2)\\(\beta_1, B_1), (\beta_2, B_2)\end{matrix}; \lambda t\right) = \sum_{k=0}^{\infty} (-\lambda)^k \frac{\Gamma(\alpha_1+2kA_1)\Gamma(\alpha_2+2kA_2)}{\Gamma(\beta_1+2kB_1)\Gamma(\beta_2+2kB_2)} \frac{t^{2k}}{(2k)!}, \tag{4.552}
$$

$$
{}_2sinh_2\left(\begin{matrix}(\alpha_1, A_1), (\alpha_2, A_2)\\(\beta_1, B_1), (\beta_2, B_2)\end{matrix}; \lambda t\right) = \sum_{k=0}^{\infty} \lambda^k \frac{\Gamma(\alpha_1+A_1(2k+1))\Gamma(\alpha_2+A_2(2k+1))}{\Gamma(\beta_1+B_1(2k+1))\Gamma(\beta_2+B_2(2k+1))} \frac{t^{2k+1}}{(2k+1)!}, \tag{4.553}
$$

and

$$
{}_2cosh_2\left(\begin{matrix}(\alpha_1, A_1), (\alpha_2, A_2)\\(\beta_1, B_1), (\beta_2, B_2)\end{matrix}; \lambda t\right) = \sum_{k=0}^{\infty} \lambda^k \frac{\Gamma(\alpha_1+2kA_1)\Gamma(\alpha_2+2kA_2)}{\Gamma(\beta_1+2kB_1)\Gamma(\beta_2+2kB_2)} \frac{t^{2k}}{(2k)!}, \tag{4.554}
$$

which leads to

$$
{}_2sin_2\left(\begin{matrix}(\alpha_1, A_1), (\beta_1, B_1)\\(\beta_1, B_1), (\beta_2, B_2)\end{matrix}; t\right) = {}_1sin_1\left(\begin{matrix}(\alpha_1, A_1)\\(\beta_2, B_2)\end{matrix}; t\right), \tag{4.555}
$$

$$
{}_2sin_2\left(\begin{matrix}(\alpha_1, A_1), (\beta_1, B_1)\\(\beta_1, B_1), (\beta_2, B_2)\end{matrix}; t\right) = {}_1sin_1\left(\begin{matrix}(\alpha_1, A_1)\\(\beta_2, B_2)\end{matrix}; t\right), \tag{4.556}
$$

$$
{}_2sinh_2\left(\begin{matrix}(\alpha_1, A_1), (\beta_1, B_1)\\(\beta_1, B_1), (\beta_2, B_2)\end{matrix}; t\right) = {}_1sinh_1\left(\begin{matrix}(\alpha_1, A_1)\\(\beta_2, B_2)\end{matrix}; t\right), \tag{4.557}
$$

and

$$
{}_2cosh_2\left(\begin{matrix}(\alpha_1, A_1), (\beta_1, B_1)\\(\beta_1, B_1), (\beta_2, B_2)\end{matrix}; t\right) = {}_1cosh_1\left(\begin{matrix}(\alpha_1, A_1)\\(\beta_2, B_2)\end{matrix}; t\right), \tag{4.558}
$$

where $j, k \in \mathbb{N}$, $\alpha_1, \beta_1, \alpha_2, \beta_2 \in \mathbb{C}$, and $A_1, B_1, A_1, B_1 \in \mathbb{R}_+$.

Theorem 4.42 *Let $\varpi, \gamma \in \mathbb{C}$, $Re(\varpi) > 0$, $Re(\gamma) > 0$, $\lambda \in \mathbb{C}$, $\vartheta > 0$, $\beta \in \mathbb{C}$, and $\alpha \in \mathbb{R}_+$. Then, we have*

$$\begin{aligned} &\frac{1}{\Gamma(\varpi)} \int_0^t (t-\tau)^{\varpi} \tau^{\gamma-1} {}_1\varphi_1 \left(\begin{matrix} (1,1) \\ (\beta,\alpha) \end{matrix} ; \lambda \tau^{\vartheta} \right) d\tau \\ &= t^{\gamma+\varpi-1} {}_2\varphi_2 \left(\begin{matrix} (1,1),(\gamma,\vartheta) \\ (\beta,\alpha),(\gamma+\varpi,\vartheta) \end{matrix} ; \lambda t^{\vartheta} \right). \end{aligned} \tag{4.559}$$

Proof The result is the special case of Chap. 2. □

Property 4.16 Let $\varpi, \gamma \in \mathbb{C}$, $Re(\varpi) > 0$, $Re(\gamma) > 0$, $\lambda \in \mathbb{C}$, $\vartheta > 0$, $\beta \in \mathbb{C}$, and $\alpha \in \mathbb{R}_+$. Then, we have

$$\frac{1}{\Gamma(\varpi)} \int_0^1 (1-\tau)^{\varpi} \tau^{\gamma-1} {}_1\varphi_1 \left(\begin{matrix} (1,1) \\ (\beta,\alpha) \end{matrix} ; \lambda \tau^{\vartheta} \right) d\tau = {}_2\varphi_2 \left(\begin{matrix} (1,1),(\gamma,\vartheta) \\ (\beta,\alpha),(\gamma+\varpi,\vartheta) \end{matrix} ; \lambda \right). \tag{4.560}$$

Proof When $t = 1$ in (4.559), we obtain the result. □

Theorem 4.43 *Let $\varpi, \gamma \in \mathbb{C}$, $Re(\varpi) > 0$, $Re(\gamma) > 0$, $\lambda \in \mathbb{C}$, $\vartheta > 0$, $\beta \in \mathbb{C}$, and $\alpha \in \mathbb{R}_+$.*

Then, we have

$$\begin{aligned} &\int_{-1}^1 (1-\tau)^{\varpi} (1+\tau)^{\gamma-1} {}_1\varphi_1 \left(\begin{matrix} (1,1) \\ (\beta,\alpha) \end{matrix} ; \lambda (1+\tau)^{\vartheta} \right) d\tau \\ &= \int_{-1}^1 (1+\tau)^{\varpi} (1-\tau)^{\gamma-1} {}_1\varphi_1 \left(\begin{matrix} (1,1) \\ (\beta,\alpha) \end{matrix} ; \lambda (1-\tau)^{\vartheta} \right) d\tau \end{aligned} \tag{4.561}$$

and

$$\begin{aligned} &\frac{1}{\Gamma(\varpi)} \int_0^2 (2-\tau)^{\varpi} \tau^{\gamma-1} {}_1\varphi_1 \left(\begin{matrix} (1,1) \\ (\beta,\alpha) \end{matrix} ; \lambda \tau^{\vartheta} \right) d\tau \\ &= \frac{1}{\Gamma(\varpi)} \int_{-1}^1 (1-\tau)^{\varpi} (1+\tau)^{\gamma-1} {}_1\varphi_1 \left(\begin{matrix} (1,1) \\ (\beta,\alpha) \end{matrix} ; \lambda (1+\tau)^{\vartheta} \right) d\tau \\ &= \frac{1}{\Gamma(\varpi)} \int_{-1}^1 (1+\tau)^{\varpi} (1-\tau)^{\gamma-1} {}_1\varphi_1 \left(\begin{matrix} (1,1) \\ (\beta,\alpha) \end{matrix} ; \lambda (1-\tau)^{\vartheta} \right) d\tau \\ &= 2^{\gamma+\varpi-1} {}_2\varphi_2 \left(\begin{matrix} (1,1),(\gamma,\vartheta) \\ (\beta,\alpha),(\gamma+\varpi,\vartheta) \end{matrix} ; \lambda 2^{\vartheta} \right). \end{aligned} \tag{4.562}$$

Proof When $t = 2$ in (4.559), we obtain the result. □

Theorem 4.44 *Let $\varpi, \gamma \in \mathbb{C}$, $Re(\varpi) > 0$, $Re(\gamma) > 0$, $\lambda \in \mathbb{C}$, $\vartheta > 0$, $\beta \in \mathbb{C}$, and $\alpha \in \mathbb{R}_{+}$. Then, we have*

$$\begin{aligned}&\frac{1}{\Gamma(\varpi)}\int_0^t (t-\tau)^{\varpi}\,\tau^{\gamma-1}presin_{\alpha,\beta}\left(\lambda\tau^{\vartheta}\right)d\tau\\&= t^{\gamma+\varpi-1}\,{}_2sin_2\left(\begin{matrix}(1,1),(\gamma,\vartheta)\\(\beta,\alpha),(\gamma+\varpi,\vartheta)\end{matrix};\lambda t^{\vartheta}\right)\end{aligned}\tag{4.563}$$

and

$$\begin{aligned}&\frac{1}{\Gamma(\varpi)}\int_0^t (t-\tau)^{\varpi}\,\tau^{\gamma-1}precos_{\alpha,\beta}\left(\lambda\tau^{\vartheta}\right)d\tau\\&= t^{\gamma+\varpi-1}\,{}_2cos_2\left(\begin{matrix}(1,1),(\gamma,\vartheta)\\(\beta,\alpha),(\gamma+\varpi,\vartheta)\end{matrix};\lambda t^{\vartheta}\right).\end{aligned}\tag{4.564}$$

Proof Making use of (4.559), (4.518), and (4.519), we have

$$\begin{aligned}&\frac{1}{\Gamma(\varpi)}\int_0^t (t-\tau)^{\varpi}\,\tau^{\gamma-1}\,{}_1\varphi_1\left(\begin{matrix}(1,1)\\(\beta,\alpha)\end{matrix};i\lambda\tau^{\vartheta}\right)d\tau\\&= t^{\gamma+\varpi-1}\,{}_2\varphi_2\left(\begin{matrix}(1,1),(\gamma,\vartheta)\\(\beta,\alpha),(\gamma+\varpi,\vartheta)\end{matrix};i\lambda t^{\vartheta}\right)\end{aligned}\tag{4.565}$$

and

$$\begin{aligned}&\frac{1}{\Gamma(\varpi)}\int_0^t (t-\tau)^{\varpi}\,\tau^{\gamma-1}\,{}_1\varphi_1\left(\begin{matrix}(1,1)\\(\beta,\alpha)\end{matrix};-i\lambda\tau^{\vartheta}\right)d\tau\\&= t^{\gamma+\varpi-1}\,{}_2\varphi_2\left(\begin{matrix}(1,1),(\gamma,\vartheta)\\(\beta,\alpha),(\gamma+\varpi,\vartheta)\end{matrix};-i\lambda t^{\vartheta}\right),\end{aligned}\tag{4.566}$$

which reduces the desired results. □

Corollary 4.2 *Let $\varpi, \gamma \in \mathbb{C}$, $Re(\varpi) > 0$, $Re(\gamma) > 0$, $\lambda \in \mathbb{C}$, $\vartheta > 0$, $\beta \in \mathbb{C}$, and $\alpha \in \mathbb{R}_{+}$. Then, we have*

$$\begin{aligned}&\frac{1}{\Gamma(\varpi)}\int_0^1 (1-\tau)^{\varpi}\,\tau^{\gamma-1}presin_{\alpha,\beta}\left(\lambda\tau^{\vartheta}\right)d\tau\\&= {}_2sin_2\left(\begin{matrix}(1,1),(\gamma,\vartheta)\\(\beta,\alpha),(\gamma+\varpi,\vartheta)\end{matrix};\lambda\right)\end{aligned}\tag{4.567}$$

and

$$\frac{1}{\Gamma(\varpi)}\int_0^1 (1-\tau)^{\varpi}\,\tau^{\gamma-1} precos_{\alpha,\beta}\left(\lambda\tau^{\vartheta}\right)d\tau = {}_2cos_2\left(\begin{matrix}(1,1),(\gamma,\vartheta)\\(\beta,\alpha),(\gamma+\varpi,\vartheta)\end{matrix};\lambda\right). \tag{4.568}$$

Proof Taking $t = 1$ in (4.563) and (4.564), we directly have the results. □

Theorem 4.45 *Let $\varpi, \gamma \in \mathbb{C}$, $Re(\varpi) > 0$, $Re(\gamma) > 0$, $\lambda \in \mathbb{C}$, $\vartheta > 0$, $\beta \in \mathbb{C}$, and $\alpha \in \mathbb{R}_+$.*

Then, we have

$$\int_{-1}^{1} (1-\tau)^{\varpi}(1+\tau)^{\gamma-1}\, presin_{\alpha,\beta}\left(\lambda(1+\tau)^{\vartheta}\right)d\tau = \int_{-1}^{1} (1+\tau)^{\varpi}(1-\tau)^{\gamma-1}\, presin_{\alpha,\beta}\left(\lambda(1-\tau)^{\vartheta}\right)d\tau \tag{4.569}$$

and

$$\begin{aligned}&\frac{1}{\Gamma(\varpi)}\int_0^2 (2-\tau)^{\varpi}\,\tau^{\gamma-1}\, presin_{\alpha,\beta}\left(\lambda\tau^{\vartheta}\right)d\tau\\ &= \frac{1}{\Gamma(\varpi)}\int_{-1}^{1} (1-\tau)^{\varpi}(1+\tau)^{\gamma-1}\, presin_{\alpha,\beta}\left(\lambda(1+\tau)^{\vartheta}\right)d\tau\\ &= \frac{1}{\Gamma(\varpi)}\int_{-1}^{1} (1+\tau)^{\varpi}(1-\tau)^{\gamma-1}\, presin_{\alpha,\beta}\left(\lambda(1-\tau)^{\vartheta}\right)d\tau\\ &= 2^{\gamma+\varpi-1}\,{}_2sin_2\left(\begin{matrix}(1,1),(\gamma,\vartheta)\\(\beta,\alpha),(\gamma+\varpi,\vartheta)\end{matrix};\lambda 2^{\vartheta}\right).\end{aligned} \tag{4.570}$$

Proof When $t = 2$ in (4.563), we obtain the result. □

Theorem 4.46 *Let $\varpi, \gamma \in \mathbb{C}$, $Re(\varpi) > 0$, $Re(\gamma) > 0$, $\lambda \in \mathbb{C}$, $\vartheta > 0$, $\beta \in \mathbb{C}$, and $\alpha \in \mathbb{R}_+$.*

Then, we have

$$\int_{-1}^{1} (1-\tau)^{\varpi}(1+\tau)^{\gamma-1}\, precos_{\alpha,\beta}\left(\lambda(1+\tau)^{\vartheta}\right)d\tau = \int_{-1}^{1} (1+\tau)^{\varpi}(1-\tau)^{\gamma-1}\, precos_{\alpha,\beta}\left(\lambda(1-\tau)^{\vartheta}\right)d\tau \tag{4.571}$$

and

$$\begin{aligned}
&\frac{1}{\Gamma(\varpi)}\int_{0}^{2}(2-\tau)^{\varpi}\,\tau^{\gamma-1}precos_{\alpha,\beta}\left(\lambda\tau^{\vartheta}\right)d\tau\\
&=\frac{1}{\Gamma(\varpi)}\int_{-1}^{1}(1-\tau)^{\varpi}\,(1+\tau)^{\gamma-1}\ precos_{\alpha,\beta}\left(\lambda\,(1+\tau)^{\vartheta}\right)d\tau\\
&=\frac{1}{\Gamma(\varpi)}\int_{-1}^{1}(1+\tau)^{\varpi}\,(1-\tau)^{\gamma-1}\ precos_{\alpha,\beta}\left(\lambda\,(1-\tau)^{\vartheta}\right)d\tau\\
&=2^{\gamma+\varpi-1}\,{}_2cos_2\left(\begin{matrix}(1,1),(\gamma,\vartheta)\\(\beta,\alpha),(\gamma+\varpi,\vartheta)\end{matrix};\lambda 2^{\vartheta}\right).
\end{aligned}\tag{4.572}$$

Proof When $t=2$ in (4.564), we obtain the result. □

Theorem 4.47 *Let $\varpi,\gamma\in\mathbb{C}$, $Re(\varpi)>0$, $Re(\gamma)>0$, $\lambda\in\mathbb{C}$, $\vartheta>0$, $\beta\in\mathbb{C}$, and $\alpha\in\mathbb{R}_{+}$. Then, we have*

$$\begin{aligned}
&\frac{1}{\Gamma(\varpi)}\int_{0}^{t}(t-\tau)^{\varpi}\,\tau^{\gamma-1}presinh_{\alpha,\beta}\left(\lambda\tau^{\vartheta}\right)d\tau\\
&=t^{\gamma+\varpi-1}\,{}_2sinh_2\left(\begin{matrix}(1,1),(\gamma,\vartheta)\\(\beta,\alpha),(\gamma+\varpi,\vartheta)\end{matrix};\lambda t^{\vartheta}\right)
\end{aligned}\tag{4.573}$$

and

$$\begin{aligned}
&\frac{1}{\Gamma(\varpi)}\int_{0}^{t}(t-\tau)^{\varpi}\,\tau^{\gamma-1}precosh_{\alpha,\beta}\left(\lambda\tau^{\vartheta}\right)d\tau\\
&=t^{\gamma+\varpi-1}\,{}_2cosh_2\left(\begin{matrix}(1,1),(\gamma,\vartheta)\\(\beta,\alpha),(\gamma+\varpi,\vartheta)\end{matrix};\lambda t^{\vartheta}\right).
\end{aligned}\tag{4.574}$$

Proof Making use of (4.559), we have

$$\begin{aligned}
&\frac{1}{\Gamma(\varpi)}\int_{0}^{t}(t-\tau)^{\varpi}\,\tau^{\gamma-1}\,{}_1\varphi_1\left(\begin{matrix}(1,1)\\(\beta,\alpha)\end{matrix};\lambda\tau^{\vartheta}\right)d\tau\\
&=t^{\gamma+\varpi-1}\,{}_2\varphi_2\left(\begin{matrix}(1,1),(\gamma,\vartheta)\\(\beta,\alpha),(\gamma+\varpi,\vartheta)\end{matrix};\lambda t^{\vartheta}\right)
\end{aligned}\tag{4.575}$$

and

$$\begin{aligned}
&\frac{1}{\Gamma(\varpi)}\int_{0}^{t}(t-\tau)^{\varpi}\,\tau^{\gamma-1}\,{}_1\varphi_1\left(\begin{matrix}(1,1)\\(\beta,\alpha)\end{matrix};-\lambda\tau^{\vartheta}\right)d\tau\\
&=t^{\gamma+\varpi-1}\,{}_2\varphi_2\left(\begin{matrix}(1,1),(\gamma,\vartheta)\\(\beta,\alpha),(\gamma+\varpi,\vartheta)\end{matrix};-\lambda t^{\vartheta}\right),
\end{aligned}\tag{4.576}$$

which reduces the results. □

Corollary 4.3 *Let $\varpi, \gamma \in \mathbb{C}$, $Re(\varpi) > 0$, $Re(\gamma) > 0$, $\lambda \in \mathbb{C}$, $\vartheta > 0$, $\beta \in \mathbb{C}$, and $\alpha \in \mathbb{R}_+$. Then, we have*

$$\begin{aligned}&\frac{1}{\Gamma(\varpi)}\int_0^1 (1-\tau)^{\varpi}\,\tau^{\gamma-1} presinh_{\alpha,\beta}\left(\lambda\tau^{\vartheta}\right)d\tau\\&= {}_2sinh_2\left(\begin{matrix}(1,1),(\gamma,\vartheta)\\(\beta,\alpha),(\gamma+\varpi,\vartheta)\end{matrix};\lambda\right)\end{aligned} \tag{4.577}$$

and

$$\begin{aligned}&\frac{1}{\Gamma(\varpi)}\int_0^1 (1-\tau)^{\varpi}\,\tau^{\gamma-1} precosh_{\alpha,\beta}\left(\lambda\tau^{\vartheta}\right)d\tau\\&= {}_2cosh_2\left(\begin{matrix}(1,1),(\gamma,\vartheta)\\(\beta,\alpha),(\gamma+\varpi,\vartheta)\end{matrix};\lambda\right).\end{aligned} \tag{4.578}$$

Proof Taking $t = 1$ in (4.573) and (4.574), we directly have the results. □

Theorem 4.48 *Let $\varpi, \gamma \in \mathbb{C}$, $Re(\varpi) > 0$, $Re(\gamma) > 0$, $\lambda \in \mathbb{C}$, $\vartheta > 0$, $\beta \in \mathbb{C}$, and $\alpha \in \mathbb{R}_+$.*

Then, we have

$$\begin{aligned}&\int_{-1}^1 (1-\tau)^{\varpi}\,(1+\tau)^{\gamma-1}\, presinh_{\alpha,\beta}\left(\lambda(1+\tau)^{\vartheta}\right)d\tau\\&= \int_{-1}^1 (1+\tau)^{\varpi}\,(1-\tau)^{\gamma-1}\, presinh_{\alpha,\beta}\left(\lambda(1-\tau)^{\vartheta}\right)d\tau\end{aligned} \tag{4.579}$$

and

$$\begin{aligned}&\frac{1}{\Gamma(\varpi)}\int_0^2 (2-\tau)^{\varpi}\,\tau^{\gamma-1} presinh_{\alpha,\beta}\left(\lambda\tau^{\vartheta}\right)d\tau\\&= \frac{1}{\Gamma(\varpi)}\int_{-1}^1 (1-\tau)^{\varpi}\,(1+\tau)^{\gamma-1}\, presinh_{\alpha,\beta}\left(\lambda(1+\tau)^{\vartheta}\right)d\tau\\&= \frac{1}{\Gamma(\varpi)}\int_{-1}^1 (1+\tau)^{\varpi}\,(1-\tau)^{\gamma-1}\, presinh_{\alpha,\beta}\left(\lambda(1-\tau)^{\vartheta}\right)d\tau\\&= 2^{\gamma+\varpi-1}\,{}_2sinh_2\left(\begin{matrix}(1,1),(\gamma,\vartheta)\\(\beta,\alpha),(\gamma+\varpi,\vartheta)\end{matrix};\lambda 2^{\vartheta}\right).\end{aligned} \tag{4.580}$$

Proof When $t = 2$ in (4.573), we obtain the result. □

Theorem 4.49 *Let $\varpi, \gamma \in \mathbb{C}$, $Re(\varpi) > 0$, $Re(\gamma) > 0$, $\lambda \in \mathbb{C}$, $\vartheta > 0$, $\beta \in \mathbb{C}$, and $\alpha \in \mathbb{R}_+$.*

Then, we have

$$\begin{aligned}&\int_{-1}^{1}(1-\tau)^{\varpi}(1+\tau)^{\gamma-1}\,precosh_{\alpha,\beta}\left(\lambda(1+\tau)^{\vartheta}\right)d\tau\\&\quad=\int_{-1}^{1}(1+\tau)^{\varpi}(1-\tau)^{\gamma-1}\,precosh_{\alpha,\beta}\left(\lambda(1-\tau)^{\vartheta}\right)d\tau\end{aligned}\tag{4.581}$$

and

$$\begin{aligned}&\frac{1}{\Gamma(\varpi)}\int_{0}^{2}(2-\tau)^{\varpi}\tau^{\gamma-1}precosh_{\alpha,\beta}\left(\lambda\tau^{\vartheta}\right)d\tau\\&=\frac{1}{\Gamma(\varpi)}\int_{-1}^{1}(1-\tau)^{\varpi}(1+\tau)^{\gamma-1}\,precosh_{\alpha,\beta}\left(\lambda(1+\tau)^{\vartheta}\right)d\tau\\&=\frac{1}{\Gamma(\varpi)}\int_{-1}^{1}(1+\tau)^{\varpi}(1-\tau)^{\gamma-1}\,precosh_{\alpha,\beta}\left(\lambda(1-\tau)^{\vartheta}\right)d\tau\\&=2^{\gamma+\varpi-1}\,{}_2cosh_2\left(\begin{matrix}(1,1),(\gamma,\vartheta)\\(\beta,\alpha),(\gamma+\varpi,\vartheta)\end{matrix};\lambda2^{\vartheta}\right).\end{aligned}\tag{4.582}$$

Proof When $t=2$ in (4.574), we obtain the result. □

Theorem 4.50 *Let $\varpi,\gamma\in\mathbb{C}$, $Re(\varpi)>0$, $Re(\gamma)>0$, $\lambda\in\mathbb{C}$, $\beta\in\mathbb{C}$, and $\alpha\in\mathbb{R}_+$. Then, we have*

$$\begin{aligned}&\frac{1}{\Gamma(\varpi)}\int_{0}^{t}(t-\tau)^{\varpi}\tau^{\gamma-1}\,{}_1\varphi_1\left(\begin{matrix}(1,1)\\(\beta,\alpha)\end{matrix};\lambda\tau\right)d\tau\\&\quad=t^{\gamma+\varpi-1}\,{}_2\varphi_2\left(\begin{matrix}(1,1),(\gamma,1)\\(\beta,\alpha),(\gamma+\varpi,1)\end{matrix};\lambda t\right).\end{aligned}\tag{4.583}$$

Proof Taking $\vartheta=1$ in (4.559), we get the result. □

Property 4.17 Let $\varpi,\gamma\in\mathbb{C}$, $Re(\varpi)>0$, $Re(\gamma)>0$, $\lambda\in\mathbb{C}$, $\beta\in\mathbb{C}$, and $\alpha\in\mathbb{R}_+$. Then, we have

$$\frac{1}{\Gamma(\varpi)}\int_{0}^{1}(1-\tau)^{\varpi}\tau^{\gamma-1}\,{}_1\varphi_1\left(\begin{matrix}(1,1)\\(\beta,\alpha)\end{matrix};\lambda\tau\right)d\tau={}_2\varphi_2\left(\begin{matrix}(1,1),(\gamma,1)\\(\beta,\alpha),(\gamma+\varpi,1)\end{matrix};\lambda\right).\tag{4.584}$$

Proof Taking $\vartheta=1$ in (4.560), we get the result. □

Theorem 4.51 *Let $\varpi, \gamma \in \mathbb{C}$, $Re(\varpi) > 0$, $Re(\gamma) > 0$, $\lambda \in \mathbb{C}$, $\beta \in \mathbb{C}$, and $\alpha \in \mathbb{R}_+$. Then, we have*

$$\begin{aligned} &\int_{-1}^{1} (1-\tau)^{\varpi} (1+\tau)^{\gamma-1} {}_1\varphi_1\left(\begin{matrix}(1,1)\\(\beta,\alpha)\end{matrix}; \lambda(1+\tau)\right) d\tau \\ &= \int_{-1}^{1} (1+\tau)^{\varpi} (1-\tau)^{\gamma-1} {}_1\varphi_1\left(\begin{matrix}(1,1)\\(\beta,\alpha)\end{matrix}; \lambda(1-\tau)\right) d\tau \end{aligned} \tag{4.585}$$

and

$$\begin{aligned} &\frac{1}{\Gamma(\varpi)} \int_{0}^{2} (2-\tau)^{\varpi} \tau^{\gamma-1} {}_1\varphi_1\left(\begin{matrix}(1,1)\\(\beta,\alpha)\end{matrix}; \lambda\tau\right) d\tau \\ &= \frac{1}{\Gamma(\varpi)} \int_{-1}^{1} (1-\tau)^{\varpi} (1+\tau)^{\gamma-1} {}_1\varphi_1\left(\begin{matrix}(1,1)\\(\beta,\alpha)\end{matrix}; \lambda(1+\tau)\right) d\tau \\ &= \frac{1}{\Gamma(\varpi)} \int_{-1}^{1} (1+\tau)^{\varpi} (1-\tau)^{\gamma-1} {}_1\varphi_1\left(\begin{matrix}(1,1)\\(\beta,\alpha)\end{matrix}; \lambda(1-\tau)\right) d\tau \\ &= 2^{\gamma+\varpi-1} {}_2\varphi_2\left(\begin{matrix}(1,1),(\gamma,1)\\(\beta,\alpha),(\gamma+\varpi,1)\end{matrix}; 2\lambda\right). \end{aligned} \tag{4.586}$$

Proof Taking $\vartheta = 1$ in (4.560) and (4.562), we get the results. □

Theorem 4.52 *Let $\varpi, \gamma \in \mathbb{C}$, $Re(\varpi) > 0$, $Re(\gamma) > 0$, $\lambda \in \mathbb{C}$, $\beta \in \mathbb{C}$, and $\alpha \in \mathbb{R}_+$. Then, we have*

$$\begin{aligned} &\frac{1}{\Gamma(\varpi)} \int_{0}^{t} (t-\tau)^{\varpi} \tau^{\gamma-1} presin_{\alpha,\beta}(\lambda\tau)\, d\tau \\ &= t^{\gamma+\varpi-1} {}_2sin_2\left(\begin{matrix}(1,1),(\gamma,1)\\(\beta,\alpha),(\gamma+\varpi,1)\end{matrix}; \lambda t\right) \end{aligned} \tag{4.587}$$

and

$$\begin{aligned} &\frac{1}{\Gamma(\varpi)} \int_{0}^{t} (t-\tau)^{\varpi} \tau^{\gamma-1} precos_{\alpha,\beta}(\lambda\tau)\, d\tau \\ &= t^{\gamma+\varpi-1} {}_2cos_2\left(\begin{matrix}(1,1),(\gamma,1)\\(\beta,\alpha),(\gamma+\varpi,1)\end{matrix}; \lambda t\right). \end{aligned} \tag{4.588}$$

Proof Taking $\vartheta = 1$ in (4.563) and (4.564), we get the results. □

Corollary 4.4 *Let $\varpi, \gamma \in \mathbb{C}$, $Re(\varpi) > 0$, $Re(\gamma) > 0$, $\lambda \in \mathbb{C}$, $\beta \in \mathbb{C}$, and $\alpha \in \mathbb{R}_+$. Then, we have*

$$\begin{aligned} &\frac{1}{\Gamma(\varpi)} \int_{0}^{1} (1-\tau)^{\varpi} \tau^{\gamma-1} presin_{\alpha,\beta}(\lambda\tau)\, d\tau \\ &= {}_2sin_2\left(\begin{matrix}(1,1),(\gamma,1)\\(\beta,\alpha),(\gamma+\varpi,1)\end{matrix}; \lambda\right) \end{aligned} \tag{4.589}$$

and

$$\begin{aligned} &\frac{1}{\Gamma(\varpi)}\int\limits_0^1 (1-\tau)^{\varpi}\,\tau^{\gamma-1} precos_{\alpha,\beta}(\lambda\tau)\,d\tau \\ &\quad = {}_2cos_2\left(\begin{matrix}(1,1),(\gamma,1)\\(\beta,\alpha),(\gamma+\varpi,1)\end{matrix};\lambda\right). \end{aligned} \tag{4.590}$$

Proof Taking $\vartheta = 1$ in (4.567) and (4.568), we get the results. □

Theorem 4.53 *Let $\varpi, \gamma \in \mathbb{C}$, $Re(\varpi) > 0$, $Re(\gamma) > 0$, $\lambda \in \mathbb{C}$, $\beta \in \mathbb{C}$, and $\alpha \in \mathbb{R}_+$. Then, we have*

$$\begin{aligned} &\int\limits_{-1}^1 (1-\tau)^{\varpi}(1+\tau)^{\gamma-1} presin_{\alpha,\beta}(\lambda(1+\tau))\,d\tau \\ &\quad = \int\limits_{-1}^1 (1+\tau)^{\varpi}(1-\tau)^{\gamma-1} presin_{\alpha,\beta}(\lambda(1-\tau))\,d\tau \end{aligned} \tag{4.591}$$

and

$$\begin{aligned} &\frac{1}{\Gamma(\varpi)}\int\limits_0^2 (2-\tau)^{\varpi}\,\tau^{\gamma-1} presin_{\alpha,\beta}(\lambda\tau)\,d\tau \\ &\quad = \frac{1}{\Gamma(\varpi)}\int\limits_{-1}^1 (1-\tau)^{\varpi}(1+\tau)^{\gamma-1} presin_{\alpha,\beta}(\lambda(1+\tau))\,d\tau \\ &\quad = \frac{1}{\Gamma(\varpi)}\int\limits_{-1}^1 (1+\tau)^{\varpi}(1-\tau)^{\gamma-1} presin_{\alpha,\beta}(\lambda(1-\tau))\,d\tau \\ &\quad = 2^{\gamma+\varpi-1}\,{}_2sin_2\left(\begin{matrix}(1,1),(\gamma,1)\\(\beta,\alpha),(\gamma+\varpi,1)\end{matrix};2\lambda\right). \end{aligned} \tag{4.592}$$

Proof Taking $\vartheta = 1$ in (4.569) and (4.570), we get the results. □

Theorem 4.54 *Let $\varpi, \gamma \in \mathbb{C}$, $Re(\varpi) > 0$, $Re(\gamma) > 0$, $\lambda \in \mathbb{C}$, $\beta \in \mathbb{C}$, and $\alpha \in \mathbb{R}_+$. Then, we have*

$$\begin{aligned} &\int\limits_{-1}^1 (1-\tau)^{\varpi}(1+\tau)^{\gamma-1} precos_{\alpha,\beta}(\lambda(1+\tau))\,d\tau \\ &\quad = \int\limits_{-1}^1 (1+\tau)^{\varpi}(1-\tau)^{\gamma-1} precos_{\alpha,\beta}(\lambda(1-\tau))\,d\tau \end{aligned} \tag{4.593}$$

and

$$\begin{aligned}
&\frac{1}{\Gamma(\varpi)}\int_0^2 (2-\tau)^{\varpi}\,\tau^{\gamma-1} precos_{\alpha,\beta}(\lambda\tau)\,d\tau \\
&= \frac{1}{\Gamma(\varpi)}\int_{-1}^1 (1-\tau)^{\varpi}\,(1+\tau)^{\gamma-1}\ precos_{\alpha,\beta}(\lambda(1+\tau))\,d\tau \\
&= \frac{1}{\Gamma(\varpi)}\int_{-1}^1 (1+\tau)^{\varpi}\,(1-\tau)^{\gamma-1}\ precos_{\alpha,\beta}(\lambda(1-\tau))\,d\tau \\
&= 2^{\gamma+\varpi-1}\ {}_2cos_2\left(\begin{matrix}(1,1),(\gamma,1)\\(\beta,\alpha),(\gamma+\varpi,1)\end{matrix};2\lambda\right).
\end{aligned} \tag{4.594}$$

Proof Taking $\vartheta = 1$ in (4.571) and (4.572), we get the results. □

Theorem 4.55 *Let $\varpi, \gamma \in \mathbb{C}$, $Re(\varpi) > 0$, $Re(\gamma) > 0$, $\lambda \in \mathbb{C}$, $\beta \in \mathbb{C}$, and $\alpha \in \mathbb{R}_+$. Then, we have*

$$\begin{aligned}
&\frac{1}{\Gamma(\varpi)}\int_0^t (t-\tau)^{\varpi}\,\tau^{\gamma-1} presinh_{\alpha,\beta}(\lambda\tau)\,d\tau \\
&= t^{\gamma+\varpi-1}\ {}_2sinh_2\left(\begin{matrix}(1,1),(\gamma,1)\\(\beta,\alpha),(\gamma+\varpi,1)\end{matrix};\lambda t\right)
\end{aligned} \tag{4.595}$$

and

$$\begin{aligned}
&\frac{1}{\Gamma(\varpi)}\int_0^t (t-\tau)^{\varpi}\,\tau^{\gamma-1} precosh_{\alpha,\beta}(\lambda\tau)\,d\tau \\
&= t^{\gamma+\varpi-1}\ {}_2cosh_2\left(\begin{matrix}(1,1),(\gamma,1)\\(\beta,\alpha),(\gamma+\varpi,1)\end{matrix};\lambda t\right).
\end{aligned} \tag{4.596}$$

Proof Taking $\vartheta = 1$ in (4.573) and (4.574), we get the results. □

Corollary 4.5 *Let $\varpi, \gamma \in \mathbb{C}$, $Re(\varpi) > 0$, $Re(\gamma) > 0$, $\lambda \in \mathbb{C}$, $\beta \in \mathbb{C}$, and $\alpha \in \mathbb{R}_+$. Then, we have*

$$\begin{aligned}
&\frac{1}{\Gamma(\varpi)}\int_0^1 (1-\tau)^{\varpi}\,\tau^{\gamma-1} presinh_{\alpha,\beta}(\lambda\tau)\,d\tau \\
&= {}_2sinh_2\left(\begin{matrix}(1,1),(\gamma,1)\\(\beta,\alpha),(\gamma+\varpi,1)\end{matrix};\lambda\right)
\end{aligned} \tag{4.597}$$

and

$$\begin{aligned}
&\frac{1}{\Gamma(\varpi)}\int_0^1 (1-\tau)^{\varpi}\,\tau^{\gamma-1} precosh_{\alpha,\beta}(\lambda\tau)\,d\tau \\
&= {}_2cosh_2\left(\begin{matrix}(1,1),(\gamma,1)\\(\beta,\alpha),(\gamma+\varpi,1)\end{matrix};\lambda\right).
\end{aligned} \tag{4.598}$$

Proof Taking $\vartheta = 1$ in (4.577) and (4.578), we get the results. □

Theorem 4.56 *Let $\varpi, \gamma \in \mathbb{C}$, $Re(\varpi) > 0$, $Re(\gamma) > 0$, $\lambda \in \mathbb{C}$, $\beta \in \mathbb{C}$, and $\alpha \in \mathbb{R}_+$. Then, we have*

$$\begin{aligned}&\int_{-1}^{1} (1-\tau)^{\varpi} (1+\tau)^{\gamma-1}\, presinh_{\alpha,\beta} (\lambda (1+\tau))\, d\tau \\ &\quad = \int_{-1}^{1} (1+\tau)^{\varpi} (1-\tau)^{\gamma-1}\, presinh_{\alpha,\beta} (\lambda (1-\tau))\, d\tau\end{aligned} \tag{4.599}$$

and

$$\begin{aligned}&\frac{1}{\Gamma(\varpi)} \int_{0}^{2} (2-\tau)^{\varpi} \tau^{\gamma-1} presinh_{\alpha,\beta} (\lambda\tau)\, d\tau \\ &= \frac{1}{\Gamma(\varpi)} \int_{-1}^{1} (1-\tau)^{\varpi} (1+\tau)^{\gamma-1}\, presinh_{\alpha,\beta} (\lambda (1+\tau))\, d\tau \\ &= \frac{1}{\Gamma(\varpi)} \int_{-1}^{1} (1+\tau)^{\varpi} (1-\tau)^{\gamma-1}\, presinh_{\alpha,\beta} (\lambda (1-\tau))\, d\tau \\ &= 2^{\gamma+\varpi-1}\, {}_2sinh_2 \left(\begin{matrix} (1,1), (\gamma,1) \\ (\beta,\alpha), (\gamma+\varpi,1) \end{matrix} ; 2\lambda \right).\end{aligned} \tag{4.600}$$

Proof Taking $\vartheta = 1$ in (4.579) and (4.580), we get the results. □

Theorem 4.57 *Let $\varpi, \gamma \in \mathbb{C}$, $Re(\varpi) > 0$, $Re(\gamma) > 0$, $\lambda \in \mathbb{C}$, $\beta \in \mathbb{C}$, and $\alpha \in \mathbb{R}_+$. Then, we have*

$$\begin{aligned}&\int_{-1}^{1} (1-\tau)^{\varpi} (1+\tau)^{\gamma-1}\, precosh_{\alpha,\beta} (\lambda (1+\tau))\, d\tau \\ &\quad = \int_{-1}^{1} (1+\tau)^{\varpi} (1-\tau)^{\gamma-1}\, precosh_{\alpha,\beta} (\lambda (1-\tau))\, d\tau\end{aligned} \tag{4.601}$$

and

$$\begin{aligned}&\frac{1}{\Gamma(\varpi)} \int_{0}^{2} (2-\tau)^{\varpi} \tau^{\gamma-1} precosh_{\alpha,\beta} (\lambda\tau)\, d\tau \\ &= \frac{1}{\Gamma(\varpi)} \int_{-1}^{1} (1-\tau)^{\varpi} (1+\tau)^{\gamma-1}\, precosh_{\alpha,\beta} (\lambda (1+\tau))\, d\tau \\ &= \frac{1}{\Gamma(\varpi)} \int_{-1}^{1} (1+\tau)^{\varpi} (1-\tau)^{\gamma-1}\, precosh_{\alpha,\beta} (\lambda (1-\tau))\, d\tau \\ &= 2^{\gamma+\varpi-1}\, {}_2cosh_2 \left(\begin{matrix} (1,1), (\gamma,1) \\ (\beta,\alpha), (\gamma+\varpi,1) \end{matrix} ; 2\lambda \right).\end{aligned} \tag{4.602}$$

Proof Taking $\vartheta = 1$ in (4.581) and (4.582), we get the results. □

Theorem 4.58 *Let* $\varpi \in \mathbb{C}$, $Re(\varpi) > 0$, $\lambda \in \mathbb{C}$, $\beta \in \mathbb{C}$, *and* $\alpha \in \mathbb{R}_{+}$. *Then, we have*

$$\frac{1}{\Gamma(\varpi)} \int_0^t (t-\tau)^{\varpi} \tau^{\beta-1} {}_1\varphi_1 \left(\begin{matrix} (1,1) \\ (\beta,\alpha) \end{matrix} ; \lambda\tau^{\alpha} \right) d\tau$$
$$= t^{\beta+\varpi-1} {}_1\varphi_1 \left(\begin{matrix} (1,1) \\ (\beta+\varpi,\alpha) \end{matrix} ; \lambda t^{\alpha} \right). \tag{4.603}$$

Proof With the aid of (4.559) and (4.507), we have

$$\tau^{\beta-1} {}_1\varphi_1 \left(\begin{matrix} (1,1) \\ (\beta,\alpha) \end{matrix} ; \lambda\tau^{\alpha} \right) = \tau^{\beta-1} E_{\alpha,\beta}\left(\lambda\tau^{\alpha}\right) \tag{4.604}$$

such that we have the desired result, where

$$t^{\beta+\varpi-1} {}_2\varphi_2 \left(\begin{matrix} (1,1),(\beta,\alpha) \\ (\beta,\alpha),(\beta+\varpi,\alpha) \end{matrix} ; \lambda t^{\alpha} \right) = t^{\beta+\varpi-1} {}_1\varphi_1 \left(\begin{matrix} (1,1) \\ (\beta+\varpi,\alpha) \end{matrix} ; \lambda t^{\alpha} \right).$$

□

Theorem 4.59 *Let* $\varpi \in \mathbb{C}$, $Re(\varpi) > 0$, $\lambda \in \mathbb{C}$, $\beta \in \mathbb{C}$, *and* $\alpha \in \mathbb{R}_{+}$. *Then, we have*

$$\frac{1}{\Gamma(\varpi)} \int_0^t (t-\tau)^{\varpi}\, supersin_{\alpha,\beta}\left(\lambda\tau^{\alpha}\right) d\tau$$
$$= t^{\beta+\varpi-1} {}_1sin_1 \left(\begin{matrix} (1,1) \\ (\beta+\varpi,\alpha) \end{matrix} ; \lambda t^{\alpha} \right). \tag{4.605}$$

Proof With the aid of (4.510) and (4.603), we have the result. □

Theorem 4.60 *Let* $\varpi \in \mathbb{C}$, $Re(\varpi) > 0$, $\lambda \in \mathbb{C}$, $\beta \in \mathbb{C}$, *and* $\alpha \in \mathbb{R}_{+}$. *Then, we have*

$$\frac{1}{\Gamma(\varpi)} \int_0^t (t-\tau)^{\varpi}\, supercos_{\alpha,\beta}\left(\lambda\tau^{\alpha}\right) d\tau = t^{\beta+\varpi-1} {}_1cos_1 \left(\begin{matrix} (1,1) \\ (\beta+\varpi,\alpha) \end{matrix} ; \lambda t^{\alpha} \right). \tag{4.606}$$

Proof With the aid of (4.511) and (4.603), we have the result. □

Theorem 4.61 *Let $\varpi \in \mathbb{C}$, $Re(\varpi) > 0$, $\lambda \in \mathbb{C}$, $\beta \in \mathbb{C}$, and $\alpha \in \mathbb{R}_+$. Then, we have*

$$\frac{1}{\Gamma(\varpi)} \int_0^t (t-\tau)^{\varpi}\, supersinh_{\alpha,\beta}\left(\lambda\tau^{\alpha}\right) d\tau = t^{\beta+\varpi-1}\ {}_1sinh_1\left(\begin{matrix}(1,1)\\(\beta+\varpi,\alpha)\end{matrix};\lambda t^{\alpha}\right). \tag{4.607}$$

Proof With the aid of (4.512) and (4.603), we have the result. □

Theorem 4.62 *Let $\varpi \in \mathbb{C}$, $Re(\varpi) > 0$, $\lambda \in \mathbb{C}$, $\beta \in \mathbb{C}$, and $\alpha \in \mathbb{R}_+$. Then, we have*

$$\frac{1}{\Gamma(\varpi)} \int_0^t (t-\tau)^{\varpi}\, supercosh_{\alpha,\beta}\left(\lambda\tau^{\alpha}\right) d\tau = t^{\beta+\varpi-1}\ {}_1cosh_1\left(\begin{matrix}(1,1)\\(\beta+\varpi,\alpha)\end{matrix};\lambda t^{\alpha}\right). \tag{4.608}$$

Proof With the aid of (4.513) and (4.603), we have the result. □

Theorem 4.63 *Let $\varpi \in \mathbb{C}$, $Re(\varpi) > 0$, $\lambda \in \mathbb{C}$, $\beta \in \mathbb{C}$, and $\alpha \in \mathbb{R}_+$. Then, we have*

$$\frac{1}{\Gamma(\varpi)} \int_0^1 (1-\tau)^{\varpi}\, \tau^{\beta-1}\ {}_1\varphi_1\left(\begin{matrix}(1,1)\\(\beta,\alpha)\end{matrix};\lambda\tau^{\alpha}\right) d\tau = {}_1\varphi_1\left(\begin{matrix}(1,1)\\(\beta+\varpi,\alpha)\end{matrix};\lambda\right). \tag{4.609}$$

Proof Taking $t = 1$ in (4.559) and (4.507), we have the desired result. □

Theorem 4.64 *Let $\varpi \in \mathbb{C}$, $Re(\varpi) > 0$, $\lambda \in \mathbb{C}$, $\beta \in \mathbb{C}$, and $\alpha \in \mathbb{R}_+$. Then, we have*

$$\frac{1}{\Gamma(\varpi)} \int_0^1 (1-\tau)^{\varpi}\, supersin_{\alpha,\beta}\left(\lambda\tau^{\alpha}\right) d\tau = {}_1sin_1\left(\begin{matrix}(1,1)\\(\beta+\varpi,\alpha)\end{matrix};\lambda\right). \tag{4.610}$$

Proof Setting $t = 1$ in (4.560) and (4.603), we have the result. □

Theorem 4.65 *Let $\varpi \in \mathbb{C}$, $Re(\varpi) > 0$, $\lambda \in \mathbb{C}$, $\beta \in \mathbb{C}$, and $\alpha \in \mathbb{R}_+$. Then, we have*

$$\frac{1}{\Gamma(\varpi)} \int_0^1 (1-\tau)^{\varpi}\, supercos_{\alpha,\beta}\left(\lambda\tau^{\alpha}\right) d\tau = {}_1cos_1\left(\begin{matrix}(1,1)\\(\beta+\varpi,\alpha)\end{matrix};\lambda\right). \tag{4.611}$$

Proof Setting $t = 1$ in (4.511) and (4.603), we have the result. □

Theorem 4.66 *Let $\varpi \in \mathbb{C}$, $Re(\varpi) > 0$, $\lambda \in \mathbb{C}$, $\beta \in \mathbb{C}$, and $\alpha \in \mathbb{R}_+$. Then, we have*

$$\frac{1}{\Gamma(\varpi)} \int_0^1 (1-\tau)^{\varpi} \, supersinh_{\alpha,\beta}\left(\lambda\tau^{\alpha}\right) d\tau = {}_1sinh_1\left(\begin{matrix}(1,1)\\(\beta+\varpi,\alpha)\end{matrix};\lambda\right). \tag{4.612}$$

Proof Taking $t = 1$ in (4.512) and (4.603), we have the result. □

Theorem 4.67 *Let $\varpi \in \mathbb{C}$, $Re(\varpi) > 0$, $\lambda \in \mathbb{C}$, $\beta \in \mathbb{C}$, and $\alpha \in \mathbb{R}_+$. Then, we have*

$$\frac{1}{\Gamma(\varpi)} \int_0^1 (1-\tau)^{\varpi} \, supercosh_{\alpha,\beta}\left(\lambda\tau^{\alpha}\right) d\tau = {}_1cosh_1\left(\begin{matrix}(1,1)\\(\beta+\varpi,\alpha)\end{matrix};\lambda\right). \tag{4.613}$$

Proof Setting $t = 1$ in (4.513) and (4.603), we have the result. □

Theorem 4.68 *Let $\varpi \in \mathbb{C}$, $Re(\varpi) > 0$, $\lambda \in \mathbb{C}$, $\beta \in \mathbb{C}$, and $\alpha \in \mathbb{R}_+$. Then, we have*

$$\begin{aligned}&\int_{-1}^{1} (1-\tau)^{\varpi} (\tau+1)^{\beta-1} \, {}_1\varphi_1\left(\begin{matrix}(1,1)\\(\beta,\alpha)\end{matrix};\lambda(\tau+1)^{\alpha}\right) d\tau \\ &\quad = \int_{-1}^{1} (1+\tau)^{\varpi} (1-\tau)^{\beta-1} \, {}_1\varphi_1\left(\begin{matrix}(1,1)\\(\beta,\alpha)\end{matrix};\lambda(1-\tau)^{\alpha}\right) d\tau\end{aligned} \tag{4.614}$$

and

$$\begin{aligned}&\frac{1}{\Gamma(\varpi)} \int_0^2 (2-\tau)^{\varpi} \tau^{\beta-1} \, {}_1\varphi_1\left(\begin{matrix}(1,1)\\(\beta,\alpha)\end{matrix};\lambda\tau^{\alpha}\right) d\tau \\ &= \frac{1}{\Gamma(\varpi)} \int_{-1}^{1} (1-\tau)^{\varpi} (\tau+1)^{\beta-1} \, {}_1\varphi_1\left(\begin{matrix}(1,1)\\(\beta,\alpha)\end{matrix};\lambda(\tau+1)^{\alpha}\right) d\tau \\ &= \frac{1}{\Gamma(\varpi)} \int_{-1}^{1} (1+\tau)^{\varpi} (1-\tau)^{\beta-1} \, {}_1\varphi_1\left(\begin{matrix}(1,1)\\(\beta,\alpha)\end{matrix};\lambda(1-\tau)^{\alpha}\right) d\tau \\ &= 2^{\beta+\varpi-1} \, {}_1\varphi_1\left(\begin{matrix}(1,1)\\(\beta+\varpi,\alpha)\end{matrix};\lambda 2^{\alpha}\right).\end{aligned} \tag{4.615}$$

Proof Setting $t = 2$ in (4.559) and (4.507), we have

$$\begin{aligned}
&\frac{1}{\Gamma(\varpi)}\int_0^2 (2-\tau)^{\varpi}\,\tau^{\beta-1}\;{}_1\varphi_1\left(\begin{matrix}(1,1)\\(\beta,\alpha)\end{matrix};\lambda\tau^{\alpha}\right)d\tau\\
&=\frac{1}{\Gamma(\varpi)}\int_{-1}^1 (1-\tau)^{\varpi}\,(\tau+1)^{\beta-1}\;{}_1\varphi_1\left(\begin{matrix}(1,1)\\(\beta,\alpha)\end{matrix};\lambda\,(\tau+1)^{\alpha}\right)d\tau\\
&=\frac{1}{\Gamma(\varpi)}\int_{-1}^1 (1+\tau)^{\varpi}\,(1-\tau)^{\beta-1}\;{}_1\varphi_1\left(\begin{matrix}(1,1)\\(\beta,\alpha)\end{matrix};\lambda\,(1-\tau)^{\alpha}\right)d\tau\\
&=2^{\beta+\varpi-1}\;{}_1\varphi_1\left(\begin{matrix}(1,1)\\(\beta+\varpi,\alpha)\end{matrix};\lambda 2^{\alpha}\right),
\end{aligned}\tag{4.616}$$

which reduces that

$$\begin{aligned}
&\frac{1}{\Gamma(\varpi)}\int_{-1}^1 (1-\tau)^{\varpi}\,(\tau+1)^{\beta-1}\;{}_1\varphi_1\left(\begin{matrix}(1,1)\\(\beta,\alpha)\end{matrix};\lambda\,(\tau+1)^{\alpha}\right)d\tau\\
&=\int_{-1}^1 (1+\tau)^{\varpi}\,(1-\tau)^{\beta-1}\;{}_1\varphi_1\left(\begin{matrix}(1,1)\\(\beta,\alpha)\end{matrix};\lambda\,(1-\tau)^{\alpha}\right)d\tau.
\end{aligned}\tag{4.617}$$

□

Theorem 4.69 *Let $\varpi \in \mathbb{C}$, $Re(\varpi) > 0$, $\lambda \in \mathbb{C}$, $\beta \in \mathbb{C}$, and $\alpha \in \mathbb{R}_+$. Then, we have*

$$\begin{aligned}
&\int_{-1}^{1} (1-\tau)^{\varpi}\, supersin_{\alpha,\beta}\left(\lambda\,(\tau+1)^{\alpha}\right)d\tau\\
&\qquad=\int_{-1}^{1} (1+\tau)^{\varpi}\, supersin_{\alpha,\beta}\left(\lambda\,(1-\tau)^{\alpha}\right)d\tau
\end{aligned}\tag{4.618}$$

and

$$\begin{aligned}
&\frac{1}{\Gamma(\varpi)}\int_0^2 (2-\tau)^{\varpi}\, supersin_{\alpha,\beta}\left(\lambda\tau^{\alpha}\right)d\tau\\
&=\frac{1}{\Gamma(\varpi)}\int_{-1}^1 (1-\tau)^{\varpi}\, supersin_{\alpha,\beta}\left(\lambda\,(\tau+1)^{\alpha}\right)d\tau\\
&=\frac{1}{\Gamma(\varpi)}\int_{-1}^1 (1+\tau)^{\varpi}\, supersin_{\alpha,\beta}\left(\lambda\,(1-\tau)^{\alpha}\right)d\tau\\
&=2^{\beta+\varpi-1}\;{}_1sin_1\left(\begin{matrix}(1,1),()\\(\beta+\varpi,\alpha)\end{matrix};\lambda 2^{\alpha}\right).
\end{aligned}\tag{4.619}$$

Proof With the aid of (4.510), (4.614), and (4.615), we have the results. □

Theorem 4.70 *Let $\varpi \in \mathbb{C}$, $Re(\varpi) > 0$, $\lambda \in \mathbb{C}$, $\beta \in \mathbb{C}$, and $\alpha \in \mathbb{R}_+$. Then, we have*

$$\int_{-1}^{1} (1-\tau)^{\varpi}\, supercos_{\alpha,\beta}\left(\lambda(\tau+1)^{\alpha}\right) d\tau$$

$$= \int_{-1}^{1} (1+\tau)^{\varpi}\, supercos_{\alpha,\beta}\left(\lambda(1-\tau)^{\alpha}\right) d\tau \tag{4.620}$$

and

$$\begin{aligned} &\frac{1}{\Gamma(\varpi)} \int_{0}^{2} (2-\tau)^{\varpi}\, supercos_{\alpha,\beta}(\lambda\tau^{\alpha})\, d\tau \\ &= \frac{1}{\Gamma(\varpi)} \int_{-1}^{1} (1-\tau)^{\varpi}\, supercos_{\alpha,\beta}(\lambda(\tau+1)^{\alpha})\, d\tau \\ &= \frac{1}{\Gamma(\varpi)} \int_{-1}^{1} (1+\tau)^{\varpi}\, supercos_{\alpha,\beta}(\lambda(1-\tau)^{\alpha})\, d\tau \\ &= 2^{\beta+\varpi-1}\; {}_1cos_1\left(\begin{matrix}(1,1)\\(\beta+\varpi,\alpha)\end{matrix};\lambda 2^{\alpha}\right). \end{aligned} \tag{4.621}$$

Proof With the aid of (4.511), (4.614), and (4.615), we have the results. □

Theorem 4.71 *Let $\varpi \in \mathbb{C}$, $Re(\varpi) > 0$, $\lambda \in \mathbb{C}$, $\beta \in \mathbb{C}$, and $\alpha \in \mathbb{R}_+$. Then, we have*

$$\int_{-1}^{1} (1-\tau)^{\varpi}\, supersinh_{\alpha,\beta}\left(\lambda(\tau+1)^{\alpha}\right) d\tau$$

$$= \int_{-1}^{1} (1+\tau)^{\varpi}\, supersinh_{\alpha,\beta}\left(\lambda(1-\tau)^{\alpha}\right) d\tau \tag{4.622}$$

and

$$\begin{aligned} &\frac{1}{\Gamma(\varpi)} \int_{0}^{2} (2-\tau)^{\varpi}\, supersinh_{\alpha,\beta}(\lambda\tau^{\alpha})\, d\tau \\ &= \frac{1}{\Gamma(\varpi)} \int_{-1}^{1} (1-\tau)^{\varpi}\, supersinh_{\alpha,\beta}(\lambda(\tau+1)^{\alpha})\, d\tau \\ &= \frac{1}{\Gamma(\varpi)} \int_{-1}^{1} (1+\tau)^{\varpi}\, supersinh_{\alpha,\beta}(\lambda(1-\tau)^{\alpha})\, d\tau \\ &= 2^{\beta+\varpi-1}\; {}_1sinh_1\left(\begin{matrix}(1,1)\\(\beta+\varpi,\alpha)\end{matrix};\lambda 2^{\alpha}\right). \end{aligned} \tag{4.623}$$

Proof With the aid of (4.512) , (4.614), and (4.615), we have the results. □

Theorem 4.72 *Let* $\varpi \in \mathbb{C}$, $Re(\varpi) > 0$, $\lambda \in \mathbb{C}$, $\beta \in \mathbb{C}$, *and* $\alpha \in \mathbb{R}_+$. *Then, we have*

$$\int_{-1}^{1} (1-\tau)^{\varpi}\, supercosh_{\alpha,\beta}\left(\lambda(\tau+1)^{\alpha}\right) d\tau = \int_{-1}^{1} (1+\tau)^{\varpi}\, supercosh_{\alpha,\beta}\left(\lambda(1-\tau)^{\alpha}\right) d\tau \tag{4.624}$$

and

$$\begin{aligned} &\frac{1}{\Gamma(\varpi)} \int_{0}^{2} (2-\tau)^{\varpi}\, supercosh_{\alpha,\beta}(\lambda\tau^{\alpha})\, d\tau \\ &= \frac{1}{\Gamma(\varpi)} \int_{-1}^{1} (1-\tau)^{\varpi}\, supercosh_{\alpha,\beta}(\lambda(\tau+1)^{\alpha})\, d\tau \\ &= \frac{1}{\Gamma(\varpi)} \int_{-1}^{1} (1+\tau)^{\varpi}\, supercosh_{\alpha,\beta}(\lambda(1-\tau)^{\alpha})\, d\tau \\ &= 2^{\beta+\varpi-1}\, {}_1cosh_1\left(\begin{matrix}(1,1)\\(\beta+\varpi,\alpha)\end{matrix};\lambda 2^{\alpha}\right). \end{aligned} \tag{4.625}$$

Proof With the aid of (4.513) , (4.614), and (4.615), we have the results. □

Theorem 4.73 *Let* $\varpi \in \mathbb{C}$, $Re(\varpi) > 0$, $\lambda \in \mathbb{C}$, $\vartheta > 0$, $\beta \in \mathbb{C}$, *and* $\alpha \in \mathbb{R}_+$. *Then, we have*

$$\begin{aligned} &\frac{1}{\Gamma(\varpi)} \int_{0}^{t} (t-\tau)^{\varpi}\, \tau^{\beta-1}\, {}_1\varphi_1\left(\begin{matrix}(1,1)\\(\beta,\alpha)\end{matrix};\lambda\tau^{\vartheta}\right) d\tau \\ &= t^{\beta+\varpi-1}\, {}_2\varphi_2\left(\begin{matrix}(1,1),(\beta,\vartheta)\\(\beta,\alpha),(\beta+\varpi,\vartheta)\end{matrix};\lambda t^{\vartheta}\right). \end{aligned} \tag{4.626}$$

Proof With the use of (4.559), we have the result. □

Theorem 4.74 *Let* $\varpi \in \mathbb{C}$, $Re(\varpi) > 0$, $\lambda \in \mathbb{C}$, $\vartheta > 0$, $\beta \in \mathbb{C}$, *and* $\alpha \in \mathbb{R}_+$. *Then, we have*

$$\frac{1}{\Gamma(\varpi)} \int_{0}^{t} (t-\tau)^{\varpi}\, Gresin_{\alpha,\beta}\left(\lambda\tau^{\vartheta}\right) d\tau = t^{\beta+\varpi-1}\, {}_2sin_2\left(\begin{matrix}(1,1),(\beta,\vartheta)\\(\beta,\alpha),(\beta+\varpi,\vartheta)\end{matrix};\lambda t^{\vartheta}\right). \tag{4.627}$$

Proof With the use of (4.559) and (4.526), we have the result. □

Theorem 4.75 *Let $\varpi \in \mathbb{C}$, $Re(\varpi) > 0$, $\lambda \in \mathbb{C}$, $\vartheta > 0$, $\beta \in \mathbb{C}$, and $\alpha \in \mathbb{R}_+$. Then, we have*

$$\frac{1}{\Gamma(\varpi)} \int_0^t (t-\tau)^{\varpi} \, Grecos_{\alpha,\beta}\left(\lambda \tau^{\vartheta}\right) d\tau$$

$$= t^{\beta+\varpi-1} \; {}_2cos_2\left(\begin{matrix} (1,1), (\beta,\vartheta) \\ (\beta,\alpha), (\beta+\varpi,\vartheta) \end{matrix} ; \lambda t^{\vartheta} \right). \tag{4.628}$$

Proof With the use of (4.559) and (4.527), we have the result. □

Theorem 4.76 *Let $\varpi \in \mathbb{C}$, $Re(\varpi) > 0$, $\lambda \in \mathbb{C}$, $\vartheta > 0$, $\beta \in \mathbb{C}$, and $\alpha \in \mathbb{R}_+$. Then, we have*

$$\frac{1}{\Gamma(\varpi)} \int_0^t (t-\tau)^{\varpi} \, Gresinh_{\alpha,\beta}\left(\lambda \tau^{\vartheta}\right) d\tau$$

$$= t^{\beta+\varpi-1} \; {}_2sinh_2\left(\begin{matrix} (1,1), (\beta,\vartheta) \\ (\beta,\alpha), (\beta+\varpi,\vartheta) \end{matrix} ; \lambda t^{\vartheta} \right). \tag{4.629}$$

Proof With the use of (4.559) and (4.528), we have the result. □

Theorem 4.77 *Let $\varpi \in \mathbb{C}$, $Re(\varpi) > 0$, $\lambda \in \mathbb{C}$, $\vartheta > 0$, $\beta \in \mathbb{C}$, and $\alpha \in \mathbb{R}_+$. Then, we have*

$$\frac{1}{\Gamma(\varpi)} \int_0^t (t-\tau)^{\varpi} \, Grecosh_{\alpha,\beta}\left(\lambda \tau^{\vartheta}\right) d\tau$$

$$= t^{\beta+\varpi-1} \; {}_2cosh_2\left(\begin{matrix} (1,1), (\beta,\vartheta) \\ (\beta,\alpha), (\beta+\varpi,\vartheta) \end{matrix} ; \lambda t^{\vartheta} \right). \tag{4.630}$$

Proof With the use of (4.559) and (4.529), we have the result. □

Theorem 4.78 *Let $\varpi \in \mathbb{C}$, $Re(\varpi) > 0$, $\lambda \in \mathbb{C}$, $\vartheta > 0$, $\beta \in \mathbb{C}$, and $\alpha \in \mathbb{R}_+$. Then, we have*

$$\frac{1}{\Gamma(\varpi)} \int_0^1 (1-\tau)^{\varpi} \, \tau^{\beta-1} \; {}_1\varphi_1\left(\begin{matrix} (1,1) \\ (\beta,\alpha) \end{matrix} ; \lambda \tau^{\vartheta} \right) d\tau = {}_2\varphi_2\left(\begin{matrix} (1,1), (\beta,\vartheta) \\ (\beta,\alpha), (\beta+\varpi,\vartheta) \end{matrix} ; \lambda \right). \tag{4.631}$$

Proof Taking $t = 1$ in (4.559), we have the result. □

Theorem 4.79 *Let $\varpi \in \mathbb{C}$, $Re(\varpi) > 0$, $\lambda \in \mathbb{C}$, $\vartheta > 0$, $\beta \in \mathbb{C}$, and $\alpha \in \mathbb{R}_+$. Then, we have*

$$\frac{1}{\Gamma(\varpi)} \int_0^1 (1-\tau)^{\varpi} \, Gresin_{\alpha,\beta}\left(\lambda\tau^{\vartheta}\right) d\tau = {}_2sin_2\left(\begin{matrix} (1,1), (\beta,\vartheta) \\ (\beta,\alpha), (\beta+\varpi,\vartheta) \end{matrix} ; \lambda \right). \tag{4.632}$$

Proof Taking $t = 1$ in (4.631) and (4.526), we have the result. □

Theorem 4.80 *Let $\varpi \in \mathbb{C}$, $Re(\varpi) > 0$, $\lambda \in \mathbb{C}$, $\vartheta > 0$, $\beta \in \mathbb{C}$, and $\alpha \in \mathbb{R}_+$. Then, we have*

$$\frac{1}{\Gamma(\varpi)} \int_0^1 (1-\tau)^{\varpi} \, Grecos_{\alpha,\beta}\left(\lambda\tau^{\vartheta}\right) d\tau = {}_2cos_2\left(\begin{matrix} (1,1), (\beta,\vartheta) \\ (\beta,\alpha), (\beta+\varpi,\vartheta) \end{matrix} ; \lambda \right). \tag{4.633}$$

Proof Taking $t = 1$ in (4.631) and (4.527), we have the result. □

Theorem 4.81 *Let $\varpi \in \mathbb{C}$, $Re(\varpi) > 0$, $\lambda \in \mathbb{C}$, $\vartheta > 0$, $\beta \in \mathbb{C}$, and $\alpha \in \mathbb{R}_+$. Then, we have*

$$\frac{1}{\Gamma(\varpi)} \int_0^1 (1-\tau)^{\varpi} \, Gresinh_{\alpha,\beta}\left(\lambda\tau^{\vartheta}\right) d\tau = {}_2sinh_2\left(\begin{matrix} (1,1), (\beta,\vartheta) \\ (\beta,\alpha), (\beta+\varpi,\vartheta) \end{matrix} ; \lambda \right). \tag{4.634}$$

Proof Taking $t = 1$ in (4.631) and (4.528), we have the result. □

Theorem 4.82 *Let $\varpi \in \mathbb{C}$, $Re(\varpi) > 0$, $\lambda \in \mathbb{C}$, $\vartheta > 0$, $\beta \in \mathbb{C}$, and $\alpha \in \mathbb{R}_+$. Then, we have*

$$\frac{1}{\Gamma(\varpi)} \int_0^1 (1-\tau)^{\varpi} \, Grecosh_{\alpha,\beta}\left(\lambda\tau^{\vartheta}\right) d\tau = {}_2cosh_2\left(\begin{matrix} (1,1), (\beta,\vartheta) \\ (\beta,\alpha), (\beta+\varpi,\vartheta) \end{matrix} ; \lambda \right). \tag{4.635}$$

Proof Taking $t = 1$ in (4.631) and (4.529), we have the result. □

Theorem 4.83 *Let $\varpi \in \mathbb{C}$, $Re(\varpi) > 0$, $\lambda \in \mathbb{C}$, $\vartheta > 0$, $\beta \in \mathbb{C}$, and $\alpha \in \mathbb{R}_+$. Then, we have*

$$\begin{aligned} &\int_{-1}^{1} (1-\tau)^{\varpi} (1+\tau)^{\beta-1} \, {}_1\varphi_1\left(\begin{matrix} (1,1) \\ (\beta,\alpha) \end{matrix} ; \lambda(1+\tau)^{\vartheta} \right) d\tau \\ &= \int_{-1}^{1} (1+\tau)^{\varpi} (1-\tau)^{\beta-1} \, {}_1\varphi_1\left(\begin{matrix} (1,1) \\ (\beta,\alpha) \end{matrix} ; \lambda(1-\tau)^{\vartheta} \right) d\tau \end{aligned} \tag{4.636}$$

and

$$\begin{aligned}
&\frac{1}{\Gamma(\varpi)} \int_0^2 (2-\tau)^{\varpi}\, \tau^{\beta-1}\ {}_1\varphi_1 \left(\begin{matrix} (1,1) \\ (\beta,\alpha) \end{matrix} ; \lambda \tau^{\vartheta} \right) d\tau \\
&= \frac{1}{\Gamma(\varpi)} \int_{-1}^1 (1-\tau)^{\varpi} (1+\tau)^{\beta-1}\ {}_1\varphi_1 \left(\begin{matrix} (1,1) \\ (\beta,\alpha) \end{matrix} ; \lambda (1+\tau)^{\vartheta} \right) d\tau \\
&= \frac{1}{\Gamma(\varpi)} \int_{-1}^1 (1+\tau)^{\varpi} (1-\tau)^{\beta-1}\ {}_1\varphi_1 \left(\begin{matrix} (1,1) \\ (\beta,\alpha) \end{matrix} ; \lambda (1-\tau)^{\vartheta} \right) d\tau \\
&= 2^{\beta+\varpi-1}\ {}_2\varphi_2 \left(\begin{matrix} (1,1)\,, (\beta,\vartheta) \\ (\beta,\alpha)\,, (\beta+\varpi,\vartheta) \end{matrix} ; \lambda 2^{\vartheta} \right).
\end{aligned} \tag{4.637}$$

Proof Taking $t = 2$ in (4.559), we have the results. □

Theorem 4.84 *Let* $\varpi \in \mathbb{C}$, $Re(\varpi) > 0$, $\lambda \in \mathbb{C}$, $\vartheta > 0$, $\beta \in \mathbb{C}$, *and* $\alpha \in \mathbb{R}_+$. *Then, we have*

$$\int_{-1}^1 (1-\tau)^{\varpi}\, Gresin_{\alpha,\beta}\left(\lambda (1+\tau)^{\vartheta}\right) d\tau = \int_{-1}^1 (1+\tau)^{\varpi}\, Gresin_{\alpha,\beta}\left(\lambda (1-\tau)^{\vartheta}\right) d\tau \tag{4.638}$$

and

$$\begin{aligned}
&\frac{1}{\Gamma(\varpi)} \int_0^2 (2-\tau)^{\varpi}\, Gresin_{\alpha,\beta}\left(\lambda \tau^{\vartheta}\right) d\tau \\
&= \frac{1}{\Gamma(\varpi)} \int_{-1}^1 (1-\tau)^{\varpi}\, Gresin_{\alpha,\beta}\left(\lambda (1+\tau)^{\vartheta}\right) d\tau \\
&= \frac{1}{\Gamma(\varpi)} \int_{-1}^1 (1+\tau)^{\varpi}\, Gresin_{\alpha,\beta}\left(\lambda (1-\tau)^{\vartheta}\right) d\tau \\
&= 2^{\beta+\varpi-1}\ {}_2sin_2 \left(\begin{matrix} (1,1)\,, (\beta,\vartheta) \\ (\beta,\alpha)\,, (\beta+\varpi,\vartheta) \end{matrix} ; \lambda 2^{\vartheta} \right).
\end{aligned} \tag{4.639}$$

Proof Taking $t = 2$ in (4.636) and (4.637), we have the results. □

Theorem 4.85 *Let* $\varpi \in \mathbb{C}$, $Re(\varpi) > 0$, $\lambda \in \mathbb{C}$, $\vartheta > 0$, $\beta \in \mathbb{C}$, *and* $\alpha \in \mathbb{R}_+$. *Then, we have*

$$\int_{-1}^1 (1-\tau)^{\varpi}\, Grecos_{\alpha,\beta}\left(\lambda (1+\tau)^{\vartheta}\right) d\tau = \int_{-1}^1 (1+\tau)^{\varpi}\, Grecos_{\alpha,\beta}\left(\lambda (1-\tau)^{\vartheta}\right) d\tau \tag{4.640}$$

and

$$\begin{aligned}
&\frac{1}{\Gamma(\varpi)}\int_{0}^{2}(2-\tau)^{\varpi}\, Grecos_{\alpha,\beta}\left(\lambda\tau^{\vartheta}\right)d\tau\\
&=\frac{1}{\Gamma(\varpi)}\int_{-1}^{1}(1-\tau)^{\varpi}\, Grecos_{\alpha,\beta}\left(\lambda\left(1+\tau\right)^{\vartheta}\right)d\tau\\
&=\frac{1}{\Gamma(\varpi)}\int_{-1}^{1}(1+\tau)^{\varpi}\, Grecos_{\alpha,\beta}\left(\lambda\left(1-\tau\right)^{\vartheta}\right)d\tau\\
&=2^{\beta+\varpi-1}\,{}_{2}cos_{2}\left(\begin{matrix}(1,1),(\beta,\vartheta)\\(\beta,\alpha),(\beta+\varpi,\vartheta)\end{matrix};\lambda 2^{\vartheta}\right).
\end{aligned} \tag{4.641}$$

Proof Taking $t=2$ in (4.636) and (4.637), we have the results. □

Theorem 4.86 *Let $\varpi\in\mathbb{C}$, $Re(\varpi)>0$, $\lambda\in\mathbb{C}$, $\vartheta>0$, $\beta\in\mathbb{C}$, and $\alpha\in\mathbb{R}_{+}$. Then, we have*

$$\begin{aligned}
&\int_{-1}^{1}(1-\tau)^{\varpi}\, Gresinh_{\alpha,\beta}\left(\lambda\left(1+\tau\right)^{\vartheta}\right)d\tau\\
&\qquad=\int_{-1}^{1}(1+\tau)^{\varpi}\, Gresinh_{\alpha,\beta}\left(\lambda\left(1-\tau\right)^{\vartheta}\right)d\tau
\end{aligned} \tag{4.642}$$

and

$$\begin{aligned}
&\frac{1}{\Gamma(\varpi)}\int_{0}^{2}(2-\tau)^{\varpi}\, Gresinh_{\alpha,\beta}\left(\lambda\tau^{\vartheta}\right)d\tau\\
&=\frac{1}{\Gamma(\varpi)}\int_{-1}^{1}(1-\tau)^{\varpi}\, Gresinh_{\alpha,\beta}\left(\lambda\left(1+\tau\right)^{\vartheta}\right)d\tau\\
&=\frac{1}{\Gamma(\varpi)}\int_{-1}^{1}(1+\tau)^{\varpi}\, Gresinh_{\alpha,\beta}\left(\lambda\left(1-\tau\right)^{\vartheta}\right)d\tau\\
&=2^{\beta+\varpi-1}\,{}_{2}sinh_{2}\left(\begin{matrix}(1,1),(\beta,\vartheta)\\(\beta,\alpha),(\beta+\varpi,\vartheta)\end{matrix};\lambda 2^{\vartheta}\right).
\end{aligned} \tag{4.643}$$

Proof Taking $t=2$ in (4.636) and (4.637), we have the results. □

Theorem 4.87 *Let $\varpi\in\mathbb{C}$, $Re(\varpi)>0$, $\lambda\in\mathbb{C}$, $\vartheta>0$, $\beta\in\mathbb{C}$, and $\alpha\in\mathbb{R}_{+}$. Then, we have*

$$\int_{-1}^{1}(1-\tau)^{\varpi}\, Grecosh_{\alpha,\beta}\left(\lambda\left(1+\tau\right)^{\vartheta}\right)d\tau=\int_{-1}^{1}(1+\tau)^{\varpi}\, Grecosh_{\alpha,\beta}\left(\lambda\left(1-\tau\right)^{\vartheta}\right)d\tau \tag{4.644}$$

and

$$\begin{aligned}
&\frac{1}{\Gamma(\varpi)}\int_0^2 (2-\tau)^{\varpi}\, Grecosh_{\alpha,\beta}\left(\lambda\tau^{\vartheta}\right)d\tau \\
&= \frac{1}{\Gamma(\varpi)}\int_{-1}^1 (1-\tau)^{\varpi}\, Grecosh_{\alpha,\beta}\left(\lambda\left(1+\tau\right)^{\vartheta}\right)d\tau \\
&= \frac{1}{\Gamma(\varpi)}\int_{-1}^1 (1+\tau)^{\varpi}\, Grecosh_{\alpha,\beta}\left(\lambda\left(1-\tau\right)^{\vartheta}\right)d\tau \\
&= 2^{\beta+\varpi-1}\,{}_2cosh_2\left(\begin{matrix}(1,1),(\beta,\vartheta)\\(\beta,\alpha),(\beta+\varpi,\vartheta)\end{matrix};\lambda 2^{\vartheta}\right).
\end{aligned} \tag{4.645}$$

Proof Taking $t = 2$ in (4.636) and (4.637), we have the results. □

Theorem 4.88 *Let $\varpi \in \mathbb{C}$, $Re(\varpi) > 0$, $\lambda \in \mathbb{C}$, $\beta \in \mathbb{C}$, and $\alpha \in \mathbb{R}_+$. Then, we have*

$$\begin{aligned}
&\frac{1}{\Gamma(\varpi)}\int_0^t (t-\tau)^{\varpi}\,\tau^{\beta-1}\,{}_1\varphi_1\left(\begin{matrix}(1,1)\\(\beta,\alpha)\end{matrix};\lambda\tau\right)d\tau \\
&= t^{\beta+\varpi-1}\,{}_2\varphi_2\left(\begin{matrix}(1,1),(\beta,1)\\(\beta,\alpha),(\beta+\varpi,1)\end{matrix};\lambda t\right).
\end{aligned} \tag{4.646}$$

Proof On setting $\vartheta = 1$ in (4.626), we have the result. □

Theorem 4.89 *Let $\varpi \in \mathbb{C}$, $Re(\varpi) > 0$, $\lambda \in \mathbb{C}$, $\beta \in \mathbb{C}$, and $\alpha \in \mathbb{R}_+$. Then, we have*

$$\frac{1}{\Gamma(\varpi)}\int_0^t (t-\tau)^{\varpi}\, Tresin_{\alpha,\beta}(\lambda\tau)\,d\tau = t^{\beta+\varpi-1}\,{}_2sin_2\left(\begin{matrix}(1,1),(\beta,1)\\(\beta,\alpha),(\beta+\varpi,1)\end{matrix};\lambda t\right). \tag{4.647}$$

Proof Taking $\vartheta = 1$ in (4.627), we have the result. □

Theorem 4.90 *Let $\varpi \in \mathbb{C}$, $Re(\varpi) > 0$, $\lambda \in \mathbb{C}$, $\beta \in \mathbb{C}$, and $\alpha \in \mathbb{R}_+$. Then, we have*

$$\frac{1}{\Gamma(\varpi)}\int_0^t (t-\tau)^{\varpi}\, Trecos_{\alpha,\beta}(\lambda\tau)\,d\tau = t^{\beta+\varpi-1}\,{}_2cos_2\left(\begin{matrix}(1,1),(\beta,1)\\(\beta,\alpha),(\beta+\varpi,1)\end{matrix};\lambda t\right). \tag{4.648}$$

Proof Setting $\vartheta = 1$ in (4.628), we have the result. □

Theorem 4.91 *Let $\varpi \in \mathbb{C}$, $Re(\varpi) > 0$, $\lambda \in \mathbb{C}$, $\beta \in \mathbb{C}$, and $\alpha \in \mathbb{R}_+$. Then, we have*

$$\frac{1}{\Gamma(\varpi)} \int_0^t (t-\tau)^{\varpi}\, Tresinh_{\alpha,\beta}(\lambda\tau)\, d\tau = t^{\beta+\varpi-1}\, {}_2sinh_2\left(\begin{matrix}(1,1),(\beta,1)\\(\beta,\alpha),(\beta+\varpi,1)\end{matrix};\lambda t\right). \tag{4.649}$$

Proof Taking $\vartheta = 1$ in (4.629), we have the result. □

Theorem 4.92 *Let $\varpi \in \mathbb{C}$, $Re(\varpi) > 0$, $\lambda \in \mathbb{C}$, $\beta \in \mathbb{C}$, and $\alpha \in \mathbb{R}_+$. Then, we have*

$$\frac{1}{\Gamma(\varpi)} \int_0^t (t-\tau)^{\varpi}\, Trecosh_{\alpha,\beta}(\lambda\tau)\, d\tau$$
$$= t^{\beta+\varpi-1}\, {}_2cosh_2\left(\begin{matrix}(1,1),(\beta,1)\\(\beta,\alpha),(\beta+\varpi,1)\end{matrix};\lambda t\right). \tag{4.650}$$

Proof Taking $\vartheta = 1$ in (4.630), we have the result. □

Theorem 4.93 *Let $\varpi \in \mathbb{C}$, $Re(\varpi) > 0$, $\lambda \in \mathbb{C}$, $\beta \in \mathbb{C}$, and $\alpha \in \mathbb{R}_+$. Then, we have*

$$\frac{1}{\Gamma(\varpi)} \int_0^1 (1-\tau)^{\varpi}\, \tau^{\beta-1}\, {}_1\varphi_1\left(\begin{matrix}(1,1)\\(\beta,\alpha)\end{matrix};\lambda\tau\right) d\tau = {}_2\varphi_2\left(\begin{matrix}(1,1),(\beta,1)\\(\beta,\alpha),(\beta+\varpi,1)\end{matrix};\lambda\right). \tag{4.651}$$

Proof Taking $\vartheta = 1$ in (4.631), we have the result. □

Theorem 4.94 *Let $\varpi \in \mathbb{C}$, $Re(\varpi) > 0$, $\lambda \in \mathbb{C}$, $\beta \in \mathbb{C}$, and $\alpha \in \mathbb{R}_+$. Then, we have*

$$\frac{1}{\Gamma(\varpi)} \int_0^1 (1-\tau)^{\varpi}\, Tresin_{\alpha,\beta}(\lambda\tau)\, d\tau = {}_2sin_2\left(\begin{matrix}(1,1),(\beta,1)\\(\beta,\alpha),(\beta+\varpi,1)\end{matrix};\lambda\right). \tag{4.652}$$

Proof Taking $\vartheta = 1$ in (4.632), we have the result. □

Theorem 4.95 *Let $\varpi \in \mathbb{C}$, $Re(\varpi) > 0$, $\lambda \in \mathbb{C}$, $\beta \in \mathbb{C}$, and $\alpha \in \mathbb{R}_+$. Then, we have*

$$\frac{1}{\Gamma(\varpi)} \int_0^1 (1-\tau)^{\varpi}\, Trecos_{\alpha,\beta}(\lambda\tau)\, d\tau = {}_2cos_2\left(\begin{matrix}(1,1),(\beta,1)\\(\beta,\alpha),(\beta+\varpi,1)\end{matrix};\lambda\right). \tag{4.653}$$

Proof Taking $\vartheta = 1$ in (4.633), we have the result. □

Theorem 4.96 *Let $\varpi \in \mathbb{C}$, $Re(\varpi) > 0$, $\lambda \in \mathbb{C}$, $\beta \in \mathbb{C}$, and $\alpha \in \mathbb{R}_+$. Then, we have*

$$\frac{1}{\Gamma(\varpi)} \int_0^1 (1-\tau)^{\varpi}\, Tresinh_{\alpha,\beta}(\lambda\tau)\, d\tau = {}_2sinh_2\left(\begin{matrix} (1,1),(\beta,1) \\ (\beta,\alpha),(\beta+\varpi,1) \end{matrix} ; \lambda \right). \tag{4.654}$$

Proof Taking $\vartheta = 1$ in (4.634), we have the result. □

Theorem 4.97 *Let $\varpi \in \mathbb{C}$, $Re(\varpi) > 0$, $\lambda \in \mathbb{C}$, $\beta \in \mathbb{C}$, and $\alpha \in \mathbb{R}_+$. Then, we have*

$$\frac{1}{\Gamma(\varpi)} \int_0^1 (1-\tau)^{\varpi}\, Trecosh_{\alpha,\beta}(\lambda\tau)\, d\tau = {}_2cosh_2\left(\begin{matrix} (1,1),(\beta,1) \\ (\beta,\alpha),(\beta+\varpi,1) \end{matrix} ; \lambda \right). \tag{4.655}$$

Proof Taking $\vartheta = 1$ in (4.636), we have the result. □

Theorem 4.98 *Let $\varpi \in \mathbb{C}$, $Re(\varpi) > 0$, $\lambda \in \mathbb{C}$, $\beta \in \mathbb{C}$, and $\alpha \in \mathbb{R}_+$. Then, we have*

$$\begin{aligned} &\int_{-1}^{1} (1-\tau)^{\varpi} (1+\tau)^{\beta-1}\, {}_1\varphi_1\left(\begin{matrix} (1,1) \\ (\beta,\alpha) \end{matrix} ; \lambda(1+\tau) \right) d\tau \\ &= \int_{-1}^{1} (1+\tau)^{\varpi} (1-\tau)^{\beta-1}\, {}_1\varphi_1\left(\begin{matrix} (1,1) \\ (\beta,\alpha) \end{matrix} ; \lambda(1-\tau) \right) d\tau \end{aligned} \tag{4.656}$$

and

$$\begin{aligned} &\frac{1}{\Gamma(\varpi)} \int_0^2 (2-\tau)^{\varpi} \tau^{\beta-1}\, {}_1\varphi_1\left(\begin{matrix} (1,1) \\ (\beta,\alpha) \end{matrix} ; \lambda\tau \right) d\tau \\ &= \frac{1}{\Gamma(\varpi)} \int_{-1}^{1} (1-\tau)^{\varpi} (1+\tau)^{\beta-1}\, {}_1\varphi_1\left(\begin{matrix} (1,1) \\ (\beta,\alpha) \end{matrix} ; \lambda(1+\tau) \right) d\tau \\ &= \frac{1}{\Gamma(\varpi)} \int_{-1}^{1} (1+\tau)^{\varpi} (1-\tau)^{\beta-1}\, {}_1\varphi_1\left(\begin{matrix} (1,1) \\ (\beta,\alpha) \end{matrix} ; \lambda(1-\tau) \right) d\tau \\ &= 2^{\beta+\varpi-1}\, {}_2\varphi_2\left(\begin{matrix} (1,1),(\beta,1) \\ (\beta,\alpha),(\beta+\varpi,1) \end{matrix} ; 2\lambda \right). \end{aligned} \tag{4.657}$$

Proof Taking $\vartheta = 1$ in (4.636) and (4.637), we have the result. □

Theorem 4.99 *Let $\varpi \in \mathbb{C}$, $Re(\varpi) > 0$, $\lambda \in \mathbb{C}$, $\beta \in \mathbb{C}$, and $\alpha \in \mathbb{R}_+$. Then, we have*

$$\int_{-1}^{1} (1-\tau)^{\varpi}\, Tresin_{\alpha,\beta}(\lambda(1+\tau))\, d\tau = \int_{-1}^{1} (1+\tau)^{\varpi}\, Tresin_{\alpha,\beta}(\lambda(1-\tau))\, d\tau \tag{4.658}$$

and

$$\begin{aligned} &\frac{1}{\Gamma(\varpi)} \int_{0}^{2} (2-\tau)^{\varpi}\, Tresin_{\alpha,\beta}(\lambda\tau)\, d\tau \\ &= \frac{1}{\Gamma(\varpi)} \int_{-1}^{1} (1-\tau)^{\varpi}\, Tresin_{\alpha,\beta}(\lambda(1+\tau))\, d\tau \\ &= \frac{1}{\Gamma(\varpi)} \int_{-1}^{1} (1+\tau)^{\varpi}\, Tresin_{\alpha,\beta}(\lambda(1-\tau))\, d\tau \\ &= 2^{\beta+\varpi-1}\, {}_2sin_2\left(\begin{matrix} (1,1),(\beta,1) \\ (\beta,\alpha),(\beta+\varpi,1) \end{matrix} ; 2\lambda \right). \end{aligned} \tag{4.659}$$

Proof Taking $\vartheta = 1$ in (4.638) and (4.639), we have the result. □

Theorem 4.100 *Let $\varpi \in \mathbb{C}$, $Re(\varpi) > 0$, $\lambda \in \mathbb{C}$, $\beta \in \mathbb{C}$, and $\alpha \in \mathbb{R}_+$. Then, we have*

$$\int_{-1}^{1} (1-\tau)^{\varpi}\, Trecos_{\alpha,\beta}(\lambda(1+\tau))\, d\tau = \int_{-1}^{1} (1+\tau)^{\varpi}\, Trecos_{\alpha,\beta}(\lambda(1-\tau))\, d\tau \tag{4.660}$$

and

$$\begin{aligned} &\frac{1}{\Gamma(\varpi)} \int_{0}^{2} (2-\tau)^{\varpi}\, Trecos_{\alpha,\beta}(\lambda\tau)\, d\tau \\ &= \frac{1}{\Gamma(\varpi)} \int_{-1}^{1} (1-\tau)^{\varpi}\, Trecos_{\alpha,\beta}(\lambda(1+\tau))\, d\tau \\ &= \frac{1}{\Gamma(\varpi)} \int_{-1}^{1} (1+\tau)^{\varpi}\, Trecos_{\alpha,\beta}(\lambda(1-\tau))\, d\tau \\ &= 2^{\beta+\varpi-1}\, {}_2cos_2\left(\begin{matrix} (1,1),(\beta,1) \\ (\beta,\alpha),(\beta+\varpi,1) \end{matrix} ; 2\lambda \right). \end{aligned} \tag{4.661}$$

Proof Taking $\vartheta = 1$ in (4.640) and (4.641), we have the result. □

Theorem 4.101 *Let $\varpi \in \mathbb{C}$, $Re(\varpi) > 0$, $\lambda \in \mathbb{C}$, $\beta \in \mathbb{C}$, and $\alpha \in \mathbb{R}_{+}$. Then, we have*

$$\int_{-1}^{1} (1-\tau)^{\varpi}\, Tresinh_{\alpha,\beta}(\lambda(1+\tau))\, d\tau = \int_{-1}^{1} (1+\tau)^{\varpi}\, Tresinh_{\alpha,\beta}(\lambda(1-\tau))\, d\tau \tag{4.662}$$

and

$$\begin{aligned} &\frac{1}{\Gamma(\varpi)} \int_{0}^{2} (2-\tau)^{\varpi}\, Tresinh_{\alpha,\beta}(\lambda\tau)\, d\tau \\ &= \frac{1}{\Gamma(\varpi)} \int_{-1}^{1} (1-\tau)^{\varpi}\, Tresinh_{\alpha,\beta}(\lambda(1+\tau))\, d\tau \\ &= \frac{1}{\Gamma(\varpi)} \int_{-1}^{1} (1+\tau)^{\varpi}\, Tresinh_{\alpha,\beta}(\lambda(1-\tau))\, d\tau \\ &= 2^{\beta+\varpi-1}\, {}_2sinh_2\left(\begin{matrix} (1,1),(\beta,1) \\ (\beta,\alpha),(\beta+\varpi,1) \end{matrix} ; 2\lambda \right). \end{aligned} \tag{4.663}$$

Proof Taking $\vartheta = 1$ in (4.642) and (4.643), we have the result. □

Theorem 4.102 *Let $\varpi \in \mathbb{C}$, $Re(\varpi) > 0$, $\lambda \in \mathbb{C}$, $\beta \in \mathbb{C}$, and $\alpha \in \mathbb{R}_{+}$. Then, we have*

$$\int_{-1}^{1} (1-\tau)^{\varpi}\, Trecosh_{\alpha,\beta}(\lambda(1+\tau))\, d\tau = \int_{-1}^{1} (1+\tau)^{\varpi}\, Trecosh_{\alpha,\beta}(\lambda(1-\tau))\, d\tau \tag{4.664}$$

and

$$\begin{aligned} &\frac{1}{\Gamma(\varpi)} \int_{0}^{2} (2-\tau)^{\varpi}\, Trecosh_{\alpha,\beta}(\lambda\tau)\, d\tau \\ &= \frac{1}{\Gamma(\varpi)} \int_{-1}^{1} (1-\tau)^{\varpi}\, Trecosh_{\alpha,\beta}(\lambda(1+\tau))\, d\tau \\ &= \frac{1}{\Gamma(\varpi)} \int_{-1}^{1} (1+\tau)^{\varpi}\, Trecosh_{\alpha,\beta}(\lambda(1-\tau))\, d\tau \\ &= 2^{\beta+\varpi-1}\, {}_2cosh_2\left(\begin{matrix} (1,1),(\beta,1) \\ (\beta,\alpha),(\beta+\varpi,1) \end{matrix} ; 2\lambda \right). \end{aligned} \tag{4.665}$$

Proof Taking $\vartheta = 1$ in (4.644) and (4.645), we have the result. □

4.3 The Truncated Wiman Functions and Related Functions

In this section, we propose the truncated Wiman functions and some truncated functions related to the Wiman functions.

Definition 4.98 The truncated Wiman function is defined as

$$\left\{E_{\alpha,\beta}\right\}^{n}(t)=\sum_{\kappa=0}^{n}\frac{t^{\kappa}}{\Gamma(\kappa\alpha+\beta)}, \tag{4.666}$$

where $t,\alpha,\beta\in\mathbb{C}$, $n,\kappa\in\mathbb{N}_0$, $Re(\alpha)>0$, and $Re(\beta)>0$.

The special cases for the truncated Wiman function are given as follows:

$$\left\{E_{\alpha,\beta}\right\}^{n}\left(t^{\alpha}\right)=\sum_{\kappa=0}^{n}\frac{t^{\kappa\alpha}}{\Gamma(\kappa\alpha+\beta)} \tag{4.667}$$

or

$$\left\{E_{\alpha,\beta}\right\}^{n}\left(\lambda t^{\gamma}\right)=\sum_{\kappa=0}^{n}\frac{\lambda^{\kappa}t^{\kappa\gamma}}{\Gamma(\kappa\alpha+\beta)} \tag{4.668}$$

where $t,\alpha,\beta,\gamma\in\mathbb{C}$, $n,\kappa\in\mathbb{N}_0$, $Re(\alpha)>0$, and $Re(\beta)>0$.

Definition 4.99 The truncated supersine via Wiman function is defined as

$$\begin{aligned}&\left\{supersin_{\alpha,\beta}\right\}^{n}\left(t^{\alpha}\right)\\&=\frac{t^{\beta-1}}{2i}\left(\left\{E_{\alpha,\beta}\right\}^{n}\left(it^{\alpha}\right)-\left\{E_{\alpha,\beta}\right\}^{n}\left(-it^{\alpha}\right)\right)\\&=\sum_{\kappa=0}^{n}\frac{(-1)^{n}t^{(2\kappa+1)\alpha}}{\Gamma((2\kappa+1)\alpha+\beta)},\end{aligned} \tag{4.669}$$

where $t,\alpha,\beta\in\mathbb{C}$, $n,\kappa\in\mathbb{N}_0$, $Re(\alpha)>0$, and $Re(\beta)>0$.

So, we present

$$\begin{aligned}&\left\{supersin_{\alpha,\beta}\right\}^{0}\left(t^{\alpha}\right)\\&=\frac{t^{\beta-1}}{2}\left(\left\{E_{\alpha,\beta}\right\}^{0}\left(it^{\alpha}\right)-\left\{E_{\alpha,\beta}\right\}^{0}\left(-it^{\alpha}\right)\right)\\&=\frac{t^{\beta+\alpha-1}}{\Gamma(\alpha+\beta)},\end{aligned} \tag{4.670}$$

$$\begin{aligned}&\left\{supersin_{\alpha,\beta}\right\}^{1}\left(t^{\alpha}\right)\\&=\frac{t^{\beta-1}}{2}\left(\left\{E_{\alpha,\beta}\right\}^{1}\left(it^{\alpha}\right)-\left\{E_{\alpha,\beta}\right\}^{1}\left(-it^{\alpha}\right)\right)\\&=t^{\beta-1}\sum_{\kappa=0}^{1}\frac{(-1)^{\kappa}t^{(2\kappa+1)\alpha}}{\Gamma((2\kappa+1)\alpha+\beta)},\end{aligned} \tag{4.671}$$

and

$$\begin{aligned}&\left\{supersinh_{\alpha,\beta}\right\}^2(t^\alpha)\\&\quad=\frac{t^{\beta-1}}{2}\left(\left\{E_{\alpha,\beta}\right\}^2(t^\alpha)-\left\{E_{\alpha,\beta}\right\}^2(-t^\alpha)\right)\\&\quad=t^{\beta-1}\sum_{\kappa=0}^{2}\frac{(-1)^\kappa t^{(2\kappa+1)\alpha}}{\Gamma((2\kappa+1)\alpha+\beta)},\end{aligned}\tag{4.672}$$

where $t, \alpha, \beta \in \mathbb{C}$, $Re(\alpha) > 0$, and $Re(\beta) > 0$.

Definition 4.100 The truncated supercosine via Wiman function is defined as

$$\begin{aligned}&\left\{supercos_{\alpha,\beta}\right\}^n(t^\alpha)\\&\quad=\frac{t^{\beta-1}}{2}\left(\left\{E_{\alpha,\beta}\right\}^n(it^\alpha)+\left\{E_{\alpha,\beta}\right\}^n(-it^\alpha)\right)\\&\quad=\sum_{\kappa=0}^{n}\frac{(-1)^n t^{2\kappa\alpha}}{\Gamma(2\kappa\alpha+\beta)},\end{aligned}\tag{4.673}$$

where $t, \alpha, \beta \in \mathbb{C}$, $n, \kappa \in \mathbb{N}_0$, $Re(\alpha) > 0$, and $Re(\beta) > 0$.

Thus, we give

$$\begin{aligned}&\left\{supercos_{\alpha,\beta}\right\}^0(t^\alpha)\\&\quad=\frac{t^{\beta-1}}{2}\left(\left\{E_{\alpha,\beta}\right\}^0(it^\alpha)+\left\{E_{\alpha,\beta}\right\}^0(-it^\alpha)\right)\\&\quad=\frac{t^{\beta-1}}{\Gamma(\beta)},\end{aligned}\tag{4.674}$$

$$\begin{aligned}&\left\{supercos_{\alpha,\beta}\right\}^1(t^\alpha)\\&\quad=\frac{t^{\beta-1}}{2}\left(\left\{E_{\alpha,\beta}\right\}^1(it^\alpha)+\left\{E_{\alpha,\beta}\right\}^1(-it^\alpha)\right)\\&\quad=t^{\beta-1}\sum_{\kappa=0}^{1}\frac{(-1)^\kappa t^{2\kappa\alpha}}{\Gamma(2\kappa\alpha+\beta)},\end{aligned}\tag{4.675}$$

and

$$\begin{aligned}&\left\{supercos_{\alpha,\beta}\right\}^2(t^\alpha)\\&\quad=\frac{t^{\beta-1}}{2}\left(\left\{E_{\alpha,\beta}\right\}^2(it^\alpha)+\left\{E_{\alpha,\beta}\right\}^2(-it^\alpha)\right)\\&\quad=t^{\beta-1}\sum_{\kappa=0}^{2}\frac{(-1)^\kappa t^{2\kappa\alpha}}{\Gamma(2\kappa\alpha+\beta)},\end{aligned}\tag{4.676}$$

where $t, \alpha, \beta \in \mathbb{C}$, $Re(\alpha) > 0$, and $Re(\beta) > 0$.

Definition 4.101 The truncated hyperbolic supersine via Wiman function is defined as

$$\begin{aligned}&\left\{supersinh_{\alpha,\beta}\right\}^{n}(t^{\alpha})\\&=\tfrac{t^{\beta-1}}{2}\left(\left\{E_{\alpha,\beta}\right\}^{n}(t^{\alpha})-\left\{E_{\alpha,\beta}\right\}^{n}(-t^{\alpha})\right)\\&=t^{\beta-1}\sum_{\kappa=0}^{n}\tfrac{t^{(2\kappa+1)\alpha}}{\Gamma((2\kappa+1)\alpha+\beta)},\end{aligned} \tag{4.677}$$

where $t,\alpha,\beta\in\mathbb{C}$, $n,\kappa\in\mathbb{N}_0$, $Re\,(\alpha)>0$, and $Re\,(\beta)>0$.

So, we present

$$\begin{aligned}&\left\{supersinh_{\alpha,\beta}\right\}^{0}(t^{\alpha})\\&=\tfrac{t^{\beta-1}}{2}\left(\left\{E_{\alpha,\beta}\right\}^{0}(t^{\alpha})-\left\{E_{\alpha,\beta}\right\}^{0}(-t^{\alpha})\right)\\&=\tfrac{t^{\beta+\alpha-1}}{\Gamma(\alpha+\beta)},\end{aligned} \tag{4.678}$$

$$\begin{aligned}&\left\{supersinh_{\alpha,\beta}\right\}^{1}(t^{\alpha})\\&=\tfrac{t^{\beta-1}}{2}\left(\left\{E_{\alpha,\beta}\right\}^{1}(t^{\alpha})-\left\{E_{\alpha,\beta}\right\}^{1}(-t^{\alpha})\right)\\&=t^{\beta-1}\sum_{\kappa=0}^{1}\tfrac{t^{(2\kappa+1)\alpha}}{\Gamma((2\kappa+1)\alpha+\beta)},\end{aligned} \tag{4.679}$$

and

$$\begin{aligned}&\left\{supersinh_{\alpha,\beta}\right\}^{2}(t^{\alpha})\\&=\tfrac{t^{\beta-1}}{2}\left(\left\{E_{\alpha,\beta}\right\}^{2}(t^{\alpha})-\left\{E_{\alpha,\beta}\right\}^{2}(-t^{\alpha})\right)\\&=t^{\beta-1}\sum_{\kappa=0}^{2}\tfrac{t^{(2\kappa+1)\alpha}}{\Gamma((2\kappa+1)\alpha+\beta)},\end{aligned} \tag{4.680}$$

where $t,\alpha,\beta\in\mathbb{C}$, $Re\,(\alpha)>0$, and $Re\,(\beta)>0$.

Definition 4.102 The truncated hyperbolic supercosine via Wiman function is defined as

$$\begin{aligned}&\left\{supercosh_{\alpha,\beta}\right\}^{n}(t^{\alpha})\\&=\tfrac{t^{\beta-1}}{2}\left(\left\{E_{\alpha,\beta}\right\}^{n}(t^{\alpha})+\left\{E_{\alpha,\beta}\right\}^{n}(-t^{\alpha})\right)\\&=t^{\beta-1}\sum_{\kappa=0}^{n}\tfrac{t^{2\kappa\alpha}}{\Gamma(2\kappa\alpha+\beta)},\end{aligned} \tag{4.681}$$

where $t,\alpha,\beta\in\mathbb{C}$, $n,\kappa\in\mathbb{N}_0$, $Re\,(\alpha)>0$, and $Re\,(\beta)>0$.

Therefore, we have that

$$\begin{aligned}&\left\{supercosh_{\alpha,\beta}\right\}^{0}(t^{\alpha})\\&\quad=\tfrac{t^{\beta-1}}{2}\left(\left\{E_{\alpha,\beta}\right\}^{0}(t^{\alpha})+\left\{E_{\alpha,\beta}\right\}^{0}(-t^{\alpha})\right)\\&\quad=\tfrac{t^{\beta-1}}{\Gamma(\beta)},\end{aligned}\tag{4.682}$$

$$\begin{aligned}&\left\{supercosh_{\alpha,\beta}\right\}^{1}(t^{\alpha})\\&\quad=\tfrac{t^{\beta-1}}{2}\left(\left\{E_{\alpha,\beta}\right\}^{1}(t^{\alpha})+\left\{E_{\alpha,\beta}\right\}^{1}(-t^{\alpha})\right)\\&\quad=t^{\beta-1}\sum_{\kappa=0}^{1}\tfrac{t^{2\kappa\alpha}}{\Gamma(2\kappa\alpha+\beta)},\end{aligned}\tag{4.683}$$

and

$$\begin{aligned}&\left\{supercosh_{\alpha,\beta}\right\}^{2}(t^{\alpha})\\&\quad=\tfrac{t^{\beta-1}}{2}\left(\left\{E_{\alpha,\beta}\right\}^{2}(t^{\alpha})+\left\{E_{\alpha,\beta}\right\}^{2}(-t^{\alpha})\right)\\&\quad=t^{\beta-1}\sum_{\kappa=0}^{2}\tfrac{t^{2\kappa\alpha}}{\Gamma(2\kappa\alpha+\beta)},\end{aligned}\tag{4.684}$$

where $t, \alpha, \beta \in \mathbb{C}$, $Re(\alpha) > 0$, and $Re(\beta) > 0$.

Here, we obtain that

$$\left\{supersin_{\alpha,\beta}\right\}^{n}(\lambda t^{\alpha})=\frac{t^{\beta-1}}{2i}\left(\left\{E_{\alpha,\beta}\right\}^{n}(i\lambda t^{\alpha})-\left\{E_{\alpha,\beta}\right\}^{n}(-i\lambda t^{\alpha})\right),\tag{4.685}$$

$$\left\{supercos_{\alpha,\beta}\right\}^{n}(\lambda t^{\alpha})=\frac{t^{\beta-1}}{2}\left(\left\{E_{\alpha,\beta}\right\}^{n}(i\lambda t^{\alpha})+\left\{E_{\alpha,\beta}\right\}^{n}(-i\lambda t^{\alpha})\right),\tag{4.686}$$

$$\left\{supersinh_{\alpha,\beta}\right\}^{n}(\lambda t^{\alpha})=\frac{t^{\beta-1}}{2}\left(\left\{E_{\alpha,\beta}\right\}^{n}(\lambda t^{\alpha})-\left\{E_{\alpha,\beta}\right\}^{n}(-\lambda t^{\alpha})\right),\tag{4.687}$$

and

$$\left\{supercosh_{\alpha,\beta}\right\}^{n}(\lambda t^{\alpha})=\frac{t^{\beta-1}}{2}\left(\left\{E_{\alpha,\beta}\right\}^{n}(\lambda t^{\alpha})+\left\{E_{\alpha,\beta}\right\}^{n}(-\lambda t^{\alpha})\right),\tag{4.688}$$

where $t, \alpha, \beta, \lambda \in \mathbb{C}$, $n, \kappa \in \mathbb{N}_0$, $Re(\alpha) > 0$, and $Re(\beta) > 0$.

Definition 4.103 The truncated presine via Wiman function is defined as

$$\begin{aligned}&\left\{presin_{\alpha,\beta}\right\}^{n}(t)\\&\quad=\tfrac{1}{2i}\left(\left\{E_{\alpha,\beta}\right\}^{n}(it)-\left\{E_{\alpha,\beta}\right\}^{n}(-it)\right)\\&\quad=\sum_{\kappa=0}^{n}\tfrac{(-1)^{n}t^{2\kappa+1}}{\Gamma((2\kappa+1)\alpha+\beta)},\end{aligned}\tag{4.689}$$

where $t, \alpha, \beta \in \mathbb{C}$, $n, \kappa \in \mathbb{N}_0$, $Re(\alpha) > 0$, and $Re(\beta) > 0$.

So, we have that

$$\begin{aligned}
&\left\{presin_{\alpha,\beta}\right\}^{0}(t)\\
&= \tfrac{1}{2}\left(\left\{E_{\alpha,\beta}\right\}^{0}(it) - \left\{E_{\alpha,\beta}\right\}^{0}(-it)\right)\\
&= \tfrac{t}{\Gamma(1+\beta)},
\end{aligned} \tag{4.690}$$

$$\begin{aligned}
&\left\{presin_{\alpha,\beta}\right\}^{1}(t)\\
&= \tfrac{1}{2}\left(\left\{E_{\alpha,\beta}\right\}^{1}(it) - \left\{E_{\alpha,\beta}\right\}^{1}(-it)\right)\\
&= \sum_{\kappa=0}^{1} \tfrac{(-1)^{\kappa} t^{2\kappa+1}}{\Gamma((2\kappa+1)\alpha+\beta)},
\end{aligned} \tag{4.691}$$

and

$$\begin{aligned}
&\left\{presinh_{\alpha,\beta}\right\}^{2}(t)\\
&= \tfrac{1}{2}\left(\left\{E_{\alpha,\beta}\right\}^{2}(it) - \left\{E_{\alpha,\beta}\right\}^{2}(-it)\right)\\
&= \sum_{\kappa=0}^{2} \tfrac{(-1)^{\kappa} t^{2\kappa+1}}{\Gamma((2\kappa+1)\alpha+\beta)},
\end{aligned} \tag{4.692}$$

where $t, \alpha, \beta \in \mathbb{C}$, $Re(\alpha) > 0$, and $Re(\beta) > 0$.

Definition 4.104 The truncated precosine via Wiman function is defined as

$$\begin{aligned}
&\left\{precos_{\alpha,\beta}\right\}^{n}(t)\\
&= \tfrac{1}{2}\left(\left\{E_{\alpha,\beta}\right\}^{n}(it) + \left\{E_{\alpha,\beta}\right\}^{n}(-it)\right)\\
&= \sum_{\kappa=0}^{n} \tfrac{(-1)^{n} t^{2\kappa}}{\Gamma(2\kappa\alpha+\beta)},
\end{aligned} \tag{4.693}$$

where $t, \alpha, \beta \in \mathbb{C}$, $n, \kappa \in \mathbb{N}_0$, $Re(\alpha) > 0$, and $Re(\beta) > 0$.

Therefore, we have that

$$\begin{aligned}
&\left\{precos_{\alpha,\beta}\right\}^{2}(t)\\
&= \tfrac{1}{2}\left(\left\{E_{\alpha,\beta}\right\}^{2}(it) + \left\{E_{\alpha,\beta}\right\}^{2}(-it)\right)\\
&= \tfrac{1}{\Gamma(\beta)},
\end{aligned} \tag{4.694}$$

$$\begin{aligned}
&\left\{precos_{\alpha,\beta}\right\}^{1}(t)\\
&= \tfrac{1}{2}\left(\left\{E_{\alpha,\beta}\right\}^{1}(it) + \left\{E_{\alpha,\beta}\right\}^{1}(-it)\right)\\
&= \sum_{\kappa=0}^{1} \tfrac{(-1)^{\kappa} t^{2\kappa}}{\Gamma(2\kappa\alpha+\beta)},
\end{aligned} \tag{4.695}$$

and

$$\begin{aligned}&\left\{precos_{\alpha,\beta}\right\}^2(t)\\&\quad=\tfrac{1}{2}\left(\left\{E_{\alpha,\beta}\right\}^2(it)+\left\{E_{\alpha,\beta}\right\}^2(-it)\right)\\&\quad=\sum_{\kappa=0}^{2}\tfrac{(-1)^{\kappa}t^{2\kappa}}{\Gamma(2\kappa\alpha+\beta)},\end{aligned} \tag{4.696}$$

where $t, \alpha, \beta \in \mathbb{C}$, $Re(\alpha) > 0$, and $Re(\beta) > 0$.

Definition 4.105 The truncated hyperbolic presine via Wiman function is defined as

$$\begin{aligned}&\left\{presinh_{\alpha,\beta}\right\}^n(t)\\&\quad=\tfrac{1}{2}\left(\left\{E_{\alpha,\beta}\right\}^n(t)-\left\{E_{\alpha,\beta}\right\}^n(-t)\right)\\&\quad=\sum_{\kappa=0}^{n}\tfrac{t^{2\kappa+1}}{\Gamma((2\kappa+1)\alpha+\beta)},\end{aligned} \tag{4.697}$$

where $t, \alpha, \beta \in \mathbb{C}$, $n, \kappa \in \mathbb{N}_0$, $Re(\alpha) > 0$, and $Re(\beta) > 0$.

So, we have that

$$\begin{aligned}&\left\{presinh_{\alpha,\beta}\right\}^0(t)\\&\quad=\tfrac{1}{2}\left(\left\{E_{\alpha,\beta}\right\}^0(t)-\left\{E_{\alpha,\beta}\right\}^0(-t)\right)\\&\quad=\tfrac{t}{\Gamma(1+\beta)},\end{aligned} \tag{4.698}$$

$$\begin{aligned}&\left\{presinh_{\alpha,\beta}\right\}^1(t)\\&\quad=\tfrac{1}{2}\left(\left\{E_{\alpha,\beta}\right\}^1(t)-\left\{E_{\alpha,\beta}\right\}^1(-t)\right)\\&\quad=\sum_{\kappa=0}^{1}\tfrac{t^{2\kappa+1}}{\Gamma((2\kappa+1)\alpha+\beta)},\end{aligned} \tag{4.699}$$

and

$$\begin{aligned}&\left\{presinh_{\alpha,\beta}\right\}^2(t)\\&\quad=\tfrac{1}{2}\left(\left\{E_{\alpha,\beta}\right\}^2(t)-\left\{E_{\alpha,\beta}\right\}^2(-t)\right)\\&\quad=\sum_{\kappa=0}^{2}\tfrac{t^{2\kappa+1}}{\Gamma((2\kappa+1)\alpha+\beta)},\end{aligned} \tag{4.700}$$

where $t, \alpha, \beta \in \mathbb{C}$, $Re(\alpha) > 0$, and $Re(\beta) > 0$.

Definition 4.106 The truncated hyperbolic precosine via Wiman function is defined as

$$\begin{aligned}&\left\{precosh_{\alpha,\beta}\right\}^{n}(t)\\&=\tfrac{1}{2}\left(\left\{E_{\alpha,\beta}\right\}^{n}(t)+\left\{E_{\alpha,\beta}\right\}^{n}(-t)\right)\\&=\sum_{\kappa=0}^{n}\tfrac{t^{2\kappa}}{\Gamma(2\kappa\alpha+\beta)},\end{aligned} \tag{4.701}$$

where $t,\alpha,\beta\in\mathbb{C}$, $n,\kappa\in\mathbb{N}_0$, $Re(\alpha)>0$, and $Re(\beta)>0$.

Thus, we present

$$\begin{aligned}&\left\{precosh_{\alpha,\beta}\right\}^{0}(t)\\&=\tfrac{1}{2}\left(\left\{E_{\alpha,\beta}\right\}^{0}(t)+\left\{E_{\alpha,\beta}\right\}^{0}(-t)\right)\\&=\tfrac{1}{\Gamma(\beta)},\end{aligned} \tag{4.702}$$

$$\begin{aligned}&\left\{precosh_{\alpha,\beta}\right\}^{1}(t)\\&=\tfrac{1}{2}\left(\left\{E_{\alpha,\beta}\right\}^{1}(t)+\left\{E_{\alpha,\beta}\right\}^{1}(-t)\right)\\&=\sum_{\kappa=0}^{1}\tfrac{t^{2\kappa}}{\Gamma(2\kappa\alpha+\beta)},\end{aligned} \tag{4.703}$$

and

$$\begin{aligned}&\left\{precosh_{\alpha,\beta}\right\}^{2}(t)\\&=\tfrac{1}{2}\left(\left\{E_{\alpha,\beta}\right\}^{2}(t)+\left\{E_{\alpha,\beta}\right\}^{2}(-t)\right)\\&=\sum_{\kappa=0}^{2}\tfrac{t^{2\kappa}}{\Gamma(2\kappa\alpha+\beta)},\end{aligned} \tag{4.704}$$

where $t,\alpha,\beta\in\mathbb{C}$, $n,\kappa\in\mathbb{N}_0$, $Re(\alpha)>0$, and $Re(\beta)>0$.

So, it is clear that

$$\left\{presin_{\alpha,\beta}\right\}^{n}(\lambda t)=\frac{1}{2i}\left(\left\{E_{\alpha,\beta}\right\}^{n}(i\lambda t)-\left\{E_{\alpha,\beta}\right\}^{n}(-i\lambda t)\right), \tag{4.705}$$

$$\left\{precos_{\alpha,\beta}\right\}^{n}(\lambda t)=\frac{1}{2}\left(\left\{E_{\alpha,\beta}\right\}^{n}(i\lambda t)+\left\{E_{\alpha,\beta}\right\}^{n}(-i\lambda t)\right), \tag{4.706}$$

$$\left\{presinh_{\alpha,\beta}\right\}^{n}(\lambda t)=\frac{1}{2}\left(\left\{E_{\alpha,\beta}\right\}^{n}(\lambda t)-\left\{E_{\alpha,\beta}\right\}^{n}(-\lambda t)\right), \tag{4.707}$$

and

$$\left\{precosh_{\alpha,\beta}\right\}^n(\lambda t)=\frac{1}{2}\left(\left\{E_{\alpha,\beta}\right\}^n(\lambda t)+\left\{E_{\alpha,\beta}\right\}^n(-\lambda t)\right) \quad (4.708)$$

where $t, \alpha, \beta, \lambda \in \mathbb{C}$, $n, \kappa \in \mathbb{N}_0$, $Re(\alpha) > 0$, and $Re(\beta) > 0$.

Definition 4.107 The truncated presine via Wiman function is defined as

$$\begin{aligned}&\left\{presin_{\alpha,\beta}\right\}^n(t^\gamma)\\&=\tfrac{1}{2i}\left(\left\{E_{\alpha,\beta}\right\}^n(it^\gamma)-\left\{E_{\alpha,\beta}\right\}^n(-it^\gamma)\right)\\&=\sum_{\kappa=0}^{n}\tfrac{(-1)^n t^{(2\kappa+1)\gamma}}{\Gamma((2\kappa+1)\alpha+\beta)},\end{aligned} \quad (4.709)$$

where $t, \alpha, \beta, \gamma \in \mathbb{C}$, $n, \kappa \in \mathbb{N}_0$, $Re(\alpha) > 0$, and $Re(\beta) > 0$.

So, we have that

$$\begin{aligned}&\left\{presin_{\alpha,\beta}\right\}^0(t^\gamma)\\&=\tfrac{1}{2}\left(\left\{E_{\alpha,\beta}\right\}^0(it^\gamma)-\left\{E_{\alpha,\beta}\right\}^0(-it^\gamma)\right)\\&=\tfrac{t^\gamma}{\Gamma(1+\beta)},\end{aligned} \quad (4.710)$$

$$\begin{aligned}&\left\{presin_{\alpha,\beta}\right\}^1(t^\gamma)\\&=\tfrac{1}{2}\left(\left\{E_{\alpha,\beta}\right\}^1(it^\gamma)-\left\{E_{\alpha,\beta}\right\}^1(-it^\gamma)\right)\\&=\sum_{\kappa=0}^{1}\tfrac{(-1)^\kappa t^{(2\kappa+1)\gamma}}{\Gamma((2\kappa+1)\alpha+\beta)},\end{aligned} \quad (4.711)$$

and

$$\begin{aligned}&\left\{presinh_{\alpha,\beta}\right\}^2(t^\gamma)\\&=\tfrac{1}{2}\left(\left\{E_{\alpha,\beta}\right\}^2(it^\gamma)-\left\{E_{\alpha,\beta}\right\}^2(-it^\gamma)\right)\\&=\sum_{\kappa=0}^{2}\tfrac{(-1)^\kappa t^{(2\kappa+1)\gamma}}{\Gamma((2\kappa+1)\alpha+\beta)},\end{aligned} \quad (4.712)$$

where $t, \alpha, \beta, \gamma \in \mathbb{C}$, $Re(\alpha) > 0$, and $Re(\beta) > 0$.

Definition 4.108 The truncated precosine via Wiman function is defined as

$$\begin{aligned}&\left\{precos_{\alpha,\beta}\right\}^n(t^\gamma)\\&=\tfrac{1}{2}\left(\left\{E_{\alpha,\beta}\right\}^n(it^\gamma)+\left\{E_{\alpha,\beta}\right\}^n(-it^\gamma)\right)\\&=\sum_{\kappa=0}^{n}\tfrac{(-1)^n t^{2\kappa\gamma}}{\Gamma(2\kappa\alpha+\beta)},\end{aligned} \quad (4.713)$$

where $t, \alpha, \beta, \gamma \in \mathbb{C}$, $n, \kappa \in \mathbb{N}_0$, $Re(\alpha) > 0$, and $Re(\beta) > 0$.

Therefore, we have that

$$\begin{aligned}&\left\{precos_{\alpha,\beta}\right\}^{2}(t^{\gamma})\\&=\tfrac{1}{2}\left(\left\{E_{\alpha,\beta}\right\}^{2}(it^{\gamma})+\left\{E_{\alpha,\beta}\right\}^{2}(-it^{\gamma})\right)\\&=\tfrac{1}{\Gamma(\beta)},\end{aligned}\tag{4.714}$$

$$\begin{aligned}&\left\{precos_{\alpha,\beta}\right\}^{1}(t^{\gamma})\\&=\tfrac{1}{2}\left(\left\{E_{\alpha,\beta}\right\}^{1}(it^{\gamma})+\left\{E_{\alpha,\beta}\right\}^{1}(-it^{\gamma})\right)\\&=\sum_{\kappa=0}^{1}\tfrac{(-1)^{\kappa}t^{2\kappa\gamma}}{\Gamma(2\kappa\alpha+\beta)},\end{aligned}\tag{4.715}$$

and

$$\begin{aligned}&\left\{precos_{\alpha,\beta}\right\}^{2}(t^{\gamma})\\&=\tfrac{1}{2}\left(\left\{E_{\alpha,\beta}\right\}^{2}(it^{\gamma})+\left\{E_{\alpha,\beta}\right\}^{2}(-it^{\gamma})\right)\\&=\sum_{\kappa=0}^{2}\tfrac{(-1)^{\kappa}t^{2\kappa\gamma}}{\Gamma(2\kappa\alpha+\beta)},\end{aligned}\tag{4.716}$$

where $t, \alpha, \beta, \gamma \in \mathbb{C}$, $Re(\alpha) > 0$, and $Re(\beta) > 0$.

Definition 4.109 The truncated hyperbolic presine via Wiman function is defined as

$$\begin{aligned}&\left\{presinh_{\alpha,\beta}\right\}^{n}(t^{\gamma})\\&=\tfrac{1}{2}\left(\left\{E_{\alpha,\beta}\right\}^{n}(t^{\gamma})-\left\{E_{\alpha,\beta}\right\}^{n}(-t^{\gamma})\right)\\&=\sum_{\kappa=0}^{n}\tfrac{t^{(2\kappa+1)\gamma}}{\Gamma((2\kappa+1)\alpha+\beta)},\end{aligned}\tag{4.717}$$

where $t, \alpha, \beta, \gamma \in \mathbb{C}$, $n, \kappa \in \mathbb{N}_0$, $Re(\alpha) > 0$, and $Re(\beta) > 0$.

So, we have that

$$\begin{aligned}&\left\{presinh_{\alpha,\beta}\right\}^{0}(t^{\gamma})\\&=\tfrac{1}{2}\left(\left\{E_{\alpha,\beta}\right\}^{0}(t^{\gamma})-\left\{E_{\alpha,\beta}\right\}^{0}(-t^{\gamma})\right)\\&=\tfrac{t^{\gamma}}{\Gamma(1+\beta)},\end{aligned}\tag{4.718}$$

$$\begin{aligned}&\left\{presinh_{\alpha,\beta}\right\}^{1}(t^{\gamma})\\&=\tfrac{1}{2}\left(\left\{E_{\alpha,\beta}\right\}^{1}(t^{\gamma})-\left\{E_{\alpha,\beta}\right\}^{1}(-t^{\gamma})\right)\\&=\sum_{\kappa=0}^{1}\tfrac{t^{(2\kappa+1)\gamma}}{\Gamma((2\kappa+1)\alpha+\beta)},\end{aligned}\tag{4.719}$$

and

$$\begin{aligned}\{presinh_{\alpha,\beta}\}^2(t^\gamma) \\ &= \tfrac{1}{2}\left(\{E_{\alpha,\beta}\}^2(t^\gamma) - \{E_{\alpha,\beta}\}^2(-t^\gamma)\right) \\ &= \sum_{\kappa=0}^{2} \frac{t^{(2\kappa+1)\gamma}}{\Gamma((2\kappa+1)\alpha+\beta)},\end{aligned} \tag{4.720}$$

where $t, \alpha, \beta, \gamma \in \mathbb{C}$, $Re(\alpha) > 0$, and $Re(\beta) > 0$.

Definition 4.110 The truncated hyperbolic precosine via Wiman function is defined as

$$\begin{aligned}\{precosh_{\alpha,\beta}\}^n(t^\gamma) \\ &= \tfrac{1}{2}\left(\{E_{\alpha,\beta}\}^n(t^\gamma) + \{E_{\alpha,\beta}\}^n(-t^\gamma)\right) \\ &= \sum_{\kappa=0}^{n} \frac{t^{2\kappa\gamma}}{\Gamma(2\kappa\alpha+\beta)},\end{aligned} \tag{4.721}$$

where $t, \alpha, \beta, \gamma \in \mathbb{C}$, $n, \kappa \in \mathbb{N}_0$, $Re(\alpha) > 0$, and $Re(\beta) > 0$.

Thus, we present

$$\begin{aligned}\{precosh_{\alpha,\beta}\}^0(t^\gamma) \\ &= \tfrac{1}{2}\left(\{E_{\alpha,\beta}\}^0(t^\gamma) + \{E_{\alpha,\beta}\}^0(-t^\gamma)\right) \\ &= \frac{1}{\Gamma(\beta)},\end{aligned} \tag{4.722}$$

$$\begin{aligned}\{precosh_{\alpha,\beta}\}^1(t^\gamma) \\ &= \tfrac{1}{2}\left(\{E_{\alpha,\beta}\}^1(t^\gamma) + \{E_{\alpha,\beta}\}^1(-t^\gamma)\right) \\ &= \sum_{\kappa=0}^{1} \frac{t^{2\kappa\gamma}}{\Gamma(2\kappa\alpha+\beta)},\end{aligned} \tag{4.723}$$

and

$$\begin{aligned}\{precosh_{\alpha,\beta}\}^2(t^\gamma) \\ &= \tfrac{1}{2}\left(\{E_{\alpha,\beta}\}^2(t^\gamma) + \{E_{\alpha,\beta}\}^2(-t^\gamma)\right) \\ &= \sum_{\kappa=0}^{2} \frac{t^{2\kappa\gamma}}{\Gamma(2\kappa\alpha+\beta)},\end{aligned} \tag{4.724}$$

where $t, \alpha, \beta, \gamma \in \mathbb{C}$, $n, \kappa \in \mathbb{N}_0$, $Re(\alpha) > 0$, and $Re(\beta) > 0$.

So, it is clear that

$$\{presin_{\alpha,\beta}\}^n(\lambda t^\gamma) = \frac{1}{2i}\left(\{E_{\alpha,\beta}\}^n(i\lambda t^\gamma) - \{E_{\alpha,\beta}\}^n(-i\lambda t^\gamma)\right), \tag{4.725}$$

$$\{precos_{\alpha,\beta}\}^n(\lambda t^\gamma) = \frac{1}{2}\left(\{E_{\alpha,\beta}\}^n(i\lambda t^\gamma) + \{E_{\alpha,\beta}\}^n(-i\lambda t^\gamma)\right), \tag{4.726}$$

$$\left\{presinh_{\alpha,\beta}\right\}^n\left(\lambda t^\gamma\right)=\frac{1}{2}\left(\left\{E_{\alpha,\beta}\right\}^n\left(\lambda t^\gamma\right)-\left\{E_{\alpha,\beta}\right\}^n\left(-\lambda t^\gamma\right)\right), \tag{4.727}$$

and

$$\left\{precosh_{\alpha,\beta}\right\}^n\left(\lambda t^\gamma\right)=\frac{1}{2}\left(\left\{E_{\alpha,\beta}\right\}^n\left(\lambda t^\gamma\right)+\left\{E_{\alpha,\beta}\right\}^n\left(-\lambda t^\gamma\right)\right) \tag{4.728}$$

where $t, \alpha, \beta, \gamma, \lambda \in \mathbb{C}$, $n, \kappa \in \mathbb{N}_0$, $Re(\alpha) > 0$, and $Re(\beta) > 0$.

Definition 4.111 The truncated presine via Wiman function is defined as

$$\begin{aligned}&\left\{Tresin_{\alpha,\beta}\right\}^n(t)\\&=\frac{t^{\beta-1}}{2i}\left(\left\{E_{\alpha,\beta}\right\}^n(it)-\left\{E_{\alpha,\beta}\right\}^n(-it)\right)\\&=t^{\beta-1}\sum_{\kappa=0}^{n}\frac{(-1)^\kappa t^{2\kappa+1}}{\Gamma((2\kappa+1)\alpha+\beta)},\end{aligned} \tag{4.729}$$

where $t, \alpha, \beta \in \mathbb{C}$, $n, \kappa \in \mathbb{N}_0$, $Re(\alpha) > 0$, and $Re(\beta) > 0$.

So, we have

$$\begin{aligned}&\left\{Tresin_{\alpha,\beta}\right\}^0(t)\\&=\frac{t^{\beta-1}}{2}\left(\left\{E_{\alpha,\beta}\right\}^0(it)-\left\{E_{\alpha,\beta}\right\}^0(-it)\right)\\&=\frac{t^\beta}{\Gamma(1+\beta)},\end{aligned} \tag{4.730}$$

$$\begin{aligned}&\left\{Tresin_{\alpha,\beta}\right\}^1(t)\\&=\frac{t^{\beta-1}}{2}\left(\left\{E_{\alpha,\beta}\right\}^1(it)-\left\{E_{\alpha,\beta}\right\}^1(-it)\right)\\&=t^{\beta-1}\sum_{\kappa=0}^{1}\frac{(-1)^\kappa t^{2\kappa+1}}{\Gamma((2\kappa+1)\alpha+\beta)},\end{aligned} \tag{4.731}$$

and

$$\begin{aligned}&\left\{Tresin_{\alpha,\beta}\right\}^2(t)\\&=\frac{t^{\beta-1}}{2}\left(\left\{E_{\alpha,\beta}\right\}^2(it)-\left\{E_{\alpha,\beta}\right\}^2(-it)\right)\\&=t^{\beta-1}\sum_{\kappa=0}^{2}\frac{(-1)^\kappa t^{2\kappa+1}}{\Gamma((2\kappa+1)\alpha+\beta)},\end{aligned} \tag{4.732}$$

where $t, \alpha, \beta \in \mathbb{C}$, $Re(\alpha) > 0$, and $Re(\beta) > 0$.

Definition 4.112 The truncated precosine via Wiman function is defined as

$$\begin{aligned}&\left\{Trecos_{\alpha,\beta}\right\}^n(t)\\&=\tfrac{t^{\beta-1}}{2}\left(\left\{E_{\alpha,\beta}\right\}^n(it)+\left\{E_{\alpha,\beta}\right\}^n(-it)\right)\\&=t^{\beta-1}\sum_{\kappa=0}^{n}\tfrac{(-1)^n t^{2\kappa}}{\Gamma(2\kappa\alpha+\beta)},\end{aligned}\tag{4.733}$$

where $t, \alpha, \beta \in \mathbb{C}$, $n, \kappa \in \mathbb{N}_0$, $Re(\alpha) > 0$, and $Re(\beta) > 0$.

Thus, we give

$$\begin{aligned}&\left\{Trecos_{\alpha,\beta}\right\}^0(t)\\&=\tfrac{t^{\beta-1}}{2}\left(\left\{E_{\alpha,\beta}\right\}^0(it)+\left\{E_{\alpha,\beta}\right\}^0(-it)\right)\\&=\tfrac{t^{\beta-1}}{\Gamma(\beta)},\end{aligned}\tag{4.734}$$

$$\begin{aligned}&\left\{Trecos_{\alpha,\beta}\right\}^1(t)\\&=\tfrac{t^{\beta-1}}{2}\left(\left\{E_{\alpha,\beta}\right\}^1(it)+\left\{E_{\alpha,\beta}\right\}^1(-it)\right)\\&=t^{\beta-1}\sum_{\kappa=0}^{1}\tfrac{(-1)^\kappa t^{2\kappa}}{\Gamma(2\kappa\alpha+\beta)},\end{aligned}\tag{4.735}$$

and

$$\begin{aligned}&\left\{Trecos_{\alpha,\beta}\right\}^2(t)\\&=\tfrac{t^{\beta-1}}{2}\left(\left\{E_{\alpha,\beta}\right\}^2(t)+\left\{E_{\alpha,\beta}\right\}^2(-t)\right)\\&=t^{\beta-1}\sum_{\kappa=0}^{2}\tfrac{(-1)^\kappa t^{2\kappa}}{\Gamma(2\kappa\alpha+\beta)},\end{aligned}\tag{4.736}$$

where $t, \alpha, \beta \in \mathbb{C}$, $Re(\alpha) > 0$, and $Re(\beta) > 0$.

Definition 4.113 The truncated hyperbolic presine via Wiman function is defined as

$$\begin{aligned}&\left\{Tresinh_{\alpha,\beta}\right\}^n(t)\\&=\tfrac{t^{\beta-1}}{2}\left(\left\{E_{\alpha,\beta}\right\}^n(t)-\left\{E_{\alpha,\beta}\right\}^n(-t)\right)\\&=t^{\beta-1}\sum_{\kappa=0}^{n}\tfrac{t^{2\kappa+1}}{\Gamma((2\kappa+1)\alpha+\beta)},\end{aligned}\tag{4.737}$$

where $t, \alpha, \beta \in \mathbb{C}$, $n, \kappa \in \mathbb{N}_0$, $Re(\alpha) > 0$, and $Re(\beta) > 0$.

So, we have

$$\begin{aligned} &\{Tresinh_{\alpha,\beta}\}^0(t)\\ &= \tfrac{t^{\beta-1}}{2}\left(\{E_{\alpha,\beta}\}^0(t) - \{E_{\alpha,\beta}\}^0(-t)\right)\\ &= \tfrac{t^{\beta}}{\Gamma(1+\beta)}, \end{aligned} \tag{4.738}$$

$$\begin{aligned} &\{Tresinh_{\alpha,\beta}\}^1(t)\\ &= \tfrac{t^{\beta-1}}{2}\left(\{E_{\alpha,\beta}\}^1(t) - \{E_{\alpha,\beta}\}^1(-t)\right)\\ &= t^{\beta-1}\sum_{\kappa=0}^{1}\tfrac{t^{2\kappa+1}}{\Gamma((2\kappa+1)\alpha+\beta)}, \end{aligned} \tag{4.739}$$

and

$$\begin{aligned} &\{Tresinh_{\alpha,\beta}\}^2(t)\\ &= \tfrac{t^{\beta-1}}{2}\left(\{E_{\alpha,\beta}\}^2(t) - \{E_{\alpha,\beta}\}^2(-t)\right)\\ &= t^{\beta-1}\sum_{\kappa=0}^{2}\tfrac{t^{2\kappa+1}}{\Gamma((2\kappa+1)\alpha+\beta)}, \end{aligned} \tag{4.740}$$

where $t, \alpha, \beta \in \mathbb{C}$, $Re(\alpha) > 0$, and $Re(\beta) > 0$.

Definition 4.114 The truncated hyperbolic precosine via Wiman function is defined as

$$\begin{aligned} &\{Trecosh_{\alpha,\beta}\}^n(t)\\ &= \tfrac{t^{\beta-1}}{2}\left(\{E_{\alpha,\beta}\}^n(t) + \{E_{\alpha,\beta}\}^n(-t)\right)\\ &= t^{\beta-1}\sum_{\kappa=0}^{n}\tfrac{t^{2\kappa}}{\Gamma(2\kappa\alpha+\beta)}, \end{aligned} \tag{4.741}$$

where $t, \alpha, \beta \in \mathbb{C}$, $n, \kappa \in \mathbb{N}_0$, $Re(\alpha) > 0$, and $Re(\beta) > 0$.

Thus,

$$\begin{aligned} &\{Trecosh_{\alpha,\beta}\}^0(t)\\ &= \tfrac{t^{\beta-1}}{2}\left(\{E_{\alpha,\beta}\}^0(t) + \{E_{\alpha,\beta}\}^0(-t)\right)\\ &= \tfrac{t^{\beta-1}}{\Gamma(\alpha+\beta)}, \end{aligned} \tag{4.742}$$

$$\begin{aligned} &\{Trecosh_{\alpha,\beta}\}^1(t)\\ &= \tfrac{t^{\beta-1}}{2}\left(\{E_{\alpha,\beta}\}^1(t) + \{E_{\alpha,\beta}\}^1(-t)\right)\\ &= t^{\beta-1}\sum_{\kappa=0}^{1}\tfrac{t^{2\kappa}}{\Gamma(2\kappa\alpha+\beta)}, \end{aligned} \tag{4.743}$$

and

$$\begin{aligned}&\left\{Trecosh_{\alpha,\beta}\right\}^{2}(t)\\&\quad=\tfrac{t^{\beta-1}}{2}\left(\left\{E_{\alpha,\beta}\right\}^{2}(t)+\left\{E_{\alpha,\beta}\right\}^{2}(-t)\right)\\&\quad=t^{\beta-1}\sum_{\kappa=0}^{2}\tfrac{t^{2\kappa}}{\Gamma(2\kappa\alpha+\beta)},\end{aligned}\tag{4.744}$$

where $t, \alpha, \beta \in \mathbb{C}$, $Re(\alpha) > 0$, and $Re(\beta) > 0$.

So, we have that

$$\left\{Tresin_{\alpha,\beta}\right\}^{n}(\lambda t)=\frac{t^{\beta-1}}{2i}\left(\left\{E_{\alpha,\beta}\right\}^{n}(i\lambda t)-\left\{E_{\alpha,\beta}\right\}^{n}(-i\lambda t)\right),\tag{4.745}$$

$$\left\{Trecos_{\alpha,\beta}\right\}^{n}(\lambda t)=\frac{t^{\beta-1}}{2}\left(\left\{E_{\alpha,\beta}\right\}^{n}(i\lambda t)+\left\{E_{\alpha,\beta}\right\}^{n}(-i\lambda t)\right),\tag{4.746}$$

$$\left\{Tresinh_{\alpha,\beta}\right\}^{n}(\lambda t)=\frac{t^{\beta-1}}{2}\left(\left\{E_{\alpha,\beta}\right\}^{n}(\lambda t)-\left\{E_{\alpha,\beta}\right\}^{n}(-\lambda t)\right),\tag{4.747}$$

and

$$\left\{Trecosh_{\alpha,\beta}\right\}^{n}(\lambda t)=\frac{t^{\beta-1}}{2}\left(\left\{E_{\alpha,\beta}\right\}^{n}(\lambda t)+\left\{E_{\alpha,\beta}\right\}^{n}(-\lambda t)\right),\tag{4.748}$$

where $t, \alpha, \beta, \lambda \in \mathbb{C}$, $n, \kappa \in \mathbb{N}_0$, $Re(\alpha) > 0$, and $Re(\beta) > 0$.

Definition 4.115 The truncated presine via Wiman function is defined as

$$\begin{aligned}&\left\{Gresin_{\alpha,\beta}\right\}^{n}(t^{\gamma})\\&\quad=\tfrac{t^{\beta-1}}{2i}\left(\left\{E_{\alpha,\beta}\right\}^{n}(it^{\gamma})-\left\{E_{\alpha,\beta}\right\}^{n}(-it^{\gamma})\right)\\&\quad=t^{\beta-1}\sum_{\kappa=0}^{n}\tfrac{(-1)^{\kappa}t^{(2\kappa+1)\gamma}}{\Gamma((2\kappa+1)\alpha+\beta)},\end{aligned}\tag{4.749}$$

where $t, \alpha, \beta, \gamma \in \mathbb{C}$, $n, \kappa \in \mathbb{N}_0$, $Re(\alpha) > 0$, and $Re(\beta) > 0$.

So, we have

$$\begin{aligned}&\left\{Gresin_{\alpha,\beta}\right\}^{0}(t^{\gamma})\\&\quad=\tfrac{t^{\beta-1}}{2}\left(\left\{E_{\alpha,\beta}\right\}^{0}(it^{\gamma})-\left\{E_{\alpha,\beta}\right\}^{0}(-it^{\gamma})\right)\\&\quad=\tfrac{t^{\beta+\gamma-1}}{\Gamma(\alpha+\beta)},\end{aligned}\tag{4.750}$$

$$\begin{aligned}&\left\{Gresin_{\alpha,\beta}\right\}^{1}(t^{\gamma})\\&\quad=\tfrac{t^{\beta-1}}{2}\left(\left\{E_{\alpha,\beta}\right\}^{1}(it^{\gamma})-\left\{E_{\alpha,\beta}\right\}^{1}(-it^{\gamma})\right)\\&\quad=t^{\beta-1}\sum_{\kappa=0}^{1}\tfrac{(-1)^{\kappa}t^{(2\kappa+1)\gamma}}{\Gamma((2\kappa+1)\alpha+\beta)},\end{aligned}\tag{4.751}$$

and

$$\begin{aligned}&\left\{Gresin_{\alpha,\beta}\right\}^{2}(t^{\gamma})\\&\quad=\tfrac{t^{\beta-1}}{2}\left(\left\{E_{\alpha,\beta}\right\}^{2}(it^{\gamma})-\left\{E_{\alpha,\beta}\right\}^{2}(-it^{\gamma})\right)\\&\quad=t^{\beta-1}\sum_{\kappa=0}^{2}\tfrac{(-1)^{\kappa}t^{(2\kappa+1)\gamma}}{\Gamma((2\kappa+1)\alpha+\beta)},\end{aligned}\tag{4.752}$$

where $t, \alpha, \beta, \gamma \in \mathbb{C}$, $Re(\alpha) > 0$, and $Re(\beta) > 0$.

Definition 4.116 The truncated precosine via Wiman function is defined as

$$\begin{aligned}&\left\{Grecos_{\alpha,\beta}\right\}^{n}(t^{\gamma})\\&\quad=\tfrac{t^{\beta-1}}{2}\left(\left\{E_{\alpha,\beta}\right\}^{n}(it^{\gamma})+\left\{E_{\alpha,\beta}\right\}^{n}(-it^{\gamma})\right)\\&\quad=t^{\beta-1}\sum_{\kappa=0}^{n}\tfrac{(-1)^{n}t^{2\kappa\gamma}}{\Gamma(2\kappa\alpha+\beta)},\end{aligned}\tag{4.753}$$

where $t, \alpha, \beta, \gamma \in \mathbb{C}$, $n, \kappa \in \mathbb{N}_0$, $Re(\alpha) > 0$, and $Re(\beta) > 0$.

Thus, we give

$$\begin{aligned}&\left\{Grecos_{\alpha,\beta}\right\}^{0}(t^{\gamma})\\&\quad=\tfrac{t^{\beta-1}}{2}\left(\left\{E_{\alpha,\beta}\right\}^{0}(it^{\gamma})+\left\{E_{\alpha,\beta}\right\}^{0}(-it^{\gamma})\right)\\&\quad=\tfrac{t^{\beta-1}}{\Gamma(\alpha+\beta)},\end{aligned}\tag{4.754}$$

$$\begin{aligned}&\left\{Grecos_{\alpha,\beta}\right\}^{1}(t^{\gamma})\\&\quad=\tfrac{t^{\beta-1}}{2}\left(\left\{E_{\alpha,\beta}\right\}^{1}(it^{\gamma})+\left\{E_{\alpha,\beta}\right\}^{1}(-it^{\gamma})\right)\\&\quad=t^{\beta-1}\sum_{\kappa=0}^{1}\tfrac{(-1)^{\kappa}t^{2\kappa\gamma}}{\Gamma(2\kappa\alpha+\beta)},\end{aligned}\tag{4.755}$$

and

$$\begin{aligned}&\left\{Grecos_{\alpha,\beta}\right\}^{2}(t^{\gamma})\\&\quad=\tfrac{t^{\beta-1}}{2}\left(\left\{E_{\alpha,\beta}\right\}^{2}(t^{\gamma})+\left\{E_{\alpha,\beta}\right\}^{2}(-t^{\gamma})\right)\\&\quad=t^{\beta-1}\sum_{\kappa=0}^{2}\tfrac{(-1)^{\kappa}t^{2\kappa\gamma}}{\Gamma(2\kappa\alpha+\beta)},\end{aligned}\tag{4.756}$$

where $t, \alpha, \beta, \gamma \in \mathbb{C}$, $Re(\alpha) > 0$, and $Re(\beta) > 0$.

Definition 4.117 The truncated hyperbolic presine via Wiman function is defined as

$$\begin{aligned}&\left\{Gresinh_{\alpha,\beta}\right\}^n(t^\gamma)\\&=\frac{t^{\beta-1}}{2}\left(\left\{E_{\alpha,\beta}\right\}^n(t^\gamma)-\left\{E_{\alpha,\beta}\right\}^n(-t^\gamma)\right)\\&=t^{\beta-1}\sum_{\kappa=0}^{n}\frac{t^{(2\kappa+1)\gamma}}{\Gamma((2\kappa+1)\alpha+\beta)},\end{aligned}\tag{4.757}$$

where $t, \alpha, \beta, \gamma \in \mathbb{C}$, $n, \kappa \in \mathbb{N}_0$, $Re(\alpha) > 0$, and $Re(\beta) > 0$.

So, we have

$$\begin{aligned}&\left\{Gresinh_{\alpha,\beta}\right\}^0(t^\gamma)\\&=\frac{t^{\beta-1}}{2}\left(\left\{E_{\alpha,\beta}\right\}^0(t^\gamma)-\left\{E_{\alpha,\beta}\right\}^0(-t^\gamma)\right)\\&=\frac{t^{\beta+\gamma-1}}{\Gamma(\alpha+\beta)},\end{aligned}\tag{4.758}$$

$$\begin{aligned}&\left\{Gresinh_{\alpha,\beta}\right\}^1(t^\gamma)\\&=\frac{t^{\beta-1}}{2}\left(\left\{E_{\alpha,\beta}\right\}^1(t^\gamma)-\left\{E_{\alpha,\beta}\right\}^1(-t^\gamma)\right)\\&=t^{\beta-1}\sum_{\kappa=0}^{1}\frac{t^{(2\kappa+1)\gamma}}{\Gamma((2\kappa+1)\alpha+\beta)},\end{aligned}\tag{4.759}$$

and

$$\begin{aligned}&\left\{Gresinh_{\alpha,\beta}\right\}^2(t^\gamma)\\&=\frac{t^{\beta-1}}{2}\left(\left\{E_{\alpha,\beta}\right\}^2(t^\gamma)-\left\{E_{\alpha,\beta}\right\}^2(-t^\gamma)\right)\\&=t^{\beta-1}\sum_{\kappa=0}^{2}\frac{t^{(2\kappa+1)\gamma}}{\Gamma((2\kappa+1)\alpha+\beta)},\end{aligned}\tag{4.760}$$

where $t, \alpha, \beta, \gamma \in \mathbb{C}$, $Re(\alpha) > 0$, and $Re(\beta) > 0$.

Definition 4.118 The truncated hyperbolic precosine via Wiman function is defined as

$$\begin{aligned}&\left\{Grecosh_{\alpha,\beta}\right\}^n(t^\gamma)\\&=\frac{t^{\beta-1}}{2}\left(\left\{E_{\alpha,\beta}\right\}^n(t^\gamma)+\left\{E_{\alpha,\beta}\right\}^n(-t^\gamma)\right)\\&=t^{\beta-1}\sum_{\kappa=0}^{n}\frac{t^{2\kappa\gamma}}{\Gamma(2\kappa\alpha+\beta)},\end{aligned}\tag{4.761}$$

where $t, \alpha, \beta, \gamma \in \mathbb{C}$, $n, \kappa \in \mathbb{N}_0$, $Re(\alpha) > 0$, and $Re(\beta) > 0$.

Thus,

$$\begin{aligned}&\left\{Grecosh_{\alpha,\beta}\right\}^{0}(t^{\gamma})\\&=\tfrac{t^{\beta-1}}{2}\left(\left\{E_{\alpha,\beta}\right\}^{0}(t^{\gamma})+\left\{E_{\alpha,\beta}\right\}^{0}(-t^{\gamma})\right)\\&=\tfrac{t^{\beta-1}}{\Gamma(\alpha+\beta)},\end{aligned}\tag{4.762}$$

$$\begin{aligned}&\left\{Grecosh_{\alpha,\beta}\right\}^{1}(t^{\gamma})\\&=\tfrac{t^{\beta-1}}{2}\left(\left\{E_{\alpha,\beta}\right\}^{1}(t^{\gamma})+\left\{E_{\alpha,\beta}\right\}^{1}(-t^{\gamma})\right)\\&=t^{\beta-1}\sum_{\kappa=0}^{1}\tfrac{t^{2\kappa\gamma}}{\Gamma(2\kappa\alpha+\beta)},\end{aligned}\tag{4.763}$$

and

$$\begin{aligned}&\left\{Grecosh_{\alpha,\beta}\right\}^{2}(t^{\gamma})\\&=\tfrac{t^{\beta-1}}{2}\left(\left\{E_{\alpha,\beta}\right\}^{2}(t^{\gamma})+\left\{E_{\alpha,\beta}\right\}^{2}(-t^{\gamma})\right)\\&=t^{\beta-1}\sum_{\kappa=0}^{2}\tfrac{t^{2\kappa\gamma}}{\Gamma(2\kappa\alpha+\beta)},\end{aligned}\tag{4.764}$$

where $t, \alpha, \beta, \gamma \in \mathbb{C}$, $Re(\alpha) > 0$, and $Re(\beta) > 0$.

So, we have that

$$\left\{Gresin_{\alpha,\beta}\right\}^{n}(\lambda t^{\gamma})=\frac{t^{\beta-1}}{2i}\left(\left\{E_{\alpha,\beta}\right\}^{n}(i\lambda t^{\gamma})-\left\{E_{\alpha,\beta}\right\}^{n}(-i\lambda t^{\gamma})\right),\tag{4.765}$$

$$\left\{Grecos_{\alpha,\beta}\right\}^{n}(\lambda t^{\gamma})=\frac{t^{\beta-1}}{2}\left(\left\{E_{\alpha,\beta}\right\}^{n}(i\lambda t^{\gamma})+\left\{E_{\alpha,\beta}\right\}^{n}(-i\lambda t^{\gamma})\right),\tag{4.766}$$

$$\left\{Gresinh_{\alpha,\beta}\right\}^{n}(\lambda t^{\gamma})=\frac{t^{\beta-1}}{2}\left(\left\{E_{\alpha,\beta}\right\}^{n}(\lambda t^{\gamma})-\left\{E_{\alpha,\beta}\right\}^{n}(-\lambda t^{\gamma})\right),\tag{4.767}$$

and

$$\left\{Grecosh_{\alpha,\beta}\right\}^{n}(\lambda t^{\gamma})=\frac{t^{\beta-1}}{2}\left(\left\{E_{\alpha,\beta}\right\}^{n}(\lambda t^{\gamma})+\left\{E_{\alpha,\beta}\right\}^{n}(-\lambda t^{\gamma})\right),\tag{4.768}$$

where $t, \alpha, \beta, \gamma, \lambda \in \mathbb{C}$, $n \in \mathbb{N}_0$, $Re(\alpha) > 0$, and $Re(\beta) > 0$.

For the more results related to the Wiman function, see Pskhu [249], Popov and Sedletskii [250], Sedletskii [251], Saxena et al. [252], Kilbas et al. [253], Saxena and Saigo [254], Gorenflo and Mainardi [255], Gorenflo et al. [256], Sixdeniers et al. [257], Metzler [258], and so on.

4.4 General Fractional Derivatives Within the Wiman Kernel

In this section, we introduce the general fractional derivative within the Wiman kernel discovered by Yang [92].

4.4.1 General Fractional Derivatives Within the Wiman Kernel

Definition 4.119 Let $1 \geq \alpha \geq 0$ and $\upsilon, \lambda \in \mathbb{R}$.

The left-sided general fractional derivative within the Wiman kernel is defined as [92]

$$\begin{aligned} {}_{GW}^{RL}D_{a+}^{\alpha,\upsilon,\lambda} f(t) &= \left(\frac{d}{dt}{}_{GWO}^{RL}I_{a+}^{\alpha,\upsilon,\lambda} f\right)(t) \\ &= \frac{d}{dt}\int_a^t (t-\tau)^{\upsilon-1} E_{\alpha,\upsilon}\left(-\lambda(t-\tau)^{\alpha}\right) f(\tau)\, d\tau, \end{aligned} \tag{4.769}$$

where

$${}_{GWO}^{RL}I_{a+}^{\alpha,\upsilon,\lambda} f(t) = \int_a^t (t-\tau)^{\upsilon-1} E_{\alpha,\upsilon}\left(-\lambda(t-\tau)^{\alpha}\right) f(\tau)\, d\tau, \tag{4.770}$$

and the right-sided general fractional derivative within the Wiman kernel as

$$\begin{aligned} {}_{GW}^{RL}D_{b-}^{\alpha,\upsilon,\lambda} f(t) &= \left(-\frac{d}{dt}{}_{GWO}^{RL}I_{b-}^{\alpha,\upsilon,\lambda} f\right)(t) \\ &= -\frac{d}{dt}\int_t^b (\tau-t)^{\upsilon-1} E_{\alpha,\upsilon}\left(-\lambda(\tau-t)^{\alpha}\right) f(\tau)\, d\tau, \end{aligned} \tag{4.771}$$

where

$${}_{GWO}^{RL}I_{b-}^{\alpha,\upsilon,\lambda} f(t) = \int_t^b (\tau-t)^{\upsilon-1} E_{\alpha,\upsilon}\left(-\lambda(\tau-t)^{\alpha}\right) f(\tau)\, d\tau. \tag{4.772}$$

The work was discovered by Yang in 2019 [92].

Definition 4.120 Let $\alpha \geq 0$, $\kappa = [\alpha] + 1$, and $\upsilon, \lambda \in \mathbb{R}$.

The left-sided general fractional derivative within the Wiman kernel is defined as [92]

$$ {}_{GW}^{RL}D_{a+}^{\alpha,\upsilon,\kappa,\lambda} f(t) = \frac{d^{\kappa}}{dt^{\kappa}} \int_{a}^{t} (t-\tau)^{\upsilon-1} E_{\alpha,\upsilon}\left(-\lambda (t-\tau)^{\alpha}\right) f(\tau)\, d\tau \tag{4.773} $$

and the right-sided general fractional derivative within the Wiman kernel as

$$ {}_{GW}^{RL}D_{b-}^{\alpha,\upsilon,\kappa,\lambda} f(t) = (-1)^{\kappa} \frac{d^{\kappa}}{dt^{\kappa}} \int_{t}^{b} (\tau-t)^{\upsilon-1} E_{\alpha,\upsilon}\left(-\lambda (\tau-t)^{\alpha}\right) f(\tau)\, d\tau. \tag{4.774} $$

The work was discovered by Yang in 2019 [92].

Definition 4.121 Let $1 \geq \alpha \geq 0$ and $\upsilon, \lambda \in \mathbb{R}$.

The left-sided general fractional derivative within the Wiman kernel is defined as [92]

$$ {}_{GW}^{LSC}D_{a+}^{\alpha,\upsilon,\lambda} f(t) = \int_{a}^{t} (t-\tau)^{\upsilon-1} E_{\alpha,\upsilon}\left(-\lambda (t-\tau)^{\alpha}\right) f^{(1)}(\tau)\, d\tau \tag{4.775} $$

and the right-sided general fractional derivative within the Wiman kernel as

$$ {}_{GW}^{LSC}D_{b-}^{\alpha,\upsilon,\lambda} f(t) = -\int_{t}^{b} (\tau-t)^{\upsilon-1} E_{\alpha,\upsilon}\left(-\lambda (\tau-t)^{\alpha}\right) f^{(1)}(\tau)\, d\tau, \tag{4.776} $$

respectively.

The work was discovered by Yang in 2019 [92].

Definition 4.122 Let $\alpha \geq 0$, $\kappa = [\alpha] + 1$, and $\upsilon, \lambda \in \mathbb{R}$.

The left-sided general fractional derivative within the Wiman kernel is defined as [92]

$$ {}_{GW}^{LSC}D_{a+}^{\alpha,\upsilon,\kappa,\lambda} f(t) = \int_{a}^{t} (t-\tau)^{\upsilon-1} E_{\alpha,\upsilon}\left(-\lambda (t-\tau)^{\alpha}\right) f^{(\kappa)}(\tau)\, d\tau \tag{4.777} $$

and the right-sided general fractional derivative within the Wiman kernel as

$$ {}_{GW}^{LSC}D_{b-}^{\alpha,\upsilon,\kappa,\lambda} f(t) = (-1)^{\kappa} \int_{t}^{b} (\tau - t)^{\upsilon-1} E_{\alpha,\upsilon}\left(-\lambda(\tau - t)^{\alpha}\right) f^{(\kappa)}(\tau)\, d\tau, \tag{4.778} $$

respectively.

The work was discovered by Yang in 2019 [92].

4.4.2 *Hilfer-Type General Fractional Derivatives with the Wiman Kernel*

In this section, we present the Hilfer-type general fractional derivatives with the Wiman kernel, which are discovered by Yang, Gao, and Ju in 2020 [92].

Definition 4.123 Let $0 < \alpha < 1, 0 \leq \beta \leq 1$, and $\upsilon, \lambda \in \mathbb{R}$.

The left-sided Hilfer-type general fractional derivative with the Wiman kernel is defined as

$$ {}_{HGW}D_{a+}^{\alpha,\upsilon,\lambda} f(t) = \left({}_{GWO}^{RL}I_{a+}^{\beta(1-\alpha),\upsilon,\lambda} \frac{d}{dt} {}_{GWO}^{RL}I_{a+}^{(1-\beta)(1-\alpha),\upsilon,\lambda} \right) f(t), \tag{4.779} $$

where

$$ {}_{GWO}^{RL}I_{a+}^{\beta(1-\alpha),\upsilon,\lambda} f(t) = \int_{a}^{t} (t - \tau)^{\upsilon-1} E_{\beta(1-\alpha),\upsilon}\left(-\lambda(t - \tau)^{\beta(1-\alpha)}\right) f(\tau)\, d\tau, \tag{4.780} $$

and

$$ {}_{GWO}^{RL}I_{a+}^{(1-\beta)(1-\alpha),\upsilon,\lambda} f(t) = \int_{a}^{t} (t - \tau)^{\upsilon-1} E_{(1-\beta)(1-\alpha),\upsilon}\left(-\lambda(t - \tau)^{(1-\beta)(1-\alpha)}\right) f(\tau)\, d\tau, \tag{4.781} $$

and the right-sided Hilfer-type general fractional derivative with the Wiman kernel as

$$ {}_{HGW}D_{b-}^{\alpha,\upsilon,\lambda} f(t) = \left({}_{GWO}^{RL}I_{b-}^{\beta(1-\alpha),\upsilon,\lambda} \frac{d}{dt} {}_{GWO}^{RL}I_{b-}^{(1-\beta)(1-\alpha),\upsilon,\lambda} \right) f(t), \tag{4.782} $$

where

$$ {}_{GWO}^{RL}I_{b-}^{\beta(1-\alpha),\upsilon,\lambda} f(t) = \int_t^b (\tau - t)^{\upsilon-1} E_{\beta(1-\alpha),\upsilon}\left(-\lambda(\tau - t)^{\beta(1-\alpha)}\right) f(\tau)\, d\tau, \tag{4.783} $$

and

$$ \begin{aligned} &{}_{GWO}^{RL}I_{b-}^{(1-\beta)(1-\alpha),\upsilon,\lambda} f(t) \\ &= \int_t^b (\tau - t)^{\upsilon-1} E_{(1-\beta)(1-\alpha),\upsilon}\left(-\lambda(\tau - t)^{(1-\beta)(1-\alpha)}\right) f(\tau)\, d\tau. \end{aligned} \tag{4.784} $$

The work was discovered by Yang, Gao, and Ju in 2020 [92].

Property 4.18 Suppose that $0 < \alpha < 1$, $0 \le \beta \le 1$, and $\upsilon, \lambda \in \mathbb{R}$, then we have [92]

$$ \begin{aligned} &L\left\{ {}_{HGW}D_{0+}^{\alpha,\upsilon,\lambda} f(t) \right\}(s) \\ &= s^{-1}\left(1 + \lambda s^{-\beta(1-\alpha)}\right)^{-1}\left(s^{1-\upsilon}\left(1+\lambda s^{-\alpha}\right)^{-1} f(s) - {}_{GWO}^{RL}I_{0+}^{(1-\beta)(1-\alpha),\upsilon,\lambda} f(0)\right). \end{aligned} \tag{4.785} $$

Proof

$$ \begin{aligned} &L\left\{ {}_{HGW}D_{0+}^{\alpha,\upsilon,\lambda} f(t)\, f(t) \right\}(s) \\ &= L\left\{ \left({}_{GWO}^{RL}I_{0+}^{\beta(1-\alpha),\upsilon,\lambda} \frac{d}{dt} {}_{GWO}^{RL}I_{0+}^{(1-\beta)(1-\alpha),\upsilon,\lambda} \right) f(t) \right\} \\ &= s^{-\upsilon}\left(1+\lambda s^{-\alpha}\right)^{-1}\left(s \times s^{-\upsilon}\left(1+\lambda s^{-\alpha}\right)^{-1} f(s) - {}_{GWO}^{RL}I_{0+}^{(1-\beta)(1-\alpha),\upsilon,\lambda} f(0)\right) \\ &= s^{-1}\left(1 + \lambda s^{-\beta(1-\alpha)}\right)^{-1}\left(s^{1-\upsilon}\left(1+\lambda s^{-\alpha}\right)^{-1} f(s) - {}_{GWO}^{RL}I_{0+}^{(1-\beta)(1-\alpha),\upsilon,\lambda} f(0)\right). \end{aligned} \tag{4.786} $$

Thus, we present the proof. □

The work was discovered by Yang, Gao, and Ju in 2020 [92].

Definition 4.124 Let $0 < \alpha < 1$, $0 \le \beta \le 1$, and $\upsilon, \lambda \in \mathbb{R}$.

The left-sided Hilfer-type general fractional derivative with the Wiman kernel is defined as [92]

$$ {}_{HGW}D_{+}^{\alpha,\upsilon,\lambda} f(t) = \left({}_{GWO}^{RL}I_{+}^{\beta(1-\alpha),\upsilon,\lambda} \frac{d}{dt} {}_{GWO}^{RL}I_{+}^{(1-\beta)(1-\alpha),\upsilon,\lambda} f \right)(t), \tag{4.787} $$

where

$$ {}_{GWO}^{RL}I_{+}^{\beta(1-\alpha),\upsilon,\lambda} f(t) = \int_{-\infty}^{t} (t-\tau)^{\upsilon-1} E_{\beta(1-\alpha),\upsilon}\left(-\lambda (t-\tau)^{\beta(1-\alpha)}\right) f(\tau)\, d\tau, \tag{4.788} $$

and

$$ {}_{GWO}^{RL}I_{+}^{(1-\beta)(1-\alpha),\upsilon,\lambda} f(t) = \int_{-\infty}^{t} (t-\tau)^{\upsilon-1} E_{(1-\beta)(1-\alpha),\upsilon}\left(-\lambda (t-\tau)^{(1-\beta)(1-\alpha)}\right) f(\tau)\, d\tau, \tag{4.789} $$

and the right-sided Hilfer-type general fractional derivative with the Wiman kernel as

$$ {}_{HGW}D_{-}^{\alpha,\upsilon,\lambda} f(t) = \left({}_{GWO}^{RL}I_{-}^{\beta(1-\alpha),\upsilon,\lambda} \frac{d}{dt} {}_{GWO}^{RL}I_{-}^{(1-\beta)(1-\alpha),\upsilon,\lambda}\right) f(t), \tag{4.790} $$

where ${}_{GWO}^{RL}I_{-}^{(1-\beta)(1-\alpha),\upsilon,\lambda} f(t) \in AC_1(-\infty, \infty)$,

$$ {}_{GWO}^{RL}I_{-}^{\beta(1-\alpha),\upsilon,\lambda} f(t) = \int_{t}^{\infty} (\tau-t)^{\upsilon-1} E_{\beta(1-\alpha),\upsilon}\left(-\lambda (\tau-t)^{\beta(1-\alpha)}\right) f(\tau)\, d\tau, \tag{4.791} $$

and

$$ {}_{GWO}^{RL}I_{-}^{(1-\beta)(1-\alpha),\upsilon,\lambda} f(t) = \int_{t}^{\infty} (\tau-t)^{\upsilon-1} E_{(1-\beta)(1-\alpha),\upsilon}\left(-\lambda (\tau-t)^{(1-\beta)(1-\alpha)}\right) f(\tau)\, d\tau. \tag{4.792} $$

The work was discovered by Yang, Gao, and Ju in 2020 [92].

4.4.3 *General Fractional Derivatives with Respect to Another Function via Wiman Function*

Definition 4.125 Let $h^{(1)}(t) > 0$, $1 \geq \alpha \geq 0$, and $\upsilon, \lambda \in \mathbb{R}$.

The left-sided general fractional derivative with respect to another function via Wiman kernel is defined as [92]

$$
\begin{aligned}
&{}_{GW}^{RL}D_{a+,h}^{\alpha,\upsilon,\lambda} f(t) \\
&= \left(\left(\frac{1}{h^{(1)}(t)} \frac{d}{dt} \right) {}_{GWO}^{RL}I_{a+,h}^{\alpha,\upsilon,\lambda} f \right)(t) \\
&= \left(\frac{1}{h^{(1)}(t)} \frac{d}{dt} \right) \int_a^t (h(t) - h(\tau))^{\upsilon-1} E_{\alpha,\upsilon} \left(-\lambda (h(t) - h(\tau))^{\alpha} \right) h^{(1)}(\tau) f(\tau) d\tau,
\end{aligned}
\tag{4.793}
$$

where

$$
{}_{GWO}^{RL}I_{a+,h}^{\alpha,\upsilon,\lambda} f(t) = \int_a^t (h(t) - h(\tau))^{\upsilon-1} E_{\alpha,\upsilon} \left(-\lambda (h(t) - h(\tau))^{\alpha} \right) h^{(1)}(\tau) f(\tau) d\tau,
\tag{4.794}
$$

and the right-sided general fractional derivative with respect to another function via Wiman kernel as

$$
\begin{aligned}
&{}_{GW}^{RL}D_{b-,h}^{\alpha,\upsilon,\lambda} f(t) \\
&= \left(\left(-\frac{1}{h^{(1)}(t)} \frac{d}{dt} \right) {}_{GWO}^{RL}I_{b-,h}^{\alpha,\upsilon,\lambda} f \right)(t) \\
&= \left(\frac{-1}{h^{(1)}(t)} \frac{d}{dt} \right) \int_t^b (h(\tau) - h(t))^{\upsilon-1} E_{\alpha,\upsilon} \left(-\lambda (h(\tau) - h(t))^{\alpha} \right) h^{(1)}(\tau) f(\tau) d\tau,
\end{aligned}
\tag{4.795}
$$

where

$$
{}_{GWO}^{RL}I_{b-,h}^{\alpha,\upsilon,\lambda} f(t) = \int_t^b (h(\tau) - h(t))^{\upsilon-1} E_{\alpha,\upsilon} \left(-\lambda (h(\tau) - h(t))^{\alpha} \right) h^{(1)}(\tau) f(\tau) d\tau.
\tag{4.796}
$$

The work was discovered by Yang, Gao, and Ju in 2020 [92].

Definition 4.126 Let $h^{(1)}(t) > 0$, $1 \geq \alpha \geq 0$, and $\upsilon, \lambda \in \mathbb{R}$.

The left-sided general fractional derivative with respect to another function via Wiman kernel is defined as [92]

$$
\begin{aligned}
&{}_{GW}^{RL}D_{+,h}^{\alpha,\upsilon,\lambda} f(t) \\
&= \left(\left(\frac{1}{h^{(1)}(t)} \frac{d}{dt} \right) {}_{GWO}^{RL}I_{+,h}^{\alpha,\upsilon,\lambda} f \right)(t) \\
&= \left(\frac{1}{h^{(1)}(t)} \frac{d}{dt} \right) \int_{-\infty}^t (h(t) - h(\tau))^{\upsilon-1} E_{\alpha,\upsilon} \left(-\lambda (h(t) - h(\tau))^{\alpha} \right) h^{(1)}(\tau) f(\tau) d\tau,
\end{aligned}
\tag{4.797}
$$

where

$$
{}_{GWO}^{RL}I_{+,h}^{\alpha,\upsilon,\lambda} f(t) = \int_{-\infty}^{t} (h(t) - h(\tau))^{\upsilon-1} E_{\alpha,\upsilon}\left(-\lambda (h(t) - h(\tau))^{\alpha}\right) h^{(1)}(\tau) f(\tau) d\tau, \tag{4.798}
$$

and the right-sided general fractional derivative with respect to another function via Wiman kernel as

$$
\begin{aligned}
&{}_{GW}^{RL}D_{-,h}^{\alpha,\upsilon,\lambda} f(t) \\
&= \left(\left(-\frac{1}{h^{(1)}(t)}\frac{d}{dt}\right) {}_{GWO}^{RL}I_{-,h}^{\alpha,\upsilon,\lambda} f\right)(t) \\
&= \left(\frac{-1}{h^{(1)}(t)}\frac{d}{dt}\right) \int_{t}^{\infty} (h(\tau) - h(t))^{\upsilon-1} E_{\alpha,\upsilon}(-\lambda (h(\tau) - h(t))^{\alpha}) h^{(1)}(\tau) f(\tau) d\tau,
\end{aligned} \tag{4.799}
$$

where

$$
{}_{GWO}^{RL}I_{-,h}^{\alpha,\upsilon,\lambda} f(t) = \int_{t}^{\infty} (h(\tau) - h(t))^{\upsilon-1} E_{\alpha,\upsilon}\left(-\lambda (h(\tau) - h(t))^{\alpha}\right) h^{(1)}(\tau) f(\tau) d\tau. \tag{4.800}
$$

The work was discovered by Yang, Gao, and Ju in 2020 [92].

Definition 4.127 Let $h^{(1)}(t) > 0$, $1 \geq \alpha \geq 0$, and $\upsilon, \lambda \in \mathbb{R}$.

The left-sided general fractional derivative with respect to another function via Wiman kernel is defined as [92]

$$
\begin{aligned}
&{}_{GW}^{LSC}D_{a+,h}^{\alpha,\upsilon,\lambda} f(t) \\
&= \int_{a}^{t} (h(t) - h(\tau))^{\upsilon-1} E_{\alpha,\upsilon}(-\lambda (h(t) - h(\tau))^{\alpha}) h^{(1)}(\tau) \left(\frac{1}{h^{(1)}(\tau)}\frac{d}{d\tau}\right) f(\tau) d\tau
\end{aligned} \tag{4.801}
$$

and the right-sided general fractional derivative with respect to another function via Wiman kernel as

$$
\begin{aligned}
&{}_{GW}^{LSC}D_{b-,h}^{\alpha,\upsilon,\lambda} f(t) \\
&= -\int_{t}^{b} (h(\tau) - h(t))^{\upsilon-1} E_{\alpha,\upsilon}(-\lambda (h(\tau) - h(t))^{\alpha}) h^{(1)}(\tau) \left(\frac{1}{h^{(1)}(\tau)}\frac{d}{d\tau}\right) f(\tau) d\tau,
\end{aligned} \tag{4.802}
$$

respectively.

The work was discovered by Yang, Gao, and Ju in 2020 [92].

Definition 4.128 Let $h^{(1)}(t) > 0$, $1 \geq \alpha \geq 0$, and $\upsilon, \lambda \in \mathbb{R}$.

The left-sided general fractional derivative with respect to another function via Wiman kernel is defined as [92]

$$
\begin{aligned}
&{}_{GW}^{LSC}D_{+,h}^{\alpha,\upsilon,\lambda} f(t) \\
&= \int_{-\infty}^{t} (h(t) - h(\tau))^{\upsilon-1} E_{\alpha,\upsilon}\left(-\lambda (h(t) - h(\tau))^{\alpha}\right) h^{(1)}(\tau) \left(\frac{1}{h^{(1)}(\tau)} \frac{d}{d\tau}\right) f(\tau)\, d\tau
\end{aligned}
\tag{4.803}
$$

and the right-sided general fractional derivative with respect to another function via Wiman kernel as

$$
\begin{aligned}
&{}_{GW}^{LSC}D_{-,h}^{\alpha,\upsilon,\lambda} f(t) \\
&= -\int_{t}^{\infty} (h(\tau) - h(t))^{\upsilon-1} E_{\alpha,\upsilon}\left(-\lambda (h(\tau) - h(t))^{\alpha}\right) h^{(1)}(\tau) \left(\frac{1}{h^{(1)}(\tau)} \frac{d}{d\tau}\right) f(\tau)\, d\tau,
\end{aligned}
\tag{4.804}
$$

respectively.

The work was discovered by Yang, Gao, and Ju in 2020 [92].

Definition 4.129 Let $h^{(1)}(t) > 0$, $0 < \alpha < 1$, $0 \leq \beta \leq 1$, and $\upsilon, \lambda \in \mathbb{R}$.

The left-sided Hilfer-type general fractional derivative with respect to another function via Wiman kernel is defined as [92]

$$
{}_{HGW}D_{a+,h}^{\alpha,\upsilon,\lambda} f(t) = \left({}_{GWO}^{RL}I_{a+,h}^{\beta(1-\alpha),\upsilon,\lambda} \left(\frac{1}{h^{(1)}(t)} \frac{d}{dt}\right) {}_{GWO}^{RL}I_{a+,h}^{(1-\beta)(1-\alpha),\upsilon,\lambda}\right) f(t),
\tag{4.805}
$$

where

$$
\begin{aligned}
&{}_{GWO}^{RL}I_{a+,h}^{\beta(1-\alpha),\upsilon,\lambda} f(t) \\
&= \int_{a}^{t} (h(t) - h(\tau))^{\upsilon-1} E_{\beta(1-\alpha),\upsilon}\left(-\lambda (h(t) - h(\tau))^{\beta(1-\alpha)}\right) h^{(1)}(\tau) f(\tau)\, d\tau,
\end{aligned}
\tag{4.806}
$$

and

$$
\begin{aligned}
&{}_{GWO}^{RL}I_{a+,h}^{(1-\beta)(1-\alpha),\upsilon,\lambda} f(t) \\
&= \int_{a}^{t} (h(t) - h(\tau))^{\upsilon-1} E_{(1-\beta)(1-\alpha),\upsilon}\left(-\lambda (h(t) - h(\tau))^{(1-\beta)(1-\alpha)}\right) h^{(1)}(\tau) f(\tau)\, d\tau,
\end{aligned}
\tag{4.807}
$$

and the right-sided Hilfer-type general fractional derivative with respect to another function via Wiman kernel as

$$ {}_{HGW}D_{b-,h}^{\alpha,\upsilon,\lambda} f(t) = \left({}_{GWO}^{RL}I_{b-,h}^{\beta(1-\alpha),\upsilon,\lambda} \left(\frac{1}{h^{(1)}(t)} \frac{d}{dt} \right) {}_{GWO}^{RL}I_{b-,h}^{(1-\beta)(1-\alpha),\upsilon,\lambda} \right) f(t), \tag{4.808} $$

where

$$ \begin{aligned} & {}_{GWO}^{RL}I_{b-}^{\beta(1-\alpha),\upsilon,\lambda} f(t) \\ & = \int_t^b (h(\tau) - h(t))^{\upsilon-1} E_{\beta(1-\alpha),\upsilon} \left(-\lambda (h(\tau) - h(t))^{\beta(1-\alpha)} \right) h^{(1)}(\tau) f(\tau) d\tau, \end{aligned} \tag{4.809} $$

and

$$ \begin{aligned} & {}_{GWO}^{RL}I_{b-,h}^{(1-\beta)(1-\alpha),\upsilon,\lambda} f(t) \\ & = \int_t^b (h(\tau) - h(t))^{\upsilon-1} E_{(1-\beta)(1-\alpha),\upsilon} \left(-\lambda (h(\tau) - h(t))^{(1-\beta)(1-\alpha)} \right) h^{(1)}(\tau) f(\tau) d\tau. \end{aligned} \tag{4.810} $$

The work was discovered by Yang, Gao, and Ju in 2020 [92].

Definition 4.130 Let $h^{(1)}(t) > 0$, $0 < \alpha < 1$, $0 \leq \beta \leq 1$, and $\upsilon, \lambda \in \mathbb{R}$.

The left-sided Hilfer-type general fractional derivative with respect to another function via Wiman kernel is defined as [92]

$$ {}_{HGW}D_{+,h}^{\alpha,\upsilon,\lambda} f(t) = \left({}_{GWO}^{RL}I_{+,h}^{\beta(1-\alpha),\upsilon,\lambda} \left(\frac{1}{h^{(1)}(t)} \frac{d}{dt} \right) {}_{GWO}^{RL}I_{+,h}^{(1-\beta)(1-\alpha),\upsilon,\lambda} \right) f(t), \tag{4.811} $$

where

$$ \begin{aligned} & {}_{GWO}^{RL}I_{+,h}^{\beta(1-\alpha),\upsilon,\lambda} f(t) \\ & = \int_{-\infty}^t (h(t) - h(\tau))^{\upsilon-1} E_{\beta(1-\alpha),\upsilon} \left(-\lambda (h(t) - h(\tau))^{\beta(1-\alpha)} \right) h^{(1)}(\tau) f(\tau) d\tau, \end{aligned} \tag{4.812} $$

and

$$ \begin{aligned} & {}_{GWO}^{RL}I_{+,h}^{(1-\beta)(1-\alpha),\upsilon,\lambda} f(t) \\ & = \int_{-\infty}^t (h(t) - h(\tau))^{\upsilon-1} E_{(1-\beta)(1-\alpha),\upsilon} \left(-\lambda (h(t) - h(\tau))^{(1-\beta)(1-\alpha)} \right) h^{(1)}(\tau) f(\tau) d\tau, \end{aligned} \tag{4.813} $$

and the right-sided Hilfer-type general fractional derivative with respect to another function via Wiman kernel as

$$_{HGW}D_{-,h}^{\alpha,\upsilon,\lambda} f(t) = \left({}_{GWO}^{RL}I_{-,h}^{\beta(1-\alpha),\upsilon,\lambda} \left(\frac{1}{h^{(1)}(t)} \frac{d}{dt} \right) {}_{GWO}^{RL}I_{-,h}^{(1-\beta)(1-\alpha),\upsilon,\lambda} \right) f(t), \tag{4.814}$$

where

$$\begin{aligned} &{}_{GWO}^{RL}I_{-}^{\beta(1-\alpha),\upsilon,\lambda} f(t) \\ &= \int_t^\infty (h(\tau) - h(t))^{\upsilon-1} E_{\beta(1-\alpha),\upsilon}\left(-\lambda (h(\tau) - h(t))^{\beta(1-\alpha)}\right) h^{(1)}(\tau) f(\tau) d\tau, \end{aligned} \tag{4.815}$$

and

$$\begin{aligned} &{}_{GWO}^{RL}I_{-,h}^{(1-\beta)(1-\alpha),\upsilon,\lambda} f(t) \\ &= \int_t^\infty (h(\tau) - h(t))^{\upsilon-1} E_{(1-\beta)(1-\alpha),\upsilon}\left(-\lambda (h(\tau) - h(t))^{(1-\beta)(1-\alpha)}\right) h^{(1)}(\tau) f(\tau) d\tau. \end{aligned} \tag{4.816}$$

The work was discovered by Yang, Gao, and Ju in 2020 [92].
For more details for the comments on the new operators, see [92].

4.5 Applications

In this section, we present the mathematical models based on the general fractional derivatives with the Wiman kernel and give the obtained examples.

Model (Newtonian-Like Dashpot Element)
The Newtonian-like dashpot element with the Wiman kernel is presented as follows [92]:

$$\sigma(\tau) = \gamma {}_{GW}^{RL}D_{0+}^{\alpha,\upsilon,\lambda} \varepsilon(\tau), \tag{4.817}$$

where γ is the viscosity of the material, $\varepsilon(\tau)$ is the strain, $\sigma(\tau)$ is the stress, and τ is the time.

The result was proposed in [92].

Model (Maxwell-Like Model)

The Maxwell-like model with the Wiman kernel can be written as follows [92]:

$$ {}_{GW}^{RL}D_{0+}^{\alpha,\upsilon,\lambda}\varepsilon(\tau) = \frac{\sigma(\tau)}{\gamma} + \frac{{}_{GW}^{RL}D_{0+}^{\alpha,\upsilon,\lambda}\sigma(\tau)}{\zeta}, \tag{4.818} $$

where γ is the viscosity of the material and ζ is the elastic modulus of the material.

The result was proposed in 2020 [92].

Model (Kelvin–Voigt-Like Model)

The Kelvin–Voigt-like model with the Wiman kernel can be given as follows [92]:

$$ \sigma(\tau) = \gamma\varepsilon(\tau) + \zeta\, {}_{GW}^{RL}D_{0+}^{\alpha,\upsilon,\lambda}\varepsilon(\tau), \tag{4.819} $$

where γ is the viscosity of the material and ζ is the elastic modulus of the material.

The result was proposed in 2020 [92].

Model (Newtonian-Like Dashpot Element)

The Newtonian-like dashpot element with the Wiman kernel is given as follows [92]:

$$ \sigma(\tau) = \gamma\, {}_{GW}^{LSC}D_{0+}^{\alpha,\upsilon,\lambda}\varepsilon(\tau), \tag{4.820} $$

where γ is the viscosity of the material, $\varepsilon(\tau)$ is the strain, $\sigma(\tau)$ is the stress, and τ is the time.

The result was proposed in 2020 [92].

Model (Maxwell-Like Element)

The Maxwell-like model with the Wiman kernel can be written as follows [92]:

$$ {}_{GW}^{LSC}D_{0+}^{\alpha,\upsilon,\lambda}\varepsilon(\tau) = \frac{\sigma(\tau)}{\gamma} + \frac{{}_{GW}^{LSC}D_{0+}^{\alpha,\upsilon,\lambda}\sigma(\tau)}{\zeta}, \tag{4.821} $$

where γ is the viscosity of the material and ζ is the elastic modulus of the material.

The result was proposed in 2020 [92].

Model (Kelvin–Voigt-Like Model)

The Kelvin–Voigt-like model with the Wiman kernel can be given as follows [92]:

$$ \sigma(\tau) = \gamma\varepsilon(\tau) + \zeta\, {}_{Ml}^{LSC}D_{0+}^{\alpha,\lambda}\varepsilon(\tau), \tag{4.822} $$

where γ is the viscosity of the material and ζ is the elastic modulus of the material.

The result was proposed in 2020 [92].

Model (The Anomalous Diffusion Model)

The anomalous diffusion model containing the general fractional derivative time with the Wiman kernel was [259]

$$ {}_{GW}^{LSC}D_{0+}^{\alpha,\upsilon,\lambda}\Xi\left(x,\tau\right)=u\frac{\partial^{2}\Xi\left(x,\tau\right)}{\partial x^{2}}, \tag{4.823}$$

where u is a real constant reflecting the magnitudes of the diffusion capacity.

The result was proposed by Liang et al. in 2018 [259].

Chapter 5
Prabhakar, Supertrigonometric, and Superhyperbolic Functions

Abstract In this chapter, we introduce the Prabhakar function and propose the supertrigonometric and superhyperbolic functions via Prabhakar function. The Laplace transforms for the new special functions and integral representations for the supertrigonometric and superhyperbolic functions are discussed in detail. The truncated Prabhakar, supertrigonometric, and superhyperbolic functions are proposed, and the general fractional calculus involving the Prabhakar function is also considered. Finally, the integral equations and mathematical models related to Prabhakar function are presented.

5.1 The Prabhakar Function: History, Definitions, Properties, and Theorems

We now introduce the history, definitions, properties, and theorems for the Prabhakar function.

5.1.1 The Prabhakar Function

In 1971, Tilak Raj Prabhakar, an Indian mathematician, worked on the generalized the Mittag-Leffler function, which is called the three-parametric Mittag-Leffler function, Mittag-Leffler type function, or Prabhakar function [260].

Definition 5.1 The Prabhakar function is defined as [260]

$$E_{\alpha,\beta}^{\gamma}(t)=\sum_{k=0}^{\infty}\frac{(\gamma)_k}{\Gamma(k\alpha+\beta)}\frac{t^{\kappa}}{k!}, \tag{5.1}$$

where $\alpha, \beta, \gamma, t \in \mathbb{C}$, $Re(\alpha) > 0$, and $Re(\beta) > 0$.

X.-J. Yang, *Theory and Applications of Special Functions for Scientists and Engineers*, https://doi.org/10.1007/978-981-33-6334-2_5

The Prabhakar function is defined as [260]

$$E_{\alpha,\beta}^{\gamma}\left(t^{\alpha}\right)=\sum_{k=0}^{\infty}\frac{(\gamma)_k}{\Gamma\left(k\alpha+\beta\right)}\frac{t^{k\alpha}}{k!}, \tag{5.2}$$

where $\alpha, \beta, \gamma, t \in \mathbb{C}$, $Re\left(\alpha\right) > 0$, and $Re\left(\beta\right) > 0$.

The Prabhakar function is defined as [210]

$$E_{\alpha,\beta}^{\gamma}\left(t^{\varsigma}\right)=\sum_{k=0}^{\infty}\frac{(\gamma)_k}{\Gamma\left(k\alpha+\beta\right)}\frac{t^{k\varsigma}}{k!}, \tag{5.3}$$

where $\alpha, \beta, \gamma, \varsigma, t \in \mathbb{C}$, $Re\left(\alpha\right) > 0$, and $Re\left(\beta\right) > 0$.

There are some special cases as follows:

$$E_{1,\beta}^{\gamma}\left(t\right)=\sum_{k=0}^{\infty}\frac{(\gamma)_k}{\Gamma\left(k+\beta\right)}\frac{t^{\kappa}}{k!}, \tag{5.4}$$

$$E_{1,\beta}^{\gamma}\left(t^{\alpha}\right)=\sum_{k=0}^{\infty}\frac{(\gamma)_k}{\Gamma\left(k+\beta\right)}\frac{t^{k\alpha}}{k!}, \tag{5.5}$$

$$E_{1,\beta}^{\gamma}\left(t^{\varsigma}\right)=\sum_{k=0}^{\infty}\frac{(\gamma)_k}{\Gamma\left(k+\beta\right)}\frac{t^{k\varsigma}}{k!}, \tag{5.6}$$

$$E_{\alpha,1}^{\gamma}\left(t\right)=\sum_{k=0}^{\infty}\frac{(\gamma)_k}{\Gamma\left(k\alpha+1\right)}\frac{t^{\kappa}}{k!}, \tag{5.7}$$

$$E_{\alpha,1}^{\gamma}\left(t^{\alpha}\right)=\sum_{k=0}^{\infty}\frac{(\gamma)_k}{\Gamma\left(k\alpha+1\right)}\frac{t^{k\alpha}}{k!}, \tag{5.8}$$

$$E_{\alpha,1}^{\gamma}\left(t^{\varsigma}\right)=\sum_{k=0}^{\infty}\frac{(\gamma)_k}{\Gamma\left(k\alpha+1\right)}\frac{t^{k\varsigma}}{k!}, \tag{5.9}$$

$$E_{1,1}^{\gamma}\left(t\right)=\sum_{k=0}^{\infty}\frac{(\gamma)_k}{\Gamma\left(k+1\right)}\frac{t^{\kappa}}{k!}, \tag{5.10}$$

$$E_{1,1}^{\gamma}\left(t^{\alpha}\right)=\sum_{k=0}^{\infty}\frac{(\gamma)_k}{\Gamma\left(k+1\right)}\frac{t^{k\alpha}}{k!}, \tag{5.11}$$

$$E_{1,1}^{\gamma}\left(t^{\varsigma}\right)=\sum_{k=0}^{\infty}\frac{(\gamma)_k}{\Gamma\left(k+1\right)}\frac{t^{k\varsigma}}{k!}, \tag{5.12}$$

$$E_{\alpha,\beta}^{1}\left(t\right)=E_{\alpha,\beta}\left(t\right)=\sum_{k=0}^{\infty}\frac{t^{\kappa}}{\Gamma\left(k\alpha+\beta\right)}, \tag{5.13}$$

$$E_{\alpha,\beta}^{1}\left(t^{\alpha}\right)=E_{\alpha,\beta}\left(t^{\alpha}\right)=\sum_{k=0}^{\infty}\frac{t^{\kappa\alpha}}{\Gamma\left(k\alpha+\beta\right)}, \tag{5.14}$$

$$E_{\alpha,\beta}^{1}\left(t^{\varsigma}\right)=E_{\alpha,\beta}\left(t^{\varsigma}\right)=\sum_{k=0}^{\infty}\frac{t^{k\varsigma}}{\Gamma\left(k\alpha+\beta\right)}, \tag{5.15}$$

$$E_{\alpha,1}^{1}\left(t\right)=E_{\alpha,1}\left(t\right)=E_{\alpha}\left(t\right)=\sum_{k=0}^{\infty}\frac{t^{\kappa}}{\Gamma\left(k\alpha+1\right)}, \tag{5.16}$$

$$E_{\alpha,1}^{1}\left(t^{\alpha}\right)=E_{\alpha,1}\left(t^{\alpha}\right)=E_{\alpha}\left(t^{\alpha}\right)=\sum_{k=0}^{\infty}\frac{t^{\kappa\alpha}}{\Gamma\left(k\alpha+1\right)}, \tag{5.17}$$

and

$$E_{\alpha,1}^{1}\left(t^{\varsigma}\right)=E_{\alpha,1}\left(t^{\varsigma}\right)=E_{\alpha,1}\left(t^{\varsigma}\right)=\sum_{k=0}^{\infty}\frac{t^{k\varsigma}}{\Gamma\left(k\alpha+1\right)}, \tag{5.18}$$

where $\alpha, \beta, \gamma, \varsigma, t \in \mathbb{C}$, $Re\left(\alpha\right) > 0$, and $Re\left(\beta\right) > 0$.

Moreover,

$$E_{1,\beta}^{\gamma}\left(-t\right)=\sum_{k=0}^{\infty}\frac{(-1)^k\left(\gamma\right)_k}{\Gamma\left(k+\beta\right)}\frac{t^{\kappa}}{k!}, \tag{5.19}$$

$$E_{1,\beta}^{\gamma}\left(-t^{\alpha}\right)=\sum_{k=0}^{\infty}\frac{(-1)^k\left(\gamma\right)_k}{\Gamma\left(k+\beta\right)}\frac{t^{k\alpha}}{k!}, \tag{5.20}$$

$$E_{1,\beta}^{\gamma}\left(-t^{\varsigma}\right)=\sum_{k=0}^{\infty}\frac{(-1)^k\left(\gamma\right)_k}{\Gamma\left(k+\beta\right)}\frac{t^{k\varsigma}}{k!}, \tag{5.21}$$

$$E_{\alpha,1}^{\gamma}\left(-t\right)=\sum_{k=0}^{\infty}\frac{(-1)^k\left(\gamma\right)_k}{\Gamma\left(k\alpha+1\right)}\frac{t^{\kappa}}{k!}, \tag{5.22}$$

$$E_{\alpha,1}^{\gamma}\left(-t^{\alpha}\right)=\sum_{k=0}^{\infty}\frac{(-1)^{k}\,(\gamma)_{k}}{\Gamma\,(k\alpha+1)}\frac{t^{k\alpha}}{k!}, \tag{5.23}$$

$$E_{\alpha,1}^{\gamma}\left(-t^{\varsigma}\right)=\sum_{k=0}^{\infty}\frac{(-1)^{k}\,(\gamma)_{k}}{\Gamma\,(k\alpha+1)}\frac{t^{k\varsigma}}{k!}, \tag{5.24}$$

$$E_{1,1}^{\gamma}\left(-t\right)=\sum_{k=0}^{\infty}\frac{(-1)^{k}\,(\gamma)_{k}}{\Gamma\,(k+1)}\frac{t^{\kappa}}{k!}, \tag{5.25}$$

$$E_{1,1}^{\gamma}\left(-t^{\alpha}\right)=\sum_{k=0}^{\infty}\frac{(-1)^{k}\,(\gamma)_{k}}{\Gamma\,(k+1)}\frac{t^{k\alpha}}{k!}, \tag{5.26}$$

$$E_{1,1}^{\gamma}\left(-t^{\varsigma}\right)=\sum_{k=0}^{\infty}\frac{(-1)^{k}\,(\gamma)_{k}}{\Gamma\,(k+1)}\frac{t^{k\varsigma}}{k!}, \tag{5.27}$$

$$E_{\alpha,\beta}^{1}\left(-t\right)=E_{\alpha,\beta}\left(-t\right)=\sum_{k=0}^{\infty}(-1)^{k}\frac{t^{\kappa}}{\Gamma\,(k\alpha+\beta)}, \tag{5.28}$$

$$E_{\alpha,\beta}^{1}\left(-t^{\alpha}\right)=E_{\alpha,\beta}\left(-t^{\alpha}\right)=\sum_{k=0}^{\infty}\frac{(-1)^{k}\,t^{\kappa\alpha}}{\Gamma\,(k\alpha+\beta)}, \tag{5.29}$$

$$E_{\alpha,\beta}^{1}\left(-t^{\varsigma}\right)=E_{\alpha,\beta}\left(-t^{\varsigma}\right)=\sum_{k=0}^{\infty}(-1)^{k}\frac{t^{k\varsigma}}{\Gamma\,(k\alpha+\beta)}, \tag{5.30}$$

$$E_{\alpha,1}^{1}\left(-t\right)=E_{\alpha,1}\left(-t\right)=E_{\alpha}\left(-t\right)=\sum_{k=0}^{\infty}\frac{(-1)^{k}\,t^{\kappa}}{\Gamma\,(k\alpha+1)}, \tag{5.31}$$

$$E_{\alpha,1}^{1}\left(-t^{\alpha}\right)=E_{\alpha,1}\left(-t^{\alpha}\right)=E_{\alpha}\left(-t^{\alpha}\right)=\sum_{k=0}^{\infty}(-1)^{k}\frac{t^{\kappa\alpha}}{\Gamma\,(k\alpha+1)}, \tag{5.32}$$

and

$$E_{\alpha,1}^{1}\left(-t^{\varsigma}\right)=E_{\alpha,1}\left(-t^{\varsigma}\right)=E_{\alpha,1}\left(-t^{\varsigma}\right)=\sum_{k=0}^{\infty}\frac{(-1)^{k}\,t^{k\varsigma}}{\Gamma\,(k\alpha+1)}, \tag{5.33}$$

where α, β, γ, ς, $t\in\mathbb{C}$, $Re\,(\alpha)>0$, and $Re\,(\beta)>0$.

The relationship between the Prabhakar function and Wright hypergeometric function is given as follows [261]:

$$E^{\gamma}_{\alpha,\beta}(t) = \frac{1}{\Gamma(\gamma)}\, {}_1\varphi_1\left(\begin{matrix}(\gamma,1)\\(\beta,\alpha)\end{matrix};t\right), \tag{5.34}$$

where $\beta, \gamma, t \in \mathbb{C}$, $\alpha > 0$, and $Re(\beta) > 0$.

Thus, we have that

$$E^{\gamma}_{\alpha,\beta}(t) = \sum_{k=0}^{\infty} \frac{(\gamma)_k}{\Gamma(k\alpha+\beta)} \frac{t^{\kappa}}{k!} = \frac{1}{\Gamma(\gamma)}\, {}_1\varphi_1\left(\begin{matrix}(\gamma,1)\\(\beta,\alpha)\end{matrix};t\right), \tag{5.35}$$

$$E^{\gamma}_{\alpha,\beta}\left(t^{\alpha}\right) = \sum_{k=0}^{\infty} \frac{(\gamma)_k}{\Gamma(k\alpha+\beta)} \frac{t^{k\alpha}}{k!} = \frac{1}{\Gamma(\gamma)}\, {}_1\varphi_1\left(\begin{matrix}(\gamma,1)\\(\beta,\alpha)\end{matrix};t^{\alpha}\right), \tag{5.36}$$

$$E^{\gamma}_{\alpha,\beta}\left(t^{\varsigma}\right) = \sum_{k=0}^{\infty} \frac{(\gamma)_k}{\Gamma(k\alpha+\beta)} \frac{t^{k\varsigma}}{k!} = \frac{1}{\Gamma(\gamma)}\, {}_1\varphi_1\left(\begin{matrix}(\gamma,1)\\(\beta,\alpha)\end{matrix};t^{\varsigma}\right), \tag{5.37}$$

$$E^{\gamma}_{1,\beta}(t) = \sum_{k=0}^{\infty} \frac{(\gamma)_k}{\Gamma(k+\beta)} \frac{t^{\kappa}}{k!} = \frac{1}{\Gamma(\gamma)}\, {}_1\varphi_1\left(\begin{matrix}(\gamma,1)\\(\beta,1)\end{matrix};t\right), \tag{5.38}$$

$$E^{\gamma}_{1,\beta}\left(t^{\alpha}\right) = \sum_{k=0}^{\infty} \frac{(\gamma)_k}{\Gamma(k+\beta)} \frac{t^{k\alpha}}{k!} = \frac{1}{\Gamma(\gamma)}\, {}_1\varphi_1\left(\begin{matrix}(\gamma,1)\\(\beta,1)\end{matrix};t^{\alpha}\right), \tag{5.39}$$

$$E^{\gamma}_{1,\beta}\left(t^{\varsigma}\right) = \sum_{k=0}^{\infty} \frac{(\gamma)_k}{\Gamma(k+\beta)} \frac{t^{k\varsigma}}{k!} = \frac{1}{\Gamma(\gamma)}\, {}_1\varphi_1\left(\begin{matrix}(\gamma,1)\\(\beta,1)\end{matrix};t^{\varsigma}\right), \tag{5.40}$$

$$E^{\gamma}_{\alpha,1}(t) = \sum_{k=0}^{\infty} \frac{(\gamma)_k}{\Gamma(k\alpha+1)} \frac{t^{\kappa}}{k!} = \frac{1}{\Gamma(\gamma)}\, {}_1\varphi_1\left(\begin{matrix}(\gamma,1)\\(1,\alpha)\end{matrix};t\right), \tag{5.41}$$

$$E^{\gamma}_{\alpha,1}\left(t^{\alpha}\right) = \sum_{k=0}^{\infty} \frac{(\gamma)_k}{\Gamma(k\alpha+1)} \frac{t^{k\alpha}}{k!} = \frac{1}{\Gamma(\gamma)}\, {}_1\varphi_1\left(\begin{matrix}(\gamma,1)\\(1,\alpha)\end{matrix};t^{\alpha}\right), \tag{5.42}$$

$$E^{\gamma}_{\alpha,1}\left(t^{\varsigma}\right) = \sum_{k=0}^{\infty} \frac{(\gamma)_k}{\Gamma(k\alpha+1)} \frac{t^{k\varsigma}}{k!} = \frac{1}{\Gamma(\gamma)}\, {}_1\varphi_1\left(\begin{matrix}(\gamma,1)\\(1,\alpha)\end{matrix};t^{\varsigma}\right), \tag{5.43}$$

$$E^{\gamma}_{1,1}(t) = \sum_{k=0}^{\infty} \frac{(\gamma)_k}{\Gamma(k+1)} \frac{t^{\kappa}}{k!} = \frac{1}{\Gamma(\gamma)}\, {}_1\varphi_1\left(\begin{matrix}(\gamma,1)\\(1,1)\end{matrix};t\right), \tag{5.44}$$

$$E^{\gamma}_{1,1}\left(t^{\alpha}\right)=\sum_{k=0}^{\infty}\frac{(\gamma)_k}{\Gamma\left(k+1\right)}\frac{t^{k\alpha}}{k!}=\frac{1}{\Gamma\left(\gamma\right)}\,{}_1\varphi_1\left(\begin{matrix}(\gamma,1)\\(1,1)\end{matrix};t^{\alpha}\right), \tag{5.45}$$

$$E^{\gamma}_{1,1}\left(t^{\varsigma}\right)=\sum_{k=0}^{\infty}\frac{(\gamma)_k}{\Gamma\left(k+1\right)}\frac{t^{k\varsigma}}{k!}=\frac{1}{\Gamma\left(\gamma\right)}\,{}_1\varphi_1\left(\begin{matrix}(\gamma,1)\\(1,1)\end{matrix};t^{\varsigma}\right), \tag{5.46}$$

where $\alpha, \beta, \gamma, \varsigma, t \in \mathbb{C}$, $Re\left(\alpha\right) > 0$, and $Re\left(\beta\right) > 0$.

Similarly,

$$E^{\gamma}_{\alpha,\beta}\left(-t\right)=\sum_{k=0}^{\infty}\frac{(-1)^k\,(\gamma)_k}{\Gamma\left(k\alpha+\beta\right)}\frac{t^{\kappa}}{k!}=\frac{1}{\Gamma\left(\gamma\right)}\,{}_1\varphi_1\left(\begin{matrix}(\gamma,1)\\(\beta,\alpha)\end{matrix};-t\right), \tag{5.47}$$

$$E^{\gamma}_{\alpha,\beta}\left(-t^{\alpha}\right)=\sum_{k=0}^{\infty}\frac{(-1)^k\,(\gamma)_k}{\Gamma\left(k\alpha+\beta\right)}\frac{t^{k\alpha}}{k!}=\frac{1}{\Gamma\left(\gamma\right)}\,{}_1\varphi_1\left(\begin{matrix}(\gamma,1)\\(\beta,\alpha)\end{matrix};-t^{\alpha}\right), \tag{5.48}$$

$$E^{\gamma}_{\alpha,\beta}\left(-t^{\varsigma}\right)=\sum_{k=0}^{\infty}\frac{(-1)^k\,(\gamma)_k}{\Gamma\left(k\alpha+\beta\right)}\frac{t^{k\varsigma}}{k!}=\frac{1}{\Gamma\left(\gamma\right)}\,{}_1\varphi_1\left(\begin{matrix}(\gamma,1)\\(\beta,\alpha)\end{matrix};-t^{\varsigma}\right), \tag{5.49}$$

$$E^{\gamma}_{1,\beta}\left(-t\right)=\sum_{k=0}^{\infty}\frac{(-1)^k\,(\gamma)_k}{\Gamma\left(k+\beta\right)}\frac{t^{\kappa}}{k!}=\frac{1}{\Gamma\left(\gamma\right)}\,{}_1\varphi_1\left(\begin{matrix}(\gamma,1)\\(\beta,1)\end{matrix};-t\right), \tag{5.50}$$

$$E^{\gamma}_{1,\beta}\left(-t^{\alpha}\right)=\sum_{k=0}^{\infty}\frac{(-1)^k\,(\gamma)_k}{\Gamma\left(k+\beta\right)}\frac{t^{k\alpha}}{k!}=\frac{1}{\Gamma\left(\gamma\right)}\,{}_1\varphi_1\left(\begin{matrix}(\gamma,1)\\(\beta,1)\end{matrix};-t^{\alpha}\right), \tag{5.51}$$

$$E^{\gamma}_{1,\beta}\left(-t^{\varsigma}\right)=\sum_{k=0}^{\infty}\frac{(-1)^k\,(\gamma)_k}{\Gamma\left(k+\beta\right)}\frac{t^{k\varsigma}}{k!}=\frac{1}{\Gamma\left(\gamma\right)}\,{}_1\varphi_1\left(\begin{matrix}(\gamma,1)\\(\beta,1)\end{matrix};-t^{\varsigma}\right), \tag{5.52}$$

$$E^{\gamma}_{\alpha,1}\left(-t\right)=\sum_{k=0}^{\infty}\frac{(-1)^k\,(\gamma)_k}{\Gamma\left(k\alpha+1\right)}\frac{t^{\kappa}}{k!}=\frac{1}{\Gamma\left(\gamma\right)}\,{}_1\varphi_1\left(\begin{matrix}(\gamma,1)\\(1,\alpha)\end{matrix};-t\right), \tag{5.53}$$

$$E^{\gamma}_{\alpha,1}\left(-t^{\alpha}\right)=\sum_{k=0}^{\infty}\frac{(-1)^k\,(\gamma)_k}{\Gamma\left(k\alpha+1\right)}\frac{t^{k\alpha}}{k!}=\frac{1}{\Gamma\left(\gamma\right)}\,{}_1\varphi_1\left(\begin{matrix}(\gamma,1)\\(1,\alpha)\end{matrix};-t^{\alpha}\right), \tag{5.54}$$

$$E^{\gamma}_{\alpha,1}\left(-t^{\varsigma}\right)=\sum_{k=0}^{\infty}\frac{(-1)^k\,(\gamma)_k}{\Gamma\left(k\alpha+1\right)}\frac{t^{k\varsigma}}{k!}=\frac{1}{\Gamma\left(\gamma\right)}\,{}_1\varphi_1\left(\begin{matrix}(\gamma,1)\\(1,\alpha)\end{matrix};-t^{\varsigma}\right), \tag{5.55}$$

$$E^{\gamma}_{1,1}(-t)=\sum_{k=0}^{\infty}\frac{(-1)^k(\gamma)_k}{\Gamma(k+1)}\frac{t^{\kappa}}{k!}=\frac{1}{\Gamma(\gamma)}{}_1\varphi_1\left(\begin{matrix}(\gamma,1)\\(1,1)\end{matrix};-t\right), \tag{5.56}$$

$$E^{\gamma}_{1,1}(-t^{\alpha})=\sum_{k=0}^{\infty}\frac{(-1)^k(\gamma)_k}{\Gamma(k+1)}\frac{t^{k\alpha}}{k!}=\frac{1}{\Gamma(\gamma)}{}_1\varphi_1\left(\begin{matrix}(\gamma,1)\\(1,1)\end{matrix};-t^{\alpha}\right), \tag{5.57}$$

$$E^{\gamma}_{1,1}(-t^{\varsigma})=\sum_{k=0}^{\infty}\frac{(-1)^k(\gamma)_k}{\Gamma(k+1)}\frac{t^{k\varsigma}}{k!}=\frac{1}{\Gamma(\gamma)}{}_1\varphi_1\left(\begin{matrix}(\gamma,1)\\(1,1)\end{matrix};-t^{\varsigma}\right), \tag{5.58}$$

where $\alpha, \beta, \gamma, \varsigma, t \in \mathbb{C}$, $Re(\alpha) > 0$, and $Re(\beta) > 0$.

So, we present that

$$E^{1}_{\alpha,\beta}(t)=E_{\alpha,\beta}(t)=\sum_{k=0}^{\infty}\frac{t^{\kappa}}{\Gamma(k\alpha+\beta)}={}_1\varphi_1\left(\begin{matrix}(1,1)\\(\beta,\alpha)\end{matrix};t\right), \tag{5.59}$$

$$E^{1}_{\alpha,\beta}(t^{\alpha})=E_{\alpha,\beta}(t^{\alpha})=\sum_{k=0}^{\infty}\frac{t^{\kappa\alpha}}{\Gamma(k\alpha+\beta)}={}_1\varphi_1\left(\begin{matrix}(1,1)\\(\beta,\alpha)\end{matrix};t^{\alpha}\right), \tag{5.60}$$

$$E^{1}_{\alpha,\beta}(t^{\varsigma})=E_{\alpha,\beta}(t^{\varsigma})=\sum_{k=0}^{\infty}\frac{t^{k\varsigma}}{\Gamma(k\alpha+\beta)}={}_1\varphi_1\left(\begin{matrix}(1,1)\\(\beta,\alpha)\end{matrix};t^{\varsigma}\right), \tag{5.61}$$

$$E^{1}_{\alpha,1}(t)=E_{\alpha,1}(t)=E_{\alpha}(t)=\sum_{k=0}^{\infty}\frac{t^{\kappa}}{\Gamma(k\alpha+1)}={}_1\varphi_1\left(\begin{matrix}(1,1)\\(1,\alpha)\end{matrix};t\right), \tag{5.62}$$

$$E^{1}_{\alpha,1}(t^{\alpha})=E_{\alpha,1}(t^{\alpha})=E_{\alpha}(t^{\alpha})=\sum_{k=0}^{\infty}\frac{t^{\kappa\alpha}}{\Gamma(k\alpha+1)}={}_1\varphi_1\left(\begin{matrix}(1,1)\\(1,\alpha)\end{matrix};t^{\alpha}\right), \tag{5.63}$$

and

$$E^{1}_{\alpha,1}(t^{\varsigma})=E_{\alpha,1}(t^{\varsigma})=E_{\alpha,1}(t^{\varsigma})=\sum_{k=0}^{\infty}\frac{t^{k\varsigma}}{\Gamma(k\alpha+1)}={}_1\varphi_1\left(\begin{matrix}(1,1)\\(1,\alpha)\end{matrix};t^{\varsigma}\right), \tag{5.64}$$

where $\alpha, \beta, \varsigma, t \in \mathbb{C}$, $Re(\alpha) > 0$, and $Re(\beta) > 0$.

Similarly, we get

$$E^{1}_{\alpha,\beta}(-t)=E_{\alpha,\beta}(-t)=\sum_{k=0}^{\infty}\frac{(-1)^k t^{\kappa}}{\Gamma(k\alpha+\beta)}={}_1\varphi_1\left(\begin{matrix}(1,1)\\(\beta,\alpha)\end{matrix};-t\right), \tag{5.65}$$

$$E^{1}_{\alpha,\beta}\left(-t^{\alpha}\right)=E_{\alpha,\beta}\left(-t^{\alpha}\right)=\sum_{k=0}^{\infty}\frac{(-1)^{k}\,t^{\kappa\alpha}}{\Gamma\left(k\alpha+\beta\right)}={}_{1}\varphi_{1}\left(\begin{matrix}(1,1)\\(\beta,\alpha)\end{matrix};-t^{\alpha}\right), \tag{5.66}$$

$$E^{1}_{\alpha,\beta}\left(-t^{\varsigma}\right)=E_{\alpha,\beta}\left(-t^{\varsigma}\right)=\sum_{k=0}^{\infty}\frac{(-1)^{k}\,t^{k\varsigma}}{\Gamma\left(k\alpha+\beta\right)}={}_{1}\varphi_{1}\left(\begin{matrix}(1,1)\\(\beta,\alpha)\end{matrix};-t^{\varsigma}\right), \tag{5.67}$$

$$E^{1}_{\alpha,1}\left(-t\right)=E_{\alpha,1}\left(-t\right)=E_{\alpha}\left(-t\right)=\sum_{k=0}^{\infty}\frac{(-1)^{k}\,t^{\kappa}}{\Gamma\left(k\alpha+1\right)}={}_{1}\varphi_{1}\left(\begin{matrix}(1,1)\\(1,\alpha)\end{matrix};-t\right), \tag{5.68}$$

$$E^{1}_{\alpha,1}\left(-t^{\alpha}\right)=E_{\alpha,1}\left(-t^{\alpha}\right)=E_{\alpha}\left(-t^{\alpha}\right)=\sum_{k=0}^{\infty}\frac{(-1)^{k}\,t^{\kappa\alpha}}{\Gamma\left(k\alpha+1\right)}={}_{1}\varphi_{1}\left(\begin{matrix}(1,1)\\(1,\alpha)\end{matrix};-t^{\alpha}\right), \tag{5.69}$$

and

$$E^{1}_{\alpha,1}\left(-t^{\varsigma}\right)=E_{\alpha,1}\left(-t^{\varsigma}\right)=E_{\alpha,1}\left(-t^{\varsigma}\right)=\sum_{k=0}^{\infty}\frac{(-1)^{k}\,t^{k\varsigma}}{\Gamma\left(k\alpha+1\right)}={}_{1}\varphi_{1}\left(\begin{matrix}(1,1)\\(1,\alpha)\end{matrix};-t^{\varsigma}\right), \tag{5.70}$$

where α, β, ς, $t\in\mathbb{C}$, $Re\,(\alpha)>0$, and $Re\,(\beta)>0$.

Here, the family of Gaussian-like functions can be written as follows:

$$E^{\gamma}_{\alpha,\beta}\left(t^{2}\right)=\sum_{k=0}^{\infty}\frac{(\gamma)_{k}}{\Gamma\left(k\alpha+\beta\right)}\frac{t^{2\kappa}}{k!}=\frac{1}{\Gamma\left(\gamma\right)}\,{}_{1}\varphi_{1}\left(\begin{matrix}(\gamma,1)\\(\beta,\alpha)\end{matrix};t^{2}\right), \tag{5.71}$$

$$E^{\gamma}_{\alpha,\beta}\left(t^{2\alpha}\right)=\sum_{k=0}^{\infty}\frac{(\gamma)_{k}}{\Gamma\left(k\alpha+\beta\right)}\frac{t^{2k\alpha}}{k!}=\frac{1}{\Gamma\left(\gamma\right)}\,{}_{1}\varphi_{1}\left(\begin{matrix}(\gamma,1)\\(\beta,\alpha)\end{matrix};t^{2\alpha}\right), \tag{5.72}$$

$$E^{\gamma}_{\alpha,\beta}\left(t^{2\varsigma}\right)=\sum_{k=0}^{\infty}\frac{(\gamma)_{k}}{\Gamma\left(k\alpha+\beta\right)}\frac{t^{2k\varsigma}}{k!}=\frac{1}{\Gamma\left(\gamma\right)}\,{}_{1}\varphi_{1}\left(\begin{matrix}(\gamma,1)\\(\beta,\alpha)\end{matrix};t^{2\varsigma}\right), \tag{5.73}$$

$$E^{\gamma}_{1,\beta}\left(t^{2}\right)=\sum_{k=0}^{\infty}\frac{(\gamma)_{k}}{\Gamma\left(k+\beta\right)}\frac{t^{2\kappa}}{k!}=\frac{1}{\Gamma\left(\gamma\right)}\,{}_{1}\varphi_{1}\left(\begin{matrix}(\gamma,1)\\(\beta,1)\end{matrix};t^{2}\right), \tag{5.74}$$

$$E^{\gamma}_{1,\beta}\left(t^{2\alpha}\right)=\sum_{k=0}^{\infty}\frac{(\gamma)_{k}}{\Gamma\left(k+\beta\right)}\frac{t^{2k\alpha}}{k!}=\frac{1}{\Gamma\left(\gamma\right)}\,{}_{1}\varphi_{1}\left(\begin{matrix}(\gamma,1)\\(\beta,1)\end{matrix};t^{2\alpha}\right), \tag{5.75}$$

$$E_{1,\beta}^{\gamma}\left(t^{2\varsigma}\right)=\sum_{k=0}^{\infty}\frac{(\gamma)_k}{\Gamma\left(k+\beta\right)}\frac{t^{2k\varsigma}}{k!}=\frac{1}{\Gamma\left(\gamma\right)}{}_1\varphi_1\left(\begin{matrix}(\gamma,1)\\(\beta,1)\end{matrix};t^{2\varsigma}\right), \tag{5.76}$$

$$E_{\alpha,1}^{\gamma}\left(t^{2}\right)=\sum_{k=0}^{\infty}\frac{(\gamma)_k}{\Gamma\left(k\alpha+1\right)}\frac{t^{2\kappa}}{k!}=\frac{1}{\Gamma\left(\gamma\right)}{}_1\varphi_1\left(\begin{matrix}(\gamma,1)\\(1,\alpha)\end{matrix};t^{2}\right), \tag{5.77}$$

$$E_{\alpha,1}^{\gamma}\left(t^{2\alpha}\right)=\sum_{k=0}^{\infty}\frac{(\gamma)_k}{\Gamma\left(k\alpha+1\right)}\frac{t^{2k\alpha}}{k!}=\frac{1}{\Gamma\left(\gamma\right)}{}_1\varphi_1\left(\begin{matrix}(\gamma,1)\\(1,\alpha)\end{matrix};t^{2\alpha}\right), \tag{5.78}$$

$$E_{\alpha,1}^{\gamma}\left(t^{2\varsigma}\right)=\sum_{k=0}^{\infty}\frac{(\gamma)_k}{\Gamma\left(k\alpha+1\right)}\frac{t^{2k\varsigma}}{k!}=\frac{1}{\Gamma\left(\gamma\right)}{}_1\varphi_1\left(\begin{matrix}(\gamma,1)\\(1,\alpha)\end{matrix};t^{2\varsigma}\right), \tag{5.79}$$

$$E_{1,1}^{\gamma}\left(t^{2}\right)=\sum_{k=0}^{\infty}\frac{(\gamma)_k}{\Gamma\left(k+1\right)}\frac{t^{2\kappa}}{k!}=\frac{1}{\Gamma\left(\gamma\right)}{}_1\varphi_1\left(\begin{matrix}(\gamma,1)\\(1,1)\end{matrix};t^{2}\right), \tag{5.80}$$

$$E_{1,1}^{\gamma}\left(t^{2\alpha}\right)=\sum_{k=0}^{\infty}\frac{(\gamma)_k}{\Gamma\left(k+1\right)}\frac{t^{2k\alpha}}{k!}=\frac{1}{\Gamma\left(\gamma\right)}{}_1\varphi_1\left(\begin{matrix}(\gamma,1)\\(1,1)\end{matrix};t^{2\alpha}\right), \tag{5.81}$$

$$E_{1,1}^{\gamma}\left(t^{2\varsigma}\right)=\sum_{k=0}^{\infty}\frac{(\gamma)_k}{\Gamma\left(k+1\right)}\frac{t^{2k\varsigma}}{k!}=\frac{1}{\Gamma\left(\gamma\right)}{}_1\varphi_1\left(\begin{matrix}(\gamma,1)\\(1,1)\end{matrix};t^{2\varsigma}\right), \tag{5.82}$$

$$E_{\alpha,\beta}^{1}\left(t^{2}\right)=E_{\alpha,\beta}\left(t^{2}\right)=\sum_{k=0}^{\infty}\frac{t^{2\kappa}}{\Gamma\left(k\alpha+\beta\right)}={}_1\varphi_1\left(\begin{matrix}(1,1)\\(\beta,\alpha)\end{matrix};t^{2}\right), \tag{5.83}$$

$$E_{\alpha,\beta}^{1}\left(t^{2\alpha}\right)=E_{\alpha,\beta}\left(t^{2\alpha}\right)=\sum_{k=0}^{\infty}\frac{t^{2\kappa\alpha}}{\Gamma\left(k\alpha+\beta\right)}={}_1\varphi_1\left(\begin{matrix}(1,1)\\(\beta,\alpha)\end{matrix};t^{2\alpha}\right), \tag{5.84}$$

$$E_{\alpha,\beta}^{1}\left(t^{2\varsigma}\right)=E_{\alpha,\beta}\left(t^{2\varsigma}\right)=\sum_{k=0}^{\infty}\frac{t^{2k\varsigma}}{\Gamma\left(k\alpha+\beta\right)}={}_1\varphi_1\left(\begin{matrix}(1,1)\\(\beta,\alpha)\end{matrix};t^{2\varsigma}\right), \tag{5.85}$$

$$E_{\alpha,1}^{1}\left(t^{2}\right)=E_{\alpha,1}\left(t^{2}\right)=E_{\alpha}\left(t^{2}\right)=\sum_{k=0}^{\infty}\frac{t^{2\kappa}}{\Gamma\left(k\alpha+1\right)}={}_1\varphi_1\left(\begin{matrix}(1,1)\\(1,\alpha)\end{matrix};t^{2}\right), \tag{5.86}$$

$$E_{\alpha,1}^{1}\left(t^{2\alpha}\right)=E_{\alpha,1}\left(t^{2\alpha}\right)=E_{\alpha}\left(t^{2\alpha}\right)=\sum_{k=0}^{\infty}\frac{t^{2\kappa\alpha}}{\Gamma\left(k\alpha+1\right)}={}_1\varphi_1\left(\begin{matrix}(1,1)\\(1,\alpha)\end{matrix};t^{2\alpha}\right), \tag{5.87}$$

and

$$E^{1}_{\alpha,1}\left(t^{2\varsigma}\right)=E_{\alpha,1}\left(t^{2\varsigma}\right)=E_{\alpha,1}\left(t^{2\varsigma}\right)=\sum_{k=0}^{\infty}\frac{t^{2k\varsigma}}{\Gamma\left(k\alpha+1\right)}={}_1\varphi_1\left(\begin{matrix}(1,1)\\(1,\alpha)\end{matrix};t^{2\varsigma}\right), \tag{5.88}$$

where $\alpha,\beta,\gamma,\varsigma,t\in\mathbb{C}$, $Re\left(\alpha\right)>0$, and $Re\left(\beta\right)>0$.

In a similar way, we have

$$E^{\gamma}_{\alpha,\beta}\left(-t^{2}\right)=\sum_{k=0}^{\infty}\frac{(-1)^k(\gamma)_k}{\Gamma\left(k\alpha+\beta\right)}\frac{t^{2\kappa}}{k!}=\frac{1}{\Gamma\left(\gamma\right)}{}_1\varphi_1\left(\begin{matrix}(\gamma,1)\\(\beta,\alpha)\end{matrix};-t^{2}\right), \tag{5.89}$$

$$E^{\gamma}_{\alpha,\beta}\left(-t^{2\alpha}\right)=\sum_{k=0}^{\infty}\frac{(-1)^k(\gamma)_k}{\Gamma\left(k\alpha+\beta\right)}\frac{t^{2k\alpha}}{k!}=\frac{1}{\Gamma\left(\gamma\right)}{}_1\varphi_1\left(\begin{matrix}(\gamma,1)\\(\beta,\alpha)\end{matrix};-t^{2\alpha}\right), \tag{5.90}$$

$$E^{\gamma}_{\alpha,\beta}\left(-t^{2\varsigma}\right)=\sum_{k=0}^{\infty}\frac{(-1)^k(\gamma)_k}{\Gamma\left(k\alpha+\beta\right)}\frac{t^{2k\varsigma}}{k!}=\frac{1}{\Gamma\left(\gamma\right)}{}_1\varphi_1\left(\begin{matrix}(\gamma,1)\\(\beta,\alpha)\end{matrix};-t^{2\varsigma}\right), \tag{5.91}$$

$$E^{\gamma}_{1,\beta}\left(-t^{2}\right)=\sum_{k=0}^{\infty}\frac{(-1)^k(\gamma)_k}{\Gamma\left(k+\beta\right)}\frac{t^{2\kappa}}{k!}=\frac{1}{\Gamma\left(\gamma\right)}{}_1\varphi_1\left(\begin{matrix}(\gamma,1)\\(\beta,1)\end{matrix};-t^{2}\right), \tag{5.92}$$

$$E^{\gamma}_{1,\beta}\left(-t^{2\alpha}\right)=\sum_{k=0}^{\infty}\frac{(-1)^k(\gamma)_k}{\Gamma\left(k+\beta\right)}\frac{t^{2k\alpha}}{k!}=\frac{1}{\Gamma\left(\gamma\right)}{}_1\varphi_1\left(\begin{matrix}(\gamma,1)\\(\beta,1)\end{matrix};-t^{2\alpha}\right), \tag{5.93}$$

$$E^{\gamma}_{1,\beta}\left(-t^{2\varsigma}\right)=\sum_{k=0}^{\infty}\frac{(-1)^k(\gamma)_k}{\Gamma\left(k+\beta\right)}\frac{t^{2k\varsigma}}{k!}=\frac{1}{\Gamma\left(\gamma\right)}{}_1\varphi_1\left(\begin{matrix}(\gamma,1)\\(\beta,1)\end{matrix};-t^{2\varsigma}\right), \tag{5.94}$$

$$E^{\gamma}_{\alpha,1}\left(-t^{2}\right)=\sum_{k=0}^{\infty}\frac{(-1)^k(\gamma)_k}{\Gamma\left(k\alpha+1\right)}\frac{t^{2\kappa}}{k!}=\frac{1}{\Gamma\left(\gamma\right)}{}_1\varphi_1\left(\begin{matrix}(\gamma,1)\\(1,\alpha)\end{matrix};-t^{2}\right), \tag{5.95}$$

$$E^{\gamma}_{\alpha,1}\left(-t^{2\alpha}\right)=\sum_{k=0}^{\infty}\frac{(-1)^k(\gamma)_k}{\Gamma\left(k\alpha+1\right)}\frac{t^{2k\alpha}}{k!}=\frac{1}{\Gamma\left(\gamma\right)}{}_1\varphi_1\left(\begin{matrix}(\gamma,1)\\(1,\alpha)\end{matrix};-t^{2\alpha}\right), \tag{5.96}$$

$$E^{\gamma}_{\alpha,1}\left(-t^{2\varsigma}\right)=\sum_{k=0}^{\infty}\frac{(-1)^k(\gamma)_k}{\Gamma\left(k\alpha+1\right)}\frac{-t^{2k\varsigma}}{k!}=\frac{1}{\Gamma\left(\gamma\right)}{}_1\varphi_1\left(\begin{matrix}(\gamma,1)\\(1,\alpha)\end{matrix};-t^{2\varsigma}\right), \tag{5.97}$$

$$E^{\gamma}_{1,1}\left(-t^{2}\right)=\sum_{k=0}^{\infty}\frac{(-1)^k(\gamma)_k}{\Gamma\left(k+1\right)}\frac{t^{2\kappa}}{k!}=\frac{1}{\Gamma\left(\gamma\right)}{}_1\varphi_1\left(\begin{matrix}(\gamma,1)\\(1,1)\end{matrix};-t^{2}\right), \tag{5.98}$$

$$E_{1,1}^{\gamma}\left(-t^{2\alpha}\right)=\sum_{k=0}^{\infty}\frac{(-1)^k\,(\gamma)_k}{\Gamma\,(k+1)}\frac{t^{2k\alpha}}{k!}=\frac{1}{\Gamma\,(\gamma)}\,{}_1\varphi_1\left(\begin{matrix}(\gamma,1)\\(1,1)\end{matrix};-t^{2\alpha}\right), \tag{5.99}$$

$$E_{1,1}^{\gamma}\left(-t^{2\varsigma}\right)=\sum_{k=0}^{\infty}\frac{(-1)^k\,(\gamma)_k}{\Gamma\,(k+1)}\frac{t^{2k\varsigma}}{k!}=\frac{1}{\Gamma\,(\gamma)}\,{}_1\varphi_1\left(\begin{matrix}(\gamma,1)\\(1,1)\end{matrix};-t^{2\varsigma}\right), \tag{5.100}$$

$$E_{\alpha,\beta}^{1}\left(-t^{2}\right)=E_{\alpha,\beta}\left(-t^{2}\right)=\sum_{k=0}^{\infty}\frac{(-1)^k\,t^{2\kappa}}{\Gamma\,(k\alpha+\beta)}={}_1\varphi_1\left(\begin{matrix}(1,1)\\(\beta,\alpha)\end{matrix};-t^{2}\right), \tag{5.101}$$

$$E_{\alpha,\beta}^{1}\left(-t^{2\alpha}\right)=E_{\alpha,\beta}\left(-t^{2\alpha}\right)=\sum_{k=0}^{\infty}\frac{(-1)^k\,t^{2\kappa\alpha}}{\Gamma\,(k\alpha+\beta)}={}_1\varphi_1\left(\begin{matrix}(1,1)\\(\beta,\alpha)\end{matrix};-t^{2\alpha}\right), \tag{5.102}$$

$$E_{\alpha,\beta}^{1}\left(-t^{2\varsigma}\right)=E_{\alpha,\beta}\left(-t^{2\varsigma}\right)=\sum_{k=0}^{\infty}\frac{(-1)^k\,t^{2k\varsigma}}{\Gamma\,(k\alpha+\beta)}={}_1\varphi_1\left(\begin{matrix}(1,1)\\(\beta,\alpha)\end{matrix};-t^{2\varsigma}\right), \tag{5.103}$$

$$E_{\alpha,1}^{1}\left(-t^{2}\right)=E_{\alpha,1}\left(-t^{2}\right)=E_{\alpha}\left(-t^{2}\right)=\sum_{k=0}^{\infty}\frac{(-1)^k\,t^{2\kappa}}{\Gamma\,(k\alpha+1)}={}_1\varphi_1\left(\begin{matrix}(1,1)\\(1,\alpha)\end{matrix};-t^{2}\right), \tag{5.104}$$

$$E_{\alpha,1}^{1}\left(-t^{2\alpha}\right)=E_{\alpha,1}\left(-t^{2\alpha}\right)=E_{\alpha}\left(t^{2\alpha}\right)=\sum_{k=0}^{\infty}\frac{(-1)^k\,t^{2\kappa\alpha}}{\Gamma\,(k\alpha+1)}={}_1\varphi_1\left(\begin{matrix}(1,1)\\(1,\alpha)\end{matrix};-t^{2\alpha}\right), \tag{5.105}$$

and

$$E_{\alpha,1}^{1}\left(-t^{2\varsigma}\right)=E_{\alpha,1}\left(-t^{2\varsigma}\right)=E_{\alpha,1}\left(-t^{2\varsigma}\right)=\sum_{k=0}^{\infty}\frac{(-1)^k\,t^{2k\varsigma}}{\Gamma\,(k\alpha+1)}={}_1\varphi_1\left(\begin{matrix}(1,1)\\(1,\alpha)\end{matrix};-t^{2\varsigma}\right), \tag{5.106}$$

where $\alpha, \beta, \gamma, \varsigma, t \in \mathbb{C}$, $Re\,(\alpha) > 0$, and $Re\,(\beta) > 0$.

On the other hand, we have the binomial series [210]:

$$(1+t)^{-\varsigma}=\sum_{k=0}^{\infty}\frac{\Gamma\,(1-\varsigma)}{\Gamma\,(1-\varsigma-k)}\frac{t^k}{k!}=\sum_{k=0}^{\infty}(-1)^k\frac{\Gamma\,(k+\varsigma)}{\Gamma\,(\varsigma)}\frac{t^k}{k!}, \tag{5.107}$$

which leads to

$$
\begin{aligned}
&(1+t)^{-\varsigma} \\
&={}_1\varphi_1\left(\begin{matrix}(1-\varsigma,0)\\(1-\varsigma,-1)\end{matrix};t\right) \\
&=\Gamma(1-\varsigma)\times{}_0\varphi_1\left(\begin{matrix}-\\(1-\varsigma,-1)\end{matrix};t\right) \\
&={}_1\varphi_1\left(\begin{matrix}(\varsigma,1)\\(\varsigma,0)\end{matrix};-t\right) \\
&=\frac{1}{\Gamma(\varsigma)}\times{}_1\varphi_0\left(\begin{matrix}(\varsigma,1)\\-\end{matrix};-t\right),
\end{aligned}
\tag{5.108}
$$

where $\varsigma, t \in \mathbb{R}$ and $|t| < 1$.

Property 5.1 Let $\alpha, \beta, \gamma, t \in \mathbb{C}$, $k \in \mathbb{N}$, $Re(\alpha) > 0$, and $Re(\beta) > 0$. Then we have [260, 261]

$$
\frac{\partial^k}{\partial t^k}\left[t^{\beta-1}E_{\alpha,\beta}^{\gamma}\left(\lambda t^{\alpha}\right)\right]=t^{\beta-k-1}E_{\alpha,\beta-k}^{\gamma}\left(\lambda t^{\alpha}\right),
\tag{5.109}
$$

and

$$
\int_0^t t^{\beta-1}E_{\alpha,\beta}^{\gamma}\left(\lambda t^{\alpha}\right)dt=t^{\beta}E_{\alpha,\beta+1}^{\gamma}\left(\lambda t^{\alpha}\right),
\tag{5.110}
$$

where

$$
t^{\beta-1}E_{\alpha,\beta}^{\gamma}\left(\lambda t^{\alpha}\right)=\sum_{k=0}^{\infty}\frac{(\gamma)_k}{\Gamma(k\alpha+\beta)}\frac{\lambda^k t^{k\alpha+\beta-1}}{k!}=\frac{t^{\beta-1}}{\Gamma(\gamma)}{}_1\varphi_1\left(\begin{matrix}(\gamma,1)\\(\beta,\alpha)\end{matrix};\lambda t^{\alpha}\right).
\tag{5.111}
$$

The work was proposed by Prabhakar in 1971 [260] and further discussed by Gorenflo et al. in 2014 [261].

Theorem 5.1 (Prabhakar Theorem [260])
Let $\alpha, \beta, \gamma, t \in \mathbb{R}$ and $\alpha > 0$. Then $E_{\alpha,\beta}^{\gamma}(t)$ is an entire function of order $\rho = \frac{1}{\alpha}$.
The work was discovered by Prabhakar in 1971 [260] and further discussed by Gorenflo et al. in 2014 [261].

Theorem 5.2 *Let $\alpha, \beta, \gamma, \lambda, t \in \mathbb{C}$, $k \in \mathbb{N}$, $Re(\alpha) > 0$, and $Re(\beta) > 0$. Then we have [238]*

$$
\int_0^t \tau^{\varsigma-1}E_{\alpha,\beta}^{\gamma}\left(\lambda\tau^{\alpha}\right)d\tau=t^{\varsigma}E_{\alpha,\varsigma+1}^{\gamma}\left(\lambda t^{\alpha}\right).
\tag{5.112}
$$

The work was reported by Mathai and Haubold in 2008 [238].

Theorem 5.3 *Let $\alpha, \beta, \gamma, \lambda, t \in \mathbb{C}$, $k \in \mathbb{N}$, $Re(\alpha) > 0$, and $Re(\beta) > 0$. Then we have [238]*

$$\frac{1}{\Gamma(\varsigma)} \int_0^t (t-\tau)^{\varsigma-1} \tau^{\beta-1} E^{\gamma}_{\alpha,\beta}\left(\lambda \tau^{\alpha}\right) d\tau = t^{\beta+\varsigma-1} E^{\gamma}_{\alpha,\beta+\varsigma}\left(\lambda t^{\alpha}\right). \tag{5.113}$$

The work was reported by Mathai and Haubold in 2008 [238].

Property 5.2 Let $\alpha, \beta, \gamma, \lambda, t \in \mathbb{C}$, $k \in \mathbb{N}$, $Re(\alpha) > 0$, and $Re(\beta) > 0$. Then we have

$$\frac{1}{\Gamma(\varsigma)} \int_0^1 (1-\tau)^{\varsigma-1} \tau^{\beta-1} E^{\gamma}_{\alpha,\beta}\left(\lambda \tau^{\alpha}\right) d\tau = E^{\gamma}_{\alpha,\beta+\varsigma}(\lambda). \tag{5.114}$$

Proof Taking $t = 1$ in (5.113), we obtain the results. □

Property 5.3 Let $\alpha, \beta, \gamma, \lambda, t \in \mathbb{C}$, $k \in \mathbb{N}$, $Re(\alpha) > 0$, and $Re(\beta) > 0$. Then we have

$$\begin{aligned} &\int_{-1}^{1} (1-\tau)^{\varsigma-1} (\tau+1)^{\beta-1} E^{\gamma}_{\alpha,\beta}\left(\lambda(\tau+1)^{\alpha}\right) d\tau \\ &= \int_{-1}^{1} (1+\tau)^{\varsigma-1} (1-\tau)^{\beta-1} E^{\gamma}_{\alpha,\beta}\left(\lambda(1-\tau)^{\alpha}\right) d\tau \end{aligned} \tag{5.115}$$

and

$$\begin{aligned} &\frac{1}{\Gamma(\varsigma)} \int_0^2 (2-\tau)^{\varsigma-1} \tau^{\beta-1} E^{\gamma}_{\alpha,\beta}\left(\lambda \tau^{\alpha}\right) d\tau \\ &= \frac{1}{\Gamma(\varsigma)} \int_{-1}^{1} (1-\tau)^{\varsigma-1} (\tau+1)^{\beta-1} E^{\gamma}_{\alpha,\beta}\left(\lambda(\tau+1)^{\alpha}\right) d\tau \\ &= 2^{\beta+\varsigma-1} E^{\gamma}_{\alpha,\beta+\varsigma}\left(\lambda 2^{\alpha}\right). \end{aligned} \tag{5.116}$$

Proof Taking $t = 2$ in (5.114), we have

$$\begin{aligned} &\frac{1}{\Gamma(\varsigma)} \int_0^2 (2-\tau)^{\varsigma-1} \tau^{\beta-1} E^{\gamma}_{\alpha,\beta}\left(\lambda \tau^{\alpha}\right) d\tau \\ &= \frac{1}{\Gamma(\varsigma)} \int_{-1}^{1} (1-\tau)^{\varsigma-1} (\tau+1)^{\beta-1} E^{\gamma}_{\alpha,\beta}\left(\lambda(\tau+1)^{\alpha}\right) d\tau \\ &= \frac{1}{\Gamma(\varsigma)} \int_{-1}^{1} (1+\tau)^{\varsigma-1} (1-\tau)^{\beta-1} E^{\gamma}_{\alpha,\beta}\left(\lambda(1-\tau)^{\alpha}\right) d\tau \\ &= 2^{\beta+\varsigma-1} E^{\gamma}_{\alpha,\beta+\varsigma}\left(\lambda 2^{\alpha}\right), \end{aligned} \tag{5.117}$$

which reduces to

$$\begin{aligned}&\int_{-1}^{1}(1-\tau)^{\varsigma-1}(\tau+1)^{\beta-1}E_{\alpha,\beta}^{\gamma}(\lambda(\tau+1)^{\alpha})\,d\tau\\&\quad=\int_{-1}^{1}(1+\tau)^{\varsigma-1}(1-\tau)^{\beta-1}E_{\alpha,\beta}^{\gamma}(\lambda(1-\tau)^{\alpha})\,d\tau.\end{aligned}\tag{5.118}$$

Thus, we finish the proof. □

Theorem 5.4 (Laplace Transform)
Let $\alpha,\beta,\gamma,\lambda,t\in\mathbb{C}$, $Re(\alpha)>0$, and $Re(\beta)>0$.
Then the Laplace transform of the function $t^{\beta-1}E_{\alpha,\beta}^{\gamma}(\lambda t^{\alpha})$ is given as follows:

$$L\left[t^{\beta-1}E_{\alpha,\beta}^{\gamma}\left(\lambda t^{\alpha}\right)\right]=\int_{0}^{\infty}e^{-st}t^{\beta-1}E_{\alpha,\beta}^{\gamma}\left(\lambda t^{\alpha}\right)dt=s^{\alpha\gamma-\beta}\left(s^{\alpha}-\lambda\right)^{-\gamma}.\tag{5.119}$$

The work was discovered by Prabhakar in 1971 [260] and further discussed by Gorenflo et al. in 2014 [261].

Theorem 5.5 (Laplace Transform)
Let $\alpha,\beta,\gamma,\lambda,\varsigma,\vartheta\in\mathbb{C}$, $Re(\alpha)>0$, and $Re(\beta)>0$.
Then the Laplace transform of the function $t^{\varsigma-1}E_{\alpha,\beta}^{\gamma}\left(\lambda t^{\vartheta}\right)$ is given as follows:

$$\begin{aligned}L\left[t^{\varsigma-1}E_{\alpha,\beta}^{\gamma}\left(\lambda t^{\vartheta}\right)\right]&=\int_{0}^{\infty}e^{-st}t^{\varsigma-1}E_{\alpha,\beta}^{\gamma}\left(\lambda t^{\vartheta}\right)dt\\&=\frac{s^{-\varsigma}}{\Gamma(\gamma)}\,{}_2\varphi_1\left(\begin{matrix}(\gamma,1),(\varsigma,\vartheta)\\(\beta,\alpha)\end{matrix};\frac{\lambda}{s^{\vartheta}}\right).\end{aligned}\tag{5.120}$$

The work was presented by Mathai and Haubold in 2008 [238] and further discussed by Gorenflo et al. in 2014 [210].

Property 5.4 (Laplace Transform)
Let $\alpha,\beta,\gamma,\lambda,\varsigma,\vartheta\in\mathbb{C}$, $Re(\alpha)>0$, and $Re(\beta)>0$.
Then the Laplace transform of the function $t^{\varsigma-1}E_{\alpha,\beta}^{\gamma}(\lambda t)$ is given as follows:

$$L\left[t^{\varsigma-1}E_{\alpha,\beta}^{\gamma}(\lambda t)\right]=\int_{0}^{\infty}e^{-st}t^{\varsigma-1}E_{\alpha,\beta}^{\gamma}(\lambda t)\,dt=\frac{s^{-\varsigma}}{\Gamma(\gamma)}\,{}_2\varphi_1\left(\begin{matrix}(\gamma,1),(\varsigma,1)\\(\beta,\alpha)\end{matrix};\frac{\lambda}{s}\right).\tag{5.121}$$

Proof Taking $\vartheta=1$ in (5.120), we obtain the result. □

Theorem 5.6 (Mellin Transform)
Let $\alpha, \beta, \gamma, \lambda \in \mathbb{C}$, $Re(\alpha) > 0$, and $Re(\beta) > 0$.
Then the Mellin transform of the function $E^{\gamma}_{\alpha,\beta}(\lambda t)$ is given as follows:

$$M\left[E^{\gamma}_{\alpha,\beta}(-\lambda t)\right] = \int\limits_0^\infty t^{\varpi-1} E^{\gamma}_{\alpha,\beta}(-\lambda t)\,dt = \frac{\Gamma(\varpi)\,\Gamma(\gamma-\varpi)}{\Gamma(\gamma)\,\Gamma(\beta-\varpi\alpha)}\lambda^{-\varpi}. \tag{5.122}$$

The work was proposed by Mathai and Haubold in 2008 [238] and further discussed by Gorenflo et al. in 2014 [210].

Theorem 5.7 (Mellin Transform)
Let $\alpha, \beta, \gamma, \lambda, t \in \mathbb{C}$, $\varsigma > 0$, $Re(\alpha) > 0$, and $Re(\beta) > 0$.
Then the Mellin transform of the function $E^{\gamma}_{\alpha,\beta}(\lambda t)$ is given as follows:

$$\int\limits_0^\infty t^{\varsigma\varpi-1} E^{\gamma}_{\alpha,\beta}(-\lambda t^{\varsigma})dt = \frac{\Gamma(\varpi)\,\Gamma(\gamma-\varpi)}{\Gamma(\gamma)\,\Gamma(\beta-\varpi\alpha)}\frac{\lambda^{-\varpi}}{\varsigma}. \tag{5.123}$$

Proof Taking the place of t by t^{ς} in (5.122), we have the result. □

It is not difficult to show that we may give

$$\begin{aligned}\int\limits_0^\infty t^{2\varpi-1} E^{\gamma}_{\alpha,\beta}(-\lambda t^2)dt &= \tfrac{1}{\Gamma(\gamma)} \int\limits_0^\infty t^{2\varpi-1}\,{}_1\varphi_1\left(\begin{matrix}(\gamma,1)\\(\beta,\alpha)\end{matrix}; -\lambda t^2\right)dt\\ &= \tfrac{\Gamma(\varpi)\Gamma(\gamma-\varpi)}{\Gamma(\gamma)\Gamma(\beta-\varpi\alpha)}\tfrac{\lambda^{-\varpi}}{2},\end{aligned} \tag{5.124}$$

$$\begin{aligned}\int\limits_0^\infty t^{2\varsigma\varpi-1} E^{\gamma}_{\alpha,\beta}(-\lambda t^{2\varsigma})dt &= \tfrac{1}{\Gamma(\gamma)} \int\limits_0^\infty t^{2\varsigma\varpi-1}\,{}_1\varphi_1\left(\begin{matrix}(\gamma,1)\\(\beta,\alpha)\end{matrix}; -\lambda t^{2\varsigma}\right)dt\\ &= \tfrac{\Gamma(\varpi)\Gamma(\gamma-\varpi)}{\Gamma(\gamma)\Gamma(\beta-\varpi\alpha)}\tfrac{\lambda^{-\varpi}}{2\varsigma},\end{aligned} \tag{5.125}$$

$$\begin{aligned}\int\limits_0^\infty t^{\alpha\varpi-1} E^{\gamma}_{\alpha,\beta}(-\lambda t^{\alpha})dt &= \tfrac{1}{\Gamma(\gamma)} \int\limits_0^\infty t^{\alpha\varpi-1}\,{}_1\varphi_1\left(\begin{matrix}(\gamma,1)\\(\beta,\alpha)\end{matrix}; -\lambda t^{\alpha}\right)dt\\ &= \tfrac{\Gamma(\varpi)\Gamma(\gamma-\varpi)}{\Gamma(\gamma)\Gamma(\beta-\varpi\alpha)}\tfrac{\lambda^{-\varpi}}{\alpha},\end{aligned} \tag{5.126}$$

$$\begin{aligned}\int\limits_0^\infty t^{2\alpha\varpi-1} E^{\gamma}_{\alpha,\beta}(-\lambda t^{2\alpha})dt &= \tfrac{1}{\Gamma(\gamma)} \int\limits_0^\infty t^{2\alpha\varpi-1}\,{}_1\varphi_1\left(\begin{matrix}(\gamma,1)\\(\beta,\alpha)\end{matrix}; -\lambda t^{2\alpha}\right)dt\\ &= \tfrac{\Gamma(\varpi)\Gamma(\gamma-\varpi)}{\Gamma(\gamma)\Gamma(\beta-\varpi\alpha)}\tfrac{\lambda^{-\varpi}}{2\alpha},\end{aligned} \tag{5.127}$$

$$\int_0^\infty t^{2\varpi-1} E_{\alpha,\beta}\left(-\lambda t^2\right)dt = \int_0^\infty t^{2\varpi-1}\, {}_1\varphi_1\left(\begin{matrix}(1,1)\\(\beta,\alpha)\end{matrix}; -\lambda t^2\right)dt = \frac{\Gamma(\varpi)\Gamma(1-\varpi)}{\Gamma(\beta-\varpi\alpha)}\frac{\lambda^{-\varpi}}{2}, \tag{5.128}$$

$$\int_0^\infty t^{2\varsigma\varpi-1} E_{\alpha,\beta}\left(-\lambda t^{2\varsigma}\right)dt = \int_0^\infty t^{2\varsigma\varpi-1}\, {}_1\varphi_1\left(\begin{matrix}(1,1)\\(\beta,\alpha)\end{matrix}; -\lambda t^{2\varsigma}\right)dt = \frac{\Gamma(\varpi)\Gamma(1-\varpi)}{\Gamma(\beta-\varpi\alpha)}\frac{\lambda^{-\varpi}}{2\varsigma}, \tag{5.129}$$

$$\int_0^\infty t^{\alpha\varpi-1} E_{\alpha,\beta}\left(-\lambda t^{\alpha}\right)dt = \int_0^\infty t^{\alpha\varpi-1}\, {}_1\varphi_1\left(\begin{matrix}(1,1)\\(\beta,\alpha)\end{matrix}; -\lambda t^{\alpha}\right)dt = \frac{\Gamma(\varpi)\Gamma(1-\varpi)}{\Gamma(\beta-\varpi\alpha)}\frac{\lambda^{-\varpi}}{\alpha}, \tag{5.130}$$

$$\int_0^\infty t^{2\alpha\varpi-1} E_{\alpha,\beta}\left(-\lambda t^{2\alpha}\right)dt = \int_0^\infty t^{2\alpha\varpi-1}\, {}_1\varphi_1\left(\begin{matrix}(1,1)\\(\beta,\alpha)\end{matrix}; -\lambda t^{2\alpha}\right)dt = \frac{\Gamma(\varpi)\Gamma(1-\varpi)}{\Gamma(\beta-\varpi\alpha)}\frac{\lambda^{-\varpi}}{2\alpha}, \tag{5.131}$$

$$\int_0^\infty t^{2\varpi-1} E_{\alpha}\left(-\lambda t^2\right)dt = \int_0^\infty t^{2\varpi-1}\, {}_1\varphi_1\left(\begin{matrix}(1,1)\\(1,\alpha)\end{matrix}; -\lambda t^2\right)dt = \frac{\Gamma(\varpi)\Gamma(1-\varpi)}{\Gamma(1-\varpi\alpha)}\frac{\lambda^{-\varpi}}{2}, \tag{5.132}$$

$$\int_0^\infty t^{2\varsigma\varpi-1} E_{\alpha}\left(-\lambda t^{2\varsigma}\right)dt = \int_0^\infty t^{2\varsigma\varpi-1}\, {}_1\varphi_1\left(\begin{matrix}(1,1)\\(1,\alpha)\end{matrix}; -\lambda t^{2\varsigma}\right)dt = \frac{\Gamma(\varpi)\Gamma(1-\varpi)}{\Gamma(1-\varpi\alpha)}\frac{\lambda^{-\varpi}}{2\varsigma}, \tag{5.133}$$

$$\int_0^\infty t^{\alpha\varpi-1} E_{\alpha}\left(-\lambda t^{\alpha}\right)dt = \int_0^\infty t^{\alpha\varpi-1}\, {}_1\varphi_1\left(\begin{matrix}(1,1)\\(1,\alpha)\end{matrix}; -\lambda t^{\alpha}\right)dt = \frac{\Gamma(\varpi)\Gamma(1-\varpi)}{\Gamma(1-\varpi\alpha)}\frac{\lambda^{-\varpi}}{\alpha}, \tag{5.134}$$

$$\int_0^\infty t^{2\alpha\varpi-1} E_{\alpha}\left(-\lambda t^{2\alpha}\right)dt = \int_0^\infty t^{2\alpha\varpi-1}\, {}_1\varphi_1\left(\begin{matrix}(1,1)\\(1,\alpha)\end{matrix}; -\lambda t^{2\alpha}\right)dt = \frac{\Gamma(\varpi)\Gamma(1-\varpi)}{\Gamma(1-\varpi\alpha)}\frac{\lambda^{-\varpi}}{2\alpha}, \tag{5.135}$$

where $\alpha, \beta, \gamma, \lambda, \varpi \in \mathbb{C}$, $\varsigma > 0$, $Re(\alpha) > 0$, and $Re(\beta) > 0$.

Here, we have

$$E_{\alpha,\beta}^{\gamma}(\lambda t) = \sum_{k=0}^{\infty} \frac{(\gamma)_k}{\Gamma(k\alpha+\beta)}\frac{\lambda^k t^\kappa}{k!} = \frac{1}{\Gamma(\gamma)}\, {}_1\varphi_1\left(\begin{matrix}(\gamma,1)\\(\beta,\alpha)\end{matrix}; \lambda t\right), \tag{5.136}$$

$$E_{\alpha,\beta}^{\gamma}\left(\lambda t^{\alpha}\right)=\sum_{k=0}^{\infty}\frac{(\gamma)_k}{\Gamma(k\alpha+\beta)}\frac{\lambda^k t^{k\alpha}}{k!}=\frac{1}{\Gamma(\gamma)}\,{}_1\varphi_1\left(\begin{matrix}(\gamma,1)\\(\beta,\alpha)\end{matrix};\lambda t^{\alpha}\right),\tag{5.137}$$

$$E_{\alpha,\beta}^{\gamma}\left(\lambda t^{\varsigma}\right)=\sum_{k=0}^{\infty}\frac{(\gamma)_k}{\Gamma(k\alpha+\beta)}\frac{\lambda^k t^{k\varsigma}}{k!}=\frac{1}{\Gamma(\gamma)}\,{}_1\varphi_1\left(\begin{matrix}(\gamma,1)\\(\beta,\alpha)\end{matrix};\lambda t^{\varsigma}\right),\tag{5.138}$$

$$E_{\alpha,\beta}^{\gamma}(-\lambda t)=\sum_{k=0}^{\infty}\frac{(\gamma)_k}{\Gamma(k\alpha+\beta)}\frac{(-\lambda)^k t^{\kappa}}{k!}=\frac{1}{\Gamma(\gamma)}\,{}_1\varphi_1\left(\begin{matrix}(\gamma,1)\\(\beta,\alpha)\end{matrix};-\lambda t\right),\tag{5.139}$$

$$E_{\alpha,\beta}^{\gamma}\left(-\lambda t^{\alpha}\right)=\sum_{k=0}^{\infty}\frac{(\gamma)_k}{\Gamma(k\alpha+\beta)}\frac{(-\lambda)^k t^{k\alpha}}{k!}=\frac{1}{\Gamma(\gamma)}\,{}_1\varphi_1\left(\begin{matrix}(\gamma,1)\\(\beta,\alpha)\end{matrix};-\lambda t^{\alpha}\right),\tag{5.140}$$

$$E_{\alpha,\beta}^{\gamma}\left(-\lambda t^{\varsigma}\right)=\sum_{k=0}^{\infty}\frac{(\gamma)_k}{\Gamma(k\alpha+\beta)}\frac{(-\lambda)^k t^{k\varsigma}}{k!}=\frac{1}{\Gamma(\gamma)}\,{}_1\varphi_1\left(\begin{matrix}(\gamma,1)\\(\beta,\alpha)\end{matrix};-\lambda t^{\varsigma}\right),\tag{5.141}$$

$$E_{\alpha,\beta}^{\gamma}(i\lambda t)=\sum_{k=0}^{\infty}\frac{(\gamma)_k}{\Gamma(k\alpha+\beta)}\frac{(i\lambda)^{\kappa} t^{\kappa}}{k!}=\frac{1}{\Gamma(\gamma)}\,{}_1\varphi_1\left(\begin{matrix}(\gamma,1)\\(\beta,\alpha)\end{matrix};i\lambda t\right),\tag{5.142}$$

$$E_{\alpha,\beta}^{\gamma}\left(i\lambda t^{\alpha}\right)=\sum_{k=0}^{\infty}\frac{(\gamma)_k}{\Gamma(k\alpha+\beta)}\frac{(i\lambda)^k t^{k\alpha}}{k!}=\frac{1}{\Gamma(\gamma)}\,{}_1\varphi_1\left(\begin{matrix}(\gamma,1)\\(\beta,\alpha)\end{matrix};i\lambda t^{\alpha}\right),\tag{5.143}$$

$$E_{\alpha,\beta}^{\gamma}\left(i\lambda t^{\varsigma}\right)=\sum_{k=0}^{\infty}\frac{(\gamma)_k}{\Gamma(k\alpha+\beta)}\frac{(i\lambda)\,t^{k\varsigma}}{k!}=\frac{1}{\Gamma(\gamma)}\,{}_1\varphi_1\left(\begin{matrix}(\gamma,1)\\(\beta,\alpha)\end{matrix};i\lambda t^{\varsigma}\right),\tag{5.144}$$

$$E_{\alpha,\beta}^{\gamma}(-i\lambda t)=\sum_{k=0}^{\infty}\frac{(\gamma)_k}{\Gamma(k\alpha+\beta)}\frac{(-i\lambda)^{\kappa} t^{\kappa}}{k!}=\frac{1}{\Gamma(\gamma)}\,{}_1\varphi_1\left(\begin{matrix}(\gamma,1)\\(\beta,\alpha)\end{matrix};-i\lambda t\right),\tag{5.145}$$

$$E_{\alpha,\beta}^{\gamma}\left(-i\lambda t^{\alpha}\right)=\sum_{k=0}^{\infty}\frac{(\gamma)_k}{\Gamma(k\alpha+\beta)}\frac{(-i\lambda)^k t^{k\alpha}}{k!}=\frac{1}{\Gamma(\gamma)}\,{}_1\varphi_1\left(\begin{matrix}(\gamma,1)\\(\beta,\alpha)\end{matrix};-i\lambda t^{\alpha}\right),\tag{5.146}$$

$$E_{\alpha,\beta}^{\gamma}\left(-i\lambda t^{\varsigma}\right)=\sum_{k=0}^{\infty}\frac{(\gamma)_k}{\Gamma\left(k\alpha+\beta\right)}\frac{(-i\lambda)\,t^{k\varsigma}}{k!}=\frac{1}{\Gamma\left(\gamma\right)}\,{}_1\varphi_1\left(\begin{matrix}(\gamma,1)\\(\beta,\alpha)\end{matrix};-i\lambda t^{\varsigma}\right),\tag{5.147}$$

where $\alpha, \beta, \gamma, \lambda \in \mathbb{C}$, $i=\sqrt{-1}$, $\varsigma > 0$, $Re\,(\alpha) > 0$, and $Re\,(\beta) > 0$.

Theorem 5.8 *Let $\alpha, \beta, \gamma, \lambda \in \mathbb{C}$, $Re\,(\alpha) > 0$, and $Re\,(\beta) > 0$.*
Then,

$$\begin{aligned}&\int_{-\infty}^{\infty}|t|^{\varpi-1}E_{\alpha,\beta}^{\gamma}\left(-\lambda\,|t|\right)dt\\&=\frac{1}{\Gamma(\gamma)}\int_{-\infty}^{\infty}|t|^{\varpi-1}\,{}_1\varphi_1\left(\begin{matrix}(\gamma,1)\\(\beta,\alpha)\end{matrix};-\lambda\,|t|\right)dt\\&=\frac{2\Gamma(\varpi)\Gamma(\gamma-\varpi)}{\Gamma(\gamma)\Gamma(\beta-\varpi\alpha)}\lambda^{-\varpi}.\end{aligned}\tag{5.148}$$

Proof With the aid of (5.122), we have the result. □

In particular,

$$\begin{aligned}\int_{-\infty}^{\infty}|t|^{2\varpi-1}\,E_{\alpha,\beta}^{\gamma}\left(-\lambda t^2\right)dt&=\frac{1}{\Gamma(\gamma)}\int_{-\infty}^{\infty}|t|^{2\varpi-1}\,{}_1\varphi_1\left(\begin{matrix}(\gamma,1)\\(\beta,\alpha)\end{matrix};-\lambda t^2\right)dt\\&=\frac{\Gamma(\varpi)\Gamma(\gamma-\varpi)\lambda^{-\varpi}}{\Gamma(\gamma)\Gamma(\beta-\varpi\alpha)},\end{aligned}\tag{5.149}$$

$$\begin{aligned}\int_{-\infty}^{\infty}|t|^{2\varsigma\varpi-1}\,E_{\alpha,\beta}^{\gamma}\left(-\lambda\,|t|^{2\varsigma}\right)dt&=\frac{1}{\Gamma(\gamma)}\int_{-\infty}^{\infty}|t|^{2\varsigma\varpi-1}\,{}_1\varphi_1\left(\begin{matrix}(\gamma,1)\\(\beta,\alpha)\end{matrix};-\lambda\,|t|^{2\varsigma}\right)dt\\&=\frac{\Gamma(\varpi)\Gamma(\gamma-\varpi)}{\Gamma(\gamma)\Gamma(\beta-\varpi\alpha)}\frac{\lambda^{-\varpi}}{\varsigma},\end{aligned}\tag{5.150}$$

$$\begin{aligned}\int_{-\infty}^{\infty}|t|^{\alpha\varpi-1}\,E_{\alpha,\beta}^{\gamma}\left(-\lambda\,|t|^{\alpha}\right)dt&=\frac{1}{\Gamma(\gamma)}\int_{-\infty}^{\infty}|t|^{\alpha\varpi-1}\,{}_1\varphi_1\left(\begin{matrix}(\gamma,1)\\(\beta,\alpha)\end{matrix};-\lambda\,|t|^{\alpha}\right)dt\\&=\frac{\Gamma(\varpi)\Gamma(\gamma-\varpi)}{\Gamma(\gamma)\Gamma(\beta-\varpi\alpha)}\frac{2}{\alpha}\lambda^{-\varpi},\end{aligned}\tag{5.151}$$

$$\begin{aligned}\int_{-\infty}^{\infty}|t|^{2\alpha\varpi-1}\,E_{\alpha,\beta}^{\gamma}\left(-\lambda\,|t|^{2\alpha}\right)dt&=\frac{1}{\Gamma(\gamma)}\int_{-\infty}^{\infty}|t|^{2\alpha\varpi-1}\,{}_1\varphi_1\left(\begin{matrix}(\gamma,1)\\(\beta,\alpha)\end{matrix};-\lambda\,|t|^{2\alpha}\right)dt\\&=\frac{\Gamma(\varpi)\Gamma(\gamma-\varpi)}{\Gamma(\gamma)\Gamma(\beta-\varpi\alpha)}\frac{\lambda^{-\varpi}}{\alpha},\end{aligned}\tag{5.152}$$

$$\int_{-\infty}^{\infty} |t|^{2\varpi-1} E_{\alpha,\beta}\left(-\lambda |t|^2\right)dt = \int_{-\infty}^{\infty} |t|^{2\varpi-1} {}_1\varphi_1\left(\begin{matrix}(1,1)\\(\beta,\alpha)\end{matrix}; -\lambda t^2\right)dt = \frac{\Gamma(\varpi)\Gamma(1-\varpi)}{\Gamma(\beta-\varpi\alpha)}\lambda^{-\varpi}, \tag{5.153}$$

$$\int_{-\infty}^{\infty} |t|^{2\varsigma\varpi-1} E_{\alpha,\beta}\left(-\lambda |t|^{2\varsigma}\right)dt = \int_{-\infty}^{\infty} |t|^{2\varsigma\varpi-1} {}_1\varphi_1\left(\begin{matrix}(1,1)\\(\beta,\alpha)\end{matrix}; -\lambda |t|^{2\varsigma}\right)dt = \frac{\Gamma(\varpi)\Gamma(1-\varpi)}{\Gamma(\beta-\varpi\alpha)}\frac{\lambda^{-\varpi}}{\varsigma}, \tag{5.154}$$

$$\int_{-\infty}^{\infty} |t|^{\alpha\varpi-1} E_{\alpha,\beta}\left(-\lambda |t|^{\alpha}\right)dt = \int_{-\infty}^{\infty} |t|^{\alpha\varpi-1} {}_1\varphi_1\left(\begin{matrix}(1,1)\\(\beta,\alpha)\end{matrix}; -\lambda |t|^{\alpha}\right)dt = \frac{\Gamma(\varpi)\Gamma(1-\varpi)}{\Gamma(\beta-\varpi\alpha)}\frac{2}{\alpha}\lambda^{-\varpi}, \tag{5.155}$$

$$\int_{-\infty}^{\infty} |t|^{2\alpha\varpi-1} E_{\alpha,\beta}\left(-\lambda |t|^{2\alpha}\right)dt = \int_{-\infty}^{\infty} |t|^{2\alpha\varpi-1} {}_1\varphi_1\left(\begin{matrix}(1,1)\\(\beta,\alpha)\end{matrix}; -\lambda |t|^{2\alpha}\right)dt = \frac{\Gamma(\varpi)\Gamma(1-\varpi)}{\Gamma(\beta-\varpi\alpha)}\frac{\lambda^{-\varpi}}{\alpha}, \tag{5.156}$$

$$\int_{-\infty}^{\infty} |t|^{2\varpi-1} E_{\alpha}\left(-\lambda |t|^2\right)dt = \int_{-\infty}^{\infty} |t|^{2\varpi-1} {}_1\varphi_1\left(\begin{matrix}(1,1)\\(1,\alpha)\end{matrix}; -\lambda t^2\right)dt = \frac{\Gamma(\varpi)\Gamma(1-\varpi)}{\Gamma(1-\varpi\alpha)}\lambda^{-\varpi}, \tag{5.157}$$

$$\int_{-\infty}^{\infty} |t|^{2\varsigma\varpi-1} E_{\alpha}\left(-\lambda |t|^{2\varsigma}\right)dt = \int_{-\infty}^{\infty} |t|^{2\varsigma\varpi-1} {}_1\varphi_1\left(\begin{matrix}(1,1)\\(1,\alpha)\end{matrix}; -\lambda |t|^{2\varsigma}\right)dt = \frac{\Gamma(\varpi)\Gamma(1-\varpi)}{\Gamma(1-\varpi\alpha)}\frac{\lambda^{-\varpi}}{\varsigma}, \tag{5.158}$$

$$\int_{-\infty}^{\infty} |t|^{\alpha\varpi-1} E_{\alpha}\left(-\lambda |t|^{\alpha}\right)dt = \int_{-\infty}^{\infty} |t|^{\alpha\varpi-1} {}_1\varphi_1\left(\begin{matrix}(1,1)\\(1,\alpha)\end{matrix}; -\lambda |t|^{\alpha}\right)dt = \frac{\Gamma(\varpi)\Gamma(1-\varpi)}{\Gamma(1-\varpi\alpha)}\frac{2}{\alpha}\lambda^{-\varpi}, \tag{5.159}$$

$$\int_{-\infty}^{\infty} |t|^{2\alpha\varpi-1} E_{\alpha}\left(-\lambda |t|^{2\alpha}\right)dt = \int_{-\infty}^{\infty} |t|^{2\alpha\varpi-1} {}_1\varphi_1\left(\begin{matrix}(1,1)\\(1,\alpha)\end{matrix}; -\lambda |t|^{2\alpha}\right)dt = \frac{\Gamma(\varpi)\Gamma(1-\varpi)}{\Gamma(1-\varpi\alpha)}\frac{\lambda^{-\varpi}}{\alpha}, \tag{5.160}$$

where $\alpha, \beta, \gamma, \lambda, \varpi \in \mathbb{C}$, $\varsigma > 0$, $Re(\alpha) > 0$, and $Re(\beta) > 0$.

Here, we the binomial series can be written as [210]

$$(1+t)^{\varsigma} = \sum_{k=0}^{\infty} \frac{\Gamma(1+\varsigma)}{\Gamma(1+\varsigma-k)} \frac{t^k}{k!} = \sum_{k=0}^{\infty} (-1)^k \frac{\Gamma(k-\varsigma)}{\Gamma(-\varsigma)} \frac{t^k}{k!}, \tag{5.161}$$

which leads to

$$\begin{aligned}(1+t)^{\varsigma} &= {}_1\varphi_1\left(\begin{matrix}(1+\varsigma,0)\\(1+\varsigma,-1)\end{matrix};t\right)\\ &= \Gamma(1+\varsigma) \times {}_0\varphi_1\left(\begin{matrix}-\\(1+\varsigma,-1)\end{matrix};t\right)\\ &= {}_1\varphi_1\left(\begin{matrix}(-\varsigma,1)\\(-\varsigma,0)\end{matrix};-t\right)\\ &= \frac{1}{\Gamma(-\varsigma)} \times {}_1\varphi_0\left(\begin{matrix}(-\varsigma,1)\\-\end{matrix};-t\right),\end{aligned} \tag{5.162}$$

where $\varsigma, t \in \mathbb{R}$, $Re(\alpha) > 0$, and $|t| < 1$.

It is not difficult to show that

$$\begin{aligned}\left(1+\lambda t^2\right)^{\varsigma} &= \Gamma(1+\varsigma) \times {}_0\varphi_1\left(\begin{matrix}-\\(1+\varsigma,-1)\end{matrix};\lambda t^2\right)\\ &= \frac{1}{\Gamma(-\varsigma)} \times {}_1\varphi_0\left(\begin{matrix}(-\varsigma,1)\\-\end{matrix};-\lambda t^2\right).\end{aligned} \tag{5.163}$$

Let us recall the Mellin transform of the Wright hypergeometric function:

$$\int_0^{\infty} t^{\varpi-1}\,{}_0\varphi_1\left(\begin{matrix}-\\(\beta,\alpha)\end{matrix};-\lambda t\right)dt = \frac{\Gamma(\varpi)}{\Gamma(\beta-\alpha\varpi)}\lambda^{-\varpi}, \tag{5.164}$$

where the Wright hypergeometric function is

$${}_0\varphi_1\left(\begin{matrix}-\\(\alpha+1,\beta)\end{matrix};\lambda t\right) = \sum_{k=0}^{\infty} \frac{1}{\Gamma(\alpha+1+\beta k)} \frac{\lambda^k t^k}{k!} \tag{5.165}$$

for $\alpha, \lambda \in \mathbb{C}$, $Re(\alpha) > 0$, and $\beta \in \mathbb{R}$.

Thus, we have

$$\int_0^{\infty} t^{\varpi-1}\,{}_0\varphi_1\left(\begin{matrix}-\\(\beta,\alpha)\end{matrix};-\lambda t\right)dt = \frac{\Gamma(\varpi)}{\Gamma(\beta-\alpha\varpi)}\lambda^{-\varpi}, \tag{5.166}$$

$$\int_0^\infty t^{\varsigma\varpi-1}{}_0\varphi_1\left(\begin{matrix}-\\(\beta,\alpha)\end{matrix};-\lambda t^{\varsigma}\right)dt=\frac{\Gamma(\varpi)}{\Gamma(\beta-\alpha\varpi)}\frac{\lambda^{-\varpi}}{\varsigma}, \tag{5.167}$$

$$\int_0^\infty t^{2\varsigma\varpi-1}{}_0\varphi_1\left(\begin{matrix}-\\(\beta,\alpha)\end{matrix};-\lambda t^{2\varsigma}\right)dt=\frac{\Gamma(\varpi)}{\Gamma(\beta-\alpha\varpi)}\frac{\lambda^{-\varpi}}{2\varsigma}, \tag{5.168}$$

$$\int_0^\infty t^{2\alpha\varpi-1}{}_0\varphi_1\left(\begin{matrix}-\\(\beta,\alpha)\end{matrix};-\lambda t^{2\alpha}\right)dt=\frac{\Gamma(\varpi)}{\Gamma(\beta-\alpha\varpi)}\frac{\lambda^{-\varpi}}{2\alpha}, \tag{5.169}$$

$$\int_0^\infty t^{2\varpi-1}{}_0\varphi_1\left(\begin{matrix}-\\(\beta,\alpha)\end{matrix};-\lambda t^{2}\right)dt=\frac{\Gamma(\varpi)}{\Gamma(\beta-\alpha\varpi)}\frac{\lambda^{-\varpi}}{2}, \tag{5.170}$$

such that

$$\int_{-\infty}^\infty |t|^{\varpi-1}{}_0\varphi_1\left(\begin{matrix}-\\(\beta,\alpha)\end{matrix};-\lambda|t|\right)dt=2\lambda^{-\varpi}\frac{\Gamma(\varpi)}{\Gamma(\beta-\alpha\varpi)}, \tag{5.171}$$

$$\int_{-\infty}^\infty |t|^{\varsigma\varpi-1}{}_0\varphi_1\left(\begin{matrix}-\\(\beta,\alpha)\end{matrix};-\lambda|t|^{\varsigma}\right)dt=\frac{2}{\varsigma}\lambda^{-\varpi}\frac{\Gamma(\varpi)}{\Gamma(\beta-\alpha\varpi)}, \tag{5.172}$$

$$\int_{-\infty}^\infty |t|^{2\varsigma\varpi-1}{}_0\varphi_1\left(\begin{matrix}-\\(\beta,\alpha)\end{matrix};-\lambda|t|^{2\varsigma}\right)dt=\frac{\lambda^{-\varpi}}{\varsigma}\frac{\Gamma(\varpi)}{\Gamma(\beta-\alpha\varpi)}, \tag{5.173}$$

$$\int_{-\infty}^\infty |t|^{2\alpha\varpi-1}{}_0\varphi_1\left(\begin{matrix}-\\(\beta,\alpha)\end{matrix};-\lambda|t|^{2\alpha}\right)dt=\frac{\Gamma(\varpi)}{\Gamma(\beta-\alpha\varpi)}\frac{\lambda^{-\varpi}}{\alpha}, \tag{5.174}$$

$$\int_{-\infty}^\infty |t|^{2\varpi-1}{}_0\varphi_1\left(\begin{matrix}-\\(\beta,\alpha)\end{matrix};-\lambda t^{2}\right)dt=\lambda^{-\varpi}\frac{\Gamma(\varpi)}{\Gamma(\beta-\alpha\varpi)}, \tag{5.175}$$

where $\alpha,\lambda\in\mathbb{C}$, $Re(\varpi)>0$, $Re(\alpha)>0$, and $\beta\in\mathbb{R}$.

Thus, we get

$$\int_0^\infty t^{2\alpha+1}{}_0\varphi_1\left(\begin{matrix}-\\(\alpha+1,1)\end{matrix};-\lambda t^{2}\right)dt=\frac{\Gamma(\alpha-1)}{2}\lambda^{-\alpha+1}, \tag{5.176}$$

$$\int_0^\infty t^{2\varpi+1}\,{}_0\varphi_1\left(\begin{matrix}-\\(\alpha+1,1)\end{matrix};-\lambda t^2\right)dt=\frac{1}{2}\frac{\Gamma(\varpi-1)}{\Gamma(\alpha+2-\varpi)}\lambda^{-\varpi+1}, \tag{5.177}$$

$$\int_{-\infty}^\infty |t|^{2\alpha+1}\,{}_0\varphi_1\left(\begin{matrix}-\\(\alpha+1,1)\end{matrix};-\lambda|t|^2\right)dt=\Gamma(\alpha-1)\,\lambda^{-\alpha+1}, \tag{5.178}$$

and

$$\int_{-\infty}^\infty |t|^{2\varpi+1}\,{}_0\varphi_1\left(\begin{matrix}-\\(\alpha+1,1)\end{matrix};-\lambda|t|^2\right)dt=\frac{\Gamma(\varpi-1)}{\Gamma(\alpha+2-\varpi)}\lambda^{-\varpi+1}, \tag{5.179}$$

where $\alpha,\lambda\in\mathbb{C}$, $Re(\varpi)>1$, and $Re(\alpha)>0$.

Let us recall the special case for the Wright's generalized hypergeometric function:

Let $\alpha,\gamma\in\mathbb{C}$, $Re(\alpha)>0$, $Re(\gamma)>0$, $\lambda\in\mathbb{C}$, $\beta>0$, $\alpha_1,\beta_1\in\mathbb{C}$, and $A_1,B_1\in\mathbb{R}_+$. Then we have

$$\begin{aligned}&\frac{1}{\Gamma(\alpha)}\int_0^t(t-\tau)^\alpha\,\tau^{\gamma-1}\,{}_1\varphi_1\left(\begin{matrix}(\alpha_1,A_1)\\(\beta_1,B_1)\end{matrix};\lambda\tau^\beta\right)d\tau\\&=t^{\gamma+\alpha-1}\,{}_2\varphi_2\left(\begin{matrix}(\alpha_1,A_1),(\gamma,\beta)\\(\beta_1,B_1),(\gamma+\alpha,\beta)\end{matrix};\lambda t^\beta\right),\end{aligned} \tag{5.180}$$

where

$${}_1\varphi_1\left(\begin{matrix}(\alpha_1,A_1)\\(\beta_1,B_1)\end{matrix};\lambda\tau^\beta\right)=\sum_{k=0}^\infty\frac{\Gamma(\alpha_1+A_1k)}{\Gamma(\beta_1+B_1k)}\frac{\lambda^k\tau^{k\beta}}{k!}$$

and

$${}_2\varphi_2\left(\begin{matrix}(\alpha_1,A_1),(\gamma,\beta)\\(\beta_1,B_1),(\gamma+\alpha,\beta)\end{matrix};\lambda\tau^\beta\right)=\sum_{k=0}^\infty\frac{\Gamma(\alpha_1+A_1k)\,\Gamma(\gamma+\beta k)}{\Gamma(\beta_1+B_1k)\,\Gamma(\gamma+\alpha+\beta k)}\frac{\lambda^k\tau^{k\beta}}{k!}.$$

From (5.180), we directly deduce the following result:

Let $\sigma,\varsigma\in\mathbb{C}$, $\lambda\in\mathbb{C}$, $\vartheta>0$, $\beta\in\mathbb{C}$, $\alpha\in\mathbb{R}_+$, $Re(\sigma)>0$, and $Re(\varsigma)>0$. Then we have

$$\begin{aligned}&\frac{1}{\Gamma(\sigma)}\int_0^t(t-\tau)^\sigma\,\tau^{\varsigma-1}\,{}_1\varphi_1\left(\begin{matrix}(\gamma,1)\\(\beta,\alpha)\end{matrix};\lambda\tau^\vartheta\right)d\tau\\&=t^{\varsigma+\sigma-1}\,{}_2\varphi_2\left(\begin{matrix}(\gamma,1),(\varsigma,\vartheta)\\(\beta,\alpha),(\varsigma+\sigma,\vartheta)\end{matrix};\lambda t^\vartheta\right).\end{aligned} \tag{5.181}$$

Theorem 5.9 *Let $\sigma, \varsigma \in \mathbb{C}$, $\lambda \in \mathbb{C}$, $\vartheta > 0$, $\beta \in \mathbb{C}$, $\alpha \in \mathbb{R}_+$, $Re(\sigma) > 0$, and $Re(\varsigma) > 0$. Then we have*

$$\begin{aligned}&\frac{\Gamma(\gamma)}{\Gamma(\sigma)}\int_0^t (t-\tau)^{\sigma}\,\tau^{\varsigma-1}E_{\alpha,\beta}^{\gamma}\left(\lambda\tau^{\vartheta}\right)d\tau\\&=t^{\varsigma+\sigma-1}\,{}_2\varphi_2\left(\begin{matrix}(\gamma,1),(\varsigma,\vartheta)\\(\beta,\alpha),(\varsigma+\sigma,\vartheta)\end{matrix};\lambda t^{\vartheta}\right).\end{aligned}\tag{5.182}$$

Proof Making use of (5.34) and (5.181), we obtain the result. □

Theorem 5.10 *Let $\sigma, \varsigma \in \mathbb{C}$, $\lambda \in \mathbb{C}$, $\vartheta > 0$, $\beta \in \mathbb{C}$, $\alpha \in \mathbb{R}_+$, $Re(\sigma) > 0$, and $Re(\varsigma) > 0$. Then we have*

$$\frac{\Gamma(\gamma)}{\Gamma(\sigma)}\int_0^1 (1-\tau)^{\sigma}\,\tau^{\varsigma-1}E_{\alpha,\beta}^{\gamma}\left(\lambda\tau^{\vartheta}\right)d\tau={}_2\varphi_2\left(\begin{matrix}(\gamma,1),(\varsigma,\vartheta)\\(\beta,\alpha),(\varsigma+\sigma,\vartheta)\end{matrix};\lambda\right).\tag{5.183}$$

Proof Putting $t = 1$ in (5.182), we obtain the result. □

Theorem 5.11 *Let $\sigma, \varsigma \in \mathbb{C}$, $\lambda \in \mathbb{C}$, $\vartheta > 0$, $\beta \in \mathbb{C}$, $\alpha \in \mathbb{R}_+$, $Re(\sigma) > 0$, and $Re(\varsigma) > 0$. Then we have*

$$\begin{aligned}&\int_{-1}^1 (1-\tau)^{\sigma}\,(1+\tau)^{\varsigma-1}\,E_{\alpha,\beta}^{\gamma}\left(\lambda\,(1+\tau)^{\vartheta}\right)d\tau\\&=\int_{-1}^1 (1+\tau)^{\sigma}\,(1-\tau)^{\varsigma-1}\,E_{\alpha,\beta}^{\gamma}\left(\lambda\,(1-\tau)^{\vartheta}\right)d\tau\end{aligned}\tag{5.184}$$

and

$$\begin{aligned}&\frac{\Gamma(\gamma)}{\Gamma(\sigma)}\int_0^2 (2-\tau)^{\sigma}\,\tau^{\varsigma-1}E_{\alpha,\beta}^{\gamma}\left(\lambda\tau^{\vartheta}\right)d\tau\\&=\frac{\Gamma(\gamma)}{\Gamma(\sigma)}\int_{-1}^1 (1-\tau)^{\sigma}\,(1+\tau)^{\varsigma-1}\,E_{\alpha,\beta}^{\gamma}\left(\lambda\,(1+\tau)^{\vartheta}\right)d\tau\\&=\frac{\Gamma(\gamma)}{\Gamma(\sigma)}\int_{-1}^1 (1+\tau)^{\sigma}\,(1-\tau)^{\varsigma-1}\,E_{\alpha,\beta}^{\gamma}\left(\lambda\,(1-\tau)^{\vartheta}\right)d\tau\\&=2^{\varsigma+\sigma-1}\,{}_2\varphi_2\left(\begin{matrix}(\gamma,1),(\varsigma,\vartheta)\\(\beta,\alpha),(\varsigma+\sigma,\vartheta)\end{matrix};\lambda 2^{\vartheta}\right).\end{aligned}\tag{5.185}$$

Proof Setting $t = 2$ in (5.182), we obtain the results. □

Theorem 5.12 *Let $\sigma, \beta \in \mathbb{C}$, $\lambda \in \mathbb{C}$, $\vartheta > 0$, $\alpha \in \mathbb{R}_+$, $Re(\sigma) > 0$, and $Re(\beta) > 0$. Then we have*

$$\frac{\Gamma(\gamma)}{\Gamma(\sigma)} \int_0^t (t-\tau)^{\sigma} \tau^{\beta-1} E_{\alpha,\beta}^{\gamma}\left(\lambda \tau^{\vartheta}\right) d\tau = t^{\beta+\sigma-1} {}_2\varphi_2 \left(\begin{matrix} (\gamma,1),(\beta,\vartheta) \\ (\beta,\alpha),(\beta+\sigma,\vartheta) \end{matrix} ; \lambda t^{\vartheta} \right). \tag{5.186}$$

Proof Putting $\varsigma = \beta$ in (5.182), we obtain the result. □

Theorem 5.13 *Let $\sigma, \beta \in \mathbb{C}$, $\lambda \in \mathbb{C}$, $\vartheta > 0$, $\alpha \in \mathbb{R}_+$, $Re(\sigma) > 0$, and $Re(\beta) > 0$. Then we have*

$$\frac{\Gamma(\gamma)}{\Gamma(\sigma)} \int_0^1 (1-\tau)^{\sigma} \tau^{\beta-1} E_{\alpha,\beta}^{\gamma}\left(\lambda \tau^{\vartheta}\right) d\tau = {}_2\varphi_2 \left(\begin{matrix} (\gamma,1),(\beta,\vartheta) \\ (\beta,\alpha),(\beta+\sigma,\vartheta) \end{matrix} ; \lambda \right). \tag{5.187}$$

Proof On setting $\varsigma = \beta$ in (5.183), we obtain the result. □

Theorem 5.14 *Let $\sigma, \beta \in \mathbb{C}$, $\lambda \in \mathbb{C}$, $\vartheta > 0$, $\alpha \in \mathbb{R}_+$, $Re(\sigma) > 0$, and $Re(\beta) > 0$. Then we have*

$$\begin{aligned} &\int_{-1}^{1} (1-\tau)^{\sigma} (1+\tau)^{\beta-1} E_{\alpha,\beta}^{\gamma}\left(\lambda (1+\tau)^{\vartheta}\right) d\tau \\ &\quad = \int_{-1}^{1} (1+\tau)^{\sigma} (1-\tau)^{\beta-1} E_{\alpha,\beta}^{\gamma}\left(\lambda (1-\tau)^{\vartheta}\right) d\tau \end{aligned} \tag{5.188}$$

and

$$\begin{aligned} &\frac{\Gamma(\gamma)}{\Gamma(\sigma)} \int_0^2 (2-\tau)^{\sigma} \tau^{\beta-1} E_{\alpha,\beta}^{\gamma}\left(\lambda \tau^{\vartheta}\right) d\tau \\ &= \frac{\Gamma(\gamma)}{\Gamma(\sigma)} \int_{-1}^{1} (1-\tau)^{\sigma} (1+\tau)^{\beta-1} E_{\alpha,\beta}^{\gamma}\left(\lambda (1+\tau)^{\vartheta}\right) d\tau \\ &= \frac{\Gamma(\gamma)}{\Gamma(\sigma)} \int_{-1}^{1} (1+\tau)^{\sigma} (1-\tau)^{\beta-1} E_{\alpha,\beta}^{\gamma}\left(\lambda (1-\tau)^{\vartheta}\right) d\tau \\ &= 2^{\beta+\sigma-1} {}_2\varphi_2 \left(\begin{matrix} (\gamma,1),(\beta,\vartheta) \\ (\beta,\alpha),(\beta+\sigma,\vartheta) \end{matrix} ; \lambda 2^{\vartheta} \right). \end{aligned} \tag{5.189}$$

Proof Putting $\varsigma = \beta$ in (5.184) and (5.185), we obtain the results. □

Theorem 5.15 *Let $\beta \in \mathbb{C}$, $\lambda \in \mathbb{C}$, $\alpha \in \mathbb{R}_+$, $Re(\sigma) > 0$, and $Re(\beta) > 0$. Then we have*

$$\frac{\Gamma(\gamma)}{\Gamma(\sigma)} \int_0^t (t-\tau)^{\sigma} \tau^{\beta-1} E_{\alpha,\beta}^{\gamma}\left(\lambda \tau^{\alpha}\right) d\tau = t^{\beta+\sigma-1} {}_1\varphi_1 \left(\begin{matrix} (\gamma,1) \\ (\beta+\sigma,\alpha) \end{matrix} ; \lambda t^{\alpha} \right). \tag{5.190}$$

Proof Putting $\vartheta = \alpha$ in (5.186), we obtain the result. □

Theorem 5.16 *Let $\beta \in \mathbb{C}$, $\lambda \in \mathbb{C}$, $\alpha \in \mathbb{R}_+$, $Re(\sigma) > 0$, and $Re(\beta) > 0$. Then we have*

$$\frac{\Gamma(\gamma)}{\Gamma(\sigma)} \int_0^1 (1-\tau)^{\sigma} \tau^{\beta-1} E^{\gamma}_{\alpha,\beta}\left(\lambda\tau^{\alpha}\right) d\tau = {}_1\varphi_1\left(\begin{matrix}(\gamma,1)\\(\beta+\sigma,\vartheta)\end{matrix};\lambda\right). \tag{5.191}$$

Proof On setting $\vartheta = \alpha$ in (5.187), we obtain the result. □

Theorem 5.17 *Let $\beta \in \mathbb{C}$, $\lambda \in \mathbb{C}$, $\alpha \in \mathbb{R}_+$, $Re(\sigma) > 0$, and $Re(\beta) > 0$. Then we have*

$$\begin{aligned}&\int_{-1}^{1} (1-\tau)^{\sigma} (1+\tau)^{\beta-1} E^{\gamma}_{\alpha,\beta}\left(\lambda (1+\tau)^{\alpha}\right) d\tau\\&\quad= \int_{-1}^{1} (1+\tau)^{\sigma} (1-\tau)^{\beta-1} E^{\gamma}_{\alpha,\beta}\left(\lambda (1-\tau)^{\alpha}\right) d\tau\end{aligned} \tag{5.192}$$

and

$$\begin{aligned}&\frac{\Gamma(\gamma)}{\Gamma(\sigma)} \int_0^2 (2-\tau)^{\sigma} \tau^{\beta-1} E^{\gamma}_{\alpha,\beta}\left(\lambda\tau^{\alpha}\right) d\tau\\&\quad= \frac{\Gamma(\gamma)}{\Gamma(\sigma)} \int_{-1}^{1} (1-\tau)^{\sigma} (1+\tau)^{\beta-1} E^{\gamma}_{\alpha,\beta}\left(\lambda (1+\tau)^{\alpha}\right) d\tau\\&\quad= \frac{\Gamma(\gamma)}{\Gamma(\sigma)} \int_{-1}^{1} (1+\tau)^{\sigma} (1-\tau)^{\beta-1} E^{\gamma}_{\alpha,\beta}\left(\lambda (1-\tau)^{\alpha}\right) d\tau\\&\quad= 2^{\beta+\sigma-1}\, {}_1\varphi_1\left(\begin{matrix}(\gamma,1)\\(\beta+\sigma,\alpha)\end{matrix};\lambda 2^{\alpha}\right).\end{aligned} \tag{5.193}$$

Proof Putting $\vartheta = \alpha$ in (5.188) and (5.189), we obtain the results. □

5.1.2 The Supertrigonometric Functions via Prabhakar Function

In this section, we introduce the supertrigonometric functions via Prabhakar function.

Definition 5.2 The supersine via Prabhakar function is defined as

$$\begin{aligned} supersin_{\alpha,\beta}^{\gamma}(t^{\alpha}) &= \frac{t^{\beta-1}}{2i}\left(E_{\alpha,\beta}^{\gamma}(it^{\alpha}) - E_{\alpha,\beta}^{\gamma}(-it^{\alpha})\right) \\ &= \sum_{k=0}^{\infty}\frac{(-1)^k(\gamma)_{2k+1}}{\Gamma((2k+1)\alpha+\beta)}\frac{t^{(2k+1)\alpha+\beta-1}}{(2k+1)!}, \end{aligned} \tag{5.194}$$

where $i=\sqrt{-1}$, $\alpha,\beta,\gamma,t\in\mathbb{C}$, $Re(\alpha)>0$, and $Re(\beta)>0$.

The result was discovered by Yang in 2019 [48].

Definition 5.3 The supercosine via Prabhakar function is defined as

$$\begin{aligned} supercos_{\alpha,\beta}^{\gamma}(t^{\alpha}) &= \frac{t^{\beta-1}}{2}\left(E_{\alpha,\beta}^{\gamma}(it^{\alpha}) + E_{\alpha,\beta}^{\gamma}(-it^{\alpha})\right) \\ &= \sum_{k=0}^{\infty}\frac{(-1)^k(\gamma)_{2k}}{\Gamma(2k\alpha+\beta)}\frac{t^{2k\alpha+\beta-1}}{(2k)!}, \end{aligned} \tag{5.195}$$

where $i=\sqrt{-1}$, $\alpha,\beta,\gamma,t\in\mathbb{C}$, $Re(\alpha)>0$, and $Re(\beta)>0$.

The result was discovered by Yang in 2019 [48].

Definition 5.4 The supertangent via Prabhakar function is defined as

$$supertan_{\alpha,\beta}^{\gamma}\left(t^{\alpha}\right) = \frac{supersin_{\alpha,\beta}^{\gamma}(t^{\alpha})}{supercos_{\alpha,\beta}^{\gamma}(t^{\alpha})}\left(supercos_{\alpha,\beta}^{\gamma}\left(t^{\alpha}\right)\neq 0\right), \tag{5.196}$$

where $\alpha,\beta,\gamma,t\in\mathbb{C}$, $Re(\alpha)>0$, and $Re(\beta)>0$.

Definition 5.5 The supercotangent via Prabhakar function is defined as

$$supercotan_{\alpha,\beta}^{\gamma}\left(t^{\alpha}\right) = \frac{supercos_{\alpha,\beta}^{\gamma}(t^{\alpha})}{supersin_{\alpha,\beta}^{\gamma}(t^{\alpha})}\left(supersin_{\alpha,\beta}^{\gamma}\left(t^{\alpha}\right)\neq 0\right), \tag{5.197}$$

where $\alpha,\beta,\gamma,t\in\mathbb{C}$, $Re(\alpha)>0$, and $Re(\beta)>0$.

Definition 5.6 The supercosecant via Prabhakar function is defined as

$$supercsc_{\alpha,\beta}^{\gamma}\left(t^{\alpha}\right) = \frac{1}{supersin_{\alpha,\beta}^{\gamma}(t^{\alpha})}\left(supersin_{\alpha,\beta}^{\gamma}\left(t^{\alpha}\right)\neq 0\right), \tag{5.198}$$

where $\alpha,\beta,\gamma,t\in\mathbb{C}$, $Re(\alpha)>0$, and $Re(\beta)>0$.

Definition 5.7 The supersecant via Prabhakar function is defined as [48]

$$supersec_{\alpha,\beta}^{\gamma}\left(t^{\alpha}\right) = \frac{1}{supercos_{\alpha,\beta}^{\gamma}(t^{\alpha})}\left(supercos_{\alpha,\beta}^{\gamma}\left(t^{\alpha}\right)\neq 0\right), \tag{5.199}$$

where $\alpha,\beta,\gamma,t\in\mathbb{C}$, $Re(\alpha)>0$, and $Re(\beta)>0$.

Generally,

$$\begin{aligned} supersin_{\alpha,\beta}^{\gamma}(\lambda t^{\alpha}) &= \tfrac{t^{\beta-1}}{2i}\left(E_{\alpha,\beta}^{\gamma}(i\lambda t^{\alpha}) - E_{\alpha,\beta}^{\gamma}(-i\lambda t^{\alpha})\right) \\ &= \tfrac{t^{\beta-1}}{2i} \times \tfrac{1}{\Gamma(\gamma)} \times {}_1sin_1\left(\begin{matrix}(\gamma,1) \\ (\beta,\alpha)\end{matrix}; \lambda t^{\alpha}\right) \\ &= \sum_{k=0}^{\infty} \tfrac{(-1)^k \lambda^{2k+1} (\gamma)_{2k+1} t^{(2n+1)\alpha+\beta-1}}{\Gamma((2k+1)\alpha+\beta)}, \end{aligned} \tag{5.200}$$

$$\begin{aligned} supercos_{\alpha,\beta}^{\gamma}(\lambda t^{\alpha}) &= \tfrac{t^{\beta-1}}{2}\left(E_{\alpha,\beta}^{\gamma}(i\lambda t^{\alpha}) + E_{\alpha,\beta}^{\gamma}(-i\lambda t^{\alpha})\right) \\ &= \tfrac{t^{\beta-1}}{2} \times \tfrac{1}{\Gamma(\gamma)} \times {}_1cos_1\left(\begin{matrix}(\gamma,1) \\ (\beta,\alpha)\end{matrix}; \lambda t^{\alpha}\right), \\ &= \sum_{k=0}^{\infty} \tfrac{(-1)^k \lambda^{2k} (\gamma)_{2k} t^{2k\alpha+\beta-1}}{\Gamma(2k\alpha+\beta)}, \end{aligned} \tag{5.201}$$

$$supertan_{\alpha,\beta}^{\gamma}\left(\lambda t^{\alpha}\right) = \frac{supersin_{\alpha,\beta}^{\gamma}(\lambda t^{\alpha})}{supercos_{\alpha,\beta}^{\gamma}(\lambda t^{\alpha})} \left(supercos_{\alpha,\beta}^{\gamma}\left(\lambda t^{\alpha}\right) \neq 0\right), \tag{5.202}$$

$$supercotan_{\alpha,\beta}^{\gamma}\left(\lambda t^{\alpha}\right) = \frac{supercos_{\alpha,\beta}^{\gamma}(\lambda t^{\alpha})}{supersin_{\alpha,\beta}^{\gamma}(\lambda t^{\alpha})} \left(supersin_{\alpha,\beta}^{\gamma}\left(\lambda t^{\alpha}\right) \neq 0\right), \tag{5.203}$$

$$supercsc_{\alpha,\beta}^{\gamma}\left(\lambda t^{\alpha}\right) = \frac{1}{supersin_{\alpha,\beta}^{\gamma}(\lambda t^{\alpha})} \left(supersin_{\alpha,\beta}^{\gamma}\left(\lambda t^{\alpha}\right) \neq 0\right), \tag{5.204}$$

$$supersec_{\alpha,\beta}^{\gamma}\left(\lambda t^{\alpha}\right) = \frac{1}{supercos_{\alpha,\beta}^{\gamma}(\lambda t^{\alpha})} \left(supercos_{\alpha,\beta}^{\gamma}\left(\lambda t^{\alpha}\right) \neq 0\right), \tag{5.205}$$

where $\alpha, \beta, \gamma, \lambda, t \in \mathbb{C}$, $Re(\alpha) > 0$, and $Re(\beta) > 0$.

The results were discovered by Yang in 2019 [48].

Based on the result, we may give the following.

Theorem 5.18 *Let $\alpha, \beta, \gamma, t \in \mathbb{C}$, $Re(\alpha) > 0$, and $Re(\beta) > 0$. Then we have [48]*

$$t^{\beta-1} E_{\alpha,\beta}^{\gamma}\left(it^{\alpha}\right) = supercos_{\alpha,\beta}^{\gamma}\left(t^{\alpha}\right) + i supersin_{\alpha,\beta}^{\gamma}\left(t^{\alpha}\right) \tag{5.206}$$

and

$$t^{\beta-1} E_{\alpha,\beta}^{\gamma}\left(-it^{\alpha}\right) = supercos_{\alpha,\beta}^{\gamma}\left(t^{\alpha}\right) - i supersin_{\alpha,\beta}^{\gamma}\left(t^{\alpha}\right). \tag{5.207}$$

The result was discovered by Yang in 2019 [48].

Generally,

$$t^{\beta-1}E^{\gamma}_{\alpha,\beta}\left(i\lambda t^{\alpha}\right) = supercos^{\gamma}_{\alpha,\beta}\left(\lambda t^{\alpha}\right) + isupersin^{\gamma}_{\alpha,\beta}\left(\lambda t^{\alpha}\right) \tag{5.208}$$

and

$$t^{\beta-1}E^{\gamma}_{\alpha,\beta}\left(-i\lambda t^{\alpha}\right) = supercos^{\gamma}_{\alpha,\beta}\left(\lambda t^{\alpha}\right) - isupersin^{\gamma}_{\alpha,\beta}\left(\lambda t^{\alpha}\right), \tag{5.209}$$

where $\alpha, \beta, \gamma, \lambda, t \in \mathbb{C}$, $Re(\alpha) > 0$, *and* $Re(\beta) > 0$.

Property 5.5 Let $\alpha, \beta, \gamma, t \in \mathbb{C}$, $Re(\alpha) > 0$, and $Re(\beta) > 0$. Then we have

$$supersin^{\gamma}_{\alpha,\beta}\left(-t^{\alpha}\right) = -supersin^{\gamma}_{\alpha,\beta}\left(t^{\alpha}\right), \tag{5.210}$$

$$supercos^{\gamma}_{\alpha,\beta}\left(-t^{\alpha}\right) = supercos^{\gamma}_{\alpha,\beta}\left(t^{\alpha}\right), \tag{5.211}$$

$$supertan^{\gamma}_{\alpha,\beta}\left(-t^{\alpha}\right) = -\, supertan^{\gamma}_{\alpha,\beta}\left(t^{\alpha}\right), \tag{5.212}$$

$$supercotan^{\gamma}_{\alpha,\beta}\left(-t^{\alpha}\right) = -\, supercotan^{\gamma}_{\alpha,\beta}\left(t^{\alpha}\right), \tag{5.213}$$

$$supercsc^{\gamma}_{\alpha,\beta}\left(-t^{\alpha}\right) = -supercsc^{\gamma}_{\alpha,\beta}\left(t^{\alpha}\right), \tag{5.214}$$

and

$$supersec^{\gamma}_{\alpha,\beta}\left(-t^{\alpha}\right) = supersec^{\gamma}_{\alpha,\beta}\left(t^{\alpha}\right). \tag{5.215}$$

Theorem 5.19 *Let* $\alpha, \beta, \gamma, \lambda, t \in \mathbb{C}$, $Re(\alpha) > 0$, *and* $Re(\beta) > 0$. *Then the Laplace transform of the function* $Supersin^{\gamma}_{\alpha,\beta}(t^{\alpha})$ *is given as follows [48]:*

$$\begin{aligned}
&L\left\{Supersin^{\gamma}_{\alpha,\beta}\left(t^{\alpha}\right)\right\} \\
&= \sum_{k=0}^{\infty}\binom{-\gamma}{2k+1}\frac{(-1)^{k+1}}{s^{(2k+1)\alpha+\beta}} \\
&= s^{-\beta}\sum_{k=0}^{\infty}\binom{-\gamma}{2k+1} i^{2k+2}\frac{1}{s^{(2k+1)\alpha}} \\
&= is^{-\beta}\sum_{k=0}^{\infty}\binom{-\gamma}{2k+1}\left(\frac{i}{s^{\alpha}}\right)^{2k+1} \\
&= \frac{is^{-\beta}}{2}\left[\left(1+\frac{i}{s^{\alpha}}\right)^{-\gamma} - \left(1-\frac{i}{s^{\alpha}}\right)^{-\gamma}\right] \\
&= \frac{is^{-\beta}}{2}\frac{\left(1-is^{-\alpha}\right)^{\gamma} - \left(1+is^{-\alpha}\right)^{\gamma}}{\left(1+s^{-2\alpha}\right)^{\gamma}} \\
&= \frac{s^{\alpha\gamma-\beta}\left[\left(s^{\alpha}-i\right)^{-\gamma} - \left(s^{\alpha}+i\right)^{-\gamma}\right]}{2i},
\end{aligned} \tag{5.216}$$

where $\left|s^{-2\alpha}\right| < 1$.

The result was discovered by Yang in 2019 [48].
More generally,

$$\begin{aligned}
&L\left\{Supersin_{\alpha,\beta}^{\gamma}\left(\lambda t^{\alpha}\right)\right\}\\
&=\sum_{k=0}^{\infty}\binom{-\gamma}{2k+1}\frac{(-1)^{k+1}\lambda^{2k+1}}{s^{(2k+1)\alpha+\beta}}\\
&=s^{-\beta}\sum_{k=0}^{\infty}\binom{-\gamma}{2k+1}i^{2k+2}\frac{\lambda^{2k+1}}{s^{(2k+1)\alpha}}\\
&=is^{-\beta}\sum_{\kappa=0}^{\infty}\binom{-\gamma}{2k+1}\left(\frac{i\lambda}{s^{\alpha}}\right)^{2k+1}\\
&=\frac{is^{-\beta}}{2}\left[\left(1+\frac{i\lambda}{s^{\alpha}}\right)^{-\gamma}-\left(1-\frac{i\lambda}{s^{\alpha}}\right)^{-\gamma}\right]\\
&=\frac{is^{-\beta}}{2}\frac{\left(1-i\lambda s^{-\alpha}\right)^{\gamma}-\left(1+i\lambda s^{-\alpha}\right)^{\gamma}}{\left(1+\lambda^{2}s^{-2\alpha}\right)^{\gamma}}\\
&=\frac{s^{\alpha\gamma-\beta}\left[\left(s^{\alpha}-i\lambda\right)^{-\gamma}-\left(s^{\alpha}+i\lambda\right)^{-\gamma}\right]}{2i},
\end{aligned}\tag{5.217}$$

where $\left|\lambda^{2}s^{-2\alpha}\right|<1$.

The result was discovered by Yang in 2019 [48].

Theorem 5.20 *Let $\alpha, \beta, \gamma, \lambda, t \in \mathbb{C}$, $Re(\alpha) > 0$, and $Re(\beta) > 0$. Then the Laplace transform of the function $Supercos_{\alpha,\beta}^{\gamma}(\lambda t^{\alpha})$ is given as follows [48]:*

$$\begin{aligned}
&L\left\{Supercos_{\alpha,\beta}^{\gamma}\left(t^{\alpha}\right)\right\}\\
&=\sum_{k=0}^{\infty}\binom{-\gamma}{2\kappa}\frac{(-1)^{k}}{s^{2k\alpha+\beta}}\\
&=s^{-\beta}\sum_{k=0}^{\infty}\binom{-\gamma}{2k}\left(\frac{i}{s^{\alpha}}\right)^{2k}\\
&=\frac{s^{-\beta}}{2}\left[\left(1+\frac{i}{s^{\alpha}}\right)^{-\gamma}+\left(1-\frac{i}{s^{\alpha}}\right)^{-\gamma}\right]\\
&=\frac{s^{-\beta}}{2}\frac{\left(1+is^{-\alpha}\right)^{\gamma}+\left(1-is^{-\alpha}\right)^{\gamma}}{\left(1+s^{-2\alpha}\right)^{\gamma}}\\
&=\frac{s^{\alpha\gamma-\beta}\left[\left(s^{\alpha}-i\right)^{-\gamma}+\left(s^{\alpha}+i\right)^{-\gamma}\right]}{2i},
\end{aligned}\tag{5.218}$$

where $\left|s^{-2\alpha}\right|<1$.

The result was discovered by Yang in 2019 [48].

More generally,

$$
\begin{aligned}
&L\left\{Supercos_{\alpha,\beta}^{\gamma}\left(\lambda t^{\alpha}\right)\right\}\\
&=\sum_{k=0}^{\infty}\binom{-\gamma}{2\kappa}\frac{(-1)^{k}\lambda^{2k}}{s^{2k\alpha+\beta}}\\
&=s^{-\beta}\sum_{k=0}^{\infty}\binom{-\gamma}{2k}\left(\frac{i\lambda}{s^{\alpha}}\right)^{2k}\\
&=\frac{s^{-\beta}}{2}\left[\left(1+\frac{i\lambda}{s^{\alpha}}\right)^{-\gamma}+\left(1-\frac{i\lambda}{s^{\alpha}}\right)^{-\gamma}\right]\\
&=\frac{s^{-\beta}}{2}\frac{\left(1+i\lambda s^{-\alpha}\right)^{\gamma}+\left(1-i\lambda s^{-\alpha}\right)^{\gamma}}{\left(1+\lambda^{2}s^{-2\alpha}\right)^{\gamma}}\\
&=\frac{s^{\alpha\gamma-\beta}\left[\left(s^{\alpha}-i\lambda\right)^{-\gamma}+\left(s^{\alpha}+i\lambda\right)^{-\gamma}\right]}{2},
\end{aligned}
\tag{5.219}
$$

where $\left|\lambda^{2}s^{-2\alpha}\right|<1$.

The result was discovered by Yang in 2019 [48].

5.1.3 The Superhyperbolic Functions via Prabhakar Function

In the section, we present the concept of the superhyperbolic functions via Prabhakar function.

Definition 5.8 The superhyperbolic supersine via Prabhakar function is defined as [48]

$$
\begin{aligned}
supersinh_{\alpha,\beta}^{\gamma}\left(t^{\alpha}\right)&=\frac{t^{\beta-1}}{2}\left(E_{\alpha,\beta}^{\gamma}\left(t^{\alpha}\right)-E_{\alpha,\beta}^{\gamma}\left(-t^{\alpha}\right)\right)\\
&=\sum_{k=0}^{\infty}\frac{(\gamma)_{2k+1}t^{(2k+1)\alpha+\beta-1}}{\Gamma((2k+1)\alpha+\beta)},
\end{aligned}
\tag{5.220}
$$

where $\alpha,\beta,t\in\mathbb{C}$, $Re\left(\alpha\right)>0$, and $Re\left(\beta\right)>0$.

The result was discovered by Yang in 2019 [48].

Definition 5.9 The superhyperbolic supercosine via Prabhakar function is defined as [48]

$$
\begin{aligned}
supercosh_{\alpha,\beta}^{\gamma}\left(t^{\alpha}\right)&=\frac{t^{\beta-1}}{2}\left(E_{\alpha,\beta}^{\gamma}\left(t^{\alpha}\right)+E_{\alpha,\beta}^{\gamma}\left(-t^{\alpha}\right)\right)\\
&=\sum_{n=0}^{\infty}\frac{(\gamma)_{2k}t^{2k\alpha+\beta-1}}{\Gamma(2k\alpha+\beta)},
\end{aligned}
\tag{5.221}
$$

where $\alpha,\beta,t\in\mathbb{C}$, $Re\left(\alpha\right)>0$, and $Re\left(\beta\right)>0$.

The result was discovered by Yang in 2019 [48].

Definition 5.10 The superhyperbolic supertangent via Prabhakar function is defined as

$$supertanh^{\gamma}_{\alpha,\beta}\left(t^{\alpha}\right) = \frac{supersinh^{\gamma}_{\alpha,\beta}\left(t^{\alpha}\right)}{supercosh^{\gamma}_{\alpha,\beta}\left(t^{\alpha}\right)}\left(supercosh^{\gamma}_{\alpha,\beta}\left(t^{\alpha}\right) \neq 0\right), \tag{5.222}$$

where $\alpha, \beta, \gamma, t \in \mathbb{C}$, $Re\left(\alpha\right) > 0$, and $Re\left(\beta\right) > 0$.

Definition 5.11 The superhyperbolic supercotangent via Prabhakar function is defined as

$$supercotanh^{\gamma}_{\alpha,\beta}\left(t^{\alpha}\right) = \frac{supercosh^{\gamma}_{\alpha,\beta}\left(t^{\alpha}\right)}{supersinh^{\gamma}_{\alpha,\beta}\left(t^{\alpha}\right)}\left(supersinh^{\gamma}_{\alpha,\beta}\left(t^{\alpha}\right) \neq 0\right), \tag{5.223}$$

where $\alpha, \beta, \gamma, t \in \mathbb{C}$, $Re\left(\alpha\right) > 0$, and $Re\left(\beta\right) > 0$.

Definition 5.12 The superhyperbolic supercosecant via Prabhakar function is defined as [48]

$$supercsch^{\gamma}_{\alpha,\beta}\left(t^{\alpha}\right) = \frac{1}{supersinh^{\gamma}_{\alpha,\beta}\left(t^{\alpha}\right)}\left(supersinh^{\gamma}_{\alpha,\beta}\left(t^{\alpha}\right) \neq 0\right), \tag{5.224}$$

where $\alpha, \beta, \gamma, t \in \mathbb{C}$, $Re\left(\alpha\right) > 0$, and $Re\left(\beta\right) > 0$.

The result was discovered by Yang in 2019 [48].

Definition 5.13 The superhyperbolic supersecant via Prabhakar function is defined as [48]

$$supersech^{\gamma}_{\alpha,\beta}\left(t^{\alpha}\right) = \frac{1}{supercosh^{\gamma}_{\alpha,\beta}\left(t^{\alpha}\right)}\left(supercosh^{\gamma}_{\alpha,\beta}\left(t^{\alpha}\right) \neq 0\right), \tag{5.225}$$

where $\alpha, \beta, \gamma, t \in \mathbb{C}$, $Re\left(\alpha\right) > 0$, and $Re\left(\beta\right) > 0$.

The result was discovered by Yang in 2019 [48].

Generally,

$$\begin{aligned} supersinh^{\gamma}_{\alpha,\beta}\left(\lambda t^{\alpha}\right) &= \tfrac{t^{\beta-1}}{2i}\left(E^{\gamma}_{\alpha,\beta}\left(\lambda t^{\alpha}\right) - E^{\gamma}_{\alpha,\beta}\left(-\lambda t^{\alpha}\right)\right) \\ &= \tfrac{t^{\beta-1}}{2} \times \tfrac{1}{\Gamma(\gamma)} \times {}_1sinh_1\left(\begin{matrix}(\gamma,1)\\(\beta,\alpha)\end{matrix}; \lambda t^{\alpha}\right) \\ &= \sum_{n=0}^{\infty} \tfrac{\lambda^{2k+1}(\gamma)_{2k+1}t^{(2n+1)\alpha+\beta-1}}{\Gamma((2n+1)\alpha+\beta)}, \end{aligned} \tag{5.226}$$

$$\begin{aligned} supercosh^{\gamma}_{\alpha,\beta}(\lambda t^{\alpha}) &= \tfrac{t^{\beta-1}}{2}\left(E^{\gamma}_{\alpha,\beta}(\lambda t^{\alpha}) + E^{\gamma}_{\alpha,\beta}(-\lambda t^{\alpha})\right) \\ &= \tfrac{t^{\beta-1}}{2} \times \tfrac{1}{\Gamma(\gamma)} \times {}_1cosh_1\left(\begin{matrix}(\gamma,1)\\(\beta,\alpha)\end{matrix};\lambda t^{\alpha}\right), \\ &= \sum_{k=0}^{\infty} \tfrac{\lambda^{2k}(\gamma)_{2k}t^{2k\alpha+\beta-1}}{\Gamma(2k\alpha+\beta)}, \end{aligned} \tag{5.227}$$

$$supertanh^{\gamma}_{\alpha,\beta}\left(\lambda t^{\alpha}\right) = \frac{supersinh^{\gamma}_{\alpha,\beta}(\lambda t^{\alpha})}{supercosh^{\gamma}_{\alpha,\beta}(\lambda t^{\alpha})}\left(supercosh^{\gamma}_{\alpha,\beta}\left(\lambda t^{\alpha}\right) \neq 0\right), \tag{5.228}$$

$$supercotanh^{\gamma}_{\alpha,\beta}\left(\lambda t^{\alpha}\right) = \frac{supercosh^{\gamma}_{\alpha,\beta}(\lambda t^{\alpha})}{supersinh^{\gamma}_{\alpha,\beta}(\lambda t^{\alpha})}\left(supersinh^{\gamma}_{\alpha,\beta}\left(\lambda t^{\alpha}\right) \neq 0\right), \tag{5.229}$$

$$supercsch^{\gamma}_{\alpha,\beta}\left(\lambda t^{\alpha}\right) = \frac{1}{supersinh^{\gamma}_{\alpha,\beta}(\lambda t^{\alpha})}\left(supersinh^{\gamma}_{\alpha,\beta}\left(\lambda t^{\alpha}\right) \neq 0\right), \tag{5.230}$$

$$supersech^{\gamma}_{\alpha,\beta}\left(\lambda t^{\alpha}\right) = \frac{1}{supercosh^{\gamma}_{\alpha,\beta}(\lambda t^{\alpha})}\left(supercosh^{\gamma}_{\alpha,\beta}\left(\lambda t^{\alpha}\right) \neq 0\right), \tag{5.231}$$

where $\alpha, \beta, \gamma, \lambda, t \in \mathbb{C}$, $Re(\alpha) > 0$, and $Re(\beta) > 0$.

Thus, we have the following result.

Theorem 5.21 *Let $\alpha, \beta, \gamma, t \in \mathbb{C}$, $Re(\alpha) > 0$, and $Re(\beta) > 0$. Then we have [48]*

$$t^{\beta-1}E^{\gamma}_{\alpha,\beta}\left(t^{\alpha}\right) = supercosh^{\gamma}_{\alpha,\beta}\left(t^{\alpha}\right) + supersinh^{\gamma}_{\alpha,\beta}\left(t^{\alpha}\right) \tag{5.232}$$

and

$$t^{\beta-1}E^{\gamma}_{\alpha,\beta}\left(-t^{\alpha}\right) = supercosh^{\gamma}_{\alpha,\beta}\left(t^{\alpha}\right) - supersinh^{\gamma}_{\alpha,\beta}\left(t^{\alpha}\right). \tag{5.233}$$

The result was discovered by Yang in 2019 [48].

Generally,

$$t^{\beta-1}E^{\gamma}_{\alpha,\beta}\left(\lambda t^{\alpha}\right) = supercosh^{\gamma}_{\alpha,\beta}\left(\lambda t^{\alpha}\right) + supersinh^{\gamma}_{\alpha,\beta}\left(\lambda t^{\alpha}\right) \tag{5.234}$$

and

$$t^{\beta-1}E^{\gamma}_{\alpha,\beta}\left(-\lambda t^{\alpha}\right)=supercosh^{\gamma}_{\alpha,\beta}\left(\lambda t^{\alpha}\right)-supersinh^{\gamma}_{\alpha,\beta}\left(\lambda t^{\alpha}\right), \tag{5.235}$$

where $\alpha,\beta,\gamma,\lambda,t\in\mathbb{C}$, $Re(\alpha)>0$, *and* $Re(\beta)>0$.

Property 5.6 Let $\alpha,\beta,\gamma,t\in\mathbb{C}$, $Re(\alpha)>0$, and $Re(\beta)>0$. Then we have

$$supersinh^{\gamma}_{\alpha,\beta}\left(-t^{\alpha}\right)=-supersinh^{\gamma}_{\alpha,\beta}\left(t^{\alpha}\right), \tag{5.236}$$

$$supercosh^{\gamma}_{\alpha,\beta}\left(-t^{\alpha}\right)=supercosh^{\gamma}_{\alpha,\beta}\left(t^{\alpha}\right), \tag{5.237}$$

$$supertanh^{\gamma}_{\alpha,\beta}\left(-t^{\alpha}\right)=-\,supertanh^{\gamma}_{\alpha,\beta}\left(t^{\alpha}\right), \tag{5.238}$$

$$supercotanh^{\gamma}_{\alpha,\beta}\left(-t^{\alpha}\right)=-\,supercotanh^{\gamma}_{\alpha,\beta}\left(t^{\alpha}\right), \tag{5.239}$$

$$supercsch^{\gamma}_{\alpha,\beta}\left(-t^{\alpha}\right)=-supercsch^{\gamma}_{\alpha,\beta}\left(t^{\alpha}\right), \tag{5.240}$$

and

$$supersech^{\gamma}_{\alpha,\beta}\left(-t^{\alpha}\right)=supersech^{\gamma}_{\alpha,\beta}\left(t^{\alpha}\right). \tag{5.241}$$

Theorem 5.22 *Let* $\alpha,\beta,\gamma,\lambda,t\in\mathbb{C}$, $Re(\alpha)>0$, *and* $Re(\beta)>0$. *Then the Laplace transform of the function* $Supersinh^{\gamma}_{\alpha,\beta}(t^{\alpha})$ *is given as follows [48]:*

$$\begin{aligned}
&L\left\{Supersinh^{\gamma}_{\alpha,\beta}(t^{\alpha})\right\}\\
&=\sum_{k=0}^{\infty}\binom{-\gamma}{2k+1}\frac{1}{s^{(2k+1)\alpha+\beta}}\\
&=s^{-\beta}\sum_{k=0}^{\infty}\binom{-\gamma}{2k+1}\frac{1}{s^{(2k+1)\alpha}}\\
&=s^{-\beta}\sum_{k=0}^{\infty}\binom{-\gamma}{2k+1}\left(\frac{1}{s^{\alpha}}\right)^{2k+1}\\
&=\frac{s^{-\beta}}{2}\left[\left(1+\frac{1}{s^{\alpha}}\right)^{-\gamma}-\left(1-\frac{1}{s^{\alpha}}\right)^{-\gamma}\right]\\
&=\frac{s^{-\beta}}{2}\frac{(1-s^{-\alpha})^{\gamma}-(1+s^{-\alpha})^{\gamma}}{(1-s^{-2\alpha})^{\gamma}}\\
&=\frac{s^{\alpha\gamma-\beta}\left[(s^{\alpha}-1)^{-\gamma}-(s^{\alpha}+1)^{-\gamma}\right]}{2},
\end{aligned} \tag{5.242}$$

where $\left|s^{-2\alpha}\right|<1$.

The result was discovered by Yang in 2019 [48].

More generally,

$$
\begin{aligned}
&L\left\{Supersinh_{\alpha,\beta}^{\gamma}\left(\lambda t^{\alpha}\right)\right\} \\
&=\sum_{k=0}^{\infty}\binom{-\gamma}{2k+1}\frac{\lambda^{2k+1}}{s^{(2k+1)\alpha+\beta}} \\
&=s^{-\beta}\sum_{k=0}^{\infty}\binom{-\gamma}{2k+1}\frac{\lambda^{2k+1}}{s^{(2k+1)\alpha}} \\
&=s^{-\beta}\sum_{\kappa=0}^{\infty}\binom{-\gamma}{2k+1}\left(\frac{\lambda}{s^{\alpha}}\right)^{2k+1} \\
&=\frac{s^{-\beta}}{2}\left[\left(1+\frac{\lambda}{s^{\alpha}}\right)^{-\gamma}-\left(1-\frac{\lambda}{s^{\alpha}}\right)^{-\gamma}\right] \\
&=\frac{s^{-\beta}}{2}\frac{\left(1-\lambda s^{-\alpha}\right)^{\gamma}-\left(1+\lambda s^{-\alpha}\right)^{\gamma}}{\left(1-\lambda^{2}s^{-2\alpha}\right)^{\gamma}}, \\
&=\frac{s^{\alpha\gamma-\beta}\left[\left(s^{\alpha}-\lambda\right)^{-\gamma}-\left(s^{\alpha}+\lambda\right)^{-\gamma}\right]}{2},
\end{aligned}
\tag{5.243}
$$

where $\left|\lambda^{2}s^{-2\alpha}\right|<1$.

The result was discovered by Yang in 2019 [48].

Theorem 5.23 *Let* $\alpha,\beta,\gamma,\lambda,t\in\mathbb{C}$, $Re(\alpha)>0$, *and* $Re(\beta)>0$. *Then the Laplace transform of the function* $Supercosh_{\alpha,\beta}^{\gamma}\left(\lambda t^{\alpha}\right)$ *is given as follows [48]:*

$$
\begin{aligned}
&L\left\{Supercosh_{\alpha,\beta}^{\gamma}\left(t^{\alpha}\right)\right\} \\
&=\sum_{k=0}^{\infty}\binom{-\gamma}{2\kappa}\frac{1}{s^{2k\alpha+\beta}} \\
&=s^{-\beta}\sum_{k=0}^{\infty}\binom{-\gamma}{2k}\left(\frac{1}{s^{\alpha}}\right)^{2k} \\
&=\frac{s^{-\beta}}{2}\left[\left(1+\frac{1}{s^{\alpha}}\right)^{-\gamma}+\left(1-\frac{1}{s^{\alpha}}\right)^{-\gamma}\right] \\
&=\frac{s^{-\beta}}{2}\frac{\left(1+s^{-\alpha}\right)^{\gamma}+\left(1-s^{-\alpha}\right)^{\gamma}}{\left(1-s^{-2\alpha}\right)^{\gamma}} \\
&=\frac{s^{\alpha\gamma-\beta}\left[\left(s^{\alpha}-1\right)^{-\gamma}+\left(s^{\alpha}+1\right)^{-\gamma}\right]}{2},
\end{aligned}
\tag{5.244}
$$

where $\left|s^{-2\alpha}\right|<1$.

The result was discovered by Yang in 2019 [48].

More generally,

$$\begin{aligned}
&L\left\{Supercosh_{\alpha,\beta}^{\gamma}\left(\lambda t^{\alpha}\right)\right\}\\
&=\sum_{k=0}^{\infty}\binom{-\gamma}{2\kappa}\frac{\lambda^{2k}}{s^{2k\alpha+\beta}}\\
&=s^{-\beta}\sum_{k=0}^{\infty}\binom{-\gamma}{2k}\left(\frac{\lambda}{s^{\alpha}}\right)^{2k}\\
&=\frac{s^{-\beta}}{2}\left[\left(1+\frac{\lambda}{s^{\alpha}}\right)^{-\gamma}+\left(1-\frac{\lambda}{s^{\alpha}}\right)^{-\gamma}\right]\\
&=\frac{s^{-\beta}}{2}\frac{\left(1+\lambda s^{-\alpha}\right)^{\gamma}+\left(1-\lambda s^{-\alpha}\right)^{\gamma}}{\left(1-\lambda^{2}s^{-2\alpha}\right)^{\gamma}}\\
&=\frac{s^{\alpha\gamma-\beta}\left[\left(s^{\alpha}-\lambda\right)^{-\gamma}+\left(s^{\alpha}+\lambda\right)^{-\gamma}\right]}{2},
\end{aligned} \tag{5.245}$$

where $\left|\lambda^{2}s^{-2\alpha}\right|<1$.

The result was discovered by Yang in 2019 [48].

5.1.4 The Pre-Supertrigonometric Functions via Prabhakar Function

In this section, we introduce the pre-supertrigonometric functions via Prabhakar function.

Definition 5.14 The presine via Prabhakar function is defined by

$$\begin{aligned}
presin_{\alpha,\beta}^{\gamma}(t)&=\frac{1}{2i}\left(E_{\alpha,\beta}^{\gamma}(it)-E_{\alpha,\beta}^{\gamma}(-it)\right)\\
&=\sum_{k=0}^{\infty}\frac{(-1)^{k}(\gamma)_{2k+1}t^{2k+1}}{\Gamma((2k+1)\alpha+\beta)},
\end{aligned} \tag{5.246}$$

where $\alpha,\beta,\gamma,t\in\mathbb{C}$, $Re(\alpha)>0$, and $Re(\beta)>0$.

Definition 5.15 The precosine via Prabhakar function is defined by

$$\begin{aligned}
precos_{\alpha,\beta}^{\gamma}(t)&=\frac{1}{2}\left(E_{\alpha,\beta}^{\gamma}(it)+E_{\alpha,\beta}^{\gamma}(-it)\right)\\
&=\sum_{k=0}^{\infty}\frac{(-1)^{k}(\gamma)_{2k}t^{2k}}{\Gamma(2k\alpha+\beta)},
\end{aligned} \tag{5.247}$$

where $\alpha,\beta,\gamma,t\in\mathbb{C}$, $Re(\alpha)>0$, and $Re(\beta)>0$.

Definition 5.16 The pretangent via Prabhakar function is defined by

$$pretan_{\alpha,\beta}^{\gamma}(t)=\frac{presin_{\alpha,\beta}^{\gamma}(t)}{precos_{\alpha,\beta}^{\gamma}(t)}\left(precos_{\alpha,\beta}^{\gamma}(t)\neq 0\right), \tag{5.248}$$

where $\alpha,\beta,\gamma,t\in\mathbb{C}$, $Re(\alpha)>0$, and $Re(\beta)>0$.

Definition 5.17 The precotangent via Prabhakar function is defined by

$$precotan^{\gamma}_{\alpha,\beta}(t) = \frac{precos^{\gamma}_{\alpha,\beta}(t)}{presin^{\gamma}_{\alpha,\beta}(t)} \left(presin^{\gamma}_{\alpha,\beta}(t) \neq 0\right), \tag{5.249}$$

where $\alpha, \beta, \gamma, t \in \mathbb{C}$, $Re(\alpha) > 0$, and $Re(\beta) > 0$.

Definition 5.18 The precosecant via Prabhakar function is defined by

$$precsc^{\gamma}_{\alpha,\beta}(t) = \frac{1}{presin^{\gamma}_{\alpha,\beta}(t)} \left(presin^{\gamma}_{\alpha,\beta}(t) \neq 0\right), \tag{5.250}$$

where $\alpha, \beta, \gamma, t \in \mathbb{C}$, $Re(\alpha) > 0$, and $Re(\beta) > 0$.

Definition 5.19 The presecant via Prabhakar function is defined by

$$presec^{\gamma}_{\alpha,\beta}(t) = \frac{1}{precos^{\gamma}_{\alpha,\beta}(t)} \left(precos^{\gamma}_{\alpha,\beta}(t) \neq 0\right), \tag{5.251}$$

where $\alpha, \beta, \gamma, t \in \mathbb{C}$, $Re(\alpha) > 0$, and $Re(\beta) > 0$.

Thus,

$$\begin{aligned} presin^{\gamma}_{\alpha,\beta}(\lambda t) &= \tfrac{1}{2i}\left(E^{\gamma}_{\alpha,\beta}(i\lambda t) - E^{\gamma}_{\alpha,\beta}(-i\lambda t)\right) \\ &= \tfrac{1}{2i} \times \tfrac{1}{\Gamma(\gamma)} \times {}_1sin_1\left(\begin{matrix}(\gamma, 1) \\ (\beta, \alpha)\end{matrix}; \lambda t\right) \\ &= \sum_{k=0}^{\infty} \tfrac{(-1)^k \lambda^{2k+1} (\gamma)_{2k+1} t^{2k+1}}{\Gamma((2k+1)\alpha+\beta)}, \end{aligned} \tag{5.252}$$

$$\begin{aligned} precos^{\gamma}_{\alpha,\beta}(\lambda t) &= \tfrac{1}{2}\left(E^{\gamma}_{\alpha,\beta}(i\lambda t) + E^{\gamma}_{\alpha,\beta}(-i\lambda t)\right) \\ &= \tfrac{1}{2\Gamma(\gamma)} \times {}_1cos_1\left(\begin{matrix}(\gamma, 1) \\ (\beta, \alpha)\end{matrix}; \lambda t\right) \\ &= \sum_{k=0}^{\infty} \tfrac{(-1)^k \lambda^{2k} (\gamma)_{2k} t^{2k}}{\Gamma(2k\alpha+\beta)}, \end{aligned} \tag{5.253}$$

$$pretan^{\gamma}_{\alpha,\beta}(\lambda t) = \frac{presin^{\gamma}_{\alpha,\beta}(\lambda t)}{precos^{\gamma}_{\alpha,\beta}(\lambda t)} \left(precos^{\gamma}_{\alpha,\beta}(\lambda t) \neq 0\right), \tag{5.254}$$

$$precotan^{\gamma}_{\alpha,\beta}(\lambda t) = \frac{precos^{\gamma}_{\alpha,\beta}(\lambda t)}{presin^{\gamma}_{\alpha,\beta}(\lambda t)} \left(presin^{\gamma}_{\alpha,\beta}(\lambda t) \neq 0\right), \tag{5.255}$$

$$precsc^{\gamma}_{\alpha,\beta}(\lambda t) = \frac{1}{presin^{\gamma}_{\alpha,\beta}(\lambda t)} \left(presin^{\gamma}_{\alpha,\beta}(\lambda t) \neq 0\right), \tag{5.256}$$

and

$$presec_{\alpha,\beta}^{\gamma}(\lambda t) = \frac{1}{precos_{\alpha,\beta}^{\gamma}(\lambda t)} \left(precos_{\alpha,\beta}^{\gamma}(\lambda t) \neq 0\right), \tag{5.257}$$

where $\alpha, \beta, \gamma, \lambda, t \in \mathbb{C}$, $Re(\alpha) > 0$, and $Re(\beta) > 0$.

Based on the result, we may have the following theorem:

Theorem 5.24 *Let $\alpha, \beta, \gamma, t \in \mathbb{C}$, $Re(\alpha) > 0$, and $Re(\beta) > 0$. Then we have*

$$E_{\alpha,\beta}^{\gamma}(it) = precos_{\alpha,\beta}^{\gamma}(t) + ipresin_{\alpha,\beta}^{\gamma}(t) \tag{5.258}$$

and

$$E_{\alpha,\beta}^{\gamma}(-it) = precos_{\alpha,\beta}^{\gamma}(t) - ipresin_{\alpha,\beta}^{\gamma}(t). \tag{5.259}$$

Generally,

$$E_{\alpha,\beta}^{\gamma}(i\lambda t) = precos_{\alpha,\beta}^{\gamma}(\lambda t) + ipresin_{\alpha,\beta}^{\gamma}(\lambda t) \tag{5.260}$$

and

$$E_{\alpha,\beta}^{\gamma}(-i\lambda t) = precos_{\alpha,\beta}^{\gamma}(\lambda t) - ipresin_{\alpha,\beta}^{\gamma}(\lambda t), \tag{5.261}$$

where $\alpha, \beta, \gamma, \lambda, t \in \mathbb{C}$, $Re(\alpha) > 0$, and $Re(\beta) > 0$.

Property 5.7 Let $\alpha, \beta, \gamma, t \in \mathbb{C}$, $Re(\alpha) > 0$, and $Re(\beta) > 0$. Then we have

$$presin_{\alpha,\beta}^{\gamma}(-t) = -presin_{\alpha,\beta}^{\gamma}(t), \tag{5.262}$$

$$precos_{\alpha,\beta}^{\gamma}(-t) = precos_{\alpha,\beta}^{\gamma}(t), \tag{5.263}$$

$$pretan_{\alpha,\beta}^{\gamma}(-t) = -\ pretan_{\alpha,\beta}^{\gamma}(t), \tag{5.264}$$

$$precotan_{\alpha,\beta}^{\gamma}(-t) = -\ precotan_{\alpha,\beta}^{\gamma}(t), \tag{5.265}$$

$$precsc_{\alpha,\beta}^{\gamma}(-t) = -precsc_{\alpha,\beta}^{\gamma}(t), \tag{5.266}$$

and

$$presec_{\alpha,\beta}^{\gamma}(-t) = presec_{\alpha,\beta}^{\gamma}(t). \tag{5.267}$$

Theorem 5.25 *Let $\alpha, \beta, \gamma, t \in \mathbb{C}$, $Re(\alpha) > 0$, and $Re(\beta) > 0$. Then the Laplace transform of the function $presin^{\gamma}_{\alpha,\beta}(t)$ is given as follows:*

$$L\left\{presin^{\gamma}_{\alpha,\beta}(t)\right\} = \frac{s^{-1}}{\Gamma(\gamma)}\, {}_2sin_1\left(\begin{matrix}(\gamma,1),(1,1)\\(\beta,\alpha)\end{matrix};\frac{1}{s}\right). \tag{5.268}$$

Proof Let us consider

$$L\left[t^{\varsigma-1}E^{\gamma}_{\alpha,\beta}(\lambda t)\right] = \int_0^{\infty} e^{-st}t^{\varsigma-1}E^{\gamma}_{\alpha,\beta}(\lambda t)\,dt = \frac{s^{-\varsigma}}{\Gamma(\gamma)}\, {}_2\varphi_1\left(\begin{matrix}(\gamma,1),(\varsigma,1)\\(\beta,\alpha)\end{matrix};\frac{\lambda}{s}\right), \tag{5.269}$$

which leads to, for $\varsigma = 1$,

$$L\left[E^{\gamma}_{\alpha,\beta}(\lambda t)\right] = \int_0^{\infty} e^{-st}E^{\gamma}_{\alpha,\beta}(\lambda t)\,dt = \frac{s^{-1}}{\Gamma(\gamma)}\, {}_2\varphi_1\left(\begin{matrix}(\gamma,1),(1,1)\\(\beta,\alpha)\end{matrix};\frac{\lambda}{s}\right). \tag{5.270}$$

It follows that

$$\begin{aligned} &L\left\{presin^{\gamma}_{\alpha,\beta}(t)\right\} \\ &= \tfrac{1}{2i}\left[L\left\{E^{\gamma}_{\alpha,\beta}(it)\right\} - L\left\{E^{\gamma}_{\alpha,\beta}(-it)\right\}\right] \\ &= \tfrac{1}{2i}\tfrac{s^{-1}}{\Gamma(\gamma)}\left[{}_2\varphi_1\left(\begin{matrix}(\gamma,1),(1,1)\\(\beta,\alpha)\end{matrix};\tfrac{i}{s}\right) - {}_2\varphi_1\left(\begin{matrix}(\gamma,1),(1,1)\\(\beta,\alpha)\end{matrix};-\tfrac{i}{s}\right)\right] \\ &= \tfrac{s^{-1}}{\Gamma(\gamma)}\, {}_2sin_1\left(\begin{matrix}(\gamma,1),(1,1)\\(\beta,\alpha)\end{matrix};\tfrac{1}{s}\right). \end{aligned} \tag{5.271}$$

More generally, for $\alpha, \beta, \gamma, \lambda, t \in \mathbb{C}$, $Re(\alpha) > 0$, and $Re(\beta) > 0$,

$$\begin{aligned} &L\left\{presin^{\gamma}_{\alpha,\beta}(\lambda t)\right\} \\ &= \tfrac{1}{2i}\left[L\left\{E^{\gamma}_{\alpha,\beta}(i\lambda t)\right\} - L\left\{E^{\gamma}_{\alpha,\beta}(-i\lambda t)\right\}\right] \\ &= \tfrac{1}{2i}\tfrac{s^{-1}}{\Gamma(\gamma)}\left[{}_2\varphi_1\left(\begin{matrix}(\gamma,1),(1,1)\\(\beta,\alpha)\end{matrix};\tfrac{i\lambda}{s}\right) - {}_2\varphi_1\left(\begin{matrix}(\gamma,1),(1,1)\\(\beta,\alpha)\end{matrix};-\tfrac{i\lambda}{s}\right)\right] \\ &= \tfrac{s^{-1}}{\Gamma(\gamma)}\, {}_2sin_1\left(\begin{matrix}(\gamma,1),(1,1)\\(\beta,\alpha)\end{matrix};\tfrac{\lambda}{s}\right). \end{aligned} \tag{5.272}$$

□

Theorem 5.26 *Let $\alpha, \beta, \gamma, t \in \mathbb{C}$, $Re(\alpha) > 0$, and $Re(\beta) > 0$. Then the Laplace transform of the function $precos^{\gamma}_{\alpha,\beta}(t)$ is given as follows:*

$$L\left\{precos^{\gamma}_{\alpha,\beta}(t)\right\} = \frac{s^{-1}}{\Gamma(\gamma)}\, {}_2cos_1\left(\begin{matrix}(\gamma,1),(1,1)\\(\beta,\alpha)\end{matrix};\frac{1}{s}\right). \tag{5.273}$$

Proof In view of

$$L\left[E^{\gamma}_{\alpha,\beta}(\lambda t)\right] = \int_0^{\infty} e^{-st} E^{\gamma}_{\alpha,\beta}(\lambda t)\, dt = \frac{s^{-1}}{\Gamma(\gamma)}\, {}_2\varphi_1\left(\begin{matrix}(\gamma,1),(1,1)\\(\beta,\alpha)\end{matrix};\frac{\lambda}{s}\right), \tag{5.274}$$

we have

$$\begin{aligned} &L\left\{precos^{\gamma}_{\alpha,\beta}(t)\right\} \\ &= \tfrac{1}{2}\left[L\left\{E^{\gamma}_{\alpha,\beta}(it)\right\} + L\left\{E^{\gamma}_{\alpha,\beta}(-it)\right\}\right] \\ &= \tfrac{1}{2}\tfrac{s^{-1}}{\Gamma(\gamma)}\left[{}_2\varphi_1\left(\begin{matrix}(\gamma,1),(1,1)\\(\beta,\alpha)\end{matrix};\tfrac{i}{s}\right) + {}_2\varphi_1\left(\begin{matrix}(\gamma,1),(1,1)\\(\beta,\alpha)\end{matrix};-\tfrac{i}{s}\right)\right] \\ &= \tfrac{s^{-1}}{\Gamma(\gamma)}\, {}_2cos_1\left(\begin{matrix}(\gamma,1),(1,1)\\(\beta,\alpha)\end{matrix};\tfrac{1}{s}\right). \end{aligned} \tag{5.275}$$

More generally, for $\alpha, \beta, \gamma, \lambda, t \in \mathbb{C}$, $Re(\alpha) > 0$, and $Re(\beta) > 0$,

$$\begin{aligned} &L\left\{precos^{\gamma}_{\alpha,\beta}(\lambda t)\right\} \\ &= \tfrac{1}{2}\left[L\left\{E^{\gamma}_{\alpha,\beta}(i\lambda t)\right\} + L\left\{E^{\gamma}_{\alpha,\beta}(-i\lambda t)\right\}\right] \\ &= \tfrac{1}{2}\tfrac{s^{-1}}{\Gamma(\gamma)}\left[{}_2\varphi_1\left(\begin{matrix}(\gamma,1),(1,1)\\(\beta,\alpha)\end{matrix};\tfrac{i\lambda}{s}\right) + {}_2\varphi_1\left(\begin{matrix}(\gamma,1),(1,1)\\(\beta,\alpha)\end{matrix};-\tfrac{i\lambda}{s}\right)\right] \\ &= \tfrac{s^{-1}}{\Gamma(\gamma)}\, {}_2cos_1\left(\begin{matrix}(\gamma,1),(1,1)\\(\beta,\alpha)\end{matrix};\tfrac{\lambda}{s}\right). \end{aligned} \tag{5.276}$$

□

5.1.5 The Pre-Superhyperbolic Functions via Prabhakar Function

In this section, we introduce the pre-superhyperbolic functions via Prabhakar function.

Definition 5.20 The hyperbolic presine via Prabhakar function is defined as

$$\begin{aligned} presinh^{\gamma}_{\alpha,\beta}(t) &= \tfrac{1}{2}\left(E^{\gamma}_{\alpha,\beta}(t) - E^{\gamma}_{\alpha,\beta}(-t)\right) \\ &= \sum_{k=0}^{\infty} \tfrac{(\gamma)_{2k+1} t^{2k+1}}{\Gamma((2k+1)\alpha+\beta)}, \end{aligned} \tag{5.277}$$

where $\alpha, \beta, t \in \mathbb{C}$, $Re(\alpha) > 0$, and $Re(\beta) > 0$.

Definition 5.21 The hyperbolic precosine via Prabhakar function is defined as

$$\begin{aligned} precosh^{\gamma}_{\alpha,\beta}(t) &= \tfrac{1}{2}\left(E^{\gamma}_{\alpha,\beta}(t) + E^{\gamma}_{\alpha,\beta}(-t)\right) \\ &= \sum_{k=0}^{\infty} \tfrac{(\gamma)_{2k} t^{2k}}{\Gamma(2k\alpha+\beta)}, \end{aligned} \tag{5.278}$$

where $\alpha, \beta, t \in \mathbb{C}$, $Re(\alpha) > 0$, and $Re(\beta) > 0$.

Definition 5.22 The hyperbolic pretangent via Prabhakar function is defined by

$$pretanh^{\gamma}_{\alpha,\beta}(t) = \frac{presinh^{\gamma}_{\alpha,\beta}(t)}{precosh^{\gamma}_{\alpha,\beta}(t)} \left(precosh^{\gamma}_{\alpha,\beta}(t) \neq 0\right), \tag{5.279}$$

where $\alpha, \beta, \gamma, t \in \mathbb{C}$, $Re(\alpha) > 0$, and $Re(\beta) > 0$.

Definition 5.23 The hyperbolic precotangent via Prabhakar function is defined by

$$precotanh^{\gamma}_{\alpha,\beta}(t) = \frac{precosh^{\gamma}_{\alpha,\beta}(t)}{presinh^{\gamma}_{\alpha,\beta}(t)} \left(presinh^{\gamma}_{\alpha,\beta}(t) \neq 0\right), \tag{5.280}$$

where $\alpha, \beta, \gamma, t \in \mathbb{C}$, $Re(\alpha) > 0$, and $Re(\beta) > 0$.

Definition 5.24 The hyperbolic precosecant via Prabhakar function is defined by

$$precsch^{\gamma}_{\alpha,\beta}(t) = \frac{1}{presinh^{\gamma}_{\alpha,\beta}(t)} \left(presinh^{\gamma}_{\alpha,\beta}(t) \neq 0\right), \tag{5.281}$$

where $\alpha, \beta, \gamma, t \in \mathbb{C}$, $Re(\alpha) > 0$, and $Re(\beta) > 0$.

Definition 5.25 The hyperbolic presecant via Prabhakar function is defined by

$$presech^{\gamma}_{\alpha,\beta}(t) = \frac{1}{precosh^{\gamma}_{\alpha,\beta}(t)} \left(precosh^{\gamma}_{\alpha,\beta}(t) \neq 0\right), \tag{5.282}$$

where $\alpha, \beta, \gamma, t \in \mathbb{C}$, $Re(\alpha) > 0$, and $Re(\beta) > 0$.

Thus,

$$\begin{aligned} presinh^{\gamma}_{\alpha,\beta}(\lambda t) &= \frac{1}{2}\left(E^{\gamma}_{\alpha,\beta}(\lambda t) - E^{\gamma}_{\alpha,\beta}(-\lambda t)\right) \\ &= \frac{1}{2} \times \frac{1}{\Gamma(\gamma)} \times {}_1sinh_1\left(\begin{matrix}(\gamma,1) \\ (\beta,\alpha)\end{matrix}; \lambda t\right) \\ &= \sum_{k=0}^{\infty} \frac{\lambda^{2k+1}(\gamma)_{2k+1}t^{2k+1}}{\Gamma((2k+1)\alpha+\beta)}, \end{aligned} \tag{5.283}$$

$$\begin{aligned} precosh^{\gamma}_{\alpha,\beta}(\lambda t) &= \frac{1}{2}\left(E^{\gamma}_{\alpha,\beta}(\lambda t) + E^{\gamma}_{\alpha,\beta}(-\lambda t)\right) \\ &= \frac{1}{2\Gamma(\gamma)} \times {}_1cosh_1\left(\begin{matrix}(\gamma,1) \\ (\beta,\alpha)\end{matrix}; \lambda t\right) \\ &= \sum_{k=0}^{\infty} \frac{\lambda^{2k}(\gamma)_{2k}t^{2k}}{\Gamma(2k\alpha+\beta)}, \end{aligned} \tag{5.284}$$

$$pretanh^{\gamma}_{\alpha,\beta}(\lambda t) = \frac{presinh^{\gamma}_{\alpha,\beta}(\lambda t)}{precosh^{\gamma}_{\alpha,\beta}(\lambda t)} \left(precosh^{\gamma}_{\alpha,\beta}(\lambda t) \neq 0\right), \tag{5.285}$$

$$precotanh^{\gamma}_{\alpha,\beta}(\lambda t) = \frac{precosh^{\gamma}_{\alpha,\beta}(\lambda t)}{presinh^{\gamma}_{\alpha,\beta}(\lambda t)} \left(presinh^{\gamma}_{\alpha,\beta}(\lambda t) \neq 0\right), \tag{5.286}$$

$$precsch^{\gamma}_{\alpha,\beta}(\lambda t) = \frac{1}{presinh^{\gamma}_{\alpha,\beta}(\lambda t)} \left(presinh^{\gamma}_{\alpha,\beta}(\lambda t) \neq 0\right), \tag{5.287}$$

and

$$presech^{\gamma}_{\alpha,\beta}(\lambda t) = \frac{1}{precosh^{\gamma}_{\alpha,\beta}(\lambda t)} \left(precosh^{\gamma}_{\alpha,\beta}(\lambda t) \neq 0\right), \tag{5.288}$$

where $\alpha, \beta, \gamma, \lambda, t \in \mathbb{C}$, $Re(\alpha) > 0$, and $Re(\beta) > 0$.

Based on the result, we may have the following theorem.

Theorem 5.27 *Let $\alpha, \beta, \gamma, t \in \mathbb{C}$, $Re(\alpha) > 0$, and $Re(\beta) > 0$. Then we have*

$$E^{\gamma}_{\alpha,\beta}(t) = precosh^{\gamma}_{\alpha,\beta}(t) + presinh^{\gamma}_{\alpha,\beta}(t) \tag{5.289}$$

and

$$E^{\gamma}_{\alpha,\beta}(-t) = precosh^{\gamma}_{\alpha,\beta}(t) - presinh^{\gamma}_{\alpha,\beta}(t). \tag{5.290}$$

Generally,

$$E^{\gamma}_{\alpha,\beta}(it) = precosh^{\gamma}_{\alpha,\beta}(\lambda t) + presinh^{\gamma}_{\alpha,\beta}(\lambda t) \tag{5.291}$$

and

$$E^{\gamma}_{\alpha,\beta}(-\lambda t) = precos^{\gamma}_{\alpha,\beta}(\lambda t) - presinh^{\gamma}_{\alpha,\beta}(\lambda t), \tag{5.292}$$

where $\alpha, \beta, \gamma, \lambda, t \in \mathbb{C}$, $Re(\alpha) > 0$, *and* $Re(\beta) > 0$.

Property 5.8 Let $\alpha, \beta, \gamma, t \in \mathbb{C}$, $Re(\alpha) > 0$, and $Re(\beta) > 0$. Then we have

$$presinh^{\gamma}_{\alpha,\beta}(-t) = -presinh^{\gamma}_{\alpha,\beta}(t), \tag{5.293}$$

$$precosh^{\gamma}_{\alpha,\beta}(-t) = precosh^{\gamma}_{\alpha,\beta}(t), \tag{5.294}$$

$$pretanh^{\gamma}_{\alpha,\beta}(-t) = -\ pretanh^{\gamma}_{\alpha,\beta}(t), \tag{5.295}$$

$$precotanh^{\gamma}_{\alpha,\beta}(-t) = -\ precotanh^{\gamma}_{\alpha,\beta}(t), \tag{5.296}$$

$$precsch^{\gamma}_{\alpha,\beta}(-t) = -precsch^{\gamma}_{\alpha,\beta}(t), \tag{5.297}$$

and

$$presech^{\gamma}_{\alpha,\beta}(-t) = presech^{\gamma}_{\alpha,\beta}(t). \tag{5.298}$$

Theorem 5.28 *Let* $\alpha, \beta, \gamma, t \in \mathbb{C}$, $Re(\alpha) > 0$, *and* $Re(\beta) > 0$. *Then the Laplace transform of the function* $presinh^{\gamma}_{\alpha,\beta}(t)$ *is given as follows:*

$$L\left\{presinh^{\gamma}_{\alpha,\beta}(t)\right\} = \frac{s^{-1}}{\Gamma(\gamma)}\,{}_2sinh_1\left(\begin{matrix}(\gamma,1),(1,1)\\(\beta,\alpha)\end{matrix};\frac{1}{s}\right). \tag{5.299}$$

Proof Let us consider that

$$L\left[E^{\gamma}_{\alpha,\beta}(\lambda t)\right] = \int_0^{\infty} e^{-st} E^{\gamma}_{\alpha,\beta}(\lambda t)\,dt = \frac{s^{-1}}{\Gamma(\gamma)}\,{}_2\varphi_1\left(\begin{matrix}(\gamma,1),(1,1)\\(\beta,\alpha)\end{matrix};\frac{\lambda}{s}\right). \tag{5.300}$$

Then,

$$\begin{aligned}
&L\left\{presinh^{\gamma}_{\alpha,\beta}(t)\right\}\\
&= \tfrac{1}{2}\left[L\left\{E^{\gamma}_{\alpha,\beta}(t)\right\} - L\left\{E^{\gamma}_{\alpha,\beta}(-t)\right\}\right]\\
&= \tfrac{1}{2}\tfrac{s^{-1}}{\Gamma(\gamma)}\left[{}_2\varphi_1\left(\begin{matrix}(\gamma,1),(1,1)\\(\beta,\alpha)\end{matrix};\tfrac{1}{s}\right) - {}_2\varphi_1\left(\begin{matrix}(\gamma,1),(1,1)\\(\beta,\alpha)\end{matrix};-\tfrac{1}{s}\right)\right]\\
&= \tfrac{s^{-1}}{\Gamma(\gamma)}\,{}_2sinh_1\left(\begin{matrix}(\gamma,1),(1,1)\\(\beta,\alpha)\end{matrix};\tfrac{1}{s}\right).
\end{aligned} \tag{5.301}$$

More generally, for $\alpha, \beta, \gamma, \lambda, t \in \mathbb{C}$, $Re(\alpha) > 0$, and $Re(\beta) > 0$,

$$\begin{aligned} & L\left\{presinh_{\alpha,\beta}^{\gamma}(\lambda t)\right\} \\ & = \frac{1}{2}\left[L\left\{E_{\alpha,\beta}^{\gamma}(\lambda t)\right\} - L\left\{E_{\alpha,\beta}^{\gamma}(-\lambda t)\right\}\right] \\ & = \frac{1}{2}\frac{s^{-1}}{\Gamma(\gamma)}\left[{}_2\varphi_1\left(\begin{matrix}(\gamma,1),(1,1)\\(\beta,\alpha)\end{matrix};\frac{\lambda}{s}\right) - {}_2\varphi_1\left(\begin{matrix}(\gamma,1),(1,1)\\(\beta,\alpha)\end{matrix};-\frac{\lambda}{s}\right)\right] \\ & = \frac{s^{-1}}{\Gamma(\gamma)}\,{}_2sinh_1\left(\begin{matrix}(\gamma,1),(1,1)\\(\beta,\alpha)\end{matrix};\frac{\lambda}{s}\right). \end{aligned} \tag{5.302}$$

□

Theorem 5.29 *Let $\alpha, \beta, \gamma, t \in \mathbb{C}$, $Re(\alpha) > 0$, and $Re(\beta) > 0$. Then the Laplace transform of the function $precosh_{\alpha,\beta}^{\gamma}(t)$ is given as follows:*

$$L\left\{precosh_{\alpha,\beta}^{\gamma}(t)\right\} = \frac{s^{-1}}{\Gamma(\gamma)}\,{}_2cosh_1\left(\begin{matrix}(\gamma,1),(1,1)\\(\beta,\alpha)\end{matrix};\frac{1}{s}\right). \tag{5.303}$$

Proof With the use of the integral representation:

$$L\left[E_{\alpha,\beta}^{\gamma}(\lambda t)\right] = \int_0^{\infty} e^{-st}E_{\alpha,\beta}^{\gamma}(\lambda t)\,dt = \frac{s^{-1}}{\Gamma(\gamma)}\,{}_2\varphi_1\left(\begin{matrix}(\gamma,1),(1,1)\\(\beta,\alpha)\end{matrix};\frac{\lambda}{s}\right), \tag{5.304}$$

we have

$$\begin{aligned} & L\left\{precosh_{\alpha,\beta}^{\gamma}(t)\right\} \\ & = \frac{1}{2}\left[L\left\{E_{\alpha,\beta}^{\gamma}(t)\right\} + L\left\{E_{\alpha,\beta}^{\gamma}(-t)\right\}\right] \\ & = \frac{1}{2}\frac{s^{-1}}{\Gamma(\gamma)}\left[{}_2\varphi_1\left(\begin{matrix}(\gamma,1),(1,1)\\(\beta,\alpha)\end{matrix};\frac{1}{s}\right) + {}_2\varphi_1\left(\begin{matrix}(\gamma,1),(1,1)\\(\beta,\alpha)\end{matrix};-\frac{1}{s}\right)\right] \\ & = \frac{s^{-1}}{\Gamma(\gamma)}\,{}_2cosh_1\left(\begin{matrix}(\gamma,1),(1,1)\\(\beta,\alpha)\end{matrix};\frac{1}{s}\right). \end{aligned} \tag{5.305}$$

More generally, for $\alpha, \beta, \gamma, \lambda, t \in \mathbb{C}$, $Re(\alpha) > 0$, and $Re(\beta) > 0$,

$$\begin{aligned} & L\left\{precosh_{\alpha,\beta}^{\gamma}(\lambda t)\right\} \\ & = \frac{1}{2}\left[L\left\{E_{\alpha,\beta}^{\gamma}(\lambda t)\right\} + L\left\{E_{\alpha,\beta}^{\gamma}(-\lambda t)\right\}\right] \\ & = \frac{1}{2}\frac{s^{-1}}{\Gamma(\gamma)}\left[{}_2\varphi_1\left(\begin{matrix}(\gamma,1),(1,1)\\(\beta,\alpha)\end{matrix};\frac{\lambda}{s}\right) + {}_2\varphi_1\left(\begin{matrix}(\gamma,1),(1,1)\\(\beta,\alpha)\end{matrix};-\frac{\lambda}{s}\right)\right] \\ & = \frac{s^{-1}}{\Gamma(\gamma)}\,{}_2cosh_1\left(\begin{matrix}(\gamma,1),(1,1)\\(\beta,\alpha)\end{matrix};\frac{\lambda}{s}\right). \end{aligned} \tag{5.306}$$

□

5.1.6 The Pre-Supertrigonometric Functions with Power Law via Prabhakar Function

In this section, we give the pre-supertrigonometric functions with power law via Prabhakar function.

Definition 5.26 The presine with power law via Prabhakar function is defined as

$$\begin{aligned} presin^{\gamma}_{\alpha,\beta}\left(t^{\varsigma}\right) &= \frac{1}{2i}\left(E^{\gamma}_{\alpha,\beta}\left(it^{\varsigma}\right) - E^{\gamma}_{\alpha,\beta}\left(-it^{\varsigma}\right)\right) \\ &= \sum_{k=0}^{\infty} \frac{(-1)^k (\gamma)_{2k+1} t^{(2k+1)\varsigma}}{\Gamma((2k+1)\alpha+\beta)}, \end{aligned} \tag{5.307}$$

where $\alpha, \beta, \gamma, \varsigma, t \in \mathbb{C}$, $Re\,(\alpha) > 0$, and $Re\,(\beta) > 0$.

Definition 5.27 The precosine with power law via Prabhakar function is defined as

$$\begin{aligned} precos^{\gamma}_{\alpha,\beta}\left(t^{\varsigma}\right) &= \frac{1}{2}\left(E^{\gamma}_{\alpha,\beta}\left(it^{\varsigma}\right) + E^{\gamma}_{\alpha,\beta}\left(-it^{\varsigma}\right)\right) \\ &= \sum_{k=0}^{\infty} \frac{(-1)^k (\gamma)_{2k} t^{2k\varsigma}}{\Gamma(2k\alpha+\beta)}, \end{aligned} \tag{5.308}$$

where $\alpha, \beta, \gamma, \varsigma, t \in \mathbb{C}$, $Re\,(\alpha) > 0$, and $Re\,(\beta) > 0$.

Definition 5.28 The pretangent with power law via Prabhakar function is defined as

$$pretan^{\gamma}_{\alpha,\beta}\left(t^{\varsigma}\right) = \frac{presin^{\gamma}_{\alpha,\beta}\left(t^{\varsigma}\right)}{precos^{\gamma}_{\alpha,\beta}\left(t^{\varsigma}\right)} \left(precos^{\gamma}_{\alpha,\beta}\left(t^{\varsigma}\right) \neq 0\right), \tag{5.309}$$

where $\alpha, \beta, \gamma, \varsigma, t \in \mathbb{C}$, $Re\,(\alpha) > 0$, and $Re\,(\beta) > 0$.

Definition 5.29 The pretangent with power law via Prabhakar function is defined as

$$pretan^{\gamma}_{\alpha,\beta}\left(t^{\varsigma}\right) = \frac{presin^{\gamma}_{\alpha,\beta}\left(t^{\varsigma}\right)}{precos^{\gamma}_{\alpha,\beta}\left(t^{\varsigma}\right)} \left(precos^{\gamma}_{\alpha,\beta}\left(t^{\varsigma}\right) \neq 0\right), \tag{5.310}$$

where $\alpha, \beta, \gamma, \varsigma, t \in \mathbb{C}$, $Re\,(\alpha) > 0$, and $Re\,(\beta) > 0$.

Definition 5.30 The precotangent with power law via Prabhakar function is defined as

$$precotan^{\gamma}_{\alpha,\beta}\left(t^{\varsigma}\right) = \frac{precos^{\gamma}_{\alpha,\beta}\left(t^{\varsigma}\right)}{presin^{\gamma}_{\alpha,\beta}\left(t^{\varsigma}\right)} \left(presin^{\gamma}_{\alpha,\beta}\left(t^{\varsigma}\right) \neq 0\right), \tag{5.311}$$

where $\alpha, \beta, \gamma, \varsigma, t \in \mathbb{C}$, $Re\,(\alpha) > 0$, and $Re\,(\beta) > 0$.

Definition 5.31 The precosecant with power law via Prabhakar function is defined as

$$precsc^{\gamma}_{\alpha,\beta}\left(t^{\varsigma}\right) = \frac{1}{presin^{\gamma}_{\alpha,\beta}\left(t^{\varsigma}\right)} \left(presin^{\gamma}_{\alpha,\beta}\left(t^{\varsigma}\right) \neq 0\right), \tag{5.312}$$

where $\alpha, \beta, \gamma, \varsigma, t \in \mathbb{C}$, $Re(\alpha) > 0$, and $Re(\beta) > 0$.

Definition 5.32 The presecant with power law via Prabhakar function is defined as

$$presec^{\gamma}_{\alpha,\beta}\left(t^{\varsigma}\right) = \frac{1}{precos^{\gamma}_{\alpha,\beta}\left(t^{\varsigma}\right)} \left(precos^{\gamma}_{\alpha,\beta}\left(t^{\varsigma}\right) \neq 0\right), \tag{5.313}$$

where $\alpha, \beta, \gamma, \varsigma, t \in \mathbb{C}$, $Re(\alpha) > 0$, and $Re(\beta) > 0$.

So, we have

$$\begin{aligned} presin^{\gamma}_{\alpha,\beta}(\lambda t^{\varsigma}) &= \tfrac{1}{2i}\left(E^{\gamma}_{\alpha,\beta}(i\lambda t^{\varsigma}) - E^{\gamma}_{\alpha,\beta}(-i\lambda t^{\varsigma})\right) \\ &= \sum_{k=0}^{\infty} \tfrac{(-1)^k \lambda^{2k+1} (\gamma)_{2k+1} t^{(2k+1)\varsigma}}{\Gamma((2k+1)\alpha+\beta)}, \end{aligned} \tag{5.314}$$

$$\begin{aligned} precos^{\gamma}_{\alpha,\beta}(\lambda t^{\varsigma}) &= \tfrac{1}{2}\left(E^{\gamma}_{\alpha,\beta}(i\lambda t^{\varsigma}) + E^{\gamma}_{\alpha,\beta}(-i\lambda t^{\varsigma})\right) \\ &= \sum_{k=0}^{\infty} \tfrac{(-1)^k \lambda^{2k} (\gamma)_{2k} t^{2k\varsigma}}{\Gamma(2k\alpha+\beta)}, \end{aligned} \tag{5.315}$$

$$pretan^{\gamma}_{\alpha,\beta}\left(\lambda t^{\varsigma}\right) = \frac{presin^{\gamma}_{\alpha,\beta}\left(\lambda t^{\varsigma}\right)}{precos^{\gamma}_{\alpha,\beta}\left(\lambda t^{\varsigma}\right)} \left(precos^{\gamma}_{\alpha,\beta}\left(\lambda t^{\varsigma}\right) \neq 0\right), \tag{5.316}$$

$$precotan^{\gamma}_{\alpha,\beta}\left(\lambda t^{\varsigma}\right) = \frac{precos^{\gamma}_{\alpha,\beta}\left(\lambda t^{\varsigma}\right)}{presin^{\gamma}_{\alpha,\beta}\left(\lambda t^{\varsigma}\right)} \left(presin^{\gamma}_{\alpha,\beta}\left(\lambda t^{\varsigma}\right) \neq 0\right), \tag{5.317}$$

$$precsc^{\gamma}_{\alpha,\beta}\left(\lambda t^{\varsigma}\right) = \frac{1}{presin^{\gamma}_{\alpha,\beta}\left(\lambda t^{\varsigma}\right)} \left(presin^{\gamma}_{\alpha,\beta}\left(\lambda t^{\varsigma}\right) \neq 0\right), \tag{5.318}$$

and

$$presec^{\gamma}_{\alpha,\beta}\left(\lambda t^{\varsigma}\right) = \frac{1}{precos^{\gamma}_{\alpha,\beta}\left(\lambda t^{\varsigma}\right)} \left(precos^{\gamma}_{\alpha,\beta}\left(\lambda t^{\varsigma}\right) \neq 0\right), \tag{5.319}$$

where $\alpha, \beta, \gamma, \varsigma, t, \lambda \in \mathbb{C}$, $Re(\alpha) > 0$, and $Re(\beta) > 0$.

Making use of the results, we may present the following theorem.

Theorem 5.30 *Let $\alpha, \beta, \gamma, \varsigma, t \in \mathbb{C}$, $Re(\alpha) > 0$, and $Re(\beta) > 0$. Then we have*

$$E^{\gamma}_{\alpha,\beta}\left(it^{\varsigma}\right) = precos^{\gamma}_{\alpha,\beta}\left(t^{\varsigma}\right) + ipresin^{\gamma}_{\alpha,\beta}\left(t^{\varsigma}\right) \tag{5.320}$$

and

$$E^{\gamma}_{\alpha,\beta}\left(-it^{\varsigma}\right) = precos^{\gamma}_{\alpha,\beta}\left(t^{\varsigma}\right) - ipresin^{\gamma}_{\alpha,\beta}\left(t^{\varsigma}\right). \tag{5.321}$$

Generally,

$$E^{\gamma}_{\alpha,\beta}\left(i\lambda t^{\varsigma}\right) = precos^{\gamma}_{\alpha,\beta}\left(\lambda t^{\varsigma}\right) + ipresin^{\gamma}_{\alpha,\beta}\left(\lambda t^{\varsigma}\right) \tag{5.322}$$

and

$$E^{\gamma}_{\alpha,\beta}\left(-i\lambda t^{\varsigma}\right) = precos^{\gamma}_{\alpha,\beta}\left(\lambda t^{\varsigma}\right) - ipresin^{\gamma}_{\alpha,\beta}\left(\lambda t^{\varsigma}\right), \tag{5.323}$$

where $\alpha, \beta, \gamma, \varsigma, t, \lambda \in \mathbb{C}$, $Re(\alpha) > 0$, and $Re(\beta) > 0$.

Property 5.9 Let $\alpha, \beta, \gamma, \varsigma, t \in \mathbb{C}$, $Re(\alpha) > 0$, and $Re(\beta) > 0$. Then we have

$$presin^{\gamma}_{\alpha,\beta}\left(-t^{\varsigma}\right) = -presin^{\gamma}_{\alpha,\beta}\left(t^{\varsigma}\right), \tag{5.324}$$

$$precos^{\gamma}_{\alpha,\beta}\left(-t^{\varsigma}\right) = precos^{\gamma}_{\alpha,\beta}\left(t^{\varsigma}\right), \tag{5.325}$$

$$pretan^{\gamma}_{\alpha,\beta}\left(-t^{\varsigma}\right) = -\ pretan^{\gamma}_{\alpha,\beta}\left(t^{\varsigma}\right), \tag{5.326}$$

$$precotan^{\gamma}_{\alpha,\beta}\left(-t^{\varsigma}\right) = -\ precotan^{\gamma}_{\alpha,\beta}\left(t^{\varsigma}\right), \tag{5.327}$$

$$precsc^{\gamma}_{\alpha,\beta}\left(-t^{\varsigma}\right) = -precsc^{\gamma}_{\alpha,\beta}\left(t^{\varsigma}\right), \tag{5.328}$$

and

$$presec^{\gamma}_{\alpha,\beta}\left(-t^{\varsigma}\right) = presec^{\gamma}_{\alpha,\beta}\left(t^{\varsigma}\right). \tag{5.329}$$

Theorem 5.31 *Let $\alpha, \beta, \gamma, \varsigma, t \in \mathbb{C}\alpha, \beta, \varsigma, t \in \mathbb{C}$, $Re(\alpha) > 0$, and $Re(\beta) > 0$. Then the Laplace transform of the function $presin^{\gamma}_{\alpha,\beta}(t^{\varsigma})$ is given as follows:*

$$L\left\{presin^{\gamma}_{\alpha,\beta}\left(t^{\varsigma}\right)\right\} = \frac{s^{-1}}{\Gamma(\gamma)}\,{}_2sin_1\left(\begin{matrix}(\gamma,1),(1,\varsigma)\\(\beta,\alpha)\end{matrix};\frac{1}{s^{\varsigma}}\right). \tag{5.330}$$

Proof Let us recall

$$L\left[t^{\varsigma-1}E^{\gamma}_{\alpha,\beta}\left(\lambda t^{\vartheta}\right)\right]=\int_0^{\infty}e^{-st}t^{\varsigma-1}E^{\gamma}_{\alpha,\beta}\left(\lambda t^{\vartheta}\right)dt \\ =\frac{s^{-\varsigma}}{\Gamma(\gamma)}\,{}_2\varphi_1\left(\begin{matrix}(\gamma,1),(\varsigma,\vartheta)\\(\beta,\alpha)\end{matrix};\frac{\lambda}{s^{\vartheta}}\right), \tag{5.331}$$

which leads to, for $\varsigma=1$,

$$L\left[E^{\gamma}_{\alpha,\beta}\left(\lambda t^{\vartheta}\right)\right]=\int_0^{\infty}e^{-st}E^{\gamma}_{\alpha,\beta}\left(\lambda t^{\vartheta}\right)dt=\frac{s^{-1}}{\Gamma(\gamma)}\,{}_2\varphi_1\left(\begin{matrix}(\gamma,1),(1,\vartheta)\\(\beta,\alpha)\end{matrix};\frac{\lambda}{s^{\vartheta}}\right). \tag{5.332}$$

It follows that

$$L\left\{presin^{\gamma}_{\alpha,\beta}\left(t^{\varsigma}\right)\right\} \\ =\frac{1}{2i}\left[L\left\{E^{\gamma}_{\alpha,\beta}\left(it^{\varsigma}\right)\right\}-L\left\{E^{\gamma}_{\alpha,\beta}\left(-it^{\varsigma}\right)\right\}\right] \\ =\frac{1}{2i}\frac{s^{-1}}{\Gamma(\gamma)}\left[{}_2\varphi_1\left(\begin{matrix}(\gamma,1),(1,\varsigma)\\(\beta,\alpha)\end{matrix};\frac{i}{s^{\varsigma}}\right)-{}_2\varphi_1\left(\begin{matrix}(\gamma,1),(1,\varsigma)\\(\beta,\alpha)\end{matrix};-\frac{i}{s^{\varsigma}}\right)\right] \\ =\frac{s^{-1}}{\Gamma(\gamma)}\,{}_2sin_1\left(\begin{matrix}(\gamma,1),(1,\varsigma)\\(\beta,\alpha)\end{matrix};\frac{1}{s^{\varsigma}}\right). \tag{5.333}$$

More generally,

$$L\left\{presin^{\gamma}_{\alpha,\beta}\left(\lambda t^{\varsigma}\right)\right\} \\ =\frac{1}{2i}\left[L\left\{E^{\gamma}_{\alpha,\beta}\left(i\lambda t^{\varsigma}\right)\right\}-L\left\{E^{\gamma}_{\alpha,\beta}\left(-i\lambda t^{\varsigma}\right)\right\}\right] \\ =\frac{1}{2i}\frac{s^{-1}}{\Gamma(\gamma)}\left[{}_2\varphi_1\left(\begin{matrix}(\gamma,1),(1,\varsigma)\\(\beta,\alpha)\end{matrix};\frac{i\lambda}{s^{\varsigma}}\right)-{}_2\varphi_1\left(\begin{matrix}(\gamma,1),(1,\vartheta)\\(\beta,\alpha)\end{matrix};-\frac{i\lambda}{s^{\varsigma}}\right)\right] \\ =\frac{s^{-1}}{\Gamma(\gamma)}\,{}_2sin_1\left(\begin{matrix}(\gamma,1),(1,\varsigma)\\(\beta,\alpha)\end{matrix};\frac{\lambda}{s^{\varsigma}}\right), \tag{5.334}$$

where $\alpha,\beta,\gamma,\varsigma,t\in\mathbb{C}$, $Re(\alpha)>0$, and $Re(\beta)>0$. □

Theorem 5.32 *Let $\alpha,\beta,\gamma,\varsigma,t\in\mathbb{C}$, $Re(\alpha)>0$, and $Re(\beta)>0$. Then the Laplace transform of the function $precos^{\gamma}_{\alpha,\beta}(t^{\varsigma})$ is given as follows:*

$$L\left\{precos^{\gamma}_{\alpha,\beta}\left(t^{\varsigma}\right)\right\}=\frac{s^{-1}}{\Gamma(\gamma)}\,{}_2cos_1\left(\begin{matrix}(\gamma,1),(1,\varsigma)\\(\beta,\alpha)\end{matrix};\frac{1}{s^{\varsigma}}\right). \tag{5.335}$$

Proof Making use of the integral representation:

$$L\left[E_{\alpha,\beta}^{\gamma}\left(\lambda t^{\varsigma}\right)\right]=\int\limits_{0}^{\infty}e^{-st}E_{\alpha,\beta}^{\gamma}\left(\lambda t^{\varsigma}\right)dt=\frac{s^{-1}}{\Gamma\left(\gamma\right)}{}_{2}\varphi_{1}\left(\begin{matrix}(\gamma,1),(1,\varsigma)\\(\beta,\alpha)\end{matrix};\frac{\lambda}{s^{\varsigma}}\right),\tag{5.336}$$

we have

$$\begin{aligned}&L\left\{precos_{\alpha,\beta}^{\gamma}\left(t^{\varsigma}\right)\right\}\\&=\tfrac{1}{2}\left[L\left\{E_{\alpha,\beta}^{\gamma}\left(it^{\varsigma}\right)\right\}+L\left\{E_{\alpha,\beta}^{\gamma}\left(-it^{\varsigma}\right)\right\}\right]\\&=\tfrac{1}{2}\tfrac{s^{-1}}{\Gamma(\gamma)}\left[{}_{2}\varphi_{1}\left(\begin{matrix}(\gamma,1),(1,\varsigma)\\(\beta,\alpha)\end{matrix};\tfrac{i}{s^{\varsigma}}\right)+{}_{2}\varphi_{1}\left(\begin{matrix}(\gamma,1),(1,\varsigma)\\(\beta,\alpha)\end{matrix};-\tfrac{i}{s^{\varsigma}}\right)\right]\\&=\tfrac{s^{-1}}{\Gamma(\gamma)}{}_{2}cos_{1}\left(\begin{matrix}(\gamma,1),(1,\varsigma)\\(\beta,\alpha)\end{matrix};\tfrac{1}{s^{\varsigma}}\right).\end{aligned}\tag{5.337}$$

More generally,

$$\begin{aligned}&L\left\{precos_{\alpha,\beta}^{\gamma}\left(\lambda t^{\varsigma}\right)\right\}\\&=\tfrac{1}{2}\left[L\left\{E_{\alpha,\beta}^{\gamma}\left(i\lambda t^{\varsigma}\right)\right\}+L\left\{E_{\alpha,\beta}^{\gamma}\left(-i\lambda t^{\varsigma}\right)\right\}\right]\\&=\tfrac{1}{2}\tfrac{s^{-1}}{\Gamma(\gamma)}\left[{}_{2}\varphi_{1}\left(\begin{matrix}(\gamma,1),(1,\varsigma)\\(\beta,\alpha)\end{matrix};\tfrac{i\lambda}{s^{\varsigma}}\right)+{}_{2}\varphi_{1}\left(\begin{matrix}(\gamma,1),(1,\varsigma)\\(\beta,\alpha)\end{matrix};-\tfrac{i\lambda}{s^{\varsigma}}\right)\right]\\&=\tfrac{s^{-1}}{\Gamma(\gamma)}{}_{2}cos_{1}\left(\begin{matrix}(\gamma,1),(1,\varsigma)\\(\beta,\alpha)\end{matrix};\tfrac{\lambda}{s^{\varsigma}}\right)\end{aligned}\tag{5.338}$$

where $\alpha,\beta,\gamma,\varsigma,t\in\mathbb{C}$, $Re\left(\alpha\right)>0$, and $Re\left(\beta\right)>0$. □

5.1.7 The Pre-Superhyperbolic Functions with Power Law via Prabhakar Function

In this section, we give the pre-superhyperbolic functions with power law via Prabhakar function.

Definition 5.33 The hyperbolic presine with power law via Prabhakar function is defined as

$$\begin{aligned}presinh_{\alpha,\beta}^{\gamma}\left(t^{\varsigma}\right)&=\tfrac{1}{2}\left(E_{\alpha,\beta}^{\gamma}\left(t^{\varsigma}\right)-E_{\alpha,\beta}^{\gamma}\left(-t^{\varsigma}\right)\right)\\&=\sum_{k=0}^{\infty}\tfrac{(\gamma)_{2k+1}t^{(2k+1)\varsigma}}{\Gamma((2k+1)\alpha+\beta)},\end{aligned}\tag{5.339}$$

where $\alpha,\beta,t,\gamma\in\mathbb{C}$, $Re\left(\alpha\right)>0$, and $Re\left(\beta\right)>0$.

Definition 5.34 The hyperbolic precosine with power law via Prabhakar function is defined as

$$\begin{aligned} precosh^{\gamma}_{\alpha,\beta}\left(t^{\varsigma}\right) &= \tfrac{1}{2}\left(E^{\gamma}_{\alpha,\beta}\left(t^{\varsigma}\right) + E^{\gamma}_{\alpha,\beta}\left(-t^{\varsigma}\right)\right) \\ &= \sum_{k=0}^{\infty} \tfrac{(\gamma)_{2k} t^{2k\varsigma}}{\Gamma(2k\alpha+\beta)}, \end{aligned} \tag{5.340}$$

where $\alpha, \beta, t, \gamma \in \mathbb{C}$, $Re\,(\alpha) > 0$, and $Re\,(\beta) > 0$.

Definition 5.35 The hyperbolic pretangent with power law via Prabhakar function is defined as

$$pretanh^{\gamma}_{\alpha,\beta}\left(t^{\varsigma}\right) = \frac{presinh^{\gamma}_{\alpha,\beta}\left(t^{\varsigma}\right)}{precosh^{\gamma}_{\alpha,\beta}\left(t^{\varsigma}\right)} \left(precosh^{\gamma}_{\alpha,\beta}\left(t^{\varsigma}\right) \neq 0\right), \tag{5.341}$$

where $\alpha, \beta, \gamma, \varsigma, t \in \mathbb{C}$, $Re\,(\alpha) > 0$, and $Re\,(\beta) > 0$.

Definition 5.36 The hyperbolic pretangent with power law via Prabhakar function is defined as

$$pretanh^{\gamma}_{\alpha,\beta}\left(t^{\varsigma}\right) = \frac{presinh^{\gamma}_{\alpha,\beta}\left(t^{\varsigma}\right)}{precosh^{\gamma}_{\alpha,\beta}\left(t^{\varsigma}\right)} \left(precosh^{\gamma}_{\alpha,\beta}\left(t^{\varsigma}\right) \neq 0\right), \tag{5.342}$$

where $\alpha, \beta, \gamma, \varsigma, t \in \mathbb{C}$, $Re\,(\alpha) > 0$, and $Re\,(\beta) > 0$.

Definition 5.37 The hyperbolic precotangent with power law via Prabhakar function is defined as

$$precotanh^{\gamma}_{\alpha,\beta}\left(t^{\varsigma}\right) = \frac{precosh^{\gamma}_{\alpha,\beta}\left(t^{\varsigma}\right)}{presinh^{\gamma}_{\alpha,\beta}\left(t^{\varsigma}\right)} \left(presinh^{\gamma}_{\alpha,\beta}\left(t^{\varsigma}\right) \neq 0\right), \tag{5.343}$$

where $\alpha, \beta, \gamma, \varsigma, t \in \mathbb{C}$, $Re\,(\alpha) > 0$, and $Re\,(\beta) > 0$.

Definition 5.38 The hyperbolic precosecant with power law via Prabhakar function is defined as

$$precsch^{\gamma}_{\alpha,\beta}\left(t^{\varsigma}\right) = \frac{1}{presinh^{\gamma}_{\alpha,\beta}\left(t^{\varsigma}\right)} \left(presinh^{\gamma}_{\alpha,\beta}\left(t^{\varsigma}\right) \neq 0\right), \tag{5.344}$$

where $\alpha, \beta, \gamma, \varsigma, t \in \mathbb{C}$, $Re\,(\alpha) > 0$, and $Re\,(\beta) > 0$.

Definition 5.39 The hyperbolic presecant with power law via Prabhakar function is defined as

$$presech^{\gamma}_{\alpha,\beta}\left(t^{\varsigma}\right) = \frac{1}{precosh^{\gamma}_{\alpha,\beta}\left(t^{\varsigma}\right)} \left(precosh^{\gamma}_{\alpha,\beta}\left(t^{\varsigma}\right) \neq 0\right), \tag{5.345}$$

where $\alpha, \beta, \gamma, \varsigma, t \in \mathbb{C}$, $Re\,(\alpha) > 0$, and $Re\,(\beta) > 0$.

So, we have

$$presinh^{\gamma}_{\alpha,\beta}(\lambda t^{\varsigma}) = \frac{1}{2}\left(E^{\gamma}_{\alpha,\beta}(\lambda t^{\varsigma}) - E^{\gamma}_{\alpha,\beta}(-\lambda t^{\varsigma})\right) = \sum_{k=0}^{\infty} \frac{\lambda^{2k+1}(\gamma)_{2k+1} t^{(2k+1)\varsigma}}{\Gamma((2k+1)\alpha+\beta)}, \tag{5.346}$$

$$precosh^{\gamma}_{\alpha,\beta}(\lambda t^{\varsigma}) = \frac{1}{2}\left(E^{\gamma}_{\alpha,\beta}(\lambda t^{\varsigma}) + E^{\gamma}_{\alpha,\beta}(-\lambda t^{\varsigma})\right) = \sum_{k=0}^{\infty} \frac{\lambda^{2k}(\gamma)_{2k} t^{2k\varsigma}}{\Gamma(2k\alpha+\beta)}, \tag{5.347}$$

$$pretanh^{\gamma}_{\alpha,\beta}\left(\lambda t^{\varsigma}\right) = \frac{presinh^{\gamma}_{\alpha,\beta}(\lambda t^{\varsigma})}{precosh^{\gamma}_{\alpha,\beta}(\lambda t^{\varsigma})}\left(precosh^{\gamma}_{\alpha,\beta}\left(\lambda t^{\varsigma}\right) \neq 0\right), \tag{5.348}$$

$$precotanh^{\gamma}_{\alpha,\beta}\left(\lambda t^{\varsigma}\right) = \frac{precosh^{\gamma}_{\alpha,\beta}(\lambda t^{\varsigma})}{presinh^{\gamma}_{\alpha,\beta}(\lambda t^{\varsigma})}\left(presinh^{\gamma}_{\alpha,\beta}\left(\lambda t^{\varsigma}\right) \neq 0\right), \tag{5.349}$$

$$precsch^{\gamma}_{\alpha,\beta}\left(\lambda t^{\varsigma}\right) = \frac{1}{presinh^{\gamma}_{\alpha,\beta}(\lambda t^{\varsigma})}\left(presinh^{\gamma}_{\alpha,\beta}\left(\lambda t^{\varsigma}\right) \neq 0\right), \tag{5.350}$$

and

$$presech^{\gamma}_{\alpha,\beta}\left(\lambda t^{\varsigma}\right) = \frac{1}{precosh^{\gamma}_{\alpha,\beta}(\lambda t^{\varsigma})}\left(precosh^{\gamma}_{\alpha,\beta}\left(\lambda t^{\varsigma}\right) \neq 0\right), \tag{5.351}$$

where $\alpha, \beta, \gamma, \varsigma, t, \lambda \in \mathbb{C}$, $Re(\alpha) > 0$, and $Re(\beta) > 0$.

With the use of the results, we may have the following theorem.

Theorem 5.33 *Let $\alpha, \beta, \gamma, \varsigma, t \in \mathbb{C}$, $Re(\alpha) > 0$, and $Re(\beta) > 0$. Then we have*

$$E^{\gamma}_{\alpha,\beta}\left(t^{\varsigma}\right) = precosh^{\gamma}_{\alpha,\beta}\left(t^{\varsigma}\right) + presinh^{\gamma}_{\alpha,\beta}\left(t^{\varsigma}\right) \tag{5.352}$$

and

$$E^{\gamma}_{\alpha,\beta}\left(-t^{\varsigma}\right) = precosh^{\gamma}_{\alpha,\beta}\left(t^{\varsigma}\right) - presinh^{\gamma}_{\alpha,\beta}\left(t^{\varsigma}\right). \tag{5.353}$$

Generally,

$$E^{\gamma}_{\alpha,\beta}\left(\lambda t^{\varsigma}\right) = precosh^{\gamma}_{\alpha,\beta}\left(\lambda t^{\varsigma}\right) + presinh^{\gamma}_{\alpha,\beta}\left(\lambda t^{\varsigma}\right) \tag{5.354}$$

and

$$E^{\gamma}_{\alpha,\beta}\left(-\lambda t^{\varsigma}\right) = precosh^{\gamma}_{\alpha,\beta}\left(\lambda t^{\varsigma}\right) - presinh^{\gamma}_{\alpha,\beta}\left(\lambda t^{\varsigma}\right), \tag{5.355}$$

where $\alpha, \beta, \gamma, \varsigma, t, \lambda \in \mathbb{C}$, $Re(\alpha) > 0$, *and* $Re(\beta) > 0$.

Property 5.10 Let $\alpha, \beta, \varsigma, t, \gamma \in \mathbb{C}$, $Re(\alpha) > 0$, and $Re(\beta) > 0$. Then we have

$$presinh^{\gamma}_{\alpha,\beta}\left(-t^{\varsigma}\right) = -presinh^{\gamma}_{\alpha,\beta}\left(t^{\varsigma}\right), \tag{5.356}$$

$$precosh^{\gamma}_{\alpha,\beta}\left(-t^{\varsigma}\right) = precosh^{\gamma}_{\alpha,\beta}\left(t^{\varsigma}\right), \tag{5.357}$$

$$pretanh^{\gamma}_{\alpha,\beta}\left(-t^{\varsigma}\right) = -\,pretanh^{\gamma}_{\alpha,\beta}\left(t^{\varsigma}\right), \tag{5.358}$$

$$precotanh^{\gamma}_{\alpha,\beta}\left(-t^{\varsigma}\right) = -\,precotanh^{\gamma}_{\alpha,\beta}\left(t^{\varsigma}\right), \tag{5.359}$$

$$precsch^{\gamma}_{\alpha,\beta}\left(-t^{\varsigma}\right) = -precsch^{\gamma}_{\alpha,\beta}\left(t^{\varsigma}\right), \tag{5.360}$$

and

$$presech^{\gamma}_{\alpha,\beta}\left(-t^{\varsigma}\right) = presech^{\gamma}_{\alpha,\beta}\left(t^{\varsigma}\right). \tag{5.361}$$

Theorem 5.34 *Let* $\alpha, \beta, \gamma, \varsigma, t \in \mathbb{C}\alpha, \beta, \varsigma, t \in \mathbb{C}$, $Re(\alpha) > 0$, *and* $Re(\beta) > 0$. *Then the Laplace transform of the function* $presinh^{\gamma}_{\alpha,\beta}(t^{\varsigma})$ *is given as follows:*

$$L\left\{presinh^{\gamma}_{\alpha,\beta}\left(t^{\varsigma}\right)\right\} = \frac{s^{-1}}{\Gamma(\gamma)}\, {}_2sinh_1\left(\begin{matrix}(\gamma,1),(1,\varsigma)\\(\beta,\alpha)\end{matrix};\frac{1}{s^{\varsigma}}\right). \tag{5.362}$$

Proof Let us consider that

$$\begin{aligned} L\left[E^{\gamma}_{\alpha,\beta}(\lambda t^{\varsigma})\right] &= \int_0^{\infty} e^{-st} E^{\gamma}_{\alpha,\beta}(\lambda t^{\varsigma})\,dt \\ &= \frac{s^{-1}}{\Gamma(\gamma)}\, {}_2\varphi_1\left(\begin{matrix}(\gamma,1),(1,\varsigma)\\(\beta,\alpha)\end{matrix};\frac{\lambda}{s^{\varsigma}}\right). \end{aligned} \tag{5.363}$$

Then,

$$\begin{aligned} &L\left\{presinh^{\gamma}_{\alpha,\beta}(t^{\varsigma})\right\} \\ &= \frac{1}{2}\left[L\left\{E^{\gamma}_{\alpha,\beta}(t^{\varsigma})\right\} - L\left\{E^{\gamma}_{\alpha,\beta}(-t^{\varsigma})\right\}\right] \\ &= \frac{1}{2}\frac{s^{-1}}{\Gamma(\gamma)}\left[{}_2\varphi_1\left(\begin{matrix}(\gamma,1),(1,\varsigma)\\(\beta,\alpha)\end{matrix};\frac{1}{s^{\varsigma}}\right) - {}_2\varphi_1\left(\begin{matrix}(\gamma,1),(1,\varsigma)\\(\beta,\alpha)\end{matrix};-\frac{1}{s^{\varsigma}}\right)\right] \\ &= \frac{s^{-1}}{\Gamma(\gamma)}\, {}_2sinh_1\left(\begin{matrix}(\gamma,1),(1,\varsigma)\\(\beta,\alpha)\end{matrix};\frac{1}{s^{\varsigma}}\right). \end{aligned} \tag{5.364}$$

More generally,

$$\begin{aligned} &L\left\{presinh^{\gamma}_{\alpha,\beta}\left(\lambda t^{\varsigma}\right)\right\} \\ &= \tfrac{1}{2}\left[L\left\{E^{\gamma}_{\alpha,\beta}\left(\lambda t^{\varsigma}\right)\right\} - L\left\{E^{\gamma}_{\alpha,\beta}\left(-\lambda t^{\varsigma}\right)\right\}\right] \\ &= \tfrac{1}{2}\tfrac{s^{-1}}{\Gamma(\gamma)}\left[{}_2\varphi_1\left(\begin{matrix}(\gamma,1),(1,\varsigma)\\(\beta,\alpha)\end{matrix};\tfrac{\lambda}{s^{\varsigma}}\right) - {}_2\varphi_1\left(\begin{matrix}(\gamma,1),(1,\vartheta)\\(\beta,\alpha)\end{matrix};-\tfrac{\lambda}{s^{\varsigma}}\right)\right] \\ &= \tfrac{s^{-1}}{\Gamma(\gamma)}\,{}_2sinh_1\left(\begin{matrix}(\gamma,1),(1,\varsigma)\\(\beta,\alpha)\end{matrix};\tfrac{\lambda}{s^{\varsigma}}\right), \end{aligned} \tag{5.365}$$

where $\alpha, \beta, \gamma, \varsigma, t \in \mathbb{C}$, $Re(\alpha) > 0$, and $Re(\beta) > 0$. □

Theorem 5.35 *Let $\alpha, \beta, \gamma, \varsigma, t \in \mathbb{C}$, $Re(\alpha) > 0$, and $Re(\beta) > 0$. Then the Laplace transform of the function $precos^{\gamma}_{\alpha,\beta}(t^{\varsigma})$ is given as follows:*

$$L\left\{precosh^{\gamma}_{\alpha,\beta}\left(t^{\varsigma}\right)\right\} = \frac{s^{-1}}{\Gamma(\gamma)}\,{}_2cosh_1\left(\begin{matrix}(\gamma,1),(1,\varsigma)\\(\beta,\alpha)\end{matrix};\frac{1}{s^{\varsigma}}\right). \tag{5.366}$$

Proof With the use of the integral representation:

$$\begin{aligned} L\left[E^{\gamma}_{\alpha,\beta}\left(\lambda t^{\varsigma}\right)\right] &= \int_0^{\infty} e^{-st} E^{\gamma}_{\alpha,\beta}\left(\lambda t^{\varsigma}\right) dt \\ &= \tfrac{s^{-1}}{\Gamma(\gamma)}\,{}_2\varphi_1\left(\begin{matrix}(\gamma,1),(1,\varsigma)\\(\beta,\alpha)\end{matrix};\tfrac{\lambda}{s^{\varsigma}}\right), \end{aligned} \tag{5.367}$$

we have

$$\begin{aligned} &L\left\{precosh^{\gamma}_{\alpha,\beta}\left(t^{\varsigma}\right)\right\} \\ &= \tfrac{1}{2}\left[L\left\{E^{\gamma}_{\alpha,\beta}\left(t^{\varsigma}\right)\right\} + L\left\{E^{\gamma}_{\alpha,\beta}\left(-t^{\varsigma}\right)\right\}\right] \\ &= \tfrac{1}{2}\tfrac{s^{-1}}{\Gamma(\gamma)}\left[{}_2\varphi_1\left(\begin{matrix}(\gamma,1),(1,\varsigma)\\(\beta,\alpha)\end{matrix};\tfrac{1}{s^{\varsigma}}\right) + {}_2\varphi_1\left(\begin{matrix}(\gamma,1),(1,\varsigma)\\(\beta,\alpha)\end{matrix};-\tfrac{1}{s^{\varsigma}}\right)\right] \\ &= \tfrac{s^{-1}}{\Gamma(\gamma)}\,{}_2cosh_1\left(\begin{matrix}(\gamma,1),(1,\varsigma)\\(\beta,\alpha)\end{matrix};\tfrac{1}{s^{\varsigma}}\right). \end{aligned} \tag{5.368}$$

More generally,

$$\begin{aligned} &L\left\{precosh^{\gamma}_{\alpha,\beta}\left(\lambda t^{\varsigma}\right)\right\} \\ &= \tfrac{1}{2}\left[L\left\{E^{\gamma}_{\alpha,\beta}\left(\lambda t^{\varsigma}\right)\right\} + L\left\{E^{\gamma}_{\alpha,\beta}\left(-\lambda t^{\varsigma}\right)\right\}\right] \\ &= \tfrac{1}{2}\tfrac{s^{-1}}{\Gamma(\gamma)}\left[{}_2\varphi_1\left(\begin{matrix}(\gamma,1),(1,\varsigma)\\(\beta,\alpha)\end{matrix};\tfrac{\lambda}{s^{\varsigma}}\right) + {}_2\varphi_1\left(\begin{matrix}(\gamma,1),(1,\varsigma)\\(\beta,\alpha)\end{matrix};-\tfrac{\lambda}{s^{\varsigma}}\right)\right] \\ &= \tfrac{s^{-1}}{\Gamma(\gamma)}\,{}_2cosh_1\left(\begin{matrix}(\gamma,1),(1,\varsigma)\\(\beta,\alpha)\end{matrix};\tfrac{\lambda}{s^{\varsigma}}\right), \end{aligned} \tag{5.369}$$

where $\alpha, \beta, \gamma, \varsigma, t \in \mathbb{C}$, $Re(\alpha) > 0$, and $Re(\beta) > 0$. □

5.1.8 The Pre-Supertrigonometric Functions with the Parameter via Prabhakar Function

In this section, we present the pre-supertrigonometric functions with the parameter via Prabhakar function.

Definition 5.40 The presine with the parameter via Prabhakar function is defined as

$$\begin{aligned} Tresin_{\alpha,\beta}^{\gamma}(t) &= \frac{t^{\beta-1}}{2i}\left(E_{\alpha,\beta}^{\gamma}(it) - E_{\alpha,\beta}^{\gamma}(-it)\right) \\ &= t^{\beta-1}\sum_{k=0}^{\infty}\frac{(-1)^k(\gamma)_{2k+1}t^{2k+1}}{\Gamma((2k+1)\alpha+\beta)}, \end{aligned} \tag{5.370}$$

where $\alpha, \beta, \gamma, t \in \mathbb{C}$, $Re(\alpha) > 0$, and $Re(\beta) > 0$.

Definition 5.41 The precosine with the parameter via Prabhakar function is defined as

$$\begin{aligned} Trecos_{\alpha,\beta}^{\gamma}(t) &= \frac{t^{\beta-1}}{2}\left(E_{\alpha,\beta}^{\gamma}(it) + E_{\alpha,\beta}^{\gamma}(-it)\right) \\ &= t^{\beta-1}\sum_{k=0}^{\infty}\frac{(-1)^k(\gamma)_{2k}t^{2k}}{\Gamma(2k\alpha+\beta)}, \end{aligned} \tag{5.371}$$

where $\alpha, \beta, \gamma, t \in \mathbb{C}$, $Re(\alpha) > 0$, and $Re(\beta) > 0$.

Definition 5.42 The pretangent with the parameter via Prabhakar function is defined as

$$Tretan_{\alpha,\beta}^{\gamma}(t) = \frac{Tresin_{\alpha,\beta}^{\gamma}(t)}{Trecos_{\alpha,\beta}^{\gamma}(t)}\left(Trecos_{\alpha,\beta}^{\gamma}(t) \neq 0\right), \tag{5.372}$$

where $\alpha, \beta, \gamma, t \in \mathbb{C}\alpha, \beta, \gamma, t \in \mathbb{C}$, $Re(\alpha) > 0$, and $Re(\beta) > 0$.

Definition 5.43 The pretangent with the parameter via Prabhakar function is defined as

$$Tretan_{\alpha,\beta}^{\gamma}(t) = \frac{Tresin_{\alpha,\beta}^{\gamma}(t)}{Trecos_{\alpha,\beta}^{\gamma}(t)}\left(Trecos_{\alpha,\beta}^{\gamma}(t) \neq 0\right), \tag{5.373}$$

where $\alpha, \beta, \gamma, t \in \mathbb{C}$, $Re(\alpha) > 0$, and $Re(\beta) > 0$.

Definition 5.44 The precotangent with the parameter via Prabhakar function is defined as

$$Trecotan_{\alpha,\beta}^{\gamma}(t) = \frac{Trecos_{\alpha,\beta}^{\gamma}(t)}{Tresin_{\alpha,\beta}^{\gamma}(t)}\left(Tresin_{\alpha,\beta}^{\gamma}(t) \neq 0\right), \tag{5.374}$$

where $\alpha, \beta, \gamma, t \in \mathbb{C}$, $Re(\alpha) > 0$, and $Re(\beta) > 0$.

Definition 5.45 The precosecant with the parameter via Prabhakar function is defined as

$$Trecsc^{\gamma}_{\alpha,\beta}(t) = \frac{1}{Tresin^{\gamma}_{\alpha,\beta}(t)} \left(Tresin^{\gamma}_{\alpha,\beta}(t) \neq 0\right), \tag{5.375}$$

where $\alpha, \beta, \gamma, t \in \mathbb{C}$, $Re(\alpha) > 0$, and $Re(\beta) > 0$.

Definition 5.46 The presecant with the parameter via Prabhakar function is defined as

$$Tresec^{\gamma}_{\alpha,\beta}(t) = \frac{1}{Trecos^{\gamma}_{\alpha,\beta}(t)} \left(Trecos^{\gamma}_{\alpha,\beta}(t) \neq 0\right), \tag{5.376}$$

where $\alpha, \beta, \gamma, t \in \mathbb{C}$, $Re(\alpha) > 0$, and $Re(\beta) > 0$.

Therefore, we have that

$$\begin{aligned} Tresin^{\gamma}_{\alpha,\beta}(\lambda t) &= \tfrac{t^{\beta-1}}{2i}\left(E^{\gamma}_{\alpha,\beta}(i\lambda t) - E^{\gamma}_{\alpha,\beta}(-i\lambda t)\right) \\ &= t^{\beta-1}\sum_{k=0}^{\infty}\tfrac{(-1)^k\lambda^{2k+1}(\gamma)_{2k+1}t^{2k+1}}{\Gamma((2k+1)\alpha+\beta)}, \end{aligned} \tag{5.377}$$

$$\begin{aligned} Trecos^{\gamma}_{\alpha,\beta}(\lambda t) &= \tfrac{t^{\beta-1}}{2}\left(E^{\gamma}_{\alpha,\beta}(i\lambda t) + E^{\gamma}_{\alpha,\beta}(-i\lambda t)\right) \\ &= t^{\beta-1}\sum_{k=0}^{\infty}\tfrac{(-1)^k\lambda^{2k}(\gamma)_{2k}t^{2k}}{\Gamma(2k\alpha+\beta)}, \end{aligned} \tag{5.378}$$

$$Tretan^{\gamma}_{\alpha,\beta}(\lambda t) = \frac{Tresin^{\gamma}_{\alpha,\beta}(\lambda t)}{Trecos^{\gamma}_{\alpha,\beta}(\lambda t)} \left(Trecos^{\gamma}_{\alpha,\beta}(\lambda t) \neq 0\right), \tag{5.379}$$

$$Trecotan^{\gamma}_{\alpha,\beta}(\lambda t) = \frac{Trecos^{\gamma}_{\alpha,\beta}(\lambda t)}{Tresin^{\gamma}_{\alpha,\beta}(\lambda t)} \left(Tresin^{\gamma}_{\alpha,\beta}(\lambda t) \neq 0\right), \tag{5.380}$$

$$Trecsc^{\gamma}_{\alpha,\beta}(\lambda t) = \frac{1}{Tresin^{\gamma}_{\alpha,\beta}(\lambda t)} \left(Tresin^{\gamma}_{\alpha,\beta}(\lambda t) \neq 0\right), \tag{5.381}$$

and

$$Tresec^{\gamma}_{\alpha,\beta}(\lambda t) = \frac{1}{Trecos^{\gamma}_{\alpha,\beta}(\lambda t)} \left(Trecos^{\gamma}_{\alpha,\beta}(\lambda t) \neq 0\right), \tag{5.382}$$

where $\alpha, \beta, \gamma, \lambda, t \in \mathbb{C}$, $Re(\alpha) > 0$, and $Re(\beta) > 0$.

With the aid of the results, we may have the following theorem.

Theorem 5.36 *Let $\alpha, \beta, \gamma, t \in \mathbb{C}$, $Re(\alpha) > 0$, and $Re(\beta) > 0$. Then there exist*

$$t^{\beta-1} E^{\gamma}_{\alpha,\beta}(it) = Trecos^{\gamma}_{\alpha,\beta}(t) + i Tresin^{\gamma}_{\alpha,\beta}(t) \tag{5.383}$$

and

$$t^{\beta-1} E^{\gamma}_{\alpha,\beta}(-it) = Trecos^{\gamma}_{\alpha,\beta}(t) - i Tresin^{\gamma}_{\alpha,\beta}(t). \tag{5.384}$$

Generally, we have

$$t^{\beta-1} E^{\gamma}_{\alpha,\beta}(i\lambda t) = Trecos^{\gamma}_{\alpha,\beta}(\lambda t) + i Tresin^{\gamma}_{\alpha,\beta}(\lambda t) \tag{5.385}$$

and

$$t^{\beta-1} E^{\gamma}_{\alpha,\beta}(-i\lambda t) = Trecos^{\gamma}_{\alpha,\beta}(\lambda t) - i Tresin^{\gamma}_{\alpha,\beta}(\lambda t), \tag{5.386}$$

where $\alpha, \beta, \gamma, t, \lambda \in \mathbb{C}$, $Re(\alpha) > 0$, and $Re(\beta) > 0$.

Property 5.11 Let $\alpha, \beta, \gamma, t \in \mathbb{C}$, $Re(\alpha) > 0$, and $Re(\beta) > 0$. Then we have

$$Tresin^{\gamma}_{\alpha,\beta}(-t) = -Tresin^{\gamma}_{\alpha,\beta}(t), \tag{5.387}$$

$$Trecos^{\gamma}_{\alpha,\beta}(-t) = Trecos^{\gamma}_{\alpha,\beta}(t), \tag{5.388}$$

$$Tretan^{\gamma}_{\alpha,\beta}(-t) = - Tretan^{\gamma}_{\alpha,\beta}(t), \tag{5.389}$$

$$Trecotan^{\gamma}_{\alpha,\beta}(-t) = - Trecotan^{\gamma}_{\alpha,\beta}(t), \tag{5.390}$$

$$Trecsc^{\gamma}_{\alpha,\beta}(-t) = -Trecsc^{\gamma}_{\alpha,\beta}(t), \tag{5.391}$$

and

$$Tresec^{\gamma}_{\alpha,\beta}(-t) = Tresec^{\gamma}_{\alpha,\beta}(t). \tag{5.392}$$

Theorem 5.37 *Let $\alpha, \beta, \gamma, t \in \mathbb{C}$, $Re(\alpha) > 0$, and $Re(\beta) > 0$. Then the Laplace transform of the function $Tresin^{\gamma}_{\alpha,\beta}(t)$ is given as follows:*

$$L\left\{Tresin^{\gamma}_{\alpha,\beta}(t)\right\} = \frac{s^{-\beta}}{\Gamma(\gamma)} {}_2sin_1\left(\begin{matrix}(\gamma,1),(\beta,1)\\(\beta,\alpha)\end{matrix};\frac{1}{s}\right). \tag{5.393}$$

Proof Let us consider

$$L\left[t^{\varsigma-1}E_{\alpha,\beta}^{\gamma}(\lambda t)\right]=\int_0^\infty e^{-st}t^{\varsigma-1}E_{\alpha,\beta}^{\gamma}(\lambda t)\,dt=\frac{s^{-\varsigma}}{\Gamma(\gamma)}\,{}_2\varphi_1\left(\begin{matrix}(\gamma,1),(\varsigma,1)\\(\beta,\alpha)\end{matrix};\frac{\lambda}{s}\right),\tag{5.394}$$

which yields, for $\varsigma=\beta$,

$$L\left[t^{\beta-1}E_{\alpha,\beta}^{\gamma}(\lambda t)\right]=\int_0^\infty e^{-st}t^{\beta-1}E_{\alpha,\beta}^{\gamma}(\lambda t)\,dt=\frac{s^{-\beta}}{\Gamma(\gamma)}\,{}_2\varphi_1\left(\begin{matrix}(\gamma,1),(\beta,1)\\(\beta,\alpha)\end{matrix};\frac{\lambda}{s}\right).\tag{5.395}$$

Thus,

$$\begin{aligned}&L\left\{Tresin_{\alpha,\beta}^{\gamma}(t)\right\}\\&=\tfrac{1}{2i}\left[L\left\{t^{\beta-1}E_{\alpha,\beta}^{\gamma}(it)\right\}-L\left\{t^{\beta-1}E_{\alpha,\beta}^{\gamma}(-it)\right\}\right]\\&=\tfrac{1}{2i}\tfrac{s^{-\beta}}{\Gamma(\gamma)}\left[{}_2\varphi_1\left(\begin{matrix}(\gamma,1),(\beta,1)\\(\beta,\alpha)\end{matrix};\tfrac{i}{s}\right)-{}_2\varphi_1\left(\begin{matrix}(\gamma,1),(\beta,1)\\(\beta,\alpha)\end{matrix};-\tfrac{i}{s}\right)\right]\\&=\tfrac{s^{-\beta}}{\Gamma(\gamma)}\,{}_2sin_1\left(\begin{matrix}(\gamma,1),(\beta,1)\\(\beta,\alpha)\end{matrix};\tfrac{1}{s}\right).\end{aligned}\tag{5.396}$$

More generally, for $\alpha,\beta,\gamma,\lambda,t\in\mathbb{C}$, $Re(\alpha)>0$, and $Re(\beta)>0$,

$$\begin{aligned}&L\left\{Tresin_{\alpha,\beta}^{\gamma}(\lambda t)\right\}\\&=\tfrac{1}{2i}\left[L\left\{t^{\beta-1}E_{\alpha,\beta}^{\gamma}(i\lambda t)\right\}-L\left\{t^{\beta-1}E_{\alpha,\beta}^{\gamma}(-i\lambda t)\right\}\right]\\&=\tfrac{1}{2i}\tfrac{s^{-\beta}}{\Gamma(\gamma)}\left[{}_2\varphi_1\left(\begin{matrix}(\gamma,1),(\beta,1)\\(\beta,\alpha)\end{matrix};\tfrac{i\lambda}{s}\right)-{}_2\varphi_1\left(\begin{matrix}(\gamma,1),(\beta,1)\\(\beta,\alpha)\end{matrix};-\tfrac{i\lambda}{s}\right)\right]\\&=\tfrac{s^{-\beta}}{\Gamma(\gamma)}\,{}_2sin_1\left(\begin{matrix}(\gamma,1),(\beta,1)\\(\beta,\alpha)\end{matrix};\tfrac{\lambda}{s}\right).\end{aligned}\tag{5.397}$$

□

Theorem 5.38 *Let $\alpha,\beta,\gamma,t\in\mathbb{C}$, $Re(\alpha)>0$, and $Re(\beta)>0$. Then the Laplace transform of the function $precos_{\alpha,\beta}^{\gamma}(t)$ is given as follows:*

$$L\left\{Trecos_{\alpha,\beta}^{\gamma}(t)\right\}=\frac{s^{-\beta}}{\Gamma(\gamma)}\,{}_2cos_1\left(\begin{matrix}(\gamma,1),(\beta,1)\\(\beta,\alpha)\end{matrix};\frac{1}{s}\right).\tag{5.398}$$

Proof Making use of the relation:

$$L\left[t^{\beta-1}E^{\gamma}_{\alpha,\beta}(\lambda t)\right] = \int\limits_0^{\infty} e^{-st}t^{\beta-1}E^{\gamma}_{\alpha,\beta}(\lambda t)\,dt = \frac{s^{-\beta}}{\Gamma(\gamma)}\,{}_2\varphi_1\left(\begin{matrix}(\gamma,1),(\beta,1)\\(\beta,\alpha)\end{matrix};\frac{\lambda}{s}\right), \tag{5.399}$$

we have

$$\begin{aligned}
&L\left\{Trecos^{\gamma}_{\alpha,\beta}(t)\right\}\\
&= \tfrac{1}{2}\left[L\left\{t^{\beta-1}E^{\gamma}_{\alpha,\beta}(it)\right\} + L\left\{t^{\beta-1}E^{\gamma}_{\alpha,\beta}(-it)\right\}\right]\\
&= \tfrac{1}{2}\tfrac{s^{-\beta}}{\Gamma(\gamma)}\left[{}_2\varphi_1\left(\begin{matrix}(\gamma,1),(\beta,1)\\(\beta,\alpha)\end{matrix};\tfrac{i}{s}\right) + {}_2\varphi_1\left(\begin{matrix}(\gamma,1),(\beta,1)\\(\beta,\alpha)\end{matrix};-\tfrac{i}{s}\right)\right]\\
&= \tfrac{s^{-\beta}}{\Gamma(\gamma)}\,{}_2cos_1\left(\begin{matrix}(\gamma,1),(\beta,1)\\(\beta,\alpha)\end{matrix};\tfrac{1}{s}\right).
\end{aligned} \tag{5.400}$$

More generally, for $\alpha,\beta,\gamma,\lambda,t\in\mathbb{C}$, $Re(\alpha)>0$, and $Re(\beta)>0$,

$$\begin{aligned}
&L\left\{Trecos^{\gamma}_{\alpha,\beta}(\lambda t)\right\}\\
&= \tfrac{1}{2}\left[L\left\{t^{\beta-1}E^{\gamma}_{\alpha,\beta}(i\lambda t)\right\} + L\left\{t^{\beta-1}E^{\gamma}_{\alpha,\beta}(-i\lambda t)\right\}\right]\\
&= \tfrac{1}{2}\tfrac{s^{-\beta}}{\Gamma(\gamma)}\left[{}_2\varphi_1\left(\begin{matrix}(\gamma,1),(\beta,1)\\(\beta,\alpha)\end{matrix};\tfrac{i\lambda}{s}\right) + {}_2\varphi_1\left(\begin{matrix}(\gamma,1),(\beta,1)\\(\beta,\alpha)\end{matrix};-\tfrac{i\lambda}{s}\right)\right]\\
&= \tfrac{s^{-\beta}}{\Gamma(\gamma)}\,{}_2cos_1\left(\begin{matrix}(\gamma,1),(\beta,1)\\(\beta,\alpha)\end{matrix};\tfrac{\lambda}{s}\right).
\end{aligned} \tag{5.401}$$

□

5.1.9 *The Pre-Superhyperbolic Functions with the Parameter via Prabhakar Function*

In this section, we give the pre-superhyperbolic functions with the parameter via Prabhakar function.

Definition 5.47 The hyperbolic presine with the parameter via Prabhakar function is defined as

$$\begin{aligned}
Tresinh^{\gamma}_{\alpha,\beta}(t) &= \tfrac{t^{\beta-1}}{2}\left(E^{\gamma}_{\alpha,\beta}(t) - E^{\gamma}_{\alpha,\beta}(-t)\right)\\
&= t^{\beta-1}\sum_{k=0}^{\infty}\tfrac{(\gamma)_{2k+1}t^{2k+1}}{\Gamma((2k+1)\alpha+\beta)},
\end{aligned} \tag{5.402}$$

where $\alpha,\beta,\gamma,t\in\mathbb{C}$, $Re(\alpha)>0$, and $Re(\beta)>0$.

Definition 5.48 The hyperbolic precosine with the parameter via Prabhakar function is defined as

$$\begin{aligned} Trecosh_{\alpha,\beta}^{\gamma}(t) &= \tfrac{t^{\beta-1}}{2}\left(E_{\alpha,\beta}^{\gamma}(t) + E_{\alpha,\beta}^{\gamma}(-t)\right) \\ &= t^{\beta-1}\sum_{k=0}^{\infty}\tfrac{(\gamma)_{2k}t^{2k}}{\Gamma(2k\alpha+\beta)}, \end{aligned} \tag{5.403}$$

where $\alpha, \beta, \gamma, t \in \mathbb{C}$, $Re(\alpha) > 0$, and $Re(\beta) > 0$.

Definition 5.49 The hyperbolic pretangent with the parameter via Prabhakar function is defined as

$$Tretanh_{\alpha,\beta}^{\gamma}(t) = \frac{Tresinh_{\alpha,\beta}^{\gamma}(t)}{Trecosh_{\alpha,\beta}^{\gamma}(t)}\left(Trecosh_{\alpha,\beta}^{\gamma}(t) \neq 0\right), \tag{5.404}$$

where $\alpha, \beta, \gamma, t \in \mathbb{C}$, $Re(\alpha) > 0$, and $Re(\beta) > 0$.

Definition 5.50 The hyperbolic precotangent with the parameter via Prabhakar function is defined as

$$Trecotanh_{\alpha,\beta}^{\gamma}(t) = \frac{Trecosh_{\alpha,\beta}^{\gamma}(t)}{Tresinh_{\alpha,\beta}^{\gamma}(t)}\left(Tresinh_{\alpha,\beta}^{\gamma}(t) \neq 0\right), \tag{5.405}$$

where $\alpha, \beta, \gamma, t \in \mathbb{C}$, $Re(\alpha) > 0$, and $Re(\beta) > 0$.

Definition 5.51 The hyperbolic precosecant with the parameter via Prabhakar function is defined as

$$Trecsch_{\alpha,\beta}^{\gamma}(t) = \frac{1}{Tresinh_{\alpha,\beta}^{\gamma}(t)}\left(Tresinh_{\alpha,\beta}^{\gamma}(t) \neq 0\right), \tag{5.406}$$

where $\alpha, \beta, \gamma, t \in \mathbb{C}$, $Re(\alpha) > 0$, and $Re(\beta) > 0$.

Definition 5.52 The hyperbolic presecant with the parameter via Prabhakar function is defined as

$$Tresech_{\alpha,\beta}^{\gamma}(t) = \frac{1}{Trecosh_{\alpha,\beta}^{\gamma}(t)}\left(Trecosh_{\alpha,\beta}^{\gamma}(t) \neq 0\right), \tag{5.407}$$

where $\alpha, \beta, \gamma, t \in \mathbb{C}$, $Re(\alpha) > 0$, and $Re(\beta) > 0$.

Therefore, we have that

$$\begin{aligned} Tresinh_{\alpha,\beta}^{\gamma}(t) &= \tfrac{t^{\beta-1}}{2}\left(E_{\alpha,\beta}^{\gamma}(t) - E_{\alpha,\beta}^{\gamma}(-t)\right) \\ &= t^{\beta-1}\sum_{k=0}^{\infty}\tfrac{(\gamma)_{2k+1}t^{2k+1}}{\Gamma((2k+1)\alpha+\beta)}, \end{aligned} \tag{5.408}$$

$$
\begin{aligned}
Trecosh_{\alpha,\beta}^{\gamma}(t) &= \tfrac{t^{\beta-1}}{2}\left(E_{\alpha,\beta}^{\gamma}(t) + E_{\alpha,\beta}^{\gamma}(-t)\right) \\
&= t^{\beta-1}\sum_{k=0}^{\infty}\tfrac{(\gamma)_{2k}t^{2k}}{\Gamma(2k\alpha+\beta)},
\end{aligned}
\tag{5.409}
$$

$$
Tretanh_{\alpha,\beta}^{\gamma}(t) = \frac{Tresinh_{\alpha,\beta}^{\gamma}(t)}{Trecosh_{\alpha,\beta}^{\gamma}(t)}\left(Trecosh_{\alpha,\beta}^{\gamma}(t) \neq 0\right), \tag{5.410}
$$

$$
Trecotanh_{\alpha,\beta}^{\gamma}(t) = \frac{Trecosh_{\alpha,\beta}^{\gamma}(t)}{Tresinh_{\alpha,\beta}^{\gamma}(t)}\left(Tresinh_{\alpha,\beta}^{\gamma}(t) \neq 0\right), \tag{5.411}
$$

$$
Trecsch_{\alpha,\beta}^{\gamma}(t) = \frac{1}{Tresinh_{\alpha,\beta}^{\gamma}(t)}\left(Tresinh_{\alpha,\beta}^{\gamma}(t) \neq 0\right), \tag{5.412}
$$

and

$$
Tresech_{\alpha,\beta}^{\gamma}(t) = \frac{1}{Trecosh_{\alpha,\beta}^{\gamma}(t)}\left(Trecosh_{\alpha,\beta}^{\gamma}(t) \neq 0\right), \tag{5.413}
$$

where $\alpha, \beta, \gamma, \lambda, t \in \mathbb{C}$, $Re(\alpha) > 0$, and $Re(\beta) > 0$.

With the aid of the results, we may have the following theorem.

Theorem 5.39 *Let $\alpha, \beta, \gamma, t \in \mathbb{C}$, $Re(\alpha) > 0$, and $Re(\beta) > 0$. Then there exist*

$$
t^{\beta-1}E_{\alpha,\beta}^{\gamma}(t) = Trecosh_{\alpha,\beta}^{\gamma}(t) + Tresinh_{\alpha,\beta}^{\gamma}(t) \tag{5.414}
$$

and

$$
t^{\beta-1}E_{\alpha,\beta}^{\gamma}(-t) = Trecosh_{\alpha,\beta}^{\gamma}(t) - Tresinh_{\alpha,\beta}^{\gamma}(t). \tag{5.415}
$$

Generally, we have

$$
t^{\beta-1}E_{\alpha,\beta}^{\gamma}(\lambda t) = Trecosh_{\alpha,\beta}^{\gamma}(\lambda t) + Tresinh_{\alpha,\beta}^{\gamma}(\lambda t) \tag{5.416}
$$

and

$$
t^{\beta-1}E_{\alpha,\beta}^{\gamma}(-\lambda t) = Trecosh_{\alpha,\beta}^{\gamma}(\lambda t) - Tresinh_{\alpha,\beta}^{\gamma}(\lambda t), \tag{5.417}
$$

where $\alpha, \beta, \gamma, t, \lambda \in \mathbb{C}$, $Re(\alpha) > 0$, and $Re(\beta) > 0$.

Property 5.12 Let $\alpha, \beta, \gamma, t \in \mathbb{C}$, $Re(\alpha) > 0$, and $Re(\beta) > 0$. Then we have

$$Tresinh^{\gamma}_{\alpha,\beta}(-t) = -Tresinh^{\gamma}_{\alpha,\beta}(t), \tag{5.418}$$

$$Trecosh^{\gamma}_{\alpha,\beta}(-t) = Trecosh^{\gamma}_{\alpha,\beta}(t), \tag{5.419}$$

$$Tretanh^{\gamma}_{\alpha,\beta}(-t) = -\,Tretanh^{\gamma}_{\alpha,\beta}(t), \tag{5.420}$$

$$Trecotanh^{\gamma}_{\alpha,\beta}(-t) = -\,Trecotanh^{\gamma}_{\alpha,\beta}(t), \tag{5.421}$$

$$Trecsch^{\gamma}_{\alpha,\beta}(-t) = -Trecsch^{\gamma}_{\alpha,\beta}(t), \tag{5.422}$$

and

$$Tresech^{\gamma}_{\alpha,\beta}(-t) = Tresech^{\gamma}_{\alpha,\beta}(t). \tag{5.423}$$

Theorem 5.40 *Let* $\alpha, \beta, \gamma, t \in \mathbb{C}$, $Re(\alpha) > 0$, *and* $Re(\beta) > 0$. *Then the Laplace transform of the function* $Tresinh^{\gamma}_{\alpha,\beta}(t)$ *is given as follows:*

$$L\left\{Tresinh^{\gamma}_{\alpha,\beta}(t)\right\} = \frac{s^{-\beta}}{\Gamma(\gamma)}\, {}_2sinh_1\left(\begin{matrix}(\gamma,1),(\beta,1)\\(\beta,\alpha)\end{matrix};\frac{1}{s}\right). \tag{5.424}$$

Proof Let us consider

$$L\left[t^{\beta-1}E^{\gamma}_{\alpha,\beta}(\lambda t)\right] = \int_0^{\infty} e^{-st}t^{\beta-1}E^{\gamma}_{\alpha,\beta}(\lambda t)\,dt = \frac{s^{-\beta}}{\Gamma(\gamma)}\, {}_2\varphi_1\left(\begin{matrix}(\gamma,1),(\beta,1)\\(\beta,\alpha)\end{matrix};\frac{\lambda}{s}\right). \tag{5.425}$$

Thus,

$$\begin{aligned}
&L\left\{Tresinh^{\gamma}_{\alpha,\beta}(t)\right\}\\
&= \tfrac{1}{2}\left[L\left\{t^{\beta-1}E^{\gamma}_{\alpha,\beta}(t)\right\} - L\left\{t^{\beta-1}E^{\gamma}_{\alpha,\beta}(-t)\right\}\right]\\
&= \tfrac{1}{2}\tfrac{s^{-\beta}}{\Gamma(\gamma)}\left[{}_2\varphi_1\left(\begin{matrix}(\gamma,1),(\beta,1)\\(\beta,\alpha)\end{matrix};\tfrac{1}{s}\right) - {}_2\varphi_1\left(\begin{matrix}(\gamma,1),(\beta,1)\\(\beta,\alpha)\end{matrix};-\tfrac{1}{s}\right)\right]\\
&= \tfrac{s^{-\beta}}{\Gamma(\gamma)}\, {}_2sinh_1\left(\begin{matrix}(\gamma,1),(\beta,1)\\(\beta,\alpha)\end{matrix};\tfrac{1}{s}\right).
\end{aligned} \tag{5.426}$$

More generally, for $\alpha, \beta, \gamma, \lambda, t \in \mathbb{C}$, $Re(\alpha) > 0$, and $Re(\beta) > 0$,

$$\begin{aligned}
&L\left\{Tresinh^{\gamma}_{\alpha,\beta}(\lambda t)\right\}\\
&= \frac{1}{2}\left[L\left\{t^{\beta-1}E^{\gamma}_{\alpha,\beta}(\lambda t)\right\} - L\left\{t^{\beta-1}E^{\gamma}_{\alpha,\beta}(-\lambda t)\right\}\right]\\
&= \frac{1}{2}\frac{s^{-\beta}}{\Gamma(\gamma)}\left[{}_2\varphi_1\left(\begin{matrix}(\gamma,1),(\beta,1)\\(\beta,\alpha)\end{matrix};\frac{\lambda}{s}\right) - {}_2\varphi_1\left(\begin{matrix}(\gamma,1),(\beta,1)\\(\beta,\alpha)\end{matrix};-\frac{\lambda}{s}\right)\right]\\
&= \frac{s^{-\beta}}{\Gamma(\gamma)}\,{}_2sinh_1\left(\begin{matrix}(\gamma,1),(\beta,1)\\(\beta,\alpha)\end{matrix};\frac{\lambda}{s}\right).
\end{aligned} \tag{5.427}$$

□

Theorem 5.41 *Let $\alpha, \beta, \gamma, t \in \mathbb{C}$, $Re(\alpha) > 0$, and $Re(\beta) > 0$. Then the Laplace transform of the function $precos^{\gamma}_{\alpha,\beta}(t)$ is given as follows:*

$$L\left\{Trecosh^{\gamma}_{\alpha,\beta}(t)\right\} = \frac{s^{-\beta}}{\Gamma(\gamma)}\,{}_2cosh_1\left(\begin{matrix}(\gamma,1),(\beta,1)\\(\beta,\alpha)\end{matrix};\frac{1}{s}\right). \tag{5.428}$$

Proof In view of

$$L\left[t^{\beta-1}E^{\gamma}_{\alpha,\beta}(\lambda t)\right] = \int_0^{\infty} e^{-st}t^{\beta-1}E^{\gamma}_{\alpha,\beta}(\lambda t)\,dt = \frac{s^{-\beta}}{\Gamma(\gamma)}\,{}_2\varphi_1\left(\begin{matrix}(\gamma,1),(\beta,1)\\(\beta,\alpha)\end{matrix};\frac{\lambda}{s}\right), \tag{5.429}$$

we have that

$$\begin{aligned}
&L\left\{Trecosh^{\gamma}_{\alpha,\beta}(t)\right\}\\
&= \frac{1}{2}\left[L\left\{t^{\beta-1}E^{\gamma}_{\alpha,\beta}(t)\right\} + L\left\{t^{\beta-1}E^{\gamma}_{\alpha,\beta}(-t)\right\}\right]\\
&= \frac{1}{2}\frac{s^{-\beta}}{\Gamma(\gamma)}\left[{}_2\varphi_1\left(\begin{matrix}(\gamma,1),(\beta,1)\\(\beta,\alpha)\end{matrix};\frac{1}{s}\right) + {}_2\varphi_1\left(\begin{matrix}(\gamma,1),(\beta,1)\\(\beta,\alpha)\end{matrix};-\frac{1}{s}\right)\right]\\
&= \frac{s^{-\beta}}{\Gamma(\gamma)}\,{}_2cosh_1\left(\begin{matrix}(\gamma,1),(\beta,1)\\(\beta,\alpha)\end{matrix};\frac{1}{s}\right).
\end{aligned} \tag{5.430}$$

More generally, for $\alpha, \beta, \gamma, \lambda, t \in \mathbb{C}$, $Re(\alpha) > 0$, and $Re(\beta) > 0$,

$$\begin{aligned}
&L\left\{Trecosh\{^{\gamma}_{\alpha,\beta}(\lambda t)\right\}\\
&= \frac{1}{2}\left[L\left\{t^{\beta-1}E^{\gamma}_{\alpha,\beta}(\lambda t)\right\} + L\left\{t^{\beta-1}E^{\gamma}_{\alpha,\beta}(-\lambda t)\right\}\right]\\
&= \frac{1}{2}\frac{s^{-\beta}}{\Gamma(\gamma)}\left[{}_2\varphi_1\left(\begin{matrix}(\gamma,1),(\beta,1)\\(\beta,\alpha)\end{matrix};\frac{\lambda}{s}\right) + {}_2\varphi_1\left(\begin{matrix}(\gamma,1),(\beta,1)\\(\beta,\alpha)\end{matrix};-\frac{\lambda}{s}\right)\right]\\
&= \frac{s^{-\beta}}{\Gamma(\gamma)}\,{}_2cosh_1\left(\begin{matrix}(\gamma,1),(\beta,1)\\(\beta,\alpha)\end{matrix};\frac{\lambda}{s}\right).
\end{aligned} \tag{5.431}$$

□

5.1.10 The Pre-Supertrigonometric Functions with the Power Law and Parameter via Prabhakar Function

In this section, we present the pre-supertrigonometric functions with the power law and parameter via Prabhakar function.

Definition 5.53 The presupersine with the power law and parameter via Prabhakar function is defined as

$$\begin{aligned} Gresin_{\alpha,\beta}^{\gamma}(t^{\varsigma}) &= \frac{t^{\beta-1}}{2i}\left(E_{\alpha,\beta}^{\gamma}(it^{\varsigma}) - E_{\alpha,\beta}^{\gamma}(-it^{\varsigma})\right) \\ &= t^{\beta-1}\sum_{k=0}^{\infty}\frac{(-1)^k(\gamma)_{2k+1}t^{(2k+1)\varsigma}}{\Gamma((2k+1)\alpha+\beta)}, \end{aligned} \tag{5.432}$$

where $\alpha, \beta, \gamma, \varsigma, t \in \mathbb{C}$, $Re(\alpha) > 0$, and $Re(\beta) > 0$.

Definition 5.54 The precosine with the power law and parameter via Prabhakar function is defined as

$$\begin{aligned} Grecos_{\alpha,\beta}^{\gamma}(t^{\varsigma}) &= \frac{t^{\beta-1}}{2}\left(E_{\alpha,\beta}^{\gamma}(it^{\varsigma}) + E_{\alpha,\beta}^{\gamma}(-it^{\varsigma})\right) \\ &= t^{\beta-1}\sum_{k=0}^{\infty}\frac{(-1)^k(\gamma)_{2k}t^{2k\varsigma}}{\Gamma(2k\alpha+\beta)}, \end{aligned} \tag{5.433}$$

where $\alpha, \beta, \gamma, \varsigma, t \in \mathbb{C}$, $Re(\alpha) > 0$, and $Re(\beta) > 0$.

Definition 5.55 The pretangent with the power law and parameter via Prabhakar function is defined as

$$Gretan_{\alpha,\beta}^{\gamma}\left(t^{\varsigma}\right) = \frac{Gresin_{\alpha,\beta}^{\gamma}(t^{\varsigma})}{Grecos_{\alpha,\beta}^{\gamma}(t^{\varsigma})}\left(Grecos_{\alpha,\beta}^{\gamma}\left(t^{\varsigma}\right) \neq 0\right), \tag{5.434}$$

where $\alpha, \beta, \gamma, \varsigma, t \in \mathbb{C}$, $Re(\alpha) > 0$, and $Re(\beta) > 0$.

Definition 5.56 The pretangent with the power law and parameter via Prabhakar function is defined as

$$Gretan_{\alpha,\beta}^{\gamma}\left(t^{\varsigma}\right) = \frac{Gresin_{\alpha,\beta}^{\gamma}(t^{\varsigma})}{Grecos_{\alpha,\beta}^{\gamma}(t^{\varsigma})}\left(Grecos_{\alpha,\beta}^{\gamma}\left(t^{\varsigma}\right) \neq 0\right), \tag{5.435}$$

where $\alpha, \beta, \gamma, \varsigma, t \in \mathbb{C}$, $Re(\alpha) > 0$, and $Re(\beta) > 0$.

Definition 5.57 The precotangent with the power law and parameter via Prabhakar function is defined as

$$Grecotan_{\alpha,\beta}^{\gamma}\left(t^{\varsigma}\right) = \frac{Grecos_{\alpha,\beta}^{\gamma}(t^{\varsigma})}{Gresin_{\alpha,\beta}^{\gamma}(t^{\varsigma})}\left(Gresin_{\alpha,\beta}^{\gamma}\left(t^{\varsigma}\right) \neq 0\right), \tag{5.436}$$

where $\alpha, \beta, \gamma, \varsigma, t \in \mathbb{C}$, $Re(\alpha) > 0$, and $Re(\beta) > 0$.

Definition 5.58 The precosecant with the power law and parameter via Prabhakar function is defined as

$$Grecsc^{\gamma}_{\alpha,\beta}\left(t^{\varsigma}\right) = \frac{1}{Gresin^{\gamma}_{\alpha,\beta}\left(t^{\varsigma}\right)} \left(Gresin^{\gamma}_{\alpha,\beta}\left(t^{\varsigma}\right) \neq 0\right), \tag{5.437}$$

where $\alpha, \beta, \gamma, \varsigma, t \in \mathbb{C}$, $Re\,(\alpha) > 0$, and $Re\,(\beta) > 0$.

Definition 5.59 The presecant with the power law and parameter via Prabhakar function is defined as

$$Gresec^{\gamma}_{\alpha,\beta}\left(t^{\varsigma}\right) = \frac{1}{Grecos^{\gamma}_{\alpha,\beta}\left(t^{\varsigma}\right)} \left(Grecos^{\gamma}_{\alpha,\beta}\left(t^{\varsigma}\right) \neq 0\right), \tag{5.438}$$

where $\alpha, \beta, \gamma, \varsigma, t \in \mathbb{C}$, $Re\,(\alpha) > 0$, and $Re\,(\beta) > 0$.

Thus,

$$\begin{aligned} Gresin^{\gamma}_{\alpha,\beta}\left(\lambda t^{\varsigma}\right) &= \tfrac{t^{\beta-1}}{2i}\left(E^{\gamma}_{\alpha,\beta}\left(i\lambda t^{\varsigma}\right) - E^{\gamma}_{\alpha,\beta}\left(-i\lambda t^{\varsigma}\right)\right) \\ &= t^{\beta-1}\sum_{k=0}^{\infty} \tfrac{(-1)^k \lambda^{2k+1} (\gamma)_{2k+1} t^{(2k+1)\varsigma}}{\Gamma((2k+1)\alpha+\beta)}, \end{aligned} \tag{5.439}$$

$$\begin{aligned} Grecos^{\gamma}_{\alpha,\beta}\left(\lambda t^{\varsigma}\right) &= \tfrac{t^{\beta-1}}{2}\left(E^{\gamma}_{\alpha,\beta}\left(i\lambda t^{\varsigma}\right) + E^{\gamma}_{\alpha,\beta}\left(-i\lambda t^{\varsigma}\right)\right) \\ &= t^{\beta-1}\sum_{k=0}^{\infty} \tfrac{(-1)^k \lambda^{2k} (\gamma)_{2k} t^{2k\varsigma}}{\Gamma(2k\alpha+\beta)}, \end{aligned} \tag{5.440}$$

$$Gretan^{\gamma}_{\alpha,\beta}\left(\lambda t^{\varsigma}\right) = \frac{Gresin^{\gamma}_{\alpha,\beta}\left(\lambda t^{\varsigma}\right)}{Grecos^{\gamma}_{\alpha,\beta}\left(\lambda t^{\varsigma}\right)} \left(Grecos^{\gamma}_{\alpha,\beta}\left(\lambda t^{\varsigma}\right) \neq 0\right), \tag{5.441}$$

$$Grecotan^{\gamma}_{\alpha,\beta}\left(\lambda t^{\varsigma}\right) = \frac{Grecos^{\gamma}_{\alpha,\beta}\left(\lambda t^{\varsigma}\right)}{Gresin^{\gamma}_{\alpha,\beta}\left(\lambda t^{\varsigma}\right)} \left(Gresin^{\gamma}_{\alpha,\beta}\left(\lambda t^{\varsigma}\right) \neq 0\right), \tag{5.442}$$

$$Grecsc^{\gamma}_{\alpha,\beta}\left(\lambda t^{\varsigma}\right) = \frac{1}{Gresin^{\gamma}_{\alpha,\beta}\left(\lambda t^{\varsigma}\right)} \left(Gresin^{\gamma}_{\alpha,\beta}\left(\lambda t^{\varsigma}\right) \neq 0\right), \tag{5.443}$$

and

$$Gresec^{\gamma}_{\alpha,\beta}\left(\lambda t^{\varsigma}\right) = \frac{1}{Grecos^{\gamma}_{\alpha,\beta}\left(\lambda t^{\varsigma}\right)} \left(Grecos^{\gamma}_{\alpha,\beta}\left(\lambda t^{\varsigma}\right) \neq 0\right), \tag{5.444}$$

where $\alpha, \beta, \gamma, \varsigma, t, \lambda \in \mathbb{C}$, $Re\,(\alpha) > 0$, and $Re\,(\beta) > 0$.

With the use of the results, we may give the following theorem.

Theorem 5.42 *Let $\alpha, \beta, \gamma, \varsigma, t \in \mathbb{C}$, $Re(\alpha) > 0$, and $Re(\beta) > 0$. Then we have*

$$t^{\beta-1} E^{\gamma}_{\alpha,\beta}\left(it^{\varsigma}\right) = Grecos^{\gamma}_{\alpha,\beta}\left(t^{\varsigma}\right) + iGresin^{\gamma}_{\alpha,\beta}\left(t^{\varsigma}\right) \tag{5.445}$$

and

$$t^{\beta-1} E^{\gamma}_{\alpha,\beta}\left(-it^{\varsigma}\right) = Grecos^{\gamma}_{\alpha,\beta}\left(t^{\varsigma}\right) - iGresin^{\gamma}_{\alpha,\beta}\left(t^{\varsigma}\right). \tag{5.446}$$

Generally,

$$t^{\beta-1} E^{\gamma}_{\alpha,\beta}\left(i\lambda t^{\varsigma}\right) = Grecos^{\gamma}_{\alpha,\beta}\left(\lambda t^{\varsigma}\right) + iGresin^{\gamma}_{\alpha,\beta}\left(\lambda t^{\varsigma}\right) \tag{5.447}$$

and

$$t^{\beta-1} E^{\gamma}_{\alpha,\beta}\left(-i\lambda t^{\varsigma}\right) = Grecos^{\gamma}_{\alpha,\beta}\left(\lambda t^{\varsigma}\right) - iGresin^{\gamma}_{\alpha,\beta}\left(\lambda t^{\varsigma}\right), \tag{5.448}$$

where $\alpha, \beta, \gamma, \varsigma, t, \lambda \in \mathbb{C}$, $Re(\alpha) > 0$, and $Re(\beta) > 0$.

Property 5.13 Let $\alpha, \beta, \gamma, \varsigma, t \in \mathbb{C}$, $Re(\alpha) > 0$, and $Re(\beta) > 0$. Then we have

$$Gresin^{\gamma}_{\alpha,\beta}\left(-t^{\varsigma}\right) = -Gresin^{\gamma}_{\alpha,\beta}\left(t^{\varsigma}\right), \tag{5.449}$$

$$Grecos^{\gamma}_{\alpha,\beta}\left(-t^{\varsigma}\right) = Grecos^{\gamma}_{\alpha,\beta}\left(t^{\varsigma}\right), \tag{5.450}$$

$$Gretan^{\gamma}_{\alpha,\beta}\left(-t^{\varsigma}\right) = -\ Gretan^{\gamma}_{\alpha,\beta}\left(t^{\varsigma}\right), \tag{5.451}$$

$$Grecotan^{\gamma}_{\alpha,\beta}\left(-t^{\varsigma}\right) = -\ Grecotan^{\gamma}_{\alpha,\beta}\left(t^{\varsigma}\right), \tag{5.452}$$

$$Grecsc^{\gamma}_{\alpha,\beta}\left(-t^{\varsigma}\right) = -Grecsc^{\gamma}_{\alpha,\beta}\left(t^{\varsigma}\right), \tag{5.453}$$

and

$$Gresec^{\gamma}_{\alpha,\beta}\left(-t^{\varsigma}\right) = Gresec^{\gamma}_{\alpha,\beta}\left(t^{\varsigma}\right). \tag{5.454}$$

Theorem 5.43 *Let $\alpha, \beta, \gamma, \varsigma, t \in \mathbb{C}\alpha, \beta, \varsigma, t \in \mathbb{C}$, $Re(\alpha) > 0$, and $Re(\beta) > 0$. Then the Laplace transform of the function $Gresin^{\gamma}_{\alpha,\beta}(t^{\varsigma})$ is given as follows:*

$$L\left\{Gresin^{\gamma}_{\alpha,\beta}\left(t^{\varsigma}\right)\right\} = \frac{s^{-\beta}}{\Gamma(\gamma)}\, {}_2sin_1\left(\begin{matrix}(\gamma, 1), (\beta, \varsigma) \\ (\beta, \alpha)\end{matrix}; \frac{1}{s^{\varsigma}}\right). \tag{5.455}$$

Proof Let us recall

$$\begin{aligned} L\left[t^{\varsigma-1}E^{\gamma}_{\alpha,\beta}\left(\lambda t^{\vartheta}\right)\right] &= \int_0^{\infty} e^{-st}t^{\varsigma-1}E^{\gamma}_{\alpha,\beta}\left(\lambda t^{\vartheta}\right)dt \\ &= \frac{s^{-\varsigma}}{\Gamma(\gamma)}\,{}_2\varphi_1\left(\begin{matrix}(\gamma,1),(\varsigma,\vartheta)\\(\beta,\alpha)\end{matrix};\frac{\lambda}{s^{\vartheta}}\right), \end{aligned} \tag{5.456}$$

which leads to, for $\varsigma = \beta$,

$$\begin{aligned} L\left[t^{\beta-1}E^{\gamma}_{\alpha,\beta}\left(\lambda t^{\vartheta}\right)\right] &= \int_0^{\infty} e^{-st}t^{\beta-1}E^{\gamma}_{\alpha,\beta}\left(\lambda t^{\vartheta}\right)dt \\ &= \frac{s^{-\beta}}{\Gamma(\gamma)}\,{}_2\varphi_1\left(\begin{matrix}(\gamma,1),(\beta,\vartheta)\\(\beta,\alpha)\end{matrix};\frac{\lambda}{s^{\vartheta}}\right). \end{aligned} \tag{5.457}$$

Therefore, we may give

$$\begin{aligned} &L\left\{Gresin^{\gamma}_{\alpha,\beta}\left(t^{\varsigma}\right)\right\} \\ &= \frac{1}{2i}\left[L\left\{t^{\beta-1}E^{\gamma}_{\alpha,\beta}\left(it^{\varsigma}\right)\right\} - L\left\{t^{\beta-1}E^{\gamma}_{\alpha,\beta}\left(-it^{\varsigma}\right)\right\}\right] \\ &= \frac{1}{2i}\frac{s^{-\beta}}{\Gamma(\gamma)}\left[{}_2\varphi_1\left(\begin{matrix}(\gamma,1),(\beta,\varsigma)\\(\beta,\alpha)\end{matrix};\frac{i}{s^{\varsigma}}\right) - {}_2\varphi_1\left(\begin{matrix}(\gamma,1),(\beta,\varsigma)\\(\beta,\alpha)\end{matrix};-\frac{i}{s^{\varsigma}}\right)\right] \\ &= \frac{s^{-\beta}}{\Gamma(\gamma)}\,{}_2sin_1\left(\begin{matrix}(\gamma,1),(\beta,\varsigma)\\(\beta,\alpha)\end{matrix};\frac{1}{s^{\varsigma}}\right). \end{aligned} \tag{5.458}$$

More generally,

$$\begin{aligned} &L\left\{Gresin^{\gamma}_{\alpha,\beta}\left(\lambda t^{\varsigma}\right)\right\} \\ &= \frac{1}{2i}\left[L\left\{t^{\beta-1}E^{\gamma}_{\alpha,\beta}\left(i\lambda t^{\varsigma}\right)\right\} - L\left\{t^{\beta-1}E^{\gamma}_{\alpha,\beta}\left(-i\lambda t^{\varsigma}\right)\right\}\right] \\ &= \frac{1}{2i}\frac{s^{-\beta}}{\Gamma(\gamma)}\left[{}_2\varphi_1\left(\begin{matrix}(\gamma,1),(\beta,\varsigma)\\(\beta,\alpha)\end{matrix};\frac{i\lambda}{s^{\varsigma}}\right) - {}_2\varphi_1\left(\begin{matrix}(\gamma,1),(\beta,\varsigma)\\(\beta,\alpha)\end{matrix};-\frac{i\lambda}{s^{\varsigma}}\right)\right] \\ &= \frac{s^{-\beta}}{\Gamma(\gamma)}\,{}_2sin_1\left(\begin{matrix}(\gamma,1),(\beta,\varsigma)\\(\beta,\alpha)\end{matrix};\frac{\lambda}{s^{\varsigma}}\right), \end{aligned} \tag{5.459}$$

where $\alpha, \beta, \gamma, \varsigma, t \in \mathbb{C}$, $Re\,(\alpha) > 0$, and $Re\,(\beta) > 0$. □

Theorem 5.44 *Let* $\alpha, \beta, \gamma, \varsigma, t \in \mathbb{C}$, $Re\,(\alpha) > 0$, *and* $Re\,(\beta) > 0$. *Then the Laplace transform of the function* $Grecos^{\gamma}_{\alpha,\beta}\left(t^{\varsigma}\right)$ *is given as follows:*

$$L\left\{Grecos^{\gamma}_{\alpha,\beta}\left(t^{\varsigma}\right)\right\} = \frac{s^{-\beta}}{\Gamma\left(\gamma\right)}\,{}_2cos_1\left(\begin{matrix}(\gamma,1),(\beta,\varsigma)\\(\beta,\alpha)\end{matrix};\frac{1}{s^{\varsigma}}\right). \tag{5.460}$$

Proof In view of the integral representation:

$$L\left[t^{\beta-1}E^{\gamma}_{\alpha,\beta}\left(\lambda t^{\vartheta}\right)\right] = \int_0^{\infty} e^{-st}t^{\beta-1}E^{\gamma}_{\alpha,\beta}\left(\lambda t^{\varsigma}\right)dt = \frac{s^{-\beta}}{\Gamma(\gamma)}\,{}_2\varphi_1\left(\begin{matrix}(\gamma,1),(\beta,\varsigma)\\(\beta,\alpha)\end{matrix};\frac{\lambda}{s^{\varsigma}}\right), \tag{5.461}$$

we have that

$$\begin{aligned}
&L\left\{Grecos^{\gamma}_{\alpha,\beta}\left(t^{\varsigma}\right)\right\}\\
&= \frac{1}{2}\left[L\left\{t^{\beta-1}E^{\gamma}_{\alpha,\beta}\left(it^{\varsigma}\right)\right\} + L\left\{t^{\beta-1}E^{\gamma}_{\alpha,\beta}\left(-it^{\varsigma}\right)\right\}\right]\\
&= \frac{1}{2}\frac{s^{-\beta}}{\Gamma(\gamma)}\left[{}_2\varphi_1\left(\begin{matrix}(\gamma,1),(\beta,\varsigma)\\(\beta,\alpha)\end{matrix};\frac{i}{s^{\varsigma}}\right) + {}_2\varphi_1\left(\begin{matrix}(\gamma,1),(\beta,\varsigma)\\(\beta,\alpha)\end{matrix};-\frac{i}{s^{\varsigma}}\right)\right]\\
&= \frac{s^{-\beta}}{\Gamma(\gamma)}\,{}_2cos_1\left(\begin{matrix}(\gamma,1),(\beta,\varsigma)\\(\beta,\alpha)\end{matrix};\frac{1}{s^{\varsigma}}\right).
\end{aligned} \tag{5.462}$$

More generally, we have that

$$\begin{aligned}
&L\left\{precos^{\gamma}_{\alpha,\beta}\left(\lambda t^{\varsigma}\right)\right\}\\
&= \frac{1}{2}\left[L\left\{t^{\beta-1}E^{\gamma}_{\alpha,\beta}\left(i\lambda t^{\varsigma}\right)\right\} + L\left\{t^{\beta-1}E^{\gamma}_{\alpha,\beta}\left(-i\lambda t^{\varsigma}\right)\right\}\right]\\
&= \frac{1}{2}\frac{s^{-\beta}}{\Gamma(\gamma)}\left[{}_2\varphi_1\left(\begin{matrix}(\gamma,1),(\beta,\varsigma)\\(\beta,\alpha)\end{matrix};\frac{i\lambda}{s^{\varsigma}}\right) + {}_2\varphi_1\left(\begin{matrix}(\gamma,1),(\beta,\varsigma)\\(\beta,\alpha)\end{matrix};-\frac{i\lambda}{s^{\varsigma}}\right)\right]\\
&= \frac{s^{-\beta}}{\Gamma(\gamma)}\,{}_2cos_1\left(\begin{matrix}(\gamma,1),(\beta,\varsigma)\\(\beta,\alpha)\end{matrix};\frac{\lambda}{s^{\varsigma}}\right)
\end{aligned} \tag{5.463}$$

where $\alpha, \beta, \gamma, \varsigma, t \in \mathbb{C}$, $Re\left(\alpha\right) > 0$, and $Re\left(\beta\right) > 0$. □

5.1.11 The Pre-Superhyperbolic Functions with the Power Law and Parameter via Prabhakar Function

In this section, we propose the pre-superhyperbolic functions with the power law and parameter via Prabhakar function.

Definition 5.60 The hyperbolic presine with the power law and parameter via Prabhakar function is defined as

$$\begin{aligned}
Gresinh^{\gamma}_{\alpha,\beta}\left(t^{\varsigma}\right) &= \frac{t^{\beta-1}}{2}\left(E^{\gamma}_{\alpha,\beta}\left(t^{\varsigma}\right) - E^{\gamma}_{\alpha,\beta}\left(-t^{\varsigma}\right)\right)\\
&= t^{\beta-1}\sum_{k=0}^{\infty}\frac{(\gamma)_{2k+1}t^{(2k+1)\varsigma}}{\Gamma((2k+1)\alpha+\beta)},
\end{aligned} \tag{5.464}$$

where $\alpha, \beta, \gamma, \varsigma, t \in \mathbb{C}$, $Re\left(\alpha\right) > 0$, and $Re\left(\beta\right) > 0$.

Definition 5.61 The hyperbolic precosine with the power law and parameter via Prabhakar function is defined as

$$\begin{aligned} Grecosh^{\gamma}_{\alpha,\beta}\left(t^{\varsigma}\right) &= \tfrac{t^{\beta-1}}{2}\left(E^{\gamma}_{\alpha,\beta}\left(t^{\varsigma}\right)+E^{\gamma}_{\alpha,\beta}\left(-t^{\varsigma}\right)\right) \\ &= t^{\beta-1}\sum_{k=0}^{\infty}\tfrac{(\gamma)_{2k}t^{2k\varsigma}}{\Gamma(2k\alpha+\beta)}, \end{aligned} \tag{5.465}$$

where $\alpha, \beta, \gamma, \varsigma, t \in \mathbb{C}$, $Re\,(\alpha) > 0$, and $Re\,(\beta) > 0$.

Definition 5.62 The hyperbolic pretangent with the power law and parameter via Prabhakar function is defined as

$$Gretanh^{\gamma}_{\alpha,\beta}\left(t^{\varsigma}\right) = \frac{Gresinh^{\gamma}_{\alpha,\beta}\left(t^{\varsigma}\right)}{Grecosh^{\gamma}_{\alpha,\beta}\left(t^{\varsigma}\right)}\left(Grecosh^{\gamma}_{\alpha,\beta}\left(t^{\varsigma}\right)\neq 0\right), \tag{5.466}$$

where $\alpha, \beta, \gamma, \varsigma, t \in \mathbb{C}$, $Re\,(\alpha) > 0$, and $Re\,(\beta) > 0$.

Definition 5.63 The hyperbolic precotangent with the power law and parameter via Prabhakar function is defined as

$$Grecotanh^{\gamma}_{\alpha,\beta}\left(t^{\varsigma}\right) = \frac{Grecosh^{\gamma}_{\alpha,\beta}\left(t^{\varsigma}\right)}{Gresinh^{\gamma}_{\alpha,\beta}\left(t^{\varsigma}\right)}\left(Gresinh^{\gamma}_{\alpha,\beta}\left(t^{\varsigma}\right)\neq 0\right), \tag{5.467}$$

where $\alpha, \beta, \gamma, \varsigma, t \in \mathbb{C}$, $Re\,(\alpha) > 0$, and $Re\,(\beta) > 0$.

Definition 5.64 The hyperbolic precosecant with the power law and parameter via Prabhakar function is defined as

$$Grecsch^{\gamma}_{\alpha,\beta}\left(t^{\varsigma}\right) = \frac{1}{Gresinh^{\gamma}_{\alpha,\beta}\left(t^{\varsigma}\right)}\left(Gresinh^{\gamma}_{\alpha,\beta}\left(t^{\varsigma}\right)\neq 0\right), \tag{5.468}$$

where $\alpha, \beta, \gamma, \varsigma, t \in \mathbb{C}$, $Re\,(\alpha) > 0$, and $Re\,(\beta) > 0$.

Definition 5.65 The hyperbolic presecant with the power law and parameter via Prabhakar function is defined as

$$Gresech^{\gamma}_{\alpha,\beta}\left(t^{\varsigma}\right) = \frac{1}{Grecosh^{\gamma}_{\alpha,\beta}\left(t^{\varsigma}\right)}\left(Grecosh^{\gamma}_{\alpha,\beta}\left(t^{\varsigma}\right)\neq 0\right), \tag{5.469}$$

where $\alpha, \beta, \gamma, \varsigma, t \in \mathbb{C}$, $Re\,(\alpha) > 0$, and $Re\,(\beta) > 0$.

Therefore, we get

$$\begin{aligned} Gresinh^{\gamma}_{\alpha,\beta}\left(\lambda t^{\varsigma}\right) &= \tfrac{t^{\beta-1}}{2}\left(E^{\gamma}_{\alpha,\beta}\left(\lambda t^{\varsigma}\right)-E^{\gamma}_{\alpha,\beta}\left(-\lambda t^{\varsigma}\right)\right) \\ &= t^{\beta-1}\sum_{k=0}^{\infty}\frac{\lambda^{2k+1}(\gamma)_{2k+1}t^{(2k+1)\varsigma}}{\Gamma((2k+1)\alpha+\beta)}, \end{aligned} \tag{5.470}$$

$$\begin{aligned} Grecosh^{\gamma}_{\alpha,\beta}(\lambda t^{\varsigma}) &= \frac{t^{\beta-1}}{2}\left(E^{\gamma}_{\alpha,\beta}(\lambda t^{\varsigma}) + E^{\gamma}_{\alpha,\beta}(-\lambda t^{\varsigma})\right) \\ &= t^{\beta-1}\sum_{k=0}^{\infty}\frac{\lambda^{2k}(\gamma)_{2k}t^{2k\varsigma}}{\Gamma(2k\alpha+\beta)}, \end{aligned} \tag{5.471}$$

$$Gretanh^{\gamma}_{\alpha,\beta}\left(\lambda t^{\varsigma}\right) = \frac{Gresinh^{\gamma}_{\alpha,\beta}(\lambda t^{\varsigma})}{Grecosh^{\gamma}_{\alpha,\beta}(\lambda t^{\varsigma})}\left(Grecosh^{\gamma}_{\alpha,\beta}\left(\lambda t^{\varsigma}\right) \neq 0\right), \tag{5.472}$$

$$Grecotanh^{\gamma}_{\alpha,\beta}\left(\lambda t^{\varsigma}\right) = \frac{Grecosh^{\gamma}_{\alpha,\beta}(\lambda t^{\varsigma})}{Gresinh^{\gamma}_{\alpha,\beta}(\lambda t^{\varsigma})}\left(Gresinh^{\gamma}_{\alpha,\beta}\left(\lambda t^{\varsigma}\right) \neq 0\right), \tag{5.473}$$

$$Grecsch^{\gamma}_{\alpha,\beta}\left(\lambda t^{\varsigma}\right) = \frac{1}{Gresinh^{\gamma}_{\alpha,\beta}(\lambda t^{\varsigma})}\left(Gresinh^{\gamma}_{\alpha,\beta}\left(\lambda t^{\varsigma}\right) \neq 0\right), \tag{5.474}$$

and

$$Gresech^{\gamma}_{\alpha,\beta}\left(\lambda t^{\varsigma}\right) = \frac{1}{Grecosh^{\gamma}_{\alpha,\beta}(\lambda t^{\varsigma})}\left(Grecosh^{\gamma}_{\alpha,\beta}\left(\lambda t^{\varsigma}\right) \neq 0\right), \tag{5.475}$$

where $\alpha, \beta, \gamma, \varsigma, t, \lambda \in \mathbb{C}$, $Re(\alpha) > 0$, and $Re(\beta) > 0$.

With the use of the results, we may give the following theorem.

Theorem 5.45 *Let $\alpha, \beta, \gamma, \varsigma, t \in \mathbb{C}$, $Re(\alpha) > 0$, and $Re(\beta) > 0$. Then we have*

$$t^{\beta-1}E^{\gamma}_{\alpha,\beta}\left(t^{\varsigma}\right) = Grecosh^{\gamma}_{\alpha,\beta}\left(t^{\varsigma}\right) + Gresinh^{\gamma}_{\alpha,\beta}\left(t^{\varsigma}\right) \tag{5.476}$$

and

$$t^{\beta-1}E^{\gamma}_{\alpha,\beta}\left(-t^{\varsigma}\right) = Grecosh^{\gamma}_{\alpha,\beta}\left(t^{\varsigma}\right) - Gresinh^{\gamma}_{\alpha,\beta}\left(t^{\varsigma}\right). \tag{5.477}$$

Generally,

$$t^{\beta-1}E^{\gamma}_{\alpha,\beta}\left(\lambda t^{\varsigma}\right) = Grecosh^{\gamma}_{\alpha,\beta}\left(\lambda t^{\varsigma}\right) + Gresinh^{\gamma}_{\alpha,\beta}\left(\lambda t^{\varsigma}\right) \tag{5.478}$$

and

$$t^{\beta-1}E^{\gamma}_{\alpha,\beta}\left(-\lambda t^{\varsigma}\right) = Grecosh^{\gamma}_{\alpha,\beta}\left(\lambda t^{\varsigma}\right) - Gresinh^{\gamma}_{\alpha,\beta}\left(\lambda t^{\varsigma}\right), \tag{5.479}$$

where $\alpha, \beta, \gamma, \varsigma, t, \lambda \in \mathbb{C}$, $Re(\alpha) > 0$, and $Re(\beta) > 0$.

Property 5.14 Let $\alpha, \beta, \varsigma, t, \gamma \in \mathbb{C}$, $Re(\alpha) > 0$, and $Re(\beta) > 0$. Then we have

$$Gresinh^{\gamma}_{\alpha,\beta}\left(-t^{\varsigma}\right) = -Gresinh^{\gamma}_{\alpha,\beta}\left(t^{\varsigma}\right), \tag{5.480}$$

$$Grecosh^{\gamma}_{\alpha,\beta}\left(-t^{\varsigma}\right) = Grecosh^{\gamma}_{\alpha,\beta}\left(t^{\varsigma}\right), \tag{5.481}$$

$$Gretanh^{\gamma}_{\alpha,\beta}\left(-t^{\varsigma}\right) = -\ Gretanh^{\gamma}_{\alpha,\beta}\left(t^{\varsigma}\right), \tag{5.482}$$

$$Grecotanh^{\gamma}_{\alpha,\beta}\left(-t^{\varsigma}\right) = -\ Grecotanh^{\gamma}_{\alpha,\beta}\left(t^{\varsigma}\right), \tag{5.483}$$

$$Grecsch^{\gamma}_{\alpha,\beta}\left(-t^{\varsigma}\right) = -Grecsch^{\gamma}_{\alpha,\beta}\left(t^{\varsigma}\right), \tag{5.484}$$

and

$$Gresech^{\gamma}_{\alpha,\beta}\left(-t^{\varsigma}\right) = Gresech^{\gamma}_{\alpha,\beta}\left(t^{\varsigma}\right). \tag{5.485}$$

Theorem 5.46 *Let $\alpha, \beta, \gamma, \varsigma, t \in \mathbb{C}\alpha, \beta, \varsigma, t \in \mathbb{C}$, $Re(\alpha) > 0$, and $Re(\beta) > 0$. Then the Laplace transform of the function $Gresinh^{\gamma}_{\alpha,\beta}(t^{\varsigma})$ is given as follows:*

$$L\left\{Gresinh^{\gamma}_{\alpha,\beta}\left(t^{\varsigma}\right)\right\} = \frac{s^{-\beta}}{\Gamma(\gamma)}\,{}_2sinh_1\left(\begin{matrix}(\gamma,1),(\beta,\varsigma)\\(\beta,\alpha)\end{matrix};\frac{1}{s^{\varsigma}}\right). \tag{5.486}$$

Proof Let us consider that

$$\begin{aligned} L\left[t^{\beta-1}E^{\gamma}_{\alpha,\beta}\left(\lambda t^{\varsigma}\right)\right] &= \int_0^{\infty} e^{-st}t^{\beta-1}E^{\gamma}_{\alpha,\beta}\left(\lambda t^{\varsigma}\right)dt \\ &= \frac{s^{-\beta}}{\Gamma(\gamma)}\,{}_2\varphi_1\left(\begin{matrix}(\gamma,1),(\beta,\varsigma)\\(\beta,\alpha)\end{matrix};\frac{\lambda}{s^{\varsigma}}\right). \end{aligned} \tag{5.487}$$

Then,

$$\begin{aligned} &L\left\{Gresinh^{\gamma}_{\alpha,\beta}\left(t^{\varsigma}\right)\right\} \\ &= \frac{1}{2}\left[L\left\{t^{\beta-1}E^{\gamma}_{\alpha,\beta}\left(t^{\varsigma}\right)\right\} - L\left\{t^{\beta-1}E^{\gamma}_{\alpha,\beta}\left(-t^{\varsigma}\right)\right\}\right] \\ &= \frac{1}{2}\frac{s^{-\beta}}{\Gamma(\gamma)}\left[{}_2\varphi_1\left(\begin{matrix}(\gamma,1),(\beta,\varsigma)\\(\beta,\alpha)\end{matrix};\frac{1}{s^{\varsigma}}\right) - {}_2\varphi_1\left(\begin{matrix}(\gamma,1),(\beta,\varsigma)\\(\beta,\alpha)\end{matrix};-\frac{1}{s^{\varsigma}}\right)\right] \\ &= \frac{s^{-\beta}}{\Gamma(\gamma)}\,{}_2sinh_1\left(\begin{matrix}(\gamma,1),(\beta,\varsigma)\\(\beta,\alpha)\end{matrix};\frac{1}{s^{\varsigma}}\right). \end{aligned} \tag{5.488}$$

More generally,

$$\begin{aligned} &L\left\{Gresinh^{\gamma}_{\alpha,\beta}\left(\lambda t^{\varsigma}\right)\right\} \\ &= \frac{1}{2}\left[L\left\{t^{\beta-1}E^{\gamma}_{\alpha,\beta}\left(\lambda t^{\varsigma}\right)\right\} - L\left\{t^{\beta-1}E^{\gamma}_{\alpha,\beta}\left(-\lambda t^{\varsigma}\right)\right\}\right] \\ &= \frac{1}{2}\frac{s^{-1}}{\Gamma(\gamma)}\left[{}_2\varphi_1\left(\begin{matrix}(\gamma,1),(1,\varsigma)\\(\beta,\alpha)\end{matrix};\frac{\lambda}{s^{\varsigma}}\right) - {}_2\varphi_1\left(\begin{matrix}(\gamma,1),(1,\vartheta)\\(\beta,\alpha)\end{matrix};-\frac{\lambda}{s^{\varsigma}}\right)\right] \\ &= \frac{s^{-1}}{\Gamma(\gamma)}\,{}_2sinh_1\left(\begin{matrix}(\gamma,1),(1,\varsigma)\\(\beta,\alpha)\end{matrix};\frac{\lambda}{s^{\varsigma}}\right), \end{aligned} \tag{5.489}$$

where $\alpha, \beta, \gamma, \varsigma, t \in \mathbb{C}$, $Re(\alpha) > 0$, and $Re(\beta) > 0$. □

Theorem 5.47 *Let $\alpha, \beta, \gamma, \varsigma, t \in \mathbb{C}$, $Re(\alpha) > 0$, and $Re(\beta) > 0$. Then the Laplace transform of the function $precos^{\gamma}_{\alpha,\beta}(t^{\varsigma})$ is given as follows:*

$$L\left\{Grecosh^{\gamma}_{\alpha,\beta}\left(t^{\varsigma}\right)\right\} = \frac{s^{-\beta}}{\Gamma(\gamma)}\,{}_2cosh_1\left(\begin{matrix}(\gamma,1),(\beta,\varsigma)\\(\beta,\alpha)\end{matrix};\frac{1}{s^{\varsigma}}\right). \tag{5.490}$$

Proof With the use of the integral representation:

$$\begin{aligned} L\left[t^{\beta-1}E^{\gamma}_{\alpha,\beta}\left(\lambda t^{\varsigma}\right)\right] &= \int_0^{\infty} e^{-st}t^{\beta-1}E^{\gamma}_{\alpha,\beta}\left(\lambda t^{\varsigma}\right)dt \\ &= \frac{s^{-\beta}}{\Gamma(\gamma)}\,{}_2\varphi_1\left(\begin{matrix}(\gamma,1),(\beta,\varsigma)\\(\beta,\alpha)\end{matrix};\frac{\lambda}{s^{\varsigma}}\right), \end{aligned} \tag{5.491}$$

we have

$$\begin{aligned} &L\left\{Grecosh^{\gamma}_{\alpha,\beta}\left(t^{\varsigma}\right)\right\} \\ &= \frac{1}{2}\left[L\left\{t^{\beta-1}E^{\gamma}_{\alpha,\beta}\left(t^{\varsigma}\right)\right\} + L\left\{t^{\beta-1}E^{\gamma}_{\alpha,\beta}\left(-t^{\varsigma}\right)\right\}\right] \\ &= \frac{1}{2}\frac{s^{-\beta}}{\Gamma(\gamma)}\left[{}_2\varphi_1\left(\begin{matrix}(\gamma,1),(\beta,\varsigma)\\(\beta,\alpha)\end{matrix};\frac{1}{s^{\varsigma}}\right) + {}_2\varphi_1\left(\begin{matrix}(\gamma,1),(\beta,\varsigma)\\(\beta,\alpha)\end{matrix};-\frac{1}{s^{\varsigma}}\right)\right] \\ &= \frac{s^{-\beta}}{\Gamma(\gamma)}\,{}_2cosh_1\left(\begin{matrix}(\gamma,1),(\beta,\varsigma)\\(\beta,\alpha)\end{matrix};\frac{1}{s^{\varsigma}}\right). \end{aligned} \tag{5.492}$$

More generally,

$$\begin{aligned} &L\left\{Grecosh^{\gamma}_{\alpha,\beta}\left(\lambda t^{\varsigma}\right)\right\} \\ &= \frac{1}{2}\left[L\left\{t^{\beta-1}E^{\gamma}_{\alpha,\beta}\left(\lambda t^{\varsigma}\right)\right\} + L\left\{t^{\beta-1}E^{\gamma}_{\alpha,\beta}\left(-\lambda t^{\varsigma}\right)\right\}\right] \\ &= \frac{1}{2}\frac{s^{-\beta}}{\Gamma(\gamma)}\left[{}_2\varphi_1\left(\begin{matrix}(\gamma,1),(\beta,\varsigma)\\(\beta,\alpha)\end{matrix};\frac{\lambda}{s^{\varsigma}}\right) + {}_2\varphi_1\left(\begin{matrix}(\gamma,1),(\beta,\varsigma)\\(\beta,\alpha)\end{matrix};-\frac{\lambda}{s^{\varsigma}}\right)\right] \\ &= \frac{s^{-\beta}}{\Gamma(\gamma)}\,{}_2cosh_1\left(\begin{matrix}(\gamma,1),(\beta,\varsigma)\\(\beta,\alpha)\end{matrix};\frac{\lambda}{s^{\varsigma}}\right), \end{aligned} \tag{5.493}$$

where $\alpha, \beta, \gamma, \varsigma, t \in \mathbb{C}$, $Re(\alpha) > 0$, and $Re(\beta) > 0$. □

5.2 The Integral Representations for Special Functions Related to Prabhakar Function

In this section, we present the integral representations for special functions related to Prabhakar function.

Let us consider that

$$\begin{aligned}supersin_{\alpha,\beta}^{\gamma}(t^{\alpha}) &= \frac{t^{\beta-1}}{2i}\left(E_{\alpha,\beta}^{\gamma}(it^{\alpha}) - E_{\alpha,\beta}^{\gamma}(-it^{\alpha})\right) \\ &= \frac{t^{\beta-1}}{\Gamma(\gamma)} \times {}_1sin_1\left(\begin{matrix}(\gamma,1)\\(\beta,\alpha)\end{matrix}; t^{\alpha}\right) = \sum_{k=0}^{\infty}\frac{(-1)^k(\gamma)_{2k+1}}{\Gamma((2k+1)\alpha+\beta)}\frac{t^{(2k+1)\alpha+\beta-1}}{(2k+1)!},\end{aligned} \tag{5.494}$$

$$\begin{aligned}supercos_{\alpha,\beta}^{\gamma}(t^{\alpha}) &= \frac{t^{\beta-1}}{2}\left(E_{\alpha,\beta}^{\gamma}(it^{\alpha}) + E_{\alpha,\beta}^{\gamma}(-it^{\alpha})\right) \\ &= \frac{t^{\beta-1}}{\Gamma(\gamma)} \times {}_1cos_1\left(\begin{matrix}(\gamma,1)\\(\beta,\alpha)\end{matrix}; t^{\alpha}\right) = \sum_{k=0}^{\infty}\frac{(-1)^k(\gamma)_{2k}}{\Gamma(2k\alpha+\beta)}\frac{t^{2k\alpha+\beta-1}}{(2k)!},\end{aligned} \tag{5.495}$$

$$\begin{aligned}supersinh_{\alpha,\beta}^{\gamma}(t^{\alpha}) &= \frac{t^{\beta-1}}{2}\left(E_{\alpha,\beta}^{\gamma}(t^{\alpha}) - E_{\alpha,\beta}^{\gamma}(-t^{\alpha})\right) \\ &= \frac{t^{\beta-1}}{\Gamma(\gamma)} \times {}_1sinh_1\left(\begin{matrix}(\gamma,1)\\(\beta,\alpha)\end{matrix}; t^{\alpha}\right) = \sum_{k=0}^{\infty}\frac{(\gamma)_{2k+1}t^{(2k+1)\alpha+\beta-1}}{\Gamma((2k+1)\alpha+\beta)},\end{aligned} \tag{5.496}$$

$$\begin{aligned}supercosh_{\alpha,\beta}^{\gamma}(t^{\alpha}) &= \frac{t^{\beta-1}}{2}\left(E_{\alpha,\beta}^{\gamma}(t^{\alpha}) + E_{\alpha,\beta}^{\gamma}(-t^{\alpha})\right) \\ &= \frac{t^{\beta-1}}{\Gamma(\gamma)} \times {}_1cosh_1\left(\begin{matrix}(\gamma,1)\\(\beta,\alpha)\end{matrix}; t^{\alpha}\right) = \sum_{n=0}^{\infty}\frac{(\gamma)_{2k}t^{2k\alpha+\beta-1}}{\Gamma(2k\alpha+\beta)},\end{aligned} \tag{5.497}$$

$$\begin{aligned}presin_{\alpha,\beta}^{\gamma}(t) &= \frac{1}{2i}\left(E_{\alpha,\beta}^{\gamma}(it) - E_{\alpha,\beta}^{\gamma}(-it)\right) \\ &= \frac{1}{\Gamma(\gamma)} \times {}_1sin_1\left(\begin{matrix}(\gamma,1)\\(\beta,\alpha)\end{matrix}; t\right) = \sum_{k=0}^{\infty}\frac{(-1)^k(\gamma)_{2k+1}t^{2k+1}}{\Gamma((2k+1)\alpha+\beta)},\end{aligned} \tag{5.498}$$

$$\begin{aligned}precos_{\alpha,\beta}^{\gamma}(t) &= \frac{1}{2}\left(E_{\alpha,\beta}^{\gamma}(it) + E_{\alpha,\beta}^{\gamma}(-it)\right) \\ &= \frac{1}{\Gamma(\gamma)} \times {}_1cos_1\left(\begin{matrix}(\gamma,1)\\(\beta,\alpha)\end{matrix}; t\right) = \sum_{k=0}^{\infty}\frac{(-1)^k(\gamma)_{2k}t^{2k}}{\Gamma(2k\alpha+\beta)},\end{aligned} \tag{5.499}$$

$$\begin{aligned}presinh_{\alpha,\beta}^{\gamma}(t) &= \frac{1}{2}\left(E_{\alpha,\beta}^{\gamma}(t) - E_{\alpha,\beta}^{\gamma}(-t)\right) \\ &= \frac{1}{\Gamma(\gamma)} \times {}_1sinh_1\left(\begin{matrix}(\gamma,1)\\(\beta,\alpha)\end{matrix}; t\right) = \sum_{k=0}^{\infty}\frac{(\gamma)_{2k+1}t^{2k+1}}{\Gamma((2k+1)\alpha+\beta)},\end{aligned} \tag{5.500}$$

$$\begin{aligned}precosh_{\alpha,\beta}^{\gamma}(t) &= \frac{1}{2}\left(E_{\alpha,\beta}^{\gamma}(t) + E_{\alpha,\beta}^{\gamma}(-t)\right) \\ &= \frac{1}{\Gamma(\gamma)} \times {}_1cosh_1\left(\begin{matrix}(\gamma,1)\\(\beta,\alpha)\end{matrix}; t\right) = \sum_{k=0}^{\infty}\frac{(\gamma)_{2k}t^{2k}}{\Gamma(2k\alpha+\beta)},\end{aligned} \tag{5.501}$$

$$\begin{aligned}presin_{\alpha,\beta}^{\gamma}(t^{\varsigma}) &= \frac{1}{2i}\left(E_{\alpha,\beta}^{\gamma}(it^{\varsigma}) - E_{\alpha,\beta}^{\gamma}(-it^{\varsigma})\right) \\ &= \frac{1}{\Gamma(\gamma)} \times {}_1sin_1\left(\begin{matrix}(\gamma,1)\\(\beta,\alpha)\end{matrix}; t^{\varsigma}\right) = \sum_{k=0}^{\infty}\frac{(-1)^k(\gamma)_{2k+1}t^{(2k+1)\varsigma}}{\Gamma((2k+1)\alpha+\beta)},\end{aligned} \tag{5.502}$$

$$precos_{\alpha,\beta}^{\gamma}(t^{\varsigma}) = \frac{1}{2}\left(E_{\alpha,\beta}^{\gamma}(it^{\varsigma}) + E_{\alpha,\beta}^{\gamma}(-it^{\varsigma})\right) = \frac{1}{\Gamma(\gamma)} \times {}_1cos_1\left(\begin{matrix}(\gamma,1)\\(\beta,\alpha)\end{matrix}; t^{\varsigma}\right) = \sum_{k=0}^{\infty}\frac{(-1)^k(\gamma)_{2k}t^{2k\varsigma}}{\Gamma(2k\alpha+\beta)}, \tag{5.503}$$

$$presinh_{\alpha,\beta}^{\gamma}(t^{\varsigma}) = \frac{1}{2}\left(E_{\alpha,\beta}^{\gamma}(t^{\varsigma}) - E_{\alpha,\beta}^{\gamma}(-t^{\varsigma})\right) = \frac{1}{\Gamma(\gamma)} \times {}_1sinh_1\left(\begin{matrix}(\gamma,1)\\(\beta,\alpha)\end{matrix}; t^{\varsigma}\right) = \sum_{k=0}^{\infty}\frac{(\gamma)_{2k+1}t^{(2k+1)\varsigma}}{\Gamma((2k+1)\alpha+\beta)}, \tag{5.504}$$

$$precosh_{\alpha,\beta}^{\gamma}(t^{\varsigma}) = \frac{1}{2}\left(E_{\alpha,\beta}^{\gamma}(t^{\varsigma}) + E_{\alpha,\beta}^{\gamma}(-t^{\varsigma})\right) = \frac{1}{\Gamma(\gamma)} \times {}_1cosh_1\left(\begin{matrix}(\gamma,1)\\(\beta,\alpha)\end{matrix}; t^{\varsigma}\right) = \sum_{k=0}^{\infty}\frac{(\gamma)_{2k}t^{2k\varsigma}}{\Gamma(2k\alpha+\beta)}, \tag{5.505}$$

$$Tresin_{\alpha,\beta}^{\gamma}(t) = \frac{t^{\beta-1}}{2i}\left(E_{\alpha,\beta}^{\gamma}(it) - E_{\alpha,\beta}^{\gamma}(-it)\right) = \frac{t^{\beta-1}}{\Gamma(\gamma)} \times {}_1sin_1\left(\begin{matrix}(\gamma,1)\\(\beta,\alpha)\end{matrix}; t\right) = t^{\beta-1}\sum_{k=0}^{\infty}\frac{(-1)^k(\gamma)_{2k+1}t^{2k+1}}{\Gamma((2k+1)\alpha+\beta)}, \tag{5.506}$$

$$Trecos_{\alpha,\beta}^{\gamma}(t) = \frac{t^{\beta-1}}{2}\left(E_{\alpha,\beta}^{\gamma}(it) + E_{\alpha,\beta}^{\gamma}(-it)\right) = \frac{t^{\beta-1}}{\Gamma(\gamma)} \times {}_1cos_1\left(\begin{matrix}(\gamma,1)\\(\beta,\alpha)\end{matrix}; t\right) = t^{\beta-1}\sum_{k=0}^{\infty}\frac{(-1)^k(\gamma)_{2k}t^{2k}}{\Gamma(2k\alpha+\beta)}, \tag{5.507}$$

$$Tresinh_{\alpha,\beta}^{\gamma}(t) = \frac{t^{\beta-1}}{2}\left(E_{\alpha,\beta}^{\gamma}(t) - E_{\alpha,\beta}^{\gamma}(-t)\right) = \frac{t^{\beta-1}}{\Gamma(\gamma)} \times {}_1sinh_1\left(\begin{matrix}(\gamma,1)\\(\beta,\alpha)\end{matrix}; t\right) = t^{\beta-1}\sum_{k=0}^{\infty}\frac{(\gamma)_{2k+1}t^{2k+1}}{\Gamma((2k+1)\alpha+\beta)}, \tag{5.508}$$

$$Trecosh_{\alpha,\beta}^{\gamma}(t) = \frac{t^{\beta-1}}{2}\left(E_{\alpha,\beta}^{\gamma}(t) + E_{\alpha,\beta}^{\gamma}(-t)\right) = \frac{t^{\beta-1}}{\Gamma(\gamma)} \times {}_1cosh_1\left(\begin{matrix}(\gamma,1)\\(\beta,\alpha)\end{matrix}; t\right) = t^{\beta-1}\sum_{k=0}^{\infty}\frac{(\gamma)_{2k}t^{2k}}{\Gamma(2k\alpha+\beta)}, \tag{5.509}$$

$$Gresin_{\alpha,\beta}^{\gamma}(t^{\varsigma}) = \frac{t^{\beta-1}}{2i}\left(E_{\alpha,\beta}^{\gamma}(it^{\varsigma}) - E_{\alpha,\beta}^{\gamma}(-it^{\varsigma})\right) = \frac{t^{\beta-1}}{\Gamma(\gamma)} \times {}_1sin_1\left(\begin{matrix}(\gamma,1)\\(\beta,\alpha)\end{matrix}; t^{\varsigma}\right) = t^{\beta-1}\sum_{k=0}^{\infty}\frac{(-1)^k(\gamma)_{2k+1}t^{(2k+1)\varsigma}}{\Gamma((2k+1)\alpha+\beta)}, \tag{5.510}$$

$$Grecos_{\alpha,\beta}^{\gamma}(t^{\varsigma}) = \frac{t^{\beta-1}}{2}\left(E_{\alpha,\beta}^{\gamma}(it^{\varsigma}) + E_{\alpha,\beta}^{\gamma}(-it^{\varsigma})\right) = \frac{t^{\beta-1}}{\Gamma(\gamma)} \times {}_1cos_1\left(\begin{matrix}(\gamma,1)\\(\beta,\alpha)\end{matrix}; t^{\varsigma}\right) = t^{\beta-1}\sum_{k=0}^{\infty}\frac{(-1)^k(\gamma)_{2k}t^{2k\varsigma}}{\Gamma(2k\alpha+\beta)}, \tag{5.511}$$

$$\begin{aligned} Gresinh^{\gamma}_{\alpha,\beta}(t^{\varsigma}) &= \frac{t^{\beta-1}}{2}\left(E^{\gamma}_{\alpha,\beta}(t^{\varsigma}) - E^{\gamma}_{\alpha,\beta}(-t^{\varsigma})\right) \\ &= \frac{t^{\beta-1}}{\Gamma(\gamma)} \times {}_1sinh_1\left(\begin{matrix}(\gamma,1)\\(\beta,\alpha)\end{matrix}; t^{\varsigma}\right) = t^{\beta-1}\sum_{k=0}^{\infty}\frac{(\gamma)_{2k+1}t^{(2k+1)\varsigma}}{\Gamma((2k+1)\alpha+\beta)}, \end{aligned} \tag{5.512}$$

and

$$\begin{aligned} Grecosh^{\gamma}_{\alpha,\beta}(t^{\varsigma}) &= \frac{t^{\beta-1}}{2}\left(E^{\gamma}_{\alpha,\beta}(t^{\varsigma}) + E^{\gamma}_{\alpha,\beta}(-t^{\varsigma})\right) \\ &= \frac{t^{\beta-1}}{\Gamma(\gamma)} \times {}_1cosh_1\left(\begin{matrix}(\gamma,1)\\(\beta,\alpha)\end{matrix}; t^{\varsigma}\right) = t^{\beta-1}\sum_{k=0}^{\infty}\frac{(\gamma)_{2k}t^{2k\varsigma}}{\Gamma(2k\alpha+\beta)}, \end{aligned} \tag{5.513}$$

where $\alpha, \beta, \gamma, \varsigma, t \in \mathbb{C}$, $Re(\alpha) > 0$, and $Re(\beta) > 0$.

Let us recall the special cases for the Wright's generalized hypergeometric function:

Let $\alpha, \gamma \in \mathbb{C}$, $Re(\alpha) > 0$, $Re(\gamma) > 0$, $\lambda \in \mathbb{C}$, $\beta > 0$, $\alpha_1, \beta_1 \in \mathbb{C}$, and $A_1, B_1 \in \mathbb{R}_+$. Then we have

$$\begin{aligned} &\frac{1}{\Gamma(\alpha)}\int_0^t (t-\tau)^{\alpha}\tau^{\gamma-1}\,{}_1sin_1\left(\begin{matrix}(\alpha_1,A_1)\\(\beta_1,B_1)\end{matrix}; \lambda\tau^{\beta}\right)d\tau \\ &= t^{\gamma+\alpha-1}\,{}_2sin_2\left(\begin{matrix}(\alpha_1,A_1),(\gamma,\beta)\\(\beta_1,B_1),(\gamma+\alpha,\beta)\end{matrix}; \lambda t^{\beta}\right), \end{aligned} \tag{5.514}$$

where

$${}_1sin_1\left(\begin{matrix}(\alpha_1,A_1)\\(\beta_1,B_1)\end{matrix}; \lambda\tau^{\beta}\right) = \sum_{k=0}^{\infty}\frac{\Gamma(\alpha_1+A_1(2k+1))}{\Gamma(\beta_1+B_1(2k+1))}\frac{(-1)^k\lambda^{2k+1}\tau^{\beta(2k+1)}}{(2k+1)!}$$

and

$$\begin{aligned} &{}_2sin_2\left(\begin{matrix}(\alpha_1,A_1),(\gamma,\beta)\\(\beta_1,B_1),(\gamma+\alpha,\beta)\end{matrix}; \lambda\tau^{\beta}\right) \\ &= \sum_{k=0}^{\infty}\frac{\Gamma(\alpha_1+A_1(2k+1))\Gamma(\gamma+\beta(2k+1))}{\Gamma(\beta_1+B_1(2k+1))\Gamma(\gamma+\alpha+\beta(2k+1))}\frac{(-1)^k\lambda^{2k+1}\tau^{\beta(2k+1)}}{(2k+1)!}. \end{aligned}$$

Let $\alpha, \gamma \in \mathbb{C}$, $Re(\alpha) > 0$, $Re(\gamma) > 0$, $\lambda \in \mathbb{C}$, $\beta > 0$, $\alpha_1, \beta_1 \in \mathbb{C}$, and $A_1, B_1 \in \mathbb{R}_+$. Then we have

$$\begin{aligned} &\frac{1}{\Gamma(\alpha)}\int_0^t (t-\tau)^{\alpha}\tau^{\gamma-1}\,{}_1cos_1\left(\begin{matrix}(\alpha_1,A_1)\\(\beta_1,B_1)\end{matrix}; \lambda\tau^{\beta}\right)d\tau \\ &= t^{\gamma+\alpha-1}\,{}_2cos_2\left(\begin{matrix}(\alpha_1,A_1),(\gamma,\beta)\\(\beta_1,B_1),(\gamma+\alpha,\beta)\end{matrix}; \lambda t^{\beta}\right), \end{aligned} \tag{5.515}$$

where

$$ {}_1cos_1\left(\begin{matrix}(\alpha_1, A_1)\\(\beta_1, B_1)\end{matrix}; \lambda\tau^{\beta}\right) = \sum_{k=0}^{\infty} \frac{\Gamma\left(\alpha_1 + 2kA_1\right)}{\Gamma\left(\beta_1 + 2kB_1\right)} \frac{(-1)^k \lambda^{2k}\tau^{2k\beta}}{(2k)!} $$

and

$$ \begin{aligned} &{}_2cos_2\left(\begin{matrix}(\alpha_1, A_1), (\gamma, \beta)\\(\beta_1, B_1), (\gamma+\alpha, \beta)\end{matrix}; \lambda\tau^{\beta}\right) \\ &= \sum_{k=0}^{\infty} \frac{\Gamma(\alpha_1+2kA_1)\Gamma(\gamma+2k\beta)}{\Gamma(\beta_1+2kB_1)\Gamma(\gamma+\alpha+2k\beta)} \frac{(-1)^k\lambda^{2k}\tau^{2k\beta}}{(2k)!}. \end{aligned} $$

Let $\alpha, \gamma \in \mathbb{C}$, $Re(\alpha) > 0$, $Re(\gamma) > 0$, $\lambda \in \mathbb{C}$, $\beta > 0$, $\alpha_1, \beta_1 \in \mathbb{C}$, and $A_1, B_1 \in \mathbb{R}_+$. Then we have

$$ \begin{aligned} &\frac{1}{\Gamma(\alpha)}\int_0^t (t-\tau)^{\alpha}\tau^{\gamma-1}\, {}_1sinh_1\left(\begin{matrix}(\alpha_1, A_1)\\(\beta_1, B_1)\end{matrix}; \lambda\tau^{\beta}\right) d\tau \\ &= t^{\gamma+\alpha-1}\, {}_2sinh_2\left(\begin{matrix}(\alpha_1, A_1), (\gamma, \beta)\\(\beta_1, B_1), (\gamma+\alpha, \beta)\end{matrix}; \lambda t^{\beta}\right), \end{aligned} \tag{5.516} $$

where

$$ {}_1sinh_1\left(\begin{matrix}(\alpha_1, A_1)\\(\beta_1, B_1)\end{matrix}; \lambda\tau^{\beta}\right) = \sum_{k=0}^{\infty} \frac{\Gamma\left(\alpha_1 + A_1\left(2k+1\right)\right)}{\Gamma\left(\beta_1 + B_1\left(2k+1\right)\right)} \frac{\lambda^{2k+1}\tau^{\beta(2k+1)}}{(2k+1)!} $$

and

$$ \begin{aligned} &{}_2sinh_2\left(\begin{matrix}(\alpha_1, A_1), (\gamma, \beta)\\(\beta_1, B_1), (\gamma+\alpha, \beta)\end{matrix}; \lambda\tau^{\beta}\right) \\ &= \sum_{k=0}^{\infty} \frac{\Gamma(\alpha_1+A_1(2k+1))\Gamma(\gamma+\beta(2k+1))}{\Gamma(\beta_1+B_1(2k+1))\Gamma(\gamma+\alpha+\beta(2k+1))} \frac{\lambda^{2k+1}\tau^{\beta(2k+1)}}{(2k+1)!}. \end{aligned} $$

Let $\alpha, \gamma \in \mathbb{C}$, $Re(\alpha) > 0$, $Re(\gamma) > 0$, $\lambda \in \mathbb{C}$, $\beta > 0$, $\alpha_1, \beta_1 \in \mathbb{C}$, and $A_1, B_1 \in \mathbb{R}_+$. Then we have

$$ \begin{aligned} &\frac{1}{\Gamma(\alpha)}\int_0^t (t-\tau)^{\alpha}\tau^{\gamma-1}\, {}_1cosh_1\left(\begin{matrix}(\alpha_1, A_1)\\(\beta_1, B_1)\end{matrix}; \lambda\tau^{\beta}\right) d\tau \\ &= t^{\gamma+\alpha-1}\, {}_2cosh_2\left(\begin{matrix}(\alpha_1, A_1), (\gamma, \beta)\\(\beta_1, B_1), (\gamma+\alpha, \beta)\end{matrix}; \lambda t^{\beta}\right), \end{aligned} \tag{5.517} $$

where

$$ {}_1cosh_1\left(\begin{matrix}(\alpha_1, A_1)\\(\beta_1, B_1)\end{matrix}; \lambda\tau^{\beta}\right) = \sum_{k=0}^{\infty} \frac{\Gamma(\alpha_1 + 2kA_1)}{\Gamma(\beta_1 + 2kB_1)} \frac{\lambda^{2k}\tau^{2k\beta}}{(2k)!} $$

and

$$ {}_2cosh_2\left(\begin{matrix}(\alpha_1, A_1), (\gamma, \beta)\\(\beta_1, B_1), (\gamma+\alpha, \beta)\end{matrix}; \lambda\tau^{\beta}\right) = \sum_{k=0}^{\infty} \frac{\Gamma(\alpha_1+2kA_1)\Gamma(\gamma+2k\beta)}{\Gamma(\beta_1+2kB_1)\Gamma(\gamma+\alpha+2k\beta)} \frac{\lambda^{2k}\tau^{2k\beta}}{(2k)!}. $$

Thus, we have the following results.

Theorem 5.48 *Let* $\sigma, \varsigma \in \mathbb{C}$, $\lambda \in \mathbb{C}$, $\vartheta > 0$, $\alpha_1, \beta_1 \in \mathbb{C}$, $A_1, B_1 \in \mathbb{R}_+$, *and* $Re(\varsigma) > 0$. *Then we have*

$$ \begin{aligned} &\frac{1}{\Gamma(\sigma)} \int_0^t (t-\tau)^{\sigma} \tau^{\varsigma-1} {}_1sin_1\left(\begin{matrix}(\alpha_1, A_1)\\(\beta_1, B_1)\end{matrix}; \lambda\tau^{\vartheta}\right) d\tau \\ &= t^{\varsigma+\sigma-1} {}_2sin_2\left(\begin{matrix}(\alpha_1, A_1), (\varsigma, \vartheta)\\(\beta_1, B_1), (\varsigma+\sigma, \vartheta)\end{matrix}; \lambda t^{\vartheta}\right). \end{aligned} \tag{5.518} $$

Proof Making use of (5.514), we have the result. □

Theorem 5.49 *Let* $\sigma, \varsigma \in \mathbb{C}$, $\lambda \in \mathbb{C}$, $\vartheta > 0$, $\alpha_1, \beta_1 \in \mathbb{C}$, $A_1, B_1 \in \mathbb{R}_+$, *and* $Re(\varsigma) > 0$. *Then we have*

$$ \begin{aligned} &\frac{1}{\Gamma(\sigma)} \int_0^t (t-\tau)^{\sigma} \tau^{\varsigma-1} {}_1sin_1\left(\begin{matrix}(\alpha_1, A_1)\\(\beta_1, B_1)\end{matrix}; \lambda\tau^{\vartheta}\right) d\tau \\ &= t^{\varsigma+\sigma-1} {}_2sin_2\left(\begin{matrix}(\alpha_1, A_1), (\varsigma, \vartheta)\\(\beta_1, B_1), (\varsigma+\sigma, \vartheta)\end{matrix}; \lambda t^{\vartheta}\right). \end{aligned} \tag{5.519} $$

Proof In view of (5.515), we obtain the result. □

Theorem 5.50 *Let* $\sigma, \varsigma \in \mathbb{C}$, $\lambda \in \mathbb{C}$, $\vartheta > 0$, $\alpha_1, \beta_1 \in \mathbb{C}$, $A_1, B_1 \in \mathbb{R}_+$, *and* $Re(\varsigma) > 0$. *Then we have*

$$ \begin{aligned} &\frac{1}{\Gamma(\alpha)} \int_0^t (t-\tau)^{\alpha} \tau^{\gamma-1} {}_1sinh_1\left(\begin{matrix}(\alpha_1, A_1)\\(\beta_1, B_1)\end{matrix}; \lambda\tau^{\beta}\right) d\tau \\ &= t^{\gamma+\alpha-1} {}_2sinh_2\left(\begin{matrix}(\alpha_1, A_1), (\gamma, \beta)\\(\beta_1, B_1), (\gamma+\alpha, \beta)\end{matrix}; \lambda t^{\beta}\right). \end{aligned} \tag{5.520} $$

Proof With the aid of (5.516), we get the result. □

Theorem 5.51 *Let* $\sigma, \varsigma \in \mathbb{C}$, $\lambda \in \mathbb{C}$, $\vartheta > 0$, $\alpha_1, \beta_1 \in \mathbb{C}$, $A_1, B_1 \in \mathbb{R}_+$, *and* $Re(\varsigma) > 0$. *Then we have*

$$\begin{aligned} &\frac{1}{\Gamma(\alpha)} \int_0^t (t-\tau)^\alpha \tau^{\gamma-1} {}_1cosh_1 \left(\begin{matrix} (\alpha_1, A_1) \\ (\beta_1, B_1) \end{matrix} ; \lambda\tau^\beta \right) d\tau \\ &= t^{\gamma+\alpha-1} {}_2cosh_2 \left(\begin{matrix} (\alpha_1, A_1), (\gamma, \beta) \\ (\beta_1, B_1), (\gamma+\alpha, \beta) \end{matrix} ; \lambda t^\beta \right). \end{aligned} \tag{5.521}$$

Proof With the use of (5.517), we present the result. □

Theorem 5.52 *Let* $\sigma, \varsigma \in \mathbb{C}$, $\lambda \in \mathbb{C}$, $\vartheta > 0$, $\beta \in \mathbb{C}$, $\alpha \in \mathbb{R}_+$, $Re(\sigma) > 0$, *and* $Re(\varsigma) > 0$. *Then we have*

$$\begin{aligned} &\frac{\Gamma(\gamma)}{\Gamma(\sigma)} \int_0^t (t-\tau)^\sigma \tau^{\varsigma-1} presin_{\alpha,\beta}^\gamma \left(\lambda\tau^\vartheta\right) d\tau \\ &= t^{\varsigma+\sigma-1} {}_2sin_2 \left(\begin{matrix} (\gamma, 1), (\varsigma, \vartheta) \\ (\beta, \alpha), (\varsigma+\sigma, \vartheta) \end{matrix} ; \lambda t^\vartheta \right). \end{aligned} \tag{5.522}$$

Proof With the use of (5.518) and (5.499), we have the result. □

Theorem 5.53 *Let* $\sigma, \varsigma \in \mathbb{C}$, $\lambda \in \mathbb{C}$, $\vartheta > 0$, $\beta \in \mathbb{C}$, $\alpha \in \mathbb{R}_+$, $Re(\sigma) > 0$, *and* $Re(\varsigma) > 0$. *Then we have*

$$\begin{aligned} &\frac{\Gamma(\gamma)}{\Gamma(\sigma)} \int_0^t (t-\tau)^\sigma \tau^{\varsigma-1} precos_{\alpha,\beta}^\gamma \left(\lambda\tau^\vartheta\right) d\tau \\ &= t^{\varsigma+\sigma-1} {}_2cos_2 \left(\begin{matrix} (\gamma, 1), (\varsigma, \vartheta) \\ (\beta, \alpha), (\varsigma+\sigma, \vartheta) \end{matrix} ; \lambda t^\vartheta \right). \end{aligned} \tag{5.523}$$

Proof With the aid of (5.519) and (5.500), we obtain the result. □

Theorem 5.54 *Let* $\sigma, \varsigma \in \mathbb{C}$, $\lambda \in \mathbb{C}$, $\vartheta > 0$, $\beta \in \mathbb{C}$, $\alpha \in \mathbb{R}_+$, $Re(\sigma) > 0$, *and* $Re(\varsigma) > 0$. *Then we have*

$$\begin{aligned} &\frac{\Gamma(\gamma)}{\Gamma(\sigma)} \int_0^t (t-\tau)^\sigma \tau^{\varsigma-1} presinh_{\alpha,\beta}^\gamma \left(\lambda\tau^\vartheta\right) d\tau \\ &= t^{\varsigma+\sigma-1} {}_2sinh_2 \left(\begin{matrix} (\gamma, 1), (\varsigma, \vartheta) \\ (\beta, \alpha), (\varsigma+\sigma, \vartheta) \end{matrix} ; \lambda t^\vartheta \right). \end{aligned} \tag{5.524}$$

Proof Making the use of (5.520) and (5.501), we get the result. □

Theorem 5.55 *Let $\sigma, \varsigma \in \mathbb{C}$, $\lambda \in \mathbb{C}$, $\vartheta > 0$, $\beta \in \mathbb{C}$, $\alpha \in \mathbb{R}_+$, $Re(\sigma) > 0$, and $Re(\varsigma) > 0$. Then we have*

$$\begin{aligned}&\frac{\Gamma(\gamma)}{\Gamma(\sigma)}\int_0^t (t-\tau)^{\sigma}\,\tau^{\varsigma-1}\,precosh_{\alpha,\beta}^{\gamma}\left(\lambda\tau^{\vartheta}\right)d\tau\\&= t^{\varsigma+\sigma-1}\,{}_2cosh_2\left(\begin{matrix}(\gamma,1),(\varsigma,\vartheta)\\(\beta,\alpha),(\varsigma+\sigma,\vartheta)\end{matrix};\lambda t^{\vartheta}\right).\end{aligned} \tag{5.525}$$

Proof With the aid of (5.522) and (5.502), we have the results. □

Theorem 5.56 *Let $\sigma, \varsigma \in \mathbb{C}$, $\lambda \in \mathbb{C}$, $\vartheta > 0$, $\beta \in \mathbb{C}$, $\alpha \in \mathbb{R}_+$, $Re(\sigma) > 0$, and $Re(\varsigma) > 0$. Then we have*

$$\frac{\Gamma(\gamma)}{\Gamma(\sigma)}\int_0^1 (1-\tau)^{\sigma}\,\tau^{\varsigma-1}\,presin_{\alpha,\beta}^{\gamma}\left(\lambda\tau^{\vartheta}\right)d\tau = {}_2sin_2\left(\begin{matrix}(\gamma,1),(\varsigma,\vartheta)\\(\beta,\alpha),(\varsigma+\sigma,\vartheta)\end{matrix};\lambda\right). \tag{5.526}$$

Proof Putting $t = 1$ in (5.523), we have the result. □

Theorem 5.57 *Let $\sigma, \varsigma \in \mathbb{C}$, $\lambda \in \mathbb{C}$, $\vartheta > 0$, $\beta \in \mathbb{C}$, $\alpha \in \mathbb{R}_+$, $Re(\sigma) > 0$, and $Re(\varsigma) > 0$. Then we have*

$$\frac{\Gamma(\gamma)}{\Gamma(\sigma)}\int_0^1 (1-\tau)^{\sigma}\,\tau^{\varsigma-1}\,precos_{\alpha,\beta}^{\gamma}\left(\lambda\tau^{\vartheta}\right)d\tau = {}_2cos_2\left(\begin{matrix}(\gamma,1),(\varsigma,\vartheta)\\(\beta,\alpha),(\varsigma+\sigma,\vartheta)\end{matrix};\lambda\right). \tag{5.527}$$

Proof Putting $t = 1$ in (5.525), we get the result. □

Theorem 5.58 *Let $\sigma, \varsigma \in \mathbb{C}$, $\lambda \in \mathbb{C}$, $\vartheta > 0$, $\beta \in \mathbb{C}$, $\alpha \in \mathbb{R}_+$, $Re(\sigma) > 0$, and $Re(\varsigma) > 0$. Then we have*

$$\frac{\Gamma(\gamma)}{\Gamma(\sigma)}\int_0^1 (1-\tau)^{\sigma}\,\tau^{\varsigma-1}\,presinh_{\alpha,\beta}^{\gamma}\left(\lambda\tau^{\vartheta}\right)d\tau = {}_2sinh_2\left(\begin{matrix}(\gamma,1),(\varsigma,\vartheta)\\(\beta,\alpha),(\varsigma+\sigma,\vartheta)\end{matrix};\lambda\right). \tag{5.528}$$

Proof Putting $t = 1$ in (5.526), we obtain the result. □

Theorem 5.59 *Let $\sigma, \varsigma \in \mathbb{C}$, $\lambda \in \mathbb{C}$, $\vartheta > 0$, $\beta \in \mathbb{C}$, $\alpha \in \mathbb{R}_+$, $Re(\sigma) > 0$, and $Re(\varsigma) > 0$. Then we have*

$$\frac{\Gamma(\gamma)}{\Gamma(\sigma)}\int_0^1 (1-\tau)^{\sigma}\,\tau^{\varsigma-1}\,precosh_{\alpha,\beta}^{\gamma}\left(\lambda\tau^{\vartheta}\right)d\tau = {}_2cosh_2\left(\begin{matrix}(\gamma,1),(\varsigma,\vartheta)\\(\beta,\alpha),(\varsigma+\sigma,\vartheta)\end{matrix};\lambda\right). \tag{5.529}$$

Proof Putting $t = 1$ in (5.527), we get the result. □

Theorem 5.60 *Let* $\sigma, \varsigma \in \mathbb{C}$, $\lambda \in \mathbb{C}$, $\vartheta > 0$, $\beta \in \mathbb{C}$, $\alpha \in \mathbb{R}_+$, $Re(\sigma) > 0$, *and* $Re(\varsigma) > 0$. *Then we have*

$$\begin{aligned}
&\frac{\Gamma(\gamma)}{\Gamma(\sigma)} \int_0^2 (2-\tau)^{\sigma} \tau^{\varsigma-1} presin^{\gamma}_{\alpha,\beta}\left(\lambda\tau^{\vartheta}\right) d\tau \\
&= \frac{\Gamma(\gamma)}{\Gamma(\sigma)} \int_{-1}^{1} (1-\tau)^{\sigma} (1+\tau)^{\varsigma-1} presin^{\gamma}_{\alpha,\beta}\left(\lambda (1+\tau)^{\vartheta}\right) d\tau \qquad (5.530) \\
&= 2^{\varsigma+\sigma-1} {}_2sin_2\left(\begin{matrix} (\gamma,1),(\varsigma,\vartheta) \\ (\beta,\alpha),(\varsigma+\sigma,\vartheta) \end{matrix} ; \lambda 2^{\vartheta} \right).
\end{aligned}$$

Proof Setting $t = 2$ in (5.523), we have the result. □

Theorem 5.61 *Let* $\sigma, \varsigma \in \mathbb{C}$, $\lambda \in \mathbb{C}$, $\vartheta > 0$, $\beta \in \mathbb{C}$, $\alpha \in \mathbb{R}_+$, $Re(\sigma) > 0$, *and* $Re(\varsigma) > 0$. *Then we have*

$$\begin{aligned}
&\frac{\Gamma(\gamma)}{\Gamma(\sigma)} \int_0^2 (2-\tau)^{\sigma} \tau^{\varsigma-1} precos^{\gamma}_{\alpha,\beta}\left(\lambda\tau^{\vartheta}\right) d\tau \\
&= \frac{\Gamma(\gamma)}{\Gamma(\sigma)} \int_{-1}^{1} (1-\tau)^{\sigma} (1+\tau)^{\varsigma-1} precos^{\gamma}_{\alpha,\beta}\left(\lambda (1+\tau)^{\vartheta}\right) d\tau \qquad (5.531) \\
&= 2^{\varsigma+\sigma-1} {}_2cos_2\left(\begin{matrix} (\gamma,1),(\varsigma,\vartheta) \\ (\beta,\alpha),(\varsigma+\sigma,\vartheta) \end{matrix} ; \lambda 2^{\vartheta} \right).
\end{aligned}$$

Proof Setting $t = 2$ in (5.524), we obtain the result. □

Theorem 5.62 *Let* $\sigma, \varsigma \in \mathbb{C}$, $\lambda \in \mathbb{C}$, $\vartheta > 0$, $\beta \in \mathbb{C}$, $\alpha \in \mathbb{R}_+$, $Re(\sigma) > 0$, *and* $Re(\varsigma) > 0$. *Then we have*

$$\begin{aligned}
&\frac{\Gamma(\gamma)}{\Gamma(\sigma)} \int_0^2 (2-\tau)^{\sigma} \tau^{\varsigma-1} presinh^{\gamma}_{\alpha,\beta}\left(\lambda\tau^{\vartheta}\right) d\tau \\
&= \frac{\Gamma(\gamma)}{\Gamma(\sigma)} \int_{-1}^{1} (1-\tau)^{\sigma} (1+\tau)^{\varsigma-1} presinh^{\gamma}_{\alpha,\beta}\left(\lambda (1+\tau)^{\vartheta}\right) d\tau \qquad (5.532) \\
&= 2^{\varsigma+\sigma-1} {}_2sinh_2\left(\begin{matrix} (\gamma,1),(\varsigma,\vartheta) \\ (\beta,\alpha),(\varsigma+\sigma,\vartheta) \end{matrix} ; \lambda 2^{\vartheta} \right).
\end{aligned}$$

Proof Setting $t = 2$ in (5.525), we get the result. □

Theorem 5.63 *Let $\sigma, \varsigma \in \mathbb{C}$, $\lambda \in \mathbb{C}$, $\vartheta > 0$, $\beta \in \mathbb{C}$, $\alpha \in \mathbb{R}_+$, $Re(\sigma) > 0$, and $Re(\varsigma) > 0$. Then we have*

$$\frac{\Gamma(\gamma)}{\Gamma(\sigma)} \int_0^2 (2-\tau)^\sigma \tau^{\varsigma-1} precosh_{\alpha,\beta}^{\gamma}\left(\lambda \tau^\vartheta\right) d\tau$$
$$= \frac{\Gamma(\gamma)}{\Gamma(\sigma)} \int_{-1}^1 (1-\tau)^\sigma (1+\tau)^{\varsigma-1} precosh_{\alpha,\beta}^{\gamma}\left(\lambda (1+\tau)^\vartheta\right) d\tau \tag{5.533}$$
$$= 2^{\varsigma+\sigma-1} {}_2cosh_2 \left(\begin{matrix} (\gamma,1),(\varsigma,\vartheta) \\ (\beta,\alpha),(\varsigma+\sigma,\vartheta) \end{matrix} ; \lambda 2^\vartheta \right).$$

Proof Setting $t = 2$ in (5.526), we obtain the result. □

Theorem 5.64 *Let $\sigma, \varsigma \in \mathbb{C}$, $\lambda \in \mathbb{C}$, $\beta \in \mathbb{C}$, $\alpha \in \mathbb{R}_+$, $Re(\sigma) > 0$, and $Re(\varsigma) > 0$. Then we have*

$$\frac{\Gamma(\gamma)}{\Gamma(\sigma)} \int_0^t (t-\tau)^\sigma \tau^{\varsigma-1} presin_{\alpha,\beta}^{\gamma}(\lambda \tau) d\tau$$
$$= t^{\varsigma+\sigma-1} {}_2sin_2 \left(\begin{matrix} (\gamma,1),(\varsigma,1) \\ (\beta,\alpha),(\varsigma+\sigma,1) \end{matrix} ; \lambda t \right). \tag{5.534}$$

Proof Taking $\vartheta = 1$ in (5.522), we have the result. □

Theorem 5.65 *Let $\sigma, \varsigma \in \mathbb{C}$, $\lambda \in \mathbb{C}$, $\beta \in \mathbb{C}$, $\alpha \in \mathbb{R}_+$, $Re(\sigma) > 0$, and $Re(\varsigma) > 0$. Then we have*

$$\frac{\Gamma(\gamma)}{\Gamma(\sigma)} \int_0^t (t-\tau)^\sigma \tau^{\varsigma-1} precos_{\alpha,\beta}^{\gamma}(\lambda \tau) d\tau$$
$$= t^{\varsigma+\sigma-1} {}_2cos_2 \left(\begin{matrix} (\gamma,1),(\varsigma,1) \\ (\beta,\alpha),(\varsigma+\sigma,1) \end{matrix} ; \lambda t \right). \tag{5.535}$$

Proof Taking $\vartheta = 1$ in (5.523), we obtain the result. □

Theorem 5.66 *Let $\sigma, \varsigma \in \mathbb{C}$, $\lambda \in \mathbb{C}$, $\beta \in \mathbb{C}$, $\alpha \in \mathbb{R}_+$, $Re(\sigma) > 0$, and $Re(\varsigma) > 0$. Then we have*

$$\frac{\Gamma(\gamma)}{\Gamma(\sigma)} \int_0^t (t-\tau)^\sigma \tau^{\varsigma-1} presinh_{\alpha,\beta}^{\gamma}(\lambda \tau) d\tau$$
$$= t^{\varsigma+\sigma-1} {}_2sinh_2 \left(\begin{matrix} (\gamma,1),(\varsigma,1) \\ (\beta,\alpha),(\varsigma+\sigma,1) \end{matrix} ; \lambda t \right). \tag{5.536}$$

Proof Taking $\vartheta = 1$ in (5.524), we get the result. □

Theorem 5.67 *Let $\sigma, \varsigma \in \mathbb{C}$, $\lambda \in \mathbb{C}$, $\beta \in \mathbb{C}$, $\alpha \in \mathbb{R}_+$, $Re(\sigma) > 0$, and $Re(\varsigma) > 0$. Then we have*

$$\begin{aligned}&\frac{\Gamma(\gamma)}{\Gamma(\sigma)} \int_0^t (t-\tau)^{\sigma} \tau^{\varsigma-1} precosh_{\alpha,\beta}^{\gamma}(\lambda\tau)\, d\tau \\ &= t^{\varsigma+\sigma-1} {}_2cosh_2 \left(\begin{matrix} (\gamma, 1), (\varsigma, 1) \\ (\beta, \alpha), (\varsigma+\sigma, 1) \end{matrix} ; \lambda t \right).\end{aligned} \tag{5.537}$$

Proof Taking $\vartheta = 1$ in (5.525), we have the result. □

Theorem 5.68 *Let $\sigma, \varsigma \in \mathbb{C}$, $\lambda \in \mathbb{C}$, $\beta \in \mathbb{C}$, $\alpha \in \mathbb{R}_+$, $Re(\sigma) > 0$, and $Re(\varsigma) > 0$. Then we have*

$$\begin{aligned}&\frac{\Gamma(\gamma)}{\Gamma(\sigma)} \int_0^1 (1-\tau)^{\sigma} \tau^{\varsigma-1} presin_{\alpha,\beta}^{\gamma}(\lambda\tau)\, d\tau \\ &= {}_2sin_2 \left(\begin{matrix} (\gamma, 1), (\varsigma, 1) \\ (\beta, \alpha), (\varsigma+\sigma, 1) \end{matrix} ; \lambda \right).\end{aligned} \tag{5.538}$$

Proof Putting $\vartheta = 1$ in (5.526), we have the result. □

Theorem 5.69 *Let $\sigma, \varsigma \in \mathbb{C}$, $\lambda \in \mathbb{C}$, $\beta \in \mathbb{C}$, $\alpha \in \mathbb{R}_+$, $Re(\sigma) > 0$, and $Re(\varsigma) > 0$. Then we have*

$$\frac{\Gamma(\gamma)}{\Gamma(\sigma)} \int_0^1 (1-\tau)^{\sigma} \tau^{\varsigma-1} precos_{\alpha,\beta}^{\gamma}(\lambda\tau)\, d\tau = {}_2cos_2 \left(\begin{matrix} (\gamma, 1), (\varsigma, 1) \\ (\beta, \alpha), (\varsigma+\sigma, 1) \end{matrix} ; \lambda \right). \tag{5.539}$$

Proof Putting $\vartheta = 1$ in (5.527), we get the result. □

Theorem 5.70 *Let $\sigma, \varsigma \in \mathbb{C}$, $\lambda \in \mathbb{C}$, $\beta \in \mathbb{C}$, $\alpha \in \mathbb{R}_+$, $Re(\sigma) > 0$, and $Re(\varsigma) > 0$. Then we have*

$$\frac{\Gamma(\gamma)}{\Gamma(\sigma)} \int_0^1 (1-\tau)^{\sigma} \tau^{\varsigma-1} presinh_{\alpha,\beta}^{\gamma}(\lambda\tau)\, d\tau = {}_2sinh_2 \left(\begin{matrix} (\gamma, 1), (\varsigma, 1) \\ (\beta, \alpha), (\varsigma+\sigma, 1) \end{matrix} ; \lambda \right). \tag{5.540}$$

Proof Putting $\vartheta = 1$ in (5.528), we obtain the result. □

Theorem 5.71 *Let $\sigma, \varsigma \in \mathbb{C}$, $\lambda \in \mathbb{C}$, $\beta \in \mathbb{C}$, $\alpha \in \mathbb{R}_+$, $Re(\sigma) > 0$, and $Re(\varsigma) > 0$. Then we have*

$$\frac{\Gamma(\gamma)}{\Gamma(\sigma)} \int_0^1 (1-\tau)^{\sigma} \tau^{\varsigma-1} precosh_{\alpha,\beta}^{\gamma}(\lambda\tau)\, d\tau = {}_2cosh_2 \left(\begin{matrix} (\gamma, 1), (\varsigma, 1) \\ (\beta, \alpha), (\varsigma+\sigma, 1) \end{matrix} ; \lambda \right). \tag{5.541}$$

Proof Putting $\vartheta = 1$ in (5.529), we get the result. □

Theorem 5.72 *Let* $\sigma, \varsigma \in \mathbb{C}$, $\lambda \in \mathbb{C}$, $\beta \in \mathbb{C}$, $\alpha \in \mathbb{R}_+$, $Re(\sigma) > 0$, *and* $Re(\varsigma) > 0$. *Then we have*

$$\begin{aligned}
&\frac{\Gamma(\gamma)}{\Gamma(\sigma)} \int_0^2 (2-\tau)^\sigma \tau^{\varsigma-1} presin_{\alpha,\beta}^{\gamma}(\lambda\tau)\, d\tau \\
&= \frac{\Gamma(\gamma)}{\Gamma(\sigma)} \int_{-1}^{1} (1-\tau)^\sigma (1+\tau)^{\varsigma-1}\, presin_{\alpha,\beta}^{\gamma}(\lambda(1+\tau))\, d\tau \qquad (5.542)\\
&= 2^{\varsigma+\sigma-1}\, {}_2sin_2\left(\begin{matrix} (\gamma,1),(\varsigma,1) \\ (\beta,\alpha),(\varsigma+\sigma,1) \end{matrix} ; 2\lambda \right).
\end{aligned}$$

Proof Setting $\vartheta = 1$ in (5.530), we have the result. □

Theorem 5.73 *Let* $\sigma, \varsigma \in \mathbb{C}$, $\lambda \in \mathbb{C}$, $\beta \in \mathbb{C}$, $\alpha \in \mathbb{R}_+$, $Re(\sigma) > 0$, *and* $Re(\varsigma) > 0$. *Then we have*

$$\begin{aligned}
&\frac{\Gamma(\gamma)}{\Gamma(\sigma)} \int_0^2 (2-\tau)^\sigma \tau^{\varsigma-1} precos_{\alpha,\beta}^{\gamma}(\lambda\tau)\, d\tau \\
&= \frac{\Gamma(\gamma)}{\Gamma(\sigma)} \int_{-1}^{1} (1-\tau)^\sigma (1+\tau)^{\varsigma-1}\, precos_{\alpha,\beta}^{\gamma}(\lambda(1+\tau))\, d\tau \qquad (5.543)\\
&= 2^{\varsigma+\sigma-1}\, {}_2cos_2\left(\begin{matrix} (\gamma,1),(\varsigma,1) \\ (\beta,\alpha),(\varsigma+\sigma,1) \end{matrix} ; 2\lambda \right).
\end{aligned}$$

Proof Setting $\vartheta = 1$ in (5.531), we obtain the result. □

Theorem 5.74 *Let* $\sigma, \varsigma \in \mathbb{C}$, $\lambda \in \mathbb{C}$, $\beta \in \mathbb{C}$, $\alpha \in \mathbb{R}_+$, $Re(\sigma) > 0$, *and* $Re(\varsigma) > 0$. *Then we have*

$$\begin{aligned}
&\frac{\Gamma(\gamma)}{\Gamma(\sigma)} \int_0^2 (2-\tau)^\sigma \tau^{\varsigma-1} presinh_{\alpha,\beta}^{\gamma}(\lambda\tau)\, d\tau \\
&= \frac{\Gamma(\gamma)}{\Gamma(\sigma)} \int_{-1}^{1} (1-\tau)^\sigma (1+\tau)^{\varsigma-1}\, presinh_{\alpha,\beta}^{\gamma}(\lambda(1+\tau))\, d\tau \qquad (5.544)\\
&= 2^{\varsigma+\sigma-1}\, {}_2sinh_2\left(\begin{matrix} (\gamma,1),(\varsigma,1) \\ (\beta,\alpha),(\varsigma+\sigma,1) \end{matrix} ; 2\lambda \right).
\end{aligned}$$

Proof Setting $\vartheta = 1$ in (5.532), we get the result. □

Theorem 5.75 *Let $\sigma, \varsigma \in \mathbb{C}$, $\lambda \in \mathbb{C}$, $\beta \in \mathbb{C}$, $\alpha \in \mathbb{R}_+$, $Re(\sigma) > 0$, and $Re(\varsigma) > 0$. Then we have*

$$\begin{aligned}
&\frac{\Gamma(\gamma)}{\Gamma(\sigma)} \int_0^2 (2-\tau)^{\sigma} \tau^{\varsigma-1} precosh^{\gamma}_{\alpha,\beta}(\lambda\tau)\, d\tau \\
&= \frac{\Gamma(\gamma)}{\Gamma(\sigma)} \int_{-1}^1 (1-\tau)^{\sigma} (1+\tau)^{\varsigma-1} precosh^{\gamma}_{\alpha,\beta}(\lambda(1+\tau))\, d\tau \\
&= 2^{\varsigma+\sigma-1} {}_2cosh_2 \left(\begin{matrix} (\gamma, 1), (\varsigma, 1) \\ (\beta, \alpha), (\varsigma+\sigma, 1) \end{matrix} ; 2\lambda \right).
\end{aligned} \tag{5.545}$$

Proof Setting $\vartheta = 1$ in (5.533), we obtain the result. □

Theorem 5.76 *Let $\sigma \in \mathbb{C}$, $\lambda \in \mathbb{C}$, $\vartheta > 0$, $\alpha \in \mathbb{R}_+$, $Re(\sigma) > 0$, and $Re(\beta) > 0$. Then we have*

$$\begin{aligned}
&\frac{\Gamma(\gamma)}{\Gamma(\sigma)} \int_0^t (t-\tau)^{\sigma} Gresin^{\gamma}_{\alpha,\beta}\left(\lambda\tau^{\vartheta}\right) d\tau \\
&= t^{\beta+\sigma-1} {}_2sin_2 \left(\begin{matrix} (\gamma, 1), (\beta, \vartheta) \\ (\beta, \alpha), (\beta+\sigma, \vartheta) \end{matrix} ; \lambda t^{\vartheta} \right).
\end{aligned} \tag{5.546}$$

Proof Setting $\varsigma = \beta$ in (5.522), we have the result. □

Theorem 5.77 *Let $\sigma \in \mathbb{C}$, $\lambda \in \mathbb{C}$, $\vartheta > 0$, $\alpha \in \mathbb{R}_+$, $Re(\sigma) > 0$, and $Re(\beta) > 0$. Then we have*

$$\begin{aligned}
&\frac{\Gamma(\gamma)}{\Gamma(\sigma)} \int_0^t (t-\tau)^{\sigma} Grecos^{\gamma}_{\alpha,\beta}\left(\lambda\tau^{\vartheta}\right) d\tau \\
&= t^{\beta+\sigma-1} {}_2cos_2 \left(\begin{matrix} (\gamma, 1), (\beta, \vartheta) \\ (\beta, \alpha), (\beta+\sigma, \vartheta) \end{matrix} ; \lambda t^{\vartheta} \right).
\end{aligned} \tag{5.547}$$

Proof Taking $\varsigma = \beta$ in (5.523), we obtain the result. □

Theorem 5.78 *Let $\sigma \in \mathbb{C}$, $\lambda \in \mathbb{C}$, $\vartheta > 0$, $\alpha \in \mathbb{R}_+$, $Re(\sigma) > 0$, and $Re(\beta) > 0$. Then we have*

$$\begin{aligned}
&\frac{\Gamma(\gamma)}{\Gamma(\sigma)} \int_0^t (t-\tau)^{\sigma} Gresinh^{\gamma}_{\alpha,\beta}\left(\lambda\tau^{\vartheta}\right) d\tau \\
&= t^{\beta+\sigma-1} {}_2sinh_2 \left(\begin{matrix} (\gamma, 1), (\beta, \vartheta) \\ (\beta, \alpha), (\beta+\sigma, \vartheta) \end{matrix} ; \lambda t^{\vartheta} \right).
\end{aligned} \tag{5.548}$$

Proof Putting $\varsigma = \beta$ in (5.524), we get the result. □

Theorem 5.79 *Let $\sigma \in \mathbb{C}$, $\lambda \in \mathbb{C}$, $\vartheta > 0$, $\alpha \in \mathbb{R}_{+}$, $Re(\sigma) > 0$, and $Re(\beta) > 0$. Then we have*

$$\frac{\Gamma(\gamma)}{\Gamma(\sigma)} \int_{0}^{t} (t-\tau)^{\sigma}\, Grecosh_{\alpha,\beta}^{\gamma}\left(\lambda\tau^{\vartheta}\right) d\tau = t^{\beta+\sigma-1}\, {}_{2}cosh_{2}\left(\begin{matrix} (\gamma,1),(\beta,\vartheta) \\ (\beta,\alpha),(\beta+\sigma,\vartheta) \end{matrix}; \lambda t^{\vartheta}\right). \tag{5.549}$$

Proof Setting $\varsigma = \beta$ in (5.525), we have the result. □

Theorem 5.80 *Let $\sigma \in \mathbb{C}$, $\lambda \in \mathbb{C}$, $\vartheta > 0$, $\alpha \in \mathbb{R}_{+}$, $Re(\sigma) > 0$, and $Re(\beta) > 0$. Then we have*

$$\frac{\Gamma(\gamma)}{\Gamma(\sigma)} \int_{0}^{1} (1-\tau)^{\sigma}\, Gresin_{\alpha,\beta}^{\gamma}\left(\lambda\tau^{\vartheta}\right) d\tau = {}_{2}sin_{2}\left(\begin{matrix} (\gamma,1),(\beta,\vartheta) \\ (\beta,\alpha),(\beta+\sigma,\vartheta) \end{matrix}; \lambda\right). \tag{5.550}$$

Proof Taking $\varsigma = \beta$ in (5.526), we have the result. □

Theorem 5.81 *Let $\sigma \in \mathbb{C}$, $\lambda \in \mathbb{C}$, $\vartheta > 0$, $\alpha \in \mathbb{R}_{+}$, $Re(\sigma) > 0$, and $Re(\beta) > 0$. Then we have*

$$\frac{\Gamma(\gamma)}{\Gamma(\sigma)} \int_{0}^{1} (1-\tau)^{\sigma}\, Grecos_{\alpha,\beta}^{\gamma}\left(\lambda\tau^{\vartheta}\right) d\tau = {}_{2}cos_{2}\left(\begin{matrix} (\gamma,1),(\beta,\vartheta) \\ (\beta,\alpha),(\beta+\sigma,\vartheta) \end{matrix}; \lambda\right). \tag{5.551}$$

Proof Putting $\varsigma = \beta$ in (5.527), we get the result. □

Theorem 5.82 *Let $\sigma \in \mathbb{C}$, $\lambda \in \mathbb{C}$, $\vartheta > 0$, $\alpha \in \mathbb{R}_{+}$, $Re(\sigma) > 0$, and $Re(\beta) > 0$. Then we have*

$$\frac{\Gamma(\gamma)}{\Gamma(\sigma)} \int_{0}^{1} (1-\tau)^{\sigma}\, Gresinh_{\alpha,\beta}^{\gamma}\left(\lambda\tau^{\vartheta}\right) d\tau = {}_{2}sinh_{2}\left(\begin{matrix} (\gamma,1),(\beta,\vartheta) \\ (\beta,\alpha),(\beta+\sigma,\vartheta) \end{matrix}; \lambda\right). \tag{5.552}$$

Proof Making $\varsigma = \beta$ in (5.528), we obtain the result. □

Theorem 5.83 *Let $\sigma \in \mathbb{C}$, $\lambda \in \mathbb{C}$, $\vartheta > 0$, $\alpha \in \mathbb{R}_{+}$, $Re(\sigma) > 0$, and $Re(\beta) > 0$. Then we have*

$$\frac{\Gamma(\gamma)}{\Gamma(\sigma)} \int_{0}^{1} (1-\tau)^{\sigma}\, Grecosh_{\alpha,\beta}^{\gamma}\left(\lambda\tau^{\vartheta}\right) d\tau = {}_{2}cosh_{2}\left(\begin{matrix} (\gamma,1),(\beta,\vartheta) \\ (\beta,\alpha),(\beta+\sigma,\vartheta) \end{matrix}; \lambda\right). \tag{5.553}$$

Proof Putting $\varsigma = \beta$ in (5.529), we get the result. □

Theorem 5.84 *Let* $\sigma \in \mathbb{C}$, $\lambda \in \mathbb{C}$, $\vartheta > 0$, $\alpha \in \mathbb{R}_+$, $Re(\sigma) > 0$, *and* $Re(\beta) > 0$. *Then we have*

$$\begin{aligned} &\frac{\Gamma(\gamma)}{\Gamma(\sigma)} \int_0^2 (2-\tau)^\sigma \, Gresin_{\alpha,\beta}^\gamma \left(\lambda \tau^\vartheta\right) d\tau \\ &= \frac{\Gamma(\gamma)}{\Gamma(\sigma)} \int_{-1}^1 (1-\tau)^\sigma \, Gresin_{\alpha,\beta}^\gamma \left(\lambda (1+\tau)^\vartheta\right) d\tau \\ &= 2^{\beta+\sigma-1} {}_2sin_2 \left(\begin{matrix} (\gamma, 1), (\beta, \vartheta) \\ (\beta, \alpha), (\beta+\sigma, \vartheta) \end{matrix} ; \lambda 2^\vartheta \right). \end{aligned} \tag{5.554}$$

Proof Setting $\varsigma = \beta$ in (5.530), we have the result. □

Theorem 5.85 *Let* $\sigma \in \mathbb{C}$, $\lambda \in \mathbb{C}$, $\vartheta > 0$, $\alpha \in \mathbb{R}_+$, $Re(\sigma) > 0$, *and* $Re(\beta) > 0$. *Then we have*

$$\begin{aligned} &\frac{\Gamma(\gamma)}{\Gamma(\sigma)} \int_0^2 (2-\tau)^\sigma \, Grecos_{\alpha,\beta}^\gamma \left(\lambda \tau^\vartheta\right) d\tau \\ &= \frac{\Gamma(\gamma)}{\Gamma(\sigma)} \int_{-1}^1 (1-\tau)^\sigma \, Grecos_{\alpha,\beta}^\gamma \left(\lambda (1+\tau)^\vartheta\right) d\tau \\ &= 2^{\beta+\sigma-1} {}_2cos_2 \left(\begin{matrix} (\gamma, 1), (\beta, \vartheta) \\ (\beta, \alpha), (\beta+\sigma, \vartheta) \end{matrix} ; \lambda 2^\vartheta \right). \end{aligned} \tag{5.555}$$

Proof Putting $\varsigma = \beta$ in (5.531), we obtain the result. □

Theorem 5.86 *Let* $\sigma \in \mathbb{C}$, $\lambda \in \mathbb{C}$, $\vartheta > 0$, $\alpha \in \mathbb{R}_+$, $Re(\sigma) > 0$, *and* $Re(\beta) > 0$. *Then we have*

$$\begin{aligned} &\frac{\Gamma(\gamma)}{\Gamma(\sigma)} \int_0^2 (2-\tau)^\sigma \, Gresinh_{\alpha,\beta}^\gamma \left(\lambda \tau^\vartheta\right) d\tau \\ &= \frac{\Gamma(\gamma)}{\Gamma(\sigma)} \int_{-1}^1 (1-\tau)^\sigma \, Gresinh_{\alpha,\beta}^\gamma \left(\lambda (1+\tau)^\vartheta\right) d\tau \\ &= 2^{\beta+\sigma-1} {}_2sinh_2 \left(\begin{matrix} (\gamma, 1), (\beta, \vartheta) \\ (\beta, \alpha), (\beta+\sigma, \vartheta) \end{matrix} ; \lambda 2^\vartheta \right). \end{aligned} \tag{5.556}$$

Proof Substituting $\varsigma = \beta$ into (5.532), we get the result. □

Theorem 5.87 *Let* $\sigma \in \mathbb{C}$, $\lambda \in \mathbb{C}$, $\vartheta > 0$, $\alpha \in \mathbb{R}_+$, $Re(\sigma) > 0$, *and* $Re(\beta) > 0$. *Then we have*

$$\begin{aligned} &\frac{\Gamma(\gamma)}{\Gamma(\sigma)} \int_0^2 (2-\tau)^\sigma \, Grecosh_{\alpha,\beta}^\gamma \left(\lambda \tau^\vartheta\right) d\tau \\ &= \frac{\Gamma(\gamma)}{\Gamma(\sigma)} \int_{-1}^1 (1-\tau)^\sigma \, Grecosh_{\alpha,\beta}^\gamma \left(\lambda (1+\tau)^\vartheta\right) d\tau \\ &= 2^{\beta+\sigma-1} {}_2cosh_2 \left(\begin{matrix} (\gamma, 1), (\beta, \vartheta) \\ (\beta, \alpha), (\beta+\sigma, \vartheta) \end{matrix} ; \lambda 2^\vartheta \right). \end{aligned} \tag{5.557}$$

Proof Taking $\varsigma = \beta$ in (5.533), we obtain the result. □

Theorem 5.88 *Let* $\sigma \in \mathbb{C}$, $\lambda \in \mathbb{C}$, $\alpha \in \mathbb{R}_+$, $Re(\sigma) > 0$, *and* $Re(\beta) > 0$. *Then we have*

$$\frac{\Gamma(\gamma)}{\Gamma(\sigma)} \int_0^t (t-\tau)^\sigma \, Tresin^{\gamma}_{\alpha,\beta}(\lambda\tau)\, d\tau = t^{\beta+\sigma-1}\, {}_2sin_2\left(\begin{matrix} (\gamma,1),(\beta,1) \\ (\beta,\alpha),(\beta+\sigma,1) \end{matrix}; \lambda t \right). \tag{5.558}$$

Proof Setting $\vartheta = 1$ in (5.546), we have the result. □

Theorem 5.89 *Let* $\sigma \in \mathbb{C}$, $\lambda \in \mathbb{C}$, $\alpha \in \mathbb{R}_+$, $Re(\sigma) > 0$, *and* $Re(\beta) > 0$. *Then we have*

$$\frac{\Gamma(\gamma)}{\Gamma(\sigma)} \int_0^t (t-\tau)^\sigma \, Trecos^{\gamma}_{\alpha,\beta}(\lambda\tau)\, d\tau = t^{\beta+\sigma-1}\, {}_2cos_2\left(\begin{matrix} (\gamma,1),(\beta,1) \\ (\beta,\alpha),(\beta+\sigma,1) \end{matrix}; \lambda t \right). \tag{5.559}$$

Proof Taking $\vartheta = 1$ in (5.547), we obtain the result. □

Theorem 5.90 *Let* $\sigma \in \mathbb{C}$, $\lambda \in \mathbb{C}$, $\alpha \in \mathbb{R}_+$, $Re(\sigma) > 0$, *and* $Re(\beta) > 0$. *Then we have*

$$\begin{aligned} &\frac{\Gamma(\gamma)}{\Gamma(\sigma)} \int_0^t (t-\tau)^\sigma \, Tresinh^{\gamma}_{\alpha,\beta}(\lambda\tau)\, d\tau \\ &= t^{\beta+\sigma-1}\, {}_2sinh_2\left(\begin{matrix} (\gamma,1),(\beta,1) \\ (\beta,\alpha),(\beta+\sigma,1) \end{matrix}; \lambda t \right). \end{aligned} \tag{5.560}$$

Proof Putting $\vartheta = 1$ in (5.548), we get the result. □

Theorem 5.91 *Let* $\sigma \in \mathbb{C}$, $\lambda \in \mathbb{C}$, $\alpha \in \mathbb{R}_+$, $Re(\sigma) > 0$, *and* $Re(\beta) > 0$. *Then we have*

$$\begin{aligned} &\frac{\Gamma(\gamma)}{\Gamma(\sigma)} \int_0^t (t-\tau)^\sigma \, Trecosh^{\gamma}_{\alpha,\beta}(\lambda\tau)\, d\tau \\ &= t^{\beta+\sigma-1}\, {}_2cosh_2\left(\begin{matrix} (\gamma,1),(\beta,1) \\ (\beta,\alpha),(\beta+\sigma,1) \end{matrix}; \lambda t \right). \end{aligned} \tag{5.561}$$

Proof Setting $\vartheta = 1$ in (5.549), we have the result. □

Theorem 5.92 *Let* $\sigma \in \mathbb{C}$, $\lambda \in \mathbb{C}$, $\alpha \in \mathbb{R}_+$, $Re(\sigma) > 0$, *and* $Re(\beta) > 0$. *Then we have*

$$\frac{\Gamma(\gamma)}{\Gamma(\sigma)} \int_0^1 (1-\tau)^\sigma \, Tresin^{\gamma}_{\alpha,\beta}(\lambda\tau)\, d\tau = {}_2sin_2\left(\begin{matrix} (\gamma,1),(\beta,1) \\ (\beta,\alpha),(\beta+\sigma,1) \end{matrix}; \lambda \right). \tag{5.562}$$

Proof Taking $\vartheta = 1$ in (5.550), we have the result. □

Theorem 5.93 *Let $\sigma \in \mathbb{C}$, $\lambda \in \mathbb{C}$, $\alpha \in \mathbb{R}_+$, $Re(\sigma) > 0$, and $Re(\beta) > 0$. Then we have*

$$\frac{\Gamma(\gamma)}{\Gamma(\sigma)} \int_0^1 (1-\tau)^{\sigma}\, Trecos_{\alpha,\beta}^{\gamma}(\lambda\tau)\, d\tau = {}_2cos_2\left(\begin{matrix} (\gamma,1),(\beta,1) \\ (\beta,\alpha),(\beta+\sigma,1) \end{matrix} ; \lambda \right). \tag{5.563}$$

Proof Putting $\vartheta = 1$ in (5.551), we get the result. □

Theorem 5.94 *Let $\sigma \in \mathbb{C}$, $\lambda \in \mathbb{C}$, $\alpha \in \mathbb{R}_+$, $Re(\sigma) > 0$, and $Re(\beta) > 0$. Then we have*

$$\frac{\Gamma(\gamma)}{\Gamma(\sigma)} \int_0^1 (1-\tau)^{\sigma}\, Tresinh_{\alpha,\beta}^{\gamma}(\lambda\tau)\, d\tau = {}_2sinh_2\left(\begin{matrix} (\gamma,1),(\beta,1) \\ (\beta,\alpha),(\beta+\sigma,1) \end{matrix} ; \lambda \right). \tag{5.564}$$

Proof Making $\vartheta = 1$ in (5.552), we obtain the result. □

Theorem 5.95 *Let $\sigma \in \mathbb{C}$, $\lambda \in \mathbb{C}$, $\alpha \in \mathbb{R}_+$, $Re(\sigma) > 0$, and $Re(\beta) > 0$. Then we have*

$$\frac{\Gamma(\gamma)}{\Gamma(\sigma)} \int_0^1 (1-\tau)^{\sigma}\, Trecosh_{\alpha,\beta}^{\gamma}(\lambda\tau)\, d\tau = {}_2cosh_2\left(\begin{matrix} (\gamma,1),(\beta,1) \\ (\beta,\alpha),(\beta+\sigma,1) \end{matrix} ; \lambda \right). \tag{5.565}$$

Proof Putting $\vartheta = 1$ in (5.553), we get the result. □

Theorem 5.96 *Let $\sigma \in \mathbb{C}$, $\lambda \in \mathbb{C}$, $\alpha \in \mathbb{R}_+$, $Re(\sigma) > 0$, and $Re(\beta) > 0$. Then we have*

$$\begin{aligned} &\frac{\Gamma(\gamma)}{\Gamma(\sigma)} \int_0^2 (2-\tau)^{\sigma}\, Tresin_{\alpha,\beta}^{\gamma}(\lambda\tau)\, d\tau \\ &= \frac{\Gamma(\gamma)}{\Gamma(\sigma)} \int_{-1}^1 (1-\tau)^{\sigma}\, Tresin_{\alpha,\beta}^{\gamma}(\lambda(1+\tau))\, d\tau \\ &= 2^{\beta+\sigma-1}\, {}_2sin_2\left(\begin{matrix} (\gamma,1),(\beta,1) \\ (\beta,\alpha),(\beta+\sigma,1) \end{matrix} ; 2\lambda \right). \end{aligned} \tag{5.566}$$

Proof Setting $\vartheta = 1$ in (5.554), we have the result. □

Theorem 5.97 *Let $\sigma \in \mathbb{C}$, $\lambda \in \mathbb{C}$, $\alpha \in \mathbb{R}_+$, $Re(\sigma) > 0$, and $Re(\beta) > 0$. Then we have*

$$\begin{aligned}
&\frac{\Gamma(\gamma)}{\Gamma(\sigma)} \int_0^2 (2-\tau)^\sigma \, Trecos_{\alpha,\beta}^{\gamma}(\lambda\tau)\, d\tau \\
&= \frac{\Gamma(\gamma)}{\Gamma(\sigma)} \int_{-1}^1 (1-\tau)^\sigma \, Trecos_{\alpha,\beta}^{\gamma}(\lambda(1+\tau))\, d\tau \\
&= 2^{\beta+\sigma-1} {}_2cos_2\left(\begin{matrix} (\gamma,1),(\beta,1) \\ (\beta,\alpha),(\beta+\sigma,1) \end{matrix} ; 2\lambda \right).
\end{aligned} \tag{5.567}$$

Proof Putting $\vartheta = 1$ in (5.555), we obtain the result. □

Theorem 5.98 *Let $\sigma \in \mathbb{C}$, $\lambda \in \mathbb{C}$, $\alpha \in \mathbb{R}_+$, $Re(\sigma) > 0$, and $Re(\beta) > 0$. Then we have*

$$\begin{aligned}
&\frac{\Gamma(\gamma)}{\Gamma(\sigma)} \int_0^2 (2-\tau)^\sigma \, Tresinh_{\alpha,\beta}^{\gamma}(\lambda\tau)\, d\tau \\
&= \frac{\Gamma(\gamma)}{\Gamma(\sigma)} \int_{-1}^1 (1-\tau)^\sigma \, Tresinh_{\alpha,\beta}^{\gamma}(\lambda(1+\tau))\, d\tau \\
&= 2^{\beta+\sigma-1} {}_2sinh_2\left(\begin{matrix} (\gamma,1),(\beta,1) \\ (\beta,\alpha),(\beta+\sigma,1) \end{matrix} ; 2\lambda \right).
\end{aligned} \tag{5.568}$$

Proof Substituting $\vartheta = 1$ into (5.556), we get the result. □

Theorem 5.99 *Let $\sigma \in \mathbb{C}$, $\lambda \in \mathbb{C}$, $\alpha \in \mathbb{R}_+$, $Re(\sigma) > 0$, and $Re(\beta) > 0$. Then we have*

$$\begin{aligned}
&\frac{\Gamma(\gamma)}{\Gamma(\sigma)} \int_0^2 (2-\tau)^\sigma \, Trecosh_{\alpha,\beta}^{\gamma}(\lambda\tau)\, d\tau \\
&= \frac{\Gamma(\gamma)}{\Gamma(\sigma)} \int_{-1}^1 (1-\tau)^\sigma \, Trecosh_{\alpha,\beta}^{\gamma}(\lambda(1+\tau))\, d\tau \\
&= 2^{\beta+\sigma-1} {}_2cosh_2\left(\begin{matrix} (\gamma,1),(\beta,1) \\ (\beta,\alpha),(\beta+\sigma,1) \end{matrix} ; 2\lambda \right).
\end{aligned} \tag{5.569}$$

Proof Taking $\vartheta = 1$ in (5.557), we obtain the result. □

Theorem 5.100 *Let $\sigma \in \mathbb{C}$, $\lambda \in \mathbb{C}$, $\vartheta > 0$, $\beta \in \mathbb{C}$, $\alpha \in \mathbb{R}_+$, and $Re(\sigma) > 0$. Then we have*

$$\frac{\Gamma(\gamma)}{\Gamma(\sigma)} \int_0^t (t-\tau)^\sigma \, presin_{\alpha,\beta}^{\gamma}\left(\lambda\tau^\vartheta\right) d\tau = t^\sigma \, {}_2sin_2\left(\begin{matrix} (\gamma,1),(1,\vartheta) \\ (\beta,\alpha),(1+\sigma,\vartheta) \end{matrix} ; \lambda t^\vartheta \right). \tag{5.570}$$

Proof Putting $\varsigma = 1$ in (5.522), we have the result. □

Theorem 5.101 *Let* $\sigma \in \mathbb{C}$, $\lambda \in \mathbb{C}$, $\vartheta > 0$, $\beta \in \mathbb{C}$, $\alpha \in \mathbb{R}_{+}$, *and* $Re(\sigma) > 0$. *Then we have*

$$\frac{\Gamma(\gamma)}{\Gamma(\sigma)} \int_{0}^{t} (t-\tau)^{\sigma}\, precos^{\gamma}_{\alpha,\beta}\left(\lambda\tau^{\vartheta}\right) d\tau = t^{\sigma}\, {}_{2}cos_{2}\left(\begin{matrix} (\gamma,1),(1,\vartheta) \\ (\beta,\alpha),(1+\sigma,\vartheta) \end{matrix};\lambda t^{\vartheta}\right). \tag{5.571}$$

Proof Setting $\varsigma = 1$ in (5.523), we obtain the result. □

Theorem 5.102 *Let* $\sigma \in \mathbb{C}$, $\lambda \in \mathbb{C}$, $\vartheta > 0$, $\beta \in \mathbb{C}$, $\alpha \in \mathbb{R}_{+}$, *and* $Re(\sigma) > 0$. *Then we have*

$$\begin{aligned} &\frac{\Gamma(\gamma)}{\Gamma(\sigma)} \int_{0}^{t} (t-\tau)^{\sigma}\, presinh^{\gamma}_{\alpha,\beta}\left(\lambda\tau^{\vartheta}\right) d\tau \\ &= t^{\sigma}\, {}_{2}sinh_{2}\left(\begin{matrix} (\gamma,1),(1,\vartheta) \\ (\beta,\alpha),(1+\sigma,\vartheta) \end{matrix};\lambda t^{\vartheta}\right). \end{aligned} \tag{5.572}$$

Proof Making $\varsigma = 1$ in (5.524), we get the result. □

Theorem 5.103 *Let* $\sigma \in \mathbb{C}$, $\lambda \in \mathbb{C}$, $\vartheta > 0$, $\beta \in \mathbb{C}$, $\alpha \in \mathbb{R}_{+}$, *and* $Re(\sigma) > 0$. *Then we have*

$$\begin{aligned} &\frac{\Gamma(\gamma)}{\Gamma(\sigma)} \int_{0}^{t} (t-\tau)^{\sigma}\, precosh^{\gamma}_{\alpha,\beta}\left(\lambda\tau^{\vartheta}\right) d\tau \\ &= t^{\sigma}\, {}_{2}cosh_{2}\left(\begin{matrix} (\gamma,1),(1,\vartheta) \\ (\beta,\alpha),(1+\sigma,\vartheta) \end{matrix};\lambda t^{\vartheta}\right). \end{aligned} \tag{5.573}$$

Proof Substituting $\varsigma = 1$ into (5.525), we have the result. □

Theorem 5.104 *Let* $\sigma \in \mathbb{C}$, $\lambda \in \mathbb{C}$, $\vartheta > 0$, $\beta \in \mathbb{C}$, $\alpha \in \mathbb{R}_{+}$, *and* $Re(\sigma) > 0$. *Then we have*

$$\begin{aligned} &\frac{\Gamma(\gamma)}{\Gamma(\sigma)} \int_{0}^{1} (1-\tau)^{\sigma}\, presin^{\gamma}_{\alpha,\beta}\left(\lambda\tau^{\vartheta}\right) d\tau \\ &= {}_{2}sin_{2}\left(\begin{matrix} (\gamma,1),(1,\vartheta) \\ (\beta,\alpha),(1+\sigma,\vartheta) \end{matrix};\lambda\right). \end{aligned} \tag{5.574}$$

Proof Putting $\varsigma = 1$ in (5.526), we have the result. □

Theorem 5.105 *Let* $\sigma \in \mathbb{C}$, $\lambda \in \mathbb{C}$, $\vartheta > 0$, $\beta \in \mathbb{C}$, $\alpha \in \mathbb{R}_{+}$, *and* $Re(\sigma) > 0$. *Then we have*

$$\frac{\Gamma(\gamma)}{\Gamma(\sigma)} \int_{0}^{1} (1-\tau)^{\sigma}\, precos^{\gamma}_{\alpha,\beta}\left(\lambda\tau^{\vartheta}\right) d\tau = {}_{2}cos_{2}\left(\begin{matrix} (\gamma,1),(1,\vartheta) \\ (\beta,\alpha),(1+\sigma,\vartheta) \end{matrix};\lambda\right). \tag{5.575}$$

Proof Taking $\varsigma = 1$ in (5.527), we get the result. □

Theorem 5.106 *Let $\sigma \in \mathbb{C}$, $\lambda \in \mathbb{C}$, $\vartheta > 0$, $\beta \in \mathbb{C}$, $\alpha \in \mathbb{R}_+$, and $Re(\sigma) > 0$. Then we have*

$$\frac{\Gamma(\gamma)}{\Gamma(\sigma)} \int_0^1 (1-\tau)^\sigma \, presinh_{\alpha,\beta}^\gamma \left(\lambda \tau^\vartheta\right) d\tau = {}_2sinh_2 \left(\begin{matrix} (\gamma, 1), (1, \vartheta) \\ (\beta, \alpha), (1+\sigma, \vartheta) \end{matrix} ; \lambda \right). \tag{5.576}$$

Proof Putting $\varsigma = 1$ in (5.528), we obtain the result. □

Theorem 5.107 *Let $\sigma \in \mathbb{C}$, $\lambda \in \mathbb{C}$, $\vartheta > 0$, $\beta \in \mathbb{C}$, $\alpha \in \mathbb{R}_+$, and $Re(\sigma) > 0$. Then we have*

$$\frac{\Gamma(\gamma)}{\Gamma(\sigma)} \int_0^1 (1-\tau)^\sigma \, precosh_{\alpha,\beta}^\gamma \left(\lambda \tau^\vartheta\right) d\tau = {}_2cosh_2 \left(\begin{matrix} (\gamma, 1), (1, \vartheta) \\ (\beta, \alpha), (1+\sigma, \vartheta) \end{matrix} ; \lambda \right). \tag{5.577}$$

Proof Making $\varsigma = 1$ in (5.529), we get the result. □

Theorem 5.108 *Let $\sigma \in \mathbb{C}$, $\lambda \in \mathbb{C}$, $\vartheta > 0$, $\beta \in \mathbb{C}$, $\alpha \in \mathbb{R}_+$, and $Re(\sigma) > 0$. Then we have*

$$\begin{aligned} &\tfrac{\Gamma(\gamma)}{\Gamma(\sigma)} \int_0^2 (2-\tau)^\sigma \, presin_{\alpha,\beta}^\gamma \left(\lambda \tau^\vartheta\right) d\tau \\ &= \tfrac{\Gamma(\gamma)}{\Gamma(\sigma)} \int_{-1}^1 (1-\tau)^\sigma \, presin_{\alpha,\beta}^\gamma \left(\lambda (1+\tau)^\vartheta\right) d\tau \\ &= 2^\sigma \, {}_2sin_2 \left(\begin{matrix} (\gamma, 1), (1, \vartheta) \\ (\beta, \alpha), (1+\sigma, \vartheta) \end{matrix} ; \lambda 2^\vartheta \right). \end{aligned} \tag{5.578}$$

Proof Putting $\varsigma = 1$ in (5.530), we have the result. □

Theorem 5.109 *Let $\sigma \in \mathbb{C}$, $\lambda \in \mathbb{C}$, $\vartheta > 0$, $\beta \in \mathbb{C}$, $\alpha \in \mathbb{R}_+$, and $Re(\sigma) > 0$. Then we have*

$$\begin{aligned} &\tfrac{\Gamma(\gamma)}{\Gamma(\sigma)} \int_0^2 (2-\tau)^\sigma \, precos_{\alpha,\beta}^\gamma \left(\lambda \tau^\vartheta\right) d\tau \\ &= \tfrac{\Gamma(\gamma)}{\Gamma(\sigma)} \int_{-1}^1 (1-\tau)^\sigma \, precos_{\alpha,\beta}^\gamma \left(\lambda (1+\tau)^\vartheta\right) d\tau \\ &= 2^\sigma \, {}_2cos_2 \left(\begin{matrix} (\gamma, 1), (1, \vartheta) \\ (\beta, \alpha), (1+\sigma, \vartheta) \end{matrix} ; \lambda 2^\vartheta \right). \end{aligned} \tag{5.579}$$

Proof Setting $\varsigma = 1$ in (5.531), we obtain the result. □

Theorem 5.110 *Let* $\sigma \in \mathbb{C}$, $\lambda \in \mathbb{C}$, $\vartheta > 0$, $\beta \in \mathbb{C}$, $\alpha \in \mathbb{R}_{+}$, *and* $Re(\sigma) > 0$. *Then we have*

$$\begin{aligned} &\frac{\Gamma(\gamma)}{\Gamma(\sigma)} \int_0^2 (2-\tau)^{\sigma} \, presinh^{\gamma}_{\alpha,\beta} \left(\lambda \tau^{\vartheta}\right) d\tau \\ &= \frac{\Gamma(\gamma)}{\Gamma(\sigma)} \int_{-1}^{1} (1-\tau)^{\sigma} \, presinh^{\gamma}_{\alpha,\beta} \left(\lambda (1+\tau)^{\vartheta}\right) d\tau \\ &= 2^{\sigma} \, {}_2sinh_2 \left(\begin{matrix} (\gamma,1),(1,\vartheta) \\ (\beta,\alpha),(1+\sigma,\vartheta) \end{matrix} ; \lambda 2^{\vartheta} \right). \end{aligned} \tag{5.580}$$

Proof Putting $\varsigma = 1$ in (5.532), we get the result. □

Theorem 5.111 *Let* $\sigma \in \mathbb{C}$, $\lambda \in \mathbb{C}$, $\vartheta > 0$, $\beta \in \mathbb{C}$, $\alpha \in \mathbb{R}_{+}$, *and* $Re(\sigma) > 0$. *Then we have*

$$\begin{aligned} &\frac{\Gamma(\gamma)}{\Gamma(\sigma)} \int_0^2 (2-\tau)^{\sigma} \, precosh^{\gamma}_{\alpha,\beta} \left(\lambda \tau^{\vartheta}\right) d\tau \\ &= \frac{\Gamma(\gamma)}{\Gamma(\sigma)} \int_{-1}^{1} (1-\tau)^{\sigma} \, precosh^{\gamma}_{\alpha,\beta} \left(\lambda (1+\tau)^{\vartheta}\right) d\tau \\ &= 2^{\sigma} \, {}_2cosh_2 \left(\begin{matrix} (\gamma,1),(1,\vartheta) \\ (\beta,\alpha),(1+\sigma,\vartheta) \end{matrix} ; \lambda 2^{\vartheta} \right). \end{aligned} \tag{5.581}$$

Proof Making $\varsigma = 1$ in (5.533), we obtain the result. □

Theorem 5.112 *Let* $\sigma \in \mathbb{C}$, $\lambda \in \mathbb{C}$, $\vartheta > 0$, $\beta \in \mathbb{C}$, $\alpha \in \mathbb{R}_{+}$, *and* $Re(\sigma) > 0$. *Then we have*

$$\frac{\Gamma(\gamma)}{\Gamma(\sigma)} \int_0^t (t-\tau)^{\sigma} \, presin^{\gamma}_{\alpha,\beta} (\lambda \tau) \, d\tau = t^{\sigma} \, {}_2sin_2 \left(\begin{matrix} (\gamma,1),(1,1) \\ (\beta,\alpha),(1+\sigma,1) \end{matrix} ; \lambda t \right). \tag{5.582}$$

Proof Putting $\vartheta = 1$ in (5.570), we have the result. □

Theorem 5.113 *Let* $\sigma \in \mathbb{C}$, $\lambda \in \mathbb{C}$, $\vartheta > 0$, $\beta \in \mathbb{C}$, $\alpha \in \mathbb{R}_{+}$, *and* $Re(\sigma) > 0$. *Then we have*

$$\frac{\Gamma(\gamma)}{\Gamma(\sigma)} \int_0^t (t-\tau)^{\sigma} \, precos^{\gamma}_{\alpha,\beta} (\lambda \tau) \, d\tau = t^{\sigma} \, {}_2cos_2 \left(\begin{matrix} (\gamma,1),(1,1) \\ (\beta,\alpha),(1+\sigma,1) \end{matrix} ; \lambda t \right). \tag{5.583}$$

Proof Setting $\vartheta = 1$ in (5.571), we obtain the result. □

Theorem 5.114 *Let $\sigma \in \mathbb{C}$, $\lambda \in \mathbb{C}$, $\vartheta > 0$, $\beta \in \mathbb{C}$, $\alpha \in \mathbb{R}_+$, and $Re(\sigma) > 0$. Then we have*

$$\frac{\Gamma(\gamma)}{\Gamma(\sigma)} \int_0^t (t-\tau)^\sigma \; presinh_{\alpha,\beta}^{\gamma}(\lambda\tau)\, d\tau = t^\sigma \; {}_2sinh_2\left(\begin{matrix} (\gamma, 1), (1, 1) \\ (\beta, \alpha), (1+\sigma, 1) \end{matrix} ; \lambda t \right). \tag{5.584}$$

Proof Making $\vartheta = 1$ in (5.572), we get the result. □

Theorem 5.115 *Let $\sigma \in \mathbb{C}$, $\lambda \in \mathbb{C}$, $\vartheta > 0$, $\beta \in \mathbb{C}$, $\alpha \in \mathbb{R}_+$, and $Re(\sigma) > 0$. Then we have*

$$\frac{\Gamma(\gamma)}{\Gamma(\sigma)} \int_0^t (t-\tau)^\sigma \; precosh_{\alpha,\beta}^{\gamma}(\lambda\tau)\, d\tau = t^\sigma \; {}_2cosh_2\left(\begin{matrix} (\gamma, 1), (1, 1) \\ (\beta, \alpha), (1+\sigma, 1) \end{matrix} ; \lambda t \right). \tag{5.585}$$

Proof Substituting $\vartheta = 1$ into (5.573), we have the result. □

Theorem 5.116 *Let $\sigma \in \mathbb{C}$, $\lambda \in \mathbb{C}$, $\vartheta > 0$, $\beta \in \mathbb{C}$, $\alpha \in \mathbb{R}_+$, and $Re(\sigma) > 0$. Then we have*

$$\frac{\Gamma(\gamma)}{\Gamma(\sigma)} \int_0^1 (1-\tau)^\sigma \; presin_{\alpha,\beta}^{\gamma}(\lambda\tau)\, d\tau = {}_2sin_2\left(\begin{matrix} (\gamma, 1), (1, 1) \\ (\beta, \alpha), (1+\sigma, 1) \end{matrix} ; \lambda \right). \tag{5.586}$$

Proof Putting $\vartheta = 1$ in (5.574), we have the result. □

Theorem 5.117 *Let $\sigma \in \mathbb{C}$, $\lambda \in \mathbb{C}$, $\vartheta > 0$, $\beta \in \mathbb{C}$, $\alpha \in \mathbb{R}_+$, and $Re(\sigma) > 0$. Then we have*

$$\frac{\Gamma(\gamma)}{\Gamma(\sigma)} \int_0^1 (1-\tau)^\sigma \; precos_{\alpha,\beta}^{\gamma}(\lambda\tau)\, d\tau = {}_2cos_2\left(\begin{matrix} (\gamma, 1), (1, 1) \\ (\beta, \alpha), (1+\sigma, 1) \end{matrix} ; \lambda \right). \tag{5.587}$$

Proof Taking $\vartheta = 1$ in (5.575), we get the result. □

Theorem 5.118 *Let $\sigma \in \mathbb{C}$, $\lambda \in \mathbb{C}$, $\vartheta > 0$, $\beta \in \mathbb{C}$, $\alpha \in \mathbb{R}_+$, and $Re(\sigma) > 0$. Then we have*

$$\frac{\Gamma(\gamma)}{\Gamma(\sigma)} \int_0^1 (1-\tau)^\sigma \; presinh_{\alpha,\beta}^{\gamma}(\lambda\tau)\, d\tau = {}_2sinh_2\left(\begin{matrix} (\gamma, 1), (1, 1) \\ (\beta, \alpha), (1+\sigma, 1) \end{matrix} ; \lambda \right). \tag{5.588}$$

Proof Putting $\vartheta = 1$ in (5.576), we obtain the result. □

Theorem 5.119 *Let $\sigma \in \mathbb{C}$, $\lambda \in \mathbb{C}$, $\vartheta > 0$, $\beta \in \mathbb{C}$, $\alpha \in \mathbb{R}_{+}$, and $Re(\sigma) > 0$. Then we have*

$$\frac{\Gamma(\gamma)}{\Gamma(\sigma)} \int_{0}^{1} (1-\tau)^{\sigma}\, precosh_{\alpha,\beta}^{\gamma}(\lambda\tau)\, d\tau = {}_{2}cosh_{2}\left(\begin{matrix} (\gamma,1),(1,1) \\ (\beta,\alpha),(1+\sigma,1) \end{matrix} ; \lambda \right). \tag{5.589}$$

Proof Making $\vartheta = 1$ in (5.577), we get the result. □

Theorem 5.120 *Let $\sigma \in \mathbb{C}$, $\lambda \in \mathbb{C}$, $\vartheta > 0$, $\beta \in \mathbb{C}$, $\alpha \in \mathbb{R}_{+}$, and $Re(\sigma) > 0$. Then we have*

$$\begin{aligned} &\frac{\Gamma(\gamma)}{\Gamma(\sigma)} \int_{0}^{2} (2-\tau)^{\sigma}\, presin_{\alpha,\beta}^{\gamma}(\lambda\tau)\, d\tau \\ &= \frac{\Gamma(\gamma)}{\Gamma(\sigma)} \int_{-1}^{1} (1-\tau)^{\sigma}\, presin_{\alpha,\beta}^{\gamma}(\lambda(1+\tau))\, d\tau \\ &= 2^{\sigma}\, {}_{2}sin_{2}\left(\begin{matrix} (\gamma,1),(1,1) \\ (\beta,\alpha),(1+\sigma,1) \end{matrix} ; 2\lambda \right). \end{aligned} \tag{5.590}$$

Proof Putting $\vartheta = 1$ in (5.578), we have the result. □

Theorem 5.121 *Let $\sigma \in \mathbb{C}$, $\lambda \in \mathbb{C}$, $\vartheta > 0$, $\beta \in \mathbb{C}$, $\alpha \in \mathbb{R}_{+}$, and $Re(\sigma) > 0$. Then we have*

$$\begin{aligned} &\frac{\Gamma(\gamma)}{\Gamma(\sigma)} \int_{0}^{2} (2-\tau)^{\sigma}\, precos_{\alpha,\beta}^{\gamma}(\lambda\tau)\, d\tau \\ &= \frac{\Gamma(\gamma)}{\Gamma(\sigma)} \int_{-1}^{1} (1-\tau)^{\sigma}\, precos_{\alpha,\beta}^{\gamma}(\lambda(1+\tau))\, d\tau \\ &= 2^{\sigma}\, {}_{2}cos_{2}\left(\begin{matrix} (\gamma,1),(1,1) \\ (\beta,\alpha),(1+\sigma,1) \end{matrix} ; 2\lambda \right). \end{aligned} \tag{5.591}$$

Proof Setting $\vartheta = 1$ in (5.579), we obtain the result. □

Theorem 5.122 *Let $\sigma \in \mathbb{C}$, $\lambda \in \mathbb{C}$, $\vartheta > 0$, $\beta \in \mathbb{C}$, $\alpha \in \mathbb{R}_{+}$, and $Re(\sigma) > 0$. Then we have*

$$\begin{aligned} &\frac{\Gamma(\gamma)}{\Gamma(\sigma)} \int_{0}^{2} (2-\tau)^{\sigma}\, presinh_{\alpha,\beta}^{\gamma}(\lambda\tau)\, d\tau \\ &= \frac{\Gamma(\gamma)}{\Gamma(\sigma)} \int_{-1}^{1} (1-\tau)^{\sigma}\, presinh_{\alpha,\beta}^{\gamma}(\lambda(1+\tau))\, d\tau \\ &= 2^{\sigma}\, {}_{2}sinh_{2}\left(\begin{matrix} (\gamma,1),(1,1) \\ (\beta,\alpha),(1+\sigma,1) \end{matrix} ; 2\lambda \right). \end{aligned} \tag{5.592}$$

Proof Putting $\vartheta = 1$ in (5.580), we get the result. □

Theorem 5.123 *Let* $\sigma \in \mathbb{C}$, $\lambda \in \mathbb{C}$, $\vartheta > 0$, $\beta \in \mathbb{C}$, $\alpha \in \mathbb{R}_{+}$, *and* $Re(\sigma) > 0$. *Then we have*

$$\begin{aligned} &\frac{\Gamma(\gamma)}{\Gamma(\sigma)} \int_0^2 (2-\tau)^{\sigma}\, precosh^{\gamma}_{\alpha,\beta}(\lambda\tau)\, d\tau \\ &= \frac{\Gamma(\gamma)}{\Gamma(\sigma)} \int_{-1}^{1} (1-\tau)^{\sigma}\, precosh^{\gamma}_{\alpha,\beta}(\lambda(1+\tau))\, d\tau \\ &= 2^{\sigma}\, {}_2cosh_2\left(\begin{matrix} (\gamma,1), (1,1) \\ (\beta,\alpha), (1+\sigma,1) \end{matrix} ; 2\lambda \right). \end{aligned} \tag{5.593}$$

Proof Making $\vartheta = 1$ in (5.581), we obtain the result. □

Theorem 5.124 *Let* $\sigma \in \mathbb{C}$, $\lambda \in \mathbb{C}$, $\alpha \in \mathbb{R}_{+}$, $Re(\sigma) > 0$, *and* $Re(\beta) > 0$. *Then we have*

$$\frac{\Gamma(\gamma)}{\Gamma(\sigma)} \int_0^t (t-\tau)^{\sigma}\, supersin^{\gamma}_{\alpha,\beta}\left(\lambda\tau^{\alpha}\right) d\tau = t^{\beta+\sigma-1}\, {}_1sin_1\left(\begin{matrix} (\gamma,1) \\ (\beta+\sigma,\alpha) \end{matrix} ; \lambda t^{\alpha} \right). \tag{5.594}$$

Proof Setting $\vartheta = \alpha$ in (5.546), we have the result. □

Theorem 5.125 *Let* $\sigma \in \mathbb{C}$, $\lambda \in \mathbb{C}$, $\alpha \in \mathbb{R}_{+}$, $Re(\sigma) > 0$, *and* $Re(\beta) > 0$. *Then we have*

$$\frac{\Gamma(\gamma)}{\Gamma(\sigma)} \int_0^t (t-\tau)^{\sigma}\, supercos^{\gamma}_{\alpha,\beta}\left(\lambda\tau^{\alpha}\right) d\tau = t^{\beta+\sigma-1}\, {}_1cos_1\left(\begin{matrix} (\gamma,1) \\ (\beta+\sigma,\alpha) \end{matrix} ; \lambda t^{\alpha} \right). \tag{5.595}$$

Proof Taking $\vartheta = \alpha$ in (5.547), we obtain the result. □

Theorem 5.126 *Let* $\sigma \in \mathbb{C}$, $\lambda \in \mathbb{C}$, $\alpha \in \mathbb{R}_{+}$, $Re(\sigma) > 0$, *and* $Re(\beta) > 0$. *Then we have*

$$\begin{aligned} &\frac{\Gamma(\gamma)}{\Gamma(\sigma)} \int_0^t (t-\tau)^{\sigma}\, supersinh^{\gamma}_{\alpha,\beta}(\lambda\tau^{\alpha})\, d\tau \\ &= t^{\beta+\sigma-1}\, {}_1sinh_1\left(\begin{matrix} (\gamma,1) \\ (\beta+\sigma,\alpha) \end{matrix} ; \lambda t^{\alpha} \right). \end{aligned} \tag{5.596}$$

Proof Putting $\vartheta = \alpha$ in (5.548), we get the result. □

Theorem 5.127 *Let $\sigma \in \mathbb{C}$, $\lambda \in \mathbb{C}$, $\alpha \in \mathbb{R}_+$, $Re(\sigma) > 0$, and $Re(\beta) > 0$. Then we have*

$$\frac{\Gamma(\gamma)}{\Gamma(\sigma)} \int_0^t (t-\tau)^\sigma \, supercosh^\gamma_{\alpha,\beta}(\lambda \tau^\alpha)\, d\tau = t^{\beta+\sigma-1} \, {}_1cosh_1\left(\begin{matrix} (\gamma, 1) \\ (\beta+\sigma, \alpha) \end{matrix} ; \lambda t^\alpha \right). \tag{5.597}$$

Proof Setting $\vartheta = \alpha$ in (5.549), we have the result. □

Theorem 5.128 *Let $\sigma \in \mathbb{C}$, $\lambda \in \mathbb{C}$, $\alpha \in \mathbb{R}_+$, $Re(\sigma) > 0$, and $Re(\beta) > 0$. Then we have*

$$\frac{\Gamma(\gamma)}{\Gamma(\sigma)} \int_0^1 (1-\tau)^\sigma \, supersin^\gamma_{\alpha,\beta}(\lambda \tau^\alpha)\, d\tau = {}_1sin_1\left(\begin{matrix} (\gamma, 1) \\ (\beta+\sigma, \alpha) \end{matrix} ; \lambda \right). \tag{5.598}$$

Proof Taking $\vartheta = \alpha$ in (5.550), we have the result. □

Theorem 5.129 *Let $\sigma \in \mathbb{C}$, $\lambda \in \mathbb{C}$, $\alpha \in \mathbb{R}_+$, $Re(\sigma) > 0$, and $Re(\beta) > 0$. Then we have*

$$\frac{\Gamma(\gamma)}{\Gamma(\sigma)} \int_0^1 (1-\tau)^\sigma \, supercos^\gamma_{\alpha,\beta}(\lambda \tau^\alpha)\, d\tau = {}_1cos_1\left(\begin{matrix} (\gamma, 1) \\ (\beta+\sigma, \alpha) \end{matrix} ; \lambda \right). \tag{5.599}$$

Proof Putting $\vartheta = \alpha$ in (5.551), we get the result. □

Theorem 5.130 *Let $\sigma \in \mathbb{C}$, $\lambda \in \mathbb{C}$, $\alpha \in \mathbb{R}_+$, $Re(\sigma) > 0$, and $Re(\beta) > 0$. Then we have*

$$\frac{\Gamma(\gamma)}{\Gamma(\sigma)} \int_0^1 (1-\tau)^\sigma \, supersinh^\gamma_{\alpha,\beta}(\lambda \tau^\alpha)\, d\tau = {}_1sinh_1\left(\begin{matrix} (\gamma, 1) \\ (\beta+\sigma, \alpha) \end{matrix} ; \lambda \right). \tag{5.600}$$

Proof Making $\vartheta = \alpha$ in (5.552), we obtain the result. □

Theorem 5.131 *Let $\sigma \in \mathbb{C}$, $\lambda \in \mathbb{C}$, $\alpha \in \mathbb{R}_+$, $Re(\sigma) > 0$, and $Re(\beta) > 0$. Then we have*

$$\frac{\Gamma(\gamma)}{\Gamma(\sigma)} \int_0^1 (1-\tau)^\sigma \, supercosh^\gamma_{\alpha,\beta}(\lambda \tau^\alpha)\, d\tau = {}_1cosh_1\left(\begin{matrix} (\gamma, 1) \\ (\beta+\sigma, \alpha) \end{matrix} ; \lambda \right). \tag{5.601}$$

Proof Putting $\vartheta = \alpha$ in (5.553), we get the result. □

Theorem 5.132 *Let $\sigma \in \mathbb{C}$, $\lambda \in \mathbb{C}$, $\alpha \in \mathbb{R}_+$, $Re(\sigma) > 0$, and $Re(\beta) > 0$. Then we have*

$$\begin{aligned}
&\frac{\Gamma(\gamma)}{\Gamma(\sigma)} \int_0^2 (2-\tau)^\sigma \, supersin^\gamma_{\alpha,\beta}(\lambda\tau^\alpha)\, d\tau \\
&= \frac{\Gamma(\gamma)}{\Gamma(\sigma)} \int_{-1}^1 (1-\tau)^\sigma \, supersin^\gamma_{\alpha,\beta}(\lambda(1+\tau)^\alpha)\, d\tau \\
&= 2^{\beta+\sigma-1} \, {}_1sin_1\left(\begin{matrix} (\gamma, 1) \\ (\beta+\sigma, \alpha) \end{matrix} ; \lambda 2^\alpha \right).
\end{aligned} \tag{5.602}$$

Proof Setting $\vartheta = \alpha$ in (5.554), we have the result. □

Theorem 5.133 *Let $\sigma \in \mathbb{C}$, $\lambda \in \mathbb{C}$, $\alpha \in \mathbb{R}_+$, $Re(\sigma) > 0$, and $Re(\beta) > 0$. Then we have*

$$\begin{aligned}
&\frac{\Gamma(\gamma)}{\Gamma(\sigma)} \int_0^2 (2-\tau)^\sigma \, supercos^\gamma_{\alpha,\beta}(\lambda\tau^\alpha)\, d\tau \\
&= \frac{\Gamma(\gamma)}{\Gamma(\sigma)} \int_{-1}^1 (1-\tau)^\sigma \, supercos^\gamma_{\alpha,\beta}(\lambda(1+\tau)^\alpha)\, d\tau \\
&= 2^{\beta+\sigma-1} \, {}_1cos_1\left(\begin{matrix} (\gamma, 1) \\ (\beta+\sigma, \alpha) \end{matrix} ; \lambda 2^\alpha \right).
\end{aligned} \tag{5.603}$$

Proof Putting $\vartheta = \alpha$ in (5.555), we obtain the result. □

Theorem 5.134 *Let $\sigma \in \mathbb{C}$, $\lambda \in \mathbb{C}$, $\alpha \in \mathbb{R}_+$, $Re(\sigma) > 0$, and $Re(\beta) > 0$. Then we have*

$$\begin{aligned}
&\frac{\Gamma(\gamma)}{\Gamma(\sigma)} \int_0^2 (2-\tau)^\sigma \, supersinh^\gamma_{\alpha,\beta}(\lambda\tau^\alpha)\, d\tau \\
&= \frac{\Gamma(\gamma)}{\Gamma(\sigma)} \int_{-1}^1 (1-\tau)^\sigma \, supersinh^\gamma_{\alpha,\beta}(\lambda(1+\tau)^\alpha)\, d\tau \\
&= 2^{\beta+\sigma-1} \, {}_1sinh_1\left(\begin{matrix} (\gamma, 1) \\ (\beta+\sigma, \alpha) \end{matrix} ; \lambda 2^\alpha \right).
\end{aligned} \tag{5.604}$$

Proof Substituting $\vartheta = \alpha$ into (5.556), we get the result. □

Theorem 5.135 *Let $\sigma \in \mathbb{C}$, $\lambda \in \mathbb{C}$, $\alpha \in \mathbb{R}_+$, $Re(\sigma) > 0$, and $Re(\beta) > 0$. Then we have*

$$\begin{aligned}
&\frac{\Gamma(\gamma)}{\Gamma(\sigma)} \int_0^2 (2-\tau)^\sigma \, supercosh^\gamma_{\alpha,\beta}(\lambda\tau^\alpha)\, d\tau \\
&= \frac{\Gamma(\gamma)}{\Gamma(\sigma)} \int_{-1}^1 (1-\tau)^\sigma \, supercosh^\gamma_{\alpha,\beta}(\lambda(1+\tau)^\alpha)\, d\tau \\
&= 2^{\beta+\sigma-1} \, {}_1cosh_1\left(\begin{matrix} (\gamma, 1) \\ (\beta+\sigma, \alpha) \end{matrix} ; \lambda 2^\alpha \right).
\end{aligned} \tag{5.605}$$

Proof Taking $\vartheta = \alpha$ in (5.557), we obtain the result. □

5.3 The Truncated Prabhakar Functions and Related Functions

In this section, we present the truncated Prabhakar functions and some functions related to the Prabhakar functions.

5.3.1 The Truncated Prabhakar Functions

In this section, we introduce the truncated Prabhakar functions.

Definition 5.66 The truncated Prabhakar function is defined as

$$\left\{E_{\alpha,\beta}^{\gamma}\right\}^{n}(\lambda t) = \sum_{\kappa=0}^{n} \frac{(\gamma)_k}{\Gamma(k\alpha+\beta)} \frac{\lambda^{\kappa} t^{\kappa}}{k!}, \tag{5.606}$$

where $t, \gamma, \alpha, \beta \in \mathbb{C}$, $n, \kappa \in \mathbb{N}_0$, $Re(\alpha) > 0$, and $Re(\beta) > 0$.

The special cases for the truncated Prabhakar function can be written as follows:

$$\left\{E_{\alpha,\beta}^{\gamma}\right\}^{n}(\lambda t^{\alpha}) = \sum_{\kappa=0}^{n} \frac{(\gamma)_k}{\Gamma(k\alpha+\beta)} \frac{\lambda^{\kappa} t^{\kappa\alpha}}{k!} \tag{5.607}$$

or

$$\left\{E_{\alpha,\beta}^{\gamma}\right\}^{n}(\lambda t^{\varsigma}) = \sum_{\kappa=0}^{n} \frac{(\gamma)_k}{\Gamma(k\alpha+\beta)} \frac{\lambda^{\kappa} t^{\kappa\varsigma}}{k!} \tag{5.608}$$

where $t, \alpha, \beta, \gamma, \varsigma, \lambda \in \mathbb{C}$, $n, \kappa \in \mathbb{N}_0$, $Re(\alpha) > 0$, and $Re(\beta) > 0$.

Definition 5.67 The truncated supersine via Prabhakar function is defined as

$$\begin{aligned}\left\{supersin_{\alpha,\beta}^{\gamma}\right\}^{n}(t^{\alpha}) &= \frac{t^{\beta-1}}{2i}\left(\left\{E_{\alpha,\beta}^{\gamma}\right\}^{n}(it^{\alpha}) - \left\{E_{\alpha,\beta}^{\gamma}\right\}^{n}(-it^{\alpha})\right)\\ &= \sum_{k=0}^{n} \frac{(-1)^k(\gamma)_{2k+1}}{\Gamma((2k+1)\alpha+\beta)} \frac{t^{(2k+1)\alpha+\beta-1}}{(2k+1)!},\end{aligned} \tag{5.609}$$

where $t, \alpha, \beta \in \mathbb{C}$, $n, \kappa \in \mathbb{N}_0$, $Re(\alpha) > 0$, and $Re(\beta) > 0$.

Therefore,

$$\begin{aligned}\left\{supersin_{\alpha,\beta}^{\gamma}\right\}^{0}(t^{\alpha}) &= \frac{t^{\beta-1}}{2}\left(\left\{E_{\alpha,\beta}^{\gamma}\right\}^{0}(it^{\alpha}) - \left\{E_{\alpha,\beta}^{\gamma}\right\}^{0}(-it^{\alpha})\right)\\ &= \frac{\Gamma(\gamma+1)}{\Gamma(\alpha+\beta)}t^{\alpha+\beta-1},\end{aligned} \tag{5.610}$$

$$\begin{aligned}\left\{supersin_{\alpha,\beta}^{\gamma}\right\}^{1}(t^{\alpha}) &= \frac{t^{\beta-1}}{2}\left(\left\{E_{\alpha,\beta}^{\gamma}\right\}^{1}(it^{\alpha}) - \left\{E_{\alpha,\beta}^{\gamma}\right\}^{1}(-it^{\alpha})\right)\\ &= \sum_{k=0}^{1}\frac{(-1)^{k}(\gamma)_{2k+1}}{\Gamma((2k+1)\alpha+\beta)}\frac{t^{(2k+1)\alpha+\beta-1}}{(2k+1)!},\end{aligned} \tag{5.611}$$

and

$$\begin{aligned}\left\{supersinh_{\alpha,\beta}^{\gamma}\right\}^{2}(t^{\alpha}) &= \frac{t^{\beta-1}}{2}\left(\left\{E_{\alpha,\beta}^{\gamma}\right\}^{2}(t^{\alpha}) - \left\{E_{\alpha,\beta}^{\gamma}\right\}^{2}(-t^{\alpha})\right)\\ &= \sum_{k=0}^{2}\frac{(-1)^{k}(\gamma)_{2k+1}}{\Gamma((2k+1)\alpha+\beta)}\frac{t^{(2k+1)\alpha+\beta-1}}{(2k+1)!},\end{aligned} \tag{5.612}$$

where $t, \alpha, \beta \in \mathbb{C}$, $Re(\alpha) > 0$, and $Re(\beta) > 0$.

Definition 5.68 The truncated supercosine via Prabhakar function is defined as

$$\begin{aligned}\left\{supercos_{\alpha,\beta}^{\gamma}\right\}^{n}(t^{\alpha}) &= \frac{t^{\beta-1}}{2}\left(\left\{E_{\alpha,\beta}^{\gamma}\right\}^{n}(it^{\alpha}) + \left\{E_{\alpha,\beta}^{\gamma}\right\}^{n}(-it^{\alpha})\right)\\ &= \sum_{k=0}^{n}\frac{(-1)^{k}(\gamma)_{2k}}{\Gamma(2k\alpha+\beta)}\frac{t^{2k\alpha+\beta-1}}{(2k)!},\end{aligned} \tag{5.613}$$

where $t, \alpha, \beta \in \mathbb{C}$, $n, \kappa \in \mathbb{N}_0$, $Re(\alpha) > 0$, and $Re(\beta) > 0$.

Thus, we have that

$$\begin{aligned}\left\{supercos_{\alpha,\beta}^{\gamma}\right\}^{0}(t^{\alpha}) &= \frac{t^{\beta-1}}{2}\left(\left\{E_{\alpha,\beta}^{\gamma}\right\}^{0}(it^{\alpha}) + \left\{E_{\alpha,\beta}^{\gamma}\right\}^{0}(-it^{\alpha})\right)\\ &= \frac{t^{\beta-1}}{\Gamma(\beta)},\end{aligned} \tag{5.614}$$

$$\begin{aligned}\left\{supercos_{\alpha,\beta}^{\gamma}\right\}^{1}(t^{\alpha}) &= \frac{t^{\beta-1}}{2}\left(\left\{E_{\alpha,\beta}^{\gamma}\right\}^{1}(it^{\alpha}) + \left\{E_{\alpha,\beta}^{\gamma}\right\}^{1}(-it^{\alpha})\right)\\ &= \sum_{k=0}^{1}\frac{(-1)^{k}(\gamma)_{2k}}{\Gamma(2k\alpha+\beta)}\frac{t^{2k\alpha+\beta-1}}{(2k)!},\end{aligned} \tag{5.615}$$

and

$$\begin{aligned}\left\{supercos_{\alpha,\beta}^{\gamma}\right\}^{2}(t^{\alpha}) &= \frac{t^{\beta-1}}{2}\left(\left\{E_{\alpha,\beta}^{\gamma}\right\}^{2}(it^{\alpha}) + \left\{E_{\alpha,\beta}^{\gamma}\right\}^{2}(-it^{\alpha})\right)\\ &= \sum_{k=0}^{2}\frac{(-1)^{k}(\gamma)_{2k}}{\Gamma(2k\alpha+\beta)}\frac{t^{2k\alpha+\beta-1}}{(2k)!},\end{aligned} \tag{5.616}$$

where $t, \alpha, \beta \in \mathbb{C}$, $Re(\alpha) > 0$, and $Re(\beta) > 0$.

Definition 5.69 The truncated hyperbolic supersine via Prabhakar function is defined as

$$\begin{aligned}\left\{supersinh_{\alpha,\beta}^{\gamma}\right\}^{n}(t^{\alpha}) &= \frac{t^{\beta-1}}{2}\left(\left\{E_{\alpha,\beta}^{\gamma}\right\}^{n}(t^{\alpha}) - \left\{E_{\alpha,\beta}^{\gamma}\right\}^{n}(-t^{\alpha})\right)\\ &= \sum_{k=0}^{n}\frac{(\gamma)_{2k+1}t^{(2k+1)\alpha+\beta-1}}{\Gamma((2k+1)\alpha+\beta)},\end{aligned} \tag{5.617}$$

where $t, \alpha, \beta \in \mathbb{C}$, $n, \kappa \in \mathbb{N}_0$, $Re(\alpha) > 0$, and $Re(\beta) > 0$.

Thus, we have that

$$\begin{aligned}\left\{supersinh_{\alpha,\beta}^{\gamma}\right\}^{0}(t^{\alpha}) &= \frac{t^{\beta-1}}{2}\left(\left\{E_{\alpha,\beta}^{\gamma}\right\}^{0}(t^{\alpha}) - \left\{E_{\alpha,\beta}^{\gamma}\right\}^{0}(-t^{\alpha})\right)\\ &= \frac{\Gamma(\gamma+1)}{\Gamma(\alpha+\beta)}t^{\alpha+\beta-1},\end{aligned} \tag{5.618}$$

$$\begin{aligned}\left\{supersinh_{\alpha,\beta}^{\gamma}\right\}^{1}(t^{\alpha}) &= \frac{t^{\beta-1}}{2}\left(\left\{E_{\alpha,\beta}^{\gamma}\right\}^{1}(t^{\alpha}) - \left\{E_{\alpha,\beta}^{\gamma}\right\}^{1}(-t^{\alpha})\right)\\ &= \sum_{k=0}^{1}\frac{(\gamma)_{2k+1}t^{(2k+1)\alpha+\beta-1}}{\Gamma((2k+1)\alpha+\beta)},\end{aligned} \tag{5.619}$$

and

$$\begin{aligned}\left\{supersinh_{\alpha,\beta}^{\gamma}\right\}^{2}(t^{\alpha}) &= \frac{t^{\beta-1}}{2}\left(\left\{E_{\alpha,\beta}^{\gamma}\right\}^{2}(t^{\alpha}) - \left\{E_{\alpha,\beta}^{\gamma}\right\}^{2}(-t^{\alpha})\right)\\ &= \sum_{k=0}^{2}\frac{(\gamma)_{2k+1}t^{(2k+1)\alpha+\beta-1}}{\Gamma((2k+1)\alpha+\beta)},\end{aligned} \tag{5.620}$$

where $t, \alpha, \beta \in \mathbb{C}$, $Re(\alpha) > 0$, and $Re(\beta) > 0$.

Definition 5.70 The truncated hyperbolic supercosine via Prabhakar function is defined as

$$\begin{aligned}&\left\{supercosh_{\alpha,\beta}^{\gamma}\right\}^{n}(t^{\alpha})\\ &= \frac{t^{\beta-1}}{2}\left(\left\{E_{\alpha,\beta}^{\gamma}\right\}^{n}(t^{\alpha}) + \left\{E_{\alpha,\beta}^{\gamma}\right\}^{n}(-t^{\alpha})\right)\\ &= \sum_{n=0}^{\infty}\frac{(\gamma)_{2k}t^{2k\alpha+\beta-1}}{\Gamma(2k\alpha+\beta)},\end{aligned} \tag{5.621}$$

where $t, \alpha, \beta \in \mathbb{C}$, $n, \kappa \in \mathbb{N}_0$, $Re(\alpha) > 0$, and $Re(\beta) > 0$.

Therefore,

$$\begin{aligned}\left\{supercosh_{\alpha,\beta}^{\gamma}\right\}^{0}(t^{\alpha}) &= \frac{t^{\beta-1}}{2}\left(\left\{E_{\alpha,\beta}^{\gamma}\right\}^{0}(t^{\alpha}) + \left\{E_{\alpha,\beta}^{\gamma}\right\}^{0}(-t^{\alpha})\right)\\ &= \frac{t^{\beta-1}}{\Gamma(\beta)},\end{aligned} \tag{5.622}$$

$$\begin{aligned}\left\{supercosh^{\gamma}_{\alpha,\beta}\right\}^1(t^\alpha) &= \tfrac{t^{\beta-1}}{2}\left(\left\{E^{\gamma}_{\alpha,\beta}\right\}^1(t^\alpha)+\left\{E^{\gamma}_{\alpha,\beta}\right\}^1(-t^\alpha)\right)\\ &= \sum_{n=0}^{1}\tfrac{(\gamma)_{2k}t^{2k\alpha+\beta-1}}{\Gamma(2k\alpha+\beta)},\end{aligned} \tag{5.623}$$

and

$$\begin{aligned}\left\{supercosh^{\gamma}_{\alpha,\beta}\right\}^2(t^\alpha) &= \tfrac{t^{\beta-1}}{2}\left(\left\{E^{\gamma}_{\alpha,\beta}\right\}^2(t^\alpha)+\left\{E^{\gamma}_{\alpha,\beta}\right\}^2(-t^\alpha)\right)\\ &= \sum_{n=0}^{2}\tfrac{(\gamma)_{2k}t^{2k\alpha+\beta-1}}{\Gamma(2k\alpha+\beta)},\end{aligned} \tag{5.624}$$

where $t, \alpha, \beta \in \mathbb{C}$, $Re(\alpha) > 0$, and $Re(\beta) > 0$.

So, we have that

$$\left\{supersin^{\gamma}_{\alpha,\beta}\right\}^n(\lambda t^\alpha) = \frac{t^{\beta-1}}{2i}\left(\left\{E^{\gamma}_{\alpha,\beta}\right\}^n(i\lambda t^\alpha)-\left\{E^{\gamma}_{\alpha,\beta}\right\}^n(-i\lambda t^\alpha)\right), \tag{5.625}$$

$$\left\{supercos^{\gamma}_{\alpha,\beta}\right\}^n(\lambda t^\alpha) = \frac{t^{\beta-1}}{2}\left(\left\{E^{\gamma}_{\alpha,\beta}\right\}^n(i\lambda t^\alpha)+\left\{E^{\gamma}_{\alpha,\beta}\right\}^n(-i\lambda t^\alpha)\right), \tag{5.626}$$

$$\left\{supersinh^{\gamma}_{\alpha,\beta}\right\}^n(\lambda t^\alpha) = \frac{t^{\beta-1}}{2}\left(\left\{E^{\gamma}_{\alpha,\beta}\right\}^n(\lambda t^\alpha)-\left\{E^{\gamma}_{\alpha,\beta}\right\}^n(-\lambda t^\alpha)\right), \tag{5.627}$$

and

$$\left\{supercosh^{\gamma}_{\alpha,\beta}\right\}^n(\lambda t^\alpha) = \frac{t^{\beta-1}}{2}\left(\left\{E^{\gamma}_{\alpha,\beta}\right\}^n(\lambda t^\alpha)+\left\{E^{\gamma}_{\alpha,\beta}\right\}^n(-\lambda t^\alpha)\right), \tag{5.628}$$

where $t, \alpha, \beta, \lambda \in \mathbb{C}$, $n, \kappa \in \mathbb{N}_0$, $Re(\alpha) > 0$, and $Re(\beta) > 0$.

Definition 5.71 The truncated presine via Prabhakar function is defined as

$$\begin{aligned}\left\{presin^{\gamma}_{\alpha,\beta}\right\}^n(t) &= \tfrac{1}{2i}\left(\left\{E^{\gamma}_{\alpha,\beta}\right\}^n(it)-\left\{E^{\gamma}_{\alpha,\beta}\right\}^n(-it)\right)\\ &= \sum_{k=0}^{n}\tfrac{(-1)^k(\gamma)_{2k+1}t^{2k+1}}{\Gamma((2k+1)\alpha+\beta)},\end{aligned} \tag{5.629}$$

where $t, \alpha, \beta, \gamma \in \mathbb{C}$, $n, \kappa \in \mathbb{N}_0$, $Re(\alpha) > 0$, and $Re(\beta) > 0$.

Thus, we have that

$$\begin{aligned}\left\{presin^{\gamma}_{\alpha,\beta}\right\}^0(t) &= \tfrac{1}{2}\left(\left\{E^{\gamma}_{\alpha,\beta}\right\}^0(it)-\left\{E^{\gamma}_{\alpha,\beta}\right\}^0(-it)\right)\\ &= \tfrac{\Gamma(\gamma+1)}{\Gamma(\alpha+\beta)}t,\end{aligned} \tag{5.630}$$

$$\left\{presin_{\alpha,\beta}^{\gamma}\right\}^{1}(t) = \frac{1}{2}\left(\left\{E_{\alpha,\beta}^{\gamma}\right\}^{1}(it) - \left\{E_{\alpha,\beta}^{\gamma}\right\}^{1}(-it)\right) = \sum_{k=0}^{1} \frac{(-1)^k (\gamma)_{2k+1} t^{2k+1}}{\Gamma((2k+1)\alpha+\beta)}, \tag{5.631}$$

and

$$\left\{presin_{\alpha,\beta}^{\gamma}\right\}^{2}(t) = \frac{1}{2}\left(\left\{E_{\alpha,\beta}^{\gamma}\right\}^{2}(it) - \left\{E_{\alpha,\beta}^{\gamma}\right\}^{2}(-it)\right) = \sum_{k=0}^{2} \frac{(-1)^k (\gamma)_{2k+1} t^{2k+1}}{\Gamma((2k+1)\alpha+\beta)}, \tag{5.632}$$

where $t, \alpha, \beta, \gamma \in \mathbb{C}$, $Re(\alpha) > 0$, and $Re(\beta) > 0$.

Definition 5.72 The truncated precosine via Prabhakar function is defined as

$$\left\{precos_{\alpha,\beta}^{\gamma}\right\}^{n}(t) = \frac{1}{2}\left(\left\{E_{\alpha,\beta}^{\gamma}\right\}^{n}(it) + \left\{E_{\alpha,\beta}^{\gamma}\right\}^{n}(-it)\right) = \sum_{k=0}^{n} \frac{(-1)^k (\gamma)_{2k} t^{2k}}{\Gamma(2k\alpha+\beta)}, \tag{5.633}$$

where $t, \alpha, \beta, \gamma \in \mathbb{C}$, $n, \kappa \in \mathbb{N}_0$, $Re(\alpha) > 0$, and $Re(\beta) > 0$.

Therefore,

$$\left\{precos_{\alpha,\beta}^{\gamma}\right\}^{0}(t) = \frac{1}{2}\left(\left\{E_{\alpha,\beta}^{\gamma}\right\}^{0}(it) + \left\{E_{\alpha,\beta}^{\gamma}\right\}^{0}(-it)\right) = \frac{1}{\Gamma(\beta)}, \tag{5.634}$$

$$\left\{precos_{\alpha,\beta}^{\gamma}\right\}^{1}(t) = \frac{1}{2}\left(\left\{E_{\alpha,\beta}^{\gamma}\right\}^{1}(it) + \left\{E_{\alpha,\beta}^{\gamma}\right\}^{1}(-it)\right) = \sum_{k=0}^{1} \frac{(-1)^k (\gamma)_{2k} t^{2k}}{\Gamma(2k\alpha+\beta)}, \tag{5.635}$$

and

$$\left\{precos_{\alpha,\beta}^{\gamma}\right\}^{2}(t) = \frac{1}{2}\left(\left\{E_{\alpha,\beta}^{\gamma}\right\}^{2}(it) + \left\{E_{\alpha,\beta}^{\gamma}\right\}^{2}(-it)\right) = \sum_{k=0}^{2} \frac{(-1)^k (\gamma)_{2k} t^{2k}}{\Gamma(2k\alpha+\beta)}, \tag{5.636}$$

where $t, \alpha, \beta, \gamma \in \mathbb{C}$, $Re(\alpha) > 0$, and $Re(\beta) > 0$.

Definition 5.73 The truncated hyperbolic presine via Prabhakar function is defined as

$$\begin{aligned}\left\{presinh_{\alpha,\beta}^{\gamma}\right\}^{n}(t) &= \frac{1}{2}\left(\left\{E_{\alpha,\beta}^{\gamma}\right\}^{n}(t) - \left\{E_{\alpha,\beta}^{\gamma}\right\}^{n}(-t)\right) \\ &= \sum_{k=0}^{\infty}\frac{(\gamma)_{2k+1}t^{2k+1}}{\Gamma((2k+1)\alpha+\beta)},\end{aligned} \tag{5.637}$$

where $t, \alpha, \beta, \gamma \in \mathbb{C}$, $n, \kappa \in \mathbb{N}_0$, $Re(\alpha) > 0$, and $Re(\beta) > 0$.

Thus, we get

$$\begin{aligned}\left\{presinh_{\alpha,\beta}^{\gamma}\right\}^{0}(t) &= \frac{1}{2}\left(\left\{E_{\alpha,\beta}^{\gamma}\right\}^{0}(t) - \left\{E_{\alpha,\beta}^{\gamma}\right\}^{0}(-t)\right) \\ &= \frac{\Gamma(\gamma+1)}{\Gamma(\alpha+\beta)}t,\end{aligned} \tag{5.638}$$

$$\begin{aligned}\left\{presinh_{\alpha,\beta}^{\gamma}\right\}^{1}(t) &= \frac{1}{2}\left(\left\{E_{\alpha,\beta}^{\gamma}\right\}^{1}(t) - \left\{E_{\alpha,\beta}^{\gamma}\right\}^{1}(-t)\right) \\ &= \sum_{k=0}^{1}\frac{(\gamma)_{2k+1}t^{2k+1}}{\Gamma((2k+1)\alpha+\beta)},\end{aligned} \tag{5.639}$$

and

$$\begin{aligned}\left\{presinh_{\alpha,\beta}^{\gamma}\right\}^{2}(t) &= \frac{1}{2}\left(\left\{E_{\alpha,\beta}^{\gamma}\right\}^{2}(t) - \left\{E_{\alpha,\beta}^{\gamma}\right\}^{2}(-t)\right) \\ &= \sum_{k=0}^{2}\frac{(\gamma)_{2k+1}t^{2k+1}}{\Gamma((2k+1)\alpha+\beta)},\end{aligned} \tag{5.640}$$

where $t, \alpha, \beta, \gamma \in \mathbb{C}$, $Re(\alpha) > 0$, and $Re(\beta) > 0$.

Definition 5.74 The truncated hyperbolic precosine via Prabhakar function is defined as

$$\begin{aligned}\left\{precosh_{\alpha,\beta}^{\gamma}\right\}^{n}(t) &= \frac{1}{2}\left(\left\{E_{\alpha,\beta}^{\gamma}\right\}^{n}(t) + \left\{E_{\alpha,\beta}^{\gamma}\right\}^{n}(-t)\right) \\ &= \sum_{k=0}^{n}\frac{(\gamma)_{2k}t^{2k}}{\Gamma(2k\alpha+\beta)},\end{aligned} \tag{5.641}$$

where $t, \alpha, \beta, \gamma \in \mathbb{C}$, $n, \kappa \in \mathbb{N}_0$, $Re(\alpha) > 0$, and $Re(\beta) > 0$.

Therefore, we get

$$\begin{aligned}\left\{precosh_{\alpha,\beta}^{\gamma}\right\}^{0}(t) &= \frac{1}{2}\left(\left\{E_{\alpha,\beta}^{\gamma}\right\}^{0}(t) + \left\{E_{\alpha,\beta}^{\gamma}\right\}^{0}(-t)\right) \\ &= \frac{1}{\Gamma(\beta)},\end{aligned} \tag{5.642}$$

$$\begin{aligned}\left\{precosh^{\gamma}_{\alpha,\beta}\right\}^{1}(t) &= \tfrac{1}{2}\left(\left\{E^{\gamma}_{\alpha,\beta}\right\}^{1}(t)+\left\{E^{\gamma}_{\alpha,\beta}\right\}^{1}(-t)\right)\\ &= \sum_{k=0}^{1}\tfrac{(\gamma)_{2k}t^{2k}}{\Gamma(2k\alpha+\beta)},\end{aligned} \tag{5.643}$$

and

$$\begin{aligned}\left\{precosh^{\gamma}_{\alpha,\beta}\right\}^{2}(t) &= \tfrac{1}{2}\left(\left\{E^{\gamma}_{\alpha,\beta}\right\}^{2}(t)+\left\{E^{\gamma}_{\alpha,\beta}\right\}^{2}(-t)\right)\\ &= \sum_{k=0}^{2}\tfrac{(\gamma)_{2k}t^{2k}}{\Gamma(2k\alpha+\beta)},\end{aligned} \tag{5.644}$$

where $t,\alpha,\beta,\gamma\in\mathbb{C}$, $n,\kappa\in\mathbb{N}_0$, $Re(\alpha)>0$, and $Re(\beta)>0$.

Thus, we show that

$$\left\{presin^{\gamma}_{\alpha,\beta}\right\}^{n}(\lambda t)=\frac{1}{2i}\left(\left\{E^{\gamma}_{\alpha,\beta}\right\}^{n}(i\lambda t)-\left\{E^{\gamma}_{\alpha,\beta}\right\}^{n}(-i\lambda t)\right), \tag{5.645}$$

$$\left\{precos^{\gamma}_{\alpha,\beta}\right\}^{n}(\lambda t)=\frac{1}{2}\left(\left\{E^{\gamma}_{\alpha,\beta}\right\}^{n}(i\lambda t)+\left\{E^{\gamma}_{\alpha,\beta}\right\}^{n}(-i\lambda t)\right), \tag{5.646}$$

$$\left\{presinh^{\gamma}_{\alpha,\beta}\right\}^{n}(\lambda t)=\frac{1}{2}\left(\left\{E^{\gamma}_{\alpha,\beta}\right\}^{n}(\lambda t)-\left\{E^{\gamma}_{\alpha,\beta}\right\}^{n}(-\lambda t)\right), \tag{5.647}$$

and

$$\left\{precosh^{\gamma}_{\alpha,\beta}\right\}^{n}(\lambda t)=\frac{1}{2}\left(\left\{E^{\gamma}_{\alpha,\beta}\right\}^{n}(\lambda t)+\left\{E^{\gamma}_{\alpha,\beta}\right\}^{n}(-\lambda t)\right), \tag{5.648}$$

where $t,\alpha,\beta,\gamma,\lambda\in\mathbb{C}$, $n,\kappa\in\mathbb{N}_0$, $Re(\alpha)>0$, and $Re(\beta)>0$.

Definition 5.75 The truncated presine via Prabhakar function is defined as

$$\begin{aligned}\left\{presin^{\gamma}_{\alpha,\beta}\right\}^{n}(t^{\varsigma}) &= \tfrac{1}{2i}\left(\left\{E^{\gamma}_{\alpha,\beta}\right\}^{n}(it^{\varsigma})-\left\{E^{\gamma}_{\alpha,\beta}\right\}^{n}(-it^{\varsigma})\right)\\ &= \sum_{k=0}^{n}\tfrac{(-1)^{k}(\gamma)_{2k+1}t^{(2k+1)\varsigma}}{\Gamma((2k+1)\alpha+\beta)},\end{aligned} \tag{5.649}$$

where $t,\alpha,\beta,\gamma,\varsigma,\lambda\in\mathbb{C}$, $n,\kappa\in\mathbb{N}_0$, $Re(\alpha)>0$, and $Re(\beta)>0$.

Therefore, we show that

$$\begin{aligned}\left\{presin^{\gamma}_{\alpha,\beta}\right\}^{0}(t^{\varsigma}) &= \tfrac{1}{2}\left(\left\{E^{\gamma}_{\alpha,\beta}\right\}^{0}(it^{\varsigma})-\left\{E^{\gamma}_{\alpha,\beta}\right\}^{0}(-it^{\varsigma})\right)\\ &= \tfrac{\Gamma(\gamma+1)}{\Gamma(\alpha+\beta)}t^{\varsigma},\end{aligned} \tag{5.650}$$

$$\left\{presin_{\alpha,\beta}^{\gamma}\right\}^{1}(t^{\varsigma}) = \frac{1}{2}\left(\left\{E_{\alpha,\beta}^{\gamma}\right\}^{1}(it^{\varsigma}) - \left\{E_{\alpha,\beta}^{\gamma}\right\}^{1}(-it^{\varsigma})\right) = \sum_{k=0}^{1}\frac{(-1)^{k}(\gamma)_{2k+1}t^{(2k+1)\varsigma}}{\Gamma((2k+1)\alpha+\beta)}, \tag{5.651}$$

and

$$\left\{presinh_{\alpha,\beta}^{\gamma}\right\}^{2}(t^{\varsigma}) = \frac{1}{2}\left(\left\{E_{\alpha,\beta}^{\gamma}\right\}^{2}(it^{\varsigma}) - \left\{E_{\alpha,\beta}^{\gamma}\right\}^{2}(-it^{\varsigma})\right) = \sum_{k=0}^{2}\frac{(-1)^{k}(\gamma)_{2k+1}t^{(2k+1)\varsigma}}{\Gamma((2k+1)\alpha+\beta)}, \tag{5.652}$$

where $t, \alpha, \beta, \gamma, \varsigma, \lambda \in \mathbb{C}$, $Re(\alpha) > 0$, and $Re(\beta) > 0$.

Definition 5.76 The truncated precosine via Wiman function is defined as

$$\left\{precos_{\alpha,\beta}^{\gamma}\right\}^{n}(t^{\varsigma}) = \frac{1}{2}\left(\left\{E_{\alpha,\beta}^{\gamma}\right\}^{n}(it^{\varsigma}) + \left\{E_{\alpha,\beta}^{\gamma}\right\}^{n}(-it^{\varsigma})\right) = \sum_{k=0}^{n}\frac{(-1)^{k}(\gamma)_{2k}t^{2k\varsigma}}{\Gamma(2k\alpha+\beta)}, \tag{5.653}$$

where $t, \alpha, \beta, \gamma, \varsigma, \lambda \in \mathbb{C}$, $n, \kappa \in \mathbb{N}_0$, $Re(\alpha) > 0$, and $Re(\beta) > 0$.

Therefore,

$$\left\{precos_{\alpha,\beta}^{\gamma}\right\}^{2}(t^{\varsigma}) = \frac{1}{2}\left(\left\{E_{\alpha,\beta}^{\gamma}\right\}^{2}(it^{\varsigma}) + \left\{E_{\alpha,\beta}^{\gamma}\right\}^{2}(-it^{\varsigma})\right) = \frac{1}{\Gamma(\beta)}, \tag{5.654}$$

$$\left\{precos_{\alpha,\beta}^{\gamma}\right\}^{1}(t^{\varsigma}) = \frac{1}{2}\left(\left\{E_{\alpha,\beta}^{\gamma}\right\}^{1}(it^{\varsigma}) + \left\{E_{\alpha,\beta}^{\gamma}\right\}^{1}(-it^{\varsigma})\right) = \sum_{k=0}^{1}\frac{(-1)^{k}(\gamma)_{2k}t^{2k\varsigma}}{\Gamma(2k\alpha+\beta)}, \tag{5.655}$$

$$\left\{precos_{\alpha,\beta}^{\gamma}\right\}^{2}(t^{\varsigma}) = \frac{1}{2}\left(\left\{E_{\alpha,\beta}^{\gamma}\right\}^{2}(it^{\varsigma}) + \left\{E_{\alpha,\beta}^{\gamma}\right\}^{2}(-it^{\varsigma})\right) = \sum_{k=0}^{2}\frac{(-1)^{k}(\gamma)_{2k}t^{2k\varsigma}}{\Gamma(2k\alpha+\beta)}, \tag{5.656}$$

where $t, \alpha, \beta, \gamma, \varsigma, \lambda \in \mathbb{C}$, $Re(\alpha) > 0$, and $Re(\beta) > 0$.

Definition 5.77 The truncated hyperbolic presine via Prabhakar function is defined as

$$\begin{aligned}\left\{presinh_{\alpha,\beta}^{\gamma}\right\}^{n}(t^{\varsigma}) &= \frac{1}{2}\left(\left\{E_{\alpha,\beta}^{\gamma}\right\}^{n}(t^{\varsigma}) - \left\{E_{\alpha,\beta}^{\gamma}\right\}^{n}(-t^{\varsigma})\right)\\ &= \sum_{k=0}^{n}\frac{(\gamma)_{2k+1}t^{(2k+1)\varsigma}}{\Gamma((2k+1)\alpha+\beta)},\end{aligned} \tag{5.657}$$

where $t, \alpha, \beta, \gamma, \varsigma, \lambda \in \mathbb{C}$, $n, \kappa \in \mathbb{N}_0$, $Re\,(\alpha) > 0$, and $Re\,(\beta) > 0$.

Thus, we have that

$$\begin{aligned}\left\{presinh_{\alpha,\beta}^{\gamma}\right\}^{0}(t^{\varsigma}) &= \frac{1}{2}\left(\left\{E_{\alpha,\beta}^{\gamma}\right\}^{0}(t^{\varsigma}) - \left\{E_{\alpha,\beta}^{\gamma}\right\}^{0}(-t^{\varsigma})\right)\\ &= \frac{\Gamma(\gamma+1)}{\Gamma(\alpha+\beta)}t^{\varsigma},\end{aligned} \tag{5.658}$$

$$\begin{aligned}\left\{presinh_{\alpha,\beta}^{\gamma}\right\}^{1}(t^{\varsigma}) &= \frac{1}{2}\left(\left\{E_{\alpha,\beta}^{\gamma}\right\}^{1}(t^{\varsigma}) - \left\{E_{\alpha,\beta}^{\gamma}\right\}^{1}(-t^{\varsigma})\right)\\ &= \sum_{k=0}^{1}\frac{(-1)^{k}(\gamma)_{2k+1}t^{(2k+1)\varsigma}}{\Gamma((2k+1)\alpha+\beta)},\end{aligned} \tag{5.659}$$

and

$$\begin{aligned}\left\{presinh_{\alpha,\beta}^{\gamma}\right\}^{2}(t^{\varsigma}) &= \frac{1}{2}\left(\left\{E_{\alpha,\beta}^{\gamma}\right\}^{2}(t^{\varsigma}) - \left\{E_{\alpha,\beta}^{\gamma}\right\}^{2}(-t^{\varsigma})\right)\\ &= \sum_{k=0}^{2}\frac{(-1)^{k}(\gamma)_{2k+1}t^{(2k+1)\varsigma}}{\Gamma((2k+1)\alpha+\beta)},\end{aligned} \tag{5.660}$$

where $t, \alpha, \beta, \gamma, \varsigma, \lambda \in \mathbb{C}$, $Re\,(\alpha) > 0$, and $Re\,(\beta) > 0$.

Definition 5.78 The truncated hyperbolic precosine via Prabhakar function is defined as

$$\begin{aligned}\left\{precosh_{\alpha,\beta}^{\gamma}\right\}^{n}(t^{\varsigma}) &= \frac{1}{2}\left(\left\{E_{\alpha,\beta}^{\gamma}\right\}^{n}(t^{\varsigma}) + \left\{E_{\alpha,\beta}^{\gamma}\right\}^{n}(-t^{\varsigma})\right)\\ &= \sum_{k=0}^{\infty}\frac{(\gamma)_{2k}t^{2k\varsigma}}{\Gamma(2k\alpha+\beta)},\end{aligned} \tag{5.661}$$

where $t, \alpha, \beta, \gamma, \varsigma, \lambda \in \mathbb{C}$, $n, \kappa \in \mathbb{N}_0$, $Re\,(\alpha) > 0$, and $Re\,(\beta) > 0$.

Thus,

$$\begin{aligned}\left\{precosh_{\alpha,\beta}^{\gamma}\right\}^{0}(t^{\varsigma}) &= \frac{1}{2}\left(\left\{E_{\alpha,\beta}^{\gamma}\right\}^{0}(t^{\varsigma}) + \left\{E_{\alpha,\beta}^{\gamma}\right\}^{0}(-t^{\varsigma})\right)\\ &= \frac{1}{\Gamma(\beta)},\end{aligned} \tag{5.662}$$

$$\begin{aligned}\left\{precosh_{\alpha,\beta}^{\gamma}\right\}^{1}(t^{\varsigma}) &= \frac{1}{2}\left(\left\{E_{\alpha,\beta}^{\gamma}\right\}^{1}(t^{\varsigma}) + \left\{E_{\alpha,\beta}^{\gamma}\right\}^{1}(-t^{\varsigma})\right) \\ &= \sum_{k=0}^{1}\frac{(\gamma)_{2k}t^{2k\varsigma}}{\Gamma(2k\alpha+\beta)},\end{aligned} \tag{5.663}$$

$$\begin{aligned}\left\{precosh_{\alpha,\beta}^{\gamma}\right\}^{2}(t^{\varsigma}) &= \frac{1}{2}\left(\left\{E_{\alpha,\beta}^{\gamma}\right\}^{2}(t^{\varsigma}) + \left\{E_{\alpha,\beta}^{\gamma}\right\}^{2}(-t^{\varsigma})\right) \\ &= \sum_{k=0}^{2}\frac{(\gamma)_{2k}t^{2k\varsigma}}{\Gamma(2k\alpha+\beta)},\end{aligned} \tag{5.664}$$

where $t, \alpha, \beta, \gamma, \varsigma, \lambda \in \mathbb{C}$, $n, \kappa \in \mathbb{N}_0$, $Re(\alpha) > 0$, and $Re(\beta) > 0$.

Thus, we have that

$$\left\{presin_{\alpha,\beta}^{\gamma}\right\}^{n}(\lambda t^{\varsigma}) = \frac{1}{2i}\left(\left\{E_{\alpha,\beta}^{\gamma}\right\}^{n}(i\lambda t^{\varsigma}) - \left\{E_{\alpha,\beta}^{\gamma}\right\}^{n}(-i\lambda t^{\varsigma})\right), \tag{5.665}$$

$$\left\{precos_{\alpha,\beta}^{\gamma}\right\}^{n}(\lambda t^{\varsigma}) = \frac{1}{2}\left(\left\{E_{\alpha,\beta}^{\gamma}\right\}^{n}(i\lambda t^{\varsigma}) + \left\{E_{\alpha,\beta}^{\gamma}\right\}^{n}(-i\lambda t^{\varsigma})\right), \tag{5.666}$$

$$\left\{presinh_{\alpha,\beta}^{\gamma}\right\}^{n}(\lambda t^{\varsigma}) = \frac{1}{2}\left(\left\{E_{\alpha,\beta}^{\gamma}\right\}^{n}(\lambda t^{\varsigma}) - \left\{E_{\alpha,\beta}^{\gamma}\right\}^{n}(-\lambda t^{\varsigma})\right), \tag{5.667}$$

and

$$\left\{precosh_{\alpha,\beta}^{\gamma}\right\}^{n}(\lambda t^{\varsigma}) = \frac{1}{2}\left(\left\{E_{\alpha,\beta}^{\gamma}\right\}^{n}(\lambda t^{\varsigma}) + \left\{E_{\alpha,\beta}^{\gamma}\right\}^{n}(-\lambda t^{\varsigma})\right), \tag{5.668}$$

where $t, \alpha, \beta, \gamma, \varsigma, \lambda \in \mathbb{C}$, $n, \kappa \in \mathbb{N}_0$, $Re(\alpha) > 0$, and $Re(\beta) > 0$.

Definition 5.79 The truncated presine via Prabhakar is defined as

$$\begin{aligned}\left\{Tresin_{\alpha,\beta}^{\gamma}\right\}^{n}(t) &= \frac{t^{\beta-1}}{2i}\left(\left\{E_{\alpha,\beta}^{\gamma}\right\}^{n}(it) - \left\{E_{\alpha,\beta}^{\gamma}\right\}^{n}(-it)\right) \\ &= t^{\beta-1}\sum_{k=0}^{n}\frac{(-1)^{k}(\gamma)_{2k+1}t^{2k+1}}{\Gamma((2k+1)\alpha+\beta)},\end{aligned} \tag{5.669}$$

where $t, \alpha, \beta, \gamma \in \mathbb{C}$, $n, \kappa \in \mathbb{N}_0$, $Re(\alpha) > 0$, and $Re(\beta) > 0$.

So, we have

$$\begin{aligned}\left\{Tresin_{\alpha,\beta}^{\gamma}\right\}^{0}(t) &= \frac{t^{\beta-1}}{2}\left(\left\{E_{\alpha,\beta}^{\gamma}\right\}^{0}(it) - \left\{E_{\alpha,\beta}^{\gamma}\right\}^{0}(-it)\right) \\ &= \frac{\Gamma(\gamma+1)}{\Gamma(\alpha+\beta)}t^{\beta},\end{aligned} \tag{5.670}$$

$$\begin{aligned} \left\{Tresin^{\gamma}_{\alpha,\beta}\right\}^{1}(t) &= \frac{t^{\beta-1}}{2}\left(\left\{E^{\gamma}_{\alpha,\beta}\right\}^{1}(it) - \left\{E^{\gamma}_{\alpha,\beta}\right\}^{1}(-it)\right) \\ &= t^{\beta-1}\sum_{k=0}^{1}\frac{(-1)^{k}(\gamma)_{2k+1}t^{2k+1}}{\Gamma((2k+1)\alpha+\beta)}, \end{aligned} \tag{5.671}$$

and

$$\begin{aligned} \left\{Tresin^{\gamma}_{\alpha,\beta}\right\}^{2}(t) &= \frac{t^{\beta-1}}{2}\left(\left\{E^{\gamma}_{\alpha,\beta}\right\}^{2}(it) - \left\{E^{\gamma}_{\alpha,\beta}\right\}^{2}(-it)\right) \\ &= t^{\beta-1}\sum_{k=0}^{2}\frac{(-1)^{k}(\gamma)_{2k+1}t^{2k+1}}{\Gamma((2k+1)\alpha+\beta)}, \end{aligned} \tag{5.672}$$

where $t, \alpha, \beta, \gamma \in \mathbb{C}$, $Re\,(\alpha) > 0$, and $Re\,(\beta) > 0$.

Definition 5.80 The truncated precosine via Prabhakar function is defined as

$$\begin{aligned} \left\{Trecos^{\gamma}_{\alpha,\beta}\right\}^{n}(t) &= \frac{t^{\beta-1}}{2}\left(\left\{E^{\gamma}_{\alpha,\beta}\right\}^{n}(it) + \left\{E^{\gamma}_{\alpha,\beta}\right\}^{n}(-it)\right) \\ &= t^{\beta-1}\sum_{k=0}^{n}\frac{(-1)^{k}(\gamma)_{2k}t^{2k}}{\Gamma(2k\alpha+\beta)}, \end{aligned} \tag{5.673}$$

where $t, \alpha, \beta, \gamma \in \mathbb{C}$, $n, \kappa \in \mathbb{N}_0$, $Re\,(\alpha) > 0$, and $Re\,(\beta) > 0$.

Thus, we show that

$$\begin{aligned} \left\{Trecos^{\gamma}_{\alpha,\beta}\right\}^{0}(t) &= \frac{t^{\beta-1}}{2}\left(\left\{E^{\gamma}_{\alpha,\beta}\right\}^{0}(it) + \left\{E^{\gamma}_{\alpha,\beta}\right\}^{0}(-it)\right) \\ &= \frac{t^{\beta-1}}{\Gamma(\beta)}, \end{aligned} \tag{5.674}$$

$$\begin{aligned} \left\{Trecos^{\gamma}_{\alpha,\beta}\right\}^{1}(t) &= \frac{t^{\beta-1}}{2}\left(\left\{E^{\gamma}_{\alpha,\beta}\right\}^{1}(it) + \left\{E^{\gamma}_{\alpha,\beta}\right\}^{1}(-it)\right) \\ &= t^{\beta-1}\sum_{k=0}^{1}\frac{(-1)^{k}(\gamma)_{2k}t^{2k}}{\Gamma(2k\alpha+\beta)}, \end{aligned} \tag{5.675}$$

and

$$\begin{aligned} \left\{Trecos^{\gamma}_{\alpha,\beta}\right\}^{2}(t) &= \frac{t^{\beta-1}}{2}\left(\left\{E^{\gamma}_{\alpha,\beta}\right\}^{2}(t) + \left\{E^{\gamma}_{\alpha,\beta}\right\}^{2}(-t)\right) \\ &= t^{\beta-1}\sum_{k=0}^{2}\frac{(-1)^{k}(\gamma)_{2k}t^{2k}}{\Gamma(2k\alpha+\beta)}, \end{aligned} \tag{5.676}$$

where $t, \alpha, \beta, \gamma \in \mathbb{C}$, $Re\,(\alpha) > 0$, and $Re\,(\beta) > 0$.

Definition 5.81 The truncated hyperbolic presine via Prabhakar function is defined as

$$\begin{aligned}\left\{Tresinh^{\gamma}_{\alpha,\beta}\right\}^{n}(t) &= \frac{t^{\beta-1}}{2}\left(\left\{E^{\gamma}_{\alpha,\beta}\right\}^{n}(t) - \left\{E^{\gamma}_{\alpha,\beta}\right\}^{n}(-t)\right)\\ &= t^{\beta-1}\sum_{k=0}^{\infty}\frac{(\gamma)_{2k+1}t^{2k+1}}{\Gamma((2k+1)\alpha+\beta)},\end{aligned} \tag{5.677}$$

where $t, \alpha, \beta, \gamma \in \mathbb{C}$, $n, \kappa \in \mathbb{N}_0$, $Re(\alpha) > 0$, and $Re(\beta) > 0$.

Thus, we get

$$\begin{aligned}\left\{Tresinh^{\gamma}_{\alpha,\beta}\right\}^{0}(t) &= \frac{t^{\beta-1}}{2}\left(\left\{E^{\gamma}_{\alpha,\beta}\right\}^{0}(t) - \left\{E^{\gamma}_{\alpha,\beta}\right\}^{0}(-t)\right)\\ &= \frac{\Gamma(\gamma+1)}{\Gamma(\alpha+\beta)}t^{\beta},\end{aligned} \tag{5.678}$$

$$\begin{aligned}\left\{Tresinh^{\gamma}_{\alpha,\beta}\right\}^{1}(t) &= \frac{t^{\beta-1}}{2}\left(\left\{E^{\gamma}_{\alpha,\beta}\right\}^{1}(t) - \left\{E^{\gamma}_{\alpha,\beta}\right\}^{1}(-t)\right)\\ &= t^{\beta-1}\sum_{k=0}^{1}\frac{(\gamma)_{2k+1}t^{2k+1}}{\Gamma((2k+1)\alpha+\beta)},\end{aligned} \tag{5.679}$$

and

$$\begin{aligned}\left\{Tresinh^{\gamma}_{\alpha,\beta}\right\}^{2}(t) &= \frac{t^{\beta-1}}{2}\left(\left\{E^{\gamma}_{\alpha,\beta}\right\}^{2}(t) - \left\{E^{\gamma}_{\alpha,\beta}\right\}^{2}(-t)\right)\\ &= t^{\beta-1}\sum_{k=0}^{2}\frac{(\gamma)_{2k+1}t^{2k+1}}{\Gamma((2k+1)\alpha+\beta)},\end{aligned} \tag{5.680}$$

where $t, \alpha, \beta, \gamma \in \mathbb{C}$, $Re(\alpha) > 0$, and $Re(\beta) > 0$.

Definition 5.82 The truncated hyperbolic precosine via Prabhakar function is defined as

$$\begin{aligned}\left\{Trecosh^{\gamma}_{\alpha,\beta}\right\}^{n}(t) &= \frac{t^{\beta-1}}{2}\left(\left\{E^{\gamma}_{\alpha,\beta}\right\}^{n}(t) + \left\{E^{\gamma}_{\alpha,\beta}\right\}^{n}(-t)\right)\\ &= t^{\beta-1}\sum_{k=0}^{n}\frac{(\gamma)_{2k}t^{2k}}{\Gamma(2k\alpha+\beta)},\end{aligned} \tag{5.681}$$

where $t, \alpha, \beta, \gamma \in \mathbb{C}$, $n, \kappa \in \mathbb{N}_0$, $Re(\alpha) > 0$, and $Re(\beta) > 0$.

Thus, we show that

$$\begin{aligned}\left\{Trecosh^{\gamma}_{\alpha,\beta}\right\}^{0}(t) &= \frac{t^{\beta-1}}{2}\left(\left\{E^{\gamma}_{\alpha,\beta}\right\}^{0}(t) + \left\{E^{\gamma}_{\alpha,\beta}\right\}^{0}(-t)\right)\\ &= \frac{t^{\beta-1}}{\Gamma(\beta)},\end{aligned} \tag{5.682}$$

$$\begin{aligned}\left\{Trecosh^{\gamma}_{\alpha,\beta}\right\}^{1}(t) &= \tfrac{t^{\beta-1}}{2}\left(\left\{E^{\gamma}_{\alpha,\beta}\right\}^{1}(t)+\left\{E^{\gamma}_{\alpha,\beta}\right\}^{1}(-t)\right)\\ &= t^{\beta-1}\sum_{k=0}^{1}\tfrac{(\gamma)_{2k}t^{2k}}{\Gamma(2k\alpha+\beta)},\end{aligned} \tag{5.683}$$

and

$$\begin{aligned}\left\{Trecosh^{\gamma}_{\alpha,\beta}\right\}^{2}(t) &= \tfrac{t^{\beta-1}}{2}\left(\left\{E^{\gamma}_{\alpha,\beta}\right\}^{2}(t)+\left\{E^{\gamma}_{\alpha,\beta}\right\}^{2}(-t)\right)\\ &= t^{\beta-1}\sum_{k=0}^{2}\tfrac{(\gamma)_{2k}t^{2k}}{\Gamma(2k\alpha+\beta)},\end{aligned} \tag{5.684}$$

where $t, \alpha, \beta, \gamma \in \mathbb{C}$, $Re(\alpha) > 0$, and $Re(\beta) > 0$.

Thus,

$$\left\{Tresin^{\gamma}_{\alpha,\beta}\right\}^{n}(\lambda t) = \frac{t^{\beta-1}}{2i}\left(\left\{E^{\gamma}_{\alpha,\beta}\right\}^{n}(i\lambda t)-\left\{E^{\gamma}_{\alpha,\beta}\right\}^{n}(-i\lambda t)\right), \tag{5.685}$$

$$\left\{Trecos^{\gamma}_{\alpha,\beta}\right\}^{n}(\lambda t) = \frac{t^{\beta-1}}{2}\left(\left\{E^{\gamma}_{\alpha,\beta}\right\}^{n}(i\lambda t)+\left\{E^{\gamma}_{\alpha,\beta}\right\}^{n}(-i\lambda t)\right), \tag{5.686}$$

$$\left\{Tresinh^{\gamma}_{\alpha,\beta}\right\}^{n}(\lambda t) = \frac{t^{\beta-1}}{2}\left(\left\{E^{\gamma}_{\alpha,\beta}\right\}^{n}(\lambda t)-\left\{E^{\gamma}_{\alpha,\beta}\right\}^{n}(-\lambda t)\right), \tag{5.687}$$

and

$$\left\{Trecosh^{\gamma}_{\alpha,\beta}\right\}^{n}(\lambda t) = \frac{t^{\beta-1}}{2}\left(\left\{E^{\gamma}_{\alpha,\beta}\right\}^{n}(\lambda t)+\left\{E^{\gamma}_{\alpha,\beta}\right\}^{n}(-\lambda t)\right), \tag{5.688}$$

where $t, \alpha, \beta, \gamma \in \mathbb{C}$, $n, \kappa \in \mathbb{N}_0$, $Re(\alpha) > 0$, and $Re(\beta) > 0$.

Definition 5.83 The truncated presine via Prabhakar function is defined as

$$\begin{aligned}\left\{Gresin^{\gamma}_{\alpha,\beta}\right\}^{n}(t^{\varsigma}) &= \tfrac{t^{\beta-1}}{2i}\left(\left\{E^{\gamma}_{\alpha,\beta}\right\}^{n}(it^{\varsigma})-\left\{E^{\gamma}_{\alpha,\beta}\right\}^{n}(-it^{\varsigma})\right)\\ &= t^{\beta-1}\sum_{k=0}^{n}\frac{(-1)^{k}(\gamma)_{2k+1}t^{(2k+1)\varsigma}}{\Gamma((2k+1)\alpha+\beta)},\end{aligned} \tag{5.689}$$

where $t, \alpha, \beta, \gamma, \varsigma \in \mathbb{C}$, $n, \kappa \in \mathbb{N}_0$, $Re(\alpha) > 0$, and $Re(\beta) > 0$.

So,

$$\begin{aligned}\left\{Gresin^{\gamma}_{\alpha,\beta}\right\}^{0}(t^{\varsigma}) &= \tfrac{t^{\beta-1}}{2}\left(\left\{E^{\gamma}_{\alpha,\beta}\right\}^{0}(it^{\varsigma})-\left\{E^{\gamma}_{\alpha,\beta}\right\}^{0}(-it^{\varsigma})\right)\\ &= \tfrac{\Gamma(\gamma+1)}{\Gamma(\alpha+\beta)}t^{\varsigma+\beta-1},\end{aligned} \tag{5.690}$$

$$\left\{Gresin^{\gamma}_{\alpha,\beta}\right\}^{1}(t^{\gamma}) = \frac{t^{\beta-1}}{2}\left(\left\{E^{\gamma}_{\alpha,\beta}\right\}^{1}(it^{\gamma}) - \left\{E^{\gamma}_{\alpha,\beta}\right\}^{1}(-it^{\gamma})\right)$$
$$= t^{\beta-1}\sum_{k=0}^{1}\frac{(-1)^k(\gamma)_{2k+1}t^{(2k+1)\varsigma}}{\Gamma((2k+1)\alpha+\beta)}, \tag{5.691}$$

and

$$\left\{Gresin^{\gamma}_{\alpha,\beta}\right\}^{2}(t^{\varsigma}) = \frac{t^{\beta-1}}{2}\left(\left\{E^{\gamma}_{\alpha,\beta}\right\}^{2}(it^{\varsigma}) - \left\{E^{\gamma}_{\alpha,\beta}\right\}^{2}(-it^{\varsigma})\right)$$
$$= t^{\beta-1}\sum_{k=0}^{2}\frac{(-1)^k(\gamma)_{2k+1}t^{(2k+1)\varsigma}}{\Gamma((2k+1)\alpha+\beta)}, \tag{5.692}$$

where $t, \alpha, \beta, \gamma, \varsigma \in \mathbb{C}$, $Re(\alpha) > 0$, and $Re(\beta) > 0$.

Definition 5.84 The truncated precosine via Prabhakar function is defined as

$$\left\{Grecos^{\gamma}_{\alpha,\beta}\right\}^{n}(t^{\varsigma}) = \frac{t^{\beta-1}}{2}\left(\left\{E^{\gamma}_{\alpha,\beta}\right\}^{n}(it^{\varsigma}) + \left\{E^{\gamma}_{\alpha,\beta}\right\}^{n}(-it^{\varsigma})\right)$$
$$= t^{\beta-1}\sum_{k=0}^{n}\frac{(-1)^k(\gamma)_{2k}t^{2k\varsigma}}{\Gamma(2k\alpha+\beta)}, \tag{5.693}$$

where $t, \alpha, \beta, \gamma, \varsigma \in \mathbb{C}$, $n, \kappa \in \mathbb{N}_0$, $Re(\alpha) > 0$, and $Re(\beta) > 0$.

Thus, we show that

$$\left\{Grecos^{\gamma}_{\alpha,\beta}\right\}^{0}(t^{\varsigma}) = \frac{t^{\beta-1}}{2}\left(\left\{E^{\gamma}_{\alpha,\beta}\right\}^{0}(it^{\varsigma}) + \left\{E^{\gamma}_{\alpha,\beta}\right\}^{0}(-it^{\varsigma})\right)$$
$$= \frac{t^{\beta-1}}{\Gamma(\beta)}, \tag{5.694}$$

$$\left\{Grecos^{\gamma}_{\alpha,\beta}\right\}^{1}(t^{\varsigma}) = \frac{t^{\beta-1}}{2}\left(\left\{E^{\gamma}_{\alpha,\beta}\right\}^{1}(it^{\varsigma}) + \left\{E^{\gamma}_{\alpha,\beta}\right\}^{1}(-it^{\varsigma})\right)$$
$$= t^{\beta-1}\sum_{k=0}^{1}\frac{(-1)^k(\gamma)_{2k}t^{2k\varsigma}}{\Gamma(2k\alpha+\beta)}, \tag{5.695}$$

and

$$\left\{Grecos^{\gamma}_{\alpha,\beta}\right\}^{2}(t^{\varsigma}) = \frac{t^{\beta-1}}{2}\left(\left\{E^{\gamma}_{\alpha,\beta}\right\}^{2}(t^{\varsigma}) + \left\{E^{\gamma}_{\alpha,\beta}\right\}^{2}(-t^{\varsigma})\right)$$
$$= t^{\beta-1}\sum_{k=0}^{2}\frac{(-1)^k(\gamma)_{2k}t^{2k\varsigma}}{\Gamma(2k\alpha+\beta)}, \tag{5.696}$$

where $t, \alpha, \beta, \gamma, \varsigma \in \mathbb{C}$, $Re(\alpha) > 0$, and $Re(\beta) > 0$.

Definition 5.85 The truncated hyperbolic presine via Prabhakar function is defined as

$$\begin{aligned}\left\{Gresinh_{\alpha,\beta}^{\gamma}\right\}^{n}(t^{\varsigma}) &= \frac{t^{\beta-1}}{2}\left(\left\{E_{\alpha,\beta}^{\gamma}\right\}^{n}(t^{\varsigma}) - \left\{E_{\alpha,\beta}^{\gamma}\right\}^{n}(-t^{\varsigma})\right)\\ &= t^{\beta-1}\sum_{k=0}^{n}\frac{(\gamma)_{2k+1}t^{(2k+1)\varsigma}}{\Gamma((2k+1)\alpha+\beta)},\end{aligned} \tag{5.697}$$

where $t, \alpha, \beta, \gamma, \varsigma \in \mathbb{C}$, $n, \kappa \in \mathbb{N}_0$, $Re(\alpha) > 0$, and $Re(\beta) > 0$.

So,

$$\begin{aligned}\left\{Gresinh_{\alpha,\beta}^{\gamma}\right\}^{0}(t^{\varsigma}) &= \frac{t^{\beta-1}}{2}\left(\left\{E_{\alpha,\beta}^{\gamma}\right\}^{0}(t^{\varsigma}) - \left\{E_{\alpha,\beta}^{\gamma}\right\}^{0}(-t^{\varsigma})\right)\\ &= \frac{\Gamma(\gamma+1)}{\Gamma(\alpha+\beta)}t^{\varsigma+\beta-1},\end{aligned} \tag{5.698}$$

$$\begin{aligned}\left\{Gresinh_{\alpha,\beta}^{\gamma}\right\}^{1}(t^{\varsigma}) &= \frac{t^{\beta-1}}{2}\left(\left\{E_{\alpha,\beta}^{\gamma}\right\}^{1}(t^{\varsigma}) - \left\{E_{\alpha,\beta}^{\gamma}\right\}^{1}(-t^{\varsigma})\right)\\ &= t^{\beta-1}\sum_{k=0}^{1}\frac{(\gamma)_{2k+1}t^{(2k+1)\varsigma}}{\Gamma((2k+1)\alpha+\beta)},\end{aligned} \tag{5.699}$$

and

$$\begin{aligned}\left\{Gresinh_{\alpha,\beta}^{\gamma}\right\}^{2}(t^{\varsigma}) &= \frac{t^{\beta-1}}{2}\left(\left\{E_{\alpha,\beta}^{\gamma}\right\}^{2}(t^{\varsigma}) - \left\{E_{\alpha,\beta}^{\gamma}\right\}^{2}(-t^{\varsigma})\right)\\ &= t^{\beta-1}\sum_{k=0}^{2}\frac{(\gamma)_{2k+1}t^{(2k+1)\varsigma}}{\Gamma((2k+1)\alpha+\beta)},\end{aligned} \tag{5.700}$$

where $t, \alpha, \beta, \gamma, \varsigma \in \mathbb{C}$, $Re(\alpha) > 0$, and $Re(\beta) > 0$.

Definition 5.86 The truncated hyperbolic precosine via Prabhakar function is defined as

$$\begin{aligned}\left\{Grecosh_{\alpha,\beta}^{\gamma}\right\}^{n}(t^{\varsigma}) &= \frac{t^{\beta-1}}{2}\left(\left\{E_{\alpha,\beta}^{\gamma}\right\}^{n}(t^{\varsigma}) + \left\{E_{\alpha,\beta}^{\gamma}\right\}^{n}(-t^{\varsigma})\right)\\ &= t^{\beta-1}\sum_{k=0}^{n}\frac{(\gamma)_{2k}t^{2k\varsigma}}{\Gamma(2k\alpha+\beta)},\end{aligned} \tag{5.701}$$

where $t, \alpha, \beta, \gamma, \varsigma \in \mathbb{C}$, $n, \kappa \in \mathbb{N}_0$, $Re(\alpha) > 0$, and $Re(\beta) > 0$.

Thus,

$$\begin{aligned}\left\{Grecosh_{\alpha,\beta}^{\gamma}\right\}^{0}(t^{\varsigma}) &= \frac{t^{\beta-1}}{2}\left(\left\{E_{\alpha,\beta}^{\gamma}\right\}^{0}(t^{\varsigma}) + \left\{E_{\alpha,\beta}^{\gamma}\right\}^{0}(-t^{\varsigma})\right)\\ &= \frac{t^{\beta-1}}{\Gamma(\beta)},\end{aligned} \tag{5.702}$$

$$\begin{aligned}\left\{Grecosh^{\gamma}_{\alpha,\beta}\right\}^{1}(t^{\varsigma}) &= \tfrac{t^{\beta-1}}{2}\left(\left\{E^{\gamma}_{\alpha,\beta}\right\}^{1}(t^{\varsigma}) + \left\{E^{\gamma}_{\alpha,\beta}\right\}^{1}(-t^{\varsigma})\right)\\ &= t^{\beta-1}\sum_{k=0}^{1}\tfrac{(\gamma)_{2k}t^{2k\varsigma}}{\Gamma(2k\alpha+\beta)},\end{aligned} \tag{5.703}$$

and

$$\begin{aligned}\left\{Grecosh^{\gamma}_{\alpha,\beta}\right\}^{2}(t^{\varsigma}) &= \tfrac{t^{\beta-1}}{2}\left(\left\{E^{\gamma}_{\alpha,\beta}\right\}^{2}(t^{\varsigma}) + \left\{E^{\gamma}_{\alpha,\beta}\right\}^{2}(-t^{\varsigma})\right)\\ &= t^{\beta-1}\sum_{k=0}^{2}\tfrac{(\gamma)_{2k}t^{2k\varsigma}}{\Gamma(2k\alpha+\beta)},\end{aligned} \tag{5.704}$$

where $t, \alpha, \beta, \gamma, \varsigma \in \mathbb{C}$, $Re(\alpha) > 0$, and $Re(\beta) > 0$.

Therefore, we have that

$$\left\{Gresin^{\gamma}_{\alpha,\beta}\right\}^{n}(\lambda t^{\varsigma}) = \frac{t^{\beta-1}}{2i}\left(\left\{E^{\gamma}_{\alpha,\beta}\right\}^{n}(i\lambda t^{\varsigma}) - \left\{E^{\gamma}_{\alpha,\beta}\right\}^{n}(-i\lambda t^{\varsigma})\right), \tag{5.705}$$

$$\left\{Grecos^{\gamma}_{\alpha,\beta}\right\}^{n}(\lambda t^{\varsigma}) = \frac{t^{\beta-1}}{2}\left(\left\{E^{\gamma}_{\alpha,\beta}\right\}^{n}(i\lambda t^{\varsigma}) + \left\{E^{\gamma}_{\alpha,\beta}\right\}^{n}(-i\lambda t^{\varsigma})\right), \tag{5.706}$$

$$\left\{Gresinh^{\gamma}_{\alpha,\beta}\right\}^{n}(\lambda t^{\varsigma}) = \frac{t^{\beta-1}}{2}\left(\left\{E^{\gamma}_{\alpha,\beta}\right\}^{n}(\lambda t^{\varsigma}) - \left\{E^{\gamma}_{\alpha,\beta}\right\}^{n}(-\lambda t^{\varsigma})\right), \tag{5.707}$$

and

$$\left\{Grecosh^{\gamma}_{\alpha,\beta}\right\}^{n}(\lambda t^{\varsigma}) = \frac{t^{\beta-1}}{2}\left(\left\{E^{\gamma}_{\alpha,\beta}\right\}^{n}(\lambda t^{\varsigma}) + \left\{E^{\gamma}_{\alpha,\beta}\right\}^{n}(-\lambda t^{\varsigma})\right), \tag{5.708}$$

where $t, \alpha, \beta, \gamma, \varsigma, \lambda \in \mathbb{C}$, $n \in \mathbb{N}_0$, $Re(\alpha) > 0$, and $Re(\beta) > 0$.

5.3.2 Other Special Functions Related to Prabhakar Function

In this section, we present the other special functions related to Prabhakar function.

Here, we have that

$$\begin{aligned}supersin^{\gamma}_{1,\beta}(t) &= \tfrac{t^{\beta-1}}{2i}\left(E^{\gamma}_{1,\beta}(it) - E^{\gamma}_{1,\beta}(-it)\right)\\ &= \tfrac{t^{\beta-1}}{\Gamma(\gamma)} \times {}_1sin_1\left(\begin{matrix}(\gamma,1)\\(\beta,1)\end{matrix};t\right)\\ &= \sum_{k=0}^{\infty}\tfrac{(-1)^k(\gamma)_{2k+1}}{\Gamma(2k+1+\beta)}\tfrac{t^{2k+\beta}}{(2k+1)!},\end{aligned} \tag{5.709}$$

$$\begin{aligned} supercos^{\gamma}_{1,\beta}(t) &= \frac{t^{\beta-1}}{2}\left(E^{\gamma}_{1,\beta}(it) + E^{\gamma}_{1,\beta}(-it)\right) \\ &= \frac{t^{\beta-1}}{\Gamma(\gamma)} \times {}_1cos_1\left(\begin{matrix}(\gamma,1)\\(\beta,1)\end{matrix}; t\right) \\ &= \sum_{k=0}^{\infty} \frac{(-1)^k(\gamma)_{2k}}{\Gamma(2k+\beta)} \frac{t^{2k+\beta-1}}{(2k)!}, \end{aligned} \tag{5.710}$$

$$\begin{aligned} supersinh^{\gamma}_{1,\beta}(t) &= \frac{t^{\beta-1}}{2}\left(E^{\gamma}_{1,\beta}(t) - E^{\gamma}_{1,\beta}(-t)\right) \\ &= \frac{t^{\beta-1}}{\Gamma(\gamma)} \times {}_1sinh_1\left(\begin{matrix}(\gamma,1)\\(\beta,1)\end{matrix}; t\right) \\ &= \sum_{k=0}^{\infty} \frac{(\gamma)_{2k+1}t^{2k+\beta}}{\Gamma(2k+1+\beta)}, \end{aligned} \tag{5.711}$$

$$\begin{aligned} supercosh^{\gamma}_{1,\beta}(t) &= \frac{t^{\beta-1}}{2}\left(E^{\gamma}_{1,\beta}(t) + E^{\gamma}_{1,\beta}(-t)\right) \\ &= \frac{t^{\beta-1}}{\Gamma(\gamma)} \times {}_1cosh_1\left(\begin{matrix}(\gamma,1)\\(\beta,1)\end{matrix}; t\right) \\ &= \sum_{n=0}^{\infty} \frac{(\gamma)_{2k}t^{2k+\beta-1}}{\Gamma(2k+\beta)}, \end{aligned} \tag{5.712}$$

$$\begin{aligned} presin^{\gamma}_{1,\beta}(t) &= \frac{1}{2i}\left(E^{\gamma}_{1,\beta}(it) - E^{\gamma}_{1,\beta}(-it)\right) \\ &= \frac{1}{\Gamma(\gamma)} \times {}_1sin_1\left(\begin{matrix}(\gamma,1)\\(\beta,1)\end{matrix}; t\right) \\ &= \sum_{k=0}^{\infty} \frac{(-1)^k(\gamma)_{2k+1}t^{2k+1}}{\Gamma(2k+1+\beta)}, \end{aligned} \tag{5.713}$$

$$\begin{aligned} precos^{\gamma}_{1,\beta}(t) &= \frac{1}{2}\left(E^{\gamma}_{1,\beta}(it) + E^{\gamma}_{1,\beta}(-it)\right) \\ &= \frac{1}{\Gamma(\gamma)} \times {}_1cos_1\left(\begin{matrix}(\gamma,1)\\(\beta,1)\end{matrix}; t\right) \\ &= \sum_{k=0}^{\infty} \frac{(-1)^k(\gamma)_{2k}t^{2k}}{\Gamma(2k+\beta)}, \end{aligned} \tag{5.714}$$

$$\begin{aligned} presinh^{\gamma}_{1,\beta}(t) &= \frac{1}{2}\left(E^{\gamma}_{1,\beta}(t) - E^{\gamma}_{1,\beta}(-t)\right) \\ &= \frac{1}{\Gamma(\gamma)} \times {}_1sinh_1\left(\begin{matrix}(\gamma,1)\\(\beta,1)\end{matrix}; t\right) \\ &= \sum_{k=0}^{\infty} \frac{(\gamma)_{2k+1}t^{2k+1}}{\Gamma(2k+1+\beta)}, \end{aligned} \tag{5.715}$$

$$\begin{aligned} precosh^{\gamma}_{1,\beta}(t) &= \tfrac{1}{2}\left(E^{\gamma}_{1,\beta}(t) + E^{\gamma}_{1,\beta}(-t)\right) \\ &= \tfrac{1}{\Gamma(\gamma)} \times {}_1cosh_1\left(\begin{matrix}(\gamma,1)\\(\beta,1)\end{matrix};t\right) \\ &= \sum_{k=0}^{\infty} \tfrac{(\gamma)_{2k}t^{2k}}{\Gamma(2k+\beta)}, \end{aligned} \tag{5.716}$$

$$\begin{aligned} presin^{\gamma}_{1,\beta}(t^{\varsigma}) &= \tfrac{1}{2i}\left(E^{\gamma}_{1,\beta}(it^{\varsigma}) - E^{\gamma}_{1,\beta}(-it^{\varsigma})\right) \\ &= \tfrac{1}{\Gamma(\gamma)} \times {}_1sin_1\left(\begin{matrix}(\gamma,1)\\(\beta,1)\end{matrix};t^{\varsigma}\right) \\ &= \sum_{k=0}^{\infty} \tfrac{(-1)^k(\gamma)_{2k+1}t^{(2k+1)\varsigma}}{\Gamma(2k+1+\beta)}, \end{aligned} \tag{5.717}$$

$$\begin{aligned} precos^{\gamma}_{1,\beta}(t^{\varsigma}) &= \tfrac{1}{2}\left(E^{\gamma}_{1,\beta}(it^{\varsigma}) + E^{\gamma}_{1,\beta}(-it^{\varsigma})\right) \\ &= \tfrac{1}{\Gamma(\gamma)} \times {}_1cos_1\left(\begin{matrix}(\gamma,1)\\(\beta,1)\end{matrix};t^{\varsigma}\right) \\ &= \sum_{k=0}^{\infty} \tfrac{(-1)^k(\gamma)_{2k}t^{2k\varsigma}}{\Gamma(2k+\beta)}, \end{aligned} \tag{5.718}$$

$$\begin{aligned} presinh^{\gamma}_{1,\beta}(t^{\varsigma}) &= \tfrac{1}{2}\left(E^{\gamma}_{1,\beta}(t^{\varsigma}) - E^{\gamma}_{1,\beta}(-t^{\varsigma})\right) \\ &= \tfrac{1}{\Gamma(\gamma)} \times {}_1sinh_1\left(\begin{matrix}(\gamma,1)\\(\beta,1)\end{matrix};t^{\varsigma}\right) \\ &= \sum_{k=0}^{\infty} \tfrac{(\gamma)_{2k+1}t^{(2k+1)\varsigma}}{\Gamma(2k+1+\beta)}, \end{aligned} \tag{5.719}$$

$$\begin{aligned} precosh^{\gamma}_{1,\beta}(t^{\varsigma}) &= \tfrac{1}{2}\left(E^{\gamma}_{1,\beta}(t^{\varsigma}) + E^{\gamma}_{1,\beta}(-t^{\varsigma})\right) \\ &= \tfrac{1}{\Gamma(\gamma)} \times {}_1cosh_1\left(\begin{matrix}(\gamma,1)\\(\beta,1)\end{matrix};t^{\varsigma}\right) \\ &= \sum_{k=0}^{\infty} \tfrac{(\gamma)_{2k}t^{2k\varsigma}}{\Gamma(2k+\beta)}, \end{aligned} \tag{5.720}$$

$$\begin{aligned} Gresin^{\gamma}_{1,\beta}(t^{\varsigma}) &= \tfrac{t^{\beta-1}}{2i}\left(E^{\gamma}_{1,\beta}(it^{\varsigma}) - E^{\gamma}_{1,\beta}(-it^{\varsigma})\right) \\ &= \tfrac{t^{\beta-1}}{\Gamma(\gamma)} \times {}_1sin_1\left(\begin{matrix}(\gamma,1)\\(\beta,1)\end{matrix};t^{\varsigma}\right) \\ &= t^{\beta-1}\sum_{k=0}^{\infty} \tfrac{(-1)^k(\gamma)_{2k+1}t^{(2k+1)\varsigma}}{\Gamma(2k+1+\beta)}, \end{aligned} \tag{5.721}$$

$$\begin{aligned} Grecos^{\gamma}_{1,\beta}(t^{\varsigma}) &= \frac{t^{\beta-1}}{2}\left(E^{\gamma}_{1,\beta}(it^{\varsigma}) + E^{\gamma}_{1,\beta}(-it^{\varsigma})\right) \\ &= \frac{t^{\beta-1}}{\Gamma(\gamma)} \times {}_1cos_1\left(\begin{matrix}(\gamma,1)\\(\beta,1)\end{matrix}; t^{\varsigma}\right) \\ &= t^{\beta-1}\sum_{k=0}^{\infty}\frac{(-1)^k(\gamma)_{2k}t^{2k\varsigma}}{\Gamma(2k+\beta)}, \end{aligned} \tag{5.722}$$

$$\begin{aligned} Gresinh^{\gamma}_{1,\beta}(t^{\varsigma}) &= \frac{t^{\beta-1}}{2}\left(E^{\gamma}_{1,\beta}(t^{\varsigma}) - E^{\gamma}_{1,\beta}(-t^{\varsigma})\right) \\ &= \frac{t^{\beta-1}}{\Gamma(\gamma)} \times {}_1sinh_1\left(\begin{matrix}(\gamma,1)\\(\beta,1)\end{matrix}; t^{\varsigma}\right) \\ &= t^{\beta-1}\sum_{k=0}^{\infty}\frac{(\gamma)_{2k+1}t^{(2k+1)\varsigma}}{\Gamma(2k+1+\beta)}, \end{aligned} \tag{5.723}$$

$$\begin{aligned} Grecosh^{\gamma}_{1,\beta}(t^{\varsigma}) &= \frac{t^{\beta-1}}{2}\left(E^{\gamma}_{1,\beta}(t^{\varsigma}) + E^{\gamma}_{1,\beta}(-t^{\varsigma})\right) \\ &= \frac{t^{\beta-1}}{\Gamma(\gamma)} \times {}_1cosh_1\left(\begin{matrix}(\gamma,1)\\(\beta,1)\end{matrix}; t^{\varsigma}\right) \\ &= t^{\beta-1}\sum_{k=0}^{\infty}\frac{(\gamma)_{2k}t^{2k\varsigma}}{\Gamma(2k+\beta)}, \end{aligned} \tag{5.724}$$

$$\begin{aligned} supersin^{\gamma}_{\alpha,1}(t^{\alpha}) &= \frac{1}{2i}\left(E^{\gamma}_{\alpha,1}(it^{\alpha}) - E^{\gamma}_{\alpha,1}(-it^{\alpha})\right) \\ &= \frac{1}{\Gamma(\gamma)} \times {}_1sin_1\left(\begin{matrix}(\gamma,1)\\(1,\alpha)\end{matrix}; t^{\alpha}\right) \\ &= \sum_{k=0}^{\infty}\frac{(-1)^k(\gamma)_{2k+1}}{\Gamma((2k+1)\alpha+1)}\frac{t^{(2k+1)\alpha}}{(2k+1)!}, \end{aligned} \tag{5.725}$$

$$\begin{aligned} supercos^{\gamma}_{\alpha,1}(t^{\alpha}) &= \frac{1}{2}\left(E^{\gamma}_{\alpha,1}(it^{\alpha}) + E^{\gamma}_{\alpha,1}(-it^{\alpha})\right) \\ &= \frac{1}{\Gamma(\gamma)} \times {}_1cos_1\left(\begin{matrix}(\gamma,1)\\(\beta,\alpha)\end{matrix}; t^{\alpha}\right) \\ &= \sum_{k=0}^{\infty}\frac{(-1)^k(\gamma)_{2k}}{\Gamma(2k\alpha+1)}\frac{t^{2k\alpha}}{(2k)!}, \end{aligned} \tag{5.726}$$

$$\begin{aligned} supersinh^{\gamma}_{\alpha,1}(t^{\alpha}) &= \frac{1}{2}\left(E^{\gamma}_{\alpha,1}(t^{\alpha}) - E^{\gamma}_{\alpha,1}(-t^{\alpha})\right) \\ &= \frac{1}{\Gamma(\gamma)} \times {}_1sinh_1\left(\begin{matrix}(\gamma,1)\\(1,\alpha)\end{matrix}; t^{\alpha}\right) \\ &= \sum_{k=0}^{\infty}\frac{(\gamma)_{2k+1}t^{(2k+1)\alpha}}{\Gamma((2k+1)\alpha+1)}, \end{aligned} \tag{5.727}$$

$$\begin{aligned} supercosh^{\gamma}_{\alpha,1}(t^{\alpha}) &= \frac{1}{2}\left(E^{\gamma}_{\alpha,1}(t^{\alpha}) + E^{\gamma}_{\alpha,1}(-t^{\alpha})\right) \\ &= \frac{1}{\Gamma(\gamma)} \times {}_1cosh_1\left(\begin{matrix}(\gamma,1)\\(1,\alpha)\end{matrix}; t^{\alpha}\right) \\ &= \sum_{n=0}^{\infty} \frac{(\gamma)_{2k} t^{2k\alpha}}{\Gamma(2k\alpha+1)}, \end{aligned} \tag{5.728}$$

$$\begin{aligned} presin^{\gamma}_{\alpha,1}(t) &= \frac{1}{2i}\left(E^{\gamma}_{\alpha,1}(it) - E^{\gamma}_{\alpha,1}(-it)\right) \\ &= \frac{1}{\Gamma(\gamma)} \times {}_1sin_1\left(\begin{matrix}(\gamma,1)\\(1,\alpha)\end{matrix}; t\right) \\ &= \sum_{k=0}^{\infty} \frac{(-1)^k(\gamma)_{2k+1} t^{2k+1}}{\Gamma((2k+1)\alpha+1)}, \end{aligned} \tag{5.729}$$

$$\begin{aligned} precos^{\gamma}_{\alpha,1}(t) &= \frac{1}{2}\left(E^{\gamma}_{\alpha,1}(it) + E^{\gamma}_{\alpha,1}(-it)\right) \\ &= \frac{1}{\Gamma(\gamma)} \times {}_1cos_1\left(\begin{matrix}(\gamma,1)\\(1,\alpha)\end{matrix}; t\right) \\ &= \sum_{k=0}^{\infty} \frac{(-1)^k(\gamma)_{2k} t^{2k}}{\Gamma(2k\alpha+1)}, \end{aligned} \tag{5.730}$$

$$\begin{aligned} presinh^{\gamma}_{\alpha,1}(t) &= \frac{1}{2}\left(E^{\gamma}_{\alpha,1}(t) - E^{\gamma}_{\alpha,1}(-t)\right) \\ &= \frac{1}{\Gamma(\gamma)} \times {}_1sinh_1\left(\begin{matrix}(\gamma,1)\\(1,\alpha)\end{matrix}; t\right) \\ &= \sum_{k=0}^{\infty} \frac{(\gamma)_{2k+1} t^{2k+1}}{\Gamma((2k+1)\alpha+1)}, \end{aligned} \tag{5.731}$$

$$\begin{aligned} precosh^{\gamma}_{\alpha,1}(t) &= \frac{1}{2}\left(E^{\gamma}_{\alpha,1}(t) + E^{\gamma}_{\alpha,1}(-t)\right) \\ &= \frac{1}{\Gamma(\gamma)} \times {}_1cosh_1\left(\begin{matrix}(\gamma,1)\\(1,\alpha)\end{matrix}; t\right) \\ &= \sum_{k=0}^{\infty} \frac{(\gamma)_{2k} t^{2k}}{\Gamma(2k\alpha+1)}, \end{aligned} \tag{5.732}$$

$$\begin{aligned} presin^{\gamma}_{\alpha,1}(t^{\varsigma}) &= \frac{1}{2i}\left(E^{\gamma}_{\alpha,1}(it^{\varsigma}) - E^{\gamma}_{\alpha,1}(-it^{\varsigma})\right) \\ &= \frac{1}{\Gamma(\gamma)} \times {}_1sin_1\left(\begin{matrix}(\gamma,1)\\(1,\alpha)\end{matrix}; t^{\varsigma}\right) \\ &= \sum_{k=0}^{\infty} \frac{(-1)^k(\gamma)_{2k+1} t^{(2k+1)\varsigma}}{\Gamma((2k+1)\alpha+1)}, \end{aligned} \tag{5.733}$$

$$
\begin{aligned}
precos^{\gamma}_{\alpha,1}(t^{\varsigma}) &= \tfrac{1}{2}\left(E^{\gamma}_{\alpha,1}(it^{\varsigma}) + E^{\gamma}_{\alpha,1}(-it^{\varsigma})\right) \\
&= \tfrac{1}{\Gamma(\gamma)} \times {}_1cos_1\left(\begin{matrix}(\gamma,1)\\(1,\alpha)\end{matrix};t^{\varsigma}\right) \\
&= \sum_{k=0}^{\infty} \tfrac{(-1)^k(\gamma)_{2k}t^{2k\varsigma}}{\Gamma(2k\alpha+1)},
\end{aligned}
\tag{5.734}
$$

$$
\begin{aligned}
presinh^{\gamma}_{\alpha,1}(t^{\varsigma}) &= \tfrac{1}{2}\left(E^{\gamma}_{\alpha,1}(t^{\varsigma}) - E^{\gamma}_{\alpha,1}(-t^{\varsigma})\right) \\
&= \tfrac{1}{\Gamma(\gamma)} \times {}_1sinh_1\left(\begin{matrix}(\gamma,1)\\(1,\alpha)\end{matrix};t^{\varsigma}\right) \\
&= \sum_{k=0}^{\infty} \tfrac{(\gamma)_{2k+1}t^{(2k+1)\varsigma}}{\Gamma((2k+1)\alpha+1)},
\end{aligned}
\tag{5.735}
$$

$$
\begin{aligned}
precosh^{\gamma}_{\alpha,1}(t^{\varsigma}) &= \tfrac{1}{2}\left(E^{\gamma}_{\alpha,1}(t^{\varsigma}) + E^{\gamma}_{\alpha,1}(-t^{\varsigma})\right) \\
&= \tfrac{1}{\Gamma(\gamma)} \times {}_1cosh_1\left(\begin{matrix}(\gamma,1)\\(1,\alpha)\end{matrix};t^{\varsigma}\right) \\
&= \sum_{k=0}^{\infty} \tfrac{(\gamma)_{2k}t^{2k\varsigma}}{\Gamma(2k\alpha+1)},
\end{aligned}
\tag{5.736}
$$

$$
\begin{aligned}
supersin^{\gamma}_{1,1}(t) &= \tfrac{1}{2i}\left(E^{\gamma}_{1,1}(it) - E^{\gamma}_{1,1}(-it)\right) \\
&= \tfrac{1}{\Gamma(\gamma)} \times {}_1sin_1\left(\begin{matrix}(\gamma,1)\\(1,1)\end{matrix};t\right) \\
&= \sum_{k=0}^{\infty} \tfrac{(-1)^k(\gamma)_{2k+1}}{\Gamma(2k+2)} \tfrac{t^{2k+1}}{(2k+1)!},
\end{aligned}
\tag{5.737}
$$

$$
\begin{aligned}
supercos^{\gamma}_{1,1}(t) &= \tfrac{1}{2}\left(E^{\gamma}_{1,1}(it) + E^{\gamma}_{1,1}(-it)\right) \\
&= \tfrac{1}{\Gamma(\gamma)} \times {}_1cos_1\left(\begin{matrix}(\gamma,1)\\(1,1)\end{matrix};t\right) \\
&= \sum_{k=0}^{\infty} \tfrac{(-1)^k(\gamma)_{2k}}{\Gamma(2k+1)} \tfrac{t^{2k}}{(2k)!},
\end{aligned}
\tag{5.738}
$$

$$
\begin{aligned}
supersinh^{\gamma}_{1,1}(t) &= \tfrac{1}{2}\left(E^{\gamma}_{1,1}(t) - E^{\gamma}_{1,1}(-t)\right) \\
&= \tfrac{1}{\Gamma(\gamma)} \times {}_1sinh_1\left(\begin{matrix}(\gamma,1)\\(1,1)\end{matrix};t\right) \\
&= \sum_{k=0}^{\infty} \tfrac{(\gamma)_{2k+1}t^{2k+1}}{\Gamma(2k+2)},
\end{aligned}
\tag{5.739}
$$

$$\begin{aligned} supercosh^{\gamma}_{1,1}(t) &= \frac{1}{2}\left(E^{\gamma}_{1,1}(t) + E^{\gamma}_{1,1}(-t)\right) \\ &= \frac{t^{\beta-1}}{\Gamma(\gamma)} \times {}_1cosh_1\left(\begin{matrix}(\gamma,1)\\(\beta,1)\end{matrix};t\right) \\ &= \sum_{n=0}^{\infty}\frac{(\gamma)_{2k}t^{2k}}{\Gamma(2k+1)}, \end{aligned} \tag{5.740}$$

$$\begin{aligned} presin^{\gamma}_{1,1}(t^{\varsigma}) &= \frac{1}{2i}\left(E^{\gamma}_{1,1}(it^{\varsigma}) - E^{\gamma}_{1,1}(-it^{\varsigma})\right) \\ &= \frac{1}{\Gamma(\gamma)} \times {}_1sin_1\left(\begin{matrix}(\gamma,1)\\(1,1)\end{matrix};t^{\varsigma}\right) \\ &= \sum_{k=0}^{\infty}\frac{(-1)^k(\gamma)_{2k+1}t^{(2k+1)\varsigma}}{\Gamma(2k+2)}, \end{aligned} \tag{5.741}$$

$$\begin{aligned} precos^{\gamma}_{1,1}(t^{\varsigma}) &= \frac{1}{2}\left(E^{\gamma}_{1,1}(it^{\varsigma}) + E^{\gamma}_{1,1}(-it^{\varsigma})\right) \\ &= \frac{1}{\Gamma(\gamma)} \times {}_1cos_1\left(\begin{matrix}(\gamma,1)\\(1,1)\end{matrix};t^{\varsigma}\right) \\ &= \sum_{k=0}^{\infty}\frac{(-1)^k(\gamma)_{2k}t^{2k\varsigma}}{\Gamma(2k+1)}, \end{aligned} \tag{5.742}$$

$$\begin{aligned} presinh^{\gamma}_{1,1}(t^{\varsigma}) &= \frac{1}{2}\left(E^{\gamma}_{1,1}(t^{\varsigma}) - E^{\gamma}_{1,1}(-t^{\varsigma})\right) \\ &= \frac{1}{\Gamma(\gamma)} \times {}_1sinh_1\left(\begin{matrix}(\gamma,1)\\(1,1)\end{matrix};t^{\varsigma}\right) \\ &= \sum_{k=0}^{\infty}\frac{(\gamma)_{2k+1}t^{(2k+1)\varsigma}}{\Gamma(2k+2)}, \end{aligned} \tag{5.743}$$

$$\begin{aligned} precosh^{\gamma}_{1,1}(t^{\varsigma}) &= \frac{1}{2}\left(E^{\gamma}_{1,1}(t^{\varsigma}) + E^{\gamma}_{1,1}(-t^{\varsigma})\right) \\ &= \frac{1}{\Gamma(\gamma)} \times {}_1cosh_1\left(\begin{matrix}(\gamma,1)\\(1,1)\end{matrix};t^{\varsigma}\right) \\ &= \sum_{k=0}^{\infty}\frac{(\gamma)_{2k}t^{2k\varsigma}}{\Gamma(2k+1)}, \end{aligned} \tag{5.744}$$

$$\begin{aligned} supersin^{1}_{1,\beta}(t) &= \frac{t^{\beta-1}}{2i}\left(E^{1}_{1,\beta}(it) - E^{1}_{1,\beta}(-it)\right) \\ &= t^{\beta-1} \times {}_1sin_1\left(\begin{matrix}(1,1)\\(\beta,1)\end{matrix};t\right), \end{aligned} \tag{5.745}$$

$$\begin{aligned} supercos^{1}_{1,\beta}(t) &= \frac{t^{\beta-1}}{2}\left(E^{1}_{1,\beta}(it) + E^{1}_{1,\beta}(-it)\right) \\ &= t^{\beta-1} \times {}_1cos_1\left(\begin{matrix}(1,1)\\(\beta,1)\end{matrix};t\right), \end{aligned} \tag{5.746}$$

$$\begin{aligned} supersinh^{1}_{1,\beta}(t) &= \tfrac{t^{\beta-1}}{2}\left(E^{1}_{1,\beta}(t) - E^{1}_{1,\beta}(-t)\right) \\ &= t^{\beta-1} \times {}_1sinh_1\left(\begin{matrix}(1,1)\\(\beta,1)\end{matrix};t\right), \end{aligned} \tag{5.747}$$

$$\begin{aligned} supercosh^{1}_{1,\beta}(t) &= \tfrac{t^{\beta-1}}{2}\left(E^{1}_{1,\beta}(t) + E^{1}_{1,\beta}(-t)\right) \\ &= t^{\beta-1} \times {}_1cosh_1\left(\begin{matrix}(1,1)\\(\beta,1)\end{matrix};t\right), \end{aligned} \tag{5.748}$$

$$\begin{aligned} presin^{1}_{1,\beta}(t) &= \tfrac{1}{2i}\left(E^{1}_{1,\beta}(it) - E^{1}_{1,\beta}(-it)\right) \\ &= {}_1sin_1\left(\begin{matrix}(1,1)\\(\beta,1)\end{matrix};t\right), \end{aligned} \tag{5.749}$$

$$\begin{aligned} precos^{1}_{1,\beta}(t) &= \tfrac{1}{2}\left(E^{1}_{1,\beta}(it) + E^{1}_{1,\beta}(-it)\right) \\ &= {}_1cos_1\left(\begin{matrix}(1,1)\\(\beta,1)\end{matrix};t\right), \end{aligned} \tag{5.750}$$

$$\begin{aligned} presinh^{1}_{1,\beta}(t) &= \tfrac{1}{2}\left(E^{1}_{1,\beta}(t) - E^{1}_{1,\beta}(-t)\right) \\ &= {}_1sinh_1\left(\begin{matrix}(1,1)\\(\beta,1)\end{matrix};t\right), \end{aligned} \tag{5.751}$$

$$\begin{aligned} precosh^{1}_{1,\beta}(t) &= \tfrac{1}{2}\left(E^{1}_{1,\beta}(t) + E^{1}_{1,\beta}(-t)\right) \\ &= {}_1cosh_1\left(\begin{matrix}(1,1)\\(\beta,1)\end{matrix};t\right), \end{aligned} \tag{5.752}$$

$$\begin{aligned} presin^{1}_{1,\beta}(t^{\varsigma}) &= \tfrac{1}{2i}\left(E^{1}_{1,\beta}(it^{\varsigma}) - E^{1}_{1,\beta}(-it^{\varsigma})\right) \\ &= {}_1sin_1\left(\begin{matrix}(1,1)\\(\beta,1)\end{matrix};t^{\varsigma}\right), \end{aligned} \tag{5.753}$$

$$\begin{aligned} precos^{1}_{1,\beta}(t^{\varsigma}) &= \tfrac{1}{2}\left(E^{1}_{1,\beta}(it^{\varsigma}) + E^{1}_{1,\beta}(-it^{\varsigma})\right) \\ &= {}_1cos_1\left(\begin{matrix}(1,1)\\(\beta,1)\end{matrix};t^{\varsigma}\right), \end{aligned} \tag{5.754}$$

$$\begin{aligned} presinh^{1}_{1,\beta}(t^{\varsigma}) &= \tfrac{1}{2}\left(E^{1}_{1,\beta}(t^{\varsigma}) - E^{1}_{1,\beta}(-t^{\varsigma})\right) \\ &= {}_1sinh_1\left(\begin{matrix}(1,1)\\(\beta,1)\end{matrix};t^{\varsigma}\right), \end{aligned} \tag{5.755}$$

$$precosh_{1,\beta}^{1}(t^{\varsigma}) = \frac{1}{2}\left(E_{1,\beta}^{1}(t^{\varsigma}) + E_{1,\beta}^{1}(-t^{\varsigma})\right) = {}_1cosh_1\left(\begin{matrix}(1,1)\\(\beta,1)\end{matrix}; t^{\varsigma}\right), \tag{5.756}$$

$$Gresin_{1,\beta}^{1}(t^{\varsigma}) = \frac{t^{\beta-1}}{2i}\left(E_{1,\beta}^{1}(it^{\varsigma}) - E_{1,\beta}^{1}(-it^{\varsigma})\right) = t^{\beta-1} \times {}_1sin_1\left(\begin{matrix}(1,1)\\(\beta,1)\end{matrix}; t^{\varsigma}\right), \tag{5.757}$$

$$Grecos_{1,\beta}^{1}(t^{\varsigma}) = \frac{t^{\beta-1}}{2}\left(E_{1,\beta}^{1}(it^{\varsigma}) + E_{1,\beta}^{1}(-it^{\varsigma})\right) = t^{\beta-1} \times {}_1cos_1\left(\begin{matrix}(1,1)\\(\beta,1)\end{matrix}; t^{\varsigma}\right), \tag{5.758}$$

$$Gresinh_{1,\beta}^{1}(t^{\varsigma}) = \frac{t^{\beta-1}}{2}\left(E_{1,\beta}^{1}(t^{\varsigma}) - E_{1,\beta}^{1}(-t^{\varsigma})\right) = t^{\beta-1} \times {}_1sinh_1\left(\begin{matrix}(1,1)\\(\beta,1)\end{matrix}; t^{\varsigma}\right), \tag{5.759}$$

and

$$Grecosh_{1,\beta}^{1}(t^{\varsigma}) = \frac{t^{\beta-1}}{2}\left(E_{1,\beta}^{1}(t^{\varsigma}) + E_{1,\beta}^{1}(-t^{\varsigma})\right) = t^{\beta-1} \times {}_1cosh_1\left(\begin{matrix}(1,1)\\(\beta,1)\end{matrix}; t^{\varsigma}\right), \tag{5.760}$$

where $\alpha, \beta, \gamma, \varsigma, t \in \mathbb{C}$, $Re(\alpha) > 0$, and $Re(\beta) > 0$.

5.4 General Fractional Calculus Operators via Prabhakar Function

In this section, we introduce the general fractional calculus operators via Prabhakar function and applications related to Prabhakar function.

5.4.1 Kilbas–Saigo–Saxena Derivative via Prabhakar Function

Let $1 \geq \alpha \geq 0$ and $\lambda, \upsilon, \varphi \in \mathbb{R}$.

The left-sided Kilbas–Saigo–Saxena derivative via Prabhakar function is defined as [92, 253]

$$\begin{aligned} & {}_{GP}^{RL}D_{a+}^{\alpha,\upsilon,\varphi,\lambda} f(t) \\ &= \frac{d}{dt}\left({}_{GP}I_{a+}^{\alpha,\upsilon,\varphi,\lambda} f(t)\right) \\ &= \frac{d}{dt}\int_a^t (t-\tau)^{\upsilon-1} E_{\alpha,\upsilon}^{\varphi}\left(-\lambda(t-\tau)^{\alpha}\right) f(\tau)\,d\tau, \end{aligned} \tag{5.761}$$

and the right-sided Kilbas–Saigo–Saxena derivative via Prabhakar function as [92]

$$\begin{aligned} & {}_{GP}^{RL}D_{b-}^{\alpha,\upsilon,\varphi,\lambda} f(t) \\ &= -\frac{d}{dt}\left({}_{GP}I_{b-}^{\alpha,\upsilon,\varphi,\lambda} f(t)\right) \\ &= -\frac{d}{dt}\int_t^b (\tau-t)^{\upsilon-1} E_{\alpha,\upsilon}^{\varphi}\left(-\lambda(\tau-t)^{\alpha}\right) f(\tau)\,d\tau, \end{aligned} \tag{5.762}$$

where the left-sided Prabhakar general fractional integral is [92, 260]

$${}_{GP}I_{a+}^{\alpha,\upsilon,\varphi,\lambda} f(t) = \int_a^t (t-\tau)^{\upsilon-1} E_{\alpha,\upsilon}^{\varphi}\left(-\lambda(t-\tau)^{\alpha}\right) f(\tau)\,d\tau, \tag{5.763}$$

and the right-sided Prabhakar general fractional integral is [92]

$${}_{GP}I_{b-}^{\alpha,\upsilon,\varphi,\lambda} f(t) = \int_t^b (\tau-t)^{\upsilon-1} E_{\alpha,\upsilon}^{\varphi}\left(-\lambda(\tau-t)^{\alpha}\right) f(\tau)\,d\tau, \tag{5.764}$$

respectively.

The result was discovered by Kilbas, Saigo, and Saxena in 2004 [253] and further discussed in 2019 by Yang [92].

5.4.2 *Garra–Gorenflo–Polito–Tomovski Derivative via Prabhakar Function*

Let $1 \geq \alpha \geq 0$ and $\lambda, \upsilon, \varphi \in \mathbb{R}$.

The left-sided Garra–Gorenflo–Polito–Tomovski derivative via Prabhakar function is defined as [92]

$$\begin{aligned} & {}_{GP}^{LS}D_{a+}^{\alpha,\upsilon,\varphi,\lambda} f(t) \\ &= {}_{GP}I_{a+}^{\alpha,\upsilon,\varphi,\lambda} f^{(1)}(t) \\ &= \int_a^t (t-\tau)^{\upsilon-1} E_{\alpha,\upsilon}^{\varphi}\left(-\lambda(t-\tau)^{\alpha}\right) f^{(1)}(\tau)\,d\tau, \end{aligned} \tag{5.765}$$

and the right-sided Garra–Gorenflo–Polito–Tomovski derivative via Prabhakar function as [92]

$$\begin{aligned} & {}_{GP}^{LS}D_{b-}^{\alpha,\upsilon,\varphi,\lambda} f(t) \\ & = {}_{GP}I_{b-}^{\alpha,\upsilon,\varphi,\lambda} \left(-f^{(1)}(t)\right) \\ & = -\int_t^b (\tau - t)^{\upsilon-1} E_{\alpha,\upsilon}^{\varphi} (-\lambda (\tau - t)^{\alpha}) f^{(1)}(\tau)\, d\tau, \end{aligned} \tag{5.766}$$

where the left-sided Prabhakar integral is [92, 260]

$${}_{GP}I_{a+}^{\alpha,\upsilon,\varphi,\lambda} f(t) = \int_a^t (t - \tau)^{\upsilon-1} E_{\alpha,\upsilon}^{\varphi} \left(-\lambda (t - \tau)^{\alpha}\right) f(\tau)\, d\tau, \tag{5.767}$$

and the right-sided Prabhakar integral is

$${}_{GP}I_{b-}^{\alpha,\upsilon,\varphi,\lambda} f(t) = \int_t^b (\tau - t)^{\upsilon-1} E_{\alpha,\upsilon}^{\varphi} \left(-\lambda (\tau - t)^{\alpha}\right) f(\tau)\, d\tau, \tag{5.768}$$

respectively.

The result was discovered by Garra, Gorenflo, Polito, and Tomovski in 2014 [92, 262].

5.4.3 *Prabhakar-Type Integrals*

Definition 5.87 Let $1 \geq \alpha \geq 0$ and $\lambda, \upsilon, \varphi \in \mathbb{R}$.

The left-sided Prabhakar-type general fractional integral is defined as [92]

$${}_{GP}^{R}I_{a+}^{\alpha,\varphi,\upsilon,\lambda} f(t) = \int_a^t (t - \tau)^{-\upsilon} E_{\alpha,1-\upsilon}^{-\varphi} \left(-\lambda (t - \tau)^{\alpha}\right) f(\tau)\, d\tau, \tag{5.769}$$

and the right-sided Prabhakar-type general fractional integral as

$${}_{GP}^{R}I_{b-}^{\alpha,\varphi,\upsilon,\lambda} f(t) = \int_t^b (\tau - t)^{-\upsilon} E_{\alpha,1-\upsilon}^{-\varphi} \left(-\lambda (\tau - t)^{\alpha}\right) f(\tau)\, d\tau, \tag{5.770}$$

respectively.

The result is the special case proposed by Prabhakar in 1971 [260].

Theorem 5.136 *If* $0 < \alpha < 1$ *and* $\lambda, \upsilon, \varphi \in \mathbb{R}$, *then we have [92]*

$$L\left\{{}_{GP}^{R}I_{0+}^{\alpha,\upsilon,\varphi,\lambda} f(t)\right\} = s^{\upsilon-1}\left(1+\lambda s^{-\alpha}\right)^{\varphi} f(s). \tag{5.771}$$

The result was discovered by Yang in 2019 [92].

Theorem 5.137 *If* $0 < \alpha < 1$ *and* $\lambda, \upsilon, \varphi \in \mathbb{R}$, *then we have [92]*

$$L\left\{{}_{GP}^{LSC}D_{0+}^{\alpha,\upsilon,\varphi,\lambda} f(t)\right\} = s^{-\upsilon}\left(1+\lambda s^{-\alpha}\right)^{-\varphi}\left(sf(s) - f(0)\right). \tag{5.772}$$

The result was discovered by Yang in 2019 [92].

Theorem 5.138 *If* $0 < \alpha < 1$ *and* $\lambda, \upsilon, \varphi \in \mathbb{R}$, *then we have [25]*

$$L\left\{{}_{GP}^{RL}D_{0+}^{\alpha,\upsilon,\varphi,\lambda} f(t)\right\} = \left(1 + \lambda s^{-\alpha}\right)^{-1} f(s) - {}_{GP}^{R}I_{0+}^{\alpha,\upsilon,\varphi,\lambda} f(0). \tag{5.773}$$

The result was discovered by Yang in 2019 [92].

Theorem 5.139 *Let* $0 < \alpha < 1$ *and* $\lambda, \upsilon, \varphi \in \mathbb{R}$. *Then we have*

$${}_{GP}^{RL}D_{a+}^{\alpha,\upsilon,\varphi,\lambda}\left({}_{GP}^{R}I_{a+}^{\alpha,\varphi,\upsilon,\lambda} f(t)\right) = f(t) \tag{5.774}$$

and

$${}_{GP}^{RL}D_{b-}^{\alpha,\upsilon,\varphi,\lambda}\left({}_{GP}^{R}I_{b-}^{\alpha,\varphi,\upsilon,\lambda} f(t)\right) = f(t). \tag{5.775}$$

Proof With the Laplace transform, we obtain the result. □

For the class of the Prabhakar fractional calculus, see Giusti [263], Garra and Garrappa [264], Colombaro et al. [265], and Alidousti [266].

5.4.4 Kilbas–Saigo–Saxena-Type Derivative via Prabhakar Function

Let $1 \geq \alpha \geq 0$ and $\lambda, \upsilon, \varphi \in \mathbb{R}$.

The left-sided Kilbas–Saigo–Saxena-type derivative via Prabhakar function is defined as [92]

$$\begin{aligned}
&{}_{GP}^{RL}D_{+}^{\alpha,\upsilon,\varphi,\lambda} f(t)\\
&= \frac{d}{dt}\left({}_{GP}I_{+}^{\alpha,\upsilon,\varphi,\lambda} f(t)\right)\\
&= \frac{d}{dt}\int_{-\infty}^{t} (t-\tau)^{\upsilon-1} E_{\alpha,\upsilon}^{\varphi}\left(-\lambda (t-\tau)^{\alpha}\right) f(\tau)\, d\tau,
\end{aligned} \tag{5.776}$$

and the right-sided Kilbas–Saigo–Saxena-type derivative via Prabhakar function as [92]

$$\begin{aligned}
&{}_{GP}^{RL}D_{-}^{\alpha,\upsilon,\varphi,\lambda} f(t) \\
&= -\frac{d}{dt}\left({}_{GP}I_{-}^{\alpha,\upsilon,\varphi,\lambda} f(t)\right) \\
&= -\frac{d}{dt}\int_{t}^{\infty} (\tau - t)^{\upsilon-1} E_{\alpha,\upsilon}^{\varphi}(-\lambda(\tau - t)^{\alpha}) f(\tau)\, d\tau,
\end{aligned} \tag{5.777}$$

where the left-sided Prabhakar general fractional integral is

$${}_{GP}I_{+}^{\alpha,\upsilon,\varphi,\lambda} f(t) = \int_{-\infty}^{t} (t - \tau)^{\upsilon-1} E_{\alpha,\upsilon}^{\varphi}\left(-\lambda(t - \tau)^{\alpha}\right) f(\tau)\, d\tau, \tag{5.778}$$

and the right-sided Prabhakar general fractional integral is

$${}_{GP}I_{-}^{\alpha,\upsilon,\varphi,\lambda} f(t) = \int_{t}^{\infty} (\tau - t)^{\upsilon-1} E_{\alpha,\upsilon}^{\varphi}\left(-\lambda(\tau - t)^{\alpha}\right) f(\tau)\, d\tau, \tag{5.779}$$

respectively.

The result was discovered by Yang in 2019 [92].

5.4.5 Garra–Gorenflo–Polito–Tomovski-Type Derivative via Prabhakar Function

Let $1 \geq \alpha \geq 0$ and $\lambda, \upsilon, \varphi \in \mathbb{R}$.

The left-sided Garra–Gorenflo–Polito–Tomovski-type derivative via Prabhakar function is defined as [92]

$$\begin{aligned}
&{}_{GP}^{LS}D_{+}^{\alpha,\upsilon,\varphi,\lambda} f(t) \\
&= {}_{GP}I_{+}^{\alpha,\upsilon,\varphi,\lambda} f^{(1)}(t) \\
&= \int_{-\infty}^{t} (t - \tau)^{\upsilon-1} E_{\alpha,\upsilon}^{\varphi}(-\lambda(t - \tau)^{\alpha}) f^{(1)}(\tau)\, d\tau,
\end{aligned} \tag{5.780}$$

and the right-sided Garra–Gorenflo–Polito–Tomovski-type derivative via Prabhakar function as [92]

$$\begin{aligned}
&{}_{GP}^{LS}D_{-}^{\alpha,\upsilon,\varphi,\lambda} f(t) \\
&= {}_{GP}I_{-}^{\alpha,\upsilon,\varphi,\lambda}\left(-f^{(1)}(t)\right) \\
&= -\int_{t}^{\infty} (\tau - t)^{\upsilon-1} E_{\alpha,\upsilon}^{\varphi}(-\lambda(\tau - t)^{\alpha}) f^{(1)}(\tau)\, d\tau,
\end{aligned} \tag{5.781}$$

where the left-sided Prabhakar general fractional integral is

$$ {}_{GP}I_{+}^{\alpha,\upsilon,\varphi,\lambda} f(t) = \int_{-\infty}^{t} (t-\tau)^{\upsilon-1} E_{\alpha,\upsilon}^{\varphi}\left(-\lambda (t-\tau)^{\alpha}\right) f(\tau)\, d\tau, \tag{5.782} $$

and the right-sided Prabhakar general fractional integral is

$$ {}_{GP}I_{-}^{\alpha,\upsilon,\varphi,\lambda} f(t) = \int_{t}^{\infty} (\tau-t)^{\upsilon-1} E_{\alpha,\upsilon}^{\varphi}\left(-\lambda (\tau-t)^{\alpha}\right) f(\tau)\, d\tau, \tag{5.783} $$

respectively.

The result was discovered by Yang in 2019 [92].

5.4.6 Prabhakar-Type Integrals

We now present the Prabhakar-type integrals via Prabhakar function.

Definition 5.88 Let $1 \geq \alpha \geq 0$ and $\lambda, \upsilon, \varphi \in \mathbb{R}$.

The left-sided Prabhakar-type integral is defined as [92]

$$ {}_{GP}^{R}I_{+}^{\alpha,\varphi,\upsilon,\lambda} f(t) = \int_{-\infty}^{t} (t-\tau)^{-\upsilon} E_{\alpha,1-\upsilon}^{-\varphi}\left(-\lambda (t-\tau)^{\alpha}\right) f(\tau)\, d\tau, \tag{5.784} $$

and the right-sided Prabhakar-type integral as

$$ {}_{GP}^{R}I_{-}^{\alpha,\varphi,\upsilon,\lambda} f(t) = \int_{t}^{\infty} (\tau-t)^{-\upsilon} E_{\alpha,1-\upsilon}^{-\varphi}\left(-\lambda (\tau-t)^{\alpha}\right) f(\tau)\, d\tau, \tag{5.785} $$

respectively.

The result was discovered by Yang in 2019 [92].

Theorem 5.140 *Let $0 < \alpha < 1$ and $\lambda, \upsilon, \varphi \in \mathbb{R}$. Then we have*

$$ {}_{GP}^{RL}D_{+}^{\alpha,\upsilon,\varphi,\lambda} \left({}_{GP}^{R}I_{+}^{\alpha,\varphi,\upsilon,\lambda} f(t) \right) = f(t) \tag{5.786} $$

and

$$ {}_{GP}^{RL}D_{+}^{\alpha,\upsilon,\varphi,\lambda} \left({}_{GP}^{R}I_{-}^{\alpha,\varphi,\upsilon,\lambda} f(t) \right) = f(t). \tag{5.787} $$

5.4.7 Hilfer-Type Derivative via Prabhakar Function

Based on the Hilfer derivative, Kilbas–Saigo–Saxena-type derivative via Prabhakar function, and Garra–Gorenflo–Polito–Tomovski-type derivative via Prabhakar function, the Hilfer-type derivative via Prabhakar function was proposed in 2019 by Yang [92].

Definition 5.89 Let $0 < \alpha < 1$ and $0 \le \beta \le 1$.

The left-sided Hilfer-type derivative via Prabhakar function is defined as [92]

$$ {}_{HGP}D_{a+}^{\alpha,\upsilon,\varphi,\lambda} f(t) = \left({}_{GP}I_{a+}^{\beta(1-\alpha),\upsilon,\varphi,\lambda} \frac{d}{dt} {}_{GP}I_{a+}^{(1-\beta)(1-\alpha),\upsilon,\varphi,\lambda} \right) f(t), \tag{5.788} $$

where

$$ {}_{GP}I_{a+}^{\beta(1-\alpha),\upsilon,\varphi,\lambda} f(t) = \int_a^t (t-\tau)^{\upsilon-1} E_{\beta(1-\alpha),\upsilon}^{\varphi} \left(-\lambda (t-\tau)^{\beta(1-\alpha)} \right) f(\tau)\, d\tau, \tag{5.789} $$

and

$$ \begin{aligned} & {}_{GP}I_{a+}^{(1-\beta)(1-\alpha),\upsilon,\varphi,\lambda} f(t) \\ & = \int_a^t (t-\tau)^{\upsilon-1} E_{(1-\beta)(1-\alpha),\upsilon}^{\varphi} \left(-\lambda (t-\tau)^{(1-\beta)(1-\alpha)} \right) f(\tau)\, d\tau, \end{aligned} \tag{5.790} $$

and the right-sided Hilfer-type derivative via Prabhakar function as

$$ {}_{HGP}D_{b-}^{\alpha,\upsilon,\varphi,\lambda} f(t) = \left({}_{GP}I_{b-}^{\beta(1-\alpha),\upsilon,\varphi,\lambda} \frac{d}{dt} {}_{GP}I_{b-}^{(1-\beta)(1-\alpha),\upsilon,\varphi,\lambda} \right) f(t), \tag{5.791} $$

where

$$ {}_{GP}I_{b-}^{\beta(1-\alpha),\upsilon,\varphi,\lambda} f(t) = \int_t^b (\tau-t)^{\upsilon-1} E_{\beta(1-\alpha),\upsilon}^{\varphi} \left(-\lambda (\tau-t)^{\beta(1-\alpha)} \right) f(\tau)\, d\tau, \tag{5.792} $$

and

$$ \begin{aligned} & {}_{GP}I_{b-}^{(1-\beta)(1-\alpha),\upsilon,\varphi,\lambda} f(t) \\ & = \int_t^b (\tau-t)^{\upsilon-1} E_{(1-\beta)(1-\alpha),\upsilon}^{\varphi} \left(-\lambda (\tau-t)^{(1-\beta)(1-\alpha)} \right) f(\tau)\, d\tau. \end{aligned} \tag{5.793} $$

The result was discovered by Yang in 2019 [92].

Theorem 5.141 *Let* $0 < \alpha < 1$ *and* $0 \leq \beta \leq 1$. *Then we have*

$$L\left\{{}_{HGP}D_{0+}^{\alpha,\upsilon,\varphi,\lambda} f(t)\right\}(s) \\ = s^{-1}\left(1+\lambda s^{-\beta(1-\alpha)}\right)^{-1}\left(s^{1-\upsilon}\left(1+\lambda s^{-\alpha}\right)^{-\varphi} f(s) - {}_{GP}I_{0+}^{(1-\beta)(1-\alpha),\upsilon,\varphi,\lambda} f(0)\right). \tag{5.794}$$

Definition 5.90 Let $0 < \alpha < 1$ and $0 \leq \beta \leq 1$.

The left-sided Hilfer-type derivative via Prabhakar function is defined as [92]

$${}_{HGP}D_{+}^{\alpha,\upsilon,\varphi,\lambda} f(t) = \left({}_{GP}I_{+}^{\beta(1-\alpha),\upsilon,\varphi,\lambda} \frac{d}{dt} {}_{GP}I_{+}^{(1-\beta)(1-\alpha),\upsilon,\varphi,\lambda} f\right)(t), \tag{5.795}$$

where

$${}_{GP}I_{+}^{\beta(1-\alpha),\upsilon,\varphi,\lambda} f(t) = \int_{-\infty}^{t} (t-\tau)^{\upsilon-1} E_{\beta(1-\alpha),\upsilon}^{\varphi}\left(-\lambda(t-\tau)^{\beta(1-\alpha)}\right) f(\tau)\, d\tau, \tag{5.796}$$

and

$${}_{GP}I_{+}^{(1-\beta)(1-\alpha),\upsilon,\varphi,\lambda} f(t) \\ = \int_{-\infty}^{t} (t-\tau)^{\upsilon-1} E_{(1-\beta)(1-\alpha),\upsilon}^{\varphi}\left(-\lambda(t-\tau)^{(1-\beta)(1-\alpha)}\right) f(\tau)\, d\tau, \tag{5.797}$$

and the right-sided Hilfer-type derivative via Prabhakar function as

$${}_{HGP}D_{-}^{\alpha,\upsilon,\varphi,\lambda} f(t) = \left({}_{GP}I_{-}^{\beta(1-\alpha),\upsilon,\varphi,\lambda} \frac{d}{dt} {}_{GP}I_{-}^{(1-\beta)(1-\alpha),\upsilon,\varphi,\lambda}\right) f(t), \tag{5.798}$$

where

$${}_{GP}I_{-}^{\beta(1-\alpha),\upsilon,\varphi,\lambda} f(t) = \int_{t}^{\infty} (\tau-t)^{\upsilon-1} E_{\beta(1-\alpha),\upsilon}^{\varphi}\left(-\lambda(\tau-t)^{\beta(1-\alpha)}\right) f(\tau)\, d\tau, \tag{5.799}$$

and

$${}_{GP}I_{-}^{(1-\beta)(1-\alpha),\upsilon,\varphi,\lambda} f(t) \\ = \int_{t}^{\infty} (\tau-t)^{\upsilon-1} E_{(1-\beta)(1-\alpha),\upsilon}^{\varphi}\left(-\lambda(\tau-t)^{(1-\beta)(1-\alpha)}\right) f(\tau)\, d\tau. \tag{5.800}$$

The result was discovered by Yang in 2019 [92].

5.4.8 Kilbas–Saigo–Saxena-Type Derivative with Respect to Another Function

Let $h^{(1)}(t) > 0$, $1 \geq \alpha \geq 0$, and $\lambda, \upsilon, \varphi \in \mathbb{R}$.

The left-sided Kilbas–Saigo–Saxena-type derivative with respect to another function is defined as [92]

$$\begin{aligned} {}_{GP}^{RL}D_{a+,h}^{\alpha,\upsilon,\varphi,\lambda} f(t) &= \left(\left(\frac{1}{h^{(1)}(t)}\frac{d}{dt}\right) {}_{GP}I_{a+,h}^{\alpha,\upsilon,\varphi,\lambda} f\right)(t) \\ &= \left(\frac{1}{h^{(1)}(t)}\frac{d}{dt}\right)\int_a^t (h(t)-h(\tau))^{\upsilon-1} E_{\alpha,\upsilon}^{\varphi}\left(-\lambda(h(t)-h(\tau))^{\alpha}\right) h^{(1)}(\tau) f(\tau)\,d\tau, \end{aligned} \tag{5.801}$$

and the right-sided Kilbas–Saigo–Saxena-type derivative with respect to another function as

$$\begin{aligned} {}_{GP}^{RL}D_{b-,h}^{\alpha,\upsilon,\varphi,\lambda} f(t) &= \left(\left(\frac{1}{h^{(1)}(t)}\frac{d}{dt}\right) {}_{GP}I_{b-,h}^{\alpha,\upsilon,\varphi,\lambda} f\right)(t) \\ &= \left(\frac{1}{h^{(1)}(t)}\frac{d}{dt}\right)\int_t^b (h(\tau)-h(t))^{\upsilon-1} E_{\alpha,\upsilon}^{\varphi}\left(-\lambda(h(\tau)-h(t))^{\alpha}\right) h^{(1)}(\tau) f(\tau)\,d\tau, \end{aligned} \tag{5.802}$$

where the left-sided Prabhakar-type integral with respect to another function is [92]

$${}_{GP}I_{a+,h}^{\alpha,\upsilon,\varphi,\lambda} f(t) = \int_a^t (h(t)-h(\tau))^{\upsilon-1} E_{\alpha,\upsilon}^{\varphi}\left(-\lambda(h(t)-h(\tau))^{\alpha}\right) h^{(1)}(\tau) f(\tau)\,d\tau, \tag{5.803}$$

and the right-sided Prabhakar-type integral with respect to another function is

$${}_{GP}I_{b-,h}^{\alpha,\upsilon,\varphi,\lambda} f(t) = \int_t^b (h(\tau)-h(t))^{\upsilon-1} E_{\alpha,\upsilon}^{\varphi}\left(-\lambda(h(\tau)-h(t))^{\alpha}\right) h^{(1)}(\tau) f(\tau)\,d\tau, \tag{5.804}$$

respectively.

The result was discovered by Yang in 2019 [92].

5.4.9 Garra–Gorenflo–Polito–Tomovski-Type Derivative with Respect to Another Function

Definition 5.91 Let $h^{(1)}(t) > 0$, $1 \geq \alpha \geq 0$, and $\lambda, \upsilon, \varphi \in \mathbb{R}$.

The left-sided Garra–Gorenflo–Polito–Tomovski-type derivative with respect to another function is defined as [92]

$$
\begin{aligned}
&{}_{GP}^{LS}D_{a+,h}^{\alpha,\upsilon,\varphi,\lambda} f(t)\\
&= {}_{GP}I_{a+,h}^{\alpha,\upsilon,\varphi,\lambda} \left(\frac{1}{h^{(1)}(t)}\frac{d}{dt}\right) f(t)\\
&= \int_a^t (h(t) - h(\tau))^{\upsilon-1} E_{\alpha,\upsilon}^{\varphi}\left(-\lambda (h(t) - h(\tau))^{\alpha}\right) h^{(1)}(\tau) \left(\frac{1}{h^{(1)}(\tau)}\frac{d}{d\tau}\right) f(\tau)\, d\tau,
\end{aligned}
$$

and the right-sided Garra–Gorenflo–Polito–Tomovski-type derivative as [92]

$$
\begin{aligned}
&{}_{GP}^{LS}D_{b-,h}^{\alpha,\upsilon,\varphi,\lambda} f(t)\\
&= {}_{GP}I_{b-,h}^{\alpha,\upsilon,\varphi,\lambda} \left(\frac{-1}{h^{(1)}(t)}\frac{d}{dt}\right) f(t)\\
&= \int_t^b (h(\tau) - h(t))^{\upsilon-1} E_{\alpha,\upsilon}^{\varphi}\left(-\lambda (h(\tau) - h(t))^{\alpha}\right) h^{(1)}(\tau) \left(\frac{-1}{h^{(1)}(\tau)}\frac{d}{d\tau}\right) f(\tau)\, d\tau,
\end{aligned}
\tag{5.805}
$$

where the left-sided Prabhakar-type integral with respect to another function is [92]

$$
{}_{GP}I_{a+,h}^{\alpha,\upsilon,\varphi,\lambda} f(t) = \int_a^t (h(t) - h(\tau))^{\upsilon-1} E_{\alpha,\upsilon}^{\varphi}\left(-\lambda (h(t) - h(\tau))^{\alpha}\right) h^{(1)}(\tau) f(\tau)\, d\tau,
\tag{5.806}
$$

and the right-sided Prabhakar-type integral with respect to another function is

$$
{}_{GP}I_{b-,h}^{\alpha,\upsilon,\varphi,\lambda} f(t) = \int_t^b (h(\tau) - h(t))^{\upsilon-1} E_{\alpha,\upsilon}^{\varphi}\left(-\lambda (h(\tau) - h(t))^{\alpha}\right) h^{(1)}(\tau) f(\tau)\, d\tau,
$$

respectively.

The result was discovered by Yang in 2019 [92].

5.4.10 Prabhakar-Type Integrals with Respect to Another Function

Definition 5.92 Let $h^{(1)}(t) > 0$, $1 \geq \alpha \geq 0$, and $\lambda, \upsilon, \varphi \in \mathbb{R}$.

The left-sided Prabhakar-type general fractional integral with respect to another function is defined as [92]

$$ {}_{GP}^{R}I_{a+,h}^{\alpha,\varphi,\upsilon,\lambda} f(t) = \int_{a}^{t} (h(t) - h(\tau))^{-\upsilon} E_{\alpha,1-\upsilon}^{-\varphi} \left(-\lambda (h(t) - h(\tau))^{\alpha}\right) h^{(1)}(\tau) f(\tau) d\tau, \tag{5.807} $$

and the right-sided Prabhakar-type integral with respect to another function as

$$ {}_{GP}^{R}I_{b-,h}^{\alpha,\varphi,\upsilon,\lambda} f(t) = \int_{t}^{b} (h(\tau) - h(t))^{-\upsilon} E_{\alpha,1-\upsilon}^{-\varphi} \left(-\lambda (h(\tau) - h(t))^{\alpha}\right) h^{(1)}(\tau) f(\tau) d\tau, \tag{5.808} $$

respectively.

The result was discovered by Yang in 2019 [92].

Theorem 5.142 *Let $h^{(1)}(t) > 0$, $0 < \alpha < 1$, and $\lambda, \upsilon, \varphi \in \mathbb{R}$. Then we have*

$$ {}_{GP}^{RL}D_{a+,h}^{\alpha,\upsilon,\varphi,\lambda} \left({}_{GP}^{R}I_{a+,h}^{\alpha,\varphi,\upsilon,\lambda} f(t) \right) = f(t) \tag{5.809} $$

and

$$ {}_{GP}^{RL}D_{b-,h}^{\alpha,\upsilon,\varphi,\lambda} \left({}_{GP}^{R}I_{b-,h}^{\alpha,\varphi,\upsilon,\lambda} f(t) \right) = f(t). \tag{5.810} $$

5.4.11 Kilbas–Saigo–Saxena-Type Derivative with Respect to Another Function

Definition 5.93 Let $h^{(1)}(t) > 0$, $1 \geq \alpha \geq 0$, and $\lambda, \upsilon, \varphi \in \mathbb{R}$.

The left-sided Kilbas–Saigo–Saxena-type derivative with respect to another function is defined as [92]

$$ \begin{aligned} &{}_{GP}^{RL}D_{+,h}^{\alpha,\upsilon,\varphi,\lambda} f(t) \\ &= \left(\frac{1}{h^{(1)}(t)}\frac{d}{dt}\right) \left({}_{GP}I_{+,h}^{\alpha,\upsilon,\varphi,\lambda} f(t) \right) \\ &= \left(\frac{1}{h^{(1)}(t)}\frac{d}{dt}\right) \int_{-\infty}^{t} (h(t) - h(\tau))^{\upsilon-1} E_{\alpha,\upsilon}^{\varphi} (-\lambda (h(t) - h(\tau))^{\alpha}) h^{(1)}(\tau) f(\tau) d\tau, \end{aligned} \tag{5.811} $$

and the right-sided Kilbas–Saigo–Saxena-type derivative with respect to another function as

$$
\begin{aligned}
&{}^{RL}_{GP}D^{\alpha,\upsilon,\varphi,\lambda}_{-,h} f(t) \\
&= \left(\frac{-1}{h^{(1)}(t)}\frac{d}{dt}\right)\left({}_{GP}I^{\alpha,\upsilon,\varphi,\lambda}_{-,h} f(t)\right) \\
&= \left(\frac{-1}{h^{(1)}(t)}\frac{d}{dt}\right)\int_{t}^{\infty} (h(\tau)-h(t))^{\upsilon-1} E^{\varphi}_{\alpha,\upsilon}\left(-\lambda (h(\tau)-h(t))^{\alpha}\right) h^{(1)}(\tau) f(\tau)\, d\tau,
\end{aligned}
\tag{5.812}
$$

where the left-sided Prabhakar-type integral with respect to another function is

$$
{}_{GP}I^{\alpha,\upsilon,\varphi,\lambda}_{+,h} f(t) = \int_{-\infty}^{t} (h(t)-h(\tau))^{\upsilon-1} E^{\varphi}_{\alpha,\upsilon}\left(-\lambda (h(t)-h(\tau))^{\alpha}\right) h^{(1)}(\tau) f(\tau)\, d\tau,
\tag{5.813}
$$

and the right-sided Prabhakar-type integral with respect to another function is

$$
{}_{GP}I^{\alpha,\upsilon,\varphi,\lambda}_{-,h} f(t) = \int_{t}^{\infty} (h(\tau)-h(t))^{\upsilon-1} E^{\varphi}_{\alpha,\upsilon}\left(-\lambda (h(\tau)-h(t))^{\alpha}\right) h^{(1)}(\tau) f(\tau)\, d\tau,
\tag{5.814}
$$

respectively.

The result was discovered by Yang in 2019 [92].

5.4.12 Garra–Gorenflo–Polito–Tomovski-Type Derivative with Respect to Another Function

Definition 5.94 Let $h^{(1)}(t) > 0$, $1 \geq \alpha \geq 0$, and $\lambda, \upsilon, \varphi \in \mathbb{R}$.

The left-sided Garra–Gorenflo–Polito–Tomovski-type derivative with respect to another function is defined as [92]

$$
\begin{aligned}
&{}^{LS}_{GP}D^{\alpha,\upsilon,\varphi,\lambda}_{+,h} f(t) \\
&= {}_{GP}I^{\alpha,\upsilon,\varphi,\lambda}_{+,h}\left(\frac{1}{h^{(1)}(t)}\frac{d}{dt}\right) f(t) \\
&= \int_{-\infty}^{t} (h(t)-h(\tau))^{\upsilon-1} E^{\varphi}_{\alpha,\upsilon}\left(-\lambda (h(t)-h(\tau))^{\alpha}\right) h^{(1)}(\tau) \left(\frac{1}{h^{(1)}(\tau)}\frac{d}{d\tau}\right) f(\tau)\, d\tau,
\end{aligned}
\tag{5.815}
$$

and the right-sided Garra–Gorenflo–Polito–Tomovski-type derivative with respect to another function as

$$\begin{aligned}
&{}_{GP}^{LS}D_{-,h}^{\alpha,\upsilon,\varphi,\lambda} f(t) \\
&= {}_{GP}I_{-,h}^{\alpha,\upsilon,\varphi,\lambda} \left(\frac{-1}{h^{(1)}(t)}\frac{d}{dt}\right) f(t) \\
&= \int_t^{\infty} (h(\tau) - h(t))^{\upsilon-1} E_{\alpha,\upsilon}^{\varphi} (-\lambda (h(\tau) - h(t))^{\alpha}) h^{(1)}(\tau) \left(\frac{-1}{h^{(1)}(\tau)}\frac{d}{d\tau}\right) f(\tau) d\tau,
\end{aligned} \tag{5.816}$$

where the left-sided Prabhakar-type integral with respect to another function is

$${}_{GP}I_{+,h}^{\alpha,\upsilon,\varphi,\lambda} f(t) = \int_{-\infty}^{t} (h(t) - h(\tau))^{\upsilon-1} E_{\alpha,\upsilon}^{\varphi} \left(-\lambda (h(t) - h(\tau))^{\alpha}\right) h^{(1)}(\tau) f(\tau) d\tau, \tag{5.817}$$

and the right-sided Prabhakar-type integral with respect to another function is

$${}_{GP}I_{-,h}^{\alpha,\upsilon,\varphi,\lambda} f(t) = \int_{t}^{\infty} (h(\tau) - h(t))^{\upsilon-1} E_{\alpha,\upsilon}^{\varphi} \left(-\lambda (h(\tau) - h(t))^{\alpha}\right) h^{(1)}(\tau) f(\tau) d\tau, \tag{5.818}$$

respectively.

The result was discovered by Yang in 2019 [92].

5.4.13 Prabhakar-Type Integrals with Respect to Another Function

Let $h^{(1)}(t) > 0$, $1 \geq \alpha > 0$, and $\lambda, \upsilon, \varphi \in \mathbb{R}$.

The left-sided Prabhakar-type integral with respect to another function is defined as [92]

$${}_{GP}^{R}I_{+,h}^{\alpha,\varphi,\upsilon,\lambda} f(t) = \int_{-\infty}^{t} (h(t) - h(\tau))^{-\upsilon} E_{\alpha,1-\upsilon}^{-\varphi} \left(-\lambda (h(t) - h(\tau))^{\alpha}\right) h^{(1)}(\tau) f(\tau) d\tau, \tag{5.819}$$

and the right-sided Prabhakar-type integral with respect to another function as

$$ {}^{R}_{GP}I^{\alpha,\varphi,\upsilon,\lambda}_{-,h} f(t) = \int_{t}^{\infty} (h(\tau) - h(t))^{-\upsilon} E^{-\varphi}_{\alpha,1-\upsilon} \left(-\lambda (h(\tau) - h(t))^{\alpha}\right) h^{(1)}(\tau) f(\tau) \, d\tau, \tag{5.820} $$

respectively.

The result was discovered by Yang in 2019 [92].

Theorem 5.143 *Let $h^{(1)}(t) > 0$, $0 < \alpha < 1$, and $\lambda, \upsilon, \varphi \in \mathbb{R}$. Then we have*

$$ {}^{RL}_{GP}D^{\alpha,\upsilon,\varphi,\lambda}_{+,h} \left({}^{R}_{GP}I^{\alpha,\varphi,\upsilon,\lambda}_{+,h} f(t)\right) = f(t) \tag{5.821} $$

and

$$ {}^{RL}_{GP}D^{\alpha,\upsilon,\varphi,\lambda}_{-,h} \left({}^{R}_{GP}I^{\alpha,\varphi,\upsilon,\lambda}_{-,h} f(t)\right) = f(t). \tag{5.822} $$

5.4.14 *Hilfer-Type Derivative with Respect to Another Function via Prabhakar Function*

Definition 5.95 Let $h^{(1)}(t) > 0$, $0 < \alpha < 1$, and $0 \le \beta \le 1$.

The left-sided Hilfer-type derivative with respect to another function via Prabhakar function is defined as [92]

$$ {}_{HGP}D^{\alpha,\upsilon,\varphi,\lambda}_{a+,h} f(t) = \left({}_{GP}I^{\beta(1-\alpha),\upsilon,\varphi,\lambda}_{a+,h} \left(\frac{1}{h^{(1)}(t)} \frac{d}{dt}\right) {}_{GP}I^{(1-\beta)(1-\alpha),\upsilon,\varphi,\lambda}_{a+,h}\right) f(t), \tag{5.823} $$

where

$$ \begin{aligned} &{}_{GP}I^{\beta(1-\alpha),\upsilon,\varphi,\lambda}_{a+,h} f(t) \\ &= \int_{a}^{t} (h(t) - h(\tau))^{\upsilon-1} E^{\varphi}_{\beta(1-\alpha),\upsilon} \left(-\lambda (h(t) - h(\tau))^{\beta(1-\alpha)}\right) h^{(1)}(\tau) f(\tau) \, d\tau, \end{aligned} \tag{5.824} $$

and

$$ \begin{aligned} &{}_{GP}I^{(1-\beta)(1-\alpha),\upsilon,\varphi,\lambda}_{a+,h} f(t) \\ &= \int_{a}^{t} (h(t) - h(\tau))^{\upsilon-1} E^{\varphi}_{(1-\beta)(1-\alpha),\upsilon} \left(-\lambda (h(t) - h(\tau))^{(1-\beta)(1-\alpha)}\right) h^{(1)}(\tau) f(\tau) \, d\tau, \end{aligned} \tag{5.825} $$

and the right-sided Hilfer-type derivative with respect to another function via Prabhakar function as

$$ {}_{HGP}D_{b-,h}^{\alpha,\upsilon,\varphi,\lambda} f(t) = \left({}_{GP}I_{b-,h}^{\beta(1-\alpha),\upsilon,\varphi,\lambda} \left(\frac{-1}{h^{(1)}(t)} \frac{d}{dt} \right) {}_{GP}I_{b-,h}^{(1-\beta)(1-\alpha),\upsilon,\varphi,\lambda} \right) f(t), \tag{5.826} $$

where

$$ {}_{GP}I_{b-,h}^{\beta(1-\alpha),\upsilon,\varphi,\lambda} f(t) = \int_t^b (h(\tau) - h(t))^{\upsilon-1} E_{\beta(1-\alpha),\upsilon}^{\varphi} \left(-\lambda (h(\tau) - h(t))^{\beta(1-\alpha)} \right) h^{(1)}(\tau) f(\tau) d\tau, \tag{5.827} $$

and

$$ {}_{GP}I_{b-,h}^{(1-\beta)(1-\alpha),\upsilon,\varphi,\lambda} f(t) = \int_t^b (h(\tau) - h(t))^{\upsilon-1} E_{(1-\beta)(1-\alpha),\upsilon}^{\varphi} \left(-\lambda (h(\tau) - h(t))^{(1-\beta)(1-\alpha)} \right) h^{(1)}(\tau) f(\tau) d\tau. \tag{5.828} $$

The result was discovered by Yang in 2019 [92].

Definition 5.96 Let $h^{(1)}(t) > 0$, $0 < \alpha < 1$, and $0 \le \beta \le 1$.

The left-sided Hilfer-type derivative with respect to another function via Prabhakar function is defined as [92]

$$ {}_{HGP}D_{+}^{\alpha,\upsilon,\varphi,\lambda} f(t) = \left({}_{GP}I_{+}^{\beta(1-\alpha),\upsilon,\varphi,\lambda} \left(\frac{1}{h^{(1)}(t)} \frac{d}{dt} \right) {}_{GP}I_{+}^{(1-\beta)(1-\alpha),\upsilon,\varphi,\lambda} f \right)(t), \tag{5.829} $$

where

$$ {}_{GP}I_{+,h}^{\beta(1-\alpha),\upsilon,\varphi,\lambda} f(t) = \int_{-\infty}^t (h(t) - h(\tau))^{\upsilon-1} E_{\beta(1-\alpha),\upsilon}^{\varphi} \left(-\lambda (h(t) - h(\tau))^{\beta(1-\alpha)} \right) h^{(1)}(\tau) f(\tau) d\tau, \tag{5.830} $$

and

$$ {}_{GP}I_{+,h}^{(1-\beta)(1-\alpha),\upsilon,\varphi,\lambda} f(t) = \int_{-\infty}^t (h(t) - h(\tau))^{\upsilon-1} E_{(1-\beta)(1-\alpha),\upsilon}^{\varphi} \left(-\lambda (h(t) - h(\tau))^{(1-\beta)(1-\alpha)} \right) h^{(1)}(\tau) f(\tau) d\tau, \tag{5.831} $$

and the right-sided Hilfer-type derivative with respect to another function via Prabhakar function as

$$ {}_{HGP}D_{-,h}^{\alpha,\upsilon,\varphi,\lambda} f(t) = \left({}_{GP}I_{-,h}^{\beta(1-\alpha),\upsilon,\varphi,\lambda} \left(\frac{-1}{h^{(1)}(t)} \frac{d}{dt} \right) {}_{GP}I_{-,h}^{(1-\beta)(1-\alpha),\upsilon,\varphi,\lambda} \right) f(t), \tag{5.832} $$

where

$$ {}_{GP}I_{-,h}^{\beta(1-\alpha),\upsilon,\varphi,\lambda} f(t) $$
$$ = \int_t^{\infty} (h(\tau) - h(t))^{\upsilon-1} E_{\beta(1-\alpha),\upsilon}^{\varphi} \left(-\lambda (h(\tau) - h(t))^{\beta(1-\alpha)} \right) h^{(1)}(\tau) f(\tau) d\tau, \tag{5.833} $$

and

$$ {}_{GP}I_{-,h}^{(1-\beta)(1-\alpha),\upsilon,\varphi,\lambda} f(t) $$
$$ = \int_t^{\infty} (h(\tau) - h(t))^{\upsilon-1} E_{(1-\beta)(1-\alpha),\upsilon}^{\varphi} \left(-\lambda (h(\tau) - h(t))^{(1-\beta)(1-\alpha)} \right) h^{(1)}(\tau) f(\tau) d\tau. \tag{5.834} $$

The result was discovered by Yang in 2019 [92].

5.5 Applications

5.5.1 *The Integral Equations in the Kernel of New Special Functions*

In this section, we consider the Volterra-type and Fredholm-type integral equations in the kernels of the Prabhakar function and supersine and supercosine via Prabhakar function.

Example 5.1 The Volterra-type integral equations via Prabhakar function can be written as follows:

$$ \rho \int_a^t (t-\tau)^{\beta-1} E_{\alpha,\beta}^{\gamma} \left(\lambda (t-\tau)^{\alpha} \right) f(\tau) d\tau + f(t) = g(t), \tag{5.835} $$

and

$$ \rho \int_a^t (t-\tau)^{\beta-1} E_{\alpha,\beta}^{\gamma} \left(\lambda (t-\tau)^{\alpha} \right) f(\tau) d\tau = g(t), \tag{5.836} $$

where ρ and λ are the constants.

Example 5.2 The Volterra-type integral equations via Prabhakar function are given as follows:

$$\rho \int_0^t (t-\tau)^{\beta-1} E_{\alpha,\beta}^{\gamma}\left(\lambda (t-\tau)^{\alpha}\right) f(\tau)\, d\tau + f(t) = g(t), \tag{5.837}$$

and

$$\rho \int_0^t (t-\tau)^{\beta-1} E_{\alpha,\beta}^{\gamma}\left(\lambda (t-\tau)^{\alpha}\right) f(\tau)\, d\tau = g(t), \tag{5.838}$$

where ρ and λ are the constants.

Example 5.3 The Fredholm-type integral equations via Prabhakar function read as follows:

$$\rho \int_0^\infty (t-\tau)^{\beta-1} E_{\alpha,\beta}^{\gamma}\left(\lambda (t-\tau)^{\alpha}\right) f(\tau)\, d\tau + f(t) = g(t), \tag{5.839}$$

and

$$\rho \int_0^\infty (t-\tau)^{\beta-1} E_{\alpha,\beta}^{\gamma}\left(\lambda (t-\tau)^{\alpha}\right) f(\tau)\, d\tau = g(t), \tag{5.840}$$

where ρ and λ are the constants.

Example 5.4 The Volterra-type integral equations within the supersine kernel via Prabhakar function can be given as follows:

$$\rho \int_a^t supersin_{\alpha,\beta}^{\gamma}\left(\lambda (t-\tau)^{\alpha}\right) f(\tau)\, d\tau + f(t) = g(t), \tag{5.841}$$

and

$$\rho \int_a^t supersin_{\alpha,\beta}^{\gamma}\left(\lambda (t-\tau)^{\alpha}\right) f(\tau)\, d\tau = g(t), \tag{5.842}$$

where ρ and λ are the constants.

Example 5.5 The Volterra-type integral equations within the supersine kernel via Prabhakar function are given as follows:

$$\rho \int_0^t supersin_{\alpha,\beta}^{\gamma}\left(\lambda (t-\tau)^{\alpha}\right) f(\tau)\, d\tau + f(t) = g(t), \tag{5.843}$$

and

$$\rho \int_0^t supersin_{\alpha,\beta}^{\gamma}\left(\lambda (t-\tau)^{\alpha}\right) f(\tau)\, d\tau = g(t), \tag{5.844}$$

where ρ and λ are the constants.

Example 5.6 The Fredholm-type integral equations in the supersine kernel via Prabhakar function are presented as follows:

$$\rho \int_0^\infty supersin_{\alpha,\beta}^{\gamma}\left(\lambda (t-\tau)^{\alpha}\right) f(\tau)\, d\tau + f(t) = g(t), \tag{5.845}$$

and

$$\rho \int_0^\infty supersin_{\alpha,\beta}^{\gamma}\left(\lambda (t-\tau)^{\alpha}\right) f(\tau)\, d\tau = g(t), \tag{5.846}$$

where ρ and λ are the constants.

Example 5.7 The Volterra-type integral equations within the supercosine kernel via Prabhakar function can be given as follows:

$$\rho \int_a^t supercos_{\alpha,\beta}^{\gamma}\left(\lambda (t-\tau)^{\alpha}\right) f(\tau)\, d\tau + f(t) = g(t), \tag{5.847}$$

and

$$\rho \int_a^t supercos_{\alpha,\beta}^{\gamma}\left(\lambda (t-\tau)^{\alpha}\right) f(\tau)\, d\tau = g(t), \tag{5.848}$$

where ρ and λ are the constants.

Example 5.8 The Volterra-type integral equations within the supercosine kernel via Prabhakar function are given as follows:

$$\rho \int_{0}^{t} supercos_{\alpha,\beta}^{\gamma}\left(\lambda\left(t-\tau\right)^{\alpha}\right) f\left(\tau\right) d\tau + f\left(t\right) = g\left(t\right), \tag{5.849}$$

and

$$\rho \int_{0}^{t} supercos_{\alpha,\beta}^{\gamma}\left(\lambda\left(t-\tau\right)^{\alpha}\right) f\left(\tau\right) d\tau = g\left(t\right), \tag{5.850}$$

where ρ and λ are the constants.

Example 5.9 The Fredholm-type integral equations in the supercosine kernel via Prabhakar function are presented as follows:

$$\rho \int_{0}^{\infty} supercos_{\alpha,\beta}^{\gamma}\left(\lambda\left(t-\tau\right)^{\alpha}\right) f\left(\tau\right) d\tau + f\left(t\right) = g\left(t\right), \tag{5.851}$$

and

$$\rho \int_{0}^{\infty} supercos_{\alpha,\beta}^{\gamma}\left(\lambda\left(t-\tau\right)^{\alpha}\right) f\left(\tau\right) d\tau = g\left(t\right), \tag{5.852}$$

where ρ and λ are the constants.

Example 5.10 The Volterra-type integral equations within the superhyperbolic supersine kernel via Prabhakar function can be given as follows:

$$\rho \int_{a}^{t} supersinh_{\alpha,\beta}^{\gamma}\left(\lambda\left(t-\tau\right)^{\alpha}\right) f\left(\tau\right) d\tau + f\left(t\right) = g\left(t\right), \tag{5.853}$$

and

$$\rho \int_{a}^{t} supersinh_{\alpha,\beta}^{\gamma}\left(\lambda\left(t-\tau\right)^{\alpha}\right) f\left(\tau\right) d\tau = g\left(t\right), \tag{5.854}$$

where ρ and λ are the constants.

Example 5.11 The Volterra-type integral equations within the superhyperbolic supersine kernel via Prabhakar function are given as follows:

$$\rho \int_{0}^{t} supersinh_{\alpha,\beta}^{\gamma}\left(\lambda\left(t-\tau\right)^{\alpha}\right) f\left(\tau\right) d\tau + f\left(t\right) = g\left(t\right), \tag{5.855}$$

and

$$\rho \int_{0}^{t} supersinh_{\alpha,\beta}^{\gamma}\left(\lambda\left(t-\tau\right)^{\alpha}\right) f\left(\tau\right) d\tau = g\left(t\right), \tag{5.856}$$

where ρ and λ are the constants.

Example 5.12 The Fredholm-type integral equations within the superhyperbolic supersine kernel via Prabhakar function are presented as follows:

$$\rho \int_{0}^{\infty} supersinh_{\alpha,\beta}^{\gamma}\left(\lambda\left(t-\tau\right)^{\alpha}\right) f\left(\tau\right) d\tau + f\left(t\right) = g\left(t\right), \tag{5.857}$$

and

$$\rho \int_{0}^{\infty} supersinh_{\alpha,\beta}^{\gamma}\left(\lambda\left(t-\tau\right)^{\alpha}\right) f\left(\tau\right) d\tau = g\left(t\right), \tag{5.858}$$

where ρ and λ are the constants.

Example 5.13 The Volterra-type integral equations within the superhyperbolic supercosine kernel via Prabhakar function can be given as follows:

$$\rho \int_{a}^{t} supercosh_{\alpha,\beta}^{\gamma}\left(\lambda\left(t-\tau\right)^{\alpha}\right) f\left(\tau\right) d\tau + f\left(t\right) = g\left(t\right), \tag{5.859}$$

and

$$\rho \int_{a}^{t} supercosh_{\alpha,\beta}^{\gamma}\left(\lambda\left(t-\tau\right)^{\alpha}\right) f\left(\tau\right) d\tau = g\left(t\right), \tag{5.860}$$

where ρ and λ are the constants.

Example 5.14 The Volterra-type integral equations within the superhyperbolic supercosine kernel via Prabhakar function are given as follows:

$$\rho \int_{0}^{t} supercosh_{\alpha,\beta}^{\gamma}\left(\lambda (t-\tau)^{\alpha}\right) f(\tau)\, d\tau + f(t) = g(t), \tag{5.861}$$

and

$$\rho \int_{0}^{t} supercosh_{\alpha,\beta}^{\gamma}\left(\lambda (t-\tau)^{\alpha}\right) f(\tau)\, d\tau = g(t), \tag{5.862}$$

where ρ and λ are the constants.

Example 5.15 The Fredholm-type integral equations in the superhyperbolic supercosine kernel via Prabhakar function are presented as follows:

$$\rho \int_{0}^{\infty} supercosh_{\alpha,\beta}^{\gamma}\left(\lambda (t-\tau)^{\alpha}\right) f(\tau)\, d\tau + f(t) = g(t), \tag{5.863}$$

and

$$\rho \int_{0}^{\infty} supercosh_{\alpha,\beta}^{\gamma}\left(\lambda (t-\tau)^{\alpha}\right) f(\tau)\, d\tau = g(t), \tag{5.864}$$

where ρ and λ are the constants.

Example 5.16 The Volterra-type integral equations via Prabhakar function can be written as follows:

$$\rho \int_{a}^{t} (t-\tau)^{\beta-1} E_{\alpha,\beta}^{\gamma}\left(\lambda (t-\tau)^{\vartheta}\right) f(\tau)\, d\tau + f(t) = g(t), \tag{5.865}$$

and

$$\rho \int_{a}^{t} (t-\tau)^{\beta-1} E_{\alpha,\beta}^{\gamma}\left(\lambda (t-\tau)^{\vartheta}\right) f(\tau)\, d\tau = g(t), \tag{5.866}$$

where ρ and λ are the constants.

Example 5.17 The Volterra-type integral equations via Prabhakar function are given as follows:

$$\rho\int_0^t (t-\tau)^{\beta-1}E_{\alpha,\beta}^{\gamma}\left(\lambda(t-\tau)^{\vartheta}\right)f(\tau)\,d\tau+f(t)=g(t), \tag{5.867}$$

and

$$\rho\int_0^t (t-\tau)^{\beta-1}E_{\alpha,\beta}^{\gamma}\left(\lambda(t-\tau)^{\vartheta}\right)f(\tau)\,d\tau=g(t), \tag{5.868}$$

where ρ and λ are the constants.

Example 5.18 The Fredholm-type integral equations via Prabhakar function read as follows:

$$\rho\int_0^\infty (t-\tau)^{\beta-1}E_{\alpha,\beta}^{\gamma}\left(\lambda(t-\tau)^{\vartheta}\right)f(\tau)\,d\tau+f(t)=g(t), \tag{5.869}$$

and

$$\rho\int_0^\infty (t-\tau)^{\beta-1}E_{\alpha,\beta}^{\gamma}\left(\lambda(t-\tau)^{\vartheta}\right)f(\tau)\,d\tau=g(t), \tag{5.870}$$

where ρ and λ are the constants.

Example 5.19 The Volterra-type integral equations within the presupersine kernel via Prabhakar function can be given as follows:

$$\rho\int_a^t Gresin_{\alpha,\beta}^{\gamma}\left(\lambda(t-\tau)^{\vartheta}\right)f(\tau)\,d\tau+f(t)=g(t), \tag{5.871}$$

and

$$\rho\int_a^t Gresin_{\alpha,\beta}^{\gamma}\left(\lambda(t-\tau)^{\vartheta}\right)f(\tau)\,d\tau=g(t), \tag{5.872}$$

where ρ and λ are the constants.

Example 5.20 The Volterra-type integral equations within the presupersine kernel via Prabhakar function are given as follows:

$$\rho \int_{0}^{t} Gresin_{\alpha,\beta}^{\gamma}\left(\lambda (t-\tau)^{\vartheta}\right) f(\tau)\, d\tau + f(t) = g(t), \tag{5.873}$$

and

$$\rho \int_{0}^{t} Gresin_{\alpha,\beta}^{\gamma}\left(\lambda (t-\tau)^{\vartheta}\right) f(\tau)\, d\tau = g(t), \tag{5.874}$$

where ρ and λ are the constants.

Example 5.21 The Fredholm-type integral equations in the presupersine kernel via Prabhakar function are presented as follows:

$$\rho \int_{0}^{\infty} Gresin_{\alpha,\beta}^{\gamma}\left(\lambda (t-\tau)^{\vartheta}\right) f(\tau)\, d\tau + f(t) = g(t), \tag{5.875}$$

and

$$\rho \int_{0}^{\infty} Gresin_{\alpha,\beta}^{\gamma}\left(\lambda (t-\tau)^{\vartheta}\right) f(\tau)\, d\tau = g(t), \tag{5.876}$$

where ρ and λ are the constants.

Example 5.22 The Volterra-type integral equations within the precosine kernel via Prabhakar function can be given as follows:

$$\rho \int_{a}^{t} Grecos_{\alpha,\beta}^{\gamma}\left(\lambda (t-\tau)^{\vartheta}\right) f(\tau)\, d\tau + f(t) = g(t), \tag{5.877}$$

and

$$\rho \int_{a}^{t} Grecos_{\alpha,\beta}^{\gamma}\left(\lambda (t-\tau)^{\vartheta}\right) f(\tau)\, d\tau = g(t), \tag{5.878}$$

where ρ and λ are the constants.

Example 5.23 The Volterra-type integral equations within the precosine kernel via Prabhakar function are given as follows:

$$\rho \int_0^t Grecos_{\alpha,\beta}^{\gamma}\left(\lambda\left(t-\tau\right)^{\vartheta}\right) f\left(\tau\right) d\tau + f\left(t\right) = g\left(t\right), \tag{5.879}$$

and

$$\rho \int_0^t Grecos_{\alpha,\beta}^{\gamma}\left(\lambda\left(t-\tau\right)^{\vartheta}\right) f\left(\tau\right) d\tau = g\left(t\right), \tag{5.880}$$

where ρ and λ are the constants.

Example 5.24 The Fredholm-type integral equations in the precosine kernel via Prabhakar function are presented as follows:

$$\rho \int_0^\infty Grecos_{\alpha,\beta}^{\gamma}\left(\lambda\left(t-\tau\right)^{\vartheta}\right) f\left(\tau\right) d\tau + f\left(t\right) = g\left(t\right), \tag{5.881}$$

and

$$\rho \int_0^\infty Grecos_{\alpha,\beta}^{\gamma}\left(\lambda\left(t-\tau\right)^{\vartheta}\right) f\left(\tau\right) d\tau = g\left(t\right), \tag{5.882}$$

where ρ and λ are the constants.

Example 5.25 The Volterra-type integral equations within the superhyperbolic sine kernel via Prabhakar function can be given as follows:

$$\rho \int_a^t Gresinh_{\alpha,\beta}^{\gamma}\left(\lambda\left(t-\tau\right)^{\vartheta}\right) f\left(\tau\right) d\tau + f\left(t\right) = g\left(t\right), \tag{5.883}$$

and

$$\rho \int_a^t Gresinh_{\alpha,\beta}^{\gamma}\left(\lambda\left(t-\tau\right)^{\vartheta}\right) f\left(\tau\right) d\tau = g\left(t\right), \tag{5.884}$$

where ρ and λ are the constants.

Example 5.26 The Volterra-type integral equations within the superhyperbolic presine kernel via Prabhakar function are given as follows:

$$\rho\int_{0}^{t} Gresinh_{\alpha,\beta}^{\gamma}\left(\lambda\left(t-\tau\right)^{\vartheta}\right)f\left(\tau\right)d\tau+f\left(t\right)=g\left(t\right), \tag{5.885}$$

and

$$\rho\int_{0}^{t} Gresinh_{\alpha,\beta}^{\gamma}\left(\lambda\left(t-\tau\right)^{\vartheta}\right)f\left(\tau\right)d\tau=g\left(t\right), \tag{5.886}$$

where ρ and λ are the constants.

Example 5.27 The Fredholm-type integral equations in the superhyperbolic presine kernel via Prabhakar function are presented as follows:

$$\rho\int_{0}^{\infty} Gresinh_{\alpha,\beta}^{\gamma}\left(\lambda\left(t-\tau\right)^{\vartheta}\right)f\left(\tau\right)d\tau+f\left(t\right)=g\left(t\right), \tag{5.887}$$

and

$$\rho\int_{0}^{\infty} Gresinh_{\alpha,\beta}^{\gamma}\left(\lambda\left(t-\tau\right)^{\vartheta}\right)f\left(\tau\right)d\tau=g\left(t\right), \tag{5.888}$$

where ρ and λ are the constants.

Example 5.28 The Volterra-type integral equations within the superhyperbolic precosine kernel via Prabhakar function can be given as follows:

$$\rho\int_{a}^{t} Grecosh_{\alpha,\beta}^{\gamma}\left(\lambda\left(t-\tau\right)^{\vartheta}\right)f\left(\tau\right)d\tau+f\left(t\right)=g\left(t\right), \tag{5.889}$$

and

$$\rho\int_{a}^{t} Grecosh_{\alpha,\beta}^{\gamma}\left(\lambda\left(t-\tau\right)^{\vartheta}\right)f\left(\tau\right)d\tau=g\left(t\right), \tag{5.890}$$

where ρ and λ are the constants.

Example 5.29 The Volterra-type integral equations within the superhyperbolic precosine kernel via Prabhakar function are given as follows:

$$\rho \int_{0}^{t} Grecosh_{\alpha,\beta}^{\gamma}\left(\lambda\left(t-\tau\right)^{\vartheta}\right) f\left(\tau\right) d\tau + f\left(t\right) = g\left(t\right), \tag{5.891}$$

and

$$\rho \int_{0}^{t} Grecosh_{\alpha,\beta}^{\gamma}\left(\lambda\left(t-\tau\right)^{\vartheta}\right) f\left(\tau\right) d\tau = g\left(t\right), \tag{5.892}$$

where ρ and λ are the constants.

Example 5.30 The Fredholm-type integral equations in the superhyperbolic precosine kernel via Prabhakar function are presented as follows:

$$\rho \int_{0}^{\infty} Grecosh_{\alpha,\beta}^{\gamma}\left(\lambda\left(t-\tau\right)^{\vartheta}\right) f\left(\tau\right) d\tau + f\left(t\right) = g\left(t\right), \tag{5.893}$$

and

$$\rho \int_{0}^{\infty} Grecosh_{\alpha,\beta}^{\gamma}\left(\lambda\left(t-\tau\right)^{\vartheta}\right) f\left(\tau\right) d\tau = g\left(t\right), \tag{5.894}$$

where ρ and λ are the constants.

Example 5.31 The Volterra-type integral equations via Prabhakar function can be written as follows:

$$\rho \int_{a}^{t} E_{\alpha,\beta}^{\gamma}\left(\lambda\left(t-\tau\right)^{\vartheta}\right) f\left(\tau\right) d\tau + f\left(t\right) = g\left(t\right), \tag{5.895}$$

and

$$\rho \int_{a}^{t} E_{\alpha,\beta}^{\gamma}\left(\lambda\left(t-\tau\right)^{\vartheta}\right) f\left(\tau\right) d\tau = g\left(t\right), \tag{5.896}$$

where ρ and λ are the constants.

Example 5.32 The Volterra-type integral equations via Prabhakar function are given as follows:

$$\rho \int_0^t E_{\alpha,\beta}^{\gamma} \left(\lambda (t - \tau)^{\vartheta} \right) f(\tau) \, d\tau + f(t) = g(t), \tag{5.897}$$

and

$$\rho \int_0^t E_{\alpha,\beta}^{\gamma} \left(\lambda (t - \tau)^{\vartheta} \right) f(\tau) \, d\tau = g(t), \tag{5.898}$$

where ρ and λ are the constants.

Example 5.33 The Fredholm-type integral equations via Prabhakar function read as follows:

$$\rho \int_0^\infty E_{\alpha,\beta}^{\gamma} \left(\lambda (t - \tau)^{\vartheta} \right) f(\tau) \, d\tau + f(t) = g(t), \tag{5.899}$$

and

$$\rho \int_0^\infty E_{\alpha,\beta}^{\gamma} \left(\lambda (t - \tau)^{\vartheta} \right) f(\tau) \, d\tau = g(t), \tag{5.900}$$

where ρ and λ are the constants.

5.5.2 Anomalous Viscoelasticity and Diffusion

In this section, we present the mathematical models based on the general fractional derivatives with the Prabhakar kernel and give the obtained examples.

Example 5.34 (Newtonian-Like Dashpot Element)

The Newtonian-like dashpot element with the Prabhakar kernel is presented as follows [92]:

$$\sigma(\tau) = \gamma \, {}_{GW}^{RL}D_{0+}^{\alpha,\upsilon,\varphi,\lambda} \varepsilon(\tau), \tag{5.901}$$

where γ is the viscosity of the material,$\varepsilon(\tau)$ is the strain, $\sigma(\tau)$ is the stress, and τ is the time.

The result was proposed in [92, 267].

Example 5.35 (Maxwell-Like Model)

The Maxwell-like model with the Prabhakar kernel can be written as follows [92]:

$$ {}_{GW}^{RL}D_{0+}^{\alpha,\upsilon,\varphi,\lambda}\varepsilon(\tau) = \frac{\sigma(\tau)}{\gamma} + \frac{{}_{GW}^{RL}D_{0+}^{\alpha,\upsilon,\varphi,\lambda}\sigma(\tau)}{\zeta}, \tag{5.902} $$

where γ is the viscosity of the material and ζ is the elastic modulus of the material.

The result was proposed in 2018 [92, 267].

Example 5.36 (Kelvin–Voigt-Like Model)

The Kelvin–Voigt-like model with the Prabhakar kernel can be given as follows [92]

$$ \sigma(\tau) = \gamma\varepsilon(\tau) + \zeta\,{}_{GW}^{RL}D_{0+}^{\alpha,\upsilon,\varphi,\lambda}\varepsilon(\tau), \tag{5.903} $$

where γ is the viscosity of the material and ζ is the elastic modulus of the material.

The result was proposed in 2019 [92].

Example 5.37 (Newtonian-Like Dashpot Element)

The Newtonian-like dashpot element with the Prabhakar kernel is given as follows [92]:

$$ \sigma(\tau) = \gamma\,{}_{GW}^{LSC}D_{0+}^{\alpha,\upsilon,\varphi,\lambda}\varepsilon(\tau), \tag{5.904} $$

where γ is the viscosity of the material, $\varepsilon(\tau)$ is the strain, $\sigma(\tau)$ is the stress, and τ is the time.

The result was proposed in 2019 [92].

Example 5.38 (Maxwell-Like Element)

The Maxwell-like model with the Prabhakar kernel can be written as follows [92]:

$$ {}_{GW}^{LSC}D_{0+}^{\alpha,\upsilon,\varphi,\lambda}\varepsilon(\tau) = \frac{\sigma(\tau)}{\gamma} + \frac{{}_{GW}^{LSC}D_{0+}^{\alpha,\upsilon,\varphi,\lambda}\sigma(\tau)}{\zeta}, \tag{5.905} $$

where γ is the viscosity of the material and ζ is the elastic modulus of the material.

The result was proposed in 2020 [92].

Example 5.39 (Kelvin–Voigt-Like Model)

The Kelvin–Voigt-like model with the Prabhakar kernel can be given as follows [92]:

$$ \sigma(\tau) = \gamma\varepsilon(\tau) + \zeta\,{}_{Ml}^{LSC}D_{0+}^{\alpha,\upsilon,\varphi,\lambda}\varepsilon(\tau), \tag{5.906} $$

where γ is the viscosity of the material and ζ is the elastic modulus of the material.

The result was proposed in 2020 [92].

Example 5.40 (The Anomalous Diffusion Model)
The anomalous diffusion model containing the general fractional derivative time with the Prabhakar kernel was [268]

$$
{}_{GW}^{LSC}D_{0+}^{\alpha,\upsilon,\varphi,\lambda}\Xi\left(x,\tau\right)=u\frac{\partial^{2}\Xi\left(x,\tau\right)}{\partial x^{2}}, \tag{5.907}
$$

where u is a real constant reflecting the magnitudes of the diffusion capacity.
The result was proposed by Yang et al. in 2018 [2].

Chapter 6
Other Special Functions Related to Mittag-Leffler Function

Abstract In this chapter, we introduce the special functions related to Mittag-Leffler function containing the Sonine functions, Rabotnov fractional exponential function, Miller–Ross function, and Lorenzo–Hartley functions. The Laplace and Mellin transforms of them are discussed, and the integral representations for the supertrigonometric and superhyperbolic functions are also presented in detail. The alternative forms associated with the Mittag-Leffler function and Wright hypergeometric function are given.

6.1 The Sonine Functions: History, Definitions, and Properties

In 1884, Nikolay Yakovlevich Sonin (Sonine), a Russian mathematician who worked on special functions and numerical integration, proposed the Sonine functions [269], which are called the Sonine functions of first type and Sonine functions of second type, respectively.

6.1.1 The Sonine Functions of First Type

Definition 6.1 The Sonine function of first type is defined as [269]

$$\begin{aligned} S_p(t) &= t^{-p}\sum_{k=0}^{\infty}\frac{t^k}{\Gamma(k+1)\,\Gamma(k-p+1)} \\ &= \sum_{k=0}^{\infty}\frac{t^{k-p}}{\Gamma(k+1)\,\Gamma(k-p+1)}, \end{aligned} \tag{6.1}$$

where $p, t \in \mathbb{C}$.

X.-J. Yang, *Theory and Applications of Special Functions for Scientists and Engineers*, https://doi.org/10.1007/978-981-33-6334-2_6

The result discovered by in Sonine in 1884 [269] is called the Sonine function of first type and further reported in 2020 by Yang [92].

In this case, the pairs of the Sonine function were represented as [269]

$$\begin{aligned} S_p(\lambda, t) &= t^{-p} \sum_{k=0}^{\infty} \frac{(\lambda t)^k}{\Gamma(k+1)\Gamma(k-p+1)} \\ &= \sum_{k=0}^{\infty} \frac{\lambda^k t^{k-p}}{\Gamma(k+1)\Gamma(k-p+1)} \end{aligned} \tag{6.2}$$

and

$$\begin{aligned} S_q(-\lambda, t) &= t^{-q} \sum_{k=0}^{\infty} \frac{(-\lambda t)^k}{\Gamma(k+1)\Gamma(k-q+1)} \\ &= \sum_{k=0}^{\infty} \frac{(-\lambda)^k t^{k-q}}{\Gamma(k+1)\Gamma(k-q+1)}, \end{aligned} \tag{6.3}$$

where $p \in \mathbb{C}$, $q \in \mathbb{C}$, and λ is a constant.

The Laplace transforms of (6.2) and (6.3) can be written as follows:

$$\begin{aligned} L\{S_p(\lambda, t)\} &= L\left\{t^{-p} \sum_{k=0}^{\infty} \frac{(\lambda t)^k}{\Gamma(k+1)\Gamma(k-p+1)}\right\} \\ &= L\left\{\sum_{k=0}^{\infty} \frac{\lambda^k t^{k-p}}{\Gamma(k+1)\Gamma(k-p+1)}\right\} \\ &= \sum_{k=0}^{\infty} \frac{\lambda^k s^{-[(k-p)+1]}}{\Gamma(k+1)} \\ &= s^{p-1} \sum_{k=0}^{\infty} \frac{\left(\frac{\lambda}{s}\right)^k}{\Gamma(k+1)} \\ &= s^{p-1} e^{\frac{\lambda}{s}} \ (|\lambda/s| < 1) \end{aligned} \tag{6.4}$$

and

$$
\begin{aligned}
L\left\{S_q(-\lambda,t)\right\} &= L\left\{t^{-q}\sum_{k=0}^{\infty}\frac{(-\lambda t)^k}{\Gamma(k+1)\,\Gamma(k-q+1)}\right\} \\
&= L\left\{\sum_{k=0}^{\infty}\frac{(-\lambda)^k\,t^{k-q}}{\Gamma(k+1)\,\Gamma(k-q+1)}\right\} \\
&= \sum_{k=0}^{\infty}\frac{(-\lambda)^k\,s^{-[(k-q)+1]}}{\Gamma(k+1)} \\
&= s^{q-1}\sum_{n=0}^{\infty}\frac{\left(-\frac{\lambda}{s}\right)^k}{\Gamma(k+1)} \\
&= s^{q-1}e^{-\frac{\lambda}{s}}\;(|\lambda/s|<1),
\end{aligned}
\tag{6.5}
$$

respectively, where $p \in \mathbb{C}$, $q \in \mathbb{C}$, and λ is a constant.

Therefore, we have the following property:

Property 6.1 Let $p \in \mathbb{C}$ and $|1/s| < 1$. Then we have

$$
L\left\{S_p(t)\right\} = s^{q-1}e^{-\frac{1}{s}}. \tag{6.6}
$$

Thus, we have

$$
\widehat{S}_p(\lambda,t) = S_{p-1}(\lambda,t) = t^{1-p}\sum_{k=0}^{\infty}\frac{(\lambda t)^k}{\Gamma(k+1)\,\Gamma(k-p+2)} \tag{6.7}
$$

and

$$
\widehat{S}_q(-\lambda,t) = \widehat{S}_{q-1}(-\lambda,t) = t^{1-q}\sum_{k=0}^{\infty}\frac{(-\lambda t)^k}{\Gamma(k+1)\,\Gamma(k-q+2)}, \tag{6.8}
$$

which lead to

$$
L\left\{\widehat{S}_p(\lambda,t)\right\} = s^{p}\sum_{k=0}^{\infty}\frac{\left(\frac{\lambda}{s}\right)^k}{\Gamma(k+1)} = s^{p}e^{\frac{\lambda}{s}}\;(|\lambda/s|<1) \tag{6.9}
$$

and

$$
L\left\{\widehat{S}_q(-\lambda,t)\right\} = s^{q}\sum_{n=0}^{\infty}\frac{\left(\frac{\lambda}{s}\right)^k}{\Gamma(k+1)} = s^{q}e^{-\frac{\lambda}{s}}\;(|\lambda/s|<1). \tag{6.10}
$$

Thus,

$$S_p(\lambda, t) = t^{-p} \sum_{k=0}^{\infty} \frac{(\lambda t)^k}{\Gamma(k+1)\,\Gamma(k-p+1)} = t^{-p}{}_0\varphi_1\left(\begin{matrix} - \\ (-p+1,1) \end{matrix}; \lambda t\right) \tag{6.11}$$

and

$$S_q(-\lambda, t) = t^{-q} \sum_{k=0}^{\infty} \frac{(-\lambda t)^k}{\Gamma(k+1)\,\Gamma(k-q+1)} = t^{-q}{}_0\varphi_1\left(\begin{matrix} - \\ (-p+1,1) \end{matrix}; -\lambda t\right), \tag{6.12}$$

$$\begin{aligned}\widehat{S}_p(\lambda, t) = S_{p-1}(\lambda, t) &= t^{1-p} \sum_{k=0}^{\infty} \frac{(\lambda t)^k}{\Gamma(k+1)\,\Gamma(k-p+2)} \\ &= t^{1-p}{}_0\varphi_1\left(\begin{matrix} - \\ (-p+2,1) \end{matrix}; \lambda t\right)\end{aligned} \tag{6.13}$$

and

$$\begin{aligned}\widehat{S}_q(-\lambda, t) = S_{q-1}(-\lambda, t) &= t^{1-q} \sum_{k=0}^{\infty} \frac{(-\lambda t)^k}{\Gamma(k+1)\,\Gamma(k-q+2)} \\ &= t^{1-q}{}_0\varphi_1\left(\begin{matrix} - \\ (-q+2,1) \end{matrix}; -\lambda t\right).\end{aligned} \tag{6.14}$$

The Mellin transform of the Sonine function of first type can be presented as follows:

$$\begin{aligned}M\left[S_q(-\lambda, t)\right] &= \int_0^{\infty} t^{\varpi-1} S_q(-\lambda, t)\, dt \\ &= \int_0^{\infty} t^{\varpi-1} t^{-q}{}_0\varphi_1\left(\begin{matrix} - \\ (-p+1,1) \end{matrix}; -\lambda t\right) dt \\ &= \frac{\Gamma(\varpi - q)\,\lambda^{q-\varpi}}{\Gamma(1-\varpi)},\end{aligned} \tag{6.15}$$

where $1 > Re(\varpi) > Re(q)$ and $\lambda \in \mathbb{R}_+$.

Thus, we obtain the following property:

Property 6.2 Let $1 > Re(\varpi) > Re(q)$. Then we have

$$M\left[S_q(t)\right] = \frac{\Gamma(\varpi - q)}{\Gamma(1-\varpi)}. \tag{6.16}$$

Thus, we have the following property:

Property 6.3 Let $1 > Re(\varpi) > Re(q)$. Then we have

$$S_q(t) = \frac{1}{2\pi i}\int_{-i\infty}^{i\infty} \frac{\Gamma(\varpi - q)}{\Gamma(1-\varpi)} (-t)^{\varpi-1} d\varpi, \tag{6.17}$$

where $|\arg(t)| < \pi$.

Thus, we have

$$M\left[\widehat{S}_q(-\lambda, t)\right] = \int_0^\infty t^{\varpi-1}\widehat{S}_q(-\lambda, t)\, dt = \frac{\Gamma(\varpi - q + 1)\lambda^{q-1-\varpi}}{\Gamma(1-\varpi)}. \tag{6.18}$$

It is not difficult to show the following theorem:

Theorem 6.1 *Let where* $1 > Re(\varpi) > Re(q)$ *and* $\lambda \in \mathbb{R}_+$. *Then we have*

$$\int_{-\infty}^\infty |t|^{\varpi-q-1}{}_0\varphi_1\left(\begin{matrix} - \\ (-p+1,1)\end{matrix}; -\lambda|t|\right) dt = \int_{-\infty}^\infty |t|^{\varpi-1} S_q(-\lambda, |t|)\, dt$$
$$= \frac{2\Gamma(\varpi - q)\lambda^{q-\varpi}}{\Gamma(1-\varpi)}. \tag{6.19}$$

Proof With the aid of (6.15), we directly deduce the result. □

Theorem 6.2 *Let* $1 > Re(\varpi) > Re(q)$ *and* $\lambda, \varsigma \in \mathbb{R}_+$. *Then we have*

$$\int_0^\infty t^{\varsigma\varpi-1} S_q(-\lambda, t^\varsigma)\, dt = \int_0^\infty t^{\varsigma\varpi-1}t^{-q\varsigma}{}_0\varphi_1\left(\begin{matrix} - \\ (-p+1,1)\end{matrix}; -\lambda t^\varsigma\right) dt$$
$$= \frac{\Gamma(\varpi - q)\lambda^{q-\varpi}}{\varsigma\Gamma(1-\varpi)}. \tag{6.20}$$

Proof With use of (6.15), we have

$$\int_0^\infty t^{\varsigma\varpi-1} S_q(-\lambda, t^\varsigma)\, dt = \int_0^\infty t^{\varsigma\varpi-1}t^{-q\varsigma}{}_0\varphi_1\left(\begin{matrix} - \\ (-p+1,1)\end{matrix}; -\lambda t^\varsigma\right) dt$$
$$= \int_0^\infty t^{\varsigma(\varpi-q)-1}{}_0\varphi_1\left(\begin{matrix} - \\ (-p+1,1)\end{matrix}; -\lambda t^\varsigma\right) dt$$
$$= \frac{\Gamma(\varpi - q)\lambda^{q-\varpi}}{\varsigma\Gamma(1-\varpi)}, \tag{6.21}$$

which is the desired result. □

Theorem 6.3 *Let* $1 > Re(\varpi) > Re(q)$ *and* $\lambda, \varsigma \in \mathbb{R}_+$. *Then we have*

$$\int_{-\infty}^{\infty} |t|^{\varsigma\varpi-1} S_q\left(-\lambda, |t|^{\varsigma}\right) dt = \int_{-\infty}^{\infty} |t|^{\varsigma(\varpi-q)-1} {}_0\varphi_1\left(\begin{matrix} - \\ (-p+1,1) \end{matrix}; -\lambda |t|^{\varsigma}\right) dt$$
$$= \frac{2\Gamma(\varpi-q)\lambda^{q-\varpi}}{\varsigma\Gamma(1-\varpi)}. \tag{6.22}$$

Proof With use of (6.21), we obtain the result. □

Theorem 6.4 *Let* $p \in \mathbb{C}$, $|\lambda/s| < 1$ *and* $\lambda \in \mathbb{C}$. *Then we have*

$$\int_0^{\infty} S_p(\lambda, t) e^{-\rho t} dt = \int_0^{\infty} t^{-p} {}_0\varphi_1\left(\begin{matrix} - \\ (-p+1,1) \end{matrix}; \lambda t\right) e^{-\rho t} dt = s^{p-1} e^{\frac{\lambda}{\rho}}. \tag{6.23}$$

Proof In view of (6.4), we deduce the result. □

Theorem 6.5 *Let* $p \in \mathbb{C}$, $|\lambda/s| < 1$, *and* $\lambda \in \mathbb{C}$. *Then we have*

$$\int_{-\infty}^{\infty} S_p(\lambda, |t|) e^{-\rho|t|} dt = \int_{-\infty}^{\infty} |t|^{-p} e^{-\rho|t|} {}_0\varphi_1\left(\begin{matrix} - \\ (-p+1,1) \end{matrix}; \lambda |t|\right) dt$$
$$= 2s^{p-1} e^{\frac{\lambda}{\rho}}. \tag{6.24}$$

Proof In view of (6.23), we deduce the result. □

Let us recall the special cases for the Wright's generalized hypergeometric function:

Let $\alpha, \gamma \in \mathbb{C}$, $Re(\alpha) > 0$, $Re(\gamma) > 0$, $\lambda \in \mathbb{C}$, $\beta > 0$, $\beta_1 \in \mathbb{C}$, and $B_1 \in \mathbb{R}_+$. Then we have

$$\frac{1}{\Gamma(\alpha)} \int_0^t (t-\tau)^{\alpha} \tau^{\gamma-1} {}_0\varphi_1\left(\begin{matrix} - \\ (\beta_1, B_1) \end{matrix}; \lambda\tau^{\beta}\right) d\tau$$
$$= t^{\gamma+\alpha-1} {}_1\varphi_2\left(\begin{matrix} (\gamma, \beta) \\ (\beta_1, B_1), (\gamma+\alpha, \beta) \end{matrix}; \lambda t^{\beta}\right), \tag{6.25}$$

where

$${}_0\varphi_1\left(\begin{matrix} - \\ (\beta_1, B_1) \end{matrix}; \lambda\tau^{\beta}\right) = \sum_{k=0}^{\infty} \frac{1}{\Gamma(\beta_1 + B_1 k)} \frac{\lambda^k \tau^{k\beta}}{k!}$$

and

$${}_1\varphi_2\left(\begin{matrix} (\gamma, \beta) \\ (\beta_1, B_1), (\gamma+\alpha, \beta) \end{matrix}; \lambda\tau^{\beta}\right) = \sum_{k=0}^{\infty} \frac{\Gamma(\gamma+\beta k)}{\Gamma(\beta_1 + B_1 k)\Gamma(\gamma+\alpha+\beta k)} \frac{\lambda^k \tau^{k\beta}}{k!}.$$

From (6.25), we may give the following case:

Let $\alpha, \gamma \in \mathbb{C}$, $Re(\alpha) > 0$, $Re(\gamma) > 0$, $\lambda \in \mathbb{C}$, $\beta > 0$, and $p \in \mathbb{C}$. Then we have

$$\frac{1}{\Gamma(\alpha)} \int_0^t (t-\tau)^{\alpha} \tau^{\gamma-1} {}_0\varphi_1 \left(\begin{matrix} - \\ (-p+1,1) \end{matrix} ; \lambda \tau^{\beta} \right) d\tau$$
$$= t^{\gamma+\alpha-1} {}_1\varphi_2 \left(\begin{matrix} (\gamma, \beta) \\ (-p+1,1), (\gamma+\alpha, \beta) \end{matrix} ; \lambda t^{\beta} \right). \tag{6.26}$$

Theorem 6.6 *Let* $\alpha \in \mathbb{C}$, $Re(\alpha) > 0$, $p \in \mathbb{C}$, $\lambda \in \mathbb{C}$, $\beta > 0$, *and* $0 < Re(p) < 1$. *Then we have*

$$\frac{1}{\Gamma(\alpha)} \int_0^t (t-\tau)^{\alpha} \tau^{-p} {}_0\varphi_1 \left(\begin{matrix} - \\ (-p+1,1) \end{matrix} ; \lambda \tau^{\beta} \right) d\tau$$
$$= t^{\alpha-p} {}_1\varphi_2 \left(\begin{matrix} (1-p, \beta) \\ (-p+1,1), (1-p+\alpha, \beta) \end{matrix} ; \lambda t^{\beta} \right). \tag{6.27}$$

Proof Making use of (6.26), we present the result. □

Theorem 6.7 *Let* $\alpha \in \mathbb{C}$, $Re(\alpha) > 0$, $p \in \mathbb{C}$, $\lambda \in \mathbb{C}$, *and* $0 < Re(p) < 1$. *Then we have*

$$\frac{1}{\Gamma(\alpha)} \int_0^t (t-\tau)^{\alpha} S_p(\lambda, \tau) \, d\tau = t^{\alpha-p} {}_0\varphi_1 \left(\begin{matrix} - \\ (1-p+\alpha,1) \end{matrix} ; \lambda t \right). \tag{6.28}$$

Proof Taking $\beta = 1$ in (6.27), we present the result. □

Theorem 6.8 *Let* $\alpha \in \mathbb{C}$, $Re(\alpha) > 0$, $p \in \mathbb{C}$, $\lambda \in \mathbb{C}$, *and* $0 < Re(p) < 1$. *Then we have*

$$\frac{1}{\Gamma(\alpha)} \int_0^1 (1-\tau)^{\alpha} S_p(\lambda, \tau) \, d\tau = {}_0\varphi_1 \left(\begin{matrix} - \\ (1-p+\alpha,1) \end{matrix} ; \lambda \right). \tag{6.29}$$

Proof Taking $t = 1$ in (6.28), we obtain the result. □

Theorem 6.9 *Let* $\alpha \in \mathbb{C}$, $Re(\alpha) > 0$, $p \in \mathbb{C}$, $\lambda \in \mathbb{C}$, *and* $0 < Re(p) < 1$. *Then we have*

$$\int_{-1}^1 (1-\tau)^{\alpha} S_p(\lambda, 1+\tau) \, d\tau = \int_{-1}^1 (1+\tau)^{\alpha} S_p(\lambda, 1-\tau) \, d\tau \tag{6.30}$$

and

$$\begin{aligned}
&\frac{1}{\Gamma(\alpha)}\int_0^2 (2-\tau)^{\alpha} S_p(\lambda,\tau)\,d\tau \\
&= \frac{1}{\Gamma(\alpha)}\int_{-1}^1 (1-\tau)^{\alpha} S_p(\lambda,1+\tau)\,d\tau \\
&= 2^{\alpha-p}\,{}_0\varphi_1\left(\begin{matrix} - \\ (1-p+\alpha,1)\end{matrix};2\lambda\right). \qquad (6.31)
\end{aligned}$$

Proof Taking $\beta = 1$ in (6.28), we give the result. □

Here, we consider that

$$\widetilde{S}_p\left(\lambda,t^{\varsigma}\right) = t^{-p}\sum_{k=0}^{\infty}\frac{\lambda^k t^{k\varsigma}}{\Gamma(k+1)\,\Gamma(k-p+1)} = t^{-p}{}_0\varphi_1\left(\begin{matrix} - \\ (-p+1,1)\end{matrix};\lambda t^{\varsigma}\right), \qquad (6.32)$$

which yields that

$$\begin{aligned}
\widetilde{S}_p\left(-\lambda,t^{\varsigma}\right) &= t^{-p}\sum_{k=0}^{\infty}\frac{(-\lambda)^k t^{k\varsigma}}{\Gamma(k+1)\,\Gamma(k-p+1)} \\
&= t^{-p}{}_0\varphi_1\left(\begin{matrix} - \\ (-p+1,1)\end{matrix};-\lambda t^{\varsigma}\right), \qquad (6.33)
\end{aligned}$$

which yields that

$$\widetilde{S}_p\left(\lambda,t^{2\varsigma}\right) = t^{-p}\sum_{k=0}^{\infty}\frac{\lambda^k t^{2k\varsigma}}{\Gamma(k+1)\,\Gamma(k-p+1)} = t^{-p}{}_0\varphi_1\left(\begin{matrix} - \\ (-p+1,1)\end{matrix};\lambda t^{2\varsigma}\right), \qquad (6.34)$$

$$\begin{aligned}
\widetilde{S}_p\left(-\lambda,t^{2\varsigma}\right) &= t^{-p}\sum_{k=0}^{\infty}\frac{(-\lambda)^k t^{2k\varsigma}}{\Gamma(k+1)\,\Gamma(k-p+1)} \\
&= t^{-p}{}_0\varphi_1\left(\begin{matrix} - \\ (-p+1,1)\end{matrix};-\lambda t^{2\varsigma}\right), \qquad (6.35)
\end{aligned}$$

$$\widetilde{S}_p\left(\lambda,t^{2}\right) = t^{-p}\sum_{k=0}^{\infty}\frac{\lambda^k t^{2k}}{\Gamma(k+1)\,\Gamma(k-p+1)} = t^{-p}{}_0\varphi_1\left(\begin{matrix} - \\ (-p+1,1)\end{matrix};\lambda t^{2}\right), \qquad (6.36)$$

and

$$\widetilde{S}_p\left(-\lambda, t^2\right) = t^{-p} \sum_{k=0}^{\infty} \frac{(-\lambda)^k t^{2k}}{\Gamma(k+1)\Gamma(k-p+1)}$$

$$= t^{-p} {}_0\varphi_1\left(\frac{-}{(-p+1,1)}; -\lambda t^2\right), \tag{6.37}$$

where $p \in \mathbb{C}$, $\lambda \in \mathbb{C}$, and $\varsigma > 0$.

Here,

$$L\left\{\widetilde{S}_p\left(\lambda, t^\varsigma\right)\right\} = L\left\{t^{-p} \sum_{k=0}^{\infty} \frac{\lambda^k t^{k\varsigma}}{\Gamma(k+1)\Gamma(k-p+1)}\right\}$$

$$= L\left\{t^{-p} {}_0\varphi_1\left(\frac{-}{(-p+1,1)}; \lambda t^\varsigma\right)\right\}$$

$$= \sum_{k=0}^{\infty} \frac{\Gamma(k\varsigma - p + 1)\lambda^k s^{-[k\varsigma-p+1]}}{\Gamma(k+1)\Gamma(k-p+1)}$$

$$= s^{p-1} \sum_{k=0}^{\infty} \frac{\Gamma(k\varsigma - p + 1)}{\Gamma(k+1)\Gamma(k-p+1)} \left(\frac{\lambda}{s^\varsigma}\right)^k$$

$$= s^{p-1} {}_1\varphi_1\left(\frac{(-p+1,\varsigma)}{(-p+1,1)}; \frac{\lambda}{s^\varsigma}\right) \left(\left|\lambda/s^\varsigma\right| < 1\right), \tag{6.38}$$

which implies that

$$L\left\{\widetilde{S}_p\left(\lambda, t^\varsigma\right)\right\} = s^{p-1} {}_1\varphi_1\left(\frac{(-p+1,\varsigma)}{(-p+1,1)}; \frac{\lambda}{s^\varsigma}\right) \left(\left|\lambda/s^\varsigma\right| < 1\right), \tag{6.39}$$

$$L\left\{\widetilde{S}_p\left(-\lambda, t^\varsigma\right)\right\} = s^{p-1} {}_1\varphi_1\left(\frac{(-p+1,\varsigma)}{(-p+1,1)}; -\frac{\lambda}{s^\varsigma}\right) \left(\left|\lambda/s^\varsigma\right| < 1\right), \tag{6.40}$$

$$L\left\{\widetilde{S}_p\left(\lambda, t^{2\varsigma}\right)\right\} = s^{p-1} {}_1\varphi_1\left(\frac{(-p+1,2\varsigma)}{(-p+1,1)}; \frac{\lambda}{s^{2\varsigma}}\right) \left(\left|\lambda/s^{2\varsigma}\right| < 1\right), \tag{6.41}$$

$$L\left\{\widetilde{S}_p\left(-\lambda, t^{2\varsigma}\right)\right\} = s^{p-1} {}_1\varphi_1\left(\frac{(-p+1,2\varsigma)}{(-p+1,1)}; -\frac{\lambda}{s^{2\varsigma}}\right) \left(\left|\lambda/s^{2\varsigma}\right| < 1\right), \tag{6.42}$$

$$L\left\{\widetilde{S}_p\left(\lambda, t^2\right)\right\} = s^{p-1} {}_1\varphi_1\left(\frac{(-p+1,2)}{(-p+1,1)}; \frac{\lambda}{s^2}\right) \left(\left|\lambda/s^2\right| < 1\right), \tag{6.43}$$

and

$$L\left\{\widetilde{S}_p\left(-\lambda, t^2\right)\right\} = s^{p-1}\,{}_1\varphi_1\left(\begin{matrix}(-p+1,2)\\(-p+1,1)\end{matrix}; -\frac{\lambda}{s^2}\right)\left(\left|\lambda/s^2\right| < 1\right), \tag{6.44}$$

where $p \in \mathbb{C}$, $\lambda \in \mathbb{C}$, and $\varsigma > 0$.

The similar cases of the Sonine functions can be considered as follows:

Let us recall the binomial series (see Chap. 5)

$$\begin{aligned}(1+\lambda t)^{\varsigma} &= \sum_{k=0}^{\infty} \frac{\Gamma(1+\varsigma)}{\Gamma(1+\varsigma-k)} \frac{(\lambda t)^k}{k!}\\ &= \Gamma(1+\varsigma)\,{}_0\varphi_1\left(\begin{matrix}-\\(1+\varsigma,-1)\end{matrix}; \lambda t\right)\end{aligned} \tag{6.45}$$

and

$$\begin{aligned}(1+\lambda t)^{-\varsigma} &= \sum_{k=0}^{\infty} \frac{\Gamma(1-\varsigma)}{\Gamma(1-\varsigma-k)} \frac{(\lambda t)^k}{k!}\\ &= \Gamma(1-\varsigma)\,{}_0\varphi_1\left(\begin{matrix}-\\(1-\varsigma,-1)\end{matrix}; \lambda t\right),\end{aligned} \tag{6.46}$$

where $|t| < 1$, $\lambda \in \mathbb{C}$, and $\varsigma > 0$.

In this case, we have that

$$\begin{aligned}t^{-q}(1+\lambda t)^{\varsigma} &= t^{-q}\sum_{k=0}^{\infty} \frac{\Gamma(1+\varsigma)}{\Gamma(1+\varsigma-k)} \frac{(\lambda t)^k}{k!}\\ &= \Gamma(1+\varsigma)\,t^{-q}\,{}_0\varphi_1\left(\begin{matrix}-\\(1+\varsigma,-1)\end{matrix}; \lambda t\right),\end{aligned} \tag{6.47}$$

$$\begin{aligned}t^{-q}(1+\lambda t)^{-\varsigma} &= t^{-q}\sum_{k=0}^{\infty} \frac{\Gamma(1-\varsigma)}{\Gamma(1-\varsigma-k)} \frac{(\lambda t)^k}{k!}\\ &= \Gamma(1-\varsigma)\,t^{-q}\,{}_0\varphi_1\left(\begin{matrix}-\\(1-\varsigma,-1)\end{matrix}; \lambda t\right),\end{aligned} \tag{6.48}$$

$$\begin{aligned}t^{-q}\left(1+\lambda t^{\vartheta}\right)^{\varsigma} &= t^{-q}\sum_{k=0}^{\infty} \frac{\Gamma(1+\varsigma)}{\Gamma(1+\varsigma-k)} \frac{\left(\lambda t^{\vartheta}\right)^k}{k!}\\ &= \Gamma(1+\varsigma)\,t^{-q}\,{}_0\varphi_1\left(\begin{matrix}-\\(1+\varsigma,-1)\end{matrix}; \lambda t^{\vartheta}\right),\end{aligned} \tag{6.49}$$

and

$$t^{-q}\left(1+\lambda t^{\vartheta}\right)^{-\varsigma} = t^{-q}\sum_{k=0}^{\infty}\frac{\Gamma(1-\varsigma)}{\Gamma(1-\varsigma-k)}\frac{\left(\lambda t^{\vartheta}\right)^{k}}{k!}$$

$$= \Gamma(1-\varsigma)\,t^{-q}{}_0\varphi_1\left(\begin{matrix}-\\(1-\varsigma,-1)\end{matrix};\lambda t^{\vartheta}\right), \tag{6.50}$$

which yield

$$t^{-q}\left(1+\lambda t^{2}\right)^{\varsigma} = t^{-q}\sum_{k=0}^{\infty}\frac{\Gamma(1+\varsigma)}{\Gamma(1+\varsigma-k)}\frac{\lambda^{k}t^{2k}}{k!}$$

$$= \Gamma(1+\varsigma)\,t^{-q}{}_0\varphi_1\left(\begin{matrix}-\\(1+\varsigma,-1)\end{matrix};\lambda t^{2}\right), \tag{6.51}$$

$$t^{-q}\left(1+\lambda t^{2}\right)^{-\varsigma} = t^{-q}\sum_{k=0}^{\infty}\frac{\Gamma(1-\varsigma)}{\Gamma(1-\varsigma-k)}\frac{\lambda^{k}t^{2k}}{k!}$$

$$= \Gamma(1-\varsigma)\,t^{-q}{}_0\varphi_1\left(\begin{matrix}-\\(1-\varsigma,-1)\end{matrix};\lambda t^{2}\right), \tag{6.52}$$

$$t^{-q}\left(1+\lambda t^{2\vartheta}\right)^{\varsigma} = t^{-q}\sum_{k=0}^{\infty}\frac{\Gamma(1+\varsigma)}{\Gamma(1+\varsigma-k)}\frac{\lambda^{k}t^{2\vartheta k}}{k!}$$

$$= \Gamma(1+\varsigma)\,t^{-q}{}_0\varphi_1\left(\begin{matrix}-\\(1+\varsigma,-1)\end{matrix};\lambda t^{2\vartheta}\right) \tag{6.53}$$

and

$$t^{-q}\left(1+\lambda t^{2\vartheta}\right)^{-\varsigma} = t^{-q}\sum_{k=0}^{\infty}\frac{\Gamma(1-\varsigma)}{\Gamma(1-\varsigma-k)}\frac{\lambda^{k}t^{2\vartheta k}}{k!}$$

$$= \Gamma(1-\varsigma)\,t^{-q}{}_0\varphi_1\left(\begin{matrix}-\\(1-\varsigma,-1)\end{matrix};\lambda t^{2\vartheta}\right), \tag{6.54}$$

where $|t| < 1$, $q > 0$, $\lambda \in \mathbb{C}$, $\vartheta > 0$, and $\varsigma > 0$.

Therefore,

$$t^{-q}(1+\lambda t)^{q} = t^{-q}\sum_{k=0}^{\infty}\frac{\Gamma(1+q)}{\Gamma(1+q-k)}\frac{\lambda^{k}t^{k}}{k!}$$

$$= \Gamma(1+q)\,t^{-q}{}_0\varphi_1\left(\begin{matrix}-\\(1+q,-1)\end{matrix};\lambda t\right), \tag{6.55}$$

$$t^{-q}\left(1+\lambda t\right)^{-q}=t^{-q}\sum_{k=0}^{\infty}\frac{\Gamma\left(1-q\right)}{\Gamma\left(1-q-k\right)}\frac{\lambda^{k}t^{k}}{k!}$$
$$=\Gamma\left(1-q\right)t^{-q}{}_{0}\varphi_{1}\left(\begin{matrix}-\\(1-q,-1)\end{matrix};\lambda t\right),\tag{6.56}$$

$$t^{-q}\left(1+\lambda t^{\vartheta}\right)^{q}=t^{-q}\sum_{k=0}^{\infty}\frac{\Gamma\left(1+q\right)}{\Gamma\left(1+q-k\right)}\frac{\lambda^{k}t^{\vartheta k}}{k!}$$
$$=\Gamma\left(1+q\right)t^{-q}{}_{0}\varphi_{1}\left(\begin{matrix}-\\(1+q,-1)\end{matrix};\lambda t^{\vartheta}\right),\tag{6.57}$$

$$t^{-q}\left(1+\lambda t^{\vartheta}\right)^{-q}=t^{-q}\sum_{k=0}^{\infty}\frac{\Gamma\left(1-q\right)}{\Gamma\left(1-q-k\right)}\frac{\lambda^{k}t^{\vartheta k}}{k!}$$
$$=\Gamma\left(1-q\right)t^{-q}{}_{0}\varphi_{1}\left(\begin{matrix}-\\(1-q,-1)\end{matrix};\lambda t^{\vartheta}\right),\tag{6.58}$$

$$t^{-q}\left(1+\lambda t^{2}\right)^{q}=t^{-q}\sum_{k=0}^{\infty}\frac{\Gamma\left(1+q\right)}{\Gamma\left(1+q-k\right)}\frac{\lambda^{k}t^{2k}}{k!}$$
$$=\Gamma\left(1+q\right)t^{-q}{}_{0}\varphi_{1}\left(\begin{matrix}-\\(1+q,-1)\end{matrix};\lambda t^{2}\right),\tag{6.59}$$

$$t^{-q}\left(1+\lambda t^{2}\right)^{-q}=t^{-q}\sum_{k=0}^{\infty}\frac{\Gamma\left(1-q\right)}{\Gamma\left(1-q-k\right)}\frac{\lambda^{k}t^{2k}}{k!}$$
$$=\Gamma\left(1-q\right)t^{-q}{}_{0}\varphi_{1}\left(\begin{matrix}-\\(1-q,-1)\end{matrix};\lambda t^{2}\right),\tag{6.60}$$

$$t^{-q}\left(1+\lambda t^{2\vartheta}\right)^{q}=t^{-q}\sum_{k=0}^{\infty}\frac{\Gamma\left(1+q\right)}{\Gamma\left(1+q-k\right)}\frac{\lambda^{k}t^{2\vartheta k}}{k!}$$
$$=\Gamma\left(1+q\right)t^{-q}{}_{0}\varphi_{1}\left(\begin{matrix}-\\(1+q,-1)\end{matrix};\lambda t^{2\vartheta}\right)\tag{6.61}$$

and

$$t^{-q}\left(1+\lambda t^{2\vartheta}\right)^{-q}=t^{-q}\sum_{k=0}^{\infty}\frac{\Gamma\left(1-q\right)}{\Gamma\left(1-q-k\right)}\frac{\lambda^{k}t^{2\vartheta k}}{k!}$$
$$=\Gamma\left(1-q\right)t^{-q}{}_{0}\varphi_{1}\left(\begin{matrix}-\\(1-q,-1)\end{matrix};\lambda t^{2\vartheta}\right),\tag{6.62}$$

where $|t|<1$, $q>0$, $\lambda\in\mathbb{C}$, and $\vartheta>0$.

Here, we consider that

$$L\left\{t^{-q}\left(1+\lambda t^{\vartheta}\right)^{\varsigma}\right\} = L\left\{\Gamma\left(1+\varsigma\right)t^{-q}{}_0\varphi_1\left(\begin{matrix} - \\ (1+\varsigma,-1)\end{matrix};\lambda t^{\vartheta}\right)\right\}$$
$$= L\left\{t^{-q}\sum_{k=0}^{\infty}\frac{\Gamma\left(1+\varsigma\right)}{\Gamma\left(1+\varsigma-k\right)}\frac{\left(\lambda t^{\vartheta}\right)^{k}}{k!}\right\}$$
$$= \sum_{k=0}^{\infty}\frac{\Gamma\left(1+\varsigma\right)}{\Gamma\left(1+\varsigma-k\right)}\frac{\lambda^{k}}{k!}s^{-(\vartheta k-q+1)}$$
$$= s^{q-1}\sum_{k=0}^{\infty}\frac{\Gamma\left(1+\varsigma\right)\Gamma\left(k\vartheta-q+1\right)}{\Gamma\left(1+\varsigma-k\right)}\frac{\left(\frac{\lambda}{s^{\vartheta}}\right)^{k}}{k!}$$
$$= \Gamma\left(1+\varsigma\right)s^{q-1}{}_1\varphi_1\left(\begin{matrix}(-q+1,\vartheta)\\(\varsigma+1,-1)\end{matrix};\frac{\lambda}{s^{\vartheta}}\right)\left(\left|\lambda/s^{\vartheta}\right|<1\right), \tag{6.63}$$

which leads to

$$L\left\{t^{-q}\left(1+\lambda t\right)^{q}\right\} = \Gamma\left(1+q\right)s^{q-1}{}_1\varphi_1\left(\begin{matrix}(-q+1,1)\\(q+1,-1)\end{matrix};\frac{\lambda}{s}\right)\left(\left|\lambda/s\right|<1\right), \tag{6.64}$$

$$L\left\{t^{-q}\left(1+\lambda t\right)^{-q}\right\} = \Gamma\left(1-q\right)s^{q-1}{}_1\varphi_1\left(\begin{matrix}(-q+1,1)\\(-q+1,-1)\end{matrix};\frac{\lambda}{s}\right)\left(\left|\lambda/s\right|<1\right), \tag{6.65}$$

$$L\left\{t^{-q}\left(1+\lambda t^{\vartheta}\right)^{q}\right\} = \Gamma\left(1+q\right)s^{q-1}{}_1\varphi_1\left(\begin{matrix}(-q+1,\vartheta)\\(q+1,-1)\end{matrix};\frac{\lambda}{s^{\vartheta}}\right)\left(\left|\lambda/s^{\vartheta}\right|<1\right), \tag{6.66}$$

$$L\left\{t^{-q}\left(1+\lambda t^{\vartheta}\right)^{-q}\right\} = \Gamma\left(1-q\right)s^{q-1}{}_1\varphi_1\left(\begin{matrix}(-q+1,\vartheta)\\(-q+1,-1)\end{matrix};\frac{\lambda}{s^{\vartheta}}\right)\left(\left|\lambda/s^{\vartheta}\right|<1\right), \tag{6.67}$$

$$L\left\{t^{-q}\left(1+\lambda t^{2}\right)^{q}\right\} = \Gamma\left(1+q\right)s^{q-1}{}_1\varphi_1\left(\begin{matrix}(-q+1,2)\\(q+1,-1)\end{matrix};\frac{\lambda}{s^{2}}\right)\left(\left|\lambda/s^{2}\right|<1\right), \tag{6.68}$$

$$L\left\{t^{-q}\left(1+\lambda t^{2}\right)^{-q}\right\} = \Gamma\left(1-q\right)s^{q-1}{}_1\varphi_1\left(\begin{matrix}(-q+1,2)\\(-q+1,-1)\end{matrix};\frac{\lambda}{s^{2}}\right)\left(\left|\lambda/s^{2}\right|<1\right), \tag{6.69}$$

$$L\left\{t^{-q}\left(1+\lambda t^{2\vartheta}\right)^{q}\right\} = \Gamma\left(1+q\right)s^{q-1}{}_1\varphi_1\left(\begin{matrix}(-q+1,2\vartheta)\\(q+1,-1)\end{matrix};\frac{\lambda}{s^{2\vartheta}}\right)\left(\left|\lambda/s^{2\vartheta}\right|<1\right), \tag{6.70}$$

and

$$L\left\{t^{-q}\left(1+\lambda t^{2\vartheta}\right)^{-q}\right\}=\Gamma(1-q)\,s^{q-1}\,{}_1\varphi_1\left(\begin{matrix}(-q+1,2\vartheta)\\(-q+1,-1)\end{matrix};\frac{\lambda}{s^{2\vartheta}}\right)$$
$$\times\left(\left|\lambda/s^{2\vartheta}\right|<1\right), \tag{6.71}$$

where $q>0$, $\lambda\in\mathbb{C}$, and $\vartheta>0$.

6.1.2 *The Supertrigonometric Functions via Sonine Function of First Type*

In this part, we give the supertrigonometric functions via Sonine function of first type.

Let us consider that

$$S_p(i\lambda,t)=t^{-p}\sum_{k=0}^{\infty}\frac{(i\lambda t)^k}{\Gamma(k+1)\,\Gamma(k-p+1)}=t^{-p}{}_0\varphi_1\left(\begin{matrix}-\\(-p+1,1)\end{matrix};i\lambda t\right), \tag{6.72}$$

and

$$S_p(-i\lambda,t)=t^{-p}\sum_{k=0}^{\infty}\frac{(-i\lambda t)^k}{\Gamma(k+1)\,\Gamma(k-p+1)}=t^{-p}{}_0\varphi_1\left(\begin{matrix}-\\(-p+1,1)\end{matrix};-i\lambda t\right). \tag{6.73}$$

We may structure the following special functions:

Definition 6.2 The supersine via Sonine function of first type is defined as

$$Ssin_p(t)=\frac{1}{2i}\left(S_p(i,t)-S_p(-i,t)\right)$$
$$=t^{-p}\sum_{k=0}^{\infty}\frac{(-1)^k\,t^{2k+1}}{\Gamma(2k+2)\,\Gamma(2k-p+2)}, \tag{6.74}$$

where $p,t\in\mathbb{C}$.

Definition 6.3 The supercosine via Sonine function of first type is defined as

$$
\begin{aligned}
Scos_p(t) &= \frac{1}{2}\left(S_p(i,t) + S_p(-i,t)\right) \\
&= t^{-p}\sum_{k=0}^{\infty}\frac{(-1)^k t^{2k}}{\Gamma(2k+1)\,\Gamma(2k-p+1)},
\end{aligned} \tag{6.75}
$$

where $p, t \in \mathbb{C}$.

Definition 6.4 The supertangent via Sonine function of first type is defined as

$$
Stan_p(t) = \frac{Ssin_p(t)}{Scos_p(t)}\left(Scos_p(t) \neq 0\right), \tag{6.76}
$$

where $p, t \in \mathbb{C}$.

Definition 6.5 The supercotangent via Sonine function of first type is defined as

$$
Scotan_p(t) = \frac{Scos_p(t)}{Ssin_p(t)}, \left(Ssin_p(t) \neq 0\right), \tag{6.77}
$$

where $p, t \in \mathbb{C}$.

Definition 6.6 The supercosecant via Sonine function of first type is defined as

$$
Scsc_p(t) = \frac{1}{Ssin_p(t)}, \left(Ssin_p(t) \neq 0\right), \tag{6.78}
$$

where $p, t \in \mathbb{C}$.

Definition 6.7 The supersecant via Sonine function of first type is defined as

$$
Ssec_p(t) = \frac{1}{Scos_p(t)}\left(Scos_p(t) \neq 0\right), \tag{6.79}
$$

where $p, t \in \mathbb{C}$.

Thus,

$$
\begin{aligned}
Ssin_p(\lambda,t) &= \frac{1}{2i}\left(S_p(i\lambda,t) - S_p(-i\lambda,t)\right) \\
&= t^{-p}\sum_{k=0}^{\infty}\frac{(-1)^k (\lambda t)^{2k+1}}{\Gamma(2k+2)\,\Gamma(2k-p+2)},
\end{aligned} \tag{6.80}
$$

$$Scos_p(\lambda, t) = \frac{1}{2}\left(S_p(i\lambda, t) + S_p(-i\lambda, t)\right)$$
$$= t^{-p}\sum_{k=0}^{\infty}\frac{(-1)^k(\lambda t)^{2k}}{\Gamma(2k+1)\Gamma(2k-p+1)}, \tag{6.81}$$

$$Stan_p(\lambda, t) = \frac{Ssin_p(\lambda, t)}{Scos_p(\lambda, t)}\left(Scos_p(\lambda, t) \neq 0\right), \tag{6.82}$$

$$Scotan_p(\lambda, t) = \frac{Scos_p(\lambda, t)}{Ssin_p(\lambda, t)}, \left(Ssin_p(\lambda, t) \neq 0\right), \tag{6.83}$$

$$Scsc_p(\lambda, t) = \frac{1}{Ssin_p(\lambda, t)}, \left(Ssin_p(\lambda, t) \neq 0\right) \tag{6.84}$$

and

$$Ssec_p(\lambda, t) = \frac{1}{Scos_p(\lambda, t)}\left(Scos_p(\lambda, t) \neq 0\right), \tag{6.85}$$

where $p, \lambda, t \in \mathbb{C}$.

Theorem 6.10 *Let $p, \lambda, t \in \mathbb{C}$. Then we have*

$$S_p(i\lambda, t) = Scos_p(\lambda, t) + iSsin_p(\lambda, t). \tag{6.86}$$

Proof Making use of (6.80) and (6.81), we obtain the result. □

6.1.3 *The Superhyperbolic Functions via Sonine Function of First Type*

In this part, we propose the superhyperbolic function via Sonine functions of first type.

Similarly, we consider that

$$S_p(\lambda, t) = t^{-p}\sum_{k=0}^{\infty}\frac{(\lambda t)^k}{\Gamma(k+1)\Gamma(k-p+1)} = t^{-p}{}_0\varphi_1\left(\begin{matrix} - \\ (-p+1, 1)\end{matrix}; \lambda t\right) \tag{6.87}$$

and

$$S_p(-\lambda, t) = t^{-p} \sum_{k=0}^{\infty} \frac{(-\lambda t)^k}{\Gamma(k+1)\Gamma(k-p+1)} = t^{-p}{}_0\varphi_1\left(\begin{matrix} - \\ (-p+1,1) \end{matrix}; -\lambda t\right). \tag{6.88}$$

We may define the following special functions:

Definition 6.8 The hyperbolic supersine via Sonine function of first type is defined as

$$\begin{aligned} Ssinh_p(t) &= \frac{1}{2}\left(S_p(1,t) - S_p(-1,t)\right) \\ &= t^{-p} \sum_{k=0}^{\infty} \frac{t^{2k+1}}{\Gamma(2k+2)\Gamma(2k-p+2)}, \end{aligned} \tag{6.89}$$

where $p, t \in \mathbb{C}$.

Definition 6.9 The hyperbolic supercosine via Sonine function of first type is defined as

$$\begin{aligned} Scosh_p(t) &= \frac{1}{2}\left(S_p(1,t) + S_p(-1,t)\right) \\ &= t^{-p} \sum_{k=0}^{\infty} \frac{t^{2k}}{\Gamma(2k+1)\Gamma(2k-p+1)}, \end{aligned} \tag{6.90}$$

where $p, t \in \mathbb{C}$.

Definition 6.10 The hyperbolic supertangent via Sonine function of first type is defined as

$$Stanh_p(t) = \frac{Ssinh_p(t)}{Scosh_p(t)} \left(Scosh_p(t) \neq 0\right), \tag{6.91}$$

where $p, t \in \mathbb{C}$.

Definition 6.11 The hyperbolic supercotangent via Sonine function of first type is defined as

$$Scotanh_p(t) = \frac{Scosh_p(t)}{Ssinh_p(t)}, \left(Ssinh_p(t) \neq 0\right), \tag{6.92}$$

where $p, t \in \mathbb{C}$.

Definition 6.12 The hyperbolic supercosecant via Sonine function of first type is defined as

$$Scsch_p(t) = \frac{1}{Ssinh_p(t)}, \left(Ssinh_p(t) \neq 0\right), \tag{6.93}$$

where $p, t \in \mathbb{C}$.

Definition 6.13 The hyperbolic supersecant via Sonine function of first type is defined as

$$Ssech_p(t) = \frac{1}{Scosh_p(t)} \left(Scosh_p(t) \neq 0\right), \tag{6.94}$$

where $p, t \in \mathbb{C}$.

Thus,

$$\begin{aligned} Ssinh_p(\lambda, t) &= \frac{1}{2}\left(S_p(\lambda, t) - S_p(-\lambda, t)\right) \\ &= t^{-p} \sum_{k=0}^{\infty} \frac{(\lambda t)^{2k+1}}{\Gamma(2k+2)\,\Gamma(2k-p+2)}, \end{aligned} \tag{6.95}$$

$$\begin{aligned} Scosh_p(\lambda, t) &= \frac{1}{2}\left(S_p(\lambda, t) + S_p(-\lambda, t)\right) \\ &= t^{-p} \sum_{k=0}^{\infty} \frac{(\lambda t)^{2k}}{\Gamma(2k+1)\,\Gamma(2k-p+1)}, \end{aligned} \tag{6.96}$$

$$Stanh_p(\lambda, t) = \frac{Ssinh_p(\lambda, t)}{Scosh_p(\lambda, t)} \left(Scosh_p(\lambda, t) \neq 0\right), \tag{6.97}$$

$$Scotanh_p(\lambda, t) = \frac{Scosh_p(\lambda, t)}{Ssinh_p(\lambda, t)}, \left(Ssinh_p(\lambda, t) \neq 0\right), \tag{6.98}$$

$$Scsch_p(\lambda, t) = \frac{1}{Ssinh_p(\lambda, t)}, \left(Ssinh_p(\lambda, t) \neq 0\right) \tag{6.99}$$

and

$$Ssech_p(\lambda, t) = \frac{1}{Scosh_p(\lambda, t)} \left(Scosh_p(\lambda, t) \neq 0\right), \tag{6.100}$$

where $p, \lambda, t \in \mathbb{C}$.

6.1.4 The Integral Representations of the Supertrigonometric and Superhyperbolic Functions

In this part, we present the properties for the supertrigonometric and superhyperbolic functions.

Property 6.4 Let $\alpha \in \mathbb{C}$, $Re(\alpha) > 0$, $p \in \mathbb{C}$, $\lambda \in \mathbb{C}$, and $0 < Re(p) < 1$. Then we have

$$\frac{1}{\Gamma(\alpha)} \int_0^t (t-\tau)^\alpha \, Ssin_p(\lambda, \tau) \, d\tau = t^{\alpha-p} \, {}_0sin_1 \left(\begin{matrix} - \\ (1-p+\alpha, 1) \end{matrix} ; \lambda t \right) = Ssin_{\alpha-p}(\lambda, t), \tag{6.101}$$

$$\frac{1}{\Gamma(\alpha)} \int_0^t (t-\tau)^\alpha \, Scos_p(\lambda, \tau) \, d\tau = t^{\alpha-p} \, {}_0cos_1 \left(\begin{matrix} - \\ (1-p+\alpha, 1) \end{matrix} ; \lambda t \right) = Scos_{\alpha-p}(\lambda, t), \tag{6.102}$$

$$\frac{1}{\Gamma(\alpha)} \int_0^t (t-\tau)^\alpha \, Ssinh_p(\lambda, \tau) \, d\tau = t^{\alpha-p} \, {}_0sinh_1 \left(\begin{matrix} - \\ (1-p+\alpha, 1) \end{matrix} ; \lambda t \right) = Ssinh_{\alpha-p}(\lambda, t) \tag{6.103}$$

and

$$\frac{1}{\Gamma(\alpha)} \int_0^t (t-\tau)^\alpha \, Scosh_p(\lambda, \tau) \, d\tau = t^{\alpha-p} \, {}_0cosh_1 \left(\begin{matrix} - \\ (1-p+\alpha, 1) \end{matrix} ; \lambda t \right) = Scosh_{\alpha-p}(\lambda, t). \tag{6.104}$$

Proof Making use of (6.28), we obtain

$$\frac{1}{\Gamma(\alpha)} \int_0^t (t-\tau)^\alpha \, S_p(i\lambda, \tau) \, d\tau = t^{\alpha-p} \, {}_0\varphi_1 \left(\begin{matrix} - \\ (1-p+\alpha, 1) \end{matrix} ; i\lambda t \right) = S_{\alpha-p}(i\lambda, t), \tag{6.105}$$

$$\frac{1}{\Gamma(\alpha)} \int_0^t (t-\tau)^\alpha \, S_p(-i\lambda, \tau) \, d\tau = t^{\alpha-p} \, {}_0\varphi_1 \left(\begin{matrix} - \\ (1-p+\alpha, 1) \end{matrix} ; -i\lambda t \right) = S_{\alpha-p}(-i\lambda, t) \tag{6.106}$$

$$\frac{1}{\Gamma(\alpha)} \int_0^t (t-\tau)^\alpha \, S_p(\lambda, \tau) \, d\tau = t^{\alpha-p} \, {}_0\varphi_1 \left(\begin{matrix} - \\ (1-p+\alpha, 1) \end{matrix} ; \lambda t \right) = S_{\alpha-p}(\lambda, t) \tag{6.107}$$

and

$$\frac{1}{\Gamma(\alpha)}\int_0^t (t-\tau)^{\alpha} S_p(-\lambda,\tau)\,d\tau = t^{\alpha-p}\,{}_0\varphi_1\left(\begin{matrix} - \\ (1-p+\alpha,1) \end{matrix};-\lambda t\right)$$
$$= S_{\alpha-p}(-\lambda,t), \tag{6.108}$$

which lead to the desired results. □

Property 6.5 Let $\alpha \in \mathbb{C}$, $Re(\alpha) > 0$, $p \in \mathbb{C}$, $\lambda \in \mathbb{C}$, and $0 < Re(p) < 1$. Then we have

$$\frac{1}{\Gamma(\alpha)}\int_0^1 (1-\tau)^{\alpha} Ssin_p(\lambda,\tau)\,d\tau = {}_0sin_1\left(\begin{matrix} - \\ (1-p+\alpha,1) \end{matrix};\lambda\right)$$
$$= Ssin_{\alpha-p}(\lambda,1), \tag{6.109}$$

$$\frac{1}{\Gamma(\alpha)}\int_0^1 (1-\tau)^{\alpha} Scos_p(\lambda,\tau)\,d\tau = {}_0cos_1\left(\begin{matrix} - \\ (1-p+\alpha,1) \end{matrix};\lambda\right)$$
$$= Scos_{\alpha-p}(\lambda,1), \tag{6.110}$$

$$\frac{1}{\Gamma(\alpha)}\int_0^1 (1-\tau)^{\alpha} Ssinh_p(\lambda,\tau)\,d\tau = {}_0sinh_1\left(\begin{matrix} - \\ (1-p+\alpha,1) \end{matrix};\lambda\right)$$
$$= Ssinh_{\alpha-p}(\lambda,1) \tag{6.111}$$

and

$$\frac{1}{\Gamma(\alpha)}\int_0^1 (1-\tau)^{\alpha} Scosh_p(\lambda,\tau)\,d\tau = {}_0cosh_1\left(\begin{matrix} - \\ (1-p+\alpha,1) \end{matrix};\lambda\right)$$
$$= Scosh_{\alpha-p}(\lambda,1). \tag{6.112}$$

Proof Putting $t = 1$ in (6.101), (6.102), (6.103), and (6.104), respectively, we have the results. □

Property 6.6 Let $\alpha \in \mathbb{C}$, $Re(\alpha) > 0$, $p \in \mathbb{C}$, $\lambda \in \mathbb{C}$, and $0 < Re(p) < 1$. Then we have

$$\frac{1}{\Gamma(\alpha)}\int_0^2 (2-\tau)^{\alpha} Ssin_p(\lambda,\tau)\,d\tau = \frac{1}{\Gamma(\alpha)}\int_{-1}^1 (1-\tau)^{\alpha} Ssin_p(\lambda,1+\tau)\,d\tau$$
$$= 2^{\alpha-p}\,{}_0sin_1\left(\begin{matrix} - \\ (1-p+\alpha,1) \end{matrix};2\lambda\right)$$
$$= Ssin_{\alpha-p}(\lambda,2), \tag{6.113}$$

$$\frac{1}{\Gamma(\alpha)}\int_0^2 (2-\tau)^{\alpha}\, Scos_p(\lambda,\tau)\,d\tau = \frac{1}{\Gamma(\alpha)}\int_{-1}^1 (1-\tau)^{\alpha}\, Scos_p(\lambda,1+\tau)\,d\tau$$
$$= 2^{\alpha-p}\,{}_0cos_1\left(\begin{matrix} - \\ (1-p+\alpha,1)\end{matrix};2\lambda\right)$$
$$= Scos_{\alpha-p}(\lambda,2)\,, \tag{6.114}$$

$$\frac{1}{\Gamma(\alpha)}\int_0^2 (2-\tau)^{\alpha}\, Ssinh_p(\lambda,\tau)\,d\tau = \frac{1}{\Gamma(\alpha)}\int_{-1}^1 (1-\tau)^{\alpha}\, Ssinh_p(\lambda,1+\tau)\,d\tau$$
$$= 2^{\alpha-p}\,{}_0sinh_1\left(\begin{matrix} - \\ (1-p+\alpha,1)\end{matrix};2\lambda\right)$$
$$= Ssinh_{\alpha-p}(\lambda,2)\,, \tag{6.115}$$

and

$$\frac{1}{\Gamma(\alpha)}\int_0^2 (2-\tau)^{\alpha}\, Scosh_p(\lambda,\tau)\,d\tau = \frac{1}{\Gamma(\alpha)}\int_{-1}^1 (1-\tau)^{\alpha}\, Scosh_p(\lambda,1+\tau)\,d\tau$$
$$= 2^{\alpha-p}\,{}_0cosh_1\left(\begin{matrix} - \\ (1-p+\alpha,1)\end{matrix};2\lambda\right)$$
$$= Scosh_{\alpha-p}(\lambda,2)\,. \tag{6.116}$$

Proof Putting $t = 2$ in (6.101), (6.102), (6.103), and (6.104), respectively, we have the results. □

Making use of (6.4), we give the following property:

Property 6.7 Let $|\lambda/s| < 1$ and $p, \lambda \in \mathbb{C}$. Then we have

$$L\left\{Ssin_p(\lambda,t)\right\} = s^{p-1} sin\left(\frac{\lambda}{s}\right), \tag{6.117}$$

$$L\left\{Scos_p(\lambda,t)\right\} = s^{p-1} cos\left(\frac{\lambda}{s}\right), \tag{6.118}$$

$$L\left\{Ssinh_p(\lambda,t)\right\} = s^{p-1} sinh\left(\frac{\lambda}{s}\right), \tag{6.119}$$

and

$$L\left\{Scosh_p(\lambda,t)\right\} = s^{p-1} cosh\left(\frac{\lambda}{s}\right). \tag{6.120}$$

Proof In view of the integral representations

$$L\left\{S_p\left(i\lambda, t\right)\right\} = s^{p-1}e^{i\frac{\lambda}{s}} \tag{6.121}$$

$$L\left\{S_p\left(-i\lambda, t\right)\right\} = s^{p-1}e^{-i\frac{\lambda}{s}}, \tag{6.122}$$

$$L\left\{S_p\left(\lambda, t\right)\right\} = s^{p-1}e^{\frac{\lambda}{s}}, \tag{6.123}$$

and

$$L\left\{S_p\left(-\lambda, t\right)\right\} = s^{p-1}e^{-\frac{\lambda}{s}}, \tag{6.124}$$

we have

$$\begin{aligned} L\left\{Ssin_p\left(\lambda, t\right)\right\} &= \frac{1}{2i}\left(s^{p-1}e^{i\frac{\lambda}{s}} - s^{p-1}e^{i\frac{\lambda}{s}}\right) = \frac{s^{p-1}}{2i}\left(e^{i\frac{\lambda}{s}} - e^{-i\frac{\lambda}{s}}\right) \\ &= s^{p-1}sin\left(\frac{\lambda}{s}\right), \end{aligned} \tag{6.125}$$

$$\begin{aligned} L\left\{Scos_p\left(\lambda, t\right)\right\} &= \frac{1}{2}\left(s^{p-1}e^{i\frac{\lambda}{s}} + s^{p-1}e^{i\frac{\lambda}{s}}\right) = \frac{s^{p-1}}{2}\left(e^{i\frac{\lambda}{s}} + e^{-i\frac{\lambda}{s}}\right) \\ &= s^{p-1}cos\left(\frac{\lambda}{s}\right), \end{aligned} \tag{6.126}$$

$$\begin{aligned} L\left\{Ssinh_p\left(\lambda, t\right)\right\} &= \frac{1}{2}\left(s^{p-1}e^{\frac{\lambda}{s}} - s^{p-1}e^{-\frac{\lambda}{s}}\right) = \frac{s^{p-1}}{2}\left(e^{\frac{\lambda}{s}} - e^{-\frac{\lambda}{s}}\right) \\ &= s^{p-1}sinh\left(\frac{\lambda}{s}\right) \end{aligned} \tag{6.127}$$

and

$$\begin{aligned} L\left\{Scosh_p\left(\lambda, t\right)\right\} &= \frac{1}{2}\left(s^{p-1}e^{\frac{\lambda}{s}} + s^{p-1}e^{-\frac{\lambda}{s}}\right) = \frac{s^{p-1}}{2}\left(e^{\frac{\lambda}{s}} + e^{-\frac{\lambda}{s}}\right) \\ &= s^{p-1}cosh\left(\frac{\lambda}{s}\right). \end{aligned} \tag{6.128}$$

□

6.1.5 The Sonine Functions of Second Type

Definition 6.14 The Sonine function of second type is defined as [269]

$$\begin{aligned} & T_p^{\xi}(t) \\ & = t^{-p} \sum_{k=0}^{\infty} \frac{(\xi)_k t^k}{\Gamma(k+1)\Gamma(k-p+1)} \\ & = \sum_{k=0}^{\infty} \frac{(\xi)_k t^{k-p}}{\Gamma(k+1)\Gamma(k-p+1)}, \end{aligned} \tag{6.129}$$

where $p, \xi, t \in \mathbb{C}$.

The result, discovered by Sonine in 1884 [269], is called the Sonine function of second type.

At the present, we have [269]

$$T_p^{\xi}(\lambda, t) = t^{-p} \sum_{k=0}^{\infty} \frac{(\xi)_k}{\Gamma(k-p+1)} \frac{(\lambda t)^k}{k!}, \tag{6.130}$$

and

$$T_p^{\xi}(-\lambda, t) = t^{-p} \sum_{k=0}^{\infty} \frac{(\xi)_k}{\Gamma(k-p+1)} \frac{(-\lambda t)^k}{k!}, \tag{6.131}$$

which lead to

$$\begin{aligned} & T_p^{\xi}(\lambda, t) \\ & = t^{-p} \sum_{k=0}^{\infty} \frac{(\xi)_k}{\Gamma(k-p+1)} \frac{(\lambda t)^k}{k!} \\ & = \sum_{k=0}^{\infty} \frac{(\xi)_k \lambda^k t^{k-p}}{\Gamma(k+1)\Gamma(k-p+1)} \\ & = \frac{1}{\Gamma(\xi+1)} \sum_{k=0}^{\infty} \frac{\Gamma(\xi+k)\lambda^k t^{k-p}}{\Gamma(k+1)\Gamma(k-p+1)} \\ & = \frac{t^{-p}}{\Gamma(\xi+1)} {}_1\varphi_1\left(\begin{matrix} (\xi, 1) \\ (-p+1, 1) \end{matrix}; \lambda t \right), \end{aligned} \tag{6.132}$$

and

$$
\begin{aligned}
& T_p^{\xi}(-\lambda, t) \\
& = t^{-p} \sum_{k=0}^{\infty} \frac{(\xi)_k}{\Gamma(k-p+1)} \frac{(-\lambda t)^k}{k!} \\
& = \sum_{k=0}^{\infty} \frac{(\xi)_k (-\lambda)^k t^{k-p}}{\Gamma(k+1) \Gamma(k-p+1)} \\
& = \frac{1}{\Gamma(\xi+1)} \sum_{k=0}^{\infty} \frac{\Gamma(\xi+k) (-\lambda)^k t^{k-p}}{\Gamma(k+1) \Gamma(k-p+1)} \\
& = \frac{t^{-p}}{\Gamma(\xi+1)} {}_1\varphi_1\left(\begin{matrix}(\xi, 1) \\ (-p+1, 1)\end{matrix}; -\lambda t\right),
\end{aligned} \tag{6.133}
$$

where $\lambda, p, \xi, t \in \mathbb{C}$.

Here, we consider the Laplace transform of the Sonine function of second type

$$
\begin{aligned}
L\left\{T_p^{\xi}(\lambda, t)\right\} & = L\left\{\frac{1}{\Gamma(\xi+1)} \sum_{k=0}^{\infty} \frac{\Gamma(\xi+k) \lambda^k t^{k-p}}{\Gamma(k+1) \Gamma(k-p+1)}\right\} \\
& = \frac{s^{p-1}}{\Gamma(\xi+1)} \sum_{k=0}^{\infty} \frac{\Gamma(\xi+k)}{k!} \left(\frac{\lambda}{s}\right)^k \\
& = \frac{s^{p-1}}{\Gamma(\xi+1)} {}_1\varphi_0\left(\begin{matrix}(\xi, 1) \\ -\end{matrix}; \frac{\lambda}{s}\right) \quad (|\lambda/s| < 1),
\end{aligned} \tag{6.134}
$$

and

$$
\begin{aligned}
L\left\{T_p^{\xi}(-\lambda, t)\right\} & = L\left\{\frac{1}{\Gamma(\xi+1)} \sum_{k=0}^{\infty} \frac{\Gamma(\xi+k) (-\lambda)^k t^{k-p}}{\Gamma(k+1) \Gamma(k-p+1)}\right\} \\
& = \frac{s^{p-1}}{\Gamma(\xi+1)} \sum_{k=0}^{\infty} \frac{\Gamma(\xi+k)}{k!} \left(-\frac{\lambda}{s}\right)^k \\
& = \frac{s^{p-1}}{\Gamma(\xi+1)} {}_1\varphi_0\left(\begin{matrix}(\xi, 1) \\ -\end{matrix}; -\frac{\lambda}{s}\right) \quad (|\lambda/s| < 1).
\end{aligned} \tag{6.135}
$$

Thus, we obtain the following property:

Property 6.8 Let $|1/s| < 1$ and $\lambda, p, \xi, t \in \mathbb{C}$. Then we have

$$L\left\{T_p^{\xi}(t)\right\} = \frac{s^{p-1}}{\Gamma(\xi+1)}\,{}_1\varphi_0\left(\begin{matrix}(\xi,1)\\ -\end{matrix};-\tfrac{1}{s}\right). \tag{6.136}$$

The Mellin transform of the Sonine function of second type can be given as follows:

$$\begin{aligned} M\left[T_p^{\xi}(-\lambda,t)\right] &= \int_0^{\infty} t^{\varpi-1} T_p^{\xi}(-\lambda,t)\,dt \\ &= \int_0^{\infty} t^{\varpi-1}\frac{t^{-p}}{\Gamma(\xi+1)}\,{}_1\varphi_1\left(\begin{matrix}(\xi,1)\\ (-p+1,1)\end{matrix};-\lambda t\right)dt \\ &= \frac{1}{\Gamma(\xi+1)}\int_0^{\infty} t^{\varpi-p-1}\,{}_1\varphi_1\left(\begin{matrix}(\xi,1)\\ (-p+1,1)\end{matrix};-\lambda t\right)dt \\ &= \frac{\Gamma(\varpi-q)\,\Gamma(\xi-\varpi+q)}{\Gamma(\xi+1)\,\Gamma(1-\varpi)}\lambda^{q-\varpi}, \end{aligned} \tag{6.137}$$

where $1 > Re(\varpi) > Re(q)$ and $\lambda \in \mathbb{R}_+$.

Thus, we have the following property:

Property 6.9 Let $1 > Re(\varpi) > Re(q)$. Then we have

$$M\left[T_p^{\xi}(t)\right] = \frac{\Gamma(\varpi-q)\,\Gamma(\xi-\varpi+q)}{\Gamma(\xi+1)\,\Gamma(1-\varpi)}. \tag{6.138}$$

Thus, we have the following property:

Property 6.10 Let $1 > Re(\varpi) > Re(q)$. Then we have

$$T_p^{\xi}(t) = \frac{1}{2\pi i}\int_{-i\infty}^{i\infty}\frac{\Gamma(\varpi-q)\,\Gamma(\xi-\varpi+q)}{\Gamma(\xi+1)\,\Gamma(1-\varpi)}(-t)^{\varpi-1}\,d\varpi, \tag{6.139}$$

where $|\arg(t)| < \pi$.

Let us recall the following result:

Let $\alpha, \gamma, \xi \in \mathbb{C}$, $Re(\alpha) > 0$, $Re(\gamma) > 0$, $\lambda \in \mathbb{C}$, $\beta > 0$, and $p \in \mathbb{C}$. Then we have

$$\begin{aligned} &\frac{1}{\Gamma(\alpha)}\int_0^t (t-\tau)^{\alpha}\,\tau^{\gamma-1}\,{}_1\varphi_1\left(\begin{matrix}(\xi,1)\\ (-p+1,1)\end{matrix};\lambda\tau^{\beta}\right)d\tau \\ &= t^{\gamma+\alpha-1}\,{}_2\varphi_2\left(\begin{matrix}(\xi,1),(\gamma,\beta)\\ (-p+1,1),(\gamma+\alpha,\beta)\end{matrix};\lambda t^{\beta}\right), \end{aligned} \tag{6.140}$$

which leads to

$$\frac{1}{\Gamma(\alpha)}\int_0^t (t-\tau)^{\alpha}\,\tau^{-p}\,{}_1\varphi_1\left(\begin{matrix}(\xi,1)\\(-p+1,1)\end{matrix};\lambda\tau^{\beta}\right)d\tau$$
$$=t^{\gamma+\alpha-1}\,{}_2\varphi_2\left(\begin{matrix}(\xi,1),(-p+1,\beta)\\(-p+1,1),(-p+1+\alpha,\beta)\end{matrix};\lambda t^{\beta}\right) \tag{6.141}$$

and

$$\frac{1}{\Gamma(\alpha)}\int_0^t (t-\tau)^{\alpha}\,\tau^{-p}\,{}_1\varphi_1\left(\begin{matrix}(\xi,1)\\(-p+1,1)\end{matrix};\lambda\tau\right)d\tau$$
$$=t^{-p+\alpha}\,{}_1\varphi_1\left(\begin{matrix}(\xi,1)\\(-p+1+\alpha,1)\end{matrix};\lambda t\right). \tag{6.142}$$

Thus, we have the following property:

Property 6.11 Let $\alpha,\xi\in\mathbb{C}$, $Re(\alpha)>0$, $\lambda\in\mathbb{C}$, and $p\in\mathbb{C}$. Then we have

$$\frac{1}{\Gamma(\alpha)}\int_0^t (t-\tau)^{\alpha}\,T_p^{\xi}(\lambda,\tau)\,d\tau=\frac{t^{-p+\alpha}}{\Gamma(\xi+1)}\,{}_1\varphi_1\left(\begin{matrix}(\xi,1)\\(-p+1+\alpha,1)\end{matrix};\lambda t\right)$$
$$=T_{\alpha-p}^{\xi}(\lambda,t). \tag{6.143}$$

Proof Making use of (6.130) and (6.143), we get the result. □

Property 6.12 Let $\alpha,\xi\in\mathbb{C}$, $Re(\alpha)>0$, $\lambda\in\mathbb{C}$, and $p\in\mathbb{C}$. Then we have

$$\frac{1}{\Gamma(\alpha)}\int_0^1 (1-\tau)^{\alpha}\,T_p^{\xi}(\lambda,\tau)\,d\tau=\frac{1}{\Gamma(\xi+1)}\,{}_1\varphi_1\left(\begin{matrix}(\xi,1)\\(-p+1+\alpha,1)\end{matrix};\lambda\right)$$
$$=T_{\alpha-p}^{\xi}(\lambda,1). \tag{6.144}$$

Proof Setting $t=1$ in (6.143), we get the result. □

Property 6.13 Let $\alpha,\xi\in\mathbb{C}$, $Re(\alpha)>0$, $\lambda\in\mathbb{C}$, and $p\in\mathbb{C}$. Then we have

$$\frac{1}{\Gamma(\alpha)}\int_0^2 (2-\tau)^{\alpha}\,T_p^{\xi}(\lambda,\tau)\,d\tau=\frac{1}{\Gamma(\alpha)}\int_{-1}^1 (1-\tau)^{\alpha}\,T_p^{\xi}(\lambda,1+\tau)\,d\tau$$
$$=\frac{1}{\Gamma(\alpha)}\int_{-1}^1 (1+\tau)^{\alpha}\,T_p^{\xi}(\lambda,1-\tau)\,d\tau$$
$$=\frac{2^{-p+\alpha}}{\Gamma(\xi+1)}\,{}_1\varphi_1\left(\begin{matrix}(\xi,1)\\(-p+1+\alpha,1)\end{matrix};2\lambda\right)$$
$$=T_{\alpha-p}^{\xi}(\lambda,2). \tag{6.145}$$

6.1.6 The Supertrigonometric Functions via Sonine Function of Second Type

In this part, we present the supertrigonometric functions via Sonine function of second type.

Now, we consider that

$$T_p^{\xi}(i\lambda, t) = t^{-p} \sum_{k=0}^{\infty} \frac{(\xi)_k}{\Gamma(k-p+1)} \frac{(i\lambda t)^k}{k!} = \frac{t^{-p}}{\Gamma(\xi+1)} {}_1\varphi_1\left(\begin{matrix} (\xi, 1) \\ (-p+1, 1) \end{matrix}; i\lambda t \right) \tag{6.146}$$

and

$$\begin{aligned} T_p^{\xi}(-i\lambda, t) &= t^{-p} \sum_{k=0}^{\infty} \frac{(\xi)_k}{\Gamma(k-p+1)} \frac{(-i\lambda t)^k}{k!} \\ &= \frac{t^{-p}}{\Gamma(\xi+1)} {}_1\varphi_1\left(\begin{matrix} (\xi, 1) \\ (-p+1, 1) \end{matrix}; -i\lambda t \right). \end{aligned} \tag{6.147}$$

We may structure the following special functions:

Definition 6.15 The supersine via Sonine function of second type is defined as

$$T sin_p^{\xi}(t) = \frac{1}{2i}\left(T_p^{\xi}(i, t) - T_p^{\xi}(-i, t)\right) = t^{-p} \sum_{k=0}^{\infty} \frac{(\xi)_{2k+1}}{\Gamma(2k-p+2)} \frac{(-1)^k t^{2k+1}}{(2k+1)!}, \tag{6.148}$$

where $p, \xi, t \in \mathbb{C}$.

Definition 6.16 The supercosine via Sonine function of second type is defined as

$$T cos_p^{\xi}(t) = \frac{1}{2}\left(T_p^{\xi}(i, t) + T_p^{\xi}(-i, t)\right) = t^{-p} \sum_{k=0}^{\infty} \frac{(\xi)_{2k}}{\Gamma(2k-p+1)} \frac{(-1)^k t^{2k}}{(2k)!}, \tag{6.149}$$

where $p, \xi, t \in \mathbb{C}$.

Definition 6.17 The supertangent via Sonine function of second type is defined as

$$T tan_p^{\xi}(t) = \frac{T sin_p^{\xi}(t)}{T cos_p^{\xi}(t)} \left(T cos_p^{\xi}(t) \neq 0\right), \tag{6.150}$$

where $p, \xi, t \in \mathbb{C}$.

Definition 6.18 The supercotangent via Sonine function of second type is defined as

$$Tcotan_p^{\xi}(t) = \frac{Tcos_p^{\xi}(t)}{Tsin_p^{\xi}(t)}, \left(Tsin_p^{\xi}(t) \neq 0\right), \tag{6.151}$$

where $p, \xi, t \in \mathbb{C}$.

Definition 6.19 The supercosecant via Sonine function of second type is defined as

$$Tcsc_p^{\xi}(t) = \frac{1}{Tsin_p^{\xi}(t)}, \left(Tsin_p^{\xi}(t) \neq 0\right), \tag{6.152}$$

where $p, \xi, t \in \mathbb{C}$.

Definition 6.20 The supersecant via Sonine function of second type is defined as

$$Tsec_p^{\xi}(t) = \frac{1}{Tcos_p^{\xi}(t)} \left(Tcos_p^{\xi}(t) \neq 0\right), \tag{6.153}$$

where $p, \xi, t \in \mathbb{C}$.

Therefore, we have

$$\begin{aligned} Tsin_p^{\xi}(\lambda, t) &= \frac{1}{2i}\left(T_p^{\xi}(i\lambda, t) - T_p^{\xi}(-i\lambda, t)\right) \\ &= t^{-p}\sum_{k=0}^{\infty} \frac{(\xi)_{2k+1}}{\Gamma(2k-p+2)} \frac{(-1)^k (\lambda t)^{2k+1}}{(2k+1)!}, \end{aligned} \tag{6.154}$$

$$\begin{aligned} Tcos_p^{\xi}(\lambda, t) &= \frac{1}{2}\left(T_p^{\xi}(i\lambda, t) + T_p^{\xi}(-i\lambda, t)\right) \\ &= t^{-p}\sum_{k=0}^{\infty} \frac{(\xi)_{2k}}{\Gamma(2k-p+1)} \frac{(-1)^k (\lambda t)^{2k}}{(2k)!}, \end{aligned} \tag{6.155}$$

$$Ttan_p^{\xi}(\lambda, t) = \frac{Tsin_p^{\xi}(\lambda, t)}{Tcos_p^{\xi}(\lambda, t)} \left(Tcos_p^{\xi}(\lambda, t) \neq 0\right), \tag{6.156}$$

$$Tcotan_p^{\xi}(\lambda, t) = \frac{Tcos_p^{\xi}(\lambda, t)}{Tsin_p^{\xi}(\lambda, t)}, \left(Tsin_p^{\xi}(\lambda, t) \neq 0\right), \tag{6.157}$$

$$Tcsc_p^{\xi}(\lambda, t) = \frac{1}{Tsin_p^{\xi}(\lambda, t)}, \left(Tsin_p^{\xi}(\lambda, t) \neq 0\right), \tag{6.158}$$

and

$$Tsec_p^{\xi}(\lambda, t) = \frac{1}{Tcos_p^{\xi}(\lambda, t)} \left(Tcos_p^{\xi}(\lambda, t) \neq 0\right), \tag{6.159}$$

where $p, \xi, \lambda, t \in \mathbb{C}$.

Property 6.14 Let $p, \xi, \lambda, t \in \mathbb{C}$. Then we have

$$T_p^{\xi}(i\lambda, t) = Tsin_p^{\xi}(\lambda, t) + iTcos_p^{\xi}(\lambda, t) \tag{6.160}$$

and

$$T_p^{\xi}(-i\lambda, t) = Tsin_p^{\xi}(\lambda, t) - iTcos_p^{\xi}(\lambda, t). \tag{6.161}$$

6.1.7 The Superhyperbolic Functions via Sonine Function of Second Type

In this part, we present the superhyperbolic functions via Sonine function of second type.

Now, we consider that

$$T_p^{\xi}(\lambda, t) = t^{-p} \sum_{k=0}^{\infty} \frac{(\xi)_k}{\Gamma(k-p+1)} \frac{(\lambda t)^k}{k!} = \frac{t^{-p}}{\Gamma(\xi+1)} {}_1\varphi_1 \left(\begin{matrix} (\xi, 1) \\ (-p+1, 1) \end{matrix} ; \lambda t \right) \tag{6.162}$$

and

$$\begin{aligned} T_p^{\xi}(-\lambda, t) &= t^{-p} \sum_{k=0}^{\infty} \frac{(\xi)_k}{\Gamma(k-p+1)} \frac{(-\lambda t)^k}{k!} \\ &= \frac{t^{-p}}{\Gamma(\xi+1)} {}_1\varphi_1 \left(\begin{matrix} (\xi, 1) \\ (-p+1, 1) \end{matrix} ; -\lambda t \right). \end{aligned} \tag{6.163}$$

We can derive the following special functions:

Definition 6.21 The hyperbolic supersine via Sonine function of second type is defined as

$$Tsinh_p^{\xi}(t) = \frac{1}{2}\left(T_p^{\xi}(1, t) - T_p^{\xi}(-1, t)\right) = t^{-p} \sum_{k=0}^{\infty} \frac{(\xi)_{2k+1}}{\Gamma(2k-p+2)} \frac{t^{2k+1}}{(2k+1)!}, \tag{6.164}$$

where $p, \xi, t \in \mathbb{C}$.

Definition 6.22 The hyperbolic supercosine via Sonine function of second type is defined as

$$Tcosh_p^{\xi}(t) = \frac{1}{2}\left(T_p^{\xi}(1,t) + T_p^{\xi}(-1,t)\right) = t^{-p}\sum_{k=0}^{\infty}\frac{(\xi)_{2k}}{\Gamma(2k-p+1)}\frac{t^{2k}}{(2k)!}, \tag{6.165}$$

where $p, \xi, t \in \mathbb{C}$.

Definition 6.23 The hyperbolic supertangent via Sonine function of second type is defined as

$$Ttanh_p^{\xi}(t) = \frac{Tsinh_p^{\xi}(t)}{Tcosh_p^{\xi}(t)}\left(Tcosh_p^{\xi}(t) \neq 0\right), \tag{6.166}$$

where $p, \xi, t \in \mathbb{C}$.

Definition 6.24 The hyperbolic supercotangent via Sonine function of second type is defined as

$$Tcotanh_p^{\xi}(t) = \frac{Tcosh_p^{\xi}(t)}{Tsinh_p^{\xi}(t)}, \left(Tsinh_p^{\xi}(t) \neq 0\right), \tag{6.167}$$

where $p, \xi, t \in \mathbb{C}$.

Definition 6.25 The hyperbolic supercosecant via Sonine function of second type is defined as

$$Tcsch_p^{\xi}(t) = \frac{1}{Tsinh_p^{\xi}(t)}, \left(Tsinh_p^{\xi}(t) \neq 0\right), \tag{6.168}$$

where $p, \xi, t \in \mathbb{C}$.

Definition 6.26 The hyperbolic supersecant via Sonine function of second type is defined as

$$Tsech_p^{\xi}(t) = \frac{1}{Tcosh_p^{\xi}(t)}\left(Tcosh_p^{\xi}(t) \neq 0\right), \tag{6.169}$$

where $p, \xi, t \in \mathbb{C}$.

Therefore, we have

$$Tsinh_p^{\xi}(\lambda, t) = \frac{1}{2}\left(T_p^{\xi}(\lambda, t) - T_p^{\xi}(-\lambda, t)\right)$$
$$= t^{-p}\sum_{k=0}^{\infty}\frac{(\xi)_{2k+1}}{\Gamma(2k-p+2)}\frac{(\lambda t)^{2k+1}}{(2k+1)!}, \tag{6.170}$$

$$Tcosh_p^{\xi}(\lambda, t) = \frac{1}{2}\left(T_p^{\xi}(\lambda, t) + T_p^{\xi}(-\lambda, t)\right)$$
$$= t^{-p}\sum_{k=0}^{\infty}\frac{(\xi)_{2k}}{\Gamma(2k-p+1)}\frac{(\lambda t)^{2k}}{(2k)!}, \tag{6.171}$$

$$Ttanh_p^{\xi}(\lambda, t) = \frac{Tsinh_p^{\xi}(\lambda, t)}{Tcosh_p^{\xi}(\lambda, t)}\left(Tcosh_p^{\xi}(\lambda, t) \neq 0\right), \tag{6.172}$$

$$Tcotanh_p^{\xi}(\lambda, t) = \frac{Tcosh_p^{\xi}(\lambda, t)}{Tsinh_p^{\xi}(\lambda, t)}, \left(Tsinh_p^{\xi}(\lambda, t) \neq 0\right), \tag{6.173}$$

$$Tcsch_p^{\xi}(\lambda, t) = \frac{1}{Tsinh_p^{\xi}(\lambda, t)}, \left(Tsinh_p^{\xi}(\lambda, t) \neq 0\right), \tag{6.174}$$

and

$$Tsech_p^{\xi}(\lambda, t) = \frac{1}{Tcosh_p^{\xi}(\lambda, t)}\left(Tcosh_p^{\xi}(\lambda, t) \neq 0\right), \tag{6.175}$$

where $p, \xi, \lambda, t \in \mathbb{C}$.

Property 6.15 Let $p, \xi, \lambda, t \in \mathbb{C}$. Then we have

$$T_p^{\xi}(\lambda, t) = Tsinh_p^{\xi}(\lambda, t) + Tcosh_p^{\xi}(\lambda, t) \tag{6.176}$$

and

$$T_p^{\xi}(-\lambda, t) = Tsinh_p^{\xi}(\lambda, t) - Tcosh_p^{\xi}(\lambda, t). \tag{6.177}$$

6.1.8 The Sonine Function of Third Type

Definition 6.27 The Sonine function of third type is defined as [269]

$$\mathbb{V}_\alpha(t) = t^\alpha \sum_{k=0}^{\infty} \frac{t^k}{\Gamma(k+\alpha)} = \sum_{k=0}^{\infty} \frac{t^{k+\alpha}}{\Gamma(k+\alpha)}, \tag{6.178}$$

where $\alpha, t \in \mathbb{C}$.

The Sonine function of third type can be given as [269]

$$\mathbb{V}_\alpha(\lambda, t) = t^\alpha \sum_{k=0}^{\infty} \frac{\lambda^k t^k}{\Gamma(k+\alpha)} = \sum_{k=0}^{\infty} \frac{\lambda^k t^{k+\alpha}}{\Gamma(k+\alpha)}, \tag{6.179}$$

where $\alpha, t \in \mathbb{C}$.

It is shown that [269]

$$\mathbb{V}_\alpha(\lambda, t) = t^\alpha \sum_{k=0}^{\infty} \frac{(\lambda t)^k}{\Gamma(k+\alpha)} = t^\alpha E_{1,\alpha}(\lambda t). \tag{6.180}$$

In fact, we may give

$$\mathbb{V}_\alpha(\lambda, t) = t^\alpha \sum_{k=0}^{\infty} \frac{(\lambda t)^k}{\Gamma(k+\alpha)} = t^\alpha E_{1,\alpha}(\lambda t) = t^\alpha \, {}_1\varphi_1\left(\begin{matrix}(1,1) \\ (\alpha,1)\end{matrix}; \lambda t\right), \tag{6.181}$$

which leads to

$$\mathbb{V}_\alpha(t) = \mathbb{V}_\alpha(1, t).$$

The Laplace transform of (6.179) reads

$$L\{\mathbb{V}_\alpha(\lambda, t)\} = L\left\{t^\alpha \sum_{k=0}^{\infty} \frac{(\lambda t)^k}{\Gamma(k+\alpha)}\right\} = s^{-\alpha} \sum_{k=0}^{\infty} (k+\alpha)\left(\frac{\lambda}{s}\right)^k. \tag{6.182}$$

Property 6.16 Let $\alpha, t \in \mathbb{C}$. Then we have

$$L\{\mathbb{V}_\alpha(t)\} = s^{-\alpha} \sum_{k=0}^{\infty} (k+\alpha)\left(\frac{\lambda}{s}\right)^k. \tag{6.183}$$

The Mellin transform of (6.179) is written as follows:

$$M\left[\mathbb{V}_\alpha(-\lambda,t)\right]=\int_0^\infty t^{\varpi+\alpha-1}E_{1,\alpha}(-\lambda t)\,dt$$
$$=\frac{\Gamma(\varpi+\alpha-1)\Gamma(2-\varpi-\alpha)}{\Gamma(1-\varpi)}\lambda^{1-(\varpi+\alpha)}, \tag{6.184}$$

where $\alpha,\varpi\in\mathbb{C}$, $\lambda\in\mathbb{C}$, $Re(\alpha)>0$, and $0<Re(\varpi)<1$.

Property 6.17 Let $\alpha\in\mathbb{C}$, $\lambda\in\mathbb{C}$, $Re(\alpha)>0$, and $0<Re(\varpi)<1$. Then we have

$$\int_{-\infty}^{\infty}|t|^{\varpi-1}\,\mathbb{V}_\alpha(\lambda,|t|)\,dt=2\lambda^{1-(\varpi+\alpha)}\frac{\Gamma(\varpi+\alpha-1)\Gamma(2-\varpi-\alpha)}{\Gamma(1-\varpi)}. \tag{6.185}$$

Property 6.18 Let $\alpha\in\mathbb{C}$, $\lambda\in\mathbb{C}$, $Re(\alpha)>0$, and $0<Re(\varpi)<1$. Then we have

$$\mathbb{V}_\alpha(t)=\frac{1}{2\pi i}\int_{-i\infty}^{i\infty}\frac{\Gamma(\varpi+\alpha)\Gamma(1-\varpi-\alpha)}{\Gamma(1-\varpi)}(-t)^{-(\varpi+\alpha)}\,d\varpi. \tag{6.186}$$

Proof With the aid of (6.184), we deduce the result. □

Let us consider the following case (found in Chap. 5):
Let $\sigma\in\mathbb{C}$, $\lambda\in\mathbb{C}$, $\alpha\in\mathbb{R}_+$, and $Re(\sigma)>0$. Then we have

$$\frac{1}{\Gamma(\sigma)}\int_0^t(t-\tau)^\sigma\,\mathbb{V}_\alpha(\lambda,\tau)\,d\tau=t^{\alpha+\sigma}\,{}_2\varphi_2\left(\begin{matrix}(1,1),(\alpha+1,1)\\(\alpha,1),(\alpha+\sigma+1,1)\end{matrix};\lambda t\right). \tag{6.187}$$

Proof From (5.181) and (6.179), we have

$$\frac{1}{\Gamma(\sigma)}\int_0^t(t-\tau)^\sigma\,\mathbb{V}_\alpha(\lambda,\tau)\,d\tau=\frac{1}{\Gamma(\sigma)}\int_0^t(t-\tau)^\sigma\,\tau^\alpha\,{}_1\varphi_1\left(\begin{matrix}(1,1)\\(\alpha,1)\end{matrix};\lambda\tau\right)d\tau$$
$$=t^{\alpha+\sigma}\,{}_2\varphi_2\left(\begin{matrix}(1,1),(\alpha+1,1)\\(\alpha,1),(\alpha+\sigma+1,1)\end{matrix};\lambda t\right). \tag{6.188}$$

□

Theorem 6.11 *Let $\sigma\in\mathbb{C}$, $\lambda\in\mathbb{C}$, $\alpha\in\mathbb{R}_+$, and $Re(\sigma)>0$. Then we have*

$$\frac{1}{\Gamma(\sigma)}\int_0^1(1-\tau)^\sigma\,\mathbb{V}_\alpha(\lambda,\tau)\,d\tau={}_2\varphi_2\left(\begin{matrix}(1,1),(\alpha+1,1)\\(\alpha,1),(\alpha+\sigma+1,1)\end{matrix};\lambda\right). \tag{6.189}$$

Proof Taking $t=1$ in (6.187), we deduce the result. □

Theorem 6.12 *Let $\sigma \in \mathbb{C}$, $\lambda \in \mathbb{C}$, $\alpha \in \mathbb{R}_+$, and $Re(\sigma) > 0$. Then we have*

$$\frac{1}{\Gamma(\sigma)} \int_0^2 (2-\tau)^\sigma \, \mathbb{V}_\alpha(\lambda,\tau)\, d\tau = \frac{1}{\Gamma(\sigma)} \int_{-1}^1 (1-\tau)^\sigma \, \mathbb{V}_\alpha(\lambda, 1+\tau)\, d\tau$$
$$= \frac{1}{\Gamma(\sigma)} \int_{-1}^1 (1+\tau)^\sigma \, \mathbb{V}_\alpha(\lambda, 1-\tau)\, d\tau$$
$$= 2^{\alpha+\sigma} \, {}_2\varphi_2\left(\begin{matrix} (1,1), (\alpha+1,1) \\ (\alpha,1), (\alpha+\sigma+1,1) \end{matrix} ; 2\lambda \right). \tag{6.190}$$

Proof Putting $t = 2$ in (6.187), we have the result. □

6.1.9 The Supertrigonometric Functions via Sonine Function of Third Type

In this part, we propose the supertrigonometric functions via Sonine function of third type.

We now consider that

$$\mathbb{V}_\alpha(i\lambda, t) = t^\alpha \sum_{k=0}^\infty \frac{(i\lambda t)^k}{\Gamma(k+\alpha)} = t^\alpha E_{1,\alpha}(i\lambda t) = t^\alpha \, {}_1\varphi_1\left(\begin{matrix} (1,1) \\ (\alpha,1) \end{matrix} ; i\lambda t \right) \tag{6.191}$$

and

$$\mathbb{V}_\alpha(-i\lambda, t) = t^\alpha \sum_{k=0}^\infty \frac{(-i\lambda t)^k}{\Gamma(k+\alpha)} = t^\alpha E_{1,\alpha}(-i\lambda t) = t^\alpha \, {}_1\varphi_1\left(\begin{matrix} (1,1) \\ (\alpha,1) \end{matrix} ; -i\lambda t \right). \tag{6.192}$$

Making use of (6.191) and (6.192), we present the following special functions:

Definition 6.28 The supersine via Sonine function of third type is defined as

$$\mathbb{V}sin_\alpha(t) = \frac{1}{2i}\left(\mathbb{V}_\alpha(i,t) - \mathbb{V}_\alpha(-i,t)\right)$$
$$= t^\alpha \sum_{k=0}^\infty \frac{(-1)^k \, t^{2k+1}}{\Gamma(2k+1+\alpha)}, \tag{6.193}$$

where $\alpha, t \in \mathbb{C}$.

Definition 6.29 The supercosine via Sonine function of third type is defined as

$$\mathbb{V}cos_{\alpha}(t) = \frac{1}{2}\left(\mathbb{V}_{\alpha}(i,t) + \mathbb{V}_{\alpha}(-i,t)\right) = t^{\alpha}\sum_{k=0}^{\infty}\frac{(-1)^{k}t^{2k}}{\Gamma(2k+\alpha)}, \tag{6.194}$$

where $\alpha, t \in \mathbb{C}$.

Definition 6.30 The supertangent via Sonine function of third type is defined as

$$\mathbb{V}tan_{\alpha}(t) = \frac{\mathbb{V}sin_{\alpha}(t)}{\mathbb{V}cos_{\alpha}(t)}\left(\mathbb{V}cos_{\alpha}(t) \neq 0\right), \tag{6.195}$$

where $\alpha, t \in \mathbb{C}$.

Definition 6.31 The supercotangent via Sonine function of third type is defined as

$$\mathbb{V}cotan_{\alpha}(t) = \frac{\mathbb{V}cos_{\alpha}(t)}{\mathbb{V}sin_{\alpha}(t)}, \left(\mathbb{V}sin_{\alpha}(t) \neq 0\right), \tag{6.196}$$

where $\alpha, t \in \mathbb{C}$.

Definition 6.32 The supercosecant via Sonine function of third type is defined as

$$\mathbb{V}csc_{\alpha}(t) = \frac{1}{\mathbb{V}sin_{\alpha}(t)}, \left(\mathbb{V}sin_{\alpha}(t) \neq 0\right), \tag{6.197}$$

where $\alpha, t \in \mathbb{C}$.

Definition 6.33 The supersecant via Sonine function of third type is defined as

$$\mathbb{V}sec_{\alpha}(t) = \frac{1}{\mathbb{V}cos_{\alpha}(t)}\left(\mathbb{V}cos_{\alpha}(t) \neq 0\right), \tag{6.198}$$

where $\alpha, t \in \mathbb{C}$.

Thus,

$$\begin{aligned}\mathbb{V}sin_{\alpha}(\lambda,t) &= \frac{1}{2i}\left(\mathbb{V}_{\alpha}(i\lambda,t) - \mathbb{V}_{\alpha}(-i\lambda,t)\right)\\ &= t^{\alpha}\sum_{k=0}^{\infty}\frac{(-1)^{k}(\lambda t)^{2k+1}}{\Gamma(2k+1+\alpha)},\end{aligned} \tag{6.199}$$

$$\begin{aligned}\mathbb{V}cos_{\alpha}(\lambda,t) &= \frac{1}{2}\left(\mathbb{V}_{\alpha}(i\lambda,t) + \mathbb{V}_{\alpha}(-i\lambda,t)\right)\\ &= t^{\alpha}\sum_{k=0}^{\infty}\frac{(-1)^{k}(\lambda t)^{2k}}{\Gamma(2k+\alpha)},\end{aligned} \tag{6.200}$$

$$\mathbb{V}tan_{\alpha}(\lambda,t)=\frac{\mathbb{V}sin_{\alpha}(\lambda,t)}{\mathbb{V}cos_{\alpha}(\lambda,t)}\left(\mathbb{V}cos_{\alpha}(\lambda,t)\neq 0\right), \tag{6.201}$$

$$\mathbb{V}cotan_{\alpha}(\lambda,t)=\frac{\mathbb{V}cos_{\alpha}(\lambda,t)}{\mathbb{V}sin_{\alpha}(\lambda,t)},\left(\mathbb{V}sin_{\alpha}(\lambda,t)\neq 0\right), \tag{6.202}$$

$$\mathbb{V}csc_{\alpha}(\lambda,t)=\frac{1}{\mathbb{V}sin_{\alpha}(\lambda,t)},\left(\mathbb{V}sin_{\alpha}(\lambda,t)\neq 0\right) \tag{6.203}$$

and

$$\mathbb{V}sec_{\alpha}(\lambda,t)=\frac{1}{\mathbb{V}cos_{\alpha}(\lambda,t)}\left(\mathbb{V}cos_{\alpha}(\lambda,t)\neq 0\right), \tag{6.204}$$

where $\alpha,\lambda,t\in\mathbb{C}$.

Theorem 6.13 *Let $\alpha,\lambda,t\in\mathbb{C}$. Then we have*

$$\mathbb{V}_{\alpha}(i\lambda,t)=\mathbb{V}cos_{\alpha}(\lambda,t)+i\mathbb{V}sin_{\alpha}(\lambda,t) \tag{6.205}$$

and

$$\mathbb{V}_{\alpha}(-i\lambda,t)=\mathbb{V}cos_{\alpha}(\lambda,t)-i\mathbb{V}sin_{\alpha}(\lambda,t). \tag{6.206}$$

Proof In view of (6.199) and (6.200), we give the result. □

In particular,

$$\mathbb{V}_{\alpha}(t)=\mathbb{V}cos_{\alpha}(t)+i\mathbb{V}sin_{\alpha}(t) \tag{6.207}$$

and

$$\mathbb{V}_{\alpha}(-i,t)=\mathbb{V}cos_{\alpha}(t)-i\mathbb{V}sin_{\alpha}(t), \tag{6.208}$$

where $\alpha,t\in\mathbb{C}$.

6.1.10 The Superhyperbolic Functions via Sonine Function of Third Type

In this part, we propose the superhyperbolic functions via Sonine function of third type.

In a similar manner, we have that

$$\mathbb{V}_{\alpha}(\lambda, t) = t^{\alpha} \sum_{k=0}^{\infty} \frac{(\lambda t)^k}{\Gamma(k+\alpha)} = t^{\alpha} E_{1,\alpha}(\lambda t) = t^{\alpha} {}_1\varphi_1\left(\begin{matrix}(1,1)\\(\alpha,1)\end{matrix}; \lambda t\right) \tag{6.209}$$

and

$$\begin{aligned}\mathbb{M}_{\alpha}(-\lambda, t) &= t^{\alpha} \sum_{k=0}^{\infty} \frac{(-\lambda t)^k}{\Gamma(k+1+\alpha)} = t^{\alpha} E_{1,1+\alpha}(-\lambda t) \\ &= t^{\alpha} {}_1\varphi_1\left(\begin{matrix}(1,1)\\(1+\alpha,1)\end{matrix}; -\lambda t\right).\end{aligned} \tag{6.210}$$

With the use of (6.32) and (6.33), we present the following special functions:

Definition 6.34 The hyperbolic supersine via Sonine function of third type is defined as

$$\mathbb{V}sinh_{\alpha}(t) = \frac{1}{2}\left(\mathbb{V}_{\alpha}(1,t) - \mathbb{V}_{\alpha}(-1,t)\right) = t^{\alpha} \sum_{k=0}^{\infty} \frac{t^{2k+1}}{\Gamma(2k+1+\alpha)}, \tag{6.211}$$

where $\alpha, t \in \mathbb{C}$.

Definition 6.35 The hyperbolic supercosine via Sonine function of third type is defined as

$$\mathbb{V}cosh_{\alpha}(t) = \frac{1}{2}\left(\mathbb{V}_{\alpha}(1,t) + \mathbb{V}_{\alpha}(-1,t)\right) = t^{\alpha} \sum_{k=0}^{\infty} \frac{t^{2k}}{\Gamma(2k+\alpha)}, \tag{6.212}$$

where $\alpha, t \in \mathbb{C}$.

Definition 6.36 The hyperbolic supertangent via Sonine function of third type is defined as

$$\mathbb{V}tanh_{\alpha}(t) = \frac{\mathbb{V}sinh_{\alpha}(t)}{\mathbb{V}cosh_{\alpha}(t)} \left(\mathbb{V}cosh_{\alpha}(t) \neq 0\right), \tag{6.213}$$

where $\alpha, t \in \mathbb{C}$.

Definition 6.37 The hyperbolic supercotangent via Sonine function of third type is defined as

$$\mathbb{V}cotanh_{\alpha}(t) = \frac{\mathbb{V}cosh_{\alpha}(t)}{\mathbb{V}sinh_{\alpha}(t)}, \left(\mathbb{V}sinh_{\alpha}(t) \neq 0\right), \tag{6.214}$$

where $\alpha, t \in \mathbb{C}$.

Definition 6.38 The hyperbolic supercosecant via Sonine function of third type is defined as

$$\mathbb{V}csch_{\alpha}(t) = \frac{1}{\mathbb{V}sinh_{\alpha}(t)}, \left(\mathbb{V}sinh_{\alpha}(t) \neq 0\right), \tag{6.215}$$

where $\alpha, t \in \mathbb{C}$.

Definition 6.39 The hyperbolic supersecant via Sonine function of third type is defined as

$$\mathbb{V}sech_{\alpha}(t) = \frac{1}{\mathbb{V}cosh_{\alpha}(t)} \left(\mathbb{V}cosh_{\alpha}(t) \neq 0\right), \tag{6.216}$$

where $\alpha, t \in \mathbb{C}$.

Thus,

$$\begin{aligned}\mathbb{V}sinh_{\alpha}(\lambda, t) &= \frac{1}{2}\left(\mathbb{V}_{\alpha}(\lambda, t) - \mathbb{V}_{\alpha}(-\lambda, t)\right) \\ &= t^{\alpha} \sum_{k=0}^{\infty} \frac{(\lambda t)^{2k+1}}{\Gamma(2k+1+\alpha)},\end{aligned} \tag{6.217}$$

$$\begin{aligned}\mathbb{V}cosh_{\alpha}(\lambda, t) &= \frac{1}{2}\left(\mathbb{V}_{\alpha}(\lambda, t) + \mathbb{V}_{\alpha}(-\lambda, t)\right) \\ &= t^{\alpha} \sum_{k=0}^{\infty} \frac{(\lambda t)^{2k}}{\Gamma(2k+\alpha)},\end{aligned} \tag{6.218}$$

$$\mathbb{V}tanh_{\alpha}(\lambda, t) = \frac{\mathbb{V}sinh_{\alpha}(\lambda, t)}{\mathbb{V}cosh_{\alpha}(\lambda, t)} \left(\mathbb{V}cosh_{\alpha}(\lambda, t) \neq 0\right), \tag{6.219}$$

$$\mathbb{V}cotanh_{\alpha}(\lambda, t) = \frac{\mathbb{V}cosh_{\alpha}(\lambda, t)}{\mathbb{V}sinh_{\alpha}(\lambda, t)}, \left(\mathbb{V}sinh_{\alpha}(\lambda, t) \neq 0\right), \tag{6.220}$$

$$\mathbb{V}csch_{\alpha}(\lambda, t) = \frac{1}{\mathbb{V}sinh_{\alpha}(\lambda, t)}, \left(\mathbb{V}sinh_{\alpha}(\lambda, t) \neq 0\right) \tag{6.221}$$

and

$$\mathbb{V}sech_{\alpha}(\lambda, t) = \frac{1}{\mathbb{V}cosh_{\alpha}(\lambda, t)} \left(\mathbb{V}cosh_{\alpha}(\lambda, t) \neq 0\right), \tag{6.222}$$

where $\alpha, \lambda, t \in \mathbb{C}$.

Theorem 6.14 *Let $\alpha, \lambda, t \in \mathbb{C}$. Then we have*

$$\mathbb{V}_{\alpha}(\lambda, t) = \mathbb{V}cosh_{\alpha}(\lambda, t) + \mathbb{V}sinh_{\alpha}(\lambda, t) \tag{6.223}$$

and

$$\mathbb{V}_{\alpha}(-\lambda, t) = \mathbb{V}cosh_{\alpha}(\lambda, t) - \mathbb{V}sinh_{\alpha}(\lambda, t). \tag{6.224}$$

Proof In view of (6.217) and (6.218), we give the result. □

In particular,

$$\mathbb{V}_{\alpha}(t) = \mathbb{V}cosh_{\alpha}(t) + \mathbb{V}sinh_{\alpha}(t) \tag{6.225}$$

and

$$\mathbb{V}_{\alpha}(-1, t) = \mathbb{V}cosh_{\alpha}(t) - \mathbb{V}sinh_{\alpha}(t), \tag{6.226}$$

where $\alpha, t \in \mathbb{C}$.

6.1.11 The Integral Representations Related to Sonine Function of Third Type

In this part, we present the integral representations related to Sonine function of third type.

Theorem 6.15 *Let $\sigma \in \mathbb{C}$, $\lambda \in \mathbb{C}$, $\alpha \in \mathbb{R}_{+}$, and $Re(\sigma) > 0$. Then we have*

$$\frac{1}{\Gamma(\sigma)} \int_{0}^{t} (t-\tau)^{\sigma} \mathbb{V}sin_{\alpha}(\lambda, \tau)\, d\tau = t^{\alpha+\sigma}\, {}_2\sin_2\left(\begin{matrix} (1,1), (\alpha+1,1) \\ (\alpha,1), (\alpha+\sigma+1,1) \end{matrix} ; \lambda t \right), \tag{6.227}$$

$$\frac{1}{\Gamma(\sigma)} \int_{0}^{t} (t-\tau)^{\sigma} \mathbb{V}cos_{\alpha}(\lambda, \tau)\, d\tau = t^{\alpha+\sigma}\, {}_2\cos_2\left(\begin{matrix} (1,1), (\alpha+1,1) \\ (\alpha,1), (\alpha+\sigma+1,1) \end{matrix} ; \lambda t \right), \tag{6.228}$$

$$\frac{1}{\Gamma(\sigma)} \int_{0}^{t} (t-\tau)^{\sigma} \mathbb{V}sinh_{\alpha}(\lambda, \tau)\, d\tau = t^{\alpha+\sigma}\, {}_2\sinh_2\left(\begin{matrix} (1,1), (\alpha+1,1) \\ (\alpha,1), (\alpha+\sigma+1,1) \end{matrix} ; \lambda t \right) \tag{6.229}$$

and

$$\frac{1}{\Gamma(\sigma)}\int_0^t (t-\tau)^{\sigma}\,\mathbb{V}cosh_{\alpha}(\lambda,\tau)\,d\tau = t^{\alpha+\sigma}\,{}_2\cosh_2\left(\begin{matrix}(1,1),(\alpha+1,1)\\(\alpha,1),(\alpha+\sigma+1,1)\end{matrix};\lambda t\right). \tag{6.230}$$

Proof Making use of the integral representations

$$\frac{1}{\Gamma(\sigma)}\int_0^t (t-\tau)^{\sigma}\,\mathbb{V}_{\alpha}(i\lambda,\tau)\,d\tau = t^{\alpha+\sigma}\,{}_2\varphi_2\left(\begin{matrix}(1,1),(\alpha+1,1)\\(\alpha,1),(\alpha+\sigma+1,1)\end{matrix};i\lambda t\right), \tag{6.231}$$

$$\frac{1}{\Gamma(\sigma)}\int_0^t (t-\tau)^{\sigma}\,\mathbb{V}_{\alpha}(-i\lambda,\tau)\,d\tau = t^{\alpha+\sigma}\,{}_2\varphi_2\left(\begin{matrix}(1,1),(\alpha+1,1)\\(\alpha,1),(\alpha+\sigma+1,1)\end{matrix};-i\lambda t\right), \tag{6.232}$$

$$\frac{1}{\Gamma(\sigma)}\int_0^t (t-\tau)^{\sigma}\,\mathbb{V}_{\alpha}(\lambda,\tau)\,d\tau = t^{\alpha+\sigma}\,{}_2\varphi_2\left(\begin{matrix}(1,1),(\alpha+1,1)\\(\alpha,1),(\alpha+\sigma+1,1)\end{matrix};\lambda t\right) \tag{6.233}$$

and

$$\frac{1}{\Gamma(\sigma)}\int_0^t (t-\tau)^{\sigma}\,\mathbb{V}_{\alpha}(-\lambda,\tau)\,d\tau = t^{\alpha+\sigma}\,{}_2\varphi_2\left(\begin{matrix}(1,1),(\alpha+1,1)\\(\alpha,1),(\alpha+\sigma+1,1)\end{matrix};-\lambda t\right), \tag{6.234}$$

we give the desired results from (6.199), (6.200), (6.217) and (6.218). □

Theorem 6.16 *Let $\sigma\in\mathbb{C}$, $\lambda\in\mathbb{C}$, $\alpha\in\mathbb{R}_+$, and $Re(\sigma)>0$. Then we have*

$$\frac{1}{\Gamma(\sigma)}\int_0^1 (1-\tau)^{\sigma}\,\mathbb{V}sin_{\alpha}(\lambda,\tau)\,d\tau = {}_2\sin_2\left(\begin{matrix}(1,1),(\alpha+1,1)\\(\alpha,1),(\alpha+\sigma+1,1)\end{matrix};\lambda\right), \tag{6.235}$$

$$\frac{1}{\Gamma(\sigma)}\int_0^1 (1-\tau)^{\sigma}\,\mathbb{V}cos_{\alpha}(\lambda,\tau)\,d\tau = {}_2cos_2\left(\begin{matrix}(1,1),(\alpha+1,1)\\(\alpha,1),(\alpha+\sigma+1,1)\end{matrix};\lambda\right), \tag{6.236}$$

$$\frac{1}{\Gamma(\sigma)}\int_0^1 (1-\tau)^{\sigma}\,\mathbb{V}sinh_{\alpha}(\lambda,\tau)\,d\tau = {}_2\sinh_2\left(\begin{matrix}(1,1),(\alpha+1,1)\\(\alpha,1),(\alpha+\sigma+1,1)\end{matrix};\lambda\right) \tag{6.237}$$

and

$$\frac{1}{\Gamma(\sigma)}\int_0^1 (1-\tau)^{\sigma}\,\mathbb{V}cosh_{\alpha}(\lambda,\tau)\,d\tau = {}_2\cosh_2\left(\begin{matrix}(1,1),(\alpha+1,1)\\(\alpha,1),(\alpha+\sigma+1,1)\end{matrix};\lambda\right). \tag{6.238}$$

Proof Taking $t = 1$ in (6.227), (6.228), (6.229) and (6.230), we get the results. □

Theorem 6.17 *Let* $\sigma \in \mathbb{C}$, $\lambda \in \mathbb{C}$, $\alpha \in \mathbb{R}_{+}$, *and* $Re(\sigma) > 0$. *Then we have*

$$\begin{aligned}\frac{1}{\Gamma(\sigma)}\int_0^2 (2-\tau)^{\sigma}\,\mathbb{V}sin_{\alpha}(\lambda,\tau)\,d\tau &= \frac{1}{\Gamma(\sigma)}\int_{-1}^1 (1-\tau)^{\sigma}\,\mathbb{V}sin_{\alpha}(\lambda,1+\tau)\,d\tau\\ &= \frac{1}{\Gamma(\sigma)}\int_{-1}^1 (1+\tau)^{\sigma}\,\mathbb{V}sin_{\alpha}(\lambda,1-\tau)\,d\tau\\ &= 2^{\alpha+\sigma}\,{}_2\sin_2\left(\begin{matrix}(1,1),(\alpha+1,1)\\(\alpha,1),(\alpha+\sigma+1,1)\end{matrix};2\lambda\right),\end{aligned} \tag{6.239}$$

$$\begin{aligned}\frac{1}{\Gamma(\sigma)}\int_0^2 (2-\tau)^{\sigma}\,\mathbb{V}cos_{\alpha}(\lambda,\tau)\,d\tau &= \frac{1}{\Gamma(\sigma)}\int_{-1}^1 (1-\tau)^{\sigma}\,\mathbb{V}cos_{\alpha}(\lambda,1+\tau)\,d\tau\\ &= \frac{1}{\Gamma(\sigma)}\int_{-1}^1 (1+\tau)^{\sigma}\,\mathbb{V}cos_{\alpha}(\lambda,1-\tau)\,d\tau\\ &= 2^{\alpha+\sigma}\,{}_2\cos_2\left(\begin{matrix}(1,1),(\alpha+1,1)\\(\alpha,1),(\alpha+\sigma+1,1)\end{matrix};2\lambda\right),\end{aligned} \tag{6.240}$$

$$\begin{aligned}\frac{1}{\Gamma(\sigma)}\int_0^2 (2-\tau)^{\sigma}\,\mathbb{V}sinh_{\alpha}(\lambda,\tau)\,d\tau &= \frac{1}{\Gamma(\sigma)}\int_{-1}^1 (1-\tau)^{\sigma}\,\mathbb{V}sinh_{\alpha}(\lambda,1+\tau)\,d\tau\\ &= \frac{1}{\Gamma(\sigma)}\int_{-1}^1 (1+\tau)^{\sigma}\,\mathbb{V}sinh_{\alpha}(\lambda,1-\tau)\,d\tau\\ &= 2^{\alpha+\sigma}\,{}_2\sinh_2\left(\begin{matrix}(1,1),(\alpha+1,1)\\(\alpha,1),(\alpha+\sigma+1,1)\end{matrix};2\lambda\right),\end{aligned} \tag{6.241}$$

and

$$\begin{aligned}\frac{1}{\Gamma(\sigma)}\int_0^2 (2-\tau)^{\sigma}\,\mathbb{V}cosh_{\alpha}(\lambda,\tau)\,d\tau &= \frac{1}{\Gamma(\sigma)}\int_{-1}^1 (1-\tau)^{\sigma}\,\mathbb{V}cosh_{\alpha}(\lambda,1+\tau)\,d\tau\\ &= \frac{1}{\Gamma(\sigma)}\int_{-1}^1 (1+\tau)^{\sigma}\,\mathbb{V}cosh_{\alpha}(\lambda,1-\tau)\,d\tau\\ &= 2^{\alpha+\sigma}\,{}_2\cosh_2\left(\begin{matrix}(1,1),(\alpha+1,1)\\(\alpha,1),(\alpha+\sigma+1,1)\end{matrix};2\lambda\right).\end{aligned} \tag{6.242}$$

Proof Putting $t = 2$ in (6.227), (6.228), (6.229) and (6.230), we get the results. □

Property 6.19 Let $\alpha, t \in \mathbb{C}$. Then we have

$$L\{\mathbb{V}sin_{\alpha}(\lambda,\tau)\} = -i\frac{s^{-\alpha}}{2}\sum_{k=0}^{\infty}(2k+1+\alpha)\left(\frac{\lambda}{s}\right)^{2k+1}, \tag{6.243}$$

$$L\{\mathbb{V}cos_{\alpha}(\lambda,\tau)\} = -\frac{s^{-\alpha}}{2}\sum_{k=0}^{\infty}(2k+\alpha)\left(\frac{\lambda}{s}\right)^{2k}, \tag{6.244}$$

$$L\{\mathbb{V}sinh_{\alpha}(\lambda,\tau)\} = \frac{s^{-\alpha}}{2}\sum_{k=0}^{\infty}(2k+1+\alpha)\left(\frac{\lambda}{s}\right)^{2k+1} \tag{6.245}$$

and

$$L\{\mathbb{V}cosh_{\alpha}(\lambda,\tau)\} = \frac{s^{-\alpha}}{2}\sum_{k=0}^{\infty}(2k+\alpha)\left(\frac{\lambda}{s}\right)^{2k}. \tag{6.246}$$

Proof With the aid of the integral representations

$$L\{\mathbb{V}_{\alpha}(i\lambda,t)\} = \frac{s^{-\alpha}}{2}\sum_{k=0}^{\infty}(k+\alpha)\left(\frac{i\lambda}{s}\right)^{k}, \tag{6.247}$$

$$L\{\mathbb{V}_{\alpha}(-i\lambda,t)\} = \frac{s^{-\alpha}}{2}\sum_{k=0}^{\infty}(k+\alpha)\left(-\frac{i\lambda}{s}\right)^{k}, \tag{6.248}$$

$$L\{\mathbb{V}_{\alpha}(\lambda,t)\} = \frac{s^{-\alpha}}{2}\sum_{k=0}^{\infty}(k+\alpha)\left(\frac{\lambda}{s}\right)^{k} \tag{6.249}$$

and

$$L\{\mathbb{V}_{\alpha}(-\lambda,t)\} = \frac{s^{-\alpha}}{2}\sum_{k=0}^{\infty}(k+\alpha)\left(-\frac{\lambda}{s}\right)^{k}, \tag{6.250}$$

we have that

$$\begin{aligned} L\{\mathbb{V}sin_{\alpha}(\lambda,\tau)\} &= \tfrac{s^{-\alpha}}{2}\left(\sum_{k=0}^{\infty}(k+\alpha)\left(\tfrac{i\lambda}{s}\right)^{k} - \sum_{k=0}^{\infty}(k+\alpha)\left(-\tfrac{i\lambda}{s}\right)^{k}\right) \\ &= \tfrac{s^{-\alpha}}{2}\sum_{k=0}^{\infty}(k+\alpha)\left[\left(\tfrac{i\lambda}{s}\right)^{k} - \left(-\tfrac{i\lambda}{s}\right)^{k}\right] \\ &= -i\tfrac{s^{-\alpha}}{2}\sum_{k=0}^{\infty}(2k+1+\alpha)\left(\tfrac{\lambda}{s}\right)^{2k+1}, \end{aligned} \tag{6.251}$$

$$\begin{aligned}L\{\mathbb{V}cos_{\alpha}(\lambda,\tau)\} &= \tfrac{s^{-\alpha}}{2}\left(\sum_{k=0}^{\infty}(k+\alpha)\left(\tfrac{i\lambda}{s}\right)^{k}+\sum_{k=0}^{\infty}(k+\alpha)\left(-\tfrac{i\lambda}{s}\right)^{k}\right)\\ &= -\tfrac{s^{-\alpha}}{2}\sum_{k=0}^{\infty}(k+\alpha)\left[\left(\tfrac{\lambda}{s}\right)^{k}+\left(-\tfrac{\lambda}{s}\right)^{k}\right]\\ &= -\tfrac{s^{-\alpha}}{2}\sum_{k=0}^{\infty}(2k+\alpha)\left(\tfrac{\lambda}{s}\right)^{2k},\end{aligned} \tag{6.252}$$

$$\begin{aligned}L\{\mathbb{V}sinh_{\alpha}(\lambda,\tau)\} &= \tfrac{s^{-\alpha}}{2}\left(\sum_{k=0}^{\infty}(k+\alpha)\left(\tfrac{\lambda}{s}\right)^{k}-\sum_{k=0}^{\infty}(k+\alpha)\left(-\tfrac{\lambda}{s}\right)^{k}\right)\\ &= \tfrac{s^{-\alpha}}{2}\sum_{k=0}^{\infty}(k+\alpha)\left[\left(\tfrac{\lambda}{s}\right)^{k}-\left(-\tfrac{\lambda}{s}\right)^{k}\right]\\ &= \tfrac{s^{-\alpha}}{2}\sum_{k=0}^{\infty}(2k+1+\alpha)\left(\tfrac{\lambda}{s}\right)^{2k+1},\end{aligned} \tag{6.253}$$

and

$$\begin{aligned}L\{\mathbb{V}cosh_{\alpha}(\lambda,\tau)\} &= \tfrac{s^{-\alpha}}{2}\left(\sum_{k=0}^{\infty}(k+\alpha)\left(\tfrac{\lambda}{s}\right)^{k}+\sum_{k=0}^{\infty}(k+\alpha)\left(-\tfrac{\lambda}{s}\right)^{k}\right)\\ &= \tfrac{s^{-\alpha}}{2}\sum_{k=0}^{\infty}(k+\alpha)\left[\left(\tfrac{\lambda}{s}\right)^{k}+\left(-\tfrac{\lambda}{s}\right)^{k}\right]\\ &= \tfrac{s^{-\alpha}}{2}\sum_{k=0}^{\infty}(2k+\alpha)\left(\tfrac{\lambda}{s}\right)^{2k}.\end{aligned} \tag{6.254}$$

□

6.1.12 The Integral Representations for the Sonine Function of Second Type

In this part, we consider the integral representations for the Sonine function of second type.

Property 6.20 Let $|\lambda/s| < 1$ and $p, \lambda \in \mathbb{C}$. Then we have

$$L\left\{Tsin_{p}^{\xi}(\lambda,t)\right\} = \frac{s^{p-1}}{\Gamma(\xi+1)}{}_{1}sin_{0}\left(\begin{matrix}(\xi,1)\\ -\end{matrix};\frac{\lambda}{s}\right), \tag{6.255}$$

$$L\left\{Tcos_{p}^{\xi}(\lambda,t)\right\} = \frac{s^{p-1}}{\Gamma(\xi+1)}{}_{1}cos_{0}\left(\begin{matrix}(\xi,1)\\ -\end{matrix};\frac{\lambda}{s}\right), \tag{6.256}$$

$$L\left\{Tsinh_{p}^{\xi}(\lambda,t)\right\} = \frac{s^{p-1}}{\Gamma(\xi+1)}{}_{1}sinh_{0}\left(\begin{matrix}(\xi,1)\\ -\end{matrix};\frac{\lambda}{s}\right), \tag{6.257}$$

and

$$L\left\{Tcosh_p^{\xi}(\lambda,t)\right\} = \frac{s^{p-1}}{\Gamma(\xi+1)}\,{}_1cosh_0\left(\begin{matrix}(\xi,1)\\ -\end{matrix};\frac{\lambda}{s}\right). \tag{6.258}$$

Proof Making use of the integral representations

$$L\left\{T_p^{\xi}(i\lambda,t)\right\} = \frac{s^{p-1}}{\Gamma(\xi+1)}\,{}_1\varphi_0\left(\begin{matrix}(\xi,1)\\ -\end{matrix};i\tfrac{\lambda}{s}\right)\ (|\lambda/s|<1), \tag{6.259}$$

$$L\left\{T_p^{\xi}(-i\lambda,t)\right\} = \frac{s^{p-1}}{\Gamma(\xi+1)}\,{}_1\varphi_0\left(\begin{matrix}(\xi,1)\\ -\end{matrix};-i\tfrac{\lambda}{s}\right)\ (|\lambda/s|<1), \tag{6.260}$$

$$L\left\{T_p^{\xi}(\lambda,t)\right\} = \frac{s^{p-1}}{\Gamma(\xi+1)}\,{}_1\varphi_0\left(\begin{matrix}(\xi,1)\\ -\end{matrix};\tfrac{\lambda}{s}\right)\ (|\lambda/s|<1), \tag{6.261}$$

and

$$L\left\{T_p^{\xi}(-\lambda,t)\right\} = \frac{s^{p-1}}{\Gamma(\xi+1)}\,{}_1\varphi_0\left(\begin{matrix}(\xi,1)\\ -\end{matrix};-\tfrac{\lambda}{s}\right)\ (|\lambda/s|<1), \tag{6.262}$$

we have

$$\begin{aligned} L\left\{Tsin_p^{\xi}(\lambda,t)\right\} &= \frac{s^{p-1}}{\Gamma(\xi+1)}\times\frac{1}{2i}\left[{}_1\varphi_0\left(\begin{matrix}(\xi,1)\\ -\end{matrix};i\tfrac{\lambda}{s}\right) - {}_1\varphi_0\left(\begin{matrix}(\xi,1)\\ -\end{matrix};-i\tfrac{\lambda}{s}\right)\right] \\ &= \frac{s^{p-1}}{\Gamma(\xi+1)}\,{}_1sin_0\left(\begin{matrix}(\xi,1)\\ -\end{matrix};\frac{\lambda}{s}\right), \end{aligned} \tag{6.263}$$

$$\begin{aligned} L\left\{Tcos_p^{\xi}(\lambda,t)\right\} &= \frac{s^{p-1}}{\Gamma(\xi+1)}\times\frac{1}{2}\left[{}_1\varphi_0\left(\begin{matrix}(\xi,1)\\ -\end{matrix};i\tfrac{\lambda}{s}\right) + {}_1\varphi_0\left(\begin{matrix}(\xi,1)\\ -\end{matrix};-i\tfrac{\lambda}{s}\right)\right] \\ &= \frac{s^{p-1}}{\Gamma(\xi+1)}\,{}_1cos_0\left(\begin{matrix}(\xi,1)\\ -\end{matrix};\frac{\lambda}{s}\right), \end{aligned} \tag{6.264}$$

$$\begin{aligned} L\left\{Tsinh_p^{\xi}(\lambda,t)\right\} &= \frac{s^{p-1}}{\Gamma(\xi+1)}\times\frac{1}{2}\left[{}_1\varphi_0\left(\begin{matrix}(\xi,1)\\ -\end{matrix};\tfrac{\lambda}{s}\right) - {}_1\varphi_0\left(\begin{matrix}(\xi,1)\\ -\end{matrix};-\tfrac{\lambda}{s}\right)\right] \\ &= \frac{s^{p-1}}{\Gamma(\xi+1)}\,{}_1sinh_0\left(\begin{matrix}(\xi,1)\\ -\end{matrix};\frac{\lambda}{s}\right), \end{aligned} \tag{6.265}$$

and

$$L\left\{Tcosh_p^{\xi}(\lambda,t)\right\} = \frac{s^{p-1}}{\Gamma(\xi+1)} \times \frac{1}{2}\left[{}_1\varphi_0\left(\begin{matrix}(\xi,1)\\ -\end{matrix};\frac{\lambda}{s}\right) + {}_1\varphi_0\left(\begin{matrix}(\xi,1)\\ -\end{matrix};-\frac{\lambda}{s}\right)\right]$$
$$= \frac{s^{p-1}}{\Gamma(\xi+1)}\,{}_1cosh_0\left(\begin{matrix}(\xi,1)\\ -\end{matrix};\frac{\lambda}{s}\right). \tag{6.266}$$

□

Thus, we have the following properties for the new special functions:

Property 6.21 Let $\alpha, \xi \in \mathbb{C}$, $Re(\alpha) > 0$, $\lambda \in \mathbb{C}$, and $p \in \mathbb{C}$. Then we have

$$\frac{1}{\Gamma(\alpha)}\int_0^t (t-\tau)^{\alpha}\, Tsin_p^{\xi}(\lambda,\tau)\,d\tau = \frac{t^{-p+\alpha}}{\Gamma(\xi+1)}\,{}_1sin_1\left(\begin{matrix}(\xi,1)\\ (-p+1+\alpha,1)\end{matrix};\lambda t\right)$$
$$= Tsin_{\alpha-p}^{\xi}(\lambda,t), \tag{6.267}$$

$$\frac{1}{\Gamma(\alpha)}\int_0^t (t-\tau)^{\alpha}\, Tcos_p^{\xi}(\lambda,\tau)\,d\tau = \frac{t^{-p+\alpha}}{\Gamma(\xi+1)}\,{}_1cos_1\left(\begin{matrix}(\xi,1)\\ (-p+1+\alpha,1)\end{matrix};\lambda t\right)$$
$$= Tcos_{\alpha-p}^{\xi}(\lambda,t), \tag{6.268}$$

$$\frac{1}{\Gamma(\alpha)}\int_0^t (t-\tau)^{\alpha}\, Tsinh_p^{\xi}(\lambda,\tau)\,d\tau = \frac{t^{-p+\alpha}}{\Gamma(\xi+1)}\,{}_1sinh_1\left(\begin{matrix}(\xi,1)\\ (-p+1+\alpha,1)\end{matrix};\lambda t\right)$$
$$= Tsinh_{\alpha-p}^{\xi}(\lambda,t), \tag{6.269}$$

and

$$\frac{1}{\Gamma(\alpha)}\int_0^t (t-\tau)^{\alpha}\, Tcosh_p^{\xi}(\lambda,\tau)\,d\tau = \frac{t^{-p+\alpha}}{\Gamma(\xi+1)}\,{}_1cosh_1\left(\begin{matrix}(\xi,1)\\ (-p+1+\alpha,1)\end{matrix};\lambda t\right)$$
$$= Tcosh_{\alpha-p}^{\xi}(\lambda,t). \tag{6.270}$$

Proof Making use of (6.143), we have that

$$\frac{1}{\Gamma(\alpha)}\int_0^t (t-\tau)^{\alpha}\, T_p^{\xi}(i\lambda,\tau)\,d\tau = \frac{t^{-p+\alpha}}{\Gamma(\xi+1)}\,{}_1\varphi_1\left(\begin{matrix}(\xi,1)\\ (-p+1+\alpha,1)\end{matrix};i\lambda t\right)$$
$$= T_{\alpha-p}^{\xi}(i\lambda,t), \tag{6.271}$$

$$\frac{1}{\Gamma(\alpha)}\int_0^t (t-\tau)^{\alpha}\, T_p^{\xi}(-i\lambda,\tau)\,d\tau = \frac{t^{-p+\alpha}}{\Gamma(\xi+1)}\,{}_1\varphi_1\left(\begin{matrix}(\xi,1)\\ (-p+1+\alpha,1)\end{matrix};-i\lambda t\right)$$
$$= T_{\alpha-p}^{\xi}(-i\lambda,t), \tag{6.272}$$

$$\frac{1}{\Gamma(\alpha)}\int_0^t (t-\tau)^{\alpha} T_p^{\xi}(\lambda,\tau)\,d\tau = \frac{t^{-p+\alpha}}{\Gamma(\xi+1)}\,{}_1\varphi_1\left(\begin{matrix}(\xi,1)\\(-p+1+\alpha,1)\end{matrix};\lambda t\right)$$
$$= T_{\alpha-p}^{\xi}(\lambda,t), \tag{6.273}$$

and

$$\frac{1}{\Gamma(\alpha)}\int_0^t (t-\tau)^{\alpha} T_p^{\xi}(-\lambda,\tau)\,d\tau = \frac{t^{-p+\alpha}}{\Gamma(\xi+1)}\,{}_1\varphi_1\left(\begin{matrix}(\xi,1)\\(-p+1+\alpha,1)\end{matrix};-\lambda t\right)$$
$$= T_{\alpha-p}^{\xi}(-\lambda,t), \tag{6.274}$$

which reduce to the desired results.

□

Property 6.22 Let $\alpha, \xi \in \mathbb{C}$, $Re(\alpha) > 0$, $\lambda \in \mathbb{C}$, and $p \in \mathbb{C}$. Then we have

$$\frac{1}{\Gamma(\alpha)}\int_0^1 (1-\tau)^{\alpha}\, Tsin_p^{\xi}(\lambda,\tau)\,d\tau = \frac{1}{\Gamma(\xi+1)}\,{}_1sin_1\left(\begin{matrix}(\xi,1)\\(-p+1+\alpha,1)\end{matrix};\lambda\right)$$
$$= Tsin_{\alpha-p}^{\xi}(\lambda,1), \tag{6.275}$$

$$\frac{1}{\Gamma(\alpha)}\int_0^1 (1-\tau)^{\alpha}\, Tcos_p^{\xi}(\lambda,\tau)\,d\tau = \frac{1}{\Gamma(\xi+1)}\,{}_1cos_1\left(\begin{matrix}(\xi,1)\\(-p+1+\alpha,1)\end{matrix};\lambda\right)$$
$$= Tcos_{\alpha-p}^{\xi}(\lambda,1), \tag{6.276}$$

$$\frac{1}{\Gamma(\alpha)}\int_0^1 (1-\tau)^{\alpha}\, Tsinh_p^{\xi}(\lambda,\tau)\,d\tau = \frac{1}{\Gamma(\xi+1)}\,{}_1sinh_1\left(\begin{matrix}(\xi,1)\\(-p+1+\alpha,1)\end{matrix};\lambda\right)$$
$$= Tsinh_{\alpha-p}^{\xi}(\lambda,1), \tag{6.277}$$

and

$$\frac{1}{\Gamma(\alpha)}\int_0^1 (1-\tau)^{\alpha}\, Tcosh_p^{\xi}(\lambda,\tau)\,d\tau = \frac{1}{\Gamma(\xi+1)}\,{}_1cosh_1\left(\begin{matrix}(\xi,1)\\(-p+1+\alpha,1)\end{matrix};\lambda\right)$$
$$= Tcosh_{\alpha-p}^{\xi}(\lambda,1). \tag{6.278}$$

Proof Putting $t = 1$ in (6.267), (6.268), (6.269), and (6.270), we have the results.

□

Property 6.23 Let $\alpha, \xi \in \mathbb{C}$, $Re(\alpha) > 0$, $\lambda \in \mathbb{C}$, and $p \in \mathbb{C}$. Then we have

$$\begin{aligned}\frac{1}{\Gamma(\alpha)}\int_0^2 (2-\tau)^\alpha\, Tsin_p^\xi(\lambda,\tau)\,d\tau &= \frac{1}{\Gamma(\alpha)}\int_{-1}^1 (1-\tau)^\alpha\, Tsin_p^\xi(\lambda,1+\tau)\,d\tau \\ &= \frac{1}{\Gamma(\alpha)}\int_{-1}^1 (1+\tau)^\alpha\, Tsin_p^\xi(\lambda,1-\tau)\,d\tau \\ &= \frac{2^{-p+\alpha}}{\Gamma(\xi+1)}\,{}_1sin_1\left(\begin{matrix}(\xi,1)\\(-p+1+\alpha,1)\end{matrix};2\lambda\right) \\ &= Tsin_{\alpha-p}^\xi(\lambda,2),\end{aligned} \tag{6.279}$$

$$\begin{aligned}\frac{1}{\Gamma(\alpha)}\int_0^2 (2-\tau)^\alpha\, Tcos_p^\xi(\lambda,\tau)\,d\tau &= \frac{1}{\Gamma(\alpha)}\int_{-1}^1 (1-\tau)^\alpha\, Tcos_p^\xi(\lambda,1+\tau)\,d\tau \\ &= \frac{1}{\Gamma(\alpha)}\int_{-1}^1 (1+\tau)^\alpha\, Tcos_p^\xi(\lambda,1-\tau)\,d\tau \\ &= \frac{2^{-p+\alpha}}{\Gamma(\xi+1)}\,{}_1cos_1\left(\begin{matrix}(\xi,1)\\(-p+1+\alpha,1)\end{matrix};2\lambda\right) \\ &= Tcos_{\alpha-p}^\xi(\lambda,2),\end{aligned} \tag{6.280}$$

$$\begin{aligned}\frac{1}{\Gamma(\alpha)}\int_0^t (t-\tau)^\alpha\, Tsinh_p^\xi(\lambda,\tau)\,d\tau &= \frac{1}{\Gamma(\alpha)}\int_{-1}^1 (1-\tau)^\alpha\, Tsinh_p^\xi(\lambda,1+\tau)\,d\tau \\ &= \frac{1}{\Gamma(\alpha)}\int_{-1}^1 (1+\tau)^\alpha\, Tsinh_p^\xi(\lambda,1-\tau)\,d\tau \\ &= \frac{2^{-p+\alpha}}{\Gamma(\xi+1)}\,{}_1sinh_1\left(\begin{matrix}(\xi,1)\\(-p+1+\alpha,1)\end{matrix};2\lambda\right) \\ &= Tsinh_{\alpha-p}^\xi(\lambda,2)\end{aligned} \tag{6.281}$$

and

$$\begin{aligned}\frac{1}{\Gamma(\alpha)}\int_0^2 (2-\tau)^\alpha\, Tcosh_p^\xi(\lambda,\tau)\,d\tau &= \frac{1}{\Gamma(\alpha)}\int_{-1}^1 (1-\tau)^\alpha\, Tcosh_p^\xi(\lambda,1+\tau)\,d\tau \\ &= \frac{1}{\Gamma(\alpha)}\int_{-1}^1 (1+\tau)^\alpha\, Tcosh_p^\xi(\lambda,1-\tau)\,d\tau \\ &= \frac{2^{-p+\alpha}}{\Gamma(\xi+1)}\,{}_1cosh_1\left(\begin{matrix}(\xi,1)\\(-p+1+\alpha,1)\end{matrix};2\lambda\right) \\ &= Tcosh_{\alpha-p}^\xi(\lambda,2).\end{aligned} \tag{6.282}$$

Proof Setting $t = 2$ in (6.267), (6.268), (6.269), and (6.270), we have the results. □

Property 6.24 Let $\alpha, \xi \in \mathbb{C}$, $Re(\alpha) > 0$, $\lambda \in \mathbb{C}$, and $p \in \mathbb{C}$. Then we have

$$\int_0^t (t-\tau)\, Tsin_p^{\xi}(\lambda, \tau)\, d\tau = \frac{t^{-p+1}}{\Gamma(\xi+1)} {}_1sin_1\left(\begin{matrix}(\xi, 1)\\(-p+2, 1)\end{matrix}; \lambda t\right) = Tsin_{1-p}^{\xi}(\lambda, t), \tag{6.283}$$

$$\int_0^t (t-\tau)\, Tcos_p^{\xi}(\lambda, \tau)\, d\tau = \frac{t^{-p+1}}{\Gamma(\xi+1)} {}_1cos_1\left(\begin{matrix}(\xi, 1)\\(-p+2, 1)\end{matrix}; \lambda t\right) = Tcos_{1-p}^{\xi}(\lambda, t), \tag{6.284}$$

$$\int_0^t (t-\tau)\, Tsinh_p^{\xi}(\lambda, \tau)\, d\tau = \frac{t^{-p+1}}{\Gamma(\xi+1)} {}_1sinh_1\left(\begin{matrix}(\xi, 1)\\(-p+2, 1)\end{matrix}; \lambda t\right) = Tsinh_{1-p}^{\xi}(\lambda, t) \tag{6.285}$$

and

$$\int_0^t (t-\tau)\, Tcosh_p^{\xi}(\lambda, \tau)\, d\tau = \frac{t^{-p+1}}{\Gamma(\xi+1)} {}_1cosh_1\left(\begin{matrix}(\xi, 1)\\(-p+2, 1)\end{matrix}; \lambda t\right) = Tcosh_{1-p}^{\xi}(\lambda, t). \tag{6.286}$$

Proof Setting $\alpha = 1$ in (6.267), (6.268), (6.269), and (6.270), we have the results. □

6.2 The Rabotnov Fractional Exponential Function

In 1948, Yuri Nikolayevich Rabotnov, a Russian academician and Russian mechanical engineer, who worked in the theory of shells, plasticity, creep, and failure of materials, proposed the special function [270], which is called the Rabotnov function or Rabotnov fractional exponential function [92] applied in viscoelasticity.

6.2.1 The Rabotnov Fractional Exponential Function: History and Properties

Definition 6.40 The Rabotnov function is defined as [270]

$$\begin{aligned}\Re_\alpha\left(t^\alpha\right) &= t^\alpha \sum_{k=0}^{\infty} \frac{t^{k(\alpha+1)}}{\Gamma\left((k+1)(\alpha+1)\right)} \\ &= \sum_{k=0}^{\infty} \frac{t^{k(\alpha+1)+\alpha}}{\Gamma\left((k+1)(\alpha+1)\right)},\end{aligned} \tag{6.287}$$

where $t \in \mathbb{C}$.

The Rabotnov function can be written as follows [92, 270]:

$$\Re_\alpha\left(\lambda, t\right) = t^\alpha \sum_{k=0}^{\infty} \frac{\lambda^k t^{k(\alpha+1)}}{\Gamma\left((k+1)(\alpha+1)\right)}, \tag{6.288}$$

where $\lambda, t \in \mathbb{C}$.

The relation between the Rabotnov function and Mittag-Leffler function can be written as follows [238]:

$$\Re_\alpha\left(\lambda, t\right) = t^\alpha \sum_{k=0}^{\infty} \frac{\lambda^k t^{k(\alpha+1)}}{\Gamma\left((k+1)(\alpha+1)\right)} = t^\alpha E_{\alpha+1,\alpha+1}\left(\lambda t^{\alpha+1}\right), \tag{6.289}$$

where $\alpha, \lambda, t \in \mathbb{C}$.

The relation between the Rabotnov function and the generalized Wright function can be given as follows:

$$\begin{aligned}\Re_\alpha\left(\lambda, t\right) &= t^\alpha \sum_{k=0}^{\infty} \frac{\lambda^k t^{k(\alpha+1)}}{\Gamma\left((k+1)(\alpha+1)\right)} \\ &= t^\alpha E_{\alpha+1,\alpha+1}\left(\lambda t^{\alpha+1}\right) \\ &= t^\alpha\, {}_1\varphi_1\left(\begin{matrix}(1,1) \\ (\alpha+1,\alpha+1)\end{matrix}; \lambda t^{\alpha+1}\right),\end{aligned} \tag{6.290}$$

where $\alpha, \lambda, t \in \mathbb{C}$.

In this case, we present the special case of the Rabotnov function as follows:

$$\begin{aligned} R_\alpha(\lambda, t) &= t^\alpha \sum_{k=0}^{\infty} \frac{\lambda^k t^k}{\Gamma((k+1)(\alpha+1))} \\ &= t^\alpha E_{\alpha+1,\alpha+1}(\lambda t) \\ &= t^\alpha\, {}_1\varphi_1\left(\begin{matrix} (1,1) \\ (\alpha+1,\alpha+1) \end{matrix} ; \lambda t \right). \end{aligned} \tag{6.291}$$

The Laplace transform for the Rabotnov function is given as [92]

$$\begin{aligned} L\{\Re_\alpha(\lambda, t)\} &= \int_0^\infty e^{-st} \Re_\alpha(\lambda, t)\, dt \\ &= L\left\{ t^\alpha \sum_{k=0}^{\infty} \frac{\lambda^k t^{k(\alpha+1)}}{\Gamma((k+1)(\alpha+1))} \right\} \\ &= \sum_{k=0}^{\infty} \frac{\lambda^k}{s^{(k+1)(\alpha+1)}} \\ &= \frac{1}{s^{\alpha+1}} \frac{1}{1-\lambda s^{-(\alpha+1)}} \left(\left|\lambda / s^{(\alpha+1)}\right| < 1 \right) \\ &= s^{-(\alpha+1)} \left(1 - \lambda s^{-(\alpha+1)}\right)^{-1} \left(\left|\lambda / s^{(\alpha+1)}\right| < 1 \right), \end{aligned} \tag{6.292}$$

where $\alpha \in \mathbb{C}$, $\left|\lambda s^{-(\alpha+1)}\right| < 1$ and $Re(\alpha) > 0$.

Theorem 6.18 *Let $\alpha \in \mathbb{C}$, $\left|s^{-(\alpha+1)}\right| < 1$ and $Re(\alpha) > 0$.*
Then the Laplace transform of the Rabotnov function is

$$L\{\Re_\alpha(1, t)\} = \frac{1}{s^{\alpha+1}} \frac{1}{1-s^{-(\alpha+1)}} = s^{-(\alpha+1)} \left(1 - s^{-(\alpha+1)}\right)^{-1} \left(\left|s^{-(\alpha+1)}\right| < 1 \right). \tag{6.293}$$

The Mellin transform for the Rabotnov function is given as

$$M[R_\alpha(-\lambda, t)] = \int_0^\infty t^{\varpi+\alpha-1} E_{\alpha+1,\alpha+1}(-\lambda t)\, dt = \frac{\Gamma(\varpi+\alpha)\Gamma(1-\varpi-\alpha)}{\Gamma((\alpha+1)(1-\varpi-\alpha))} \lambda^{-(\varpi+\alpha)}, \tag{6.294}$$

where $\alpha \in \mathbb{C}$ and $Re(\alpha) > 0$.

Property 6.25 Let $\alpha \in \mathbb{C}$ and $Re(\alpha) > 0$. Then we have

$$M[R_\alpha(-1,t)] = \int_0^\infty t^{\varpi+\alpha-1} E_{\alpha+1,\alpha+1}(-t)\,dt = \frac{\Gamma(\varpi+\alpha)\Gamma(1-\varpi-\alpha)}{\Gamma((\alpha+1)(1-\varpi-\alpha))}. \tag{6.295}$$

Proof Making use of (6.294), we obtain the result. □

Property 6.26 Let $\alpha \in \mathbb{C}$ and $Re(\alpha) > 0$. Then we have

$$\int_{-\infty}^{\infty} |t|^{\varpi+\alpha-1} E_{\alpha+1,\alpha+1}(-\lambda|t|)\,dt = 2\lambda^{-(\varpi+\alpha)}\frac{\Gamma(\varpi+\alpha)\Gamma(1-\varpi-\alpha)}{\Gamma((\alpha+1)(1-\varpi-\alpha))}. \tag{6.296}$$

Proof With the use of (6.294), we give the result. □

Property 6.27 Let $\alpha > 0$. Then we have

$$t^\alpha E_{\alpha+1,\alpha+1}(t) = \frac{1}{2\pi i}\int_{-i\infty}^{i\infty} \frac{\Gamma(\varpi+\alpha)\Gamma(1-\varpi-\alpha)}{\Gamma((\alpha+1)(1-\varpi-\alpha))}(-t)^{\varpi-1}\,d\varpi, \tag{6.297}$$

where $|\arg(t)| < \pi$.

Proof With the use of (6.294), we have the result. □

Let us recall the special case:
Let $\alpha, \varsigma \in \mathbb{C}$, $Re(\alpha) > 0$, and $\lambda \in \mathbb{C}$. Then we have

$$\begin{aligned}&\frac{1}{\Gamma(\varsigma)}\int_0^t (t-\tau)^\varsigma t^\alpha {}_1\varphi_1\left(\begin{matrix}(1,1)\\(\alpha+1,\alpha+1)\end{matrix};\lambda\tau^{\alpha+1}\right)d\tau\\&= t^{\alpha+\varsigma} {}_1\varphi_1\left(\begin{matrix}(1,1)\\(\alpha+1+\varsigma,\alpha+1)\end{matrix};\lambda t^{\alpha+1}\right).\end{aligned} \tag{6.298}$$

Thus, we have the alternative form as follows:
Let $\alpha, \varsigma \in \mathbb{C}$, $Re(\alpha) > 0$, and $\lambda \in \mathbb{C}$. Then we have

$$\frac{1}{\Gamma(\varsigma)}\int_0^t (t-\tau)^\varsigma \Re_\alpha(\lambda,\tau)\,d\tau = t^{\alpha+\varsigma} {}_1\varphi_1\left(\begin{matrix}(1,1)\\(\alpha+1+\varsigma,\alpha+1)\end{matrix};\lambda t^{\alpha+1}\right). \tag{6.299}$$

Property 6.28 Let $\alpha, \varsigma \in \mathbb{C}$, $Re(\alpha) > 0$, and $\lambda \in \mathbb{C}$. Then we have

$$\frac{1}{\Gamma(\varsigma)}\int_0^1 (1-\tau)^\varsigma \Re_\alpha(\lambda,\tau)\,d\tau = {}_1\varphi_1\left(\begin{matrix}(1,1)\\(\alpha+1+\varsigma,\alpha+1)\end{matrix};\lambda\right). \tag{6.300}$$

Proof Putting $t = 1$ in (6.299), we obtain the result. □

Property 6.29 Let $\alpha, \varsigma \in \mathbb{C}$, $Re(\alpha) > 0$, and $\lambda \in \mathbb{C}$. Then we have

$$\begin{aligned}\frac{1}{\Gamma(\varsigma)} \int_0^2 (2-\tau)^{\varsigma} \, \mathfrak{R}_\alpha(\lambda, \tau) \, d\tau &= \frac{1}{\Gamma(\varsigma)} \int_{-1}^1 (1-\tau)^{\varsigma} \, \mathfrak{R}_\alpha(\lambda, 1+\tau) \, d\tau \\ &= \frac{1}{\Gamma(\varsigma)} \int_{-1}^1 (1+\tau)^{\varsigma} \, \mathfrak{R}_\alpha(\lambda, 1-\tau) \, d\tau \\ &= 2^{\alpha+\varsigma} \, {}_1\varphi_1\left(\begin{matrix} (1,1) \\ (\alpha+1+\varsigma, \alpha+1) \end{matrix} ; \lambda 2^{\alpha+1} \right). \end{aligned} \tag{6.301}$$

Proof Putting $t = 2$ in (6.299), we obtain the result. □

6.2.2 *The Supertrigonometric Functions via Rabotnov Function*

In this part, we propose the supertrigonometric functions via Rabotnov function.

We now consider that [92]

$$\begin{aligned}\mathfrak{R}_\alpha(i\lambda, t) &= t^\alpha \sum_{k=0}^{\infty} \frac{(i\lambda)^k \, t^{k(\alpha+1)}}{\Gamma((k+1)(\alpha+1))} \\ &= t^\alpha E_{\alpha+1,\alpha+1}\left(i\lambda t^{\alpha+1}\right) \\ &= t^\alpha \, {}_1\varphi_1\left(\begin{matrix} (1,1) \\ (\alpha+1, \alpha+1) \end{matrix} ; i\lambda t^{\alpha+1} \right) \end{aligned} \tag{6.302}$$

and

$$\begin{aligned}\mathfrak{R}_\alpha(i\lambda, t) &= t^\alpha \sum_{k=0}^{\infty} \frac{(i\lambda)^k \, t^{k(\alpha+1)}}{\Gamma((k+1)(\alpha+1))} \\ &= t^\alpha E_{\alpha+1,\alpha+1}\left(i\lambda t^{\alpha+1}\right) \\ &= t^\alpha \, {}_1\varphi_1\left(\begin{matrix} (1,1) \\ (\alpha+1, \alpha+1) \end{matrix} ; i\lambda t^{\alpha+1} \right). \end{aligned} \tag{6.303}$$

Making use of (6.302) and (6.303), we propose the following special functions:

Definition 6.41 The supersine via Rabotnov function is defined as [92]

$$\begin{aligned}\Re sin_{\alpha}(t) &= \frac{1}{2i}\left(\Re_{\alpha}(i,t) - \Re_{\alpha}(-i,t)\right) \\ &= t^{\alpha}\sum_{k=0}^{\infty}\frac{(-1)^{k}\, t^{(2k+1)(\alpha+1)}}{\Gamma\left((2k+2)(\alpha+1)\right)},\end{aligned} \tag{6.304}$$

where $\alpha, t \in \mathbb{C}$.

Definition 6.42 The supercosine via Rabotnov function is defined as [92]

$$\begin{aligned}\Re cos_{\alpha}(t) &= \frac{1}{2}\left(\Re_{\alpha}(i,t) + \Re_{\alpha}(-i,t)\right) \\ &= t^{\alpha}\sum_{k=0}^{\infty}\frac{(-1)^{k}\, t^{2k(\alpha+1)}}{\Gamma\left((2k+1)(\alpha+1)\right)},\end{aligned} \tag{6.305}$$

where $\alpha, t \in \mathbb{C}$.

Note that (6.304) and (6.305) were discovered by Yang in 2019 [92].

Definition 6.43 The supertangent via Rabotnov function is defined as

$$\Re tan_{\alpha}(t) = \frac{\Re sin_{\alpha}(t)}{\Re cos_{\alpha}(t)}\left(\Re cos_{\alpha}(t) \neq 0\right), \tag{6.306}$$

where $\alpha, t \in \mathbb{C}$.

Definition 6.44 The supercotangent via Rabotnov function is defined as

$$\Re cotan_{\alpha}(t) = \frac{\Re cos_{\alpha}(t)}{\Re sin_{\alpha}(t)}, \left(\Re sin_{\alpha}(t) \neq 0\right), \tag{6.307}$$

where $\alpha, t \in \mathbb{C}$.

Definition 6.45 The supercosecant via Rabotnov function is defined as

$$\Re csc_{\alpha}(t) = \frac{1}{\Re sin_{\alpha}(t)}, \left(\Re sin_{\alpha}(t) \neq 0\right), \tag{6.308}$$

where $\alpha, t \in \mathbb{C}$.

Definition 6.46 The supersecant via Rabotnov function is defined as

$$\Re sec_{\alpha}(t) = \frac{1}{\Re cos_{\alpha}(t)}\left(\Re cos_{\alpha}(t) \neq 0\right), \tag{6.309}$$

where $\alpha, t \in \mathbb{C}$.

Thus,

$$\Re sin_{\alpha}(\lambda, t) = \frac{1}{2i}\left(\Re_{\alpha}(i\lambda, t) - \Re_{\alpha}(-i\lambda, t)\right)$$
$$= t^{\alpha}\sum_{k=0}^{\infty}\frac{(-1)^k \lambda^{2k+1} t^{(2k+1)(\alpha+1)}}{\Gamma((2k+2)(\alpha+1))}, \tag{6.310}$$

$$\Re cos_{\alpha}(\lambda, t) = \frac{1}{2}\left(\Re_{\alpha}(i\lambda, t) + \Re_{\alpha}(-i\lambda, t)\right)$$
$$= t^{\alpha}\sum_{k=0}^{\infty}\frac{(-1)^k \lambda^{2k} t^{2k(\alpha+1)}}{\Gamma((2k+1)(\alpha+1))}, \tag{6.311}$$

$$\Re tan_{\alpha}(\lambda, t) = \frac{\Re sin_{\alpha}(\lambda, t)}{\Re cos_{\alpha}(\lambda, t)}\left(\Re cos_{\alpha}(\lambda, t) \neq 0\right), \tag{6.312}$$

$$\Re cotan_{\alpha}(\lambda, t) = \frac{\Re cos_{\alpha}(\lambda, t)}{\Re sin_{\alpha}(\lambda, t)}, \left(\Re sin_{\alpha}(\lambda, t) \neq 0\right), \tag{6.313}$$

$$\Re csc_{\alpha}(\lambda, t) = \frac{1}{\Re sin_{\alpha}(\lambda, t)}, \left(\Re sin_{\alpha}(\lambda, t) \neq 0\right), \tag{6.314}$$

and

$$\Re sec_{\alpha}(\lambda, t) = \frac{1}{\Re cos_{\alpha}(\lambda, t)}\left(\Re cos_{\alpha}(\lambda, t) \neq 0\right), \tag{6.315}$$

where $\alpha, \lambda, t \in \mathbb{C}$.

Theorem 6.19 *Let $\alpha, \lambda, t \in \mathbb{C}$. Then we have*

$$\Re_{\alpha}(i\lambda, t) = \Re cos_{\alpha}(\lambda, t) + i\Re sin_{\alpha}(\lambda, t) \tag{6.316}$$

and

$$\Re_{\alpha}(-i\lambda, t) = \Re cos_{\alpha}(\lambda, t) - i\Re sin_{\alpha}(\lambda, t). \tag{6.317}$$

Proof In view of (6.304) and (6.305), we give the result. □

6.2.3 The Superhyperbolic Functions via Rabotnov Function

In this part, we propose the superhyperbolic functions via Rabotnov function.

In a similar way, we present that [92]

$$\begin{aligned}\mathfrak{R}_{\alpha}(\lambda, t) &= t^{\alpha}\sum_{k=0}^{\infty}\frac{\lambda^{k}t^{k(\alpha+1)}}{\Gamma((k+1)(\alpha+1))}\\&= t^{\alpha}E_{\alpha+1,\alpha+1}\left(\lambda t^{\alpha+1}\right)\\&= t^{\alpha}\,{}_{1}\varphi_{1}\left(\begin{matrix}(1,1)\\(\alpha+1,\alpha+1)\end{matrix};\lambda t^{\alpha+1}\right)\end{aligned}\tag{6.318}$$

and

$$\begin{aligned}\mathfrak{R}_{\alpha}(-\lambda, t) &= t^{\alpha}\sum_{k=0}^{\infty}\frac{(-\lambda)^{k}t^{k(\alpha+1)}}{\Gamma((k+1)(\alpha+1))}\\&= t^{\alpha}E_{\alpha+1,\alpha+1}\left(-\lambda t^{\alpha+1}\right)\\&= t^{\alpha}\,{}_{1}\varphi_{1}\left(\begin{matrix}(1,1)\\(\alpha+1,\alpha+1)\end{matrix};-\lambda t^{\alpha+1}\right).\end{aligned}\tag{6.319}$$

We may give the following special functions:

Definition 6.47 The hyperbolic supersine via Rabotnov function is defined as [92]

$$\begin{aligned}\mathfrak{R}sinh_{\alpha}(t) &= \frac{1}{2}\left(\mathfrak{R}_{\alpha}(1,t)-\mathfrak{R}_{\alpha}(-1,t)\right)\\&= t^{\alpha}\sum_{k=0}^{\infty}\frac{t^{(2k+1)(\alpha+1)}}{\Gamma((2k+2)(\alpha+1))},\end{aligned}\tag{6.320}$$

where $\alpha, t \in \mathbb{C}$.

Definition 6.48 The hyperbolic supercosine via Rabotnov function is defined as [92]

$$\begin{aligned}\mathfrak{R}cosh_{\alpha}(t) &= \frac{1}{2}\left(\mathfrak{R}_{\alpha}(1,t)+\mathfrak{R}_{\alpha}(-1,t)\right)\\&= t^{\alpha}\sum_{k=0}^{\infty}\frac{t^{2k(\alpha+1)}}{\Gamma((2k+1)(\alpha+1))},\end{aligned}\tag{6.321}$$

where $\alpha, t \in \mathbb{C}$.

Note that (6.320) and (6.321) were discovered by Yang in 2019 [92].

Definition 6.49 The hyperbolic supertangent via Rabotnov function is defined as

$$\Re tanh_\alpha(t) = \frac{\Re sinh_\alpha(t)}{\Re cosh_\alpha(t)} \left(\Re cosh_\alpha(t) \neq 0\right), \tag{6.322}$$

where $\alpha, t \in \mathbb{C}$.

Definition 6.50 The hyperbolic supercotangent via Rabotnov function is defined as

$$\Re cotanh_\alpha(t) = \frac{\Re cosh_\alpha(t)}{\Re sinh_\alpha(t)}, \left(\Re sinh_\alpha(t) \neq 0\right), \tag{6.323}$$

where $\alpha, t \in \mathbb{C}$.

Definition 6.51 The hyperbolic supercosecant via Rabotnov function is defined as

$$\Re csch_\alpha(t) = \frac{1}{\Re sinh_\alpha(t)}, \left(\Re sinh_\alpha(t) \neq 0\right), \tag{6.324}$$

where $\alpha, t \in \mathbb{C}$.

Definition 6.52 The hyperbolic supersecant via Rabotnov function is defined as

$$\Re sech_\alpha(t) = \frac{1}{\Re cosh_\alpha(t)} \left(\Re cosh_\alpha(t) \neq 0\right), \tag{6.325}$$

where $\alpha, t \in \mathbb{C}$.

Thus,

$$\begin{aligned}\Re sinh_\alpha(\lambda, t) &= \frac{1}{2}\left(\Re_\alpha(\lambda, t) - \Re_\alpha(-\lambda, t)\right) \\ &= t^\alpha \sum_{k=0}^{\infty} \frac{\lambda^{2k+1} t^{(2k+1)(\alpha+1)}}{\Gamma\left((2k+2)(\alpha+1)\right)},\end{aligned} \tag{6.326}$$

$$\begin{aligned}\Re cosh_\alpha(\lambda, t) &= \frac{1}{2}\left(\Re_\alpha(\lambda, t) + \Re_\alpha(-\lambda, t)\right) \\ &= t^\alpha \sum_{k=0}^{\infty} \frac{\lambda^{2k} t^{2k(\alpha+1)}}{\Gamma\left((2k+1)(\alpha+1)\right)},\end{aligned} \tag{6.327}$$

$$\Re tanh_\alpha(\lambda, t) = \frac{\Re sinh_\alpha(\lambda, t)}{\Re cosh_\alpha(\lambda, t)} \left(\Re cosh_\alpha(\lambda, t) \neq 0\right), \tag{6.328}$$

$$\Re cotanh_{\alpha}(\lambda, t) = \frac{\Re cosh_{\alpha}(\lambda, t)}{\Re sinh_{\alpha}(\lambda, t)}, \left(\Re sinh_{\alpha}(\lambda, t) \neq 0\right), \tag{6.329}$$

$$\Re csch_{\alpha}(\lambda, t) = \frac{1}{\Re sinh_{\alpha}(\lambda, t)}, \left(\Re sinh_{\alpha}(\lambda, t) \neq 0\right), \tag{6.330}$$

and

$$\Re sech_{\alpha}(\lambda, t) = \frac{1}{\Re cosh_{\alpha}(\lambda, t)} \left(\Re cosh_{\alpha}(\lambda, t) \neq 0\right), \tag{6.331}$$

where $\alpha, \lambda, t \in \mathbb{C}$.

Theorem 6.20 *Let $\alpha, \lambda, t \in \mathbb{C}$. Then we have*

$$\Re_{\alpha}(\lambda, t) = \Re cosh_{\alpha}(\lambda, t) + \Re sinh_{\alpha}(\lambda, t) \tag{6.332}$$

and

$$\Re_{\alpha}(-\lambda, t) = \Re cosh_{\alpha}(\lambda, t) - \Re sinh_{\alpha}(\lambda, t). \tag{6.333}$$

Proof With use of (6.320) and (6.321), we give the result. □

6.2.4 The Supertrigonometric Functions via Rabotnov Type Function

In this part, we propose the supertrigonometric functions via Rabotnov type function.

From (6.291), we now consider that

$$\begin{aligned} R_{\alpha}(i\lambda, t) &= t^{\alpha} \sum_{k=0}^{\infty} \frac{(i\lambda)^{k} t^{k}}{\Gamma((k+1)(\alpha+1))} \\ &= t^{\alpha} E_{\alpha+1,\alpha+1}(i\lambda t) \\ &= t^{\alpha} {}_{1}\varphi_{1}\left(\begin{matrix}(1,1) \\ (\alpha+1,\alpha+1)\end{matrix}; i\lambda\right) \end{aligned} \tag{6.334}$$

and

$$\begin{aligned} R_{\alpha}(i\lambda, t) &= t^{\alpha} \sum_{k=0}^{\infty} \frac{(i\lambda)^{k} t^{k}}{\Gamma((k+1)(\alpha+1))} \\ &= t^{\alpha} E_{\alpha+1,\alpha+1}(i\lambda t) \\ &= t^{\alpha} {}_{1}\varphi_{1}\left(\begin{matrix}(1,1) \\ (\alpha+1,\alpha+1)\end{matrix}; i\lambda t\right). \end{aligned} \tag{6.335}$$

Making use of (6.302) and (6.303), we propose the following special functions:

Definition 6.53 The supersine via Rabotnov type function is defined as

$$Rsin_{\alpha}(t) = \frac{1}{2i}\left(R_{\alpha}(i,t) - R_{\alpha}(-i,t)\right) = t^{\alpha}\sum_{k=0}^{\infty}\frac{(-1)^k t^{2k+1}}{\Gamma((2k+2)(\alpha+1))}, \tag{6.336}$$

where $\alpha, t \in \mathbb{C}$.

Definition 6.54 The supercosine via Rabotnov type function is defined as

$$Rcos_{\alpha}(t) = \frac{1}{2}\left(R_{\alpha}(i,t) + R_{\alpha}(-i,t)\right) = t^{\alpha}\sum_{k=0}^{\infty}\frac{t^{2k}}{\Gamma((2k+1)(\alpha+1))}, \tag{6.337}$$

where $\alpha, t \in \mathbb{C}$.

Definition 6.55 The supertangent via Rabotnov type function is defined as

$$Rtan_{\alpha}(t) = \frac{Rsin_{\alpha}(t)}{Rcos_{\alpha}(t)}\left(Rcos_{\alpha}(t) \neq 0\right), \tag{6.338}$$

where $\alpha, t \in \mathbb{C}$.

Definition 6.56 The supercotangent via Rabotnov type function is defined as

$$Rcotan_{\alpha}(t) = \frac{Rcos_{\alpha}(t)}{Rsin_{\alpha}(t)}, \left(Rsin_{\alpha}(t) \neq 0\right), \tag{6.339}$$

where $\alpha, t \in \mathbb{C}$.

Definition 6.57 The supercosecant via Rabotnov type function is defined as

$$Rcsc_{\alpha}(t) = \frac{1}{Rsin_{\alpha}(t)}, \left(Rsin_{\alpha}(t) \neq 0\right), \tag{6.340}$$

where $\alpha, t \in \mathbb{C}$.

Definition 6.58 The supersecant via Rabotnov type function is defined as

$$Rsec_{\alpha}(t) = \frac{1}{Rcos_{\alpha}(t)}\left(Rcos_{\alpha}(t) \neq 0\right), \tag{6.341}$$

where $\alpha, t \in \mathbb{C}$.

Thus,

$$Rsin_\alpha(\lambda,t)=\frac{1}{2i}\left(R_\alpha(i\lambda,t)-R_\alpha(-i\lambda,t)\right)$$
$$=t^\alpha\sum_{k=0}^{\infty}\frac{(-1)^k\lambda^{2k+1}t^{2k+1}}{\Gamma((2k+2)(\alpha+1))},\tag{6.342}$$

$$Rcos_\alpha(\lambda,t)=\frac{1}{2}\left(R_\alpha(i\lambda,t)+R_\alpha(-i\lambda,t)\right)$$
$$=t^\alpha\sum_{k=0}^{\infty}\frac{(-1)^k\lambda^{2k}t^{2k}}{\Gamma((2k+1)(\alpha+1))},\tag{6.343}$$

$$Rtan_\alpha(\lambda,t)=\frac{Rsin_\alpha(\lambda,t)}{Rcos_\alpha(\lambda,t)}\left(Rcos_\alpha(\lambda,t)\neq 0\right),\tag{6.344}$$

$$Rcotan_\alpha(\lambda,t)=\frac{Rcos_\alpha(\lambda,t)}{Rsin_\alpha(\lambda,t)},\left(Rsin_\alpha(\lambda,t)\neq 0\right),\tag{6.345}$$

$$Rcsc_\alpha(\lambda,t)=\frac{1}{Rsin_\alpha(\lambda,t)},\left(Rsin_\alpha(\lambda,t)\neq 0\right),\tag{6.346}$$

and

$$Rsec_\alpha(\lambda,t)=\frac{1}{Rcos_\alpha(\lambda,t)}\left(Rcos_\alpha(\lambda,t)\neq 0\right),\tag{6.347}$$

where $\alpha,\lambda,t\in\mathbb{C}$.

Theorem 6.21 *Let $\alpha,\lambda,t\in\mathbb{C}$. Then we have*

$$R_\alpha(i\lambda,t)=Rcos_\alpha(\lambda,t)+iRsin_\alpha(\lambda,t)\tag{6.348}$$

and

$$R_\alpha(-i\lambda,t)=Rcos_\alpha(\lambda,t)-iRsin_\alpha(\lambda,t).\tag{6.349}$$

Proof With use of (6.36) and (6.337), we give the result. □

6.2.5 The Superhyperbolic Functions via Rabotnov Type Function

In this part, we present the superhyperbolic functions via Rabotnov type function.

In a similar way, we present that

$$R_{\alpha}(\lambda, t) = t^{\alpha} \sum_{k=0}^{\infty} \frac{\lambda^k t^k}{\Gamma((k+1)(\alpha+1))}$$
$$= t^{\alpha} E_{\alpha+1,\alpha+1}(\lambda t)$$
$$= t^{\alpha} {}_1\varphi_1\left(\begin{matrix}(1,1)\\(\alpha+1,\alpha+1)\end{matrix}; \lambda t\right) \tag{6.350}$$

and

$$R_{\alpha}(-\lambda, t) = t^{\alpha} \sum_{k=0}^{\infty} \frac{(-\lambda)^k t^k}{\Gamma((k+1)(\alpha+1))}$$
$$= t^{\alpha} E_{\alpha+1,\alpha+1}(-\lambda t)$$
$$= t^{\alpha} {}_1\varphi_1\left(\begin{matrix}(1,1)\\(\alpha+1,\alpha+1)\end{matrix}; -\lambda t\right). \tag{6.351}$$

We may present the following special functions:

Definition 6.59 The hyperbolic supersine via Rabotnov type function is defined as

$$Rsinh_{\alpha}(t) = \frac{1}{2}\left(R_{\alpha}(1,t) - R_{\alpha}(-1,t)\right)$$
$$= t^{\alpha} \sum_{k=0}^{\infty} \frac{t^{2k+1}}{\Gamma((2k+2)(\alpha+1))}, \tag{6.352}$$

where $\alpha, t \in \mathbb{C}$.

Definition 6.60 The hyperbolic supercosine via Rabotnov type function is defined as

$$Rcosh_{\alpha}(t) = \frac{1}{2}\left(R_{\alpha}(1,t) + R_{\alpha}(-1,t)\right)$$
$$= t^{\alpha} \sum_{k=0}^{\infty} \frac{t^{2k}}{\Gamma((2k+1)(\alpha+1))}, \tag{6.353}$$

where $\alpha, t \in \mathbb{C}$.

Definition 6.61 The hyperbolic supertangent via Rabotnov type function is defined as

$$Rtanh_{\alpha}(t) = \frac{Rsinh_{\alpha}(t)}{Rcosh_{\alpha}(t)}\left(Rcosh_{\alpha}(t) \neq 0\right), \tag{6.354}$$

where $\alpha, t \in \mathbb{C}$.

Definition 6.62 The hyperbolic supercotangent via Rabotnov type function is defined as

$$Rcotanh_{\alpha}(t) = \frac{Rcosh_{\alpha}(t)}{Rsinh_{\alpha}(t)}, (Rsinh_{\alpha}(t) \neq 0), \tag{6.355}$$

where $\alpha, t \in \mathbb{C}$.

Definition 6.63 The hyperbolic supercosecant via Rabotnov type function is defined as

$$Rcsch_{\alpha}(t) = \frac{1}{Rsinh_{\alpha}(t)}, (Rsinh_{\alpha}(t) \neq 0), \tag{6.356}$$

where $\alpha, t \in \mathbb{C}$.

Definition 6.64 The hyperbolic supersecant via Rabotnov type function is defined as

$$Rsech_{\alpha}(t) = \frac{1}{Rcosh_{\alpha}(t)} (Rcosh_{\alpha}(t) \neq 0), \tag{6.357}$$

where $\alpha, t \in \mathbb{C}$.

Thus,

$$\begin{aligned} Rsinh_{\alpha}(\lambda, t) &= \frac{1}{2}(R_{\alpha}(\lambda, t) - R_{\alpha}(-\lambda, t)) \\ &= t^{\alpha} \sum_{k=0}^{\infty} \frac{\lambda^{2k+1} t^{2k+1}}{\Gamma((2k+2)(\alpha+1))}, \end{aligned} \tag{6.358}$$

$$\begin{aligned} Rcosh_{\alpha}(\lambda, t) &= \frac{1}{2}(R_{\alpha}(\lambda, t) + R_{\alpha}(-\lambda, t)) \\ &= t^{\alpha} \sum_{k=0}^{\infty} \frac{\lambda^{2k} t^{2k}}{\Gamma((2k+1)(\alpha+1))}, \end{aligned} \tag{6.359}$$

$$Rtanh_{\alpha}(\lambda, t) = \frac{Rsinh_{\alpha}(\lambda, t)}{Rcosh_{\alpha}(\lambda, t)} (Rcosh_{\alpha}(\lambda, t) \neq 0), \tag{6.360}$$

$$Rcotanh_{\alpha}(\lambda, t) = \frac{Rcosh_{\alpha}(\lambda, t)}{Rsinh_{\alpha}(\lambda, t)}, (Rsinh_{\alpha}(\lambda, t) \neq 0), \tag{6.361}$$

$$Rcsch_{\alpha}(\lambda, t) = \frac{1}{Rsinh_{\alpha}(\lambda, t)}, (Rsinh_{\alpha}(\lambda, t) \neq 0), \tag{6.362}$$

and

$$Rsech_{\alpha}(\lambda, t) = \frac{1}{Rcosh_{\alpha}(\lambda, t)} \left(Rcosh_{\alpha}(\lambda, t) \neq 0\right), \tag{6.363}$$

where $\alpha, \lambda, t \in \mathbb{C}$.

Theorem 6.22 *Let $\alpha, \lambda, t \in \mathbb{C}$. Then we have*

$$R_{\alpha}(\lambda, t) = Rcosh_{\alpha}(\lambda, t) + Rsinh_{\alpha}(\lambda, t) \tag{6.364}$$

and

$$R_{\alpha}(-\lambda, t) = Rcosh_{\alpha}(\lambda, t) - Rsinh_{\alpha}(\lambda, t). \tag{6.365}$$

Proof Making use of (6.352) and (6.353), we get the result. □

6.2.6 The Integral Representations of the Supertrigonometric and Superhyperbolic Functions

In this part, we present the properties for the supertrigonometric and superhyperbolic functions.

Property 6.30 Let $\alpha, \varsigma \in \mathbb{C}$, $Re(\alpha) > 0$, and $\lambda \in \mathbb{C}$. Then we have

$$\frac{1}{\Gamma(\varsigma)} \int_0^t (t-\tau)^{\varsigma} \, \Re sin_{\alpha}(\lambda, \tau)\, d\tau = t^{\alpha+\varsigma} \, {}_1sin_1 \left(\begin{matrix} (1,1) \\ (\alpha+1+\varsigma, \alpha+1) \end{matrix} ; \lambda t^{\alpha+1} \right), \tag{6.366}$$

$$\frac{1}{\Gamma(\varsigma)} \int_0^t (t-\tau)^{\varsigma} \, \Re cos_{\alpha}(\lambda, \tau)\, d\tau = t^{\alpha+\varsigma} \, {}_1cos_1 \left(\begin{matrix} (1,1) \\ (\alpha+1+\varsigma, \alpha+1) \end{matrix} ; \lambda t^{\alpha+1} \right), \tag{6.367}$$

$$\frac{1}{\Gamma(\varsigma)} \int_0^t (t-\tau)^{\varsigma} \, \Re sinh_{\alpha}(\lambda, \tau)\, d\tau = t^{\alpha+\varsigma} \, {}_1sinh_1 \left(\begin{matrix} (1,1) \\ (\alpha+1+\varsigma, \alpha+1) \end{matrix} ; \lambda t^{\alpha+1} \right), \tag{6.368}$$

and

$$\frac{1}{\Gamma(\varsigma)} \int_0^t (t-\tau)^{\varsigma} \, \Re cosh_{\alpha}(\lambda, \tau)\, d\tau = t^{\alpha+\varsigma} \, {}_1cosh_1 \left(\begin{matrix} (1,1) \\ (\alpha+1+\varsigma, \alpha+1) \end{matrix} ; \lambda t^{\alpha+1} \right). \tag{6.369}$$

Proof Making use of (6.299), we obtain

$$\frac{1}{\Gamma(\varsigma)}\int_0^t (t-\tau)^{\varsigma}\,\Re_{\alpha}(i\lambda,\tau)\,d\tau = t^{\alpha+\varsigma}\;{}_1\varphi_1\left(\begin{matrix}(1,1)\\(\alpha+1+\varsigma,\alpha+1)\end{matrix};i\lambda t^{\alpha+1}\right), \tag{6.370}$$

$$\frac{1}{\Gamma(\varsigma)}\int_0^t (t-\tau)^{\varsigma}\,\Re_{\alpha}(-i\lambda,\tau)\,d\tau = t^{\alpha+\varsigma}\;{}_1\varphi_1\left(\begin{matrix}(1,1)\\(\alpha+1+\varsigma,\alpha+1)\end{matrix};-i\lambda t^{\alpha+1}\right), \tag{6.371}$$

$$\frac{1}{\Gamma(\varsigma)}\int_0^t (t-\tau)^{\varsigma}\,\Re_{\alpha}(\lambda,\tau)\,d\tau = t^{\alpha+\varsigma}\;{}_1\varphi_1\left(\begin{matrix}(1,1)\\(\alpha+1+\varsigma,\alpha+1)\end{matrix};\lambda t^{\alpha+1}\right), \tag{6.372}$$

and

$$\frac{1}{\Gamma(\varsigma)}\int_0^t (t-\tau)^{\varsigma}\,\Re_{\alpha}(-\lambda,\tau)\,d\tau = t^{\alpha+\varsigma}\;{}_1\varphi_1\left(\begin{matrix}(1,1)\\(\alpha+1+\varsigma,\alpha+1)\end{matrix};-\lambda t^{\alpha+1}\right), \tag{6.373}$$

which yield the desired results. □

Property 6.31 Let $\alpha, \varsigma \in \mathbb{C}$, $Re(\alpha) > 0$, and $\lambda \in \mathbb{C}$. Then we have

$$\frac{1}{\Gamma(\varsigma)}\int_0^1 (1-\tau)^{\varsigma}\,\Re sin_{\alpha}(\lambda,\tau)\,d\tau = {}_1sin_1\left(\begin{matrix}(1,1)\\(\alpha+1+\varsigma,\alpha+1)\end{matrix};\lambda\right), \tag{6.374}$$

$$\frac{1}{\Gamma(\varsigma)}\int_0^1 (1-\tau)^{\varsigma}\,\Re cos_{\alpha}(\lambda,\tau)\,d\tau = {}_1cos_1\left(\begin{matrix}(1,1)\\(\alpha+1+\varsigma,\alpha+1)\end{matrix};\lambda\right), \tag{6.375}$$

$$\frac{1}{\Gamma(\varsigma)}\int_0^1 (1-\tau)^{\varsigma}\,\Re sinh_{\alpha}(\lambda,\tau)\,d\tau = {}_1sinh_1\left(\begin{matrix}(1,1)\\(\alpha+1+\varsigma,\alpha+1)\end{matrix};\lambda\right), \tag{6.376}$$

and

$$\frac{1}{\Gamma(\varsigma)}\int_0^1 (1-\tau)^{\varsigma}\,\Re cosh_{\alpha}(\lambda,\tau)\,d\tau = {}_1cosh_1\left(\begin{matrix}(1,1)\\(\alpha+1+\varsigma,\alpha+1)\end{matrix};\lambda\right). \tag{6.377}$$

Proof Putting $t = 1$ in (6.366), (6.367), (6.368), and (6.369), respectively, we get the results. □

Property 6.32 Let $\alpha, \varsigma \in \mathbb{C}$, $Re(\alpha) > 0$, and $\lambda \in \mathbb{C}$. Then we have

$$\frac{1}{\Gamma(\varsigma)}\int_0^2 (2-\tau)^{\varsigma}\,\Re sin_{\alpha}(\lambda,\tau)\,d\tau = \frac{1}{\Gamma(\varsigma)}\int_{-1}^1 (1-\tau)^{\varsigma}\,\Re sin_{\alpha}(\lambda,1+\tau)\,d\tau$$
$$= \frac{1}{\Gamma(\varsigma)}\int_{-1}^1 (1+\tau)^{\varsigma}\,\Re sin_{\alpha}(\lambda,1-\tau)\,d\tau$$
$$= 2^{\alpha+\varsigma}\,{}_1sin_1\left(\begin{matrix}(1,1)\\(\alpha+1+\varsigma,\alpha+1)\end{matrix};\lambda 2^{\alpha+1}\right), \tag{6.378}$$

$$\frac{1}{\Gamma(\varsigma)}\int_0^2 (2-\tau)^{\varsigma}\,\Re cos_{\alpha}(\lambda,\tau)\,d\tau = \frac{1}{\Gamma(\varsigma)}\int_{-1}^1 (1-\tau)^{\varsigma}\,\Re cos_{\alpha}(\lambda,1+\tau)\,d\tau$$
$$= \frac{1}{\Gamma(\varsigma)}\int_{-1}^1 (1+\tau)^{\varsigma}\,\Re cos_{\alpha}(\lambda,1-\tau)\,d\tau$$
$$= 2^{\alpha+\varsigma}\,{}_1cos_1\left(\begin{matrix}(1,1)\\(\alpha+1+\varsigma,\alpha+1)\end{matrix};\lambda 2^{\alpha+1}\right), \tag{6.379}$$

$$\frac{1}{\Gamma(\varsigma)}\int_0^2 (2-\tau)^{\varsigma}\,\Re sinh_{\alpha}(\lambda,\tau)\,d\tau = \frac{1}{\Gamma(\varsigma)}\int_{-1}^1 (1-\tau)^{\varsigma}\,\Re sinh_{\alpha}(\lambda,1+\tau)\,d\tau$$
$$= \frac{1}{\Gamma(\varsigma)}\int_{-1}^1 (1+\tau)^{\varsigma}\,\Re sinh_{\alpha}(\lambda,1-\tau)\,d\tau$$
$$= 2^{\alpha+\varsigma}\,{}_1sinh_1\left(\begin{matrix}(1,1)\\(\alpha+1+\varsigma,\alpha+1)\end{matrix};\lambda 2^{\alpha+1}\right) \tag{6.380}$$

and

$$\frac{1}{\Gamma(\varsigma)}\int_0^2 (2-\tau)^{\varsigma}\,\Re cosh_{\alpha}(\lambda,\tau)\,d\tau = \frac{1}{\Gamma(\varsigma)}\int_{-1}^1 (1-\tau)^{\varsigma}\,\Re cosh_{\alpha}(\lambda,1+\tau)\,d\tau$$
$$= \frac{1}{\Gamma(\varsigma)}\int_{-1}^1 (1+\tau)^{\varsigma}\,\Re cosh_{\alpha}(\lambda,1-\tau)\,d\tau$$
$$= 2^{\alpha+\varsigma}\,{}_1cosh_1\left(\begin{matrix}(1,1)\\(\alpha+1+\varsigma,\alpha+1)\end{matrix};\lambda 2^{\alpha+1}\right). \tag{6.381}$$

Proof Putting $t = 2$ in (6.366), (6.367), (6.368), and (6.369), respectively, we have the results. □

Making use of (6.290), we give the following property:

Property 6.33 Let $|\lambda/s| < 1$ and $\alpha, \lambda \in \mathbb{C}$. Then, we have

$$L\{\Re sin_\alpha(\lambda, t)\} = \frac{\lambda s^{-2(\alpha+1)}}{1+\lambda^2 s^{-2(\alpha+1)}}, \tag{6.382}$$

$$L\{\Re cos_\alpha(\lambda, t)\} = \frac{s^{-(\alpha+1)}}{1+\lambda^2 s^{-2(\alpha+1)}}, \tag{6.383}$$

$$L\{\Re sinh_\alpha(\lambda, t)\} = \frac{\lambda s^{-2(\alpha+1)}}{1-\lambda^2 s^{-2(\alpha+1)}}, \tag{6.384}$$

and

$$L\{\Re cosh_\alpha(\lambda, t)\} = \frac{s^{-(\alpha+1)}}{1-\lambda^2 s^{-2(\alpha+1)}}. \tag{6.385}$$

Proof In view of the integral representations

$$L\{\Re_\alpha(i\lambda, t)\} = s^{-(\alpha+1)}\left(1 - i\lambda s^{-(\alpha+1)}\right)^{-1}, \tag{6.386}$$

$$L\{\Re_\alpha(-i\lambda, t)\} = s^{-(\alpha+1)}\left(1 + i\lambda s^{-(\alpha+1)}\right)^{-1}, \tag{6.387}$$

$$L\{\Re_\alpha(\lambda, t)\} = s^{-(\alpha+1)}\left(1 - \lambda s^{-(\alpha+1)}\right)^{-1}, \tag{6.388}$$

and

$$L\{\Re_\alpha(-\lambda, t)\} = s^{-(\alpha+1)}\left(1 + \lambda s^{-(\alpha+1)}\right)^{-1}, \tag{6.389}$$

we have

$$\begin{aligned} L\{\Re sin_\alpha(\lambda, t)\} &= \frac{1}{2i}\left(s^{-(\alpha+1)}\left(1 - i\lambda s^{-(\alpha+1)}\right)^{-1} - s^{-(\alpha+1)}\left(1 + i\lambda s^{-(\alpha+1)}\right)^{-1}\right) \\ &= \frac{\lambda s^{-2(\alpha+1)}}{1+\lambda^2 s^{-2(\alpha+1)}}, \end{aligned} \tag{6.390}$$

$$\begin{aligned} L\{\Re cos_\alpha(\lambda, t)\} &= \frac{1}{2}\left(s^{-(\alpha+1)}\left(1 - i\lambda s^{-(\alpha+1)}\right)^{-1} + s^{-(\alpha+1)}\left(1 + i\lambda s^{-(\alpha+1)}\right)^{-1}\right) \\ &= \frac{s^{-(\alpha+1)}}{1+\lambda^2 s^{-2(\alpha+1)}}, \end{aligned} \tag{6.391}$$

$$L\left\{\Re sinh_{\alpha}(\lambda,t)\right\}=\frac{1}{2}\left(s^{-(\alpha+1)}\left(1-\lambda s^{-(\alpha+1)}\right)^{-1}-s^{-(\alpha+1)}\left(1+\lambda s^{-(\alpha+1)}\right)^{-1}\right)$$
$$=\frac{\lambda s^{-2(\alpha+1)}}{1-\lambda^{2}s^{-2(\alpha+1)}}, \tag{6.392}$$

and

$$L\left\{\Re cosh_{\alpha}(\lambda,t)\right\}=\frac{1}{2}\left(s^{-(\alpha+1)}\left(1-\lambda s^{-(\alpha+1)}\right)^{-1}+s^{-(\alpha+1)}\left(1+\lambda s^{-(\alpha+1)}\right)^{-1}\right)$$
$$=\frac{s^{-(\alpha+1)}}{1-\lambda^{2}s^{-2(\alpha+1)}}. \tag{6.393}$$

□

6.2.7 The Gauss–Rabotnov Type Functions

From (6.318), (6.319), (6.349), and (6.351), we derive the Gauss–Rabotnov type functions.

The Gauss–Rabotnov type function can be written as follows:

$$\Re_{\alpha}\left(\lambda,t^{2}\right)=t^{2\alpha}\sum_{k=0}^{\infty}\frac{\lambda^{k}t^{2k(\alpha+1)}}{\Gamma\left((k+1)(\alpha+1)\right)}$$
$$=t^{2\alpha}E_{\alpha+1,\alpha+1}\left(\lambda t^{2(\alpha+1)}\right)$$
$$=t^{2\alpha}\,{}_{1}\varphi_{1}\left(\begin{matrix}(1,1)\\(\alpha+1,\alpha+1)\end{matrix};\lambda t^{2(\alpha+1)}\right), \tag{6.394}$$

$$\Re_{\alpha}\left(-\lambda,t^{2}\right)=t^{2\alpha}\sum_{k=0}^{\infty}\frac{(-\lambda)^{k}t^{2k(\alpha+1)}}{\Gamma\left((k+1)(\alpha+1)\right)}$$
$$=t^{2\alpha}E_{\alpha+1,\alpha+1}\left(-\lambda t^{2(\alpha+1)}\right)$$
$$=t^{2\alpha}\,{}_{1}\varphi_{1}\left(\begin{matrix}(1,1)\\(\alpha+1,\alpha+1)\end{matrix};-\lambda t^{2(\alpha+1)}\right), \tag{6.395}$$

$$R_{\alpha}\left(\lambda,t^{2}\right)=t^{\alpha}\sum_{k=0}^{\infty}\frac{\lambda^{k}t^{2k}}{\Gamma\left((k+1)(\alpha+1)\right)}$$
$$=t^{2\alpha}E_{\alpha+1,\alpha+1}\left(\lambda t^{2}\right)$$
$$=t^{2\alpha}\,{}_{1}\varphi_{1}\left(\begin{matrix}(1,1)\\(\alpha+1,\alpha+1)\end{matrix};\lambda t^{2}\right), \tag{6.396}$$

$$R_\alpha\left(-\lambda, t^2\right) = t^\alpha \sum_{k=0}^{\infty} \frac{(-\lambda)^k t^{2k}}{\Gamma((k+1)(\alpha+1))}$$
$$= t^{2\alpha} E_{\alpha+1,\alpha+1}\left(-\lambda t^2\right)$$
$$= t^{2\alpha} {}_1\varphi_1\left(\begin{matrix}(1,1)\\(\alpha+1,\alpha+1)\end{matrix}; -\lambda t^2\right), \tag{6.397}$$

$$R_\alpha\left(\lambda, t^2\right) = t^\alpha \sum_{k=0}^{\infty} \frac{\lambda^k t^{2k}}{\Gamma((k+1)(\alpha+1))}$$
$$= t^{2\alpha} E_{\alpha+1,\alpha+1}\left(\lambda t^2\right)$$
$$= t^{2\alpha} {}_1\varphi_1\left(\begin{matrix}(1,1)\\(\alpha+1,\alpha+1)\end{matrix}; \lambda t^2\right), \tag{6.398}$$

$$R_\alpha\left(-\lambda, t^2\right) = t^\alpha \sum_{k=0}^{\infty} \frac{(-\lambda)^k t^{2k}}{\Gamma((k+1)(\alpha+1))}$$
$$= t^{2\alpha} E_{\alpha+1,\alpha+1}\left(-\lambda t^2\right)$$
$$= t^{2\alpha} {}_1\varphi_1\left(\begin{matrix}(1,1)\\(\alpha+1,\alpha+1)\end{matrix}; -\lambda t^2\right), \tag{6.399}$$

$$\Re_\alpha\left(t^2\right) = t^{2\alpha} \sum_{k=0}^{\infty} \frac{t^{2k(\alpha+1)}}{\Gamma((k+1)(\alpha+1))}$$
$$= t^{2\alpha} E_{\alpha+1,\alpha+1}\left(t^{2(\alpha+1)}\right)$$
$$= t^{2\alpha} {}_1\varphi_1\left(\begin{matrix}(1,1)\\(\alpha+1,\alpha+1)\end{matrix}; t^{2(\alpha+1)}\right), \tag{6.400}$$

$$\Re_\alpha\left(-1, t^2\right) = t^{2\alpha} \sum_{k=0}^{\infty} \frac{(-1)^k t^{2k(\alpha+1)}}{\Gamma((k+1)(\alpha+1))}$$
$$= t^{2\alpha} E_{\alpha+1,\alpha+1}\left(-t^{2(\alpha+1)}\right)$$
$$= t^{2\alpha} {}_1\varphi_1\left(\begin{matrix}(1,1)\\(\alpha+1,\alpha+1)\end{matrix}; -t^{2(\alpha+1)}\right), \tag{6.401}$$

$$\begin{aligned} R_\alpha\left(t^2\right) &= t^\alpha \sum_{k=0}^{\infty} \frac{t^{2k}}{\Gamma\left((k+1)(\alpha+1)\right)} \\ &= t^{2\alpha} E_{\alpha+1,\alpha+1}\left(t^2\right) \\ &= t^{2\alpha}\, {}_1\varphi_1\left(\begin{matrix}(1,1)\\(\alpha+1,\alpha+1)\end{matrix};t^2\right), \end{aligned} \tag{6.402}$$

$$\begin{aligned} R_\alpha\left(-1,t^2\right) &= t^\alpha \sum_{k=0}^{\infty} \frac{(-1)^k t^{2k}}{\Gamma\left((k+1)(\alpha+1)\right)} \\ &= t^{2\alpha} E_{\alpha+1,\alpha+1}\left(-t^2\right) \\ &= t^{2\alpha}\, {}_1\varphi_1\left(\begin{matrix}(1,1)\\(\alpha+1,\alpha+1)\end{matrix};-t^2\right), \end{aligned} \tag{6.403}$$

$$\begin{aligned} R_\alpha\left(t^2\right) &= t^\alpha \sum_{k=0}^{\infty} \frac{t^{2k}}{\Gamma\left((k+1)(\alpha+1)\right)} \\ &= t^{2\alpha} E_{\alpha+1,\alpha+1}\left(t^2\right) \\ &= t^{2\alpha}\, {}_1\varphi_1\left(\begin{matrix}(1,1)\\(\alpha+1,\alpha+1)\end{matrix};t^2\right), \end{aligned} \tag{6.404}$$

and

$$\begin{aligned} R_\alpha\left(-1,t^2\right) &= t^\alpha \sum_{k=0}^{\infty} \frac{(-1)^k t^{2k}}{\Gamma\left((k+1)(\alpha+1)\right)} \\ &= t^{2\alpha} E_{\alpha+1,\alpha+1}\left(-t^2\right) \\ &= t^{2\alpha}\, {}_1\varphi_1\left(\begin{matrix}(1,1)\\(\alpha+1,\alpha+1)\end{matrix};-t^2\right), \end{aligned} \tag{6.405}$$

where $\alpha, \lambda \in \mathbb{C}$.

For the applications of Rabotnov function (the so-called Rabotnov fractional exponential function) in the applied sciences, see Yang et al. [271, 272], Kumar et al. [273], Rossikhin and Shitikova [274], Kiryakova [275], and Rabotnov et al. [276].

6.3 The Miller–Ross Function

In 1993, Kenneth S. Miller and Bertram Ross proposed the speical function, which is called the Miller–Ross function [92, 238, 277].

6.3.1 The Miller–Ross Function: History and Properties

Definition 6.65 The Miller–Ross function is defined as [92, 277]

$$\begin{aligned} \mathbb{M}_\alpha (t) &= t^\alpha \sum_{k=0}^{\infty} \frac{t^k}{\Gamma (k+1+\alpha)} \\ &= \sum_{k=0}^{\infty} \frac{t^{k+\alpha}}{\Gamma (k+1+\alpha)}, \end{aligned} \tag{6.406}$$

where $\alpha, t \in \mathbb{C}$.

The Miller–Ross function can be written as [92, 277]

$$\begin{aligned} \mathbb{M}_\alpha (\lambda, t) &= t^\alpha \sum_{k=0}^{\infty} \frac{\lambda^k t^k}{\Gamma (k+1+\alpha)} \\ &= \sum_{k=0}^{\infty} \frac{\lambda^k t^{k+\alpha}}{\Gamma (k+1+\alpha)}, \end{aligned} \tag{6.407}$$

where $\alpha, t \in \mathbb{C}$.

It is convenient to show that [238]

$$\begin{aligned} \mathbb{M}_\alpha (\lambda, t) &= t^\alpha \sum_{k=0}^{\infty} \frac{(\lambda t)^k}{\Gamma (k+1+\alpha)} \\ &= t^\alpha E_{1,1+\alpha} (\lambda t). \end{aligned} \tag{6.408}$$

In fact, we have

$$\begin{aligned} \mathbb{M}_\alpha (\lambda, t) &= t^\alpha \sum_{k=0}^{\infty} \frac{(\lambda t)^k}{\Gamma (k+1+\alpha)} \\ &= t^\alpha E_{1,1+\alpha} (\lambda t) \\ &= t^\alpha \, {}_1\varphi_1 \left(\begin{matrix} (1,1) \\ (1+\alpha, 1) \end{matrix} ; \lambda t \right), \end{aligned} \tag{6.409}$$

which leads to

$$\mathbb{M}_\alpha(t) = \mathbb{M}_\alpha(1, t).$$

The Laplace transform of the Miller–Ross function can be given as follows [277]:

$$\begin{aligned} L\{\mathbb{M}_\alpha(\lambda, t)\} &= L\left\{t^\alpha \sum_{k=0}^{\infty} \frac{(\lambda t)^k}{\Gamma(k+1+\alpha)}\right\} \\ &= s^{-\alpha-1} \sum_{k=0}^{\infty} \left(\frac{\lambda}{s}\right)^k \\ &= \frac{s^{-\alpha}}{s-\lambda} \; (|\lambda/s| < 1). \end{aligned} \tag{6.410}$$

Property 6.34 Let $\alpha, t \in \mathbb{C}$. Then we have [277]

$$L\{\mathbb{M}_\alpha(t)\} = \frac{s^{-\alpha}}{s-1} \; (|1/s| < 1). \tag{6.411}$$

The Mellin transform of the Miller–Ross function is given as follows:

$$\begin{aligned} M[\mathbb{M}_\alpha(-\lambda, t)] &= \int_0^\infty t^{\varpi+\alpha-1} E_{1,1+\alpha}(-\lambda t)\, dt \\ &= \frac{\Gamma(\varpi+\alpha)\Gamma(1-\varpi-\alpha)}{\Gamma(1-\varpi)} \lambda^{-(\varpi+\alpha)}, \end{aligned} \tag{6.412}$$

where $\alpha, \varpi \in \mathbb{C}$, $\lambda \in \mathbb{C}$, $Re(\alpha) > -1$, and $0 < Re(\varpi) < 1$.

Property 6.35 Let $\alpha \in \mathbb{C}$, $\lambda \in \mathbb{C}$, $Re(\alpha) > -1$, and $0 < Re(\varpi) < 1$. Then we have

$$\int_{-\infty}^{\infty} |t|^{\varpi-1} \mathbb{M}_\alpha(\lambda, |t|)\, dt = 2\lambda^{-(\varpi+\alpha)} \frac{\Gamma(\varpi+\alpha)\Gamma(1-\varpi-\alpha)}{\Gamma(1-\varpi)}. \tag{6.413}$$

Property 6.36 Let $\alpha \in \mathbb{C}$, $\lambda \in \mathbb{C}$, $Re(\alpha) > -1$, and $0 < Re(\varpi) < 1$. Then we have

$$\mathbb{M}_\alpha(t) = \frac{1}{2\pi i} \int_{-i\infty}^{i\infty} \frac{\Gamma(\varpi+\alpha)\Gamma(1-\varpi-\alpha)}{\Gamma(1-\varpi)} (-t)^{-(\varpi+\alpha)}\, d\varpi. \tag{6.414}$$

Proof Making use of (6.412), we obtain the result. □

Let us recall the following case that can be found in Chap. 5.

Let $\beta \in \mathbb{C}$, $\lambda \in \mathbb{C}$, $\alpha \in \mathbb{R}_+$, $Re(\sigma) > 0$, and $Re(\beta) > 0$. Then we have

$$\frac{\Gamma(\gamma)}{\Gamma(\sigma)} \int_0^t (t-\tau)^\sigma \tau^{\beta-1} E_{\alpha,\beta}^{\gamma}\left(\lambda\tau^\alpha\right) d\tau = t^{\beta+\sigma-1} {}_1\varphi_1 \left(\begin{matrix} (\gamma, 1) \\ (\beta+\sigma, \alpha) \end{matrix} ; \lambda t^\alpha \right). \tag{6.415}$$

Theorem 6.23 *Let $\beta \in \mathbb{C}$, $\lambda \in \mathbb{C}$, $\alpha \in \mathbb{R}_+$, $Re(\sigma) > 0$, and $Re(\beta) > 0$. Then we have*

$$\begin{aligned} \frac{1}{\Gamma(\sigma)} \int_0^t (t-\tau)^\sigma \mathbb{M}_\alpha(\lambda, \tau)\, d\tau &= t^{\alpha+\sigma} {}_1\varphi_1 \left(\begin{matrix} (1, 1) \\ (\alpha+\sigma+1, 1) \end{matrix} ; \lambda t \right) \\ &= \mathbb{M}_{\alpha+\sigma}(\lambda, t). \end{aligned} \tag{6.416}$$

Proof Taking $\gamma = 1$ and $\alpha = 1$ in (6.415) and using $\beta = 1 + \alpha$ and (6.409), we obtain

$$\begin{aligned} \frac{1}{\Gamma(\sigma)} \int_0^t (t-\tau)^\sigma \mathbb{M}_\alpha(\lambda, \tau)\, d\tau &= t^{\alpha+\sigma} {}_1\varphi_1 \left(\begin{matrix} (1, 1) \\ (\alpha+\sigma+1, 1) \end{matrix} ; \lambda t \right) \\ &= \mathbb{M}_{\alpha+\sigma}(\lambda, t). \end{aligned} \tag{6.417}$$

□

Theorem 6.24 *Let $\beta \in \mathbb{C}$, $\lambda \in \mathbb{C}$, $\alpha \in \mathbb{R}_+$, $Re(\sigma) > 0$, and $Re(\beta) > 0$. Then we have*

$$\frac{1}{\Gamma(\sigma)} \int_0^1 (1-\tau)^\sigma \mathbb{M}_\alpha(\lambda, \tau)\, d\tau = {}_1\varphi_1 \left(\begin{matrix} (1, 1) \\ (\alpha+\sigma+1, 1) \end{matrix} ; \lambda \right) = \mathbb{M}_\alpha(\lambda, 1). \tag{6.418}$$

Proof Taking $t = 1$ in (6.416), we get the result. □

Theorem 6.25 *Let $\beta \in \mathbb{C}$, $\lambda \in \mathbb{C}$, $\alpha \in \mathbb{R}_+$, $Re(\sigma) > 0$, and $Re(\beta) > 0$. Then we have*

$$\begin{aligned} \frac{1}{\Gamma(\sigma)} \int_0^2 (2-\tau)^\sigma \mathbb{M}_\alpha(\lambda, \tau)\, d\tau &= \frac{1}{\Gamma(\sigma)} \int_{-1}^1 (1-\tau)^\sigma \mathbb{M}_\alpha(\lambda, 1+\tau)\, d\tau \\ &= \frac{1}{\Gamma(\sigma)} \int_{-1}^1 (1+\tau)^\sigma \mathbb{M}_\alpha(\lambda, 1-\tau)\, d\tau \\ &= 2^{\alpha+\sigma} {}_1\varphi_1 \left(\begin{matrix} (1, 1) \\ (\alpha+\sigma+1, 1) \end{matrix} ; 2\lambda \right) \\ &= \mathbb{M}_{\alpha+\sigma}(\lambda, 2). \end{aligned} \tag{6.419}$$

Proof Taking $t = 2$ in (6.416), we deduce the result. □

6.3.2 *The Supertrigonometric Functions via Miller–Ross Function*

In this part, we propose the supertrigonometric functions via Miller–Ross function.

We now present that

$$\begin{aligned}\mathbb{M}_{\alpha}(i\lambda, t) &= t^{\alpha}\sum_{k=0}^{\infty}\frac{(i\lambda t)^{k}}{\Gamma(k+1+\alpha)}\\ &= t^{\alpha}E_{1,1+\alpha}(i\lambda t)\\ &= t^{\alpha}\,{}_{1}\varphi_{1}\left(\begin{matrix}(1,1)\\(1+\alpha,1)\end{matrix}; i\lambda t\right)\end{aligned} \tag{6.420}$$

and

$$\begin{aligned}\mathbb{M}_{\alpha}(-i\lambda, t) &= t^{\alpha}\sum_{k=0}^{\infty}\frac{(-i\lambda t)^{k}}{\Gamma(k+1+\alpha)}\\ &= t^{\alpha}E_{1,1+\alpha}(-i\lambda t)\\ &= t^{\alpha}\,{}_{1}\varphi_{1}\left(\begin{matrix}(1,1)\\(1+\alpha,1)\end{matrix}; -i\lambda t\right).\end{aligned} \tag{6.421}$$

Making use of (6.420) and (6.421), we recall the following special functions:

Definition 6.66 The Miller–Ross supersine function is defined as [92, 277]

$$\begin{aligned}\mathbb{M}sin_{\alpha}(t) &= \frac{1}{2i}\left(\mathbb{M}_{\alpha}(i,t)-\mathbb{M}_{\alpha}(-i,t)\right)\\ &= t^{\alpha}\sum_{k=0}^{\infty}\frac{(-1)^{k}t^{2k+1}}{\Gamma(2k+2+\alpha)},\end{aligned} \tag{6.422}$$

where $\alpha, t \in \mathbb{C}$.

Definition 6.67 The Miller–Ross supercosine function is defined as [92, 277]

$$\begin{aligned}\mathbb{M}cos_{\alpha}(t) &= \frac{1}{2}\left(\mathbb{M}_{\alpha}(i,t)+\mathbb{M}_{\alpha}(-i,t)\right)\\ &= t^{\alpha}\sum_{k=0}^{\infty}\frac{(-1)^{k}t^{2k}}{\Gamma(2k+1+\alpha)},\end{aligned} \tag{6.423}$$

where $\alpha, t \in \mathbb{C}$.

Definition 6.68 The supertangent via Miller–Ross function is defined as

$$\mathbb{M}tan_{\alpha}(t) = \frac{\mathbb{M}sin_{\alpha}(t)}{\mathbb{M}cos_{\alpha}(t)} \left(\mathbb{M}cos_{\alpha}(t) \neq 0\right), \tag{6.424}$$

where $\alpha, t \in \mathbb{C}$.

Definition 6.69 The supercotangent via Miller–Ross function is defined as

$$\mathbb{M}cotan_{\alpha}(t) = \frac{\mathbb{M}cos_{\alpha}(t)}{\mathbb{M}sin_{\alpha}(t)}, \left(\mathbb{M}sin_{\alpha}(t) \neq 0\right), \tag{6.425}$$

where $\alpha, t \in \mathbb{C}$.

Definition 6.70 The supercosecant via Miller–Ross function is defined as

$$\mathbb{M}csc_{\alpha}(t) = \frac{1}{\mathbb{M}sin_{\alpha}(t)}, \left(\mathbb{M}sin_{\alpha}(t) \neq 0\right), \tag{6.426}$$

where $\alpha, t \in \mathbb{C}$.

Definition 6.71 The supersecant via Miller–Ross function is defined as

$$\mathbb{M}sec_{\alpha}(t) = \frac{1}{\mathbb{M}cos_{\alpha}(t)} \left(\mathbb{M}cos_{\alpha}(t) \neq 0\right), \tag{6.427}$$

where $\alpha, t \in \mathbb{C}$.

Thus,

$$\begin{aligned}\mathbb{M}sin_{\alpha}(\lambda, t) &= \frac{1}{2i}\left(\mathbb{M}_{\alpha}(i\lambda, t) - \mathbb{M}_{\alpha}(-i\lambda, t)\right) \\ &= t^{\alpha} \sum_{k=0}^{\infty} \frac{(-1)^k (\lambda t)^{2k+1}}{\Gamma(2k+2+\alpha)},\end{aligned} \tag{6.428}$$

$$\begin{aligned}\mathbb{M}cos_{\alpha}(\lambda, t) &= \frac{1}{2}\left(\mathbb{M}_{\alpha}(i\lambda, t) + \mathbb{M}_{\alpha}(-i\lambda, t)\right) \\ &= t^{\alpha} \sum_{k=0}^{\infty} \frac{(-1)^k (\lambda t)^{2k}}{\Gamma(2k+1+\alpha)},\end{aligned} \tag{6.429}$$

$$\mathbb{M}tan_{\alpha}(\lambda, t) = \frac{\mathbb{M}sin_{\alpha}(\lambda, t)}{\mathbb{M}cos_{\alpha}(\lambda, t)} \left(\mathbb{M}cos_{\alpha}(\lambda, t) \neq 0\right), \tag{6.430}$$

$$\mathbb{M}cotan_{\alpha}(\lambda, t) = \frac{\mathbb{M}cos_{\alpha}(\lambda, t)}{\mathbb{M}sin_{\alpha}(\lambda, t)}, \left(\mathbb{M}sin_{\alpha}(\lambda, t) \neq 0\right), \tag{6.431}$$

$$\mathbb{M}csc_{\alpha}(\lambda, t) = \frac{1}{\mathbb{M}sin_{\alpha}(\lambda, t)}, \left(\mathbb{M}sin_{\alpha}(\lambda, t) \neq 0\right), \tag{6.432}$$

and

$$\mathbb{M}sec_{\alpha}(\lambda, t) = \frac{1}{\mathbb{M}cos_{\alpha}(\lambda, t)} \left(\mathbb{M}cos_{\alpha}(\lambda, t) \neq 0\right), \tag{6.433}$$

where $\alpha, \lambda, t \in \mathbb{C}$.

The formulas (6.428) and (6.429) were considered in [92, 277].

Theorem 6.26 *Let* $\alpha, \lambda, t \in \mathbb{C}$. *Then we have*

$$\mathbb{M}_{\alpha}(i\lambda, t) = \mathbb{M}cos_{\alpha}(\lambda, t) + i\mathbb{M}sin_{\alpha}(\lambda, t) \tag{6.434}$$

and

$$\mathbb{M}_{\alpha}(-i\lambda, t) = \mathbb{M}cos_{\alpha}(\lambda, t) - i\mathbb{M}sin_{\alpha}(\lambda, t). \tag{6.435}$$

Proof In view of (6.428) and (6.429), we give the result. □

In particular,

$$\mathbb{M}_{\alpha}(t) = \mathbb{M}cos_{\alpha}(t) + i\mathbb{M}sin_{\alpha}(t) \tag{6.436}$$

and

$$\mathbb{M}_{\alpha}(-i, t) = \mathbb{M}cos_{\alpha}(t) - i\mathbb{M}sin_{\alpha}(t), \tag{6.437}$$

where $\alpha, t \in \mathbb{C}$.

6.3.3 The Superhyperbolic Functions via Miller–Ross Function

In this part, we propose the superhyperbolic functions via Miller–Ross function.

In a similar manner, we consider that

$$\begin{aligned}\mathbb{M}_{\alpha}(\lambda, t) &= t^{\alpha} \sum_{k=0}^{\infty} \frac{(\lambda t)^k}{\Gamma(k+1+\alpha)} \\ &= t^{\alpha} E_{1,1+\alpha}(\lambda t) \\ &= t^{\alpha}\, {}_1\varphi_1\left(\begin{matrix}(1,1) \\ (1+\alpha,1)\end{matrix}; \lambda t\right)\end{aligned} \tag{6.438}$$

and

$$\begin{aligned}\mathbb{M}_{\alpha}(-\lambda, t) &= t^{\alpha}\sum_{k=0}^{\infty}\frac{(-\lambda t)^{k}}{\Gamma(k+1+\alpha)} \\ &= t^{\alpha}E_{1,1+\alpha}(-\lambda t) \\ &= t^{\alpha}\,{}_{1}\varphi_{1}\left(\begin{matrix}(1,1)\\(1+\alpha,1)\end{matrix};-\lambda t\right).\end{aligned} \tag{6.439}$$

With the use of (6.438) and (6.439), we recall the following special functions:

Definition 6.72 The Miller–Ross hyperbolic supersine function is defined as

$$\begin{aligned}\mathbb{M}sinh_{\alpha}(t) &= \frac{1}{2}\left(\mathbb{M}_{\alpha}(1,t) - \mathbb{M}_{\alpha}(-1,t)\right) \\ &= t^{\alpha}\sum_{k=0}^{\infty}\frac{t^{2k+1}}{\Gamma(2k+2+\alpha)},\end{aligned} \tag{6.440}$$

where $\alpha, t \in \mathbb{C}$.

Definition 6.73 The Miller–Ross hyperbolic supercosine function is defined as [92, 277]

$$\begin{aligned}\mathbb{M}cosh_{\alpha}(t) &= \frac{1}{2}\left(\mathbb{M}_{\alpha}(1,t) + \mathbb{M}_{\alpha}(-1,t)\right) \\ &= t^{\alpha}\sum_{k=0}^{\infty}\frac{t^{2k}}{\Gamma(2k+1+\alpha)},\end{aligned} \tag{6.441}$$

where $\alpha, t \in \mathbb{C}$.

Definition 6.74 The hyperbolic supertangent via Miller–Ross function is defined as [92, 277]

$$\mathbb{M}tanh_{\alpha}(t) = \frac{\mathbb{M}sinh_{\alpha}(t)}{\mathbb{M}cosh_{\alpha}(t)}\left(\mathbb{M}cosh_{\alpha}(t) \neq 0\right), \tag{6.442}$$

where $\alpha, t \in \mathbb{C}$.

Definition 6.75 The hyperbolic supercotangent via Miller–Ross function is defined as

$$\mathbb{M}cotanh_{\alpha}(t) = \frac{\mathbb{M}cosh_{\alpha}(t)}{\mathbb{M}sinh_{\alpha}(t)}, \left(\mathbb{M}sinh_{\alpha}(t) \neq 0\right), \tag{6.443}$$

where $\alpha, t \in \mathbb{C}$.

Definition 6.76 The hyperbolic supercosecant via Miller–Ross function is defined as

$$\mathbb{M}csch_{\alpha}(t) = \frac{1}{\mathbb{M}sinh_{\alpha}(t)}, \ (\mathbb{M}sinh_{\alpha}(t) \neq 0), \tag{6.444}$$

where $\alpha, t \in \mathbb{C}$.

Definition 6.77 The hyperbolic supersecant via Miller–Ross function is defined as

$$\mathbb{M}sech_{\alpha}(t) = \frac{1}{\mathbb{M}cosh_{\alpha}(t)} \ (\mathbb{M}cosh_{\alpha}(t) \neq 0), \tag{6.445}$$

where $\alpha, t \in \mathbb{C}$.

Thus,

$$\begin{aligned} \mathbb{M}sinh_{\alpha}(\lambda, t) &= \frac{1}{2i}\left(\mathbb{M}_{\alpha}(\lambda, t) - \mathbb{M}_{\alpha}(-\lambda, t)\right) \\ &= t^{\alpha} \sum_{k=0}^{\infty} \frac{(\lambda t)^{2k+1}}{\Gamma(2k+2+\alpha)}, \end{aligned} \tag{6.446}$$

$$\begin{aligned} \mathbb{M}cosh_{\alpha}(\lambda, t) &= \frac{1}{2}\left(\mathbb{M}_{\alpha}(\lambda, t) + \mathbb{M}_{\alpha}(-\lambda, t)\right) \\ &= t^{\alpha} \sum_{k=0}^{\infty} \frac{(\lambda t)^{2k}}{\Gamma(2k+1+\alpha)}, \end{aligned} \tag{6.447}$$

$$\mathbb{M}tanh_{\alpha}(\lambda, t) = \frac{\mathbb{M}sinh_{\alpha}(\lambda, t)}{\mathbb{M}cosh_{\alpha}(\lambda, t)} \ (\mathbb{M}cosh_{\alpha}(\lambda, t) \neq 0), \tag{6.448}$$

$$\mathbb{M}cotanh_{\alpha}(\lambda, t) = \frac{\mathbb{M}cosh_{\alpha}(\lambda, t)}{\mathbb{M}sinh_{\alpha}(\lambda, t)}, \ (\mathbb{M}sinh_{\alpha}(\lambda, t) \neq 0), \tag{6.449}$$

$$\mathbb{M}csch_{\alpha}(\lambda, t) = \frac{1}{\mathbb{M}sinh_{\alpha}(\lambda, t)}, \ (\mathbb{M}sinh_{\alpha}(\lambda, t) \neq 0), \tag{6.450}$$

and

$$\mathbb{M}sech_{\alpha}(\lambda, t) = \frac{1}{\mathbb{M}cosh_{\alpha}(\lambda, t)} \ (\mathbb{M}cosh_{\alpha}(\lambda, t) \neq 0), \tag{6.451}$$

where $\alpha, \lambda, t \in \mathbb{C}$.

The formulas (6.446) and (6.447) were considered in [92, 277].

Theorem 6.27 *Let $\alpha, \lambda, t \in \mathbb{C}$. Then we have*

$$\mathbb{M}_{\alpha}(\lambda, t) = \mathbb{M}cosh_{\alpha}(\lambda, t) + \mathbb{M}sinh_{\alpha}(\lambda, t) \tag{6.452}$$

and

$$\mathbb{M}_{\alpha}(-\lambda, t) = \mathbb{M}cosh_{\alpha}(\lambda, t) - \mathbb{M}sinh_{\alpha}(\lambda, t). \tag{6.453}$$

Proof In view of (6.446) and (6.447), we give the result. □

In particular,

$$\mathbb{M}_{\alpha}(t) = \mathbb{M}cosh_{\alpha}(t) + \mathbb{M}sinh_{\alpha}(t) \tag{6.454}$$

and

$$\mathbb{M}_{\alpha}(-1, t) = \mathbb{M}cosh_{\alpha}(t) - \mathbb{M}sinh_{\alpha}(t), \tag{6.455}$$

where $\alpha, t \in \mathbb{C}$.

6.3.4 The Integral Representations Related to Miller–Ross Function

In this part, we consider the integral representations related to Miller–Ross function.

Theorem 6.28 *Let $\beta \in \mathbb{C}$, $\lambda \in \mathbb{C}$, $\alpha \in \mathbb{R}_{+}$, $Re(\sigma) > 0$, and $Re(\beta) > 0$. Then we have*

$$\frac{1}{\Gamma(\sigma)} \int_0^t (t-\tau)^{\sigma} \, \mathbb{M}sin_{\alpha}(\lambda, \tau) \, d\tau = t^{\alpha+\sigma} \, {}_1sin_1\left(\begin{matrix} (1,1) \\ (\alpha+\sigma+1, 1) \end{matrix}; \lambda t \right)$$
$$= \mathbb{M}sin_{\alpha+\sigma}(\lambda, t), \tag{6.456}$$

$$\frac{1}{\Gamma(\sigma)} \int_0^t (t-\tau)^{\sigma} \, \mathbb{M}cos_{\alpha}(\lambda, \tau) \, d\tau = t^{\alpha+\sigma} \, {}_1cos_1\left(\begin{matrix} (1,1) \\ (\alpha+\sigma+1, 1) \end{matrix}; \lambda t \right)$$
$$= \mathbb{M}cos_{\alpha+\sigma}(\lambda, t), \tag{6.457}$$

$$\frac{1}{\Gamma(\sigma)} \int_0^t (t-\tau)^{\sigma} \, \mathbb{M}sinh_{\alpha}(\lambda, \tau) \, d\tau = t^{\alpha+\sigma} \, {}_1sinh_1\left(\begin{matrix} (1,1) \\ (\alpha+\sigma+1, 1) \end{matrix}; \lambda t \right)$$
$$= \mathbb{M}sinh_{\alpha+\sigma}(\lambda, t), \tag{6.458}$$

and

$$\frac{1}{\Gamma(\sigma)}\int_0^t (t-\tau)^\sigma\, \mathbb{M}cosh_\alpha(\lambda,\tau)\,d\tau = t^{\alpha+\sigma}\ {}_1cosh_1\left(\begin{matrix}(1,1)\\(\alpha+\sigma+1,1)\end{matrix};\lambda t\right)$$
$$= \mathbb{M}cosh_{\alpha+\sigma}(\lambda,t)\,. \tag{6.459}$$

Proof With the aid of the integral representions

$$\frac{1}{\Gamma(\sigma)}\int_0^t (t-\tau)^\sigma\, \mathbb{M}_\alpha(i\lambda,\tau)\,d\tau = t^{\alpha+\sigma}\ {}_1\varphi_1\left(\begin{matrix}(1,1)\\(\alpha+\sigma+1,1)\end{matrix};i\lambda t\right)$$
$$= \mathbb{M}_{\alpha+\sigma}(i\lambda,t)\,, \tag{6.460}$$

$$\frac{1}{\Gamma(\sigma)}\int_0^t (t-\tau)^\sigma\, \mathbb{M}_\alpha(-i\lambda,\tau)\,d\tau = t^{\alpha+\sigma}\ {}_1\varphi_1\left(\begin{matrix}(1,1)\\(\alpha+\sigma+1,1)\end{matrix};-i\lambda t\right)$$
$$= \mathbb{M}_{\alpha+\sigma}(-i\lambda,t)\,, \tag{6.461}$$

$$\frac{1}{\Gamma(\sigma)}\int_0^t (t-\tau)^\sigma\, \mathbb{M}_\alpha(\lambda,\tau)\,d\tau = t^{\alpha+\sigma}\ {}_1\varphi_1\left(\begin{matrix}(1,1)\\(\alpha+\sigma+1,1)\end{matrix};\lambda t\right)$$
$$= \mathbb{M}_{\alpha+\sigma}(\lambda,t)\,, \tag{6.462}$$

and

$$\frac{1}{\Gamma(\sigma)}\int_0^t (t-\tau)^\sigma\, \mathbb{M}_\alpha(-\lambda,\tau)\,d\tau = t^{\alpha+\sigma}\ {}_1\varphi_1\left(\begin{matrix}(1,1)\\(\alpha+\sigma+1,1)\end{matrix};-\lambda t\right)$$
$$= \mathbb{M}_{\alpha+\sigma}(-\lambda,t)\,, \tag{6.463}$$

we set up the desired results from (6.428), (6.429), (6.446), and (6.447). □

Theorem 6.29 *Let $\beta \in \mathbb{C}$, $\lambda \in \mathbb{C}$, $\alpha \in \mathbb{R}_+$, $Re(\sigma) > 0$, and $Re(\beta) > 0$. Then we have*

$$\frac{1}{\Gamma(\sigma)}\int_0^1 (1-\tau)^\sigma\, \mathbb{M}sin_\alpha(\lambda,\tau)\,d\tau = {}_1sin_1\left(\begin{matrix}(1,1)\\(\alpha+\sigma+1,1)\end{matrix};\lambda\right)$$
$$= \mathbb{M}sin_{\alpha+\sigma}(\lambda,1)\,, \tag{6.464}$$

$$\frac{1}{\Gamma(\sigma)}\int_0^1 (1-\tau)^\sigma\, \mathbb{M}cos_\alpha(\lambda,\tau)\,d\tau = {}_1cos_1\left(\begin{matrix}(1,1)\\(\alpha+\sigma+1,1)\end{matrix};\lambda\right)$$
$$= \mathbb{M}cos_{\alpha+\sigma}(\lambda,1)\,, \tag{6.465}$$

$$\frac{1}{\Gamma(\sigma)}\int_0^1 (1-\tau)^{\sigma}\,\mathbb{M}sinh_{\alpha}(\lambda,\tau)\,d\tau = {}_1sinh_1\left(\begin{array}{c}(1,1)\\(\alpha+\sigma+1,1)\end{array};\lambda\right)$$
$$= \mathbb{M}sinh_{\alpha+\sigma}(\lambda,1), \qquad (6.466)$$

and

$$\frac{1}{\Gamma(\sigma)}\int_0^1 (1-\tau)^{\sigma}\,\mathbb{M}cosh_{\alpha}(\lambda,\tau)\,d\tau = {}_1cosh_1\left(\begin{array}{c}(1,1)\\(\alpha+\sigma+1,1)\end{array};\lambda\right)$$
$$= \mathbb{M}cosh_{\alpha+\sigma}(\lambda,1). \qquad (6.467)$$

Proof Taking $t = 1$ in (6.456), (6.457), (6.458), and (6.459), we get the results. □

Theorem 6.30 *Let $\beta \in \mathbb{C}$, $\lambda \in \mathbb{C}$, $\alpha \in \mathbb{R}_+$, $Re(\sigma) > 0$, and $Re(\beta) > 0$. Then we have*

$$\frac{1}{\Gamma(\sigma)}\int_0^2 (2-\tau)^{\sigma}\,\mathbb{M}sin_{\alpha}(\lambda,\tau)\,d\tau = \frac{1}{\Gamma(\sigma)}\int_{-1}^1 (1-\tau)^{\sigma}\,\mathbb{M}sin_{\alpha}(\lambda,1+\tau)\,d\tau$$
$$= \frac{1}{\Gamma(\sigma)}\int_{-1}^1 (1+\tau)^{\sigma}\,\mathbb{M}sin_{\alpha}(\lambda,1-\tau)\,d\tau$$
$$= 2^{\alpha+\sigma}\,{}_1sin_1\left(\begin{array}{c}(1,1)\\(\alpha+\sigma+1,1)\end{array};2\lambda\right)$$
$$= \mathbb{M}sin_{\alpha+\sigma}(\lambda,2), \qquad (6.468)$$

$$\frac{1}{\Gamma(\sigma)}\int_0^2 (2-\tau)^{\sigma}\,\mathbb{M}cos_{\alpha}(\lambda,\tau)\,d\tau = \frac{1}{\Gamma(\sigma)}\int_{-1}^1 (1-\tau)^{\sigma}\,\mathbb{M}cos_{\alpha}(\lambda,1+\tau)\,d\tau$$
$$= \frac{1}{\Gamma(\sigma)}\int_{-1}^1 (1+\tau)^{\sigma}\,\mathbb{M}cos_{\alpha}(\lambda,1-\tau)\,d\tau$$
$$= 2^{\alpha+\sigma}\,{}_1cos_1\left(\begin{array}{c}(1,1)\\(\alpha+\sigma+1,1)\end{array};2\lambda\right)$$
$$= \mathbb{M}cos_{\alpha+\sigma}(\lambda,2), \qquad (6.469)$$

$$\frac{1}{\Gamma(\sigma)}\int_0^2 (2-\tau)^{\sigma}\,\mathbb{M}sinh_{\alpha}(\lambda,\tau)\,d\tau = \frac{1}{\Gamma(\sigma)}\int_{-1}^1 (1-\tau)^{\sigma}\,\mathbb{M}sinh_{\alpha}(\lambda,1+\tau)\,d\tau$$
$$= \frac{1}{\Gamma(\sigma)}\int_{-1}^1 (1+\tau)^{\sigma}\,\mathbb{M}sinh_{\alpha}(\lambda,1-\tau)\,d\tau$$
$$= 2^{\alpha+\sigma}\,{}_1sinh_1\left(\begin{array}{c}(1,1)\\(\alpha+\sigma+1,1)\end{array};2\lambda\right)$$
$$= \mathbb{M}sinh_{\alpha+\sigma}(\lambda,2), \qquad (6.470)$$

and

$$\frac{1}{\Gamma(\sigma)} \int_0^2 (2-\tau)^\sigma \, \mathbb{M}cosh_\alpha(\lambda, \tau)\, d\tau = \frac{1}{\Gamma(\sigma)} \int_{-1}^1 (1-\tau)^\sigma \, \mathbb{M}cosh_\alpha(\lambda, 1+\tau)\, d\tau$$

$$= \frac{1}{\Gamma(\sigma)} \int_{-1}^1 (1+\tau)^\sigma \, \mathbb{M}cosh_\alpha(\lambda, 1-\tau)\, d\tau$$

$$= 2^{\alpha+\sigma} \, {}_1cosh_1 \left(\begin{matrix} (1,1) \\ (\alpha+\sigma+1, 1) \end{matrix} ; 2\lambda \right)$$

$$= \mathbb{M}cosh_{\alpha+\sigma}(\lambda, 2). \tag{6.471}$$

Proof Taking $t = 2$ in (6.456), (6.457), (6.458), and (6.459), we get the results. □

Property 6.37 Let $\alpha, t \in \mathbb{C}$. Then we have [92, 277]

$$L\{\mathbb{M}sin_\alpha(\lambda, \tau)\} = \frac{s^{-\alpha}\lambda}{s^2+\lambda^2} \quad (|\lambda/s| < 1), \tag{6.472}$$

$$L\{\mathbb{M}cos_\alpha(\lambda, \tau)\} = \frac{s^{1-\alpha}}{s^2+\lambda^2} \quad (|\lambda/s| < 1), \tag{6.473}$$

$$L\{\mathbb{M}sinh_\alpha(\lambda, \tau)\} = \frac{s^{-\alpha}\lambda}{s^2-\lambda^2} \quad (|\lambda/s| < 1), \tag{6.474}$$

and

$$L\{\mathbb{M}cosh_\alpha(\lambda, \tau)\} = \frac{s^{1-\alpha}}{s^2-\lambda^2} \quad (|\lambda/s| < 1). \tag{6.475}$$

Proof With the aid of the integral representations

$$L\{\mathbb{M}_\alpha(i\lambda, t)\} = \frac{s^{-\alpha}}{s-i\lambda} \quad (|\lambda/s| < 1), \tag{6.476}$$

$$L\{\mathbb{M}_\alpha(-i\lambda, t)\} = \frac{s^{-\alpha}}{s+i\lambda} \quad (|\lambda/s| < 1), \tag{6.477}$$

$$L\{\mathbb{M}_\alpha(\lambda, t)\} = \frac{s^{-\alpha}}{s-\lambda} \quad (|\lambda/s| < 1), \tag{6.478}$$

and

$$L\{\mathbb{M}_\alpha(-\lambda, t)\} = \frac{s^{-\alpha}}{s+\lambda} \quad (|\lambda/s| < 1), \tag{6.479}$$

we have

$$L\{\mathbb{M}sin_{\alpha}(\lambda,\tau)\} = s^{-\alpha}\left(\frac{1}{s-i\lambda} - \frac{1}{s+i\lambda}\right)$$
$$= \frac{s^{-\alpha}\lambda}{s^2+\lambda^2}\,(|\lambda/s|<1)\,, \tag{6.480}$$

$$L\{\mathbb{M}cos_{\alpha}(\lambda,\tau)\} = s^{-\alpha}\left(\frac{1}{s-i\lambda} + \frac{1}{s+i\lambda}\right)$$
$$= \frac{s^{1-\alpha}}{s^2+\lambda^2}\,(|\lambda/s|<1)\,, \tag{6.481}$$

$$L\{\mathbb{M}sinh_{\alpha}(\lambda,\tau)\} = s^{-\alpha}\left(\frac{1}{s-\lambda} - \frac{1}{s+\lambda}\right)$$
$$= \frac{s^{-\alpha}\lambda}{s^2-\lambda^2}\,(|\lambda/s|<1) \tag{6.482}$$

and

$$L\{\mathbb{M}cosh_{\alpha}(\lambda,\tau)\} = s^{-\alpha}\left(\frac{1}{s-\lambda} + \frac{1}{s+\lambda}\right)$$
$$= \frac{s^{1-\alpha}}{s^2-\lambda^2}\,(|\lambda/s|<1)\,. \tag{6.483}$$

The formulas (6.472), (6.473), (6.474), and (6.475) were considered in [92, 277].

□

6.4 The Lorenzo–Hartley Functions

In 1998 and 2008, Carl F. Lorenzo and Tom T. Hartley proposed the special functions, which are called the Lorenzo–Hartley function of first type (one-parametric Lorenzo–Hartley function) [92, 278, 279] and Lorenzo–Hartley function of second type (two-parametric Lorenzo–Hartley function) [92, 279, 280], respectively.

6.4.1 *The Lorenzo–Hartley Function of First Type*

In this part, we consider the special functions related to the Lorenzo–Hartley function of first type.

Definition 6.78 The Lorenzo–Hartley function of first type is defined as [92, 278, 279]

$$\mathbb{L}_{\alpha}(t) = t^{\alpha-1} \sum_{k=0}^{\infty} \frac{t^{k\alpha}}{\Gamma((k+1)\alpha)}, \tag{6.484}$$

where $\alpha, t \in \mathbb{C}$.

We may consider the Lorenzo–Hartley function of first type [92, 278, 279]

$$\mathbb{L}_{\alpha}(\lambda, t) = t^{\alpha-1} \sum_{k=0}^{\infty} \frac{\lambda^k t^{k\alpha}}{\Gamma((k+1)\alpha)}, \tag{6.485}$$

which leads to

$$\mathbb{L}_{\alpha}(t) = \mathbb{L}_{\alpha}(1, t),$$

where $\alpha, \lambda, t \in \mathbb{C}$.

The alternative form of the Lorenzo–Hartley function of first type can be given as follows:

$$\begin{aligned} \mathbb{L}_{\alpha}(\lambda, t) &= t^{\alpha-1} \sum_{k=0}^{\infty} \frac{\lambda^k t^{k\alpha}}{\Gamma(k\alpha+\alpha)} \\ &= t^{\alpha-1} E_{\alpha,\alpha}\left(\lambda t^{\alpha}\right) \\ &= t^{\alpha-1} {}_1\varphi_1\left(\begin{matrix}(1,1)\\(\alpha,\alpha)\end{matrix}; \lambda t^{\alpha}\right). \end{aligned} \tag{6.486}$$

The Laplace transform of the Lorenzo–Hartley function of first type can be written as follows [92, 279]:

$$\begin{aligned} L\left\{\mathbb{L}_{\alpha}(\lambda, t)\right\} &= L\left\{t^{\alpha-1} \sum_{k=0}^{\infty} \frac{\lambda^k t^{k\alpha}}{\Gamma(k\alpha+\alpha)}\right\} \\ &= s^{-\alpha} \sum_{k=0}^{\infty} \frac{\lambda^k}{s^{k\alpha}} \\ &= \frac{1}{s^{\alpha}-\lambda} \left(\left|\frac{\lambda}{s^{\alpha}}\right| < 1\right). \end{aligned} \tag{6.487}$$

In this case, we may structure the Lorenzo–Hartley type function of first type as follows:

$$\begin{aligned}\widehat{\mathbb{L}}_{\alpha}(\lambda, t) &= t^{\alpha-1}\sum_{k=0}^{\infty}\frac{\lambda^k t^k}{\Gamma(k\alpha+\alpha)}\\ &= t^{\alpha-1}E_{\alpha,\alpha}(\lambda t)\\ &= t^{\alpha-1}{}_1\varphi_1\left(\begin{matrix}(1,1)\\(\alpha,\alpha)\end{matrix};\lambda t\right),\end{aligned} \tag{6.488}$$

which leads to

$$\widehat{\mathbb{L}}_{\alpha}(t) = \widehat{\mathbb{L}}_{\alpha}(1, t),$$

and its Mellin transform reads

$$M\left[\widehat{\mathbb{L}}_{\alpha}(-\lambda, t)\right] = \int_0^{\infty} t^{\varpi+\alpha-1}E_{\alpha,\alpha}(-\lambda t)\,dt = \frac{\Gamma(\varpi+\alpha)\Gamma(1-\varpi-\alpha)}{\Gamma(\alpha(1-\varpi-\alpha))}\lambda^{-(\varpi+\alpha)}, \tag{6.489}$$

where $\alpha, \varpi \in \mathbb{C}$, $\lambda \in \mathbb{C}$, $Re(\varpi) > 0$, $Re(\alpha) > 0$, and $Re(\alpha) + Re(\varpi) < 1$.

Property 6.38 Let $\alpha, \varpi \in \mathbb{C}$, $\lambda \in \mathbb{C}$, $Re(\varpi) > 0$, $Re(\alpha) > 0$, and $Re(\alpha) + Re(\varpi) < 1$.

$$\int_{-\infty}^{\infty} |t|^{\varpi+\alpha-1} E_{\alpha,\alpha}(-\lambda|t|)\,dt = 2\lambda^{-(\varpi+\alpha)}\frac{\Gamma(\varpi+\alpha)\Gamma(1-\varpi-\alpha)}{\Gamma(\alpha(1-\varpi-\alpha))}. \tag{6.490}$$

Proof Making use of (6.489), we derive the result. □

Property 6.39 Let $\alpha, \varpi \in \mathbb{C}$, $\lambda \in \mathbb{C}$, $Re(\varpi) > 0$, $Re(\alpha) > 0$, and $Re(\alpha) + Re(\varpi) < 1$. Then we have

$$\widehat{\mathbb{L}}_{\alpha}(t) = \frac{1}{2\pi i}\int_{-i\infty}^{i\infty}\frac{\Gamma(\varpi+\alpha)\Gamma(1-\varpi-\alpha)}{\Gamma(\alpha(1-\varpi-\alpha))}(-t)^{-\varpi}\,d\varpi. \tag{6.491}$$

Proof Making use of (6.489), we obtain the result. □

It is not difficult to show the following result:
Let $\lambda \in \mathbb{C}$, $\alpha \in \mathbb{R}_+$, and $Re(\sigma) > 0$. Then we have

$$\frac{1}{\Gamma(\sigma)}\int_0^t (t-\tau)^{\sigma}\,\mathbb{L}_{\alpha}(\lambda, \tau)\,d\tau = t^{\alpha+\sigma-1}{}_1\varphi_1\left(\begin{matrix}(1,1)\\(\alpha+\sigma,\alpha)\end{matrix};\lambda t^{\alpha}\right). \tag{6.492}$$

Property 6.40 Let $\lambda \in \mathbb{C}$, $\alpha \in \mathbb{R}_+$, and $Re(\sigma) > 0$. Then we have

$$\frac{1}{\Gamma(\sigma)}\int_0^1 (1-\tau)^\sigma \, \mathbb{L}_\alpha(\lambda,\tau)\, d\tau = {}_1\varphi_1\left(\begin{matrix}(1,1)\\(\alpha+\sigma,\alpha)\end{matrix};\lambda\right). \tag{6.493}$$

Proof Putting $t = 1$ in (6.492), we obtain the result. □

Property 6.41 Let $\lambda \in \mathbb{C}$, $\alpha \in \mathbb{R}_+$, and $Re(\sigma) > 0$. Then we have

$$\begin{aligned}\frac{1}{\Gamma(\sigma)}\int_0^2 (2-\tau)^\sigma \, \mathbb{L}_\alpha(\lambda,\tau)\, d\tau &= \frac{1}{\Gamma(\sigma)}\int_{-1}^1 (1-\tau)^\sigma \, \mathbb{L}_\alpha(\lambda,1+\tau)\, d\tau\\ &= \frac{1}{\Gamma(\sigma)}\int_{-1}^1 (1+\tau)^\sigma \, \mathbb{L}_\alpha(\lambda,1-\tau)\, d\tau\\ &= 2^{\alpha+\sigma-1}\, {}_1\varphi_1\left(\begin{matrix}(1,1)\\(\alpha+\sigma,\alpha)\end{matrix};\lambda 2^\alpha\right).\end{aligned} \tag{6.494}$$

Proof Setting $t = 2$ in (6.492), we give the result. □

6.4.2 *The Supertrigonometric Functions via Lorenzo–Hartley Function of First Type*

In this part, we propose the supertrigonometric functions via Lorenzo–Hartley function of first type.

We now consider that

$$\begin{aligned}\mathbb{L}_\alpha(i\lambda,t) &= t^{\alpha-1}\sum_{k=0}^{\infty}\frac{(i\lambda)^k t^{k\alpha}}{\Gamma(k\alpha+\alpha)}\\ &= t^{\alpha-1}E_{\alpha,\alpha}\left(i\lambda t^\alpha\right)\\ &= t^{\alpha-1}\, {}_1\varphi_1\left(\begin{matrix}(1,1)\\(\alpha,\alpha)\end{matrix};i\lambda t^\alpha\right)\end{aligned} \tag{6.495}$$

and

$$\begin{aligned}\mathbb{L}_\alpha(-i\lambda,t) &= t^{\alpha-1}\sum_{k=0}^{\infty}\frac{(-i\lambda)^k t^{k\alpha}}{\Gamma(k\alpha+\alpha)}\\ &= t^{\alpha-1}E_{\alpha,\alpha}\left(-i\lambda t^\alpha\right)\\ &= t^{\alpha-1}\, {}_1\varphi_1\left(\begin{matrix}(1,1)\\(\alpha,\alpha)\end{matrix};-i\lambda t^\alpha\right).\end{aligned} \tag{6.496}$$

In view of (6.495) and (6.496), we present the following special functions:

Definition 6.79 The supersine via Lorenzo–Hartley function of first type is defined as [92]

$$\begin{aligned} \mathbb{L}sin_{\alpha}(t) &= \frac{1}{2i}\left(\mathbb{L}_{\alpha}(i,t) - \mathbb{L}_{\alpha}(-i,t)\right) \\ &= t^{\alpha-1}\sum_{k=0}^{\infty}\frac{(-1)^k\, t^{(2k+1)\alpha}}{\Gamma(2k\alpha+2\alpha)}, \end{aligned} \tag{6.497}$$

where $\alpha, t \in \mathbb{C}$.

Definition 6.80 The supercosine via Lorenzo–Hartley function of first type is defined as [92]

$$\begin{aligned} \mathbb{L}cos_{\alpha}(t) &= \frac{1}{2}\left(\mathbb{L}_{\alpha}(i,t) + \mathbb{L}_{\alpha}(-i,t)\right) \\ &= t^{\alpha-1}\sum_{k=0}^{\infty}\frac{(-1)^k\, t^{2k\alpha}}{\Gamma(2k\alpha+\alpha)}, \end{aligned} \tag{6.498}$$

where $\alpha, t \in \mathbb{C}$.

Definition 6.81 The supertangent via Lorenzo–Hartley function of first type is defined as

$$\mathbb{L}tan_{\alpha}(t) = \frac{\mathbb{L}sin_{\alpha}(t)}{\mathbb{L}cos_{\alpha}(t)}\left(\mathbb{L}cos_{\alpha}(t) \neq 0\right), \tag{6.499}$$

where $\alpha, t \in \mathbb{C}$.

Definition 6.82 The supercotangent via Lorenzo–Hartley function of first type is defined as

$$\mathbb{L}cotan_{\alpha}(t) = \frac{\mathbb{L}cos_{\alpha}(t)}{\mathbb{L}sin_{\alpha}(t)}, \left(\mathbb{L}sin_{\alpha}(t) \neq 0\right), \tag{6.500}$$

where $\alpha, t \in \mathbb{C}$.

Definition 6.83 The supercosecant via Lorenzo–Hartley function of first type is defined as

$$\mathbb{L}csc_{\alpha}(t) = \frac{1}{\mathbb{L}sin_{\alpha}(t)}, \left(\mathbb{L}sin_{\alpha}(t) \neq 0\right), \tag{6.501}$$

where $\alpha, t \in \mathbb{C}$.

Definition 6.84 The supersecant via Lorenzo–Hartley function of first type is defined as

$$\mathbb{L}sec_{\alpha}(t) = \frac{1}{\mathbb{L}cos_{\alpha}(t)} \left(\mathbb{L}cos_{\alpha}(t) \neq 0\right), \tag{6.502}$$

where $\alpha, t \in \mathbb{C}$.

Therefore, we have that

$$\begin{aligned}\mathbb{L}sin_{\alpha}(\lambda, t) &= \frac{1}{2i}\left(\mathbb{L}_{\alpha}(i\lambda, t) - \mathbb{L}_{\alpha}(-i\lambda, t)\right) \\ &= t^{\alpha-1} \sum_{k=0}^{\infty} \frac{(-1)^k \lambda^{2k+1} t^{(2k+1)\alpha}}{\Gamma(2k\alpha + 2\alpha)},\end{aligned} \tag{6.503}$$

$$\begin{aligned}\mathbb{L}cos_{\alpha}(\lambda, t) &= \frac{1}{2}\left(\mathbb{L}_{\alpha}(i\lambda, t) + \mathbb{L}_{\alpha}(-i\lambda, t)\right) \\ &= t^{\alpha-1} \sum_{k=0}^{\infty} \frac{(-1)^k \lambda^{2k} t^{2k\alpha}}{\Gamma(2k\alpha + \alpha)},\end{aligned} \tag{6.504}$$

$$\mathbb{L}tan_{\alpha}(\lambda, t) = \frac{\mathbb{L}sin_{\alpha}(\lambda, t)}{\mathbb{L}cos_{\alpha}(\lambda, t)} \left(\mathbb{L}cos_{\alpha}(\lambda, t) \neq 0\right), \tag{6.505}$$

$$\mathbb{L}cotan_{\alpha}(\lambda, t) = \frac{\mathbb{L}cos_{\alpha}(\lambda, t)}{\mathbb{L}sin_{\alpha}(\lambda, t)}, \left(\mathbb{L}sin_{\alpha}(\lambda, t) \neq 0\right), \tag{6.506}$$

$$\mathbb{L}csc_{\alpha}(\lambda, t) = \frac{1}{\mathbb{L}sin_{\alpha}(\lambda, t)}, \left(\mathbb{L}sin_{\alpha}(\lambda, t) \neq 0\right), \tag{6.507}$$

and

$$\mathbb{L}sec_{\alpha}(\lambda, t) = \frac{1}{\mathbb{L}cos_{\alpha}(\lambda, t)} \left(\mathbb{L}cos_{\alpha}(\lambda, t) \neq 0\right), \tag{6.508}$$

where $\alpha, \lambda, t \in \mathbb{C}$.

Theorem 6.31 *Let $\alpha, \lambda, t \in \mathbb{C}$. Then we have*

$$\mathbb{L}_{\alpha}(i\lambda, t) = \mathbb{L}cos_{\alpha}(\lambda, t) + i\mathbb{L}sin_{\alpha}(\lambda, t) \tag{6.509}$$

and

$$\mathbb{L}_{\alpha}(-i\lambda, t) = \mathbb{L}cos_{\alpha}(\lambda, t) - i\mathbb{L}sin_{\alpha}(\lambda, t). \tag{6.510}$$

Proof In view of (6.506) and (6.507), we give the result. □

In particular,

$$\mathbb{L}_\alpha(t) = \mathbb{L}cos_\alpha(t) + i\mathbb{L}sin_\alpha(t) \tag{6.511}$$

and

$$\mathbb{L}_\alpha(-i,t) = \mathbb{L}cos_\alpha(t) - i\mathbb{L}sin_\alpha(t), \tag{6.512}$$

where $\alpha, t \in \mathbb{C}$.

6.4.3 *The Superhyperbolic Functions via Lorenzo–Hartley Function of First Type*

In this part, we propose the superhyperbolic functions via Lorenzo–Hartley function of first type.

We now consider that

$$\begin{aligned}\mathbb{L}_\alpha(\lambda,t) &= t^{\alpha-1}\sum_{k=0}^{\infty}\frac{\lambda^k t^{k\alpha}}{\Gamma(k\alpha+\alpha)}\\ &= t^{\alpha-1}E_{\alpha,\alpha}\left(\lambda t^\alpha\right)\\ &= t^{\alpha-1}{}_1\varphi_1\left(\begin{matrix}(1,1)\\(\alpha,\alpha)\end{matrix};\lambda t^\alpha\right)\end{aligned} \tag{6.513}$$

and

$$\begin{aligned}\mathbb{L}_\alpha(-\lambda,t) &= t^{\alpha-1}\sum_{k=0}^{\infty}\frac{(-\lambda)^k t^{k\alpha}}{\Gamma(k\alpha+\alpha)}\\ &= t^{\alpha-1}E_{\alpha,\alpha}\left(-\lambda t^\alpha\right)\\ &= t^{\alpha-1}{}_1\varphi_1\left(\begin{matrix}(1,1)\\(\alpha,\alpha)\end{matrix};-\lambda t^\alpha\right).\end{aligned} \tag{6.514}$$

With the use of (6.513) and (6.514), we recall the following special functions:

Definition 6.85 The hyperbolic supersine via Lorenzo–Hartley function of first type is defined as [92]

$$\begin{aligned}\mathbb{L}sinh_\alpha(t) &= \frac{1}{2}\left(\mathbb{L}_\alpha(1,t) - \mathbb{L}_\alpha(-1,t)\right)\\ &= t^{\alpha-1}\sum_{k=0}^{\infty}\frac{t^{(2k+1)\alpha}}{\Gamma(2k\alpha+2\alpha)},\end{aligned} \tag{6.515}$$

where $\alpha, t \in \mathbb{C}$.

Definition 6.86 The hyperbolic supercosine via Lorenzo–Hartley function of first type is defined as [92]

$$\mathbb{L}cosh_{\alpha}(t) = \frac{1}{2}\left(\mathbb{L}_{\alpha}(1,t) + \mathbb{L}_{\alpha}(-1,t)\right) = t^{\alpha-1}\sum_{k=0}^{\infty}\frac{\lambda^{2k}t^{2k\alpha}}{\Gamma(2k\alpha+\alpha)}, \tag{6.516}$$

where $\alpha, t \in \mathbb{C}$.

Definition 6.87 The hyperbolic supertangent via Lorenzo–Hartley function of first type is defined as

$$\mathbb{L}tanh_{\alpha}(t) = \frac{\mathbb{L}sinh_{\alpha}(t)}{\mathbb{L}cosh_{\alpha}(t)}\,(\mathbb{L}cosh_{\alpha}(t) \neq 0), \tag{6.517}$$

where $\alpha, t \in \mathbb{C}$.

Definition 6.88 The hyperbolic supercotangent via Lorenzo–Hartley function of first type is defined as

$$\mathbb{L}cotanh_{\alpha}(t) = \frac{\mathbb{L}cosh_{\alpha}(t)}{\mathbb{L}sinh_{\alpha}(t)}, (\mathbb{L}sinh_{\alpha}(t) \neq 0), \tag{6.518}$$

where $\alpha, t \in \mathbb{C}$.

Definition 6.89 The hyperbolic supercosecant via Lorenzo–Hartley function of first type is defined as

$$\mathbb{L}csch_{\alpha}(t) = \frac{1}{\mathbb{L}sinh_{\alpha}(t)}, (\mathbb{L}sinh_{\alpha}(t) \neq 0), \tag{6.519}$$

where $\alpha, t \in \mathbb{C}$.

Definition 6.90 The hyperbolic supersecant via Lorenzo–Hartley function of first type is defined as

$$\mathbb{L}sech_{\alpha}(t) = \frac{1}{\mathbb{L}cosh_{\alpha}(t)}\,(\mathbb{L}cosh_{\alpha}(t) \neq 0), \tag{6.520}$$

where $\alpha, t \in \mathbb{C}$.

Therefore, we have that

$$\mathbb{L}sinh_\alpha(\lambda, t) = \frac{1}{2}\left(\mathbb{L}_\alpha(\lambda, t) - \mathbb{L}_\alpha(-\lambda, t)\right)$$
$$= t^{\alpha-1}\sum_{k=0}^{\infty}\frac{\lambda^{2k+1}t^{(2k+1)\alpha}}{\Gamma(2k\alpha+2\alpha)}, \quad (6.521)$$

$$\mathbb{L}cosh_\alpha(\lambda, t) = \frac{1}{2}\left(\mathbb{L}_\alpha(\lambda, t) + \mathbb{L}_\alpha(-\lambda, t)\right)$$
$$= t^{\alpha-1}\sum_{k=0}^{\infty}\frac{\lambda^{2k}t^{2k\alpha}}{\Gamma(2k\alpha+\alpha)}, \quad (6.522)$$

$$\mathbb{L}tanh_\alpha(\lambda, t) = \frac{\mathbb{L}sinh_\alpha(\lambda, t)}{\mathbb{L}cosh_\alpha(\lambda, t)}\ \left(\mathbb{L}cosh_\alpha(\lambda, t) \neq 0\right), \quad (6.523)$$

$$\mathbb{L}cotanh_\alpha(\lambda, t) = \frac{\mathbb{L}cosh_\alpha(\lambda, t)}{\mathbb{L}sinh_\alpha(\lambda, t)}, \left(\mathbb{L}sinh_\alpha(\lambda, t) \neq 0\right), \quad (6.524)$$

$$\mathbb{L}csch_\alpha(\lambda, t) = \frac{1}{\mathbb{L}sinh_\alpha(\lambda, t)}, \left(\mathbb{L}sinh_\alpha(\lambda, t) \neq 0\right), \quad (6.525)$$

and

$$\mathbb{L}sech_\alpha(\lambda, t) = \frac{1}{\mathbb{L}cosh_\alpha(\lambda, t)}\ \left(\mathbb{L}cosh_\alpha(\lambda, t) \neq 0\right), \quad (6.526)$$

where $\alpha, \lambda, t \in \mathbb{C}$.

Theorem 6.32 *Let $\alpha, \lambda, t \in \mathbb{C}$. Then we have [92]*

$$\mathbb{L}_\alpha(\lambda, t) = \mathbb{L}cosh_\alpha(\lambda, t) + \mathbb{L}sinh_\alpha(\lambda, t) \quad (6.527)$$

and

$$\mathbb{L}_\alpha(-\lambda, t) = \mathbb{L}cosh_\alpha(\lambda, t) - \mathbb{L}sinh_\alpha(\lambda, t). \quad (6.528)$$

Proof In view of (6.521) and (6.522), we deduce the result. □

In particular,

$$\mathbb{L}_\alpha(t) = \mathbb{L}cosh_\alpha(t) + \mathbb{L}sinh_\alpha(t) \quad (6.529)$$

and

$$\mathbb{L}_\alpha(-1, t) = \mathbb{L}cosh_\alpha(t) - i\mathbb{L}sinh_\alpha(t), \tag{6.530}$$

where $\alpha, t \in \mathbb{C}$.

6.4.4 The Integral Representations for the Special Functions Related to the Lorenzo–Hartley Function of First Type

In this part, we consider the integral representations for the special functions related to the Lorenzo–Hartley function of first type.

Property 6.42 Let $\lambda \in \mathbb{C}$, $\alpha \in \mathbb{R}_+$, and $Re(\sigma) > 0$. Then we have

$$\frac{1}{\Gamma(\sigma)} \int_0^t (t-\tau)^\sigma \mathbb{L}sin_\alpha(\lambda, \tau)\, d\tau = t^{\alpha+\sigma-1}\, {}_1sin_1\left(\begin{matrix}(1,1)\\(\alpha+\sigma,\alpha)\end{matrix}; \lambda t^\alpha\right), \tag{6.531}$$

$$\frac{1}{\Gamma(\sigma)} \int_0^t (t-\tau)^\sigma \mathbb{L}cos_\alpha(\lambda, \tau)\, d\tau = t^{\alpha+\sigma-1}\, {}_1cos_1\left(\begin{matrix}(1,1)\\(\alpha+\sigma,\alpha)\end{matrix}; \lambda t^\alpha\right), \tag{6.532}$$

$$\frac{1}{\Gamma(\sigma)} \int_0^t (t-\tau)^\sigma \mathbb{L}sinh_\alpha(\lambda, \tau)\, d\tau = t^{\alpha+\sigma-1}\, {}_1sinh_1\left(\begin{matrix}(1,1)\\(\alpha+\sigma,\alpha)\end{matrix}; \lambda t^\alpha\right), \tag{6.533}$$

and

$$\frac{1}{\Gamma(\sigma)} \int_0^t (t-\tau)^\sigma \mathbb{L}cosh_\alpha(\lambda, \tau)\, d\tau = t^{\alpha+\sigma-1}\, {}_1cosh_1\left(\begin{matrix}(1,1)\\(\alpha+\sigma,\alpha)\end{matrix}; \lambda t^\alpha\right). \tag{6.534}$$

Proof With the aid of (6.492), we have

$$\frac{1}{\Gamma(\sigma)} \int_0^t (t-\tau)^\sigma \mathbb{L}_\alpha(i\lambda, \tau)\, d\tau = t^{\alpha+\sigma-1}\, {}_1\varphi_1\left(\begin{matrix}(1,1)\\(\alpha+\sigma,\alpha)\end{matrix}; i\lambda t^\alpha\right), \tag{6.535}$$

$$\frac{1}{\Gamma(\sigma)} \int_0^t (t-\tau)^\sigma \mathbb{L}_\alpha(-i\lambda, \tau)\, d\tau = t^{\alpha+\sigma-1}\, {}_1\varphi_1\left(\begin{matrix}(1,1)\\(\alpha+\sigma,\alpha)\end{matrix}; -i\lambda t^\alpha\right), \tag{6.536}$$

$$\frac{1}{\Gamma(\sigma)} \int_0^t (t-\tau)^\sigma \mathbb{L}_\alpha(\lambda, \tau)\, d\tau = t^{\alpha+\sigma-1}\, {}_1\varphi_1\left(\begin{matrix}(1,1)\\(\alpha+\sigma,\alpha)\end{matrix}; \lambda t^\alpha\right), \tag{6.537}$$

$$\frac{1}{\Gamma(\sigma)}\int_0^t (t-\tau)^\sigma\, \mathbb{L}_\alpha(-\lambda,\tau)\, d\tau = t^{\alpha+\sigma-1}\, {}_1\varphi_1\left(\begin{matrix}(1,1)\\(\alpha+\sigma,\alpha)\end{matrix};-\lambda t^\alpha\right), \tag{6.538}$$

so that

$$\begin{aligned}
&\frac{1}{\Gamma(\sigma)}\int_0^t (t-\tau)^\sigma\, \mathbb{L}sin_\alpha(\lambda,\tau)\, d\tau \\
&= \frac{1}{\Gamma(\sigma)}\int_0^t \frac{1}{2i}\left(\mathbb{L}_\alpha(i\lambda,\tau)-\mathbb{L}_\alpha(-i\lambda,\tau)\right) d\tau \\
&= t^{\alpha+\sigma-1}\left\{\frac{1}{2i}\left[{}_1\varphi_1\left(\begin{matrix}(1,1)\\(\alpha+\sigma,\alpha)\end{matrix};i\lambda t^\alpha\right)-{}_1\varphi_1\left(\begin{matrix}(1,1)\\(\alpha+\sigma,\alpha)\end{matrix};-i\lambda t^\alpha\right)\right]\right\} \\
&= t^{\alpha+\sigma-1}\, {}_1sin_1\left(\begin{matrix}(1,1)\\(\alpha+\sigma,\alpha)\end{matrix};\lambda t^\alpha\right),
\end{aligned} \tag{6.539}$$

$$\begin{aligned}
&\frac{1}{\Gamma(\sigma)}\int_0^t (t-\tau)^\sigma\, \mathbb{L}cos_\alpha(\lambda,\tau)\, d\tau \\
&= \frac{1}{\Gamma(\sigma)}\int_0^t \frac{1}{2}\left(\mathbb{L}_\alpha(i\lambda,\tau)+\mathbb{L}_\alpha(-i\lambda,\tau)\right) d\tau \\
&= t^{\alpha+\sigma-1}\left\{\frac{1}{2}\left[{}_1\varphi_1\left(\begin{matrix}(1,1)\\(\alpha+\sigma,\alpha)\end{matrix};i\lambda t^\alpha\right)+{}_1\varphi_1\left(\begin{matrix}(1,1)\\(\alpha+\sigma,\alpha)\end{matrix};-i\lambda t^\alpha\right)\right]\right\} \\
&= t^{\alpha+\sigma-1}\, {}_1cos_1\left(\begin{matrix}(1,1)\\(\alpha+\sigma,\alpha)\end{matrix};\lambda t^\alpha\right),
\end{aligned} \tag{6.540}$$

$$\begin{aligned}
&\frac{1}{\Gamma(\sigma)}\int_0^t (t-\tau)^\sigma\, \mathbb{L}sinh_\alpha(\lambda,\tau)\, d\tau \\
&= \frac{1}{\Gamma(\sigma)}\int_0^t \frac{1}{2}\left(\mathbb{L}_\alpha(\lambda,\tau)-\mathbb{L}_\alpha(-\lambda,\tau)\right) d\tau \\
&= t^{\alpha+\sigma-1}\left\{\frac{1}{2}\left[{}_1\varphi_1\left(\begin{matrix}(1,1)\\(\alpha+\sigma,\alpha)\end{matrix};\lambda t^\alpha\right)-{}_1\varphi_1\left(\begin{matrix}(1,1)\\(\alpha+\sigma,\alpha)\end{matrix};-\lambda t^\alpha\right)\right]\right\} \\
&= t^{\alpha+\sigma-1}\, {}_1sinh_1\left(\begin{matrix}(1,1)\\(\alpha+\sigma,\alpha)\end{matrix};\lambda t^\alpha\right),
\end{aligned} \tag{6.541}$$

and

$$\begin{aligned}
&\frac{1}{\Gamma(\sigma)}\int_0^t (t-\tau)^\sigma\, \mathbb{L}cosh_\alpha(\lambda,\tau)\,d\tau \\
&= \frac{1}{\Gamma(\sigma)}\int_0^t \frac{1}{2}\left(\mathbb{L}_\alpha(\lambda,\tau)+\mathbb{L}_\alpha(-\lambda,\tau)\right)d\tau \\
&= t^{\alpha+\sigma-1}\left\{\frac{1}{2}\left[{}_1\varphi_1\left(\begin{matrix}(1,1)\\(\alpha+\sigma,\alpha)\end{matrix};\lambda t^\alpha\right)+{}_1\varphi_1\left(\begin{matrix}(1,1)\\(\alpha+\sigma,\alpha)\end{matrix};-\lambda t^\alpha\right)\right]\right\} \\
&= t^{\alpha+\sigma-1}\,{}_1cosh_1\left(\begin{matrix}(1,1)\\(\alpha+\sigma,\alpha)\end{matrix};\lambda t^\alpha\right). \qquad (6.542)
\end{aligned}$$

□

Property 6.43 Let $\lambda \in \mathbb{C}$, $\alpha \in \mathbb{R}_+$, and $Re(\sigma) > 0$. Then we have

$$\frac{1}{\Gamma(\sigma)}\int_0^1 (1-\tau)^\sigma\, \mathbb{L}sin_\alpha(\lambda,\tau)\,d\tau = {}_1sin_1\left(\begin{matrix}(1,1)\\(\alpha+\sigma,\alpha)\end{matrix};\lambda\right), \qquad (6.543)$$

$$\frac{1}{\Gamma(\sigma)}\int_0^1 (1-\tau)^\sigma\, \mathbb{L}cos_\alpha(\lambda,\tau)\,d\tau = {}_1cos_1\left(\begin{matrix}(1,1)\\(\alpha+\sigma,\alpha)\end{matrix};\lambda\right), \qquad (6.544)$$

$$\frac{1}{\Gamma(\sigma)}\int_0^1 (1-\tau)^\sigma\, \mathbb{L}sinh_\alpha(\lambda,\tau)\,d\tau = {}_1sinh_1\left(\begin{matrix}(1,1)\\(\alpha+\sigma,\alpha)\end{matrix};\lambda\right), \qquad (6.545)$$

and

$$\frac{1}{\Gamma(\sigma)}\int_0^1 (1-\tau)^\sigma\, \mathbb{L}cosh_\alpha(\lambda,\tau)\,d\tau = {}_1cosh_1\left(\begin{matrix}(1,1)\\(\alpha+\sigma,\alpha)\end{matrix};\lambda\right). \qquad (6.546)$$

Proof Setting $t = 1$ in (6.531), (6.532), (6.533), and (6.534), we get the results. □

Property 6.44 Let $\lambda \in \mathbb{C}$, $\alpha \in \mathbb{R}_+$, and $Re(\sigma) > 0$. Then we have

$$\begin{aligned}
\frac{1}{\Gamma(\sigma)}\int_0^2 (2-\tau)^\sigma\, \mathbb{L}sin_\alpha(\lambda,\tau)\,d\tau &= \frac{1}{\Gamma(\sigma)}\int_{-1}^1 (1-\tau)^\sigma\, \mathbb{L}sin_\alpha(\lambda,1+\tau)\,d\tau \\
&= \frac{1}{\Gamma(\sigma)}\int_{-1}^1 (1+\tau)^\sigma\, \mathbb{L}sin_\alpha(\lambda,1-\tau)\,d\tau \\
&= 2^{\alpha+\sigma-1}\,{}_1sin_1\left(\begin{matrix}(1,1)\\(\alpha+\sigma,\alpha)\end{matrix};\lambda 2^\alpha\right), \qquad (6.547)
\end{aligned}$$

$$\frac{1}{\Gamma(\sigma)}\int_0^2 (2-\tau)^\sigma\, \mathbb{L}cos_\alpha(\lambda,\tau)\,d\tau = \frac{1}{\Gamma(\sigma)}\int_{-1}^1 (1-\tau)^\sigma\, \mathbb{L}cos_\alpha(\lambda,1+\tau)\,d\tau$$
$$= \frac{1}{\Gamma(\sigma)}\int_{-1}^1 (1+\tau)^\sigma\, \mathbb{L}cos_\alpha(\lambda,1-\tau)\,d\tau$$
$$= 2^{\alpha+\sigma-1}\,{}_1cos_1\left(\begin{matrix}(1,1)\\(\alpha+\sigma,\alpha)\end{matrix};\lambda 2^\alpha\right), \tag{6.548}$$

$$\frac{1}{\Gamma(\sigma)}\int_0^2 (2-\tau)^\sigma\, \mathbb{L}sinh_\alpha(\lambda,\tau)\,d\tau = \frac{1}{\Gamma(\sigma)}\int_{-1}^1 (1-\tau)^\sigma\, \mathbb{L}sinh_\alpha(\lambda,1+\tau)\,d\tau$$
$$= \frac{1}{\Gamma(\sigma)}\int_{-1}^1 (1+\tau)^\sigma\, \mathbb{L}sinh_\alpha(\lambda,1-\tau)\,d\tau$$
$$= 2^{\alpha+\sigma-1}\,{}_1sinh_1\left(\begin{matrix}(1,1)\\(\alpha+\sigma,\alpha)\end{matrix};\lambda 2^\alpha\right), \tag{6.549}$$

and

$$\frac{1}{\Gamma(\sigma)}\int_0^2 (2-\tau)^\sigma\, \mathbb{L}cosh_\alpha(\lambda,\tau)\,d\tau = \frac{1}{\Gamma(\sigma)}\int_{-1}^1 (1-\tau)^\sigma\, \mathbb{L}cosh_\alpha(\lambda,1+\tau)\,d\tau$$
$$= \frac{1}{\Gamma(\sigma)}\int_{-1}^1 (1+\tau)^\sigma\, \mathbb{L}cosh_\alpha(\lambda,1-\tau)\,d\tau$$
$$= 2^{\alpha+\sigma-1}\,{}_1cosh_1\left(\begin{matrix}(1,1)\\(\alpha+\sigma,\alpha)\end{matrix};\lambda 2^\alpha\right). \tag{6.550}$$

Property 6.45 Let $\lambda \in \mathbb{C}$, $\alpha \in \mathbb{R}_+$, and $Re(\sigma) > 0$. Then we have [92]

$$L\{\mathbb{L}sin_\alpha(\lambda,t)\} = \frac{\lambda}{s^{2\alpha}+\lambda^2}\left(\left|\frac{\lambda}{s^\alpha}\right|<1\right), \tag{6.551}$$

$$L\{\mathbb{L}cos_\alpha(\lambda,\tau)\} = \frac{s^\alpha}{s^{2\alpha}+\lambda^2}\left(\left|\frac{\lambda}{s^\alpha}\right|<1\right), \tag{6.552}$$

$$L\{\mathbb{L}sinh_\alpha(\lambda,t)\} = \frac{\lambda}{s^{2\alpha}-\lambda^2}\left(\left|\frac{\lambda}{s^\alpha}\right|<1\right), \tag{6.553}$$

and

$$L\left\{\mathbb{L}cosh_{\alpha}(\lambda,\tau)\right\}=\frac{s^{\alpha}}{s^{2\alpha}-\lambda^{2}}\quad\left(\left|\frac{\lambda}{s^{\alpha}}\right|<1\right). \tag{6.554}$$

Proof Making use of the integral representations

$$L\left\{\mathbb{L}_{\alpha}(i\lambda,t)\right\}=\frac{1}{s^{\alpha}-i\lambda}\quad\left(\left|\frac{\lambda}{s^{\alpha}}\right|<1\right), \tag{6.555}$$

$$L\left\{\mathbb{L}_{\alpha}(-i\lambda,t)\right\}=\frac{1}{s^{\alpha}+i\lambda}\quad\left(\left|\frac{\lambda}{s^{\alpha}}\right|<1\right), \tag{6.556}$$

$$L\left\{\mathbb{L}_{\alpha}(\lambda,t)\right\}=\frac{1}{s^{\alpha}-\lambda}\quad\left(\left|\frac{\lambda}{s^{\alpha}}\right|<1\right), \tag{6.557}$$

and

$$L\left\{\mathbb{L}_{\alpha}(-\lambda,t)\right\}=\frac{1}{s^{\alpha}+\lambda}\quad\left(\left|\frac{\lambda}{s^{\alpha}}\right|<1\right), \tag{6.558}$$

we have that

$$\begin{aligned}L\left\{\mathbb{L}sin_{\alpha}(\lambda,t)\right\}&=L\left\{\frac{1}{2i}\left(\mathbb{L}_{\alpha}(i\lambda,t)-\mathbb{L}_{\alpha}(-i\lambda,t)\right)\right\}\\&=\frac{1}{2i}\left(\frac{1}{s^{\alpha}-i\lambda}-\frac{1}{s^{\alpha}+i\lambda}\right)\\&=\frac{\lambda}{s^{2\alpha}+\lambda^{2}}\quad\left(\left|\frac{\lambda}{s^{\alpha}}\right|<1\right),\end{aligned} \tag{6.559}$$

$$\begin{aligned}L\left\{\mathbb{L}cos_{\alpha}(\lambda,\tau)\right\}&=L\left\{\frac{1}{2}\left(\mathbb{L}_{\alpha}(i\lambda,t)+\mathbb{L}_{\alpha}(-i\lambda,t)\right)\right\}\\&=\frac{1}{2}\left(\frac{1}{s^{\alpha}-i\lambda}+\frac{1}{s^{\alpha}+i\lambda}\right)\\&=\frac{s^{\alpha}}{s^{2\alpha}+\lambda^{2}}\quad\left(\left|\frac{\lambda}{s^{\alpha}}\right|<1\right),\end{aligned} \tag{6.560}$$

$$\begin{aligned}L\left\{\mathbb{L}sinh_{\alpha}(\lambda,t)\right\}&=L\left\{\frac{1}{2}\left(\mathbb{L}_{\alpha}(\lambda,t)-\mathbb{L}_{\alpha}(-\lambda,t)\right)\right\}\\&=\frac{1}{2}\left(\frac{1}{s^{\alpha}-\lambda}-\frac{1}{s^{\alpha}+\lambda}\right)\\&=\frac{\lambda}{s^{2\alpha}-\lambda^{2}}\quad\left(\left|\frac{\lambda}{s^{\alpha}}\right|<1\right),\end{aligned} \tag{6.561}$$

and

$$
\begin{aligned}
L\left\{\mathbb{L}cosh_{\alpha}(\lambda,\tau)\right\} &= L\left\{\frac{1}{2}\left(\mathbb{L}_{\alpha}(\lambda,t)+\mathbb{L}_{\alpha}(-\lambda,t)\right)\right\} \\
&= \frac{1}{2}\left(\frac{1}{s^{\alpha}-\lambda}+\frac{1}{s^{\alpha}+\lambda}\right) \\
&= \frac{s^{\alpha}}{s^{2\alpha}-\lambda^{2}}\left(\left|\frac{\lambda}{s^{\alpha}}\right|<1\right). \qquad (6.562)
\end{aligned}
$$

The results were discovered by Yang in 2019 [92]. □

6.4.5 *The Lorenzo–Hartley Function of Second Type*

In this part, we investigate the special functions related to the Lorenzo–Hartley function of second type.

Definition 6.91 The Lorenzo–Hartley function of second type is defined as [92, 279, 280]

$$
\mathbb{L}_{\alpha,\beta}(t) = t^{\alpha-\beta-1}\sum_{k=0}^{\infty}\frac{t^{k\alpha}}{\Gamma((k+1)\alpha-\beta)}, \qquad (6.563)
$$

where $\alpha,\beta,t\in\mathbb{C}$ and $Re(\alpha)>Re(\beta)$.

For $\alpha,\beta,t,\lambda\in\mathbb{C}$ and $Re(\alpha)>Re(\beta)$, we have the Lorenzo–Hartley function of second type [92, 279, 280]

$$
\mathbb{L}_{\alpha,\beta}(\lambda,t) = t^{\alpha-\beta-1}\sum_{k=0}^{\infty}\frac{\lambda^{k}t^{k\alpha}}{\Gamma((k+1)\alpha-\beta)}, \qquad (6.564)
$$

which leads to

$$
\mathbb{L}_{\alpha,\beta}(t) = \mathbb{L}_{\alpha,\beta}(1,t).
$$

The alternative form of the Lorenzo–Hartley function of second type can be given as follows:

$$
\begin{aligned}
\mathbb{L}_{\alpha,\beta}(\lambda,t) &= t^{\alpha-\beta-1}\sum_{k=0}^{\infty}\frac{\lambda^{k}t^{k\alpha}}{\Gamma((k+1)\alpha-\beta)} \\
&= t^{\alpha-\beta-1}E_{\alpha,\alpha-\beta}\left(\lambda t^{\alpha}\right) \\
&= t^{\alpha-\beta-1}{}_{1}\varphi_{1}\left(\begin{matrix}(1,1)\\(\alpha,\alpha-\beta)\end{matrix};\lambda t^{\alpha}\right), \qquad (6.565)
\end{aligned}
$$

where $\alpha,\beta,t,\lambda\in\mathbb{C}$ and $Re(\alpha)>Re(\beta)$.

The Laplace transform of the Lorenzo–Hartley function of second type can be written as follows [92, 279]:

$$\begin{aligned}L\left\{\mathbb{L}_{\alpha,\beta}(\lambda,t)\right\} &= L\left\{t^{\alpha-\beta-1}\sum_{k=0}^{\infty}\frac{\lambda^k t^{k\alpha}}{\Gamma((k+1)\alpha-\beta)}\right\}\\ &= s^{\beta-\alpha}\sum_{k=0}^{\infty}\frac{\lambda^k}{s^{k\alpha}}\\ &= \frac{s^{\beta}}{s^{\alpha}-\lambda}\left(\left|\frac{\lambda}{s^{\alpha}}\right|<1\right).\end{aligned} \tag{6.566}$$

The Lorenzo–Hartley type function of second type can be given as follows:

$$\begin{aligned}\widehat{\mathbb{L}}_{\alpha,\beta}(\lambda,t) &= t^{\alpha-\beta-1}\sum_{k=0}^{\infty}\frac{\lambda^k t^k}{\Gamma((k+1)\alpha-\beta)}\\ &= t^{\alpha-\beta-1}E_{\alpha,\alpha-\beta}(\lambda t)\\ &= t^{\alpha-\beta-1}\,{}_1\varphi_1\left(\begin{matrix}(1,1)\\(\alpha,\alpha-\beta)\end{matrix};\lambda t\right),\end{aligned} \tag{6.567}$$

which yields that

$$\widehat{\mathbb{L}}_{\alpha,\beta}(t) = \widehat{\mathbb{L}}_{\alpha,\beta}(1,t),$$

where $\alpha, \beta, t, \lambda \in \mathbb{C}$ and $Re(\alpha) > Re(\beta)$.

The Mellin transform of the Lorenzo–Hartley type function of second type is given as follows:

$$\begin{aligned}M\left[\widehat{\mathbb{L}}_{\alpha,\beta}(-\lambda,t)\right] &= \int_0^{\infty} t^{\varpi+\alpha-\beta-2}E_{\alpha,\alpha-\beta}(-\lambda t)\,dt\\ &= \frac{\Gamma(\varpi+\alpha-\beta-1)\Gamma(2+\beta-\varpi-\alpha)}{\Gamma(2\alpha-\beta-\alpha(\varpi+\alpha-\beta))}\lambda^{-(\varpi+\alpha-\beta-1)},\end{aligned} \tag{6.568}$$

where $\alpha, \beta, t, \lambda, \varpi \in \mathbb{C}$ and $Re(\alpha) > Re(\beta)$.

Property 6.46 Let $\alpha, \beta, t, \lambda, \varpi \in \mathbb{C}$ and $Re(\alpha) > Re(\beta)$. Then we have

$$\widehat{\mathbb{L}}_{\alpha,\beta}(t) = \frac{1}{2\pi i}\int_{-i\infty}^{i\infty}\frac{\Gamma(\varpi+\alpha-\beta-1)\Gamma(2+\beta-\varpi-\alpha)}{\Gamma(2\alpha-\beta-\alpha(\varpi+\alpha-\beta))}(-t)^{-\varpi}\,d\varpi. \tag{6.569}$$

Proof Making use of (6.568), we give the result. □

Property 6.47 Let $\alpha, \beta, t, \lambda, \varpi \in \mathbb{C}$ and $Re(\alpha) > Re(\beta)$. Then we have

$$\widehat{\mathbb{L}}_{\alpha,\beta}(t) = \frac{1}{2\pi i} \int_{-i\infty}^{i\infty} \frac{\Gamma(\varpi+\alpha-\beta-1)\Gamma(2+\beta-\varpi-\alpha)}{\Gamma(2\alpha-\beta-\alpha(\varpi+\alpha-\beta))} (-t)^{-\varpi} \, d\varpi . \tag{6.570}$$

Proof In view of (6.568), we give the result. □

Here we recall the following result (found in Chap. 5):
Let $\alpha, \beta, t, \sigma, \lambda \in \mathbb{C}$, $Re(\sigma) > 0$, and $Re(\alpha) > Re(\beta)$. Then we have

$$\frac{1}{\Gamma(\sigma)} \int_0^t (t-\tau)^{\sigma} \mathbb{L}_{\alpha,\beta}(\lambda, \tau) \, d\tau = t^{\alpha-\beta+\sigma-1} {}_1\varphi_1 \left(\begin{matrix} (1,1) \\ (\alpha-\beta+\sigma, \alpha) \end{matrix} ; \lambda t^{\alpha} \right). \tag{6.571}$$

Property 6.48 Let $\alpha, \beta, t, \sigma, \lambda \in \mathbb{C}$, $Re(\sigma) > 0$, and $Re(\alpha) > Re(\beta)$. Then we have

$$\frac{1}{\Gamma(\sigma)} \int_0^1 (1-\tau)^{\sigma} \mathbb{L}_{\alpha,\beta}(\lambda, \tau) \, d\tau = {}_1\varphi_1 \left(\begin{matrix} (1,1) \\ (\alpha-\beta+\sigma, \alpha) \end{matrix} ; \lambda \right). \tag{6.572}$$

Proof Putting $t = 1$ in (6.571), we obtain the result. □

Property 6.49 Let $\alpha, \beta, t, \sigma, \lambda \in \mathbb{C}$, $Re(\sigma) > 0$, and $Re(\alpha) > Re(\beta)$. Then we have

$$\begin{aligned} \frac{1}{\Gamma(\sigma)} \int_0^2 (2-\tau)^{\sigma} \mathbb{L}_{\alpha,\beta}(\lambda, \tau) \, d\tau &= \frac{1}{\Gamma(\sigma)} \int_{-1}^1 (1-\tau)^{\sigma} \mathbb{L}_{\alpha,\beta}(\lambda, 1+\tau) \, d\tau \\ &= \frac{1}{\Gamma(\sigma)} \int_{-1}^1 (1+\tau)^{\sigma} \mathbb{L}_{\alpha,\beta}(\lambda, 1-\tau) \, d\tau \\ &= 2^{\alpha-\beta+\sigma-1} {}_1\varphi_1 \left(\begin{matrix} (1,1) \\ (\alpha-\beta+\sigma, \alpha) \end{matrix} ; \lambda 2^{\alpha} \right). \end{aligned} \tag{6.573}$$

Proof Putting $t = 2$ in (6.571), we obtain the result. □

6.4.6 *The Supertrigonometric Functions via Lorenzo–Hartley Function of Second Type*

In this part, we give the supertrigonometric functions via Lorenzo–Hartley function of second type.

Let us consider

$$\begin{aligned}\mathbb{L}_{\alpha,\beta}\left(i\lambda,t\right)&=t^{\alpha-\beta-1}\sum_{k=0}^{\infty}\frac{(i\lambda)^{k}\,t^{k\alpha}}{\Gamma\left((k+1)\,\alpha-\beta\right)}\\&=t^{\alpha-\beta-1}E_{\alpha,\alpha-\beta}\left(i\lambda t^{\alpha}\right)\\&=t^{\alpha-\beta-1}\,{}_{1}\varphi_{1}\left(\begin{matrix}(1,1)\\(\alpha,\alpha-\beta)\end{matrix};i\lambda t^{\alpha}\right),\end{aligned}\tag{6.574}$$

and

$$\begin{aligned}\mathbb{L}_{\alpha,\beta}\left(-i\lambda,t\right)&=t^{\alpha-\beta-1}\sum_{k=0}^{\infty}\frac{(-i\lambda)^{k}\,t^{k\alpha}}{\Gamma\left((k+1)\,\alpha-\beta\right)}\\&=t^{\alpha-\beta-1}E_{\alpha,\alpha-\beta}\left(-i\lambda t^{\alpha}\right)\\&=t^{\alpha-\beta-1}\,{}_{1}\varphi_{1}\left(\begin{matrix}(1,1)\\(\alpha,\alpha-\beta)\end{matrix};-i\lambda t^{\alpha}\right),\end{aligned}\tag{6.575}$$

where $\alpha,\beta,t,\lambda\in\mathbb{C}$ and $Re\left(\alpha\right)>Re\left(\beta\right)$.

In view of (6.574) and (6.575), we define the special functions as follows:

Definition 6.92 The supersine via Lorenzo–Hartley function of second type is defined as [92]

$$\begin{aligned}\mathbb{L}sin_{\alpha,\beta}\left(t\right)&=\frac{1}{2i}\left(\mathbb{L}_{\alpha,\beta}\left(i,t\right)-\mathbb{L}_{\alpha,\beta}\left(-i,t\right)\right)\\&=t^{\alpha-\beta-1}\sum_{k=0}^{\infty}\frac{(-1)^{k}\,t^{(2k+1)\alpha}}{\Gamma\left((2k+2)\,\alpha-\beta\right)},\end{aligned}\tag{6.576}$$

where $\alpha,\beta,t,\lambda\in\mathbb{C}$ and $Re\left(\alpha\right)>Re\left(\beta\right)$.

Definition 6.93 The supercosine via Lorenzo–Hartley function of second type is defined as [92]

$$\begin{aligned}\mathbb{L}cos_{\alpha,\beta}\left(t\right)&=\frac{1}{2}\left(\mathbb{L}_{\alpha,\beta}\left(i,t\right)+\mathbb{L}_{\alpha,\beta}\left(-i,t\right)\right)\\&=t^{\alpha-\beta-1}\sum_{k=0}^{\infty}\frac{(-1)^{k}\,t^{2k\alpha}}{\Gamma\left((2k+1)\,\alpha-\beta\right)},\end{aligned}\tag{6.577}$$

where $\alpha,\beta,t,\lambda\in\mathbb{C}$ and $Re\left(\alpha\right)>Re\left(\beta\right)$.

Definition 6.94 The supertangent via Lorenzo–Hartley function of second type is defined as

$$\mathbb{L}tan_{\alpha,\beta}(t) = \frac{\mathbb{L}sin_{\alpha,\beta}(t)}{\mathbb{L}cos_{\alpha,\beta}(t)} \left(\mathbb{L}cos_{\alpha,\beta}(t) \neq 0\right), \tag{6.578}$$

where $\alpha, \beta, t, \lambda \in \mathbb{C}$ and $Re(\alpha) > Re(\beta)$.

Definition 6.95 The supercotangent via Lorenzo–Hartley function of second type is defined as

$$\mathbb{L}cotan_{\alpha,\beta}(t) = \frac{\mathbb{L}cos_{\alpha,\beta}(t)}{\mathbb{L}sin_{\alpha,\beta}(t)}, \left(\mathbb{L}sin_{\alpha,\beta}(t) \neq 0\right), \tag{6.579}$$

where $\alpha, \beta, t, \lambda \in \mathbb{C}$ and $Re(\alpha) > Re(\beta)$.

Definition 6.96 The supercosecant via Lorenzo–Hartley function of second type is defined as

$$\mathbb{L}csc_{\alpha,\beta}(t) = \frac{1}{\mathbb{L}sin_{\alpha,\beta}(t)}, \left(\mathbb{L}sin_{\alpha,\beta}(t) \neq 0\right), \tag{6.580}$$

where $\alpha, \beta, t, \lambda \in \mathbb{C}$ and $Re(\alpha) > Re(\beta)$.

Definition 6.97 The supersecant via Lorenzo–Hartley function of second type is defined as

$$\mathbb{L}sec_{\alpha,\beta}(t) = \frac{1}{\mathbb{L}cos_{\alpha,\beta}(t)} \left(\mathbb{L}cos_{\alpha,\beta}(t) \neq 0\right), \tag{6.581}$$

where $\alpha, \beta, t, \lambda \in \mathbb{C}$ and $Re(\alpha) > Re(\beta)$.

Therefore, it is shown that

$$\begin{aligned}\mathbb{L}sin_{\alpha,\beta}(\lambda, t) &= \frac{1}{2i}\left(\mathbb{L}_{\alpha,\beta}(i\lambda, t) - \mathbb{L}_{\alpha,\beta}(-i\lambda, t)\right)\\ &= t^{\alpha-\beta-1}\sum_{k=0}^{\infty}\frac{(-1)^k \lambda^{2k+1} t^{(2k+1)\alpha}}{\Gamma((2k+2)\alpha-\beta)},\end{aligned} \tag{6.582}$$

$$\begin{aligned}\mathbb{L}cos_{\alpha,\beta}(\lambda, t) &= \frac{1}{2}\left(\mathbb{L}_{\alpha,\beta}(i\lambda, t) + \mathbb{L}_{\alpha,\beta}(-i\lambda, t)\right)\\ &= t^{\alpha-\beta-1}\sum_{k=0}^{\infty}\frac{(-1)^k \lambda^{2k} t^{2k\alpha}}{\Gamma((2k+1)\alpha-\beta)},\end{aligned} \tag{6.583}$$

$$\mathbb{L}tan_{\alpha,\beta}(\lambda, t) = \frac{\mathbb{L}sin_{\alpha,\beta}(\lambda, t)}{\mathbb{L}cos_{\alpha,\beta}(\lambda, t)} \left(\mathbb{L}cos_{\alpha,\beta}(\lambda, t) \neq 0\right), \tag{6.584}$$

$$\mathbb{L}cotan_{\alpha,\beta}(\lambda,t) = \frac{\mathbb{L}cos_{\alpha,\beta}(\lambda,t)}{\mathbb{L}sin_{\alpha,\beta}(\lambda,t)}, \left(\mathbb{L}sin_{\alpha,\beta}(\lambda,t) \neq 0\right), \tag{6.585}$$

$$\mathbb{L}csc_{\alpha,\beta}(\lambda,t) = \frac{1}{\mathbb{L}sin_{\alpha,\beta}(\lambda,t)}, \left(\mathbb{L}sin_{\alpha,\beta}(\lambda,t) \neq 0\right), \tag{6.586}$$

and

$$\mathbb{L}sec_{\alpha,\beta}(\lambda,t) = \frac{1}{\mathbb{L}cos_{\alpha,\beta}(\lambda,t)} \left(\mathbb{L}cos_{\alpha,\beta}(\lambda,t) \neq 0\right), \tag{6.587}$$

where $\alpha, \beta, t, \lambda \in \mathbb{C}$ and $Re(\alpha) > Re(\beta)$.

The formulas (6.582) and (6.583) are discussed in [92].

Theorem 6.33 *Let $\alpha, \beta, t, \lambda \in \mathbb{C}$ and $Re(\alpha) > Re(\beta)$. Then we have [92]*

$$\mathbb{L}_{\alpha,\beta}(i\lambda,t) = \mathbb{L}cos_{\alpha,\beta}(\lambda,t) + i\mathbb{L}sin_{\alpha,\beta}(\lambda,t) \tag{6.588}$$

and

$$\mathbb{L}_{\alpha,\beta}(-i\lambda,t) = \mathbb{L}cos_{\alpha,\beta}(\lambda,t) - i\mathbb{L}sin_{\alpha,\beta}(\lambda,t). \tag{6.589}$$

Proof In view of (6.585) and (6.586), we get the result. □

In particular, it is shown that

$$\mathbb{L}_{\alpha,\beta}(t) = \mathbb{L}cos_{\alpha,\beta}(t) + i\mathbb{L}sin_{\alpha,\beta}(t) \tag{6.590}$$

and

$$\mathbb{L}_{\alpha,\beta}(-i,t) = \mathbb{L}cos_{\alpha,\beta}(t) - i\mathbb{L}sin_{\alpha,\beta}(t), \tag{6.591}$$

where $\alpha, \beta, t, \lambda \in \mathbb{C}$ and $Re(\alpha) > Re(\beta)$.

6.4.7 *The Superhyperbolic Functions via Lorenzo–Hartley Function of Second Type*

In this part, we give the superhyperbolic functions via Lorenzo–Hartley function of second type.

Let us consider

$$\begin{aligned}\mathbb{L}_{\alpha,\beta}(\lambda,t) &= t^{\alpha-\beta-1}\sum_{k=0}^{\infty}\frac{\lambda^k t^{k\alpha}}{\Gamma((k+1)\alpha-\beta)}\\ &= t^{\alpha-\beta-1}E_{\alpha,\alpha-\beta}(\lambda t^\alpha)\\ &= t^{\alpha-\beta-1}\,{}_1\varphi_1\left(\begin{matrix}(1,1)\\(\alpha,\alpha-\beta)\end{matrix};\lambda t^\alpha\right)\end{aligned}\tag{6.592}$$

and

$$\begin{aligned}\mathbb{L}_{\alpha,\beta}(-\lambda,t) &= t^{\alpha-\beta-1}\sum_{k=0}^{\infty}\frac{(-\lambda)^k t^{k\alpha}}{\Gamma((k+1)\alpha-\beta)}\\ &= t^{\alpha-\beta-1}E_{\alpha,\alpha-\beta}(-\lambda t^\alpha)\\ &= t^{\alpha-\beta-1}\,{}_1\varphi_1\left(\begin{matrix}(1,1)\\(\alpha,\alpha-\beta)\end{matrix};-\lambda t^\alpha\right),\end{aligned}\tag{6.593}$$

where $\alpha,\beta,t,\lambda\in\mathbb{C}$ and $Re(\alpha)>Re(\beta)$.

In view of (6.592) and (6.593), we present the following special functions:

Definition 6.98 The hyperbolic supersine via Lorenzo–Hartley function of second type is defined as [92]

$$\begin{aligned}\mathbb{L}sinh_{\alpha,\beta}(t) &= \frac{1}{2}\left(\mathbb{L}_{\alpha,\beta}(1,t)-\mathbb{L}_{\alpha,\beta}(-1,t)\right)\\ &= t^{\alpha-\beta-1}\sum_{k=0}^{\infty}\frac{t^{(2k+1)\alpha}}{\Gamma((2k+2)\alpha-\beta)},\end{aligned}\tag{6.594}$$

where $\alpha,\beta,t,\lambda\in\mathbb{C}$ and $Re(\alpha)>Re(\beta)$.

Definition 6.99 The hyperbolic supercosine via Lorenzo–Hartley function of second type is defined as [92]

$$\begin{aligned}\mathbb{L}cosh_{\alpha,\beta}(t) &= \frac{1}{2}\left(\mathbb{L}_{\alpha,\beta}(1,t)+\mathbb{L}_{\alpha,\beta}(-1,t)\right)\\ &= t^{\alpha-\beta-1}\sum_{k=0}^{\infty}\frac{t^{2k\alpha}}{\Gamma((2k+1)\alpha-\beta)},\end{aligned}\tag{6.595}$$

where $\alpha,\beta,t,\lambda\in\mathbb{C}$ and $Re(\alpha)>Re(\beta)$.

Definition 6.100 The hyperbolic supertangent via Lorenzo–Hartley function of second type is defined as

$$\mathbb{L}tanh_{\alpha,\beta}(t) = \frac{\mathbb{L}sinh_{\alpha,\beta}(t)}{\mathbb{L}cosh_{\alpha,\beta}(t)} \left(\mathbb{L}cosh_{\alpha,\beta}(t) \neq 0\right), \tag{6.596}$$

where $\alpha, \beta, t, \lambda \in \mathbb{C}$ and $Re(\alpha) > Re(\beta)$.

Definition 6.101 The hyperbolic supercotangent via Lorenzo–Hartley function of second type is defined as

$$\mathbb{L}cotanh_{\alpha,\beta}(t) = \frac{\mathbb{L}cosh_{\alpha,\beta}(t)}{\mathbb{L}sinh_{\alpha,\beta}(t)}, \left(\mathbb{L}sinh_{\alpha,\beta}(t) \neq 0\right), \tag{6.597}$$

where $\alpha, \beta, t, \lambda \in \mathbb{C}$ and $Re(\alpha) > Re(\beta)$.

Definition 6.102 The hyperbolic supercosecant via Lorenzo–Hartley function of second type is defined as

$$\mathbb{L}csch_{\alpha,\beta}(t) = \frac{1}{\mathbb{L}sinh_{\alpha,\beta}(t)}, \left(\mathbb{L}sinh_{\alpha,\beta}(t) \neq 0\right), \tag{6.598}$$

where $\alpha, \beta, t, \lambda \in \mathbb{C}$ and $Re(\alpha) > Re(\beta)$.

Definition 6.103 The hyperbolic supersecant via Lorenzo–Hartley function of second type is defined as

$$\mathbb{L}sech_{\alpha,\beta}(t) = \frac{1}{\mathbb{L}cosh_{\alpha,\beta}(t)} \left(\mathbb{L}cosh_{\alpha,\beta}(t) \neq 0\right), \tag{6.599}$$

where $\alpha, \beta, t, \lambda \in \mathbb{C}$ and $Re(\alpha) > Re(\beta)$.

Therefore, it is found that

$$\begin{aligned}\mathbb{L}sinh_{\alpha,\beta}(\lambda, t) &= \frac{1}{2}\left(\mathbb{L}_{\alpha,\beta}(\lambda, t) - \mathbb{L}_{\alpha,\beta}(-\lambda, t)\right) \\ &= t^{\alpha-\beta-1} \sum_{k=0}^{\infty} \frac{\lambda^{2k+1} t^{(2k+1)\alpha}}{\Gamma((2k+2)\alpha - \beta)},\end{aligned} \tag{6.600}$$

$$\begin{aligned}\mathbb{L}cosh_{\alpha,\beta}(\lambda, t) &= \frac{1}{2}\left(\mathbb{L}_{\alpha,\beta}(\lambda, t) + \mathbb{L}_{\alpha,\beta}(-\lambda, t)\right) \\ &= t^{\alpha-\beta-1} \sum_{k=0}^{\infty} \frac{\lambda^{2k} t^{2k\alpha}}{\Gamma((2k+1)\alpha - \beta)},\end{aligned} \tag{6.601}$$

$$\mathbb{L}tanh_{\alpha,\beta}(\lambda, t) = \frac{\mathbb{L}sinh_{\alpha,\beta}(\lambda, t)}{\mathbb{L}cosh_{\alpha,\beta}(\lambda, t)} \left(\mathbb{L}cosh_{\alpha,\beta}(\lambda, t) \neq 0\right), \tag{6.602}$$

$$\mathbb{L}cotanh_{\alpha,\beta}(\lambda,t)=\frac{\mathbb{L}cosh_{\alpha,\beta}(\lambda,t)}{\mathbb{L}sinh_{\alpha,\beta}(\lambda,t)},\ \left(\mathbb{L}sinh_{\alpha,\beta}(\lambda,t)\neq 0\right), \tag{6.603}$$

$$\mathbb{L}csch_{\alpha,\beta}(\lambda,t)=\frac{1}{\mathbb{L}sinh_{\alpha,\beta}(\lambda,t)},\ \left(\mathbb{L}sinh_{\alpha,\beta}(\lambda,t)\neq 0\right), \tag{6.604}$$

and

$$\mathbb{L}sech_{\alpha,\beta}(\lambda,t)=\frac{1}{\mathbb{L}cosh_{\alpha,\beta}(\lambda,t)}\ \left(\mathbb{L}cosh_{\alpha,\beta}(\lambda,t)\neq 0\right), \tag{6.605}$$

where $\alpha,\beta,t,\lambda\in\mathbb{C}$ and $Re(\alpha)>Re(\beta)$.

The formulas (6.582) and (6.583) are discussed in [92].

Theorem 6.34 *Let $\alpha,\beta,t,\lambda\in\mathbb{C}$ and $Re(\alpha)>Re(\beta)$. Then we have [92]*

$$\mathbb{L}_{\alpha,\beta}(\lambda,t)=\mathbb{L}cosh_{\alpha,\beta}(\lambda,t)+\mathbb{L}sinh_{\alpha,\beta}(\lambda,t) \tag{6.606}$$

and

$$\mathbb{L}_{\alpha,\beta}(-\lambda,t)=\mathbb{L}cosh_{\alpha,\beta}(\lambda,t)-\mathbb{L}sinh_{\alpha,\beta}(\lambda,t). \tag{6.607}$$

Proof In view of (6.600) and (6.601), we have the result. □

In particular,

$$\mathbb{L}_{\alpha,\beta}(t)=\mathbb{L}cosh_{\alpha,\beta}(t)+\mathbb{L}sinh_{\alpha,\beta}(t) \tag{6.608}$$

and

$$\mathbb{L}_{\alpha,\beta}(-1,t)=\mathbb{L}cosh_{\alpha,\beta}(t)-i\mathbb{L}sinh_{\alpha,\beta}(t), \tag{6.609}$$

where $\alpha,\beta,t\in\mathbb{C}$ and $Re(\alpha)>Re(\beta)$.

6.4.8 The Integral Representations for the Special Functions Related to the Lorenzo–Hartley Function of Second Type

In this part, we present the integral representations for the special functions related to the Lorenzo–Hartley function of second type.

Property 6.50 Let $\alpha, \beta, t, \lambda \in \mathbb{C}$, $Re(\alpha) > Re(\beta)$, and $Re(\sigma) > 0$. Then we have

$$\frac{1}{\Gamma(\sigma)} \int_0^t (t-\tau)^\sigma \, \mathbb{L}sin_{\alpha,\beta}(\lambda, \tau)\, d\tau = t^{\alpha-\beta+\sigma-1} \, {}_1sin_1 \left(\begin{matrix} (1,1) \\ (\alpha-\beta+\sigma, \alpha) \end{matrix} ; \lambda t^\alpha \right), \tag{6.610}$$

$$\frac{1}{\Gamma(\sigma)} \int_0^t (t-\tau)^\sigma \, \mathbb{L}cos_{\alpha,\beta}(\lambda, \tau)\, d\tau = t^{\alpha-\beta+\sigma-1} \, {}_1cos_1 \left(\begin{matrix} (1,1) \\ (\alpha-\beta+\sigma, \alpha) \end{matrix} ; \lambda t^\alpha \right), \tag{6.611}$$

$$\frac{1}{\Gamma(\sigma)} \int_0^t (t-\tau)^\sigma \, \mathbb{L}sinh_{\alpha,\beta}(\lambda, \tau)\, d\tau = t^{\alpha-\beta+\sigma-1} \, {}_1sinh_1 \left(\begin{matrix} (1,1) \\ (\alpha-\beta+\sigma, \alpha) \end{matrix} ; \lambda t^\alpha \right), \tag{6.612}$$

and

$$\frac{1}{\Gamma(\sigma)} \int_0^t (t-\tau)^\sigma \, \mathbb{L}cosh_{\alpha,\beta}(\lambda, \tau)\, d\tau = t^{\alpha-\beta+\sigma-1} \, {}_1cosh_1 \left(\begin{matrix} (1,1) \\ (\alpha-\beta+\sigma, \alpha) \end{matrix} ; \lambda t^\alpha \right). \tag{6.613}$$

Proof With the aid of (6.571), we have

$$\frac{1}{\Gamma(\sigma)} \int_0^t (t-\tau)^\sigma \, \mathbb{L}_{\alpha,\beta}(i\lambda, \tau)\, d\tau = t^{\alpha-\beta+\sigma-1} \, {}_1\varphi_1 \left(\begin{matrix} (1,1) \\ (\alpha-\beta+\sigma, \alpha) \end{matrix} ; i\lambda t^\alpha \right), \tag{6.614}$$

$$\frac{1}{\Gamma(\sigma)} \int_0^t (t-\tau)^\sigma \, \mathbb{L}_{\alpha,\beta}(-i\lambda, \tau)\, d\tau = t^{\alpha-\beta+\sigma-1} \, {}_1\varphi_1 \left(\begin{matrix} (1,1) \\ (\alpha-\beta+\sigma, \alpha) \end{matrix} ; -i\lambda t^\alpha \right), \tag{6.615}$$

$$\frac{1}{\Gamma(\sigma)} \int_0^t (t-\tau)^\sigma \, \mathbb{L}_{\alpha,\beta}(\lambda, \tau)\, d\tau = t^{\alpha-\beta+\sigma-1} \, {}_1\varphi_1 \left(\begin{matrix} (1,1) \\ (\alpha-\beta+\sigma, \alpha) \end{matrix} ; \lambda t^\alpha \right), \tag{6.616}$$

$$\frac{1}{\Gamma(\sigma)} \int_0^t (t-\tau)^\sigma \, \mathbb{L}_{\alpha,\beta}(-\lambda, \tau)\, d\tau = t^{\alpha-\beta+\sigma-1} \, {}_1\varphi_1 \left(\begin{matrix} (1,1) \\ (\alpha-\beta+\sigma, \alpha) \end{matrix} ; -\lambda t^\alpha \right), \tag{6.617}$$

so that

$$\frac{1}{\Gamma(\sigma)}\int_0^t (t-\tau)^{\sigma}\,\mathbb{L}sin_{\alpha,\beta}(\lambda,\tau)\,d\tau$$
$$=\frac{1}{\Gamma(\sigma)}\int_0^t \frac{1}{2i}\left(\mathbb{L}_{\alpha,\beta}(i\lambda,\tau)-\mathbb{L}_{\alpha,\beta}(-i\lambda,\tau)\right)d\tau$$
$$=t^{\alpha-\beta+\sigma-1}\left\{\frac{1}{2i}\left[{}_1\varphi_1\left(\begin{matrix}(1,1)\\(\alpha-\beta+\sigma,\alpha)\end{matrix};i\lambda t^{\alpha}\right)\right.\right.$$
$$\left.\left.-\ {}_1\varphi_1\left(\begin{matrix}(1,1)\\(\alpha-\beta+\sigma,\alpha)\end{matrix};-i\lambda t^{\alpha}\right)\right]\right\}$$
$$=t^{\alpha-\beta+\sigma-1}\,{}_1sin_1\left(\begin{matrix}(1,1)\\(\alpha-\beta+\sigma,\alpha)\end{matrix};\lambda t^{\alpha}\right), \tag{6.618}$$

$$\frac{1}{\Gamma(\sigma)}\int_0^t (t-\tau)^{\sigma}\,\mathbb{L}cos_{\alpha,\beta}(\lambda,\tau)\,d\tau$$
$$=\frac{1}{\Gamma(\sigma)}\int_0^t \frac{1}{2}\left(\mathbb{L}_{\alpha,\beta}(i\lambda,\tau)+\mathbb{L}_{\alpha,\beta}(-i\lambda,\tau)\right)d\tau$$
$$=t^{\alpha-\beta+\sigma-1}\left\{\frac{1}{2}\left[{}_1\varphi_1\left(\begin{matrix}(1,1)\\(\alpha-\beta+\sigma,\alpha)\end{matrix};i\lambda t^{\alpha}\right)\right.\right.$$
$$\left.\left.+\,{}_1\varphi_1\left(\begin{matrix}(1,1)\\(\alpha-\beta+\sigma,\alpha)\end{matrix};-i\lambda t^{\alpha}\right)\right]\right\}$$
$$=t^{\alpha-\beta+\sigma-1}\,{}_1cos_1\left(\begin{matrix}(1,1)\\(\alpha-\beta+\sigma,\alpha)\end{matrix};\lambda t^{\alpha}\right), \tag{6.619}$$

$$\frac{1}{\Gamma(\sigma)}\int_0^t (t-\tau)^{\sigma}\,\mathbb{L}sinh_{\alpha,\beta}(\lambda,\tau)\,d\tau$$
$$=\frac{1}{\Gamma(\sigma)}\int_0^t \frac{1}{2}\left(\mathbb{L}_{\alpha,\beta}(\lambda,\tau)-\mathbb{L}_{\alpha,\beta}(-\lambda,\tau)\right)d\tau$$
$$=t^{\alpha-\beta+\sigma-1}\left\{\frac{1}{2}\left[{}_1\varphi_1\left(\begin{matrix}(1,1)\\(\alpha-\beta+\sigma,\alpha)\end{matrix};\lambda t^{\alpha}\right)\right.\right.$$
$$\left.\left.-\,{}_1\varphi_1\left(\begin{matrix}(1,1)\\(\alpha-\beta+\sigma,\alpha)\end{matrix};-\lambda t^{\alpha}\right)\right]\right\}$$
$$=t^{\alpha-\beta+\sigma-1}\,{}_1sinh_1\left(\begin{matrix}(1,1)\\(\alpha-\beta+\sigma,\alpha)\end{matrix};\lambda t^{\alpha}\right), \tag{6.620}$$

and

$$\frac{1}{\Gamma(\sigma)}\int_0^t (t-\tau)^{\sigma}\,\mathbb{L}cosh_{\alpha,\beta}(\lambda,\tau)\,d\tau$$
$$=\frac{1}{\Gamma(\sigma)}\int_0^t \frac{1}{2}\left(\mathbb{L}_{\alpha,\beta}(\lambda,\tau)+\mathbb{L}_{\alpha,\beta}(-\lambda,\tau)\right)d\tau$$
$$=t^{\alpha-\beta+\sigma-1}\left\{\frac{1}{2}\left[{}_1\varphi_1\left(\begin{matrix}(1,1)\\(\alpha-\beta+\sigma,\alpha)\end{matrix};\lambda t^{\alpha}\right)\right.\right.$$
$$\left.\left.+{}_1\varphi_1\left(\begin{matrix}(1,1)\\(\alpha-\beta+\sigma,\alpha)\end{matrix};-\lambda t^{\alpha}\right)\right]\right\}$$
$$=t^{\alpha-\beta+\sigma-1}\,{}_1cosh_1\left(\begin{matrix}(1,1)\\(\alpha-\beta+\sigma,\alpha)\end{matrix};\lambda t^{\alpha}\right). \tag{6.621}$$

□

Property 6.51 Let $\lambda\in\mathbb{C}$, $\alpha\in\mathbb{R}_+$, and $Re(\sigma)>0$. Then we have

$$\frac{1}{\Gamma(\sigma)}\int_0^1 (1-\tau)^{\sigma}\,\mathbb{L}sin_{\alpha,\beta}(\lambda,\tau)\,d\tau={}_1sin_1\left(\begin{matrix}(1,1)\\(\alpha-\beta+\sigma,\alpha)\end{matrix};\lambda\right), \tag{6.622}$$

$$\frac{1}{\Gamma(\sigma)}\int_0^1 (1-\tau)^{\sigma}\,\mathbb{L}cos_{\alpha,\beta}(\lambda,\tau)\,d\tau={}_1cos_1\left(\begin{matrix}(1,1)\\(\alpha-\beta+\sigma,\alpha)\end{matrix};\lambda t^{\alpha}\right), \tag{6.623}$$

$$\frac{1}{\Gamma(\sigma)}\int_0^1 (1-\tau)^{\sigma}\,\mathbb{L}sinh_{\alpha,\beta}(\lambda,\tau)\,d\tau={}_1sinh_1\left(\begin{matrix}(1,1)\\(\alpha-\beta+\sigma,\alpha)\end{matrix};\lambda t^{\alpha}\right), \tag{6.624}$$

and

$$\frac{1}{\Gamma(\sigma)}\int_0^1 (1-\tau)^{\sigma}\,\mathbb{L}cosh_{\alpha,\beta}(\lambda,\tau)\,d\tau={}_1cosh_1\left(\begin{matrix}(1,1)\\(\alpha-\beta+\sigma,\alpha)\end{matrix};\lambda t^{\alpha}\right). \tag{6.625}$$

Proof Setting $t=1$ in (6.610), (6.611), (6.612), and (6.613), we get the results. □

Property 6.52 Let $\lambda\in\mathbb{C}$, $\alpha\in\mathbb{R}_+$, and $Re(\sigma)>0$. Then we have

$$\frac{1}{\Gamma(\sigma)}\int_0^2 (2-\tau)^{\sigma}\,\mathbb{L}sin_{\alpha,\beta}(\lambda,\tau)\,d\tau=\frac{1}{\Gamma(\sigma)}\int_{-1}^1 (1-\tau)^{\sigma}\,\mathbb{L}sin_{\alpha,\beta}(\lambda,1+\tau)\,d\tau$$
$$=\frac{1}{\Gamma(\sigma)}\int_{-1}^1 (1+\tau)^{\sigma}\,\mathbb{L}sin_{\alpha,\beta}(\lambda,1-\tau)\,d\tau$$
$$=2^{\alpha-\beta+\sigma-1}\,{}_1sin_1\left(\begin{matrix}(1,1)\\(\alpha-\beta+\sigma,\alpha)\end{matrix};\lambda 2^{\alpha}\right), \tag{6.626}$$

$$\frac{1}{\Gamma(\sigma)}\int_0^2 (2-\tau)^{\sigma}\,\mathbb{L}cos_{\alpha,\beta}(\lambda,\tau)\,d\tau = \frac{1}{\Gamma(\sigma)}\int_{-1}^{1} (1-\tau)^{\sigma}\,\mathbb{L}cos_{\alpha,\beta}(\lambda,1+\tau)\,d\tau$$
$$= \frac{1}{\Gamma(\sigma)}\int_{-1}^{1} (1+\tau)^{\sigma}\,\mathbb{L}cos_{\alpha,\beta}(\lambda,1-\tau)\,d\tau$$
$$= 2^{\alpha-\beta+\sigma-1}\,{}_1cos_1\left(\begin{matrix}(1,1)\\(\alpha-\beta+\sigma,\alpha)\end{matrix};\lambda 2^{\alpha}\right), \tag{6.627}$$

$$\frac{1}{\Gamma(\sigma)}\int_0^2 (2-\tau)^{\sigma}\,\mathbb{L}sinh_{\alpha,\beta}(\lambda,\tau)\,d\tau = \frac{1}{\Gamma(\sigma)}\int_{-1}^{1} (1-\tau)^{\sigma}\,\mathbb{L}sinh_{\alpha,\beta}(\lambda,1+\tau)\,d\tau$$
$$= \frac{1}{\Gamma(\sigma)}\int_{-1}^{1} (1+\tau)^{\sigma}\,\mathbb{L}sinh_{\alpha,\beta}(\lambda,1-\tau)\,d\tau$$
$$= 2^{\alpha-\beta+\sigma-1}\,{}_1sinh_1\left(\begin{matrix}(1,1)\\(\alpha-\beta+\sigma,\alpha)\end{matrix};\lambda 2^{\alpha}\right), \tag{6.628}$$

and

$$\frac{1}{\Gamma(\sigma)}\int_0^2 (2-\tau)^{\sigma}\,\mathbb{L}cosh_{\alpha,\beta}(\lambda,\tau)\,d\tau = \frac{1}{\Gamma(\sigma)}\int_{-1}^{1} (1-\tau)^{\sigma}\,\mathbb{L}cosh_{\alpha,\beta}(\lambda,1+\tau)\,d\tau$$
$$= \frac{1}{\Gamma(\sigma)}\int_{-1}^{1} (1+\tau)^{\sigma}\,\mathbb{L}cosh_{\alpha,\beta}(\lambda,1-\tau)\,d\tau$$
$$= 2^{\alpha-\beta+\sigma-1}\,{}_1cosh_1\left(\begin{matrix}(1,1)\\(\alpha-\beta+\sigma,\alpha)\end{matrix};\lambda t^{\alpha}\right). \tag{6.629}$$

Property 6.53 Let $\lambda \in \mathbb{C}$, $\alpha \in \mathbb{R}_+$, and $Re(\sigma) > 0$. Then we have [92]

$$L\left\{\mathbb{L}sin_{\alpha,\beta}(\lambda,t)\right\} = \frac{\lambda s^{\beta}}{s^{2\alpha}+\lambda^2}\left(\left|\frac{\lambda}{s^{\alpha}}\right| < 1\right), \tag{6.630}$$

$$L\left\{\mathbb{L}cos_{\alpha,\beta}(\lambda,\tau)\right\} = \frac{s^{\alpha+\beta}}{s^{2\alpha}+\lambda^2}\left(\left|\frac{\lambda}{s^{\alpha}}\right| < 1\right), \tag{6.631}$$

$$L\left\{\mathbb{L}sinh_{\alpha,\beta}(\lambda,t)\right\} = \frac{\lambda s^{\beta}}{s^{2\alpha}-\lambda^2}\left(\left|\frac{\lambda}{s^{\alpha}}\right| < 1\right), \tag{6.632}$$

and

$$L\left\{\mathbb{L}cosh_{\alpha,\beta}\left(\lambda,\tau\right)\right\}=\frac{s^{\alpha+\beta}}{s^{2\alpha}-\lambda^{2}}\left(\left|\frac{\lambda}{s^{\alpha}}\right|<1\right). \tag{6.633}$$

Proof Making use of the integral representations

$$L\left\{\mathbb{L}_{\alpha,\beta}\left(i\lambda,t\right)\right\}=\frac{s^{\beta}}{s^{\alpha}-i\lambda}\left(\left|\frac{\lambda}{s^{\alpha}}\right|<1\right), \tag{6.634}$$

$$L\left\{\mathbb{L}_{\alpha,\beta}\left(-i\lambda,t\right)\right\}=\frac{s^{\beta}}{s^{\alpha}+i\lambda}\left(\left|\frac{\lambda}{s^{\alpha}}\right|<1\right), \tag{6.635}$$

$$L\left\{\mathbb{L}_{\alpha,\beta}\left(\lambda,t\right)\right\}=\frac{s^{\beta}}{s^{\alpha}-\lambda}\left(\left|\frac{\lambda}{s^{\alpha}}\right|<1\right), \tag{6.636}$$

and

$$L\left\{\mathbb{L}_{\alpha,\beta}\left(-\lambda,t\right)\right\}=\frac{s^{\beta}}{s^{\alpha}+\lambda}\left(\left|\frac{\lambda}{s^{\alpha}}\right|<1\right), \tag{6.637}$$

we have that

$$\begin{aligned}L\left\{\mathbb{L}sin_{\alpha,\beta}\left(\lambda,t\right)\right\}&=L\left\{\frac{1}{2i}\left(\mathbb{L}_{\alpha,\beta}\left(i\lambda,t\right)-\mathbb{L}_{\alpha,\beta}\left(-i\lambda,t\right)\right)\right\}\\&=\frac{1}{2i}\left(\frac{s^{\beta}}{s^{\alpha}-i\lambda}-\frac{s^{\beta}}{s^{\alpha}+i\lambda}\right)\\&=\frac{\lambda s^{\beta}}{s^{2\alpha}+\lambda^{2}}\left(\left|\frac{\lambda}{s^{\alpha}}\right|<1\right),\end{aligned} \tag{6.638}$$

$$\begin{aligned}L\left\{\mathbb{L}cos_{\alpha,\beta}\left(\lambda,\tau\right)\right\}&=L\left\{\frac{1}{2}\left(\mathbb{L}_{\alpha,\beta}\left(i\lambda,t\right)+\mathbb{L}_{\alpha,\beta}\left(-i\lambda,t\right)\right)\right\}\\&=\frac{1}{2}\left(\frac{s^{\beta}}{s^{\alpha}-i\lambda}+\frac{s^{\beta}}{s^{\alpha}+i\lambda}\right)\\&=\frac{s^{\alpha+\beta}}{s^{2\alpha}+\lambda^{2}}\left(\left|\frac{\lambda}{s^{\alpha}}\right|<1\right),\end{aligned} \tag{6.639}$$

$$\begin{aligned} L\left\{\mathbb{L}sinh_{\alpha,\beta}(\lambda,t)\right\} &= L\left\{\frac{1}{2}\left(\mathbb{L}_{\alpha,\beta}(\lambda,t)-\mathbb{L}_{\alpha,\beta}(-\lambda,t)\right)\right\} \\ &= \frac{1}{2}\left(\frac{s^{\beta}}{s^{\alpha}-\lambda}-\frac{s^{\beta}}{s^{\alpha}+\lambda}\right) \\ &= \frac{\lambda s^{\beta}}{s^{2\alpha}-\lambda^{2}}\left(\left|\frac{\lambda}{s^{\alpha}}\right|<1\right), \end{aligned} \tag{6.640}$$

and

$$\begin{aligned} L\left\{\mathbb{L}cosh_{\alpha,\beta}(\lambda,\tau)\right\} &= L\left\{\frac{1}{2}\left(\mathbb{L}_{\alpha,\beta}(\lambda,t)+\mathbb{L}_{\alpha,\beta}(-\lambda,t)\right)\right\} \\ &= \frac{1}{2}\left(\frac{s^{\beta}}{s^{\alpha}-\lambda}+\frac{s^{\beta}}{s^{\alpha}+\lambda}\right) \\ &= \frac{s^{\alpha+\beta}}{s^{2\alpha}-\lambda^{2}}\left(\left|\frac{\lambda}{s^{\alpha}}\right|<1\right). \end{aligned} \tag{6.641}$$

The results were discovered by Yang in 2019 [92]. □

Chapter 7
Kohlrausch–Williams–Watts Function and Related Topics

Abstract In this chapter, we investigate the history, definitions, and properties of the Kohlrausch–Williams–Watts function. The subtrigonometric and subhyperbolic functions are addressed based on the Kohlrausch–Williams–Watts function. The subsurfaces and geometric representations are also proposed in detail. The Fourier-type series, the Fourier-like integral transforms, the Laplace-like transforms, and the Mellin-like transforms are investigated and presented with the aid of the new special functions.

7.1 The Kohlrausch–Williams–Watts Function: History, Definitions, and Properties

In 1854, Rudolph Kohlrausch proposed the stretched exponential function of integral order applied in the electric circuit [281]. In 1970, Graham Williams and David C. Watts rediscovered the stretched exponential function of fractional order considered in the field of the non-symmetrical dielectric relaxation with the characteristics of the decay function [282]. The special function is called the Kohlrausch–Williams–Watts function [283, 284] or the Kohlrausch–Williams–Watts stretched exponential function [285, 286]. The function is the well-known "*Material Function*" [287].

7.1.1 The Kohlrausch–Williams–Watts Function

Definition 7.1 The Kohlrausch–Williams–Watts function is defined as [281, 282]

$$\mathbb{K}_{\alpha}(t) = e^{\lambda t^{\alpha}}, \tag{7.1}$$

where $\alpha, \lambda, t \in \mathbb{C}$.

X.-J. Yang, *Theory and Applications of Special Functions for Scientists and Engineers*, https://doi.org/10.1007/978-981-33-6334-2_7

The series form of the Kohlrausch–Williams–Watts function can be given as follows:

$$\mathbb{K}_{\alpha}(\lambda, t) = e^{\lambda t^{\alpha}} = \sum_{k=0}^{\infty} \frac{\lambda^k t^{k\alpha}}{k!} = E_1(\lambda t^{\alpha}) = {}_0\varphi_0(-; \lambda t^{\alpha}), \tag{7.2}$$

which leads to

$$\mathbb{K}_{\alpha}(t) = \mathbb{K}_{\alpha}(1, t),$$

where $\alpha, t \in \mathbb{C}$.

Definition 7.2 The Kohlrausch–Williams–Watts type function is defined as [287]

$$\mathbb{K}_{\alpha,\beta}(\lambda, t) = t^{\beta-1} e^{\lambda t^{\alpha}} = t^{\beta-1} \sum_{k=0}^{\infty} \frac{\lambda^k t^{k\alpha}}{k!} = t^{\beta-1} E_1(\lambda t^{\alpha}) = t^{\beta-1} {}_0\varphi_0(-; \lambda t^{\alpha}), \tag{7.3}$$

which leads to

$$\mathbb{K}_{\alpha,\beta}(t) = \mathbb{K}_{\alpha,\beta}(1, t)$$

and

$$\mathbb{K}_{\alpha}(\lambda, t) = \mathbb{K}_{\alpha,1}(\lambda, t),$$

where $\alpha, \beta, t \in \mathbb{C}$.

In particular, for $\alpha = \beta$, we have

$$\mathbb{K}_{\alpha,\alpha}(\lambda, t) = t^{\alpha-1} e^{\lambda t^{\alpha}} = t^{\alpha-1} \sum_{k=0}^{\infty} \frac{\lambda^k t^{k\alpha}}{k!} = t^{\alpha-1} E_1(\lambda t^{\alpha}) = t^{\alpha-1} {}_0\varphi_0(-; \lambda t^{\alpha}), \tag{7.4}$$

where $\alpha, t \in \mathbb{C}$.

For $\alpha = \frac{1}{2}$ and $\lambda = -1$, the Friedrich function, proposed in 1991 by Friedrich, can be written as follows [287]:

$$\mathbb{K}_{\frac{1}{2},\frac{1}{2}}(\lambda, t) = t^{-\frac{1}{2}} e^{-t^{\alpha}} = t^{-\frac{1}{2}} E_1(-t^{\alpha}) = t^{-\frac{1}{2}} {}_0\varphi_0(-; -t^{\alpha}), \tag{7.5}$$

where $\alpha, t \in \mathbb{C}$.

The Laplace transform of the Kohlrausch–Williams–Watts function can be given as follows:

$$\begin{aligned} & L\{\mathbb{K}_\alpha(\lambda, t)\} \\ & = L\left\{e^{\lambda t^\alpha}\right\} \\ & = L\left\{\sum_{k=0}^{\infty} \frac{\lambda^k t^{k\alpha}}{k!}\right\} \\ & = \sum_{k=0}^{\infty} \Gamma(k\alpha+1) \frac{\left(\frac{\lambda}{s^\alpha}\right)^k}{k!} \\ & = {}_1\varphi_0\left(\begin{matrix}(1,\alpha) \\ -\end{matrix}; \frac{\lambda}{s^\alpha}\right), \end{aligned} \tag{7.6}$$

where $\alpha, \lambda \in \mathbb{C}$ and $\left|\lambda s^{-\alpha}\right| < 1$.

Similarly,

$$\begin{aligned} & L\left\{\mathbb{K}_{\alpha,\beta}(\lambda, t)\right\} \\ & = L\left\{t^{\beta-1} e^{\lambda t^\alpha}\right\} \\ & = L\left\{\sum_{k=0}^{\infty} \frac{\lambda^k t^{k\alpha+\beta-1}}{k!}\right\} \\ & = s^{1-\beta} \sum_{k=0}^{\infty} \Gamma(k\alpha+\beta) \frac{\left(\frac{\lambda}{s^\alpha}\right)^k}{k!} \\ & = s^{1-\beta}\, {}_1\varphi_0\left(\begin{matrix}(\beta,\alpha) \\ -\end{matrix}; \frac{\lambda}{s^\alpha}\right), \end{aligned} \tag{7.7}$$

where $\alpha, \beta, \lambda \in \mathbb{C}$ and $\left|\lambda s^{-\alpha}\right| < 1$.

The Mellin transform of the Kohlrausch–Williams–Watts function can be presented as follows.

Property 7.1 Let $\alpha, \lambda \in \mathbb{C}$, $Re(\lambda) \geq 0$, and $Re(\alpha) \geq 0$. Then, we have [288]

$$M\{\mathbb{K}_\alpha(-\lambda, t)\} = M\left\{e^{-\lambda t^\alpha}\right\} = \int_0^\infty t^{\varpi-1} e^{-\lambda t^\alpha} dt = \frac{\lambda^{-\frac{\varpi}{\alpha}}}{\alpha} \Gamma\left(\frac{\varpi}{\alpha}\right). \tag{7.8}$$

The result was discovered by Meijer in 1936 [288].

Some important formulas connected with the Kohlrausch–Williams–Watts function are given as follows:

$$\int_0^\infty t^{\varpi-1} e^{-\lambda t} dt = \lambda^{-\varpi} \Gamma(\varpi), \tag{7.9}$$

$$\int_0^\infty e^{-\lambda t^\alpha} dt = \lambda^{-\frac{1}{\alpha}} \Gamma\left(\frac{1}{\alpha}+1\right), \tag{7.10}$$

$$\int_0^\infty t^{\varpi-1}e^{-t^\alpha}dt = \frac{1}{\alpha}\Gamma\left(\frac{\varpi}{\alpha}\right), \tag{7.11}$$

$$\int_0^\infty e^{-t^\alpha}dt = \Gamma\left(\frac{1}{\alpha}+1\right), \tag{7.12}$$

$$\int_0^\infty t^{\varpi-1}e^{-\lambda t^2}dt = \frac{\lambda^{-\frac{\varpi}{2}}}{2}\Gamma\left(\frac{\varpi}{2}\right), \tag{7.13}$$

and

$$\int_0^\infty t^{\varpi-1}e^{-\lambda t^\alpha}dt = \lambda^{-\frac{\varpi}{\alpha}}\frac{1}{\alpha}\Gamma\left(\frac{\varpi}{\alpha}\right), \tag{7.14}$$

where $\alpha, \lambda \in \mathbb{C}$, $Re(\lambda) \geq 0$, and $Re(\alpha) \geq 0$.

The formula (7.9) was discussed in [289], (7.10), (7.11), (7.13), and (7.14) are considered in [290–292], and the detailed discussions are presented in [92].

In a similar way, we have the following result:

Property 7.2 Let $\alpha, \lambda \in \mathbb{C}$, $Re(\lambda) \geq 0$, $Re(\beta) \geq 0$, and $Re(\alpha) \geq 0$. Then, we have

$$\begin{aligned} &M\left\{\mathbb{K}_{\alpha,\beta}(-\lambda,t)\right\} \\ &= M\left\{t^{\beta-1}e^{-\lambda t^\alpha}\right\} \\ &= \int_0^\infty t^{\varpi+\beta-2}e^{-\lambda t^\alpha}dt \\ &= \frac{\lambda^{-\frac{\varpi+\beta-1}{\alpha}}}{\alpha}\Gamma\left(\frac{\varpi+\beta-1}{\alpha}\right). \end{aligned} \tag{7.15}$$

Proof From (7.8), we reduce the result. □

In particular,

$$\begin{aligned} &M\left\{\mathbb{K}_{1,\beta}(-\lambda,t)\right\} \\ &= M\left\{t^{\beta-1}e^{-\lambda t}\right\} \\ &= \int_0^\infty t^{\varpi+\beta-2}e^{-\lambda t}dt \\ &= \lambda^{1-(\varpi+\beta)}\Gamma(\varpi+\beta-1), \end{aligned} \tag{7.16}$$

$$
\begin{aligned}
&M\left\{\mathbb{K}_{2,\beta}(-\lambda, t)\right\} \\
&\quad = M\left\{t^{\beta-1} e^{-\lambda t^2}\right\} \\
&\quad = \int_0^{\infty} t^{\varpi+\beta-2} e^{-\lambda t^2} dt \\
&\quad = \frac{\lambda^{-\frac{\varpi+\beta-1}{2}}}{2} \Gamma\left(\frac{\varpi+\beta-1}{2}\right),
\end{aligned}
\tag{7.17}
$$

$$
\begin{aligned}
&M\left\{\mathbb{K}_{2\alpha,\beta}(-\lambda, t)\right\} \\
&\quad = M\left\{t^{\beta-1} e^{-\lambda t^{2\alpha}}\right\} \\
&\quad = \int_0^{\infty} t^{\varpi+\beta-2} e^{-\lambda t^{2\alpha}} dt \\
&\quad = \frac{\lambda^{-\frac{\varpi+\beta-1}{2\alpha}}}{2\alpha} \Gamma\left(\frac{\varpi+\beta-1}{2\alpha}\right),
\end{aligned}
\tag{7.18}
$$

$$
\begin{aligned}
&M\left\{\mathbb{K}_{2\alpha,\alpha}(-\lambda, t)\right\} \\
&\quad = M\left\{t^{\alpha-1} e^{-\lambda t^{2\alpha}}\right\} \\
&\quad = \int_0^{\infty} t^{\varpi+\alpha-2} e^{-\lambda t^{2\alpha}} dt \\
&\quad = \frac{\lambda^{-\frac{\varpi+\alpha-1}{2\alpha}}}{2\alpha} \Gamma\left(\frac{\varpi+\alpha-1}{2\alpha}\right),
\end{aligned}
\tag{7.19}
$$

and

$$
\begin{aligned}
&M\left\{\mathbb{K}_{\alpha,\alpha}(-\lambda, t)\right\} \\
&\quad = M\left\{t^{\alpha-1} e^{-\lambda t^{\alpha}}\right\} \\
&\quad = \int_0^{\infty} t^{\varpi+\alpha-2} e^{-\lambda t^{\alpha}} dt \\
&\quad = \frac{\lambda^{-\frac{\varpi+\alpha-1}{\alpha}}}{\alpha} \Gamma\left(\frac{\varpi+\alpha-1}{\alpha}\right).
\end{aligned}
\tag{7.20}
$$

Thus,

$$
\int_{-\infty}^{\infty} |t|^{\varpi-1} e^{-\lambda|t|^{\alpha}} dt = \frac{2}{\alpha} \lambda^{-\frac{\varpi}{\alpha}} \Gamma\left(\frac{\varpi}{\alpha}\right),
\tag{7.21}
$$

$$
\int_{-\infty}^{\infty} |t|^{\varpi-1} e^{-\lambda|t|} dt = 2\lambda^{-\varpi} \Gamma(\varpi),
\tag{7.22}
$$

$$
\int_{-\infty}^{\infty} e^{-\lambda|t|^{\alpha}} dt = \frac{2}{\lambda^{\frac{1}{\alpha}}} \Gamma\left(\frac{1}{\alpha}+1\right),
\tag{7.23}
$$

$$\int_{-\infty}^{\infty} |t|^{\varpi-1} e^{-|t|^{\alpha}} dt = \frac{2}{\alpha}\Gamma\left(\frac{\varpi}{\alpha}\right), \tag{7.24}$$

$$\int_{-\infty}^{\infty} e^{-|t|^{\alpha}} dt = 2\Gamma\left(\frac{1}{\alpha}+1\right), \tag{7.25}$$

$$\int_{-\infty}^{\infty} |t|^{\varpi-1} e^{-\lambda|t|^{2}} dt = \lambda^{-\frac{\varpi}{2}}\Gamma\left(\frac{\varpi}{2}\right), \tag{7.26}$$

$$\int_{-\infty}^{\infty} |t|^{\varpi-1} e^{-\lambda|t|^{\alpha}} dt = \frac{2}{\alpha}\lambda^{-\frac{\varpi}{\alpha}}\Gamma\left(\frac{\varpi}{\alpha}\right), \tag{7.27}$$

$$\int_{-\infty}^{\infty} |t|^{\varpi+\beta-2} e^{-\lambda|t|^{\alpha}} dt = \frac{2}{\alpha}\lambda^{-\frac{\varpi+\beta-1}{\alpha}}\Gamma\left(\frac{\varpi+\beta-1}{\alpha}\right), \tag{7.28}$$

$$\int_{-\infty}^{\infty} |t|^{\varpi+\beta-2} e^{-\lambda|t|} dt = 2\lambda^{1-(\varpi+\beta)}\Gamma\left(\varpi+\beta-1\right), \tag{7.29}$$

$$\int_{-\infty}^{\infty} |t|^{\varpi+\beta-2} e^{-\lambda|t|^{2}} dt = \lambda^{-\frac{\varpi+\beta-1}{2}}\Gamma\left(\frac{\varpi+\beta-1}{2}\right), \tag{7.30}$$

$$\int_{-\infty}^{\infty} |t|^{\varpi+\beta-2} e^{-\lambda|t|^{2\alpha}} dt = \frac{\lambda^{-\frac{\varpi+\beta-1}{2\alpha}}}{\alpha}\Gamma\left(\frac{\varpi+\beta-1}{2\alpha}\right), \tag{7.31}$$

$$\int_{-\infty}^{\infty} |t|^{\varpi+\alpha-2} e^{-\lambda|t|^{2\alpha}} dt = \frac{\lambda^{-\frac{\varpi+\alpha-1}{2\alpha}}}{\alpha}\Gamma\left(\frac{\varpi+\alpha-1}{2\alpha}\right), \tag{7.32}$$

and

$$\int_{-\infty}^{\infty} |t|^{\varpi+\alpha-2} e^{-\lambda|t|^{\alpha}} dt = \frac{2}{\alpha}\lambda^{-\frac{\varpi+\alpha-1}{\alpha}}\Gamma\left(\frac{\varpi+\alpha-1}{\alpha}\right). \tag{7.33}$$

Let us consider the following formula:

$$\frac{1}{\Gamma(\sigma)}\int_0^t (t-\tau)^{\sigma}\,\mathbb{K}_{\alpha,\beta}(\lambda,\tau)\,d\tau = t^{\beta+\sigma-1}\,{}_1\varphi_1\left(\begin{matrix}(\beta,\alpha)\\(\beta+\sigma,\alpha)\end{matrix};\lambda t^{\alpha}\right), \tag{7.34}$$

where $\alpha,\beta,\sigma,\lambda\in\mathbb{C}$, $Re(\sigma)>0$, $Re(\alpha)>0$, and $Re(\beta)>0$.

Here, it is shown that the formula (7.34) can be found in Chap. 5 as the special case.

Property 7.3 Let $\alpha,\beta,\sigma,\lambda\in\mathbb{C}$, $Re(\sigma)>0$, $Re(\alpha)>0$, and $Re(\beta)>0$. Then, we have

$$\frac{1}{\Gamma(\sigma)}\int_0^1 (1-\tau)^{\sigma}\,\mathbb{K}_{\alpha,\beta}(\lambda,\tau)\,d\tau = {}_1\varphi_1\left(\begin{matrix}(\beta,\alpha)\\(\beta+\sigma,\alpha)\end{matrix};\lambda\right). \tag{7.35}$$

Proof Putting $t=1$ in (7.34), we have the result. □

Property 7.4 Let $\alpha,\beta,\sigma,\lambda\in\mathbb{C}$, $Re(\sigma)>0$, $Re(\alpha)>0$, and $Re(\beta)>0$. Then, we have

$$\begin{aligned}&\tfrac{1}{\Gamma(\sigma)}\int_0^2 (2-\tau)^{\sigma}\,\mathbb{K}_{\alpha,\beta}(\lambda,\tau)\,d\tau\\&=\tfrac{1}{\Gamma(\sigma)}\int_{-1}^1 (1-\tau)^{\sigma}\,\mathbb{K}_{\alpha,\beta}(\lambda,1+\tau)\,d\tau\\&=\tfrac{1}{\Gamma(\sigma)}\int_{-1}^1 (1+\tau)^{\sigma}\,\mathbb{K}_{\alpha,\beta}(\lambda,1-\tau)\,d\tau\\&=2^{\beta+\sigma-1}\,{}_1\varphi_1\left(\begin{matrix}(\beta,\alpha)\\(\beta+\sigma,\alpha)\end{matrix};\lambda 2^{\alpha}\right).\end{aligned} \tag{7.36}$$

Proof Setting $t=2$ in (7.34), we obtain the result. □

Let us consider the cases of (7.34), (7.35), and (7.36) for $\beta=1$.

Property 7.5 Let $\alpha,\sigma,\lambda\in\mathbb{C}$, $Re(\sigma)>0$, and $Re(\alpha)>0$. Then, we have

$$\frac{1}{\Gamma(\sigma)}\int_0^t (t-\tau)^{\sigma}\,\mathbb{K}_{\alpha}(\lambda,\tau)\,d\tau = t^{\sigma}\,{}_1\varphi_1\left(\begin{matrix}(1,\alpha)\\(1+\sigma,\alpha)\end{matrix};\lambda t^{\alpha}\right), \tag{7.37}$$

$$\frac{1}{\Gamma(\sigma)}\int_0^1 (1-\tau)^{\sigma}\,\mathbb{K}_{\alpha}(\lambda,\tau)\,d\tau = {}_1\varphi_1\left(\begin{matrix}(1,\alpha)\\(1+\sigma,\alpha)\end{matrix};\lambda\right), \tag{7.38}$$

and

$$\begin{aligned}
&\frac{1}{\Gamma(\sigma)}\int_0^2 (2-\tau)^\sigma\, \mathbb{K}_\alpha(\lambda,\tau)\, d\tau \\
&= \frac{1}{\Gamma(\sigma)}\int_{-1}^1 (1-\tau)^\sigma\, \mathbb{K}_\alpha(\lambda,1+\tau)\, d\tau \\
&= \frac{1}{\Gamma(\sigma)}\int_{-1}^1 (1+\tau)^\sigma\, \mathbb{K}_\alpha(\lambda,1-\tau)\, d\tau \\
&= 2^\sigma\, {}_1\varphi_1\left(\begin{matrix}(1,\alpha)\\(1+\sigma,\alpha)\end{matrix};\lambda 2^\alpha\right).
\end{aligned} \tag{7.39}$$

7.1.2 The Subtrigonometric Functions via Kohlrausch–Williams–Watts Function

In this part, we propose the subtrigonometric functions via Kohlrausch–Williams–Watts function.

Now, we consider that [92]

$$\mathbb{K}_\alpha(i\lambda,t) = e^{i\lambda t^\alpha} = \sum_{k=0}^{\infty}\frac{(i\lambda)^k t^{k\alpha}}{k!} = E_1(i\lambda t^\alpha) = {}_0\varphi_0(-;i\lambda t^\alpha)\, l \tag{7.40}$$

and

$$\mathbb{K}_\alpha(-i\lambda,t) = e^{-i\lambda t^\alpha} = \sum_{k=0}^{\infty}\frac{(-i\lambda)^k\, t^{k\alpha}}{k!} = E_1\left(-i\lambda t^\alpha\right) = {}_0\varphi_0(-;-i\lambda t^\alpha). \tag{7.41}$$

In view of (7.40) and (7.41), we present the following special functions:

Definition 7.3 The subsine via Kohlrausch–Williams–Watts function is defined as [92]

$$\begin{aligned}
subsin_\alpha(t) &= \frac{1}{2i}\left(\mathbb{K}_\alpha(i,t) - \mathbb{K}_\alpha(-i,t)\right) = \sum_{k=0}^{\infty}\frac{(-1)^k t^{(2k+1)\alpha}}{\Gamma(2k+2)} \\
&= \sum_{k=0}^{\infty}\frac{(-1)^k t^{(2k+1)\alpha}}{(2k+1)!},
\end{aligned} \tag{7.42}$$

where $\alpha, t \in \mathbb{C}$.

Definition 7.4 The subcosine via Kohlrausch–Williams–Watts function is defined as [92]

$$\begin{aligned} subcos_{\alpha}(t) &= \frac{1}{2}\left(\mathbb{K}_{\alpha}(i,t) + \mathbb{K}_{\alpha}(-i,t)\right) = \sum_{k=0}^{\infty} \frac{(-1)^k t^{2k\alpha}}{\Gamma(2k+1)} \\ &= \sum_{k=0}^{\infty} \frac{(-1)^k t^{2k\alpha}}{(2k)!}, \end{aligned} \tag{7.43}$$

where $\alpha, t \in \mathbb{C}$.

The formulas (7.42) and (7.43) were discovered by Yang in 2019 [92].

Definition 7.5 The subtangent via Kohlrausch–Williams–Watts function is defined as

$$subtan_{\alpha}(t) = \frac{subsin_{\alpha}(t)}{subcos_{\alpha}(t)}\,\left(subcos_{\alpha}(t) \neq 0\right), \tag{7.44}$$

where $\alpha, t \in \mathbb{C}$.

Definition 7.6 The subcotangent via Kohlrausch–Williams–Watts function is defined as

$$subcotan_{\alpha}(t) = \frac{subcos_{\alpha}(t)}{subsin_{\alpha}(t)}, \left(subsin_{\alpha}(t) \neq 0\right), \tag{7.45}$$

where $\alpha, t \in \mathbb{C}$.

Definition 7.7 The subcosecant via Kohlrausch–Williams–Watts function is defined as

$$subcsc_{\alpha}(t) = \frac{1}{subsin_{\alpha}(t)}, \left(subsin_{\alpha}(t) \neq 0\right), \tag{7.46}$$

where $\alpha, t \in \mathbb{C}$.

Definition 7.8 The subsecant via Kohlrausch–Williams–Watts function is defined as

$$subsec_{\alpha}(t) = \frac{1}{subcos_{\alpha}(t)}\,\left(subcos_{\alpha}(t) \neq 0\right), \tag{7.47}$$

where $\alpha, t \in \mathbb{C}$.

Thus, we have that

$$
\begin{aligned}
subsin_{\alpha}(\lambda, t) &= \frac{1}{2i}\left(\mathbb{K}_{\alpha}(i\lambda, t) - \mathbb{K}_{\alpha}(-i\lambda, t)\right) \\
&= \sum_{k=0}^{\infty} \frac{(-1)^k \lambda^{2k+1} t^{(2k+1)\alpha}}{\Gamma(2k+2)} \\
&= \sum_{k=0}^{\infty} \frac{(-1)^k \lambda^{2k+1} t^{(2k+1)\alpha}}{(2k+1)!},
\end{aligned} \tag{7.48}
$$

$$
\begin{aligned}
subcos_{\alpha}(\lambda, t) &= \frac{1}{2}\left(\mathbb{K}_{\alpha}(i\lambda, t) + \mathbb{K}_{\alpha}(-i\lambda, t)\right) \\
&= \sum_{k=0}^{\infty} \frac{(-1)^k \lambda^{2k} t^{2k\alpha}}{\Gamma(2k+1)} \\
&= \sum_{k=0}^{\infty} \frac{(-1)^k \lambda^{2k} t^{2k\alpha}}{(2k)!},
\end{aligned} \tag{7.49}
$$

$$subtan_{\alpha}(\lambda, t) = \frac{subsin_{\alpha}(\lambda, t)}{subcos_{\alpha}(\lambda, t)} \left(subcos_{\alpha}(\lambda, t) \neq 0\right), \tag{7.50}$$

$$subcotan_{\alpha}(\lambda, t) = \frac{subcos_{\alpha}(\lambda, t)}{subsin_{\alpha}(\lambda, t)}, \left(subsin_{\alpha}(\lambda, t) \neq 0\right), \tag{7.51}$$

$$subcsc_{\alpha}(\lambda, t) = \frac{1}{subsin_{\alpha}(\lambda, t)}, \left(subsin_{\alpha}(\lambda, t) \neq 0\right), \tag{7.52}$$

and

$$subsec_{\alpha}(\lambda, t) = \frac{1}{subcos_{\alpha}(\lambda, t)} \left(subcos_{\alpha}(\lambda, t) \neq 0\right), \tag{7.53}$$

where $\alpha, \lambda, t \in \mathbb{C}$.

Theorem 7.1 *Let $\alpha, \lambda, t \in \mathbb{C}$. Then, we have*

$$\mathbb{K}_{\alpha}(i\lambda, t) = subcos_{\alpha}(\lambda, t) + i subsin_{\alpha}(\lambda, t) \tag{7.54}$$

and

$$\mathbb{K}_{\alpha}(-i\lambda, t) = subcos_{\alpha}(\lambda, t) - i subsin_{\alpha}(\lambda, t). \tag{7.55}$$

Proof With the aid of (7.48) and (7.49), we deduce the result. □

7.1.3 The Subhyperbolic Functions via Kohlrausch–Williams–Watts Function

In this part, we propose the subhyperbolic functions via Kohlrausch–Williams–Watts function.

In a similar way, we present that [92]

$$\mathbb{K}_{\alpha}(\lambda, t) = e^{\lambda t^{\alpha}} = \sum_{k=0}^{\infty} \frac{\lambda^{k} t^{k\alpha}}{k!} = E_1\left(\lambda t^{\alpha}\right) = {}_0\varphi_0\left(-; \lambda t^{\alpha}\right) \tag{7.56}$$

and

$$\mathbb{K}_{\alpha}(-\lambda, t) = e^{-\lambda t^{\alpha}} = \sum_{k=0}^{\infty} \frac{(-\lambda)^{k} t^{k\alpha}}{k!} = E_1\left(-\lambda t^{\alpha}\right) = {}_0\varphi_0\left(-; -\lambda t^{\alpha}\right). \tag{7.57}$$

We may give the following special functions:

Definition 7.9 The hyperbolic subsine via Kohlrausch–Williams–Watts function is defined as [92]

$$subsinh_{\alpha}(t) = \frac{1}{2}\left(\mathbb{K}_{\alpha}(1, t) - \mathbb{K}_{\alpha}(-1, t)\right) = \sum_{k=0}^{\infty} \frac{t^{(2k+1)\alpha}}{\Gamma(2k+2)} = \sum_{k=0}^{\infty} \frac{t^{(2k+1)\alpha}}{(2k+1)!}, \tag{7.58}$$

where $\alpha, t \in \mathbb{C}$.

Definition 7.10 The hyperbolic subcosine via Kohlrausch–Williams–Watts function is defined as [92]

$$subcosh_{\alpha}(t) = \frac{1}{2}\left(\mathbb{K}_{\alpha}(1, t) + \mathbb{K}_{\alpha}(-1, t)\right) = \sum_{k=0}^{\infty} \frac{t^{2k\alpha}}{\Gamma(2k+1)} = \sum_{k=0}^{\infty} \frac{t^{2k\alpha}}{(2k)!}, \tag{7.59}$$

where $\alpha, t \in \mathbb{C}$.

Note that (7.58) and (7.59) were discovered by Yang in 2019 [92].

Definition 7.11 The hyperbolic subtangent via Kohlrausch–Williams–Watts function is defined as

$$subtanh_{\alpha}(t) = \frac{subsinh_{\alpha}(t)}{subcosh_{\alpha}(t)} \left(subcosh_{\alpha}(t) \neq 0\right), \tag{7.60}$$

where $\alpha, t \in \mathbb{C}$.

Definition 7.12 The hyperbolic subcotangent via Kohlrausch–Williams–Watts function is defined as

$$subcotanh_{\alpha}(t) = \frac{subcosh_{\alpha}(t)}{sub\Re sinh_{\alpha}(t)}, (subsinh_{\alpha}(t) \neq 0), \tag{7.61}$$

where $\alpha, t \in \mathbb{C}$.

Definition 7.13 The hyperbolic subcosecant via Kohlrausch–Williams–Watts function is defined as

$$subcsch_{\alpha}(t) = \frac{1}{subsinh_{\alpha}(t)}, (subsinh_{\alpha}(t) \neq 0), \tag{7.62}$$

where $\alpha, t \in \mathbb{C}$.

Definition 7.14 The hyperbolic subsecant via Kohlrausch–Williams–Watts function is defined as

$$subsech_{\alpha}(t) = \frac{1}{subcosh_{\alpha}(t)} (subcosh_{\alpha}(t) \neq 0), \tag{7.63}$$

where $\alpha, t \in \mathbb{C}$.

Thus,

$$\begin{aligned} subsinh_{\alpha}(\lambda, t) &= \tfrac{1}{2}\left(\mathbb{K}_{\alpha}(\lambda, t) - \mathbb{K}_{\alpha}(-\lambda, t)\right) \\ &= \sum_{k=0}^{\infty} \frac{\lambda^{2k+1} t^{(2k+1)\alpha}}{\Gamma(2k+2)} \\ &= \sum_{k=0}^{\infty} \frac{\lambda^{2k+1} t^{(2k+1)\alpha}}{(2k+1)!}, \end{aligned} \tag{7.64}$$

$$\begin{aligned} subcosh_{\alpha}(\lambda, t) &= \tfrac{1}{2}\left(\mathbb{K}_{\alpha}(\lambda, t) + \mathbb{K}_{\alpha}(-\lambda, t)\right) \\ &= \sum_{k=0}^{\infty} \frac{\lambda^{2k} t^{2k\alpha}}{\Gamma(2k+1)} \\ &= \sum_{k=0}^{\infty} \frac{\lambda^{2k} t^{2k\alpha}}{(2k)!}, \end{aligned} \tag{7.65}$$

$$subtanh_{\alpha}(\lambda, t) = \frac{subsinh_{\alpha}(\lambda, t)}{subcosh_{\alpha}(\lambda, t)} (subcosh_{\alpha}(\lambda, t) \neq 0), \tag{7.66}$$

$$subcotanh_{\alpha}(\lambda, t) = \frac{subcosh_{\alpha}(\lambda, t)}{subsinh_{\alpha}(\lambda, t)}, (subsinh_{\alpha}(\lambda, t) \neq 0), \tag{7.67}$$

$$subcsch_{\alpha}(\lambda, t) = \frac{1}{subsinh_{\alpha}(\lambda, t)}, (subsinh_{\alpha}(\lambda, t) \neq 0), \tag{7.68}$$

and

$$subsech_{\alpha}(\lambda, t) = \frac{1}{subcosh_{\alpha}(\lambda, t)} \; (subcosh_{\alpha}(\lambda, t) \neq 0), \tag{7.69}$$

where $\alpha, \lambda, t \in \mathbb{C}$.

Theorem 7.2 *Let $\alpha, \lambda, t \in \mathbb{C}$. Then, we have*

$$\mathbb{K}_{\alpha}(\lambda, t) = subcosh_{\alpha}(\lambda, t) + subsinh_{\alpha}(\lambda, t) \tag{7.70}$$

and

$$\mathbb{K}_{\alpha}(-\lambda, t) = subcosh_{\alpha}(\lambda, t) - subsinh_{\alpha}(\lambda, t). \tag{7.71}$$

Proof In view of (7.64) and (7.65), we obtain the result. □

7.1.4 The Subtrigonometric Functions via Kohlrausch–Williams–Watts Type Function

In this part, we propose the subtrigonometric functions via Kohlrausch–Williams–Watts type function.

Let us consider

$$\begin{aligned} \mathbb{K}_{\alpha,\beta}(i\lambda, t) &= t^{\beta-1} e^{i\lambda t^{\alpha}} = t^{\beta-1} \sum_{k=0}^{\infty} \frac{(i\lambda)^k t^{k\alpha}}{k!} \\ &= t^{\beta-1} E_1(i\lambda t^{\alpha}) \\ &= t^{\beta-1} {}_0\varphi_0(-; i\lambda t^{\alpha}) \end{aligned} \tag{7.72}$$

and

$$\begin{aligned} \mathbb{K}_{\alpha,\beta}(-i\lambda, t) &= t^{\beta-1} e^{-i\lambda t^{\alpha}} = t^{\beta-1} \sum_{k=0}^{\infty} \frac{(-i\lambda)^k t^{k\alpha}}{k!} \\ &= t^{\beta-1} E_1(-i\lambda t^{\alpha}) \\ &= t^{\beta-1} {}_0\varphi_0(-; -i\lambda t^{\alpha}). \end{aligned} \tag{7.73}$$

In view of (7.72) and (7.73), we have the following special functions:

Definition 7.15 The subsine via Kohlrausch–Williams–Watts type function is defined as

$$\begin{aligned} subsin_{\alpha,\beta}(t) &= \frac{1}{2i}\left(\mathbb{K}_{\alpha,\beta}(i, t) - \mathbb{K}_{\alpha,\beta}(-i, t)\right) \\ &= t^{\beta-1} \sum_{k=0}^{\infty} \frac{(-1)^k t^{(2k+1)\alpha}}{\Gamma(2k+2)} \\ &= t^{\beta-1} \sum_{k=0}^{\infty} \frac{(-1)^k t^{(2k+1)\alpha}}{(2k+1)!}, \end{aligned} \tag{7.74}$$

where $\alpha, \beta, t \in \mathbb{C}$.

Definition 7.16 The subcosine via Kohlrausch–Williams–Watts type function is defined as

$$\begin{aligned} subcos_{\alpha,\beta}(t) &= \tfrac{1}{2}\left(\mathbb{K}_{\alpha,\beta}(i,t)+\mathbb{K}_{\alpha,\beta}(-i,t)\right) \\ &= t^{\beta-1}\sum_{k=0}^{\infty}\frac{(-1)^k t^{2k\alpha}}{\Gamma(2k+1)} \\ &= t^{\beta-1}\sum_{k=0}^{\infty}\frac{(-1)^k t^{2k\alpha}}{(2k)}, \end{aligned} \tag{7.75}$$

where $\alpha, \beta, t \in \mathbb{C}$.

Definition 7.17 The subtangent via Kohlrausch–Williams–Watts type function is defined as

$$subtan_{\alpha,\beta}(t) = \frac{subsin_{\alpha,\beta}(t)}{subcos_{\alpha,\beta}(t)}\left(subcos_{\alpha,\beta}(t) \neq 0\right), \tag{7.76}$$

where $\alpha, \beta, t \in \mathbb{C}$.

Definition 7.18 The subcotangent via Kohlrausch–Williams–Watts type function is defined as

$$subcotan_{\alpha,\beta}(t) = \frac{subcos_{\alpha,\beta}(t)}{subsin_{\alpha,\beta}(t)}, \left(subsin_{\alpha,\beta}(t) \neq 0\right), \tag{7.77}$$

where $\alpha, \beta, t \in \mathbb{C}$.

Definition 7.19 The subcosecant via Kohlrausch–Williams–Watts type function is defined as

$$subcsc_{\alpha,\beta}(t) = \frac{1}{subsin_{\alpha,\beta}(t)}, \left(subsin_{\alpha,\beta}(t) \neq 0\right), \tag{7.78}$$

where $\alpha, \beta, t \in \mathbb{C}$.

Definition 7.20 The subsecant via Kohlrausch–Williams–Watts type function is defined as

$$subsec_{\alpha,\beta}(t) = \frac{1}{subcos_{\alpha,\beta}(t)}\left(subcos_{\alpha,\beta}(t) \neq 0\right), \tag{7.79}$$

where $\alpha, \beta, t \in \mathbb{C}$.

Thus, we have that

$$\begin{aligned} subsin_{\alpha,\beta}(\lambda,t) &= \frac{1}{2i}\left(\mathbb{K}_{\alpha,\beta}(i\lambda,t)-\mathbb{K}_{\alpha,\beta}(-i\lambda,t)\right) \\ &= t^{\beta-1}\sum_{k=0}^{\infty}\frac{(-1)^k\lambda^{2k+1}t^{(2k+1)\alpha}}{\Gamma(2k+2)} \\ &= t^{\beta-1}\sum_{k=0}^{\infty}\frac{(-1)^k\lambda^{2k+1}t^{(2k+1)\alpha}}{(2k+1)!}, \end{aligned} \tag{7.80}$$

$$\begin{aligned} subcos_{\alpha,\beta}(\lambda,t) &= \frac{1}{2}\left(\mathbb{K}_{\alpha,\beta}(i\lambda,t)+\mathbb{K}_{\alpha,\beta}(-i\lambda,t)\right) \\ &= t^{\beta-1}\sum_{k=0}^{\infty}\frac{(-1)^k\lambda^{2k}t^{2k\alpha}}{\Gamma(2k+1)} \\ &= t^{\beta-1}\sum_{k=0}^{\infty}\frac{(-1)^k\lambda^{2k}t^{2k\alpha}}{(2k)!}, \end{aligned} \tag{7.81}$$

$$subtan_{\alpha,\beta}(\lambda,t) = \frac{subsin_{\alpha,\beta}(\lambda,t)}{subcos_{\alpha,\beta}(\lambda,t)}\left(subcos_{\alpha,\beta}(\lambda,t)\neq 0\right), \tag{7.82}$$

$$subcotan_{\alpha,\beta}(\lambda,t) = \frac{subcos_{\alpha,\beta}(\lambda,t)}{subsin_{\alpha,\beta}(\lambda,t)}, \left(subsin_{\alpha,\beta}(\lambda,t)\neq 0\right), \tag{7.83}$$

$$subcsc_{\alpha,\beta}(\lambda,t) = \frac{1}{subsin_{\alpha,\beta}(\lambda,t)}, \left(subsin_{\alpha,\beta}(\lambda,t)\neq 0\right), \tag{7.84}$$

and

$$subsec_{\alpha,\beta}(\lambda,t) = \frac{1}{subcos_{\alpha,\beta}(\lambda,t)}\left(subcos_{\alpha,\beta}(\lambda,t)\neq 0\right), \tag{7.85}$$

where $\alpha, \beta, \lambda, t \in \mathbb{C}$.

Theorem 7.3 *Let $\alpha, \beta, \lambda, t \in \mathbb{C}$. Then, we have*

$$\mathbb{K}_{\alpha,\beta}(i\lambda,t) = subcos_{\alpha,\beta}(\lambda,t) + i\,subsin_{\alpha,\beta}(\lambda,t) \tag{7.86}$$

and

$$\mathbb{K}_{\alpha,\beta}(-i\lambda,t) = subcos_{\alpha,\beta}(\lambda,t) - i\,subsin_{\alpha,\beta}(\lambda,t). \tag{7.87}$$

Proof With the aid of (7.80) and (7.81), we deduce the result. □

7.1.5 The Subhyperbolic Functions via Kohlrausch–Williams–Watts Type Function

In this part, we propose the subhyperbolic functions via Kohlrausch–Williams–Watts type function.

Similarly,

$$\mathbb{K}_{\alpha,\beta}(\lambda,t) = t^{\beta-1}e^{\lambda t^{\alpha}} = t^{\beta-1}\sum_{k=0}^{\infty}\frac{\lambda^k t^{k\alpha}}{k!} = t^{\beta-1}E_1\left(\lambda t^{\alpha}\right) = t^{\beta-1}\,{}_0\varphi_0\left(-;\lambda t^{\alpha}\right) \tag{7.88}$$

and

$$\begin{aligned}\mathbb{K}_{\alpha}(-\lambda,t) &= t^{\beta-1}e^{-\lambda t^{\alpha}} = t^{\beta-1}\sum_{k=0}^{\infty}\frac{(-\lambda)^k t^{k\alpha}}{k!} = t^{\beta-1}E_1\left(-\lambda t^{\alpha}\right)\\ &= t^{\beta-1}\,{}_0\varphi_0\left(-;-\lambda t^{\alpha}\right).\end{aligned} \tag{7.89}$$

Here, we may give the following special functions:

Definition 7.21 The hyperbolic subsine via Kohlrausch–Williams–Watts type function is defined as

$$\begin{aligned}subsinh_{\alpha,\beta}(t) &= \tfrac{1}{2}\left(\mathbb{K}_{\alpha,\beta}(1,t) - \mathbb{K}_{\alpha,\beta}(-1,t)\right)\\ &= t^{\beta-1}\sum_{k=0}^{\infty}\frac{t^{(2k+1)\alpha}}{\Gamma(2k+2)}\\ &= t^{\beta-1}\sum_{k=0}^{\infty}\frac{t^{(2k+1)\alpha}}{(2k+1)!},\end{aligned} \tag{7.90}$$

where $\alpha,\beta,t\in\mathbb{C}$.

Definition 7.22 The hyperbolic subcosine via Kohlrausch–Williams–Watts type function is defined as

$$\begin{aligned}subcosh_{\alpha,\beta}(t) &= \tfrac{1}{2}\left(\mathbb{K}_{\alpha,\beta}(1,t) + \mathbb{K}_{\alpha,\beta}(-1,t)\right)\\ &= t^{\beta-1}\sum_{k=0}^{\infty}\frac{t^{2k\alpha}}{\Gamma(2k+1)}\\ &= t^{\beta-1}\sum_{k=0}^{\infty}\frac{t^{2k\alpha}}{(2k)!},\end{aligned} \tag{7.91}$$

where $\alpha,\beta,t\in\mathbb{C}$.

Definition 7.23 The hyperbolic subtangent via Kohlrausch–Williams–Watts type function is defined as

$$subtanh_{\alpha,\beta}(t) = \frac{subsinh_{\alpha,\beta}(t)}{subcosh_{\alpha,\beta}(t)} \left(subcosh_{\alpha,\beta}(t) \neq 0\right), \tag{7.92}$$

where α, $\beta, t \in \mathbb{C}$.

Definition 7.24 The hyperbolic subcotangent via Kohlrausch–Williams–Watts type function is defined as

$$subcotanh_{\alpha,\beta}(t) = \frac{subcosh_{\alpha,\beta}(t)}{sub\Re sinh_{\alpha,\beta}(t)}, \left(subsinh_{\alpha,\beta}(t) \neq 0\right), \tag{7.93}$$

where α, $\beta, t \in \mathbb{C}$.

Definition 7.25 The hyperbolic subcosecant via Kohlrausch–Williams–Watts type function is defined as

$$subcsch_{\alpha,\beta}(t) = \frac{1}{subsinh_{\alpha,\beta}(t)}, \left(subsinh_{\alpha,\beta}(t) \neq 0\right), \tag{7.94}$$

where α, $\beta, t \in \mathbb{C}$.

Definition 7.26 The hyperbolic subsecant via Kohlrausch–Williams–Watts type function is defined as

$$subsech_{\alpha,\beta}(t) = \frac{1}{subcosh_{\alpha,\beta}(t)} \left(subcosh_{\alpha,\beta}(t) \neq 0\right), \tag{7.95}$$

where α, $\beta, t \in \mathbb{C}$.

Therefore, we have that

$$\begin{aligned} subsinh_{\alpha,\beta}(\lambda, t) &= \tfrac{1}{2}\left(\mathbb{K}_{\alpha,\beta}(\lambda, t) - \mathbb{K}_{\alpha,\beta}(-\lambda, t)\right) \\ &= t^{\beta-1} \sum_{k=0}^{\infty} \frac{\lambda^{2k+1} t^{(2k+1)\alpha}}{\Gamma(2k+2)} \\ &= t^{\beta-1} \sum_{k=0}^{\infty} \frac{\lambda^{2k+1} t^{(2k+1)\alpha}}{(2k+1)!}, \end{aligned} \tag{7.96}$$

$$\begin{aligned} subcosh_{\alpha,\beta}(\lambda, t) &= \tfrac{1}{2}\left(\mathbb{K}_{\alpha,\beta}(\lambda, t) + \mathbb{K}_{\alpha,\beta}(-\lambda, t)\right) \\ &= t^{\beta-1} \sum_{k=0}^{\infty} \frac{\lambda^{2k} t^{2k\alpha}}{\Gamma(2k+1)} \\ &= t^{\beta-1} \sum_{k=0}^{\infty} \frac{\lambda^{2k} t^{2k\alpha}}{(2k)!}, \end{aligned} \tag{7.97}$$

$$subtanh_{\alpha,\beta}(\lambda,t) = \frac{subsinh_{\alpha,\beta}(\lambda,t)}{subcosh_{\alpha,\beta}(\lambda,t)} \left(subcosh_{\alpha,\beta}(\lambda,t) \neq 0\right), \tag{7.98}$$

$$subcotanh_{\alpha,\beta}(\lambda,t) = \frac{subcosh_{\alpha,\beta}(\lambda,t)}{subsinh_{\alpha,\beta}(\lambda,t)}, \left(subsinh_{\alpha,\beta}(\lambda,t) \neq 0\right), \tag{7.99}$$

$$subcsch_{\alpha,\beta}(\lambda,t) = \frac{1}{subsinh_{\alpha,\beta}(\lambda,t)}, \left(subsinh_{\alpha,\beta}(\lambda,t) \neq 0\right), \tag{7.100}$$

and

$$subsech_{\alpha,\beta}(\lambda,t) = \frac{1}{subcosh_{\alpha,\beta}(\lambda,t)} \left(subcosh_{\alpha,\beta}(\lambda,t) \neq 0\right), \tag{7.101}$$

where $\alpha, \beta, \lambda, t \in \mathbb{C}$.

Theorem 7.4 *Let $\alpha, \beta, \lambda, t \in \mathbb{C}$. Then, we have*

$$\mathbb{K}_{\alpha,\beta}(\lambda,t) = subcosh_{\alpha,\beta}(\lambda,t) + subsinh_{\alpha,\beta}(\lambda,t) \tag{7.102}$$

and

$$\mathbb{K}_{\alpha,\beta}(-\lambda,t) = subcosh_{\alpha,\beta}(\lambda,t) - subsinh_{\alpha,\beta}(\lambda,t). \tag{7.103}$$

Proof In view of (7.96) and (7.97), we obtain the result. □

7.1.6 The Integral Representations Associated with Kohlrausch–Williams–Watts Function

In this part, we present the integral representations associated with Kohlrausch–Williams–Watts function.

Property 7.6 Let $\alpha, \beta, \sigma, \lambda \in \mathbb{C}$, $Re(\sigma) > 0$, $Re(\alpha) > 0$, and $Re(\beta) > 0$. Then, we have

$$\frac{1}{\Gamma(\sigma)} \int_0^t (t-\tau)^{\sigma} subsin_{\alpha,\beta}(\lambda,\tau)\, d\tau = t^{\beta+\sigma-1} {}_1sin_1\left(\begin{matrix} (\beta,\alpha) \\ (\beta+\sigma,\alpha) \end{matrix} ; \lambda t^{\alpha} \right), \tag{7.104}$$

$$\frac{1}{\Gamma(\sigma)} \int_0^t (t-\tau)^{\sigma} subcos_{\alpha,\beta}(\lambda,\tau)\, d\tau = t^{\beta+\sigma-1} {}_1cos_1\left(\begin{matrix} (\beta,\alpha) \\ (\beta+\sigma,\alpha) \end{matrix} ; \lambda t^{\alpha} \right), \tag{7.105}$$

$$\frac{1}{\Gamma(\sigma)}\int_0^t (t-\tau)^\sigma\, subsinh_{\alpha,\beta}(\lambda,\tau)\,d\tau = t^{\beta+\sigma-1}\,{}_1sinh_1\left(\begin{matrix}(\beta,\alpha)\\(\beta+\sigma,\alpha)\end{matrix};\lambda t^\alpha\right), \tag{7.106}$$

and

$$\frac{1}{\Gamma(\sigma)}\int_0^t (t-\tau)^\sigma\, subcosh_{\alpha,\beta}(\lambda,\tau)\,d\tau = t^{\beta+\sigma-1}\,{}_1cosh_1\left(\begin{matrix}(\beta,\alpha)\\(\beta+\sigma,\alpha)\end{matrix};\lambda t^\alpha\right). \tag{7.107}$$

Proof With the aid of (7.34), we present

$$\frac{1}{\Gamma(\sigma)}\int_0^t (t-\tau)^\sigma\, \mathbb{K}_{\alpha,\beta}(i\lambda,\tau)\,d\tau = t^{\beta+\sigma-1}\,{}_1\varphi_1\left(\begin{matrix}(\beta,\alpha)\\(\beta+\sigma,\alpha)\end{matrix};i\lambda t^\alpha\right), \tag{7.108}$$

$$\frac{1}{\Gamma(\sigma)}\int_0^t (t-\tau)^\sigma\, \mathbb{K}_{\alpha,\beta}(-i\lambda,\tau)\,d\tau = t^{\beta+\sigma-1}\,{}_1\varphi_1\left(\begin{matrix}(\beta,\alpha)\\(\beta+\sigma,\alpha)\end{matrix};-i\lambda t^\alpha\right), \tag{7.109}$$

$$\frac{1}{\Gamma(\sigma)}\int_0^t (t-\tau)^\sigma\, \mathbb{K}_{\alpha,\beta}(\lambda,\tau)\,d\tau = t^{\beta+\sigma-1}\,{}_1\varphi_1\left(\begin{matrix}(\beta,\alpha)\\(\beta+\sigma,\alpha)\end{matrix};\lambda t^\alpha\right), \tag{7.110}$$

and

$$\frac{1}{\Gamma(\sigma)}\int_0^t (t-\tau)^\sigma\, \mathbb{K}_{\alpha,\beta}(-\lambda,\tau)\,d\tau = t^{\beta+\sigma-1}\,{}_1\varphi_1\left(\begin{matrix}(\beta,\alpha)\\(\beta+\sigma,\alpha)\end{matrix};-\lambda t^\alpha\right), \tag{7.111}$$

and we have

$$\begin{aligned}&\tfrac{1}{\Gamma(\sigma)}\int_0^t (t-\tau)^\sigma\, subsin_{\alpha,\beta}(\lambda,\tau)\,d\tau\\
&= \tfrac{1}{\Gamma(\sigma)}\int_0^t (t-\tau)^\sigma \left[\tfrac{1}{2i}\left(\mathbb{K}_{\alpha,\beta}(i\lambda,\tau)-\mathbb{K}_{\alpha,\beta}(-i\lambda,\tau)\right)\right]d\tau\\
&= \tfrac{t^{\beta+\sigma-1}}{2i}\left[{}_1\varphi_1\left(\begin{matrix}(\beta,\alpha)\\(\beta+\sigma,\alpha)\end{matrix};i\lambda t^\alpha\right)-{}_1\varphi_1\left(\begin{matrix}(\beta,\alpha)\\(\beta+\sigma,\alpha)\end{matrix};-i\lambda t^\alpha\right)\right]\\
&= t^{\beta+\sigma-1}\,{}_1sin_1\left(\begin{matrix}(\beta,\alpha)\\(\beta+\sigma,\alpha)\end{matrix};\lambda t^\alpha\right),\end{aligned} \tag{7.112}$$

$$\begin{aligned}
&\frac{1}{\Gamma(\sigma)}\int_0^t (t-\tau)^\sigma \, subcos_{\alpha,\beta}(\lambda,\tau)\, d\tau \\
&= \frac{1}{\Gamma(\sigma)}\int_0^t (t-\tau)^\sigma \left[\tfrac{1}{2}\left(\mathbb{K}_{\alpha,\beta}(i\lambda,\tau)+\mathbb{K}_{\alpha,\beta}(-i\lambda,\tau)\right)\right] d\tau \\
&= \frac{t^{\beta+\sigma-1}}{2}\left[{}_1\varphi_1\left(\begin{matrix}(\beta,\alpha)\\(\beta+\sigma,\alpha)\end{matrix};i\lambda t^\alpha\right)+{}_1\varphi_1\left(\begin{matrix}(\beta,\alpha)\\(\beta+\sigma,\alpha)\end{matrix};-i\lambda t^\alpha\right)\right] \\
&= t^{\beta+\sigma-1}\,{}_1cos_1\left(\begin{matrix}(\beta,\alpha)\\(\beta+\sigma,\alpha)\end{matrix};\lambda t^\alpha\right),
\end{aligned} \tag{7.113}$$

$$\begin{aligned}
&\frac{1}{\Gamma(\sigma)}\int_0^t (t-\tau)^\sigma \, subsinh_{\alpha,\beta}(\lambda,\tau)\, d\tau \\
&= \frac{1}{\Gamma(\sigma)}\int_0^t (t-\tau)^\sigma \left[\tfrac{1}{2}\left(\mathbb{K}_{\alpha,\beta}(\lambda,\tau)-\mathbb{K}_{\alpha,\beta}(-\lambda,\tau)\right)\right] d\tau \\
&= \frac{t^{\beta+\sigma-1}}{2}\left[{}_1\varphi_1\left(\begin{matrix}(\beta,\alpha)\\(\beta+\sigma,\alpha)\end{matrix};\lambda t^\alpha\right)-{}_1\varphi_1\left(\begin{matrix}(\beta,\alpha)\\(\beta+\sigma,\alpha)\end{matrix};-\lambda t^\alpha\right)\right] \\
&= t^{\beta+\sigma-1}\,{}_1sinh_1\left(\begin{matrix}(\beta,\alpha)\\(\beta+\sigma,\alpha)\end{matrix};\lambda t^\alpha\right),
\end{aligned} \tag{7.114}$$

and

$$\begin{aligned}
&\frac{1}{\Gamma(\sigma)}\int_0^t (t-\tau)^\sigma \, subcosh_{\alpha,\beta}(\lambda,\tau)\, d\tau \\
&= \frac{1}{\Gamma(\sigma)}\int_0^t (t-\tau)^\sigma \left[\tfrac{1}{2}\left(\mathbb{K}_{\alpha,\beta}(\lambda,\tau)+\mathbb{K}_{\alpha,\beta}(-\lambda,\tau)\right)\right] d\tau \\
&= \frac{t^{\beta+\sigma-1}}{2}\left[{}_1\varphi_1\left(\begin{matrix}(\beta,\alpha)\\(\beta+\sigma,\alpha)\end{matrix};\lambda t^\alpha\right)+{}_1\varphi_1\left(\begin{matrix}(\beta,\alpha)\\(\beta+\sigma,\alpha)\end{matrix};-\lambda t^\alpha\right)\right] \\
&= t^{\beta+\sigma-1}\,{}_1cosh_1\left(\begin{matrix}(\beta,\alpha)\\(\beta+\sigma,\alpha)\end{matrix};\lambda t^\alpha\right).
\end{aligned} \tag{7.115}$$

□

Property 7.7 Let $\alpha, \beta, \sigma, \lambda \in \mathbb{C}$, $Re(\sigma) > 0$, $Re(\alpha) > 0$, and $Re(\beta) > 0$. Then, we have

$$\frac{1}{\Gamma(\sigma)}\int_0^1 (1-\tau)^\sigma \, subsin_{\alpha,\beta}(\lambda,\tau)\, d\tau = {}_1sin_1\left(\begin{matrix}(\beta,\alpha)\\(\beta+\sigma,\alpha)\end{matrix};\lambda\right), \tag{7.116}$$

$$\frac{1}{\Gamma(\sigma)}\int_0^1 (1-\tau)^\sigma \, subcos_{\alpha,\beta}(\lambda,\tau)\, d\tau = {}_1cos_1\left(\begin{matrix}(\beta,\alpha)\\(\beta+\sigma,\alpha)\end{matrix};\lambda\right), \tag{7.117}$$

$$\frac{1}{\Gamma(\sigma)}\int_0^1 (1-\tau)^{\sigma}\, subsinh_{\alpha,\beta}(\lambda,\tau)\,d\tau = {}_1sinh_1\left(\begin{matrix}(\beta,\alpha)\\(\beta+\sigma,\alpha)\end{matrix};\lambda\right), \tag{7.118}$$

and

$$\frac{1}{\Gamma(\sigma)}\int_0^1 (1-\tau)^{\sigma}\, subcosh_{\alpha,\beta}(\lambda,\tau)\,d\tau = {}_1cosh_1\left(\begin{matrix}(\beta,\alpha)\\(\beta+\sigma,\alpha)\end{matrix};\lambda\right). \tag{7.119}$$

Proof Taking $t = 1$ in (7.104), (7.105), (7.106), and (7.107), we obtain the results. □

Property 7.8 Let $\alpha, \beta, \sigma, \lambda \in \mathbb{C}$, $Re(\sigma) > 0$, $Re(\alpha) > 0$, and $Re(\beta) > 0$. Then, we have

$$\begin{aligned}\frac{1}{\Gamma(\sigma)}\int_0^2 (2-\tau)^{\sigma}\, subsin_{\alpha,\beta}(\lambda,\tau)\,d\tau &= \frac{1}{\Gamma(\sigma)}\int_{-1}^1 (1-\tau)^{\sigma}\, subsin_{\alpha,\beta}(\lambda,1+\tau)\,d\tau\\ &= \frac{1}{\Gamma(\sigma)}\int_{-1}^1 (1+\tau)^{\sigma}\, subsin_{\alpha,\beta}(\lambda,1-\tau)\,d\tau\\ &= 2^{\beta+\sigma-1}\,{}_1sin_1\left(\begin{matrix}(\beta,\alpha)\\(\beta+\sigma,\alpha)\end{matrix};\lambda 2^{\alpha}\right),\end{aligned} \tag{7.120}$$

$$\begin{aligned}\frac{1}{\Gamma(\sigma)}\int_0^2 (2-\tau)^{\sigma}\, subcos_{\alpha,\beta}(\lambda,\tau)\,d\tau &= \frac{1}{\Gamma(\sigma)}\int_{-1}^1 (1-\tau)^{\sigma}\, subcos_{\alpha,\beta}(\lambda,1+\tau)\,d\tau\\ &= \frac{1}{\Gamma(\sigma)}\int_{-1}^1 (1+\tau)^{\sigma}\, subcos_{\alpha,\beta}(\lambda,1-\tau)\,d\tau\\ &= 2^{\beta+\sigma-1}\,{}_1cos_1\left(\begin{matrix}(\beta,\alpha)\\(\beta+\sigma,\alpha)\end{matrix};\lambda 2^{\alpha}\right),\end{aligned} \tag{7.121}$$

$$\begin{aligned}\frac{1}{\Gamma(\sigma)}\int_0^2 (2-\tau)^{\sigma}\, subsinh_{\alpha,\beta}(\lambda,\tau)\,d\tau &= \frac{1}{\Gamma(\sigma)}\int_{-1}^1 (1-\tau)^{\sigma}\, subsinh_{\alpha,\beta}(\lambda,1+\tau)\,d\tau\\ &= \frac{1}{\Gamma(\sigma)}\int_{-1}^1 (1+\tau)^{\sigma}\, subsinh_{\alpha,\beta}(\lambda,1-\tau)\,d\tau\\ &= 2^{\beta+\sigma-1}\,{}_1sinh_1\left(\begin{matrix}(\beta,\alpha)\\(\beta+\sigma,\alpha)\end{matrix};\lambda 2^{\alpha}\right),\end{aligned} \tag{7.122}$$

and

$$\frac{1}{\Gamma(\sigma)}\int_{0}^{2}(2-\tau)^{\sigma}\, subcosh_{\alpha,\beta}(\lambda,\tau)\,d\tau = \frac{1}{\Gamma(\sigma)}\int_{-1}^{1}(1-\tau)^{\sigma}\, subcosh_{\alpha,\beta}(\lambda,1+\tau)\,d\tau$$
$$= \frac{1}{\Gamma(\sigma)}\int_{-1}^{1}(1+\tau)^{\sigma}\, subcosh_{\alpha,\beta}(\lambda,1-\tau)\,d\tau$$
$$= 2^{\beta+\sigma-1}\,{}_1cosh_1\left(\begin{matrix}(\beta,\alpha)\\(\beta+\sigma,\alpha)\end{matrix};\lambda 2^{\alpha}\right). \tag{7.123}$$

Proof Putting $t=2$ in (7.104), (7.105), (7.106), and (7.107), we get the results. □

Property 7.9 Let $\alpha,\sigma,\lambda\in\mathbb{C}$, $Re(\sigma)>0$, and $Re(\alpha)>0$. Then, we have

$$\frac{1}{\Gamma(\sigma)}\int_{0}^{t}(t-\tau)^{\sigma}\, subsin_{\alpha}(\lambda,\tau)\,d\tau = t^{\sigma}\,{}_1sin_1\left(\begin{matrix}(1,\alpha)\\(1+\sigma,\alpha)\end{matrix};\lambda t^{\alpha}\right), \tag{7.124}$$

$$\frac{1}{\Gamma(\sigma)}\int_{0}^{t}(t-\tau)^{\sigma}\, subcos_{\alpha}(\lambda,\tau)\,d\tau = t^{\sigma}\,{}_1cos_1\left(\begin{matrix}(1,\alpha)\\(1+\sigma,\alpha)\end{matrix};\lambda t^{\alpha}\right), \tag{7.125}$$

$$\frac{1}{\Gamma(\sigma)}\int_{0}^{t}(t-\tau)^{\sigma}\, subsinh_{\alpha}(\lambda,\tau)\,d\tau = t^{\sigma}\,{}_1sinh_1\left(\begin{matrix}(1,\alpha)\\(1+\sigma,\alpha)\end{matrix};\lambda t^{\alpha}\right), \tag{7.126}$$

and

$$\frac{1}{\Gamma(\sigma)}\int_{0}^{t}(t-\tau)^{\sigma}\, subcosh_{\alpha}(\lambda,\tau)\,d\tau = t^{\sigma}\,{}_1cosh_1\left(\begin{matrix}(1,\alpha)\\(1+\sigma,\alpha)\end{matrix};\lambda t^{\alpha}\right). \tag{7.127}$$

Proof Taking $\beta=1$ in (7.104), (7.105), (7.106), and (7.107), we deduce the results. □

Property 7.10 Let $\alpha,\sigma,\lambda\in\mathbb{C}$, $Re(\sigma)>0$, and $Re(\alpha)>0$. Then, we have

$$\frac{1}{\Gamma(\sigma)}\int_{0}^{1}(1-\tau)^{\sigma}\, subsin_{\alpha}(\lambda,\tau)\,d\tau = {}_1sin_1\left(\begin{matrix}(1,\alpha)\\(1+\sigma,\alpha)\end{matrix};\lambda\right), \tag{7.128}$$

$$\frac{1}{\Gamma(\sigma)}\int_0^1 (1-\tau)^{\sigma}\, subcos_{\alpha}(\lambda,\tau)\, d\tau = {}_1cos_1\left(\begin{matrix}(1,\alpha)\\(1+\sigma,\alpha)\end{matrix};\lambda\right), \tag{7.129}$$

$$\frac{1}{\Gamma(\sigma)}\int_0^1 (1-\tau)^{\sigma}\, subsinh_{\alpha}(\lambda,\tau)\, d\tau = {}_1sinh_1\left(\begin{matrix}(1,\alpha)\\(1+\sigma,\alpha)\end{matrix};\lambda\right), \tag{7.130}$$

and

$$\frac{1}{\Gamma(\sigma)}\int_0^1 (1-\tau)^{\sigma}\, subcosh_{\alpha}(\lambda,\tau)\, d\tau = {}_1cosh_1\left(\begin{matrix}(1,\alpha)\\(1+\sigma,\alpha)\end{matrix};\lambda\right). \tag{7.131}$$

Proof Taking $t = 1$ in (7.124), (7.125), (7.126), and (7.127), we obtain the results.

□

Property 7.11 Let $\alpha, \sigma, \lambda \in \mathbb{C}$, $Re(\sigma) > 0$, and $Re(\alpha) > 0$. Then, we have

$$\begin{aligned}
&\tfrac{1}{\Gamma(\sigma)}\int_0^2 (2-\tau)^{\sigma}\, subsin_{\alpha}(\lambda,\tau)\, d\tau\\
&= \tfrac{1}{\Gamma(\sigma)}\int_{-1}^1 (1-\tau)^{\sigma}\, subsin_{\alpha}(\lambda,1+\tau)\, d\tau\\
&= \tfrac{1}{\Gamma(\sigma)}\int_{-1}^1 (1+\tau)^{\sigma}\, subsin_{\alpha}(\lambda,1-\tau)\, d\tau\\
&= 2^{\sigma}\, {}_1sin_1\left(\begin{matrix}(1,\alpha)\\(1+\sigma,\alpha)\end{matrix};\lambda 2^{\alpha}\right),
\end{aligned} \tag{7.132}$$

$$\begin{aligned}
&\tfrac{1}{\Gamma(\sigma)}\int_0^2 (2-\tau)^{\sigma}\, subcos_{\alpha}(\lambda,\tau)\, d\tau\\
&= \tfrac{1}{\Gamma(\sigma)}\int_{-1}^1 (1-\tau)^{\sigma}\, subcos_{\alpha}(\lambda,1+\tau)\, d\tau\\
&= \tfrac{1}{\Gamma(\sigma)}\int_{-1}^1 (1+\tau)^{\sigma}\, subcos_{\alpha}(\lambda,1-\tau)\, d\tau\\
&= 2^{\sigma}\, {}_1cos_1\left(\begin{matrix}(1,\alpha)\\(1+\sigma,\alpha)\end{matrix};\lambda 2^{\alpha}\right),
\end{aligned} \tag{7.133}$$

$$\begin{aligned}
&\frac{1}{\Gamma(\sigma)}\int_0^2 (2-\tau)^{\sigma}\, subsinh_{\alpha}(\lambda,\tau)\, d\tau \\
&= \frac{1}{\Gamma(\sigma)}\int_{-1}^{1} (1-\tau)^{\sigma}\, subsinh_{\alpha}(\lambda,1+\tau)\, d\tau \\
&= \frac{1}{\Gamma(\sigma)}\int_{-1}^{1} (1+\tau)^{\sigma}\, subsinh_{\alpha}(\lambda,1-\tau)\, d\tau \\
&= 2^{\sigma}\ {}_1sinh_1\left(\begin{matrix}(1,\alpha)\\(1+\sigma,\alpha)\end{matrix};\lambda 2^{\alpha}\right),
\end{aligned} \tag{7.134}$$

and

$$\begin{aligned}
&\frac{1}{\Gamma(\sigma)}\int_0^2 (2-\tau)^{\sigma}\, subcosh_{\alpha}(\lambda,\tau)\, d\tau \\
&= \frac{1}{\Gamma(\sigma)}\int_{-1}^{1} (1-\tau)^{\sigma}\, subcosh_{\alpha}(\lambda,1+\tau)\, d\tau \\
&= \frac{1}{\Gamma(\sigma)}\int_{-1}^{1} (1+\tau)^{\sigma}\, subcosh_{\alpha}(\lambda,1-\tau)\, d\tau \\
&= 2^{\sigma}\ {}_1cosh_1\left(\begin{matrix}(1,\alpha)\\(1+\sigma,\alpha)\end{matrix};\lambda 2^{\alpha}\right).
\end{aligned} \tag{7.135}$$

Proof Taking $t = 2$ in (7.124), (7.125), (7.126), and (7.127), we obtain the results. □

Here, we have

$$subsin_{\alpha,\beta}(-\lambda,t) = -subsin_{\alpha,\beta}(\lambda,t), \tag{7.136}$$

$$subcos_{\alpha,\beta}(-\lambda,t) = subcos_{\alpha,\beta}(\lambda,t), \tag{7.137}$$

$$subtan_{\alpha,\beta}(-\lambda,t) = -subtan_{\alpha,\beta}(\lambda,t), \tag{7.138}$$

$$subcotan_{\alpha,\beta}(-\lambda,t) = -subcotan_{\alpha,\beta}(\lambda,t), \tag{7.139}$$

$$subcsc_{\alpha,\beta}(-\lambda,t) = -subcsc_{\alpha,\beta}(\lambda,t), \tag{7.140}$$

$$subsec_{\alpha,\beta}(-\lambda,t) = subsec_{\alpha,\beta}(\lambda,t), \tag{7.141}$$

$$subsinh_{\alpha,\beta}(-\lambda,t) = -subsinh_{\alpha,\beta}(\lambda,t), \tag{7.142}$$

$$subcosh_{\alpha,\beta}(-\lambda,t) = subcosh_{\alpha,\beta}(\lambda,t), \tag{7.143}$$

$$subtanh_{\alpha,\beta}(-\lambda,t) = -subtanh_{\alpha,\beta}(\lambda,t), \tag{7.144}$$

$$subcotanh_{\alpha,\beta}\left(-\lambda,t\right)=-subcotanh_{\alpha,\beta}\left(\lambda,t\right), \tag{7.145}$$

$$subcsch_{\alpha,\beta}\left(-\lambda,t\right)=-subcsch_{\alpha,\beta}\left(\lambda,t\right), \tag{7.146}$$

$$subsech_{\alpha,\beta}\left(-\lambda,t\right)=subsech_{\alpha,\beta}\left(\lambda,t\right), \tag{7.147}$$

such that

$$subsin_{\alpha}\left(-\lambda,t\right)=-subsin_{\alpha}\left(\lambda,t\right), \tag{7.148}$$

$$subcos_{\alpha}\left(-\lambda,t\right)=subcos_{\alpha}\left(\lambda,t\right), \tag{7.149}$$

$$subtan_{\alpha}\left(-\lambda,t\right)=-subtan_{\alpha}\left(\lambda,t\right), \tag{7.150}$$

$$subcotan_{\alpha}\left(-\lambda,t\right)=-subcotan_{\alpha}\left(\lambda,t\right), \tag{7.151}$$

$$subcsc_{\alpha}\left(-\lambda,t\right)=-subcsc_{\alpha}\left(\lambda,t\right), \tag{7.152}$$

$$subsec_{\alpha}\left(-\lambda,t\right)=subsec_{\alpha}\left(\lambda,t\right), \tag{7.153}$$

$$subsinh_{\alpha}\left(-\lambda,t\right)=-subsinh_{\alpha}\left(\lambda,t\right), \tag{7.154}$$

$$subcosh_{\alpha}\left(-\lambda,t\right)=subcosh_{\alpha}\left(\lambda,t\right), \tag{7.155}$$

$$subtanh_{\alpha}\left(-\lambda,t\right)=-subtanh_{\alpha}\left(\lambda,t\right), \tag{7.156}$$

$$subcotanh_{\alpha}\left(-\lambda,t\right)=-subcotanh_{\alpha}\left(\lambda,t\right), \tag{7.157}$$

$$subcsch_{\alpha}\left(-\lambda,t\right)=-subcsch_{\alpha}\left(\lambda,t\right), \tag{7.158}$$

and

$$subsech_{\alpha}\left(-\lambda,t\right)=subsech_{\alpha}\left(\lambda,t\right). \tag{7.159}$$

Moreover, for $i=\sqrt{-1}$, we have

$$subsinh_{\alpha,\beta}\left(\lambda,t\right)=-isubsin_{\alpha,\beta}\left(i\lambda,t\right), \tag{7.160}$$

$$subcosh_{\alpha,\beta}\left(\lambda,t\right)=subcos_{\alpha,\beta}\left(i\lambda,t\right), \tag{7.161}$$

$$subtanh_{\alpha,\beta}\left(\lambda,t\right)=-isubtan_{\alpha,\beta}\left(i\lambda,t\right), \tag{7.162}$$

$$subcotanh_{\alpha,\beta}\left(\lambda,t\right)=isubcotanh_{\alpha,\beta}\left(i\lambda,t\right), \tag{7.163}$$

$$subcsch_{\alpha,\beta}\left(\lambda,t\right)=isubcsc_{\alpha,\beta}\left(i\lambda,t\right), \tag{7.164}$$

and

$$subsech_{\alpha,\beta}(\lambda, t) = subsec_{\alpha,\beta}(i\lambda, t), \tag{7.165}$$

which yield that

$$-isubsin_{\alpha}(i\lambda, t) = subsinh_{\alpha}(\lambda, t), \tag{7.166}$$

$$subcos_{\alpha}(\lambda, t) = subcos_{\alpha}(i\lambda, t), \tag{7.167}$$

$$subtanh_{\alpha}(\lambda, t) = -isubtan_{\alpha}(i\lambda, t), \tag{7.168}$$

$$subcotanh_{\alpha}(\lambda, t) = isubcotan_{\alpha}(i\lambda, t), \tag{7.169}$$

$$subcsc_{\alpha}(\lambda, t) = isubcsc_{\alpha}(i\lambda, t), \tag{7.170}$$

and

$$subsech_{\alpha}(\lambda, t) = subsec_{\alpha}(i\lambda, t). \tag{7.171}$$

In particular,

$$subsin_{1,1}(\lambda, t) = sin(\lambda t), \tag{7.172}$$

$$subcos_{1,1}(\lambda, t) = cos(\lambda t), \tag{7.173}$$

$$subtan_{1,1}(\lambda, t) = tan(\lambda t), \tag{7.174}$$

$$subcotan_{1,1}(\lambda, t) = cotan(\lambda t), \tag{7.175}$$

$$subcsc_{1,1}(\lambda, t) = csc(\lambda t), \tag{7.176}$$

$$subsec_{1,1}(\lambda, t) = sec(\lambda t), \tag{7.177}$$

$$subsinh_{1,1}(\lambda, t) = sinh(\lambda t), \tag{7.178}$$

$$subcosh_{1,1}(\lambda, t) = cosh(\lambda t), \tag{7.179}$$

$$subtanh_{1,1}(\lambda, t) = tanh(\lambda, t), \tag{7.180}$$

$$subcotanh_{1,1}(\lambda, t) = cotanh(\lambda t), \tag{7.181}$$

$$subcsch_{1,1}(\lambda, t) = csch(\lambda t), \tag{7.182}$$

$$subsech_{1,1}(\lambda, t) = sech(\lambda t), \tag{7.183}$$

$$subsin_1(\lambda, t) = sin(\lambda t), \tag{7.184}$$

$$subcos_1(\lambda, t) = cos(\lambda t), \tag{7.185}$$

$$subtan_1(\lambda, t) = tan(\lambda t), \tag{7.186}$$

$$subcotan_1(\lambda, t) = cotan(\lambda t), \tag{7.187}$$

$$subcsc_1(\lambda, t) = csc(\lambda t), \tag{7.188}$$

$$subsec_1(\lambda, t) = sec(\lambda t), \tag{7.189}$$

$$subsinh_1(\lambda, t) = sinh_\alpha(\lambda t), \tag{7.190}$$

$$subcosh_1(\lambda, t) = cosh(\lambda t), \tag{7.191}$$

$$subtanh_1(\lambda, t) = tanh(\lambda t), \tag{7.192}$$

$$subcotanh_1(\lambda, t) = cotanh(\lambda t), \tag{7.193}$$

$$subcsch_1(\lambda, t) = csch(\lambda t), \tag{7.194}$$

and

$$subsech_1(\lambda, t) = sech(\lambda t). \tag{7.195}$$

In fact, it is not difficult to derive the following results:

$$subsin_\alpha(\lambda, \tau + t) = subsin_\alpha(\lambda, \tau)subcos_\alpha(\lambda, t) + subsin_\alpha(\lambda, t)subcos_\alpha(\lambda, \tau), \tag{7.196}$$

$$subsin_\alpha(\lambda, \tau - t) = subsin_\alpha(\lambda, \tau)subcos_\alpha(\lambda, t) - subsin_\alpha(\lambda, t)subcos_\alpha(\lambda, \tau), \tag{7.197}$$

$$subcos_\alpha(\lambda, \tau + t) = subcos_\alpha(\lambda, \tau)subcos_\alpha(\lambda, t) - subsin_\alpha(\lambda, t)subsin_\alpha(\lambda, \tau), \tag{7.198}$$

$$subcos_\alpha(\lambda, \tau - t) = subcos_\alpha(\lambda, \tau)subcos_\alpha(\lambda, t) + subsin_\alpha(\lambda, t)subsin_\alpha(\lambda, \tau), \tag{7.199}$$

$$\begin{aligned} subsinh_\alpha(\lambda, \tau + t) = {} & subsinh_\alpha(\lambda, \tau)\, subcosh_\alpha(\lambda, t) \\ & + subsinh_\alpha(\lambda, t)\, subcosh_\alpha(\lambda, \tau), \end{aligned} \tag{7.200}$$

$$\begin{aligned} subsinh_\alpha(\lambda, \tau - t) = {} & subsinh_\alpha(\lambda, \tau)\, subcosh_\alpha(\lambda, t) \\ & - subsinh_\alpha(\lambda, t)\, subcosh_\alpha(\lambda, \tau), \end{aligned} \tag{7.201}$$

$$\begin{aligned} subcosh_\alpha(\lambda, \tau + t) = {} & subcosh_\alpha(\lambda, \tau)\, subcosh_\alpha(\lambda, t) \\ & + subsinh_\alpha(\lambda, t)\, subsinh_\alpha(\lambda, \tau), \end{aligned} \tag{7.202}$$

and

$$subcosh_{\alpha}(\lambda, \tau - t) = subcosh_{\alpha}(\lambda, \tau)\, subcosh_{\alpha}(\lambda, t) \\ - subsinh_{\alpha}(\lambda, t)\, subsinh_{\alpha}(\lambda, \tau). \tag{7.203}$$

Thus, it is found that

$$subsin_{\alpha}^{2}(\lambda, t) + subcos_{\alpha}^{2}(\lambda, t) = 1 \tag{7.204}$$

and

$$subcosh_{\alpha}^{2}(\lambda, t) - subsinh_{\alpha}^{2}(\lambda, t) = 1. \tag{7.205}$$

The scaling-law derivatives of the special functions can be given as follows:

$${}^{SL}D_t^{(1)} subsin_{\alpha}(\lambda, t) = subcos_{\alpha}(\lambda, t), \tag{7.206}$$

$${}^{SL}D_t^{(1)} subcos_{\alpha}(\lambda, t) = -subsin_{\alpha}(\lambda, t), \tag{7.207}$$

$${}^{SL}D_t^{(1)} subtan_{\alpha}(\lambda, t) = subsec_{\alpha}^{2}(\lambda, t), \tag{7.208}$$

$${}^{SL}D_t^{(1)} subcotan_{\alpha}(\lambda, t) = -subcsc_{\alpha}^{2}(\lambda, t), \tag{7.209}$$

$${}^{SL}D_t^{(1)} subsec_{\alpha}(\lambda, t) = subtan_{\alpha}(\lambda, t) \cdot subsec_{\alpha}(\lambda, t), \tag{7.210}$$

$${}^{SL}D_t^{(1)} subcsc_{\alpha}(\lambda, t) = -subcotan_{\alpha}(\lambda, t) \cdot subcsc_{\alpha}(\lambda, t), \tag{7.211}$$

$${}^{SL}D_t^{(1)} subsinh_{\alpha}(\lambda, t) = subcosh_{\alpha}(\lambda, t), \tag{7.212}$$

$${}^{SL}D_t^{(1)} subcosh_{\alpha}(\lambda, t) = subsinh_{\alpha}(\lambda, t), \tag{7.213}$$

$${}^{SL}D_t^{(1)} subtanh_{\alpha}(\lambda, t) = subsech_{\alpha}^{2}(\lambda, t), \tag{7.214}$$

$${}^{SL}D_t^{(1)} subcotanh_{\alpha}(\lambda, t) = -subcsch_{\alpha}^{2}(\lambda, t), \tag{7.215}$$

$${}^{SL}D_t^{(1)} subsech_{\alpha}(\lambda, t) = -subtanh_{\alpha}(\lambda, t) \cdot subsech_{\alpha}(\lambda, t), \tag{7.216}$$

and

$${}^{SL}D_t^{(1)} subcsch_{\alpha}(\lambda, t) = -subcotanh_{\alpha}(\lambda, t) \cdot subcsch_{\alpha}(\lambda, t). \tag{7.217}$$

The scaling-law integrals of the special functions can be presented as follows:

$$^{SL}I_t^{(1)} subsin_\alpha(\lambda, t) = -subcos_\alpha(\lambda, t) + C, \tag{7.218}$$

$$^{SL}I_t^{(1)} subcos_\alpha(\lambda, t) = subsin_\alpha(\lambda, t) + C, \tag{7.219}$$

$$^{SL}I_t^{(1)} subtan_\alpha(\lambda, t) = \ln|subsec_\alpha(\lambda, t)| + C, \tag{7.220}$$

$$^{SL}I_t^{(1)} subcotan_\alpha(\lambda, t) = |subcsc_\alpha(\lambda, t)| + C, \tag{7.221}$$

$$^{SL}I_t^{(1)} subsec_\alpha(\lambda, t) = |subtan_\alpha(\lambda, t) + subsec_\alpha(\lambda, t)| + C, \tag{7.222}$$

$$^{SL}I_t^{(1)} subcsc_\alpha(\lambda, t) = |subcotan_\alpha(\lambda, t) - subcsc_\alpha(\lambda, t)| + C, \tag{7.223}$$

$$^{SL}I_t^{(1)} subsinh_\alpha(\lambda, t) = subcosh_\alpha(\lambda, t) + C, \tag{7.224}$$

$$^{SL}I_t^{(1)} subcosh_\alpha(\lambda, t) = subsinh_\alpha(\lambda, t) + C, \tag{7.225}$$

$$^{SL}I_t^{(1)} subtanh_\alpha(\lambda, t) = \ln(subcosh_\alpha(\lambda, t)) + C, \tag{7.226}$$

and

$$^{SL}I_t^{(1)} subcotanh_\alpha(\lambda, t) = \ln(subsinh_\alpha(\lambda, t)) + C, \tag{7.227}$$

where C is a constant.

The power-law derivatives of the special functions can be given as follows:

$$^{P}D_t^{(1)} subsin_\alpha(t) = subcos_\alpha(t), \tag{7.228}$$

$$^{P}D_t^{(1)} subcos_\alpha(t) = -subsin_\alpha(t), \tag{7.229}$$

$$^{P}D_t^{(1)} subtan_\alpha(t) = subsec_\alpha^2(t), \tag{7.230}$$

$$^{P}D_t^{(1)} subcotan_\alpha(t) = -subcsc_\alpha^2(t), \tag{7.231}$$

$$^{P}D_t^{(1)} subsec_\alpha(t) = subtan_\alpha(t) \cdot subsec_\alpha(t), \tag{7.232}$$

$$^{P}D_t^{(1)} subcsc_\alpha(t) = -subcotan_\alpha(t) \cdot subcsc_\alpha(t), \tag{7.233}$$

$$^{P}D_t^{(1)} subsinh_\alpha(t) = subcosh_\alpha(t), \tag{7.234}$$

$$^{P}D_t^{(1)} subcosh_\alpha(t) = subsinh_\alpha(t), \tag{7.235}$$

$$ {}^{P}D_t^{(1)} subtanh_\alpha(t) = subsec_\alpha^2(t), \tag{7.236}$$

$$ {}^{P}D_t^{(1)} subcotanh_\alpha(t) = -subcsch_\alpha^2(t), \tag{7.237}$$

$$ {}^{P}D_t^{(1)} subsech_\alpha(t) = -subtanh_\alpha(t) \cdot subsech_\alpha(t), \tag{7.238}$$

and

$$ {}^{P}D_t^{(1)} subcsch_\alpha(t) = -subcotanh_\alpha(t) \cdot subcsch_\alpha(t). \tag{7.239}$$

The power-law integrals of the special functions can be presented as follows:

$$ {}^{P}I_t^{(1)} subsin_\alpha(t) = -subcos_\alpha(t) + C, \tag{7.240}$$

$$ {}^{P}I_t^{(1)} subcos_\alpha(t) = subsin_\alpha(t) + C, \tag{7.241}$$

$$ {}^{P}I_t^{(1)} subtan_\alpha(t) = \ln|subsec_\alpha(t)| + C, \tag{7.242}$$

$$ {}^{P}I_t^{(1)} subcotan_\alpha(t) = |subcsc_\alpha(t)| + C, \tag{7.243}$$

$$ {}^{P}I_t^{(1)} subsec_\alpha(t) = |subtan_\alpha(t) + subsec_\alpha(t)| + C, \tag{7.244}$$

$$ {}^{P}I_t^{(1)} subcsc_\alpha(t) = |subcotan_\alpha(t) - subcsc_\alpha(t)| + C, \tag{7.245}$$

$$ {}^{P}I_t^{(1)} subsinh_\alpha(t) = subcosh_\alpha(t) + C, \tag{7.246}$$

$$ {}^{P}I_t^{(1)} subcosh_\alpha(t) = subsinh_\alpha(t) + C, \tag{7.247}$$

$$ {}^{P}I_t^{(1)} subtanh_\alpha(t) = \ln(subcosh_\alpha(t)) + C, \tag{7.248}$$

and

$$ {}^{P}I_t^{(1)} subcotanh_\alpha(t) = \ln(subsinh_\alpha(t)) + C, \tag{7.249}$$

where C is a constant.

7.1.7 The Special Functions with Complex Topology

For $\mu(t) = \kappa t^{\beta} + c$, we have that

$$subsin_{\beta}\left(\kappa t^{\beta}+c\right) = \frac{e^{i\left(\kappa t^{\beta}+c\right)} - e^{-i\left(\kappa t^{\beta}+c\right)}}{2i} = \sum_{\kappa=0}^{\infty} \frac{(-1)^{\kappa}\left(\kappa t^{\beta}+c\right)^{2\kappa+1}}{\Gamma(2\kappa+2)}, \tag{7.250}$$

$$subcos_{\beta}\left(\kappa t^{\beta}+c\right) = \frac{e^{i\left(\kappa t^{\beta}+c\right)} + e^{-i\left(\kappa t^{\beta}+c\right)}}{2} = \sum_{\kappa=0}^{\infty} \frac{(-1)^{\kappa}\left(\kappa t^{\beta}+c\right)^{2\kappa}}{\Gamma(2\kappa+1)}, \tag{7.251}$$

$$subtan_{\beta}\left(\kappa t^{\beta}+c\right) = \frac{subsin_{\beta}\left(\kappa t^{\beta}+c\right)}{subcos_{\beta}\left(\kappa t^{\beta}+c\right)}, \tag{7.252}$$

$$subcotan_{\beta}\left(\kappa t^{\beta}+c\right) = \frac{subcos_{\beta}\left(\kappa t^{\beta}+c\right)}{subsin_{\beta}\left(\kappa t^{\beta}+c\right)}, \tag{7.253}$$

$$subcsc_{\beta}\left(\kappa t^{\beta}+c\right) = \frac{1}{subsin_{\beta}\left(\kappa t^{\beta}+c\right)}, \tag{7.254}$$

and

$$subsec_{\beta}\left(\kappa t^{\beta}+c\right) = \frac{1}{subcos_{\beta}\left(\kappa t^{\beta}+c\right)}, \tag{7.255}$$

where κ is the normalization constant, β is the scaling exponent, t is the radius, and c is the moving term.

In an analogous manner, for $\mu(t) = \kappa t^{\beta} + c$, we have

$$subsinh_{\beta}\left(\kappa t^{\beta}+c\right) = \frac{e^{\kappa t^{\beta}+c} - e^{-\left(\kappa t^{\beta}+c\right)}}{2} = \sum_{\kappa=0}^{\infty} \frac{\left(\kappa t^{\beta}+c\right)^{2\kappa+1}}{\Gamma(2\kappa+2)}, \tag{7.256}$$

$$subcosh_{\beta}\left(\kappa t^{\beta}+c\right) = \frac{e^{\kappa t^{\beta}+c} + e^{-\left(\kappa t^{\beta}+c\right)}}{2} = \sum_{\kappa=0}^{\infty} \frac{\left(\kappa t^{\beta}+c\right)^{2\kappa}}{\Gamma(2\kappa+1)}, \tag{7.257}$$

$$subtanh_{\beta}\left(\kappa t^{\beta}+c\right) = \frac{subsinh_{\beta}\left(\kappa t^{\beta}+c\right)}{subcosh_{\beta}\left(\kappa t^{\beta}+c\right)}, \tag{7.258}$$

$$subcotanh_{\beta}\left(\kappa t^{\beta}+c\right) = \frac{subcosh_{\beta}\left(\kappa t^{\beta}+c\right)}{subsinh_{\beta}\left(\kappa t^{\beta}+c\right)}, \tag{7.259}$$

$$subcsch_{\beta}\left(\kappa t^{\beta}+c\right) = \frac{1}{subsinh_{\beta}\left(\kappa t^{\beta}+c\right)}, \tag{7.260}$$

and

$$subsech_{\beta}\left(\kappa t^{\beta}+c\right)=\frac{1}{subcosh_{\beta}\left(\kappa t^{\beta}+c\right)}, \tag{7.261}$$

where κ is the normalization constant, β is the scaling exponent, t is the radius, and c is the moving term.

The topology derivatives of the special functions can be given as follows:

$${}^{T}D_{t}^{(1)}subsin_{\beta}\left(\kappa t^{\beta}+c\right)=subcos_{\beta}\left(\kappa t^{\beta}+c\right), \tag{7.262}$$

$${}^{T}D_{t}^{(1)}subcos_{\beta}\left(\kappa t^{\beta}+c\right)=-subsin_{\beta}\left(\kappa t^{\beta}+c\right), \tag{7.263}$$

$${}^{T}D_{t}^{(1)}subtan_{\beta}\left(\kappa t^{\beta}+c\right)=subsec_{\beta}^{2}\left(\kappa t^{\beta}+c\right), \tag{7.264}$$

$${}^{T}D_{t}^{(1)}subcotan_{\beta}\left(\kappa t^{\beta}+c\right)=-subcsc_{\beta}^{2}\left(\kappa t^{\beta}+c\right), \tag{7.265}$$

$${}^{T}D_{t}^{(1)}subsec_{\beta}\left(\kappa t^{\beta}+c\right)=subtan_{\beta}\left(\kappa t^{\beta}+c\right)\cdot subsec_{\beta}\left(\kappa t^{\beta}+c\right), \tag{7.266}$$

$${}^{T}D_{t}^{(1)}subcsc_{\beta}\left(\kappa t^{\beta}+c\right)=-subcotan_{\beta}\left(\kappa t^{\beta}+c\right)\cdot subcsc_{\beta}\left(\kappa t^{\beta}+c\right), \tag{7.267}$$

$${}^{T}D_{t}^{(1)}subsinh_{\beta}\left(\kappa t^{\beta}+c\right)=subcosh_{\beta}\left(\kappa t^{\beta}+c\right), \tag{7.268}$$

$${}^{T}D_{t}^{(1)}subcosh_{\beta}\left(\kappa t^{\beta}+c\right)=subsinh_{\beta}\left(\kappa t^{\beta}+c\right), \tag{7.269}$$

$${}^{T}D_{t}^{(1)}subtanh_{\beta}\left(\kappa t^{\beta}+c\right)=subsech_{\beta}^{2}\left(\kappa t^{\beta}+c\right), \tag{7.270}$$

$${}^{T}D_{t}^{(1)}subcotanh_{\beta}\left(\kappa t^{\beta}+c\right)=-subcsch_{\beta}^{2}\left(\kappa t^{\beta}+c\right), \tag{7.271}$$

$${}^{T}D_{t}^{(1)}subsech_{\beta}\left(\kappa t^{\beta}+c\right)=-subtanh_{\beta}\left(\kappa t^{\beta}+c\right)\cdot subsech_{\beta}\left(\kappa t^{\beta}+c\right), \tag{7.272}$$

and

$${}^{T}D_{t}^{(1)}subcsch_{\beta}\left(\kappa t^{\beta}+c\right)=-subcotanh_{\beta}\left(\kappa t^{\beta}+c\right)\cdot subcsch_{\beta}\left(\kappa t^{\beta}+c\right). \tag{7.273}$$

The topology integrals of the special functions can be given as follows:

$${}^{T}I_{t}^{(1)}subsin_{\beta}\left(\kappa t^{\beta}+c\right)=-subcos_{\beta}\left(\kappa t^{\beta}+c\right)+C, \tag{7.274}$$

$${}^{T}I_{t}^{(1)}subcos_{\beta}\left(\kappa t^{\beta}+c\right)=subsin_{\beta}\left(\kappa t^{\beta}+c\right)+C, \tag{7.275}$$

$$ {}^{T}I_t^{(1)} subtan_\beta\left(\kappa t^\beta + c\right) = \ln\left|subsec_\beta\left(\kappa t^\beta + c\right)\right| + C, \tag{7.276} $$

$$ {}^{T}I_t^{(1)} subcotan_\beta\left(\kappa t^\beta + c\right) = \left|subcsc_\beta\left(\kappa t^\beta + c\right)\right| + C, \tag{7.277} $$

$$ {}^{T}I_t^{(1)} subsec_\beta\left(\kappa t^\beta + c\right) = \left|subtan_\beta\left(\kappa t^\beta + c\right) + subsec_\beta\left(\kappa t^\beta + c\right)\right| + C, \tag{7.278} $$

$$ {}^{T}I_t^{(1)} subcsc_\beta\left(\kappa t^\beta + c\right) = \left|subcotan_\beta\left(\kappa t^\beta + c\right) - subcsc_\beta\left(\kappa t^\beta + c\right)\right| + C, \tag{7.279} $$

$$ {}^{T}I_t^{(1)} subsinh_\beta\left(\kappa t^\beta + c\right) = subcosh_\beta\left(\kappa t^\beta + c\right) + C, \tag{7.280} $$

$$ {}^{T}I_t^{(1)} subcosh_\beta\left(\kappa t^\beta + c\right) = subsinh_\beta\left(\kappa t^\beta + c\right) + C, \tag{7.281} $$

$$ {}^{T}I_t^{(1)} subtanh_\beta\left(\kappa t^\beta + c\right) = \ln\left(subcosh_\beta\left(\kappa t^\beta + c\right)\right) + C, \tag{7.282} $$

and

$$ {}^{T}I_t^{(1)} subcotanh_\beta\left(\kappa t^\beta + c\right) = \ln\left(subsinh_\beta\left(\kappa t^\beta + c\right)\right) + C, \tag{7.283} $$

where C is a constant.

7.1.8 Subsurfaces and Geometric Representations Related to Kohlrausch–Williams–Watts Function

In this part, we plan to extend the Euler task associated with the Kohlrausch–Williams–Watts function and to propose the circle-like, ellipse-like, hyperbola-like, and parabola-like equations. The geometric representations for the subsurfaces are also given.

In 1545, the Italian mathematician, physician, gambler, and philosopher Girolamo Cardano was the first to introduce complex numbers into algebra in his *Ars Magna* [293, 294].

Cardano discovered the complex number $\mathbf{z}$, given as [293]

$$ \mathbf{z} = a + \sqrt{-b}, \tag{7.284} $$

where $a, b \in \mathbb{R}$ and $\mathbf{z} \in \mathbb{C}$.

In 1707, Abraham de Moivre [295] introduced the so-called de Moivre formulas, its proof is due to C. F. Gauss [296], and its modern notation was suggested by L. Euler [71]

The de Moivre formulas can be written as follows [297–300]:

$$ (cosx + isinx)^n = cosnx + isinnx, \tag{7.285} $$

where $n \in \mathbb{N}$.

In 1777, Leonhard Euler introduced the equation of the form [71]

$$h^2 + 1 = 0 \tag{7.286}$$

and considered that

$$e^{\pm i\theta} = cos\theta \pm isin\theta, \tag{7.287}$$

$$e^{i\pi} = 1, \tag{7.288}$$

and

$$a + ib = r\,(cos\theta + isin\theta)\,, \tag{7.289}$$

with the geometric representations

$$cos\theta = \frac{a}{r} \tag{7.290}$$

and

$$sin\theta = \frac{b}{r}, \tag{7.291}$$

where $h = i = \sqrt{-1}$ is the Euler symbol, $a, b \in \mathbb{R}$, $r = \sqrt{a^2 + b^2}$,

$$cos\theta = \sum_{n=0}^{\infty} (-1)^n \frac{\theta^{2n}}{(2n)!} \tag{7.292}$$

and

$$sin\theta = \sum_{n=0}^{\infty} (-1)^n \frac{\theta^{2n+1}}{(2n+1)!}. \tag{7.293}$$

The work was proposed in 1777 and published in 1794 [71].

Let us consider that

$$e^{it^\alpha} = subcos\left(t^\alpha\right) + isubsin\left(t^\alpha\right) \tag{7.294}$$

and

$$e^{-it^\alpha} = subcos\left(t^\alpha\right) - isubsin\left(t^\alpha\right), \tag{7.295}$$

where $i = \sqrt{-1}$, and the special functions are denoted as [92]

$$subcos\left(t^{\alpha}\right) = \sum_{n=0}^{\infty} (-1)^n \frac{t^{2n\alpha}}{(2n)!} \tag{7.296}$$

and

$$subsin\left(t^{\alpha}\right) = \sum_{n=0}^{\infty} (-1)^n \frac{t^{(2n+1)\alpha}}{(2n+1)!}. \tag{7.297}$$

If one sets $\ln r + ix$ in place of $r^{\alpha} + it^{\alpha}$, it is convenient to show that

$$\begin{aligned} & a^{\alpha} + ib^{\alpha} \\ & = e^{r^{\alpha}+it^{\alpha}} \\ & = e^{r^{\alpha}} \left(subcos\left(t^{\alpha}\right) + isubsin\left(t^{\alpha}\right)\right) \\ & = \widehat{r}^{\alpha} \left(subcos\left(t^{\alpha}\right) + isubsin\left(t^{\alpha}\right)\right), \end{aligned} \tag{7.298}$$

where

$$\widehat{r}^{\alpha} = e^{r^{\alpha}} = \sum_{n=0}^{\infty} \frac{r^{n\alpha}}{n!} \tag{7.299}$$

for $\widehat{r} > 0$ and $\widehat{r}^{\alpha} = \sqrt{a^{2\alpha} + b^{2\alpha}} > 0$.

Here, we have

$$t = subtan_{\alpha}^{-1} \left(\frac{b}{a}\right)^{\alpha}, \tag{7.300}$$

which leads to

$$t^{\alpha} = \arg tan \left(\frac{b}{a}\right)^{\alpha}, \tag{7.301}$$

where $subtan_{\alpha}^{-1}\left(subtan_{\alpha}\left(t^{\alpha}\right)\right) = t^{\alpha}$.

Let us recall the Terjanian equation of the form [301]

$$x^{2\alpha} + y^{2\alpha} = z^{2\alpha}, \tag{7.302}$$

where $x, y, z \in \mathbb{C}$ and $\alpha \in \mathbb{C}$.

It is clear that for $\widehat{r}^{2\alpha} = z^{2\alpha} \in \mathbb{R}_{+}$ and $\alpha \in \mathbb{R}$, we have

$$x^{2\alpha} + y^{2\alpha} = \widehat{r}^{2\alpha}, \tag{7.303}$$

so that the equations

$$subcos\left(t^{\alpha}\right)=\frac{x^{\alpha}}{\widehat{r}^{\alpha}} \tag{7.304}$$

and

$$subsin\left(t^{\alpha}\right)=\frac{y^{\alpha}}{\widehat{r}^{\alpha}}, \tag{7.305}$$

where the subdistance is defined as follows:

$$x^{2\alpha}+y^{2\alpha}=\widehat{r}^{2\alpha}, \tag{7.306}$$

with

$$\widehat{r}^{\alpha}=\sqrt{x^{2\alpha}+y^{2\alpha}}, \tag{7.307}$$

or equivalently,

$$\widehat{r}=\sqrt[\alpha]{x^{2\alpha}+y^{2\alpha}}. \tag{7.308}$$

Taking $\widehat{r}=1$, we have

$$\widehat{r}^{\alpha}=1 \tag{7.309}$$

such that

$$subcos\left(t^{\alpha}\right)=x^{\alpha} \tag{7.310}$$

and

$$subsin\left(t^{\alpha}\right)=y^{\alpha}, \tag{7.311}$$

where

$$x^{2\alpha}+y^{2\alpha}=1 \tag{7.312}$$

for $\alpha\in\mathbb{R}$.

It is clear that there are the equations

$$subcos\left(t^{\alpha}+2\pi\right)=subcos\left(t^{\alpha}\right) \tag{7.313}$$

and

$$subsin\left(t^{\alpha}+2\pi\right)=subsin\left(t^{\alpha}\right), \tag{7.314}$$

which leads to the following relation:

$$subcos\left(t^{\alpha}+\varphi\right)=subcos\left(t^{\alpha}\right) \tag{7.315}$$

and

$$subsin\left(t^{\alpha}+\varphi\right)=subsin\left(t^{\alpha}\right), \tag{7.316}$$

where $\varphi=2\pi$.

The outer-period of the function is defined as

$$G\left(t^{\alpha}+\varphi\right)=G\left(t^{\alpha}\right), \tag{7.317}$$

where φ is the outer-period if there is the representation

$$G\left(t+\varphi\right)=G\left(t\right). \tag{7.318}$$

More generally, the outer-period of the function is defined as

$$G\left(g\left(t\right)+\varphi\right)=G\left(g\left(t\right)\right), \tag{7.319}$$

where φ is the outer-period if there is the representation

$$G\left(t+\varphi\right)=G\left(t\right). \tag{7.320}$$

There are the facts

$$x^{\alpha}=\widehat{r}^{\alpha}subcos\left(t^{\alpha}\right) \tag{7.321}$$

and

$$y^{\alpha}=\widehat{r}^{\alpha}subsin\left(t^{\alpha}\right), \tag{7.322}$$

lead to the Terjanian equation of the form

$$x^{2\alpha}+y^{2\alpha}=z^{2\alpha}=\widehat{r}^{2\alpha}, \tag{7.323}$$

where $\alpha\in\mathbb{R}$.

So, there are the equations

$$subcos\left(t^{\alpha}\right)=\left(cos\theta\right)^{\alpha} \tag{7.324}$$

and

$$subsin\left(t^{\alpha}\right) = (sin\theta)^{\alpha} \tag{7.325}$$

since $x = a$ and $y = b$.

It is very easy to rewrite (7.324) and (7.325) as

$$(cos\theta)^{\alpha} = \left(\frac{e^{i\theta} + e^{-i\theta}}{2}\right)^{\alpha} = \frac{e^{i\alpha\theta}}{2^{\alpha}}\left(1 + e^{-i2\theta}\right)^{\alpha} \tag{7.326}$$

and

$$(sin\theta)^{\alpha} = \left(\frac{e^{i\theta} - e^{-i\theta}}{2i}\right)^{\alpha} = \frac{e^{i\alpha\theta}}{2^{\alpha}i^{\alpha}}\left(1 - e^{-i2\theta}\right)^{\alpha}. \tag{7.327}$$

Consider the special case of the Terjanian equation when $|x|^{\alpha} + |y|^{\alpha} = 1$, e.g.,

$$|x|^{\alpha} + |y|^{\alpha} = 1, \tag{7.328}$$

which yields that

$$x^{\alpha} = subcos\left(t^{\alpha}\right) \tag{7.329}$$

and

$$y^{\alpha} = subsin\left(t^{\alpha}\right), \tag{7.330}$$

or alternatively,

$$x = \sqrt[\alpha]{subcos\left(t^{\alpha}\right)} \tag{7.331}$$

and

$$y = \sqrt[\alpha]{subsin\left(t^{\alpha}\right)}, \tag{7.332}$$

where $\alpha \in \mathbb{R}$.

Making use of the expressions

$$x^{\alpha} = \widehat{a}^{\alpha} subcos\left(t^{\alpha}\right) \tag{7.333}$$

and

$$y^{\alpha} = \widehat{b}^{\alpha} subsin\left(t^{\alpha}\right), \tag{7.334}$$

we easily obtain the equation of the circle-like

$$\frac{x^{2\alpha}}{\widehat{a}^{2\alpha}} + \frac{y^{2\alpha}}{\widehat{a}^{2\alpha}} = 1, \tag{7.335}$$

the equation of the ellipse-like

$$\frac{x^{2\alpha}}{\widehat{a}^{2\alpha}} + \frac{y^{2\alpha}}{\widehat{b}^{2\alpha}} = 1, \tag{7.336}$$

the equation of the hyperbola-like

$$\frac{x^{2\alpha}}{\widehat{a}^{2\alpha}} - \frac{y^{2\alpha}}{\widehat{b}^{2\alpha}} = 1, \tag{7.337}$$

and the equation of the parabola-like

$$x^{2\alpha} = 4ay^{\alpha}, \tag{7.338}$$

where $\alpha \in \mathbb{R}$.

Example 7.1 When $\alpha = \frac{1}{2}$, we give the equation of the circle-like

$$\frac{|x|}{\widehat{a}} + \frac{|y|}{\widehat{a}} = 1, \tag{7.339}$$

the equation of the ellipse-like

$$\frac{|x|}{\widehat{a}} + \frac{|y|}{\widehat{b}} = 1, \tag{7.340}$$

the equation of the hyperbola-like

$$\frac{|x|}{\widehat{a}} - \frac{|y|}{\widehat{b}} = 1, \tag{7.341}$$

and the equation of the parabola-like

$$|x| = 4a\sqrt{|y|}, \tag{7.342}$$

where $\widehat{a}, \widehat{b} \in \mathbb{R}_{+}$.

Example 7.2 When $\alpha = 1$ and $\widehat{a}, \widehat{b} \in \mathbb{R}_{+}$, we present the equation of the ellipse [302]

$$\frac{x^2}{\widehat{a}^2} + \frac{y^2}{\widehat{b}^2} = 1, \tag{7.343}$$

the equation of the hyperbola [302]

$$\frac{x^2}{\widehat{a}^2} - \frac{y^2}{\widehat{b}^2} = 1, \tag{7.344}$$

and the equation of the parabola [302]

$$x^2 = 4ay. \tag{7.345}$$

It is not difficult to observe that the equation of the circle-like can be given as follows:

$$x^{2\alpha} + y^{2\alpha} = 1, \tag{7.346}$$

where $\alpha \in \mathbb{R}$.

Case 7.1 For $\alpha = 0.3$, we have the equation of the circle-like

$$|x|^{0.6} + |y|^{0.6} = 1 \tag{7.347}$$

such that

$$x^{\frac{3}{10}} = subcos\left(t^{\frac{3}{10}}\right) \tag{7.348}$$

and

$$y^{\frac{3}{10}} = subsin\left(t^{\frac{3}{10}}\right). \tag{7.349}$$

Case 7.2 For $\alpha = 0.9$, we have the equation of the circle-like

$$|x|^{0.9} + |y|^{0.9} = 1 \tag{7.350}$$

such that

$$x^{\frac{9}{20}} = subcos\left(t^{\frac{9}{20}}\right) \tag{7.351}$$

and

$$y^{\frac{9}{20}} = subsin\left(t^{0.9}\right). \tag{7.352}$$

Case 7.3 For $\alpha = 1.5$, we have the equation of the circle-like

$$|x|^{1.5} + |y|^{1.5} = 1 \tag{7.353}$$

such that

$$x^{\frac{3}{4}} = subcos\left(t^{\frac{3}{4}}\right) \tag{7.354}$$

and

$$y^{\frac{3}{4}} = subsin\left(t^{\frac{3}{4}}\right). \tag{7.355}$$

Case 7.4 For $\alpha = 2.5$, we have the equation of the circle-like

$$|x|^{2.5} + |y|^{2.5} = 1 \tag{7.356}$$

such that

$$x^{\frac{5}{4}} = subcos\left(t^{\frac{5}{4}}\right) \tag{7.357}$$

and

$$y^{\frac{5}{4}} = subsin\left(t^{\frac{5}{4}}\right). \tag{7.358}$$

Case 7.5 For $\alpha = 3$, we have the equation of the circle-like

$$|x|^{3} + |y|^{3} = 1 \tag{7.359}$$

such that

$$x^{\frac{3}{2}} = subcos\left(t^{\frac{3}{2}}\right) \tag{7.360}$$

and

$$y^{\frac{3}{2}} = subsin\left(t^{\frac{3}{2}}\right). \tag{7.361}$$

7.2 The Fourier-Type Series Theory via Subtrigonometric Series with Respect to Monotone Function

7.2.1 Theory of Fourier Series: History and Properties

In 1822, Jean Baptiste Joseph Fourier showed the well-known Fourier series and Fourier transform, which were applied in the theory of the heat conduction [303].

The Fourier series is defined by [303]

$$\Omega(t) = \frac{1}{2}a_0 + \sum_{\kappa=1}^{\infty} b_\kappa cos(\kappa t) + \sum_{\kappa=1}^{\infty} c_\kappa sin(\kappa t), \tag{7.362}$$

where the Fourier coefficients are as follows [303]:

$$a_0 = \frac{1}{\pi}\int_{-\pi}^{\pi} \Omega(t)\,dt, \tag{7.363}$$

$$b_\kappa = \frac{1}{\pi}\int_{-\pi}^{\pi} \Omega(t)\, cos(\kappa t)\,dt, \tag{7.364}$$

and

$$c_\kappa = \frac{1}{\pi}\int_{-\pi}^{\pi} \Omega(t)\, sin(\kappa t)\,dt \tag{7.365}$$

with Euler's formula [71]

$$e^{i\kappa t} = cos(\kappa t) + i sin(\kappa t), \tag{7.366}$$

for $\kappa \in \mathbb{N}_0$ and $i = \sqrt{-1}$, and the orthogonal relations [21]

$$\int_{-\pi}^{\pi} sin(\kappa_1 t)\, sin(\kappa_2 t)\,dt = \begin{cases} \pi\ (\kappa_1 = \kappa_2) \\ 0\ (\kappa_1 \neq \kappa_2), \end{cases} \tag{7.367}$$

$$\int_{-\pi}^{\pi} cos(\kappa_1 t)\, cos(\kappa_2 t)\,dt = \begin{cases} \pi\ (\kappa_1 = \kappa_2) \\ 0\ (\kappa_1 \neq \kappa_2), \end{cases} \tag{7.368}$$

$$\int_{-\pi}^{\pi} sin(\kappa_1 t)\, cos(\kappa_2 t)\,dt = 0, \tag{7.369}$$

$$\int_{-\pi}^{\pi} sin(\kappa_1 t)\,dt = 0, \tag{7.370}$$

and

$$\int_{-\pi}^{\pi} \cos(\kappa_1 t)\, dt = 0 \tag{7.371}$$

for any integers κ_1 and κ_2.

Generally, there are [190]

$$\int_{-\frac{L}{2}}^{\frac{L}{2}} \sin\left(\frac{2\pi\kappa_1}{L}\upsilon\right) \sin\left(\frac{2\pi\kappa_2}{L}\upsilon\right) d\upsilon = \pi\delta_{\kappa_1\kappa_2}, \tag{7.372}$$

$$\int_{-\frac{L}{2}}^{\frac{L}{2}} \cos\left(\frac{2\pi\kappa_1}{L}\upsilon\right) \cos\left(\frac{2\pi\kappa_2}{L}\upsilon\right) d\upsilon = \pi\delta_{\kappa_1\kappa_2}, \tag{7.373}$$

$$\int_{-\frac{L}{2}}^{\frac{L}{2}} \cos\left(\frac{2\pi\kappa_1}{L}\upsilon\right) \sin\left(\frac{2\pi\kappa_2}{L}\upsilon\right) d\upsilon = 0, \tag{7.374}$$

$$\int_{-\frac{L}{2}}^{\frac{L}{2}} \cos\left(\frac{2\pi\kappa_1}{L}\upsilon\right) d\upsilon = 0, \tag{7.375}$$

and

$$\int_{-\frac{L}{2}}^{\frac{L}{2}} \sin\left(\frac{2\pi\kappa_2}{L}\upsilon\right) d\upsilon = 0. \tag{7.376}$$

7.2.2 The Subtrigonometric and Subhyperbolic Functions with Respect to Monotone Function

Let $\mu(t)$ be the monotone function, e.g., $\mu^{(1)}(t) > 0$.

Definition 7.27 The function Φ_μ with the variable t is a polynomial with respect to monotone function $\mu(t)$ if

$$\Phi_\mu(t) = \sum_{\kappa=1}^{n} h_\kappa (\mu(t))^\kappa + h_0, \tag{7.377}$$

where κ is a nonnegative integer, and h_κ are the coefficients.

Definition 7.28 The exponential function with respect to monotone function $\mu(t)$ is represented as

$$e^{\mu(t)} = \sum_{\kappa=0}^{\infty} \frac{(\mu(t))^\kappa}{\Gamma(\kappa+1)}, \tag{7.378}$$

where $\Gamma(\kappa)$ is the gamma function.

In an analogous manner of Euler [3], one has

$$e^{\lambda\mu(t)} = \sum_{\kappa=0}^{\infty} \frac{\lambda^\kappa (\mu(t))^\kappa}{\Gamma(\kappa+1)}, \tag{7.379}$$

$$e^{-\lambda\mu(t)} = \sum_{\kappa=0}^{\infty} \frac{(-\lambda)^\kappa (\mu(t))^\kappa}{\Gamma(\kappa+1)}, \tag{7.380}$$

$$e^{i\lambda\mu(t)} = cos_\mu(\lambda\mu(t)) + isin_\mu(\lambda\mu(t)), \tag{7.381}$$

and

$$e^{-i\lambda\mu(t)} = cos_\mu(\lambda\mu(t)) - isin_\mu(\lambda\mu(t)), \tag{7.382}$$

where $\mu^{(1)}(t) > 0$.

7.2.3 *The Subtrigonometric Functions with Respect to Monotone Function*

Definition 7.29 The subsine function with respect to monotone function $\mu(t)$ is defined as

$$subsin_\mu(\mu(t)) = \frac{e^{i\mu(t)} - e^{-i\mu(t)}}{2i} = \sum_{\kappa=0}^{\infty} \frac{(-1)^\kappa (\mu(t))^{2\kappa+1}}{\Gamma(2\kappa+2)}, \tag{7.383}$$

where $\mu^{(1)}(t) > 0$.

Definition 7.30 The subcosine function with respect to monotone function $\mu(t)$ is defined as

$$subcos_{\mu}(\mu(t)) = \frac{e^{i\mu(t)} + e^{-i\mu(t)}}{2} = \sum_{\kappa=0}^{\infty} \frac{(-1)^{\kappa} (\mu(t))^{2\kappa}}{\Gamma(2\kappa+1)}, \tag{7.384}$$

where $\mu^{(1)}(t) > 0$.

Definition 7.31 The subtangent function with respect to monotone function $\mu(t)$ is defined as

$$subtan_{\mu}(\mu(t)) = \frac{subsin_{\mu}(\mu(t))}{subcos_{\mu}(\mu(t))}, \tag{7.385}$$

where $\mu^{(1)}(t) > 0$.

Definition 7.32 The subcotangent function with respect to monotone function $\mu(t)$ is defined as

$$subcotan_{\mu}(\mu(t)) = \frac{subcos_{\mu}(\mu(t))}{subsin_{\mu}(\mu(t))}, \tag{7.386}$$

where $\mu^{(1)}(t) > 0$.

Definition 7.33 The subcosecant function with respect to monotone function $\mu(t)$ is defined as

$$subcsc_{\mu}(\mu(t)) = \frac{1}{subsin_{\mu}(\mu(t))}, \tag{7.387}$$

where $\mu^{(1)}(t) > 0$.

Definition 7.34 The subsecant function with respect to monotone function $\mu(t)$ is defined as

$$subsec_{\mu}(\mu(t)) = \frac{1}{subcos_{\mu}(\mu(t))}, \tag{7.388}$$

where $\mu^{(1)}(t) > 0$.

The relations among the subtrigonometric functions are given as follows:

$$subsin_{\mu}(\lambda\mu(t)) = \frac{e^{i\lambda\mu(t)} - e^{-i\lambda\mu(t)}}{2i} = \sum_{\kappa=0}^{\infty} \frac{(-\lambda)^{\kappa} (\mu(t))^{2\kappa+1}}{\Gamma(2\kappa+2)}, \tag{7.389}$$

$$subcos_{\mu}\left(\lambda\mu\left(t\right)\right)=\frac{e^{i\lambda\mu(t)}+e^{-i\lambda\mu(t)}}{2}=\sum_{\kappa=0}^{\infty}\frac{\left(-\lambda\right)^{\kappa}\left(\mu\left(t\right)\right)^{2\kappa}}{\Gamma\left(2\kappa+1\right)}, \tag{7.390}$$

$$subtan_{\mu}\left(\lambda\mu\left(t\right)\right)=\frac{subsin_{\mu}\left(\lambda\mu\left(t\right)\right)}{subcos_{\mu}\left(\lambda\mu\left(t\right)\right)}=\frac{1}{i}\cdot\frac{e^{i\lambda\mu(t)}-e^{-i\lambda\mu(t)}}{e^{i\lambda\mu(t)}+e^{-i\lambda\mu(t)}}, \tag{7.391}$$

$$subcotan_{\mu}\left(\lambda\mu\left(t\right)\right)=\frac{subcos_{\mu}\left(\lambda\mu\left(t\right)\right)}{subsin_{\mu}\left(\lambda\mu\left(t\right)\right)}=i\cdot\frac{e^{i\lambda\mu(t)}+e^{-i\lambda\mu(t)}}{e^{i\lambda\mu(t)}-e^{-i\lambda\mu(t)}}, \tag{7.392}$$

$$subcsc_{\mu}\left(\lambda\mu\left(t\right)\right)=\frac{1}{subsin_{\mu}\left(\lambda\mu\left(t\right)\right)}=i\cdot\frac{2}{e^{i\lambda\mu(t)}-e^{-i\lambda\mu(t)}}, \tag{7.393}$$

$$subsec_{\mu}\left(\lambda\mu\left(t\right)\right)=\frac{1}{subcos_{\mu}\left(\lambda\mu\left(t\right)\right)}=\frac{2}{e^{i\lambda\mu(t)}+e^{-i\lambda\mu(t)}}, \tag{7.394}$$

where

$$subsin_{\mu}\left(-\mu\left(t\right)\right)=-subsin_{\mu}\left(\mu\left(t\right)\right), \tag{7.395}$$

$$subcos_{\mu}\left(-\mu\left(t\right)\right)=subcos_{\mu}\left(\mu\left(t\right)\right), \tag{7.396}$$

$$subtan_{\mu}\left(-\mu\left(t\right)\right)=-subtan_{\mu}\left(\mu\left(t\right)\right), \tag{7.397}$$

$$subcotan_{\mu}\left(-\mu\left(t\right)\right)=-subcotan_{\mu}\left(\mu\left(t\right)\right), \tag{7.398}$$

$$subcsc_{\mu}\left(-\mu\left(t\right)\right)=-subcsc_{\mu}\left(\mu\left(t\right)\right), \tag{7.399}$$

$$subsec_{\mu}\left(-\mu\left(t\right)\right)=subsec_{\mu}\left(\mu\left(t\right)\right), \tag{7.400}$$

$$subsin_{\mu}\left(\mu\left(t\right)\right)=subsin_{\mu}\left(\mu\left(t\right)+2\pi n\right), \tag{7.401}$$

$$subcos_{\mu}\left(\mu\left(t\right)\right)=subcos_{\mu}\left(\mu\left(t\right)+2\pi n\right), \tag{7.402}$$

$$subtan_{\mu}\left(\mu\left(t\right)\right)=subtan_{\mu}\left(\mu\left(t\right)+\pi n\right), \tag{7.403}$$

$$subcotan_{\mu}\left(\mu\left(t\right)\right)=subcotan_{\mu}\left(\mu\left(t\right)+\pi n\right), \tag{7.404}$$

$$subcsc_{\mu}\left(\mu\left(t\right)\right)=subcsc_{\mu}\left(\mu\left(t\right)+2\pi n\right), \tag{7.405}$$

$$subsec_{\mu}\left(\mu\left(t\right)\right)=subsec_{\mu}\left(\mu\left(t\right)+2\pi n\right), \tag{7.406}$$

$$\left(subsin_{\mu}\left(\lambda\mu\left(t\right)\right)\right)^{2}+\left(subcos_{\mu}\left(\lambda\mu\left(t\right)\right)\right)^{2}=1, \tag{7.407}$$

$$2subsin_{\mu}(\lambda\mu(t)) \cdot subcos_{\mu}(\lambda\mu(t)) = subsin_{\mu}(2\lambda\mu(t)), \tag{7.408}$$

$$subcos_{\mu}(2\lambda\mu(t)) = 2\left(subcos_{\mu}(\lambda\mu(t))\right)^2 - 1, \tag{7.409}$$

$$subcos_{\mu}(2\lambda\mu(t)) = 1 - 2\left(subsin_{\mu}(\lambda\mu(t))\right)^2, \tag{7.410}$$

$$subcos_{\mu}(2\lambda\mu(t)) = \left(subcos_{\mu}(\lambda\mu(t))\right)^2 - \left(subsin_{\mu}(\lambda\mu(t))\right)^2 \tag{7.411}$$

$$1 + \left(subtan_{\mu}(\lambda\mu(t))\right)^2 = \left(subsec_{\mu}(\lambda\mu(t))\right)^2, \tag{7.412}$$

and

$$1 + \left(subcotan_{\mu}(\lambda\mu(t))\right)^2 = \left(subcsc_{\mu}(\lambda\mu(t))\right)^2, \tag{7.413}$$

with $\lambda \in \mathbb{C}$ and $n \in \mathbb{N}_0$.

7.2.4 The Subhyperbolic Functions with Respect to Monotone Function

Definition 7.35 The hyperbolic subsine function with respect to monotone function $\mu(t)$ is defined as

$$subsinh_{\mu}(\mu(t)) = \frac{e^{\mu(t)} - e^{-\mu(t)}}{2} = \sum_{\kappa=0}^{\infty} \frac{(\mu(t))^{2\kappa+1}}{\Gamma(2\kappa+2)}, \tag{7.414}$$

where $\mu^{(1)}(t) > 0$.

Definition 7.36 The hyperbolic subcosine function with respect to monotone function $\mu(t)$ is defined as

$$subcosh_{\mu}(\mu(t)) = \frac{e^{\mu(t)} + e^{-\mu(t)}}{2} = \sum_{\kappa=0}^{\infty} \frac{(\mu(t))^{2\kappa}}{\Gamma(2\kappa+1)}, \tag{7.415}$$

where $\mu^{(1)}(t) > 0$.

Definition 7.37 The subhyperbolic subtangent function with respect to monotone function $\mu(t)$ is defined as

$$subtanh_{\mu}(\mu(t)) = \frac{subsinh_{\mu}(\mu(t))}{subcosh_{\mu}(\mu(t))}, \tag{7.416}$$

where $\mu^{(1)}(t) > 0$.

Definition 7.38 The subhyperbolic subcotangent function with respect to monotone function $\mu(t)$ is defined as

$$subcotanh_{\mu}(\mu(t)) = \frac{subcosh_{\mu}(\mu(t))}{subsinh_{\mu}(\mu(t))}, \tag{7.417}$$

where $\mu^{(1)}(t) > 0$.

Definition 7.39 The subhyperbolic subcosecant function with respect to monotone function $\mu(t)$ is defined as

$$subcsch_{\mu}(\mu(t)) = \frac{1}{subsinh_{\mu}(\mu(t))}, \tag{7.418}$$

where $\mu^{(1)}(t) > 0$.

Definition 7.40 The subhyperbolic subsecant function with respect to monotone function $\mu(t)$ is defined as

$$subsech_{\mu}(\mu(t)) = \frac{1}{subcosh_{\mu}(\mu(t))}, \tag{7.419}$$

where $\mu^{(1)}(t) > 0$.

The subhyperbolic functions with respect to monotone function $\mu(t)$ are given as follows:

$$subsinh_{\mu}(\lambda\mu(t)) = \frac{e^{\lambda\mu(t)} - e^{-\lambda\mu(t)}}{2} = \sum_{\kappa=0}^{\infty} \frac{\lambda^{\kappa}(\mu(t))^{2\kappa+1}}{\Gamma(2\kappa+2)}, \tag{7.420}$$

$$subcosh_{\mu}(\lambda\mu(t)) = \frac{e^{\lambda\mu(t)} + e^{-\lambda\mu(t)}}{2} = \sum_{\kappa=0}^{\infty} \frac{\lambda^{\kappa}(\mu(t))^{2\kappa}}{\Gamma(2\kappa+1)}, \tag{7.421}$$

$$subtanh_{\mu}(\lambda\mu(t)) = \frac{subsinh_{\mu}(\lambda\mu(t))}{subcosh_{\mu}(\lambda\mu(t))}, \tag{7.422}$$

$$subcotanh_{\mu}(\lambda\mu(t)) = \frac{subcosh_{\mu}(\lambda\mu(t))}{subsinh_{\mu}(\lambda\mu(t))}, \tag{7.423}$$

$$subcsch_{\mu}(\lambda\mu(t)) = \frac{1}{subsinh_{\mu}(\lambda\mu(t))}, \tag{7.424}$$

and

$$subsech_{\mu}\left(\lambda\mu\left(t\right)\right)=\frac{1}{subcosh_{\mu}\left(\lambda\mu\left(t\right)\right)},\tag{7.425}$$

where

$$subsinh_{\mu}\left(-\lambda\mu\left(t\right)\right)=-subsinh_{\mu}\left(\lambda\mu\left(t\right)\right),\tag{7.426}$$

$$subcosh_{\mu}\left(-\lambda\mu\left(t\right)\right)=subcosh_{\mu}\left(\lambda\mu\left(t\right)\right),\tag{7.427}$$

$$subtanh_{\mu}\left(-\lambda\mu\left(t\right)\right)=-subtanh_{\mu}\left(\lambda\mu\left(t\right)\right),\tag{7.428}$$

$$subcotanh_{\mu}\left(-\lambda\mu\left(t\right)\right)=-subcotanh_{\mu}\left(\lambda\mu\left(t\right)\right),\tag{7.429}$$

$$subcsch_{\mu}\left(-\lambda\mu\left(t\right)\right)=subcsch_{\mu}\left(\lambda\mu\left(t\right)\right),\tag{7.430}$$

$$subsech_{\mu}\left(-\lambda\mu\left(t\right)\right)=-subsech_{\mu}\left(\lambda\mu\left(t\right)\right),\tag{7.431}$$

$$\left(subcosh_{\mu}\left(\lambda\mu\left(t\right)\right)\right)^{2}-\left(subsinh_{\mu}\left(\lambda\mu\left(t\right)\right)\right)^{2}=1,\tag{7.432}$$

$$subsinh_{\mu}\left(i\mu\left(t\right)\right)=i\cdot subsin_{\mu}\left(i\mu\left(t\right)\right),\tag{7.433}$$

$$subcosh_{\mu}\left(i\mu\left(t\right)\right)=subcos_{\mu}\left(\mu\left(t\right)\right),\tag{7.434}$$

$$subsinh_{\mu}\left(2\mu\left(t\right)\right)=2subsinh_{\mu}\left(\mu\left(t\right)\right)\cdot subcosh_{\mu}\left(\mu\left(t\right)\right),\tag{7.435}$$

$$subcosh_{\mu}\left(2\mu\left(t\right)\right)=2\left(subcosh_{\mu}\left(\mu\left(t\right)\right)\right)^{2}-1,\tag{7.436}$$

and

$$subcosh_{\mu}\left(2\mu\left(t\right)\right)=2\left(subsinh_{\mu}\left(\mu\left(t\right)\right)\right)^{2}+1,\tag{7.437}$$

with $\lambda\in\mathbb{C}$.

The Leibniz derivatives of the special functions can be given as follows:

$$D_{t,\mu(\cdot)}^{(1)}subsin_{\mu}\left(\mu\left(t\right)\right)=subco_{\mu}\left(\mu\left(t\right)\right),\tag{7.438}$$

$$D_{t,\mu(\cdot)}^{(1)}subcos_{\mu}\left(\mu\left(t\right)\right)=-subsin_{\mu}\left(\mu\left(t\right)\right),\tag{7.439}$$

$$D_{t,\mu(\cdot)}^{(1)}subtan_{\mu}\left(\mu\left(t\right)\right)=subsec_{\mu}^{2}\left(\mu\left(t\right)\right),\tag{7.440}$$

$$D_{t,\mu(\cdot)}^{(1)}subcotan_{\mu}\left(\mu\left(t\right)\right)=-subcsc_{\mu}^{2}\left(\mu\left(t\right)\right),\tag{7.441}$$

$$D^{(1)}_{t,\mu(\cdot)} subsec_{\mu}(\mu(t)) = subtan_{\mu}(\mu(t)) \cdot subsec_{\mu}(\mu(t)), \tag{7.442}$$

$$D^{(1)}_{t,\mu(\cdot)} subcsc_{\mu}(\mu(t)) = -subcotan_{\mu}(\mu(t)) \cdot subcsc_{\mu}(\mu(t)), \tag{7.443}$$

$$D^{(1)}_{t,\mu(\cdot)} subsinh_{\mu}(\mu(t)) = subcosh_{\mu}(\mu(t)), \tag{7.444}$$

$$D^{(1)}_{t,\mu(\cdot)} subcosh_{\mu}(\mu(t)) = subsinh_{\mu}(\mu(t)), \tag{7.445}$$

$$D^{(1)}_{t,\mu(\cdot)} subtanh_{\mu}(\mu(t)) = subsec^2_{\mu}(\mu(t)), \tag{7.446}$$

$$D^{(1)}_{t,\mu(\cdot)} subcotanh_{\mu}(\mu(t)) = -subcsch^2_{\mu}(\mu(t)), \tag{7.447}$$

$$D^{(1)}_{t,\mu(\cdot)} subsech_{\mu}(\mu(t)) = -subtanh_{\mu}(\mu(t)) \cdot subsech_{\mu}(\mu(t)), \tag{7.448}$$

and

$$D^{(1)}_{t,\mu(\cdot)} subcsch_{\mu}(\mu(t)) = -subcotanh_{\mu}(\mu(t)) \cdot subcsch_{\mu}(\mu(t)). \tag{7.449}$$

The Riemann–Stieltjes integrals of the special functions can be presented as follows:

$$I^{(1)}_{t,\mu(\cdot)} subsin_{\mu}(\mu(t)) = -subcos_{\mu}(\mu(t)) + C, \tag{7.450}$$

$$I^{(1)}_{t,\mu(\cdot)} subcos_{\mu}(\mu(t)) = subsin_{\mu}(\mu(t)) + C, \tag{7.451}$$

$$I^{(1)}_{t,\mu(\cdot)} subtan_{\mu}(\mu(t)) = \ln\left|subsec_{\mu}(\mu(t))\right| + C, \tag{7.452}$$

$$I^{(1)}_{t,\mu(\cdot)} subcotan_{\mu}(\mu(t)) = \left|subcsc_{\mu}(\mu(t))\right| + C, \tag{7.453}$$

$$I^{(1)}_{t,\mu(\cdot)} subsec_{\mu}(\mu(t)) = \left|subtan_{\mu}(\mu(t)) + subsec_{\mu}(\mu(t))\right| + C, \tag{7.454}$$

$$I^{(1)}_{t,\mu(\cdot)} subcsc_{\mu}(\mu(t)) = \left|subcotan_{\mu}(\mu(t)) - subcsc_{\mu}(\mu(t))\right| + C, \tag{7.455}$$

$$I^{(1)}_{t,\mu(\cdot)} subsinh_{\mu}(\mu(t)) = subcosh_{\mu}(\mu(t)) + C, \tag{7.456}$$

$$I^{(1)}_{t,\mu(\cdot)} subcosh_{\mu}(\mu(t)) = subsinh_{\mu}(\mu(t)) + C, \tag{7.457}$$

$$I^{(1)}_{t,\mu(\cdot)} subtanh_{\mu}(\mu(t)) = \ln\left(subcosh_{\mu}(\mu(t))\right) + C, \tag{7.458}$$

and

$$I_{t,\mu(\cdot)}^{(1)} subcotanh_{\mu}(\mu(t)) = \ln\left(subsinh_{\mu}(\mu(t))\right) + C, \tag{7.459}$$

where C is a constant.

7.3 Theory of Subtrigonometric Series with Respect to Monotone Function

7.3.1 *The Subtrigonometric Series with Respect to Monotone Function*

To start with the subtrigonometric series with respect to monotone function, we present the theorems for the series with respect to monotone function based on the Leibniz derivatives and the Riemann–Stieltjes integrals.

Let $\left(\mu^{-1} \circ \mu\right)(t) = I(t) = t$ and $\mu^{(1)}(t) > 0$, where I is the identity operator.

Theorem 7.5 *Let $\mu^{(1)}(t) > 0$, $\mu(a) = -\pi$, and $\mu(b) = \pi$, $t \in \mathbb{R}$, $a \in \mathbb{R}$, and $b \in \mathbb{R}$. Then, there exist*

$$\int_a^b subsin_{\mu}(m\mu(t)) \cdot subsin_{\mu}(n\mu(t))\, \mu^{(1)}(t)\, dt = \pi\delta_{mn}, \tag{7.460}$$

$$\int_a^b subcos_{\mu}(m\mu(t)) \cdot subcos_{\mu}(n\mu(t))\, \mu^{(1)}(t)\, dt = \pi\delta_{mn}, \tag{7.461}$$

$$\int_a^b subsin_{\mu}(m\mu(t)) \cdot subcos_{\mu}(m\mu(t))\, \mu^{(1)}(t)\, dt = 0, \tag{7.462}$$

$$\int_a^b subsin_{\mu}(m\mu(t))\, \mu^{(1)}(t)\, dt = 0, \tag{7.463}$$

and

$$\int_a^b subcos_{\mu}(m\mu(t))\, \mu^{(1)}(t)\, dt = 0, \tag{7.464}$$

where $m \in \mathbb{N}$, $n \in \mathbb{N}$, and δ_{mn} is the Kronecker delta function [23].

Proof Putting $\upsilon = \mu(t)$, one has

$$\int_a^b subsin_\mu(m\mu(t)) \cdot subsin_\mu(n\mu(t))\, \mu^{(1)}(t)\, dt = \pi\delta_{mn}, \tag{7.465}$$

$$\int_a^b subcos_\mu(m\mu(t)) \cdot subcos_\mu(n\mu(t))\, \mu^{(1)}(t)\, dt = \pi\delta_{mn}, \tag{7.466}$$

$$\int_a^b subsin_\mu(m\mu(t)) \cdot subcos_\mu(m\mu(t))\, \mu^{(1)}(t)\, dt = 0, \tag{7.467}$$

$$\int_a^b subsin_\mu(m\mu(t))\, \mu^{(1)}(t)\, dt = 0, \tag{7.468}$$

and

$$\int_a^b subcos_\mu(m\mu(t))\, \mu^{(1)}(t)\, dt = 0 \tag{7.469}$$

due to (7.367), (7.368), (7.369), (7.370), and (7.371).

Thus, we finish the proof. □

Theorem 7.6 *Let $\Xi_\mu(t) = (\Xi \circ \mu)(t) = \Xi(\mu(t))$, $\Xi(\mu(t)) = \Xi(\mu(t) + L)$, $\mu^{(1)}(t) > 0$ $\mu(a) = -\pi$ $\mu(b) = \pi$, and $t, a, b \in \mathbb{R}$. Then, there exist the series with respect to monotone function, defined by*

$$\Xi_\mu(t) = \frac{1}{2}\psi_0 + \sum_{n=1}^{\infty} \varphi_n subcos_\mu(n\mu(t)) + \sum_{n=1}^{\infty} \psi_n subsin_\mu(n\mu(t)), \tag{7.470}$$

where $n \in \mathbb{N}$ and the coefficients ψ_0, φ_n, and ψ_n are determined by

$$\psi_0 = \frac{1}{\pi}\int_a^b \Xi_\mu(t)\, \mu^{(1)}(t)\, dt \tag{7.471}$$

$$\varphi_n = \frac{1}{\pi}\int_a^b \Xi_\mu(t)\, subcos_\mu(n\mu(t))\, \mu^{(1)}(t)\, dt, \tag{7.472}$$

and

$$\psi_n = \frac{1}{\pi} \int_a^b \Xi_\mu (t)\, subsin_\mu (n\mu (t))\, \mu^{(1)} (t)\, dt. \tag{7.473}$$

Proof Taking $\upsilon = \mu (t)$, we have that

$$\begin{aligned}
&\tfrac{1}{\pi} \int_a^b \Xi_\mu (t)\, \mu^{(1)} (t)\, dt \\
&= \tfrac{1}{\pi} \int_a^b \left(\tfrac{1}{2}\psi_0 + \sum_{n=1}^{\infty} \varphi_n subcos_\mu (n\mu (t)) + \sum_{n=1}^{\infty} \psi_n subsin_\mu (n\mu (t)) \right) \mu^{(1)} (t)\, dt \\
&= \tfrac{\psi_0}{2\pi} \int_{-\pi}^{\pi} d\upsilon \\
&= \psi_0,
\end{aligned} \tag{7.474}$$

since the formulas (7.470) and (7.471) hold.

Similarly, we present

$$\begin{aligned}
&\tfrac{1}{\pi} \int_a^b \Xi_\mu (t)\, subcos_\mu (n\mu (t))\, \mu^{(1)} (t)\, dt \\
&= \tfrac{1}{\pi} \int_a^b \left(\tfrac{1}{2}\psi_0 + \sum_{n=1}^{\infty} \varphi_n subcos_\mu (n\mu (t)) + \sum_{\kappa=1}^{\infty} \psi_n subsin_\mu (n\mu (t)) \right) \\
&\quad \times subcos_\mu (n\mu (t))\, \mu^{(1)} (t)\, dt \\
&= \tfrac{1}{\pi} \int_a^b \varphi_n subcos_\mu (n\mu (t))\, subcos_\mu (n\mu (t))\, \mu^{(1)} (t)\, dt \\
&= \tfrac{\varphi_n}{\pi} \int_{-\pi}^{\pi} cos (n\upsilon)\, cos (n\upsilon)\, d\upsilon \\
&= \varphi_n
\end{aligned} \tag{7.475}$$

and

$$\begin{aligned}
\psi_n &= \tfrac{1}{\pi} \int_a^b \Xi_\mu (t)\, subsin_\mu (n\mu (t))\, \mu^{(1)} (t)\, dt \\
&= \tfrac{1}{\pi} \int_a^b \left(\tfrac{1}{2}\psi_0 + \sum_{\kappa=1}^{\infty} \varphi_n subcos_\mu (n\mu (t)) + \sum_{\kappa=1}^{\infty} \psi_n subsin_\mu (n\mu (t)) \right) \\
&\quad \times subsin_\mu (n\mu (t))\, \mu^{(1)} (t)\, dt \\
&= \tfrac{1}{\pi} \int_a^b \psi_n subsin_\mu (n\mu (t))\, subsin_\mu (n\mu (t))\, \mu^{(1)} (t)\, dt \\
&= \tfrac{\psi_n}{\pi} \int_{-\pi}^{\pi} sin (n\upsilon)\, sin (n\upsilon)\, d\upsilon \\
&= \psi_n
\end{aligned} \tag{7.476}$$

since the formulas (7.467) and (7.468) are valid.

Thus, we give the proof. □

Theorem 7.7 *Let* $\mu^{(1)}(t) > 0$, $\mu(a) = -\pi \quad \mu(b) = \pi$, *and* $t, a, b \in \mathbb{R}$ *Then, there exists*

$$\int_a^b e^{i(m-n)\mu(t)} \mu^{(1)}(t)\, dt = 2\pi \delta_{mn}. \tag{7.477}$$

Proof Taking $\upsilon = \mu(t)$ and using [304]

$$\int_{-\pi}^{\pi} e^{i(m-n)\upsilon} d\upsilon = 2\pi \delta_{mn}, \tag{7.478}$$

we have

$$\int_a^b e^{i(m-n)\mu(t)} \mu^{(1)}(t)\, dt = \int_{-\pi}^{\pi} e^{i(m-n)\upsilon} d\upsilon = 2\pi \delta_{mn}. \tag{7.479}$$

Thus, we finish the result. □

Theorem 7.8 *Let* $\mu^{(1)}(t) > 0$, $\mu(a) = -\pi \quad \mu(b) = \pi$, *and* $t, a, b \in \mathbb{R}$.

If $\Theta_\mu(t) = (\Theta \circ \mu)(t) = \Theta(\mu(t))$ *with* $\Theta(\mu(t) + 2\pi) = \Theta(\mu(t))$, *then we have [304]*

$$\Theta_\mu(t) = \sum_{n=-\infty}^{\infty} \Pi_n e^{in\mu(t)}, \tag{7.480}$$

where

$$\Pi_n = \frac{1}{2\pi} \int_a^b \Theta_\mu(t)\, e^{-in\mu(t)} \mu^{(1)}(t)\, dt. \tag{7.481}$$

Proof Taking $\upsilon = \mu(t)$, we have

$$\begin{aligned} &\int_a^b \Theta_\mu(t)\, e^{-in\mu(t)} \mu^{(1)}(t)\, dt \\ &\quad = \int_a^b \left(\sum_{m=-\infty}^{\infty} \Pi_m e^{im\mu(t)} \right) e^{-in\mu(t)} \mu^{(1)}(t)\, dt \\ &\quad = \sum_{m=-\infty}^{\infty} \Pi_m \int_a^b e^{i(m-n)\mu(t)} \mu^{(1)}(t)\, dt \\ &\quad = \sum_{m=-\infty}^{\infty} 2\pi \Pi_m \delta_{mn} \\ &\quad = 2\pi \Pi_n, \end{aligned} \tag{7.482}$$

which implies that

$$\Pi_n = \frac{1}{2\pi} \int_a^b \Theta_\mu(t) e^{-in\mu(t)} \mu^{(1)}(t)\, dt. \tag{7.483}$$

Therefore, we finish the proof. □

As the direct result, we have the following:

Corollary 7.1 *Let $\mu^{(1)}(t) > 0$, $\mu(a) = -\frac{L}{2}$, $\mu(b) = \frac{L}{2}$, and $t, a, b \in \mathbb{R}$. Then, there exists*

$$\int_a^b e^{\frac{2\pi i(m-n)\mu(t)}{L}} \mu^{(1)}(t)\, dt = L\delta_{mn}. \tag{7.484}$$

Proof Taking $\upsilon = \mu(t)$ and using [305]

$$\int_{-\frac{L}{2}}^{\frac{L}{2}} e^{i(m-n)\upsilon} d\upsilon = L\delta_{mn}, \tag{7.485}$$

we have

$$\int_a^b e^{\frac{2\pi i(m-n)\mu(t)}{L}} \mu^{(1)}(t)\, dt = \int_{-\frac{L}{2}}^{\frac{L}{2}} e^{\frac{2\pi i(m-n)\upsilon}{L}} d\upsilon = L\delta_{mn}. \tag{7.486}$$

□

Theorem 7.9 *Let $\mu^{(1)}(t) > 0$, $\mu(a) = -\frac{L}{2}$, $\mu(b) = \frac{L}{2}$, and $t, a, b \in \mathbb{R}$.*

If $\mathbb{M}_\mu(t) = (\mathbb{M} \circ \mu)(t) = \mathbb{M}(\mu(t))$ with $\mathbb{M}(\mu(t) + L) = \mathbb{M}(\mu(t))$, then we have

$$\mathbb{M}_\mu(t) = \sum_{m=-\infty}^{\infty} \phi_m e^{\frac{2\pi i m\mu(t)}{L}}, \tag{7.487}$$

where

$$\phi_m = \frac{1}{L} \int_a^b \mathbb{M}_\mu(t) e^{-\frac{2\pi m i\mu(t)}{L}} \mu^{(1)}(t)\, dt. \tag{7.488}$$

Proof Taking $\upsilon = \mu(t)$ and using (7.486), we present

$$\begin{aligned}
&\int_a^b \mathbb{M}_\mu(t)\, e^{-\frac{2\pi n i \mu(t)}{L}} \mu^{(1)}(t)\, dt \\
&= \int_a^b \left(\sum_{m=-\infty}^{\infty} \phi_m e^{\frac{2\pi m i \mu(t)}{L}} \right) e^{-\frac{2\pi n i \mu(t)}{L}} \mu^{(1)}(t)\, dt \\
&= \sum_{m=-\infty}^{\infty} \phi_m \int_a^b e^{\frac{2\pi i (m-n) \mu(t)}{L}} \mu^{(1)}(t)\, dt \\
&= \sum_{m=-\infty}^{\infty} L m_m \delta_{mn} \\
&= L m_n,
\end{aligned} \tag{7.489}$$

which can be written as follows:

$$\phi_m = \frac{1}{L} \int_a^b \mathbb{M}_\mu(t)\, e^{-\frac{2\pi m i \mu(t)}{L}} \mu^{(1)}(t)\, dt. \tag{7.490}$$

Therefore, we present the proof. □

Theorem 7.10 *Let $\Xi(\mu(t) + L) = \Xi(\mu(t))$, $\mu^{(1)}(t) > 0$, $\mu(a) = -\frac{L}{2}$, $\mu(b) = \frac{L}{2}$, and $t, a, b \in \mathbb{R}$.*

Then, there exist the series with respect to another function, defined by

$$\Xi_\mu(t) = \frac{1}{2}\psi_0 + \sum_{\kappa=1}^{\infty} \varphi_\kappa subcos_\mu \left(\frac{2\pi\kappa}{L} \mu(t) \right) + \sum_{\kappa=1}^{\infty} \psi_\kappa subsin_\mu \left(\frac{2\pi\kappa}{L} \mu(t) \right), \tag{7.491}$$

where $\kappa \in \mathbb{N}$, and the coefficients ψ_0, φ_κ, and ψ_κ are determined by

$$\psi_0 = \frac{2}{L} \int_a^b \Xi_\mu(t)\, \mu^{(1)}(t)\, dt \tag{7.492}$$

$$\varphi_\kappa = \frac{2}{L} \int_a^b \Xi_\mu(t)\, subcos_\mu \left(\frac{2\pi\kappa}{L} \mu(t) \right) \mu^{(1)}(t)\, dt, \tag{7.493}$$

and

$$\psi_\kappa = \frac{2}{L} \int_a^b \Xi_\mu(t)\, subsin_\mu \left(\frac{2\pi\kappa}{L} \mu(t) \right) \mu^{(1)}(t)\, dt. \tag{7.494}$$

Proof Taking $\upsilon = \mu(t)$, we have

$$\begin{aligned}
&\tfrac{2}{L}\int_a^b \Xi_\mu(t)\mu^{(1)}(t)dt \\
&= \tfrac{2}{L}\int_a^b \left(\tfrac{1}{2}\psi_0 + \sum_{\kappa=1}^{\infty}\varphi_\kappa subcos_\mu\left(\tfrac{2\pi\kappa}{L}\mu(t)\right) + \sum_{\kappa=1}^{\infty}\psi_\kappa subsin_\mu\left(\tfrac{2\pi\kappa}{L}\mu(t)\right)\right)\mu^{(1)}(t)dt \\
&= \tfrac{\psi_0}{L}\int_{-\frac{L}{2}}^{\frac{L}{2}} d\upsilon \\
&= \psi_0,
\end{aligned} \tag{7.495}$$

$$\begin{aligned}
&\tfrac{2}{L}\int_a^b \Xi_\mu(t)\, subcos_\mu\left(\tfrac{2\pi\kappa}{L}\mu(t)\right)\mu^{(1)}(t)\,dt \\
&= \tfrac{2}{L}\int_a^b \left(\tfrac{1}{2}\psi_0 + \sum_{\kappa=1}^{\infty}\varphi_\kappa subcos_\mu\left(\tfrac{2\pi\kappa}{L}\mu(t)\right) + \sum_{\kappa=1}^{\infty}\psi_\kappa subsin_\mu\left(\tfrac{2\pi\kappa}{L}\mu(t)\right)\right) \\
&\times subcos_\mu\left(\tfrac{2\pi\kappa}{L}\mu(t)\right)\mu^{(1)}(t)\,dt \\
&= \tfrac{2}{L}\int_a^b \varphi_\kappa subcos_\mu\left(\tfrac{2\pi\kappa}{L}\mu(t)\right) subcos_\mu\left(\tfrac{2\pi\kappa}{L}\mu(t)\right)\mu^{(1)}(t)\,dt \\
&= \varphi_\kappa,
\end{aligned} \tag{7.496}$$

and

$$\begin{aligned}
\psi_\kappa &= \tfrac{2}{L}\int_a^b \Xi_\mu(t)\, subsin_\mu\left(\tfrac{2\pi\kappa}{L}\mu(t)\right)\mu^{(1)}(t)\,dt \\
&= \tfrac{2}{L}\int_a^b \left(\tfrac{1}{2}\psi_0 + \sum_{\kappa=1}^{\infty}\varphi_\kappa subcos_\mu\left(\tfrac{2\pi\kappa}{L}\mu(t)\right) + \sum_{\kappa=1}^{\infty}\psi_\kappa subsin_\mu\left(\tfrac{2\pi\kappa}{L}\mu(t)\right)\right) \\
&\times subsin_\mu\left(\tfrac{2\pi\kappa}{L}\mu(t)\right)\mu^{(1)}(t)dt \\
&= \tfrac{2}{L}\int_a^b \psi_\kappa subsin_\mu\left(\tfrac{2\pi\kappa}{L}\mu(t)\right) subsin_\mu\left(\tfrac{2\pi\kappa}{L}\mu(t)\right)\mu^{(1)}(t)\,dt \\
&= \psi_\kappa,
\end{aligned} \tag{7.497}$$

where

$$\int_a^b subsin_\mu\left(\frac{2\pi\kappa_2}{L}\mu(t)\right)\mu^{(1)}(t)\,dt = \int_{-\frac{L}{2}}^{\frac{L}{2}} sin\left(\frac{2\pi\kappa_2}{L}\upsilon\right)d\upsilon = 0 \tag{7.498}$$

$$\int_a^b subcos_\mu\left(\frac{2\pi\kappa_1}{L}\mu(t)\right)\mu^{(1)}(t)\,dt = \int_{-\frac{L}{2}}^{\frac{L}{2}} cos\left(\frac{2\pi\kappa_1}{L}\upsilon\right)d\upsilon = 0, \tag{7.499}$$

$$\int_a^b subcos_\mu\left(\frac{2\pi\kappa_1}{L}\mu(t)\right) subcos_\mu\left(\frac{2\pi\kappa_2}{L}\mu(t)\right)\mu^{(1)}(t)\,dt = \delta_{\kappa_1\kappa_2}, \tag{7.500}$$

$$\int_a^b subsin_\mu\left(\frac{2\pi\kappa_1}{L}\mu(t)\right) subcos_\mu\left(\frac{2\pi\kappa_2}{L}\mu(t)\right)\mu^{(1)}(t)\,dt = 0, \tag{7.501}$$

$$\int_a^b subsin_\mu\left(\frac{2\pi\kappa_1}{L}\mu(t)\right) subsin_\mu\left(\frac{2\pi\kappa_2}{L}\mu(t)\right)\mu^{(1)}(t)\,dt = \delta_{\kappa_1\kappa_2}, \tag{7.502}$$

and

$$\int_a^b subsin_\mu\left(\frac{2\pi\kappa_1}{L}\mu(t)\right) subcos_\mu\left(\frac{2\pi\kappa_2}{L}\mu(t)\right)\mu^{(1)}(t)\,dt = 0. \tag{7.503}$$

Thus, we give the proof. □

Theorem 7.11 *Let $\mu^{(1)}(t) > 0$, $\mu(a) = -\pi$ $\mu(b) = \pi$, and $t, a, b \in \mathbb{R}$*
Let $\varpi^{(1)}(\kappa) > 0$, $\kappa_1 = \varpi(c)$, $\kappa_2 = \varpi(d)$, and $\kappa, c, d \in \mathbb{N}$.
Then, there exist

$$\int_a^b subsin_\mu(\varpi(c)\mu(t)) \cdot subsin_\mu(\varpi(d)\mu(t))\,\mu^{(1)}(t)\,dt = \pi\delta_{\varpi(c)\varpi(d)}, \tag{7.504}$$

$$\int_a^b subcos_\mu(\varpi(c)\mu(t)) \cdot subcos_\mu(\varpi(d)\mu(t))\,\mu^{(1)}(t)\,dt = \pi\delta_{\varpi(c)\varpi(d)}, \tag{7.505}$$

$$\int_a^b subsin_\mu(\varpi(c)\mu(t)) \cdot subcos_\mu(\varpi(d)\mu(t))\,\mu^{(1)}(t)\,dt = 0, \tag{7.506}$$

$$\int_a^b subsin_\mu(\varpi(c)\mu(t))\,\mu^{(1)}(t)\,dt = 0, \tag{7.507}$$

and

$$\int_a^b subcos_\mu(\varpi(c)\mu(t))\,\mu^{(1)}(t)\,dt = 0, \tag{7.508}$$

where $\delta_{\varpi(c)\varpi(d)} = \delta_{\kappa_1\kappa_2}$ with $\varpi(c) = \kappa_1$ and $\varpi(d) = \kappa_2$.

Proof Putting $\upsilon = \mu(t)$, we have that

$$\begin{aligned}
&\int_a^b subsin_{\mu}(\varpi(c)\mu(t)) \cdot subsin_{\mu}(\varpi(d)\mu(t))\,\mu^{(1)}(t)\,dt \\
&\quad = \int_{-\pi}^{\pi} sin(\varpi(c)\upsilon)\,sin(\varpi(d)\upsilon)\,d\upsilon \\
&\quad = \int_{-\pi}^{\pi} sin(\kappa_1\upsilon)\,sin(\kappa_2\upsilon)\,d\upsilon \\
&\quad = \pi\delta_{\varpi(c)\varpi(d)},
\end{aligned} \tag{7.509}$$

$$\begin{aligned}
&\int_a^b subcos_{\mu}(\varpi(c)\mu(t)) \cdot subcos_{\mu}(\varpi(d)\mu(t))\,\mu^{(1)}(t)\,dt \\
&\quad = \int_{-\pi}^{\pi} cos(\varpi(c)\upsilon)\,cos(\varpi(d)\upsilon)\,d\upsilon \\
&\quad = \int_{-\pi}^{\pi} cos(\kappa_1\upsilon)\,cos(\kappa_2\upsilon)\,d\upsilon \\
&\quad = \pi\delta_{\varpi(c)\varpi(d)},
\end{aligned} \tag{7.510}$$

$$\begin{aligned}
&\int_a^b subsin_{\mu}(\varpi(c)\mu(t)) \cdot subcos_{\mu}(\varpi(d)\mu(t))\,\mu^{(1)}(t)\,dt \\
&\quad = \int_{-\pi}^{\pi} sin(\varpi(c)\upsilon)\,cos(\varpi(d)\upsilon)\,d\upsilon \\
&\quad = \int_{-\pi}^{\pi} sin(\kappa_1\upsilon)\,cos(\kappa_2\upsilon)\,d\upsilon \\
&\quad = 0,
\end{aligned} \tag{7.511}$$

$$\int_a^b subsin_{\mu}(\varpi(c)\mu(t))\,\mu^{(1)}(t)\,dt = \int_{-\pi}^{\pi} sin(\varpi(c)\upsilon)\,d\upsilon = \int_{-\pi}^{\pi} sin(\kappa_1\upsilon)\,d\upsilon = 0, \tag{7.512}$$

and

$$\int_a^b subcos_{\mu}(\varpi(c)\mu(t))\mu^{(1)}(t)dt = \int_{-\pi}^{\pi} cos(\varpi(c)\upsilon)d\upsilon = \int_{-\pi}^{\pi} cos(\kappa_1\upsilon)d\upsilon = 0. \tag{7.513}$$

Thus, we give the proof. □

Theorem 7.12 *Let* $\mu^{(1)}(t) > 0$, $\mu(a) = -\pi$, $\mu(b) = \pi$, *and* $t, a, b \in \mathbb{R}$ *Let* $\varpi^{(1)}(\kappa) > 0$, $\kappa_1 = \varpi(c)$, $\kappa_2 = \varpi(d)$, *and* $\kappa, c, d \in \mathbb{N}$.

Then, there exist the series with respect to another function, defined by

$$\begin{aligned}\Xi_{\mu}(t) = \tfrac{1}{2}\psi_0 + \sum_{\kappa=1}^{\infty} \varphi_{\varpi}(\kappa) subcos_{\mu}(\varpi(\kappa)\mu(t)) \\ + \sum_{\kappa=1}^{\infty} \psi_{\varpi}(\kappa) subsin_{\mu}(\varpi(\kappa)\mu(t)),\end{aligned} \tag{7.514}$$

where $\kappa \in \mathbb{N}$ *and the coefficients* $\psi_{\varpi}(0)$, $\varphi_{\varpi}(\kappa)$, *and* $\psi_{\varpi}(\kappa)$ *are determined by*

$$\psi_{\varpi}(0) = \frac{1}{\pi}\int_a^b \Xi_{\mu}(t)\,\mu^{(1)}(t)\,dt, \tag{7.515}$$

$$\varphi_{\varpi}(\kappa) = \frac{1}{\pi}\int_a^b \Xi_{\mu}(t)\, subcos_{\mu}(\varpi(\kappa)\,\mu(t))\,\mu^{(1)}(t)\,dt, \tag{7.516}$$

and

$$\psi_{\varpi}(\kappa) = \frac{1}{\pi}\int_a^b \Xi_{\mu}(t)\, subsin_{\mu}(\varpi(\kappa)\,\mu(t))\,\mu^{(1)}(t)\,dt. \tag{7.517}$$

Proof Setting $\mu(t) = \upsilon$, we have

$$\frac{1}{\pi}\int_a^b \Xi_{\mu}(t)\,\mu^{(1)}(t)\,dt = \frac{1}{\pi}\int_a^b \frac{\psi_{\varpi}(0)}{2}\mu^{(1)}(t)\,dt = \frac{\psi_{\varpi}(0)}{2\pi}\int_{-\pi}^{\pi} d\upsilon = \psi_{\varpi}(0), \tag{7.518}$$

$$\begin{aligned}&\tfrac{1}{\pi}\int_a^b \Xi_{\mu}(t)\, subcos_{\mu}(\varpi(\kappa)\,\mu(t))\,\mu^{(1)}(t)\,dt \\ &\quad= \tfrac{1}{\pi}\int_a^b \varphi_{\varpi}(\kappa)\, subcos_{\mu}(\varpi(\kappa)\,\mu(t))\, subcos_{\mu}(\varpi(\kappa)\,\mu(t))\,\mu^{(1)}(t)\,dt \\ &\quad= \varphi_{\varpi}(\kappa),\end{aligned} \tag{7.519}$$

and

$$\begin{aligned}\psi_{\varpi}(\kappa) &= \tfrac{1}{\pi}\int_a^b \Xi_{\mu}(t)\, subsin_{\mu}(\varpi(\kappa)\,\mu(t))\,\mu^{(1)}(t)\,dt \\ &= \tfrac{1}{\pi}\int_a^b \psi_{\varpi}(\kappa)\, subsin_{\mu}(\varpi(\kappa)\,\mu(t))\, subsin_{\mu}(\varpi(\kappa)\,\mu(t))\,\mu^{(1)}(t)\,dt \\ &= \psi_{\varpi}(\kappa).\end{aligned} \tag{7.520}$$

Thus, the proof is finished. □

Theorem 7.13 *Let* $\mu^{(1)}(t) > 0$, $\mu(a) = -\pi$, $\mu(b) = \pi$, *and* $t, a, b \in \mathbb{R}$
Let $\varpi^{(1)}(\kappa) > 0$, $\kappa_1 = \varpi(c)$, $\kappa_2 = \varpi(d)$, *and* $\kappa, c, d \in \mathbb{N}$.
If $\Theta_\mu(t) = (\Theta \circ \mu)(t) = \Theta(\mu(t))$ *with* $\Theta(\mu(t) + 2\pi) = \Theta(\mu(t))$, *then we have*

$$\Theta_\mu(t) = \sum_{\kappa=-\infty}^{\infty} m_\varpi(\kappa)\, e^{i\varpi(\kappa)\mu(t)}, \tag{7.521}$$

where

$$m_\varpi(\kappa) = \frac{1}{2\pi}\int_a^b \Theta_\mu(t)\, e^{-i\varpi(\kappa)\mu(t)}\mu^{(1)}(t)\, dt. \tag{7.522}$$

Proof Putting $\upsilon = \mu(t)$ and using

$$\int_{-\pi}^{\pi} e^{i(\varpi(\kappa)-\varpi(\xi))\upsilon} d\upsilon = 2\pi\,\delta_{\varpi(\kappa)\varpi(\xi)}, \tag{7.523}$$

we have

$$\begin{aligned}
&\int_a^b \Theta_\mu(t)\, e^{-i\varpi(\xi)\mu(t)}\mu^{(1)}(t)\, dt \\
&= \int_a^b \left(\sum_{\kappa=-\infty}^{\infty} m_\varpi(\kappa)\, e^{i\varpi(\kappa)\mu(t)}\right) e^{-i\varpi(\xi)\mu(t)}\mu^{(1)}(t)\, dt \\
&= \sum_{\kappa=-\infty}^{\infty} m_\varpi(\kappa) \int_a^b e^{i(\varpi(\kappa)-\varpi(\xi))\mu(t)}\mu^{(1)}(t)\, dt \\
&= \sum_{\kappa=-\infty}^{\infty} m_\varpi(\kappa) \int_{-\pi}^{\pi} e^{i(\varpi(\kappa)-\varpi(\xi))\upsilon} d\upsilon \\
&= \sum_{\kappa=-\infty}^{\infty} 2\pi m_\varpi(\kappa)\, \delta_{\varpi(\kappa)\varpi(\xi)} \\
&= 2\pi m_\varpi(\xi),
\end{aligned} \tag{7.524}$$

which implies that

$$m_\varpi(\kappa) = \frac{1}{2\pi}\int_a^b \Theta_\mu(t)\, e^{-i\varpi(\kappa)\mu(t)}\mu^{(1)}(t)\, dt. \tag{7.525}$$

Therefore, we complete the proof. □

Theorem 7.14 *Let* $\mu^{(1)}(t) > 0$, $\mu(a) = -\pi$, $\mu(b) = \pi$, *and* $t, a, b \in \mathbb{R}$
Let $\varpi^{(1)}(\kappa) > 0$, $\kappa_1 = \varpi(c)$, $\kappa_2 = \varpi(d)$, *and* $\kappa, c, d \in \mathbb{N}$.

If $\mathbb{M}_{\mu}(t) = (\mathbb{M} \circ \mu)(t) = \mathbb{M}(\mu(t))$ *with* $\mathbb{M}(\mu(t) + L) = \mathbb{M}(\mu(t))$, *then we have*

$$\mathbb{M}_{\mu}(t) = \sum_{\kappa=-\infty}^{\infty} \phi_{\varpi}(\kappa)\, e^{\frac{2\pi i \varpi(\kappa)\mu(t)}{L}}, \tag{7.526}$$

where

$$\phi_{\varpi}(\kappa) = \frac{1}{L} \int_{a}^{b} \mathbb{M}_{\mu}(t)\, e^{-\frac{2\pi i \varpi(\kappa)\mu(t)}{L}} \mu^{(1)}(t)\, dt. \tag{7.527}$$

Proof Taking $\upsilon = \mu(t)$ and using

$$\int_{a}^{b} e^{\frac{2\pi i(\varpi(\kappa)-\varpi(\xi))\mu(t)}{L}} \mu^{(1)}(t)\, dt = \int_{-\frac{L}{2}}^{\frac{L}{2}} e^{\frac{2\pi i(\varpi(\kappa)-\varpi(\xi))\upsilon}{L}} d\upsilon = L\delta_{\varpi(\kappa)\varpi(\xi)}, \tag{7.528}$$

we present

$$\begin{aligned} &\int_{a}^{b} \mathbb{M}_{\mu}(t)\, e^{-\frac{2\pi i \varpi(\xi)\mu(t)}{L}} \mu^{(1)}(t)\, dt \\ &= \int_{a}^{b} \left(\sum_{\kappa=-\infty}^{\infty} \phi_{\varpi}(\kappa)\, e^{\frac{2\pi i \varpi(\kappa)\mu(t)}{L}} \right) e^{-\frac{2\pi i \varpi(\xi)\mu(t)}{L}} \mu^{(1)}(t)\, dt \\ &= \sum_{\kappa=-\infty}^{\infty} \phi_{\varpi}(\kappa) \int_{a}^{b} e^{\frac{2\pi i(\varpi(\kappa)-\varpi(\xi))\mu(t)}{L}} \mu^{(1)}(t)\, dt \\ &= \sum_{\kappa=-\infty}^{\infty} \phi_{\varpi}(\kappa) \int_{-\frac{L}{2}}^{\frac{L}{2}} e^{\frac{2\pi i(\varpi(\kappa)-\varpi(\xi))\upsilon}{L}} d\upsilon \\ &= \sum_{\kappa=-\infty}^{\infty} L\phi_{\varpi}(\kappa)\, \delta_{\varpi(\kappa)\varpi(\xi)} \\ &= Lm_{\varpi(\xi)}, \end{aligned} \tag{7.529}$$

which can be written as follows:

$$\phi_{\varpi}(\kappa) = \frac{1}{L} \int_{a}^{b} \mathbb{M}_{\mu}(t)\, e^{-\frac{2\pi i \varpi(\kappa)\mu(t)}{L}} \mu^{(1)}(t)\, dt. \tag{7.530}$$

Therefore, we present the proof. □

Theorem 7.15 *Let* $\mu^{(1)}(t) > 0$, $\mu(a) = -\pi$, $\mu(b) = \pi$, *and* $t, a, b \in \mathbb{R}$
Let $\varpi^{(1)}(\kappa) > 0$, $\kappa_1 = \varpi(c)$, $\kappa_2 = \varpi(d)$, *and* $\kappa, c, d \in \mathbb{N}$.

Then, there exist

$$\int_a^b subsin_\mu(\varpi(c)\mu(t)) \cdot subsin_\mu(\varpi(d)\mu(t))\mu^{(1)}(t)\,dt = \pi\delta_{\varpi(c)\varpi(d)}, \tag{7.531}$$

$$\int_a^b subcos_\mu(\varpi(c)\mu(t)) \cdot subcos_\mu(\varpi(d)\mu(t))\mu^{(1)}(t)\,dt = \pi\delta_{\varpi(c)\varpi(d)}, \tag{7.532}$$

$$\int_a^b subsin_\mu(\varpi(c)\mu(t)) \cdot subcos_\mu(\varpi(d)\mu(t))\mu^{(1)}(t)\,dt = 0, \tag{7.533}$$

$$\int_a^b subsin_\mu(\varpi(c)\mu(t))\mu^{(1)}(t)\,dt = 0, \tag{7.534}$$

and

$$\int_a^b subcos_\mu(\varpi(c)\mu(t))\mu^{(1)}(t)\,dt = 0, \tag{7.535}$$

where $\delta_{\varpi(c)\varpi(d)} = \delta_{\kappa_1\kappa_2}$ *due to* $\varpi(c) = \kappa_1$ *and* $\varpi(d) = \kappa_2$.

Proof Taking $\upsilon = \mu(t)$, we present

$$\begin{aligned}
&\int_a^b subsin_\mu(\varpi(c)\mu(t)) \cdot subsin_\mu(\varpi(d)\mu(t))\mu^{(1)}(t)\,dt \\
&= \int_{-\pi}^{\pi} sin(\varpi(c)\upsilon)\,sin(\varpi(d)\upsilon)\,d\upsilon \\
&= \int_{-\pi}^{\pi} sin(\kappa_1\upsilon)\,sin(\kappa_2\upsilon)\,d\upsilon \\
&= \pi\delta_{\varpi(c)\varpi(d)},
\end{aligned} \tag{7.536}$$

$$\begin{aligned}
&\int_a^b subcos_\mu(\varpi(c)\mu(t)) \cdot subcos_\mu(\varpi(d)\mu(t))\mu^{(1)}(t)\,dt \\
&= \int_{-\pi}^{\pi} cos(\varpi(c)\upsilon)\,cos(\varpi(d)\upsilon)\,d\upsilon \\
&= \int_{-\pi}^{\pi} cos(\kappa_1\upsilon)\,cos(\kappa_2\upsilon)\,d\upsilon \\
&= \pi\delta_{\varpi(c)\varpi(d)},
\end{aligned} \tag{7.537}$$

$$
\begin{aligned}
&\int_a^b subsin_\mu\left(\varpi\left(c\right)\mu\left(t\right)\right)\cdot subcos_\mu\left(\varpi\left(d\right)\mu\left(t\right)\right)\mu^{(1)}\left(t\right)dt \\
&\quad=\int_{-\pi}^{\pi} sin\left(\varpi\left(c\right)\upsilon\right)cos\left(\varpi\left(d\right)\upsilon\right)d\upsilon \\
&\quad=\int_{-\pi}^{\pi} sin\left(\kappa_1\upsilon\right)cos\left(\kappa_2\upsilon\right)d\upsilon \\
&\quad=0,
\end{aligned}
\tag{7.538}
$$

$$
\int_a^b subsin_\mu\left(\varpi\left(c\right)\mu\left(t\right)\right)\mu^{(1)}\left(t\right)dt=\int_{-\pi}^{\pi} sin\left(\varpi\left(c\right)\upsilon\right)d\upsilon=\int_{-\pi}^{\pi} sin\left(\kappa_1\upsilon\right)d\upsilon=0,
\tag{7.539}
$$

and

$$
\int_a^b subcos_\mu(\varpi(c)\mu(t))\mu^{(1)}(t)dt=\int_{-\pi}^{\pi} cos(\varpi(c)\upsilon)d\upsilon=\int_{-\pi}^{\pi} cos(\kappa_1\upsilon)d\upsilon=0.
\tag{7.540}
$$

Thus, we give the proof. □

Theorem 7.16 *Let $\mu^{(1)}(t)>0$, $\mu(a)=-\pi$, $\mu(b)=\pi$, and $t,a,b\in\mathbb{R}$*
Let $\varpi^{(1)}(\kappa)>0$, $\kappa_1=\varpi(c)$, $\kappa_2=\varpi(d)$, and $\kappa,c,d\in\mathbb{N}$.
Let $\mathbb{M}_\mu(t)=(\mathbb{M}\circ\mu)(t)=\mathbb{M}(\mu(t))$ with $\mathbb{M}(\mu(t)+L)=\mathbb{M}(\mu(t))$
Then, there exist the series with respect to another function, defined by

$$
\begin{aligned}
\Xi_\mu\left(t\right)=\tfrac{1}{2}\psi_\varpi\left(0\right)&+\sum_{\kappa=1}^{\infty}\varphi_\varpi\left(\kappa\right)subcos_\mu\left(\tfrac{2\pi\varpi(\kappa)}{L}\mu\left(t\right)\right)\\
&+\sum_{\kappa=1}^{\infty}\psi_\varpi\left(\kappa\right)subsin_\mu\left(\tfrac{2\pi\varpi(\kappa)}{L}\mu\left(t\right)\right),
\end{aligned}
\tag{7.541}
$$

where $\kappa\in\mathbb{N}$ and the coefficients $\psi_\varpi(0)$, $\varphi_\varpi(\kappa)$, and $\psi_\varpi(\kappa)$ are determined by

$$
\psi_\varpi\left(0\right)=\frac{2}{L}\int_a^b \Xi_\mu\left(t\right)\mu^{(1)}\left(t\right)dt
\tag{7.542}
$$

$$
\varphi_\varpi\left(\kappa\right)=\frac{2}{L}\int_a^b \Xi_\mu\left(t\right)subcos_\mu\left(\frac{2\pi\varpi\left(\kappa\right)}{L}\mu\left(t\right)\right)\mu^{(1)}\left(t\right)dt,
\tag{7.543}
$$

and

$$
\psi_\varpi\left(\kappa\right)=\frac{2}{L}\int_a^b \Xi_\mu\left(t\right)subsin_\mu\left(\frac{2\pi\varpi\left(\kappa\right)}{L}\mu\left(t\right)\right)\mu^{(1)}\left(t\right)dt.
\tag{7.544}
$$

Proof Taking $\mu(t) = \upsilon$ and using (7.11), (7.12), (7.13), (7.14), and (7.15) [306], we have that

$$\begin{aligned}
&\frac{2}{L}\int_a^b \Xi_\mu(t)\,\mu^{(1)}(t)\,dt \\
&\quad = \frac{2}{L}\int_a^b \left(\frac{1}{2}\psi_0 + \sum_{\kappa=1}^{\infty} \varphi_\varpi(\kappa)\, subcos_\mu\left(\frac{2\pi\varpi(\kappa)}{L}\mu(t)\right)\right)\mu^{(1)}(t)\,dt \\
&\quad + \frac{2}{L}\int_a^b \left(\sum_{\kappa=1}^{\infty} \psi_\varpi(\kappa)\, subsin_\mu\left(\frac{2\pi\varpi(\kappa)}{L}\mu(t)\right)\right)\mu^{(1)}(t)\,dt \\
&\quad = \frac{\psi_\varpi(0)}{L}\int_{-\frac{L}{2}}^{\frac{L}{2}} d\upsilon \\
&\quad = \psi_\varpi(0),
\end{aligned} \tag{7.545}$$

$$\begin{aligned}
&\frac{2}{L}\int_a^b \Xi_\mu(t)\, subcos_\mu\left(\frac{2\pi\varpi(\kappa)}{L}\mu(t)\right)\mu^{(1)}(t)\,dt \\
&= \frac{1}{L}\int_a^b \psi_\varpi(0)\, subcos_\mu\left(\frac{2\pi\varpi(\kappa)}{L}\mu(t)\right)\mu^{(1)}(t)\,dt \\
&+ \frac{2}{L}\int_a^b \left(\sum_{\kappa=1}^{\infty} \varphi_\kappa subcos_\mu\left(\frac{2\pi\varpi(\kappa)}{L}\mu(t)\right)\right) subcos_\mu\left(\frac{2\pi\varpi(\kappa)}{L}\mu(t)\right)\mu^{(1)}(t)\,dt \\
&+ \frac{2}{L}\int_a^b \left(\sum_{\kappa=1}^{\infty} \psi_\varpi(\kappa)\, subsin_\mu\left(\frac{2\pi\varpi(\kappa)}{L}\mu(t)\right)\right) subcos_\mu\left(\frac{2\pi\varpi(\kappa)}{L}\mu(t)\right)\mu^{(1)}(t)\,dt \\
&= \frac{2}{L}\int_a^b \varphi_\varpi(\kappa)\, subcos_\mu\left(\frac{2\pi\varpi(\kappa)}{L}\mu(t)\right) subcos_\mu\left(\frac{2\pi\varpi(\kappa)}{L}\mu(t)\right)\mu^{(1)}(t)\,dt \\
&= \varphi_\varpi(\kappa),
\end{aligned} \tag{7.546}$$

$$\begin{aligned}
&\frac{2}{L}\int_a^b \Xi_\mu(t)\, subsin_\mu\left(\frac{2\pi\varpi(\kappa)}{L}\mu(t)\right)\mu^{(1)}(t)\,dt \\
&\quad = \frac{1}{L}\int_a^b \psi_\varpi(0)\, subsin_\mu\left(\frac{2\pi}{L}\mu(t)\right)\mu^{(1)}(t)\,dt \\
&\quad + \frac{2}{L}\int_a^b \left(\sum_{\kappa=1}^{\infty} \varphi_\varpi(\kappa)\, subcos_\mu\left(\frac{2\pi\varpi(\kappa)}{L}\mu(t)\right)\right) subsin_\mu\left(\frac{2\pi}{L}\mu(t)\right)\mu^{(1)}(t)\,dt \\
&\quad + \frac{2}{L}\int_a^b \left(\sum_{\kappa=1}^{\infty} \psi_\varpi(\kappa)\, subsin_\mu\left(\frac{2\pi\varpi(\kappa)}{L}\mu(t)\right)\right) subsin_\mu\left(\frac{2\pi}{L}\mu(t)\right)\mu^{(1)}(t)\,dt \\
&\quad = \frac{2}{L}\int_a^b \psi_\varpi(\kappa)\, subsin_\mu\left(\frac{2\pi\varpi(\kappa)}{L}\mu(t)\right) subsin_\mu\left(\frac{2\pi\varpi(\kappa)}{L}\mu(t)\right)\mu^{(1)}(t)\,dt \\
&\quad = \psi_\varpi(\kappa),
\end{aligned} \tag{7.547}$$

where

$$\int_a^b subsin_\mu\left(\frac{2\pi\varpi(\kappa)}{L}\mu(t)\right)\mu^{(1)}(t)\,dt=\int_{-\frac{L}{2}}^{\frac{L}{2}} sin\left(\frac{2\pi\varpi(\kappa)}{L}\upsilon\right)d\upsilon=0 \tag{7.548}$$

$$\int_a^b subcos_\mu\left(\frac{2\pi\varpi(\kappa)}{L}\mu(t)\right)\mu^{(1)}(t)\,dt=\int_{-\frac{L}{2}}^{\frac{L}{2}} cos\left(\frac{2\pi\varpi(\kappa)}{L}\upsilon\right)d\upsilon=0, \tag{7.549}$$

$$\begin{aligned}&\int_a^b subcos_\mu\left(\tfrac{2\pi\varpi(\kappa_2)}{L}\mu(t)\right)subcos_\mu\left(\tfrac{2\pi\varpi(\kappa_2)}{L}\mu(t)\right)\mu^{(1)}(t)\,dt\\&\quad=L\delta_{\varpi(\kappa_1)\varpi(\kappa_2)},\end{aligned} \tag{7.550}$$

and

$$\int_a^b subsin_\mu\left(\frac{2\pi\varpi(\kappa_1)}{L}\mu(t)\right)subcos_\mu\left(\frac{2\pi\varpi(\kappa_2)}{L}\mu(t)\right)\mu^{(1)}(t)\,dt=0, \tag{7.551}$$

$$\begin{aligned}&\int_a^b subsin_\mu\left(\tfrac{2\pi\varpi(\kappa_1)}{L}\mu(t)\right)subsin_\mu\left(\tfrac{2\pi\varpi(\kappa_2)}{L}\mu(t)\right)\mu^{(1)}(t)\,dt\\&\quad=L\delta_{\varpi(\kappa_1)\varpi(\kappa_2)},\end{aligned} \tag{7.552}$$

and

$$\int_a^b subsin_\mu\left(\frac{2\pi\varpi(\kappa_1)}{L}\mu(t)\right)subcos_\mu\left(\frac{2\pi\varpi(\kappa_2)}{L}\mu(t)\right)\mu^{(1)}(t)\,dt=0. \tag{7.553}$$

Thus, we give the proof. □

Corollary 7.2 *Let* $\varpi^{(1)}(\kappa)>0$, $\kappa_1=\varpi(c)$, $\kappa_2=\varpi(d)$, *and* $\kappa, c, d\in\mathbb{N}$.

The subtrigonometric polynomial $\Xi(t)=\Xi(T+t)$ *with* $T=2\pi/\varpi(\omega)$, *where* T *is the exterior period and* $\varpi(\omega)$ *is the exterior frequency, is defined as*

$$\Xi(t)=\varphi_\varpi(0)+\sum_{\kappa=1}^{\infty}\varphi_\varpi(\kappa)\,cos_\varpi(\kappa\varpi(\omega)\,t)+\sum_{\kappa=1}^{\infty}\psi_\varpi(\kappa)\,sin_\varpi(\kappa\varpi(\omega)\,t), \tag{7.554}$$

where $\kappa \in \mathbb{N}$ *and the coefficients* $\varphi_{\varpi}(0)$, $\varphi_{\varpi}(\kappa)$, *and* $\psi_{\varpi}(\kappa)$ *are determined by*

$$\varphi_{\varpi}(0) = \frac{1}{T}\int_{-\frac{T}{2}}^{\frac{T}{2}} \Xi(t)\,dt, \tag{7.555}$$

$$\varphi_{\varpi}(\kappa) = \frac{2}{T}\int_{-\frac{T}{2}}^{\frac{T}{2}} \Xi(t)\, subcos_{\varpi}(\kappa\varpi(\omega)\,t)\,dt, \tag{7.556}$$

and

$$\psi_{\varpi}(\kappa) = \frac{2}{T}\int_{-\frac{T}{2}}^{\frac{T}{2}} \Xi(t)\, subsin_{\varpi}(\kappa\varpi(\omega)\,t)\,dt. \tag{7.557}$$

Corollary 7.3 *Let* $\varpi^{(1)}(\kappa) > 0$, $\kappa_1 = \varpi(c)$, $\kappa_2 = \varpi(d)$, *and* $\kappa, c, d \in \mathbb{N}$.

If $\Theta(t+T) = \Theta(t)$ *with* $T = 2\pi/\varpi(\omega)$, *where* T *is the exterior period and* $\varpi(\omega)$ *is the exterior frequency, then we have*

$$\Theta(t) = \sum_{\kappa=-\infty}^{\infty} m_{\varpi}(\kappa)\, e^{i\kappa\varpi(\omega)t}, \tag{7.558}$$

where

$$m_{\varpi}(\kappa) = \frac{1}{T}\int_{-\frac{T}{2}}^{\frac{T}{2}} \Theta(t)\, e^{-i\kappa\varpi(\omega)t}\,dt. \tag{7.559}$$

Corollary 7.4 *Let* $\varpi^{(1)}(\kappa) > 0$, $\kappa_1 = \varpi(c)$, $\kappa_2 = \varpi(d)$, *and* $\kappa, c, d \in \mathbb{N}$.

Then, there exist the series with respect to another function, which is $\Xi(t+L) = \Xi(t)$, *defined by*

$$\begin{aligned}\Xi(t) = {} & \tfrac{1}{2}\psi_{\varpi}(0) + \sum_{\kappa=1}^{\infty} \varphi_{\varpi}(\kappa)\, subcos_{\mu}\left(\tfrac{2\pi\varpi(\kappa)}{L}t\right) \\ & + \sum_{\kappa=1}^{\infty} \psi_{\varpi}(\kappa)\, subsin_{\mu}\left(\tfrac{2\pi\varpi(\kappa)}{L}t\right),\end{aligned} \tag{7.560}$$

where $\kappa \in \mathbb{N}$ *and the coefficients* $\psi_{\varpi}(0)$, $\varphi_{\varpi}(\kappa)$, *and* $\psi_{\varpi}(\kappa)$ *are determined by*

$$\psi_{\varpi}(0) = \frac{2}{L}\int_{a}^{b} \Xi(t)\,dt, \tag{7.561}$$

$$\varphi_{\varpi}(\kappa) = \frac{2}{L}\int_{a}^{b} \Xi(t)\,subcos_{\mu}\left(\frac{2\pi\varpi(\kappa)}{L}t\right)dt, \tag{7.562}$$

and

$$\psi_{\varpi}(\kappa) = \frac{2}{L}\int_{a}^{b} \Xi(t)\,subsin_{\mu}\left(\frac{2\pi\varpi(\kappa)}{L}t\right)dt. \tag{7.563}$$

Corollary 7.5 *Let* $\varpi^{(1)}(\kappa) > 0$, $\kappa_1 = \varpi(c)$, $\kappa_2 = \varpi(d)$, *and* $\kappa, c, d \in \mathbb{N}$. *If* $\mathbb{M}(t+L) = \mathbb{M}(t)$, *then we have*

$$\mathbb{M}(t) = \sum_{\kappa=-\infty}^{\infty} \phi_{\varpi}(\kappa)\,e^{\frac{2\pi i\varpi(\kappa)t}{L}}, \tag{7.564}$$

where

$$\phi_{\varpi}(\kappa) = \frac{1}{L}\int_{a}^{b} \mathbb{M}(t)\,e^{-\frac{2\pi i\varpi(\kappa)t}{L}}dt. \tag{7.565}$$

Proof With the aid of (7.562), (7.563), (7.564), and (7.565), we obtain the result. □

7.3.2 *The Subtrigonometric Series with Respect to Scaling-Law Function*

In this part, we present the subtrigonometric series with respect to scaling-law function

Theorem 7.17 *Let* $\alpha, \lambda > 0$ *and* $(-t)^{\alpha} = -t^{\alpha}$. *Then, there exist*

$$\int_{-\sqrt[\alpha]{\frac{\pi}{\lambda}}}^{\sqrt[\alpha]{\frac{\pi}{\lambda}}} subsin_{\alpha}(\lambda m, t)\cdot subsin_{\alpha}(\lambda n, t)\,t^{\alpha-1}dt = \frac{\pi}{\lambda\alpha}\delta_{mn}, \tag{7.566}$$

$$\int_{-\sqrt[\alpha]{\frac{\pi}{\lambda}}}^{\sqrt[\alpha]{\frac{\pi}{\lambda}}} subcos_{\alpha}(\lambda m, t) \cdot subcos_{\alpha}(\lambda n, t)\, t^{\alpha-1} dt = \frac{\pi}{\lambda\alpha}\delta_{mn}, \tag{7.567}$$

$$\int_{-\sqrt[\alpha]{\frac{\pi}{\lambda}}}^{\sqrt[\alpha]{\frac{\pi}{\lambda}}} subsin_{\alpha}(\lambda m, t) \cdot subcos_{\alpha}(\lambda n, t)\, t^{\alpha-1} dt = 0, \tag{7.568}$$

$$\int_{-\sqrt[\alpha]{\frac{\pi}{\lambda}}}^{\sqrt[\alpha]{\frac{\pi}{\lambda}}} subsin_{\alpha}(\lambda m, t)\, t^{\alpha-1} dt = 0, \tag{7.569}$$

and

$$\int_{-\sqrt[\alpha]{\frac{\pi}{\lambda}}}^{\sqrt[\alpha]{\frac{\pi}{\lambda}}} subcos_{\alpha}(\lambda m, t)\, t^{\alpha-1} dt = 0, \tag{7.570}$$

where $m \in \mathbb{N}$ *and* $n \in \mathbb{N}$.

Proof Taking the place of t by λt^{α}, we obtain the result. □

Theorem 7.18 *Let* $\Xi_{\alpha}(\lambda, t) = \Xi(\lambda t^{\alpha})$, $\Xi(\lambda t^{\alpha}) = \Xi(\lambda t^{\alpha} + 2\pi)$, $\alpha, \lambda > 0$, *and* $(-t)^{\alpha} = -t^{\alpha}$.

Then, there exist the series with respect to scaling-law function

$$\Xi_{\alpha}(\lambda, t) = \frac{1}{2}\psi_0 + \sum_{n=1}^{\infty} \varphi_n subcos_{\alpha}(\lambda n, t) + \sum_{n=1}^{\infty} \psi_n subsin_{\alpha}(\lambda n, t), \tag{7.571}$$

where $n \in \mathbb{N}$ *and the coefficients* ψ_0, φ_n, *and* ψ_n *are determined by*

$$\psi_0 = \frac{\alpha\lambda}{\pi} \int_{-\sqrt[\alpha]{\frac{\pi}{\lambda}}}^{\sqrt[\alpha]{\frac{\pi}{\lambda}}} \Xi_{\alpha}(\lambda, t)\, t^{\alpha-1} dt, \tag{7.572}$$

$$\varphi_n = \frac{\alpha\lambda}{\pi} \int_{-\sqrt[\alpha]{\frac{\pi}{\lambda}}}^{\sqrt[\alpha]{\frac{\pi}{\lambda}}} \Xi_{\alpha}(\lambda, t)\, subcos_{\alpha}(\lambda n, t)\, t^{\alpha-1} dt, \tag{7.573}$$

and

$$\psi_n = \frac{\alpha\lambda}{\pi} \int_{-\sqrt[\alpha]{\frac{\pi}{\lambda}}}^{\sqrt[\alpha]{\frac{\pi}{\lambda}}} \Xi_\alpha(\lambda, t)\, subsin_\alpha(\lambda n, t)\, t^{\alpha-1} dt. \tag{7.574}$$

Proof Based on the formulas (7.568), (7.569), (7.570), (7.571), and (7.572), we show that

$$\begin{aligned}
&\tfrac{\alpha\lambda}{\pi} \int_{-\sqrt[\alpha]{\frac{\pi}{\lambda}}}^{\sqrt[\alpha]{\frac{\pi}{\lambda}}} \Xi_\alpha(\lambda, t)\, t^{\alpha-1} dt \\
&= \tfrac{\alpha\lambda}{\pi} \int_{-\sqrt[\alpha]{\pi}}^{\sqrt[\alpha]{\pi}} \left(\tfrac{1}{2}\psi_0 + \sum_{n=1}^{\infty} \varphi_n subcos_\alpha(\lambda n, t) + \sum_{n=1}^{\infty} \psi_n subsin_\alpha(\lambda n, t) \right) t^{\alpha-1} dt \\
&= \tfrac{\psi_0}{2\pi} \int_{-\pi}^{\pi} d\upsilon \\
&= \psi_0,
\end{aligned} \tag{7.575}$$

$$\begin{aligned}
&\tfrac{\alpha\lambda}{\pi} \int_{-\sqrt[\alpha]{\frac{\pi}{\lambda}}}^{\sqrt[\alpha]{\frac{\pi}{\lambda}}} \Xi_\alpha(\lambda, t)\, subcos_\alpha(\lambda n, t)\, t^{\alpha-1} dt \\
&= \tfrac{\alpha\lambda}{\pi} \int_{-\sqrt[\alpha]{\frac{\pi}{\lambda}}}^{\sqrt[\alpha]{\frac{\pi}{\lambda}}} \left(\tfrac{1}{2}\psi_0 + \sum_{n=1}^{\infty} \varphi_n subcos_\alpha(\lambda n, t) + \sum_{\kappa=1}^{\infty} \psi_n subsin_\alpha(\lambda n, t) \right) \\
&\quad \times subcos_\alpha(\lambda n, t)\, t^{\alpha-1} dt \\
&= \tfrac{\alpha\lambda}{\pi} \int_{-\sqrt[\alpha]{\frac{\pi}{\lambda}}}^{\sqrt[\alpha]{\frac{\pi}{\lambda}}} \varphi_n subcos_\alpha(\lambda n, t)\, subcos_\alpha(\lambda n, t)\, t^{\alpha-1} dt \\
&= \tfrac{\varphi_n}{\pi} \int_{-\pi}^{\pi} \cos(n\upsilon)\cos(n\upsilon)\, d\upsilon \\
&= \varphi_n,
\end{aligned} \tag{7.576}$$

and

$$\begin{aligned}
\psi_n &= \frac{\alpha\lambda}{\pi} \int_{-\sqrt[\alpha]{\frac{\pi}{\lambda}}}^{\sqrt[\alpha]{\frac{\pi}{\lambda}}} \Xi_\alpha(\lambda, t)\, subsin_\alpha(\lambda n, t)\, t^{\alpha-1} dt \\
&= \frac{\alpha\lambda}{\pi} \int_{-\sqrt[\alpha]{\frac{\pi}{\lambda}}}^{\sqrt[\alpha]{\frac{\pi}{\lambda}}} \left(\frac{1}{2}\psi_0 + \sum_{\kappa=1}^{\infty} \varphi_n subcos_\alpha(\lambda n, t) + \sum_{\kappa=1}^{\infty} \psi_n subsin_\alpha(\lambda n, t) \right) \\
&\times subsin_\alpha(\lambda n, t)\, t^{\alpha-1} dt \\
&= \frac{\alpha\lambda}{\pi} \int_{-\sqrt[\alpha]{\frac{\pi}{\lambda}}}^{\sqrt[\alpha]{\frac{\pi}{\lambda}}} \psi_n subsin_\alpha(\lambda n, t)\, subsin_\alpha(\lambda n, t)\, t^{\alpha-1} dt \\
&= \frac{\psi_n}{\pi} \int_{-\pi}^{\pi} sin(n\upsilon)\, sin(n\upsilon)\, d\upsilon \\
&= \psi_n.
\end{aligned} \tag{7.577}$$

Thus, we finish the proof. □

Let us consider the alternative form of the formula (7.573) as follows:

Theorem 7.19 *Let $\alpha, \lambda > 0$ and $(-t)^\alpha = -t^\alpha$.*
Then, there exists

$$\int_{-\sqrt[\alpha]{\frac{\pi}{\lambda}}}^{\sqrt[\alpha]{\frac{\pi}{\lambda}}} e^{i\lambda(m-n)t^\alpha} t^{\alpha-1} dt = \frac{2\pi}{\lambda\alpha}\delta_{mn}. \tag{7.578}$$

Proof Based on the definition, we have the integral representation

$$\lambda\alpha \int_{-\sqrt[\alpha]{\frac{\pi}{\lambda}}}^{\sqrt[\alpha]{\frac{\pi}{\lambda}}} e^{i\lambda(m-n)t^\alpha} t^{\alpha-1} dt = \int_{-\pi}^{\pi} e^{i(m-n)t} dt = 2\pi\delta_{mn}. \tag{7.579}$$

Thus, we give the result. □

Theorem 7.20 *Let $\Xi_\alpha(\lambda, t) = \Xi(\lambda t^\alpha)$, $\Xi(\lambda t^\alpha) = \Xi(\lambda t^\alpha + 2\pi)$, $\alpha, \lambda > 0$, and $(-t)^\alpha = -t^\alpha$.*
Then, we have

$$\Theta_\alpha(\lambda, t) = \sum_{n=-\infty}^{\infty} \Pi_n e^{in\lambda t^\alpha}, \tag{7.580}$$

where

$$\varPi_n = \frac{\lambda\alpha}{2\pi} \int_{-\sqrt[\alpha]{\frac{\pi}{\lambda}}}^{\sqrt[\alpha]{\frac{\pi}{\lambda}}} \Theta_\alpha(\lambda, t)\, e^{-in\lambda t^\alpha} t^{\alpha-1} dt. \tag{7.581}$$

Proof Based on the definition, we have that

$$\begin{aligned}
&\alpha\lambda \int_{-\sqrt[\alpha]{\frac{\pi}{\lambda}}}^{\sqrt[\alpha]{\frac{\pi}{\lambda}}} \Theta_\alpha(\lambda, t)\, e^{-in\lambda t^\alpha} t^{\alpha-1} dt \\
&= \alpha\lambda \int_{-\sqrt[\alpha]{\frac{\pi}{\lambda}}}^{\sqrt[\alpha]{\frac{\pi}{\lambda}}} \left(\sum_{m=-\infty}^{\infty} \varPi_m e^{im\lambda t^\alpha} \right) e^{-in\lambda t^\alpha} t^{\alpha-1} dt \\
&= \alpha\lambda \sum_{m=-\infty}^{\infty} \varPi_m \int_{-\sqrt[\alpha]{\frac{\pi}{\lambda}}}^{\sqrt[\alpha]{\frac{\pi}{\lambda}}} e^{i\lambda(m-n)t^\alpha} t^{\alpha-1} dt \\
&= \sum_{m=-\infty}^{\infty} 2\pi \varPi_m \delta_{mn} \\
&= 2\pi \varPi_n,
\end{aligned} \tag{7.582}$$

which implies that

$$\varPi_n = \frac{\alpha\lambda}{2\pi} \int_{-\sqrt[\alpha]{\frac{\pi}{\lambda}}}^{\sqrt[\alpha]{\frac{\pi}{\lambda}}} \Theta_\alpha(\lambda, t)\, e^{-in\lambda t^\alpha} t^{\alpha-1} dt. \tag{7.583}$$

Therefore, we finish the proof. □

Here, we have the generalized results.

Corollary 7.6 *Let $\alpha, \lambda > 0$ and $(-t)^\alpha = -t^\alpha$. Then, there exists*

$$\int_{-\sqrt[\alpha]{\frac{L}{2\lambda}}}^{\sqrt[\alpha]{\frac{L}{2\lambda}}} e^{\frac{2\pi i\lambda(m-n)t^\alpha}{L}} t^{\alpha-1} dt = \frac{L}{\alpha\lambda}\delta_{mn}. \tag{7.584}$$

Proof We now consider that

$$\int_{-\sqrt[\alpha]{\frac{L}{2\lambda}}}^{\sqrt[\alpha]{\frac{L}{2\lambda}}} e^{\frac{2\pi i\lambda(m-n)t^\alpha}{L}} t^{\alpha-1} dt = \int_{-\frac{L}{2}}^{\frac{L}{2}} e^{\frac{2\pi i(m-n)t}{L}} t^{\alpha-1} dt = \frac{L}{\alpha\lambda}\delta_{mn}. \tag{7.585}$$

□

Theorem 7.21 *Let* $\mathbb{M}_\alpha(\lambda, t) = \mathbb{M}(\lambda t^\alpha)$, $\mathbb{M}(\lambda t^\alpha) = \mathbb{M}(\lambda t^\alpha + L)$, $\alpha, \lambda > 0$, *and* $(-t)^\alpha = -t^\alpha$.

Then, we have

$$\mathbb{M}_\alpha(\lambda, t) = \sum_{m=-\infty}^{\infty} \phi_m e^{\frac{2\pi i m \lambda t^\alpha}{L}}, \tag{7.586}$$

where

$$\phi_m = \frac{\alpha\lambda}{L} \int_{-\sqrt[\alpha]{\frac{L}{2\lambda}}}^{\sqrt[\alpha]{\frac{L}{2\lambda}}} \mathbb{M}_\alpha(\lambda, t)\, e^{-\frac{2\pi m i \lambda t^\alpha}{L}} t^{\alpha-1} dt. \tag{7.587}$$

Proof Now, we consider that

$$\begin{aligned}
&\alpha\lambda \int_{-\sqrt[\alpha]{\frac{L}{2\lambda}}}^{\sqrt[\alpha]{\frac{L}{2\lambda}}} \mathbb{M}_\alpha(\lambda, t)\, e^{-\frac{2\pi n i \lambda t^\alpha}{L}} t^{\alpha-1} dt \\
&= \alpha\lambda \int_{-\sqrt[\alpha]{\frac{L}{2\lambda}}}^{\sqrt[\alpha]{\frac{L}{2\lambda}}} \left(\sum_{m=-\infty}^{\infty} \phi_m e^{\frac{2\pi m i \lambda t^\alpha}{L}} \right) e^{-\frac{2\pi n i \lambda t^\alpha}{L}} t^{\alpha-1} dt \\
&= \alpha\lambda \sum_{m=-\infty}^{\infty} \phi_m \int_{-\sqrt[\alpha]{\frac{L}{2\lambda}}}^{\sqrt[\alpha]{\frac{L}{2\lambda}}} e^{\frac{2\pi i (m-n)\lambda t^\alpha}{L}} t^{\alpha-1} dt \\
&= \sum_{m=-\infty}^{\infty} L m_m \delta_{mn} \\
&= L m_n,
\end{aligned} \tag{7.588}$$

which can be written as follows:

$$\phi_m = \frac{\alpha\lambda}{L} \int_{-\sqrt[\alpha]{\frac{L}{2\lambda}}}^{\sqrt[\alpha]{\frac{L}{2\lambda}}} \mathbb{M}_\alpha(\lambda, t)\, e^{-\frac{2\pi n i \lambda t^\alpha}{L}} t^{\alpha-1} dt. \tag{7.589}$$

Therefore, we finish the proof. □

Theorem 7.22 *Let* $\Xi_\alpha(\lambda, t) = \Xi(\lambda t^\alpha)$, $\Xi(\lambda, t^\alpha) = \Xi(\lambda t^\alpha + L)$, $\alpha, \lambda > 0$, *and* $(-t)^\alpha = -t^\alpha$.

Then, there exist the series with respect to scaling-law function

$$\Xi_\alpha(\lambda, t) = \frac{1}{2}\psi_0 + \sum_{\kappa=1}^{\infty} \varphi_\kappa subcos_\alpha\left(\frac{2\pi\lambda\kappa}{L}, t\right) + \sum_{\kappa=1}^{\infty} \psi_\kappa subsin_\alpha\left(\frac{2\pi\lambda\kappa}{L}, t\right), \tag{7.590}$$

where $\kappa \in \mathbb{N}$ *and the coefficients* ψ_0, φ_κ, *and* ψ_κ *are determined by*

$$\psi_0 = \frac{2\alpha\lambda}{L} \int_{-\sqrt[\alpha]{\frac{L}{2\lambda}}}^{\sqrt[\alpha]{\frac{L}{2\lambda}}} \Xi_\alpha(\lambda, t)\, t^{\alpha-1} dt, \tag{7.591}$$

$$\varphi_\kappa = \frac{2\alpha\lambda}{L} \int_{-\sqrt[\alpha]{\frac{L}{2\lambda}}}^{\sqrt[\alpha]{\frac{L}{2\lambda}}} \Xi_\alpha(\lambda, t)\, subcos_\alpha\left(\frac{2\pi\lambda\kappa}{L}, t\right) t^{\alpha-1} dt, \tag{7.592}$$

and

$$\psi_\kappa = \frac{2\alpha\lambda}{L} \int_{-\sqrt[\alpha]{\frac{L}{2\lambda}}}^{\sqrt[\alpha]{\frac{L}{2\lambda}}} \Xi_\alpha(\lambda, t)\, subsin_\alpha\left(\frac{2\pi\lambda\kappa}{L}, t\right) t^{\alpha-1} dt. \tag{7.593}$$

Proof We now consider that

$$\begin{aligned}
&\tfrac{2\alpha\lambda}{L} \int_{-\sqrt[\alpha]{\frac{L}{2\lambda}}}^{\sqrt[\alpha]{\frac{L}{2\lambda}}} \Xi_\alpha(\lambda, t)\, t^{\alpha-1} dt \\
&= \tfrac{\alpha\lambda}{L} \int_{-\sqrt[\alpha]{\frac{L}{2\lambda}}}^{\sqrt[\alpha]{\frac{L}{2\lambda}}} \psi_0 t^{\alpha-1} dt \\
&+ \tfrac{2\alpha\lambda}{L} \int_{-\sqrt[\alpha]{\frac{L}{2\lambda}}}^{\sqrt[\alpha]{\frac{L}{2\lambda}}} \left(\sum_{\kappa=1}^{\infty} \varphi_\kappa subcos_\alpha\left(\tfrac{2\pi\lambda\kappa}{L}, t\right)\right) t^{\alpha-1} dt \\
&+ \tfrac{2\alpha\lambda}{L} \int_{\sqrt[\alpha]{\frac{L}{2\lambda}}}^{\sqrt[\alpha]{\frac{L}{2\lambda}}} \left(\sum_{\kappa=1}^{\infty} \psi_\kappa subsin_\alpha\left(\tfrac{2\pi\lambda\kappa}{L}, t\right)\right) t^{\alpha-1} dt \\
&= \tfrac{\psi_0}{L} \int_{-\frac{L}{2}}^{\frac{L}{2}} d\upsilon \\
&= \psi_0,
\end{aligned} \tag{7.594}$$

$$\begin{aligned}
&\frac{2\alpha\lambda}{L}\int_{-\sqrt[\alpha]{\frac{L}{2\lambda}}}^{\sqrt[\alpha]{\frac{L}{2\lambda}}} \Xi_\alpha(\lambda,t)\, subcos_\alpha\left(\frac{2\pi\lambda\kappa}{L},t\right) t^{\alpha-1}dt \\
&=\frac{\alpha\lambda}{L}\int_{-\sqrt[\alpha]{\frac{L}{2\lambda}}}^{\sqrt[\alpha]{\frac{L}{2\lambda}}} \psi_0 subcos_\alpha\left(\frac{2\pi\lambda\kappa}{L},t\right) t^{\alpha-1}dt \\
&+\frac{2\alpha\lambda}{L}\int_{-\sqrt[\alpha]{\frac{L}{2\lambda}}}^{\sqrt[\alpha]{\frac{L}{2\lambda}}} \left(\sum_{\kappa=1}^{\infty}\varphi_\kappa subcos_\alpha\left(\frac{2\pi\lambda\kappa}{L},t\right)\right) subcos_\alpha\left(\frac{2\pi\lambda\kappa}{L},t\right) t^{\alpha-1}dt \\
&+\frac{2\alpha\lambda}{L}\int_{-\sqrt[\alpha]{\frac{L}{2\lambda}}}^{\sqrt[\alpha]{\frac{L}{2\lambda}}} \left(\sum_{\kappa=1}^{\infty}\psi_\kappa subsin_\alpha\left(\frac{2\pi\lambda\kappa}{L},t\right)\right) subcos_\alpha\left(\frac{2\pi\lambda\kappa}{L},t\right) t^{\alpha-1}dt \\
&=\frac{2\alpha\lambda}{L}\int_{-\sqrt[\alpha]{\frac{L}{2\lambda}}}^{\sqrt[\alpha]{\frac{L}{2\lambda}}} \varphi_\kappa subcos_\alpha\left(\frac{2\pi\lambda\kappa}{L},t\right) subcos_\alpha\left(\frac{2\pi\lambda\kappa}{L},t\right) t^{\alpha-1}dt \\
&=\varphi_\kappa,
\end{aligned} \tag{7.595}$$

$$\begin{aligned}
\psi_\kappa &= \frac{2\alpha\lambda}{L}\int_{-\sqrt[\alpha]{\frac{L}{2\lambda}}}^{\sqrt[\alpha]{\frac{L}{2\lambda}}} \Xi_\alpha(\lambda,t)\, subsin_\alpha\left(\frac{2\pi\lambda\kappa}{L},t\right) t^{\alpha-1}dt \\
&=\frac{\alpha\lambda}{L}\int_{-\sqrt[\alpha]{\frac{L}{2\lambda}}}^{\sqrt[\alpha]{\frac{L}{2\lambda}}} \psi_0 subsin_\alpha\left(\frac{2\pi\lambda\kappa}{L},t\right) t^{\alpha-1}dt \\
&+\frac{2\alpha\lambda}{L}\int_{-\sqrt[\alpha]{\frac{L}{2\lambda}}}^{\sqrt[\alpha]{\frac{L}{2\lambda}}} \left(\sum_{\kappa=1}^{\infty}\varphi_\kappa subcos_\alpha\left(\frac{2\pi\lambda\kappa}{L},t\right)\right) subsin_\alpha\left(\frac{2\pi\lambda\kappa}{L},t\right) t^{\alpha-1}dt \\
&+\frac{2\alpha\lambda}{L}\int_{-\sqrt[\alpha]{\frac{L}{2\lambda}}}^{\sqrt[\alpha]{\frac{L}{2\lambda}}} \left(\sum_{\kappa=1}^{\infty}\psi_\kappa subsin_\alpha\left(\frac{2\pi\lambda\kappa}{L},t\right)\right) subsin_\alpha\left(\frac{2\pi\lambda\kappa}{L},t\right) t^{\alpha-1}dt \\
&=\frac{2\alpha\lambda}{L}\int_{-\sqrt[\alpha]{\frac{L}{2\lambda}}}^{\sqrt[\alpha]{\frac{L}{2\lambda}}} \psi_\kappa subsin_\alpha\left(\frac{2\pi\lambda\kappa}{L},t\right) subsin_\alpha\left(\frac{2\pi\lambda\kappa}{L},t\right) t^{\alpha-1}dt \\
&=\psi_\kappa,
\end{aligned} \tag{7.596}$$

where

$$\int_{-\sqrt[\alpha]{\frac{L}{2\lambda}}}^{\sqrt[\alpha]{\frac{L}{2\lambda}}} subsin_{\alpha}\left(\frac{2\pi\lambda\kappa_2}{L},t\right)t^{\alpha-1}dt=0, \tag{7.597}$$

$$\int_{-\sqrt[\alpha]{\frac{L}{2\lambda}}}^{\sqrt[\alpha]{\frac{L}{2\lambda}}} subcos_{\alpha}\left(\frac{2\pi\lambda\kappa_1}{L},t\right)t^{\alpha-1}dt=0, \tag{7.598}$$

$$\int_{-\sqrt[\alpha]{\frac{L}{2\lambda}}}^{\sqrt[\alpha]{\frac{L}{2\lambda}}} subcos_{\alpha}\left(\frac{2\pi\lambda\kappa_1}{L},t\right) subcos_{\alpha}\left(\frac{2\pi\lambda\kappa_2}{L},t\right)t^{\alpha-1}dt=\frac{\delta_{\kappa_1\kappa_2}}{\alpha\lambda}, \tag{7.599}$$

$$\int_{-\sqrt[\alpha]{\frac{L}{2\lambda}}}^{\sqrt[\alpha]{\frac{L}{2\lambda}}} subsin_{\alpha}\left(\frac{2\pi\lambda\kappa_1}{L},t\right) subcos_{\alpha}\left(\frac{2\pi\lambda\kappa_2}{L},t\right)t^{\alpha-1}dt=0, \tag{7.600}$$

$$\int_{-\sqrt[\alpha]{\frac{L}{2\lambda}}}^{\sqrt[\alpha]{\frac{L}{2\lambda}}} subsin_{\alpha}\left(\frac{2\pi\lambda\kappa_1}{L},t\right) subsin_{\alpha}\left(\frac{2\pi\lambda\kappa_2}{L},t\right)t^{\alpha-1}dt=\frac{\delta_{\kappa_1\kappa_2}}{\alpha\lambda}, \tag{7.601}$$

and

$$\int_{-\sqrt[\alpha]{\frac{L}{2\lambda}}}^{\sqrt[\alpha]{\frac{L}{2\lambda}}} subsin_{\alpha}\left(\frac{2\pi\lambda\kappa_1}{L},t\right) subcos_{\alpha}\left(\frac{2\pi\lambda\kappa_2}{L},t\right)t^{\alpha-1}dt=0. \tag{7.602}$$

Thus, we finish the proof. □

7.3.3 Theory of Subtrigonometric Series with Respect to Complex and Power-Law Functions

In this part, we give the subtrigonometric series with respect to power-law function.

Corollary 7.7 *Let $m, n \in \mathbb{N}$, $\alpha > 0$, and $(-t)^\alpha = -t^\alpha$. Then, there exist*

$$\int_{-\sqrt[\alpha]{\pi}}^{\sqrt[\alpha]{\pi}} subsin_\alpha (m, t) \cdot subsin_\alpha (n, t)\, t^{\alpha-1} dt = \frac{\pi}{\alpha} \delta_{mn}, \tag{7.603}$$

$$\int_{-\sqrt[\alpha]{\pi}}^{\sqrt[\alpha]{\pi}} subcos_\alpha (m, t) \cdot subcos_\alpha (n, t)\, t^{\alpha-1} dt = \frac{\pi}{\alpha} \delta_{mn}, \tag{7.604}$$

$$\int_{-\sqrt[\alpha]{\pi}}^{\sqrt[\alpha]{\pi}} subsin_\alpha (m, t) \cdot subcos_\alpha (n, t)\, t^{\alpha-1} dt = 0, \tag{7.605}$$

$$\int_{-\sqrt[\alpha]{\pi}}^{\sqrt[\alpha]{\pi}} subsin_\alpha (m, t)\, t^{\alpha-1} dt = 0, \tag{7.606}$$

and

$$\int_{-\sqrt[\alpha]{\pi}}^{\sqrt[\alpha]{\pi}} subcos_\alpha (m, t)\, t^{\alpha-1} dt = 0, \tag{7.607}$$

where $m \in \mathbb{N}$ and $n \in \mathbb{N}$

Proof On setting $\lambda = 1$ in (7.568), (7.569), (7.570), (7.571), and (7.572), we obtain the result. □

Corollary 7.8 *Let $\alpha > 0$ $\Xi_\alpha (t) = \Xi (t^\alpha)$, $\Xi (t^\alpha) = \Xi (t^\alpha + 2\pi)$, $n \in \mathbb{N}$, and $(-t)^\alpha = -t^\alpha$.*

Then, we have the series with respect to power-law function

$$\Xi_\alpha (t) = \frac{1}{2}\psi_0 + \sum_{n=1}^{\infty} \varphi_n subcos_\alpha (n, t) + \sum_{n=1}^{\infty} \psi_n subsin_\alpha (n, t), \tag{7.608}$$

where the coefficients ψ_0, φ_n, *and* ψ_n *are*

$$\psi_0 = \frac{\alpha}{\pi} \int_{-\sqrt[\alpha]{\pi}}^{\sqrt[\alpha]{\pi}} \Xi_\alpha(t)\, t^{\alpha-1} dt, \tag{7.609}$$

$$\varphi_n = \frac{\alpha}{\pi} \int_{-\sqrt[\alpha]{\pi}}^{\sqrt[\alpha]{\pi}} \Xi_\alpha(t)\, subcos_\alpha(n,t)\, t^{\alpha-1} dt, \tag{7.610}$$

and

$$\psi_n = \frac{\alpha}{\pi} \int_{-\sqrt[\alpha]{\pi}}^{\sqrt[\alpha]{\pi}} \Xi_\alpha(t)\, subsin_\alpha(n,t)\, t^{\alpha-1} dt. \tag{7.611}$$

Proof From (7.577), (7.578), and (7.579), we have

$$\begin{aligned}
&\tfrac{\alpha}{\pi} \int_{-\sqrt[\alpha]{\pi}}^{\sqrt[\alpha]{\pi}} \Xi_\alpha(t)\, t^{\alpha-1} dt \\
&= \tfrac{\alpha}{\pi} \int_{-\sqrt[\alpha]{\pi}}^{\sqrt[\alpha]{\pi}} \left(\tfrac{1}{2}\psi_0 + \sum_{n=1}^{\infty} \varphi_n subcos_\alpha(n,t) + \sum_{n=1}^{\infty} \psi_n subsin_\alpha(n,t) \right) t^{\alpha-1} dt \\
&= \tfrac{\psi_0}{2\pi} \int_{-\pi}^{\pi} d\upsilon \\
&= \psi_0,
\end{aligned} \tag{7.612}$$

$$\begin{aligned}
&\tfrac{\alpha}{\pi} \int_{-\sqrt[\alpha]{\pi}}^{\sqrt[\alpha]{\pi}} \Xi_\alpha(t)\, subcos_\alpha(n,t)\, t^{\alpha-1} dt \\
&= \tfrac{\alpha}{\pi} \int_{-\sqrt[\alpha]{\pi}}^{\sqrt[\alpha]{\pi}} \left(\tfrac{1}{2}\psi_0 + \sum_{n=1}^{\infty} \varphi_n subcos_\alpha(n,t) + \sum_{\kappa=1}^{\infty} \psi_n subsin_\alpha(n,t) \right) subcos_\alpha(n,t)\, t^{\alpha-1} dt \\
&= \tfrac{\alpha}{\pi} \int_{-\sqrt[\alpha]{\pi}}^{\sqrt[\alpha]{\pi}} \varphi_n subcos_\alpha(n,t)\, subcos_\alpha(n,t)\, t^{\alpha-1} dt \\
&= \tfrac{\varphi_n}{\pi} \int_{-\pi}^{\pi} cos(n\upsilon)\, cos(n\upsilon)\, d\upsilon \\
&= \varphi_n,
\end{aligned} \tag{7.613}$$

and

$$\begin{aligned}
&\psi_n \\
&= \tfrac{1}{\pi} \int_a^b \Xi_\alpha (t)\, subsin_\alpha (n, t)\, t^{\alpha-1} dt \\
&= \tfrac{\alpha}{\pi} \int_a^b \left(\tfrac{1}{2}\psi_0 + \sum_{\kappa=1}^{\infty} \varphi_n subcos_\alpha (n, t) + \sum_{\kappa=1}^{\infty} \psi_n subsin_\alpha (n, t) \right) subsin_\alpha (n, t)\, t^{\alpha-1} dt \\
&= \tfrac{\alpha}{\pi} \int_a^b \psi_n subsin_\alpha (n, t)\, subsin_\alpha (n, t)\, t^{\alpha-1} dt \\
&= \tfrac{\psi_n}{\pi} \int_{-\pi}^{\pi} sin\,(nv)\, sin\,(nv)\, dv \\
&= \psi_n .
\end{aligned} \tag{7.614}$$

So, we finish the proof. □

Corollary 7.9 *Let* $\alpha > 0$ *and* $(-t)^\alpha = -t^\alpha$.
Then, there exists

$$\int_{-\sqrt[\alpha]{\pi}}^{\sqrt[\alpha]{\pi}} e^{i(m-n)t^\alpha} t^{\alpha-1} dt = \frac{2\pi}{\alpha} \delta_{mn}. \tag{7.615}$$

Proof Taking $\lambda = 1$ in (7.580), we have the result. □

Corollary 7.10 *Let* $\Xi_\alpha (t) = \Xi (t^\alpha)$, $\Xi (t^\alpha) = \Xi (t^\alpha + 2\pi)$, $\alpha > 0$, *and* $(-t)^\alpha = -t^\alpha$.
Then, we have

$$\Theta_\alpha (t) = \sum_{n=-\infty}^{\infty} \Pi_n e^{int^\alpha}, \tag{7.616}$$

where

$$\Pi_n = \frac{\alpha}{2\pi} \int_{-\sqrt[\alpha]{\pi}}^{\sqrt[\alpha]{\pi}} \Theta_\alpha (t)\, e^{-int^\alpha} t^{\alpha-1} dt. \tag{7.617}$$

Proof We now have that

$$\begin{aligned}
&\alpha \int_{-\sqrt[\alpha]{\pi}}^{\sqrt[\alpha]{\pi}} \Theta_\alpha(t)\, e^{-int^\alpha} t^{\alpha-1} dt \\
&= \alpha \int_{-\sqrt[\alpha]{\pi}}^{\sqrt[\alpha]{\pi}} \left(\sum_{m=-\infty}^{\infty} \Pi_m e^{imt^\alpha} \right) e^{-int^\alpha} t^{\alpha-1} dt \\
&= \alpha \sum_{m=-\infty}^{\infty} \Pi_m \int_{-\sqrt[\alpha]{\pi}}^{\sqrt[\alpha]{\pi}} e^{i(m-n)t^\alpha} t^{\alpha-1} dt \\
&= \sum_{m=-\infty}^{\infty} 2\pi \Pi_m \delta_{mn} \\
&= 2\pi \Pi_n,
\end{aligned} \tag{7.618}$$

which implies that

$$\Pi_n = \frac{\alpha}{2\pi} \int_{\sqrt[\alpha]{\pi}}^{-\sqrt[\alpha]{\pi}} \Theta_\mu(t)\, e^{-int^\alpha} t^{\alpha-1} dt. \tag{7.619}$$

So, we finish the proof. □

Corollary 7.11 *Let $\alpha > 0$ and $(-t)^\alpha = -t^\alpha$. Then, there exists*

$$\int_{-\sqrt[\alpha]{\frac{L}{2}}}^{\sqrt[\alpha]{\frac{L}{2}}} e^{\frac{2\pi i(m-n)t^\alpha}{L}} t^{\alpha-1} dt = \frac{L}{\alpha} \delta_{mn}. \tag{7.620}$$

Proof We now consider that

$$\alpha \int_{-\sqrt[\alpha]{\frac{L}{2}}}^{\sqrt[\alpha]{\frac{L}{2}}} e^{\frac{2\pi i(m-n)t^\alpha}{L}} t^{\alpha-1} dt = \int_{-\frac{L}{2}}^{\frac{L}{2}} e^{\frac{2\pi i(m-n)t}{L}} dt = L\delta_{mn}. \tag{7.621}$$

□

Corollary 7.12 *Let $\mathbb{M}_\alpha(t) = \mathbb{M}(t^\alpha)$, $\mathbb{M}(t^\alpha) = \mathbb{M}(t^\alpha + L)$, $\alpha > 0$, and $(-t)^\alpha = -t^\alpha$.*

Then, we have

$$\mathbb{M}_\alpha(t) = \sum_{m=-\infty}^{\infty} \phi_m e^{\frac{2\pi i m t^\alpha}{L}}, \tag{7.622}$$

where

$$\phi_m = \frac{\alpha}{L} \int_{-\sqrt[\alpha]{\frac{L}{2}}}^{\sqrt[\alpha]{\frac{L}{2}}} \mathbb{M}_\alpha(t)\, e^{-\frac{2\pi m i t^\alpha}{L}} t^{\alpha-1} dt. \tag{7.623}$$

Proof We now have that

$$\begin{aligned}
&\alpha \int_{-\sqrt[\alpha]{\frac{L}{2}}}^{\sqrt[\alpha]{\frac{L}{2}}} \mathbb{M}_\alpha(t)\, e^{-\frac{2\pi n i t^\alpha}{L}} t^{\alpha-1} dt \\
&= \alpha \int_{-\sqrt[\alpha]{\frac{L}{2}}}^{\sqrt[\alpha]{\frac{L}{2}}} \left(\sum_{m=-\infty}^{\infty} \phi_m e^{\frac{2\pi m i t^\alpha}{L}} \right) e^{-\frac{2\pi n i t^\alpha}{L}} t^{\alpha-1} dt \\
&= \alpha \sum_{m=-\infty}^{\infty} \phi_m \int_{-\sqrt[\alpha]{\frac{L}{2}}}^{\sqrt[\alpha]{\frac{L}{2}}} e^{\frac{2\pi i (m-n) t^\alpha}{L}} t^{\alpha-1} dt \\
&= \sum_{m=-\infty}^{\infty} L m_m \delta_{mn} \\
&= L m_n,
\end{aligned} \tag{7.624}$$

which can be given as follows:

$$\phi_m = \frac{\alpha}{L} \int_{-\sqrt[\alpha]{\frac{L}{2}}}^{\sqrt[\alpha]{\frac{L}{2}}} \mathbb{M}_\alpha(t)\, e^{-\frac{2\pi m i t^\alpha}{L}} t^{\alpha-1} dt. \tag{7.625}$$

Therefore, we complete the proof. □

Corollary 7.13 *Let* $\Xi_\alpha(t) = \Xi(t^\alpha)$, $\Xi(t^\alpha) = \Xi(t^\alpha + L)$, $\alpha > 0$ $(-t)^\alpha = -t^\alpha$, *and* $\kappa \in \mathbb{N}$.

Then, we have the series with respect to power-law function

$$\Xi_\alpha(t) = \frac{1}{2}\psi_0 + \sum_{\kappa=1}^{\infty} \varphi_\kappa subcos_\alpha\left(\frac{2\pi\kappa}{L}, t\right) + \sum_{\kappa=1}^{\infty} \psi_\kappa subsin_\alpha\left(\frac{2\pi\kappa}{L}, t\right), \tag{7.626}$$

where the coefficients ψ_0, φ_κ, *and* ψ_κ *are presented as*

$$\psi_0 = \frac{2\alpha}{L} \int_{-\sqrt[\alpha]{\frac{L}{2}}}^{\sqrt[\alpha]{\frac{L}{2}}} \Xi_\alpha (t)\, t^{\alpha-1} dt, \tag{7.627}$$

$$\varphi_\kappa = \frac{2\alpha}{L} \int_{-\sqrt[\alpha]{\frac{L}{2}}}^{\sqrt[\alpha]{\frac{L}{2}}} \Xi_\alpha (t)\, subcos_\alpha \left(\frac{2\pi\kappa}{L}, t \right) t^{\alpha-1} dt, \tag{7.628}$$

and

$$\psi_\kappa = \frac{2\alpha}{L} \int_{-\sqrt[\alpha]{\frac{L}{2}}}^{\sqrt[\alpha]{\frac{L}{2}}} \Xi_\alpha (t)\, subsin_\alpha \left(\frac{2\pi\kappa}{L}, t \right) t^{\alpha-1} dt. \tag{7.629}$$

Proof We now have that

$$\begin{aligned}
&\tfrac{2}{L} \int_{-\sqrt[\alpha]{\frac{L}{2}}}^{\sqrt[\alpha]{\frac{L}{2}}} \Xi_\alpha (t)\, t^{\alpha-1} dt \\
&= \tfrac{\alpha}{L} \int_{-\sqrt[\alpha]{\frac{L}{2}}}^{\sqrt[\alpha]{\frac{L}{2}}} \psi_0 t^{\alpha-1} dt \\
&+ \tfrac{2\alpha}{L} \int_{-\sqrt[\alpha]{\frac{L}{2}}}^{\sqrt[\alpha]{\frac{L}{2}}} \left(\sum_{\kappa=1}^{\infty} \varphi_\kappa subcos_\alpha \left(\tfrac{2\pi\kappa}{L}, t \right) \right) t^{\alpha-1} dt \\
&+ \tfrac{2\alpha}{L} \int_{\sqrt[\alpha]{\frac{L}{2}}}^{\sqrt[\alpha]{\frac{L}{2}}} \left(\sum_{\kappa=1}^{\infty} \psi_\kappa subsin_\alpha \left(\tfrac{2\pi\kappa}{L}, t \right) \right) t^{\alpha-1} dt \\
&= \tfrac{\psi_0}{L} \int_{-\frac{L}{2}}^{\frac{L}{2}} d\upsilon \\
&= \psi_0,
\end{aligned} \tag{7.630}$$

$$
\begin{aligned}
&\frac{2\alpha}{L}\int_{-\sqrt[\alpha]{\frac{L}{2}}}^{\sqrt[\alpha]{\frac{L}{2}}} \Xi_\alpha(t)\, subcos_\alpha\left(\frac{2\pi\kappa}{L}, t\right) t^{\alpha-1}dt \\
&=\frac{\alpha}{L}\int_{-\sqrt[\alpha]{\frac{L}{2}}}^{\sqrt[\alpha]{\frac{L}{2}}} \psi_0 subcos_\alpha\left(\frac{2\pi\kappa}{L}, t\right) t^{\alpha-1}dt \\
&+\frac{2\alpha}{L}\int_{-\sqrt[\alpha]{\frac{L}{2}}}^{\sqrt[\alpha]{\frac{L}{2}}} \left(\sum_{\kappa=1}^{\infty}\varphi_\kappa subcos_\alpha\left(\frac{2\pi\kappa}{L}, t\right)\right) subcos_\alpha\left(\frac{2\pi\kappa}{L}, t\right) t^{\alpha-1}dt \\
&+\frac{2\alpha}{L}\int_{-\sqrt[\alpha]{\frac{L}{2}}}^{\sqrt[\alpha]{\frac{L}{2}}} \left(\sum_{\kappa=1}^{\infty}\psi_\kappa subsin_\alpha\left(\frac{2\pi\kappa}{L}, t\right)\right) subcos_\alpha\left(\frac{2\pi\kappa}{L}, t\right) t^{\alpha-1}dt \\
&=\frac{2\alpha}{L}\int_{-\sqrt[\alpha]{\frac{L}{2}}}^{\sqrt[\alpha]{\frac{L}{2}}} \varphi_\kappa subcos_\alpha\left(\frac{2\pi\kappa}{L}, t\right) subcos_\alpha\left(\frac{2\pi\kappa}{L}, t\right) t^{\alpha-1}dt \\
&=\varphi_\kappa,
\end{aligned}
\tag{7.631}
$$

$$
\begin{aligned}
\psi_\kappa &= \frac{2\alpha}{L}\int_{-\sqrt[\alpha]{\frac{L}{2}}}^{\sqrt[\alpha]{\frac{L}{2}}} \Xi_\alpha(t)\, subsin_\alpha\left(\frac{2\pi\kappa}{L}, t\right) t^{\alpha-1}dt \\
&=\frac{\alpha}{L}\int_{-\sqrt[\alpha]{\frac{L}{2}}}^{\sqrt[\alpha]{\frac{L}{2}}} \psi_0 subsin_\alpha\left(\frac{2\pi\kappa}{L}, t\right) t^{\alpha-1}dt \\
&+\frac{2\alpha}{L}\int_{-\sqrt[\alpha]{\frac{L}{2}}}^{\sqrt[\alpha]{\frac{L}{2}}} \left(\sum_{\kappa=1}^{\infty}\varphi_\kappa subcos_\alpha\left(\frac{2\pi\kappa}{L}, t\right)\right) subsin_\alpha\left(\frac{2\pi\kappa}{L}, t\right) t^{\alpha-1}dt \\
&+\frac{2\alpha}{L}\int_{-\sqrt[\alpha]{\frac{L}{2}}}^{\sqrt[\alpha]{\frac{L}{2}}} \left(\sum_{\kappa=1}^{\infty}\psi_\kappa subsin_\alpha\left(\frac{2\pi\kappa}{L}, t\right)\right) subsin_\alpha\left(\frac{2\pi\kappa}{L}, t\right) t^{\alpha-1}dt \\
&=\frac{2\alpha}{L}\int_{-\sqrt[\alpha]{\frac{L}{2}}}^{\sqrt[\alpha]{\frac{L}{2}}} \psi_\kappa subsin_\alpha\left(\frac{2\pi\kappa}{L}, t\right) subsin_\alpha\left(\frac{2\pi\kappa}{L}, t\right) t^{\alpha-1}dt \\
&=\psi_\kappa,
\end{aligned}
\tag{7.632}
$$

where

$$\int_{-\sqrt[\alpha]{\frac{L}{2}}}^{\sqrt[\alpha]{\frac{L}{2}}} subsin_{\alpha}\left(\frac{2\pi\kappa_2}{L}, t\right) t^{\alpha-1}dt = 0, \tag{7.633}$$

$$\int_{-\sqrt[\alpha]{\frac{L}{2}}}^{\sqrt[\alpha]{\frac{L}{2}}} subcos_{\alpha}\left(\frac{2\pi\kappa_1}{L}, t\right) t^{\alpha-1}dt = 0, \tag{7.634}$$

$$\int_{-\sqrt[\alpha]{\frac{L}{2}}}^{\sqrt[\alpha]{\frac{L}{2}}} subcos_{\alpha}\left(\frac{2\pi\kappa_1}{L}, t\right) subcos_{\alpha}\left(\frac{2\pi\kappa_2}{L}, t\right) t^{\alpha-1}dt = \frac{\delta_{\kappa_1\kappa_2}}{\alpha}, \tag{7.635}$$

$$\int_{-\sqrt[\alpha]{\frac{L}{2}}}^{\sqrt[\alpha]{\frac{L}{2}}} subsin_{\alpha}\left(\frac{2\pi\kappa_1}{L}, t\right) subcos_{\alpha}\left(\frac{2\pi\kappa_2}{L}, t\right) t^{\alpha-1}dt = 0, \tag{7.636}$$

$$\int_{-\sqrt[\alpha]{\frac{L}{2}}}^{\sqrt[\alpha]{\frac{L}{2}}} subsin_{\alpha}\left(\frac{2\pi\kappa_1}{L}, t\right) subsin_{\alpha}\left(\frac{2\pi\kappa_2}{L}, t\right) t^{\alpha-1}dt = \frac{\delta_{\kappa_1\kappa_2}}{\alpha}, \tag{7.637}$$

and

$$\int_{-\sqrt[\alpha]{\frac{L}{2}}}^{\sqrt[\alpha]{\frac{L}{2}}} subsin_{\alpha}\left(\frac{2\pi\kappa_1}{L}, t\right) subcos_{\alpha}\left(\frac{2\pi\kappa_2}{L}, t\right) t^{\alpha-1}dt = 0. \tag{7.638}$$

Thus, we complete the proof. □

7.3.4 Applications

Example 7.3 Let $\Xi_{\frac{1}{3}}(t) = \Xi\left(t^{\frac{1}{3}}\right) = \frac{t}{2}$ and $\Xi\left(t^{\frac{1}{3}}\right) = \Xi\left(t^{\frac{1}{3}} + 2\pi\right)$, where $-\sqrt[\frac{1}{3}]{\pi} \le t \le \sqrt[\frac{1}{3}]{\pi}$.

We can structure the subtrigonometric series

$$\Xi_{\frac{1}{3}}(t) = \frac{1}{2}\psi_0 + \sum_{n=1}^{\infty} \varphi_n subcos_{\frac{1}{3}}(n,t) + \sum_{n=1}^{\infty} \psi_n subsin_{\frac{1}{3}}(n,t), \tag{7.639}$$

where ψ_0, φ_n, and ψ_n are the coefficients for $n \in \mathbb{N}$.

Thus, we have

$$\psi_0 = \frac{1}{3\pi}\int_{-\sqrt[\frac{1}{3}]{\pi}}^{\sqrt[\frac{1}{3}]{\pi}} \Xi_{\frac{1}{3}}(t)\, t^{-\frac{2}{3}} dt = \frac{1}{3\pi}\int_{-\sqrt[\frac{1}{3}]{\pi}}^{\sqrt[\frac{1}{3}]{\pi}} \frac{t}{2} t^{-\frac{2}{3}} dt = \frac{1}{\pi}\int_{-\pi}^{\pi} \frac{3}{2} t dt = 0, \tag{7.640}$$

$$\begin{aligned} \varphi_n &= \tfrac{1}{3\pi} \int_{-\sqrt[\frac{1}{3}]{\pi}}^{\sqrt[\frac{1}{3}]{\pi}} \Xi_{\frac{1}{3}}(t)\, subcos_{\frac{1}{3}}(n,t)\, t^{-\frac{2}{3}} dt \\ &= \tfrac{1}{3\pi} \int_{-\sqrt[\frac{1}{3}]{\pi}}^{\sqrt[\frac{1}{3}]{\pi}} \tfrac{t}{2} subcos_{\frac{1}{3}}(n,t)\, t^{-\frac{2}{3}} dt = \tfrac{1}{\pi}\int_{-\pi}^{\pi} \tfrac{3}{2} t cos(nt)\, dt = 0, \end{aligned} \tag{7.641}$$

and

$$\begin{aligned} \psi_n &= \tfrac{1}{3\pi} \int_{-\sqrt[\frac{1}{3}]{\pi}}^{\sqrt[\frac{1}{3}]{\pi}} \Xi_{\frac{1}{3}}(t)\, subsin_{\frac{1}{3}}(n,t)\, t^{-\frac{2}{3}} dt \\ &= \tfrac{1}{3\pi} \int_{-\sqrt[\frac{1}{3}]{\pi}}^{\sqrt[\frac{1}{3}]{\pi}} \tfrac{t}{2} subsin_{\frac{1}{3}}(n,t)\, t^{-\frac{2}{3}} dt \\ &= \tfrac{3}{2\pi}\int_{-\pi}^{\pi} t sin(nt)\, dt = \tfrac{3}{2}(-1)^{n-1}\tfrac{2}{n} = (-1)^{n-1}\tfrac{3}{n}. \end{aligned} \tag{7.642}$$

So, we have

$$\frac{t}{2} = \sum_{n=1}^{\infty} (-1)^{n-1} \frac{3}{n} subsin_{\frac{1}{3}}(n,t). \tag{7.643}$$

Example 7.4 Let $\Xi_3(t) = \Xi\left(t^3\right) = \frac{t}{2}$ and $\Xi\left(t^3\right) = \Xi\left(t^3 + 2\pi\right)$, where $0 < t < \pi^3$.

Now, we present

$$\psi_0 = \frac{3}{\pi}\int_0^{\pi^3} \Xi_3(t)\, t^2 dt = \frac{3}{\pi}\int_0^{\pi^3} \frac{1}{2} t^2 dt = \frac{1}{\pi}\int_0^{\pi} \frac{1}{2} dt = \frac{1}{2}, \tag{7.644}$$

$$\varphi_n = \frac{3}{\pi}\int_0^{\pi^3} \Xi_3(t)\, subcos_3(n,t)\, t^2 dt = \frac{1}{\pi}\int_0^{\pi} cos(nt)\, dt = 0, \tag{7.645}$$

and

$$\begin{aligned}\psi_n &= \tfrac{3}{\pi}\int_0^{\pi^3} \Xi_3(t)\, subsin_3(n,t)\, t^2 dt \\ &= \tfrac{3}{\pi}\int_0^{\pi^3} subsin_3(n,t)\, t^2 dt = \tfrac{1}{\pi}\int_0^{\pi} sin(nt)\, dt = \tfrac{1+(-1)^{n+1}}{n\pi}.\end{aligned} \tag{7.646}$$

Thus, we present

$$\Xi_3(t) = \sum_{n=1}^{\infty} \frac{1+(-1)^{n+1}}{n\pi} subsin_3(n,t). \tag{7.647}$$

Example 7.5 Let $\Xi_{\frac{1}{3}}(t) = \Xi\left(t^{\frac{1}{3}}\right) = t$ and $\Xi\left(t^{\frac{1}{3}}\right) = \Xi\left(t^{\frac{1}{3}}+1\right)$, where $-\sqrt[\frac{1}{3}]{\frac{1}{2}} \le t \le \sqrt[\frac{1}{3}]{\frac{1}{2}}$.

We may present the subtrigonometric series

$$\Xi_{\frac{1}{3}}(t) = \frac{1}{2}\psi_0 + \sum_{\kappa=1}^{\infty} \varphi_\kappa subcos_{\frac{1}{3}}(2\pi\kappa, t) + \sum_{\kappa=1}^{\infty} \psi_\kappa subsin_{\frac{1}{3}}(2\pi\kappa, t), \tag{7.648}$$

where ψ_0, φ_κ, and ψ_κ are the coefficients for $\kappa \in \mathbb{N}$.

Thus, we have

$$\psi_0 = \frac{2}{3}\int_{-\sqrt[\frac{1}{3}]{\frac{1}{2}}}^{\sqrt[\frac{1}{3}]{\frac{1}{2}}} \Xi_{\frac{1}{3}}(t)\, t^{-\frac{2}{3}} dt = 2\int_{-\frac{1}{2}}^{\frac{1}{2}} t^3 dt = 0, \tag{7.649}$$

$$\varphi_\kappa = \frac{2}{3}\int_{-\sqrt[\frac{1}{3}]{\frac{1}{2}}}^{\sqrt[\frac{1}{3}]{\frac{1}{2}}} \Xi_{\frac{1}{3}}(t)\, subcos_{\frac{1}{3}}(2\pi\kappa, t)\, t^{-\frac{2}{3}} dt = 2\int_{-\frac{1}{2}}^{\frac{1}{2}} t^3 cos(2\pi\kappa t)\, dt = 0, \tag{7.650}$$

and

$$\psi_\kappa = \tfrac{2}{3} \int_{-\sqrt[\frac{1}{3}]{\frac{1}{2}}}^{\sqrt[\frac{1}{3}]{\frac{1}{2}}} \Xi_{\frac{1}{3}}(t)\, subsin_{\frac{1}{3}}(2\pi\kappa, t)\, t^{-\frac{2}{3}} dt = 2 \int_{-\frac{1}{2}}^{\frac{1}{2}} t^3 sin\,(2\pi\kappa t)\, dt$$
$$= 4 \int_0^{\frac{1}{2}} t^3 sin\,(2\pi\kappa t)\, dt = (-1)^\kappa \left[\tfrac{3}{2(\pi\kappa)^3} - \tfrac{1}{4\pi\kappa}\right]. \quad (7.651)$$

So, we have

$$t = \sum_{\kappa=1}^{\infty} (-1)^\kappa \left[\frac{3}{2\,(\pi\kappa)^3} - \frac{1}{4\pi\kappa}\right] subsin_{\frac{1}{3}}(2\pi\kappa, t). \quad (7.652)$$

Example 7.6 Let $\Xi_3(t) = \Xi\left(t^3\right) = t^3$ and $\Xi\left(t^3\right) = \Xi\left(t^3 + 2\right)$, where $0 < t < 1$.
We now show that

$$\Xi_3(t) = \frac{1}{2}\psi_0 + \sum_{\kappa=1}^{\infty} \varphi_\kappa subcos_3(\pi\kappa, t) + \sum_{\kappa=1}^{\infty} \psi_\kappa subsin_3(\pi\kappa, t), \quad (7.653)$$

where ψ_0, φ_κ, and ψ_κ are the coefficients.
Here, we give

$$\psi_0 = 3 \int_0^1 \Xi_3(t)\, t^2 dt = \int_0^1 t dt = \frac{1}{2}, \quad (7.654)$$

$$\varphi_\kappa = 3 \int_0^1 \Xi_3(t)\, subcos_3(\pi\kappa, t)\, t^2 dt = \int_0^1 tcos\,(\pi\kappa t)\, dt = \frac{(-1)^\kappa - 1}{(\pi\kappa)^2}, \quad (7.655)$$

and

$$\psi_\kappa = 3 \int_0^1 \Xi_3(t)\, subsin_\alpha(\pi\kappa, t)\, t^2 dt = \int_0^1 tsin\,(\pi\kappa t)\, dt = \frac{(-1)^\kappa}{\pi\kappa}. \quad (7.656)$$

Thus, we obtain

$$t^3 = \frac{1}{4} + \sum_{\kappa=1}^{\infty} \frac{(-1)^\kappa - 1}{(\pi\kappa)^2} subcos_3(\pi\kappa, t) + \sum_{\kappa=1}^{\infty} \frac{(-1)^\kappa}{\pi\kappa} subsin_3(\pi\kappa, t). \quad (7.657)$$

Example 7.7 Let $\Xi_{\frac{1}{3}}(27,t) = \Xi\left(27t^{\frac{1}{3}}\right) = 27t^{\frac{1}{3}}$ and $\Xi\left(27t^{\frac{1}{3}}\right) = \Xi\left(27t^{\frac{1}{3}}+1\right)$, where $0 \le t \le \frac{1}{3}$.

We may give the subtrigonometric series

$$\Xi_{\frac{1}{3}}(27,t) = \frac{1}{2}\psi_0 + \sum_{\kappa=1}^{\infty}\varphi_\kappa subcos_{\frac{1}{3}}(54\pi\kappa,t) + \sum_{\kappa=1}^{\infty}\psi_\kappa subsin_{\frac{1}{3}}(54\pi\kappa,t), \tag{7.658}$$

where ψ_0, φ_κ, and ψ_κ are the coefficients for $\kappa \in \mathbb{N}$.

Thus, we may write

$$\psi_0 = 18\int_0^{\frac{1}{3}}\Xi_{\frac{1}{3}}(27,t)\,t^{-\frac{2}{3}}dt = 2\int_0^1 tdt = 2, \tag{7.659}$$

$$\begin{aligned}\varphi_\kappa &= 18\int_0^{\frac{1}{3}}\Xi_{\frac{1}{3}}(27,t)\,subcos_{\frac{1}{3}}(54\pi\kappa,t)\,t^{-\frac{2}{3}}dt = 2\int_0^1 tcos(2\pi\kappa t)\,dt \\ &= 2\int_0^1 tcos(2\pi\kappa t)\,dt = 0,\end{aligned} \tag{7.660}$$

and

$$\psi_\kappa = 18\int_0^1\Xi_{\frac{1}{3}}(27,t)\,subsin_{\frac{1}{3}}(54\pi\kappa,t)\,t^{-\frac{2}{3}}dt = 2\int_0^1 tsin(2\pi\kappa t)\,dt = \frac{1}{\pi\kappa}. \tag{7.661}$$

Thus, we get

$$27t^{\frac{1}{3}} = 1 + \sum_{\kappa=1}^{\infty}\frac{1}{\pi\kappa}subsin_{\frac{1}{3}}(54\pi\kappa,t). \tag{7.662}$$

7.4 The Fourier-Like Integral Transforms via Subtrigonometric Functions with Respect to Monotone Function

In 1822, Jean Baptiste Joseph Fourier, a French mathematician, physicist, and government administrator, proposed the well-known Fourier transform, which was applied in the theory of the heat conduction [303].

7.4.1 *Fourier Transform: History, Concepts, and Theorems*

The Fourier transform of the function $g(t)$ is defined by [303]

$$g(\omega) = F\{g(t)\} = \int_{-\infty}^{\infty} g(t)\, e^{-i\omega t} dt, \tag{7.663}$$

where the inverse Fourier transform is [303]

$$g(t) = F^{-1}\{g(\omega)\} = \frac{1}{2\pi}\int_{-\infty}^{\infty} g(\omega)\, e^{i\omega t} d\omega. \tag{7.664}$$

Theorem 7.23 *Let $g(\omega) = F\{g(t)\}$ and $g(t) = F^{-1}\{g(\omega)\}$. Then we have [303]*

$$g(\omega) = F\left\{F^{-1}\{g(\omega)\}\right\} \tag{7.665}$$

and

$$g(t) = F^{-1}\{F\{g(t)\}\}. \tag{7.666}$$

Theorem 7.24 *Let $F(\omega) = F\{f(t)\}$ and $\lim_{t\to\pm\infty} f(t) = 0$. Then we have [307]*

$$F\left\{\frac{dg(t)}{dt}\right\} = i\omega g(\omega). \tag{7.667}$$

Theorem 7.25 *Let $F(\omega) = F\{f(t)\}$ and $\lim_{t\to\infty} \int_{-\infty}^{t} g(t)\,dt = 0$. Then, we have [307]*

$$F\left\{\int_{0}^{t} g(t)\,dt\right\} = i\omega g(\omega). \tag{7.668}$$

7.4.2 *The Integral Transform Operator with Respect to Monotone Function of First Type*

Let $\mu^{(1)}(t) > 0$, $\mu(a) = -\infty$, $\mu(b) = \infty$, and $t, a, b \in \mathbb{R}$. Let $\varpi^{(1)}(\omega) > 0$, $\varpi(x) = -\infty$, $\varpi(y) = +\infty$, and $\omega, x, y \in \mathbb{R}$.

Definition 7.41 The integral transform operator with respect to monotone function of first type is defined as

$$\mathbb{Y}_{\varpi}\left[\Theta_{\mu}(t)\right]=\Theta_{\mu,\varpi}(\omega)=\int_{a}^{b}\Theta_{\mu}(t)\,e^{-i\varpi(\omega)\mu(t)}\mu^{(1)}(t)\,dt, \tag{7.669}$$

where $\mathbb{Y}_{\varpi}\left[\Theta_{\mu}(t)\right]=\Theta_{\mu,\varpi}(\omega)$ is the integral transform operator with respect to monotone function of first type.

Definition 7.42 The inverse integral transform operator with respect to monotone function of first type is defined as

$$\Theta_{\mu}(t)=\mathbb{Y}_{\varpi}{}^{-1}\left[\Theta_{\mu,\varpi}(\omega)\right]=\frac{1}{2\pi}\int_{x}^{y}\Theta_{\mu,\varpi}(\omega)\,e^{i\varpi(\omega)\mu(t)}\varpi^{(1)}(\omega)\,d\omega, \tag{7.670}$$

where $\Theta_{\mu}(t)=\mathbb{Y}_{\varpi}{}^{-1}\left[\Theta_{\mu,\varpi}(\omega)\right]$ is the inverse integral transform operator with respect to monotone function of first type.

There is the convergent condition as follows:

The integral transform operator with respect to monotone function of first type converges absolutely if

$$\left|\Theta_{\mu}(t)\right|\leq\rho e^{-\varpi(\omega)|\mu(t)|} \tag{7.671}$$

exists, where ρ is the constant.

Based on the formulas (7.671) and (7.672), we have the following:

Theorem 7.26 *Let* $\mu^{(1)}(t)>0$, $\mu(a)=-\infty$, $\mu(b)=\infty$, *and* $t,a,b\in\mathbb{R}$.
Let $\varpi^{(1)}(\omega)>0$, $\varpi(x)=-\infty$, $\varpi(y)=+\infty$, *and* $\omega,x,y\in\mathbb{R}$.
Then, we have

$$\Theta_{\mu}(t)=\left(\mathbb{Y}_{\varpi}^{-1}\circ\mathbb{Y}_{\varpi}\right)\Theta_{\mu}(t) \tag{7.672}$$

and

$$\Theta_{\mu,\varpi}(\omega)=\left(\mathbb{Y}_{\varpi}\circ\mathbb{Y}_{\varpi}^{-1}\right)\Theta_{\mu,\varpi}(\omega). \tag{7.673}$$

Proof With the aid of Eqs. (7.665) and (7.666), one has

$$\begin{aligned}\left(\mathbb{Y}_{\varpi}^{-1}\circ\mathbb{Y}_{\varpi}\right)\Theta_{\mu}(t)=&\frac{1}{2\pi}\int_{x}^{y}\left(\int_{a}^{b}\Theta_{\mu}(t)\,e^{-i\varpi(\omega)\mu(t)}\mu^{(1)}(t)\,dt\right)\\&\times e^{i\varpi(\omega)\mu(t)}\varpi^{(1)}(\omega)\,d\omega.\end{aligned} \tag{7.674}$$

Taking $\mu = \mu(t)$ and $\varpi = \varpi(\omega)$, we have

$$\left(\mathbb{Y}_{\varpi}^{-1} \circ \mathbb{Y}_{\varpi}\right) \Theta(t) = \frac{1}{2\pi} \int_{-\infty}^{\infty} \left(\int_{-\infty}^{\infty} \Theta(\mu) e^{-i\varpi\mu} d\mu \right) e^{i\varpi\mu} d\varpi. \tag{7.675}$$

It follows from Eq. (7.677) that

$$\frac{1}{2\pi} \int_{-\infty}^{\infty} \left(\int_{-\infty}^{\infty} \Theta(\mu) e^{-i\varpi\mu} d\mu \right) e^{i\varpi\mu} d\varpi = \Theta(t), \tag{7.676}$$

which leads to

$$\Theta(t) = \Theta_{\mu}(t). \tag{7.677}$$

Similarly, we give the proof of the formula (7.675).

Thus, the proof is completed. □

7.4.3 *The Integral Transform Operator with Respect to Monotone Function of Second Type*

Let $\mu^{(1)}(t) > 0$, $\mu(a) = -\infty$, $\mu(b) = \infty$, and $t, a, b \in \mathbb{R}$.

Definition 7.43 The integral transform operator with respect to monotone function of second type is defined as

$$\mathbb{Y}\left[\Theta_{\mu}(t)\right] = \Theta_{\mu}(\omega) = \int_{a}^{b} \Theta_{\mu}(t) e^{-i\omega\mu(t)} \mu^{(1)}(t) dt, \tag{7.678}$$

where $\mathbb{Y}\left[\Theta_{\mu}(t)\right] = \Theta_{\mu}(\omega)$ is the integral transform operator with respect to monotone function of second type.

Definition 7.44 The inverse integral transform operator with respect to monotone function of second type is defined as

$$\mathbb{Y}^{-1}\left[\Theta_{\mu}(\omega)\right] = \Theta_{\mu}(t) = \frac{1}{2\pi} \int_{-\infty}^{\infty} \Theta_{\varpi}(\omega) e^{i\omega\mu(t)} d\omega, \tag{7.679}$$

where $\mathbb{Y}^{-1}\left[\Theta_{\mu}(\omega)\right] = \Theta_{\mu}(t)$ is the inverse integral transform operator with respect to monotone function of second type.

The integral transform operator with respect to monotone function of second type converges absolutely if

$$\left|\Theta_{\mu}(t)\right| \leq \rho e^{-\omega|\mu(t)|} \tag{7.680}$$

exists, where ρ is the constant.

Theorem 7.27 *Let $\mu^{(1)}(t) > 0$, $\mu(a) = -\infty$, $\mu(b) = \infty$, and $t, a, b \in \mathbb{R}$. Then, we have*

$$\Theta_{\mu}(t) = \left(\mathbb{Y}^{-1} \circ \mathbb{Y}\right) \Theta_{\mu}(t) \tag{7.681}$$

and

$$\Theta_{\mu}(\omega) = \left(\mathbb{Y} \circ \mathbb{Y}^{-1}\right) \Theta_{\mu}(\omega). \tag{7.682}$$

Proof Taking $\varpi^{(1)}(\omega) = \omega$ in (7.674) and (7.675), we obtain the results. □

7.4.4 *The Integral Transform Operator with Respect to Monotone Function of Third Type*

Let $\varpi^{(1)}(\omega) > 0$, $\varpi(x) = -\infty$, $\varpi(y) = +\infty$, and $\omega, x, y \in \mathbb{R}$.

Definition 7.45 The integral transform operator with respect to monotone function of third type is defined as

$$\mathbb{Y}_{\varpi}[\Theta(t)] = \Theta_{\varpi}(\omega) = \int_{-\infty}^{\infty} \Theta_{\mu}(t) e^{-i\varpi(\omega)t} dt, \tag{7.683}$$

where $\mathbb{Y}_{\varpi}[\Theta(t)] = \Theta_{\varpi}(\omega)$ is the integral transform operator with respect to monotone function of third type.

Definition 7.46 The inverse integral transform operator with respect to monotone function of third type is defined as

$$\Theta(t) = \mathbb{Y}_{\varpi}^{-1}[\Theta_{\varpi}(\omega)] = \frac{1}{2\pi} \int_{x}^{y} \Theta_{\varpi}(\omega) e^{i\varpi(\omega)t} \varpi^{(1)}(\omega) d\omega, \tag{7.684}$$

where $\Theta(t) = \mathbb{Y}_{\varpi}^{-1}[\Theta_{\varpi}(\omega)]$ is the inverse integral transform operator with respect to monotone function of third type.

The integral transform operator with respect to monotone function of third type converges absolutely if

$$|\Theta(t)| \le \rho e^{-\varpi(\omega)|t|} \tag{7.685}$$

exists, where ρ is the constant.

Theorem 7.28 *Let $\varpi^{(1)}(\omega) > 0$, $\varpi(x) = -\infty$, $\varpi(y) = +\infty$, and $\omega, x, y \in \mathbb{R}$. Then, we have*

$$\Theta(t) = \left(\Upsilon_{\varpi}^{-1} \circ \Upsilon_{\varpi}\right)\Theta(t) \tag{7.686}$$

and

$$\Theta_{\varpi}(\omega) = \left(\Upsilon_{\varpi} \circ \Upsilon_{\varpi}^{-1}\right)\Theta_{\varpi}(\omega) \tag{7.687}$$

Proof Putting $\mu(t) = t$ in (7.674) and (7.675), we have the desired results. □

7.4.5 The Integral Transform Operator with Respect to Monotone Function of Fourth Type

Let $\mu^{(1)}(t) > 0$, $\mu(a) = -\infty$, $\mu(b) = \infty$, and $t, a, b \in \mathbb{R}$.
Let $\varpi^{(1)}(\omega) > 0$, $\varpi(x) = -\infty$, $\varpi(y) = +\infty$, and $\omega, x, y \in \mathbb{R}$.

Definition 7.47 The integral transform operator with respect to monotone function of fourth type is defined as

$$\Upsilon_{\varpi}\left[\Theta_{\mu}(t)\right] = \Theta_{\mu,\varpi}(\omega) = \int_{a}^{b} \Theta_{\mu}(t)\, e^{-2\pi i \varpi(\omega)\mu(t)} \mu^{(1)}(t)\, dt, \tag{7.688}$$

where $\Upsilon_{\varpi}\left[\Theta_{\mu}(t)\right] = \Theta_{\mu,\varpi}(\omega)$ is the integral transform operator with respect to monotone function of fourth type.

Definition 7.48 The inverse integral transform operator with respect to monotone function of fourth type is defined as

$$\Theta_{\mu}(t) = \Upsilon_{\varpi}^{-1}\left[\Theta_{\mu,\varpi}(\omega)\right] = \int_{x}^{y} \Theta_{\mu,\varpi}(\omega)\, e^{2\pi i \varpi(\omega)\mu(t)} \varpi^{(1)}(\omega)\, d\omega, \tag{7.689}$$

where $\Theta_{\mu}(t) = \Upsilon_{\varpi}^{-1}\left[\Theta_{\mu,\varpi}(\omega)\right]$ is the inverse integral transform operator with respect to monotone function of fourth type.

The integral transform operator with respect to monotone function of fourth type converges absolutely if

$$\lim_{\substack{|\mu(t)| \to 0 \\ t \in \mathbb{R}}} |\mu(t)|^r \left| \left(\Theta_\mu(t) \right)^{(l)} \right| = 0 \tag{7.690}$$

exists for $r, l \in \mathbb{N}$ and

$$\left(\Theta_\mu(t) \right)^{(l)} = \left(\frac{1}{\mu^{(1)}(t)} \frac{d}{dt} \right)^l \Theta_\mu(t). \tag{7.691}$$

Theorem 7.29 *Let $\mu^{(1)}(t) > 0$, $\mu(a) = -\infty$, $\mu(b) = \infty$, and $t, a, b \in \mathbb{R}$.*
Let $\varpi^{(1)}(\omega) > 0$, $\varpi(x) = -\infty$, $\varpi(y) = +\infty$, and $\omega, x, y \in \mathbb{R}$.
Then, we have

$$\Theta_\mu(t) = \left(\mathbb{Y}_\varpi^{-1} \circ \mathbb{Y}_\varpi \right) \Theta_\mu(t) \tag{7.692}$$

and

$$\Theta_{\mu,\varpi}(\omega) = \left(\mathbb{Y}_\varpi \circ \mathbb{Y}_\varpi^{-1} \right) \Theta_{\mu,\varpi}(\omega). \tag{7.693}$$

Proof Taking the place of $\mu(t)$ and $\varpi(\omega)$ by $\mu(t)$ and $2\pi\varpi(\omega)$ in (7.672) and (7.673), respectively, we have the desired results. □

7.4.6 The Integral Transform Operator with Respect to Monotone Function of Fifth Type

Let $\mu^{(1)}(t) > 0$, $\mu(a) = -\infty$, $\mu(b) = \infty$, and $t, a, b \in \mathbb{R}$.

Definition 7.49 The integral transform operator with respect to monotone function of fifth type is defined as

$$\mathbb{Y}\left[\Theta_\mu(t)\right] = \Theta_\mu(\omega) = \int_a^b \Theta_\mu(t)\, e^{-2\pi i \omega \mu(t)} \mu^{(1)}(t)\, dt, \tag{7.694}$$

where $\mathbb{Y}\left[\Theta_\mu(t)\right] = \Theta_\mu(\omega)$ is the integral transform operator with respect to monotone function of fifth type.

Definition 7.50 The inverse integral transform operator with respect to monotone function of fifth type is defined as

$$\Theta_\mu(t) = \Upsilon^{-1}\left[\Theta_\mu(\omega)\right] = \int_x^y \Theta_\mu(\omega)\, e^{2\pi i \omega \mu(t)} d\omega, \tag{7.695}$$

where $\Theta_\mu(t) = \Upsilon^{-1}\left[\Theta_\mu(\omega)\right]$ is the inverse integral transform operator with respect to monotone function of fifth type.

The integral transform operator with respect to monotone function of fifth type converges absolutely if

$$\lim_{\substack{|\mu(t)| \to 0 \\ t \in \mathbb{R}}} |\mu(t)|^r \left|\left(\Theta_\mu(t)\right)^{(l)}\right| = 0 \tag{7.696}$$

exists for $r, l \in \mathbb{N}$ and

$$\left(\Theta_\mu(t)\right)^{(l)} = \left(\frac{1}{\mu^{(1)}(t)}\frac{d}{dt}\right)^l \Theta_\mu(t). \tag{7.697}$$

Theorem 7.30 *Let* $\mu^{(1)}(t) > 0$, $\mu(a) = -\infty$, $\mu(b) = \infty$, *and* $t, a, b \in \mathbb{R}$.
Then, we have

$$\Theta_\mu(t) = \left(\Upsilon^{-1} \circ \Upsilon\right) \Theta_\mu(t) \tag{7.698}$$

and

$$\Theta_\mu(\omega) = \left(\Upsilon \circ \Upsilon^{-1}\right) \Theta_\mu(\omega). \tag{7.699}$$

Proof Taking $\varpi^{(1)}(\omega) = \omega$ in (7.694) and (7.695), we deduce the results. □

7.4.7 The Integral Transform Operator with Respect to Monotone Function of Sixth Type

Let $\varpi^{(1)}(\omega) > 0$, $\varpi(x) = -\infty$, $\varpi(y) = +\infty$, and $\omega, x, y \in \mathbb{R}$.

Definition 7.51 The integral transform operator with respect to monotone function of sixth type is defined as

$$\Upsilon_\varpi[\Theta(t)] = \Theta_\varpi(\omega) = \int_{-\infty}^{\infty} \Theta_\mu(t)\, e^{-2\pi i \varpi(\omega) t} dt, \tag{7.700}$$

where $\mathbb{Y}_{\varpi}\left[\Theta\left(t\right)\right] = \Theta_{\varpi}\left(\omega\right)$ is the integral transform operator with respect to monotone function of sixth type.

Definition 7.52 The inverse integral transform operator with respect to monotone function of sixth type is defined as

$$\Theta\left(t\right) = \mathbb{Y}_{\varpi}^{-1}\left[\Theta_{\varpi}\left(\omega\right)\right] = \int\limits_{x}^{y} \Theta_{\varpi}\left(\omega\right) e^{2\pi i \varpi(\omega) t} \varpi^{(1)}\left(\omega\right) d\omega, \tag{7.701}$$

where $\Theta\left(t\right) = \mathbb{Y}_{\varpi}^{-1}\left[\Theta_{\varpi}\left(\omega\right)\right]$ is the inverse integral transform operator with respect to monotone function of sixth type.

The integral transform operator with respect to monotone function of sixth type converges absolutely if

$$\lim_{\substack{|t| \to 0 \\ t \in \mathbb{R}}} |t|^{r} \left|\Theta^{(l)}\left(t\right)\right| = 0 \tag{7.702}$$

exists for $r, l \in \mathbb{N}$.

Theorem 7.31 *Let* $\varpi^{(1)}\left(\omega\right) > 0$, $\varpi\left(x\right) = -\infty$, $\varpi\left(y\right) = +\infty$, *and* $\omega, x, y \in \mathbb{R}$. *Then, we have*

$$\Theta\left(t\right) = \left(\mathbb{Y}_{\varpi}^{-1} \circ \mathbb{Y}_{\varpi}\right) \Theta\left(t\right) \tag{7.703}$$

and

$$\Theta_{\varpi}\left(\omega\right) = \left(\mathbb{Y}_{\varpi} \circ \mathbb{Y}_{\varpi}^{-1}\right) \Theta_{\varpi}\left(\omega\right). \tag{7.704}$$

Proof Putting $\mu\left(t\right) = t$ in (7.694) and (7.695), we reduce to the results. □

7.4.8 *The Integral Transform Operator with Respect to Monotone Function of Seventh Type*

Let $\mu^{(1)}\left(t\right) > 0$, $\mu\left(a\right) = -\infty$, $\mu\left(b\right) = \infty$, and $t, a, b \in \mathbb{R}$.

Let $\varpi^{(1)}\left(\omega\right) > 0$, $\varpi\left(x\right) = -\infty$, $\varpi\left(y\right) = +\infty$, and $\omega, x, y \in \mathbb{R}$.

Definition 7.53 The integral transform operator with respect to monotone function of seventh type is defined as

$$\mathbb{Y}_{\varpi}\left[\Theta_{\mu}\left(t\right)\right] = \Theta_{\mu,\varpi}\left(\omega\right) = \frac{1}{\sqrt{2\pi}} \int\limits_{a}^{b} \Theta_{\mu}\left(t\right) e^{-i\varpi(\omega)\mu(t)} \mu^{(1)}\left(t\right) dt, \tag{7.705}$$

where $\Upsilon_{\varpi}\left[\Theta_{\mu}(t)\right]=\Theta_{\mu,\varpi}(\omega)$ is the integral transform operator with respect to monotone function of seventh type.

Definition 7.54 The inverse integral transform operator with respect to monotone function of seventh type is defined as

$$\Theta_{\mu}(t)=\Upsilon_{\varpi}^{-1}\left[\Theta_{\mu,\varpi}(\omega)\right]=\frac{1}{\sqrt{2\pi}}\int_{x}^{y}\Theta_{\mu,\varpi}(\omega)\,e^{i\varpi(\omega)\mu(t)}\varpi^{(1)}(\omega)\,d\omega, \tag{7.706}$$

where $\Theta_{\mu}(t)=\Upsilon_{\varpi}^{-1}\left[\Theta_{\mu,\varpi}(\omega)\right]$ is the inverse integral transform operator with respect to monotone function of seventh type.

The integral transform operator with respect to monotone function of seventh type converges absolutely if

$$\left|\Theta_{\mu}(t)\right|\leq\rho e^{-\varpi(\omega)|\mu(t)|} \tag{7.707}$$

exists, where ρ is the constant.

Theorem 7.32 *Let* $\mu^{(1)}(t)>0$, $\mu(a)=-\infty$, $\mu(b)=\infty$, *and* $t,a,b\in\mathbb{R}$.
Let $\varpi^{(1)}(\omega)>0$, $\varpi(x)=-\infty$, $\varpi(y)=+\infty$, *and* $\omega,x,y\in\mathbb{R}$.
Then, we have

$$\Theta_{\mu}(t)=\left(\Upsilon_{\varpi}^{-1}\circ\Upsilon_{\varpi}\right)\Theta_{\mu}(t) \tag{7.708}$$

and

$$\Theta_{\mu,\varpi}(\omega)=\left(\Upsilon_{\varpi}\circ\Upsilon_{\varpi}^{-1}\right)\Theta_{\mu,\varpi}(\omega). \tag{7.709}$$

Proof From (7.12), we have

$$\begin{aligned}&\frac{1}{2\pi}\int_{x}^{y}\left(\int_{a}^{b}\Theta_{\mu}(t)\,e^{-i\varpi(\omega)\mu(t)}\mu^{(1)}(t)\,dt\right)e^{i\varpi(\omega)\mu(t)}\varpi^{(1)}(\omega)\,d\omega\\&=\frac{1}{\sqrt{2\pi}}\int_{x}^{y}\left(\frac{1}{\sqrt{2\pi}}\int_{a}^{b}\Theta_{\mu}(t)\,e^{-i\varpi(\omega)\mu(t)}\mu^{(1)}(t)\,dt\right)e^{i\varpi(\omega)\mu(t)}\varpi^{(1)}(\omega)\,d\omega,\end{aligned} \tag{7.710}$$

which leads to the desired formula (7.708). □

Similarly, we present the formula (7.709).

7.4.9 The Integral Transform Operator with Respect to Monotone Function of Eighth Type

Let $\mu^{(1)}(t) > 0$, $\mu(a) = -\infty$, $\mu(b) = \infty$, and $t, a, b \in \mathbb{R}$.

Definition 7.55 The integral transform operator with respect to monotone function of eighth type is defined as

$$\mathbb{Y}\left[\Theta_{\mu}(t)\right] = \Theta_{\mu}(\omega) = \frac{1}{\sqrt{2\pi}} \int_{a}^{b} \Theta_{\mu}(t)\, e^{-i\omega\mu(t)} \mu^{(1)}(t)\, dt, \tag{7.711}$$

where $\mathbb{Y}\left[\Theta_{\mu}(t)\right] = \Theta_{\mu}(\omega)$ is the integral transform operator with respect to monotone function of eighth type.

Definition 7.56 The inverse integral transform operator with respect to monotone function of eighth type is defined as

$$\mathbb{Y}^{-1}\left[\Theta_{\mu}(\omega)\right] = \Theta_{\mu}(t) = \frac{1}{\sqrt{2\pi}} \int_{-\infty}^{\infty} \Theta_{\varpi}(\omega)\, e^{i\omega\mu(t)} d\omega, \tag{7.712}$$

where $\mathbb{Y}^{-1}\left[\Theta_{\mu}(\omega)\right] = \Theta_{\mu}(t)$ is the inverse integral transform operator with respect to monotone function of eighth type.

The integral transform operator with respect to monotone function of eighth type converges absolutely if

$$\left|\Theta_{\mu}(t)\right| \leq \rho e^{-\omega|\mu(t)|} \tag{7.713}$$

exists, where ρ is the constant.

Theorem 7.33 *Let $\mu^{(1)}(t) > 0$, $\mu(a) = -\infty$, $\mu(b) = \infty$, and $t, a, b \in \mathbb{R}$. Then, we have*

$$\Theta_{\mu}(t) = \left(\mathbb{Y}^{-1} \circ \mathbb{Y}\right) \Theta_{\mu}(t) \tag{7.714}$$

and

$$\Theta_{\mu}(\omega) = \left(\mathbb{Y} \circ \mathbb{Y}^{-1}\right) \Theta_{\mu}(\omega). \tag{7.715}$$

Proof Taking $\varpi^{(1)}(\omega) = \omega$ in (7.710) and (7.711), we have the results. □

7.4.10 The Integral Transform Operator with Respect to Monotone Function of Ninth Type

Let $\varpi^{(1)}(\omega) > 0$, $\varpi(x) = -\infty$, $\varpi(y) = +\infty$, and $\omega, x, y \in \mathbb{R}$.

Definition 7.57 The integral transform operator with respect to monotone function of ninth type is defined as

$$\mathbb{Y}_{\varpi}[\Theta(t)] = \Theta_{\varpi}(\omega) = \frac{1}{\sqrt{2\pi}} \int_{-\infty}^{\infty} \Theta_{\mu}(t)\, e^{-i\varpi(\omega)t} dt, \tag{7.716}$$

where $\mathbb{Y}_{\varpi}[\Theta(t)] = \Theta_{\varpi}(\omega)$ is the integral transform operator with respect to monotone function of ninth type.

Definition 7.58 The inverse integral transform operator with respect to monotone function of ninth type is defined as

$$\Theta(t) = \mathbb{Y}_{\varpi}^{-1}[\Theta_{\varpi}(\omega)] = \frac{1}{\sqrt{2\pi}} \int_{x}^{y} \Theta_{\varpi}(\omega)\, e^{i\varpi(\omega)t} \varpi^{(1)}(\omega)\, d\omega, \tag{7.717}$$

where $\Theta_{\mu}(t) = \mathbb{Y}_{\varpi}^{-1}\left[\Theta_{\mu,\varpi}(\omega)\right]$ is the inverse integral transform operator with respect to monotone function of ninth type.

The integral transform operator with respect to monotone function of ninth type converges absolutely if

$$\left|\Theta_{\mu}(t)\right| \le \rho e^{-\varpi(\omega)|t|} \tag{7.718}$$

exists, where ρ is the constant.

Theorem 7.34 *Let* $\varpi^{(1)}(\omega) > 0$, $\varpi(x) = -\infty$, $\varpi(y) = +\infty$, *and* $\omega, x, y \in \mathbb{R}$. *Then, we have*

$$\Theta(t) = \left(\mathbb{Y}_{\varpi}^{-1} \circ \mathbb{Y}_{\varpi}\right) \Theta(t) \tag{7.719}$$

and

$$\Theta_{\varpi}(\omega) = \left(\mathbb{Y}_{\varpi} \circ \mathbb{Y}_{\varpi}^{-1}\right) \Theta_{\varpi}(\omega). \tag{7.720}$$

Proof Taking $\mu(t) = t$ in (7.710) and (7.711), we present the desired results. □

7.4.11 The Integral Transform Operator with Respect to Power-Law Function of Second Type

Let $\alpha > 0$ and $t^{\alpha-1} > 0$ for $t \in \mathbb{R}$.

Definition 7.59 The integral transform operator with respect to power-law function of second type is defined as

$$\mathbb{Y}\left[\Theta(t)\right] = \Theta_\alpha(\omega) = \alpha \int_{-\infty}^{\infty} \Theta(t)\, e^{-i\omega t^\alpha} t^{\alpha-1} dt, \tag{7.721}$$

where $\mathbb{Y}\left[\Theta_\mu(t)\right] = \Theta_\alpha(\omega)$ is the integral transform operator with respect to power-law function of second type.

Definition 7.60 The inverse integral transform operator with respect to power-law function of second type is defined as

$$\mathbb{Y}^{-1}\left[\Theta_\alpha(\omega)\right] = \Theta(t) = \frac{1}{2\pi} \int_{-\infty}^{\infty} \Theta_\alpha(\omega)\, e^{i\omega t^\alpha} d\omega, \tag{7.722}$$

where $\mathbb{Y}^{-1}\left[\Theta_\alpha(\omega)\right] = \Theta(t)$ is the inverse integral transform operator with respect to power-law function of second type.

The integral transform operator with respect to power-law function of second type converges absolutely if

$$|\Theta(t)| \le \rho e^{-\omega|t|^\alpha} \tag{7.723}$$

exists, where ρ is the constant, $\alpha > 0$, and $t^{\alpha-1} > 0$ for $t \in \mathbb{R}$.

Theorem 7.35 *Let $\alpha > 0$ and $t^{\alpha-1} > 0$ for $t \in \mathbb{R}$.*
Then, we have

$$\Theta(t) = \left(\mathbb{Y}^{-1} \circ \mathbb{Y}\right) \Theta(t) \tag{7.724}$$

and

$$\Theta_\alpha(\omega) = \left(\mathbb{Y} \circ \mathbb{Y}^{-1}\right) \Theta_\alpha(\omega). \tag{7.725}$$

Proof Taking $\mu(t) = t^\alpha$ in (7.683) and (7.684), we obtain the results. □

7.4.12 The Integral Transform Operator with Respect to Power-Law Function of Fifth Type

Let $\alpha > 0$ and $t^{\alpha-1} > 0$ for $t \in \mathbb{R}$.

Definition 7.61 The integral transform operator with respect to power-law function of fifth type is defined as

$$\mathbb{Y}[\Theta(t)] = \Theta_\alpha(\omega) = \int_{-\infty}^{\infty} \Theta(t) e^{-2\pi i \omega t^\alpha} t^{\alpha-1} dt, \tag{7.726}$$

where $\mathbb{Y}[\Theta(t)] = \Theta_\alpha(\omega)$ is the integral transform operator with respect to power-law function of fifth type.

Definition 7.62 The inverse integral transform operator with respect to power-law function of fifth type is defined as

$$\Theta(t) = \mathbb{Y}^{-1}[\Theta_\alpha(\omega)] = \int_{-\infty}^{\infty} \Theta_\alpha(\omega) e^{2\pi i \omega t^\alpha} d\omega, \tag{7.727}$$

where $\Theta(t) = \mathbb{Y}^{-1}[\Theta_\alpha(\omega)]$ is the inverse integral transform operator with respect to power-law function of fifth type.

Theorem 7.36 *Let $\alpha > 0$ and $t^{\alpha-1} > 0$ for $t \in \mathbb{R}$.*
Then, we have

$$\Theta(t) = \left(\mathbb{Y}^{-1} \circ \mathbb{Y}\right) \Theta(t) \tag{7.728}$$

and

$$\Theta_\alpha(\omega) = \left(\mathbb{Y} \circ \mathbb{Y}^{-1}\right) \Theta_\alpha(\omega). \tag{7.729}$$

Proof Taking $\mu(t) = t^\alpha$ in (7.700) and (7.701), we deduce the results. □

7.4.13 The Integral Transform Operator with Respect to Scaling-Law Function of Second Type

Let $\alpha, \lambda > 0$ and $t^{\alpha-1} > 0$ for $t \in \mathbb{R}$.

Definition 7.63 The integral transform operator with respect to scaling-law function of second type is defined as

$$\mathbb{Y}\left[\Theta\left(t\right)\right]=\Theta_{\alpha}\left(\lambda,\omega\right)=\alpha\lambda\int_{-\infty}^{\infty}\Theta\left(t\right)e^{-i\omega\lambda t^{\alpha}}t^{\alpha-1}dt, \tag{7.730}$$

where $\mathbb{Y}\left[\Theta\left(t\right)\right]=\Theta_{\alpha}\left(\lambda,\omega\right)$ is the integral transform operator with respect to scaling-law function of second type.

Definition 7.64 The inverse integral transform operator with respect to scaling-law function of second type is defined as

$$\mathbb{Y}^{-1}\left[\Theta_{\alpha}\left(\lambda,\omega\right)\right]=\Theta\left(t\right)=\frac{1}{2\pi}\int_{-\infty}^{\infty}\Theta_{\alpha}\left(\lambda,\omega\right)e^{i\omega\lambda t^{\alpha}}d\omega, \tag{7.731}$$

where $\mathbb{Y}^{-1}\left[\Theta_{\alpha}\left(\lambda,\omega\right)\right]=\Theta\left(t\right)$ is the inverse integral transform operator with respect to scaling-law function of second type.

The integral transform operator with respect to scaling-law function of second type converges absolutely if

$$\left|\Theta\left(t\right)\right|\leq\rho e^{-\omega\lambda|t|^{\alpha}} \tag{7.732}$$

exists, where ρ is the constant, $\alpha,\lambda>0$, and $t^{\alpha-1}>0$ for $t\in\mathbb{R}$.

Theorem 7.37 *Let $\alpha,\lambda>0$ and $t^{\alpha-1}>0$ for $t\in\mathbb{R}$.*
Then, we have

$$\Theta\left(t\right)=\left(\mathbb{Y}^{-1}\circ\mathbb{Y}\right)\Theta\left(t\right) \tag{7.733}$$

and

$$\Theta_{\alpha}\left(\lambda,\omega\right)=\left(\mathbb{Y}\circ\mathbb{Y}^{-1}\right)\Theta_{\alpha}\left(\lambda,\omega\right). \tag{7.734}$$

Proof Taking $\mu\left(t\right)=\lambda t^{\alpha}$ in (7.683) and (7.684), we obtain the results. □

7.4.14 The Integral Transform Operator with Respect to Scaling-Law Function of Fifth Type

Let $\alpha,\lambda>0$ and $t^{\alpha-1}>0$ for $t\in\mathbb{R}$.

Definition 7.65 The integral transform operator with respect to scaling-law function of fifth type is defined as

$$\mathbb{Y}\left[\Theta\left(t\right)\right]=\Theta_{\alpha}\left(\lambda,\omega\right)=\lambda\alpha\int\limits_{-\infty}^{\infty}\Theta\left(t\right)e^{-2\pi i\omega\lambda t^{\alpha}}t^{\alpha-1}dt, \tag{7.735}$$

where $\mathbb{Y}\left[\Theta\left(t\right)\right]=\Theta_{\alpha}\left(\lambda,\omega\right)$ is the integral transform operator with respect to scaling-law function of fifth type.

Definition 7.66 The inverse integral transform operator with respect to scaling-law function of fifth type is defined as

$$\Theta\left(t\right)=\mathbb{Y}^{-1}\left[\Theta_{\alpha}\left(\lambda,\omega\right)\right]=\int\limits_{-\infty}^{\infty}\Theta_{\alpha}\left(\lambda,\omega\right)e^{2\pi i\omega\lambda t^{\alpha}}d\omega, \tag{7.736}$$

where $\Theta\left(t\right)=\mathbb{Y}^{-1}\left[\Theta_{\alpha}\left(\lambda,\omega\right)\right]$ is the inverse integral transform operator with respect to scaling-law function of fifth type.

Theorem 7.38 *Let $\alpha,\lambda>0$ and $t^{\alpha-1}>0$ for $t\in\mathbb{R}$. Then we have*

$$\Theta_{\mu}\left(t\right)=\left(\mathbb{Y}^{-1}\circ\mathbb{Y}\right)\Theta_{\mu}\left(t\right) \tag{7.737}$$

and

$$\Theta_{\alpha}\left(\lambda,\omega\right)=\left(\mathbb{Y}\circ\mathbb{Y}^{-1}\right)\Theta_{\alpha}\left(\lambda,\omega\right). \tag{7.738}$$

Proof Taking $\mu\left(t\right)=\lambda t^{\alpha}$ in (7.700) and (7.701), we deduce the results. □

7.4.15 Applications

Theorem 7.39 *Let $\phi\left(t\right)=\left(\varphi\circ\mu\right)\left(t\right)=\varphi\left(\mu\left(t\right)\right)$, $\lim\limits_{\mu(t)\rightarrow\pm\infty}\varphi\left(\mu\left(t\right)\right)=0$, and $\mu^{(1)}\left(t\right)>0$.*

The integral transform of the Leibniz derivative of the function $\phi\left(t\right)$ is given as follows:

$$\int\limits_{a}^{b}D_{t,\mu(\cdot)}^{(1)}\phi\left(t\right)e^{-i\omega\mu(t)}\mu^{(1)}\left(t\right)dt=i\omega\varphi\left(\omega\right). \tag{7.739}$$

Proof Making use of the formula (7.680), we have

$$\int_a^b D_{t,\mu(\cdot)}^{(1)} \phi(t)\, e^{-i\omega\mu(t)} \mu^{(1)}(t)\, dt = \int_a^b D_{t,\mu(\cdot)}^{(1)} \phi(t)\, e^{-i\omega\mu(t)} d\mu(t), \tag{7.740}$$

which can be written as

$$\int_a^b D_{t,\mu(\cdot)}^{(1)} \phi(t)\, e^{-i\omega\mu(t)} d\mu(t) = \int_{-\infty}^{\infty} D_t^{(1)} \varphi(t)\, e^{-i\omega t} dt = i\omega\varphi(\omega), \tag{7.741}$$

where

$$\lim_{\mu(t)\to\pm\infty} \varphi(\mu(t)) = \lim_{t\to\pm\infty} \varphi(t) = 0. \tag{7.742}$$

Thus, we complete the proof. □

Theorem 7.40 *Let* $\psi(t) = (\Omega \circ \mu)(t) = \Omega(\mu(t))$, $\int_a^t \psi(t)\, \mu^{(1)}(t)\, dt = 0$, *and* $\mu^{(1)}(t) > 0$.

The integral transform of the Riemann–Stieltjes integral of the function $\psi(t)$ *is given as follows:*

$$\int_a^b {}_{-\infty}I_{t,\mu(\cdot)}^{(1)} \psi(t)\, e^{-i\omega\mu(t)} \mu^{(1)}(t)\, dt = \frac{\Omega(\omega)}{i\omega}. \tag{7.743}$$

Proof Making use of the formula (7.680), we have

$$\int_a^b {}_aI_{t,\mu(\cdot)}^{(1)} \psi(t)\, e^{-i\omega\mu(t)} \mu^{(1)}(t)\, dt = \int_a^b {}_aI_{t,\mu(\cdot)}^{(1)} \psi(t)\, e^{-i\omega\mu(t)} d\mu(t), \tag{7.744}$$

which can be presented as follows:

$$\int_a^b {}_aI_t^{(1)} \Omega(t)\, e^{-i\omega\mu(t)} d\mu(t) = \int_{-\infty}^{\infty} {}_{-\infty}I_t^{(1)} \Omega(t)\, e^{-i\omega t} dt = \frac{\Omega(\omega)}{i\omega}, \tag{7.745}$$

where

$$\int_{a}^{t} \psi(t)\,\mu^{(1)}(t)\,dt = \lim_{t\to\infty}\int_{-\infty}^{t} \Omega(t)\,dt = 0. \tag{7.746}$$

□

Theorem 7.41 *Let* $\beta, \lambda > 0$, $\phi(t) = \left(\varphi \circ \left(\lambda t^{\beta}\right)\right)(t) = \varphi\left(\lambda t^{\beta}\right)$, $\lim_{t\to\pm\infty}\varphi(t) = 0$, *and* $t^{\beta-1} > 0$.

The integral transform of the scaling-law derivative of the function $\phi(t)$ *is given as follows:*

$$\int_{-\infty}^{\infty} {}^{SL}D_t^{(1)}\phi(t)\,e^{-i\omega\lambda t^{\beta}}t^{\beta-1}dt = \frac{i\omega}{\lambda\beta}\varphi(\omega). \tag{7.747}$$

Proof Taking $\mu(t) = \lambda t^{\beta}$ in (7.741), we obtain the result. □

Theorem 7.42 *Let* $\beta, \lambda > 0$, $\psi(t) = \left(\Omega \circ \left(\lambda t^{\beta}\right)\right)(t) = \Omega\left(\lambda t^{\beta}\right)$, $\lim_{t\to\infty} {}^{SL}_{-\infty}I_t^{(1)}\psi(t) = 0$, *and* $t^{\beta-1} > 0$.

The integral transform of the scaling-law integral of the function $\psi(t)$ *is given as follows:*

$$\int_{-\infty}^{\infty} {}^{SL}_{-\infty}I_t^{(1)}\psi(t)\,e^{-i\omega\lambda t^{\beta}}t^{\beta-1}dt = \frac{\Omega(\omega)}{i\omega\lambda\beta}. \tag{7.748}$$

Proof Taking $\mu(t) = \lambda t^{\beta}$ in (7.745), we have the result. □

Theorem 7.43 *Let* $\beta > 0$, $\phi(t) = \left(\varphi \circ \left(t^{\beta}\right)\right)(t) = \varphi\left(t^{\beta}\right)$, $\lim_{t\to\pm\infty}\varphi(t) = 0$, *and* $t^{\beta-1} > 0$.

The integral transform of the power-law derivative of the function $\phi(t)$ *is given as follows:*

$$\int_{-\infty}^{\infty} {}^{P}D_t^{(1)}\phi(t)\,e^{-i\omega t^{\beta}}t^{\beta-1}dt = \frac{i\omega}{\beta}\varphi(\omega). \tag{7.749}$$

Proof Taking $\lambda = 1$ in (7.749), we obtain the result. □

Theorem 7.44 *Let* $\beta > 0$, $\psi(t) = \left(\Omega \circ \left(t^{\beta}\right)\right)(t) = \Omega\left(t^{\beta}\right)$, $\lim_{t\to\infty} {}^{P}_{-\infty}I_t^{(1)}\psi(t) = 0$, *and* $t^{\beta-1} > 0$.

The integral transform of the power-law integral of the function $\psi(t)$ is given as follows:

$$\int_{-\infty}^{\infty} {}_{-\infty}^{P}I_t^{(1)}\psi(t)\, e^{-i\omega t^{\beta}} t^{\beta-1} dt = \frac{\Omega(\omega)}{i\omega\beta}. \tag{7.750}$$

Proof Taking $\lambda = 1$ in (7.750), we have the result. □

The results are the special cases of Chap. 2.

7.5 The Laplace-Like Transforms via Subtrigonometric Functions with Respect to Monotone Function

In this section, we consider the integral transform operators based on the subtrigonometric functions with respect to monotone function.

7.5.1 Laplace Transform: History, Concepts, and Theorems

In 1812, Pierre-Simon Laplace, a French mathematician and astronomer, coined the well-known Laplace transform [88].

Definition 7.67 The Laplace transform is defined as [88]

$$F(s) = L\{f(t)\} = \int_{-\infty}^{\infty} f(t)\, e^{-st} dt, \tag{7.751}$$

where $s = \sigma + i\omega$.

The work was discovered by Laplace in 1812 [88].

Definition 7.68 The inverse Laplace transform, the Bromwich integral, is defined as [306]

$$f(t) = L^{-1}\{F(s)\} = \frac{1}{2\pi i} \int_{\sigma-i\infty}^{\sigma+i\infty} F(s)\, e^{st} ds, \tag{7.752}$$

where $0 < \sigma$.

The work was discovered by Bromwich in 1908 [306].

7.5.2 The Integral Transform Operator with Respect to Monotone Function of First Type

Let $\mu^{(1)}(t) > 0$, $\mu(a) = 0$, $\mu(b) = \infty$ and $t, a, b \in \mathbb{R}$. Let $\nu^{(1)}(u) > 0$, $s(u) = \sigma - i\nu(u)$, $s(x) = \sigma - i\infty = \sigma + i\nu(x)$, $s(y) = \sigma + i\infty = \sigma + i\nu(y)$, and $u, x, y \in \mathbb{R}$.

Definition 7.69 The integral transform operator with respect to monotone function of first type is defined as

$$\mathbb{X}_s\left[\Theta_\mu(t)\right] = \Theta_{\mu,s}(u) = \int_a^b \Theta_\mu(t)\, e^{-s(u)\mu(t)} \mu^{(1)}(t)\, dt, \tag{7.753}$$

where $\mathbb{X}_s\left[\Theta_\mu(t)\right] = \Theta_{\mu,s}(u)$ is the integral transform operator with respect to monotone function of first type.

Definition 7.70 The inverse integral transform operator with respect to monotone function of first type is defined as

$$\Theta_\mu(t) = \mathbb{X}_s^{-1}\left[\Theta_{\mu,s}(u)\right] = \frac{1}{2\pi i}\int_x^y \Theta_{\mu,s}(u)\, e^{s(u)\mu(t)} s^{(1)}(u)\, du, \tag{7.754}$$

where $\Theta_\mu(t) = \mathbb{X}_s^{-1}\left[\Theta_{\mu,s}(u)\right]$ is the inverse integral transform operator with respect to monotone function of first type.

The integral transform operator with respect to monotone function of first type converges absolutely if

$$\int_a^b \left|\Theta_\mu(t)\, e^{-s(u)\mu(t)}\right| \mu^{(1)}(t)\, dt \tag{7.755}$$

exists.

Theorem 7.45 *Let $\mu^{(1)}(t) > 0$, $\mu(a) = 0$, $\mu(b) = \infty$ and $t, a, b \in \mathbb{R}$. Let $\nu^{(1)}(u) > 0$, $s(u) = \sigma - i\nu(u)$, $s(x) = \sigma - i\infty = \sigma + i\nu(x)$, $s(y) = \sigma + i\infty = \sigma + i\nu(y)$, and $u, x, y \in \mathbb{R}$. Then, we have*

$$\Theta_\mu(t) = \left(\mathbb{X}_s^{-1} \circ \mathbb{X}_s\right)\Theta_\mu(t) \tag{7.756}$$

and

$$\Theta_{\mu,s}(u) = \left(\mathbb{X}_s \circ \mathbb{X}_s^{-1}\right)\Theta_{\mu,s}(u). \tag{7.757}$$

Proof Setting $\mu = \mu(t)$ and $s = s(u) = \sigma - i\nu(u)$ in Eq. (7.754), we have

$$\left(\mathbb{X}_s^{-1} \circ \mathbb{X}_s\right) \Theta_\mu(t) = \frac{1}{2\pi i} \int_x^y \left(\int_a^b \Theta_\mu(t) e^{-s(u)\mu(t)} \mu^{(1)}(t) dt \right) e^{s(u)\mu(t)} s^{(1)}(u) du, \tag{7.758}$$

where

$$\Theta_\mu(t) = \Theta(\mu) = \Theta(\mu(t)). \tag{7.759}$$

Similarly,

$$\begin{aligned} &\left(\mathbb{X}_s \circ \mathbb{X}_s^{-1}\right) \Theta_{\mu,s}(u) \\ &= \int_a^b \left(\frac{1}{2\pi i} \int_x^y \Theta_{\mu,s}(u) e^{s(u)\mu(t)} s^{(1)}(u) du \right) e^{-s(u)\mu(t)} \mu^{(1)}(t) dt. \end{aligned} \tag{7.760}$$

Thus, we finish the proof. □

7.5.3 *The Integral Transform Operator with Respect to Monotone Function of Second Type*

Let $\mu^{(1)}(t) > 0$, $\mu(a) = 0$, $\mu(b) = \infty$, and $t, a, b \in \mathbb{R}$

Definition 7.71 The integral transform operator with respect to monotone function of second type is defined as

$$\mathbb{X}\left[\Theta_\mu(t)\right] = \Theta_\mu(s) = \int_a^b \Theta_\mu(t) e^{-s\mu(t)} \mu^{(1)}(t) dt, \tag{7.761}$$

where $\mathbb{X}\left[\Theta_\mu(t)\right] = \Theta_\mu(s)$ is the integral transform operator with respect to monotone function of second type.

Definition 7.72 The inverse integral transform operator with respect to monotone function of second type is defined as

$$\Theta_\mu(t) = \mathbb{X}^{-1}\left[\Theta_\mu(s)\right] = \frac{1}{2\pi} \int_{\sigma - i\infty}^{\sigma + i\infty} \Theta_\mu(s) e^{s\mu(t)} ds, \tag{7.762}$$

where $\Theta_\mu(t) = \mathbb{X}^{-1}[\Theta_\mu(s)]$ is the inverse integral transform operator with respect to monotone function of second type.

The integral transform operator with respect to monotone function of second type converges absolutely if

$$\int_a^b \left|\Theta_\mu(t)\, e^{-s\mu(t)}\right| \mu^{(1)}(t)\, dt \tag{7.763}$$

exists.

Corollary 7.14 *Let $\mu^{(1)}(t) > 0$, $\mu(a) = 0$, $\mu(b) = \infty$, and $t, a, b \in \mathbb{R}$. Then, we have*

$$\Theta_\mu(t) = \left(\mathbb{X}^{-1} \circ \mathbb{X}\right) \Theta_\mu(t) \tag{7.764}$$

and

$$\Theta_\mu(s) = \left(\mathbb{X} \circ \mathbb{X}^{-1}\right) \Theta_\mu(s). \tag{7.765}$$

Proof Taking $s = \sigma - i\nu$ in (7.759) and (7.760), we have the results. □

There are the special cases:

$$\mathbb{X}\left[\Theta_\mu(t)\right] = \Theta_\mu(s) = \int_a^\infty \Theta_\mu(t)\, e^{-s\mu(t)} \mu^{(1)}(t)\, dt \tag{7.766}$$

and

$$\Theta_\mu(t) = \mathbb{X}^{-1}\left[\Theta_\mu(s)\right] = \frac{1}{2\pi} \int_{\sigma - i\infty}^{\sigma + i\infty} \Theta_\mu(s)\, e^{s\mu(t)} ds, \tag{7.767}$$

where

$$\left(\mathbb{X}^{-1} \circ \mathbb{X}\right) \Theta_\mu(t) = \frac{1}{2\pi i} \int_{\sigma - i\infty}^{\sigma + i\infty} \left(\int_a^\infty \Theta_\mu(t)\, e^{-s\mu(t)} \mu^{(1)}(t)\, dt \right) e^{s\mu(t)} ds. \tag{7.768}$$

The formula (7.765) was discovered by Fahad and coauthors [308, 309].

7.5.4 The Integral Transform Operator with Respect to Monotone Function of Third Type

Let $\nu^{(1)}(u) > 0$, $s(u) = \sigma - i\nu(u)$, $s(x) = \sigma - i\infty = \sigma + i\nu(x), s(y) = \sigma + i\infty = \sigma + i\nu(y)$, and $u, x, y \in \mathbb{R}$.

Definition 7.73 The integral transform operator with respect to monotone function of third type is defined as

$$\mathbb{X}_s[\Theta(t)] = \Theta_s(u) = \int_0^\infty \Theta(t)\, e^{-s(u)t} dt, \tag{7.769}$$

where $\mathbb{X}_s[\Theta(t)] = \Theta_s(u)$ is the integral transform operator with respect to monotone function of third type.

Definition 7.74 The inverse integral transform operator with respect to monotone function of third type is defined as

$$\Theta(t) = \mathbb{X}_s^{-1}[\Theta_s(u)] = \frac{1}{2\pi i}\int_x^y \Theta_s(u)\, e^{s(u)t} s^{(1)}(u)\, du, \tag{7.770}$$

where $\Theta(t) = \mathbb{X}_s^{-1}[\Theta_s(u)]$ is the inverse integral transform operator with respect to monotone function of third type.

The integral transform operator with respect to monotone function of third type converges absolutely if

$$\int_0^\infty \left|\Theta(t)\, e^{-s(u)t}\right| dt \tag{7.771}$$

exists.

Corollary 7.15 *Let $\nu^{(1)}(u) > 0$, $s(u) = \sigma - i\nu(u)$, $s(x) = \sigma - i\infty = \sigma + i\nu(x), s(y) = \sigma + i\infty = \sigma + i\nu(y)$, and $u, x, y \in \mathbb{R}$.*

Then, we have

$$\Theta(t) = \left(\mathbb{X}_s^{-1} \circ \mathbb{X}_s\right)\Theta(t) \tag{7.772}$$

and

$$\Theta_s(u) = \left(\mathbb{X}_s \circ \mathbb{X}_s^{-1}\right)\Theta_s(u). \tag{7.773}$$

Proof Taking $\mu(t) = t$ in (7.759) and (7.760), we have the results. □

As a special case of (7.773), there is the formula by [310]

$$f(t) = \frac{1}{2\pi i} \int_{-\infty}^{\infty} F_s(u) e^{s(u)t} s^{(1)}(u) du, \tag{7.774}$$

where $F_s(u) = F(s(u))$ and $s(u)$ is the general parabola in [310], and the integral transform operator with respect to monotone function of third type is defined as

$$F_s(u) = \int_0^{\infty} f(t) e^{-s(u)t} dt \tag{7.775}$$

with

$$f(t) = \frac{1}{2\pi i} \int_{-\infty}^{\infty} \left(\int_0^{\infty} f(t) e^{-s(u)t} dt \right) e^{s(u)t} s^{(1)}(u) du. \tag{7.776}$$

For more details of (7.777), see Talbot [311], López-Fernández and Palencia [312], and Rizzardi [313].

7.5.5 *The Integral Transform Operator with Respect to Scaling-Law Function of Second Type*

Let $\alpha, \lambda > 0$ and $t^{\alpha-1} > 0$ for $t \in \mathbb{R}_+$.

Definition 7.75 The integral transform operator with respect to scaling-law function of second type is defined as

$$\widetilde{\mathbb{X}}[\Theta_\alpha(t)] = \Theta_\alpha(s) = \lambda\alpha \int_0^{\infty} \Theta_\alpha(t) e^{-\lambda s t^\alpha} t^{\alpha-1} dt, \tag{7.777}$$

where $\widetilde{\mathbb{X}}[\Theta_\alpha(t)] = \Theta_\alpha(s)$ is the integral transform operator with respect to scaling-law function of second type.

Definition 7.76 The inverse integral transform operator with respect to scaling-law function of second type is defined as

$$\Theta_\alpha(t) = \widetilde{\mathbb{X}}^{-1}[\Theta_\alpha(s)] = \frac{1}{2\pi} \int_{\sigma-i\infty}^{\sigma+i\infty} \Theta_\alpha(s) e^{\lambda s t^\alpha} ds, \tag{7.778}$$

where $\Theta_\alpha(t) = \widetilde{\mathbb{X}}^{-1}[\Theta_\alpha(s)]$ is the inverse integral transform operator with respect to scaling-law function of second type.

The integral transform operator with respect to scaling-law function of second type converges absolutely if

$$\lambda\alpha \int_0^\infty \left|\Theta_\mu(t)\, e^{-\lambda s t^\alpha}\right| t^{\alpha-1} dt \tag{7.779}$$

exists.

Corollary 7.16 *Let $\alpha, \lambda > 0$ and $t^{\alpha-1} > 0$ for $t \in \mathbb{R}_+$. Then, we have*

$$\Theta_\alpha(t) = \left(\widetilde{\mathbb{X}}^{-1} \circ \widetilde{\mathbb{X}}\right)\Theta_\alpha(t) \tag{7.780}$$

and

$$\Theta_\alpha(s) = \left(\widetilde{\mathbb{X}} \circ \widetilde{\mathbb{X}}^{-1}\right)\Theta_\alpha(s). \tag{7.781}$$

Proof Taking $\mu(t) = \lambda t^\alpha$ in (7.775) and (7.776), we have the results. □

7.5.6 The Integral Transform Operator with Respect to Power-Law Function of Second Type

Let $\alpha > 0$ and $t^{\alpha-1} > 0$ for $t \in \mathbb{R}_+$.

Definition 7.77 The integral transform operator with respect to power-law function of second type is defined as

$$\widehat{\mathbb{X}}[\Theta_\alpha(t)] = \Theta_\alpha(s) = \alpha \int_0^\infty \Theta_\alpha(t)\, e^{-st^\alpha} t^{\alpha-1} dt, \tag{7.782}$$

where $\widehat{\mathbb{X}}[\Theta_\alpha(t)]$ is the integral transform operator with respect to power-law function of second type.

Definition 7.78 The inverse integral transform operator with respect to power-law function of second type is defined as

$$\Theta_\alpha(t) = \widehat{\mathbb{X}}^{-1}[\Theta_\alpha(s)] = \frac{1}{2\pi} \int_{\sigma-i\infty}^{\sigma+i\infty} \Theta_\alpha(s)\, e^{st^\alpha} ds, \tag{7.783}$$

where $\Theta_\alpha(t) = \widehat{\mathbb{X}}^{-1}[\Theta_\alpha(s)]$ is the inverse integral transform operator with respect to power-law function of second type.

The integral transform operator with respect to power-law function of second type converges absolutely if

$$\alpha \int_0^\infty \left| \Theta_\mu(t) e^{-st^\alpha} \right| t^{\alpha-1} dt \tag{7.784}$$

exists.

Corollary 7.17 *Let $\alpha > 0$ and $t^{\alpha-1} > 0$ for $t \in \mathbb{R}_+$. Then, we have*

$$\Theta_\alpha(t) = \left(\widehat{\mathbb{X}}^{-1} \circ \widehat{\mathbb{X}}\right) \Theta_\alpha(t) \tag{7.785}$$

and

$$\Theta_\alpha(s) = \left(\widehat{\mathbb{X}} \circ \widehat{\mathbb{X}}^{-1}\right) \Theta_\alpha(s). \tag{7.786}$$

Proof Taking $\lambda = 1$ in (7.783) and (7.784), we have the results. □

7.5.7 The Bilateral Integral Transform Operator with Respect to Monotone Function of Fourth Type

Let $\mu^{(1)}(t) > 0$, $\mu(a) = -\infty$, $\mu(b) = \infty$ and $t, a, b \in \mathbb{R}$. Let $\nu^{(1)}(u) > 0$, $s(u) = \sigma - i\nu(u)$, $s(x) = \sigma - i\infty = \sigma + i\nu(x)\, s(y) = \sigma + i\infty = \sigma + i\nu(y)$, and $u, x, y \in \mathbb{R}$.

Definition 7.79 The bilateral integral transform operator with respect to monotone function of first type is defined as

$$\mathbb{Y}_s\left[\Theta_\mu(t)\right] = \Theta_{\mu,s}(u) = \int_a^b \Theta_\mu(t) e^{-s(u)\mu(t)} \mu^{(1)}(t)\, dt, \tag{7.787}$$

where $\mathbb{Y}_s\left[\Theta_\mu(t)\right] = \Theta_{\mu,s}(u)$ is the bilateral integral transform operator with respect to monotone function of first type.

Definition 7.80 The inverse bilateral integral transform operator with respect to monotone function of first type is defined as

$$\Theta_{\mu}(t) = \Upsilon_{s}^{-1}\left[\Theta_{\mu,s}(u)\right] = \frac{1}{2\pi i}\int_{x}^{y} \Theta_{\mu,s}(u)\, e^{s(u)\mu(t)} s^{(1)}(u)\, du, \tag{7.788}$$

where $\Theta_{\mu}(t) = \Upsilon_{s}^{-1}\left[\Theta_{\mu,s}(u)\right]$ is the bilateral inverse integral transform operator with respect to monotone function of first type.

The bilateral integral transform operator with respect to monotone function of first type converges absolutely if

$$\int_{a}^{b} \left|\Theta_{\mu}(t)\, e^{-s(u)\mu(t)}\right| \mu^{(1)}(t)\, dt \tag{7.789}$$

exists.

Theorem 7.46 *Let $\mu^{(1)}(t) > 0$, $\mu(a) = -\infty$, $\mu(b) = \infty$, and $t, a, b \in \mathbb{R}$. Let $\nu^{(1)}(u) > 0$, $s(u) = \sigma - i\nu(u)$, $s(x) = \sigma - i\infty = \sigma + i\nu(x)$, $s(y) = \sigma + i\infty = \sigma + i\nu(y)$, and $u, x, y \in \mathbb{R}$. Then, we have*

$$\Theta_{\mu}(t) = \left(\Upsilon_{s}^{-1} \circ \Upsilon_{s}\right) \Theta_{\mu}(t) \tag{7.790}$$

and

$$\Theta_{\mu,s}(u) = \left(\Upsilon_{s} \circ \Upsilon_{s}^{-1}\right) \Theta_{\mu,s}(u) \tag{7.791}$$

Proof Let us consider the formulas

$$\Upsilon_{\varpi}\left[\Theta_{\mu}(t)\right] = \Theta_{\mu,\varpi}(\omega) = \int_{a}^{b} \Theta_{\mu}(t)\, e^{-i\varpi(\omega)\mu(t)} \mu^{(1)}(t)\, dt \tag{7.792}$$

and

$$\Theta_{\mu}(t) = \Upsilon_{\varpi}^{-1}\left[\Theta_{\mu,\varpi}(\omega)\right] = \frac{1}{2\pi}\int_{x}^{y} \Theta_{\mu,\varpi}(\omega)\, e^{i\varpi(\omega)\mu(t)} \varpi^{(1)}(\omega)\, d\omega, \tag{7.793}$$

where $\mu^{(1)}(t) > 0$, $\mu(a) = -\infty$, $\mu(b) = \infty$, $\varpi^{(1)}(\omega) > 0$, $\varpi(x) = -\infty$, $\varpi(y) = +\infty$, and $t, a, b, \omega, x, y \in \mathbb{R}$.

Taking $i\varpi(\omega) = s(u) = -i\nu(u)$ in (7.795) and (7.796), we have the results. □

7.5.8 *The Bilateral Integral Transform Operator with Respect to Monotone Function of Second Type*

Let $\mu^{(1)}(t) > 0$, $\mu(a) = -\infty$, $\mu(b) = \infty$, and $t, a, b \in \mathbb{R}$.

Definition 7.81 The bilateral integral transform operator with respect to monotone function of second type is defined as

$$\mathbb{Y}\left[\Theta_\mu(t)\right] = \Theta_\mu(s) = \int_a^b \Theta_\mu(t)\, e^{-s\mu(t)} \mu^{(1)}(t)\, dt, \tag{7.794}$$

where $\mathbb{Y}\left[\Theta_\mu(t)\right] = \Theta_\mu(s)$ is the bilateral integral transform operator with respect to monotone function of second type.

Definition 7.82 The inverse bilateral integral transform operator with respect to monotone function of second type is defined as

$$\Theta_\mu(t) = \mathbb{Y}^{-1}\left[\Theta_\mu(s)\right] = \frac{1}{2\pi} \int_{\sigma-i\infty}^{\sigma+i\infty} \Theta_\mu(s)\, e^{s\mu(t)} ds, \tag{7.795}$$

where $\Theta_\mu(t) = \mathbb{Y}^{-1}\left[\Theta_\mu(s)\right]$ is the inverse bilateral integral transform operator with respect to monotone function of second type.

The bilateral integral transform operator with respect to monotone function of second type converges absolutely if

$$\int_a^b \left|\Theta_\mu(t)\, e^{-s\mu(t)}\right| \mu^{(1)}(t)\, dt \tag{7.796}$$

exists.

Corollary 7.18 *Let* $\mu^{(1)}(t) > 0$, $\mu(a) = 0$, $\mu(b) = \infty$, *and* $t, a, b \in \mathbb{R}$
Then, we have

$$\Theta_\mu(t) = \left(\mathbb{Y}^{-1} \circ \mathbb{Y}\right) \Theta_\mu(t) \tag{7.797}$$

and

$$\Theta_\mu(s) = \left(\mathbb{Y} \circ \mathbb{Y}^{-1}\right) \Theta_\mu(s). \tag{7.798}$$

Proof Taking $s = \sigma - i\nu$ in (7.793) and (7.794), we deduce the results. □

7.5.9 The Bilateral Integral Transform Operator with Respect to Monotone Function of Third Type

Let $\nu^{(1)}(u) > 0$, $s(u) = \sigma - i\nu(u)$, $s(x) = \sigma - i\infty = \sigma + i\nu(x)$, $s(y) = \sigma + i\infty = \sigma + i\nu(y)$, and $u, x, y \in \mathbb{R}$

Definition 7.83 The bilateral integral transform operator with respect to monotone function of third type is defined as

$$\mathbb{Y}_s[\Theta(t)] = \Theta_s(u) = \int_{-\infty}^{\infty} \Theta(t)\, e^{-s(u)t} dt, \tag{7.799}$$

where $\mathbb{Y}_s[\Theta(t)] = \Theta_s(u)$ is the bilateral integral transform operator with respect to monotone function of third type.

Definition 7.84 The inverse bilateral integral transform operator with respect to monotone function of third type is defined as

$$\Theta(t) = \mathbb{Y}_s^{-1}[\Theta_s(u)] = \frac{1}{2\pi i}\int_{x}^{y} \Theta_s(u)\, e^{s(u)t} s^{(1)}(u)\, du, \tag{7.800}$$

where $\Theta(t) = \mathbb{Y}_s^{-1}[\Theta_s(u)]$ is the inverse bilateral integral transform operator with respect to monotone function of third type.

The bilateral integral transform operator with respect to monotone function of third type converges absolutely if

$$\int_{-\infty}^{\infty} \left|\Theta(t)\, e^{-s(u)t}\right| dt \tag{7.801}$$

exists.

Corollary 7.19 *Let* $\nu^{(1)}(u) > 0$, $s(u) = \sigma - i\nu(u)$, $s(x) = \sigma - i\infty = \sigma + i\nu(x)$, $s(y) = \sigma + i\infty = \sigma + i\nu(y)$, *and* $u, x, y \in \mathbb{R}$.

Then, we have

$$\Theta(t) = \left(\mathbb{Y}_s^{-1} \circ \mathbb{Y}_s\right)\Theta(t) \tag{7.802}$$

and

$$\Theta_s(u) = \left(\mathbb{Y}_s \circ \mathbb{Y}_s^{-1}\right)\Theta_s(u). \tag{7.803}$$

Proof Taking $\mu(t) = t$ in (7.793) and (7.794), we reduce to the desired results. □

7.5.10 The Bilateral Integral Transform Operator with Respect to Scaling-Law Function of Second Type

Let $\alpha, \lambda > 0$ and $t^{\alpha-1} > 0$ for $t \in \mathbb{R}$.

Definition 7.85 The bilateral integral transform operator with respect to scaling-law function of second type is defined as

$$\widetilde{\mathbb{Y}}\left[\Theta_\alpha(t)\right] = \Theta_\alpha(s) = \lambda\alpha \int_{-\infty}^{\infty} \Theta_\alpha(t)\, e^{-\lambda s t^\alpha} t^{\alpha-1} dt, \tag{7.804}$$

where $\widetilde{\mathbb{Y}}\left[\Theta_\alpha(t)\right] = \Theta_\alpha(s)$ is the bilateral integral transform operator with respect to scaling-law function of second type.

Definition 7.86 The inverse bilateral integral transform operator with respect to scaling-law function of second type is defined as

$$\Theta_\alpha(t) = \widetilde{\mathbb{Y}}^{-1}\left[\Theta_\alpha(s)\right] = \frac{1}{2\pi} \int_{\sigma-i\infty}^{\sigma+i\infty} \Theta_\alpha(s)\, e^{\lambda s t^\alpha} ds, \tag{7.805}$$

where $\Theta_\alpha(t) = \widetilde{\mathbb{Y}}^{-1}\left[\Theta_\alpha(s)\right]$ is the inverse bilateral integral transform operator with respect to scaling-law function of second type.

The bilateral integral transform operator with respect to scaling-law function of second type converges absolutely if

$$\lambda\alpha \int_{-\infty}^{\infty} \left|\Theta_\mu(t)\, e^{-\lambda s t^\alpha}\right| t^{\alpha-1} dt \tag{7.806}$$

exists.

Corollary 7.20 *Let $\alpha, \lambda > 0$ and $t^{\alpha-1} > 0$ for $t \in \mathbb{R}$. Then we have*

$$\Theta_\alpha(t) = \left(\widetilde{\mathbb{Y}}^{-1} \circ \widetilde{\mathbb{Y}}\right) \Theta_\alpha(t) \tag{7.807}$$

and

$$\Theta_\alpha(s) = \left(\widetilde{\mathbb{Y}} \circ \widetilde{\mathbb{Y}}^{-1}\right) \Theta_\alpha(s). \tag{7.808}$$

Proof Taking $\mu(t) = \lambda t^\alpha$ in (7.797) and (7.798), we obtain the results. □

7.5.11 The Bilateral Integral Transform Operator with Respect to Power-Law Function of Second Type

Let $\alpha > 0$ and $t^{\alpha-1} > 0$ for $t \in \mathbb{R}$.

Definition 7.87 The bilateral integral transform operator with respect to power-law function of second type is defined as

$$\widehat{\mathbb{Y}}\left[\Theta_\alpha(t)\right] = \Theta_\alpha(s) = \alpha \int\limits_{-\infty}^{\infty} \Theta_\alpha(t)\, e^{-st^\alpha} t^{\alpha-1} dt, \tag{7.809}$$

where $\widehat{\mathbb{Y}}\left[\Theta_\alpha(t)\right]$ is the bilateral integral transform operator with respect to power-law function of second type.

Definition 7.88 The inverse bilateral integral transform operator with respect to power-law function of second type is defined as

$$\Theta_\alpha(t) = \widehat{\mathbb{Y}}^{-1}\left[\Theta_\alpha(s)\right] = \frac{1}{2\pi} \int\limits_{\sigma-i\infty}^{\sigma+i\infty} \Theta_\alpha(s)\, e^{st^\alpha} ds, \tag{7.810}$$

where $\Theta_\alpha(t) = \widehat{\mathbb{Y}}^{-1}\left[\Theta_\alpha(s)\right]$ is the inverse integral transform operator with respect to power-law function of second type.

The bilateral integral transform operator with respect to power-law function of second type converges absolutely if

$$\alpha \int\limits_{-\infty}^{\infty} \left|\Theta_\mu(t)\, e^{-st^\alpha}\right| t^{\alpha-1} dt \tag{7.811}$$

exists.

Corollary 7.21 *Let $\alpha > 0$ and $t^{\alpha-1} > 0$ for $t \in \mathbb{R}$. Then, we have*

$$\Theta_\alpha(t) = \left(\widehat{\mathbb{Y}}^{-1} \circ \widehat{\mathbb{Y}}\right) \Theta_\alpha(t) \tag{7.812}$$

and

$$\Theta_\alpha(s) = \left(\widehat{\mathbb{Y}} \circ \widehat{\mathbb{Y}}^{-1}\right) \Theta_\alpha(s). \tag{7.813}$$

Proof Taking $\lambda = 1$ in (7.797) and (7.798), we have the results. □

7.5.12 Applications

Theorem 7.47 *Let* $\mu^{(1)}(t) > 0$, $\mu(a) = 0$, $\mu(b) = \infty$, $\phi(t) = (\varphi \circ \mu)(t) = \varphi(\mu(t))$, *and* $t, a, b \in \mathbb{R}$

The integral transform of the Leibniz derivative of the function $\phi(t)$ *is given as follows:*

$$\int_a^b D_{t,\mu(\cdot)}^{(1)} \phi(t)\, e^{-s\mu(t)} \mu^{(1)}(t)\, dt = s\varphi(s) - \varphi(0). \tag{7.814}$$

Proof Making use of the formula (7.772), we have

$$\int_a^b D_{t,\mu(\cdot)}^{(1)} \phi(t)\, e^{-s\mu(t)} \mu^{(1)}(t)\, dt = \int_a^b D_{t,\mu(\cdot)}^{(1)} \phi(t)\, e^{-s\mu(t)} d\mu(t), \tag{7.815}$$

which can be given as

$$\int_a^b D_{t,\mu(\cdot)}^{(1)} \phi(t)\, e^{-s\mu(t)} d\mu(t) = \int_0^\infty D_t^{(1)} \varphi(t)\, e^{-st} dt = s\varphi(s) - \varphi(0). \tag{7.816}$$

Thus, we complete the proof. □

Theorem 7.48 *Let* $\mu^{(1)}(t) > 0$, $\mu(a) = 0$, $\mu(b) = \infty$, $\psi(t) = (\Omega \circ \mu)(t) = \Omega(\mu(t))$, *and* $t, a, b \in \mathbb{R}$

The integral transform of the Riemann–Stieltjes integral of the function $\psi(t)$ *is given as follows:*

$$\int_a^b {}_aI_{t,\mu(\cdot)}^{(1)} \psi(t)\, e^{-s\mu(t)} \mu^{(1)}(t)\, dt = \frac{\Omega(s)}{s}, \tag{7.817}$$

Proof Making use of the formula (7.772), we have that

$$\int_a^b {}_aI_{t,\mu(\cdot)}^{(1)} \psi(t)\, e^{-s\mu(t)} \mu^{(1)}(t)\, dt = \int_a^b {}_aI_{t,\mu(\cdot)}^{(1)} \psi(t)\, e^{-s\mu(t)} d\mu(t), \tag{7.818}$$

which can be presented as follows:

$$\int_a^b {}_aI_t^{(1)} \Omega(t)\, e^{-s\mu(t)} d\mu(t) = \int_0^\infty {}_0I_t^{(1)} \Omega(t)\, e^{-st} dt = \frac{\Omega(s)}{s}. \tag{7.819}$$

□

Theorem 7.49 *Let* $\beta, \lambda > 0$, $\phi(t) = \left(\varphi \circ \left(\lambda t^{\beta}\right)\right)(t) = \varphi\left(\lambda t^{\beta}\right)$, *and* $t^{\beta-1} > 0$.

The integral transform of the scaling-law derivative of the function $\phi(t)$ *is given as follows:*

$$\int_0^{\infty} {}^{SL}D_t^{(1)} \phi(t)\, e^{-s\lambda t^{\beta}} t^{\beta-1} dt = \frac{1}{\lambda\beta}\left(s\varphi(s) - \varphi(0)\right). \tag{7.820}$$

Proof Taking $\mu(t) = \lambda t^{\beta}$ in (7.63), we obtain the result. □

Theorem 7.50 *Let* $\beta, \lambda > 0$, $\psi(t) = \left(\Omega \circ \left(\lambda t^{\beta}\right)\right)(t) = \Omega\left(\lambda t^{\beta}\right)$, *and* $t^{\beta-1} > 0$.

The integral transform of the scaling-law integral of the function $\psi(t)$ *is given as follows:*

$$\int_0^{\infty} {}_0^{SL}I_t^{(1)} \psi(t)\, e^{-i\omega\lambda t^{\beta}} t^{\beta-1} dt = \frac{\Omega(s)}{s\lambda\beta}. \tag{7.821}$$

Proof Taking $\mu(t) = \lambda t^{\beta}$ in (7.820), we have the result. □

Theorem 7.51 *Let* $\beta > 0$, $\phi(t) = \left(\varphi \circ \left(t^{\beta}\right)\right)(t) = \varphi\left(t^{\beta}\right)$, *and* $t^{\beta-1} > 0$.

The integral transform of the power-law derivative of the function $\phi(t)$ *is given as follows:*

$$\int_0^{\infty} {}^{P}D_t^{(1)} \phi(t)\, e^{-st^{\beta}} t^{\beta-1} dt = \frac{s}{\beta}\varphi(s). \tag{7.822}$$

Proof Taking $\lambda = 1$ in (7.823), we obtain the result. □

Theorem 7.52 *Let* $\beta > 0$, $\psi(t) = \left(\Omega \circ \left(t^{\beta}\right)\right)(t) = \Omega\left(t^{\beta}\right)$, *and* $t^{\beta-1} > 0$.

The integral transform of the power-law integral of the function $\psi(t)$ *is given as follows:*

$$\int_0^{\infty} {}_0^{P}I_t^{(1)} \psi(t)\, e^{-st^{\beta}} t^{\beta-1} dt = \frac{\Omega(s)}{s\beta}. \tag{7.823}$$

Proof Taking $\lambda = 1$ in (7.824), we obtain the result. □

Note: The more generalized results can be found in Chap. 2.

7.6 The Mellin-Like Transforms via Subtrigonometric Functions with Respect to Monotone Function

7.6.1 *Mellin Transform: History, Concepts, and Theorems*

In 1896, Robert Hjalmar Mellin, a Finnish mathematician, proposed the well-known Mellin transform [91].

Definition 7.89 The Mellin transform is defined as [91]

$$F(z) = M\{f(\zeta)\} = \int_0^\infty f(\zeta)\,\zeta^{z-1}d\zeta, \tag{7.824}$$

where $z = \sigma + i\omega$.

Definition 7.90 The inverse Mellin transform is defined as [91]

$$f(t) = M^{-1}\{F(z)\} = \frac{1}{2\pi i}\int_{\sigma-i\infty}^{\sigma+i\infty} F(z)\,\zeta^{-z}dz, \tag{7.825}$$

where $0 < a < \sigma < b$ and $|\arg\zeta| < \pi$.

7.6.2 *The Integral Transform Operator with Respect to Monotone Function of First Type*

Definition 7.91 Let $\mu^{(1)}(t) > 0$, $\mu(a) = 0$, $\mu(b) = \infty$, and $t, a, b \in \mathbb{R}$.

Let $\varpi^{(1)}(u) > 0$, $s(u) = \sigma - i\varpi(u)$, $s(x) = \sigma - i\infty = \sigma + i\varpi(x)$, $s(y) = \sigma + i\infty = \sigma + i\varpi(y)$, and $u, x, y \in \mathbb{R}$.

The integral transform operator with respect to monotone function of first type is defined as

$$\mathbb{J}_s\left[\Theta_\mu(t)\right] = \Theta_{\mu,s}(u) = \int_a^b \Theta_\mu(t)\,(\mu(t))^{s(u)-1}\,\mu^{(1)}(t)\,dt, \tag{7.826}$$

where $\mathbb{J}_s\left[\Theta_\mu(t)\right] = \Theta_{\mu,s}(u)$ is the integral transform operator with respect to monotone function of first type.

Definition 7.92 Let $0 < \sigma_1 < \sigma < \sigma_2$ and $|\arg\mu(t)| < \pi$.

The inverse integral transform operator with respect to monotone function of first type is defined as

$$\Theta_{\mu}(t) = \mathbb{J}_s^{-1}\left[\Theta_{\mu,s}(u)\right] = \frac{1}{2\pi i}\int_x^y \Theta_{\mu,s}(u)\,(\mu(t))^{-s(u)}\,s^{(1)}(u)\,du, \tag{7.827}$$

where $\Theta_{\mu}(t) = \mathbb{J}_s^{-1}\left[\Theta_{\mu,s}(u)\right]$ is the inverse integral transform operator with respect to monotone function of first type.

The integral transform operator with respect to monotone function of first type converges absolutely if

$$\int_a^b \left|\Theta_{\mu}(t)\,(\mu(t))^{s(u)-1}\right|\mu^{(1)}(t)\,dt \tag{7.828}$$

exists for $0 < \sigma_1 < \sigma < \sigma_2$ and $|\arg\mu(t)| < \pi$.

Theorem 7.53 *Let* $\mu^{(1)}(t) > 0$, $\mu(a) = 0$, $\mu(b) = \infty$, *and* $t, a, b \in \mathbb{R}$.

Let $\varpi^{(1)}(u) > 0$, $s(u) = \sigma - i\varpi(u)$, $s(x) = \sigma - i\infty = \sigma + i\varpi(x)$, $s(y) = \sigma + i\infty = \sigma + i\varpi(y)$, *and* $u, x, y \in \mathbb{R}$.

Then, we have

$$\Theta_{\mu}(t) = \left(\mathbb{J}_s^{-1}\circ\mathbb{J}_s\right)\Theta_{\mu}(t) \tag{7.829}$$

and

$$\Theta_{\mu,s}(u) = \left(\mathbb{J}_s\circ\mathbb{J}_s^{-1}\right)\Theta_{\mu,s}(u). \tag{7.830}$$

Proof From (7.829) and (7.830), we have that

$$\begin{aligned}&\left(\mathbb{J}_s^{-1}\circ\mathbb{J}_s\right)\Theta_{\mu}(t)\\&\quad= \frac{1}{2\pi i}\int_x^y\left(\int_a^b \Theta_{\mu}(t)\,(\mu(t))^{s(u)-1}\,\mu^{(1)}(t)\,dt\right)(\mu(t))^{-s(u)}\,s^{(1)}(u)\,du,\end{aligned} \tag{7.831}$$

and

$$\begin{aligned}&\left(\mathbb{J}_s\circ\mathbb{J}_s^{-1}\right)\Theta_{\mu,s}(u)\\&\quad= \int_a^b\left(\frac{1}{2\pi i}\int_x^y \Theta_{\mu,s}(u)\,(\mu(t))^{-s(u)}\,s^{(1)}(u)\,du\right)(\mu(t))^{s(u)-1}\,\mu^{(1)}(t)\,dt.\end{aligned} \tag{7.832}$$

It is the generalized version of the Mellin transform. □

7.6.3 The Integral Transform Operator with Respect to Monotone Function of Second Type

Let $\mu^{(1)}(t) > 0$, $\mu(a) = 0$, $\mu(b) = \infty$, and $t, a, b \in \mathbb{R}$.

Definition 7.93 The integral transform operator with respect to monotone function of second type is defined as

$$\mathbb{J}\left[\Theta_{\mu}(t)\right] = \Theta_{\mu}(s) = \int_{a}^{b} \Theta_{\mu}(t)(\mu(t))^{s-1}\mu^{(1)}(t)\,dt, \tag{7.833}$$

where $\mathbb{J}\left[\Theta_{\mu}(t)\right] = \Theta_{\mu}(s)$ is the integral transform operator with respect to monotone function of second type.

Definition 7.94 Let $0 < \sigma_1 < \sigma < \sigma_2$ and $|\arg \mu(t)| < \pi$.

The inverse integral transform operator with respect to monotone function of second type is defined as

$$\Theta_{\mu}(t) = \mathbb{J}^{-1}\left[\Theta_{\mu}(s)\right] = \frac{1}{2\pi i}\int_{\sigma-i\infty}^{\sigma+i\infty} \Theta_{\mu}(s)(\mu(t))^{-s}\,ds, \tag{7.834}$$

where $\Theta_{\mu}(t) = \mathbb{J}^{-1}\left[\Theta_{\mu}(s)\right]$ is the inverse integral transform operator with respect to monotone function of second type.

The integral transform operator with respect to monotone function of second type converges absolutely if

$$\int_{a}^{b} \left|\Theta_{\mu}(t)(\mu(t))^{s-1}\right|\mu^{(1)}(t)\,dt \tag{7.835}$$

exists for $0 < \sigma_1 < \sigma < \sigma_2$ and $|\arg \mu(t)| < \pi$.

Theorem 7.54 *Let $\mu^{(1)}(t) > 0$, $\mu(a) = 0$, $\mu(b) = \infty$, and $t, a, b \in \mathbb{R}$.*

Then, we have

$$\Theta_{\mu}(t) = \left(\mathbb{J}^{-1} \circ \mathbb{J}\right)\Theta_{\mu}(t) \tag{7.836}$$

and

$$\Theta_{\mu}(s) = \left(\mathbb{J}^{-1} \circ \mathbb{J}\right)\Theta_{\mu}(s). \tag{7.837}$$

Proof Taking $s = \sigma - i\varpi$ in (7.832) and (7.833), we obtain the results. □

Definition 7.95 Let $\varpi^{(1)}(u) > 0$, $s(u) = \sigma - i\varpi(u)$, $s(x) = \sigma - i\infty = \sigma + i\varpi(x)$, $s(y) = \sigma + i\infty = \sigma + i\varpi(y)$, and $u, x, y \in \mathbb{R}$.

The integral transform operator with respect to monotone function of third type is defined as

$$\mathbb{J}_s[\Theta(t)] = \Theta_s(u) = \int_0^\infty \Theta(t)\, t^{s(u)-1} dt, \tag{7.838}$$

where $\mathbb{J}_s[\Theta(t)] = \Theta_s(u)$ is the integral transform operator with respect to monotone function of third type.

Definition 7.96 Let $\varpi^{(1)}(u) > 0$, $s(u) = \sigma - i\varpi(u)$, $s(x) = \sigma - i\infty = \sigma + i\varpi(x)$, $s(y) = \sigma + i\infty = \sigma + i\varpi(y)$, and $u, x, y \in \mathbb{R}$.

Let $0 < \sigma_1 < \sigma < \sigma_2$ and $|\arg t| < \pi$.

The inverse integral transform operator with respect to monotone function of third type is defined as

$$\Theta(t) = \mathbb{J}_s^{-1}[\Theta_s(u)] = \frac{1}{2\pi i} \int_x^y \Theta_s(u)\, t^{-s(u)} s^{(1)}(u)\, du, \tag{7.839}$$

where $\Theta(t) = \mathbb{J}_s^{-1}[\Theta_s(u)]$ is the inverse integral transform operator with respect to monotone function of third type.

The integral transform operator with respect to monotone function of third type converges absolutely if

$$\int_a^b \left| \Theta(t)\, t^{s(u)-1} \right| dt \tag{7.840}$$

exists for $0 < \sigma_1 < \sigma < \sigma_2$ and $|\arg \mu(t)| < \pi$.

Theorem 7.55 *Let $\varpi^{(1)}(u) > 0$, $s(u) = \sigma - i\varpi(u)$, $s(x) = \sigma - i\infty = \sigma + i\varpi(x)$, $s(y) = \sigma + i\infty = \sigma + i\varpi(y)$, and $u, x, y \in \mathbb{R}$.*

Let $0 < \sigma_1 < \sigma < \sigma_2$ and $|\arg t| < \pi$.

Then, we have

$$\Theta(t) = \left(\mathbb{J}_s^{-1} \circ \mathbb{J}_s\right) \Theta(t) \tag{7.841}$$

and

$$\Theta_s(u) = \left(\mathbb{J}_s \circ \mathbb{J}_s^{-1}\right) \Theta_s(u). \tag{7.842}$$

Proof Taking $\mu(t) = t$ in (7.832) and (7.833), we obtain the results. □

7.6.4 The Integral Transform Operator with Respect to Scaling-Law Function of Second Type

Let $\alpha, \lambda > 0$ and $t^{\alpha-1} > 0$ for $t \in \mathbb{R}_+$.

Definition 7.97 The integral transform operator with respect to scaling-law function of second type is defined as

$$\mathbb{X}\left[\Theta_\alpha(t)\right] = \Theta_\alpha(s) = \lambda\alpha \int_0^\infty \Theta_\alpha(t)\left(\lambda t^\alpha\right)^{s-1} t^{\alpha-1} dt, \tag{7.843}$$

where $\mathbb{X}\left[\Theta_\alpha(t)\right] = \Theta_\alpha(s)$ is the integral transform operator with respect to scaling-law function of second type.

Definition 7.98 Let $0 < \sigma_1 < \sigma < \sigma_2$ and $|\arg \mu(t)| < \pi$.

The inverse integral transform operator with respect to scaling-law function of second type is defined as

$$\Theta_\alpha(t) = \mathbb{X}^{-1}\left[\Theta_\alpha(s)\right] = \frac{1}{2\pi i} \int_{\sigma-i\infty}^{\sigma+i\infty} \Theta_\alpha(s)\left(\lambda t^\alpha\right)^{-s} ds, \tag{7.844}$$

where $\Theta_\alpha(t) = \mathbb{X}^{-1}\left[\Theta_\alpha(s)\right]$ is the inverse integral transform operator with respect to scaling-law function of second type.

The integral transform operator with respect to scaling-law function of second type converges absolutely if

$$\lambda\alpha \int_0^\infty \left|\Theta_\alpha(t)\left(\lambda t^\alpha\right)^{s-1}\right| t^{\alpha-1} dt \tag{7.845}$$

exists for $0 < \sigma_1 < \sigma < \sigma_2$ and $|\arg \mu(t)| < \pi$.

Theorem 7.56 *Let $\mu^{(1)}(t) > 0$, $\mu(a) = 0$, $\mu(b) = \infty$, and $t, a, b \in \mathbb{R}$.*

Then, we have

$$\Theta_\alpha(t) = \left(\mathbb{X}^{-1} \circ \mathbb{X}\right) \Theta_\alpha(t) \tag{7.846}$$

and

$$\Theta_\alpha(s) = \left(\mathbb{X}^{-1} \circ \mathbb{X}\right) \Theta_\alpha(s). \tag{7.847}$$

Proof Taking $\mu(t) = \lambda t^\alpha$ in (7.839) and (7.840), we obtain the results. □

7.6.5 *The Integral Transform Operator with Respect to Power-Law Function of Second Type*

Let $\alpha > 0$ and $t^{\alpha-1} > 0$ for $t \in \mathbb{R}_+$.

Definition 7.99 The integral transform operator with respect to power-law function of second type is defined as

$$\widetilde{\mathbb{X}}\left[\Theta_\alpha(t)\right] = \Theta_\alpha(s) = \lambda\alpha \int_0^\infty \Theta_\alpha(t)\left(\lambda t^\alpha\right)^{s-1} t^{\alpha-1} dt, \tag{7.848}$$

where $\widetilde{\mathbb{X}}\left[\Theta_\alpha(t)\right] = \Theta_\alpha(s)$ is the integral transform operator with respect to power-law function of second type.

Definition 7.100 Let $0 < \sigma_1 < \sigma < \sigma_2$ and $|\arg\mu(t)| < \pi$.

The inverse integral transform operator with respect to power-law function of second type is defined as

$$\Theta_\alpha(t) = \widetilde{\mathbb{X}}^{-1}\left[\Theta_\alpha(s)\right] = \frac{1}{2\pi i} \int_{\sigma-i\infty}^{\sigma+i\infty} \Theta_\alpha(s)\left(\lambda t^\alpha\right)^{-s} ds, \tag{7.849}$$

where $\Theta_\alpha(t) = \widetilde{\mathbb{X}}^{-1}\left[\Theta_\alpha(s)\right]$ is the inverse integral transform operator with respect to power-law function of second type.

The integral transform operator with respect to power-law function of second type converges absolutely if

$$\lambda\alpha \int_0^\infty \left|\Theta_\alpha(t)\left(\lambda t^\alpha\right)^{s-1}\right| t^{\alpha-1} dt \tag{7.850}$$

exists for $0 < \sigma_1 < \sigma < \sigma_2$ and $|\arg\mu(t)| < \pi$.

Theorem 7.57 *Let $\alpha > 0$ and $t^{\alpha-1} > 0$ for $t \in \mathbb{R}_+$.*

Let $0 < \sigma_1 < \sigma < \sigma_2$ and $|\arg\mu(t)| < \pi$.

Then, we have

$$\Theta_\alpha(t) = \left(\widetilde{\mathbb{X}}^{-1} \circ \widetilde{\mathbb{X}}\right)\Theta_\alpha(t) \tag{7.851}$$

and

$$\Theta_\alpha(s) = \left(\widetilde{\mathbb{X}}^{-1} \circ \widetilde{\mathbb{X}}\right)\Theta_\alpha(s). \tag{7.852}$$

Proof Taking $\lambda = 1$ in (7.849) and (7.850), we obtain the results. □

References

1. Euler, L.: Letter to Goldbach, 13 Oct 1729. Correspondence math. et phys. de quelques celebres geometres du 18e siecle, publiee par Fuss, vol. 1, St. Petersburg, 1843
2. Legendre, A.M.: Exercises de calcul intégral. 2 vols. Courcier, Paris (1814)
3. Weierstrass, K.: Über die Theorie der analytischen Facultäten. J. Reine Angew. Math. **51**, 1–60 (1856)
4. Srivastava, H.M., Choi, J.: Zeta and q-Zeta Functions and Associated Series and Integrals. Elsevier, Amsterdam (2012)
5. Newman, F.W.: On Γ (a), especially when a is negative. Camb. Dubl. Math. J. **3**, 57–60 (1848)
6. Hölder, O.: Ueber die Eigenschaft der Gammafunction keiner algebraischen Differentialgleichung zu genügen. Math. Ann. **28**(1), 1–13 (1886)
7. Moore, E.H.: Concerning transcendentally transcendental functions. Math. Ann. **48**(1–2), 49–74 (1896)
8. Barnes, E.W.: The theory of gamma functions. Messenger Math. **29**, 64–128 (1900)
9. Euler, L.: De progressionibus transcendentibus sen quaroum termini generales algebrare dari nequeunt (1738). Commentarii academiae scientiarum Petropolitanae, 5, 36–57, presented to the St. Petersburg Academy on 28 Nov 1729
10. Brunel, G.E.A.: Monographie de la fonction gamma. Memoires de la societe des sciences physiques et naturelles de Bordeaux, ser. 3, vol. 3, pp. 1–184 (1886)
11. Gronwall, T.H.: An elementary exposition of the theory of the gamma function. Ann. Math. **17**(3), 124–166 (1916)
12. Olver, F.: Asymptotics and Special Functions. CRC Press, Routledge (1997)
13. Bell, W.W.: Special Functions for Scientists and Engineers. Courier Corporation, D. Van Nostrand Company, London (1968)
14. Luke, Y.L. (ed.): Special Functions and Their Approximations, vol. 1. Academic, New York (1969)
15. Bendersky, L.: Sur la fonction gamma généralisée. Acta Math. **61**, 263–322 (1933)
16. Winckler, A.: Neue Theoreme zur Lehre von den bestimmten Integralen, Sitzungs ber. Akad. Wien **21**, 389–426 (1856)
17. Schlömilch, O.: Einiges uiber die Eulerschen Integrale der zweiten Art. Art. Arch. Math. Phys. **2**(4), 167–174 (1844)
18. Newman, F.W.: On Γ (a), especially when a is negative. Camb. Dubl. Math. J. **3**, 57–60 (1848)
19. Whittaker, E.T.: A Course of Modern Analysis: An Introduction to the General Theory of Infinite Series and of Analytic Functions, with an Account of the Principal Transcendental Functions. University Press, Cambridge (1902)

X.-J. Yang, *Theory and Applications of Special Functions for Scientists and Engineers*, https://doi.org/10.1007/978-981-33-6334-2

20. Whittaker, E.T., Watson, G.N.: A Course of Modern Analysis. Cambridge University Press, Cambridge (1927); Reprinted by Cambridge University Press in 1927
21. Titchmarsh, E.C.: Introduction to the Theory of Fourier Integrals. Clarendon Press, Oxford (1948)
22. Euler, L.: Evolutio formulae integralis $\int x^{(f-1)} dx(lx)^{(m/n)}$ integratione a valore x= 0 ad x= 1 extensa. Novi Comment. Acad. Sci. Petropol. **16**, 91–139 (1771). (Published in 1772)
23. Manocha, H.S.H., Srivastava, H.M.: A Treatise on Generating Functions. Mathematics and Its Applications. Ellis Horwood Series, Chichester (1984)
24. Godefroy, M.: Théorie élémentaire des séries: limites–séries à termes constants–séries à termes variables–fonction exponentielle–fonctions circulaires–fonction gamma. Gauthier-Villars, Paris (1903)
25. Tannery, J.: Introduction à la théorie des fonctions d'une variable, vol. 1. A. Hermann, Paris (1910)
26. Andrews, G.E., Askey, R., Roy, R.: Special Functions, vol. 71. Cambridge University Press, Cambridge (1999)
27. Gauss, C.F.: Disquisitiones generales circa seriem infinitam $1+\frac{\alpha\beta}{1\cdot\gamma}x+\frac{\alpha(\alpha+1)\beta(\beta+1)}{1\cdot 2\cdot\gamma(\gamma+1)}xx+\text{etc.}$. Commentationes Societiones Regiae Scientiarum Gottingensis Recentiores, vol. 2, pp. 1–46 (1812). Reprinted in Werke **3**(2), 122–162 (1876)
28. Winckler, A.: Neue Theoreme zur Lehre von den bestimmten Integralen, Sitzungs ber. Akad. Wien **21**, 389–426 (1856)
29. Lerch, M.: Theorie funkce gamma, Vestník ceske Akademie císare Františka Josefa pro vedy, slovesnost a umení v Praze, II (1893)
30. Godefroy, M.: La fonction gamma: théorie, histoire, bibliographie. Gauthier-Villars, Paris (1901)
31. Stieltjes, T.J.: Sur le développement de log $\Gamma(a)$. J. Math. Pures Appl. **5**, 425–466 (1889)
32. Bateman, H.: Higher Transcendental Functions. California Institute of Technology Bateman Manuscript Project, vol. 1. McGraw-Hill, New York (1953), vol. 3 (1955)
33. Temme, N.M.: Special Functions: An Introduction to the Classical Functions of Mathematical Physics. Wiley, New York (2011)
34. Pochhammer, L.: Ueber Hypergeometrischen Funktionen Hoheren Ordungen. J. Reine Angew. Math. **71**(216), 316–352 (1870)
35. Yang, X.J.: General Fractional Derivatives: Theory, Methods and Applications. CRC Press, New York (2019)
36. Slater, L.J.: Generalized Hypergeometric Functions. Cambridge University Press, Cambridge (1966)
37. Rainville, E.D.: Special Functions. Macmillan, New York (1960)
38. Stirling, J.: Methodus differentialis, sive Tractatus de summatione et interpolatione serierum infinitarum, London (1730)
39. Yang, X.-J.: An Introduction to Hypergeometric Supertrigonometric and Superhyperbolic Functions. Academic, New York (2021)
40. Legendre, A.M.: Exercices, I. Courcier, Paris (1811)
41. Binet, J.P.M.: Mémoire sur les intégrales définies eulériennes et sur leur application à la théorie des suites ainsi qu'à l'évaluation des fonctions des grands nombres. J. Ec. Polytech. **16**, 123–343 (1839)
42. Hankel, H.: Dissertation, Leipzig, p. 23; see also the paper: Die Euler'schen Integrale bei unbeschränkter Variabilität des Argumentes. Voss, Leipzig (1863). Z. Angew. Math. Phys. **9**, 1–21 (1864)
43. Gudermann, C.: Additamentum ad functionis…theoriam. J. Reine Angew. Math. **29**, 209–212 (1845)
44. Chaudhry, M.A., Zubair, S.M.: Generalized incomplete gamma functions with applications. J. Comput. Appl. Math. **55**(1), 99–123 (1994)
45. Chaudhry, M.A., Zubair, S.M.: Generalized incomplete gamma functions with applications. J. Comput. Appl. Math. **55**(1), 99–123 (1994)

46. Chaudhry, M.A., Zubair, S.M.: On the decomposition of generalized incomplete gamma functions with applications to Fourier transforms. J. Comput. Appl. Math. **59**(3), 253–284 (1995)
47. Chaudhry, M.A., Zubair, S.M.: Extended incomplete gamma functions with applications. J. Math. Anal. Appl. **274**(2), 725–745 (2002)
48. Chaudhry, M.A., Qadir, A., Rafique, M., Zubair, S.M.: Extension of Euler's beta function. J. Comput. Appl. Math. **78**(1), 19–32 (1997)
49. Choi, J., Rathie, A.K., Parmar, R.K.: Extension of extended beta, hypergeometric and confluent hypergeometric functions. Honam Math. J. **36**(2), 339–367 (2014)
50. Chaudhry, M.A., Qadir, A., Srivastava, H.M., Paris, R.B.: Extended hypergeometric and confluent hypergeometric functions. Appl. Math. Comput. **159**(2), 589–602 (2004)
51. Euler, L.: Institutiones calculi integralis, vol. 12. Academia Imperialis Scientiarum, pp. 221–230 (1679)
52. Chu, C.-k.: Manuscript, 1303, China. (See J. Needham, Science and Civilization in China, vol. 3. Cambridge University Press, New York, 1959)
53. Vandermonde, A.T.: Mémoire sur des irrationnelles de différens ordres avec une application au cercle. Mémoires de l'Académie (royale) des sciences de l'Institut (imperial) de France, pp. 489–498 (1772)
54. Bateman, H.: The solution of linear differential equations by means of definite integrals. Trans. Camb. Philos. Soc. **21**, 171–196 (1909)
55. Koshliakov, N.S.: On Sonine's polynomials. Messenger Math. **55**, 152–160 (1926)
56. Bailey, W.N.: Generalized Hypergeometric Series. Camrbridge University Press, Camrbridge (1935)
57. Barnes, E.W.: A new development of the theory of the hypergeometric functions. Proc. Lond. Math. Soc. **2**(1), 141–177 (1908)
58. Sun, B.H., Yi, H.: New development in classical theory of conical shells. Adv. Mech. **19**(4), 497–519 (1989)
59. Sun, B.H.: Exact solution for the P.Seide's equation. Eng. Mech. **5**(1), 127–131 (1988)
60. Sun, B.H., Zhang, W., Yeh, K.Y., Rimrott, F.P.J.: Exact displacement solution of arbitrary degree paraboloidal shallow shell of revolution made of linear elastic materials. Int. J. Solids Struct. **33**(16), 2299–2308 (1996)
61. Sun, B.H.: The general bending problem of a shallow paraboloidal shell. Acta Mech. Sin. **21**(2), 245–249 (1989)
62. Gasper, G., Rahman, M., George, G.: Basic Hypergeometric Series, vol. 96. Cambridge University Press, Cambridge (2004)
63. Fine, N.J.: Basic Hypergeometric Series and Applications (No. 27). American Mathematical Society, Providence (1988)
64. Ono, K.: Values of Gaussian Hypergeometric Series. Trans. Am. Math. Soc. **350**(3), 1205–1223 (1998)
65. Ahlgreen, S., Ono, K.: A Gaussian hypergeometric series evaluation and Apéry number congruences. J. Reine Angew. Math. **518**, 187–212 (2000)
66. Ahlgren, S.: Gaussian hypergeometric series and combinatorial congruences. In: Symbolic Computation, Number Theory, Special Functions, Physics and Combinatorics, pp. 1–12. Springer, Boston, MA (2001)
67. Wang, Z.X., Guo, D.R.: Special Functions. World Scientific, Singapore (1989)
68. Clausen, T.: Ueber die Fälle, wenn die Reihe von der Form y= 1+...etc. ein Quadrat von der Form z= 1...etc. hat. J. Reine Angew. Math. **3**, 89–95 (1828)
69. Srivastava, H.M., Kashyap, B.R.K.: Special Functions in Queuing Theory and Related Stochastic Processes. Academic, New York (1982)
70. Askey, R.: Orthogonal Polynomials and Special Functions. Society for Industrial and Applied Mathematics, Philadelphia, PA (1975)
71. Euler, L.: Introductio in analysin infinitorum. Bousquet, Lausanne (1748)
72. Burchnall, J.L., Chaundy, T.W.: Expansions of Appell's double hypergeometric functions. Q. J. Math. **11**(1), 249–270 (1940)

73. Fasenmyer, M.C.: Some generalized hypergeometric polynomials. Bull. Am. Math. Soc. **53**(8), 806–812 (1947)
74. Nørlund, N.E.: Hypergeometric functions. Acta Math. **94**, 289–349 (1955)
75. Fields, J.L., Wimp, J.: Expansions of hypergeometric functions in hypergeometric functions. Math. Comput. **15**(76), 390–395 (1961)
76. Srivastava, H.M.: Some biorthogonal polynomials suggested by the Laguerre polynomials. Pac. J. Math. **98**(1), 235–250 (1982)
77. Weisner, L.: Generating functions for Bessel functions. Can. J. Math. **11**, 148–155 (1959)
78. Beals, R., Wong, R.: Special functions and orthogonal polynomials, vol. 153. Cambridge University Press, Cambridge (2016)
79. Burchnall, J.L., Chaundy, T.W.: The hypergeometric identities of Cayley, Orr, and Bailey. Proc. Lond. Math. Soc. **2**(1), 56–74 (1948)
80. Srivastava, H.M.: Generating functions for Jacobi and Laguerre polynomials. Proc. Am. Math. Soc. **23**(3), 590–595 (1969)
81. Srivastava, H.M., Buschman, R.G.: Some polynomials defined by generating relations. Trans. Am. Math. Soc. **205**, 360–370 (1975)
82. Srivastava, H.M.: An extension of the Hille-Hardy formula. Math. Comput. **23**, 305–311 (1969)
83. Abdul-Halim, N., Al-Salam, W.A.: Double Euler transformations of certain hypergeometric functions. Duke Math. J. **30**(1), 51–62 (1963)
84. Carlitz, L.: Some generating functions of Weisner. Duke Math. J. **28**(4), 523–529 (1961)
85. Watson, G.N.: Notes on generating functions of polynomials:(1) Laguerre polynomials. J. Lond. Math. Soc. **1**(3), 189–192 (1933)
86. Whipple, F.J.W.: A group of generalized hypergeometric series: relations between 120 allied series of the type F (a, b, ce, f). Proc. Lond. Math. Soc. **2**(1), 104–114 (1925)
87. Darling, H.B.C.: On the differential equation satisfied by the hypergeometric series of the second order. J. Lond. Math. Soc. **1**(1), 63–70 (1935)
88. Laplace, P.S.: Theorie Analytique des Probability. Courcier, Paris (1820). (First appeared in 1812)
89. Erdelyi, A., Magnus, W., Oberhettinger, F., Tricomi, F.G.: Tables of Integral Transforms, vols. I and II. McGraw-Hill, New York (1954)
90. Oberhettinger, F., Badii, L.: Tables of Laplace Transforms. Springer, Berlin (1973)
91. Mellin, H.: Über die fundamentale Wichtigkeit des Satzes von Cauchy für die Theorien der Gamma-und der hypergeometrischen Functionen. Officina Typographica Societatis Litterari Fennic, vol. 21, no. 1, pp. 1–115 (1896)
92. Yang, X.J., Gao, F., Ju, Y.: General Fractional Derivatives with Applications in Viscoelasticity. Academic, New York (2020)
93. Newton, I.S.: Method of Fluxions, 1665; Annotations from Wallis, manuscript of 1665, published in The Mathematical Papers of Isaac Newton 1, Cambridge University Press, 1967
94. Leibniz, G.W.: Nova methodus pro maximis et minimis, itemque tangentibus, qua nec irrationals quantitates moratur. Acta Eruditorum (1684)
95. Boyer, C.B.: The History of the Calculus and Its Conceptual Development. Dover, New York (1959)
96. Hairer, E., Wanner, G.: Analysis by Its History. Springer, Berlin (2008)
97. Leibniz, G.W.: Memoir using the chain rule (1676)
98. Stieltjes, T.J.: Recherches sur les fractions continues. C. R. Acad. Sci. Ser. I Math. **118**, 1401–1403 (1894)
99. Lebesgue, H.: Intégrale, longueur, aire. Ann. Mat. Pura Appl. **7**(1), 231–359 (1902)
100. Riemann, B.: Ueber die Darstellbarkeit einer Function durch eine trigonometrische Reihe. Dieterich, Göttingen (1867)
101. Carter, M., Brunt, B.V.: The Lebesgue-Stieltjes integral: a practical introduction. Am. Math. Mon. **109**(8), 357–362 (2002)
102. Carter, M., Van Brunt, B.: The Lebesgue-Stieltjes Integral. Springer, Berlin (2000)

103. Hildebrandt, T.H.: Stieltjes integrals of the Riemann type. Am. Math. Mon. **45**, 265–277 (1938)
104. Hildebrandt, T.H.: Definitions of Stieltjes integrals of the Riemann type. Am. Math. Mon. **45**(5), 265–278 (1938)
105. Hildebrandt, T.H.: Introduction to the Theory of Integration. Academic, New York (1963)
106. Hildebrandt, T.H., Schoenberg, I.J.: On linear functional operations and the moment problem for a finite interval in one or several dimensions. Ann. Math. **34**(2), 317–328 (1933)
107. Clarkson, J.A.: On double Riemann-Stieltjes integrals. Bull. Am. Math. Soc. **39**(12), 929–937 (1933)
108. Lukkarinen, J., Pakkanen, M.S.: On the positivity of Riemann–Stieltjes integrals. Bull. Aust. Math. Soc. **87**(3), 400–405 (2013)
109. Hobson, E.W.: On change of the variable in a Lebesgue integral. Proc. Lond. Math. Soc. **2**(1), 10–21 (1910)
110. Yang, X.J.: New general calculi with respect to another functions applied to describe the Newton-like dashpot models in anomalous viscoelasticity. Therm. Sci. **23**(6B), 3751–3757 (2019)
111. Yang, X.J.: New non-conventional methods for quantitative concepts of anomalous rheology. Therm. Sci. **23**(6B), 4117–4127 (2019)
112. Carmichael, R.D.: Conditions necessary and sufficient for the existence of a Stieltjes integral. Proc. Natl. Acad. Sci. **5**(12), 551–555 (1919)
113. Russell, A.M.: Necessary and sufficient conditions for the existence of a generalized Stieltjes integral. J. Aust. Math. Soc. **26**(4), 501–510 (1978)
114. Widder, D.V.: Advanced Calculus. Prentice-Hall, New York (1947)
115. Bullock, G.L.: A geometric interpretation of the Riemann-Stieltjes integral. Am. Math. Mon. **95**(5), 448–455 (1988)
116. Bliss, G.A.: A necessary and sufficient condition for the existence of a Stieltjes integral. Proc. Natl. Acad. Sci. USA **3**(11), 3, 633–637 (1917)
117. Carmichael, R.D.: Conditions necessary and sufficient for the existence of a Stieltjes integral. Proc. Natl. Acad. Sci. **5**(12), 551–555 (1919)
118. Aleksander, P., Rolewicz, S.: Remarks on the existence of the Riemann-Stieltjes integral. Colloq. Math. **5**(1), 43–52 (1958)
119. Russell, A.M.: Necessary and sufficient conditions for the existence of a generalized Stieltjes integral. J. Aust. Math. Soc. **26**(4), 501–510 (1978)
120. Nakasho, K., Narita, K., Shidama, Y.: The basic existence theorem of Riemann-Stieltjes integral. Formaliz. Math. **24**(4), 253–259 (2016)
121. Reitano, R.R.: Foundations of Quantitative Finance: 3. The Integrals of Lebesgue and (Riemann-)Stieltjes. Brandeis International Business School, Waltham (2017)
122. Bray, H.E.: Elementary properties of the Stieltjes integral. Ann. Math. **20**(3), 177–186 (1919)
123. Dushnik, B.: On the Stieltjes integral. Dissertation, University of Michigan (1931)
124. McShane, E.J.: A Riemann-Type Integral that Includes Lebesgue-Stieltjes, Bochner and Stochastic Integrals. American Mathematical Society, Providence (1983)
125. Hille, E., Tamarkin, J.D.: On moment functions. Proc. Natl. Acad. Sci. USA **19**(10), 902–908 (1933)
126. Volterra, V.: Sopra alcune questioni di inversione di integrali definiti. Ann. Mat. Pura Appl. (1867–1897) **25**(1), 139–178 (1897)
127. Fredholm, I.: Sur une classe d'équations fonctionnelles. Acta Math. **27**(1), 365–390 (1903)
128. Osler, T.J.: Leibniz rule for fractional derivatives generalized and an application to infinite series. SIAM J. Appl. Math. **18**(3), 658–674 (1970)
129. Osler, T.J.: Taylor's series generalized for fractional derivatives and applications. SIAM J. Math. Anal. **2**(1), 37–48 (1971)
130. Almeida, R.: A Caputo fractional derivative of a function with respect to another function. Commun. Nonlinear Sci. Numer. Simul. **44**, 460–481 (2017)
131. Samko, S.G., Kilbas, A.A., Marichev, O.I.: Fractional Integrals and Derivatives. Theory and Applications. Gordon and Breach, Yverdon (1993) (Russian version, 1987)

132. Yang, X.J.: Open problems for anomalous linear viscoelasticity. Zhejiang University, 10 Oct 2019
133. Abel, N.: Solution de quelques problèmes à l'aide d'intégrales définies. Oeuvres **1**, 11–27 (1881). (N.H. Abel, "Solutions de quelques problèmes à l'aide d'intégrales défines" , Oeuvres complètes, nouvelle éd., 1, pp. 11–27. Grondahl & Son, Christiania (1881) (Edition de Holmboe))
134. Liouville, J.: Memoire sur le Calcul des different idles a indices quelconques. J. Ec. Polytech. **13**(21), 71–162 (1832)
135. Riemann, B.: Versuch einer allgemeinen Auffassung der Integration und Differentiation, 14 Janvier 1847. Bernhard Riemanns Gesammelte Mathematische Werke, pp. 353–362 (1892)
136. Holmgren, H.: On differential kalkylen: Med indices af hvad natur somhelst. Kongl. Stenh Vetendap Ahd. Handl. Stockholm **5**(11), 1–83 (1865)
137. Ross, B.: The development of fractional calculus 1695–1900. Hist. Math. **4**(1), 75–89 (1977)
138. Krug, A.: Theorie der Derivationen Akad. Wiss. Wien Denkschriften Math. Naturwiss **57**, 151–228 (1890)
139. Tamarkin, J.D.: On integrable solutions of Abel's integral equation. Ann. Math. **31**(2), 219–229 (1930)
140. Sonine, N.Y.: Sur la differentiation a indice quelconque. Mat. Sb. **6**(1), 1–38 (1872)
141. Caputo, M.: Linear models of dissipation whose Q is almost frequency independent II. Geophys. J. Int. **13**(5), 529–539 (1967)
142. Smit, W., De Vries, H.: Rheological models containing fractional derivatives. Rheol. Acta **9**(4), 525–534 (1970)
143. Weyl, H.: Bemerkungen zum Begriffe des Differential quotienten gebrochener Ordnung. Vierteljal rechrift tier Ntdrforch. Geellch. Zirich **62**(1–2), 296–302 (1917)
144. Gerasimov, A.N.: A generalization of linear laws of deformation and its application to problems of internal friction. Akad. Nauk SSSR. Prikl. Mat. Meh. **12**, 251–260 (1948)
145. Dzhrbashian, M.M., Nersesyan, A.B.: Fractional derivatives and the Cauchy problem for fractional differential equations. Izv. Akad. Nauk. Armyan **3**, 3–29 (1968) (in Russian)
146. Chen, Y.W.: Entire solutions of class of differential equations of mixed type. Commun. Pure Appl. Math. **14**(3), 229–255 (1961)
147. Agrawal, O.P.: Fractional variational calculus in terms of Riesz fractional derivatives. J. Phys. A **40**(24), 6287 (2007)
148. Kokilashvili, V.M.: Anisotropic potentials and singular integrals with weak singularity on curves. Soobshch. Akad. Nad. Gruzin. SSR (1969)
149. Sato, T.: On Abel's integral equation. Mem. Coll. Sci. Kyoto **18**, 63–78 (1935)
150. Euler, L., Blanton, J.D.: Foundations of Differential Calculus, pp. 1–9. Springer, New York (2000)
151. Dirac, P.A.M.: The physical interpretation of the quantum dynamics. Proc. R. Soc. Lond. Ser. A Containing Pap. Math. Phys. Character **113**(765), 621–641 (1927)
152. Bracewell, R.N.: The Fourier Transform and Its Applications, vol. 31999. McGraw-Hill, New York (1986)
153. Yang, X.J., Gao, F., Jing, H.W.: New mathematical models in anomalous viscoelasticity from the derivative with respect to another function view point. Therm. Sci. **23**(3A), 1555–1561 (2019)
154. Chen, W.: Time-space fabric underlying anomalous diffusion. Chaos Solitons Fractals **28**(4), 923–929 (2006)
155. Mandelbrot, B.: How long is the coast of Britain? Statistical self-similarity and fractional dimension. Science **156**(3775), 636–638 (1967)
156. Richardson, L.F.: Atmospheric diffusion shown on a distance-neighbour graph. Proc. R. Soc. A Math. Phys. Eng. Sci. **11**(756), 709–737 (1926)
157. Korcak, J.: Geopolitické základy skoslovenska. Jeho kmenové oblasti. (The Geopolitic Foundations of Czechoslovakia. Its Tribal Areas). Prague, Orbis (1938)
158. Turcotte, D.L.: Fractals and Chaos in Geology and Geophysics. Cambridge University Press, Cambridge (1997)

159. Imre, A.R. , Novotny, J.: Fractals and the Korcak-law: a history and a correction. Eur. Phys. J. H **41**(1), 69–91 (2016)
160. West, G.B., Brown, J.H., Enquist, B.J.: The fourth dimension of life: fractal geometry and allometric scaling of organisms. Science **284**(5420), 1677–1679 (1999)
161. Mandelbrot, B.: The fractal geometry of nature. W. H. Freeman, New York (1982)
162. Shlesinger, M.F., Montroll, E.W.: On the Williams—Watts function of dielectric relaxation. Proc. Natl. Acad. Sci. **81**(4), 1280–1283 (1984)
163. Wright, E.M.: The asymptotic expansion of the generalized hypergeometric function. J. Lond. Math. Soc. **1**(4), 286–293 (1935)
164. Wright, E.M.: The asymptotic expansion of integral functions defined by Taylor series. Philos. Trans. R. Soc. Lond. Ser. A Math. Phys. Sci. **238**(795), 423–451 (1940)
165. Kilbas, A.A., Saigo, M., Trujillo, J.J.: On the generalized Wright function. Fract. Calc. Appl. Anal. **5**(4), 437–460 (2002)
166. Dotsenko, M.R.: On some applications of Wrights hypergeometric function. Dokl. Bolg. Akad. Nauk. **44**(6), 13–16 (1991)
167. Srivastava, H.M.: Some Fox-Wright generalized hypergeometric functions and associated families of convolution operators. Appl. Anal. Discrete Math. **1**(1), 56–71 (2007)
168. Murugusundaramoorthy, G., Raina, R.K.: On a subclass of harmonic functions associated with the Wright's generalized hypergeometric functions. Hacettepe J. Math. Stat. **38**(2), 129–136 (2009)
169. Aouf, M.K., Dziok, J.: Distortion and convolutional theorems for operators of generalized fractional calculus involving Wright function. J. Appl. Anal. **14**(2), 183–192 (2008)
170. Kilbas, A.: Fractional calculus of the generalized Wright function. Fract. Calc. Appl. Anal. **8**(2), 113–126 (2005)
171. Kilbas, A.A., Srivastava, H.M., Trujillo, J.J.: Theory and Applications of Fractional Differential Equations. Elsevier, Amsterdam (2006)
172. Wright, E.M.: On the coefficients of power series having exponential singularities. J. Lond. Math. Soc. **1**(1), 71–79 (1933)
173. Luchko, Y.: The Wright function and its applications. In: Kochubei, A., Luchko, Yu. (eds.) Handbook of Fractional Calculus with Applications, vol. 1, pp. 241–268. De Gruyter, Berlin (2019)
174. Wright, E.M.: The asymptotic expansion of the generalized Bessel function. Proc. Lond Math. Soc. **2**(1), 257–270 (1935)
175. Marichev, O.I.: Handbook of Integral Transforms of Higher Transcendental Functions: Theory and Algorithmic Tables. Horwood, Chichester (1985)
176. Kiryakova, V.S.: Generalized Fractional Calculus and Applications. CRC Press, Boca Raton (1993)
177. Srivastava, H.M., Karlsson, P.W.: Multiple Gaussian Hypergeometric Series. Halsted Press (Ellis Horwood Limited), Chichester (1985)
178. Raina, R.K.: On generalized Wright's hypergeometric functions and fractional calculus operators. East Asian J. Appl. Math. **21**(2), 191–203 (2005)
179. Virchenko, N.: On some generalizations of the functions of hypergeometric type. Fract. Calc. Appl. Anal. **2**(3), 233–244 (1999)
180. Džrbašjan, M.M.: On the integral transformations generated by the generalized Mittag-Leffler function. Izvestiya Akademii Nauk Armyanskoi SSR **13**(3), 21–63 (1960)
181. Luchko, Y.: Asymptotics of zeros of the Wright function. Zeitschrift für Analysis und ihre Anwendungen **19**(2), 583–595 (2000)
182. Luchko, Y.: On the distribution of zeros of the Wright function. Integral Transform. Spec. Funct. **11**(2), 195–200 (2001)
183. Pogány, T.K., Srivastava, H.M.: Some Mathieu-type series associated with the Fox–Wright function. Comput. Math. Appl. **57**(1), 127–140 (2009)
184. Mehrez, K.: New integral representations for the Fox–Wright functions and its applications. J. Math. Anal. Appl. **468**(2), 650–673 (2018)

185. Mathai, A.M., Haubold, H.J.: Mittag-Leffler functions to pathway model to Tsallis statistics. Integral Transform. Spec. Funct. **21**(11), 867–875 (2010)
186. Gorenflo, R., Luchko, Y., Mainardi, F.: Analytical properties and applications of the Wright function. Fract. Calc. Appl. Anal. **2**(4), 383–414 (1999)
187. Ansari, A.: Green's function of two-dimensional time-fractional diffusion equation using addition formula of Wright function. Integral Transform. Spec. Funct. **30**(4), 301–315 (2019)
188. Dziok, J., Raina, R.K.: Some results based on first order differential subordination with the Wright's generalized hypergeometric function. Rikkyo Daigaku Sugaku Zasshi **58**(2), 87–94 (2009)
189. Raina, R.K.: Certain subclasses of analytic functions with fixed argument of coefficients involving the Wright's function. Tamsui Oxf. J. Math. Sci. **22**(1), 51 (2006)
190. Stankovic, B.: On the function of EM Wright. Publ. Inst. Math. Beograd N. S. **10**, 113–124 (1970)
191. Paris, R.B.: Exponentially small expansions of the Wright function on the Stokes lines. Lith. Math. J. **54**(1), 82–105 (2014)
192. Dziok, J., Raina, R.K.: Families of analytic functions associated with the Wright generalized hypergeometric function. Demonstratio Math. **37**(3), 533–542 (2004)
193. Mainardi, F., Pagnini, G.: The Wright functions as solutions of the time-fractional diffusion equation. Appl. Math. Comput. **141**(1), 51–62 (2003)
194. Virchenko, N., Kalla, S.L., Al-Zamel, A.: Some results on a generalized hypergeometric function. Integral Transform. Spec. Funct. **12**(1), 89–100 (2001)
195. Aouf, M.K., Dziok, J.: Certain class of analytic functions associated with the Wright generalized hypergeometric function. J. Math. Appl. **30**, 23–32 (2008)
196. Raina, R.K., Sharma, P.: Harmonic univalent functions associated with Wright's generalized hypergeometric functions. Integral Transform. Spec. Funct. **22**(8), 561–572 (2011)
197. Virchenko, N.: On the generalized confluent hypergeometric function and its application. Fract. Calc. Appl. Anal. **9**(2), 101–108 (2006)
198. Dunkl, C.F.: Differential-difference operators associated to reflection groups. Trans. Am. Math. Soc. **311**(1), 167–183 (1989)
199. De Jeu, M.F.: The Dunkl transform. Invent. Math. **113**(1), 147–162 (1993)
200. Soltani, F.: Sonine transform associated to the Dunkl kernel on the real line. Symmetry Integrability Geom. Methods Appl. **4**, 092 (2008)
201. Kallel, S.: Some results on generalized Dunkl–Lipschitz spaces. Math. Nachr. **293**(2), 305–326 (2020)
202. Mittag-Leffler, G.M.: Sur la nouvelle fonction Eα(x). C. R. Acad. Sci. **137**, 554–558 (1903)
203. Bateman, H.: The polynomial of Mittag-Leffler. Proc. Natl. Acad. Sci. **26**(8), 491–496 (1940)
204. Saxena, R.K., Mathai, A.M., Haubold, H.J.: On fractional kinetic equations. Astrophys. Space Sci. **282**(1), 281–287 (2002)
205. Haubold, H.J., Mathai, A.M., Saxena, R.K.: Mittag-Leffler functions and their applications. J. Appl. Math. **2011**, 1–51 (2011)
206. Berberan-Santos, M.N.: Properties of the Mittag-Leffler relaxation function. J. Math. Chem. **38**(4), 629–635 (2005)
207. Peng, J., Li, K.: A note on property of the Mittag-Leffler function. J. Math. Anal. Appl. **370**(2), 635–638 (2010)
208. Barnes, E.W.: The asymptotic expansion of integral functions defined by Taylor's series. Philos. Trans. R. Soc. Lond. **206**, 249–297 (1906)
209. Kilbas, A.A., Koroleva, A.A.: Inversion of integral transform with an extended generalized Mittag-Leffler function. Dokl. Math. **74**(3), 805–808 (2006)
210. Gorenflo, R., Kilbas, A.A., Mainardi, F., Rogosin, S.V.: Mittag-Leffler Functions, Related Topics and Applications, vol. 2. Springer, Berlin (2014)
211. Kilbas, A.A., Saigo, M., Saxena, R.K.: Solution of Volterra integro-differential equations with generalized Mittag-Leffler function in the kernels. J. Integral Equ. Appl. **14**, 377–396 (2002)
212. Valiron, G.: Entire functions and Borel's directions. Proc. Natl. Acad. Sci. USA **20**(3), 211 (1934)

213. Hardy, G.H.: On the summability of series by Borel's and Mittag-Leffler's methods. J. Lond. Math. Soc. **1**(2), 153–157 (1934)
214. Hille, E.: The mathematical congress in Stockholm. Bull. Am. Math. Soc. **41**(1), 4–5 (1935)
215. Hille, E., Tamarkin, J.D.: On moment functions. Proc. Natl. Acad. Sci. USA **19**(10), 902 (1933)
216. Cartwright, M.L.: On functions which are regular and of finite order in an angle. Proc. Lond. Math. Soc. **2**(1), 158–179 (1935)
217. Rademacher, H.: A convergent series for the partition function p (n). Proc. Natl. Acad. Sci. USA **23**(2), 78 (1937)
218. Pollard, H.: The completely monotonic character of the Mittag-Leffler function $E_a(\{-x\})$. Bull. Am. Math. Soc. **54**(12), 1115–1117 (1948)
219. Dattoli, G., Gorska, K., Horzela, A., Licciardi, S., Pidatella, R.M.: Comments on the properties of Mittag-Leffler function. Eur. Phys. J. Spec. Top. **226**(16–18), 3427–3443 (2017)
220. Berberan-Santos, M.N.: Relation between the inverse Laplace transforms of I (t β) and I (t): application to the Mittag-Leffler and asymptotic inverse power law relaxation functions. J. Math. Chem. **38**(2), 265–270 (2005)
221. Phragmén, E.: Sur une extension d'un théorème classique de la théorie des fonctions. Acta Math. **28**, 351–368 (1904)
222. Buhl, A.: Séries analytiques: sommabilité. Gauthier-Villars, Paris (1925)
223. Bierberbach, L.: Lehrbuch der Funktionentheorie, vol. 2. Teubner, Leipzig and Berlin (1931)
224. Feller, W.: Fluctuation theory of recurrent events. Trans. Am. Math. Soc. **67**(1), 98–119 (1949)
225. Gorenflo, R., Mainardi, F.: Fractional calculus: integral and differential equations of fractional order. In: Carpinteri, A., Mainardi, F. (eds.) Fractals and Fractional Calculus in Continuum Mechanics, pp. 223–276. Springer, Wien (1997)
226. Erdelyi, A., Magnus, W., Oberhettinger, F., Tricomi, F.G.: Higher Transcendental Functions, vol. 3, McGraw-Hill, New York, NY (1955)
227. Humbert, P.: Quelques résultats relatifs à la fonction de Mittag-Leffler. C. R. Hebd. Seances Acad. Sci. **236**(15), 1467–1468 (1953)
228. Bagley, R.L.: The initial value problem for fractional order differential equations with constant coefficients. Air Force Institute of Technology Report, AFIT-TR-EN-88-1 (1988)
229. Zhang, S., Zhang, H.Q.: Fractional sub-equation method and its applications to nonlinear fractional PDEs. Phys. Lett. A **375**(7), 1069–1073 (2011)
230. Jumarie, G.: Fourier's transform of fractional order via Mittag-Leffler function and modiffied Riemann-Liouville derivative. J. Appl. Math. Inform. **26**(5–6), 1101–1121 (2008)
231. Riccati, V.: Vincentii Riccati soc. Jesu Opusculorum ad res physicas & mathematicas pertinentium Tomus primus.[-secundus] (1757)
232. Muldoon, M.E., Ungar, A.A.: Beyond sin and cos. Math. Mag. **69**(1), 3–14 (1996)
233. Ungar, A.: Generalized hyperbolic functions. Am. Math. Mon. **89**(9), 688–691 (1982)
234. Mainardi, F.: Fractional relaxation-oscillation and fractional diffusion-wave phenomena. Chaos Solitons Fractals **7**(9), 1461–1477 (1996)
235. Gorenflo, R., Mainardi, F.: Fractional oscillations and Mittag-Leffler functions. In: International Workshop on the Recent Advances in Applied Mathematics (RAAM96), State of Kuwait, Proceedings, Kuwait University, pp. 193–208 (1996)
236. Atangana, A., Baleanu, D.: New fractional derivatives with nonlocal and non-singular kernel: theory and application to heat transfer model. Therm. Sci. **20**(2), 763–769 (2016)
237. Hille, E., Tamarkin, J.D.: On the theory of linear integral equations. Ann. Math. **31**(3), 479–528 (1930)
238. Mathai, A.M., Haubold, H.J.: Special Functions for Applied Scientists. Springer, New York (2008)
239. Wiman, A.: Über den Fundamentalsatz in der Teorie der Funktionen Ea (x). Acta Math. **29**, 191–201 (1905)
240. Wiman, A.: Über din Nullstellen der Funktionen Ea (x). Acta Math. **29**, 217–234 (1905)

241. Agarwal, R.P.: A propos d'une note de M. Pierre Humbert. C. R. Acad. Sci. Ser. I Math. Paris **236**(21), 2031–2032 (1953)
242. Djrbashian, M.M.: Integral Transforms and Representations of Functions in the Complex Domain (Russian). Nauka, Moscow (1966)
243. Djrbashian, M.M.: Harmonic Analysis and Boundary Value Problems in the Complex Domain. Birkhauser, Berlin (1995)
244. Gorenflo, R., Kilbas, A.A., Rogosin, S.V.: On the generalized Mittag-Leffler type functions. Integral Transform. Spec. Funct. **7**(3–4), 215–224 (1998)
245. Srivastava, H.M.: A certain family of sub-exponential series. Int. J. Math. Educ. Sci. Technol. **25**(2), 211–216 (1994)
246. Yu, R., Zhang, H.: New function of Mittag–Leffler type and its application in the fractional diffusion-wave equation. Chaos Solitons Fractals **30**(4), 946–955 (2006)
247. Schneider, W.R.: Completely monotone generalized Mittag-Leffler functions. Expo. Math. **14**(1), 3–24 (1996)
248. Miller, K.S., Samko, S.G.: A note on the complete monotonicity of the generalized Mittag-Leffler function. Real Anal. Exch. **23**(2), 753–756 (1999)
249. Pskhu, A.V.: On the real zeros of functions of Mittag-Leffler type. Math. Notes **77**(3), 546–552 (2005)
250. Popov, A.Yu., Sedletskii, A.M.: The distribution of the zeros of a function of Mittag-Leffler type. Russ. Acad. Sci. Dokl. Math. **390**(3), 165–168 (2003)
251. Sedletskii, A.M.: Asymptotic formulas for zeros of a function of Mittag-Lefler type (Russian). Anal. Math. **20**, 117–132 (1994)
252. Saxena, R.K., Mathai, A.M., Haubold, H.J.: On generalized fractional kinetic equations. Phys. A **344**(3–4), 657–664 (2004)
253. Kilbas, A.A., Saigo, M., Saxena, R.K.: Generalized Mittag-Leffler function and generalized fractional calculus operators. Integral Transform. Spec. Funct. **15**(1), 31–49 (2004)
254. Saxena, R.K., Saigo, M.: Certain properties of fractional calculus operators associated with generalized Mittag-Leffler function. Fract. Calc. Appl. Anal. **8**(2), 141–154 (2005)
255. Gorenflo, R., Mainardi, F.: The Mittag-Leffler type function in the Riemann-Liouville fractional calculus, In: Kilbas, A A. (ed.) Boundary Value Problems, Special Functions and Fractional Calculus (Proc. Int. Conf., Minsk, 1996), Belarusian State Univ., Minsk, pp. 215–225 (1996)
256. Gorenflo, R., Luchko, Yu., Rogosin, S.V.: Mittag-Leffler type functions: notes on growth properties and distribution of zeros. Preprint No. A04-97, Freie Universitat Berlin, Serie A Mathematik, Berlin (1997)
257. Sixdeniers, J.M., Penson, K.A., Solomon, A.I.: Mittag-Leffler coherent states. J. Phys. A **32**(43), 7543 (1999)
258. Metzler, R.: Generalized Chapman-Kolmogorov equation: a unifying approach to the description of anomalous transport in external fields. Phys. Rev. E **62**(5), 6233 (2000)
259. Liang, X., Gao, F., Zhou, C.B., Wang, Z., Yang, X.J.: An anomalous diffusion model based on a new general fractional operator with the Mittag-Leffler function of Wiman type. Adv. Differ. Equ. **2018**(1), 25 (2018)
260. Prabhakar, T.R.: A singular integral equation with a generalized Mittag Leffler function in the kernel. Yokohama Math. J. **19**, 7–15 (1971)
261. Mathai, A.M., Saxena, R.K.: Generalized Hypergeometric Functions with Applications in Statistics and Physical Sciences. Lecture Notes in Mathematics, vol. 348. Springer, Berlin (1973)
262. Garra, R., Gorenflo, R., Polito, F., Tomovski, Ž.: Hilfer-Prabhakar derivatives and some applications. Appl. Math. Comput. **242**, 576–589 (2014)
263. Giusti, A.: General fractional calculus and Prabhakar's theory. Commun. Nonlinear Sci. Numer. Simul. **83**, 105114 (2020)
264. Garra, R., Garrappa, R.: The Prabhakar or three parameter Mittag–Leffler function: theory and application. Commun. Nonlinear Sci. Numer. Simul. **56**, 314–329 (2018)

265. Colombaro, I., Giusti, A., Vitali, S.: Storage and dissipation of energy in Prabhakar viscoelasticity. Mathematics **6**(2), 15 (2018)
266. Alidousti, J.: Stability region of fractional differential systems with Prabhakar derivative. J. Appl. Math. Comput. **62**(1–2), 135–155 (2020)
267. Giusti, A., Colombaro, I.: Prabhakar-like fractional viscoelasticity. Commun. Nonlinear Sci. Numer. Simul. **56**, 138–143 (2018)
268. Yang, X.J., Gao, F., Ju, Y., Zhou, H.W.: Fundamental solutions of the general fractional-order diffusion equations. Math. Methods Appl. Sci. **41**(18), 9312–9320 (2018)
269. Sonine, N.: Sur la généralisation d'une formule d'Abel. Acta Math. **4**(1), 171–176 (1884)
270. Rabotnov, Y.N.: The equilibrium of an elastic medium with after-effect. Prikl. Mat. Mekh. (J. Appl. Math. Mech.) **12**(1), 53–62. Reprinted in Fract. Calc. Appl. Anal. **17**(3), 684–696 (1948)
271. Yang, X.J., Abdel-Aty, M., Cattani, C.: A new general fractional-order derivative with Rabotnov fractional-exponential kernel applied to model the anomalous heat transfer. Therm. Sci. **23**(3 Part A), 1677–1681 (2019)
272. Yang, X.J., Ragulskis, M., Taha, T.: A new general fractional-order derivative with Rabotnov fractional-exponential kernel. Therm. Sci. **23**(6B), 3711–3718 (2019)
273. Kumar, S., Nisar, K.S., Kumar, R., Cattani, C., Samet, B.: A new Rabotnov fractional-exponential function-based fractional derivative for diffusion equation under external force. Math. Methods Appl. Sci. **43**(7), 4460–4471 (2020)
274. Rossikhin, Y.A., Shitikova, M.V.: Centennial jubilee of academician Rabotnov and contemporary handling of his fractional operator. Fract. Calc. Appl. Anal. **17**(3), 674–683 (2014)
275. Kiryakova, V.: The multi-index Mittag-Leffler functions as an important class of special functions of fractional calculus. Comput. Math. Appl. **59**(5), 1885–1895 (2010)
276. Rabotnov, Y.N., Papernik, L.K., Zvonov, E.N.: Tables of the Fractional-Exponential Function and Its Integral, Moscow (1969) (in Russian)
277. Miller, K.S., Ross, B.: An Introduction to the Fractional Calculus and Fractional Differential Equations. Wiley, New York (1993)
278. Hartley, T.T., Lorenzo, C.F.: A solution to the fundamental linear fractional order differential equation. NASA/TP-1998-208693, December 1998
279. Hartley, T.T., Lorenzo, C.F.: The Fractional Trigonometry: With Applications to Fractional Differential Equations and Science. Wiley, New York (2016)
280. Hartley, T.T., Lorenzo, C.F.: Generalized functions for the fractional calculus. Crit. Rev. Biomed. Eng. **36**(1), 39–55 (2008)
281. Kohlrausch, R.: Theorie des elektrischen rckstandes in der leidener flasche. Ann. Phys. **167**(2), 179–214 (1854)
282. Williams, G., Watts, D.C.: Non-symmetrical dielectric relaxation behaviour arising from a simple empirical decay function. Trans. Faraday Soc. **66**, 80–85 (1970)
283. Chung, S.H., Stevens, J.R.: Time-dependent correlation and the evaluation of the stretched exponential or Kohlrausch–Williams–Watts function. Am. J. Phys. **59**(11), 1024–1030 (1991)
284. Alvarez, F., Alegra, A., Colmenero, J.: Relationship between the time-domain Kohlrausch-Williams-Watts and frequency-domain Havriliak-Negami relaxation functions. Phys. Rev. B **44**(14), 7306 (1991)
285. Yoshioka, S., Tajima, S., Aso, Y., Kojima, S.: Inactivation and aggregation of β-galactosidase in lyophilized formulation described by Kohlrausch-Williams-Watts stretched exponential function. Pharm. Res. **20**(10), 1655–1660 (2003)
286. Rabani, E., Gezelter, J.D., Berne, B.J.: Direct observation of stretched-exponential relaxation in low-temperature Lennard-Jones systems using the cage correlation function. Phys. Rev. Lett. **82**(18), 3649 (1999)
287. Friedrich, C.H.R.: Relaxation and retardation functions of the Maxwell model with fractional derivatives. Rheol. Acta **30**(2), 151–158 (1991)
288. Meijer, C.S.: Uber Whittakersche bzw. Besselsche Funktionen und deren Produkte. Nieuw Archief Voor Wiskunde **18**(4) 10–39 (1936) (in German)

289. Freeden, W., Gutting, M.: Special Functions of Mathematical(Geo-)Physics. Springer, Berlin (2013)
290. Bell, W.W.: Special Functions and Their Applications, Prentice Hall, London (1965). Revised English edition, translated and edited by Richard A. Silverman (1965)
291. Mainardi, F.: Fractional Calculus and Waves in Linear Viscoelasticity: An Introduction to Mathematical Models. World Scientific, Singapore (2010)
292. Kiryakova, V.S.: Generalized Fractional Calculus and Applications. CRC Press, Boca Raton (1993)
293. Cardano, G.: Ars Magna or the Rules of Algebra. Dover, New York (1554). Translated and Edited by Witmer, T.R. (1968); with a foreword by Ore, O
294. Confalonieri, S.: The casus irreducibilis in cardano's ars magna and de regula aliza. Arch. Hist. Exact Sci. **69**(3), 257–289 (2015)
295. De Moivre, A.: Aequationum Quarundam Potestatis Tertiae, Quintae, Septimae, Nonae, & Superiorum, ad Infinitum Usque Pergendo, in Terminis Finitis, ad Instar Regularum pro Cubicis Quae Vocantur Cardani, Resolutio Analytica. Philos. Trans. **25**, 2368–2371 (1707)
296. Gauss, C.F.: Arithmetiche Geometrisches Mittel. Werke **3**, 361–432 (1799)
297. Schneider, I.: Der Mathematiker Abraham de Moivre (1667–1754). Arch. Hist. Exact Sci. **5**(3–4), 177–317 (1968)
298. De Moivre, A.: Reduction of radicals to more simple terms. In: Hutton, C., Shaw, G., Pearson, R. (eds.) Philosophical Transactions of the Royal Society of London from their commencement in 1665 to the year 1800; Abridged, vol. 8, pp. 1735–1743 (1738)
299. Green, D.R.: The historical development of complex numbers. Math. Gaz. **60**(412), 99–107 (1976)
300. Merino, O.: A short history of complex numbers. University of Rhode Island (2006)
301. Terjanian, G.: Sur l'équation $x^{2p} + y^{2p} = z^{2p}$. C. R. Acad. Sci. Paris Sér. **285**(16), A973–A975 (1977)
302. Gray, A.: Modern Differential Geometry of Curves and Surfaces with Mathematica, 2nd ed. CRC Press, Boca Raton (1997)
303. Fourier, J.: Théorie Analytique de la Chaleur (The Analytical Theory of Heat). Didot, Paris (1822). English translation by Alexander Freeman, first published in 1878, Cambridge University Press, 2009
304. Pereyra, M.C., Ward, L.A.: Harmonic Analysis: From Fourier to Wavelets, vol. 63. American Mathematical Society, Providence (2012)
305. Serov, V.: Fourier Series, Fourier Transform and Their Applications to Mathematical Physics. Springer, Berlin (2017)
306. Bromwich, T.J.I.A.: An Introduction to the Theory of Infinite Series. Macmillan, London (1908)
307. Beerends, R.J., ter Morsche, H.G., van den Berg, J.C., van de Vrie, E.M.: Fourier and Laplace Transforms. Cambridge University Press, Cambridge (2003)
308. Fahad, H.M., Rehman, M., Fernandez, A.: On Laplace transforms with respect to functions and their applications to fractional differential equations (2020). ArXiv:1907.04541v2 [math.CA] 10 Aug 2020
309. Jarad, F., Abdeljawad, T.: Generalized fractional derivatives and Laplace transform. Discrete Contin. Dyn. Syst. **13**(3), 709–722 (2020)
310. Weideman, J., Trefethen, L.: Parabolic and hyperbolic contours for computing the Bromwich integral. Math. Comput. **76**(259), 1341–1356 (2007)

311. Talbot, A.: The accurate numerical inversion of Laplace transforms. IMA J. Appl. Math. **23**(1), 97–120 (1979)
312. López-Fernández, M., Palencia, C.: On the numerical inversion of the Laplace transform of certain holomorphic mappings. Appl. Numer. Math. **51**(2–3), 289–303 (2004)
313. Rizzardi, M.: A modification of Talbot's method for the simultaneous approximation of several values of the inverse Laplace transform. ACM Trans. Math. Softw. **21**(4), 347–371 (1995)

The manufacturer's authorised representative in the EU is Springer Nature Customer Service Centre GmbH, Europaplatz 3, 69115 Heidelberg, Germany. If you have any concerns regarding our products, please contact ProductSafety@springernature.com

Printed and bound by CPI Group (UK) Ltd, Croydon, CR0 4YY
30/09/2026
02225329-0020